AutoCAD® 2000 for Architecture

Alan Jefferis
Michael Jones

Artwork by
Tereasa Jefferis

AutoCAD® 2000 for Architecture

Autodesk Press

Thomson Learning™

Africa • Australia • Canada • Denmark • Japan • Mexico • New Zealand
Phillipines • Puerto Rico • Singapore • Spain • United Kingdom • United States

NOTICE TO THE READER

TRADEMARKS

AUTODESK PRESS STAFF

Executive Director: Alar Elken
Executive Editor: Sandy Clark
Developmental Editor: John Fisher
Editorial Assistant: Allyson Powell
Executive Marketing Manager: Maura Theriault
Executive Production Manager: Mary Ellen Black
Production Coordinator: Jennifer Gaines
Art and Design Coordinator: Mary Beth Vought
Marketing Coordinator: Paula Collins
Technology Project Manager: Tom Smith
Cover design by Scott Keidong's Image Enterprises

Printed in Canada

1 2 3 4 5 6 7 8 9 10 XXX 04 03 02 01 00 99

For more information, contact

Autodesk Press
3 Columbia Circle, Box 15-015
Albany, New York USA 12212-15015;

or find us on the World Wide Web at http://www.autodeskpress.com

Library of Congress Cataloging-in-Publication Data

Jefferis, Alan.
AutoCAD 2000 for architecture / Alan Jefferis, Michael Jones.
p. cm.
ISBN 0-7668-1242-1 (pbk.)
1.Architectural drawing -- Data processing Programmed instruction.
2.Computer-aided design Programmed instruction. 3. AutoCAD (Computer file) I. Jones, Michael, 1964- . II. Title.
NA2728.J438 1999
720' . 28' 402855369—dc21 99-35162
CIP

CONTENTS

SECTION I

CHAPTER 3 Creating Drawing Aids

CHAPTER 4 Drawing and Controlling Lines

CHAPTER 5 Working with Drawing Files

CHAPTER 6 Drawing Organization

SECTION 3

CHAPTER 17 Placing Text on a Drawing

CHAPTER 18 Introduction to Dimensioning

SECTION 4

CHAPTER 20 Creating Blocks and Wblocks

CHAPTER 21 Adding Attributes to Enhance Blocks

SECTION 5

CHAPTER 27 Oblique and Isometric Drawing

INTRODUCTION

AutoCAD 2000 for Architecture is a practical, comprehensive text and workbook that is easy to use and understand. The contents are designed to teach the skills needed to master the 2D drawing commands used on most construction-related drawings. The content may be used as presented to provide an introduction to the use of AutoCAD. This text can also be used with older releases of AutoCAD. All concepts are presented using architectural examples so that each command can be correlated to a specific skill that will be required of architects, engineers, and drafters. Commands and drawing skills are presented in a sequence that is easy to understand and follow with each chapter building on the skills presented in the previous chapter. Other specific features of this text include the following:

- Explanation of how a skill will relate to office practice
- Professional drawings by architects, engineers, and residential designers
- Chapter exercises that review chapter contents as well as reinforce previous material
- Drawing chapters with drawing problems to reinforce the chapter concepts

Individual chapters will provide knowledge to master the following drawing skills:

- Drawing lines and geometric shapes
- Varying linetype and width
- Editing basic geometric shapes
- Combining skills to produce drawings such as floor and foundation plans, elevations, details, and sections
- Editing an entire drawing to produce other similar drawings
- Dimensioning methods and techniques
- Placing text within a drawing
- Creating varied text and dimensioning styles
- Creating symbols libraries to increase drawing speed and efficiency
- Providing a user-friendly environment to master computer drafting skills

In addition, the authors have written an Instructor's Guide that contains supplemental drawing exercises, which are also available on disk. For more information, instructors can contact Delmar Publishers.

INTRODUCTION TO DRAWING COMMANDS

AutoCAD 2000 is a powerful program that contains hundreds of commands to meet a variety of drawing needs. In this text, only the command options that are typically used within the construction field will be explored. Several methods are available to execute each drawing command. Data entry can executed using toolbars, shortcut menus, the keyboard, or pull-down menus. The selection method used will depend on individual preference. Each command will be presented initially through the method that provides the fastest access. For many commands the fastest access is through the keyboard, with available options used to supplement the basic procedure. Using the keyboard, the entire command sequence can be presented and many options that are available within a specific command will be displayed.

Commands, options, or values that must be typed at the keyboard are listed with **BOLDFACE** capital letters. Once the command is entered, pressing the ENTER or RETURN key is required and will be represented by the symbol (EnterKey) throughout the text. Command sequences will be presented in the following manner:

Command: **LINE** (EnterKey)

LINE

Specify first point: **4,8** (EnterKey)

Specify next point or [Undo]: **8,8** (EnterKey)

Specify next point or [Undo]: (EnterKey)

When input is required to complete a command, and the selected point can be any point, the command sequence will be presented using italics in the following manner:

Command: **LINE (**EnterKey)

LINE

Specify first point: *(Select a point.)*

Specify next point or [Undo]: *(Select a point.)*

Specify next point or [Undo]: (EnterKey)

Command:

Some commands prompt the user to select a key to conrrtinue a specific sequence. *Select any key* means strike any key on the keyboard to continue. There is no key labeled "select any key" on the keyboard, but many novice users spend a lot of time looking for it.

PREREQUISITES

Although this text can serve as an excellent reference for experienced professionals, no knowledge of architecture or construction is required to learn the command structure of AutoCAD. Some of the advanced projects require a knowledge of basic residential architectural standards.

ONLINE COMPANION

The Online Companion™ is your link to AutoCAD on the Internet. We have compiled supporting resources with links to a variety of sites. Not only can you find out about training and education, industry, and the online community, we also point to valuable archives compiled for AutoCAD users from various Web sites. In addition, the Online Companion contains pages designed specifically for users of AutoCAD 2000 for Architecture. These include an owner's page with updates, a swap bank where you can share your drawings with other AutoCAD students, and a page where you can send us your comments. You can find the Online Companion at:

http://www.autodeskpress.com/onlinecompanion.html

When you reach the Online Companion page, click on the title **AutoCAD 2000 for Architecture**.

ABOUT THE AUTHORS

Mike Jones is the CAD Systems Manager and a drafting instructor at Clackamas Community College, an Autodesk Premier Training Center in Oregon City, Oregon. Mike has introduced new users and experienced professionals to the commands and options of AutoCAD for over 16 years. His training center experience is enhanced by 19 years of professional drafting and CAD system management experience in private industry.

Alan Jefferis has been a drafting instructor at Clackamas Community College for 21 years. His professional experience includes eight years of drafting experience with structural engineers, and 19 years as a professional building designer. He is currently the owner of Residential Designs, A.I.B.D., and is coauthor of *Architectural Drafting and Design, Commercial Drafting and Detailing, Print Reading for Architectural and Construction Technology, AutoCAD for Architecture Release 12, AutoCAD for Architecture Release 13. and AutoCAD for Architecture Release 14.* All are available through Delmar Publishers.

ACKNOWLEDGMENTS

The quality of this text has been greatly enhanced by the donations of many professional offices and individuals within private and municipal agencies. My special thanks to:

Tereasa Jefferis
Stark, Jefferis Designs

Kenneth D. Smith
Kenneth D. Smith Architects
El Cajon, CA

Tom Sponsler
Director, Department of Environmental Services
City of Gresham, OR

Jessie O. Kempter
International Business Machines Corporation

F.D. Lee
Lee Engineering, Inc.
Oregon City, OR

G. Williamson Archer, A.I.A.
Archer & Archer P.A.
Meridian, MI

Ron Lee, A.I.A.
Architects Barrentine, Bates & Lee, A.I.A.
Lake Oswego, OR

Thomas J. Kuhns A.I.A
Michael & Kuhns Architects, P.C.
Portland, OR

Ned Peck,
Peck, Smiley, Ettlin Architects

Havlin G. Kemp, P.E.
Van Domelen/Looijenga/McGarrigle/Knauf Consulting Engineers
Portland, OR

David Rogencamp
kpff Consulting Engineers
Portland, OR

Walt Cheever
South Central Technical College
N. Mankato, MN

Shelly Fry
CPCC Corporate Center
Charlotte, NC

Aaron Rumple
St. Louis Community College
St. Louis, MO.

Dean Urevig
Dunwoody Instuite
Minneapolis, MN

Sylvia Sullivan
College of the Canyons

Joseph Yamello
Greater Lowell Regional Vocational
Technical School

Robert Kamenski
Lenape A.V.T.S.

Walter Eric Lawrence
Gwinnett Technical Institute

All of these people have made substantial contributions to the success of this text. We're also especially thankful for the work and efforts of John Fisher and Sandy Clark of Delmar Publishers; and Ralph Pescatore, Michael Pescatore, and Suzanne Pescatore of TKM Productions. No one has been more helpful and supportive than Janice Ann, David Alan, and Michael Aaron. Thanks for the help and support.

TO THE STUDENTS

AutoCAD 2000 for Architecture is designed for students in architectural or engineering programs. The contents are presented in an order that has been tested within the classroom. Information is based on common drafting practices from the offices of architects, engineers, and designers. To make the best use of the information within the text, students should follow these few helpful hints.

READ THE TEXT

I know it sounds so simple, but if you'll read a chapter prior to sitting at a computer, you'll find your drawing time will be much more productive. Read through the chapter and make note of things that you're not quite sure of. With the reading complete, now try to mentally work through the commands that you've been reading about.

STUDY THE EXAMPLES

One of the many features of this text are the many illustrations and examples of how to complete each command sequence. Carefully study and compare the command sequence with the illustration of how to complete the command.

PRACTICE AT A WORKSTATION

Once you've read the chapter and studied the illustrations, complete the command sequences at a computer workstation. Practice each command several times and don't be afraid to explore all of the options within a command. Because each chapter is based on the previous chapters, don't hesitate to reread past material. Above all else, don't be afraid to experiment. It's very common for new computer users to be filled with a fear of breaking something. If you save your drawings often, there is very little that can go wrong by experimenting. Relax and have fun!

SECTION 1

EXPLORING AutoCAD

CHAPTER 1

Exploring the Tools That Control AutoCAD

Whether their drawings were painted on walls, methodically scribed into clay, or etched on papyrus, people have been planning structures and equipment to better their lives. Most recently, those charged with converting a sketch into a usable drawing have progressed from T-squares, to parallel bars, to drafting arms, to tract machines, to computers. Computers are now found in most architectural and engineering firms. This probably has a lot to do with why you're reading this book. Computer-aided drafting, or CAD, has changed the marketplace greatly. To work efficiently in an architectural or engineering firm, you must be a proficient computer operator. You're about to start exploring the most popular computer-aided drafting program on the market. If you're like many new AutoCAD users, it will only take a few minutes of exploring to realize that AutoCAD offers lots of options, and the task of learning the program may seem a bit overwhelming. Relax! It's really not that bad. CAD drafting, and AutoCAD in particular, tend to become addictive. AutoCAD may seem frustrating for the first few weeks, but I think you'll love it.

An understanding of the tools you will be using will help to eliminate the natural frustration that comes with learning. This chapter will introduce key components of the hardware and the software that you will be using to create and control your drawings. Before you begin to explore AutoCAD, it's important that you understand the tools you will be working with. If you have computer experience, you can skip to Starting AutoCAD at the end of this chapter.

BENEFITS

AutoCAD has many benefits that will make its use invaluable. Among the benefits that make AutoCAD so popular are the accuracy, speed, consistency, and neatness that can be achieved with each drawing.

ACCURACY

As you will see in the next few chapters, the accuracy of drawings can be enhanced greatly by using AutoCAD. This is because you will be able to control the exact placement of objects in the drawing, enlarge the size of an object for better viewing, or use a series of drawing aids that accurately place an object exactly were it was intended to go. Figure 1.1 shows an example of an architectural drawing created using AutoCAD.

DRAWING SPEED

Speed in the office is a great asset, as long as quality can be maintained. AutoCAD drafting offers both. Although you may be able to manually draw some objects more quickly, AutoCAD allows you to make changes much more quickly than manual drafting. AutoCAD also excels in repetitive functions. For example, several different command options are available that enable you to draw 10 similar bolts almost as quickly as 1.

CONSISTENCY

Multiple objects can not only be drawn quickly, they can be drawn consistently. This consistency is a major consideration in the office setting, not just within your own drawings, but within the drawings of the entire staff. For example, with AutoCAD each employee can letter exactly the same as fellow staff members. Other important drawing qualities, such as line weight, linetype, and lettering angle, can be controlled easily by AutoCAD.

THE CHALLENGE

Of all the features of AutoCAD, perhaps its best one is that it is challenging. As you develop your manual skills, you can learn to use a drafting machine in an hour. You can develop the manual skills to get quality linework and lettering by the time you've completed your first few drawings. You will be able to draw simple objects using AutoCAD with little effort. By the end of just a few more chapters, you will be doing complex drawings. Because AutoCAD is always changing and very complex, it may take you several months before you feel totally at ease with it. You can become very

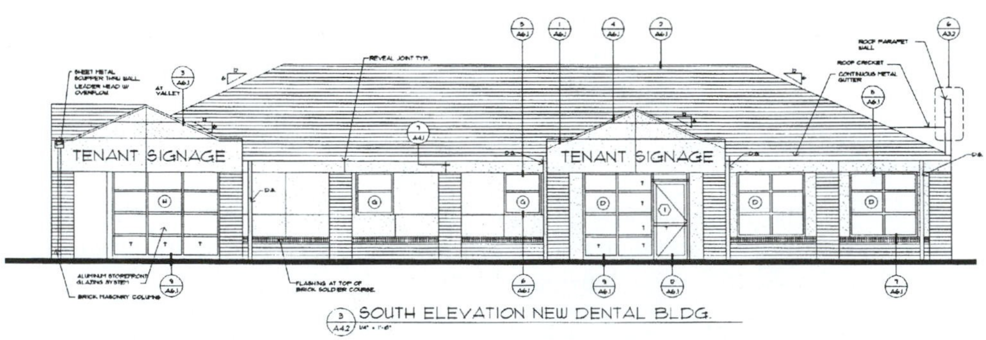

Figure 1.1 *AutoCAD and programs created by third-party vendors are used throughout the construction field to create drawings quickly and economically. (Courtesy Scott R. Beck Architect.)*

proficient with the core of the AutoCAD program in a short time and then choose to expand your knowledge as needed. Once proficient with AutoCAD, a whole world of third-party software is available for adjusting the program to meet specific needs. Figure 1.2 shows an example of a rendering completed using AutoCAD and third-party software.

HARDWARE

Not only does AutoCAD drafting offer new benefits to the architect, engineer, and designer, it also introduces new tools to complete the drawing. These tools are the hardware and software. Software is the program or instructions that make the computer perform specified tasks such as word processing or computer-aided drafting. Hardware is the equipment used to run the software. One is useless without the other. Before you can begin to draw or design, you need a basic understanding of the main components of hardware found at a computer workstation. This will be especially true if you decide to purchase or update your workstation at home to meet the demands of AutoCAD. Key workstation components include the computer, a monitor, input devices such as a mouse and keyboard, and output devices such as a printer or plotter.

Figure 1.2 *Once the basic drawing skills are mastered, add-on programs can be used to produce realistic drawings.*

COMPUTERS

Computers typically are referred to, from largest to smallest, as mainframe computers, minicomputers, and microcomputers. Mainframe computers are used by large firms, which may have several hundred employee terminals connected to the same computer. A mainframe can run several large, sophisticated programs at the same time. Your local building and planning department is probably served by a mainframe computer. The receptionist, zoning department, plans examiner, document control department, and fee collectors all can run different programs on individual terminals that are connected to the mainframe computer.

A minicomputer allows several people to use the same program at one time. An office with several drafters and engineers may provide each with a terminal served by one computer. A desktop computer or microcomputer is typically found in most educational and small office settings. It is also known as a PC (personal computer) because it is designed to be used by one person at a time. Microcomputers can be interconnected or networked for better efficiency of class or office personnel. An example of a microcomputer is shown in Figure 1.3.

Figure 1.3 *A typical work station includes the computer case that protects the motherboard, a central processing unit, CD ROM, hard drive, Zip drive, monitor, mouse, keyboard, and sound systems.*

A laptop computer is a small, portable computer that draws its power from a battery. Because of its small size, it can be easily carried, allowing productive use of time when the user must be away from an office. Laptop computers are excellent for use at a construction site because of their mobility. Many professionals are looking into the use of laptop computers to allow the transfer of information electronically from the architect or engineer to the contractor at the job site and reduce the amount of paper now required to convey ideas. No matter the size or type of computer, the key components include a systems board, central processing unit, memory, and drives.

Systems Board

The systems board contains the electronic chips and boards that perform the basic functions of the computer. It is referred to as the motherboard because other components are all connected to it. In addition to the CPU and memory chips, the motherboard contains terminals or bus slots for control cards. A control card similar to Figure 1.4 is used to control the monitor, printers, plotters, and other external devices.

Central Processing Unit

The central processing unit (CPU) is a chip located on the motherboard where the software program is processed. Once information is processed, the chip relays messages to other parts of the computer. The most common CPU is an Intel Pentium® chip. Common Intel CPUs are the Pentium®, Pentium Pro®, Pentium II® and Pentium III®. The power of a computer is also measured by the speed of the CPU in megahertz (MHz). A megahertz is a unit of frequency equal to 1 million cycles per second. The common Pentium processing speeds range from 90 to over 500MHz for

Figure 1.4 *A control card is installed onto the systems board to control each piece of equipment attached to the computer. (Courtesy Hercules Computer Tech.)*

Pentium III chips. Processors with faster speeds are constantly being introduced. If you're about to purchase your own computer, remember that faster is better. AutoCAD 2000 requires a minimum of a Pentium 133MHz machine to function.

Memory

Each computer must have memory to load, run, and save the programs it uses. Memory chips are located on the motherboard. Read only memory (ROM) and random access memory (RAM) are the two major types of memory typically specified when referring to the power of a computer.

ROM is the memory built into electronic chips on the motherboard that enables basic commands to be stored. Your disk operating system (Windows 98, 95, or NT) is stored in ROM. Each time your computer is started, or booted, as it is typically called, ROM is checked for specific commands that tell the computer what functions to perform with the needed internal components.

RAM is memory that is used as a program is being run. It provides temporary storage, meaning that any information that has not been saved is lost each time the computer is turned off. RAM should be thought of as the computer's work space. If you were a manual drafter, it would be your desktop. The bigger the desktop, the more stuff you can look at. The more complicated the task that needs to be done, the more RAM space will be required. RAM is measured in kilobytes or megabytes. Typical computers have 640K of RAM (640,000 bytes) as base memory.

The terms *extended memory* and *cache* are often used when referring to RAM memory. Extended memory enables you to run programs requiring you to use large amounts of data, which could not otherwise be run efficiently in conventional memory. Computers generally have between 16 to 128MB of RAM. The more RAM, the more programs that can be opened on your desktop at once. AutoCAD requires 32MB of RAM. Cache memory is used by your computer to achieve the highest possible data transfer rate between the main memory and the CPU. Cache memory holds just those items in memory that are for imminent use. This allows the computer to function faster than when it works in main memory. Cache memory is a very high-speed, small storage area in the CPU. Main memory is a much larger area, but information stored there takes longer to access because of the area's size. When a byte of information is requested, your CPU will check cache memory first. If the information is found there, the next request can be processed. If it's not found in cache memory, the CPU will examine main memory.

STORAGE DEVICES

Before AutoCAD can be used effectively, you'll need an understanding of how information is stored. Future chapters will deal with the process of saving drawings. This chapter will deal with where drawing files are stored. Common choices for storage include the fixed disk drive, diskettes, CDs, Zip disks, Jaz cartridges, and tape storage. Most personal computers are purchased with hard disk and floppy disk drive units already installed. Although the size and number of drives will vary, their basic functions remain the same.

Fixed Disks

Depending on your computer, you may have one or more fixed disks and one or more diskette drives. Fixed disks, which are sometimes referred to as hard drives, are mounted inside the computer. There is no physical opening in the computer case to allow access to the hard drive. This drive is accessed by electronic signals processed by the other computer components. AutoCAD is installed on your fixed disk. Fixed disk drives of 3 to 10GB have become standard equipment on most PCs. As you complete your first few drawing assignments, you'll notice that they are very small, and will require only a few kilobytes of space. You may not see the need for space measured in gigabytes now, but as with any good structure you'll be working on, the more storage space the better. Multimedia projects, audio or video files, and the proliferation of information available through the Web make large storage capacity a must for most users.

Diskettes

Diskettes or floppy disks are read in a diskette drive. Most PCs are equipped with one or more diskette drives. Figure 1.5 shows a typical arrangement of drives. Notice that the diskette drive has an access slot for inserting and retrieving diskettes, a light to indicate that the drive is in use, and a button for removing the diskette from the drive. Diskettes have a storage capacity of 1.44MB. Diskettes similar to the disk shown in

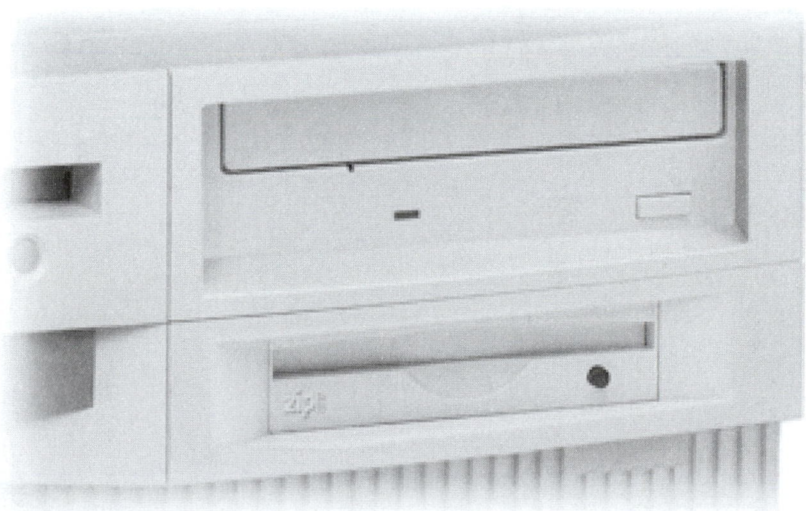

Figure 1.5 *The access slots for a diskette drive, Zip drive, and a CD ROM.*

Figure 1.6 should be used as a secondary storage device or for transferring information between computers.

Alternative Drives

Most computers come equipped with internal drives that read CDs. Drives for writing information to CDs are also available. External drives that can be moved between individual work stations are also available for reading CDs. AutoCAD and many of the programs that you will load on your computer are placed on CDs.

Zip, Jaz, or tape drives can be installed on a computer to provide virtually unlimited storage capacity. A Zip drive can be externally mounted to a computer, but is quickly becoming standard equipment on many PCs and laptops. A Zip disk looks similar to a diskette, but has storage capacities of 100 or 250MB, providing portable flexibility for saving, storing, or sharing files. The big brother of a Zip disk is a Jaz cartridge that is ideal for storing 3D drawings or other graphic files. With storage capacity of 2GB and rapid storage speeds, Jaz cartridges can also be used to provide backup for network applications. Tape systems range from 2 to 10GB capacity and provide a portable backup system for important files, operating systems, program applications, or an entire hard drive.

The final storage device to be considered in this chapter is much more than a storage method. Most schools and professional users link their computers using a local area network (LAN). A network is a combination of hardware and software that allows information to be shared between computers, and links multiple computers to a printer or plotter. A network allows multiple computers to be linked by cable so that the computers can communicate with each other. Each computer is linked to a central computer, referred to as the network server. The server is similar to a PC, but has

Figure 1.6 *The storage capacity of a 3 ½" high-density disk is 1.44 MB. (Courtesy International Business Machines Corporation.)*

a large-capacity storage space to allow each user to store on the server. Other advantages of a network will be considered later. A key benefit to be examined now is the private, secure, short-term storage areas that can be provided to each user. If you are a network user, your instructor or network administrator must provide you with a user name or account number before you can access or store information on a network. Once you have access, you will be given a folder that is located on the hard drive of the network server. Network servers should not be used for long-term storage, but they do provide an excellent storage location for active drawing projects. Many design firms place active drawing files in folders that can be accessed by anyone on the network, ensuring that each member of the design team is working on the most current file without having to pass a diskette between team members. AutoCAD has features that allow a base drawing such as a floor plan to be accessed from the network, and essentially lock the file so that no one else will access the file while it is being updated. Networks can also be configured to allow access via the Internet to consulting firms. A designer working for the structural engineer can be allowed access to the floor plan created by the architectural firm, ensuring the transfer of the most up-to-date drawing files between offices.

Your school or office administrator will determine where you will save your drawing files. Most users save their projects at about 15-minute intervals on the hard drive. At the end of a drawing session, the file is saved to the hard drive. Because of limited space, some schools do not allow storage on the hard drive. Check with your instructor before you create a folder on the hard drive. If you have access to a network, back up your project at the end of each drawing session by saving it in your folder on the network and to a floppy disk. Diskettes and fixed drives are often used for the day-to-day storage of drawing files because of their access speed. Once the drawing session or project is completed, it can be stored on the network, on a Zip or Jaz disk, or on a tape cartridge where it is less subject to damage.

Drive Identification

The various drives within your computer are designated by a letter followed by a colon. Typically, C is used to designate the fixed drive. The letter A is used to designate the diskette drive. If you have a second diskette drive, it would be accessed by the letter B. A workstation equipped with a diskette drive, CD ROM, Zip drive, and a hard drive could include the following drive letters:

A: diskette drive (3 ½")

B: diskette drive (optional)

C: hard drive (fixed disk)

D: CD ROM

E: Zip drive

F: network drive

MONITORS

The monitor is a display device that allows users to see the results of their input into the software program. Depending on the manufacturer, a color monitor with AutoCAD can display between 256 to 16 million colors on a white or black background. Monitors are specified by screen size and display ability. Most PCs come with 15", 17", 19", or 21" monitors, measured diagonally across the screen. The clarity or resolution of the monitor is influenced by the number of pixels. A pixel is a specific point on the screen that can be lit or remain unlit. Objects are displayed on the screen as pixels are lit. An 800 × 600 pixel display is typical for most 2D AutoCAD applications; a 1024 × 768 setting is typical for many 3D applications. As the number of pixels is increased, the resolution increases. Monitors are referred to as *VGA (video graphics array),* or *SVGA (super video graphics array).* SVGA is the norm for Pentium machines.

Input Devices

The two most common input devices used with AutoCAD are the mouse and the keyboard.

Mouse

A mouse is the pointing device used with Windows applications. Two common types are roller and optical. On a two-button mouse, the left button is used for the select button, and the right button to terminate a command or to select a menu to alter a command. Future chapters will introduce additional functions of each button. The mouse can be used to select commands from menus that are displayed beside, above, or inside the drawing area, or to select starting and ending points while in a command. When a command is to be selected from a menu, the text will refer to selecting a command from a menu. This is the process of going to a menu, placing the cursor on the desired option, and pressing the left button once.

Keyboard

The keyboard of your computer is similar to a typewriter keyboard. Some variation will be found between manufacturers. Across the top of the standard keys is an additional row of keys called function keys. These keys can be used as shortcuts within specific programs in place of typing out an entire command. On the right side of the keyboard you will see directional keys and a 10-key layout for math functions. An alternative to a standard keyboard is shown in Figure 1.7. Keyboards with the keys placed to meet the natural position of the hands and wrist can be a great aid in overcoming fatigue caused from long hours of typing. Even with all of the icons and toolbars provided by Windows and AutoCAD, the keyboard is the fastest and easiest method of executing many commands. This is true even if you're a two-finger typist. Throughout this text, commands will be discussed by providing examples of both icon, toolbar, and keyboard entry. For example, once you are in AutoCAD, the only typing required to start the LINE command would be to type **L** (EnterKey) at the

Figure 1.7 *The use of an ergonomic keyboard, such as the Natural Keyboard by Microsoft, can greatly ease the stress placed on the wrist and fingers. (Courtesy Microsoft Corporation.)*

command prompt. L represents the LINE command, and (EnterKey) represents pressing the enter key. The command sequence for keyboard entry for the LINE command will be displayed on the command line as:

Command: L (EnterKey)

LINE Specify first point:

OUTPUT DEVICES

Two types of output devices are printers and plotters.

Printers

The most common types of printers are laser and inkjet. Each provides a quick method of getting a printout or hardcopy of a drawing or a list of commands. Printers are typically used in an office for check prints. Laser printers can produce a very sharp graphic image allowing the user to set the density of lines to be produced. Laser printers allow for a quick, quiet print to be made, although the size is often limited. Typical sizes of hardcopy from a laser are 8 1/2" × 11" or 8 1/2" × 17", with a resolution of 300 to 600 dots per inch (dpi). A laser printer is shown in Figure 1.8. Inkjet printers offer the highest quality of prints at prices that are often below the cost of laser printers. These printers use color cartridges, and provide resolutions of 300 to 600 dpi. Inkjet printers that produce color prints at inexpensive prices are also available. Many inkjet printers are able to produce drawings using resolutions that match

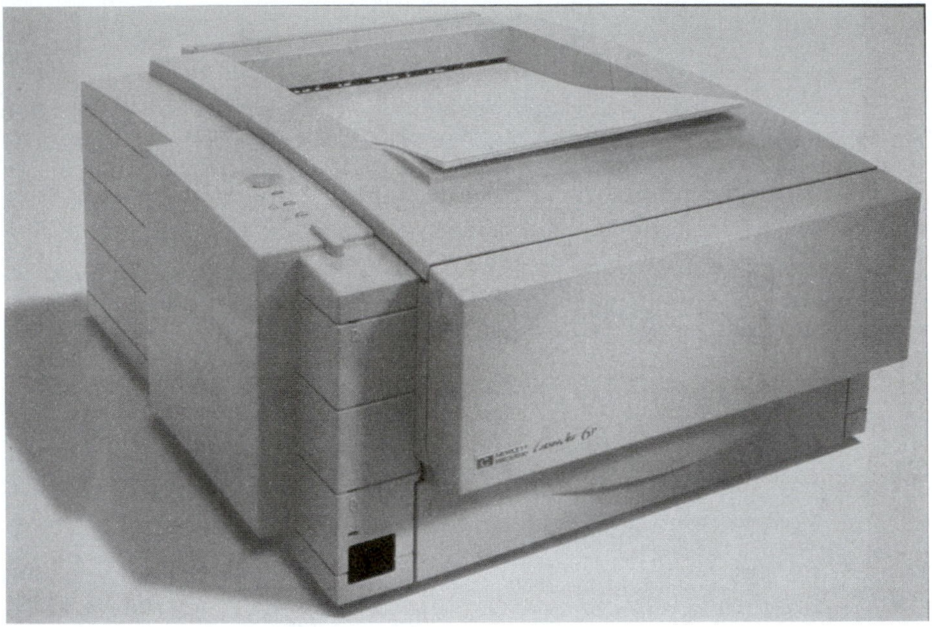

Figure 1.8 *Laser printers such as the HP LaserJet 6P produce a fast, economical method of reproducing check prints. (Courtesty Hewlett Packard Company.)*

photo-quality color images. Portable laptop printers are also available for use with laptop computers. Some notebook printers allow prints to be made by use of infrared signals with no printer cables connected to the computer.

Plotters

Plotters are typically used to get a final print of a completed drawing. As AutoCAD drawings are completed, they typically are reproduced onto vellum and then multiple copies are blueprinted. Many schools use older flatbed or drum pen plotters because of their low cost. Although pen plotters are still supported by AutoCAD, most professionals use inkjet or electrostatic plotters. Each type of plotter allows for black and white or color printing. Plotters use LED technology and a dry toner cartridge to produce plots that match the quality and speeds of laser printers in sizes up to E-size material. These plotters produce prints with a resolution of up to 600 dpi, with a speed that is as much as ten times faster than a pen plotter. Also available are plotters that feature direct image technology to produce drawings up to E-size material with up to 800 dpi resolution. Direct-image plotters require no pens, ink, or toner and are approximately twice as fast as pen plotters. They are often used on a network or to meet other high-volume plotting demands (Figure 1.9). Although not common in schools or small offices because of their expense, electrophotographic plotters are a third type of plotter. These plotters provide fast, quiet, high-resolution, full-color plots.

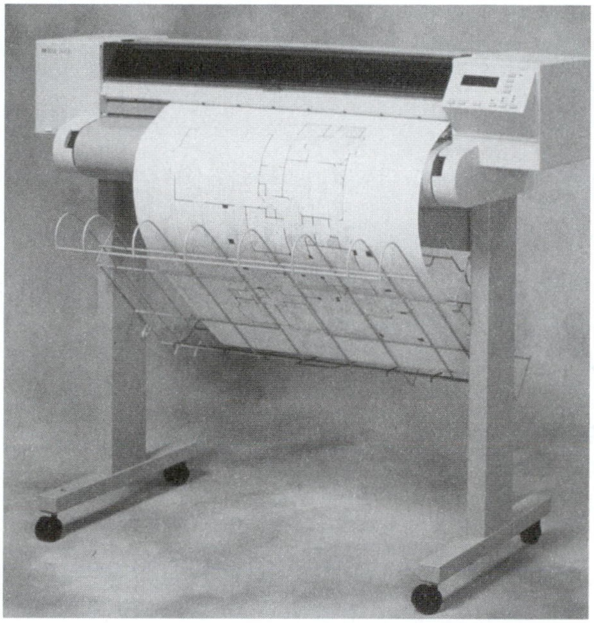

Figure 1.9 *Laser-quality plots can be made in seconds using 800 dpi resolution on plotters similar to the CalComp Solus 4. (Courtesy of CalComp Corporation.)*

ERGONOMIC CONSIDERATIONS

Although becoming a proficient AutoCAD user will offer many exciting opportunities, spending long hours at a workstation also has its pitfalls. You may start to notice problems with your eyes, back and wrist. Fortunately, most physical problems arising from computer use can be minimized or eliminated by proper workstation configuration, good work habits, and good posture.

Equipment

You may have little control over your workstation at school, but if you're setting up your own workstation, it should meet the following requirements:

- Your chair should be adjusted to a height that allows your feet to be supported on an inclined footrest or to rest on the floor. The chair should also provide a tilt adjustment so that your back is firmly supported as you work. It should also have an armrest to provide support for your wrists. The height of the chair should allow your wrist to remain level as you type.
- The monitor should be placed so that the top of the monitor is level with your line of eyesight and between 18" to 30" from your eyes. Lighting should be placed so that it will not produce a glare on the display screen.
- Use a desk with a keyboard drawer that allows your wrist to remain level.
- Use an ergonomic keyboard, a foam wrist support or forearm supports.

Work Habits

You can have a state-of-the-art workstation, but if you don't maintain good work habits, your body will wear out from fatigue. To keep your productivity high and maintain your sanity, you'll need to control your work habits and posture.

- Avoid looking at a monitor for long periods of time. To ease eyestrain, several times an hour look away from the monitor and force your eyes to focus on an object several feet away from your workstation.
- Avoid resting your hand on the mouse when you're not actually using it. When you are using the mouse, avoid resting your hand on the desk. Get in the habit of keeping your wrist level, or adjust the armrest of your chair to provide support.
- Take time to stretch your legs, neck, shoulders, back, and wrist several times each hour. In addition to reducing tension throughout your body, it also provides an excellent opportunity to get your eyes off the monitor.
- Most importantly, take time to walk away from your workstation at least once an hour. It's easy to get caught up with a deadline and convince yourself that you just don't have time for a short break. Short breaks will help keep you productive, and help to eliminate drawing errors.

AN INTRODUCTION TO CONTROLLING AUTOCAD

A workstation is neat stuff, but these gadgets can do nothing without software. Software consists of programs that are loaded into the computer to tell it what to do, such as AutoCAD. The platform coordinates the computer and the software so that each can function. The operating systems that coordinate the software programs in your computer include Microsoft Windows 98, Windows 95, or Windows NT. Although Windows 98 or Windows 95 are used on many PCs, Windows NT is used for most educational and business settings. Windows NT (Network Technology) allows multiple computers in an office to work with each other. One or more CAD operators can access a floor plan stored on the local network server. Windows NT also allows several computers to link to one plotter. It's important to remember that these platforms are just tools used to operate AutoCAD.

If you purchase a PC (personal computer), it will come loaded with Windows 98 as the controlling platform. When a computer is booted up using Windows 98, Windows 95, or Windows NT, a screen display similar to Figure 1.10 will be displayed. Methods of accessing programs will be discussed later in this chapter.

A WINDOWS SURVIVAL GUIDE

Before AutoCAD is explored, the basis of Windows must be understood. No matter which Windows platform you use to run AutoCAD, they each have common

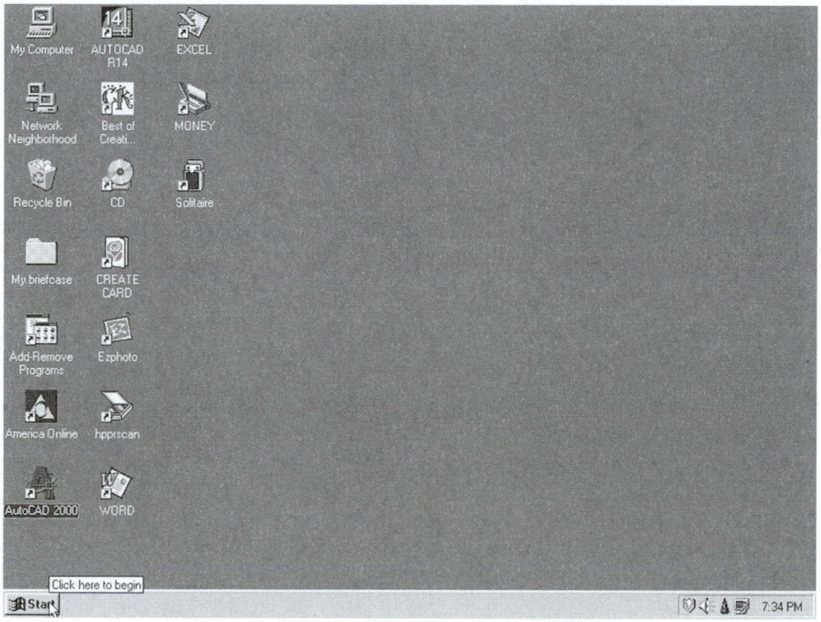

Figure 1.10 *Although the display will vary greatly depending on what programs are loaded on the machine, the program icons are displayed as Windows is entered. Icons are used throughout Windows and AutoCAD to represents specific programs, or components within a program.*

features that must be understood to function in Windows. If you have experience working with a Windows-based program, skip to the next section, which discusses entering AutoCAD. If you're a new computer user, it's important that you understand the basics of Windows—including icons, clicking, double-clicking, and dragging—before you tackle AutoCAD.

Icons

Pictures called icons are displayed to identify what program or command in a program is to be activated. Windows and AutoCAD use an abundance of icons. The icons allow greater productivity and are used in place of written commands. Icons are also used within AutoCAD to identify specific portions of the programs to be executed. Figure 1.11 shows the icons from the standard tool bar of AutoCAD 2000. Chapter 2 will discuss the use of each of the icons. Consult the Help menu of Windows or a related text to explore other options of Windows.

Clicking and Double-Clicking

Selecting an icon requires pressing the left mouse button as you would a touch-tone phone button. A single click of a mouse button is referred to as a *pick*. Menu items are typically selected with a single click. If a menu is displayed with six options, selecting one of the options with a single click will highlight that menu option. Figure 1.12

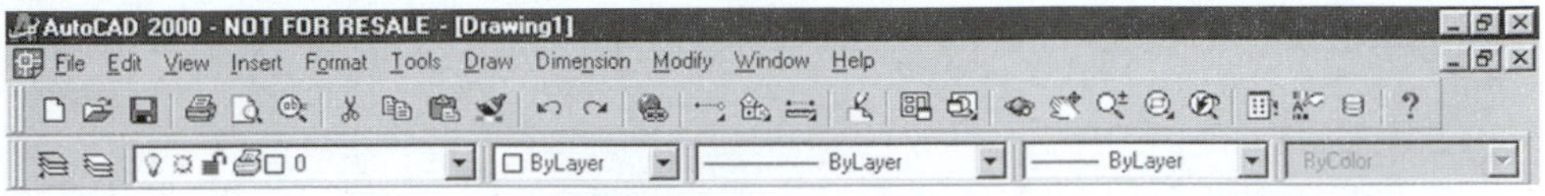

Figure 1.11 *Icons of the AutoCAD Standard toolbar can be used to perform specific functions.*

shows the Draw menu with the Line command highlighted. Clicking once will start the command sequence.

Pressing a mouse button twice in rapid succession is referred to as double-clicking. If a double-click is entered too slowly, it will be interpreted as two single clicks. A double-click is used to select a program, folder, or file for an immediate use. Double-clicking the AutoCAD 2000 icon of the original Windows display shown in Figure 1.10 will remove that display and start the program. If you find yourself unsure about when to click or double-click, start with a single click. If the desired results are not achieved with a single click, use a double-click.

Most commands and options are executed by using the left mouse button to make selections in Windows and AutoCAD. The right mouse button can produce several responses depending on what you're doing. Clicking a blank portion of the screen will

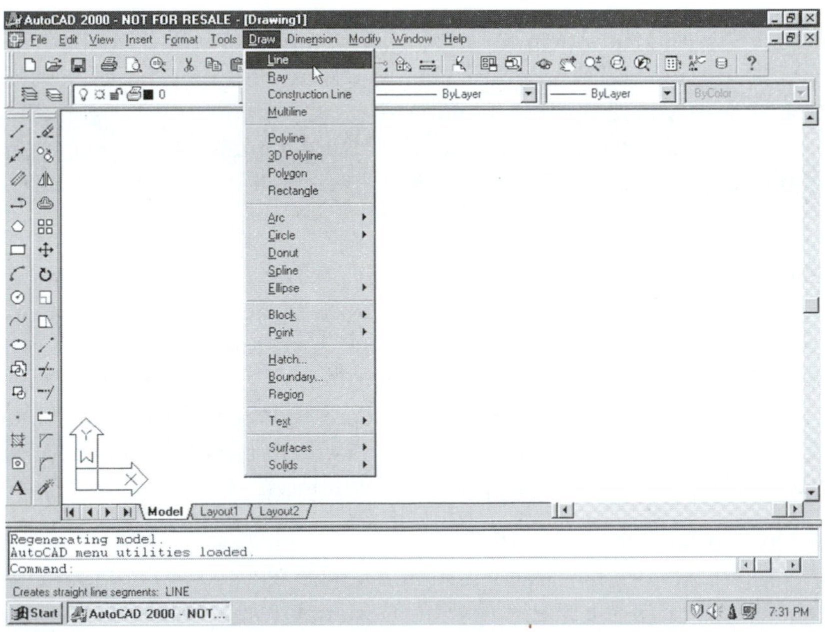

Figure 1.12 *Selecting the Draw listing from the Standard menu displays the Draw pull-down menu. As the cursor is moved to Line command, the command is highlighted for selection.*

typically reveal a hidden menu. The menu will be altered depending if you are in the middle of a command sequence, or if the right button is pressed before entering a command. Variations of the right button menus will be discussed in future chapters as each command is discussed. Clicking with the left button will remove the menu. Other uses of the right mouse button produce a similar result as the left button. The selections will be introduced with the command they affect.

STARTING AUTOCAD

Throughout this chapter you've been introduced to the tools that control AutoCAD. The balance of this chapter will introduce you to opening AutoCAD. Double-clicking the program icon in the Program menu, the Windows Explorer, or the Run menu will open AutoCAD. Each method is accessed by selecting the Start button. Double-clicking the AutoCAD 2000 icon will produce a display similar to Figure 1.13. Chapter 2 will explain each of the options of the AutoCAD Start-up menu.

USING THE PROGRAM MANAGER

Select the Windows Start button to open AutoCAD using the Program Manager. Move the cursor to the Programs icon. As the cursor is moved to the right end of Programs, a listing of each program that is loaded on the workstation will be displayed. Move the cursor to the listing for AutoCAD 2000 and the menu shown in

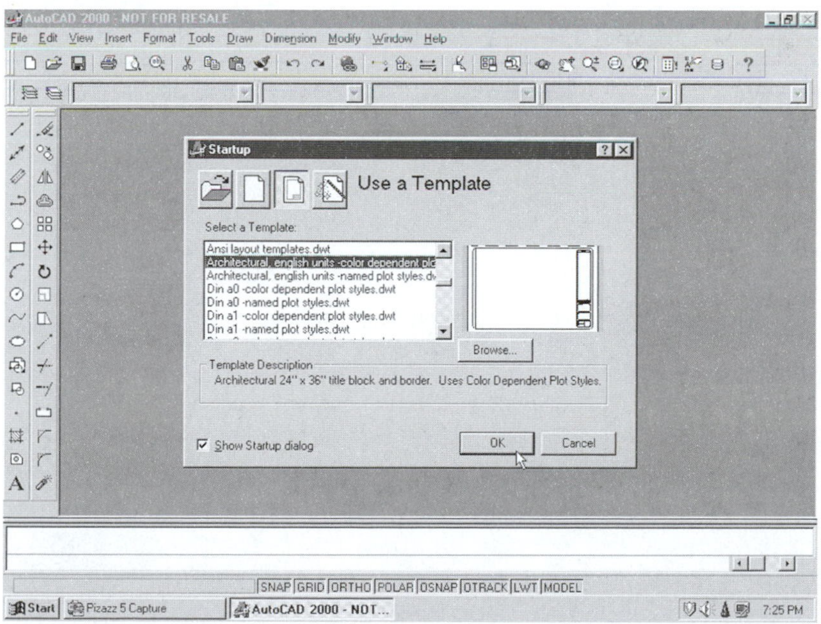

Figure 1.13 *Double-clicking the AutoCAD icon will produce the AutoCAD Startup dialog box. The dialog box is used to define how the drawing session will be started.*

Figure 1.14 will be displayed. Each of these menus can be displayed by moving the mouse, without clicking the select button. With the AutoCAD 2000 icon highlighted, click once with the left mouse button. This will produce the AutoCAD display graphics momentarily, followed by the AutoCAD drawing screen shown in Figure 1.13. If you're sitting at a work station, and have just entered AutoCAD, take a few minutes to explore the icons displayed on the screen. Each screen component will be discussed in Chapter 2, but it rarely hurts to peek. As long as you keep liquids away from your computer, you'll be pretty safe. After you've spent a little time exploring, invest some time in Chapter 2, so you can get a better idea of just how powerful AutoCAD is.

STARTING AUTOCAD IN WINDOWS NT

Starting AutoCAD in Windows NT is very similar to starting the program in Windows 98 or Windows 95. When a computer is booted up using Windows NT, the Microsoft icon is displayed. This icon is quickly replaced with a login display. Your name and password must be entered before you are allowed to proceed. Once you have logged in, double-click the AutoCAD icon. The only difference Windows NT users might encounter is that Windows NT often has a personalized start-up menu provided by the network administrator.

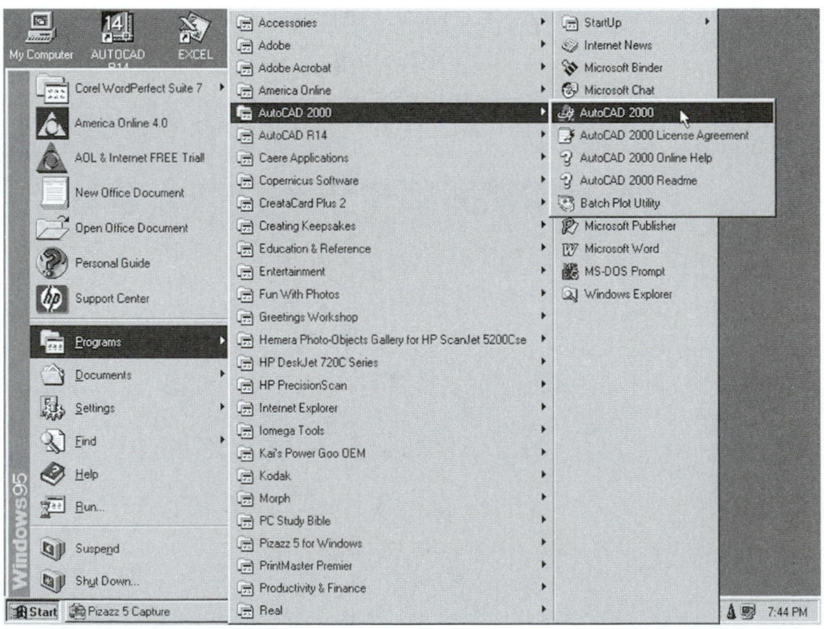

Figure 1.14 *Selecting the Start button will open the Windows StartUp menu. Move the cursor to Programs, and then to the AutoCAD 2000 listing. Click the AutoCAD 2000 listing in the Program menu to display the AutoCAD Startup menu shown in Figure 1.13.*

CHAPTER 1 QUIZ

1. List two types of memory and explain their uses.
2. List four components that comprise a computer.
3. List four advantages of AutoCAD over manual drawing.
4. What units are used to measure memory?
5. What are the components of a well-equipped workstation?
6. What are two tools that can be used to input information into the computer?
7. You've just purchased a new monitor and display card. What does the display card connect to?
8. Use three local suppliers to research the various speeds available for Pentium II computers, as well as their price range.
9. Use three local suppliers to research hard disk size and price range.
10. List two types of disk drives.
11. How is information entered onto the hard drive?
12. What letter is typically used to describe the hard drive?
13. Select the My Computer icon from the startup display. Select the Printer icon and determine which printers or plotters are configured to the current workstation.
14. Compare the cost and features of printers made by at least three different manufacturers that can be used to obtain high-quality colored prints.
15. Verify with a minimum of three local retailers the cost of setting up a work station using a 300MHz Pentium II or faster machine equipped with Windows 98, a CD ROM, multimedia kit, 512K cache memory, and 32MB of memory. Get estimates to include the following equipment:

    64 and 128MB of extended memory.

    Internal Zip drive.

    13", 15", and 17" monitors.

    56Kbps internal fax modem.

    4-gigabyte hard disk drive.

    1.44MB floppy drive
16. Explain the difference of a click and a double-click.
17. What steps would be required to boot your computer and start AutoCAD?
18. Is a click or double-click required to start AutoCAD using the Programs menu?
19. Describe key functions of the left and right mouse buttons.
20. List five menu options of the Windows Explorer menu bar.

CHAPTER 2

Exploring AutoCAD

The drawing display shown in Figure 2.1 is displayed as AutoCAD is entered. The Start-up dialog box aids in setting drawing parameters for each drawing session. Options include:

- Open a Drawing—Allows an existing drawing to be opened. This option will provide a list of the four most recently opened AutoCAD drawings.
- Start from Scratch—This option allows a new drawing to be started with the expectation that you will provide all required measurement and drawing parameters.
- Use a Template—This option will allow you to begin a drawing session using an existing template. A template contains preset measurement and drawing parameters. This option will open an existing drawing template for the creation of a new drawing.
- Use a Wizard—This option will allow a new drawing to be created with drawing parameters set by the use of either the Quick Setup wizard or the Advanced Setup wizard.

Chapter 3 will introduce each of these tools for controlling the parameters of a drawing session. For now, press the (EscKey), select the OK button or the CLOSE button to remove the Start-up dialog box. This will allow access to the default version of AutoCAD displaying a drawing area of approximately 12" × 9" with measuring units set to four-place decimal inches and produces the display shown in Figure 2.2. AutoCAD will provide either a light gray or black drawing screen, depending on how your machine has been configured. The exact size of the drawing area will also vary, depending on how the program is configured. Later chapters will explore how these settings can be set to match your preferences. The drawing area contains several important tools for controlling AutoCAD. To use AutoCAD effectively, an understanding of each of the methods of interacting with the software must be understood. This includes a knowledge of the drawing area displays, toolbars, types of menus available, common components of a dialog box, function keys, and control keys.

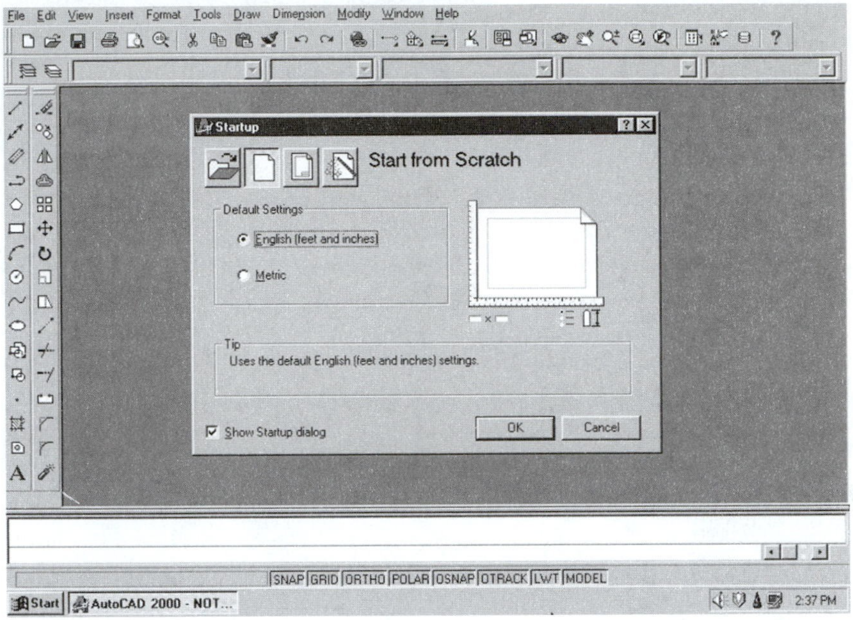

Figure 2.1 *The Start-up menu is provided as AutoCAD is entered to help establish drawing parameters.*

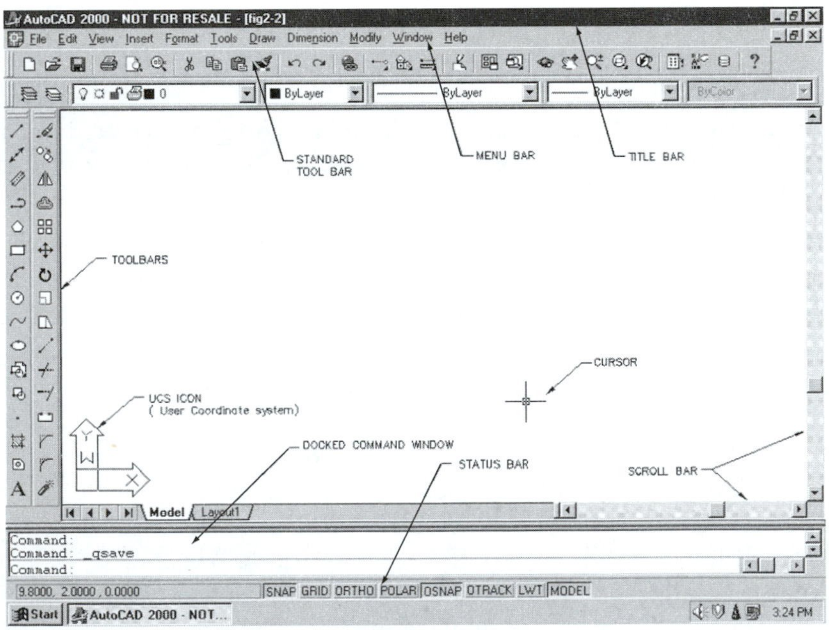

Figure 2.2 *Key components of the AutoCAD drawing screen.*

Don't let all of the different menus, boxes, and buttons confuse you. Each will become quite easy to use if you are willing to explore the program. Push a few keys, explore some options. As long as you keep liquid away from the keyboard, there is very little you can do to hurt the hardware, software, or yourself.

DRAWING AREA TOOLS AND DISPLAYS

AutoCAD offers the user many tools and displays to input information into the program, and to determine the status of the current drawing file. These tools include the drawing area, title bar, menu bar, command line, status bar, scroll bars, user coordinate icon, and cursors.

DRAWING AREA

Central to each of the AutoCAD tools is the drawing area. The drawing area is where the material that you input is displayed. The size of the drawing area will vary depending on how many toolbars or dialog boxes are displayed. In the default setting of AutoCAD, it is the white area, surrounded by gray areas. Methods of altering the colors of each portion of the display will be introduced at the end of this chapter.

Multiple Layouts

One of the features that made Microsoft Windows so popular is that multiple programs can be opened at the same time. With older versions of AutoCAD, if you wanted to view multiple drawings, you opened multiple sessions of AutoCAD. One of the many new features of AutoCAD 2000 is the ability to open multiple drawing projects in one AutoCAD session. Figure 2.3 shows an example of a drawing session with two drawings displayed. AutoCAD refers to the side-by-side arrangement as vertical tiles. As with other windows, the arrangement and size of the drawing window can be altered. The size of a window is changed by using the minimize/maximize buttons or by activating the window, and then dragging the edge to the desired size. Selecting Window from the Menu bar will provide options for altering the arrangement of each window. Figure 2.4 shows the drawings arranged in horizontal and cascading windows. The arrangement of windows can be altered at any point during the drawing session. Future chapters explore the process for opening and working with multiple files.

Model and Paper Space Layouts

New to the drawing area of AutoCAD 2000 are the Model and Layout tabs. Drawings are created in Model space at full scale. If walls are to be 6" wide, the lines that represent the walls are drawn 6" apart. If you're drawing a building that is 300' long, the drawing area can be adjusted so that the entire structure can be seen. Figure 2.5 shows a floor plan displayed in the drawing area. Paper space is used to plot drawings that have been reduced to a scale such as ¼"=1'-0". With 6"-wide walls, the lines will be ⅛" apart. Paper space allows a 300'-long building to fit on a 24" × 36" sheet of

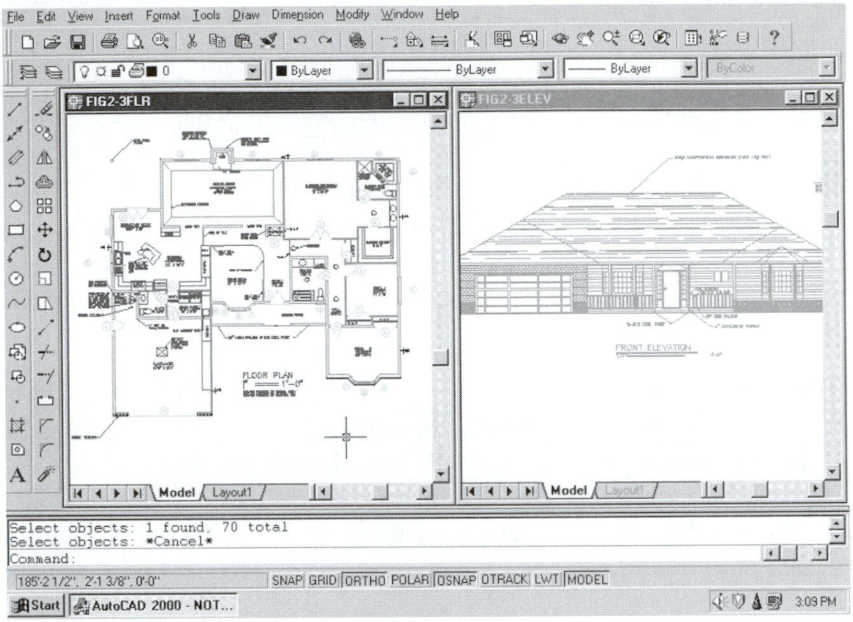

Figure 2.3 *AutoCAD 2000 allows multiple drawings to be opened. The name of each drawing is listed in the title bar of each window. The active drawing (Floor) is displayed in a blue title bar, and the inactive drawing (Elevation) has a gray title bar.*

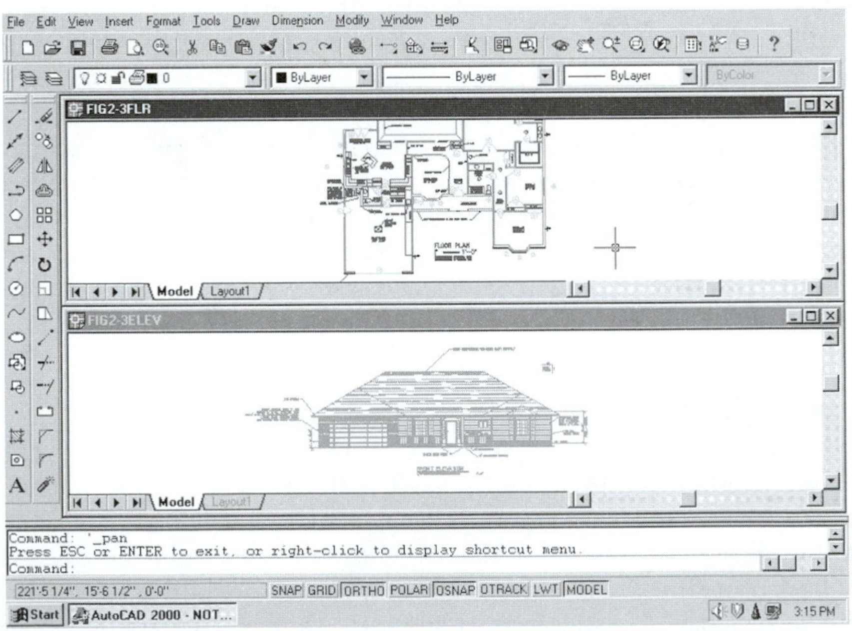

Figure 2.4a *AutoCAD drawings can be arranged in horizontal (Figure 2.4a) or cascading window arrangements (Figure 2.4b).*

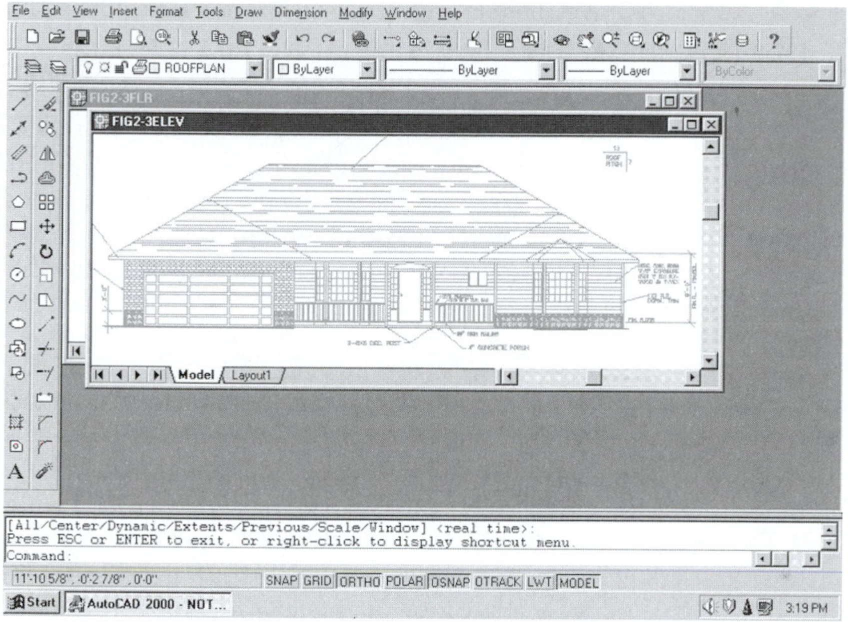

Figure 2.4b

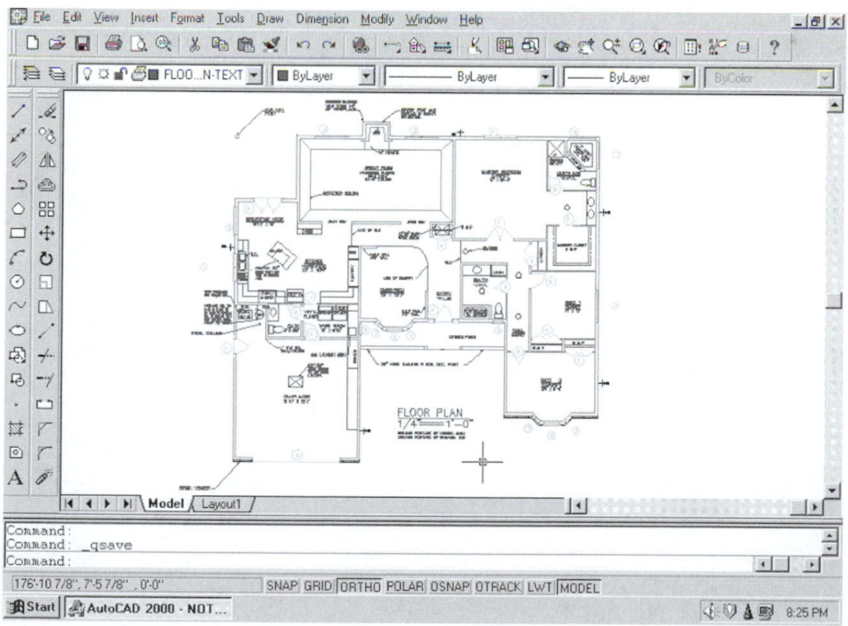

Figure 2.5 *Drawings are created using model space at full scale. Enlarging the drawing area will show the entire floor plan.*

paper. Figure 2.6 shows the plan from Figure 2.5 in paper space. This is how the drawing will appear when plotted.

AutoCAD allows both model and paper space views to be displayed. New to AutoCAD 2000 is the ability to have multiple paper space layouts in a single drawing. By combining multiple sheets into a single drawing, the number of individual drawings needed to produce and easily generate different views of a project can be consolidated. Multiple layouts can be used to aid in the layout of an object, and in the plotting of an object. Uses for model and paper space layouts will be further discussed through the balance of this text.

TITLE BAR

A title bar is located across the top of every window. It displays the icon and name of the current program as well as the name of the current file. The title bar serves as a handle to drag the window to a different location on the screen. Notice that in Figure 2.3, two drawings are opened and displayed in two windows within the AutoCAD 2000 program. In addition to the AutoCAD title bar at the top of the drawing screen, each drawing window has its own title bar. If multiple drawings are open, the title bar of the active window is highlighted with a blue bar, while inactive windows are displayed in another color. The name of each drawing session is listed in the title bar. Notice in Figure 2.3 that the active drawing is named Floor and the inactive drawing is named Elevation. At the far right end of each title bar are three

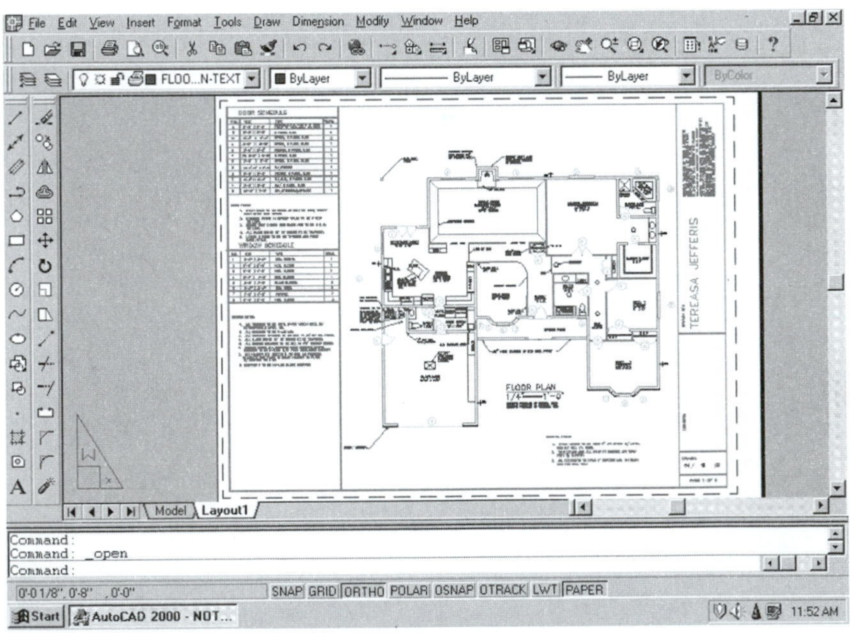

Figure 2.6 *The floor plan from Figure 2.5 is shown in paper space. Paper space scales a drawing to fit within the limits of a sheet of paper.*

buttons that can be used to control the size of the current window. These buttons are the minimize, maximize, restore, and close file buttons. Each can be seen in Figure 2.7. The minimize and close buttons will always be displayed, but only the maximize or the restore button is displayed depending on the current window setting. You'll notice that sometimes there are two rows of buttons. Although they function in a similar manner, the buttons at the end of the title bar control the display of the current program window, and the buttons at the end of the menu bar control the display of the current file.

The Minimize Button

The minimize button is the horizontal bar icon on the left side of the three buttons (See Figure 2.7a). Selecting the minimize button with the mouse pick button will remove the current program or drawing from the display screen and place the program icon in the taskbar at the bottom of the drawing area. The taskbar will be discussed later in this chapter. It's important to remember that if you select the minimize button and remove the file from the screen, the file still exists, it's just been placed out of your view. It hasn't been saved, or destroyed, just moved for convenience so you can do something else. To restore the window, use the mouse and a single click to select the program icon in the task bar.

The Maximize Button

The maximize button is the file icon in the center of the three buttons. (See Figure 2.7b.) Selecting a window corner or border of a window with the cursor allows the size of the window display to be altered. When the size of a window has been reduced, selecting the maximize button will restore the current window to its largest size. Other windows that may have been open are still open, they've just been covered by the current window. Notice that as soon as the window is maximized, the maximize icon is changed to the restore icon.

The Restore Button

The restore button shown in Figure 2.7c is displayed only when a window is maximized. Selecting the restore button allows the window to be reduced to its previous size.

Figure 2.7 *Four buttons for controlling windows are located on the right end of the title bar. Selecting the minimize button (a) will remove a window or drawing from the screen and place the title in the task bar. Selecting the maximize button (b) will increase the window to its largest possible size. The restore button (c) is only displayed when a window is maximized. Selecting the restore button will return the window to its previous size. Selecting the close button (d) will close a window and remove the program icon from the task bar.*

The Close Button

Selecting the button with the X on the far right side of the program title bar (Figure 2.7d) will close the current program. When the close button is selected while in a program such as AutoCAD, a box will be displayed with options to save or not to save the current file, or cancel the close command.

THE MENU BAR

The menu bar is located below the title bar and is used to display menus specific to the open program. The menu bar for AutoCAD can be seen in Figure 2.2. Each word of the menu bar represents a separate menu with options related to the key word. Moving the cursor to one of these words and clicking will display the corresponding pull-down menu. Figure 2.8 shows the Draw pull-down menu for AutoCAD. Some commands such as Arc have a triangle on the right side. The triangle indicates selecting this command will produce another menu of options related to that command. The options don't have to be selected with the mouse pick button. Moving the mouse to the Arc listing will produce the menu shown in Figure 2.9. Select the Format menu with a left-click of the mouse. Notice that the Plot Style listing appears different from other listings. This option is inactive and is not currently available for selection. Inactive options only become active as other options are activated.

THE COMMAND LINE

The command line displays the prompt "Command:" and allows you to enter your request into AutoCAD using the keyboard. Command entry methods will be dis-

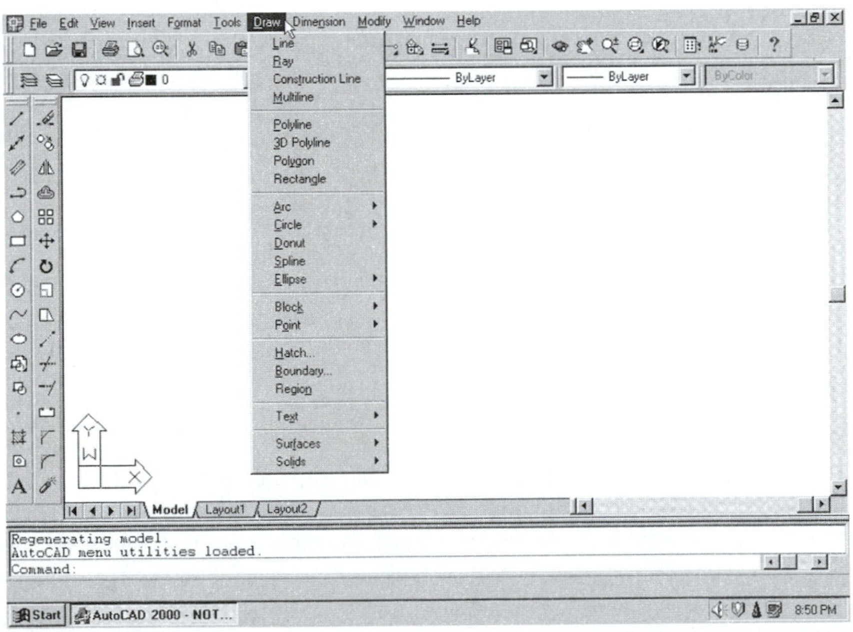

Figure 2.8 *The Draw pull-down menu.*

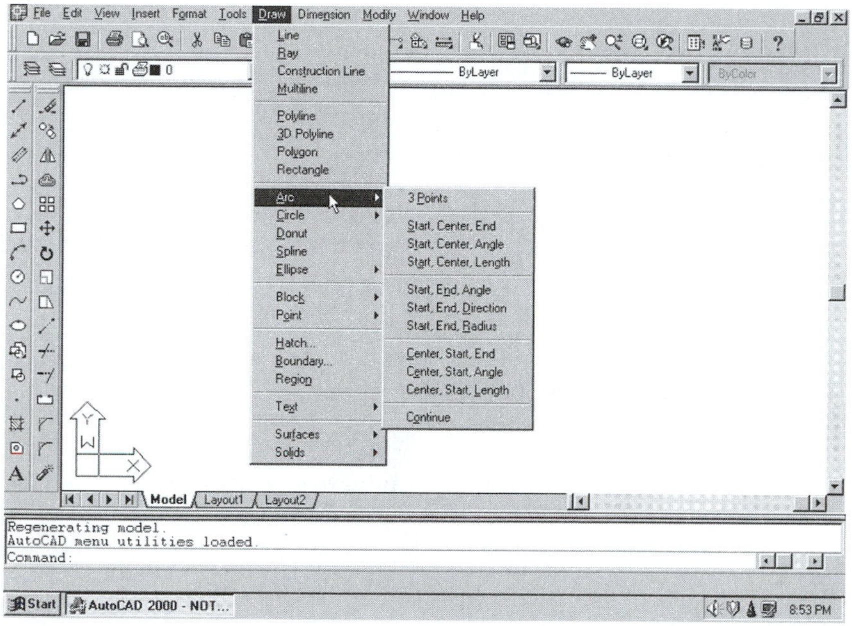

Figure 2.9 *Listings in a menu followed by a triangle will produce a flyout menu. The Arc flyout menu is displayed by highlighting Arc.*

cussed later in this chapter. As you explore AutoCAD, it's important to remember that the command line displays prompts for completing commands. Even if you enter commands by using a toolbar or pull-down menu, AutoCAD will display prompts and a command history in the command window. Users new to AutoCAD often fail to read the command prompt and miss AutoCAD's attempt to communicate with them.

STATUS BAR

The Status bar is located below the command line. Key elements of the status bar can be seen in Figure 2.10 and include the Coordinate display and toggle buttons that can be used to control a drawing aid. The drawing aids in the status bar will effect placement or display of your drawings. Options are in the ON position when the button appears to be pressed in. Buttons include:

- Snap—Toggles an invisible grid ON/OFF for placing lines at precise intervals. Snap increments can be set using the Snap and Grid tab of the Drawing Settings Dialog box. The box is displayed by selecting Drafting Settings... from the Tools pull-down menu.
- Grid—Toggles a visible grid ON/OFF for placing modular objects. Grid increments can be set using the Snap and Grid tab of the Drawing Settings Dialog box.
- Ortho—Toggles the ability to draw lines that are vertical and horizontal (Ortho ON) or at any angle (Ortho OFF).

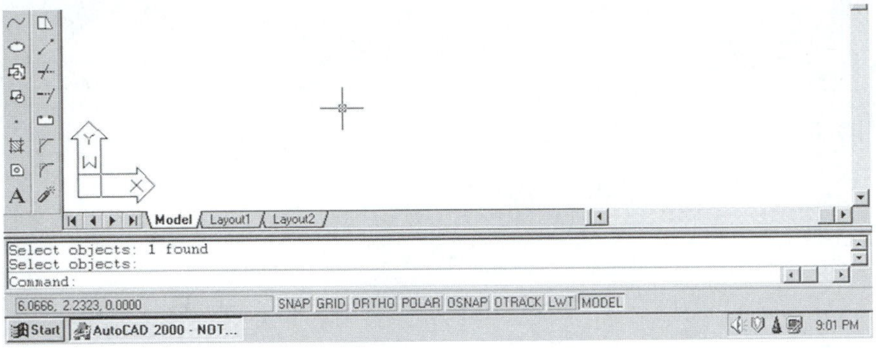

Figure 2.10 *The status bar contains the coordinate display and eight buttons for controlling drawing aids.*

- Polar—Toggles the polar tracking feature that will describe the position of the cursor in distance and angle relative to another point. By default, the tracking angle is set to a 90°-increment angle. The angle can be altered using the Polar Tracking tab of the Drawing Settings Dialog box.
- Osnap—Toggles on the ability to select drawing objects such as endpoints, intersections, or eleven other options. AutoCAD automatically snaps to the endpoints of existing lines as the cursor is moved toward those lines. Other modes can be set by selecting the desired icon from the Object Snap toolbar, or by selecting the desired mode from the Drafting Settings dialog box. Modes can also be selected by holding down the (ShiftKey) and the right mouse button at the same time.
- Otrack—Toggles the tracking feature that will allow specific angles to be tracked along a projected path based on an object endpoint to increase drafting precision.
- Lwt—Toggles the display of lineweight. When ON, assigned lineweights are displayed. When OFF, all lines are shown using the same lineweight.
- Model—Toggles between Model and Paper space.

Each button will be explained in detail in future chapters.

SCROLL BARS

The scroll bars consist of a vertical bar located on the right side of the drawing area and a horizontal bar located below the drawing area. The vertical bar consists of an up arrow at the top of the bar, a down arrow at the bottom of the bar, and a slide box that changes its position relative to the location to the start or end of the display. Scroll bars are used to alter the material viewed in the drawing area. They are also used in dialog boxes that are displayed throughout AutoCAD to alter program settings. For the drawing area, the scroll bars are one of many ways to alter the display in the drawing area. There are better, faster ways than scroll bars to alter the displayed view. In a dialog box, scroll bars offer a quick method of moving through a long list of options. Dialog boxes will be discussed later in this chapter.

THE USER COORDINATE ICON

The User Coordinate System Icon describes the current coordinate system and orientation of the drawing area. AutoCAD drawings are imposed over an invisible coordinate system based on X, Y, and Z coordinates. AutoCAD uses both a fixed world coordinate system (WCS) and a movable user coordinate system (UCS). Each method will be explained in future chapters. In its standard setting, as you draw in AutoCAD, you're working in model space. Model space can best be thought of as drawing at full scale. If a 300'-long structure is to be drawn, instead of working at a scale so the structure will fit on the screen, the area displayed on the screen is altered so that the entire building can be seen. When a drawing is to be plotted, it is plotted using paper space. Paper space is space created so that one or several drawings at one or more scales can be plotted on one sheet of paper. Model and paper space will be discussed in detail in future chapters. Figure 2.11 shows the model and paper space icons. As you explore AutoCAD, the important thing to remember is that the icon is a reminder of what mode you are currently using to create your drawing. Unless your instructor tells you otherwise, work in model space as you create a drawing, and work in paper space to plot a drawing.

CURSORS

The term *cursor* refers to the pointing device that is placed on the screen to indicate the current mouse position. As seen in Figure 2.2, the cursor for the AutoCAD drawing area is crosshairs. As the mouse is moved, the location of the crosshairs is moved. The cursor provides a method of entering the location for drawing objects. When commands such as LINE are started, the cursor is the primary method of selecting the beginning and ending points for a line.

Altering the Crosshairs size

Depending on the complexity of the drawing being completed, some operators like to adjust the size of the cursor. Crosshairs that extend across the screen will aid in pro-

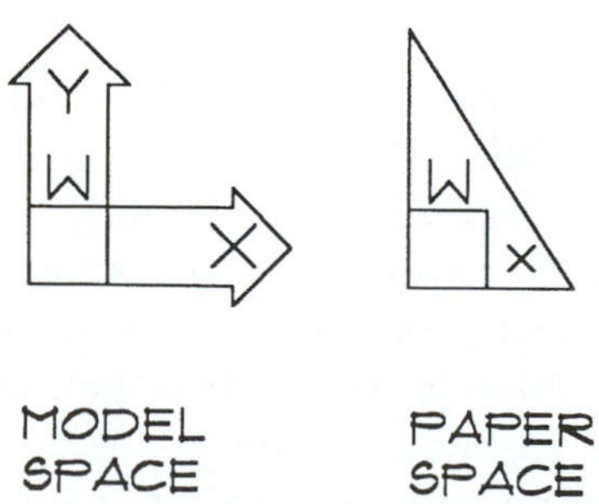

Figure 2.11 *The user coordinate system (UCS) icons for model and paper space.*

jecting objects from one view to another. Projecting heights from one elevation to another is much easier with enlarged crosshairs. The size of the cursor can be altered using either of the following methods:

- Using a dialog box:

  Select Options... from the Tools pull-down menu. This will display the Options dialog box.

  Select the Display tab. Use the slider bar in the Crosshairs size edit box to alter the cursor size.

  When adjusted, select the OK button to return to the drawing area.

- From the command line:

  Type: **CURSORSIZE** (EnterKey) at the command prompt. This will produce the prompt:

  Command: **CURSORSIZE** (EnterKey)

  Enter new value for CURSORSIZE <5>:

No matter how the command is entered, the default value is displayed as 5. The 5 represents the percentage of the screen the cursor will occupy. Select and hold the slider bar with the left mouse button and move it to the right to increase the size of the cursor. If you're working at the command line, enter the desired value followed by the (EnterKey). Changing the value to 10 will double the size of the crosshairs. Entering a value of 2.5 will decrease the size of the cursor to half of the original size. Entering a value of 100 will cause the lines forming the crosshairs to cross the entire screen.

Alternate Cursors

Just as with other Windows programs, the cursor changes shapes depending on where it is located in AutoCAD. When the cursor moves outside of the drawing area, the crosshairs turn into an arrow. It may also appear as an hourglass, pointing finger, flashing vertical line, double arrow, or four-sided arrow. Each cursor can be seen in Figure 2.12.

The Hourglass

The cursor changes to an hourglass (Figure 2.12b) each time AutoCAD or Windows is processing information. While the hourglass is present, the current program is unable to process additional information. For example, if you save a file, the hourglass will be displayed, indicating no new information can be processed. Once the command is executed, the hourglass reverts to the previous cursor, and new commands can be processed. The hourglass may also be displayed beside the arrow cursor. The arrow means you can still keep working but the hourglass is telling you that Windows is performing a task while you're doing another. Your current program may run slightly slower, but full function will be restored when Windows is done.

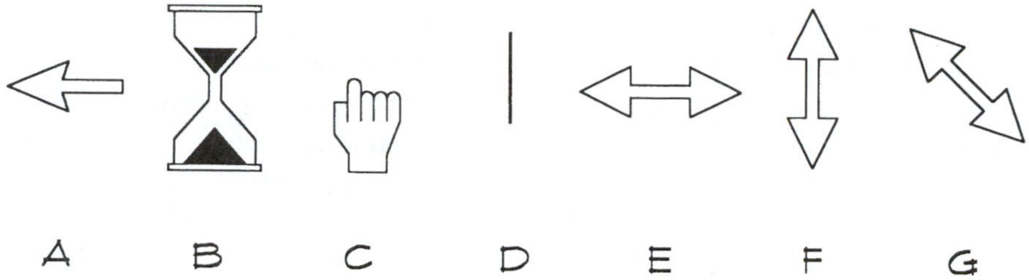

Figure 2.12 *The cursors of a Windows platform include (a) the arrow—used to make selections; (b) the hourglass— indicating a command is being processed; (c) the pointing finger—for making a selection; (d) the flashing bar—for entering text; (e) the horizontal double arrow—for altering a window size in the left/right direction; (f) the vertical double arrow—for altering a window in the top/bottom direction; and (g) the diagonal double arrow—for altering a window in two directions in one movement.*

Pointing Finger

The pointing finger (Figure 2.12c) is the politically correct symbol used to indicate that there is additional information available for a specified command or option. Clicking the mouse will display additional information related to the designated topic.

Flashing Bar

The arrow cursor (Figure 2.12d) changes to a flashing bar cursor when the cursor is moved into a portion of the screen where text input is required. Any text that is entered at the keyboard will be entered just to the left of the cursor bar.

Double Arrow

As the cursor is placed on the border of a window, it will change to a double arrow (Figure 2.12e, f, g). When placed on a side border of a window, the cursor will be a horizontal double arrow. When placed on the top or bottom border, the cursor will be a vertical arrow. If placed in a corner, it will be a diagonal double arrow. If the select button of the mouse is selected and held, the size of the window can be enlarged or decreased depending on the movement of the mouse. Using the arrow to move a side border will alter the horizontal size. Selecting the bottom border will alter the vertical size. If the pointer is placed in the corner of a window and the select button is held, the size of the window can be altered in both directions in one movement. The size of a window can be altered throughout a drawing session.

COMMAND SELECTION METHODS

AutoCAD contains more than 250 commands related to drawing or editing a drawing. Four options are available to enter commands, including selecting icons from toolbars, pull-down menus, keyboard entry, and dialog boxes. Consideration will also be given to using Function and Control keys to alter drawing settings or alter commands. Each method has its advantages. Throughout the balance of the text, commands will be presented in the order that tends to be the fastest method of entry. For most commands the fastest entry method is by selecting an icon from a toolbar or by keyboard entry. Remember that the key word is fastest, not best. As you're learning each command, take the time to explore each option. You may decide the keyboard is the fastest method of command entry. As you begin to work with AutoCAD menus, you will see common terms used repeatedly. You will need to understand these so you can progress through AutoCAD. Common terms include:

- Command—An instruction to be carried out by the computer, such as LINE or ARC. Commands will be written in capital letters throughout this text.
- Default—A value that will remain constant until a new value is entered. Default values are shown on the command line in brackets: < 3 >.
- Option—A portion of a command that requires a selection.
- Pick—To use a pointing device to make a selection.
- Enter—(EnterKey) is used to denote pressing the appropriate key (key names vary, depending on the keyboard manufacturer) after inputting a request at the command line.
- Select—Making a choice from a menu of commands or options.
- Specify—To enter a value, such as a coordinate on the command line.

TOOLBARS

A toolbar contains icons that represent commands. In its default display, the Standard and Object Properties toolbars are displayed above the drawing area just below the menu bar, and the Modify and Draw toolbars are displayed on the left side of the drawing screen. Key features of these four toolbars include:

- The Standard toolbar contains the names of frequently used buttons. In addition to common methods for controlling AutoCAD, the toolbar contains standard Microsoft buttons such as Open, Save, Print, Cut, Copy, and Paste.
- The Object Properties toolbar can be used to set the properties of objects such as color, lineweight, linetype, and the layer an object is displayed on.
- The Draw toolbar contains icons to represent each of the major drawing commands including Line, Polygon, Rectangle, Arc, Circle, Ellipse, and Point.
- The Modify toolbar contains commands that can be used to edit drawing objects such as Erase, Copy, Mirror, Offset, Array, Move, and Rotate.

Because of their ease of use, and availability, for most AutoCAD users the toolbars will become the primary method of selecting commands. AutoCAD offers 24 different toolbars that can be displayed to meet various drawing needs. Toolbars can be displayed by selecting Toolbar from the View pull-down menu, or by typing

TOOLBAR (EnterKey) at the command line. This will produce the menu shown in Figure 2.13. A similar toolbar menu can be displayed by moving the cursor to any open toolbar and right-clicking. Selecting Dimensioning will produce the Dimension toolbar shown in Figure 2.14. Each of the toolbars listed in Figure 2.13 will be discussed as specific commands are introduced throughout the text. As you get started using AutoCAD, it is more important that you feel comfortable using the toolbars that are currently displayed. To remove the Toolbars dialog box, click the close button in the upper right hand corner. The Dimension toolbar can also be removed from the screen by clicking the close button in the toolbox title bar.

As the cursor is moved from the drawing screen to a toolbar, it changes from crosshairs to the arrow cursor. The cursor can now be used to select the desired icon. You'll also notice as the cursor passes over, or rests on an icon in the toolbar, that the button takes on a 3D appearance. Move the cursor to the icon in the left corner beside the drawing window. As the cursor is placed above the icon, the title of the icon will be displayed below the cursor, similar to the display in Figure 2.15. The title is referred to as a tooltip. Tooltips provide a written display of the results that will occur if the icon is selected.

In this case, the cursor is over the Line icon. Select the Line icon with a single-click to start the LINE command. This will display the following prompt at the command line:

Command:__Specify first point:

The program is now waiting for you to select a starting point for a line. Move the cursor and select an endpoint for a line by pressing the select button of the mouse. As the endpoint is selected, the command prompt is changed to request the next point of the line. The prompt will display:

Specify next point or [undo]:

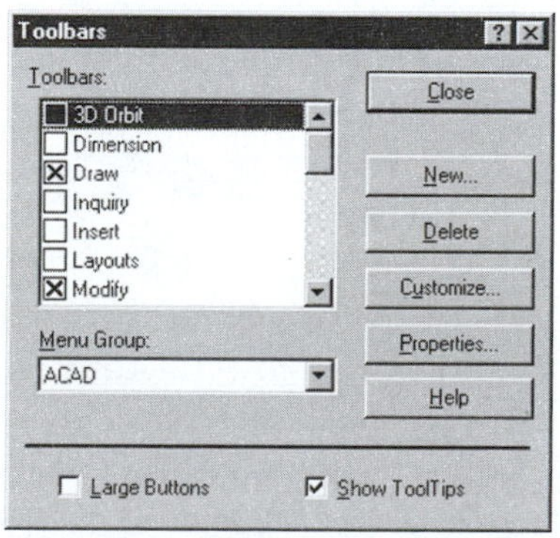

Figure 2.13 *Additional toolbars can be displayed by selecting Toolbars from the View pull-down menu.*

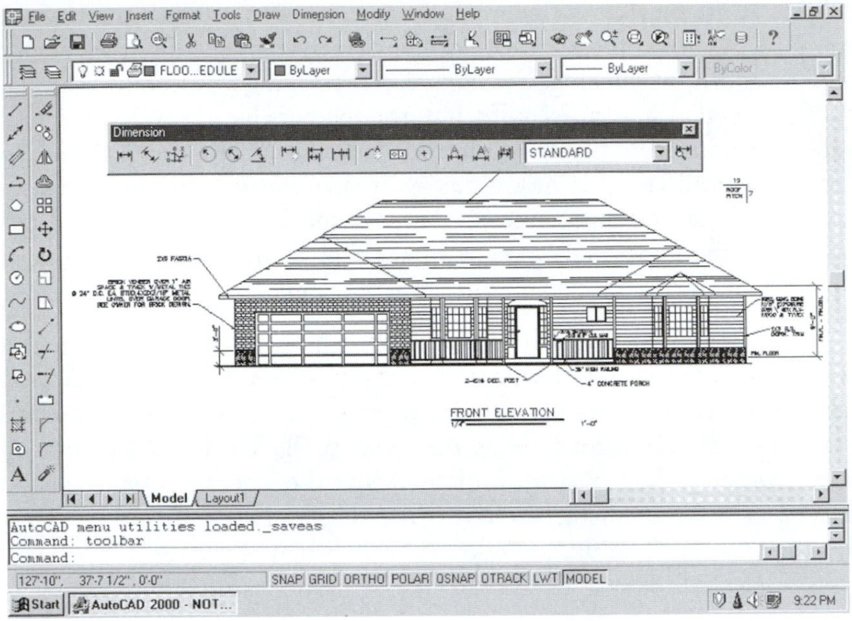

Figure 2.14 *The Dimension toolbar is displayed by making the selection started in Figure 2.13.*

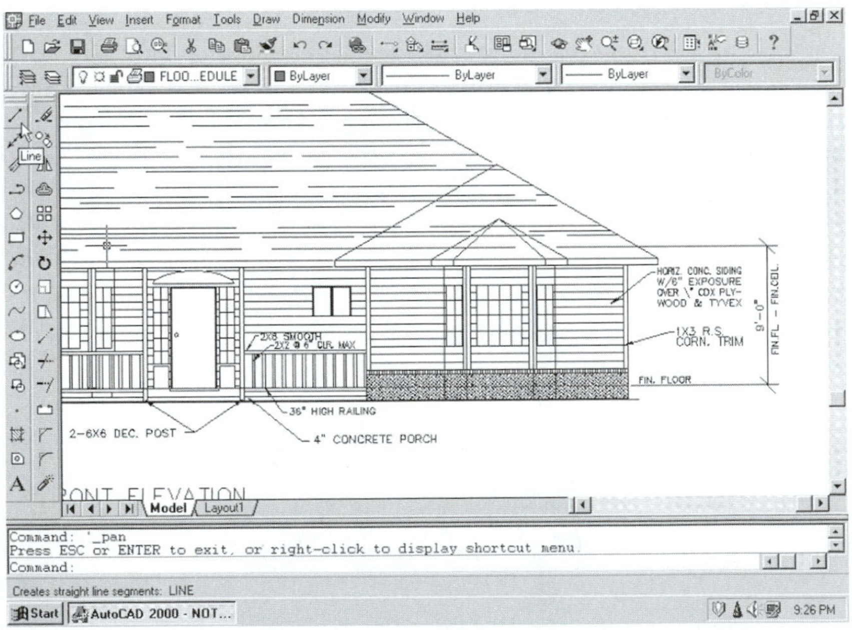

Figure 2.15 *Resting the cursor over an icon provides a written description of the icon's function called a Tooltip. Notice that below the command line a brief description is also displayed.*

AutoCAD is waiting for you to select an endpoint for a line. Move the cursor to a new location and select a point. The prompt will continue to request additional points after each new point is entered until you choose to stop drawing lines. As you draw one line, the command will continue to allow you to add to the existing line. To terminate the LINE command, press the space bar, the (EnterKey), or the right mouse key. Pressing the right mouse key will produce the menu shown in Figure 2.16. Selecting the Enter or Cancel option will end the LINE command. The other options of this menu will be explored in Chapter 4 as the LINE command is explored.

To start another line in a different location, press the (EnterKey). Pressing the (EnterKey) will automatically restart the last command sequence. Repeating this process will allow an unlimited number of line segments to be drawn without having to select the Line icon for each line. Take time to enter AutoCAD and draw several different continuous line segments. Once you've drawn several lines, end the command, and then reenter the LINE command. Now draw several individual line segments. This process can be used with each of the icons in the Draw toolbar.

Although the command will be discussed in detail in future chapters, take time to explore several of the other commands in the Draw toolbar. As you're exploring, you

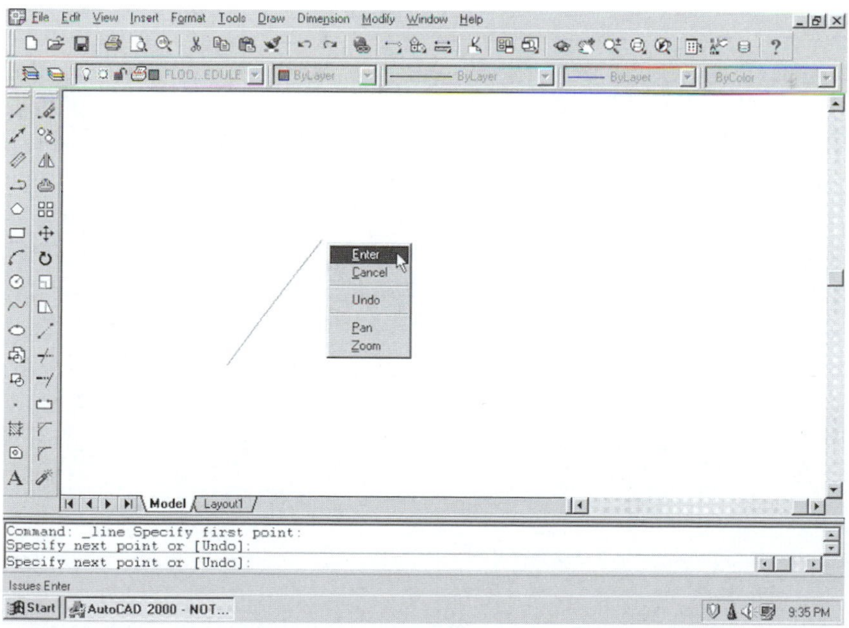

Figure 2.16 *Right-clicking while in the middle of the LINE command will produce a command-mode menu. Selecting Enter or Cancel will terminate the LINE command at the last point that was specified.*

might also want to investigate the Erase command in the Modify toolbar. Selecting the Erase icon will allow you to select lines and remove them from the drawing screen.

As you're exploring, notice that several of the icons in the Standard toolbar have a small triangle in the corner of the icon. Clicking and holding an icon with a small black triangle in the lower right corner will produce a flyout menu. A flyout menu is a submenu of the selected menu. Figure 2.17 shows the flyout menu for the Temporary Tracking Point icon.

Altering Toolbars

Toolbars are used most often to enter commands because they are extremely flexible. In addition to being able to display multiple toolbars on the screen, you can move, dock, float, resize, and alter the contents of toolbars to allow for more convenient usage.

Moving Toolbars

A toolbar can be moved to any convenient area of the drawing screen. Placing the cursor in the title bar of the menu and then selecting and holding the title bar moves the toolbar. While pressing the select button and moving the mouse, the location of the toolbar is altered.

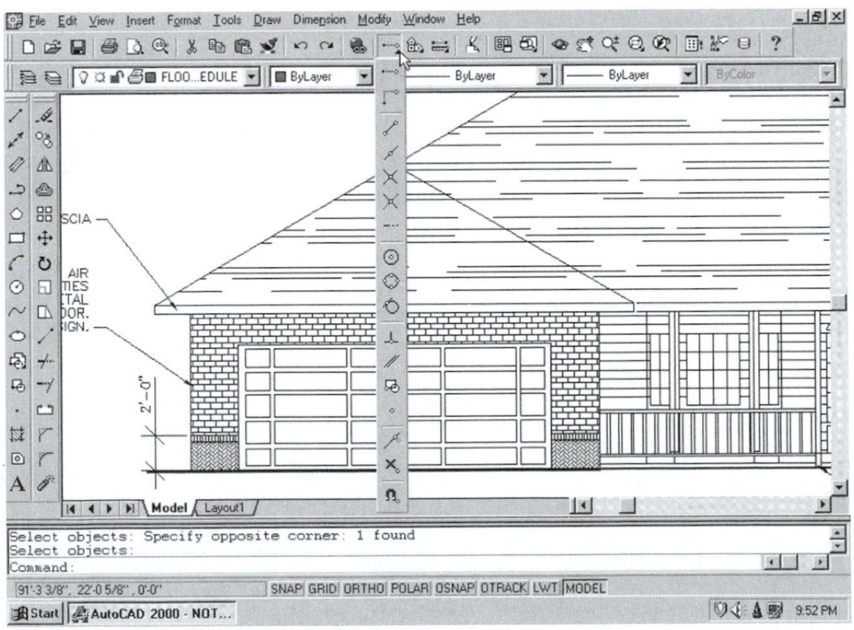

Figure 2.17 *Clicking and holding an icon that contains a triangle will produce a flyout menu.*

Floating Toolbars

A toolbar is referred to as a floating toolbar once it is moved from the border of the drawing display. Once a toolbar is floating, it can be resized, docked, or the contents altered. When the Dimension toolbar is activated (see Figure 2.14), it is placed in the middle of the drawing screen, making it a floating tool bar. Selecting the title bar and holding the select button allows the toolbar to be moved to any desired location. Move the cursor to the top border of the Draw toolbar. Press the select button and move the toolbar into the drawing area. (See Figure 2.18.)

Resizing Toolbars

If the toolbar interferes with the drawing display, the size of a toolbar can be altered. Placing the cursor anywhere on the border of the toolbar and dragging it in the direction you want the bar to be resized alters the shape of a toolbar. Figure 2.19 shows examples of the resized Draw toolbar.

Docking Toolbars

A toolbar is docked by selecting and holding the title bar and then moving the toolbar to a position above, below, or on either side of the drawing screen. Figure 2.18 shows the Draw toolbar floating. Figure 2.20 shows the Draw toolbar docked on the right side of the drawing area. An alternative to docking a toolbar is to totally remove it. A toolbar can be removed from the screen while it is floating by selecting the close but-

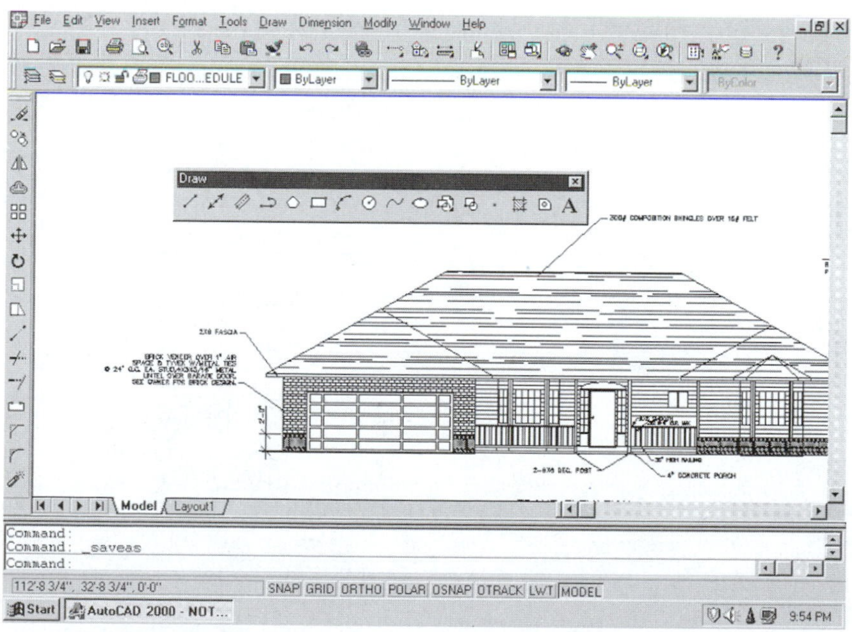

Figure 2.18 *Toolbars can be moved to any convenient location by selecting and holding the title bar and then dragging the toolbar to the desired location.*

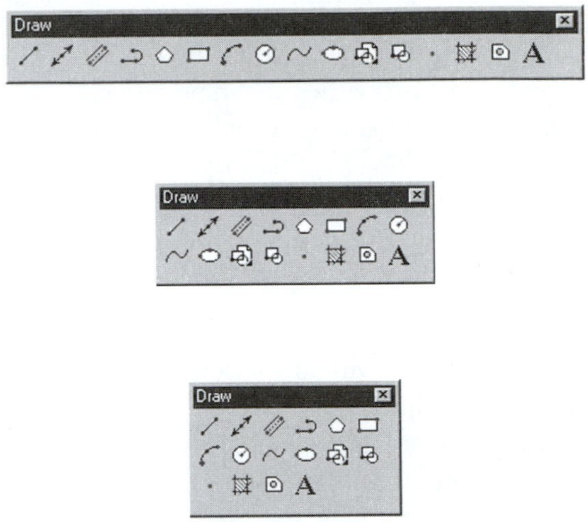

Figure 2.19 *The shape of floating toolbars can be adjusted to please the user.*

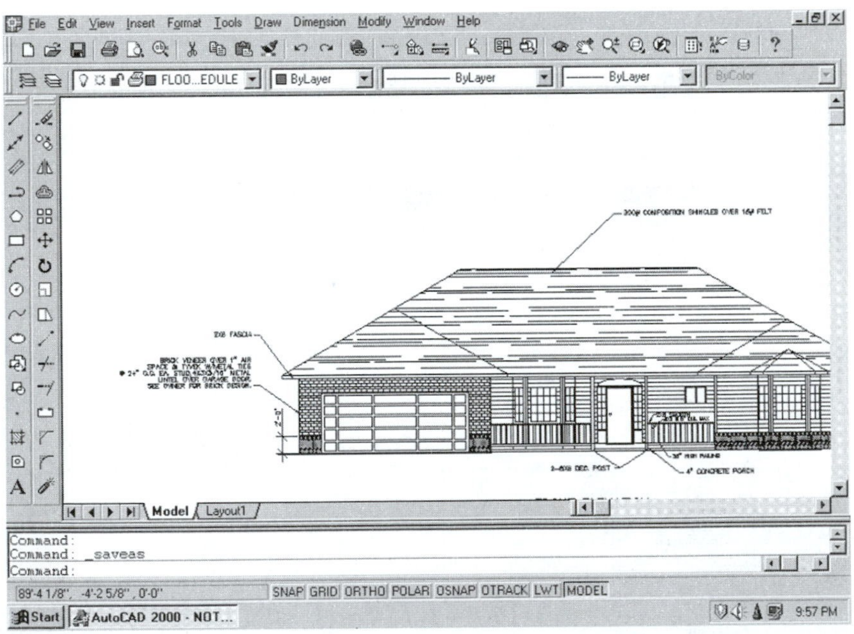

Figure 2.20 *Floating toolbars can be docked above, below, or on either side of the drawing area. The toolbar is docked by selecting and holding the Title bar, and then dragging the toolbar to the desired edge.*

ton in the upper right corner. To restore the window, select the check box of the desired toolbar from Toolbars in the View pull-down menu.

Refloating Docked Toolbars

A docked toolbar can be refloated by placing the cursor anywhere on the border of the toolbar and then dragging and dropping the box in its desired position on the drawing screen.

Altering Toolbar Content

It may seem unimportant now, but one of the nice features of AutoCAD is that it allows you to modify existing toolbars or to create a new one with your favorite icons in it. You can change the size or location of toolbar buttons. You can also add or delete buttons from a toolbar. Use the following method to modify a toolbar:

- Select Toolbar from the View pull-down menu, and then activate the desired toolbar.
- While the Toolbar menu is still displayed, select Customize. This will display the Customize menu.
- Select the desired menu that you would like displayed in the image box.
- Drag buttons right or left in the toolbar by selecting the desired icon and using the drag and drop feature of Windows.
- Add buttons to a toolbar by choosing the desired icon from the image box and then drag and drop the button in the desired location in the toolbar.
- Remove buttons from a toolbar by choosing the desired icon and then drag and drop the button outside of the toolbar.

PULL-DOWN MENUS

As the cursor is moved into the menu bar, a key method of interacting with AutoCAD is presented. The menu bar displays the names of eleven menus used to control AutoCAD. As with other window menus, selecting a specific name will produce a display of that pull-down menu.

Selecting a pull-down menu displays a complete listing of related commands. Move the cursor to the Draw menu and select Draw to display the menu shown in Figure 2.21. If you accidentally select a wrong menu, move to the desired menu. Once a menu has been opened, other menus will automatically open and close as the mouse is moved.

Once in the Draw menu, you'll notice that three different types of commands are listed. These are typical commands where the command is selected and executed, commands with a fly-out menu of related commands or options, and commands that produce a dialog box. Selecting one of the first eight options removes the menu and returns the crosshairs to the screen so that the selected command can be executed. Select Line from the menu. This will remove the menu from the screen and start the

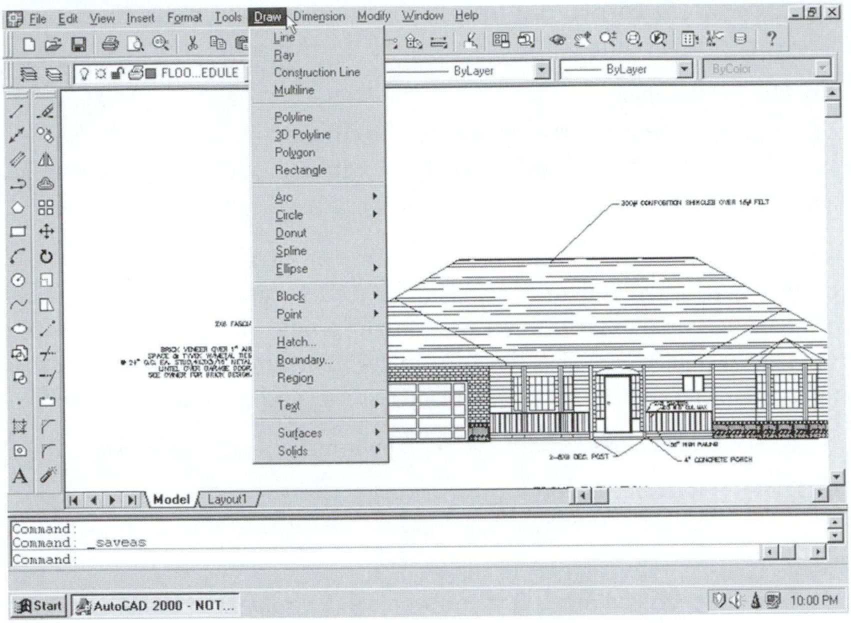

Figure 2.21 *Moving the cursor to the Menu bar and selecting Draw will display the Draw pull-down menu.*

LINE command. Notice the command line has now changed, and "_line Specify first point:" is displayed. AutoCAD now waits for you to select a point with the mouse. Draw a few lines. You'll notice that the command works the same as when you entered the command using the Line icon from the Draw toolbar.

Exit the LINE command by pressing the (EnterKey) key and return to the Draw pull-down menu. Select Arc from the Draw menu to display the menu shown in Figure 2.22. Select the 3Points option. This will remove the menu, return the crosshairs cursor, and allow the selected command option to be executed.

Once the Arc option is selected, the command sequence to draw a three-point arc is:

> Command:_arc Specify start point of arc or [CEnter]: *(Select a start point for the arc with the mouse.)*
>
> Specify second point of arc or [CEnter/ENd]: *(Select a mid point for the arc.)*
>
> Specify end point of arc: *(Select the end point of the arc.)*

As the third point of the arc is selected, the arc will be displayed in the drawing area and the command prompt awaits your next command. Open the Draw pull-down menu.

A third type of display is found in the Draw pull-down menu. Notice that two of the options are followed by three periods. Selecting an option such as Hatch... with a

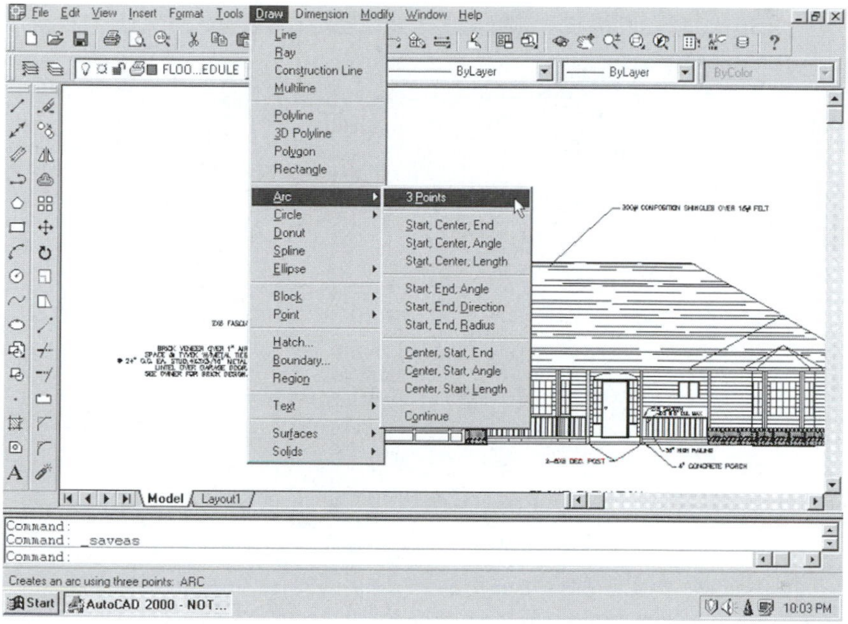

Figure 2.22 *The submenu box produced by selecting the Arc option.*

single click will produce the dialog box shown in Figure 2.23. Using dialog boxes will be introduced later in this chapter and explained throughout the balance of the text as specific commands are introduced. To exit the Boundary Hatch dialog box, move the arrow to the Cancel button, and press the pick button to return to the drawing screen.

KEYBOARD ENTRY

The command line allows you to enter your request into the AutoCAD using the keyboard. After typing the desired command, press (EnterKey) to activate the command. AutoCAD will provide prompts on the command line to complete the command. New AutoCAD users often fail to read the command prompt and miss AutoCAD's attempt to communicate with them. To enter a command by keyboard, type the name of the command followed by pressing (EnterKey). To draw a line, type the word **LINE** (EnterKey) or the command alias of **L** (EnterKey). The text will automatically be entered at the command prompt. Once the command is entered, pressing (EnterKey) will activate the specified command. Notice that once (EnterKey) is pressed, the command line display will be altered. The command sequence is:

Command: **L** (EnterKey)

LINE Specify first point:

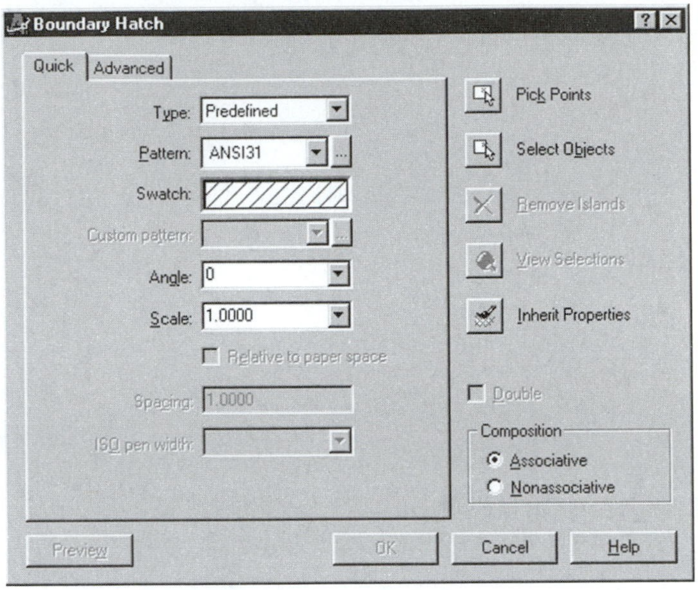

Figure 2.23 *Menu options followed by three periods will produce a dialog box. Dialog boxes allow the functions performed by certain commands to be adjusted.*

Use the cursor to select a starting point for the line just as you did when the command was started from the toolbar or pull-down menu.

Note: AutoCAD is *not* case sensitive. You can use all caps, all lower case or a combination.

Expanding the Command Line

As the program is entered, the command window will display three lines of text. Many AutoCAD users prefer to have more lines of text displayed so that an entire command sequence can be seen at one time. Additional lines of command text can be viewed by selecting the top edge of the window and dragging it to a new location. Selecting Options... from the Tools pull-down menu can also alter the size of the command window. This will produce the Options dialog box that is displayed in Figure 2.24. Select the Display tab at the top of the box. This will produce six groupings for controlling the screen display. (See Figure 2.25.)

Notice that the box in the upper left corner is titled Window Elements and allows the size of the command line to be altered. Other options for controlling the display will be discussed in future chapters. To change the command line from three to four lines use the following sequence:

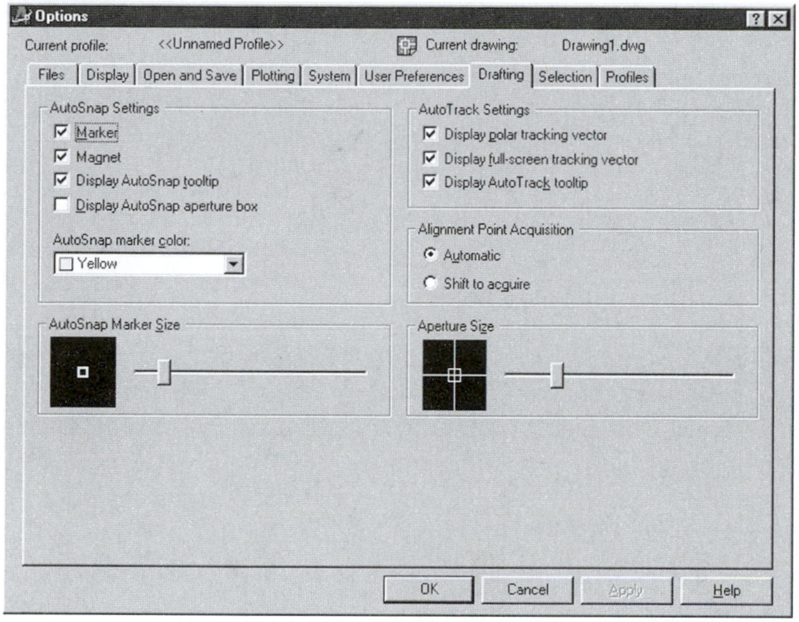

Figure 2.24 *Selecting Options… from the Tools pull-down menu will display the Options dialog box.*

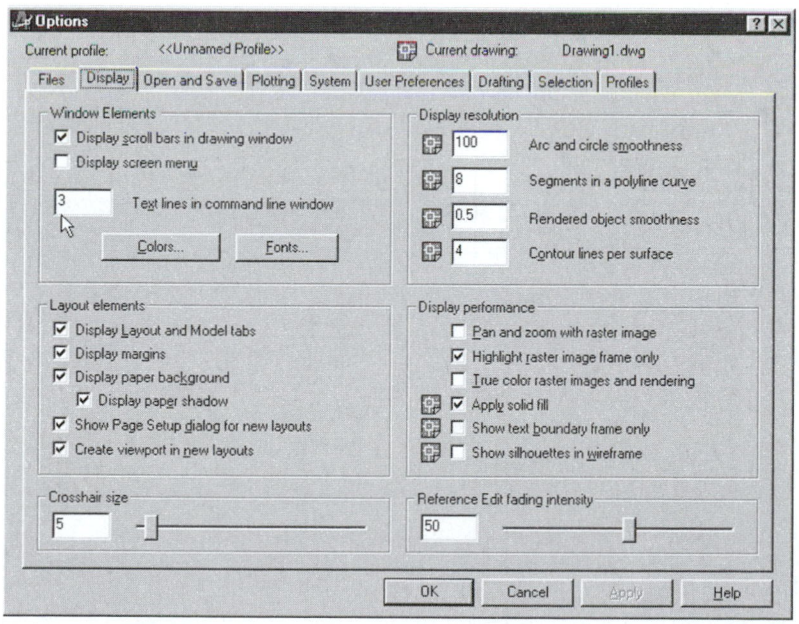

Figure 2.25 *Selecting the Display tab of the Options dialog box will allow the size of the command window to be adjusted.*

- Move the cursor to the text box for Text lines in the command line window.
- Highlight the 3 by placing the cursor beside the 3, and then moving the cursor over the 3 while pressing the pick button. This will place the 3 in a blue box, indicating that the 3 can be altered.
- Enter the desired number using the keyboard, in this case 4.
- Select the OK button at the bottom of the dialog box. This will close the Option dialog box and return the drawing area. The command line will now display four lines of text.

As the size of the window is increased, a scroll bar is added to the window to allow viewing of previous commands. Up to 100 lines of text may be displayed, but a display of over 35 lines will totally fill the drawing area. The scroll bar on the right edge of the display can be used to show additional lines of text. In addition to displaying information about the current command sequence, AutoCAD can display a list of commands used throughout the drawing. This can be done by using the (F2 Key) function key. The function keys are located across the top of the keyboard. The use of each function key will be discussed later in this chapter. Pressing the F2 key will produce a display of the AutoCAD Text window that contains the drawing history. (See Figure 2.26.) The window can be used to display all of the commands used to create the current drawing file. The display can be removed by pressing the F2 key again. Like other windows, the drawing history display can be moved to a new location or resized.

Note: You've just used Options from the Tools pull-down menu to change the size of the command window. One of the most useful tools for AutoCAD users who share a workstation is to be able to have the settings that control the work environment saved

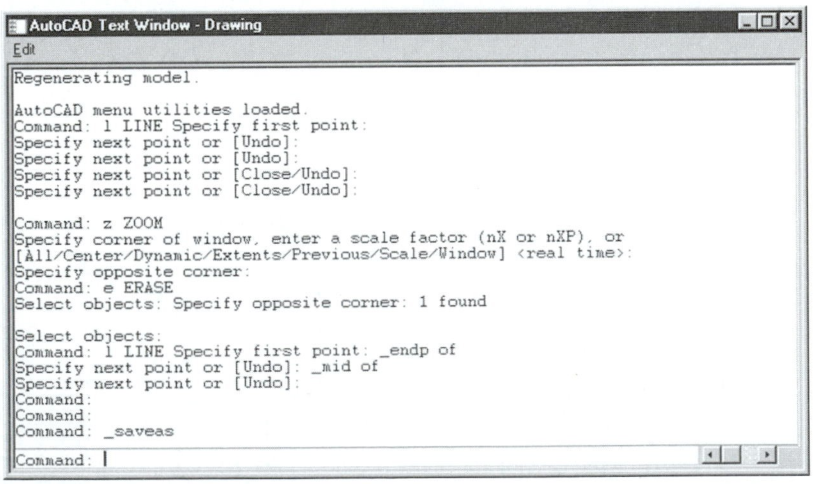

Figure 2.26 *Pressing the F2 function key of the keyboard will display the drawing history. Pressing the F2 key a second time will remove the text box.*

for future work sessions. The Options dialog box can be used to set and save user profiles for future drawing sessions. Although these options will not be introduced until Chapter 5, it is important to remember your favorite settings can be saved as you're exploring AutoCAD's options in this chapter. If you can't wait, see the Help menu.

DIALOG BOXES

Although not a menu of commands, dialog boxes provide a convenient way to adjust certain command parameters. You encountered your first dialog box when AutoCAD opened and the Startup dialog box filled the screen. You've also opened the Toolbar and Options dialog boxes. A dialog box provides options for controlling commands, options, or portions of the drawing program. Figure 2.27 is a display of the dialog box for controlling LAYERS. This dialog box can be accessed by selecting the Layer icon in the Object Properties toolbar, or by selecting the Layer… option from the Format pull-down menu. Dialog boxes can also be opened from the command line by typing a command that begins with the letters DD (dynamic dialog). The Layer Properties Manager dialog box is opened by typing **DDLMODES** (EnterKey).

Typing **DDLMODES** (EnterKey) (dynamic dialog layer modes) will display the dialog box shown in Figure 2.27. Layers will be discussed in future chapters . Don't worry about what a layer is or "DDwhat?" for now. Each will be explained as the

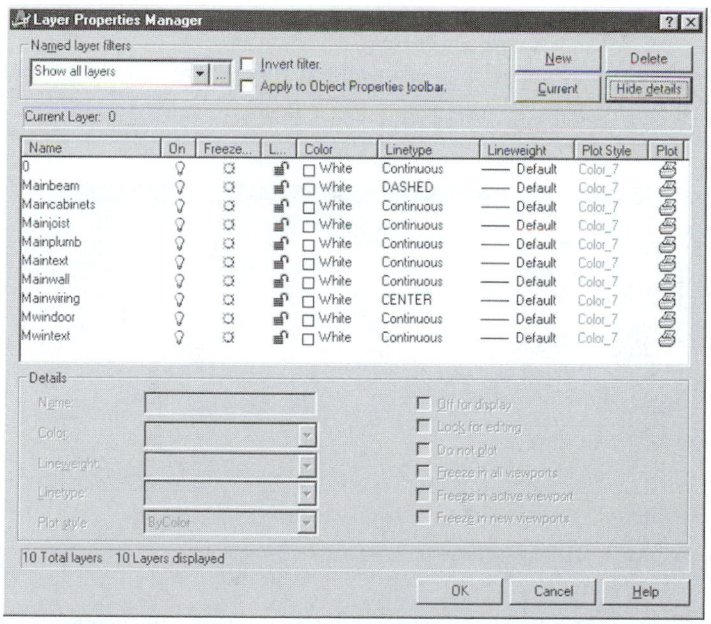

Figure 2.27 *A dialog box provides options for controlling commands, options, or portions of the drawing program. The Layer Properties Manager is displayed by selecting the Layer icon from the Object properties toolbar, by selecting Layer… from the Format pull-down menu, or by typing* ***DDLMODES*** *(EnterKey).*

skill is needed. For now, it is most important to understand the features found in dialog boxes.

WARNING: AutoCAD provides system variables to control the display of dialog boxes. If dialog boxes are not displayed, type **FILEDIA** (EnterKey) at the command line. The default value is 1. Options include:

0 — does not display dialog boxes.

1— displays file dialog boxes.

Other system variables will be introduced as the commands they effect are introduced.

As a dialog box is selected, the crosshairs of the drawing area changes to an arrow. The arrow can be used to select an option by placing the arrow in the box or button by the desired option and pressing the pick button. More specific information regarding the actual use of each dialog box will be discussed as the controlling command is introduced. Common components of a dialog box are seen in Figure 2.28 and include OK/Cancel buttons, scroll bars, buttons, radio buttons, check boxes, edit boxes, image tiles, and alerts.

OK/Cancel

As you enter a dialog box, you will notice that the cursor no longer responds to the drawing area. To return to the drawing and remove the dialog box, select either the

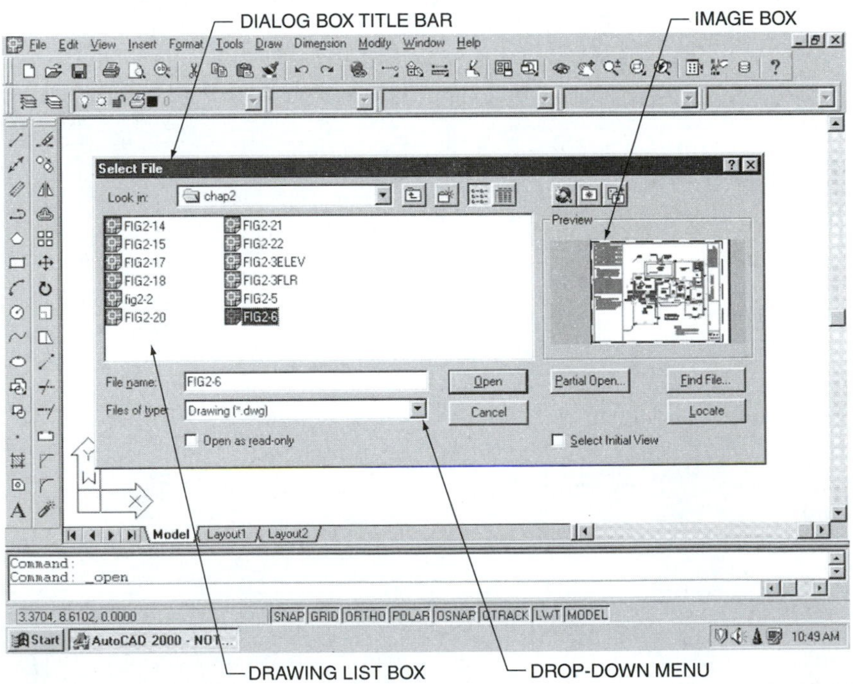

Figure 2.28 *Common components of a dialog box.*

OK or Cancel options. If you've altered the contents of a dialog box, close the box using the OK option. OK updates the dialog settings, and removes the box from the drawing area. Closing a dialog box with the Cancel option will remove the box from the screen, but will also delete any changes you may have made. Pressing the (EscKey) key has the same effect as selecting cancel.

Buttons

Several buttons often are included in a dialog box in addition to OK and Cancel. Examples are seen in Figure 2.29. The Plot dialog box can be displayed by selecting the Plot icon from the Standard toolbar. Selecting one of these buttons with the pick button will cause the selected option to be performed. Buttons with a wide border are default options. A button with an option followed by ... will display another dialog box that must be addressed before proceeding. A button with an option followed by < indicates that an action needs to be performed in the graphics area before proceeding. This might include selecting an object or point in the drawing field. When an action button is selected, the dialog box is removed temporarily to allow for selecting an object from the drawing. Selecting the Windows< button provides prompts for selecting the area to be plotted.

Radio Buttons

Radio buttons are buttons that list mutually exclusive options. Examples can be seen in Figure 2.29. Limits, Extents, Display, View, and Window are each radio buttons

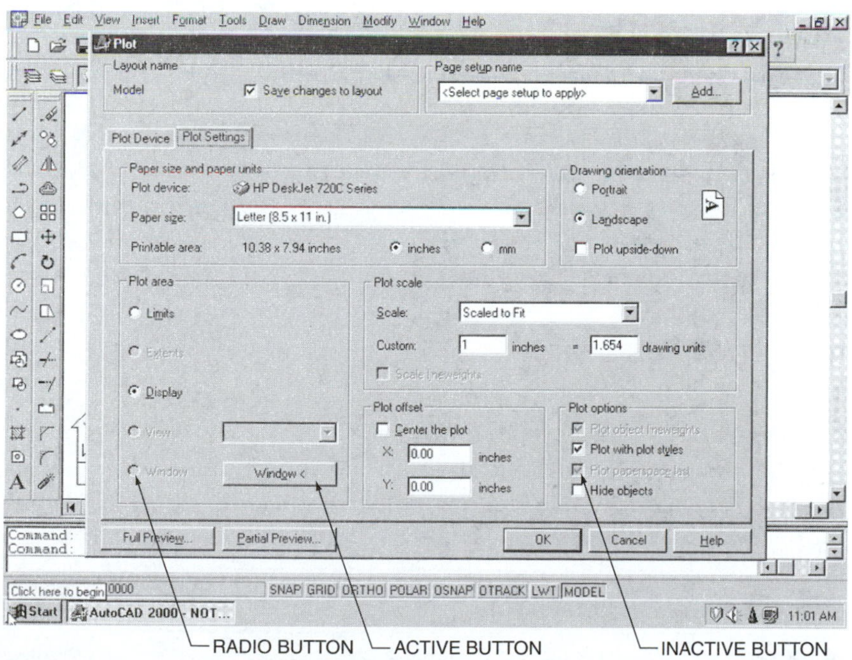

Figure 2.29 *Common buttons for controlling options in a dialog box.*

found in the Print/Plot Configuration dialog box. Each will present a different area to be plotted, but only one of the four options can be used per plot. Notice in Figure 2.29 that the View and Window options are not displayed with black text. Gray options are inactive options and can't be selected until another option has been completed. One of the active or black options may be selected by moving the cursor to the desired button and pressing the pick button.

Check Boxes

Check boxes serve as toggle switches and can be seen in Figure 2.30. Open the Drafting Settings dialog box by selecting Drafting Settings… from the Tools pull-down menu. Once open, select the Object Snap tab to produce the display shown in Figure 2.31. In the Drafting Settings… dialog box, each of the Modes is represented by a check box. Each mode is independent of the other listings. Selecting the Endpoint mode will allow the cursor to jump to the nearest endpoint of existing lines as the cursor is moved. Each of these options, as well as many others, will be presented throughout the text. For now, remember that if a box has an "x" or check in it, that option is enabled. If the box is empty, the option is off.

Edit Boxes

An edit box is a text entry box that allows information to be entered by keyboard. Figure 2.28 shows an example of an edit box. The Save Drawings As dialog box is displayed by selecting Save from the File pull-down menu. When the edit box is selected, it contains a vertical line (a cursor) that flashes slowly. The flashing cursor means AutoCAD is waiting to be told the name of the file. Typed information can

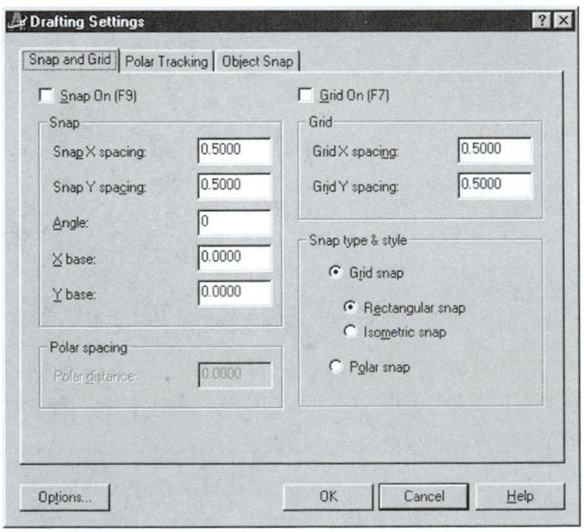

Figure 2.30 *Check boxes in a dialog box serve as toggle switches.*

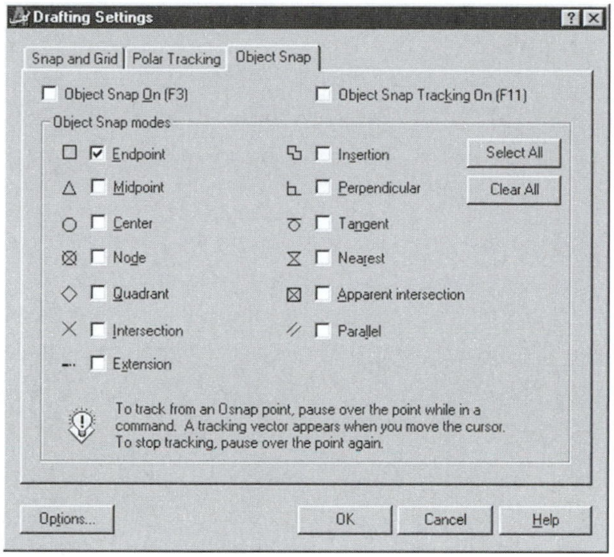

Figure 2.31 *Object snap check boxes are displayed by selecting the Object Snap tab of the Drafting Settings dialog box. In the current state, Endpoint is on, and all other options are off.*

be entered into the box, in this case the "File:" box. If errors are made in typing, the cursor can be moved using the left and right arrow keys. The text in the box is activated by pressing (EnterKey).

Edit Keys

Several keys can be used to aid input into a dialog box. These include text keys, arrow keys, and control keys. These functions include:

- Text—Symbols from the keyboard are inserted at the cursor. As a letter is inserted, the cursor and existing text move to the right. When the cursor reaches the right side limit, the cursor remains and text is moved to the left.
- Left arrow—This will move the cursor to the left without changing the existing text.
- Right arrow—This will move the cursor to the right without changing the existing text.
- Del—This will delete the character at the cursor and move all remaining text to the left. The cursor will remain stationary.
- Backspace—This will delete the character to the left of the cursor while the cursor and all remaining text is moved to the left.

Image Boxes

An image box is a portion of a dialog box that displays an image of the object or pattern that is available. Figure 2.32 shows the image box used to show the pattern that

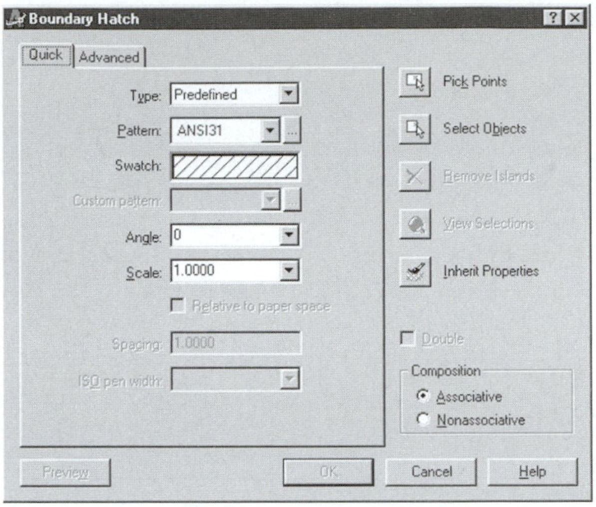

Figure 2.32 *An image box is a portion of a dialog box that displays an image of the selected item.*

will be placed if this hatching option is selected. This box can be found by selecting Hatch ... from the Draw pull-down menu.

Alerts

An alert is a message that is displayed in the lower left corner of a dialog box, or in a separate box that is superimposed over the dialog box. Alerts typically warn the user that they have made a mistake by displaying a message such as "Invalid filename." Another common type of alert is a warning that a file is about to be changed, for example:

> The specified file already exists.
>
> Do you want to replace it?
>
> Yes No Cancel

CONTEXT MENUS

Similar to most Microsoft Windows programs, AutoCAD contains menus that provide rapid access to options for tasks that are underway. These shortcut menus are available by clicking the right mouse button. As the name implies, the contents of the menu will be changed depending on when the menu is selected. If you're in the middle of a drawing command and right-click, the menu will be different than if you right-click while trying to edit a drawing. Five different shortcut menus are available while in AutoCAD. Each can be customized. The following menus are shown in their default setting in Figure 2.33:

- The Default menu is displayed when a right-click is performed with the cursor in the drawing area prior to starting a command sequence.

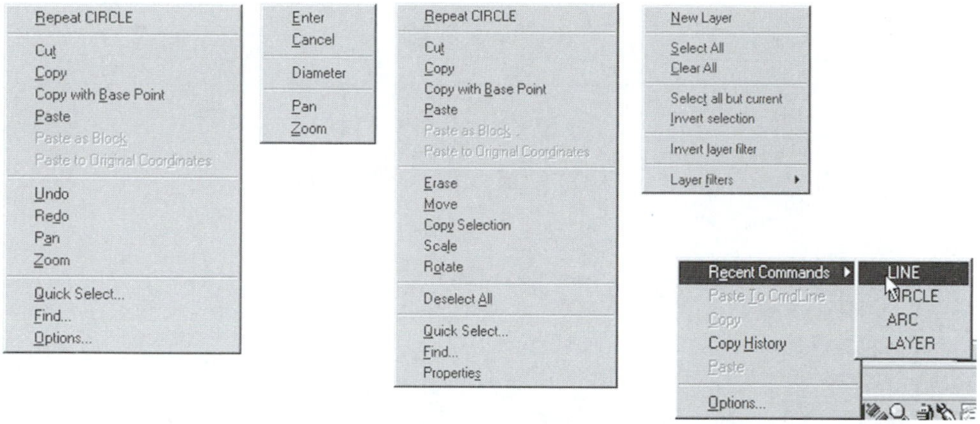

Figure 2.33 *Context menus are new to AutoCAD 2000. Five options are included in the default settings of AutoCAD.*

- The Command-mode menu appears when a right-click is performed in the middle of a command sequence. Options specific to the command will be displayed.
- The Edit-mode menu is displayed with a right-click if objects have been selected, but no command is in progress.
- The Dialog-mode menus appear if a right-click is performed while in a dialog box.
- Other menus will be displayed when right-clicking based on the cursor location. Right-clicking with the cursor over a toolbar, produces a list of all toolbars. Right-clicking while over the command line, layout tabs or status bar will each produce different context-specific menus.

FUNCTION KEYS

Twelve function keys are located across the top of the keyboard or to the left of the keyboard. The actual layout depends on the keyboard manufacturer. Nine function keys are used by AutoCAD to provide access to commands. Each is used as a toggle from ON to OFF and will be explained further in Chapter 5. Other keys can be programmed to perform specific commands. The following keys are used by AutoCAD:

- F1—Selecting this key clears half of the screen and displays the HELP menu. Using the Help options of AutoCAD will be discussed later in this chapter. Select the cancel button or the close box to remove the display from the screen.
- F2—Flip Screen. Pressing F2 in the middle of a drawing will remove the drawing from the screen and display a written list of the commands that were used to generate the drawing. Press F2 a second time and the Text window will be removed. (See Figure 2.34.)
- F3—Object Snap. Snap allows the cursor to be moved to specified points on drawing objects such as an endpoint or midpoint of a line. Future chapters will

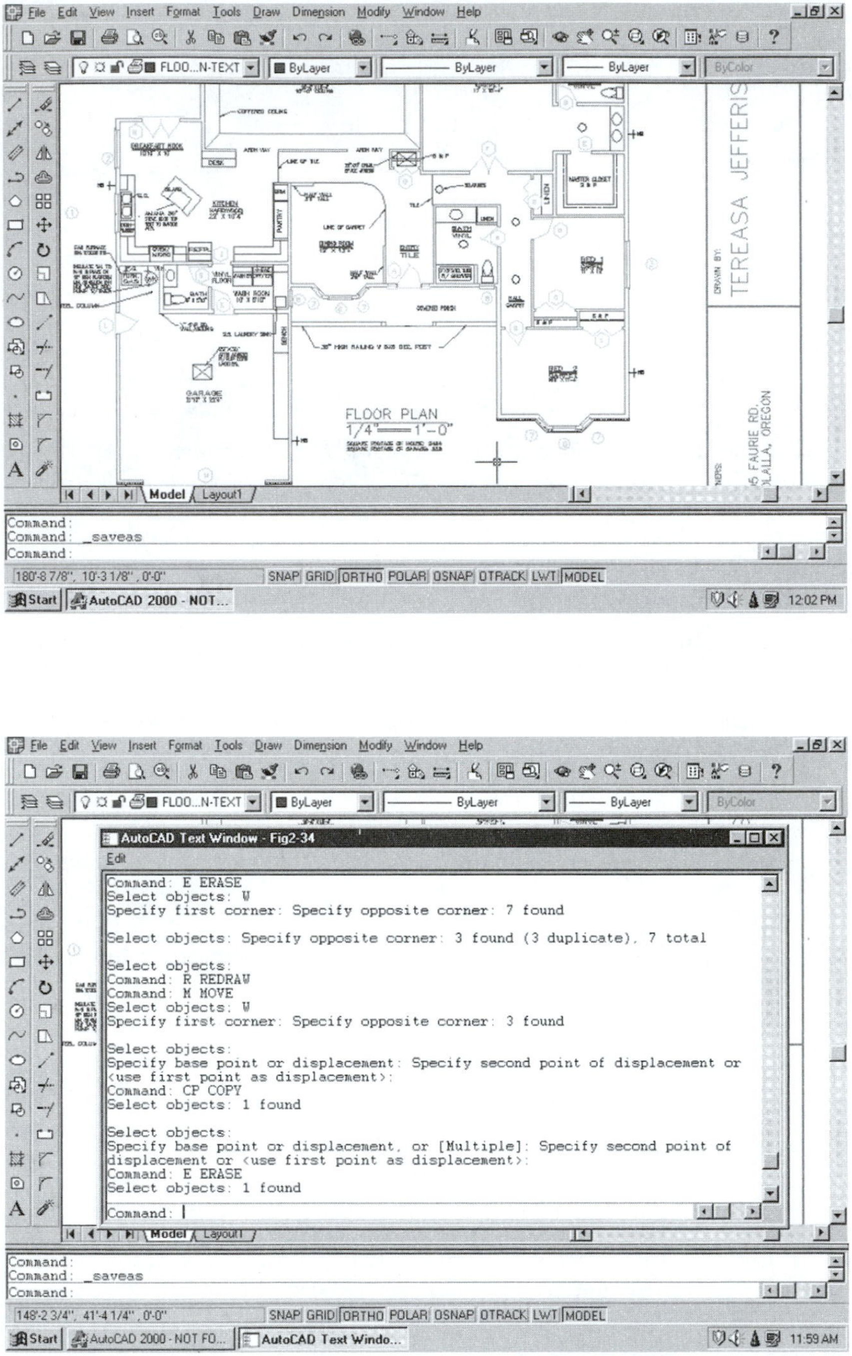

Figure 2.34 *Using function key F2 will produce a display of the commands used to produce a drawing. Pressing F2 once will produce the graphic display. Pressing F2 again will restore the previous drawing screen.*

explore how to set the object snap settings. Pressing F3 will toggle Object Snap ON/OFF.

- F4—Tablet mode. This function key serves as a toggle if a digitizing tablet is used. Some professional offices use a digitizer tablet to control command options specific to their company. A digitizing tablet can be used as an alternative to a mouse for selecting objects from screen menus and controlling the crosshairs. If you're working at a station with no tablet, you'll see the following prompt:

  CALibrate the TABLET before turning it ON <off>

If you're at a machine with no tablet, ignore the prompt and continue exploring.

- F5—Isoplane crosshairs mode. This is used when creating isometric drawings. This button toggles between the top, left, right, and bottom planes of an isometric square. These planes will be discussed as isometric drawings are explored.
- F6—Coordinate display. The coordinate display can be seen on the status line in Figure 2.2. This function key will toggle the coordinated display ON or OFF.
- F7—GRID display. In future chapters you will learn how to create a visible grid to aid in drawing layout. Once you have selected the size of the grid to be displayed, the grid can be displayed or removed by selecting the F7 key. Grids can also be toggled ON/OFF by selecting the GRID button on the status bar.
- F8—ORTHO. Ortho can be used to control the movement of the cursor. When it is ON, only vertical and horizontal lines can be drawn. When OFF, lines at any angle may be drawn. These options will be explored in Chapter 5. By switching Ortho OFF, angular features can be drawn.
- F9—SNAP mode. This key activates the ability to move the crosshairs within the drawing area at specific intervals. Snap will be explained in Chapter 5. Snap can also be toggled ON/OFF by selecting the SNAP button on the status bar.
- F10—This key toggles polar tracking ON/OFF.

Note: The key thing to remember about the function keys is that they can be used at any time throughout the life of a drawing including in the middle of a command. Although Ortho in the ON setting will make a floor plan much easier to draw, it can be quickly toggled OFF to allow for lines representing an angled wall to be drawn, and then reset to the ON position to finish other perpendicular walls.

CONTROL KEYS

Like many other software programs, AutoCAD uses control keys to perform common functions. These functions are achieved when the control key is pressed in conjunction with a specified letter key. Many of these functions duplicate the functions performed by the function keys. The control keys typically are located in the lower corners of the keyboard and may be labeled CTRL. The control key functions are:

CTRL + B—Snap mode toggle ON/OFF.

CTRL + C—Copies selection to the clipboard.

CTRL + D—Coordinate display of status line toggle ON/OFF.

CTRL + E—Crosshairs in isoplane position toggle left/top/right.

CTRL + G—Grid toggle ON/OFF.

CTRL + H—Same as using the backspace arrow.

CTRL + L—Ortho mode toggle ON/OFF.

CTRL + O—Used to open a new file.

CTRL + T—Tablet mode toggle ON/OFF.

CTRL + V—Paste clipboard locations to a specified location.

CTRL + X—Cut selected objects to the clipboard.

CTRL + Z—Undo.

EXPLORING THE SAVE COMMAND

Throughout this chapter you've explored several methods of controlling commands. The LINE command was used as methods of using toolbars, pull-down menus and keyboard entry for command entry were explored. The lines you've created thus far will be destroyed if you end your drawing session without saving them. Chapter 5 will explore how and where files can be saved. This chapter will introduce the procedure for saving a drawing file.

If you have not already done so, enter the drawing area and draw several lines. Since this is a chapter for exploring, use the Draw toolbar to draw a circle and a rectangle. You've now created a work of art that must be saved for future generations to enjoy. Probably not, but it will create a file that can be used in Chapter 3 as you explore methods for opening an existing drawing. Use the following procedure to save a file to a floppy diskette:

- Select Save from the File pull-down menu. This will display the Save As dialog box shown in Figure 2.35.
- Select the arrow in the Save in: drop-down menu. This will allow the storage destination for the file to be selected. Select the C: drive for storage on a hard drive or A: for storage on a diskette.
- Provide a name in the File name: edit box. The name of Drawing 1 is currently displayed. To alter the name, use the following procedure:

  Place the cursor in the edit box, press the pick button, and move the cursor so that Drawing 1 is now highlighted (in a blue box.).

  With the current name highlighted, enter the desired filename using the keyboard. Use a name of EXE2-1 for the filename.

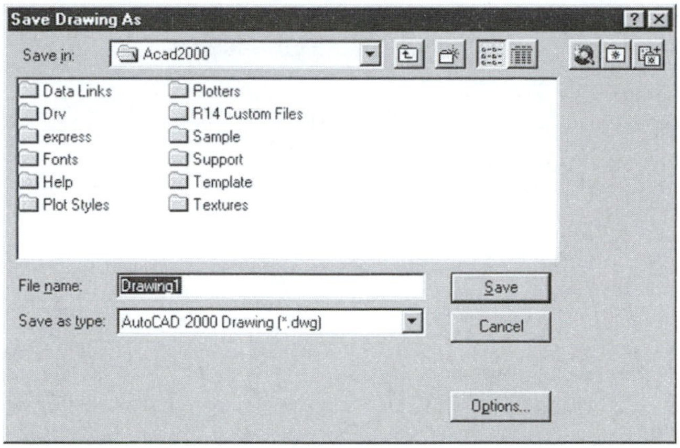

Figure 2.35 *The Save Drawing As dialog box is displayed by selecting Save from the File pull-down menu.*

- Select the SAVE button. The dialog box will be removed from the screen, the drawing will be saved, and the drawing area will be restored.

ENDING A DRAWING SESSION

You can continue to add objects to this drawing for as long as you so desire. When you want to end the drawing session, save the drawing again using the procedure that was just discussed to update the file on your diskette. To exit the drawing session, select Exit from the File pull-down menu or select the Close icon from the title bar. For now, continue in AutoCAD and explore the HELP menu.

HELP

One of the best features of AutoCAD is that it offers free advice with no ridicule. You're going to be exposed to hundreds of commands, options, and menus, and sometimes it's easy to feel lost or overwhelmed. If you can avoid panic in the first 30 seconds you'll probably do fine. Your chances for success are even better if you remember that there has been a help button in every menu you've looked at.

The fastest way to get help is to press the F1 function key. The HELP command can also be entered by selecting Help from the menu bar, selecting the ? key from the Standard toolbar, or by typing **HELP** (EnterKey) at the command prompt. Each method will produce the dialog box seen in Figure 2.36.

Selecting HELP from the Standard toolbar will also activate the Help menu. The HELP command is a transparent command. A transparent command is a command that can be completed in the middle of another command sequence. A transparent command is started by typing an apostrophe before entering the command. To get help in the middle of the LINE command, make the following entry:

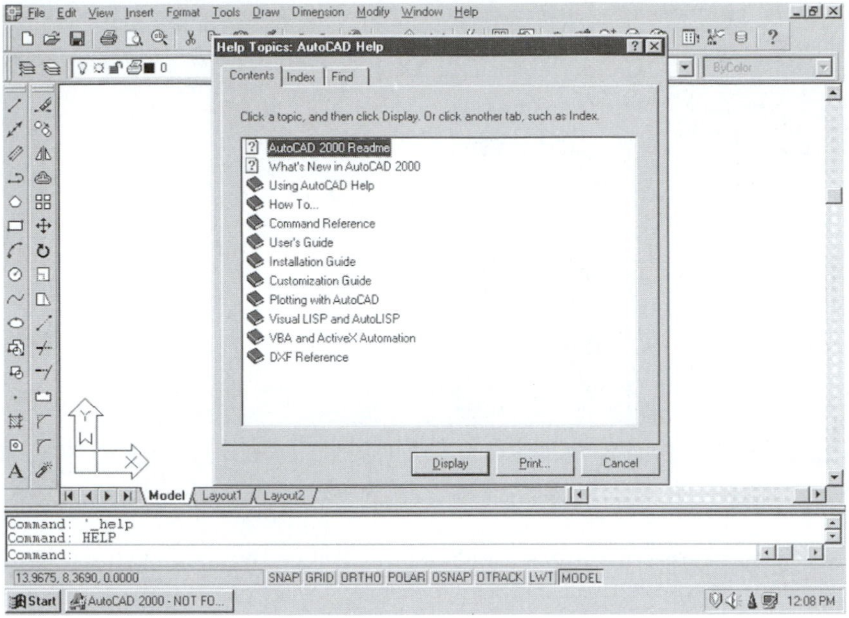

Figure 2.36 *Typing* ***HELP*** *(EnterKey) at the command line, selecting the Help icon from the Standard toolbar, or pressing the F1 key will produce the Help menu.*

Command: **L** (EnterKey)

LINE Specify first point: *(Select desired starting point.)*

Specify next point or [undo]: **'HELP** (EnterKey)

The F1 key can also be used in the middle of another command. When the HELP command is accessed in the middle of a command, the HELP command automatically provides command-related help. Asking for help in the middle of the LINE command will produce the Help display shown in Figure 2.37 that is specific to the command.

The Help menu offers help in the areas of Contents, Index, and Find. The Contents folder is opened as the Help menu is opened. If you use the Index folder and then exit Help, it will be the open folder the next time the Help is accessed. Each menu has other listings that will produce information related to that topic. In each of the menus that will be viewed, you'll find Back and Help Topics, as well as commands that are surrounded by double brackets. Selecting Back will take you back through other help menus that you've already viewed. Selecting Help Topics will return you to the original Help menu. Selecting either pair of the double brackets will sequentially display a list of topics that can be selected for a help search. Selecting the << button will display previous portions of the menu. Selecting the >> button will show the next menu listing. Most menus also provide a method of reaching a glossary of AutoCAD terms.

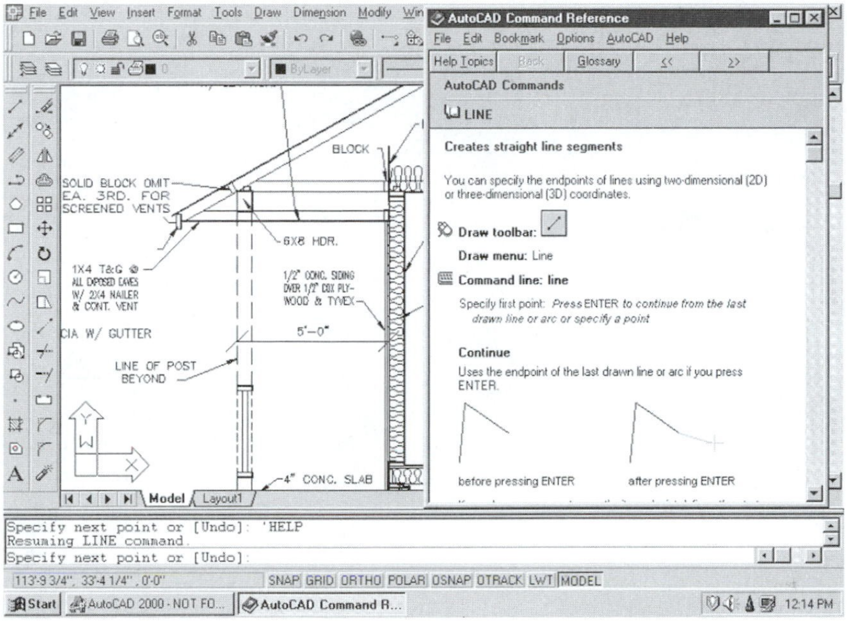

Figure 2.37 *Asking for help in the middle of a command will provide a help menu specific to the command.*

CONTENTS

The Contents folder of the Help menu can be used to obtain information about the Help program, or individual portions of the AutoCAD program. The Contents folder contains a How to Guide and a Users Guide to provide instruction on using each aspect of the program. The Command Reference portion of the folder provides information on each command used by AutoCAD. To receive information about the LINE command, enter the Help menu by selecting the F1 button or by any one of the other methods introduced in this chapter. Select Command Reference by double-clicking. This will produce the menu shown in Figure 2.38 with eleven options. Double-click the Commands listing option to display the dialog box shown in Figure 2.39. This menu provides a list of commands presented in alphabetical order and two methods of moving through the menu. At the top of the menu is a partial keyboard. Selecting a letter from the keyboard will scroll the menu display to that portion of the listing. On the right side of the display is a scroll bar that can be used to move through the list. Use the keys to move to an area of the listing, and then use the scroll bar to refine the portion of the listing to be viewed.

To find information on the LINE command, select the **L** button from the menu display. This will produce the display shown in Figure 2.40, and provide a brief description of the LINE command. A detailed description of the command can be found by

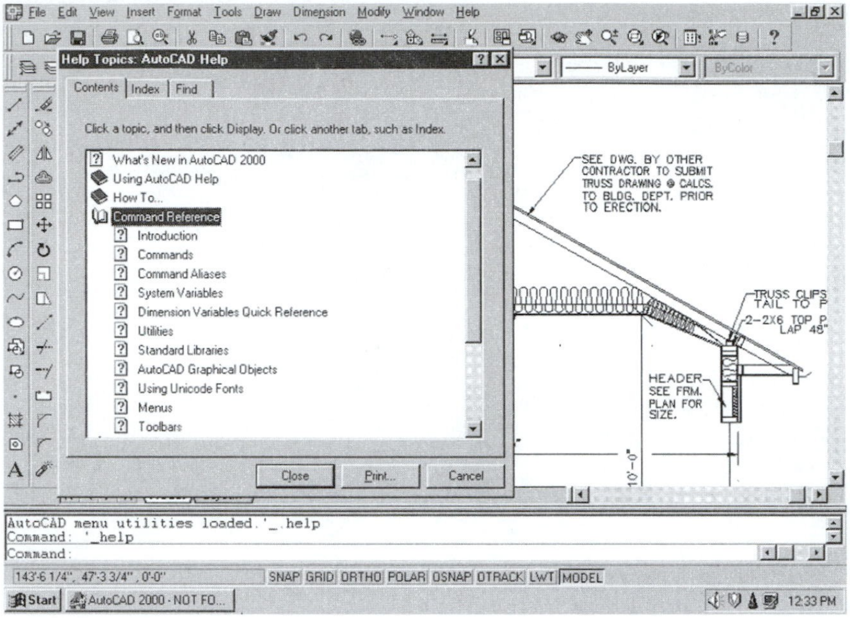

Figure 2.38 *Selecting Command Reference from the Contents folder will provide eleven options for seeking help about AutoCAD commands.*

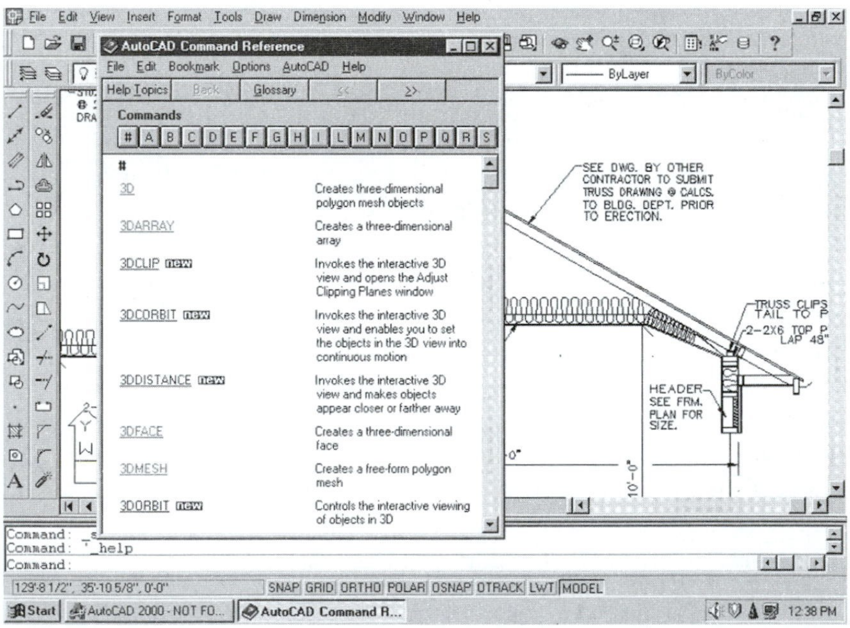

Figure 2.39 *Selecting Commands from the Command Reference listing provides access to an alphabetical listing of AutoCAD commands.*

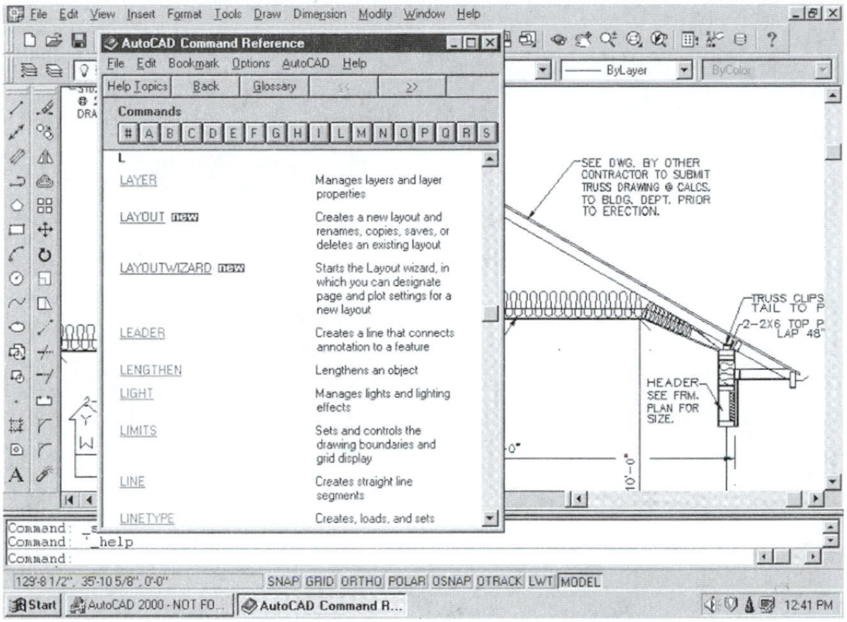

Figure 2.40 *Selecting the L button at the top of the menu will scroll to the start L commands list.*

moving the cursor to the command name and selecting the name with a single click. Figure 2.41 shows a portion of the Line command description. Select the F1 key to close the Line Display. Select the Close button in the Help title bar to close the Help window and return to the drawing screen.

Reopen the Help menu and select the System Variables option of the Command Reference folder. This will provide a listing similar to the Command listing. System variables control how a specific command will function and will be discussed as each command is introduced. Selecting Menus from Command Reference will produce a listing of each of the pull-down menus. Selecting Draw will display the listing shown in Figure 2.42. Selecting Line will display the menu shown in Figure 2.37. This menu can also be displayed by selecting the green listing of LINE on the right side of the definition.

Selecting the Toolbars option will provide a list of all AutoCAD toolbars. Selecting one of the names will produce a listing of the options contained in the toolbar. By selecting the Draw listing, the menu shown in Figure 2.43 will be displayed to provide a brief description of each command.

Selecting the Standard Library option provides a menu of each of the libraries in AutoCAD. A library is a set of predetermined symbols designed to save the user drawing time. If you're going to draw a floor plan, you'll need to draw toilets, sinks, doors, windows, as well as many other common symbols. Each of these symbols are

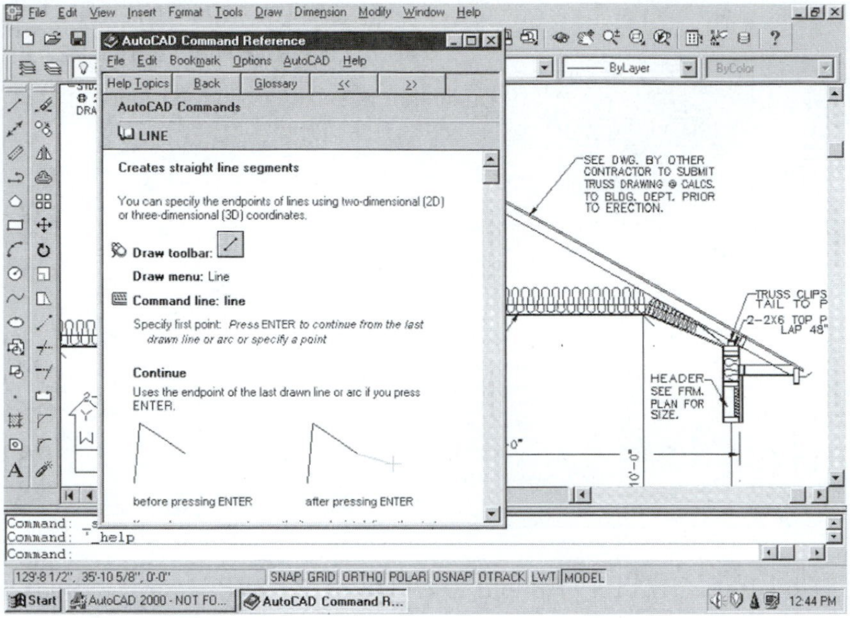

Figure 2.41 *The Help display for the LINE command.*

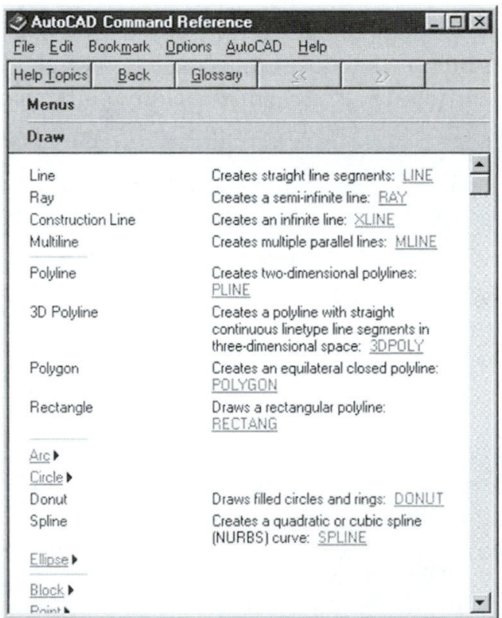

Figure 2.42 *Selecting Draw from the Menu listing will provide a list and brief description of each command in the Draw menu.*

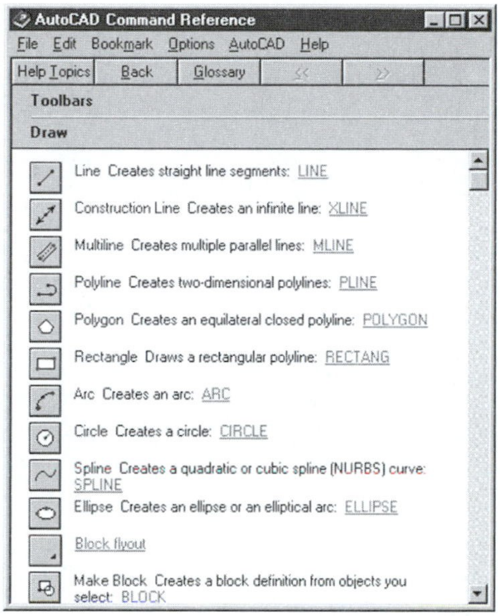

Figure 2.43 *Selecting Toolbar from the Command Reference provides a list of the toolbars. Selecting a specific toolbar, such as Draw, will display each of the icons found in that tool bar and a brief description of that command.*

examples of symbols that can be placed in a library to reduce drawing time. Many useful symbols have been created by Autodesk and can be viewed from this menu. These symbols include common linetypes (see Chapter 6), hatch patterns (see Chapter 15), and text fonts (see Chapter 17).

Index

Selecting the Index tab will produce the menu shown in Figure 2.44. This menu is like the index found at the end of a book. Look up a key term and the book index will list several pages where information can be found. With the Help index, you don't even have to know the word you're looking for. By entering the first few letters of a command or option, the help menu will display a list of commands and sources of information about that command. If time is no object, the scroll bar can be used to scan through the list until your subject is displayed. Entering "line" in the edit box will produce the display shown in Figure 2.45. This display box can be used to select information about the command. Selecting LINE command display will produce a listing of four options. Select Drawing Line Objects and press the Display button. This will display the help window shown in Figure 2.46. This window, like all other Help windows can be reduced in size and moved beside the drawing so that it serves as a tutorial for the specified command.

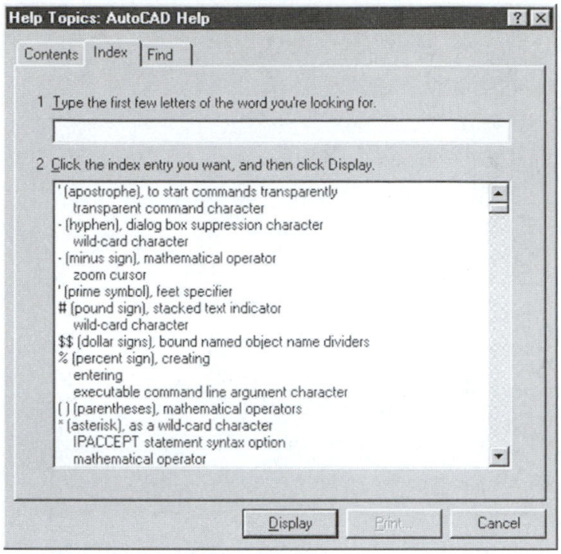

Figure 2.44 *Selecting the Index tab will produce a menu that is like the index found at the end of a book.*

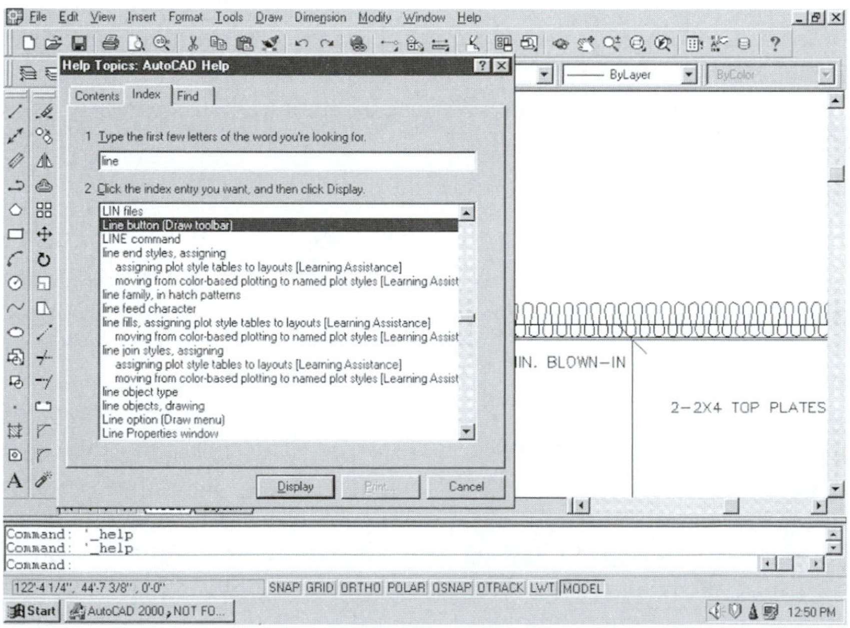

Figure 2.45 *As commands are entered in the edit box, the display options related to that command are listed.*

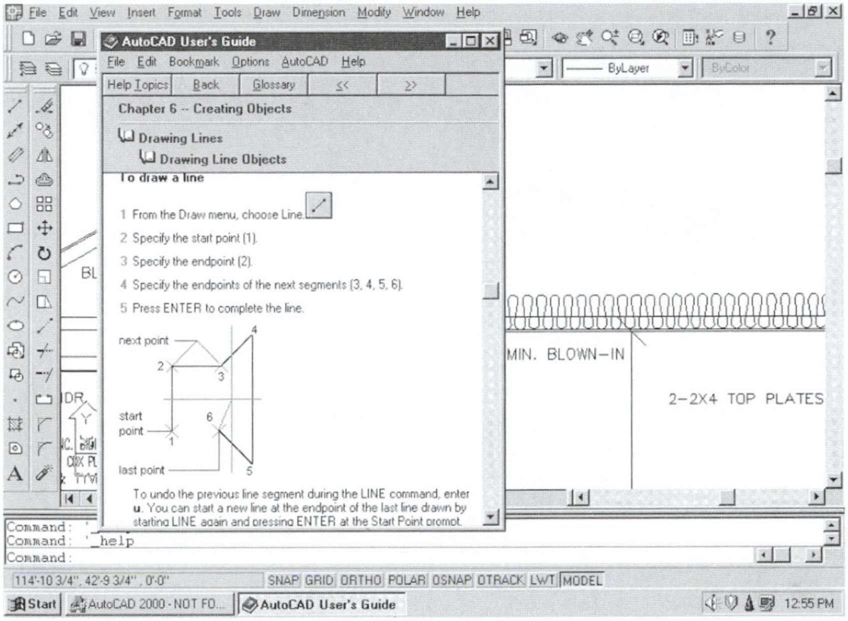

Figure 2.46 *Selecting a specific option from the line index listing provides a help box. This is the Line Display for the LINE command. Remember that this, and other windows, can be minimized and dragged into the drawing area for easy reference.*

Find

The Find option of Help is similar to an index, but it's also faster than a speeding bullet. Before the Find tab is opened, a display listing three options is presented. Select the default option, the Next button, and then the Finish button to open Find. There will be a slight delay as the dictionary is loaded into memory. Figure 2.47a shows how the Find box will appear as the folder is opened. Once active, to obtain information about the LINE command, type the letter L into the edit. The upper windows will change to display the L portion of the index and the lower window will change to display related information about the selected word. Figure 2. 47b shows the changes in the display as LINE is entered into the edit box. By selecting the word LINE in box 2 (the matching word box), the search is restricted to the LINE command, as opposed to other line options. As the search field is restricted in box two, box three is updated. Selecting Draw Menu from the topic box (box 3), the display shown in Figure 2.48 is displayed. This is a result of asking for information about the LINE command as it relates to the Draw toolbar.

AUTOCAD ASSISTANCE

In addition to the Help menu, help can be obtained through the AutoCAD Learning Assistance tutorial, and the AutoCAD Support Assistant. AutoCAD provides addi-

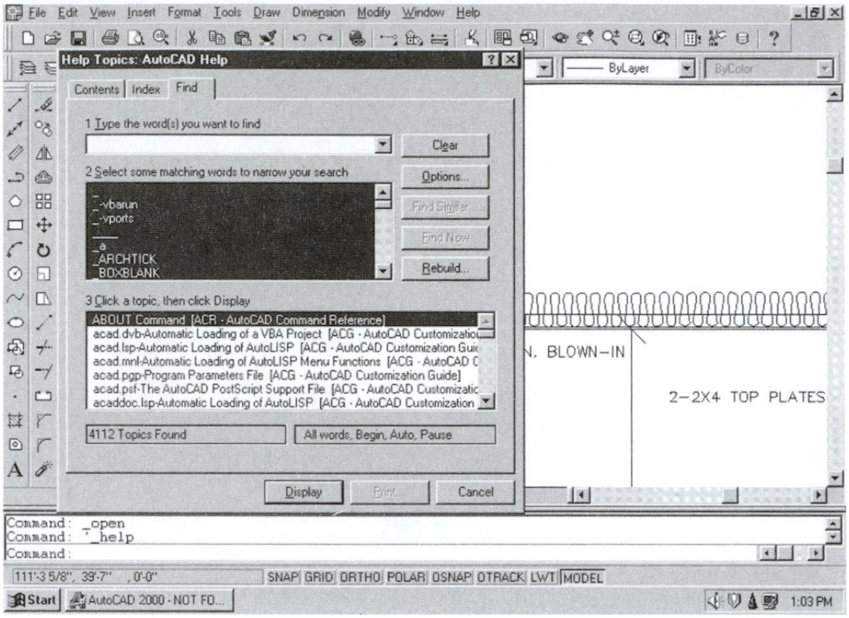

Figure 2.47a *Using the Find option at the Help menu alters the display listing as each letter is entered in the edit box.*

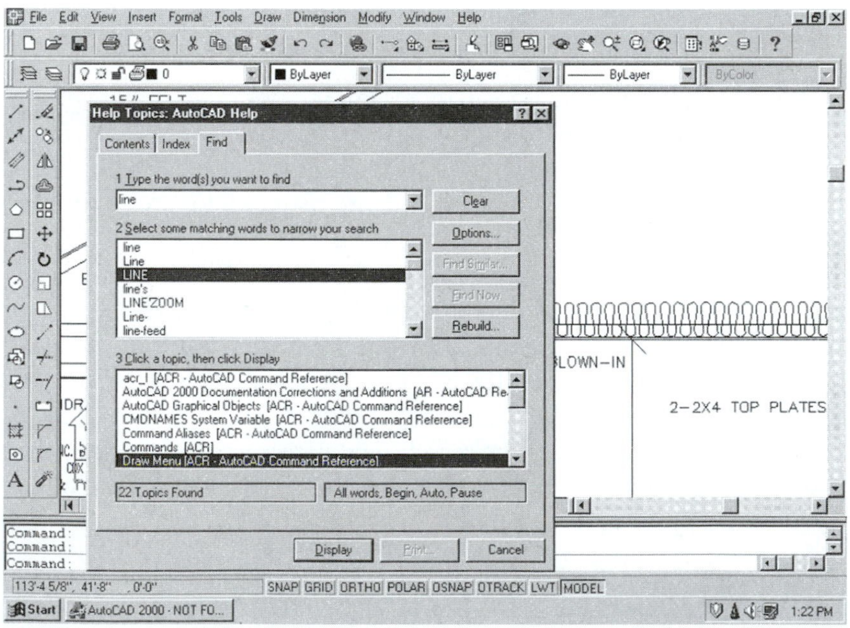

Figure 2.47b *Selecting a specific option from the line index listing provides a list of related subjects.*

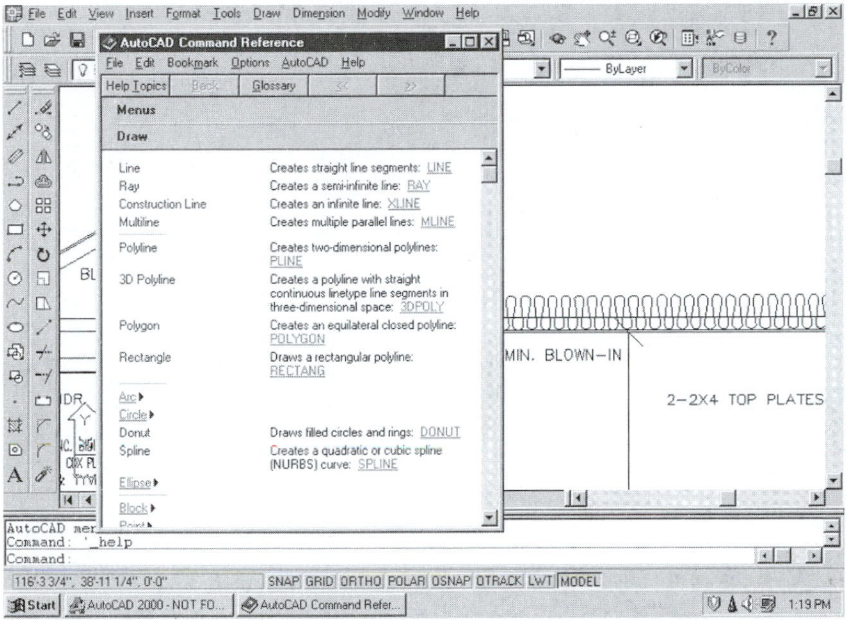

Figure 2.48 *Selecting the desired subject displays the related menu. By selecting LINE and then the Draw Toolbar [ACR-AutoCAD Command Reference] option, a display of the contents of the Draw toolbar are displayed.*

tional help through a tutorial-based learning tool named the AutoCAD 2000 Learning Assistance (ALA). The ALA is provided on a separate CD from AutoCAD and must be installed before it can be accessed throught the Help menu The ALA includes lessons for learning many of the new features of AutoCAD. The AutoCAD Support Assistant (ASA) contains technical documentation, solutions to common support issues and trouble shooting guidance for common AutoCAD error messages. AutoCAD users can periodically update the ASA through Web access. Selecting Autodesk on the Web will produce the drop-down menu shown in Figure 2.49. Selecting the AutoCAD Home Page will keep you updated on AutoCAD 2000. Selecting Autodesk Home Page will allow access to the world of Autodesk. If you're not in AutoCAD, Autodesk can be reached at www.Autodesk.com.

WHAT IT ALL MEANS

By now, you might be feeling overwhelmed by all of the bells and whistles of AutoCAD. In this chapter you've been introduced to several methods of entering a command, and explored a multitude of buttons, boxes, and controls for altering the commands and options. Each has been introduced using the LINE command. By the time you explore each of these controls in relations to the hundreds of commands, options, and icons, your brain may be on overload. The good news is that you don't

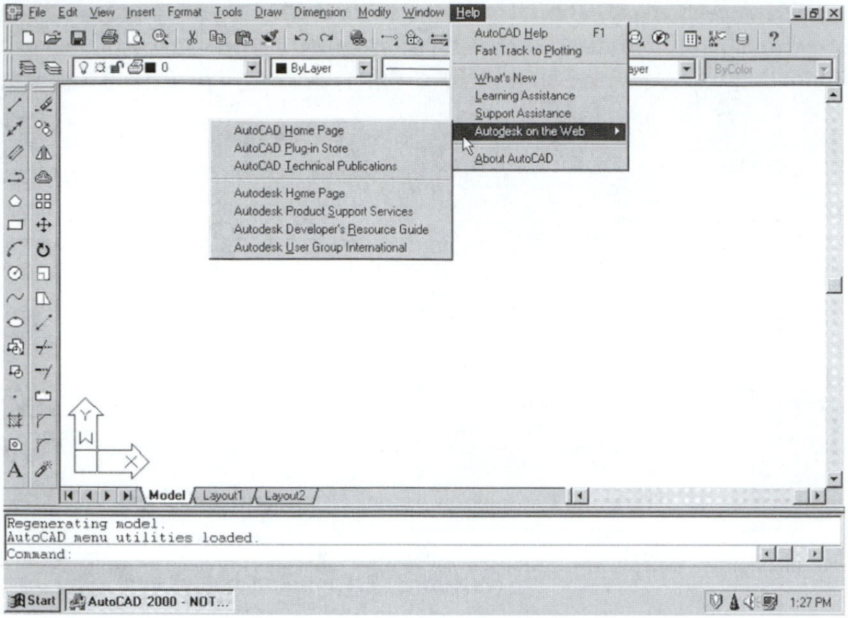

Figure 2.49 *Help on the Web can be accessed through the Help pull-down menu. Selecting Autodesk on the Web will produce the drop-down menu. Selecting the AutoCAD Home Page will keep you updated on AutoCAD 2000. Selecting Autodesk Home Page will allow access to the world of Autodesk. If you're not in AutoCAD, Autodesk can be reached at www.Autodesk.com.*

have to memorize every command, option, and control method. You only have to remember your favorites and the ones you use on a regular basis. Use the Help menu (and this book) to help keep track of options that you don't use on a regular basis. The other key point of this chapter is that there is usually several ways to do one thing. Generally, one method is as good as another, but don't limit yourself to the first method you find. The key to becoming a successful AutoCAD operator is to explore. The following quiz will require you to do just that. Many of the questions will deal with information not introduced in this chapter. This is your first real opportunity to explore AutoCAD.

CHAPTER 2 QUIZ

1. List three different methods that can be used for selecting commands and tell how they are accessed.
2. List the commands found in the Draw toolbar. What additional commands are listed in the Draw pull-down menu?
3. You've entered a dialog box with 40 listings. Describe two methods that can be used to move through the list to the end.
4. List three sources for obtaining help with AutoCAD.
5. List the four screen areas displayed after entering AutoCAD.
6. Explain the following terms:

   Default

   Pick

   Command
7. List in order the first eight commands for the DRAW pull-down menu.
8. You have just finished drawing a line and would like to draw an arc. Explain how to get the arc options using the pull-down menu.
9. What is the last arc option listed in the pull-down menu?
10. List the options in the File pull-down menu.
11. You've entered the Insert pull-down menu, but you wanted the View menu. How can you get View to be displayed?
12. List nine tabs that may be seen on the status line.
13. Is ELLIPSE, of the draw menu, a command or an option? How can you be sure?
14. An icon showing a paintbrush is in the standard toolbar. What is the icon, and what does it do?
15. List the four toolbars shown as AutoCAD is loaded.
16. What key will execute the following request?

    Display grid

    Ortho

    Snap mode
17. Someone has asked you what the last three commands were that you used on your drawing. How could you get a record if you don't remember?
18. List five items that can be found in a dialog box.

19. Search through each pull-down menu and find the location of:

    Erase

    Tablet

    Toolbars

20. What letter key would be needed when using the control key to complete the following options?

    Grid toggle

    Save

    Backspace

    Coordinate display

21. List three methods for entering the Help menu.
22. Explain the procedure to float the Draw toolbar.
23. Describe the process to display a hidden toolbar.
24. You've drawn 100 lines and would now like to quit. How can you terminate the command?
25. You've discovered after terminating the command, that you actually wanted to draw one more line. What is the quickest way to reenter the LINE command?
26. How many different types of arcs are listed in the Draw pull-down menu?
27. How many sides will be drawn if the default value is selected for a polygon?
28. What is the first and last listing in the Linetype library?
29. What is the effect of a 0 and a 1 setting on Tooltips?
30. List and describe the five context menus of AutoCAD.

CHAPTER 3

Creating Drawing Aids

By now you've become familiar with the hardware, and entered the software. It's time to start building your drawing files. In Chapter 2 you opened AutoCAD and were exposed to the Startup window. To quickly get you drawing, you selected the OK button and removed the Startup dialog box from the drawing screen. The Startup dialog box offers four alternatives to help set up a drawing file. The Open a Drawing option is used to open an existing drawing file. For efficient use of AutoCAD when you know the drawing parameters to be used, the Use a Wizard and Use a Template options offer help to organize the drawing area. The Use a Wizard offers helpful prompts to guide in drawing setup. The Wizard contains the Quick Setup and Advanced Setup options for determining how the new drawing environment will be established. Selecting the Use a Template button will allow predetermined drawing parameters to be used to control the drawing environment.

As you gain an understanding of the options that are available, you can create your own templates. Once you've created and stored your own template, the drawing parameters will be preset to meet your standards each time you start a new drawing. Chapter 4 will help you create your own template. The Start from Scratch option can be used to enter a drawing with minimal drawing controls and provide them as they are needed. The Start from Scratch option is the equivalent of grabbing a blank sheet of paper with no title block. It is best suited for users who like to plan as they go. No matter which option you chose to start a drawing session, several tools must be understood before exploring the methods to set these options.

DRAWING TOOLS

The basic drawing tools for controlling the drawing environment include UNITS, LIMITS, SNAP, GRIDS, and ORTHO. An additional tool, the Help menu, will also be explored. Help can be used at any point of a drawing session to get information about almost any aspect of a drawing.

UNITS

If you have previous manual drawing experience, you've worked with a scale similar to that shown in Figure 3.1. Many architectural and structural projects are drawn using ¼" = 1'-0". Figure 3.2 shows a floor plan drawn in architectural units and scaled down to fit into the drawing area.

One of the unique features of AutoCAD is that as you work in model space you will be drawing at full scale. The drawing will not be scaled until you change to paper space to print or plot your drawing. Figure 3.3 shows a portion of Figure 3.2 printed at ¼" = 1'-0". If you want to draw a structure that is 300' long, you will draw it full size, and then choose a scale as you plot, using AutoCAD's paper space option. This will be covered in future chapters. Units will allow you to choose which measuring device you will use to establish size. As your computer is currently set, you are measuring lines in four-place decimal inch units. This can be confirmed by looking at the coordinate display in the lower left corner of the status bar. To describe the current location of the crosshairs, two four-place decimal numbers are used. The drawing units may be set using the Quick Wizard, the Advanced Wizard, and with the methods used in Start from Scratch as the drawing is started. Each of the unit controls can be changed at any time throughout the life of the drawing. Units can be altered using the methods described at the end of this chapter.

Unit Selection

Unless you are using a template, setting the UNITS should be the first thing you do once you have entered the drawing editor. Notice that the coordinate display on the task bar shows the default value is for decimal inches. This would be fine for a drawing such as a beam connector shown in Figure 3.4. Most architectural projects are better suited to the Architectural option. This will measure objects in feet and inches.

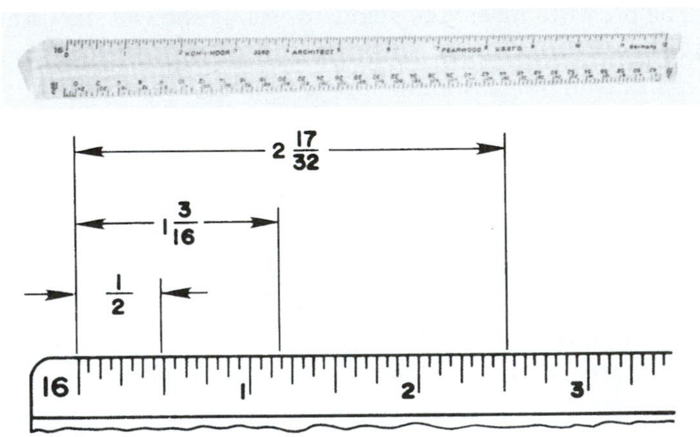

Figure 3.1 *Architectural scales have been used for years to reduce or enlarge construction components. (Photo courtesy of Koh-I-Noor Rapidograph, Inc.)*

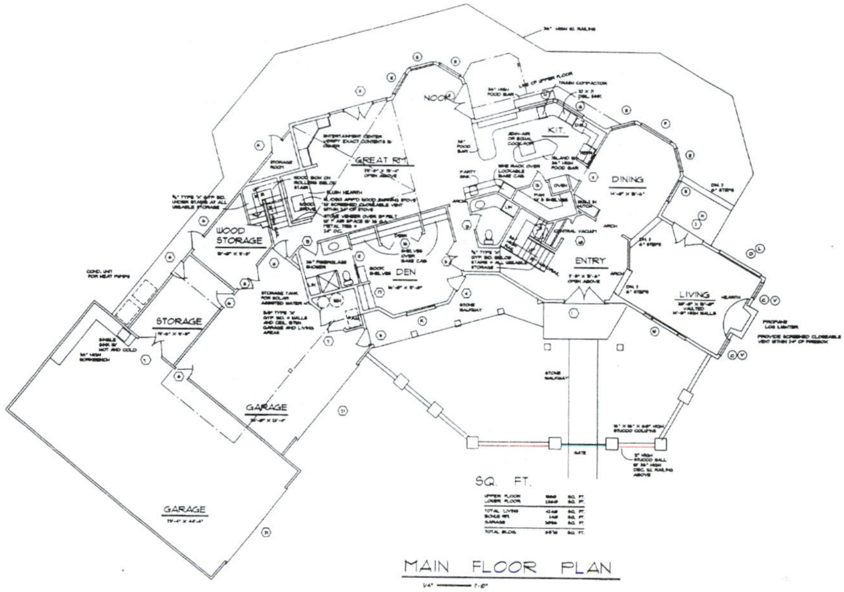

Figure 3.2 *Architectural projects such as floor plans are typically drawn manually at a scale of ¼" = 1'-0". AutoCAD will allow you to draw at full scale by altering the size of the display shown on the screen. The "scale" can then be set as the drawing is plotted.*

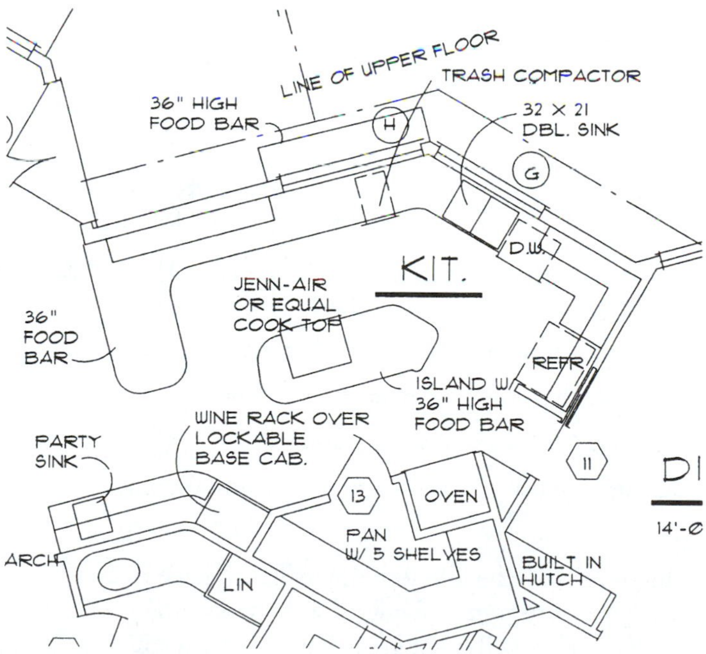

Figure 3.3 *The floor plan shown in Figure 3.2 can be enlarged by using the ZOOM command to view a specific area of a drawing or by changing the scale factor during plotting.*

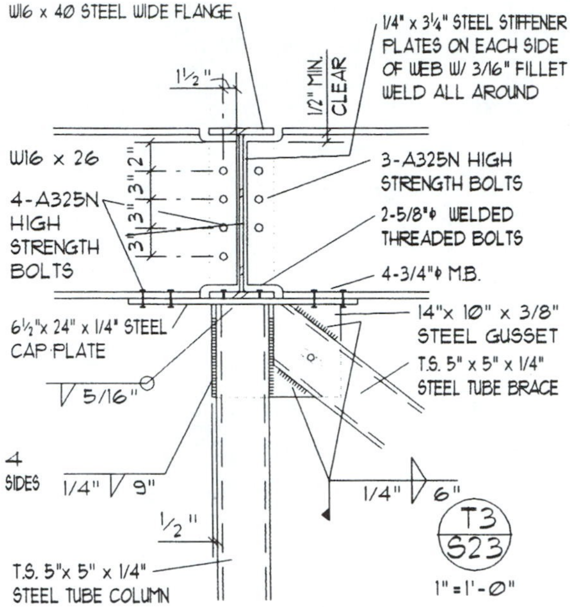

Figure 3.4 *Although some drawings can be drawn using Decimal or Engineering units, most are created using Architectural units.*

The Engineering option would be the best option for working on drawings such as site plans or street layouts, where larger tracts of land are involved, as seen in Figure 3.5.

Before proceeding, select the type of units that are appropriate for an architectural drawing such as a floor plan. This can be done using the following procedure.

- Select Units... from the Format pull-down menu. This will display the Drawing Units dialog box shown in Figure 3.6.
- Select the down arrow in the Type: drop-down list. (See Figure 3.7.)
- Select the Architectural option from the menu.
- Select the OK button to close the dialog box.
- Remember that using a template or a wizard will prompt you for this process.

Unit Accuracy

Once the drawing units have been selected, you'll also have to consider the precision that will be used to measure your drawing. For Architectural units, the default precision is set at 0'-0 1/16". This will work great for drawing a small detail, but will be of little help for drawing a 300'-long structure. Setting the unit precision in whole inches will allow you to draw lines that are an even number of inches in length. The

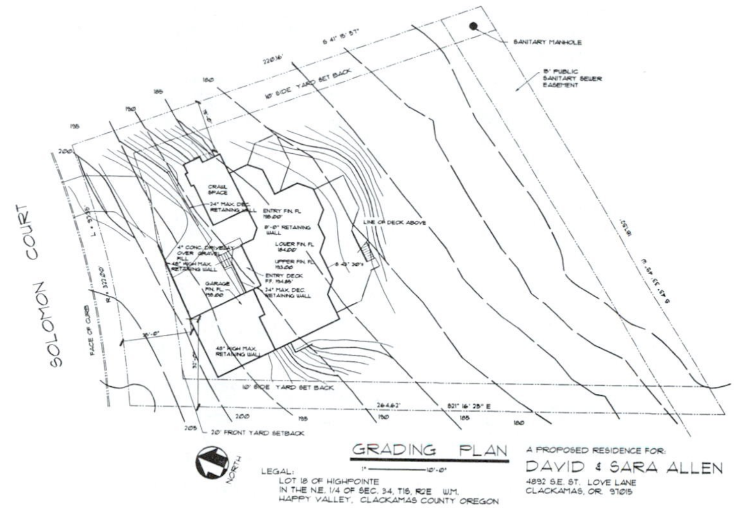

Figure 3.5 *Large parcels of land are often created using Engineering Units. (Courtesy Residential Designs.)*

drawing precision can be changed throughout the drawing, but should be kept as large as possible to facilitate drawing speed. The unit precision may be set using the Advanced Wizard or by using the following procedure.

- Select Units... from the Format pull-down menu.
- Select the down arrow in the Precision: drop-down list. (See Figure 3.6.)
- Select the desired increment of accuracy to be used. (For this example, select the 0.0 option from the menu).

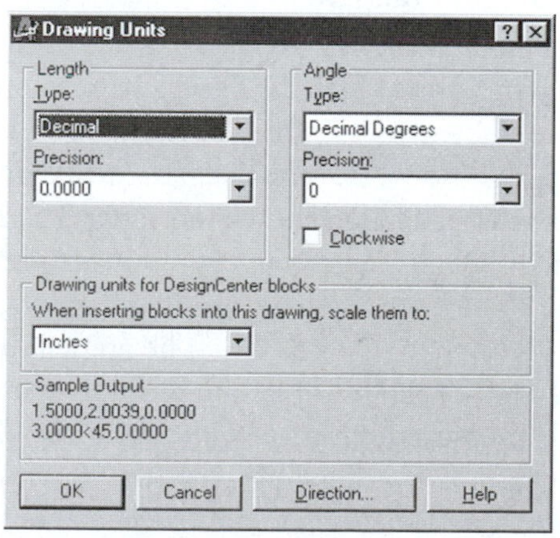

Figure 3.6 *Selecting Units... from the Format pull–down menu will display the Drawing Units dialog box.*

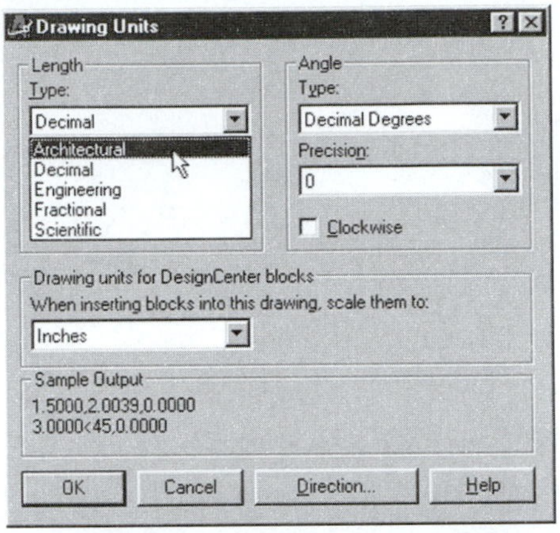

Figure 3.7 *Selecting the down arrow in the Type: drop-down list provides options for controlling the drawing units. The Architectural option will meet the needs of most drawings.*

- Open the lower edit box by picking the arrow and change Unitless to Inches.
- Select the OK button to close the dialog box.

Remember that using a template or a wizard will prompt you for this process

CONTROLLING ANGLES

If your drawing will include lines drawn at an angle other than vertical or horizontal, you'll need to decide how the angle increment will be displayed. Common angle units seen on construction drawings include angles measured in degrees/minutes/seconds and angles measured in degrees located from compass points.

Angle Measurement Type and Precision

The method of angle measurement and precision is set using similar methods that were used to set the drawing units. The Decimal Degrees option is best suited for plan views, such as those shown in Figure 3.8. This option will display angles in degrees and decimal parts of a degree. The Deg/Min/Sec option is best suited for land measurements, such as the site plan shown in Figure 3.5. This will display angles in degrees, minutes, and seconds. The Surveyor option presents surveyor's units, which also could be used to plot land, street, or sewer layouts, with angles displayed as bearings displayed as <N/S> <angle> <E/W>. The angle is based on north or south and will always be less than 90°. An example of common bearings is listed in Figure 3.9. The method of angle measurement may be set using the Advanced Wizard or be adjusted throughout the life of the drawing with the methods used in Start from Scratch. The method of angle measurement and precision is set using similar methods that were used to set the drawing units.

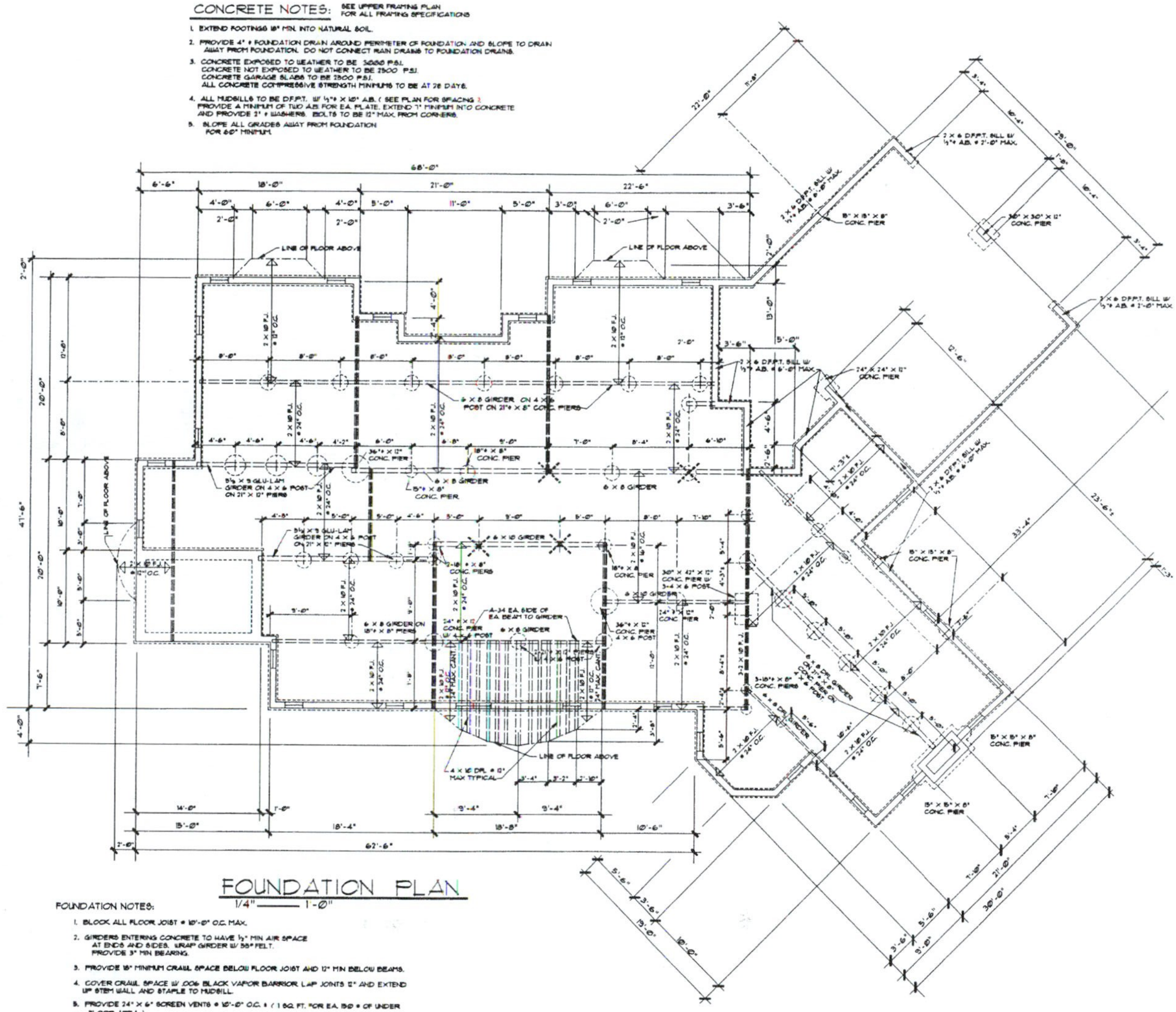

Figure 3.8 *Angles on a drawing such as this foundation plan can be measured conveniently using option 1, Decimal Degrees.*

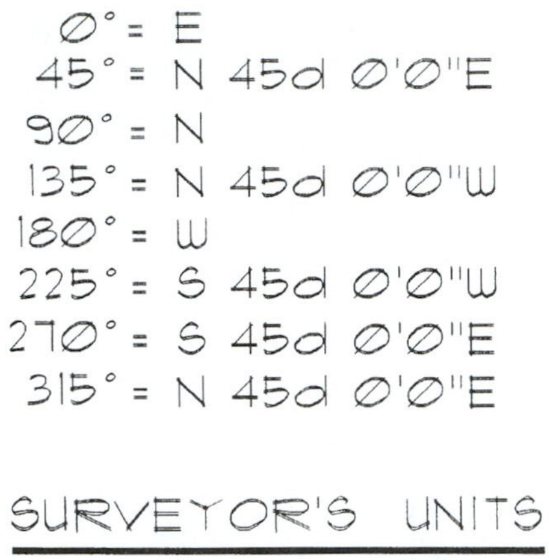

Figure 3.9 *Bearings are useful measuring units on site-related drawing. All angles are based on looking either north or south and turning either east or west.*

Angle Direction

Once the measurement units and precision are selected, you can decide the starting point for angular measurement. Selecting the Direction… button at the bottom of the Drafting Units dialog box will display the Direction Control dialog box shown in Figure 3.10. Figure 3.11 shows the visual representation of the angle menu. Standard AutoCAD layout of angles places zero degrees at 3 o'clock and then moves counterclockwise.

You may specify that angle measurement starts in any location. This will be particularly helpful if you want to work on a site plan where north does not point to the top of the paper. To place north in another location, enter the new starting angle at the prompt. The starting angle that is being specified is the location of east. Once the starting place is determined, you are given an option to decide the direction of angle

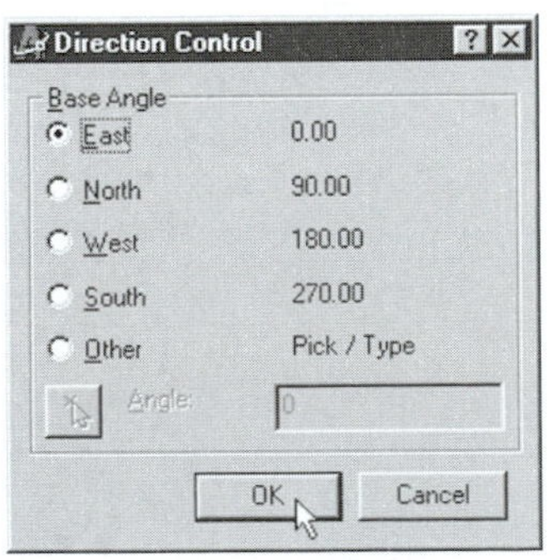

Figure 3.10 *Selecting the Direction… button at the bottom of the Drafting Units dialog box will display the Direction Control dialog box.*

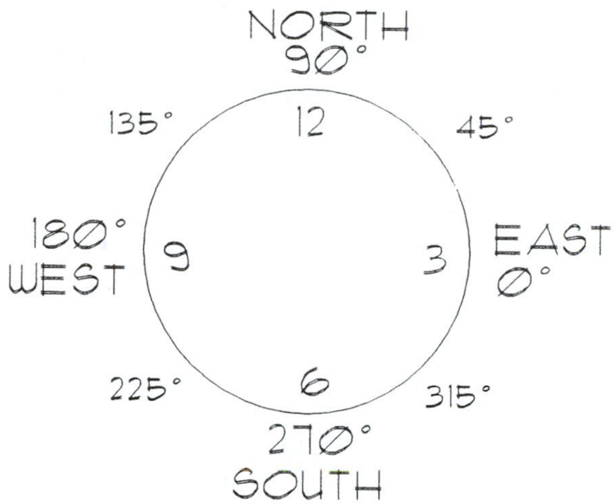

Figure 3.11 *A visual representation of the angle menu comparing compass and angular measurement.*

measurement. The default direction measures angles counterclockwise. To alter the direction to clockwise, select the Clockwise box in the Drawing Units dialog box. The angle direction may be set using the Advanced Wizard or be adjusted throughout the life of the drawing with the methods used in Start from Scratch.

AREA

Selecting the Area defines the area that future settings will control. Later in this chapter you'll be introduced to Snap and Grid. The area that is specified in the drawing setup will determine the location where these features will be displayed. The Area setting also adjusts the default settings for features such as text height and linetype scaling. The drawing area may be set using the Quick Wizard or the Advanced Wizard.

LIMITS

Setting UNITS is equal to deciding which scale to use in manual drafting. Setting the LIMITS would be equal to deciding what size drawing paper to use. As the drawing screen is configured now, it is equal to a 12" × 9" piece of paper. This is great for drawing a bolt, but very difficult for drawing an office structure. Remember, when you're working in model space, AutoCAD is drawing in real size, not at scale as in manual drafting.

Drawing limits are expressed by a pair of X, Y coordinates that represent the lower left corner, and another pair that represent the upper right corner of the drawing area. The drawing area or limits may be set using the Advanced Wizard or be adjusted throughout the life of the drawing by selecting Drawing Limits from the Format pull-down menu.

Selecting Drawing Limits

Your model space drawing limits will be determined by the size of the object to be drawn and the number of notes and dimensions to be placed. The paper space drawing limits will be determined by the size of the paper your drawing will be plotted on. Drawing limits for model space may be set arbitrarily and adjusted throughout the life of the drawing. For instance, you may start to draw a site plan with the limits of 300', 250'. Halfway through the project you may decide you need to draw the adjoining lot and may need to change the limits to 750', 475'.

Limits can also be set by the size of paper the project will be printed on. Common paper sizes include A: 8½" × 11", B: 12" × 18", C: 18" × 24", D: 24" × 36", E: 36" × 48". Remember that these sizes are in inches. If you want to draw a structure that will be printed on C-size paper at a scale of ⅛" = 1'-0", there are eight feet per inch of paper. Multiply 24" × 8 and 18" × 8 to determine the limits. This would produce limits of 192' × 144'. You also need to plan for the border that will be placed on the drawing sheet, and keep in mind that the plotter can not plot near the edge of the paper. Most plotters require between ½" to ¾" of space between the paper edge and the first printed material. This will be explained in more detail as title blocks are added to a drawing. Future chapters will introduce listings of common scales, paper size limits, and suggestions for setting up drawing templates. Once the area values have been entered, select the Next > button to proceed.

Note: A template is a sheet of paper that contains all of the drawing settings that you will need to complete a drawing. It will include the title block and border, text fonts, dimension settings, and every other setting that will be introduced throughout this text. Because most architectural drawings are created on D-size material, once you've created a D template, you'll be set for all future drawings.

GRIDS

Another helpful tool for drawing layout is GRID. This will provide the equivalent of the nonrepro blue grid that is preprinted on drafting vellum. The GRID command will produce a visible grid of dots at any desired spacing. The grid size can be adjusted throughout the life of the drawing, depending on the size of the object being designed. The grid also can be displayed or removed to aid visual clarity, but will not be produced when the drawing is printed or plotted, even if it is displayed on the screen. Figure 3.12 shows an example of a floor plan with the grid display. The current status is displayed in the Task bar. Options for grid settings include ON, OFF, Snap, and Aspect. Each option will be discussed later in this chapter.

SNAP

When working on a modular component, it is helpful to be able to control the accuracy of lines entered with the mouse. The SNAP command sets up an imaginary rectangular grid that controls placement of the drawing cursor in precise intervals.

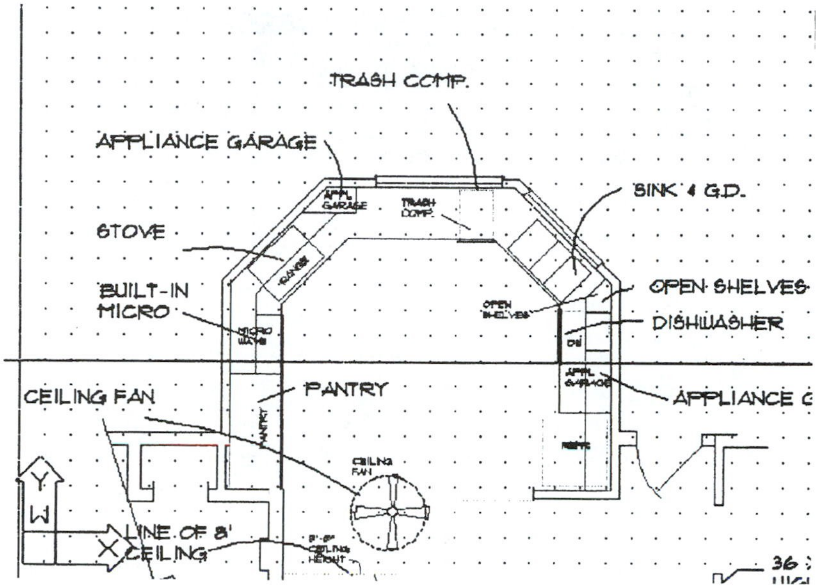

Figure 3.12 *Drawing limits of 15', 12' with a GRID value of 6".*

The distance between cursor movements is defined as snap resolution. Snap intervals can be set at any interval and may be changed throughout the life of the drawing. A change in the snap grid affects only the points entered after the change in resolution. Points entered prior to the change in resolution may no longer line up with the new snap location. The size of the grid setting and the size of the object to be drawn will effect the snap value. If the grid is set at 2', a snap value of 6" or 12" will be helpful in producing accuracy. If a detail of a beam to post is drawn, a snap of ½" might be used to accurately move through the drawing. If a 200-foot-long structure is to be drawn, a 4" or 6" snap will be more helpful than the ½" accuracy used for a detail. The snap value can be adjusted throughout the life of the drawing using the methods described in Start from Scratch. The current status is displayed in the Task bar. Snap settings include toggling between ON/OFF, Aspect, Rotate, and Style.

DEFINING THE DRAWING ENVIRONMENT

Five methods are available to set up a drawing. Four are listed in the Startup window. The fifth method is to close the Startup window, and begin drawing. This was the method used when you explored Chapter 2. It is the equivalent to taking a piece of vellum with no other tools and starting to draw. This system worked well in kindergarten but is not suitable for the construction industry. AutoCAD offers the use of the Open an Existing Drawing, Use a Wizard, Use a Template, and Start from Scratch options to organize drawings. Each can be seen in the Startup dialog box in Figure 3.13. Each has advantages for drawing setup.

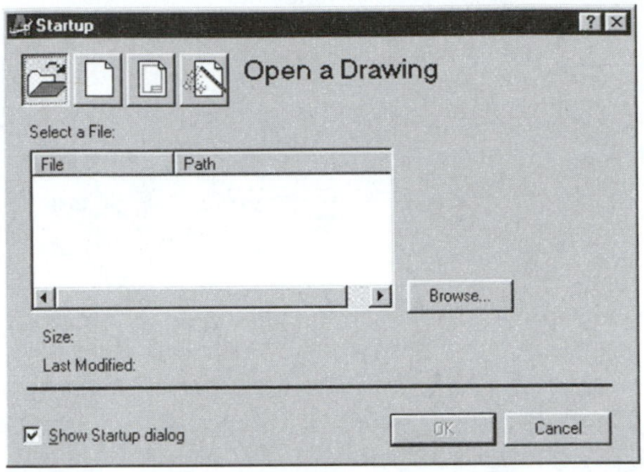

Figure 3.13 *The Startup dialog box is displayed as AutoCAD is entered to provide help in setting a drawing parameter.*

- The Open an Existing Drawing option will open a drawing file created in a previous drawing session, and allow you to edit or add to the existing drawing.
- Use the Drawing Wizards to start a drawing session when you know the scale or the paper size that will be needed to plot the project. This will allow you to use the setup assistance from AutoCAD
- Use a template if you have preset drawing parameters established that will meet the current drawing needs.
- Use the Start from Scratch for sketching, line drawings that will not be drawn to scale, or for drawings that do not require a high level of accuracy. This option also works great when you want to establish the drawing's parameters as they are needed. Although this option provides some basic drawing controls, many of the controls and variables will need to be added to increase your drawing efficiency.

Note: If the Startup menu is not displayed as you enter AutoCAD, someone has hidden it from view. When hidden, you will automatically enter AutoCAD using the last method that was used to access the drawing area. The Startup dialog box can be displayed or hidden from view by selecting Options… from the Tools pull-down menu, and then selecting the Systems tab of the Options dialog box. Use the Show Startup Dialog button in the General Options box to control the display of the Startup menu This can be a setting that you adjust and save as you create your own profile in Chapter 5.

OPEN AN EXISTING DRAWING

Much of your time in an office will be spent adding to or editing existing drawings. The default option in the Startup menu is Open a Drawing. Selecting this option will display a Startup menu similar to Figure 3.14. The Select a File: edit box will display

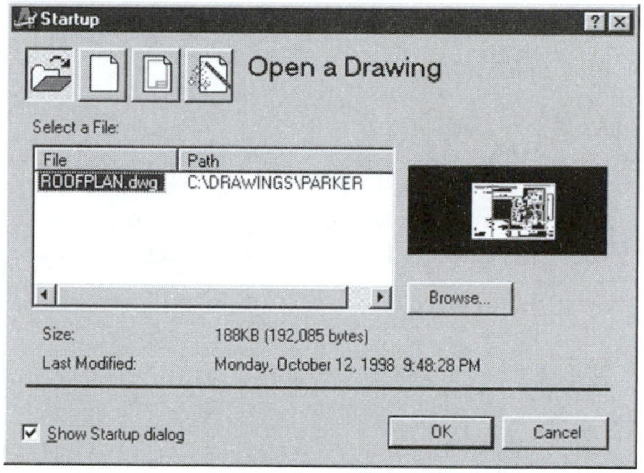

Figure 3.14 *Selecting the Open an Existing File icon will display this dialog box. The last four drawings that you have worked on will be displayed in the Select a File: edit box. Files that are not listed in the Select a File edit box can be found by selecting the Browse... button. To open an existing drawing, select the name of the desired drawings and select the OK button.*

a list of the last four drawings that have been opened. Notice that the last drawing listed is the file created in Chapter 2, titled EXE2-1. Selecting the OK button will remove the dialog box and open the drawing that was created in Chapter 2. As you develop more than four drawings, the drawing file that you want to open may not be displayed in the Select a File list box. Select the Browse... button to find an existing drawing file that is not listed in the Startup dialog box. Selecting Browse... will produce a Select File dialog box similar to Figure 3.15. The display will vary depending on the number of drives and the file contents.

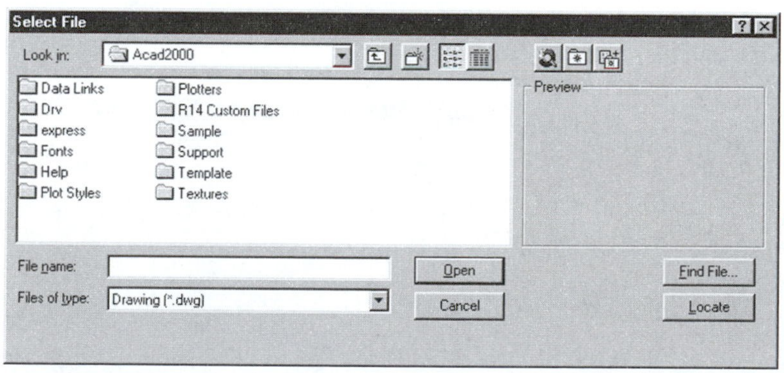

Figure 3.15 *Selecting the Browse... button will produce a listing of availabe drives. Once in the desired drive, select the desired file to be opened. An image of the selected file will be displayed in the Preview box. If this is the file you want to open, select the Open button. If this is not the file you expected, select another filename. Once you've found the desired file, select the Open button.*

The Select File dialog box can be used to list and select a drawing file from any of your storage locations. By default, files will be selected from the Acad2000 folder. Use the following steps to open an existing file that is not listed in the Select a File listing:

- Select the arrow beside the Look in: drop-down menu to select the desired drive that contains the desired drawing file to be opened. As the arrow is selected, a drop-down menu is displayed, reflecting the contents of the Windows Explorer. (See Figure 3.16.)
- Move to the desired drive just as you would using the Windows Explorer and select the file to be opened.
- Clicking the desired drive to be searched displays a list of the drawing folders contained in the selected drive.
- Double-clicking a folder produces a list of files contained in that folder. Highlighting a file created in Release 13 or 14 with a single click displays an image of the file in the Preview box.
- If this is the file you want to open, select the Open button to remove the dialog box and open the selected drawing on the drawing area, similar to Figure 3.17. If this is not the file you expected, highlight another filename. Once you've found the desired file, select the Open button to bring the file into the drawing area. A drawing can also be opened by double-clicking the filename.

Note: Remember AutoCAD allows multiple drawings to be opened. An existing drawing can also be opened in the middle of another drawing session by selecting the Open icon from the Standard toolbar, by selecting the Open icon from the File of the pull-down menu, or by typing OPEN (EnterKey) at the command prompt. Each method will produce the dialog box shown in Figure 3.15.

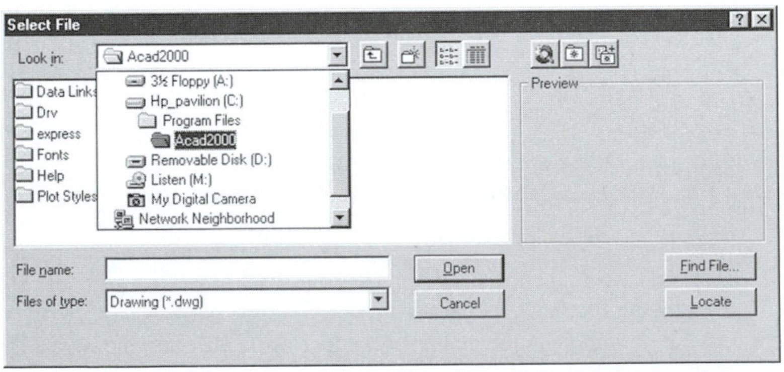

Figure 3.16 *The Look in: drop-down menu is used in a similar manner to using the Windows Explorer.*

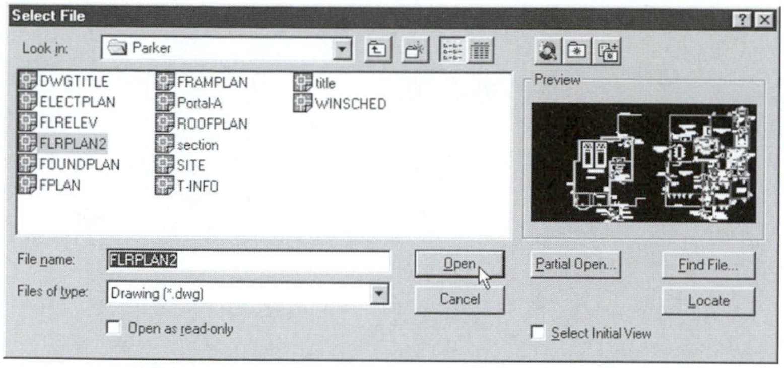

Figure 3.17a *Selecting a drawing file from the browser is a result of highlighting a drawing file and selecting the Open button.*

DRAWING SETUP USING A WIZARD

Using a wizard is an excellent method of starting a new drawing session if you know the size of paper the project will be plotted on. AutoCAD provides the Quick Startup wizard and the Advanced Setup wizard. The method chosen will depend on how much help you would like. Choosing Quick Startup will assist you in setting the drawing units and the drawing area. The Advanced setup will provide six startup prompts.

Quick Setup

The Quick Startup option produces a drawing area in model space. This option works well for drawing, but requires some adjustment for plotting. Selecting the OK button will produce the display shown in Figure 3.18a and allow the selection of the drawing units. Use the mouse to place the cursor in the desired setting and press the select button. This will place a black dot in the button, indicating the selection is activated. (See Figure 3.18b.)

Select the Next > button to continue the drawing setup. This will produce the display shown in Figure 3.19, showing a default value of 12" × 9". The drawing area should be based on what you expect to draw. For now, accept the default value and select the Finish button. This will display the drawing screen shown in Figure 3.20. Notice that the grid is confined within the drawing area of 12" × 9", even though the screen displays a larger area.

Advanced Setup

With this button active, the AutoCAD wizard will display the dialog box shown in Figure 3.21. The Advanced Setup wizard will help you set six controls related to the units of measurement, angle of measurement, direction of angle, and area. These controls open a new drawing and provide a title block that is viewed in either paper or model space, depending on the decisions made during the setup. Select OK to start the wizard.

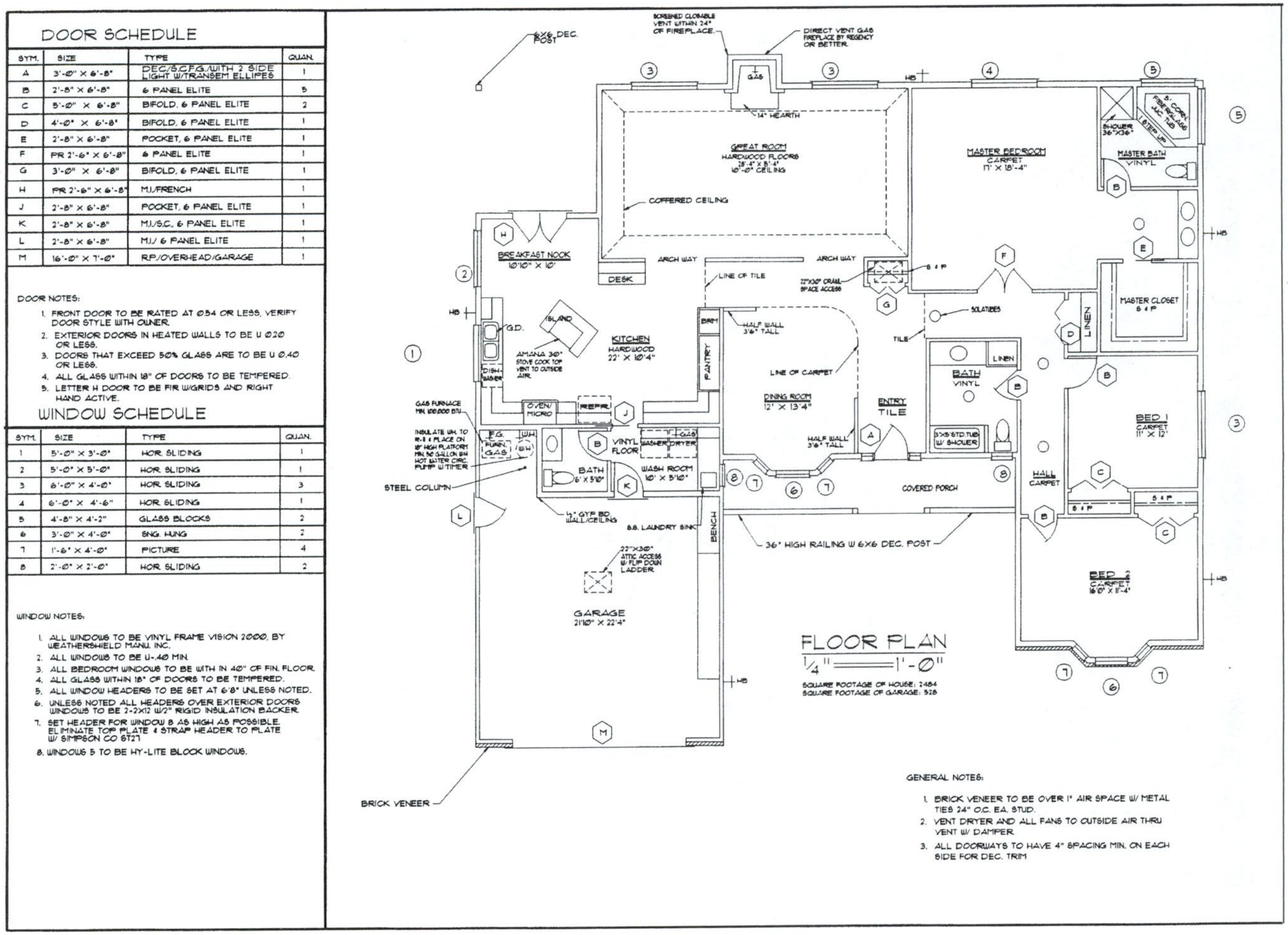

Figure 3.17b *The result of selecting the floor plan in Figure 3.17a.*

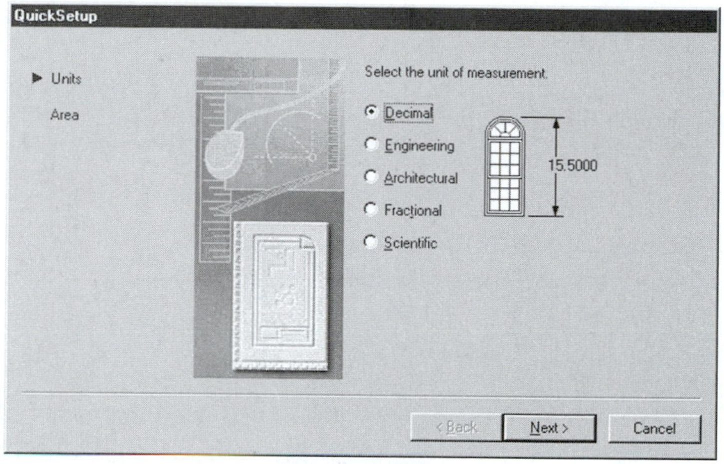

Figure 3.18a *Selecting the Use a Wizard button provides two choices for drawing setup. Choosing Quick Setup produces this display to aid in the selection of the drawing units.*

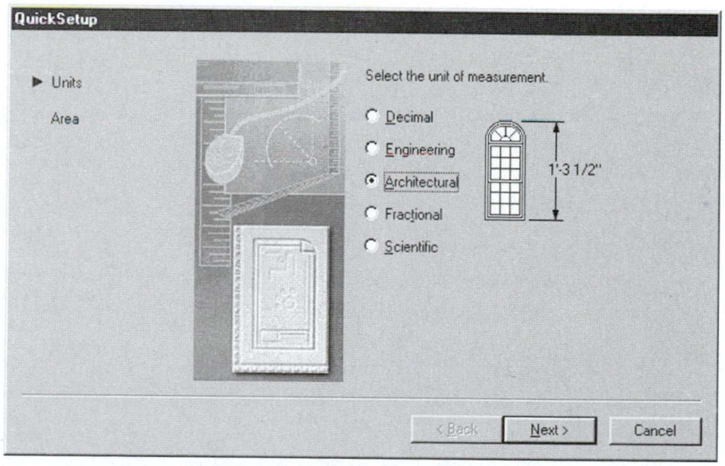

Figure 3.18b *Selecting the Architectural button will display distances measured in units of feet and inches.*

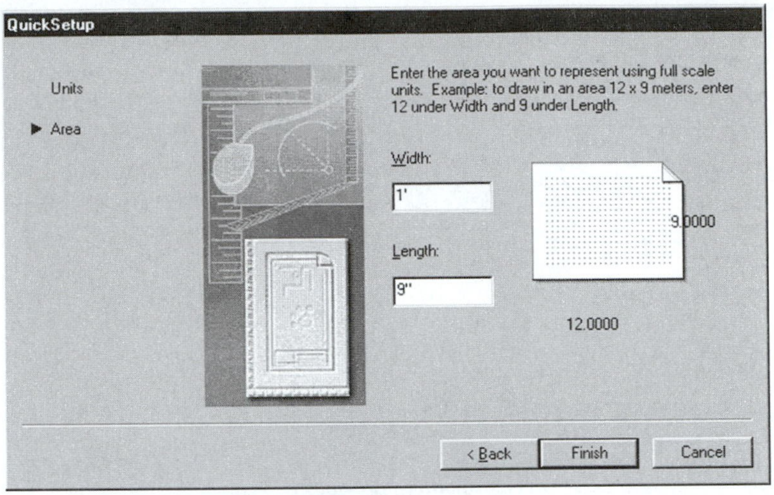

Figure 3.19 *Setting the area defines how much space is displayed on the screen. Set the area slightly larger than what you think you will need to complete the drawing.*

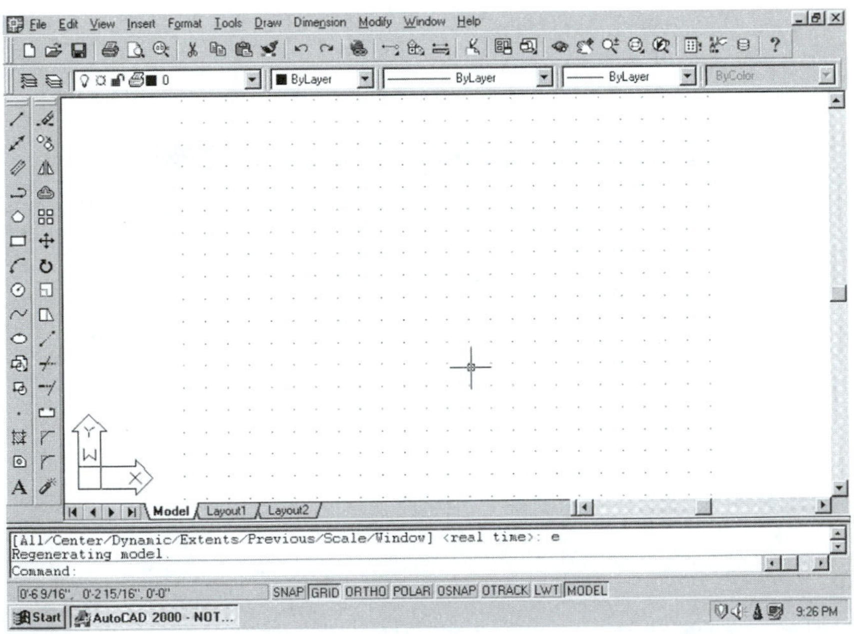

Figure 3.20 *Accepting the default value of 12" x 9" and selecting the Finish button will display the drawing screen. The grid is displayed in the drawing area by selecting the GRID button in the Status bar.*

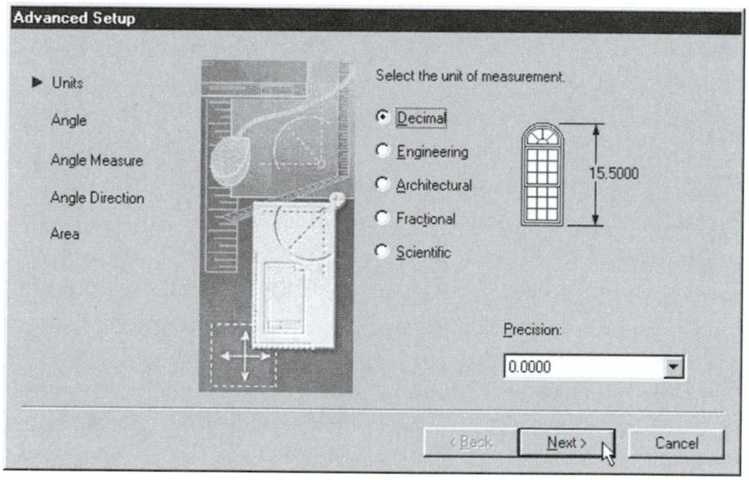

Figure 3.21 *Selecting the Advanced Setup wizard option will aid in setting six drawing aids.*

Units

The Units control folder is displayed as the Advanced Setup is opened. The default value is decimal. Choose the value that best meets the need of the drawing to be started. Figure 3.22 shows the value set to record Architectural units and displays a sample of how measurements will be displayed.

Unit Precision

The Precision edit box is located at the bottom of the Units dialog box. The precision is now set at 0'-0 1/16" for Architectural units. To change the setting, select the arrow

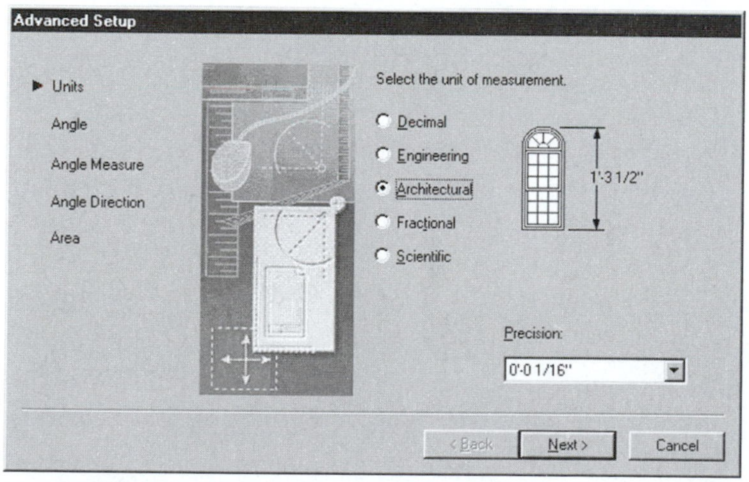

Figure 3.22 *The first step in using the Advanced Setup wizard is to set the drawing units. Notice this method allows the precision of the units to be measured.*

on the right side of the box. This displays the accuracy menu shown in Figure 3.23. Selecting the down arrow will scroll through the menu, allowing a greater degree of accuracy to be selected. Selecting the up arrow sets the accuracy in larger units. For now, highlight the 0'-0" setting, and then select the option with the mouse. This will return you to the units box and allow you to continue the drawing setup. Notice, as the unit precision was changed from 1⁄16" to 0", the display of units was also changed. The units will now display 1'-4". If you're satisfied with the units, select the Next > button to continue the drawing setup. The units can also be changed at any point during the life of the drawing. Selecting the Next> button will set the Units, exit the wizard, and place the drawing area on the screen. Select the Next> button to allow the wizard to continue.

Angles

Once the Units are selected, the folder for controlling angles will be displayed. Figure 3.24 shows the Angle display. Most architectural drawings can be completed easily using decimal angles. For now accept the default option of Decimal Degrees. The precision for measuring angles can be set using the same method used to set the unit precision. Selecting the down arrow will provide options for zero to eight decimal places, with the default of 0. Normally, two-place accuracy is sufficient for architectural drawings, while civil drawings might require a greater accuracy. Accept the default by selecting the Next > button to proceed to the next wizard.

Angle Measurement

The Advanced Setup wizard allows the direction of angle measurement to be selected. The default option is east. Options are shown in Figure 3.25. To change the starting point of angles to north, select the North radio button with the mouse select

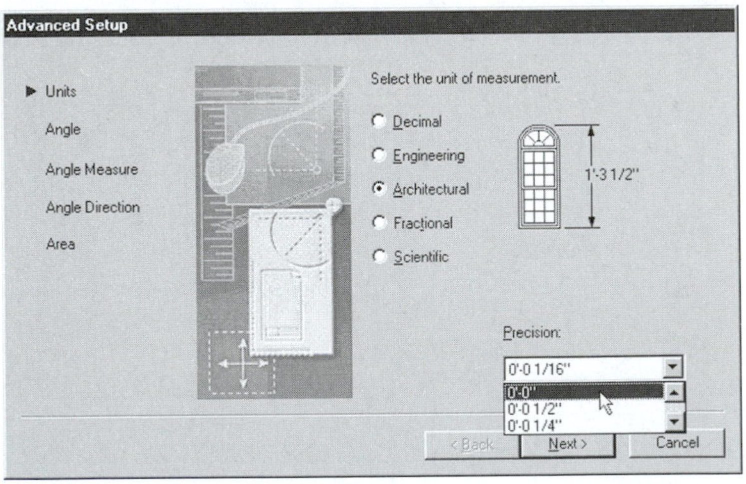

Figure 3.23 *Selecting the Precision button displays the Units menu. Accepting the default value displays units with an accuracy of 1⁄16". Selecting 0'-0" will measure units in whole inches.*

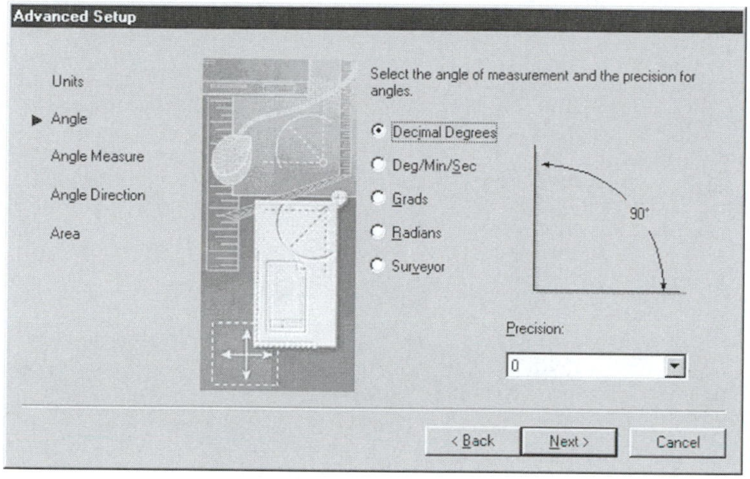

Figure 3.24 *The Advanced Setup menu allows the units for measuring angles and the precision to be selected.*

button. If you select north as the new starting direction, 0° will now be placed at north, 90° will be west, 180° will be south, and east will be at 270°. An angle other than one of the four compass points can be selected by choosing the Other button, and then entering the desired starting angle in the edit box. Selecting Other and entering a starting angle of 45° will place 0° halfway between north and east. When the desired starting point has been selected, selecting Next > will allow the wizard to continue.

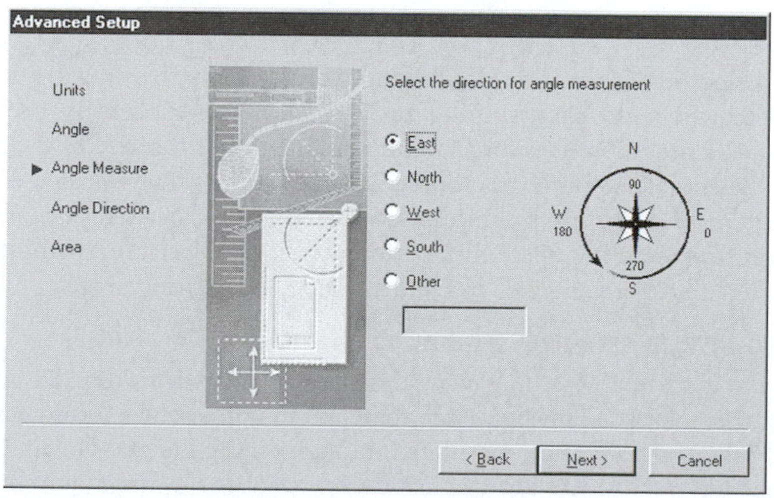

Figure 3.25 *The default starting point for measuring angles is east. One of the other three quadrants, or any point in between, can be selected by choosing Other and providing an angle.*

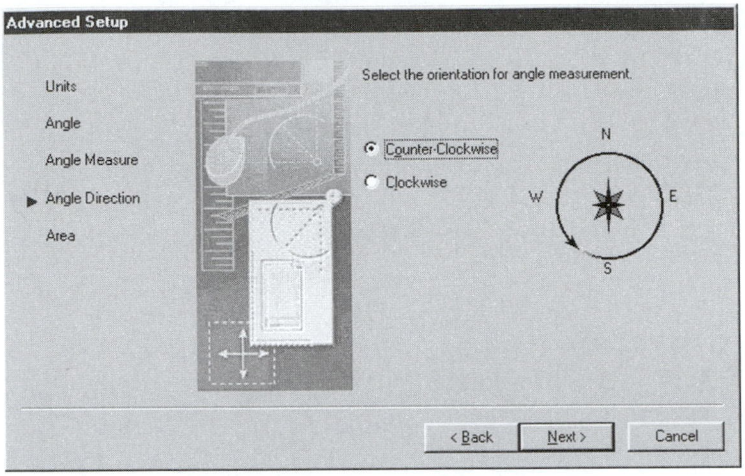

Figure 3.26 *The Advanced Setup wizard allows the direction for angle measurement to be entered. For most drawings the default setting of counterclockwise works well.*

Angle Direction

Figure 3.26 shows the Advance Setup wizard for selecting the directions that angles will be measured. The default setting for angle measurement is counterclockwise. Selecting the Clockwise button will change the direction of angle measurement. For now, accept the default. Remember that the defaults are there because they fit the needs of most users. The direction, like each of the other variables, can be altered at any time throughout the life of the drawing. Once the direction is selected, select the Next > button to proceed.

Area

The final dialog box of the Advanced Setup wizard allows the drawing area to be specified using the folder shown in Figure 3.27. In the default setting, the drawing will be 1' × 9". Enter the values for the width and length of the area of the intended drawing to be completed in the appropriate edit box. The values entered will determine the extents that the grid will be displayed in. Enter the desired value, and select the Finish button. Selecting the Finish button will close the wizard, establish each of the parameters that were entered, and return you to the drawing area.

DRAWING SETUP USING A TEMPLATE

In addition to the Setup wizard, drawing templates with predetermined drawing values can be used to start a drawing. Each template contains values for drawing units, limits, and other values that can be altered throughout the life of a drawing. A template supplied by AutoCAD can be customized to meet specific requirements and saved for future use. Select the Use a Template button from the Startup window to use a template to start a drawing session. This will produce the display shown in

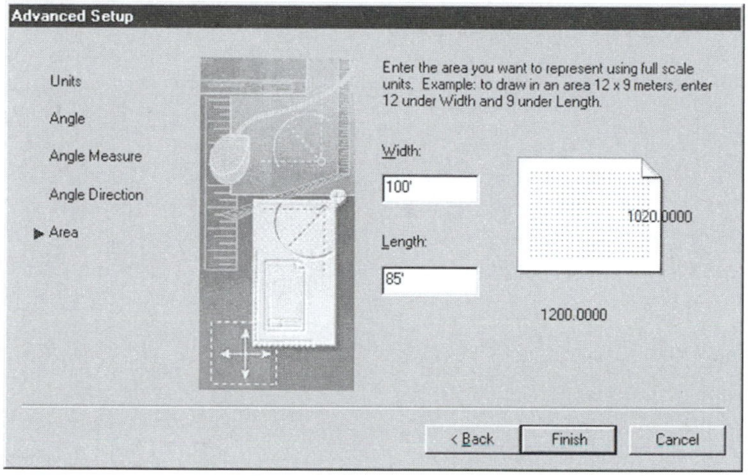

Figure 3.27 *Entering drawing units of 100' by 85' will allow the layout of a floor plan to be started.*

Figure 3.28. Accepting the default setting of AutoCAD-named plot styles.dwt and selecting the OK button will produce a display similar to Figure 3.20. The drawing area will be approximately 12" × 9" (size varies depending on monitor size) with units measured in four-place decimal inches. The limits can be adjusted as needed. Scroll through the list of templates and select Architectural, english-color dependent plot styles.dwt and the OK button. This will produce a drawing display as shown in Figure 3.29. In addition to using the templates supplied with AutoCAD, you may develop you own template to meet specific needs of a project. In addition to each of the drawing controls that were introduced in this chapter, the template could also include common text fonts, linetypes, layers, and other related features. Each of these

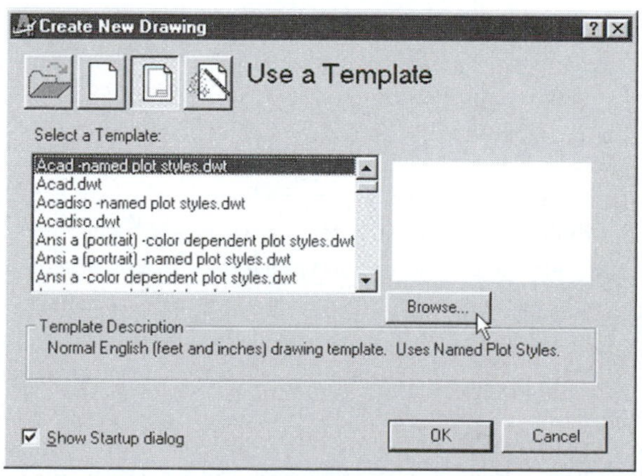

Figure 3.28 *Selecting the Use a Template button allows a drawing with preset drawing parameters to be used. Selecting Browse... allows user-created templets to be accessed.*

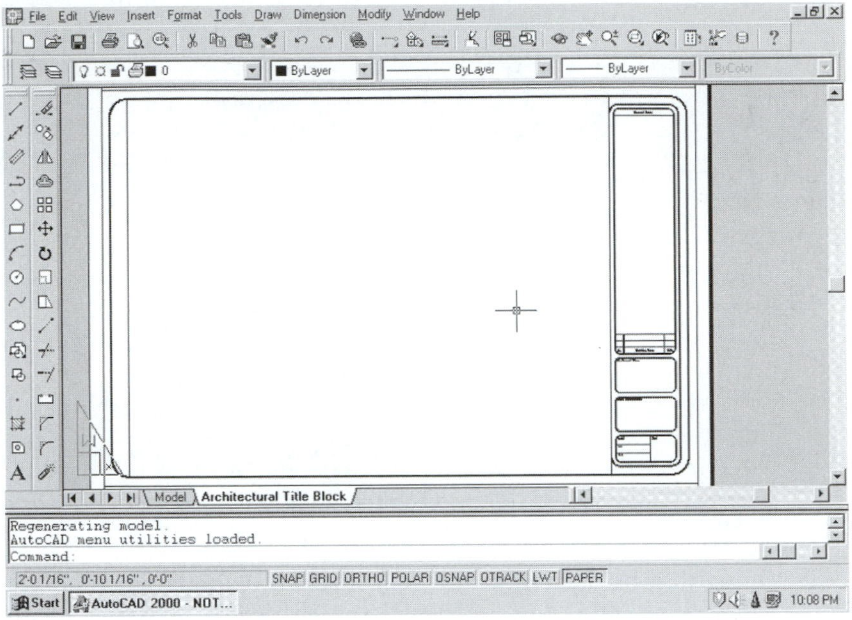

Figure 3.29 *A title block suitable for basic architectural or engineering projects. Chapter 5 will introduce options for creating and saving your own title block.*

components will be introduced throughout the text. As each is introduced, features that you find useful can be added to your template. Methods for creating and saving templates will be covered in Chapter 5. For now, if you choose to use a template to enter the drawing area, accept the default and OK.

Note: As the Architectural.dwt template is entered, you enter the drawing session in paper space. Select the Model tab at the bottom of the drawing area. This will remove the title block and produce a blank drawing screen. Now draw a few lines. Select the Architectural Title Block tab and the lines that you drew will be displayed on the template. Click the model space button in the Status bar. If your drawing does not fit into the template, select the Pan icon (the hand) from the Standard toolbar. Pan allows the drawing to be moved to the desired location. After selecting PAN, a hand will be displayed in the drawing area. Press the pick button, and move the cursor in the direction you would like to move the drawing. When you're satisfied with the location, press the (EnterKey) to exit the Pan command. If the drawing is too big to be seen in the template, use the Zoom command (the magnifying lens) from the Standard toolbar. After selecting ZOOM, a magnifying lens will be displayed in the drawing area. Press the pick button, and move the cursor down to reduce the size of the object. When you're satisfied with the location, press the (EnterKey) to exit the ZOOM command. The full power of each command will be introduced later.

STARTING FROM SCRATCH

Selecting the Start from Scratch icon from the Startup window will produce a display similar to Figure 3.20. This option allows a drawing to be started with a minimum of predetermined values. Units are displayed in four-place decimal units, with drawing limits of 12" × 9" (varies based on monitor size). Rather than being prompted by a wizard or restricted by a template, Start from Scratch allows drawing controls to be set as needed using a dialog box or from the command prompt. Keep in mind that if you enter a drawing using this option, that the units are set in four-place decimals with a 12" × 9" drawing area.

STARTING A NEW DRAWING FROM WITHIN AUTOCAD

Earlier in this chapter you explored the Startup options of AutoCAD. A new feature of AutoCAD allows a drawing to be started from inside of an AutoCAD drawing. Use one of the methods introduced earlier in this chapter to start a new drawing. Once the drawing is started, draw a triangle so that the screen resembles Figure 3.30a. A new drawing can be started without harming the current drawing. Another drawing window can be opened by selecting the New icon at the far left end of the Standard toolbar, or by selecting New... from the File pull-down menu. A new

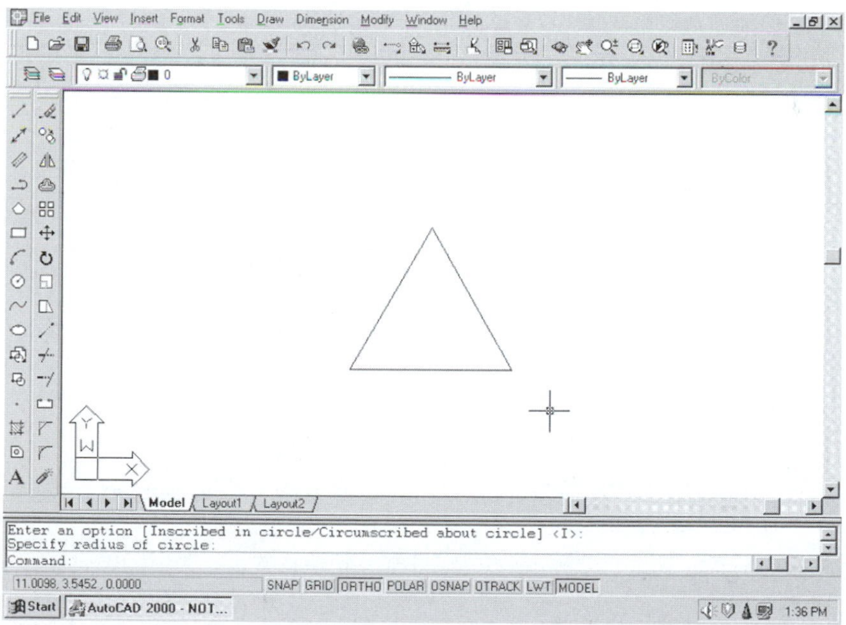

Figure 3.30a *A new feature of AutoCAD allows a drawing to be started from inside an active AutoCAD drawing. Use one of the methods introduced earlier in this chapter to start a new drawing. Once the drawing is started, draw a triangle.*

drawing can also be started by typing **NEW**(EnterKey) at the command prompt. Each method will remove the existing drawing file and begin the process of starting a new file. Start the new drawing using any one of the four Startup methods you prefer. Your original drawing will be removed from the drawing area, and a new drawing screen will be displayed. Use the LINE command to draw a polygon so that your drawing area resembles Figure 3.30b. Press the minimize button and the display will resemble Figure 3.30c. The name tabs of both of the open drawings are displayed. Several new drawings can be started, or you can open several existing files in one AutoCAD session. Although this might not seem important now, several of AutoCAD's new features will make this very important. AutoCAD allows you to transfer objects, settings, and other types of information from one drawing into another drawing. Some of AutoCAD's new features allow drawing attributes to be transferred between drawings using methods similar to those used with the Windows Explorer. Drawing features can be moved between a drawing stored on your hard drive, from a drawing on a network, or from a drawing located across the Internet. These features will be addressed throughout the balance of the text. For now, remember that two or more drawings can be worked on in the same drawing session.

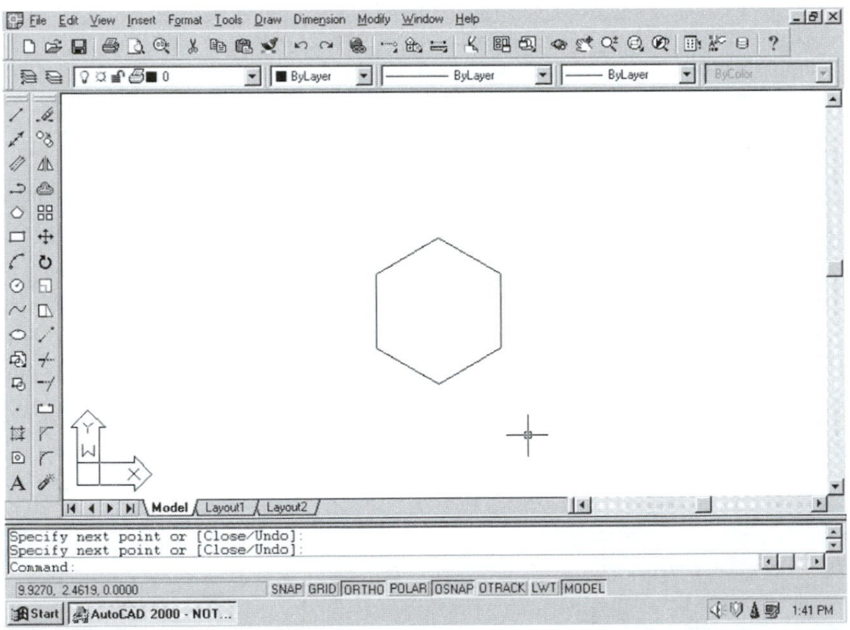

Figure 3.30b *A new drawing can be started without harming the current drawing. Another drawing window can be opened by selecting the New icon from the Standard toolbar, by selecting New... from the File pull-down menu, or by typing **NEW** (EnterKey) at the command prompt. Each method will remove the existing drawing from view and begin the process of starting a new file.*

Figure 3.30c *Press the minimize button and the current display will be minimized to display the drawing tab for all current drawings.*

Displaying Windows

As with other Windows programs, AutoCAD provides choices on how open windows will be displayed. Individual program or drawing windows within AutoCAD can be arranged in horizontal, vertical, or cascading tiles. Multiple layouts can be displayed by selecting Window from the Standard toolbar. The menu can be seen in Figure 3.31. The lower half of the menu will display a listing of the current open drawing layouts. Selecting the name of the desired view will remove the current drawing and display the selected drawing. Multiple windows are displayed by selecting one of the options from the top half of the menu. A cascading or tiled display is often used when two or more windows are to be displayed.

Cascading Windows

Selecting Cascade from the Windows menu displays the program windows similar to those shown in Figure 3.32a. Cascading windows are stacked over each other in descending order. Programs are displayed in the order that they are selected. To display one of the lower windows, click the menu bar of the desired window to move it to the top of the stack. Clicking the menu bar and holding the pick button down allows the window to be dragged anywhere on the screen.

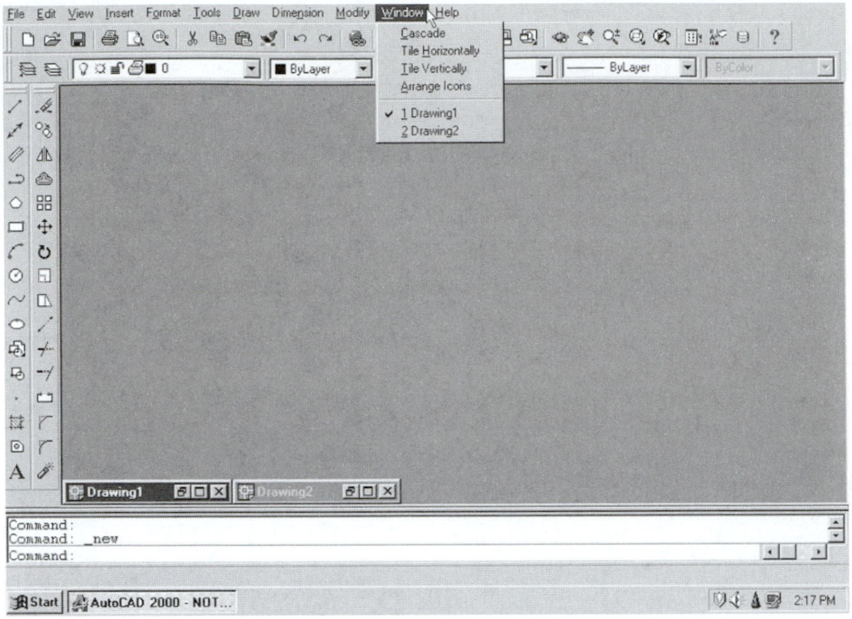

Figure 3.31 *Multiple layouts can be displayed by selecting Window from the Standard toolbar.*

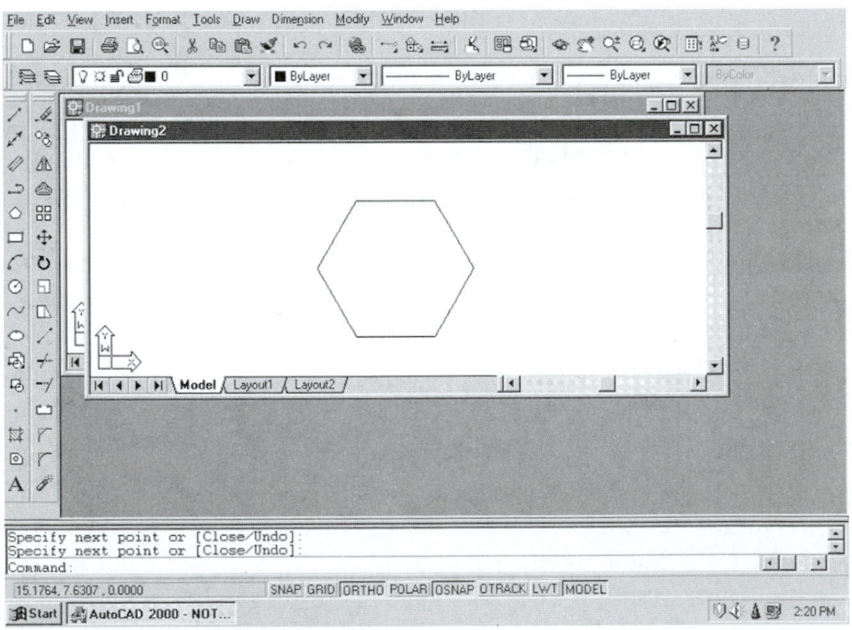

Figure 3.32a *Selecting Cascade from the Windows menu displays drawing windows that are stacked over each other in descending order.*

Tile Displays

The Tile options allow drawing windows to be arranged in horizontal (Figure 3.32b) or vertical displays (Figure 3.32c). Tiled windows permit each window to be viewed at the same time. Just as with other programs, if you're unhappy with the size of a window, the size can be altered by selecting and dragging the edge to the desired size.

WORKING WITH PARTIAL FILES

AutoCAD 2000 offers two new methods of working with drawing files. Although each method is beyond your current needs as you explore AutoCAD, you should at least be aware of the PARTIALOPEN and PARTIALOAD commands. Both commands can be used when working with large files to allow only a specific portion of the drawing to be accessed. The projects that you work on throughout this text will typically be too small to benefit from PARTIALOPEN. On extremely detailed drawings, PARTIALOPEN will allow specific information to be viewed in the drawing area. If you're working on a complex floor plan, the plan could be divided into zones, and then each zone can be saved by what AutoCAD refers to as a view. PARTIALOPEN would allow you to open a specific zone of the floor plan and display the framing information while hiding the electrical or plumbing information. See Chapter 5 for an explanation of how each command can be used to start a drawing session.

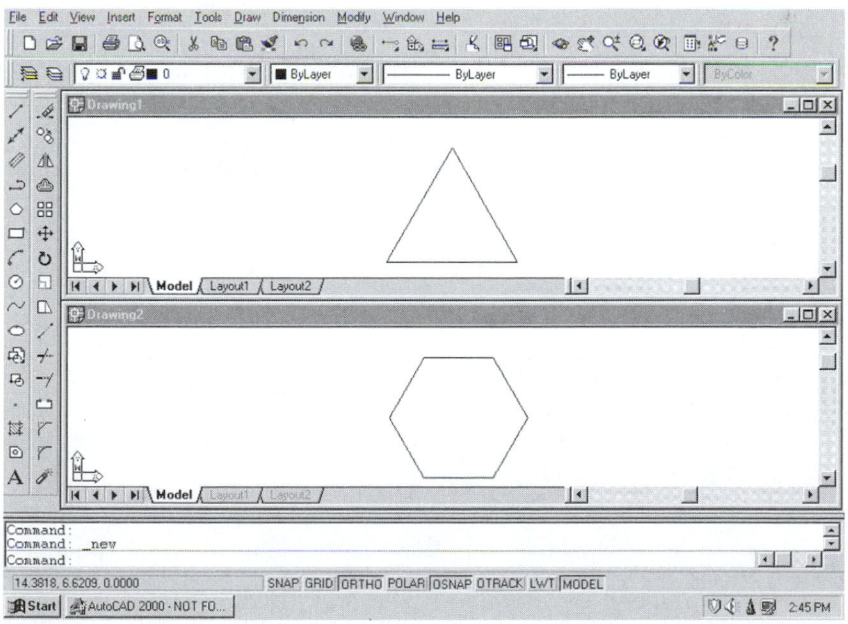

Figure 3.32b *The Tile Horizontal option displays drawing windows in horizontal rows.*

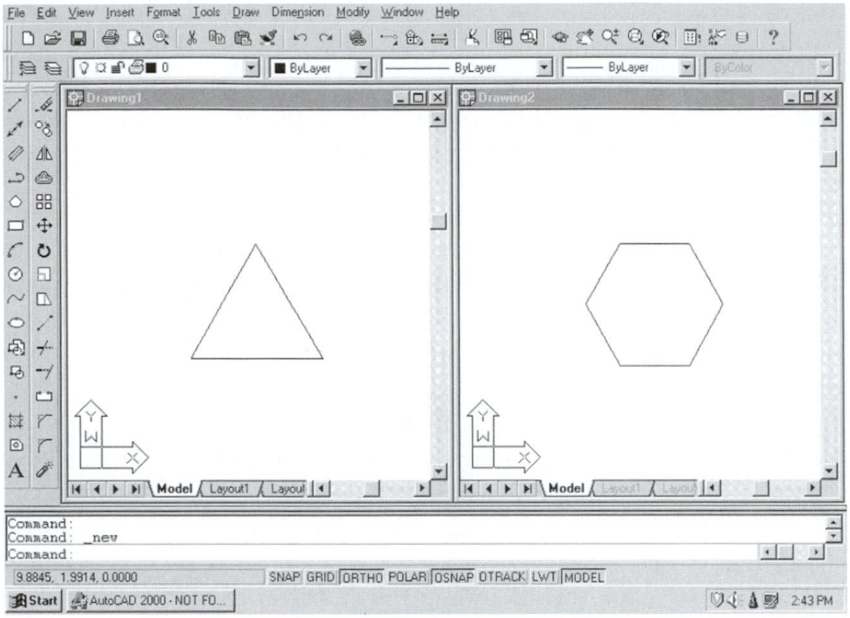

Figure 3.32c *The Tile Vertical option displays drawing windows in vertical columns*

Note: You now have several methods that can be used to start a drawing session. Just as important as how you start a drawing session, is where you select the file that you want to open from. When you're working with existing files, always use files stored on your hard drive to open the drawing session. AutoCAD needs workspace as you work on a drawing session. If the drawing and the workspace exceed the storage space on the disk, you may encounter unwanted results. If the drawing is stored on a network, you'll typically have no shortage of workspace, but your hard drive is generally faster. If the most current file is saved on a floppy or in a network folder, copy the file to the hard drive prior to opening the file.

CONTROLING DRAWING PARAMETERS FROM WITHIN A DRAWING

No matter what method is used to enter a drawing, the drawing parameters can be altered at any time. Parameters can be altered using dialog boxes, pull-down menus, or the keyboard.

SETTING UNITS

Drawing units can be changed by using the Drawing Settings dialog box. (See Figure 3.33.) This box was introduced earlier in this chapter, and will be automatically displayed if a drawing wizard is used. It's important to remember that settings can be altered at any point during the drawing session. The Drawing units dialog box can be displayed at any point of a drawing session using one of the following methods:

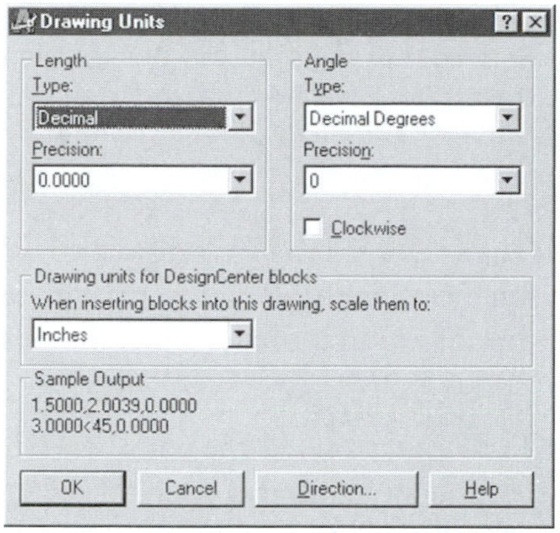

Figure 3.33 *Drawing units can be altered at any time using the Drawing Unit dialog box. The dialog box is displayed by selecting Units... from the Format pull-down menu, or by typing* **UN** *(EnterKey) at the command prompt.*

- Select Units... from the Format pull-down menu.
- Type **UN** (EnterKey) at the command line.
- The Drawing units are set using by selecting the desired style from the Type menu. Selecting the arrow on the right side of the Precision edit box will display the drop–down menu. The desired precision is selected by choosing the desired value from the precision menu. The box can also be used to select the method for measuring angles and for setting the angle precision. Angle precision accuracy is set using the same methods to change the unit precision. Selecting the Direction... button will display the Direction Control menu. (See Figure 3.34.)

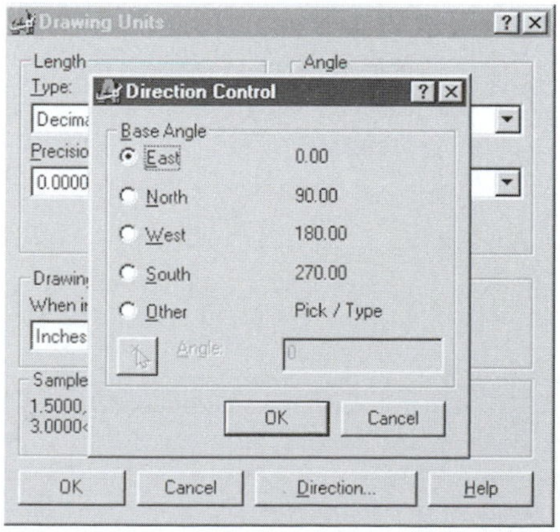

Figure 3.34 *Selecting the Direction... button in the Drawing Unit dialog box allows the orientation for angle direction to be altered.*

This menu allows for the angle starting point and the angle direction to be set, by selecting the appropriate radio button or highlighting the desired accuracy. Once selected, the accuracy will be displayed in the edit box and the menu will be closed.

SETTING LIMITS

Changing the limits alters the size of the drawing displayed on the screen. Before you change the limits, draw a square inside the drawing area, similar to Figure 3.35a. This will help you to visualize the change that occurs when limits are set.

To change the size of your "paper," select Drawing Limits from the Format pull–down menu or type **LIMITS** (EnterKey) at the command line. The drawing limits can also be altered in the middle of another command by typing **'LIMITS** (Enter-Key) at the command line. Each method of entering the command will produce the display:

Reset Model space limits:

Specify lower left corner or [ON/OFF] <0'-0",0'-0">:

The acceptable responses are ON, OFF, and (EnterKey).

ON will keep the current values (1'-0",0'-9") and activate limit checking. Limit checking will not allow objects to be drawn that are outside the limits. Think of limit

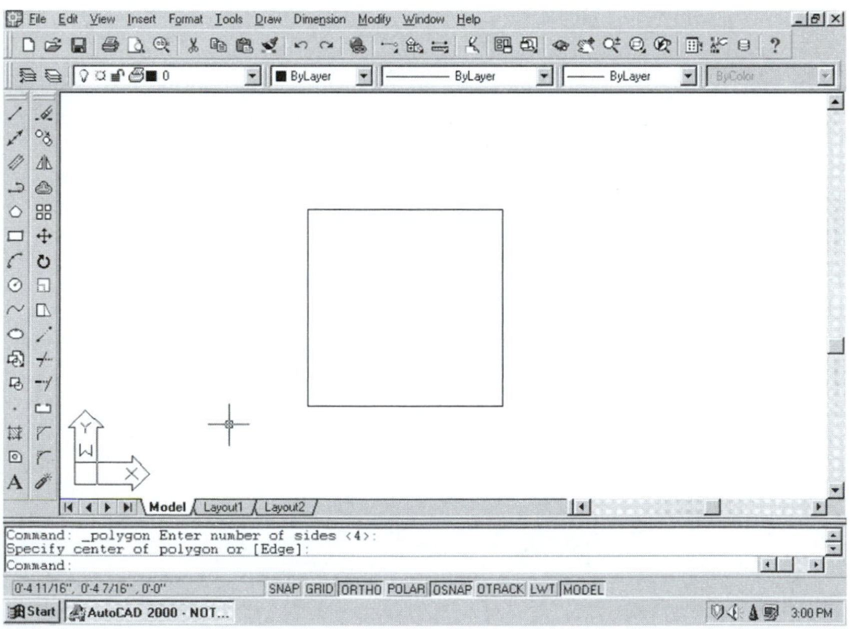

Figure 3.35a *With the drawing limits set at the default units of 1'-0", 9", an object may appear quite large.*

checking as a border on your drawing paper to keep your project on the paper. OFF will disable limit checking but retain the limit values for future use of limit check.

Pressing (EnterKey) retains the current location of the lower left drawing limit of (0.0000,0.0000). This will be the most typical response. Pressing (EnterKey) produces the prompt:

Specify upper right corner (1'-0", 0'-9"):

The default value is 1'-0" × 9", with the first number representing the horizontal dimension and the second number representing the vertical dimension. This is your opportunity to change the size of the drawing area to fit your current project. Enter 15',12' (EnterKey), and the command prompt will return with no apparent change. Type **ZOOM** (EnterKey) at the keyboard, then **A** (EnterKey) (the A represents all). Now the effect of your new limits can be seen in Figure 3.35b. Remember that the box is still the same size but it appears smaller because the paper size has increased.

SETTING GRIDS

Grid can be set by dialog box or from the command line. Once the values have been set, the grid can be toggled ON/OFF by double-clicking the Grid button in the status bar. The grid can also be toggled ON/OFF in the middle of another command by using the F7 key or by pressing the Ctrl+G keys.

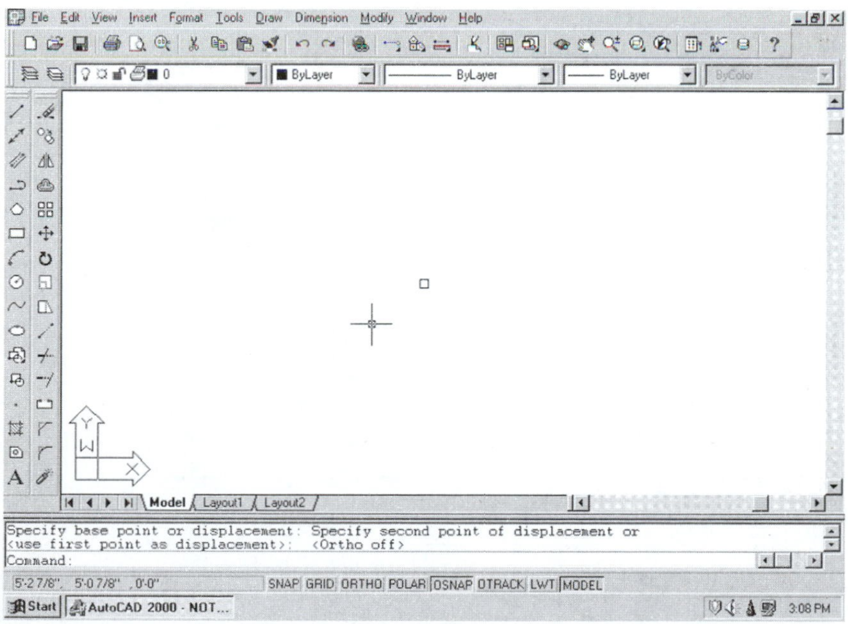

Figure 3.35b *By changing the location of the upper right-hand corner, and pressing the (EnterKey), the screen will allow viewing of a larger area. In this example, the limits are set to 15', 12'.*

Assigning Grid Values Using a Dialog Box

Grid values can be set using the Drafting Settings dialog box. The box is accessed by selecting Drafting Settings... from the Tools pull-down menu. This will produce the dialog box shown in Figure 3.36. The dialog box can also be displayed by moving the cursor to the Grid or Snap buttons on the status bar and right-clicking the Settings... option, or by typing **RM** (EnterKey) at the command prompt. Once the Drafting Settings dialog box is displayed, grids are set using the following procedure:

- Select the Snap and Grid tab.
- Enter the desired value for the X (left-to-right) grid spacing.
- Enter the desired value for the Y (top-to-bottom) grid spacing. If you want the Y value to match the X value, left-click with the mouse. Pressing the (EnterKey) will also match the X and Y values and close the dialog box. To enter a different Y value, enter the desired value and then select the OK button.

Once selected, the grid can be toggled ON/OFF by selecting the Grid button on the Status bar. The dialog box also offers methods of altering the angle used to display the grid. Enter the desired angle for the grid in the Angle: edit box. Figure 3.37 shows an example of a floor plan drawn with a grid rotated to 45 degrees. Selecting the Isometric Snap setting will establish a grid suitable for isometric drawings. Isometric drawings will be discussed later in this text.

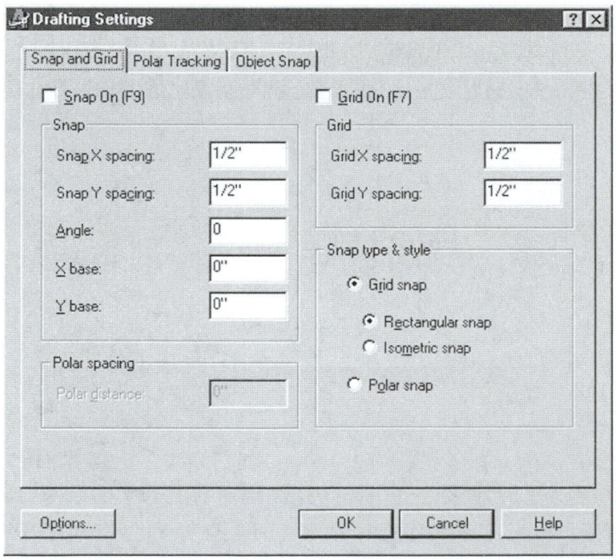

Figure 3.36 *Values for Grid can be altered by right-clicking the Grid button on the Status bar and then selecting the Settings... option. Grids can also be altered by selecting Drawing Aids... from the Tools pull-down menu or by typing* **RM** *(EnterKey) at the command prompt and then selecting the Snap and Grid tab.*

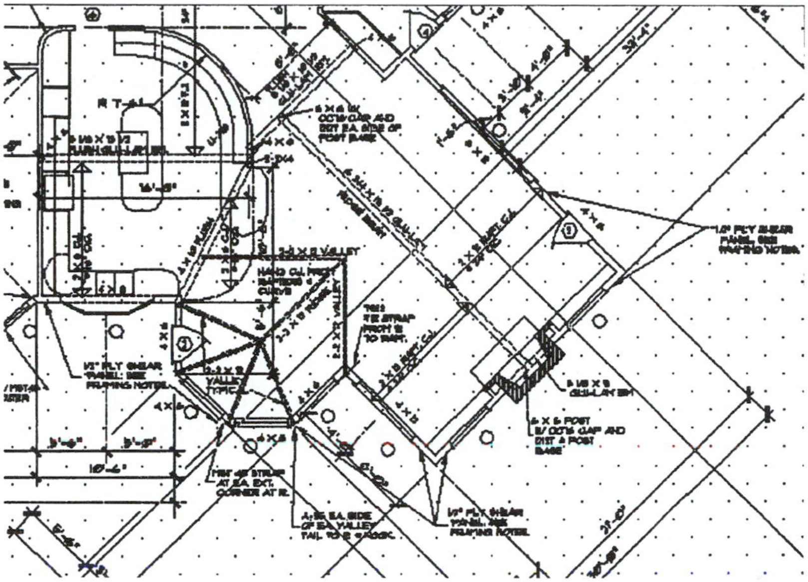

Figure 3.37 *The grid can be rotated to ease layout of an irregular shape.*

Setting Grids at the Command Line

The grid values can be established, altered, and displayed from the Command prompt. The controls for GRID are accessed at the command prompt by typing **GRID** (EnterKey). The command line will now read:

> Command: **GRID** (EnterKey)
>
> Specify grid spacing (X) or [ON/OFF/Snap/Aspect] <0'-0">:

Enter **12** (EnterKey). This will return the command prompt and display the grid field with dots at 1'-0" spacing on the drawing screen.

Spacing

Grid is typically set in units common to the object to be drawn. For instance, when drawing a foundation plan, the grid might be set at 6".

OFF

The grid may be removed by typing **GRID** (EnterKey) at the command prompt. As the menu is displayed, type **OFF** (EnterKey).

Snap

Selecting the Snap option sets the grid spacing to the current snap resolution. If the snap resolution is changed, the grid spacing will be changed as well. Snap may be set by typing S (EnterKey) at the command prompt.

> Specify grid spacing(X) or [ON/OFF/SNAP/ASPECT] <1'-0">: **S** (EnterKey)

Aspect

Aspect allows the grid to be set with a different value for the horizontal (X) and vertical (Y) directions. To do this, type **GRID** (EnterKey). The standard grid prompt is repeated:

Specify grid spacing (X) or [ON/OFF/SNAP/ASPECT] <1'-0">: **A** (EnterKey)

Type **Aspect** or **A** and the prompt will display the additional listings:

Specify the horizontal spacing (X) <1'-0">: (EnterKey)

Specify the vertical spacing (Y) <1'-0">: **24** (EnterKey)

Command:

An X may follow either of these variables to make it equal to the Snap value. Figure 3.38 shows an example of a floor plan with different X and Y values.

SETTING SNAP

Snap can be set using the Snap and Grid tab of the Drafting Settings dialog box. It can also be set by right-clicking the SNAP button on the Status bar and selecting the Settings… option, or from the command line by entering **SNAP** (EnterKey). Snap is adjusted using the same methods that were used to adjust the grid.

Values for the X and Y spacing can be entered in the appropriate edit box. Once selected, Snap can be toggled ON/OFF. Once the values have been set, snap can be

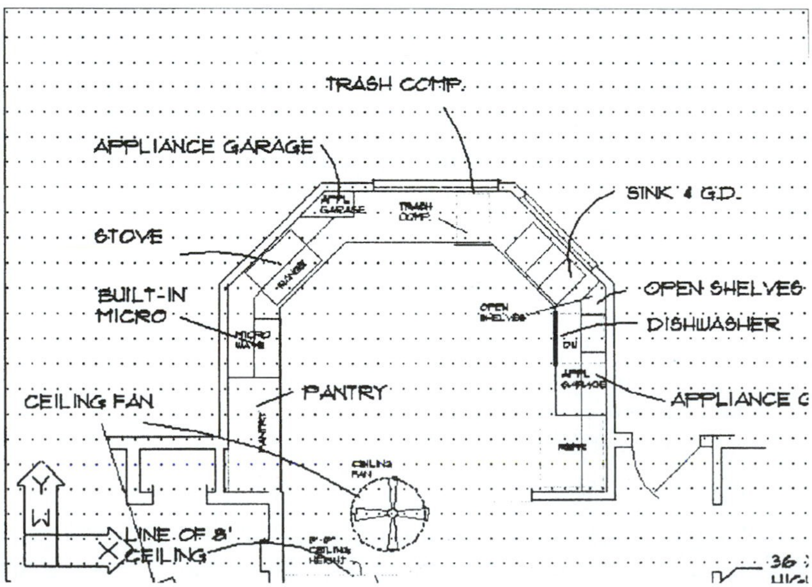

Figure 3.38 *Grid aspect allows the spacing between horizontal and vertical grid units to be altered.*

toggled ON/OFF by double-clicking the Snap button in the status bar. Snap can also be toggled ON/OFF in the middle of another command by using the F9 key or by pressing the Ctrl+B keys. Entering an angle in the Snap Angle edit box controls the display angle for Snap and Grid. Entering a value in the X or Y Base edit boxes allows the base point for the snap pattern to be established.

Setting Snap at the Command Line

The snap values can be established, altered and displayed from the command prompt. The controls for SNAP can be accessed at the Command prompt by entering **SN** (EnterKey). This will produce the prompt:

Specify snap spacing or [ON/OFF/Aspect/Rotate/Style/Type] <0'-0">:

Entering a numeric value will set the spacing of the grid for entry points. Type the number **6** (EnterKey). The command line returns and allows for drawing. Type L(EnterKey) and draw a line. You'll notice that the crosshairs jump across the screen in 6" intervals rather than moving smoothly as they had before.

Off

To terminate the use of snap, type **SN** (EnterKey). Once the menu is presented, type **OFF** (EnterKey). The command line will return and allow for free movement of the crosshairs.

On

Snap can be activated at any time by typing **SN** (EnterKey). Once the SNAP menu is presented, type **ON** (EnterKey) . Notice that the last value used is presented as the default. If you would like to activate the snap with a different spacing, just enter the desired numeric value and (EnterKey).

Aspect

Some projects are set up when a different value for the horizontal (X) and vertical (Y) directions is used. To do this, type **SN** (EnterKey). Type **A** (EnterKey) and the prompt will display the additional listings:

Specify horizontal spacing <0'-0">: **6** (EnterKey)

Specify vertical spacing <0'-0">: **12** (EnterKey)

Command:

Rotate

Because not all projects are rectangular, such as the floor plan in Figure 3.34, it may be helpful to rotate the snap grid to match the drawing. Type **SN** (EnterKey) and enter **R** (EnterKey) at the prompt. This will produce the additional display:

Specify snap spacing or [ON/OFF/Aspect/Rotate/Style/Type] <A>: **R** (EnterKey) .

Specify base point <0'-0", 0'-0">:

AutoCAD is asking for the point of the drawing where you would like the snap grid to be rotated. To select a location, move the crosshairs to the desired location in the drawing and press the pick button. This will now provide the prompt:

Base point <0'-0", 0'-0">: Rotation angle <0>:

A number between –90 and 90 may be entered. Using a negative angle will rotate the snap grid counterclockwise. Using a positive angle will rotate the snap grid clockwise. Enter a value and (EnterKey). The snap grid will be rotated around the base point just selected.

Style

The format of the snap grid is adjusted by using the style option. By entering S (EnterKey) at the snap menu, the prompt will display:

Specify snap spacing or [ON/OFF/Aspect/Rotate/Style/Type] <A>: **S** (EnterKey).

Enter Snap grid style [Standard/Isometric] <S>:

For orthographic drawings the style will not need to be set. For isometric drawings enter **I** (EnterKey).

Type

This setting toggles between polar and grid snap. Polar snaps align with polar angles relative to the Polar tracking point. Grid snaps along the grid. By entering **T** (EnterKey) at the snap menu, the prompt will display:

Specify snap type [Polar/Grid]<Grid>:

Note: Coordinating the grid and snap will provide great accuracy and speed. If the grid value is set at zero, the grid spacing automatically adjusts to the snap resolution. To specify the grid spacing as a multiple of the snap resolution, enter **X** (EnterKey) after the grid value. For instance, if the snap value were 2", entering **6X** (EnterKey) at the command prompt would produce a grid of 12".

CHAPTER 3 QUIZ

1. What command will set an invisible grid to aid layout?
2. List the five options of the units.
3. What is the default unit for angle measurement?
4. What is the default for angle direction?
5. Your boss would like a drawing of a street and sewer layout based on the surveyor's notes. Describe the process for setting up the proper method of two-place angle measurement by keyboard.
6. List three methods to toggle between GRID ON/OFF.
7. List three methods to toggle between SNAP ON/OFF.
8. Access the Users Guide. Research and briefly describe Using Commands Transparently.
9. List the first and last listings in the HELP menu and explain how to find the last listing.
10. What is the main difference between GRID and SNAP?
11. You've just entered AutoCAD and don't want to establish any drawing parameters. How can you get to the drawing screen?
12. When would Start from Scratch be a useful method of entering a drawing?
13. What does Area control?
14. A site plan must be drawn with angles measured in a format such as N42° 30'-30" E. How can this be done if you forgot to adjust the angle measurement as you set up the drawing?
15. A friend wants to draw a house, but has no idea how big it will be. What would you recommend setting for drawing limits as the drawing is started?
16. How do GRID and SNAP relate to each other?
17. What is the default method of entering AutoCAD?
18. How can you use the Advanced Wizard to start a drawing?
19. What command is used to access the Drawing Settings dialog box from the command prompt?
20. What is the effect of entering **DDUNITS** (EnterKey) at the Command prompt?

CHAPTER 4

Drawing and Controlling Lines

Now that you're able to set up drawing units and limits, basic drawing components can be explored. The main components of the drawings you will be working with will consist of lines. Chapter 2 introduced the LINE command using the Draw toolbar, the keyboard, and the Draw pull-down menu. This chapter will introduce you to methods for making effective use of the LINE command, as well as methods for locating and altering lines

LINES

The basic components of every drawing are lines, which are used to represent topography, grades, pipelines, easements, walls, roads, property lines, and an endless list of other construction materials. Figure 4.1 shows the lines required to represent a portion of a framing plan for a multilevel steel structure. Although many of these lines are represented with different line patterns or thickness, each can be drawn using the LINE command. Lines are used to represent different types of information within a drawing. Establishing and controlling varied linetypes will be introduced in future chapters.

ENTERING THE LINE COMMAND

The LINE command is accessed by toolbar, keyboard, or pull-down menu. Each method was introduced in Chapter 2. To start the command, select the Line icon from the Draw toolbar. This will produce the prompt:

Command:_line Specify first point:

Move the cursor to where you would like to start a line and press the pick button. The "first point" will become the beginning point of a line segment. As the first point is selected, the prompt displays:

Specify next point or [Undo]:

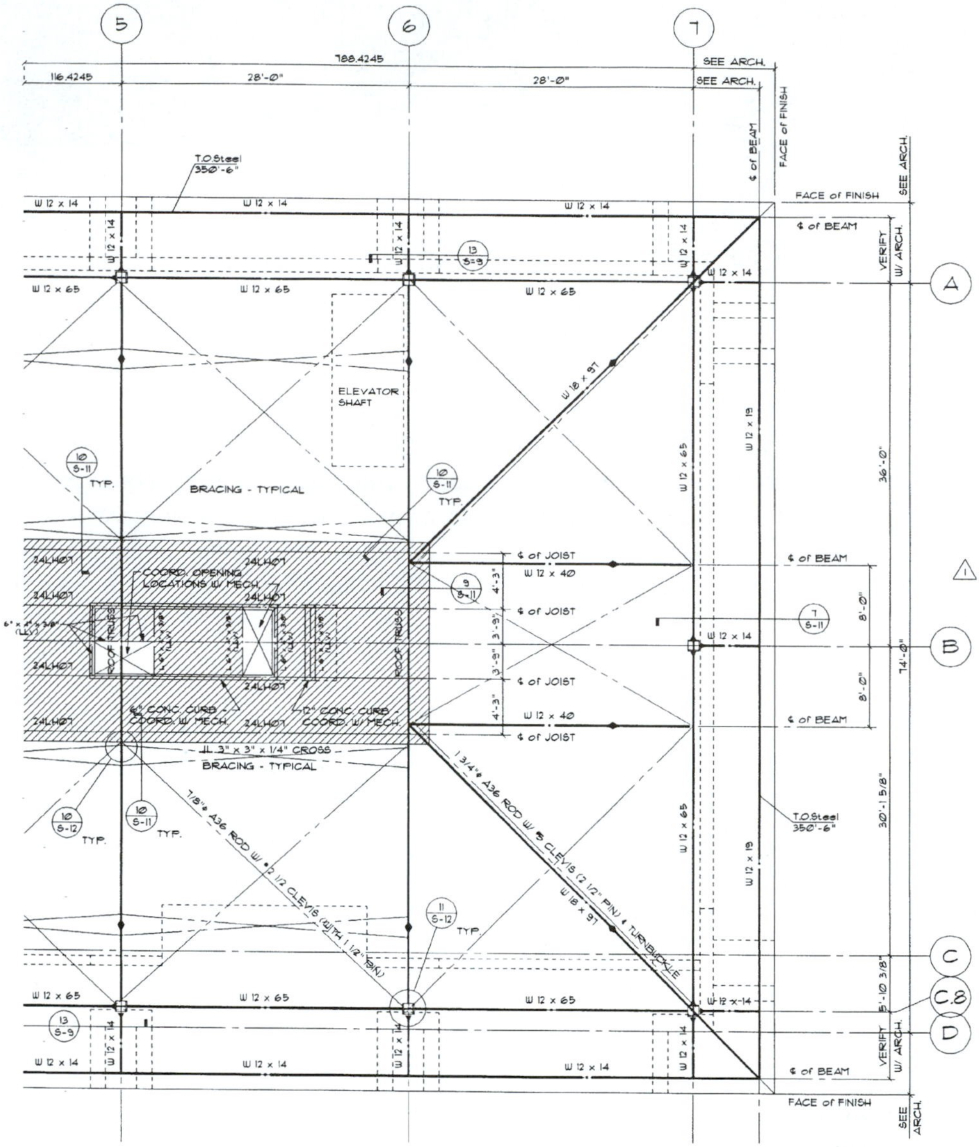

Figure 4.1 *Lines of varied widths and patterns are used on most construction drawings to represent varied materials. Methods of changing linetype and line width will be discussed in future chapters. (Courtesy Van Domelen/Looijenga/McGarrigle/Knauf Consulting Engineers.)*

Move the cursor to the desired location and press the pick button to specify the endpoint. As the cursor is moved, a rubberband line is displayed from the first point to the crosshairs location. This can be seen in Figure 4.2. The rubberband line will help you to visualize the placement of the line to be drawn and to experiment with its placement. The rubberband line will remain as the crosshairs are moved until the next point is selected or the line command is terminated. To terminate the LINE command, press the (EnterKey), the SPACEBAR, or the right mouse button. Selecting (EnterKey) will draw the indicated line and then exit the command. Pressing the SPACEBAR will terminate the command, but no line will be drawn. Right-clicking and selecting (EnterKey) will produce the same results as pressing the SPACEBAR. Right-clicking and selecting CANCEL will draw the indicated lines and terminate the command.

COMMAND LINE INPUT

A second method of entering commands is by keyboard. The LINE command is entered by typing L (EnterKey) at the command prompt. The sequence for entering the LINE command is:

Command: **L** (EnterKey)

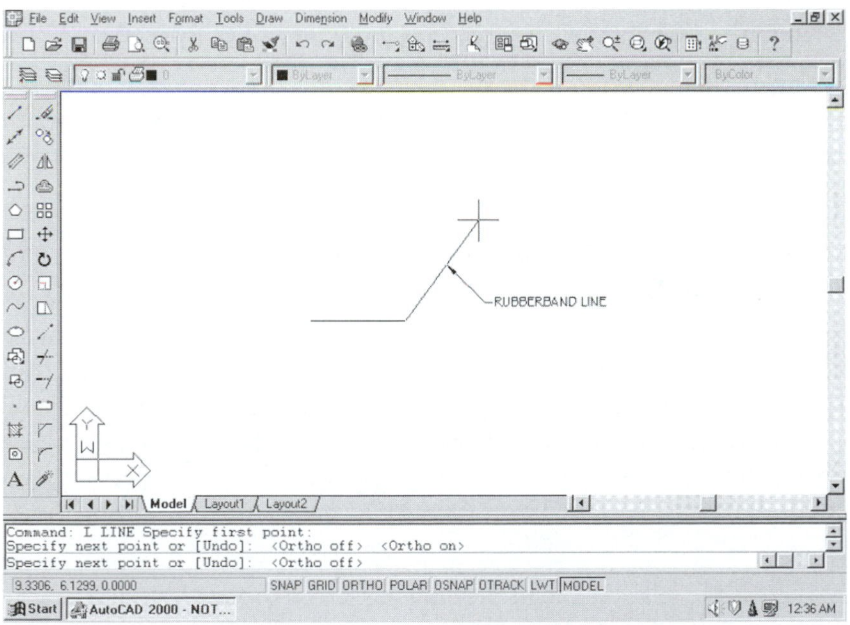

Figure 4.2 *Notice that when a "next point" is provided, a rubberband line extends from the "first point" to the present crosshairs location. This line indicates the line that will be drawn if the current crosshairs location is used.*

LINE specify first point: (*Specify an end point.*)

Specify next point or [Undo]: (*Specify an end point.*)

The LINE command functions exactly the same when entered by keyboard or by toolbar.

Pull-Down Menu Entry

A third method for entering a command is to select Line from a pull-down menu. By selecting Line from the Draw menu, the pull-down menu will disappear and the command prompt will show:

Command:_line Specify first point:

Notice that four other options for lines are presented in the Draw menu—Ray, Construction Line, Multilines, and Polylines commands. A fifth method of drawing lines is Sketch. The SKETCH command allows freehand drawing methods to be reproduced. Each of these commands will be introduced in future chapters.

No matter which method is used to start the command, a request for the "first" and "next" points are displayed at the command line, and can be entered by selecting a point with the mouse, or by keyboard. Specific coordinates can be entered by keyboard rather than selecting the points with the mouse. Methods of selecting line and point by keyboard will be discussed later in this chapter. As you start to experiment with drawing lines, you should draw objects combining all entry methods.

Note: You've entered the LINE command using three different methods. Each method works well to initially start the command. The easiest way to repeat a command is to press the (EnterKey) or the SPACE BAR; both will repeat the previous command.

LOCATING LINES USING COORDINATE ENTRY METHODS

As you start to complete drawings in this chapter, it will be important to understand the user coordinate system (UCS) of AutoCAD. The UCS is based on a Cartesian coordinate system, dividing space into four quadrants. Figure 4.3 shows the division of space based on the Cartesian coordinates. Points in these quadrants are located in turn by using their points in a horizontal (X) and vertical (Y) direction along the plane. Points are always specified by listing the X coordinate followed by the Y coordinate.

Figure 4.4 shows the relationship of the drawing screen to the Cartesian coordinates. As you begin to draw, AutoCAD will ask you to specify a "first point." As you move to the "next point," the coordinate display at the left end of the status line will reflect the crosshairs movement. To activate coordinate display, click the coordinate display button, or use the F6 key, or CTLR +D.

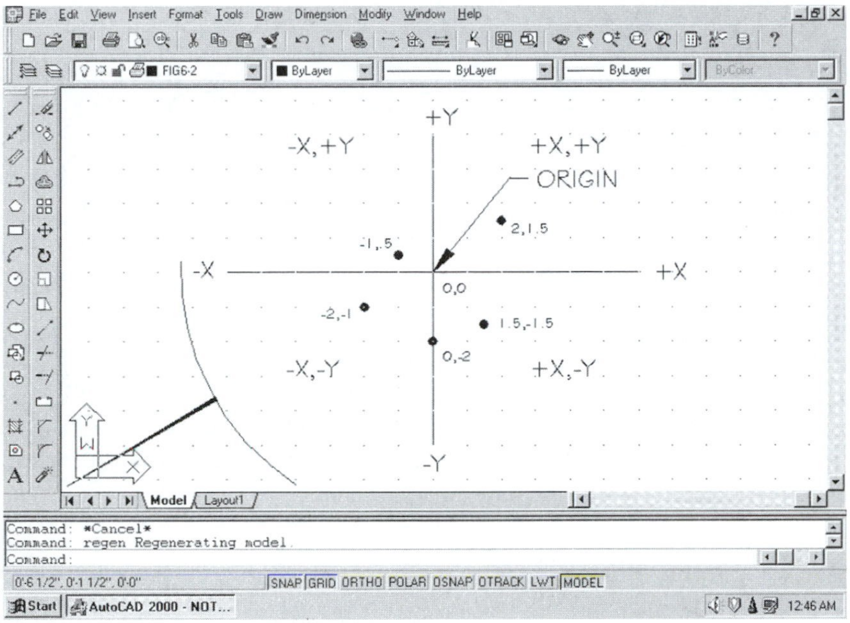

Figure 4.3 *The division of space using the Cartesian coordinate system. The intersection of the X and Y coordinates is assumed to be 0,0, which is the lower left corner of the screen. The grid has been moved to the center of the screen to represent negative X and Y locations.*

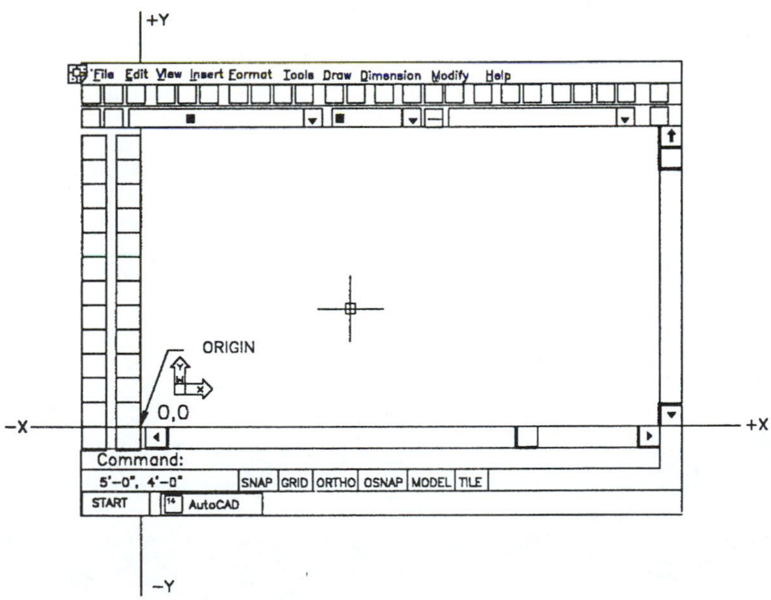

Figure 4.4 *The relationship of the drawing screen to Cartesian coordinates with the Limits set at the lower left = 0,0, and the upper right values set as two positive values.*

Because the fields of engineering and architecture tend to deal with very specific locations, a means of controlling the "to point" is often desirable. This can be done by using absolute, relative, or polar coordinates for point-entry methods.

ABSOLUTE COORDINATES

The absolute coordinate system locates all points from an origin assumed to be 0,0. The 0,0 origin point is assumed to be the same 0,0 point that the limits, snap, and grid are based on, although it can be changed. The axis for the X and Y coordinates intersects at 0,0. Each point on the screen is located by a numeric coordinate based on the original axis. Objects are drawn by entering the coordinates of X and Y separated by a comma. In the example in Figure 4.5 a line is drawn between 2,2 and 2,4. Notice the coordinates that were typed in at the command line or compare the line with the grid pattern. The next "to point" is 4,4, based on the cursor location and the current coordinate display. To complete the drawing in Figure 4.6 enter the following command sequence.

Command: **L** (EnterKey)

LINE Specify first point: **4,1.5** (EnterKey)

Specify next point or [UNDO]: **6,1.5** (EnterKey)

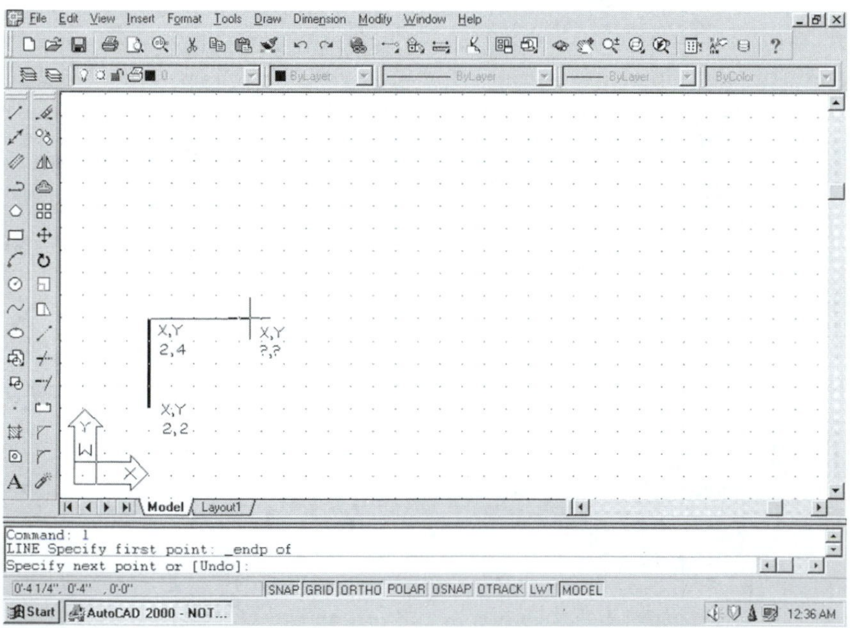

Figure 4.5 *With absolute coordinates, and the 0,0 location as the lower left corner of the drawing screen, the line drawn starts at 2" over (+X) and 2" up (+Y) and extends to a location 2" over (+X) and 4" up (+Y).*

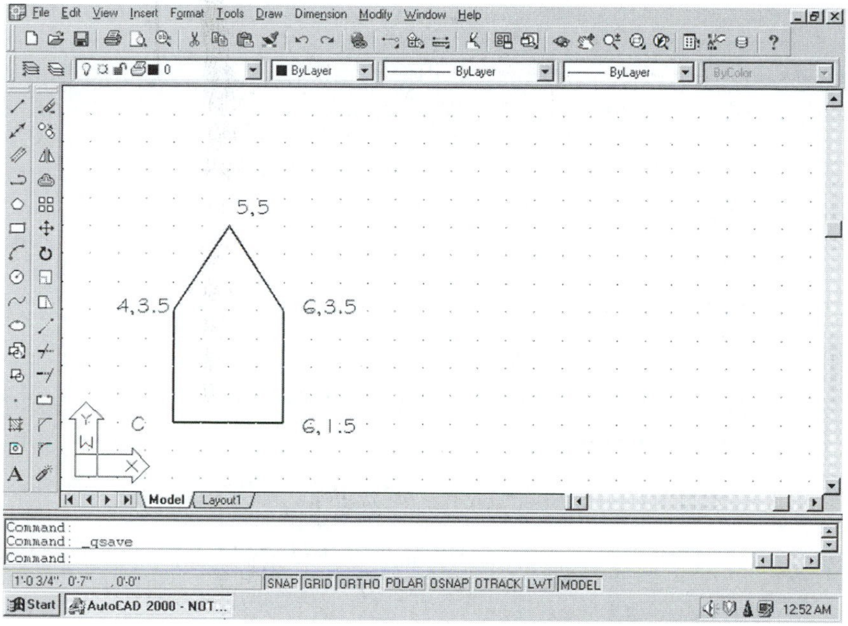

Figure 4.6 *A simple shape and the absolute coordinates required to describe it.*

Specify next point or [UNDO]: **6, 3.5** (EnterKey)

Specify next point or [UNDO]: **5,5** (EnterKey)

Specify next point or [UNDO]: **4,3.5** (EnterKey)

Specify next point or [UNDO]: **C** (EnterKey)

Command:

Notice that C (EnterKey) was entered as the last command, rather than numeric coordinates. C represents close. C (EnterKey) will end the LINE command sequence at the point where the line sequence was begun. With OSNAP activated, moving the cursor to the desired endpoint will also allow easy closure.

Because absolute coordinates are based on an origin of 0,0, they are often not convenient for most architectural and engineering projects. Many CAD users find relative coordinates more suited to a construction project.

RELATIVE COORDINATES

Relative coordinates locate "to points" based on the last point instead of the 0,0 origin. Relative coordinates are entered using a similar method as absolute coordinates. Absolute coordinates were entered X,Y. Relative coordinates are entered @ X,Y. In addition to preceding the coordinate value with the @ symbol, relative coordinates require you to be mindful of positive and negative values.

Figure 4.7 shows an object drawn using relative coordinates. Positive coordinates will move the crosshairs to the right or above the origin point. Negative coordinates will move the crosshairs down or to the left of the last entry point. The object in Figure 4.7 was drawn using the following command sequence:

Command: **L**

LINE Specify first point: **4,1.5** (EnterKey)

Specify next point or [UNDO]: **@2,0** (EnterKey)

Specify next point or [UNDO]: **@0,2** (EnterKey)

Specify next point or [UNDO]: **@–1,1.5** (EnterKey)

Specify next point or [UNDO]: **@–1,–1.5** (EnterKey)

Specify next point or [UNDO]: **@–0,–2** (EnterKey)

The first point was entered using absolute coordinates, with the following coordinate points entered by relative coordinates. The final coordinates could have been omitted and replaced with C (EnterKey) or closed using OSNAP.

POLAR COORDINATES

Polar coordinates are similar to relative coordinates. Polar coordinates use the last entry point as the origin and combine a distance and an angle. This is useful on draw-

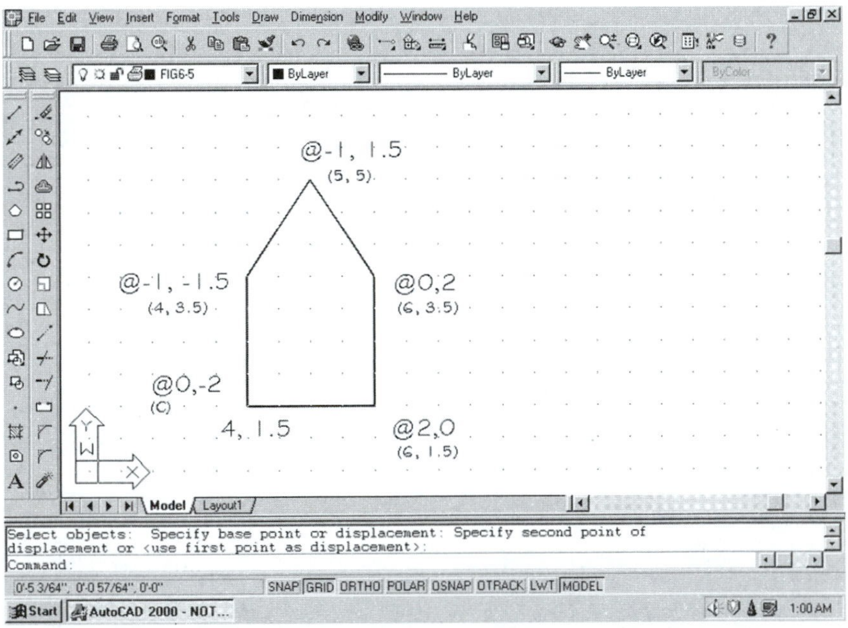

Figure 4.7 *Relative coordinates describe the location of the next point, based on the last point. Relative coordinates must be preceded by the* **@** *symbol. The value* **@2,0** *would extend a line 2" to the right with no vertical rise. Relative coordinates are a common entry choice for many projects.*

ings such as sections for drawing a roof, or site plans with angled property lines. Polar entries are entered as:

@3<27.5 (EnterKey)

This would produce a line 3" long at a 27½° angle, as shown in Figure 4.8. To draw the opposite side of this "roof truss" would require an entry of **@3<332.5** (EnterKey). Remember, polar coordinates are based on the current origin of the crosshairs. In Figure 4.9 the location of line two at 332½° is determined by subtracting 27½° from 360°.

To complete the triangle started in Figure 4.9, type **C** (EnterKey) or use the OSNAP feature of the cursor. Figure 4.10 shows an object drawn using polar coordinates. The required command sequence would be:

Command: **L** (EnterKey)

LINE Specify first point: **4,1.5** (EnterKey)

Specify next point or [UNDO]: **@2<0** (EnterKey)

Specify next point or [UNDO]: **@2<90** (EnterKey)

Specify next point or [UNDO]: **@1.8<124**(EnterKey)

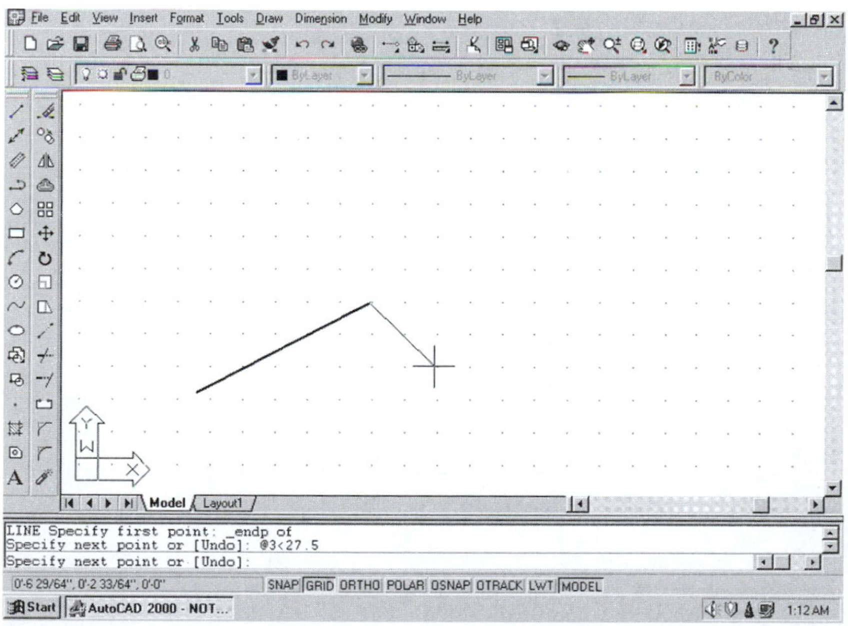

Figure 4.8 *Polar coordinates describe a point relative to the last position using a length and an angle. The @ symbol precedes the length, and the < symbol precedes the angle specification in degrees. Polar coordinates are useful in drawing angled sides of a parcel of land.*

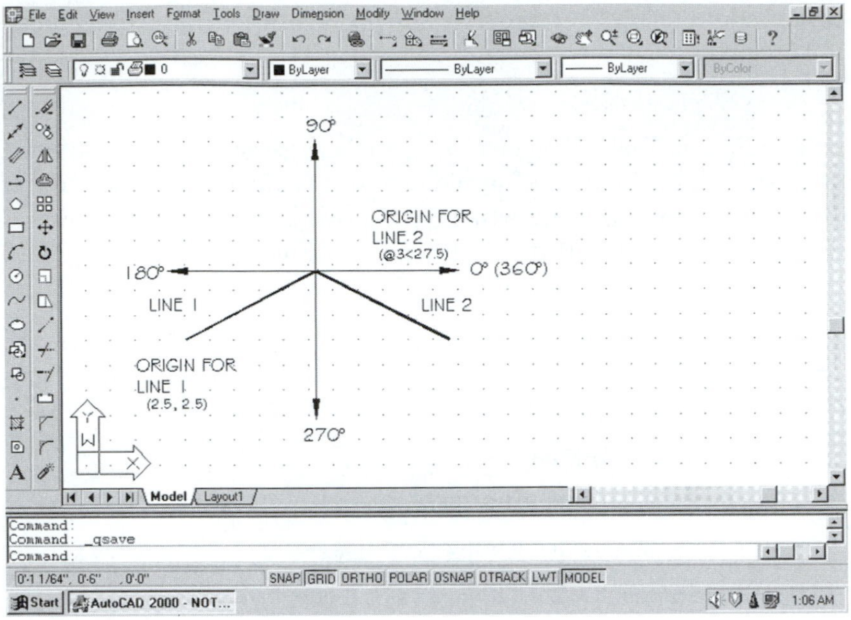

Figure 4.9 *Specifying angles using polar coordinates.*

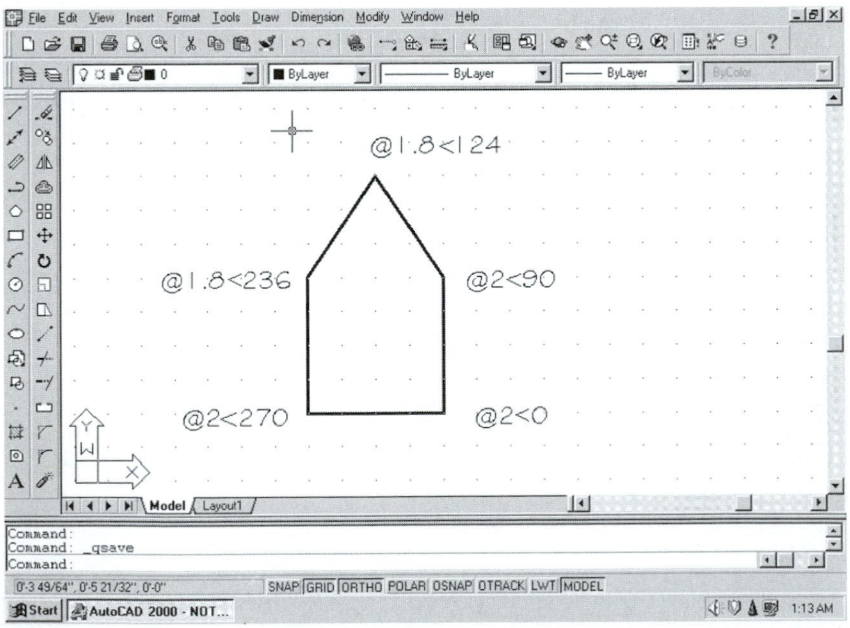

Figure 4.10 *The use of polar coordinates to establish a simple shape.*

Specify next point or [UNDO]: **@1.8<236** (EnterKey)

Specify next point or [UNDO]: **C** (EnterKey)

CONTROLLING LINES

In addition to the aids for placing lines accurately, AutoCAD offers help for placing lines easily. Line continuation, closure, and the tools of the status bar can each be used to control the placement of lines.

LINE CONTINUATION

Once the LINE command has been terminated, it can be resumed from the end of the last line segment that was drawn. This can be done even if other commands were used. The horizontal line segment drawn in Figure 4.11 was drawn prior to the circle. The circle was drawn by typing C (EnterKey) at the command prompt and selecting a center and radius point. This command will be further explored in the Chapter 7. By entering L (EnterKey) and then pressing (EnterKey) for the "first point:" the new line will begin at the end of the last line segment.

LINE CLOSURE

If a sequence of lines will form a closed polygon, the last line can be drawn without providing endpoints or coordinates for the termination point. Rather than struggling to hit the exact end point of the first line, typing C (EnterKey) at the "next point:"

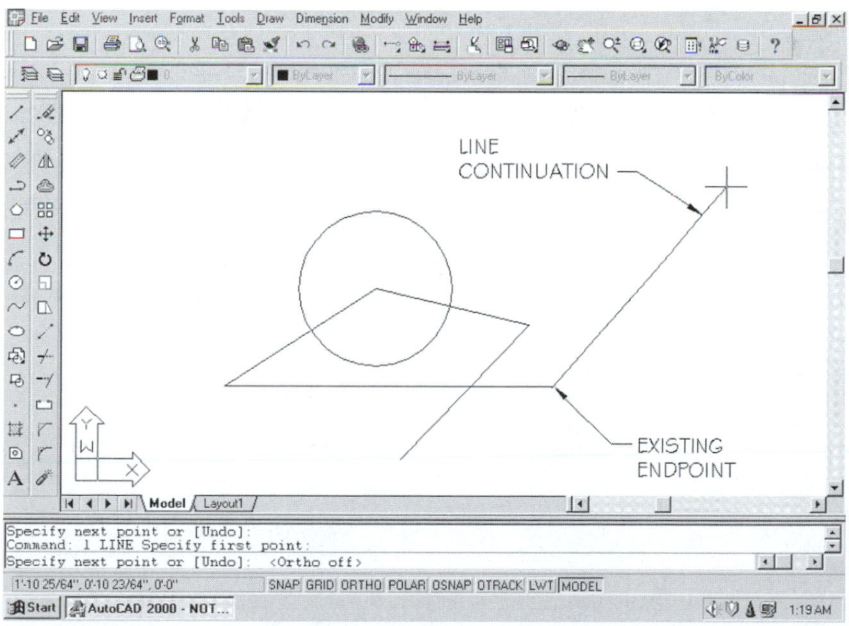

Figure 4.11 *A line can be continued from the end of the last line drawn by pressing (EnterKey) for the "first point."*

will close the polygon precisely. This can be seen in Figure 4.12 and in the following command sequence:

Command: **L** (EnterKey)

LINE Specify first point: **3,3** (EnterKey)

Specify next point or [UNDO] :**@6,0** (EnterKey)

Specify next point or [UNDO]: **@0,2** (EnterKey)

Specify next point or [UNDO]: **@–6,0** (EnterKey)

Specify next point or [UNDO]: **C** (EnterKey)

Typically, the polygons you will draw for projects will be much more complex. It's not uncommon to have to interrupt the drawing of the shape to execute another command, such as ERASE. Once the polygon has been interrupted, C (EnterKey) cannot be used. This can be seen in Figure 4.13 and in the following command sequence:

Command: **L** (EnterKey)

LINE Specify first point: **4,2** (EnterKey)

Specify next point or [UNDO]: **@2,0** (EnterKey)

Specify next point or [UNDO]: **@0,2** (EnterKey)

Specify next point or [UNDO]: **@2,2** (EnterKey)

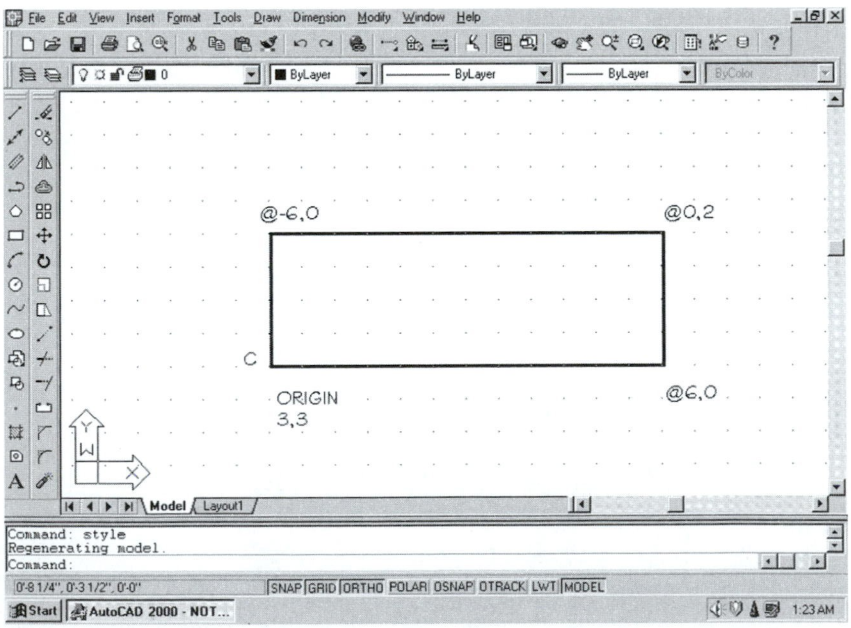

Figure 4.12 *Entering* **C** *(EnterKey) for the last "to point" will close the polygon.*

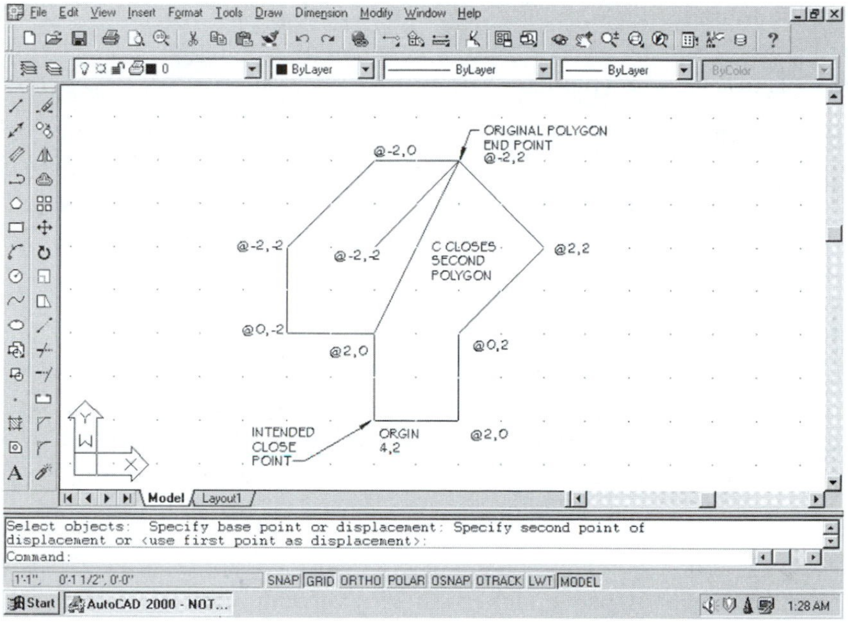

Figure 4.13 *If the drawing of the polygon has been interrupted,* **C** *(EnterKey) will terminate the line at the original endpoint rather than at the close of the total polygon. The desired close point can be easily selected with OSNAP activated by moving the cursor near the desired point. As the cursor nears the endpoint, a box will be placed around the crosshairs, indicating an endpoint will be selected.*

Specify next point or [UNDO]: **@–2,2** (EnterKey)

Specify next point or [UNDO]: **@–2,–2** (EnterKey) *(Wrong coordinate entered and a line drawn.)*

Specify next point or [UNDO]: (EnterKey)

Command: **E** (EnterKey)

ERASE

Select objects: (EnterKey)

Command: **L** (EnterKey)

Specify next point or [UNDO]: (EnterKey) *(This uses the endpoint of the last line drawn as the new endpoint.)*

Specify next point or [UNDO]: **@–2,0** (EnterKey)

Specify next point or [UNDO]: **@–2,–2** (EnterKey)

Specify next point or [UNDO]:**@0,–2** (EnterKey)

Specify next point or [UNDO]: **@2,0** (EnterKey)

To point: **C** (EnterKey)

Although the intended close is at the origin point of 4,2, the polygon will close at the start of the second sequence of lines that were drawn. Chapter 8 will offer other drawing tools to select exact points of a drawing entity as alternatives to **C** (EnterKey)

In Figure 4.13 an error was made and the line command was exited to correct the mistake. A better way to correct the mistake would have been to use the UNDO command when the "next point: **@–2,–2** (EnterKey) " was selected. This would have returned the origin point to the end of the last segment drawn (–2,–2) and the polygon could have been continued, as shown in Figure 4.14.

CONTROLLING LINES WITH THE STATUS BAR TOOLS

Each of the tools of the Status bar were introduced in Chapter 3. This discussion will help you make practical application of these tools. Snap, Grid, and Polar Tracking can each be set using the Drafting Settings dialog box. The box is accessed by selecting Drafting Settings from the Tools pull-down menu, and then selecting the desired tab.

Snap

The Snap grid may be invisible, but it can be a valuable tool for placing lines and other drawing elements. Snap should be adjusted to a modular size of the object being drawn. When working on a floor plan, a SNAP setting of 12" would be helpful for the initial layout of the walls. A setting of 4" would allow lines to be easily placed

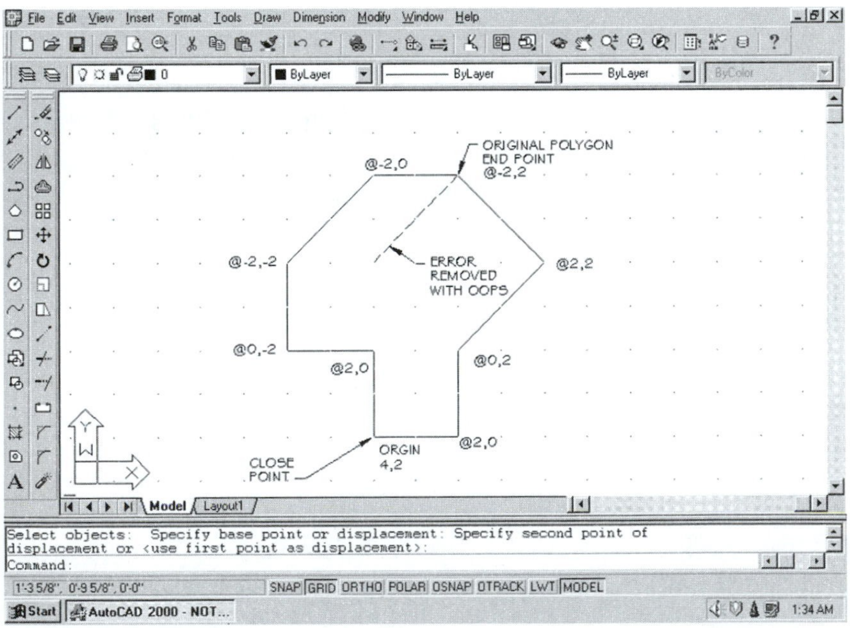

Figure 4.14 *The OOPS command should be used to remove mistakes rather than exiting the LINE command and using the ERASE command.*

as the plan is being fine tuned. SNAP will be most helpful if you adjust the value throughout the life of the drawing to meet specific needs. Keep in mind that when SNAP is On, you may have a difficult time drawing angled lines.

Grid

The visible grid should be used to plan and layout a drawing. Setting the GRID value as a multiple of the SNAP setting will greatly aid in placing drawing objects. The grid serves as a visual reference as well as an aid in placing lines. A grid of 24" and a snap of 6" would greatly aid in the layout of most plan views. The 24" spacing of the grid will provide a reference to keep track of object sizes, and the snap will provide a useful grid for placing lines.

Ortho

Most of the structures that you will draw will be composed of perpendicular lines. Enter LINE at the command prompt and draw a few lines. You'll notice that as a line is drawn from the "first point" to the "next point", a line is extended to the crosshairs and moves as the crosshairs move. This line is referred to as a rubberband line. The ORTHO command will allow you to place the end of the rubberband line in an exact horizontal or vertical position based on the current snap or grid pattern. If the snap and grid are rotated, ORTHO will respond by producing perpendicular lines based on the new grid. The residence shown in Figure 3.37 has a grid set to 45° for the right edge of the structure. As ORTHO was used on the right half of the structure, lines were automatically set to 45°, 135°, 225°, and 315°.

Polar tracking

Earlier in this chapter you were introduced to using polar coordinates to place a line. By activating Polar, a display will be given at specified intervals to aid in placing lines. The display will show the distance and the angle of the line in progress. Figure 4.15 shows an example of a polar display for placing a line. The display can be a helpful alternative to entering coordinates at the keyboard. Settings can be adjusted from the Polar Tracking tab of the Drafting Settings menu in the Tools pull-down menu. The Polar Tracking tab can be seen in Figure 4.16. Key elements of the menu include:

- Increment angle—This edit box allows the desired angle to be used with the tracker to be set. Default angles include 5, 10, 15, 18, 22.5, 30, 45, and 90 degrees. Selecting 10 will display a tracking indicator as the cursor is moved in 10° increments.
- Additional angles—This display box allows up to 10 additional angles to be added to the default list. If you're drawing an elevation with a roof at a 6/12 pitch, selecting an angle of 27 ½° would aid in the layout of the roof. The setting could be entered by:

  Select the new button

  Enter 27 ½ in the edit box

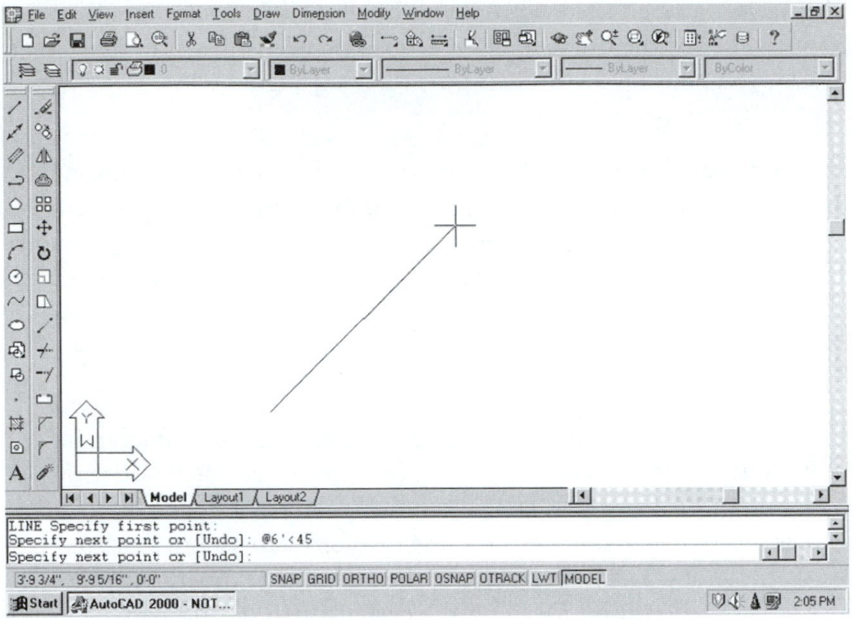

Figure 4.15 *Polar display for tracking a line. The display can be a helpful alternative to entering coordinates at the keyboard. Settings can be adjusted from the Polar Tracking tab of the Drafting Settings menu.*

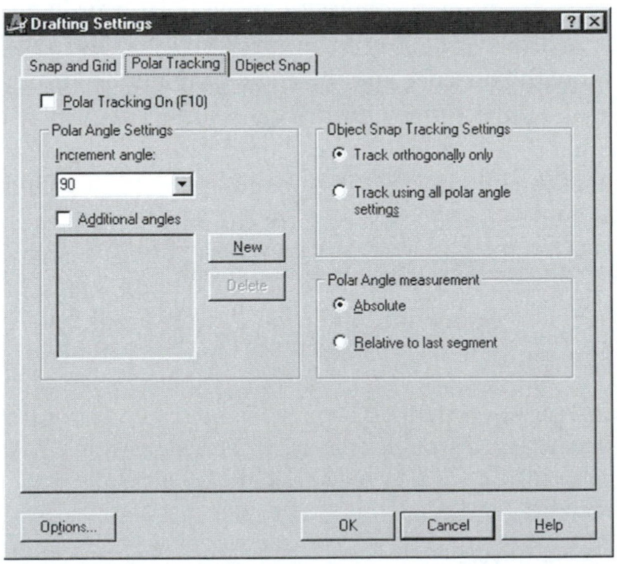

Figure 4.16 *The Polar Tracking tab of the Drafting Settings dialog box can be used to control the display used with polar tracking.*

Select the OK button.

Settings can be removed from the default list by highlighting the setting to be removed and then selecting the Delete button.

- Object Snap Tracking Settings—This option selects options for object track settings. Options include Orthogonally and Polar Angle Tracking.
- Track orthogonally only—With this option active, the object tracker displays only horizontal or vertical object snap tracking paths for acquired object snap points when object snap tracking is on.
- Track using all polar angles—This option allows the cursor to track along any polar angle tracking path when Object Snap Tracking is on.
- Polar Angle Measurements—This option sets the method to be used to measure polar angles. Options include Absolute and Relative.

  Absolute—displays polar tracking angles based on the current UCS.

  Relative to last segment—displays polar tracking angles on the most recent object.
- Options—Selecting this option will display the Options dialog box. This dialog box can be used to adjust the color and size of the object snap marker. The box also provides a convenient method of altering the size of the aperture box.

Osnap

Activating OSNAP provides one of the easiest and fastest means of accurately drawing in AutoCAD. With this option on, a marker will be placed on the crosshairs as it is moved near an endpoint, midpoint, or intersection of lines. With the marker displayed, the "next point" of a line will be placed exactly at the endpoint. Figure 4.17 shows an example of the display as the cursor is moved near the endpoint of an existing line. Chapter 8 will explore methods of altering the default settings of OSNAP. If you want to explore on your own, select the Drafting Settings from the Tools pull-down menu and select the Object Snap tab to view the options. Use the Help menu if you just can't wait.

AN INTRODUCTION TO EDITING LINES

This section will introduce three methods of removing drawing errors from a drawing file. These methods include the ERASE, OOPS, and UNDO command. The REDRAW command will also be introduced as a method of cleaning the drawing display. Chapter 9 will present methods to use these commands more effectively.

ERASE

The ERASE command will allow you to remove unwanted drawing objects. To remove objects, select the Erase icon from the Modify toolbar, select Erase from the Modify pull-down menu, or type E (EnterKey). This will display the prompt:

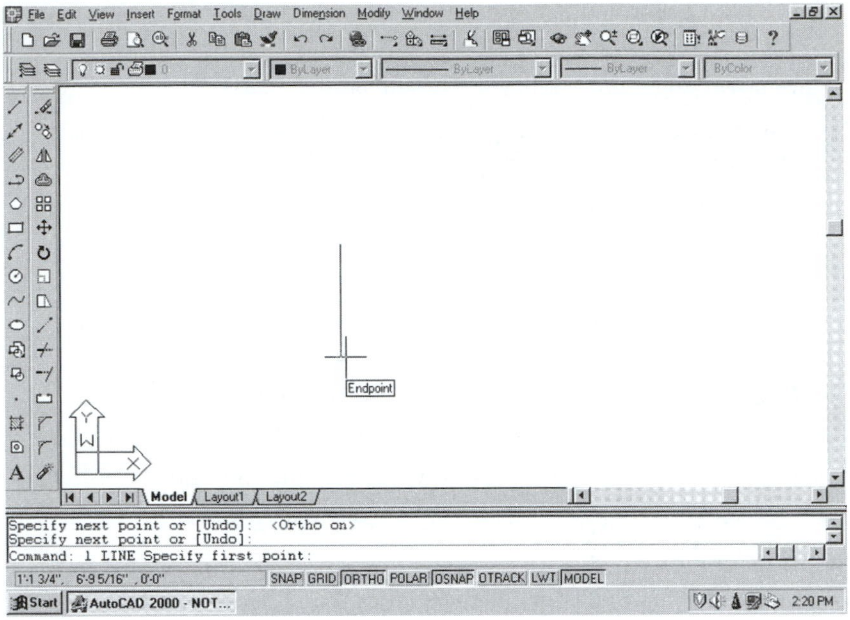

Figure 4.17 *OSNAP is an excellent method of increasing drawing accuracy. With the OSNAP button in the Status bar active, a square will be added to the crosshairs as it moves near the endpoint of a line. The "next point" will automatically snap to the endpoint of the selected line.*

Command: **E** (EnterKey)

Select objects:

As the prompt is displayed, the crosshairs turn into a pick box. The pick box can be used to select objects to be removed. Place the box on the object to be removed and left-click. In Figure 4.18 the column centerline has been selected for removal. Once selected, the selected line will now appear highlighted. An unlimited number of objects to be erased can be selected. The command line will continue to show "Select objects:" until the (EnterKey) key is selected. Pressing (EnterKey) will remove the selected lines and restore the command prompt as seen in Figure 4.19.

ERASE allows for the last drawing entity to be erased without having to select the entity. This can be done by typing E (EnterKey) L (EnterKey). The command prompt will appear as:

Command: **E** (EnterKey)

Select object: **L** (EnterKey)

1 found:

Select objects: (EnterKey)

Command:

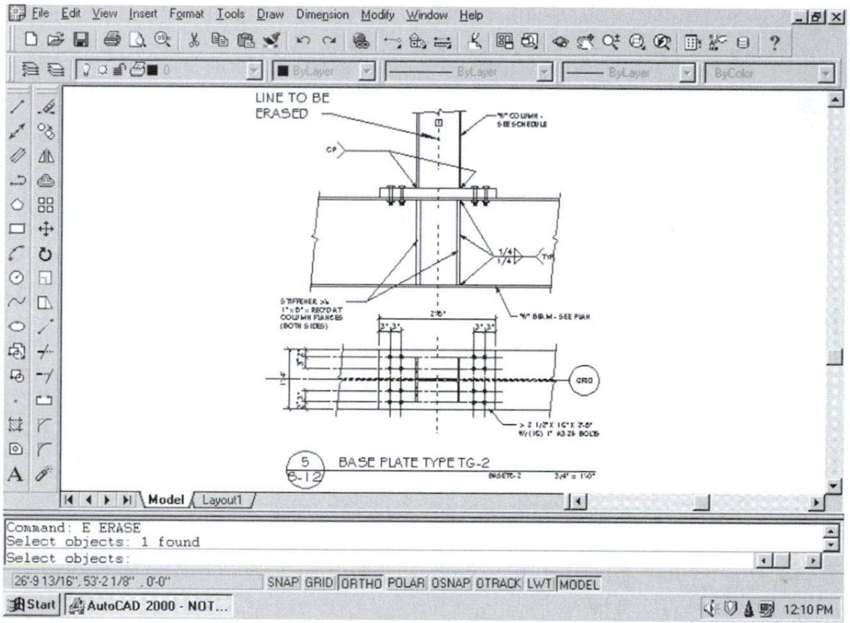

Figure 4.18 *As the ERASE command is entered, the crosshairs will be replaced by a pick box. Placing the pick box on the desired item to be erased and selecting it, will select that object for removal. Notice that the line in the column centerline has been selected for removal. Once an item has been selected for removal from the drawing, it will be displayed as short dashed lines.*
(Courtesy Van Domelen/Looijenga/McGarrigle/Knauf Consulting Engineers.)

As the command line is restored, the last entity drawn is eliminated. If you would like to remove the objects drawn with the last drawing sequence type:

> Command: **E** (EnterKey)
>
> ERASE
>
> Select objects: **L** (EnterKey)
>
> Select objects: *(Select object.)*
>
> 1 found
>
> Select objects *(Continue to select objects or (EnterKey) to end sequence.)* (EnterKey)
>
> Command:

Selecting objects to be erased can be continued indefinitely until the entire drawing is removed, but this process is not the most efficient way to remove multiple objects. Future chapters will provide in-depth coverage of ERASE and other editing options, as well as more efficient methods of selecting objects to be edited.

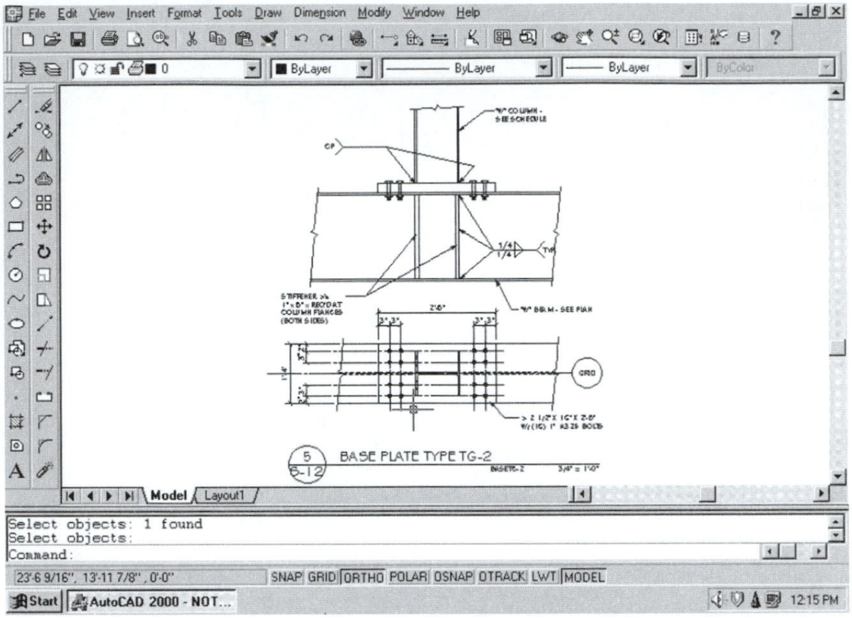

Figure 4.19 *Pressing the (EnterKey) or right-clicking will remove the selected object from the drawing screen. (Courtesy Van Domelen/Looijenga/McGarrigle/Knauf Consulting Engineers.)*

OOPS

Occasionally an object will be erased by mistake. The OOPS command will restore the object removed by a previous ERASE command. Chapter 10 will provide other options for removing elements that cannot be removed by OOPS.

UNDO

An alternative to ERASE or OOPS is the UNDO command. UNDO will remove the command that was just executed. This command will remove the most recent line and then allow you to continue with the LINE command from the previous line. This command can be used as many times as you wish, and will continue to move you backward through the drawing sequence one-step at a time. Undo can be selected from either the Standard toolbar, by keyboard, or by selecting Undo from the Edit pull-down menu. When the toolbar is used to enter the command, simply select the icon when you wish to remove the command that has just been executed. Selecting the icon in the middle of a command sequence is an excellent method to remove an unwanted line, without ending the LINE command. If the Undo icon is selected after ending the command, the entire line sequence will be removed. The Line Undo command can be entered by keyboard, in the middle of the LINE command using the following sequence:

Specify next point or [UNDO]: **@–2,–2** (EnterKey)

Specify next point or [UNDO]: **U** (EnterKey)

Specify next point or [UNDO]: **@–2,0** (EnterKey)

If you decide that you really do want the line after you have used the UNDO command, the REDO command can be used to correct your change of mind. REDO will reverse the effects of UNDO. A REDO is preformed by selecting the Redo button in the Standard toolbar or by typing **REDO** (EnterKey) at the command prompt.

CHAPTER 4 EXERCISES

Set limits, units, snap, and grid to best suit each project. Save each exercise to a floppy disk and to a place on the hard drive or network, as per your instructor's directions.

1. Draw the following outline using the coordinates that are provided.

   Point of beginning: 1,1

point b	@4.2426<45
point c	@3.1623<342
point d	@2.5<0
point e	@4.0311<120
point f	@4.2720<159
point g	@3.9051<230
point a	@4.000<270

   Save your drawing on your floppy disk as filename E-4-1.

2. Draw the following outline using the coordinates that are provided.

   Point of beginning: 2,1

point b	@2.5<N80dE
point c	@3.5<N38d23'15"E
point d	@2.5<S40d20'40"E
point e	@1,1
point f	@1,–1
point g	@4<N
point h	@–6,1
point i	@3.5<S39d7'W
point a	back to the true point of beginning

   Save your drawing on your floppy disk as filename E-4-2.

3. Draw the following outline using the coordinates that are provided. Save the drawing as filename E-4-3.

   Point of beginning: 9,9

point b	@–5,–2.5
point c	@0,–1.5
point d	@4<S39d48'W

point e	@–1.5,–2
point f	@4.5,1.5
point g	@1.5<90
point h	@2<S65dE
point i	@2.5<N30d15'15"W
point j	@5.5<S60d20'40"E
point k	@1,1
point l	@–2.5,1
point m	@–.5,–.5
point n	@3<N61d25'10"W
point o	@3.8125<E
point a	@4.5<N to the true point of beginning

4. Use the following coordinates to lay out the required shape. Save as filename E-4-4.

   Starting point: 7.75, 7.95

   @–2,0

   @0,–2

   @–1.50, 1.75

   @–.96,0

   @1.62<137°

   @.46<227°

   @1.22<310½°

   @–.92,0

   @1.62,–.50

   @2.17,2.91

   @3.5,0

   @1.62,1.62

   @1.75<308

   @3.32<80.73°

   back to the point of beginning

5. Use the dimensions on the floor plan below and draw the outline for the perimeter of the structure. No footings or beams are to be drawn. Save the drawing in a directory labeled EXCER with the filename E-4-5PLAN.

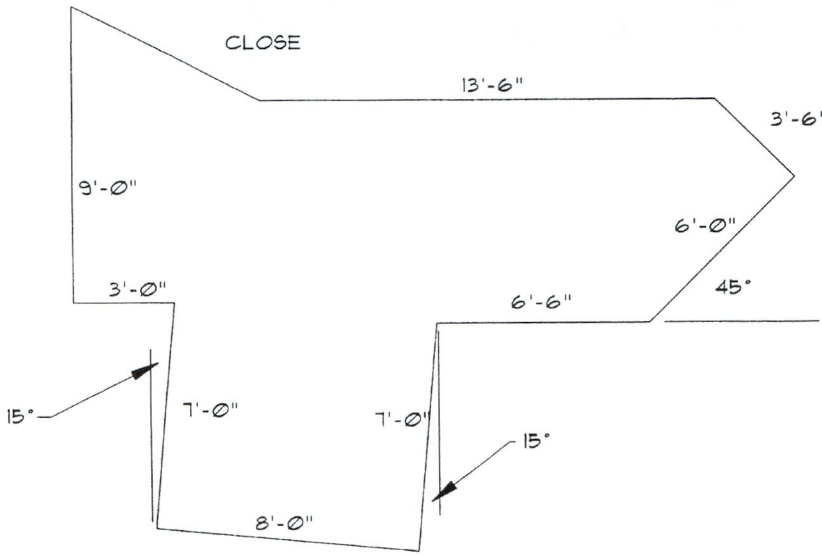

6 Use the attached drawing to draw the roofline of the structure. Assume all hips and valleys are drawn at 45° angles. Save the drawing as filename E-4-6ROOF.

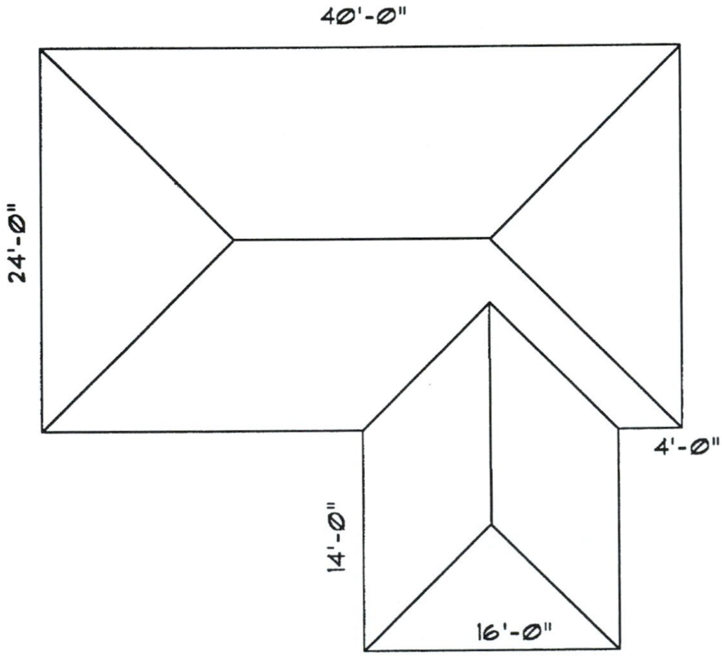

7. Use the attached drawing to draw a section of a one-story footing for a wood floor system. Assume the bottom of the footing extends 12" into the grade, and the top of the stem wall extends 8" above grade. Save the drawing as filename E-4-7-1STFTG.

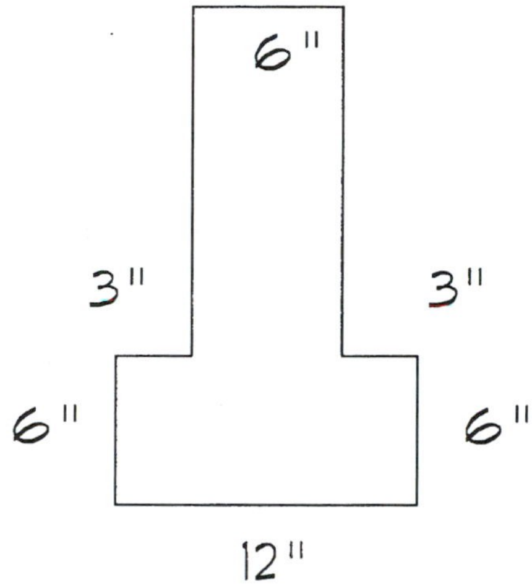

8. Use the attached drawing to draw a section of a one-story footing for an on-grade concrete floor system. Save the drawing as filename E-4-8-CONCFTG.

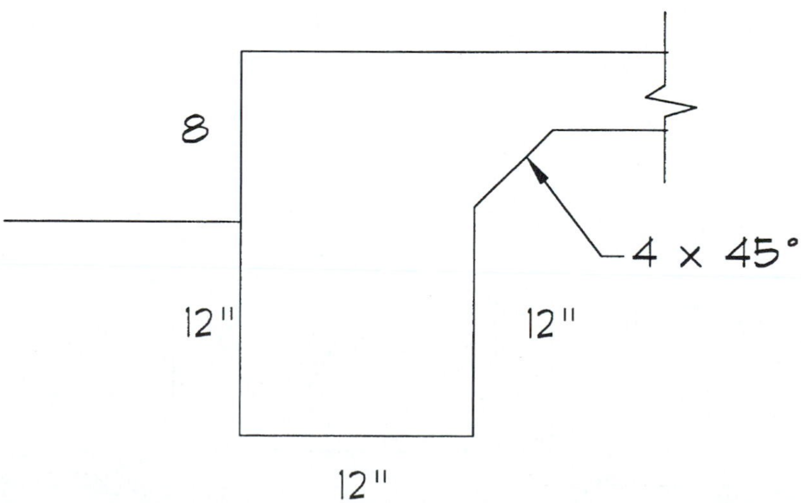

9. Use the attached drawing to complete a section through a 6"-wide concrete panel, resting on a concrete footing 24" wide × 12" deep. Support the wall panel on 1"-deep grout with beveled edges. Show a 3" × 6" pocket in the footing for reinforcing. Save the drawing as filename E-4-9CONCWALL.

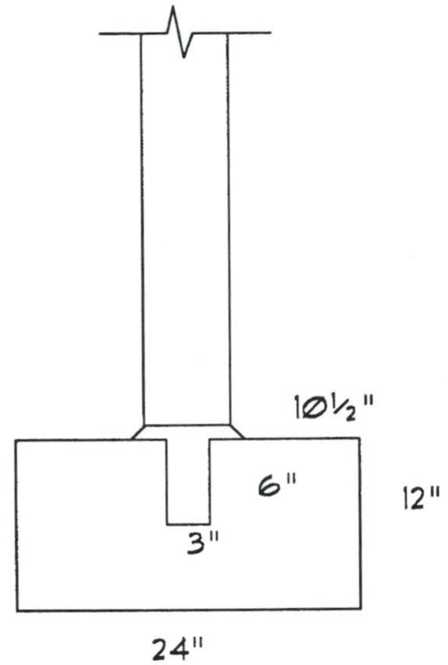

10. Use the attached drawing to complete a section view of a floor truss intersecting a 6"-wide concrete tilt-up wall. Assume the truss to be 24" deep with 2" deep top and bottom. Save the drawing as filename E-4-10TRUSSWALL.

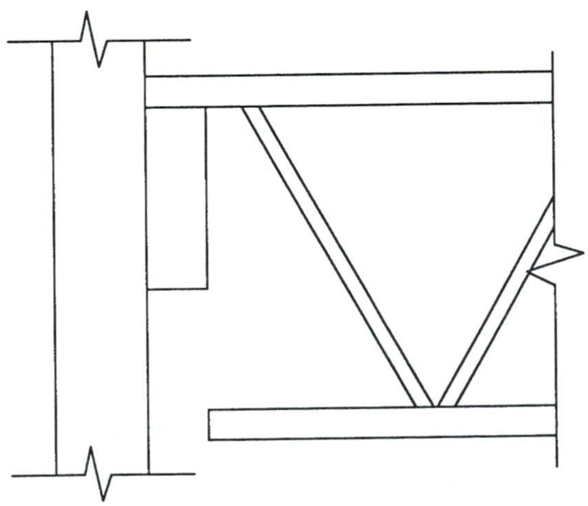

CHAPTER 4 QUIZ

1. Use the drawing below to provide absolute coordinates. Grid = .5", Snap = .25"

a. 2.5,2	e. ________	i. ________
b. ________	f. ________	j. ________
c. ________	g. ________	k. ________
d. ________	h. ________	l. ________
		a. ________

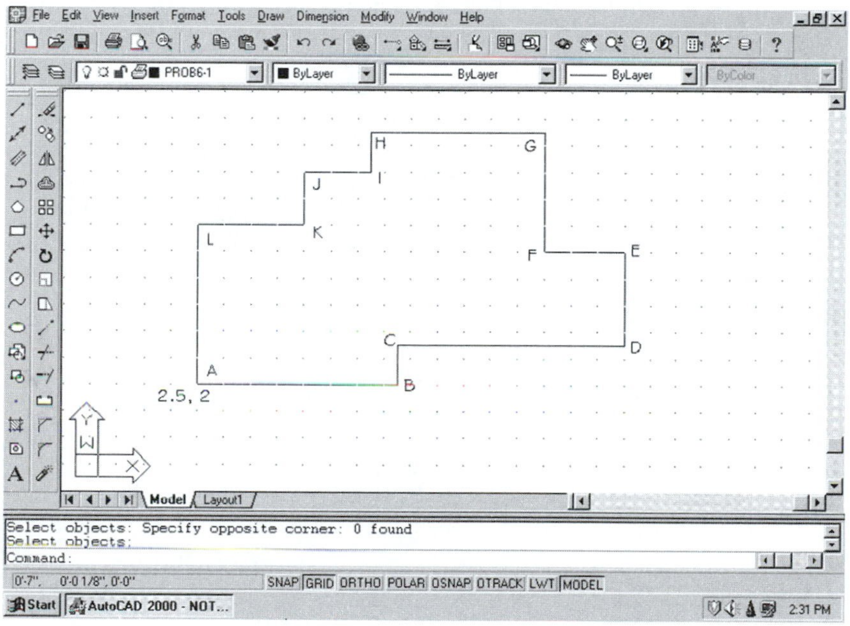

2. Use the drawing from Problem 1 to provide relative coordinates.

a. 2.5,2	e. ________	i. ________
b. ________	f. ________	j. ________
c. ________	g. ________	k. ________
d. ________	h. ________	l. ________
		a. ________

3. Use the drawing from Problem 1 to provide polar coordinates.

a. 2.5,2	e. ________	i. ________
b. ________	f. ________	j. ________
c. ________	g. ________	k. ________
d. ________	h. ________	l. ________
		a. ________

4. Use the drawing below to provide relative coordinates.

a. 1.5,1.5	e. ________	i. ________
b. ________	f. ________	j. ________
c. ________	g. ________	a. ________
d. ________	h. ________	

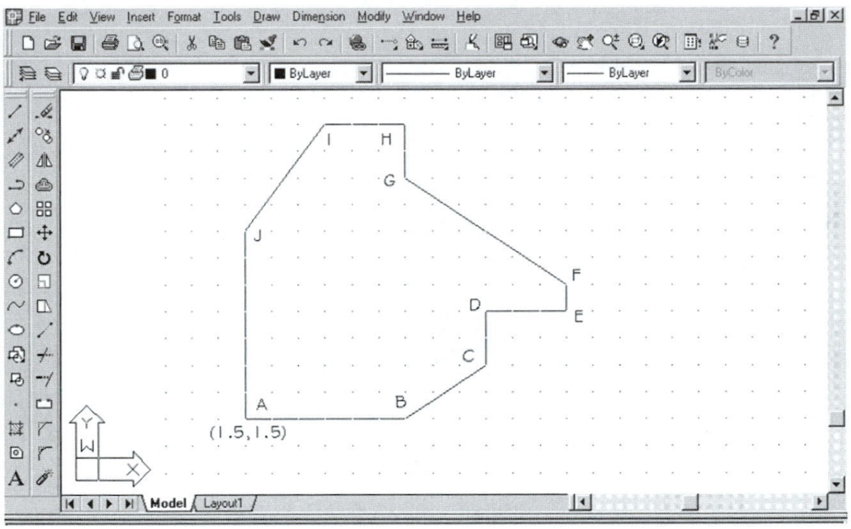

5. Use the drawing from Problem 4 to provide absolute coordinates.

a. 1.5,1.5	e. ________	i. ________
b. ________	f. ________	j. ________
c. ________	g. ________	a. ________
d. ________	h. ________	

6. Use the drawing from Problem 4 to provide polar coordinates. Use relative coordinates where insufficient information is available to determine polar entries.

a. 1.5,1.5	e. ________	i. ________
b. ________	f. ________	j. ________
c. ________	g. ________	a. ________
d. ________	h. ________	

7. What entry method uses the X,Y format to enter coordinates?
8. What entry method uses the @X<Y format to enter coordinates?
9. What entry method uses the @X,Y format to enter coordinates?
10. Several entry coordinates are listed below. Circle the coordinates that will not work.

a. 3,4	f. –3,<27	k. 27'6
b. –3,–4	g. @1'–0",<27.5	l. 25',13'6
c. 3,4	h. @ 27'6<n3615'e	m. 14'<W30N
d. @3,<45	i. @ 36'<27.5	n. 20'4,S27d36,15E
e. @–3,–5	j. N27d36'27"e,15'	o. @3'6,7'4

11. Can an angled line be drawn with snap and grid both on?
12. Describe the line that is formed between 2,0 and 4,0.
13. Describe the line that is formed between 2,0 and 4,2.
14. List two methods for activating the coordinate display.
15. What settings must be changed to use absolute, relative, and polar coordinates on the same drawing?
16. How can the LINE command be accessed from the pull-down menu?
17. List four methods for accessing the LINE command.
18. What steps are required to erase a line?
19. If a line has been erased by mistake, how can it be restored?
20. What effect does ERASE have on the cursor?
21. Describe the difference between OOPS and UNDO.
22. What effect will **C** (EnterKey) have on a line segment?
23. How does the Snap setting affect the use of the LINE command?
24. What type of coordinate entry is best suited for drawing E-4-5?
25. Use the HELP command and list four methods of accessing the ERASE command.

CHAPTER 5

Working with Drawing Files

In Chapter 2 you discovered a little about the drawing environment of AutoCAD and how to save a drawing. In Chapter 3 you were introduced to methods of entering the AutoCAD drawing area when a new drawing file is created using the wizard or a template. In Chapter 4 you started to create great works of art. This chapter will introduce you to methods for creating a template drawing, methods to save drawing properties for future use, expand your methods for saving and managing drawing files, merge existing drawings, and introduce methods of making a print.

CREATING AND SAVING DRAWING TEMPLATES

Just as manual drafters used plastic templates to make drawing faster and easier, AutoCAD provides templates to help make each new drawing that you start. Many professionals often refer to template drawings as prototype drawings. The template can be created and saved using the Startup dialog box, and then used each time a drawing requiring those parameters is needed. Several of the parameters were introduced in Chapter 3. Other useful values such as linetype, lettering size and styles, and dimensioning styles and format will be introduced in succeeding chapters. Necessary values can be added to the drawing template for future use. Once created and saved, they can be opened to form the base for a new drawing file and save time.

USING AN EXISTING TEMPLATE

AutoCAD provides several templates to ease drawing setup. The default template is the AutoCAD-named plot styles. Scroll through the list and you'll find two architectural templates. Both templates will provide a drawing screen showing a 24" × 36" sheet of paper with an architectural style border that is suitable for most construction drawings. You'll also notice that most of the templates are either "named plot styles" or "color dependent." Plot styles alter the way a plotted drawing will appear. Each object in a drawing has a plot style property. These styles can be assigned by color or by named groups. By changing the plot style, the color, linetype, and lineweight of an object can be altered on the print. This will allow objects within the drawing to be highlighted. The benefits of plotting using assigned styles will be addressed in detail in Chapter 24 as plotting is explored. For now, open the Startup menu, select the Use a Template option, and select the Architectural, english units-color dependent plot style template. This will produce the template shown in Figure 5.1. Although the title implies that this template is for architectural drawing, it should not be used without some basic changes. The limits are appropriate for most construction drawings. The term *Architectural* is applied for the shape of the title block, rather than the current settings. As you proceed through the next section and set limits, it will help to give you a better understanding of how the three areas of a drawing template can be used.

As the template is displayed, you're working in paper space (notice the paper space icon in the lower left corner) with the Architectural Title Block tab active. The paper is 24" × 36" with a drawing area of approximately 2'-5 ½" × 1'-10 ½". The size will vary depending on the unit accuracy that is set. In effect, you're now working on a blank sheet of paper. Move the cursor and you'll notice that it can be moved to any point in the drawing area. Paper space will be a great aid for plotting the drawing, but it should not be used to create a drawing.

Select the Paper button in the Status bar. The Paper button will now change to read Model, and the drawing will change to display the drawing shown in Figure 5.2. The

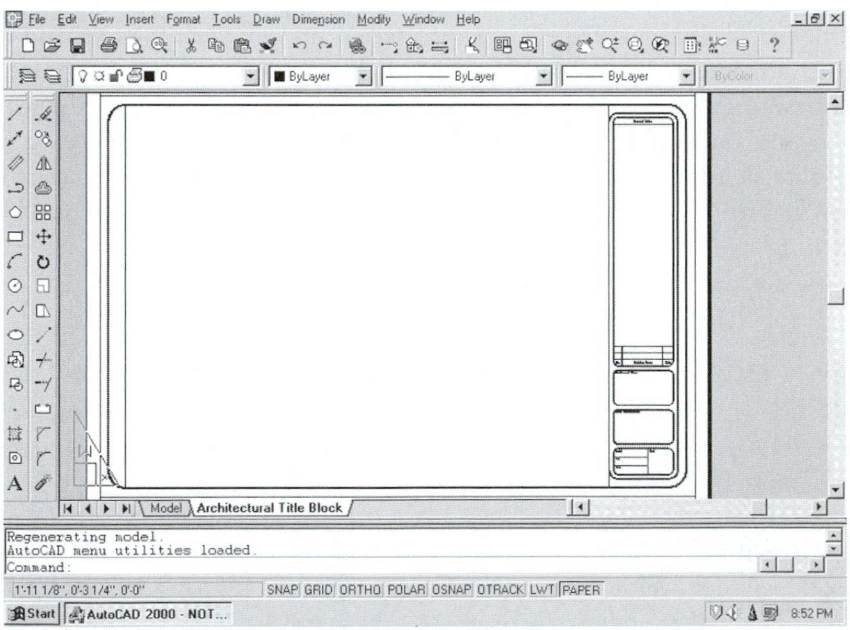

Figure 5.1 *The Architectural template can be used for a base layout that will meet most of the needs for construction related projects. The template is displayed by selecting the Use a Template option from the Startup menu and selecting the Architectural, english units option.*

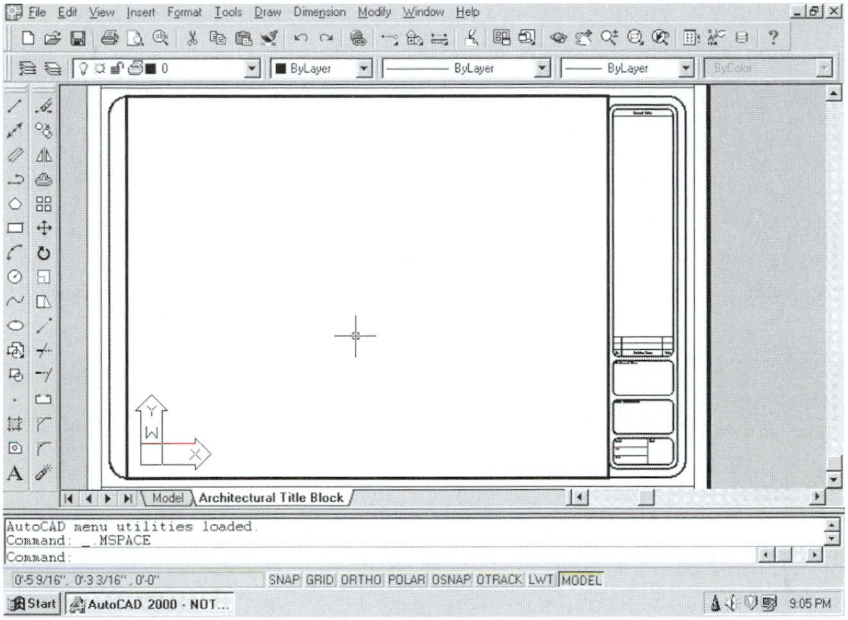

Figure 5.2 *Selecting the Paper button in the Status bar, will toggle the display to model space. Notice that the UCS icon is now inside the drawing area.*

template appears similar to Figure 5.1, but now the UCS icon is displayed within the border, and has changed to reflect model space. Move the crosshairs to the border of the template. As soon as the crosshairs touches the border, it changes to an arrow. You're not allowed to draw outside the border. You're now working with a 12" × 9" drawing area or viewport. A viewport is an area where a drawing can be displayed. Nice looking border, but not many projects will fit into the drawing area. Future chapters will introduce altering the size and number of the viewports.

Examining the Layout Tabs

Although not a part of the Architectural template, many templates contain multiple layouts. These layouts can be created to help control the plotting of layouts that contain drawings to be displayed at varied scales. To add additional layouts to the architectural template, move the cursor to the area beside the Model and Architectural Title Block tabs, and right-click. Select New Layout from the shortcut menu, and a Layout1 tab will be added to the display. Selecting the Layout1 tab will display the Page Setup dialog box shown in Figure 5.3. The dialog box is a new feature to AutoCAD and allows the plotting paramaters to be established, saved, and reused throughout the life of the drawings. These paramaters will be addresed in future chapters. Select the OK button to close the dialog box and produce a display similar to Figure 5.4. With model space active, you'll have a drawing area of 12" × 9". The size can be altered to meet the drawing needs. The important thing to consider about

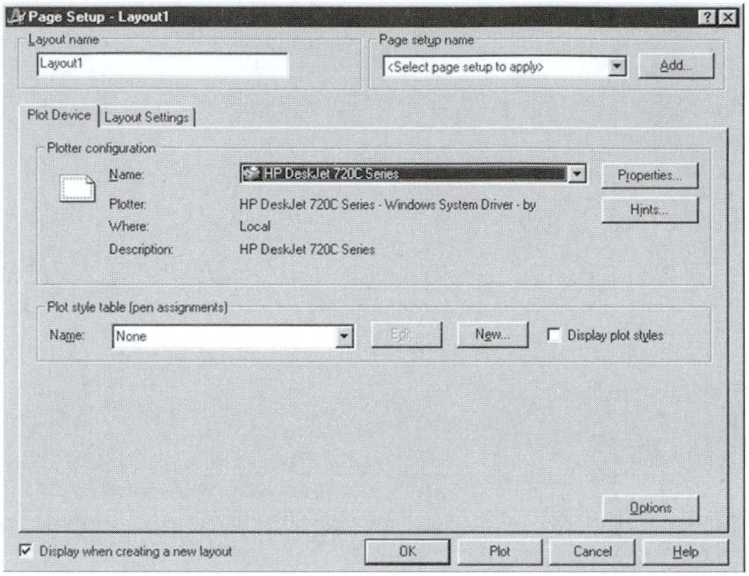

Figure 5.3 *Once in the Architectural template, place the cursor in the tab bar and right-click. Selecting New Layout from the shortcut menu will produce an additional layout titled Layout1. As the layout is created, the Page Setup dialog box will be displayed. The menu allows plotting paramaters to be established that can be saved and used throughout the life of the drawings. Additional layouts can be added to aid in the plotting of layouts containing drawings that must be displayed at varied scales. Future chapters will introduce methods for creating and using multiple layouts.*

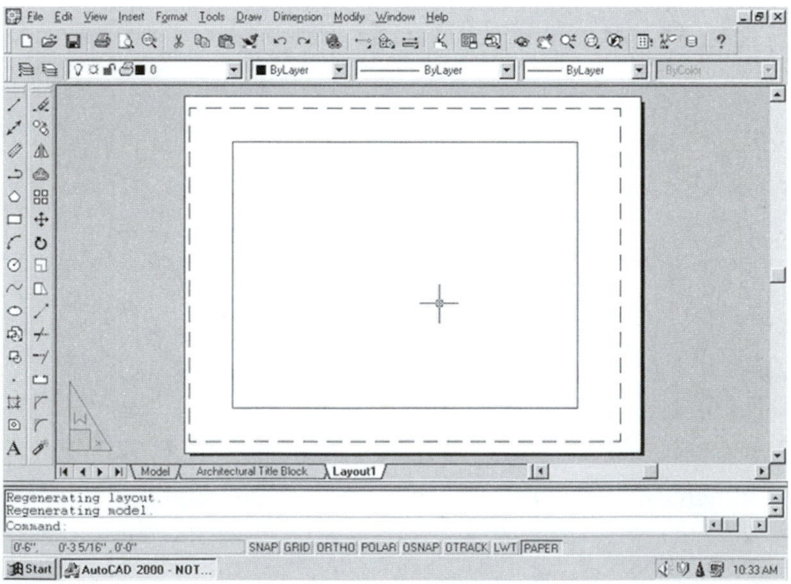

Figure 5.4 *With model space active, you'll have a drawing area of 12" × 9" using Layout1 of the ACAD template.*

Layout1 now is its use. Layouts are used to create a drawing for plotting. You'll often hear layouts referred to as a viewport. A viewport allows you to place a 300'-long building on a 36"-wide sheet of paper. Viewports will be explored in future chapters after you've had some time to gain experience with other basic AutoCAD skills.

Selecting the MODEL tab will produce a drawing area of approximately 1'-8" × 10 3/8". The size will vary depending on the monitor size. For most applications, this will be the area of the template where you will construct most of your future drawings. Model space allows you to work in "real" space. By changing the size of the space, the 300'-long building can be viewed on the screen in its entirety. The size of the drawing space is altered using the LIMITS command.

Setting the Limits of Model Space

If you haven't already done so, open the Architectural template and select the Model tab. Use the methods described in Chapter 3 to set the limits as 118' × 90'. Because many plan views in architecture are plotted at a scale of 1/4" = 1'-0", this example will establish limits that allow a plan view to fit within the drawing area of the template. This size was established because at 1/4" = 1'-0", each inch in paper space represents 4 feet in model space. Multiplying the drawing area size 29.5" (the horizontal distance between the template borders in paper space) × 4 (the number of feet per inch) and 22.5" (the vertical distance between the template borders in paper space) × 4 provides the basis for how the limits were determined. Methods for establishing drawing units for other drawing scales will be presented in this chapter. Use the following procedures to update the template to architectural standards.

- Enter **UNITS** (EnterKey) at the command prompt. Using the Drawing Units dialog box to adjust the units to the desired precision.
- Enter **LIMITS** (EnterKey) at the command prompt. Use the following sequence to alter the drawing limits:

  Command: **LIMITS** (EnterKey)

  Reset Model space limits:

  Specify lower left corner [ON/OFF]: <-0'-0",-0'-0">: (EnterKey)

  Specify upper right corner <1'-0",0'-9">: **118',90'** (EnterKey)

  Command: **Z** (EnterKey)

  Specify corner of window, enter a scale factor (nX or nXP), or

  All/Center/Dynamic/Extents/Previous/Scale/Window/<Realtime>: **A** (EnterKey)

  Command:

You should be familiar with the effects of using the LIMITS command. As you entered Z (EnterKey), you started the ZOOM command. Think of ZOOM as a telephoto lens on a camera. ZOOM alters the area of a drawing to be displayed. The

ZOOM command will be explored in detail in Chapter 9. If you enter the ZOOM command and type **A** (EnterKey), you've selected the ALL option. You've asked AutoCAD to display all of the display area. This process will alter the size of the drawing area from 29.5" × 22.5" to 118' × 90'. Now you have a drawing area that is suitable for construction drawings. Future chapters will show you how to add other elements to the template to ease drawing setup.

Comparing Model, Layout1, and the Title Block

The limits that have just been set are for model space. The model space UCS icon is displayed in the lower left corner of the drawing area, and the MODEL button in the Status bar is activated. You're working in model space. Use the LINE command to draw a box that is 90' × 50'. The drawing can be seen, but the template title block is not displayed. You're still working in the template, but in the model space, you're only displaying the space between the borders. Pressing the Layout1 tab will display the box inside the layout as shown in Figure 5.5. This is representing how the drawing will appear in the template in paper space. Any portion of the drawing that extends past the bold line (the viewport) will not be plotted. Selecting the Architectural tab displays the template with the title block and border but the box can't be seen. Select the PAPER button in the Status bar and the drawing and the template, with the title block and border, will be displayed as seen in Figure 5.6. Future chapters will introduce when the title block and Layout1 should be used.

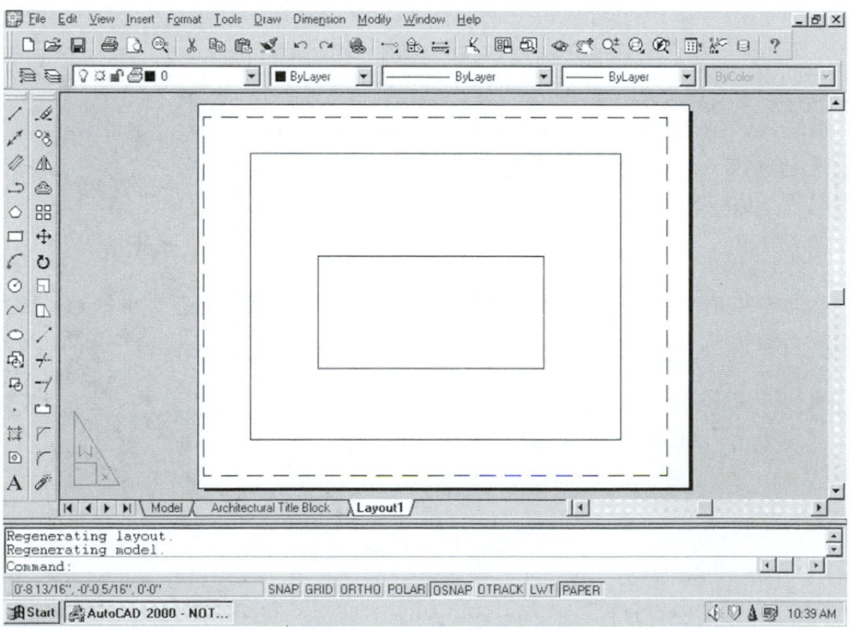

Figure 5.5 *Pressing the Layout1 tab will display the box inside the layout.*

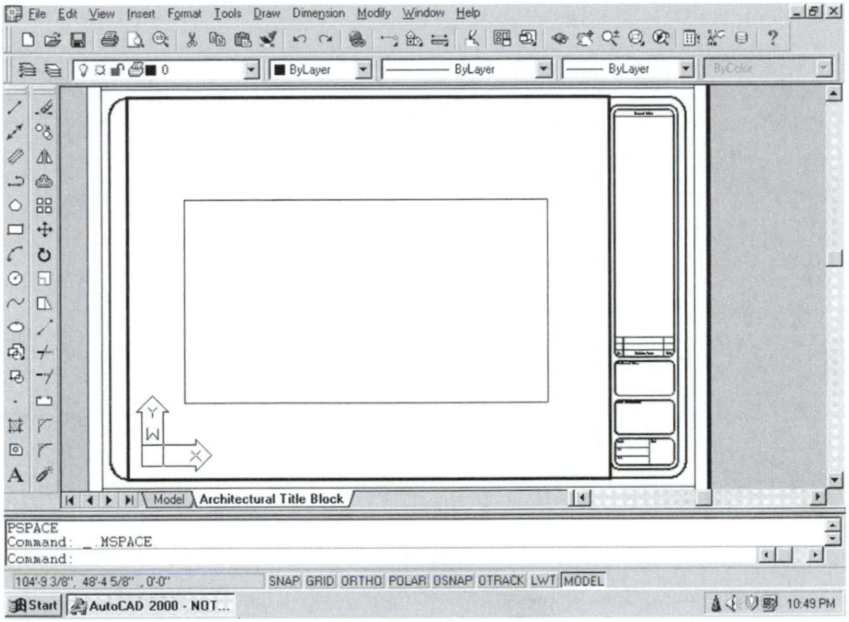

Figure 5.6 *Selecting the Architectural tab displays the template, but the box can't be seen. Select the Model button in the Status bar and the drawing and the template are displayed.*

Note: As a new AutoCAD operator this may all seem a bit overwhelming. If nothing else, keep in mind that a template is a great place to store drawing tools. It's like the junk drawer we all have in a desk. All the good stuff goes in the drawer. A template is a drawing base where you will begin to store the settings that you use over and over. No one should be setting the precision of units or the drawing limits each time a drawing is started. The second thing to remember is that you want to work in model space to create objects, and paper space to plot them. Each of the preceding chapters has reminded you of this key point. It's important. Future chapters will introduce when the title block and Layout1 should be used.

- Use templates to save time.
- Draw in model space, plot in paper space.

Creating Your Own Template

Any drawing can be saved and used as a template. A template drawing can be created using methods used with any other drawing. Open a drawing and establish the drawing units, limits, snap, and grid values. Special consideration will need to be given to the drawing limits as you establish a drawing template. In Chapter 3, as the limits were established common paper sizes were introduced. The size of paper that the completed project will be printed or plotted on must be considered as the drawing template is established. Although the drawing limits that are shown on the screen can be altered at anytime throughout the life of the drawing, establishing limits that match standard paper sizes will aid in the use of the template. To establish a template for use with drawings completed at ¼" = 1'-0", drawing values will need to be multiplied by a factor of 48 to determine the size to be used in the template. The scale factor for ¼" = 1'-0" or .25 = 12 is determined by dividing 12 by .25. If the drawing is to be placed on D-size material, limits would need to be 1,152" × 1,728" or 96' × 144'. The limits can also be determined by multiplying the desired paper size by the number of feet in 1 inch of paper space. At a scale of ¼" = 1'-0", 1 inch of paper will contain 4 feet of drawing, resulting in the same 96' × 144' limits. Other common limits for D material are shown in Figure 5.7. Once the desired settings are established, the template can be saved and recalled for future use. Methods of saving a template will be introduced later in this chapter.

SAVING USER PREFERENCES

A drawing template works well for saving drawing settings that must remain constant between each member of the drawing team. There are many features of the drawing environment that do not affect the final project, but affect the productivity of each team member. Drawing profiles allow each user to adjust and save the drawing settings to meet their personal preferences. Drawing profiles can be especially helpful if you're working at a computer that you share with several other people. If you like a black background and you share your computer with someone who likes a white background, AutoCAD allows these differences to be stored so that they can be changed quickly. Items that might be included in a profile include: Startup dialog box display, crosshairs size, display resolution of objects, intervals between automatic saves, text display format, screen colors, and plotter or printer information. Using a

DRAWING SCALE FACTORS							
ARCHITECTURAL SCALE FACTORS				ENGINEERING SCALE FACTORS			
SCALE	LIMITS		TEXT SCALE	SCALE	LIMITS		TEXT SCALE
	18 X 24	24 X 36			18 X 24	24 X 36	
1" = 1'-0"	18 X 24	24 X 36	12	1" = 1'-0"	18 X 24	24 X 36	12
3/4" = 1'-0"	24 X 32	32 X 48	16	1" = 10'-0"	180 X 240	240 X 360	120
1/2" = 1'-0"	36 X 48	48 X 72	24	1" = 100'-0"	1800 X 2400	2400 X 3600	1200
3/8" = 1'-0"	48 X 64	64 X 96	32	1" = 20'-0"	360 X 480	480 X 720	240
1/4" = 1'-0"	72 X 96	96 X 144	48	1" = 200'-0"	3600 X 4800	4800 X 7200	2400
3/16" = 1'-0"	96 X 128	128 X 192	64	1" = 30'-0"	540 X 864	864 X 1080	360
1/8" = 1'-0"	144 X 192	192 X 288	96	1" = 40'-0"	720 X 960	960 X 1440	480
3/32" = 1'-0"	192 X 256	256 X 384	128	1" = 50'-0"	900 X 1200	1200 X 1800	600
1/16" = 1'-0"	288 X 384	384 X 576	192	1" = 60'-0"	1080 X 1440	1440 X 2160	720

NOTE: ALL LIMIT SIZES ARE GIVEN IN FEET.

Figure 5.7 *The scale that a drawing will be plotted at must be considered as a template is constructed. The drawing limits will be affected by the scale to be used and the paper size.*

profile, you'll be able to adjust, save, and retrieve drawing settings the next time you use the workstation.

To set User profiles, select Options... from the Tools pull-down menu. This will display the Options dialog box with nine tabs to choose from. The tab that was used last is opened as the default display. Each of these tabs will be explored in detail as commands are introduced throughout the text. The goal of this chapter is to introduce the Options dialog box and to help you save a few of your favorite settings. As new features are introduced, they can be added to your profile. Highlights of the Options tabs include:

- Files—This tab is used to control the directories where AutoCAD will search for driver, menu, and other files. At this point of your AutoCAD career, only a few of the options need to be considered. Both points will be discussed later in this chapter.

  Display—This tab allows the display of AutoCAD to be customized. Several of the options of this tab have been discussed in previous chapters. By setting the values and saving them in a profile, they will automatically be reset as the profile is selected as the active profile. Take time to evaluate each portion of this box. If in doubt, select the help button at the bottom of the tab for an explanation. Items you should adjust include:

  Text Lines in the Command Line Window—Specifies the number of lines displayed in the docked window.

  Colors—Allows the color of each aspect of the display screen to be altered. Selecting this option will display the Color Option dialog box and allow you to set the color of each element in the drawing area.

  Crosshairs size—Allows the size of the crosshairs to be altered by entering a number that represents the percentage of the screen to be filled.

  The Display Resolution—The settings in this box affect the quality of the display of objects in the monitor. Do you want to display a circle as a circle, or as a series of straight line segments that appear to make a circle. The effects of this area will be discussed in Chapter 7.

- Open and Save—This tab will affect how and where drawing files are saved. This tab will be covered in detail later in this chapter.

  SaveAs—This portion of the menu can be used to determine the format used when saving your drawings.

  Automatic Save—This option controls the time interval between automatic saves. This options is one of the greatest safety measures you can have to protect your drawing files.

  Create Backup Copy with Each Save—This option toggles the creation of backup files. Typically, having a backup copy is a good safety feature. If your machine is short of storage space, deactivating this option will save your file over the existing copy of the file. If you decide to delete backup files, make sure you make multiple copies of your files on the hard drive, multiple floppy disks, or a Zip drive.

- Plotting—This tab controls all of the options that affect plotting. Components of this tab will be explored later in this chapter and discussed in detail in Chapter 24.
- System—This tab controls how the system will operate. Take time to examine the General Options box. These settings control general options that relate to drawing operation. Key elements to consider are:

  Single Drawing Compatibility Mode—One of the new features of AutoCAD 2000 is that you can truly work on two drawings at once. This only holds true if you have this setting in the inactive mode.

  Show Startup Dialog—This setting will control the display of the Startup menu. Most new users prefer to have the dialog box displayed until they have developed their own work habits. Keep this setting activated in your profile.

  Beep on Error in User Input—In the active setting, AutoCAD will sound an alarm (a small beep) when an invalid response is given. If you're in the middle of the LINE command, and AutoCAD is expecting a "to point" to be provided, the alarm will sound if you try and use the CIRCLE command. As a new user, keep this activated. When deactivated, many new users often get frustrated when the desired results are not achieved. The alarm often serves to bring you back to reality. The alarm should remind you to look at the command prompt so that the desired information can be provided.
- User Preferences—This tab affects options that optimize your performance in AutoCAD. One feature that experienced users of AutoCAD may want to explore is the Right-click Customization box. Some experienced users are so accustomed to using the right-click button as the (EnterKey) that they may be willing to forsake the new shortcut menus. If you're a new user, leave the menus active. Other options will be addressed in future chapters.
- Drafting—Most of the options in this box have been introduced in previous chapters, and will be covered in detail in future chapters. Options that you should address as you set your own profile include:

  AutoSnap Marker Color—This control can be used to alter the color of the snap box

  AutoSnap Marker Size—This control can be used to alter the size of the snap box. As the size of the box is increased, it gets easier to see. As the drawing complexity increases, a large snap box can lead to inaccurate point selection.

  Aperture Size—This slide bar can be used to control the size of the aperture box. The aperture box is the AutoCAD selection tool when a command is started.

- Selection—These features control how AutoCAD will select drawing objects. This box will be explored in detail in Chapter 8. A feature that you may want to set in your initial profile is the pickbox size.
- Profiles—This tab controls the creation and use of profiles. The Profile tab can be seen in Figure 5.8. The next section of this chapter will help you explore this tab and create your first profile. As you gain experience and discover new settings that you would like to include in the profile, the profile can be updated using this tab.

CREATING AND SAVING A PROFILE

To create a user profile, select Options... from the Tools pull-down menu. Select the Profiles tab to display the menu shown in Figure 5.8. Go through each of the nine tabs and set the options to meet your preferences. Once you have selected the desired settings:

- Select the Add to List option. This will display the Add Profile dialog box shown in Figure 5.9.
- Enter a profile name in the Provide Name edit box. Use a name that will quickly describe the user or drawing type.
- Provide a short description in the Provide Description edit box that will quickly describe the contents.

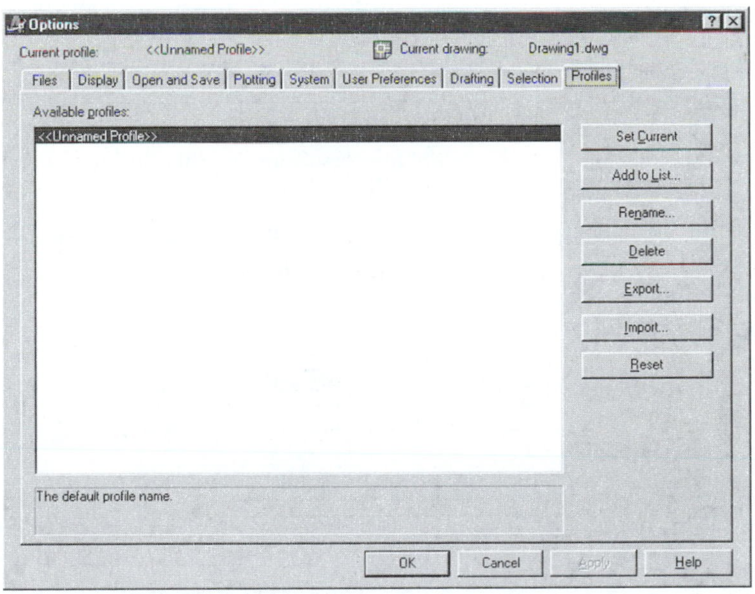

Figure 5.8 *The Profile tab controls the creation and use of profiles. As you gain experience and discover new settings that you would like to include in the profile, the profile can be updated using this tab.*

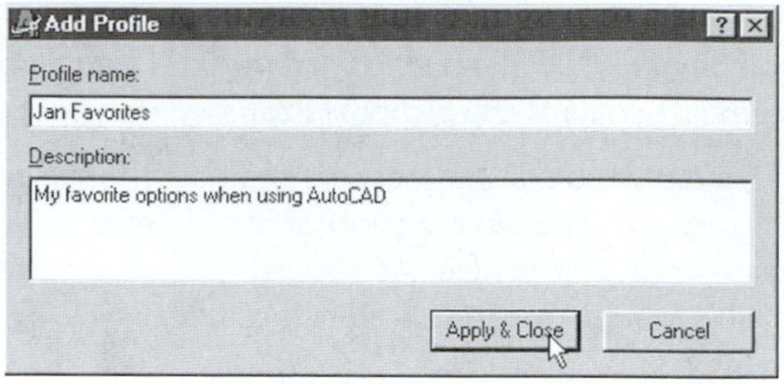

Figure 5.9 *Selecting the Add to List option will display the Add Profile dialog box. The box is used to name and describe the profile to be created.*

- Select the Apply and Close button to accept the settings. When the button is selected, the dialog box will be closed and the listing of "Jan Favorites" will be added in the Available profiles listings.
- Once more than one profile has been created, you can quickly select the desired profile from the Profile tab of the Options dialog box by highlighting the listing of the desired profile and then selecting the Set Current button. Other buttons in this menu that should be considered include:

  Rename—This option allows an existing profile to be renamed. Selecting this option will produce the Change Profile dialog box show in Figure 5.10. Alter the name as desired and then select the Apply and Close button to alter the profile name and close the dialog box.

  Delete—This option allows an existing profile to be removed from the Available Profiles list. Highlight the profile to be deleted and select the Delete button. This will produce a warning to verify that you really want to delete the profile. Selecting No will terminate the delete process. Selecting the Yes button will delete the selected profile.

  Export—Selecting this option displays the Export Profile dialog box. (See Figure 5.11.) This dialog box can be used to export a profile as a file with an

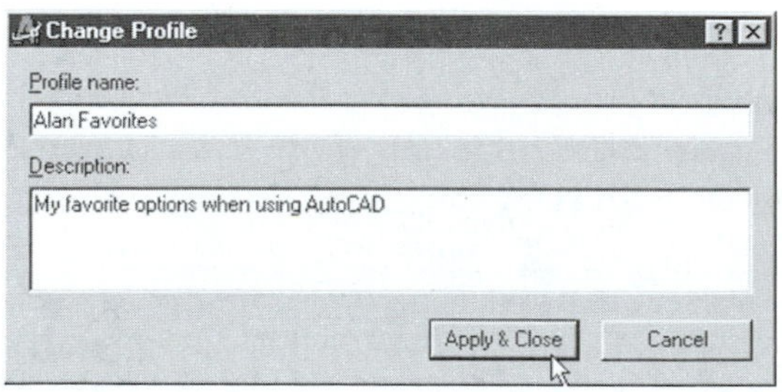

Figure 5.10 *Selecting the Rename button will produce the Change Profile dialog box. The box is used to alter the name of existing profiles.*

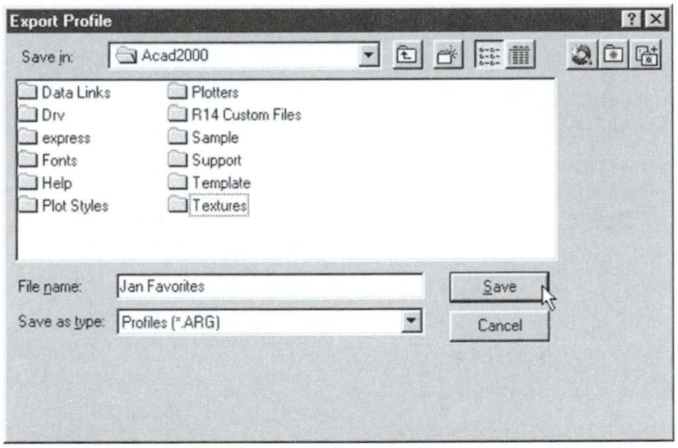

Figure 5.11 *The Export Profile dialog box can be used to export a profile as an .ARG file. The profile can be saved and then shared with another computer.*

extension of .ARG. The profile can be saved and then shared with another computer. If you're working on a network, save your profile to your secure folder. Profiles should also be saved to a floppy disk.

Import—Selecting this option will display the Import Profile dialog box. Similar to the Export Profile dialog box, the display can be used to import a profile that was created using Export.

Reset—Selecting this option will reset the computer to the AutoCAD default settings.

SAVING DRAWINGS

Chapter 2 introduced the SAVE command and Chapter 4 gave you practice using it. This chapter will help you expand the use of the command. Once you have created your drawing, it will need to be saved. This will mean saving your drawings in a folder, on the hard drive, on a floppy disk, or in your own folder on a network. Check with your instructor to see if saving on the hard disk is allowed. If so, set up a folder so that all of your drawings can be easily accessed. If not, select the letter of the appropriate drive so files will be saved to the floppy disk. Generally you should avoid saving a file to a floppy drive, except when the drawing session will be ended. Typically, a file should be saved to the hard drive throughout the drawing session to save time and avoid a possible problem caused by a full disk.

If you are learning AutoCAD in an office setting you may be saving on the hard drive, tape drives, Zip drives, as well as on floppies. Because working on the hard drive is so much faster than accessing floppies, you will be saving to the hard drive during the drawing session, and saving on the floppy and the hard drive at the end of the drawing session. Many professionals save on the hard drive and make two floppy backups, which are each stored in different places. The open drawing file should be saved to the hard drive approximately every 15 minutes. When the drawing session is done, the file should again be saved to update the existing file on the hard drive. The file should also be saved on another source, such as a floppy disk. This is especially true for students who must transport their drawing between home and school.

No matter where you save your drawing, the most important thing to remember is to save often. Saving every 15 to 20 minutes will help you avoid timely redraws if the power supply to your computer is interrupted. As you start saving your drawings you will have five options: SAVE, SAVEAS, QSAVE, QUIT, and EXPORT.

THE SAVE COMMAND

Using the SAVE command will store your drawing in a selected location but still leave it displayed on the screen. This will allow you to continue working on the drawing. If your computer should crash, only work done after the last save would be lost. To save a drawing, select the Save icon from the Standard toolbar, select Save from the File pull-down menu, or type **SAVE** (EnterKey) at the command prompt. Each will produce a similar result and produce the Save Drawing As dialog box shown in Figure 5.12a. The dialog box is only displayed the initial time the drawing is saved. The second time the drawing is saved, the SAVE will occur automatically using the name that was assigned during the initial save.

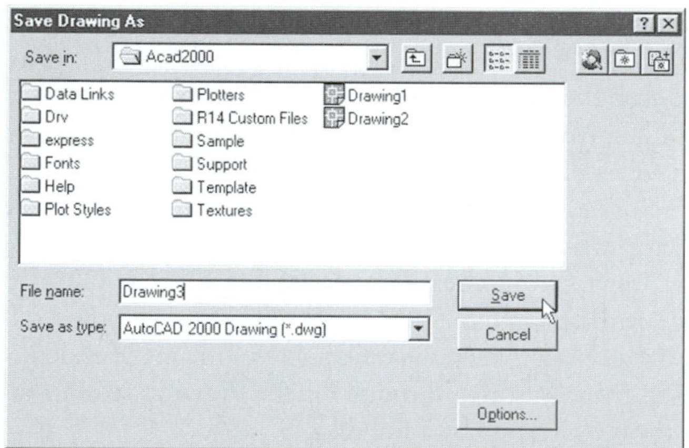

Figure 5.12a *The Save Drawing As dialog box is displayed by selecting the Save icon in the standard toolbar, selecting Save or SaveAs... from the File pull-down menu, or by typing* ***SAVE*** *(EnterKey) or* ***SAVEAS*** *(EnterKey) at the command prompt. The dialog box is only displayed the initial time the drawing is saved using the SAVE command. The second time the drawing is saved, the SAVE will occur automatically using the name that was assigned during the initial save. Current drawing file names are displayed. The name of the file to be saved must be typed in the File name: edit box.*

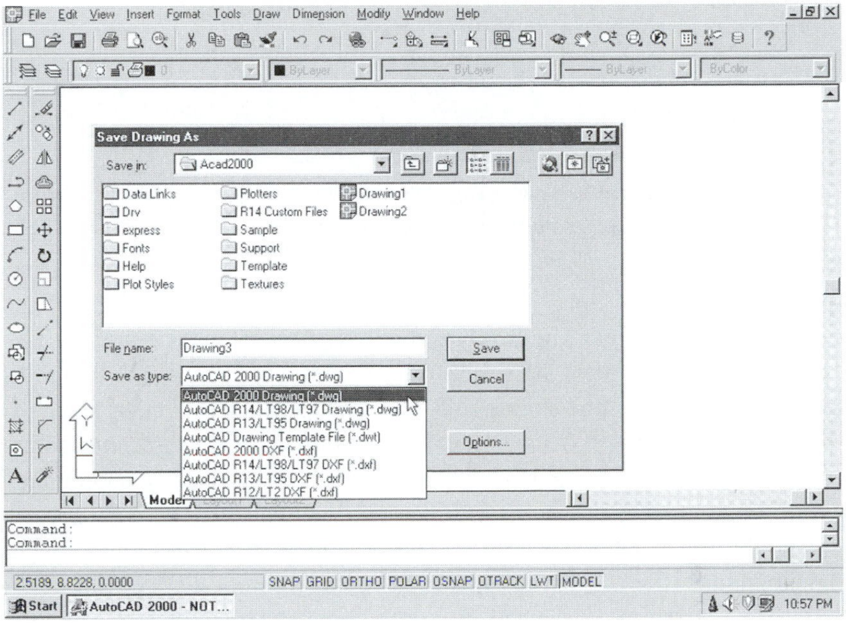

Figure 5.12b *The Save as type: edit box allows drawings to be saved and be compatable with older versions of AutoCAD.*

Selecting the Save in: edit box arrow will display the contents of the Windows Explorer. When the destination drive has been selected, the contents of the specified drive will be listed. Enter the desired file name in the File name: edit box. Use the Save as Type edit box to control the drawing type to be used. Figure 5-12b shows a menu of alternative drawing types that can be used. Once the type has been selected, press the Save button. Guidelines for file names will be given later in the chapter. The .DWG extension is added to the drawing file and it is saved to the designated drive. This will be the desired method of saving for most AutoCAD users.

Filenames

As you save drawings, you may misname a drawing. If you haven't hit the enter key, just use the backspace key to remove the unwanted name. If the name is entered, the file can be renamed using the Windows Explorer. Occasionally you will assign a drawing file a name that already has been used. In this case you will see a display similar to the one in Figure 5.13. If you proceed, you will destroy the existing file. The name will remain in the folder and your present drawing will be saved using the old filename, but the existing drawing will vanish. This is a fine way to update an old drawing, but be sure you can live without the old drawing.

Back-Up Files

Earlier you discovered that when AutoCAD saves a file it adds an extension of .DWG or .BAK. The .DWG stands for a drawing file. AutoCAD also adds .BAK as

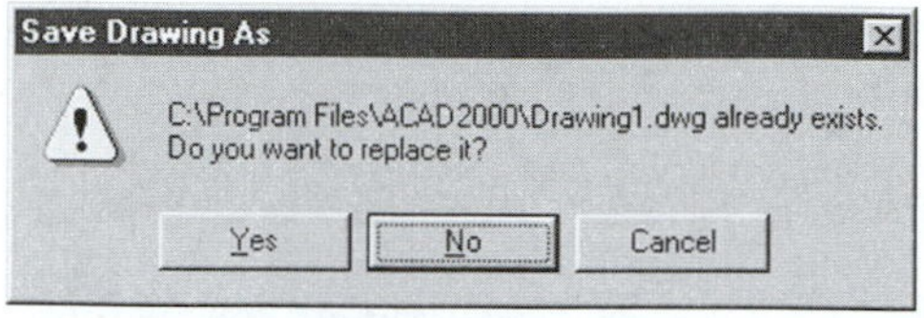

Figure 5.13 *Attempting to save a drawing using a name that has been assigned to another file will result in this alert message. If the Yes option is accepted, the drawing will be saved using the selected name, and the existing drawing will be altered.*

an extension to drawing files. The .BAK is the extension for a back-up file. These are created automatically by AutoCAD each time you end a drawing. When a drawing is terminated using SAVE, the existing .DWG file becomes the .BAK file and a new .DWG file is created. The .BAK file may not contain all of your recent changes, but it should be only 15 minutes out of date if you've been saving at proper intervals. What a motivator to keep saving your drawings! Back-up files are also great if your disk has been damaged. The .BAK file can be renamed using the Windows Explorer and then reloaded. If the .BAK file cannot be converted to a .DWG file, or if it was scrambled, this will serve as one last reminder to save your files in more than one location. If you have a limited storage space, or if you save to multiple locations, you may not want to create back-up files. If you decide you really don't want .BAK files created, the option can be deactivated by selecting the Create back-up copy with each save option on the Open and Save tab of the Options dialog box

THE SAVEAS COMMAND

The SAVEAS command is similar to the SAVE command. With SAVE the SaveAs dialog box is only displayed the first time the drawing is saved. With SaveAs the dialog box is displayed each time the SAVEAS command is accessed. SAVEAS allows the name of the file and the file type to be altered. SAVEAS can only be accessed from the File pull-down menu or from the command prompt. Enter SAVEAS (EnterKey) at the command prompt. This will produce the Save Drawing As dialog box, seen in Figure 5.12a. The default name will be the name you selected the last time the drawing was saved. If you like that name, select the Save button. If you want to change the name of the drawing file, enter the new name in the File: edit box.

Be sure that the correct drive destination is current. Note that the destination is listed in the Drives: edit box. To change the current drive, use the arrow beside the Drives: edit box to display the desired drive. Pressing the Save button will save the drawing to the specified drive. When a drawing is saved, it will be placed in the current drive. AutoCAD will assign the extension .DWG or .BAK following the filename to help identify the contents of your directory.

Another benefit of SaveAs is that it allows the file type to be altered. Examine Figure 5.12b again. In Figure 5.12b seven other alternatives for saving a file are presented.

These alternatives allow drawings created in AutoCAD 2000 to be saved so they will be compatible with different versions of AutoCAD. These options can be especially useful for students whose home machines are not current with those at school or for architectural and engineering firms with different versions of the software. To save a drawing created in AutoCAD 2000 to be used on a machine running R14, select the AutoCAD R14/LT98 Drawing (*.DWG) option. Remember that you can set the type of drawing format to be used for saving using the Open and Save tab of the Option dialog box, and then saving the setting as part of your profile.

Open a drawing session, draw a few lines, and then save the file with a name of JUNK. As you assign a name, you do not have to type the .DWG at the end of the drawing file because it will be added automatically by AutoCAD. Once the drawing is saved, add a few more lines to the drawing and save the drawing again. If you save the drawing using the SAVE command, the Save Drawing As dialog box will not be displayed. The current file of JUNK.DWG is automatically updated. If you save the drawing using the SAVEAS command, the Save Drawing As dialog box will be displayed allowing a new drawing name to be assigned to the drawing. Draw a few more lines and save the file using the SAVEAS command. When the SaveAs dialog box is displayed, change the drawing name from JUNK to TRASH, using the File Name: edit box. You will now have files named JUNK and TRASH. When you're done with the drawing session be sure to delete these files using Windows Explorer

QSAVE

QSAVE, or quick save, will probably be your most frequently used method of saving a drawing. Selecting the Save icon from the Standard menu or typing QSAVE (EnterKey) at the command prompt will save the current drawing using its given name without displaying the Saveas dialog box. If QSAVE is used on an unnamed drawing file, the Save Drawing As dialog box of Figure 5.12a will be displayed.

Note: SAVE, SAVEAS, and QSAVE will each save a file for future use. Use QSAVE for saving a drawing that you want to continue working on. It's the fastest method of saving because you only need to select the Save icon from the Standard tool bar. QSAVE saves named drawings without requesting a file name. Use SAVEAS when you want to save a drawing by giving it a new name. SAVEAS renames the current drawing and saves it to a filename. Save does the same thing as QSAVE but is slightly slower to access because you need to select both File pull-down menu and then select Save. SAVE saves the current drawing to a filename.

AUTOMATIC SAVING

By default, the existing drawing will be saved every 120 minutes. SAVETIME can be used to alter the time between saves. The time between automatic saves can be altered by using the Open and Save tab of the Options dialog box. The box is accessed by selecting Options… from the Tools pull-down menu. Use the following procedure to alter the interval between saves:

- Open the Option dialog box.
- Select the Open and Save tab.
- Highlight the current value in the Minutes between Saves edit box.
- Enter the desired value.
- Select the OK button.

The save time can also be adjusted by typing **SAVETIME** (EnterKey) at the command prompt. This will produce the prompt:

Command: **SAVETIME** (EnterKey)

New value for SAVETIME <120>: **15** (EnterKey)

AutoCAD allows a value between 1 and 120 minutes to be entered. The value entered will start the timer when a change is made to the current drawing file. The commands SAVE, SAVEAS, and QSAVE will reset the timer. With a value of 20, your drawing file will be automatically saved every 20 minutes with a name of AUTO.SV$. If this file needs to be accessed, the Windows Explorer can be used to rename the file from a .SV$ file to a .DWG file. SAVETIME can be disabled by entering a value of 0, but this could prove to be a huge mistake if your computer ever crashes.

SAVING A TEMPLATE

Earlier in this chapter you learned how to create a template. Once each of the values for the template is established, the drawing can be saved and reused whenever suitable values are used. Save the template using SAVEAS with the Drawing Template File (*.DWT) option selected. (See Figure 5.12b.) Enter the name of the drawing just as you would do with any other drawing. A name that describes the scale or paper size such as 14DSIZE or CSIZE will be helpful for sorting templates. Once the name has been entered, select the SAVE button. Provide a brief description of the template, such as "template for plan views". This description is displayed whenever you select this template from the Create New Drawing dialog box. The final step to save the template is to select the OK button. The template can now be reused as a new drawing is opened.

LEAVING AUTOCAD

Once a drawing is saved, you may want to leave the drawing and start another project. AutoCAD allows you to exit the program using the Close button or by using QUIT and CLOSE commands.

The QUIT Command

The QUIT command will exit the drawing and close AutoCAD. The QUIT command can be executed by selecting the close button from the AutoCAD title bar, or by typing QUIT (EnterKey) at the command prompt. If you attempt to leave a draw-

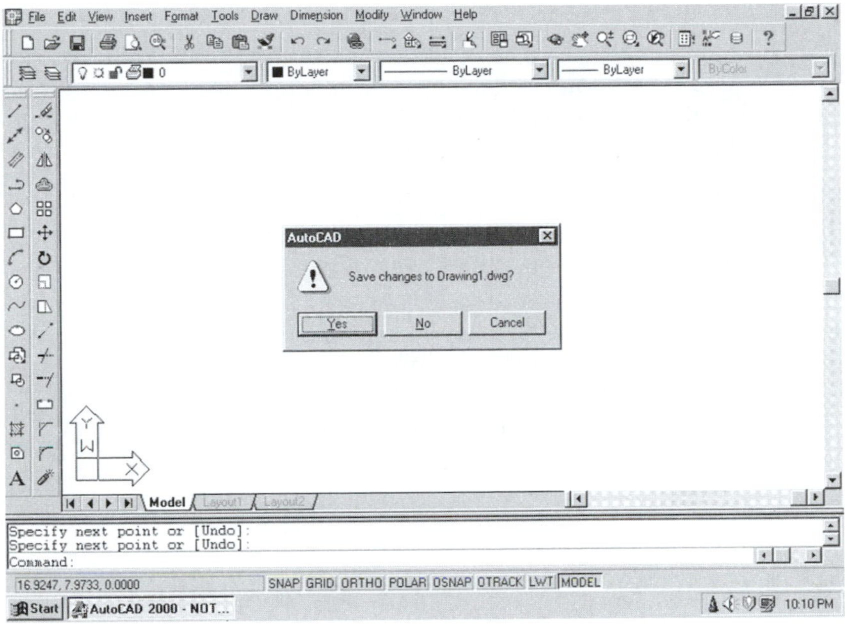

Figure 5.14 *An alert message will be displayed if you attempt to open a new drawing without saving the existing drawing.*

ing with unsaved changes, you will see the alert box shown in Figure 5.14. If you select the Yes button, AutoCAD will save the changes to the default file. Selecting No will exit the drawing, close AutoCAD, and return you to other opened programs. Any changes made since the last SAVE will be destroyed. CANCEL will terminate the QUIT command and return you to the drawing.

The CLOSE Command

The CLOSE command can be used to close the current drawings but remain in AutoCAD. The command must be entered from the command prompt by entering CLOSE (EnterKey) at the prompt. If there have been no changes, the drawing will be closed automatically. If changes have been made since it was last saved, you will be prompted to save or discard the changes.

MANAGING FILES AND FOLDERS

No office could function if the drafters took their finished drawings and threw them into a room. If a drawing was needed, some poor drafter would have to wade in and hunt through thousands of sheets of paper to find the specific project. Architecture and engineering offices typically use well-organized filing systems to keep track of each project for quick, easy retrieval.

FILES

No matter what method is used to save a drawing, information that is to be kept on your disk must be stored in some orderly manner to allow quick, easy retrieval. Disks are sorted in files, folders, and paths. Your drawings will be stored as files and will need names to distinguish them from other files. As you store your files, three-letter extensions will be added to titles to aid retrieval. For example, a drawing saved by AutoCAD might be filed as FLOOR.DWG. FLOOR is the drawing name, and .DWG is the extension. Filenames for AutoCAD can contain up to 255 characters. Other guidelines for naming files include:

- Layer names can contain letters, digits, spaces, and special characters such as $ (dollars), - (hyphen), and _ (underscore).
- Special characters used by Microsoft, Windows, or AutoCAD cannot be used.
- Do not use the following symbols:

* (asterisks)	< > (less-than and greater-than symbols)
‘ (backquotes)	? (question marks)
: (colons)	“ (quotation marks)
, (commas)	; (semicolons)
= (equal signs)	| (vertical bar)
/ \ (forward and back slashes)	

Note: If you're saving an AutoCAD 2000 drawing to be compatible with an older release, names should be restricted to 31 characters. Longer names will automatically be shortened to 31 characters and illegal characters will be replaced with an underscore (_). The original name is saved and restored when the file is opened in AutoCAD 2000. If the AutoCAD 2000 file is renamed during a drawing session in a previous release of AutoCAD, the original name will be lost.

FOLDERS

Files should be organized in folders to provide efficient storage. Directories are the equivalent of dividers in a file cabinet drawer. By default your drawings will be stored in the Acad 2000 folder. Information about files such as file size, time of creation, and the time of last file revision is automatically kept as files are stored. Folders are usually created by software programs as they are loaded onto the hard drive. Folders can also be created as needed to organize information based on client names, class requirements, or any other similar criteria. A folder can be created by using the Windows Explorer. Folder names should follow the same guidelines used to name drawing files. Names should be used that will quickly define the contents such as DRAWINGS, MAPS, REPORTS, or CLIENTS. Folders can be placed inside other folders to better organize information. Subfolders can be created and named using the same command and guidelines used to create a folder. Figure 5.15 shows an

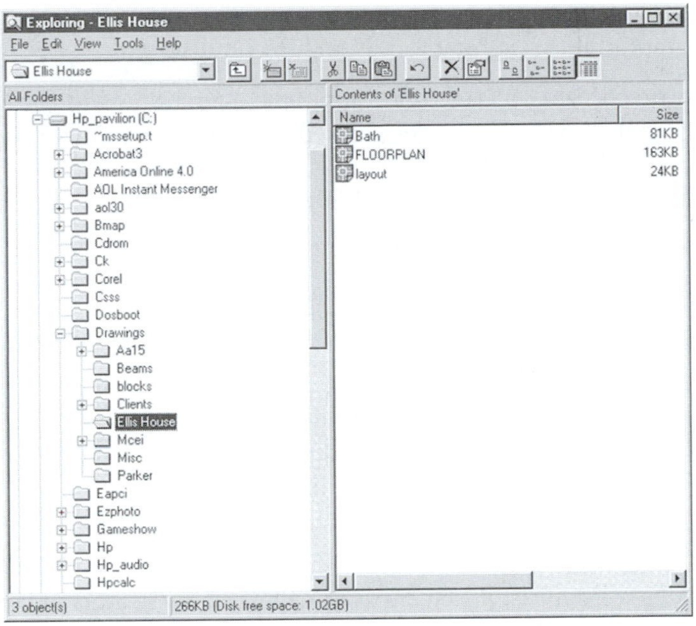

Figure 5.15 *Folders should be created to contain similar drawing files to ease retrieval. Notice in the C drive, that all drawings are stored in the DRAWING folder. The DRAWING folder is further broken down into subfolders named for clients to ease retrieval.*

example of how folders are set up within folders created by the Windows Explorer in a hierarchical method. This use of paths in saving material will help in retrieval.

OFFICE PROCEDURE

Many small residential offices keep each client in separate folders or disks and label the disk by the client name. Drawings are then saved by contents such as floor, found, elev, sects, specs, or site. Some offices assign a combination of numbers and letters to name each project. Numbers are usually assigned to represent the year the project is started, as well as a job number with letters representing the type of drawing. For example, 9953fl would represent the floor plan for the fifty-third project started in 1999. A few offices will save an entire house plan in one file. The drawings are separated by layers, which will be discussed in Chapters 6 and 14. The drawback to saving large amounts of information in one file is that more time is needed to load and process the information. Future chapters will introduce External Reference as another method of saving drawing files.

AIA Guidelines for Naming Files

Because it is rare that only one person in an office will use a file, most architectural and engineering offices use file names based on the AIA (American Institute of Architects) guidelines. The AIA guidelines provide a uniform file naming system that will be recognized by the various consultants that work on each project. Two methods of naming files are often used by architectural systems.

One common practice in architectural firms is to name drawing files by the page number it will occupy in the drawing set based on AIA page numbering guidelines.

Common Page Numbers and File Names

A0.01—Index, symbols, abbreviations, notes, and location maps

A1.01—Demolition, site plans, and temporary work

A2.01—Plans, and schedules such as room material, door, or windows and keyed drawings

A3.01—Sections and exterior elevations

A4.01—Detailed, large-scale floor plans

A5.01—Interior elevations

A6.01—Reflected ceiling plans

A7.01—Vertical circulation drawings such as stair, elevator, and escalator drawings

A8.01—Exterior details

A9.01—Interior details

When this system is followed, a drawing file named A4.01 would represent a detailed floor plan. If plans for a five-level building were to be drawn, each floor could be saved as a separate drawing file using drawing names such as:

A2.01—level one

A2.02—level two

A2.03—level three

A2.04—level four

A2.05—level five

A second method of naming files is based on a letter that represents the discipline of the drawing originator, and a two-letter code that represents drawing types common to all disciplines. A number that represents the sheet sequence number is sometimes added to the filename. Using this system, drawing files would have names such as A–FP01 or S–DT13. Some of the most common disciplines and drawing codes include:

Common Discipline Codes

A	Architectural	M	Mechanical
C	Civil	P	Plumbing

E	Electrical	Q	Equipment
F	Fire Protection	S	Structural
I	Interiors	T	Telecommunications
L	Landscape		

Common Drawing Codes

FP	Floor Plan	DT	Detail
SP	Site Plan	SH	Schedules
DP	Demolition Plan	3D	Isometric/3D drawings
QP	Equipment Plan	DG	Diagrams
XP	Existing Plan	EL	Elevation
SC	Section		

CSI Guidelines for Naming Files

Another popular method of naming construction drawing files is to use the guidelines provided by the Construction Specifications Institute (CSI). These guidelines recommend titles start with a letter to represent a discipline code, a sheet designator, and a page number. Page numbers in filenames are designated sequentially from 01 through 99. Discipline codes are the same as used by the AIA. Using the CSI format, files names such as A-501 would represent **A** is a drawing by the architectural team, **5** represents details, and **01** represents sheet 1 of the detail section.

Sheet Type Designators

0—General symbols, abbreviations, notes, and location maps

1—Plans, and schedules such as room material, door, or windows and keyed drawings

2—Exterior elevations

3—Sections

4—Detailed, large-scale floor plans

5—Details

6—Schedules and diagrams

7—User defined

8—User defined

9—3D views such as isometrics, perspectives, and photographs

STORAGE PROBLEMS

One of the great thrills for a new CAD operator is seeing the message, "DISK FULL." It's not a serious problem, but it is discouraging. As a drawing is started,

AutoCAD examines the disk to find a suitable storage space. Because AutoCAD creates space for several files, it will need to perform expected drawing commands and there may not be enough space to create the intended drawing. In the middle of a command you may find that a DISK FULL error is displayed. Your drawing will be terminated, but all work up to that point will be saved. You will typically only encounter this problem if you save to a floppy drive in the middle of a drawing session.

Occasionally a full disk will cause FATAL ERROR to flash across the screen just before your machine destroys two hours worth of drawing. Sure, you know to save every 15 minutes, but sometimes you've got to learn the hard way. The unfortunate aspect of a fatal error is that it may disrupt or destroy other files on your floppy. You'll find that you only have FATAL ERROR when you're 99 percent done with a project.

One Last Reminder: As you are working on a drawing file, save the drawing to the hard drive. Only save to a floppy when you are done with the drawing session and are closing the file. Working from a floppy drive will slow the drawing session and can lead to storage problems.

Diskette Handling Procedures

Diskettes should only serve to back up projects that have been saved on the hard disk or network. Because of limited memory in many school PCs, you may not be allowed to save on the hard disk. This will make your diskettes even more valuable. Because diskettes can be damaged easily, extreme care must be taken with them. The biggest destroyers of information on a diskette are dust, smoke, magnets, and bending.

Dust and smoke interfere with the head in the disk drive as it attempts to interpret the diskette. Magnetic fields from such common items as a stereo speaker or a telephone can scramble the information on a disk into an unusable form. In addition to destroying information, a bend may hinder retrieval of a diskette from the drive unit. Disk labels that are not affixed well to the disk also can cause the disk to jam in the drive.

Guidelines for Safe Diskette Handling

- ALWAYS keep the disk labels in the outer corners of the disk.
- ALWAYS label diskettes clearly, indicating the contents.
- ALWAYS store diskettes in a protective carrying case and in a cool, dry location.
- ALWAYS store diskettes between 50 and 125° Fahrenheit (10–52° centigrade).
- NEVER place diskettes on top of a monitor.

Protecting Disks

Diskettes can be protected so that no one can accidentally enter information or change the original contents. There is a notch in the top right-hand side of the disk when looking at it from the front. Turn the disk over and slide the black tab to its upper position. This will allow you to see through the notch and make the diskette write protected.

Correcting File Errors

No matter how carefully you follow the guidelines for handling a disk, they will occasionally become corrupted. AutoCAD provides the AUDIT and RECOVERY commands that attempt to make corrupted drawing files useable. Each are located in Drawing Utilities of the File pull-down menu. The PURGE command can be used to clean unneeded information from the drawing base and will be introduce in Chapter 16.

Examining a Drawing File with Audit

This command can be used to examine the current drawing file for errors and to correct any errors that exist. These errors are not spelling mistakes or drawing errors, but errors that occasionally are made by the software. This command will detect errors, generate extensive descriptions of the problems, and recommend actions to correct the errors. Audit can be used by selecting Audit of Drawing Utilities File in the pull-down menu, or by entering **AUDIT** (EnterKey) at the command prompt. The command sequence is:

Command: **AUDIT** (EnterKey)

Fix any errors detected? <N>

Responding with an **N** (EnterKey) will produce a report of the file and list the errors, but will not repair the errors. The final listing of the report resembles:

Total errors found 3 fixed 0

Responding **Y** (EnterKey) at the prompt will produce a report of the file and list the errors. The listing resembles:

3 Blocks audited

Pass 1 12 entities audited

Pass 2 4 entities audited

Total errors found 2 fixed 2

With the AUDITCTL system variable set to 1, an Audit report file will be created using the filename with .ADT as the extension. This file will be placed in the same directory containing the original drawing file and will list all of the functions required to fix file errors.

Recovering a Drawing

If errors are discovered that cannot be repaired by AUDIT, the RECOVER command can be used to try and salvage the file. RECOVER can be accessed by selecting Recover... from Drawing Utilities of the File pull-down menu or by typing

RECOVER (EnterKey) at the command prompt. AutoCAD will automatically attempt to repair damaged drawings as they are opened. If the drawing can't be opened, a message will be displayed that the drawing must be recovered. Once the file to be recovered has been selected, AutoCAD will begin a series of diagnostic procedures. When the process is complete, a listing of problems and the corrective measures taken by the program will be displayed. Generally the recovery process will open a damaged drawing. If the RECOVERY command fails, it's time to get out the back-up file. This is one last opportunity to remind you that no matter how fast and good computers have become, they occasionally fail. They generally only fail at times of great stress such as the end of a school term or the due date of a big project. Back-up files can greatly reduce the stress of a file that can't be recovered.

CHAPTER 5 EXERCISES

1. Start a new drawing. You will be using this exercise as a base for a drawing of a small subdivision. The drawing will be drawn at a scale of 1" = 30' and plotted on D-size paper. North will be at the top of the paper (12 o'clock). Set all applicable defaults to two-place decimals.

   What unit setting should be used for this drawing? ____________

   Set the limits at the proper value. What option number will be required? ____________

   Many of the property lines are described with angles measured in degrees, minutes, and seconds. Set the angle measurement system in the appropriate system. What default number was selected? ____________

2. Start a new drawing. Draw a square, a rectangle, and a triangle. Save the drawing using the name E-5-2.
3. Start a new drawing called EXER5-3 that has the same drawings contained on E-5-2. Add another circle and a rectangle. Save as E-5-3.
4. Edit drawing E-5-3 by adding any figure that you would like. Save file as E-5-4.
5. Get a directory of your floppy disk; list the following:

   The contents.

   The time your first exercise was finished.

   The date of your last exercise.

   The number of bytes used.

   The number of bytes free.

6. Open the ARCHITECTURAL.DWT template.

   What is the current snap resolution? X _____ Y _____

   What is the current grid setting? X _____ Y _____

   What layer will new lines be added to? ____________

   You will be using this exercise to set up a drawing for a plan view of a room that will be drawn at ¼" = 1'-0" and plotted on D-size paper. Two-place decimal angles are desired.

   Set the units at the proper value. What option number will be required? ____________ What was the default? ____________

   The smallest fraction to be worked with will be ¼". What denominator should be used? ____________

   Give the current system for measuring angles. ____________

Will the current setting allow for two-place decimals? ____________

In which direction will angles be drawn? ____________

What are the current drawing limits? ____________

What limits will be needed to achieve the listed parameters of this exercise?
____

What is the current snap value? ____________

Set the snap value to 6".

What is the current grid default value? ____________

Set the grid to 12".

List two other methods of listing the 12" value.

CHAPTER 5 QUIZ

1. You're working on the fifth level of a multilevel project. Suggest three different names that can be used to save the file.
2. What command will store the drawing on the hard disk but leave the drawing on the screen?
3. How many characters may be used for a file using AutoCAD 2000? How many characters may be used if the file is to be compatible with an R14 file?
4. What command will store the drawing on a disk and terminate the drawing?
5. List three different methods to access the SAVE command.
6. You have a drawing file labeled SECTION. You draw a roof detail and start to save it as SECTION. What will AutoCAD respond and how will your drawing be affected?
7. Give six forms of information listed when the dialog box is used to save a drawing.
8. What two extensions are assigned to a drawing file?
9. What does QUIT do differently from SAVE?
10. What is a fatal error?
11. Give two reasons for backing up a drawing in multiple locations.
12. List two reasons why you might receive an alert message while saving a drawing.
13. You've opened an existing drawing named FLOOR.DWG, made some changes, and now want to save the drawing while keeping FLOOR.DWG intact. How can this be done?
14. You've saved a file named LOST.DWG some where on your hard drive, but you're not sure where. How can you find the file?
15. Explain the difference between SAVE and SAVEAS.
16. What is the default value for SAVETIME, and is this value adequate?

CHAPTER 6

Drawing Organization

Now that you're able to set up drawing units and limits, basic drawing components can be explored. The main component of the drawings you will be working with will be lines. This chapter will introduce you to various linetypes used on construction drawings and to methods for assigning and controlling varied linetypes. As linetypes are added to a drawing, a method of managing these linetypes must also be considered. Color can be assigned to lines to help distinguish building components from each other. Most professional AutoCAD users sort linetypes by the use of layers. The LINETYPE, COLOR, and LAYER commands will be introduced in this chapter, with a detailed discussion presented in future chapters. In addition to exploring the methods of controlling lines, this chapter will explore the process of making a non-scaled check print. Plotting a drawing to scale will be explored in future chapters.

AN INTRODUCTION TO LINETYPES

Varied linetypes are use in every type of drawing and consulting firm involved in the construction process. This chapter will introduce you to common linetypes used to make construction drawings, introduce how to load linetypes to be used in a drawing session, and explain how to alter and control linetypes that are part of a drawing.

COMMON LINETYPES OF CONSTRUCTION DRAWINGS

Although each office may have its own standard, guidelines established by the American National Standards Institute (ANSI) are used throughout most of the construction industry to ensure drawing uniformity. Common line width for thick lines is 0.6mm. Use a width of 0.3mm for thin lines. Methods of altering line width will be introduced in this chapter and discussed in Chapter 24 as the drawing is printed. Common linetypes include object, hidden, center, viewing plane, cutting plane, break, phantom, extension, dimension, and leader. Examples of each type of line can be seen in Figure 6.1. Methods to create and control each linetype will be introduced in this chapter. Specific uses for each linetype will be discussed throughout the text.

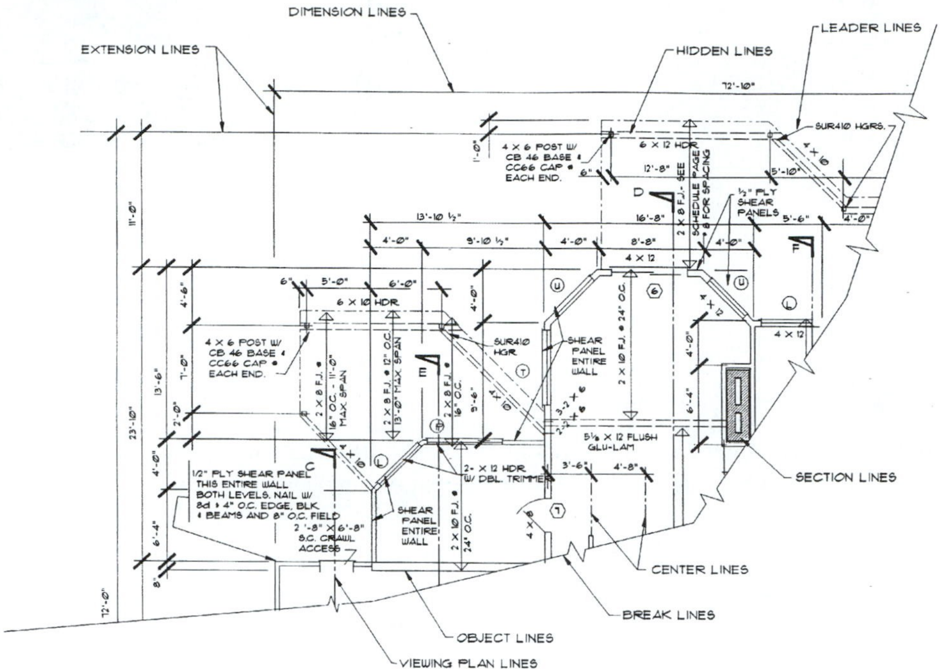

Figure 6.1 *Common linetypes used on construction drawings.*

Object Lines

Object lines are continuous lines used to describe the shape of an object or to show changes in the surface of an object. Object lines may be thick or thin depending on how they are being used. On a drawing such as a floor plan, thick object lines are used to represent walls, and thin object lines are used to provide contrast when representing doors, windows, cabinets or appliances. Figure 6.2 shows both uses of object lines. Uses of object lines include, but are not limited to:

- Site Plans

  Thick: Building footprint,

  Thin: Flatwork, street and sidewalk edges, decks, balconies, or patios.

- Floor and Framing Plans

  Thick: Walls and fireplace outlines.

  Thin: Doors, windows, cabinets, appliances, stairs, railings, plumbing fixtures, shelving, and framing members.

- Elevations

  Thick: Finish grades, outlines of objects, and shading.

  Thin: Windows, doors, trim, and wall and roof outlines.

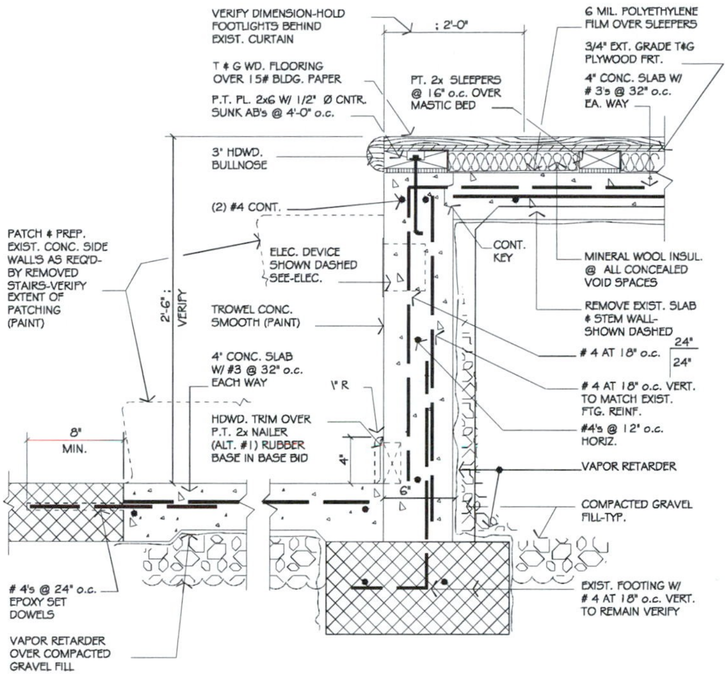

Figure 6.2 *Object lines are continuous lines used to describe the shape of an object or to show changes in the surface of an object.*

- Sections and Details

  Thick: Outline of structural material cut by the cutting plane such as beams, plates, sills, and girders; concrete shapes such as piers or footings; steel reinforcing; finish grades; and natural grades.

  Thin: Repetitive structural materials such as studs, posts, columns, joists, and rafters that lie beyond the cutting plane. Structural materials such as sheathing. Non-structural materials such as sheet rock, and materials beyond the cutting plane.

Hidden Lines

Dashed or *hidden lines* are thin lines used to represent a surface or object that is hidden from view. ANSI specifies that the lines should be ⅛" (3mm) long with $\frac{1}{16}$" (1.5mm) space between lines. The size of the line can be controlled using the LTSCALE command. AutoCAD will automatically control the spacing, based on the line length. Figure 6.3 shows an example of hidden lines. In addition to hidden features, uses of hidden lines for drawings include, but are not limited to:

- Site Plan—Easements, sewer or water lines, and structures to be removed.
- Floor Plan—Appliances such as dishwashers that are under counters, appliances such as refrigerators that are not in the contract, and knee spaces in cabinets.

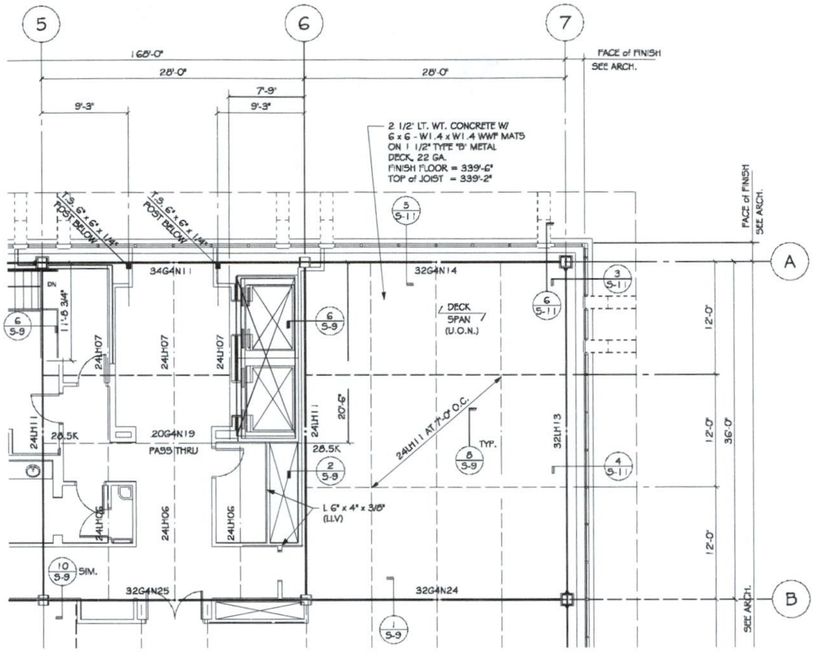

Figure 6.3 *Dashed or hidden lines are thin lines used to represent a surface or object that is hidden from view.*

- Framing Plan—Beams and headers.
- Foundation Plan—Outlines of footings, piers, and girders.
- Elevations—Outline of footings.
- Sections and Details—Interior of tube steel columns, and steel reinforcing.

Centerlines

A *centerline* is composed of thin lines that create a long-short-long pattern. The short line should be 1/8" long, the space should be 1/16", and the long segment must be between 3/4" – 1" long. Although the length of the long segment can vary from drawing to drawing, all long segments in a drawing should be consistent. AutoCAD will control the length and space of each portion of the line or you can create your own pattern using the methods presented in Chapter 14. Centerlines are used to locate the center axis of circular features such as drilled holes, bolts, columns, and piers. Figure 6.4 shows the use of centerlines on a foundation plan. Uses for centerlines include, but are not limited to:

- Site Plans—Center of easements, streets.
- Floor and Framing Plans—Upper levels or projections above the current floor, grid lines.
- Elevations—Finish floor or ceiling levels.

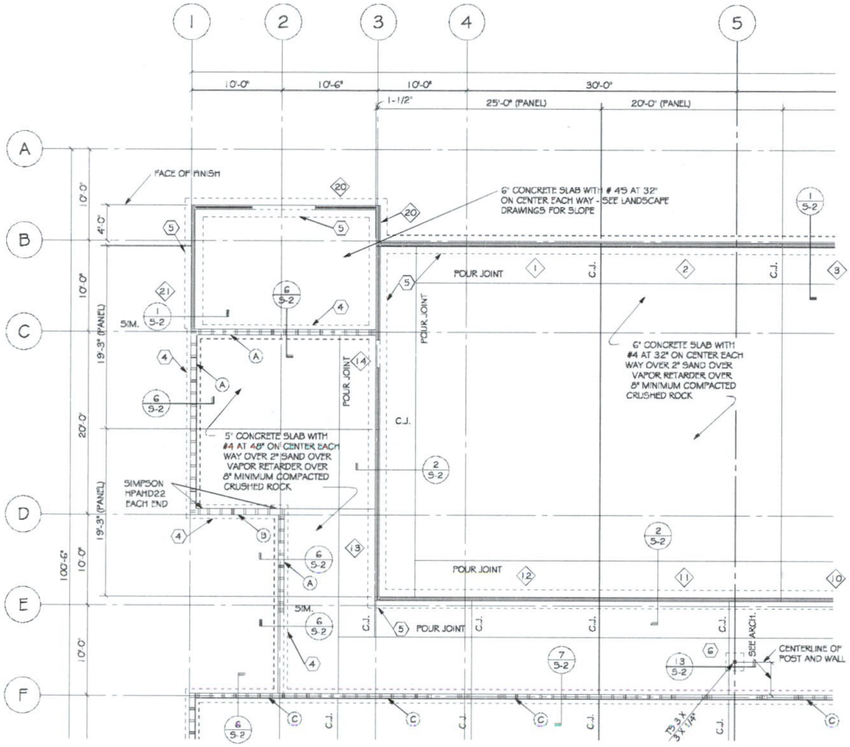

Figure 6.4 *A centerline is composed of thin lines that create a long-short-long pattern. Centerlines are used to locate the center axis of circular features such as drilled holes, bolts, columns, and piers. Centerlines are also used to show the girders on a foundation plan.*

- Sections and Details—Finish floor, ceiling of roofing levels, centers of steel columns and chevrons.

Cutting Plane Lines

A *cutting plane line* is placed on the floor or framing plan to indicate the location of a section and the direction of sight when viewing the section. Cutting planes are a long-short-short-long line pattern. Cutting plane lines should be thicker than object lines. The long line segment should be between ¾" – 1" long. The short lines should be ⅛" long, and the space between lines should be 1⁄16". Terminate the line with arrows to indicate the viewing direction. Place a letter at each end of the line using ¼"-high text to indicate the view or a detail reference bubble. Figure 6.5 shows examples of cutting planes and line terminators.

Section Lines

When an object has been sectioned to reveal its cross section, a pattern is placed to indicate the portion of the part that has been sectioned. In its simplest form, *Section lines* are thin parallel lines used to indicate what portion of the object has been cut by

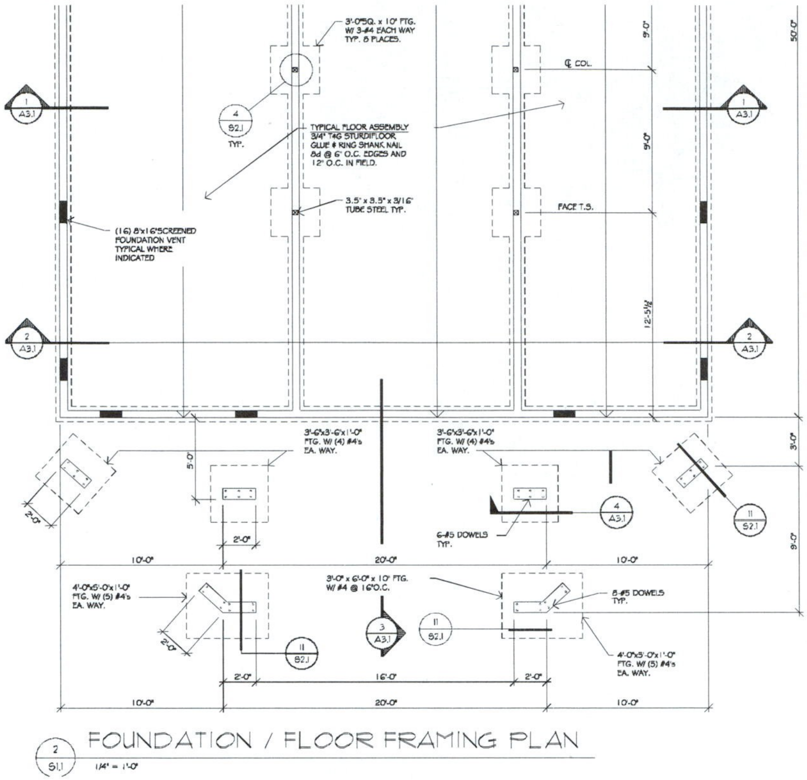

Figure 6.5 *A cutting plane line is used to indicate the location of a section and the direction of sight when viewing the section. The line is thicker than object lines and must terminate with arrows to indicate the viewing direction or a detail reference bubble.*

the cutting plane. Section lines are usually drawn at a 45° angle but should not be parallel or perpendicular to any part of the object that has been sectioned. If an angle other than 45° is used, it should be between 15° and 75°. If two adjacent parts have been sectioned, section lines should be placed at opposing angles. (See Figure 6.6.) Spacing between section lines can be altered, depending on the size of the part to be sectioned. Typically, a spacing of 1/8" should be provided between section lines.

Section lines are generally only used to represent masonry in plan or sectioned views. Most materials have their own special pattern. Figure 6.7 shows common patterns used to indicate that an object has been sectioned. AutoCAD refers to these patterns as hatch patterns. Chapter 15 will discuss representing sectioned material, guidelines for placement of section lines and hatch patterns, and methods of using the BHATCH command.

Break Lines

Break lines are used to remove unimportant portions of an object from a drawing so that it will fit into a specific space. If you consider a 10'-long steel column that sup-

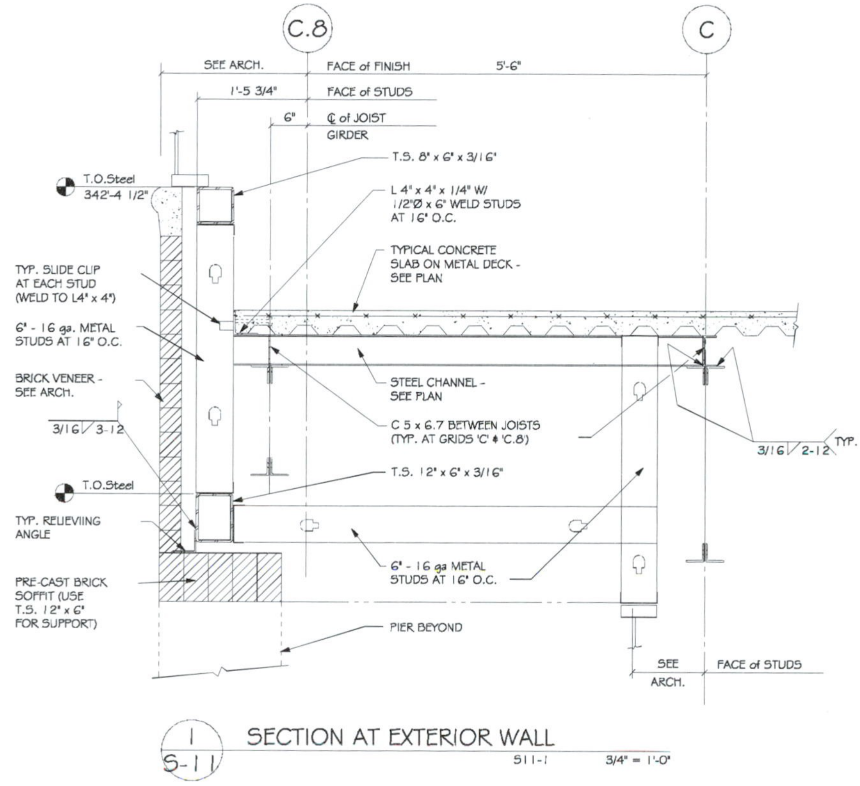

Figure 6.6 *Section lines are thin parallel lines drawn at a 45° angle to indicate what portion of the object has been cut by the cutting plane. Section lines are generally only used to represent masonry in plan or sectioned views.*

ports a beam and rests on a concrete footing, the column is the same between the beam and the bottom plate. If you had to draw a 10' steel column at full scale in a 5' space, break lines can be used to remove 5' of the column. The three common break lines used on construction drawings are the short, long, and cylindrical break lines. A short break line is a thick jagged line placed where material has been removed. See Figure 6.8a. A long break line is a thin line with a zigzag shape or inverted S shape inserted into the line at intervals. See Figure 6.8b. A cylindrical break line is a thin line resembling a backward S. (See Figure 6.8c).

Phantom Lines

Phantom lines are thin lines in a long-short-short-long pattern that are used to show motion, or to show an alternative position of a moving part. They can also be used in place of centerlines to represent upper levels or projections of a structure, or on site plans to represent easements or utilities. The long line segment should be between ¾" – 1" long, but remain a constant length within each drawing. The short lines should be ⅛" long. The space between lines should be 1⁄16".

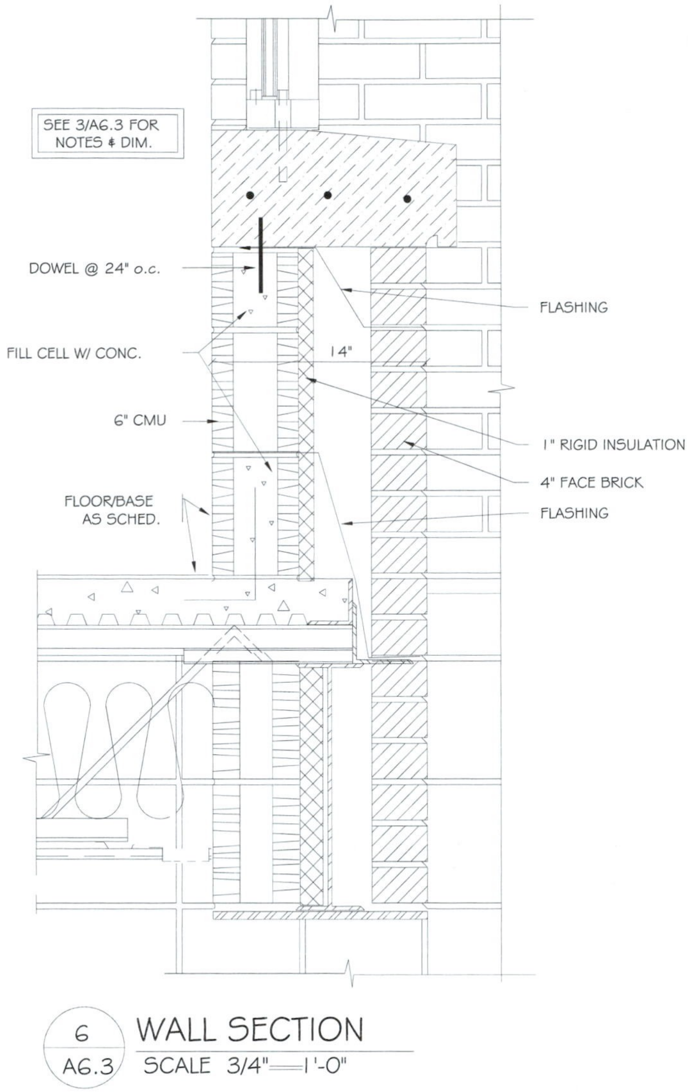

Figure 6.7 *Common patterns used to indicate that an object has been sectioned. Most construction materials have their own special pattern. AutoCAD creates these patterns using the BHATCH command.*

Extension Lines

Extension lines are thin lines used to relate dimension text to a specific surface of a part. (See Figure 6.9.) Provide a 1/16" to 1/8" space between the object being described and the start of the extension line. Extend the extension line 1/8" past the dimension line. An extension line may cross other extension lines, object lines, center, hidden, and any other line, with one exception. An extension line can never cross a dimension

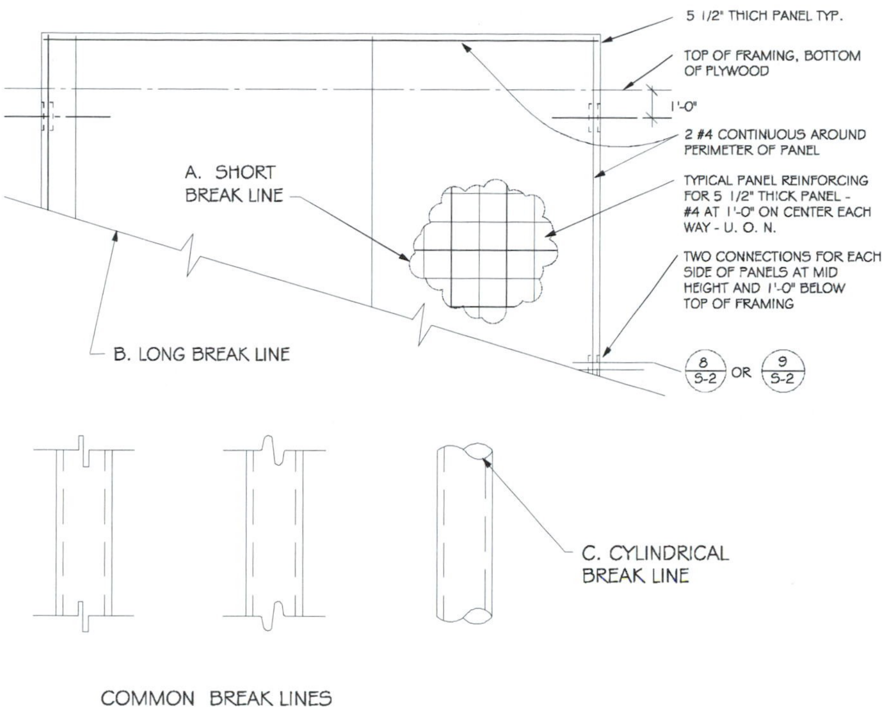

Figure 6.8 *Common break lines of construction drawings; a.) Short break line is a thick jagged line placed where material has been removed; b.) Long break line is a thin line with a zigzag shape or inverted S shape inserted into the line at uniform intervals; c.) Cylindrical break line is a thin line resembling a backward S.*

line. No exceptions! Chapters 19 and 20 will introduce other guidelines for placing extension lines.

Dimension Lines

Dimension lines are thin lines used to relate dimension text to a specific part of an object (See Figure 6.9). Dimension lines extend from one extension line to another. Dimension lines require a line terminator where the line intersects the extension line. The terminator is usually a tick mark or a solid arrowhead. Dimension text is typically ⅛" high text. Dimension text should be placed centered over the dimension line, although it may need to be placed off center to aid clarity when several lines of text are near each other. The space between the dimension text and the dimension line should be equal to half the height of the dimension text. Chapters 19 and 20 will provide specific uses of dimension lines and dimension text.

Leader Lines

Leader lines are thin lines used to relate a dimension or note to a specific portion of a drawing. A solid arrowhead should be used at the object end of the leader line. A horizontal line approximately .125" long should be used at the note end of the leader

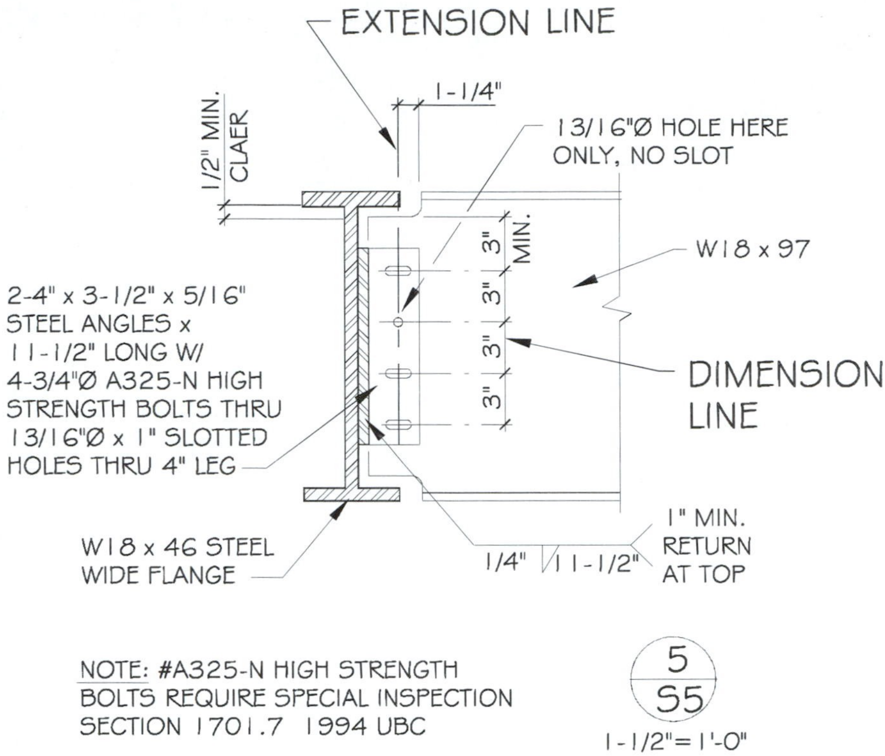

Figure 6.9 *Dimension lines are thin lines used to relate dimension text to a specific part of an object.*

line. Leader lines can be placed at any angle, but angles between 15° and 75° are most typical. Vertical and horizontal leader lines should never be used. The arrow end of the leader line should touch the edge of the part it describes. When used to describe a circle, the arrow should end at the edge of the circle, but if extended, the line would pass through the center of the circle. The leader line should extend far enough away from the object it is describing so that the attached note is a minimum of ¾" from the outer surface of the object. See Figure 6.10 and Chapters 18 and 19.

Figure 6.10 *Leader lines are thin lines used to relate a dimension or note to a specific portion of a drawing. A terminator should be used at the object end of the leader line. Leader lines should extend a minimum of ¾" from the object being described*

AN INTRODUCTION TO CREATING LINETYPES

You've been introduced to the varied types of lines that are required to produce construction drawings. AutoCAD contains many of the linetypes found in the construction industry in the Standard linetype library. These lines can be accessed by selecting Linetype... from the Format pull-down menu or by typing **LINETYPE** (EnterKey) at the command prompt. Each method will produce the Linetype Manager dialog box shown in Figure 6.11. The Linetype library is displayed by selecting the Load... button. Figure 6.12a shows the linetype display. Figure 6.12b shows a listing of the lines found in the linetype library.

LINETYPE CHARACTERISTICS

The LINETYPE command will allow you to vary the linetype pattern for each object you draw or for the entire layer. Most offices assign a specific linetype to a specific layer. As you start to assign linetypes to drawing entities, you need to remember the following two key elements regarding linetypes:

Effects

Objects that will be affected by a change of linetype include line, circle, arcs, polygons, donuts, ellipse, hatch, and 2D polylines. Any other drawing object will be displayed as a continuous line. That will be good news as you start to add dimensions or text to a drawing and you're working on a layer drawn with centerlines or another line pattern as the current linetype.

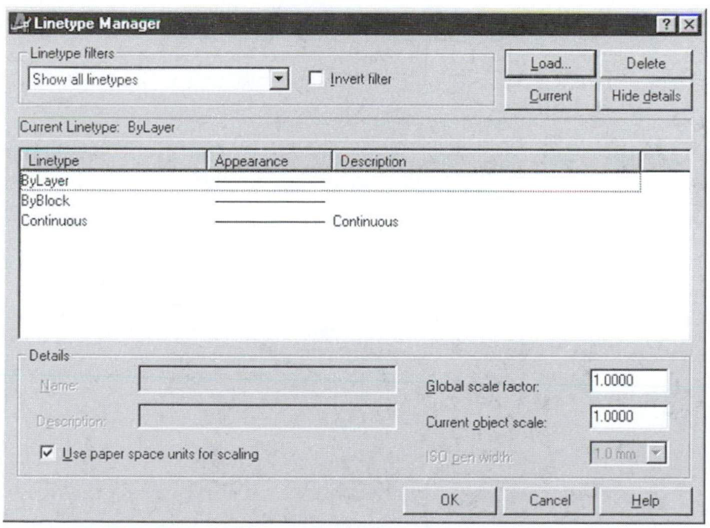

Figure 6.11 *The Linetype Manager dialog box can be used to change and load linetypes. The Linetype manager is accessed by selecting Other... from the Linetype Control edit menu of the Object Properties toolbar, or selecting Linetype... from the Format pull-down menu.*

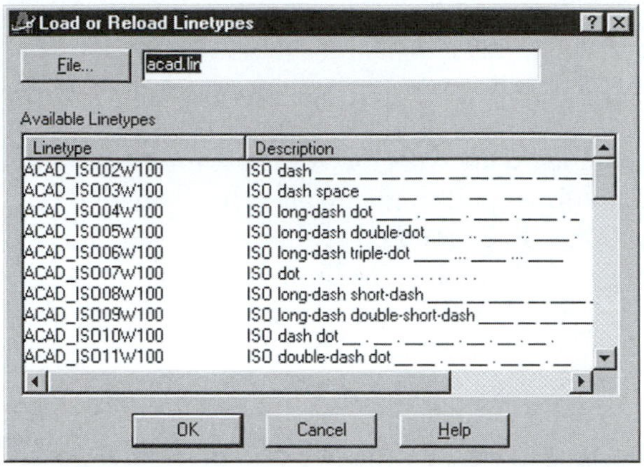

Figure 6.12a *Selecting the Load button will display the Load or Reload Linetypes menu with a display of each AutoCAD linetype. Use the scroll bar to see the balance of the menu.*

Loading

A second concept to keep in mind as you work with linetypes is that you can't use a linetype unless it has been loaded into your drawing. The linetypes of Figure 6.12b exist in a library file stored in AutoCAD. They will be loaded automatically when LINETYPE is entered as a function of the Linetype Manager dialog box.

LOADING LINETYPES INTO A DRAWING

As you enter the Linetype Manager dialog box, you'll notice that three linetypes are displayed: ByLayer, ByBlock, and Continuous. Additional linetypes are loaded by

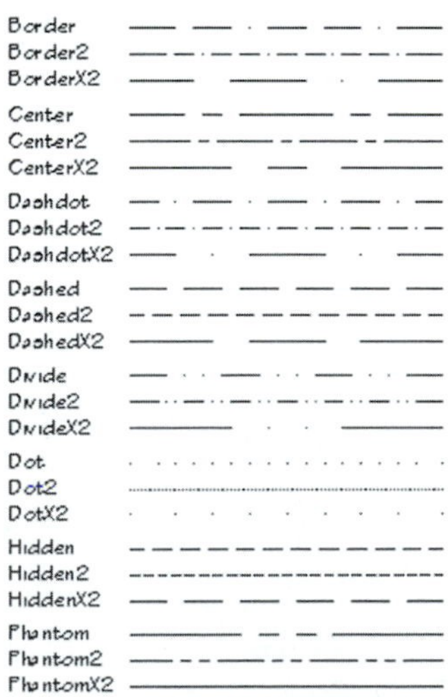

Figure 6.12b *A partial listing of linetypes that are used on construction drawings.*

using the Load... option of the Linetype manager. Use the following procedure to load the center, dashed, hidden, and phantom linetypes.

- Open the Linetype Manager dialog box by selecting Other from the Linetype Control edit box, by selecting Linetype... from the Format menu, or by typing **LINETYPE** (EnterKey) at the command prompt.
- Select the Load button. This will display the displayed in the Load or Reload Linetypes dialog box shown in Figure 6.12a.
- Select the name, the description, or the line. Either selection will highlight the indicated linetype. To load more than one linetype, press and hold the (CtrlKey) while selecting the names of the desired linetypes with the select button of the mouse.
- When you're through selecting the linetypes to be loaded into a drawing, press the OK button.

The menu will be removed, and the selected linetypes will be displayed in the Linetype Edit window of the Linetype Manager dialog box. (See Figure 6.13.) The lines are also shown in the Linetype control box of the Object Properties toolbar when the display arrow is selected. Figure 6.14 shows a listing of the linetypes currently loaded into the drawing base.

MAKING A LINETYPE CURRENT

Remember that all you've done so far is to load linetypes into the drawing. To use the linetype you must select one of the loaded linetypes as the current linetype. The fol-

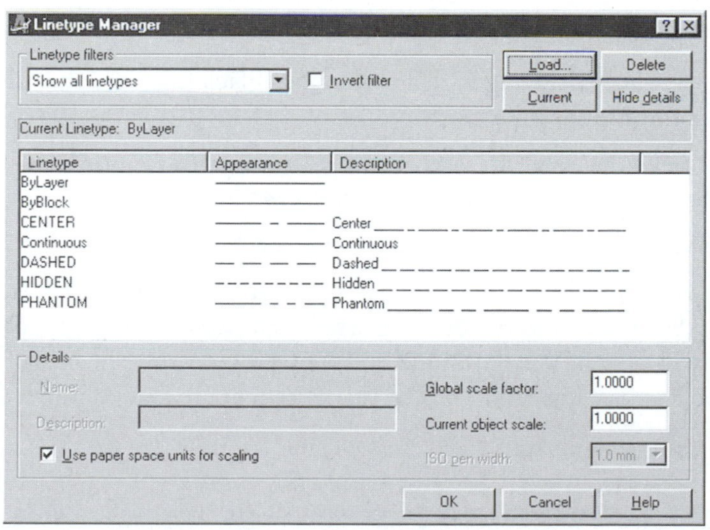

Figure 6.13 *Once a linetype has been selected to load, it will be displayed in the active linetypes edit box.*

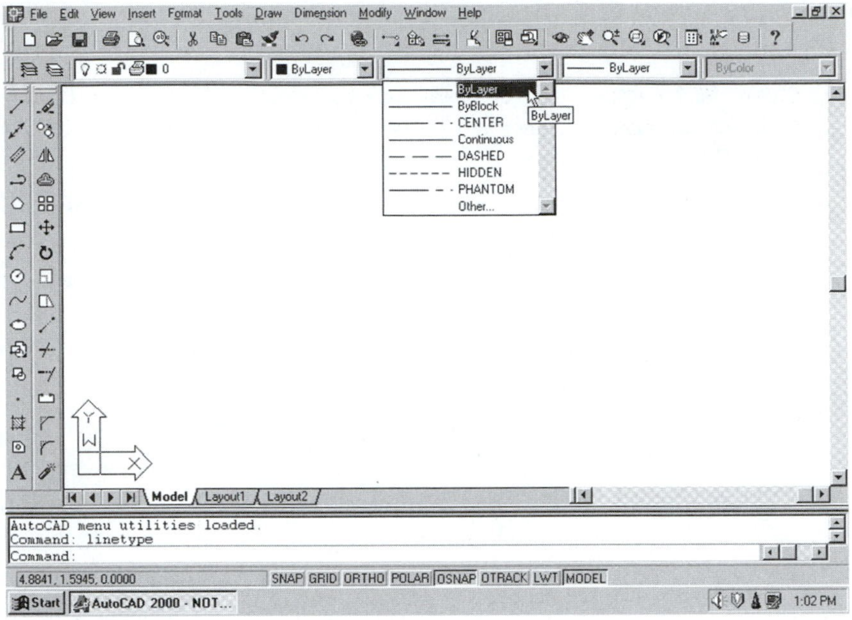

Figure 6.14 *A listing of the linetypes currently loaded into the drawing base is found in the Linetype Control edit box of the Object Properties toolbar.*

lowing process will allow you to alter the lines that are used to create a drawing. Later in the chapter you'll be introduced to methods of assigning linetypes to a specific layer. Most professionals find linetypes easier to control when they are associated to a specific layer rather than scattered throughout one layer.

By default, you've been drawing with a continuous black line. Use the following procedure to alter the current linetype.

- Select the linetype that will become the current linetype from the Linetype Manager dialog box. Select Center for now.
- Select the Current button, and then the OK button.

This process will make the selected linetype the current linetype to be used, close the dialog box, and return you to the drawing screen. Notice the name and pattern of the selected line will now be displayed in the Linetype Control edit box of the Object Properties toolbar. See Figure 6.15. The linetypes that are loaded into a drawing can

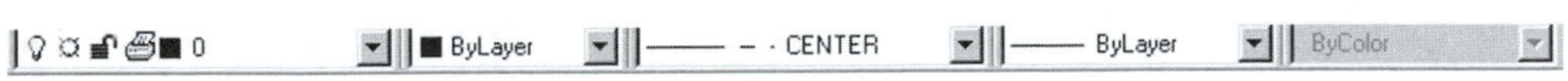

Figure 6.15 *The current linetype is displayed in the Object Properties toolbar. The Linetype Control box can be used to alter the linetypes that will be drawn.*

also be displayed by selecting Properties from the Modify pull-down menu or by typing **PROPERTIES** (EnterKey) at the command prompt. Each method will produce the Properties dialog box. Selecting the current linetype will display a menu arrow similar to Figure 6.16a. Selecting the arrow will display a listing of the current linetypes similar to Figure 6.16b.

Note: New users of AutoCAD often expect the linetype command to work miracles. Once a linetype has been made the current linetype, it will only affect drawing objects that are added after changing the linetype. Objects drawn prior to the selection will not be effected. Objects drawn prior to the selection can be transformed to a different linetype using Match Properties or the CHANGE command. Each will be discussed in Chapter 14.

ALTERING THE LINETYPE SCALE

Enter the LINE command and draw several lines. The new linetype should now be reflected on the drawing screen. If the line still appears as a continuous line, check the Object Properties toolbar to verify that you did select the current button. If the name and linetype show the desired linetype, most likely the linetype does not appear altered because you've altered the drawing limits from the original 12" × 9" screen. Use the following steps to alter the linetype scale.

- Reenter the Linetype Manager dialog box and select the Show Details button. This will produce a dialog box similar to Figure 6.17.

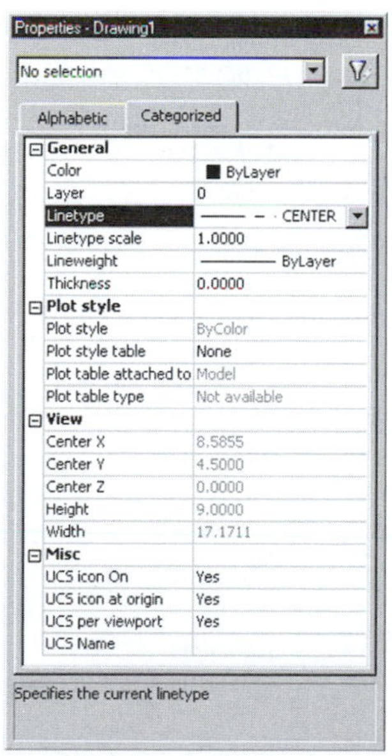

Figure 6.16a *The linetypes that are loaded into a drawing can be displayed by selecting the Properties icon from the Standard toolbar or by selecting Properties from the Modify pull-down menu.*

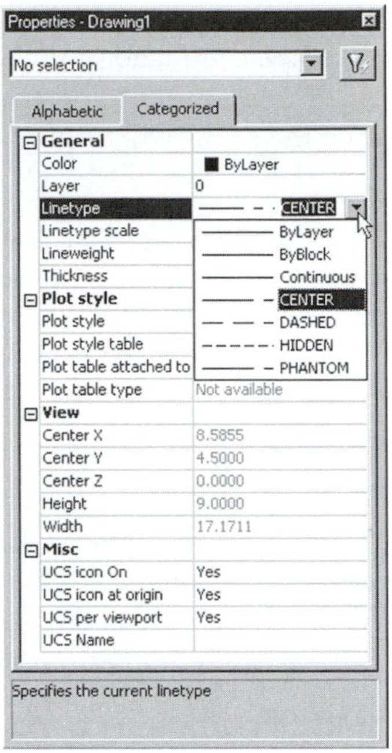

Figure 6.16b *Selecting the current linetype will display a menu arrow similar to Figure 6.16a. Selecting the arrow will display a listing of the current linetypes.*

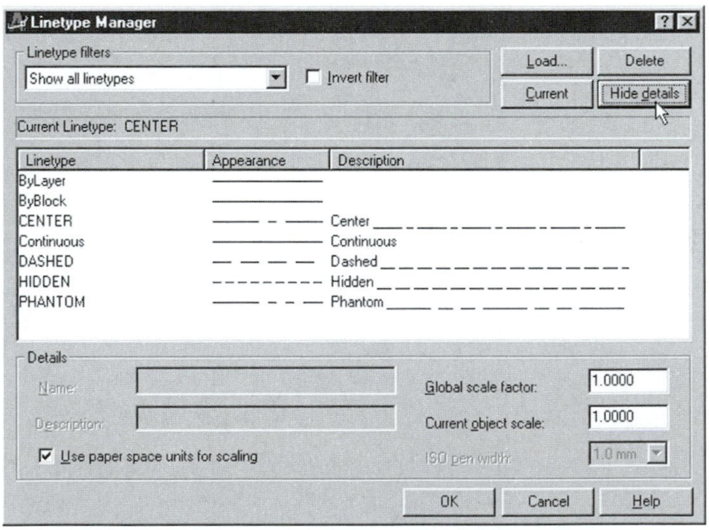

Figure 6.17 *Selecting the Show Details button of the Linetype Manager dialog box will display controls for linetype scaling. If you've altered the limits of the drawing screen from the 12" × 9", a line drawn with a centerline may appear to be drawn as continuous.*

- Alter the global scale factor. If you're working on a template set for ¼" = 1'-0" plotting, change the global scale factor to 48 and press the OK button.

As you return to the drawing screen, lines should now reflect the selected pattern as shown in Figure 6.18.

AN INTRODUCTION TO LINEWEIGHT

You've been introduced to common linetypes and line widths earlier in this chapter. Varied line widths will help you and the print reader to keep track of various components of a complex drawing. As you assign lineweight to drawings, remember that you're using thick and thin lines to add contrast not represent thickness of an object. Chapter 12 will introduce an alternative method of using a thickness to accurately represent objects.

EXPLORING LINEWEIGHTS

Lineweights can be assigned to a layer by selecting Lineweights... from the Format pull-down menu, or by typing **LWEIGHT** (EnterKey) at the command prompt. Each method will produce the Lineweight Settings dialog box. Once you've made some basic settings, lineweights can be altered using the Lineweight Control menu from the Object Properties toolbar. For now, enter the Lineweight Settings dialog

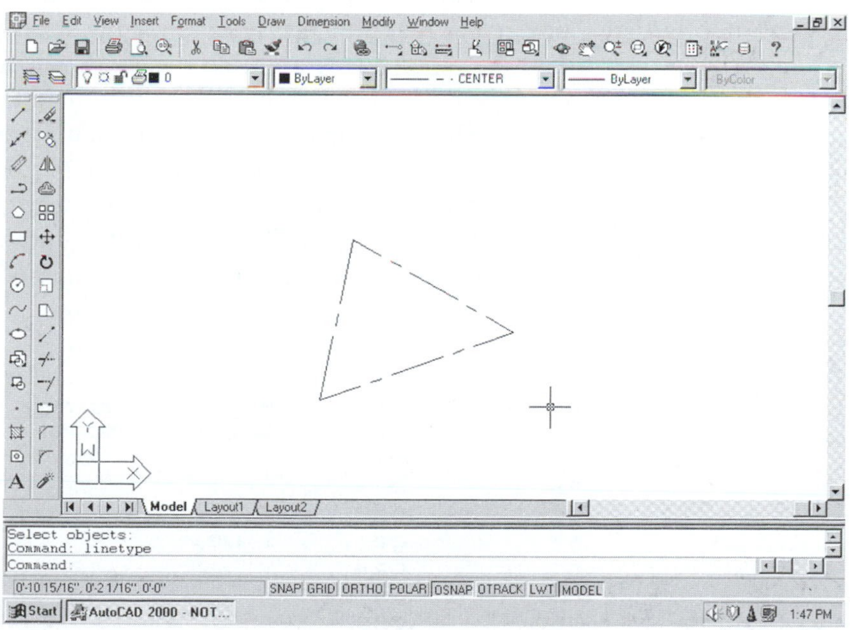

Figure 6.18 *Setting the Globel scale factor. If you're working on a drawing that will be plotted at a scale of ¼" = 1'-0", the scale factor is 48. Once the linetype scale is altered, the object will now be displayed using centerlines.*

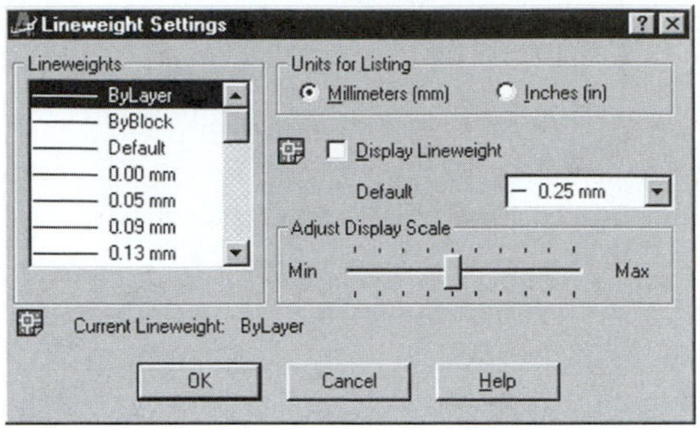

Figure 6.19 *The Lineweight Settings dialog box can be used to control drawing lineweights. The dialog box is accessed by selecting Lineweights... from the Format pull-down menu or by typing* ***LWEIGHT*** *(EnterKey) at the command prompt.*

box by selecting Lineweights… from the Format pull-down menu. This will produce a display similar to Figure 6.19. Common lineweights and their width in metric and inches, point size and pen size can be seen in Figure 6.20.

ASSIGNING LINEWEIGHTS

The Linetype Settings dialog box can be used to set the units for measuring line width, the default lineweight, the display scale, and as a display of the current settings. As you enter the dialog box, notice that the Display Lineweights box is OFF. In its current state, you can draw using varied lineweights, but the width will only be reflected on a print and not in the drawing display. For most users, this should be the first adjustment you make. Select the Display Lineweight box to make it active and varied lineweights will be displayed on the screen. Most users are also more comfortable working in inches rather than metric units. This change can be made by choosing the Inches radio button. Making this change, you'll notice that the lineweights are now displayed in inches. To assign a lineweight, scroll down the list of line-

LINEWEIGHT	MILLIMETERS	INCHES	POINTS	PEN SIZE
———	.25	.010	3/4 pt.	000
———	.30	.012		00
———	.35	.014	1 pt.	0
———	.50	.020		1
———	.53	.021	1-1/2pt.	
———	.60	.024		2
———	.80	.031		3
———	1.00	.039		3-1/2
———	1.20	.047		4

Figure 6.20 *Common lineweights and their width in metric and inches, point size and pen size.*

weights, find the desired line weight, and select the OK button. As you use the LINE command the line, any line added to the drawing will now reflect the current linetype and lineweight.

Once the unit and display settings have be set, lineweights can be altered by selecting Lineweight Control from the Object Properties toolbar. See Figure 6.21. Lineweights can also be toggled ON/OFF by selecting the Lineweight Display control button (LWT) from the Status bar. This will be helpful in the future, as you enlarge areas of the drawing using the ZOOM command.

DISPLAYING LINEWEIGHTS

In addition to activating the Display Lineweight in the Lineweight Settings dialog box, you also must consider the difference in lineweight display between model and paper space. In model space, lineweights are displayed in relation to pixels. Lineweights in paper space are displayed in the exact plotting width. For work in model space, lineweights should be selected that are proportional to their pixel value and not their real world width. Chapter 14 will explore the display scale for lineweights. For now, use the following guide for determining lineweight. Examples are all related to a floor plan.

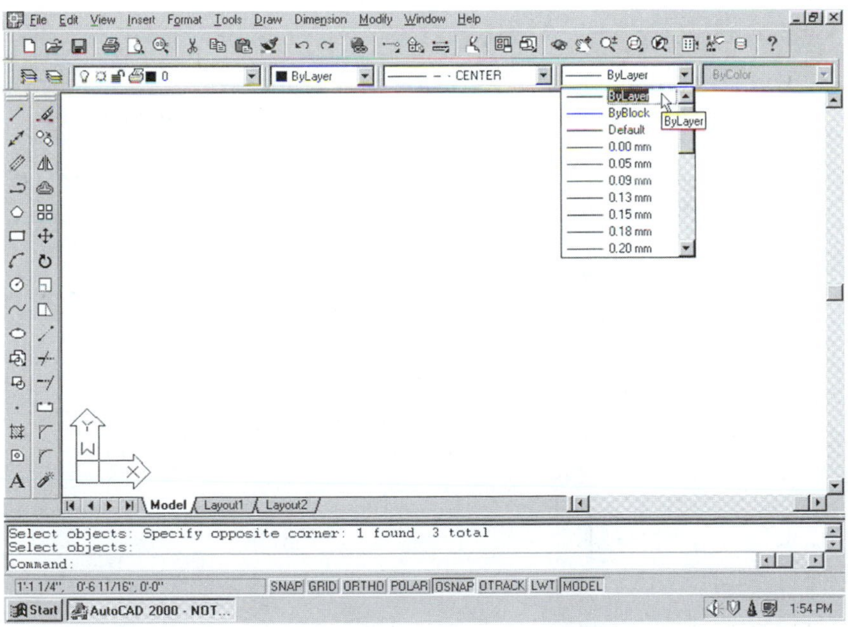

Figure 6.21 *Once the unit and display settings have been set, lineweights can also be altered by selecting Lineweight Control from the Object Properties toolbar.*

mm	In.	Description	Use
.25	.010	very, very, thin	Very detailed areas
.30	.012	very thin	Window symbols and door swings
.35	.014	thin	Cabinets and plumbing fixtures
.50	.020	moderate	Doors
.60	.24	thick	Outline of major components, such as walls
.80	.031	very thick	Underlining room titles
1.20	.047	very, very thick	Borders

AN INTRODUCTION TO COLOR

The Options dialog box was briefly introduced in Chapter 2 as a means of adjusting screen colors. Up to this point, all of your drawings have been done with black lines on a white background, or white lines on a black background. The choice of screen background can be altered by:

- Selecting Options from the Tools pull-down menu. This will display the Options dialog box.
- Select the Display tab from the Options dialog box.
- Select the Colors... button from the Windows Elements box. This will produce the AutoCAD Window Color Options dialog box shown in Figure 6.22.
- The box allows for the color of each of the drawing window displays to be altered. To change the color of the drawing area, select the desired color tile, press the Apply and Close button to close the Color box, and then press the OK button of the Options box. The dialog box will close and the screen will display the selected background color.

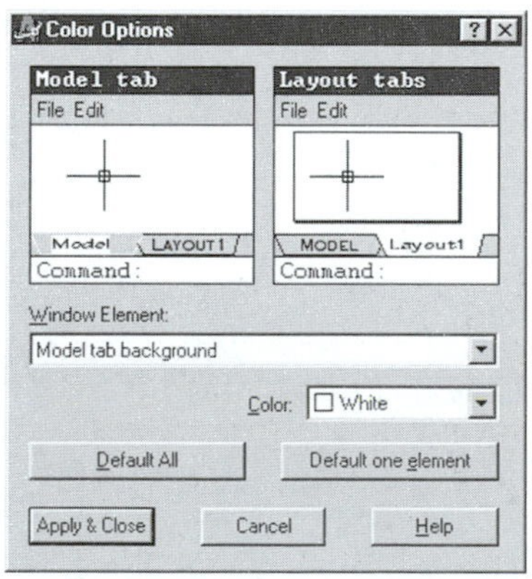

Figure 6.22 *Selecting the Colors... button from the Windows Elements box will produce the AutoCAD Window Color Options dialog box.*

AutoCAD also allows the color of each drawing object to be altered. For most AutoCAD users, a whole new world is about to be revealed. You will be able to start drawing with 256 colors. Most users will have millions of shades to chose from. The number of colors available will depend on the hardware you are working with and how the software is configured.

COLOR NUMBERS AND NAMES

Each drawing object or layer may be assigned a color. The seven basic colors used by AutoCAD are listed in Figure 6.23. AutoCAD identifies colors using numbers from 1 through 256, with the selection based on the color displayed in the color palette. Multiple objects or layers can have the same color number. Objects of the same color can be assigned to a specific pen during the plotting process. For instance, all red objects can be assigned to a thick pen, and all blue objects can be assigned to a thin pen to produce good line contrast on the plotted drawing. Plot styles can be used to control plotting and colors and will be explored later.

SETTING OBJECT COLOR

The COLOR command can be used to assign colors to an entire layer or to individual objects on a layer. Setting colors for layers is the default and will be explained as you proceed through layer options later in this chapter. The easiest method to assign color is to select the Color icon from the middle of the Object Properties toolbar. You may select the color tile, the color name, or the down arrow. Each method will display the menu shown in Figure 6.24. Selecting one of the listed colors will close the menu and return the drawing display. Now, if a line is drawn, it will be drawn using the selected color. Using the current settings, the line would be a red line drawn with a center linetype. Select the Color icon again to reopen the color menu. Notice the last menu option is Other. Selecting this option will display a menu similar to Figure 6.25. The Select Color dialog box can also be displayed by selecting Color... from the Format pull-down menu, or by typing **COLOR** (EnterKey) at the command

COLOR NUMBER	COLOR NAME
1	RED
2	YELLOW
3	GREEN
4	CYAN
5	BLUE
6	MAGENTA
7	BLACK

Figure 6.23 *The eight major colors of AutoCAD. The default color is 7. Depending on the background color, 7 will produce either a black or white line.*

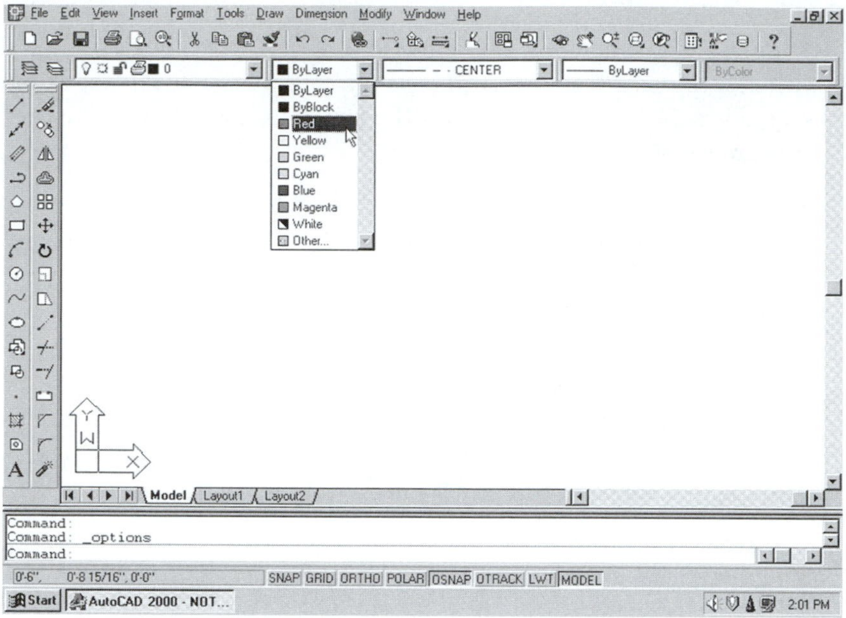

Figure 6.24 *Selecting the Color Control tile from the Object Properties toolbar will display the color menu and allow the color of the current layer to be changed.*

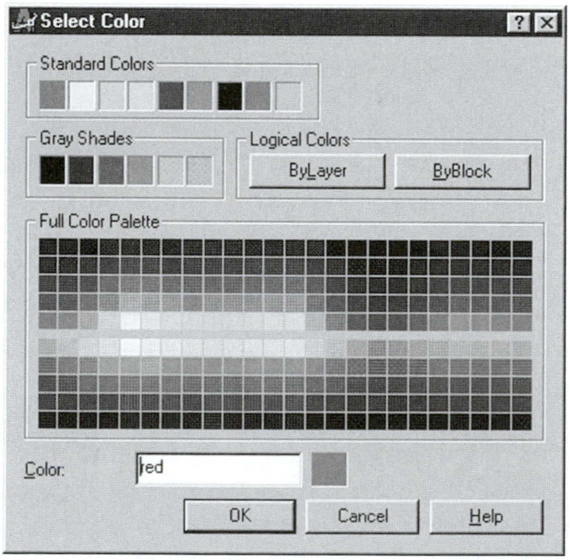

Figure 6.25 *Selecting Other... from the color menu displays the color pallet and allows a wide variety of color tints to be selected for drawing objects.*

prompt. The Select Color dialog box allows shades of the standard colors to be selected. Notice the current color is listed as red. Selecting any of the color tiles will alter the current color display tile and list the color by number. Selecting the OK button will close the display and make that the current color. Using the LINE command now will create lines using the selected color.

Notice that the Select Colors dialog box contains buttons for ByLayer and ByBlock. The ByBlock option will be discussed in Chapter 21. If ByLayer is accepted as the default, the color will be assigned once each layer has been defined. Typically, most offices set colors using the ByLayer option rather than assigning colors to individual objects.

AN INTRODUCTION TO LAYERS

The use of layers in a CAD drawing is similar to overlay drafting procedures of manual design. Overlay drawing will have a base drawing, such as the valve assembly seen in Figure 6.26, on one sheet of Mylar®, and have overlay sheets of notes and dimensions on separate sheets of Mylar®, as seen in Figure 6.27. Depending on who needed what, different sheets could be attached to the base and various prints made. Layers allow all the flexibility with none of the alignment problems of overlay drafting procedures.

Until now you have been drawing on the 0 layer. AutoCAD will allow you to create different layers, assign each layer a different name, linetype, lineweight, color, and allow some layers to be made invisible.

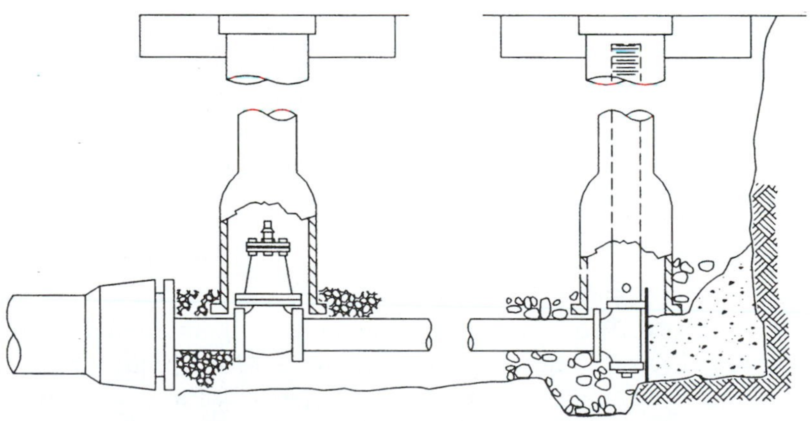

Figure 6.26 *The use of layers is similar to placing information on several sheets of paper in overlay drafting. This drawing shows the basic drawing with no supplemental information provided. (Courtesy Department of Environmental Services, City of Gresham, OR.)*

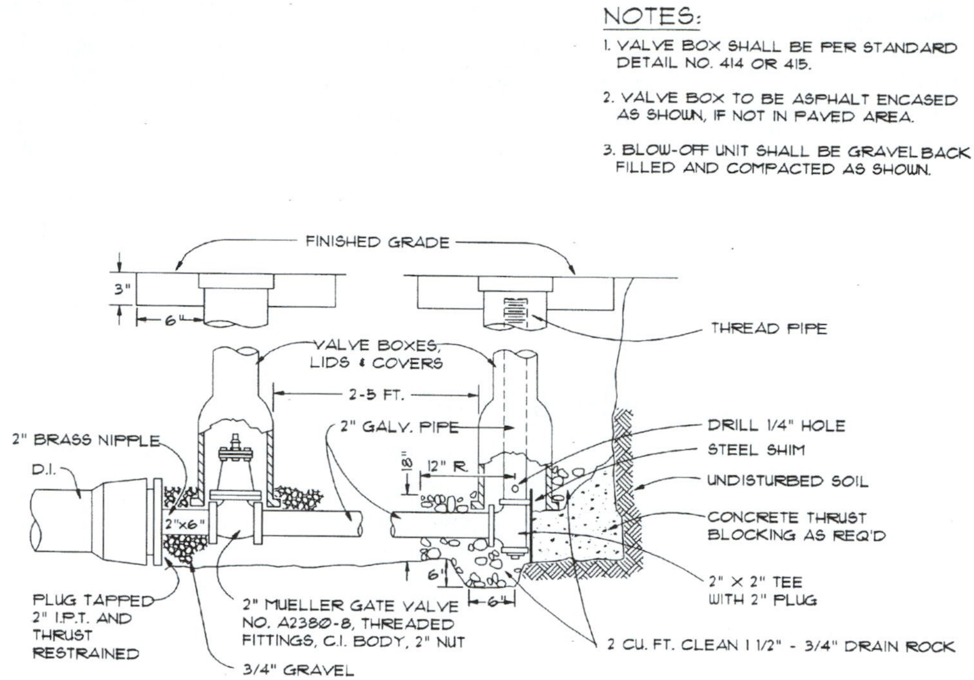

Figure 6.27 *Supplemental information added to the base drawing. (Courtesy Department of Environmental Services, City of Gresham, OR.)*

ENTERING THE LAYER COMMAND

The LAYER command can be started by selecting the Layers icon from the Object Properties toolbar, or by selecting Layer... from the Format pull-down menu. The LAYER command can also be entered by typing **LA** (EnterKey) at the command prompt. Each will display the Layer Properties Manager dialog box shown in Figure 6.28. Options for controlling layers can also be accessed.

CREATING A NEW LAYER

Once in the Layer Properties Manager dialog box, press the (EnterKey) or choose the New button. Each method will add the highlighted name of Layer 1 in the Name edit box. (See Figure 6.29a.) With Layer 1 highlighted, a new layer name can be created by keyboard. Press the select button of the mouse to accept the new layer name. If you would like to add multiple layers, press (EnterKey). Pressing (EnterKey) will add the previously changed name to the drawings and add a new layer titled Layer1. In Figure 6.29b, Layer1 is assigned the name of COOLSTUFF. Follow the guidelines from Chapter 5 for naming files when assigning layer name. Additional guidelines for naming layers will be discussed in Chapter 14. For now, keep the names

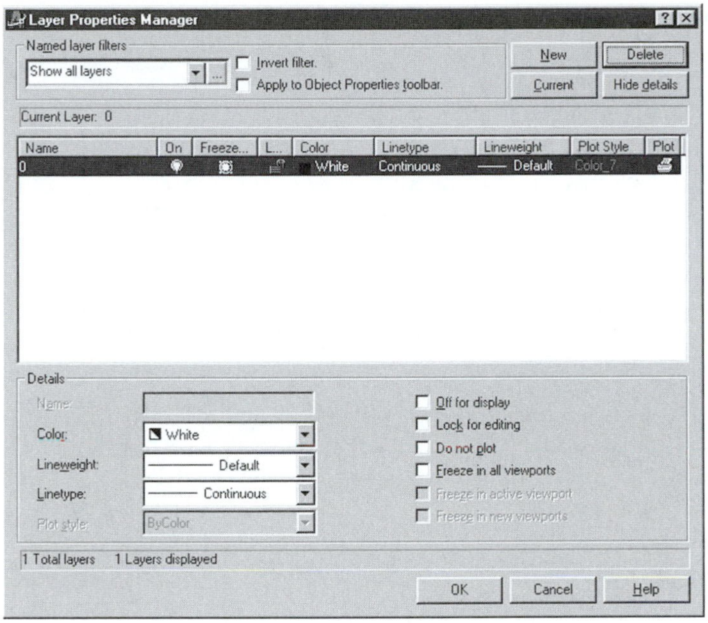

Figure 6.28 *Picking the Layer icon from the Object Properties toolbar or typing* **LA** *(EnterKey) will display the Layer Properties Manager dialog box. The box can be used to create and control the properties of each drawing layer.*

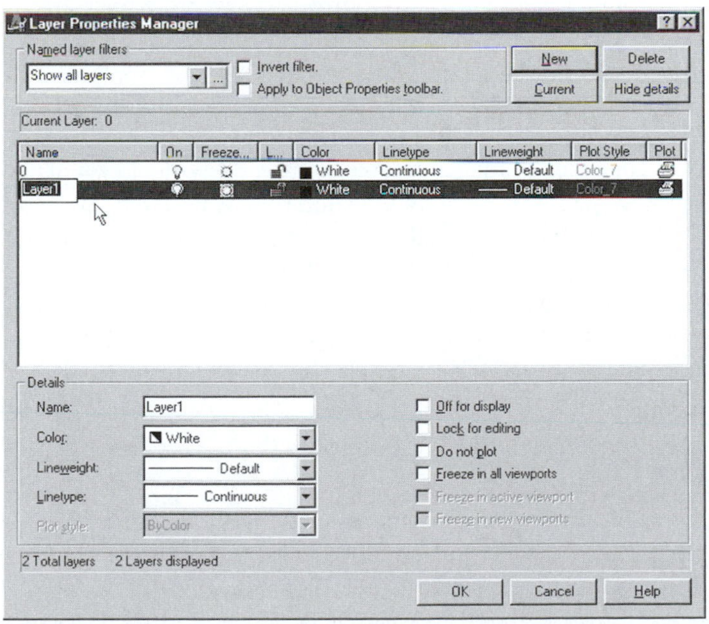

Figure 6.29a *Pressing (EnterKey) or selecting the New button will display a layer title of Layer1.*

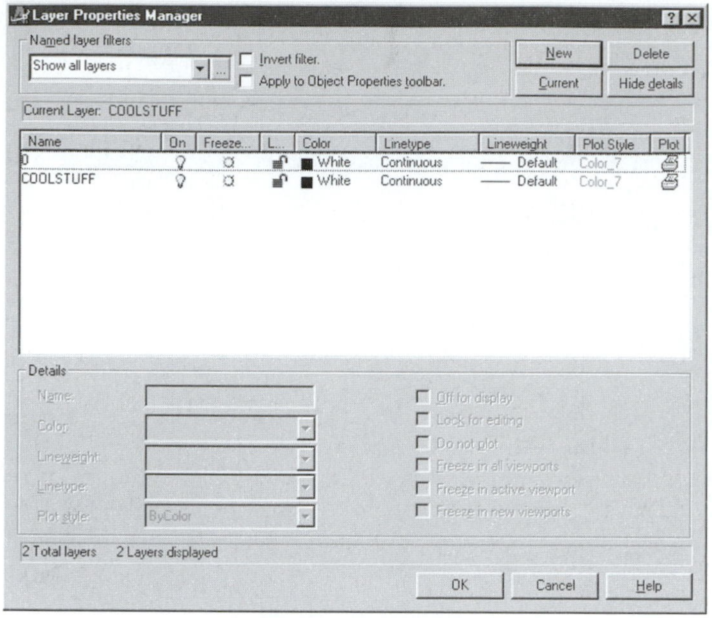

Figure 6.29b *The selected layer name of Figure 6.29a can be altered to reflect the desired layer name. Guidelines for naming layers are the same as those used to name a drawing file.*

short, simple, and descriptive, using names that will identify the contents. Once the name has been supplied, select the OK button to save the layer and return to the drawing area. Just as with Linetype, all you've done is to create a new layer. The current layer is still the default layer of 0.

MAKING A LAYER CURRENT

Close the Layer Properties Manager and reenter the drawing area and add a few lines. The lines will be the same linetype and color that was drawn before creating the COOLSTUFF layer. Reenter the Layer Properties Manager by selecting the Layers icon. The dialog box displays 0 and COOLSTUFF as the only layers in the drawing. The box surrounds the 0 layer indicating that it is the current layer. Use the following steps to make COOLSTUFF the current layer.

- Select COOLSTUFF.
- Select the Current button.
- Select OK.

Selecting the OK button will make the selected layer the current layer, close the Layer Properties Manager, and return you to the drawing area.

Note: A layer can also be made current by opening the Layer Properties Manager and double-clicking the name of the layer to be made current.

Figure 6.30a *The Layer edit box describes the properties of the current layer.*

Notice that just to the right of the Layers icon of the Object Properties toolbar is an edit bar that describes the current layer status. (See Figure 6.30a.) You'll see icons that describe the current status of the layers, with COOLSTUFF listed as the current layer. Picking the down arrow will create a display of all of the current layers similar to Figure 6.30b. Currently, your only other option is the 0 layer, but this is a very convenient method for altering the current layer. A layer can be made the current layer by selecting the down arrow, and then selecting the name of the desired layer to be current.

ALTERING LAYER QUALITIES

Reenter the Layer Properties Manager dialog box, select the COOLSTUFF layer and then select the Current button. COOLSTUFF has the same color, linetype, and lineweight qualities as the 0 layer. Any object created on the COOLSTUFF layer will appear the same if it were drawn on the 0 layer. Layers become a powerful drawing tool when varied linetypes, lineweights, and colors are assigned to specific layers. Creating a layer with thin, blue, hidden lines for appliances can be a great way of easily separating walls that are drawn with thick, continuous red lines.

Linetype qualities can be assigned to all objects created, when this layer is the current layer, by selecting the current Linetype display. Use the following steps to alter the linetype of the COOLSTUFF layer.

- Open the Layer Properties Manager dialog box.
- Highlight the layer that you wish to alter.
- With the layer highlighted, select the current linetype for that layer. This will produce the menu shown in Figure 6.31a. Only linetypes that have been loaded into the drawing will be displayed.
- Select the desired linetype to be used.
- Select the OK button to close the Select Linetype edit box.
- Select OK to close the Layer Properties Manager dialog box and return to the drawing area.

Selecting one of the linetypes will use that linetype for all drawing entities created using that layer. The lineweight and color can be altered using the same procedure.

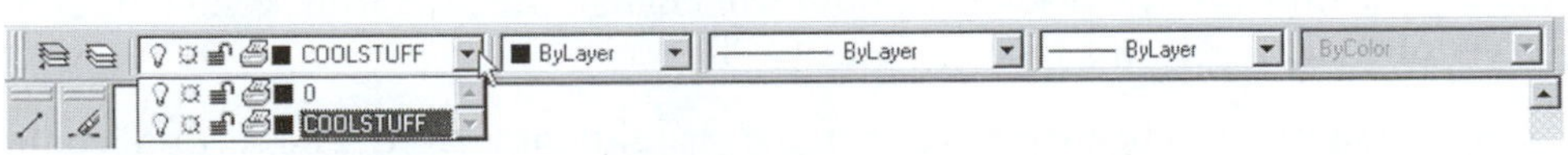

Figure 6.30b *Selecting the arrow displays the status of each drawing layer.*

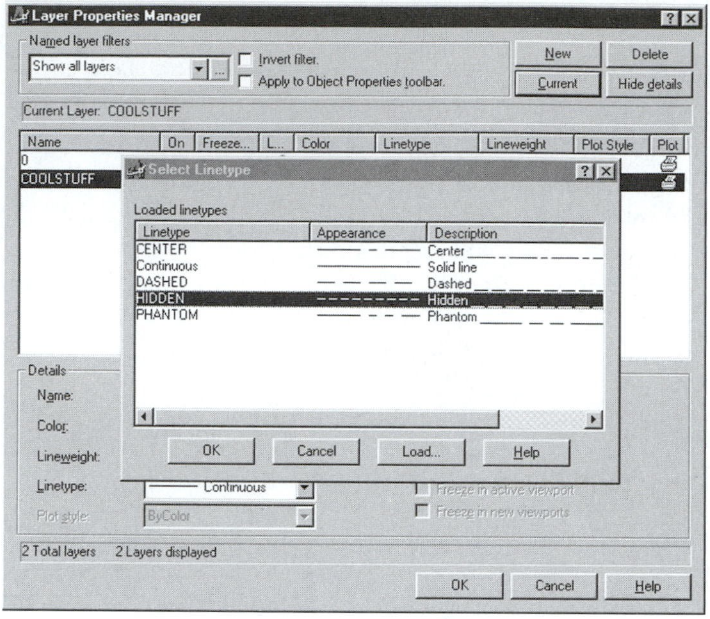

Figure 6.31a *Selecting the current linetype of Continuous will produce the Select Linetype menu. Only linetypes that have been loaded into the drawing will be displayed. Additional linetypes can be loaded into the drawing by selecting the Load button and following the guidelines presented early in this chapter.*

Selecting the color display for COOLSTUFF will produce the Select Color menu shown in Figure 6.31b. Selecting the lineweight display for COOLSTUFF will display the Lineweight menu similar to Figure 6.31c. Once the desired qualities of a layer have been assigned, select the Current button if you want to use that layer as you return to the drawing editor. Selecting the OK button will close the Layer Properties Manager dialog box and return you to the drawing screen.

Note: Assigning colors, linetypes, and lineweights will affect all objects drawn on the layer that is altered, even objects drawn prior to making the changes.

CONTROLLING THE VISIBILITY OF LAYERS

Eight icons are displayed by each layer name. These icons represent various controls for the use of each layer. The light bulb icon is a toggle for the ON/OFF setting of a layer. The drawings shown in Figure 6.26 and 6.27 are controlled by toggling the TEXT layer ON/OFF. Clicking the bulb will change the bulb from yellow to gray. The yellow bulb represents ON, the gray OFF

The Sun icon can also be used to control the visibility of a layer. The second icon is a toggle between Thaw/Freeze. When the sun is displayed, the contents of that layer

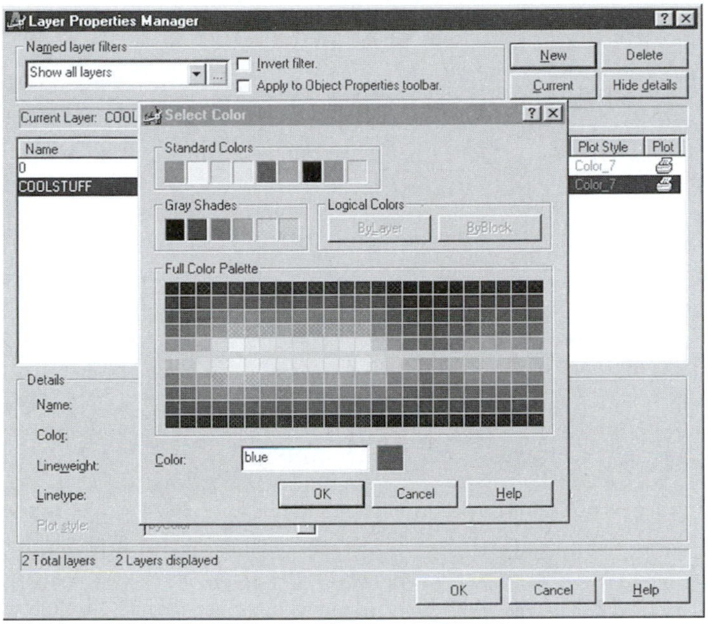

Figure 6.31b *The color of a layer can be adjusted by selecting the current color display for that line. This will display the Select Color menu allowing the color to be altered.*

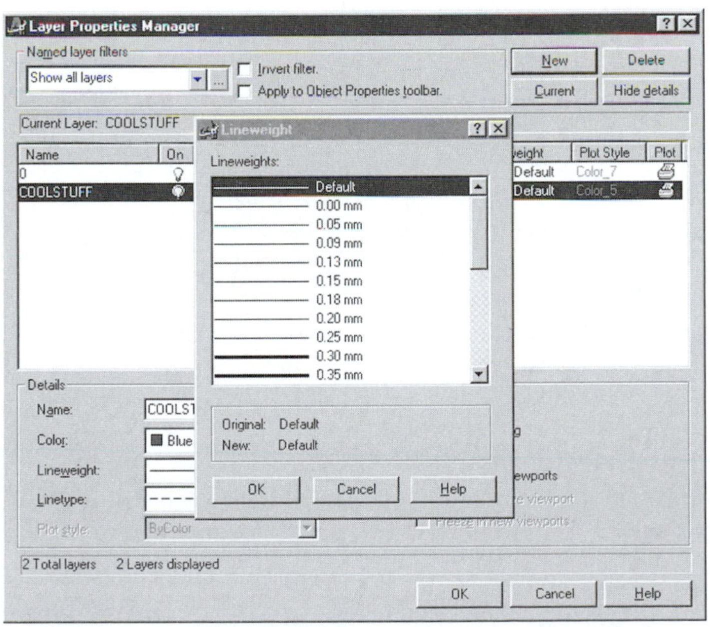

Figure 6.31c *The lineweight of a layer can be adjusted by selecting the current lineweight display for that line. This will display the Lineweight menu, allowing the line width to be altered.*

will be displayed. Selecting the sun icon will change it to a snowflake. The snowflake represents a frozen layer. The contents of a frozen layer will not be displayed. The benefits of using the ON/OFF and THAW/FREEZE options, as well as each of the other layer controls, will be discussed in Chapter 14. To remove the contents of a layer from the screen display, select the desired layer name and then select the bulb or the sun icon and the select the OK button.

The visibility of layers can also be controlled from the Layer Control box of the Object Properties toolbar. Select the down arrow to display a list of current layers and select the light bulb or sun icon of the desired layer to be altered.

AN INTRODUCTION TO PRINTING

Once you've created a drawing, you'll need some method of passing the contents of the file onto those who will put it to use. This might include sending the drawing electronically over the Internet to another firm, or sending the drawing by disk. For many, it also means making a hard copy such as a paper print or plot. No matter how technology advances, for most AutoCAD operators, a paper copy is still the most dependable method to check drawings. The plotting process has undergone many changes in AutoCAD 2000. You've already been introduced to the layout templates as you explored drawing setup. Layouts provide a preview of how the drawing will appear prior to plotting. Each layout displays a border and the drawing as it will appear at the current lineweight and scale factor. Although it will be a bit overwhelming now, AutoCAD will allow you to establish multiple plotting layouts of the same drawing project. The same layout can be plotted at different scales or configured for different plotting devices. The setting to repeat the plotting process can be saved with the drawing, providing a valuable saving of time. The purpose of this chapter is not to explore all of the plotting functions, but to help you produce a quick print of your drawing. This chapter will introduce the process of making a paper print using the system printer configured using your Windows platform. Future chapters will provide additional information on choosing additional printers, configuring a plotter, and controlling the many variables associated with plotting.

EXPLORING THE PLOTTING PROCESS

The print process can be started by selecting the Print icon from the Object Properties toolbar, by selecting Print... from the File pull-down menu, or by entering **PLOT** (EnterKey) at the command prompt. Each selection method will produce the Plot dialog box shown in Figure 6.32. The dialog box contains the Plot Device and Plot Settings tabs, as well as an option to take the Fast Track to Plotting Help tour. The Plot Device tab and the tour will not be considered in this chapter. See Chapter 24 for further discussion of altering the plotter. The Plot Settings tab contains eight major areas that must be considered to produce a print.

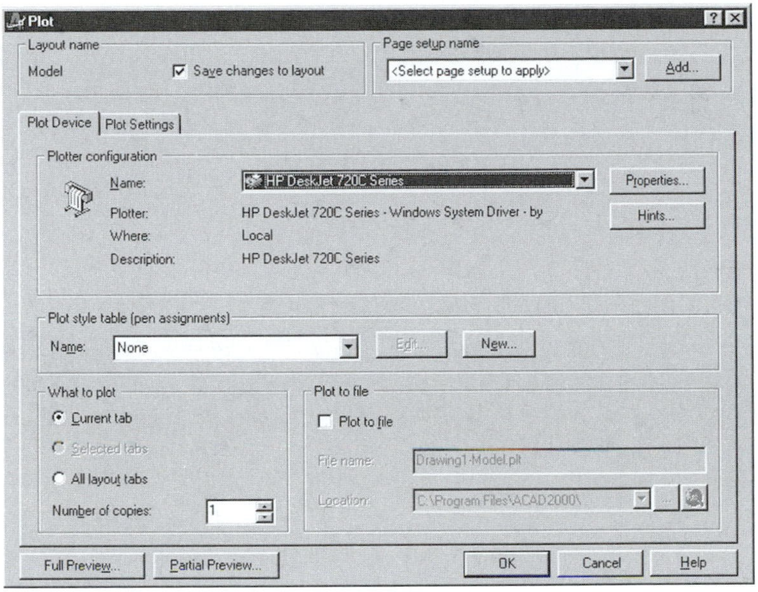

Figure 6.32 *Selecting the Plot icon from the Standard toolbar, selecting Plot from the File pull-down menu, or typing* **PLOT** *(EnterKey) at the command line will display the Plot dialog box. The dialog box can be used to control each aspect of creating a print.*

LAYOUT AND PAGE SETUP NAMES

These portions of the Plot dialog box are used to specify layout and plotting device settings. The name of the current layout (Model, Layout1, or Layout2), is displayed on the Left side. Options for listing and choosing pages setups for plotting are listed on the right side. Each option will be explored as plotting is explored in later chapters. Chapter 24 will introduce methods of creating new page setups.

PAPER SIZE AND PAPER UNITS

The Paper Size and Paper Units box displays the current printing device, the paper size the plot will be made on, and the size of the printable work area. Selecting Options… from the Tools pull-down menu and selecting the Plotting Tab from the Options dialog box can also control printer selections. For the current discussion, the default printer will be used. The box also displays the plotting units of inch and millimeter. Most construction projects will use the default setting of Inches for the plotting standard.

PLOT AREA

This selection box provides options for describing what portion of the current drawing will be printed. Because the entire drawing often does not need to be printed, five options are given to decide what portion of the drawing will be plotted.

Limits

Using this option will print all of the drawing entities that lie within the current drawing limits that were defined during the drawing setup.

Extents

Choosing this option will print the current drawing with the extent of the drawing objects as the maximum limits that will be displayed in the plot.

Display

The default setting is to print only the material that is currently displayed on the screen. If you have zoomed into a drawing, only that portion of the drawing will be plotted.

View

This option is currently not active. Methods of creating views will be introduced in Chapter 9 and the use of this option will be introduced in Chapter 23.

Window

This option will allow a specific area of a drawing to be defined for plotting. Initially the Window radio box is inactive. By selecting the Window< button, the Plot dialog box will be removed from the screen, allowing you to select what is to be printed. The command line will show the prompt:

```
Command:_plot:
Specify window for printing
Specify first corner: Specify opposite corner:
```

AutoCAD is waiting for you to define a window or box that will surround what you would like to print. To create the plotting window, move the cursor and select a point. Move the cursor to the opposite corner. As the cursor is moved, a box (the window) will be formed. Figure 6.33 shows the process for defining the window. Anything that lies within the window will be printed. Once the window is defined, the Plot dialog box will be redisplayed, allowing you to continue defining the print to be created. The Window radio button will be activated, allowing other options to be set. Each option will be discussed in future chapters.

DRAWING ORIENTATION

This selection box allows the drawing orientation to be specified. The current orientation of the box indicates the drawing will be printed in the normal orientation for construction drawing. The drawing will be read looking from the bottom to the top, and the title block will be along the right side of the print. This setting is referred to as Landscape. When the print is rotated so that it is read from the title block end, the print is referred to as a Portrait print. A portrait print is more likely to be used with laser printers. The orientation of the drawing to the paper can be adjusted by select-

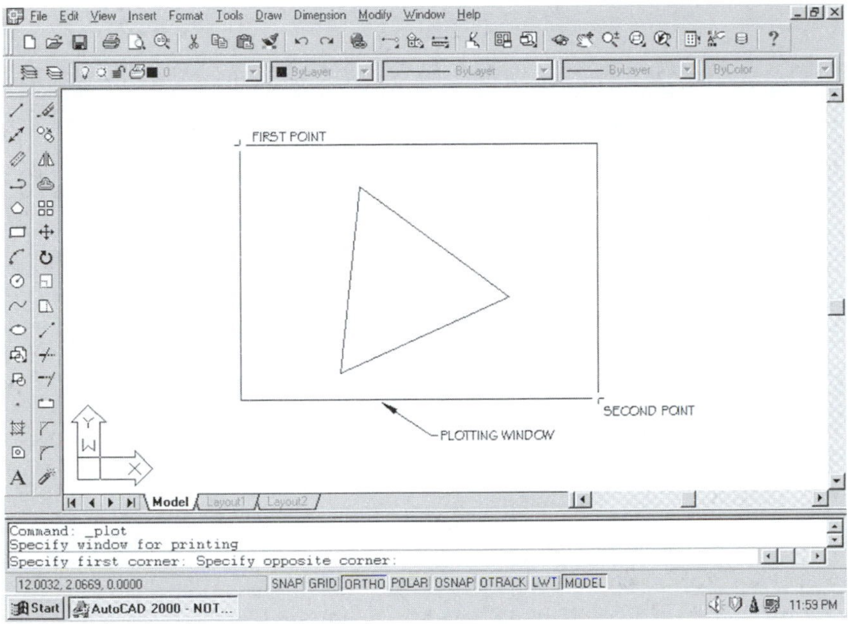

Figure 6.33 *When objects are selected to be printed using the Window option, the size of the window can be set by selecting the window corners with the cursor. Selecting the Window< button will remove the menu and return the drawing to the screen. Once the corners of the select window are located, the Plot dialog box will be displayed again so that other print options can be set.*

ing the appropriate radio button. The Plot upside-down could be useful to see a mirror image of the drawing. This could be useful when working on a floor plan for a duplex if you wanted to see the reverse image without drawing it.

PLOT SCALE

The Plot Scale box controls settings for the Scale of the plot. Its options include scale and custom.

Selecting a Scale

In the default setting, when a print is made it will be reduced in size to fit on the current printer paper. This is indicated by the Scaled to Fit display in the Scale: edit box. Unless you'll be printing a small detail, most architectural drawings can't be printed at full size on a printer. Rarely could you get the floor plan for a structure to fit on A-size material. You can, however, print the entire floor plan on A-size material if the scale is reduced. The Scaled to Fit option reduces the selected material to be printed to an appropriate size to fill the current paper. You can print portions of a drawing to scale by selecting the desired scale, and using the window selection method. To print a drawing to a scale, select the arrow beside the edit box to reveal the scale menu. (See Figure 6.34.) Select the desired scale and the menu will close to allow you to continue setting the print parameters.

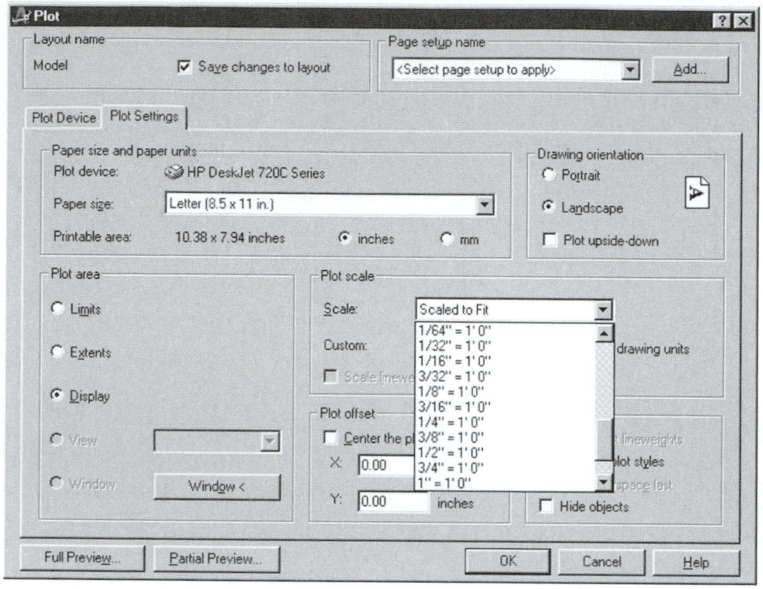

Figure 6.34 *To print a drawing to a scale, select the arrow beside the edit box to reveal the scale menu. Select the desired scale. Once selected, the menu will close to allow you to continue setting the print parameters.*

Custom Scale

You've just altered the scale from "plot to scale" to ¼" = 1'-0". The Custom: window now has been altered to reveal 1 in the inches box, and 48 in the drawing units box. These units reflect the plot scale factor of 1/48 for a drawing to be scaled to ¼"= 1'-0" when printed. The Custom: edit box allows you to print a drawing to a scale that you make up. Figure 6.35a and 6.35b compare the differences in scaling options.

Note: Keep in mind that the goal of this chapter is to help you make a quick, non-scaled, check print so that you can evaluate your work. Stick with plot to scale or one of the common architectural scales until plotting is discussed in detail.

Plot Offset

This portion of the Plot dialog box can be used to specify an offset of the plotting area from the lower left corner of the paper. The default setting for plotting origin is X = 0, Y = 0. Printers typically specify the origin from the upper right-hand corner. Specifying various locations for the origin can be a useful technique for altering the drawing location on the paper. By selecting the X or Y origin box, new values can be entered. Entering an X value of 12 and a Y value as 10 would place the new origin 12 units to the right and 10 units above the original plotting origin.

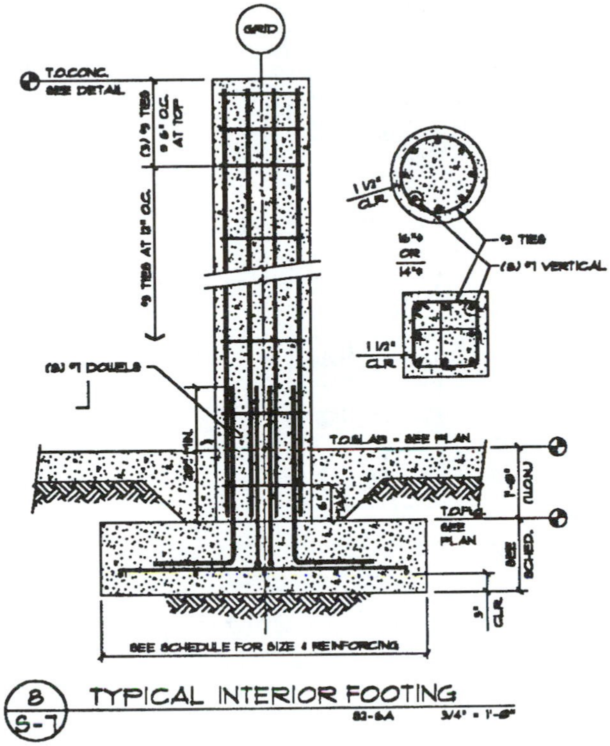

Figure 6.35a *A detail printed using the Scaled to Fit option. Most professionals use drawings printed using the Fit option for check prints prior to plotting the drawing to scale.*

Plot Options

This portion of the Plot dialog box can be used to specify options for lineweights, plot styles, and the current plot style table. By default, the Plot with lineweights and the Plot with plot styles options are active. Each option of this box will be discussed in later chapters.

PRINT PREVIEW

Selecting Preview allows the outcome of the current plot settings to be viewed prior to making the print. Two options are available for previewing the print include Partial or Full Previews.

Partial Preview

Selecting the Partial Preview... button will produce a display similar to the one seen in Figure 6.36. A white box represents the paper size and the limits of the effective printing area are specified on the screen by a blue box. A warning also will be given

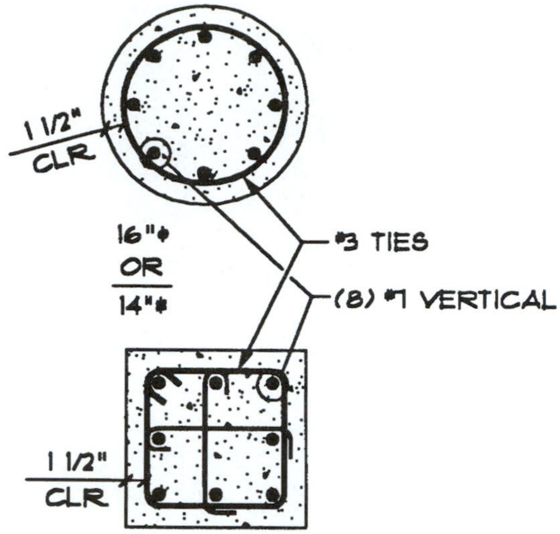

Figure 6.35b *A portion of the detail from Figure 6.35a printed using the scale of ¾"=1'-0".*

of any problems that may be encountered during the plot. Common warnings displayed in the warning box include:

Effective area too small to display

Origin forced effective area off display

Plotting area exceeds plotter maximum

A small triangular Rotation Icon is displayed in the upper left-hand corner to represent a landscape rotation. As the rotation is set to portrait, the icon is moved to the lower left-hand corner.

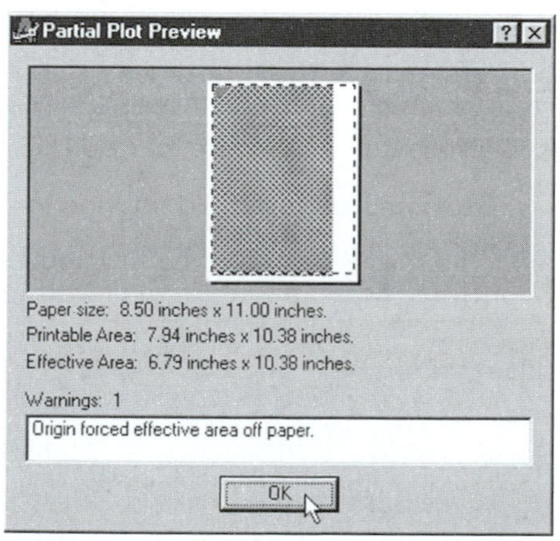

Figure 6.36 *A partial preview of a sheet of details to be printed using the scale of ¾"=1'-0".*

Full Preview

Choosing this option will produce a graphic display of the drawing as it will appear when printed. Figure 6.37 shows an example of a full preview. An outline of the paper size will be drawn on the screen, which displays the portion of the drawing to be printed. The cursor is changed to show a magnifying glass with a + and – sign beside it. This represents the ZOOM command and allows the image for viewing to be enlarged or reduced. It will not affect the plotted image. By pressing and holding the mouse pick button, the size of the image can be altered. Moving the mouse toward the top of the screen enlarges the image, and moving the image toward the bottom of the screen reduces the size of the image. Pressing the right mouse button will display the menu shown in Figure 6.38, allowing other options to be selected. Selecting the Pan option allows the image to be shifted, but the magnification of the image is not altered. Each of the options will be further discussed in Chapter 24. To continue with the print process, select the Exit option to remove the menu.

PRODUCING THE PRINT

Seven areas of the Plot dialog box have been examined. If you change your mind on the need for a print, select the Cancel box; the dialog box will be terminated, and the drawing will be redisplayed. If you are satisfied with the printing parameters that have been established, select the OK box. The dialog box will be removed from the screen and the print will be produced. Figure 6.39a and 6.39b show the results of the selected plot settings.

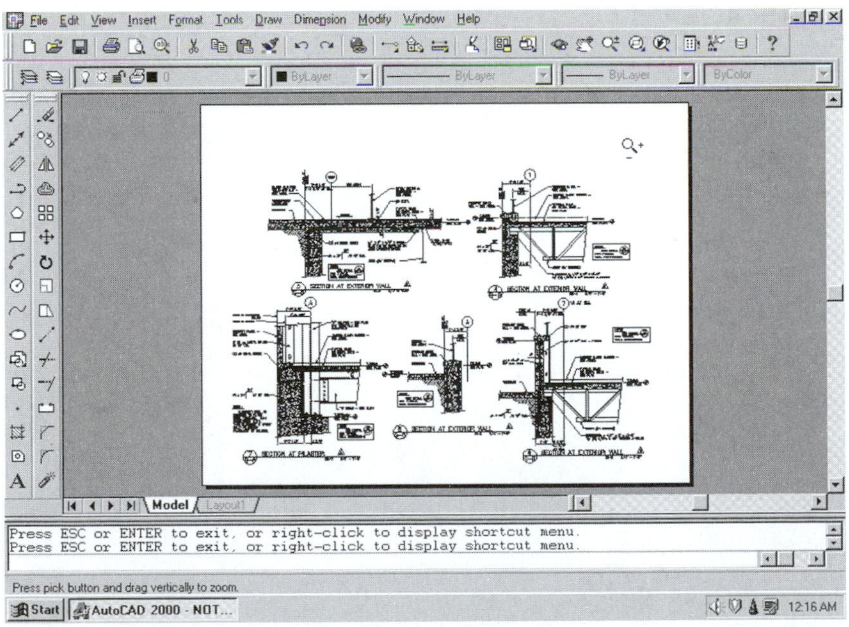

Figure 6.37 *A full preview will show the limits of the paper and the drawing to be plotted.*

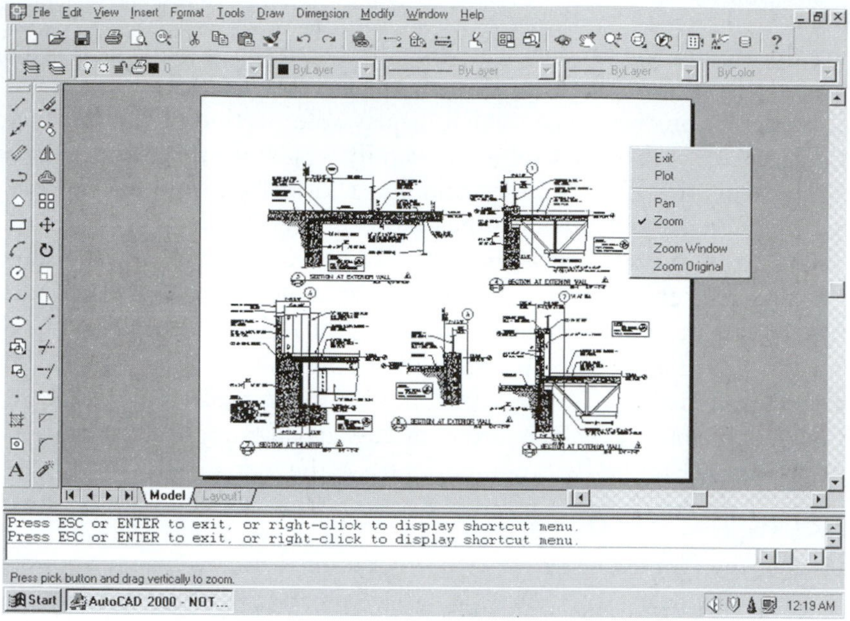

Figure 6.38 *Pressing the right mouse button while using a full preview, will display this menu. The options can be used to control the display to be printed.*

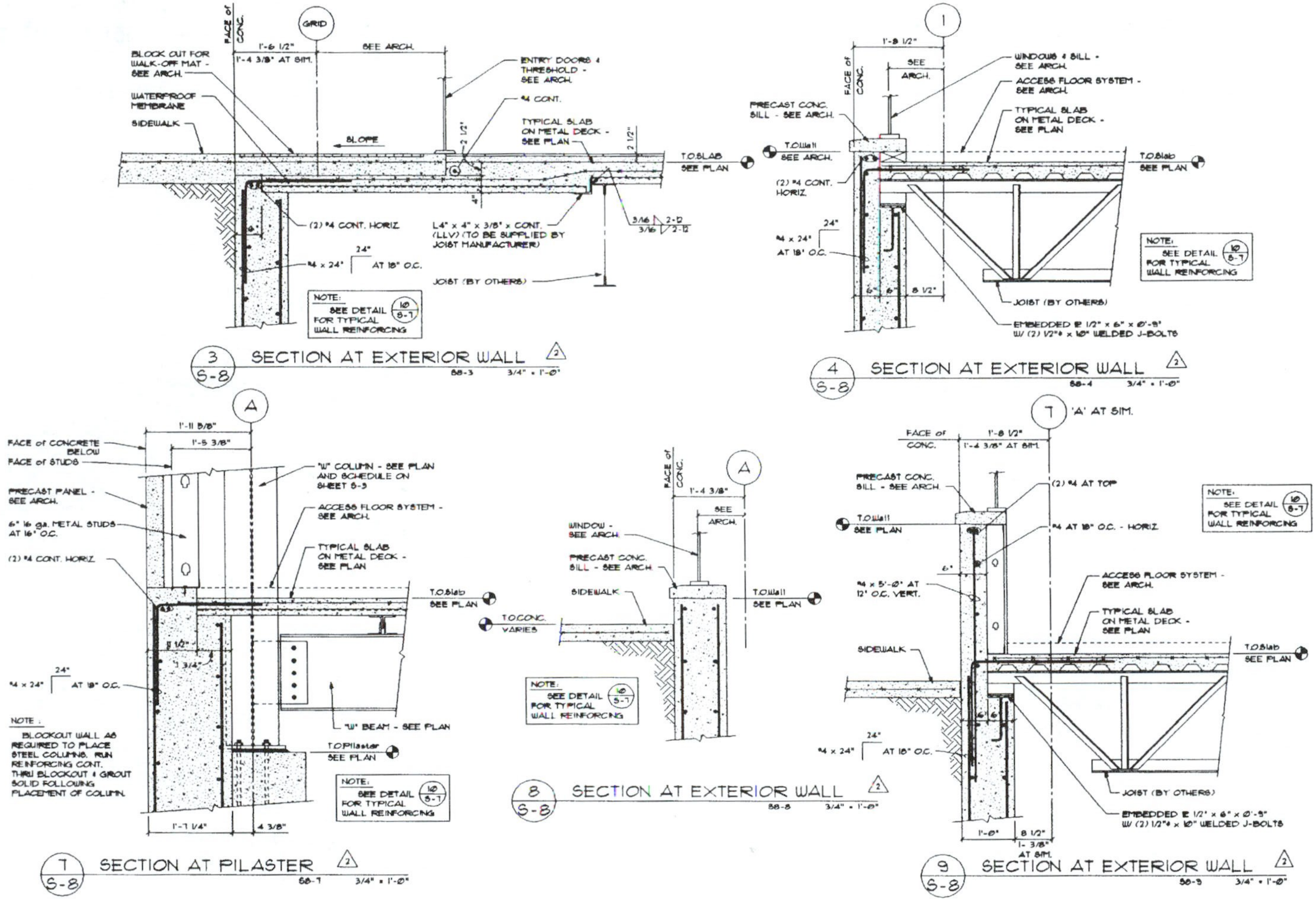

Figure 6.39a *The results of the preview shown in Figure 6.36. (Courtesy Van Domelen/Looijenga/McGarrigle/Knauf Consulting Engineers.)*

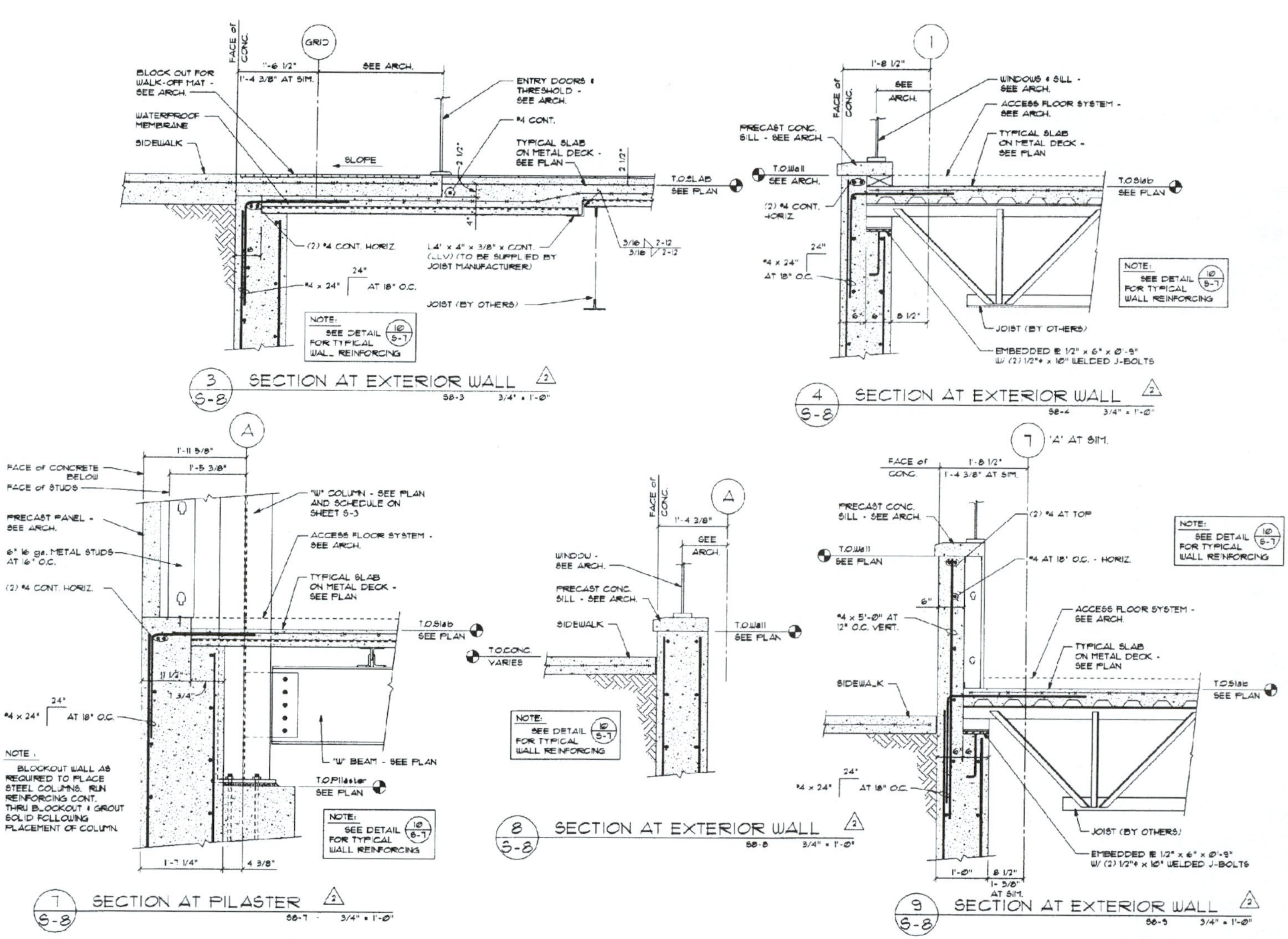

Figure 6.39b *The results of the preview shown in Figure 6.38. The End Preview button was selected and the 90° rotation was selected prior to selecting the OK button to produce the print. (Courtesy Van Domelen/Looijenga/McGarrigle/Knaf Consulting Engineers.)*

CHAPTER 6 EXERCISES

Set limits, units, snap, and grid to best suit each project. Use appropriate linetypes and lineweights that best suit each project. Dimensions, hatch patterns, and text do not need to be placed at this time. Save the following exercises in safe locations.

1. A parcel of land is to be developed. Use the following metes and bounds supplied by the surveyor and draw the site plan and the indicated utility easement. Set limits to 200',150'. Use a phantom line to represent the property line and use a centerline to represent the easement. Select different colors and layers for each linetype.

   SITE. Beginning at a point to be described as the northwesterly corner, 100.00' due south, thence 37.52' N89°37'E, thence 96.85' N58°40'52"E, thence 57.25' N3°50'40"W, and then back to the true point of beginning.

   EASEMENT. The easement is approximately an inverted L-shaped easement. The long leg of the "L" is 10'-wide; the short leg is 15' wide. Lay out the easement beginning at a point that lies on the southerly property line, 20'-0" from the southwesterly corner of the property. Thence northerly 70'-0" along the center line of said easement to a point at N7°32'15"W, thence a 15'-wide easement lying along a center line extending to the westerly property line at N50°55'04"W. Save the drawing with the filename E6-1 SITE.

2. Create a template drawing containing appropriate values for an architectural drawing. Include settings for each of the options discussed in previous chapters. Set limits and paper sizes assuming 24" × 36" plotting paper. Load linetypes and lineweights for lines that might be necessary on a plan view. Create a layer titled FLRWALLS. Save the file as a PLAN TEMPLATE drawing.

3. Use the drawing below to complete a side view of a steel beam resting on a steel plate supported on a steel column. Assume the beam to be 18" deep, and show the thickness of each plate as 7/8" thick. Provide a 10"-long support plate, and an 18"- high stiffener plate at the end of the beam. The support column is a TS 6" × 6" × 3/8". Provide appropriate break lines where needed. Save the drawing as E-6-3.

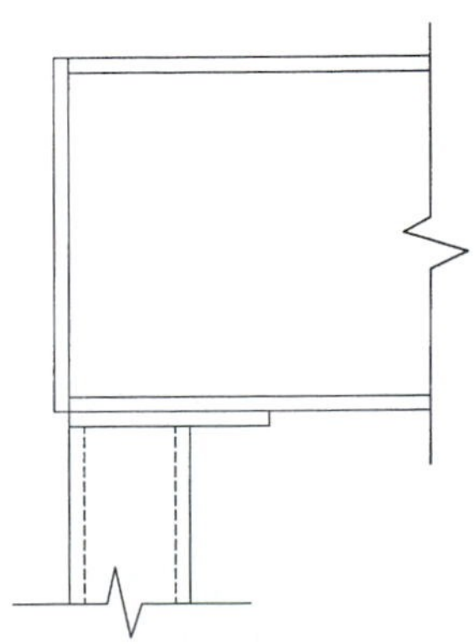

4. Use the attached drawing to complete a side view of a steel beam resting on a steel plate supported on a steel column. Assume the beam to be 16" deep, and show the thickness of each plate as ⅝" thick. Provide a 14"-long support plate. The support column is a TS 6" × 6" × ⅜" with a TS 6" × 6" × ⅜" chevron at a 45° angle. Provide centerlines for each TS. Provide appropriate break lines where needed. Save the drawing as E-6-4.

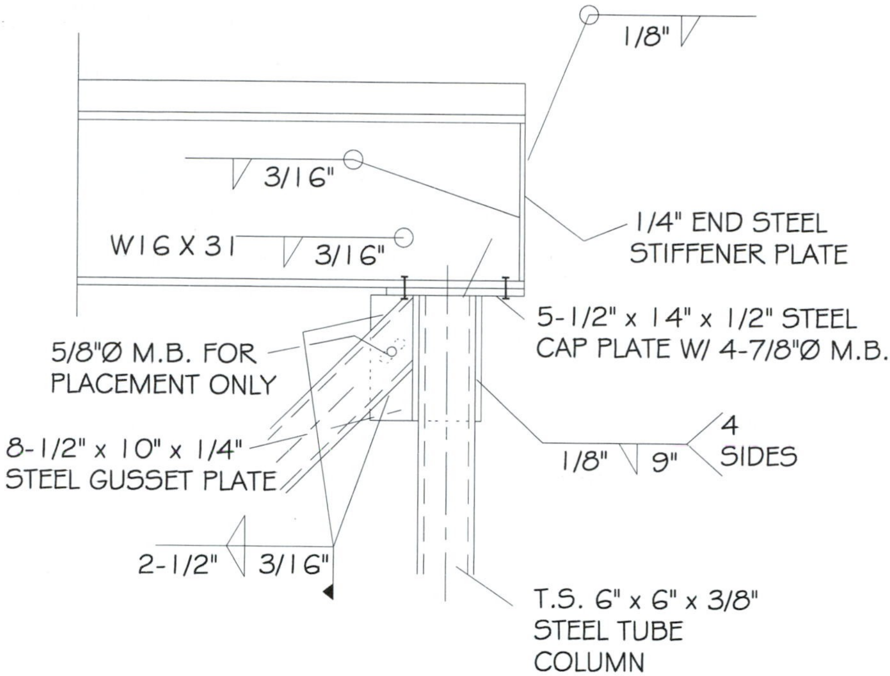

5. Use the drawing below to complete a side view of a steel column resting on a steel plate that is supported on 1" mortar over a 24" × 24" concrete pedestal. Assume the thickness of the plate as ⅝" and show ¾"×14" long anchor bolts 1-½" from each edge of the plate. Provide a 14" × 10" × ⅜" support plate. The support column is a TS 6" × 6" × ⅜" with a TS 6" × 6" × ⅜" chevron at a 45° angle. Provide centerlines for each TS. Provide appropriate break lines where needed. Save the drawing as E-6-5.

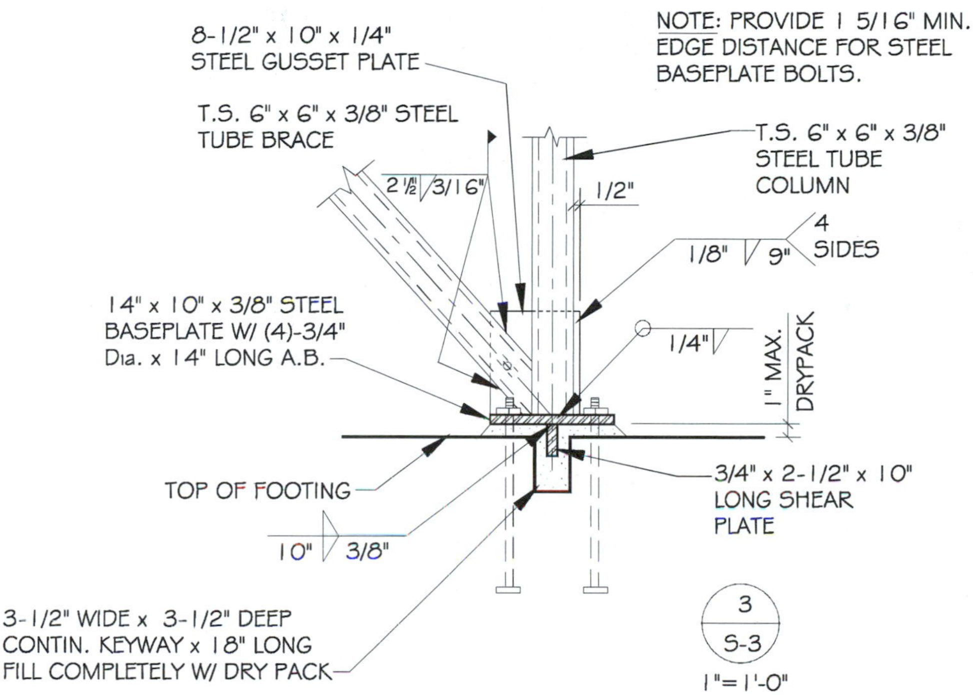

6. Use the drawing below to draw a partial foundation plan and place a detail marker. Show the outline of the footing with thin dashed lines and show the outline of the slab with a thick, continuous line. Draw a side view for a concrete slab foundation. Save the drawing as E-6-6.

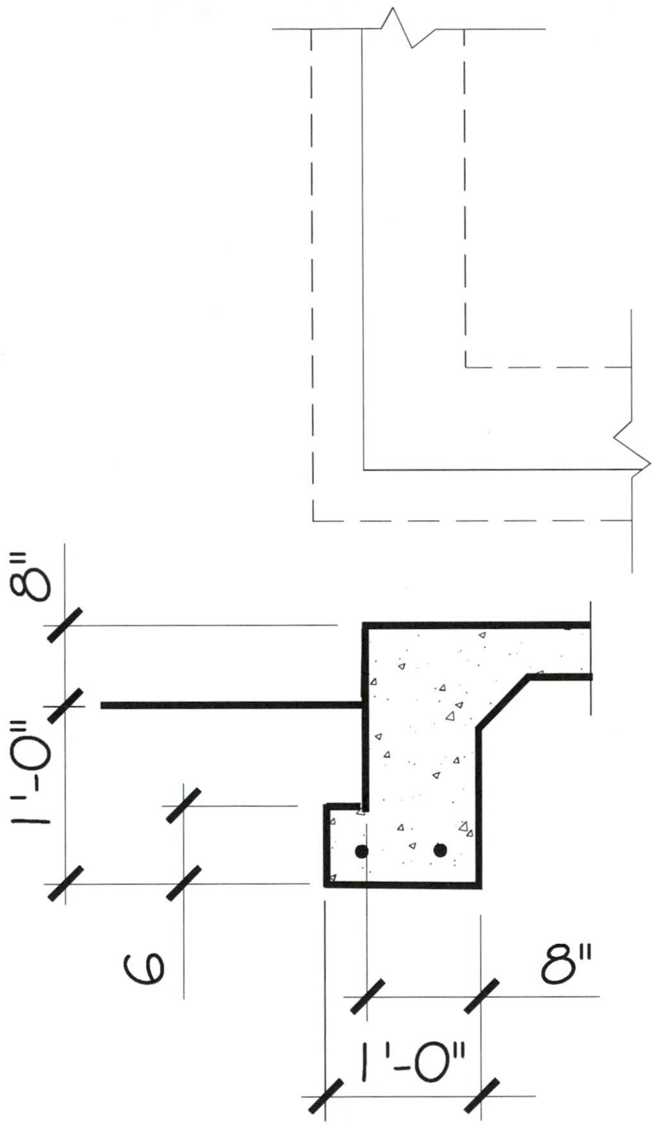

7. Use the drawing from Exercise 6 to obtain the needed sizes to draw a partial foundation plan and place a detail marker. Show the outline of the footing with thin dashed lines and show the outline of the 8"-wide stem wall with thick, continuous lines. Save the drawing as E-6-7.

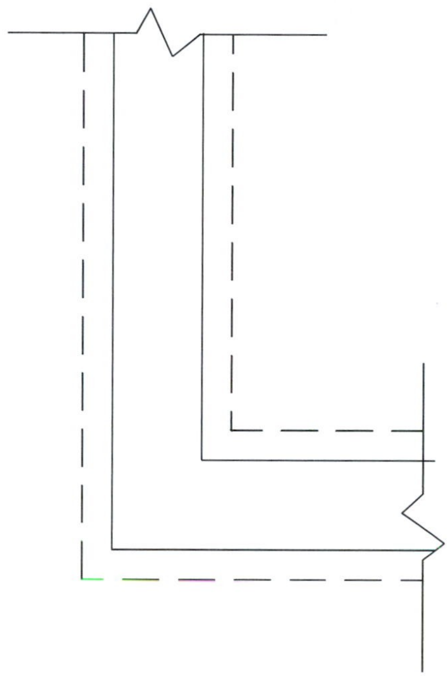

8. Use the drawing below to draw a site plan. Create your own north arrow. Save the drawings as E-6-8.

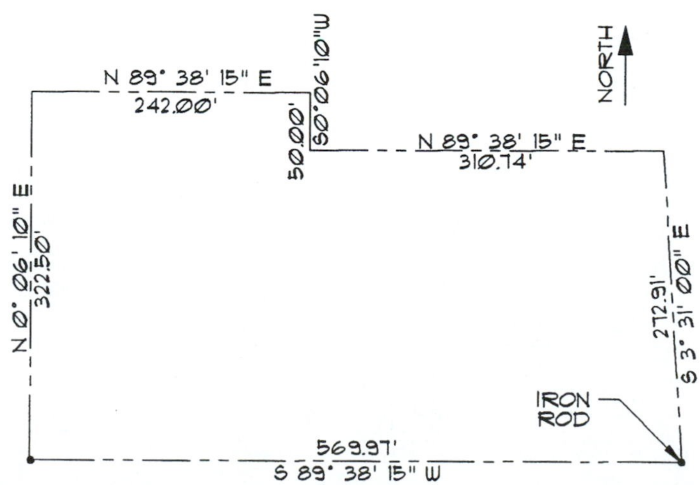

9. Use the drawing below to draw a section of a wall constructed of 8" concrete blocks. Use thin lines to represent the block. Show a #5 rebar centered in the wall using a thick dashed line. The individual blocks and hatch pattern do not need to be shown at this time. Save the drawings as E-6-9.

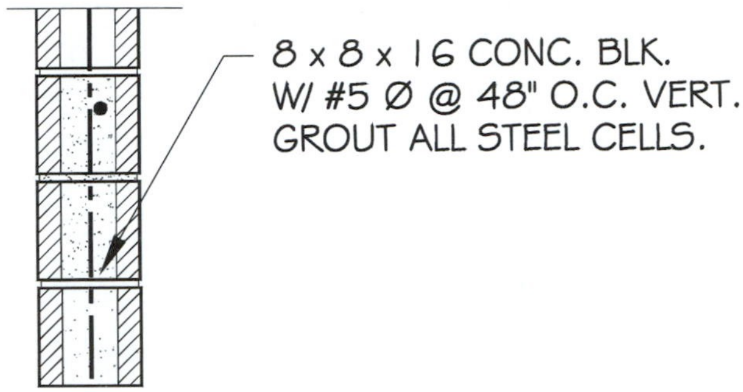

10. Use the drawing below to draw a 24' × 26' concrete wall panel with a 10' × 12' door and a 3' × 7' door. Place each door 2' from the panel edge. Show the finish floor 12" above the bottom of the panel. Use thin lines to represent the wall outline and openings. Save the drawings as E-6-10.

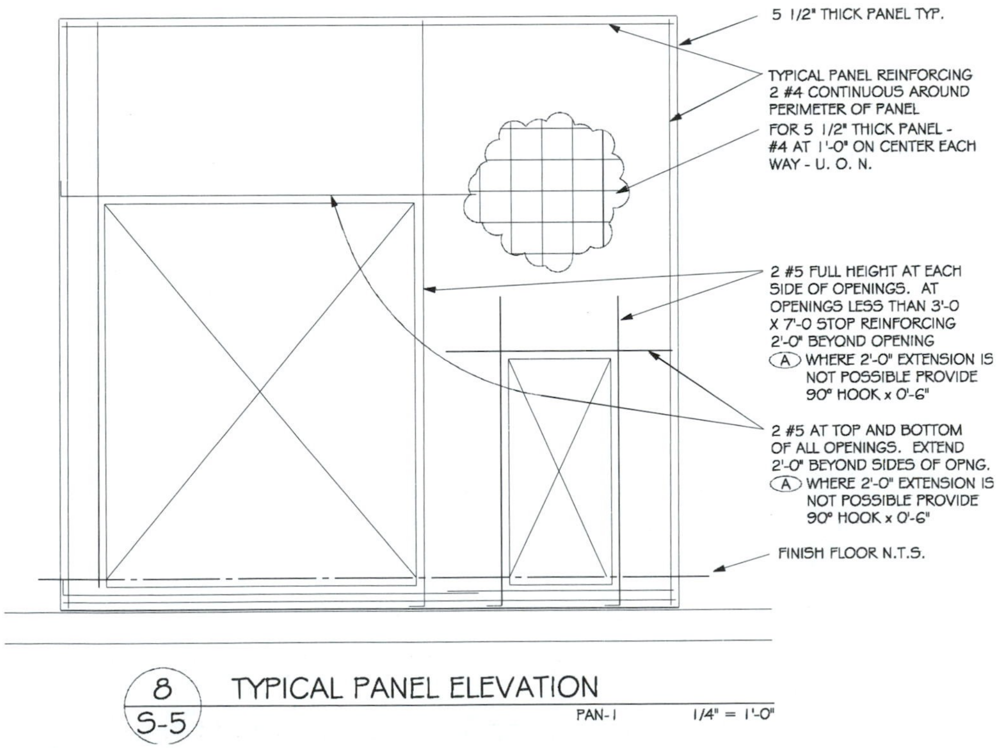

CHAPTER 6 QUIZ

1. List the methods for accessing linetype. Which is your favorite method? Which method is your least favorite method? Explain each answer.
2. List two possible uses for an object line.
3. Describe the length and thickness of each line used in a centerline pattern.
4. List examples of lines that should be thick on a floor plan.
5. Describe the length and thickness of each line used in a hidden line pattern.
6. Describe the use of a section cutting plane line.
7. A note must be placed to describe a feature. Name and describe the type of line to be used.
8. A length of 24'-6" must be placed to describe the length of a room. What kind of line is used to reference the text to the drawing? In addition to this line, what else is needed to complete the correct placement?
9. Sketch how long and short break lines should be drawn.
10. List nine types of lines often found on construction drawings.
11. Describe the differences between a section line and a section cutting plane line.
12. A portion of a beam needs to be removed so that it will fit in a drawing. Name and describe the line used where the portion of the object has been removed.
13. Describe the options for a print preview.
14. What command will allow the line pattern to be altered?
15. Explain the difference between printing using the Display and Window options.
16. List four options for controlling the visibility of a layer
17. You've just loaded five linetypes, and have entered the LINE command. Only continuous lines are being drawn. List two possible problems.
18. What is the difference between assigning a color to an object or a layer?
19. List four options for entering the Layer command.
20. What is the current layer name, and how can a new layer be named?
21. What are the limitations to assigning layer names?
22. A friend needs help to load five linetypes into a drawing. How can this be done quickly?
23. Centerline has just been made the current linetype. You return to the drawing and nothing is different? What's happened and why did it happen?
24. You've gone into the Layer Properties Manager and made centerline the current linetype. You return to the drawing? What's happened and why did it happen?
25. Use the Help menu and explore how the Global Scale Factor will affect linetypes.

26. You've gone into Lineweight Control in the Object Properties toolbar and altered the lineweight. You return to the drawing area and draw several lines, but no change has been made to the lineweight. Why is there no change, and what must be done to see varied lineweight?
27. List the suggested lineweight for thick ____, moderate ____, and thin lines ____.
28. You want to draw one red line. How can it be done?
29. You want all of your lines to be red. What is the best way to do this?
30. You want to plot a drawing at ⅜" = 1'-0". How can this be done?

CHAPTER 7

Drawing Geometric Shapes

Lines and points may be the base of all drawings, but it's difficult to imagine a large engineering or architectural project without curved features. This chapter will introduce you to drawing circles, arcs, ellipses, polygons, and donuts.

CIRCLES

Circles are found in a variety of projects throughout the construction world. Round windows, concrete piers and columns, steel tubing, and conduit lines—circular features will be prevalent throughout your drawings. Figure 7.1 uses circles to represent concrete piers for a structure. The CIRCLE command may be entered by selecting the Circle icon from the Draw toolbar, by selecting Circle from the Draw pull-down menu, or by keyboard. The Draw pull-down menu can be seen in Figure 7.2. To draw a circle, select the Circle icon from the Draw toolbar, type **C** (EnterKey) at the command prompt. Entering the CIRCLE command from the toolbar or keyboard will display the prompt:

Command: **C** (EnterKey)

CIRCLE Specify center point for circle or [3P/2P/TTR (tan tan radius)]:

AutoCAD is waiting for you to select a center point for the origin point of the circle. Once the center point is selected, you're prompted for the edge of a circle. The process can be seen in Figure 7.3.

Circles may be drawn by entering seven different combinations of information about the circle. Six of the circles are listed in the Circle pull-down menu shown in Figure 7.2. The DONUT command in this menu produces a circle with varied line width and will be discussed later in this chapter.

Notice that once the center point is selected, a circle will appear on the screen, but the size will be altered as the mouse is moved. AutoCAD refers to this as drag or rubber band. Moving the mouse and changing the circle size refers to dragging the circle into

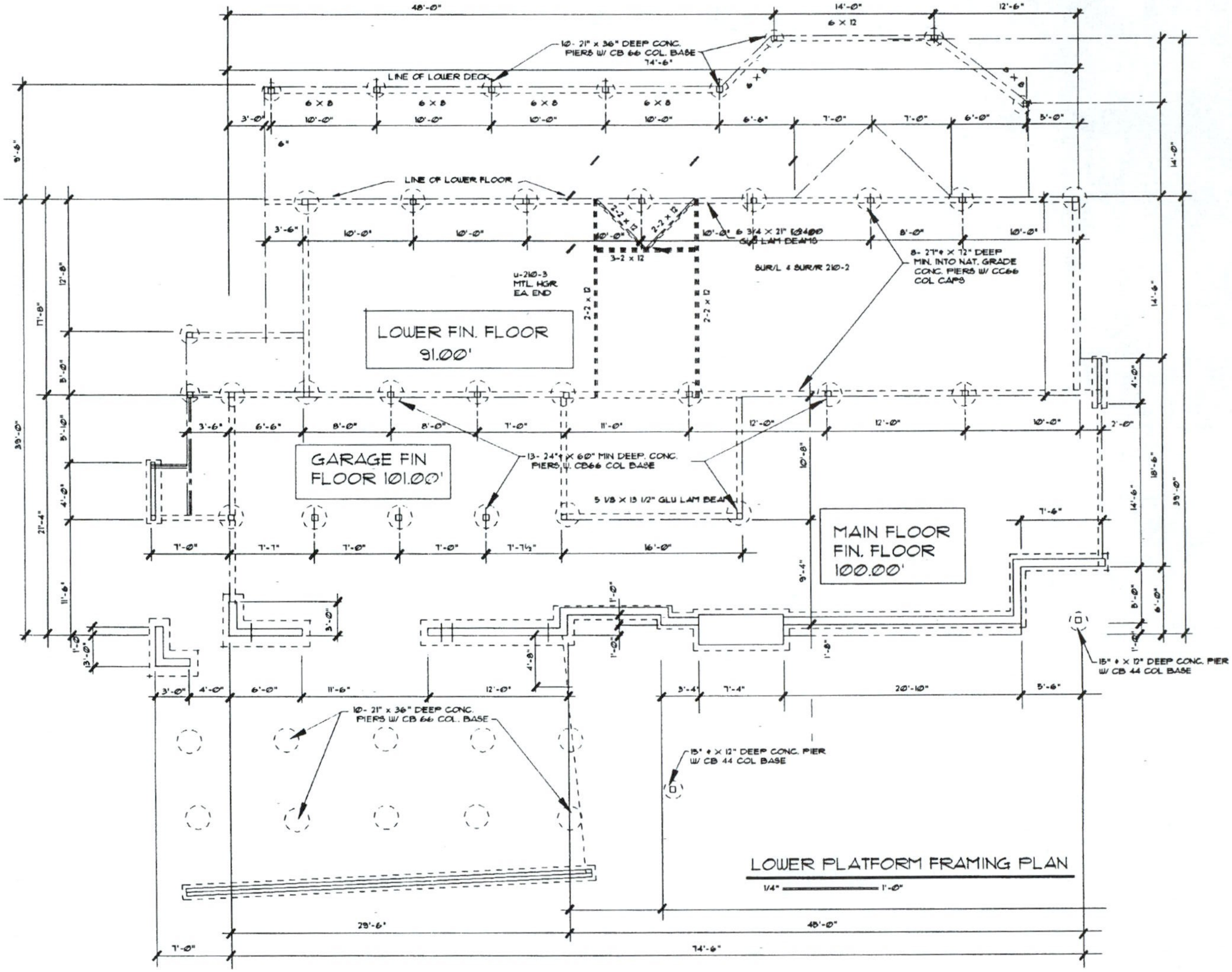

Figure 7.1 *Circles are used throughout architecture to represent many building components. (Courtesy Residential Designs.)*

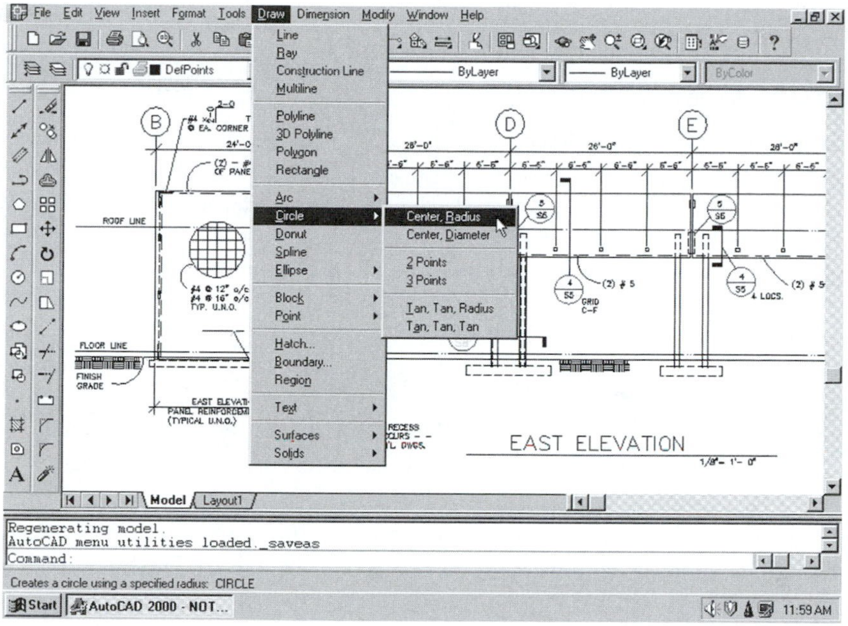

Figure 7.2 *Circles can be drawn by selecting the Circle icon from the Draw toolbar, by selecting Circle from the Draw pull-down menu, or by typing* **C** *(EnterKey) at the command prompt.*

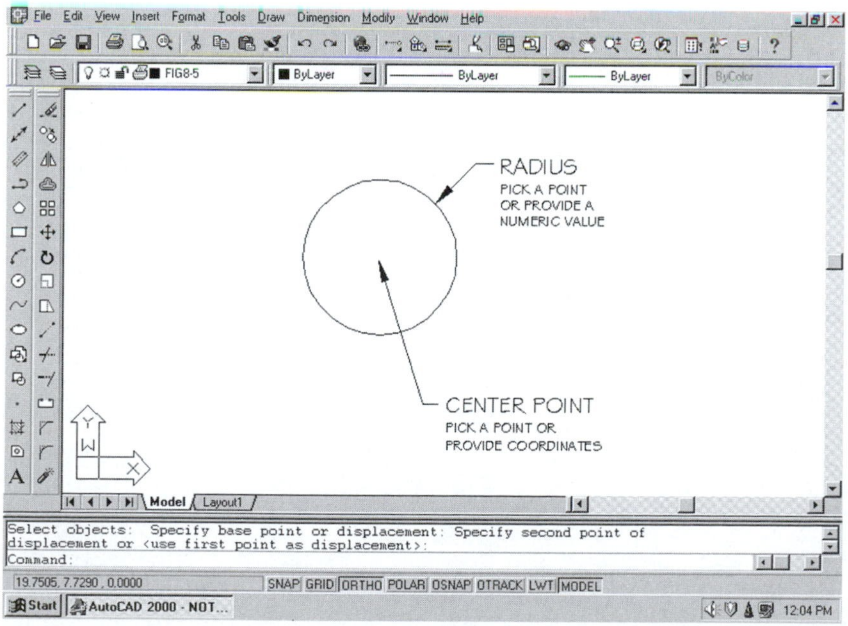

Figure 7.3 *The size of a circle can be controlled by entering a value for the radius or the diameter.*

position. This drag image is typically very helpful in determining the final locations of lines and other geometric shapes. If you find it distracting, type **DRAGMODE** (EnterKey) at the command prompt and select OFF to disable it.

Note: As you draw circular features, you may want to review the Display tab of the Options dialog box. The Arc and Circle Smoothness setting can be used to control the smoothness of the display of arcs, circles, and ellipses. The default value is 100, with a range of 1 through 20,000. A higher number will produce a smoother display, but will require longer regeneration times on larger drawings. A lower number will display circular features with a series of straight lines. The true shape of objects will be displayed on plots, regardless of the resolution.

CENTER AND RADIUS

The default method of drawing a circle is to enter the center point and a radius. As the CIRCLE is started, the center point may be entered by typing coordinates or by selecting a center point with the pointing device. In Figure 7.4 three circles have been drawn, and a fourth has been dragged into position. Notice that the rubberband line extends from the chosen center point to the radius indicated by the crosshairs. By moving the crosshairs, the size of the circle can be changed. By entering a numeric value and (EnterKey), a circle will be drawn using the desired radius. To draw a 24" diameter circle, the sequence would be:

Command: **C** (EnterKey)

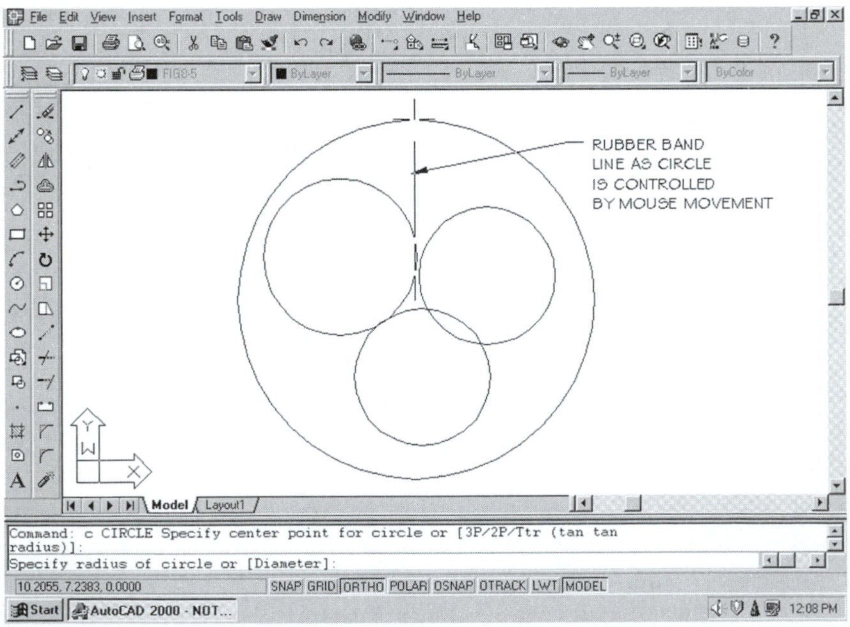

Figure 7.4 *Rather than entering a specific size, use the rubberband line to see the effect moving the cursor will have on the circle to be drawn.*

CIRCLE Specify center point for circle or [3P/2P/TTR (tan tan radius)]
(Select center point.)

Specify radius of circle or [Diameter] <0'-0">: **12** (EnterKey)

Command:

Note: To draw another circle, the command can be continued by pressing the (EnterKey), pressing the (ShiftKey), or right-clicking and then selecting Repeat CIRCLE

CENTER AND DIAMETER

Sizes of circular features such as piers, pipes, and windows typically are given as a diameter. At the circle prompt enter D (EnterKey) to switch the prompt from radius to diameter. When entered by toolbar or keyboard, the command sequence is:

Command: **C** (EnterKey)

CIRCLE Specify center point for circle or [3P/2P/TTR (tan tan radius)]
(Select center point.)

Specify radius of circle or [Diameter] <0'-0">: **D** (EnterKey)

Specify diameter of circle:<2'-0"> (*Enter a diameter value or move the crosshairs to the desired location.*)

Command:

The command sequencc is similar when selected from the pull-down menu. AutoCAD simplifies the process with the pull-down menu so that the diameter option is automatic.

Once the circle center is selected, the circle will drag across the screen but will not line up with the crosshairs. Figure 7.5 shows an example of a circle being created using the diameter. Notice that the length of the line extending from the center point is the diameter of the circle to be drawn. The line will disappear once the diameter is selected. The diameter entered will become the default setting for the next circle drawn.

TWO-POINT CIRCLES

Drawing a two-point circle can be useful if the diameter is known, but the center point is not. The two-point circle is drawn by selecting a point and then selecting a point on the opposite side of the circle, as seen in Figure 7.6. The points may be selected with the crosshairs, or coordinates may be entered. The command sequence is as follows:

Command: **C** (EnterKey)

CIRCLE Specify center point for circle or [3P/2P/TTR (tan tan radius)]: **2P** (EnterKey)

Specify first end point of circle's diameter: *(Enter coordinates or select a point.)*

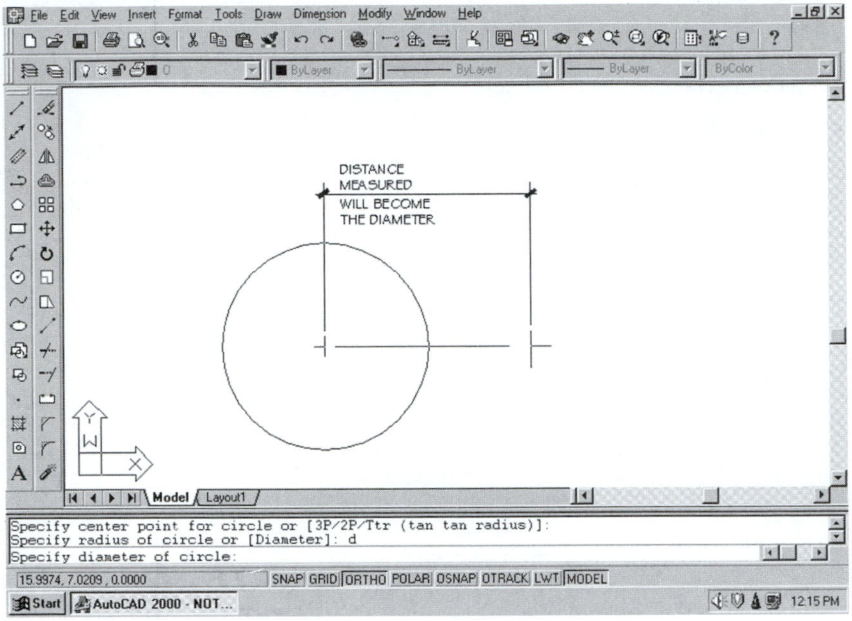

Figure 7.5 *When using the Diameter option, the distance that the cursor is moved from the center point will become the diameter of the circle.*

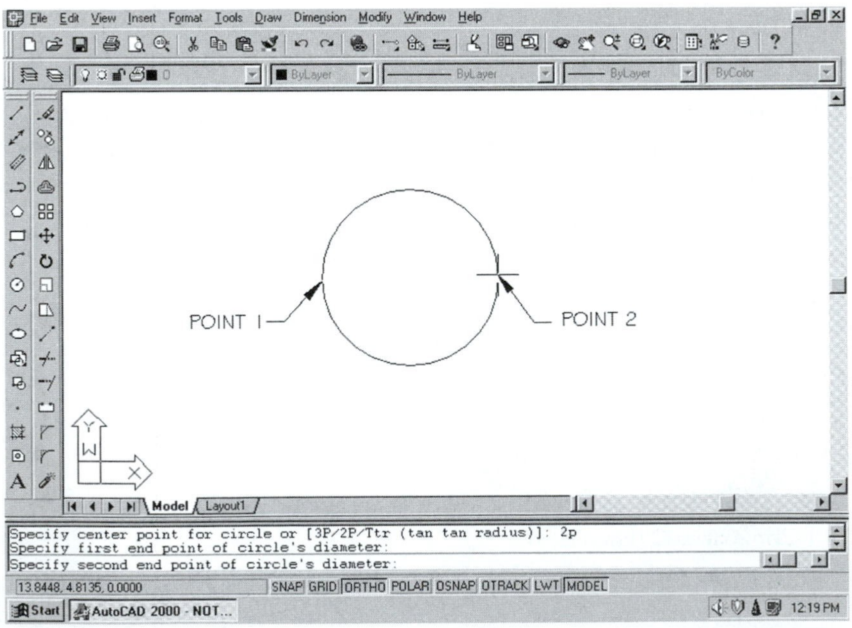

Figure 7.6 *A circle can be drawn by specifying the endpoints of a line that would be the diameter.*

Specify second end point of circle's diameter: *(Enter coordinates or select a point.)*

Command:

THREE-POINT CIRCLES

If certain points of a circle are known, a circle can be drawn by entering any three points. Figure 7.7 shows an example of a three-point circle. The points may be selected with the crosshairs, or coordinates may be entered. The process for drawing a three-point circle is:

Command: **C** (EnterKey)

CIRCLE Specify center point for circle or [3P/2P/TTR (tan tan radius)]: **3P** (EnterKey)

Specify first point on circle: *(Enter coordinates or select a point.)*

Specify second point on circle: *(Enter coordinates or select a point.)*

Specify third point on circle: *(Enter coordinates or select a point.)*

Command:

TANGENT, TANGENT, AND RADIUS

An object is tangent if it intersects another object at one and only one point. A circle may be drawn tangent to existing features by selecting two features that the circle will

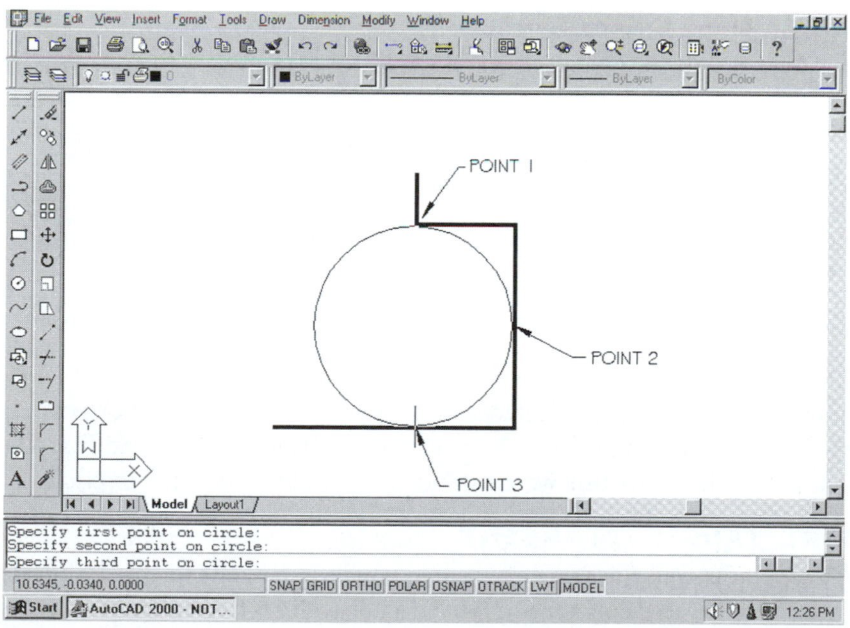

Figure 7.7 *The size of a circle can be determined by specifying three tangent points.*

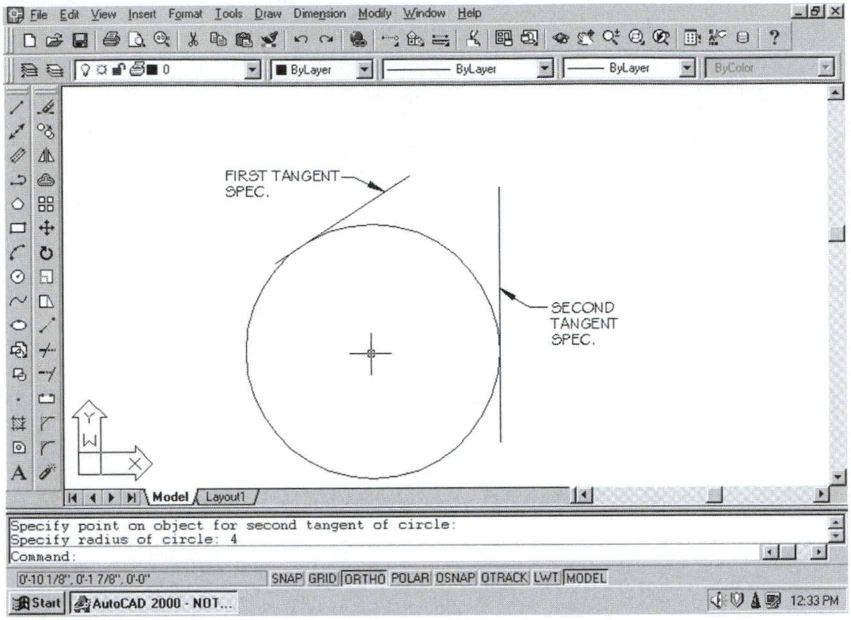

Figure 7.8 *A circle can be drawn by specifying two tangent surfaces and a radius.*

be tangent to, and then specifying a radius. This can be seen in Figure 7.8. The command sequence is:

> Command: **C** (EnterKey)
>
> CIRCLE Specify center point for circle or [3P/2P/TTR (tan tan radius)]: **TTR** (EnterKey)
>
> Specify point on object for first tangent of circle: *(Select a line or circle.)*
>
> Specify point on object for second tangent of circle: *(Select a line or circle.)*
>
> Specify radius of circle: <2'-0"> *(Enter a numeric value or select a point.)* (EnterKey)
>
> Specify second point: *(Enter a numeric value or select a point.)* (EnterKey)
>
> Command:

In Figure 7.9 each line was selected as a tangent point and a radius was entered. Because the radius is smaller than the distance between the lines, the first line does not touch the circle. If the line were extended, it would be tangent to circle.

TANGENT, TANGENT, TANGENT

This option works well for drawing a circle which is tangent to three objects. This option can only be accessed through Tan, Tan, Tan of the pull-down and screen menus. The command sequence is:

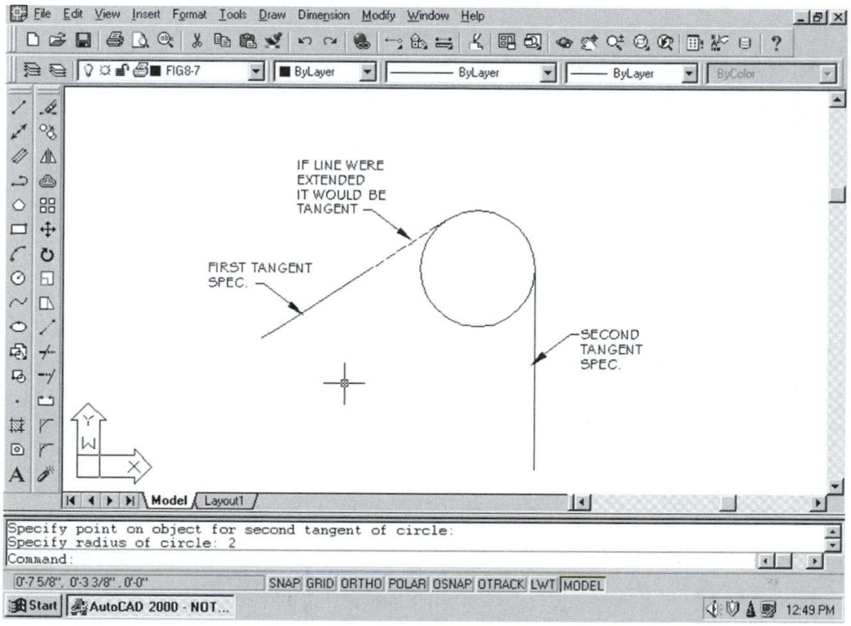

Figure 7.9 *If a radius is selected that is not large enough for the specified circle to touch each line, a circle is drawn tangent to one line and tangent to where the other line would be if it were to be extended.*

Command: _circle Specify center point for circle or [3P/2P/TTR (tan tan radius)]: _3p

Specify first point on circle: _tan to *(Select the first object the circle is to be tangent to.)*

Specify second point on circle: _tan to *(Select the second object the circle is to be tangent to.)*

Specify third point on circle: _tan to *(Select the third object the circle is to be tangent to.)*

Command:

Figure 7.10 shows the development of a circle tangent to three other circles.

CIRCLE@

This option allows a circle to be drawn using the last endpoint of the last line drawn for the center point of a new circle. The option is best used for drawing concentric circles associated with pipes and circular columns. The command sequence is:

Command: **C** (EnterKey)

CIRCLE 3P/2P/TTR/<Centerpoint>: @ (EnterKey)

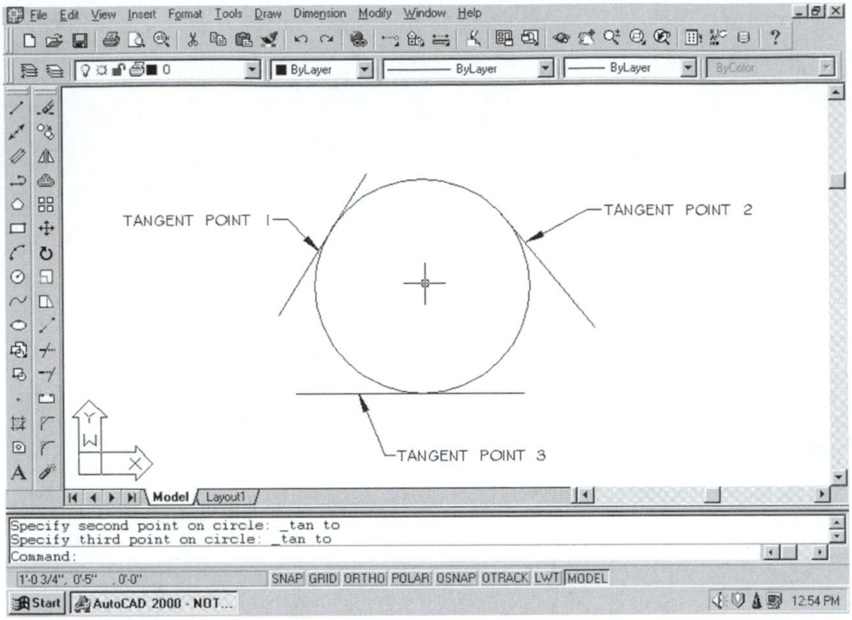

Figure 7.10 *Tan,Tan,Tan can be used to draw circles that are tangent to three lines.*

Diameter/<radius> <1.000> *(Specify desired diameter.)*

Command:

DRAWING ARCS

Arcs are used where a full circle is not required. Curved components are found throughout the structure seen in Figure 7.11. The command can be started by selecting the Arc icon from the Draw toolbar or by selecting Arc from the Draw pull-down menu. Arcs can also be drawn by typing A (EnterKey) on the command line The pull-down menu can be seen in Figure 7.12. Eleven different options are available for drawing arcs, depending on what information is available or on personal preference. Selecting Arc from the toolbar will produce the prompt:

Command: _arc Specify start point of arc or [CEnter]:

The default setting of the toolbar and keyboard is the three-point arc. With the exception of the three-point arc, each arc drawing method may be selected by typing a letter and (EnterKey) or the space bar when prompted for a point of the arc. Letters and the function of the arc they represent are:

A: included Angle

C: Center

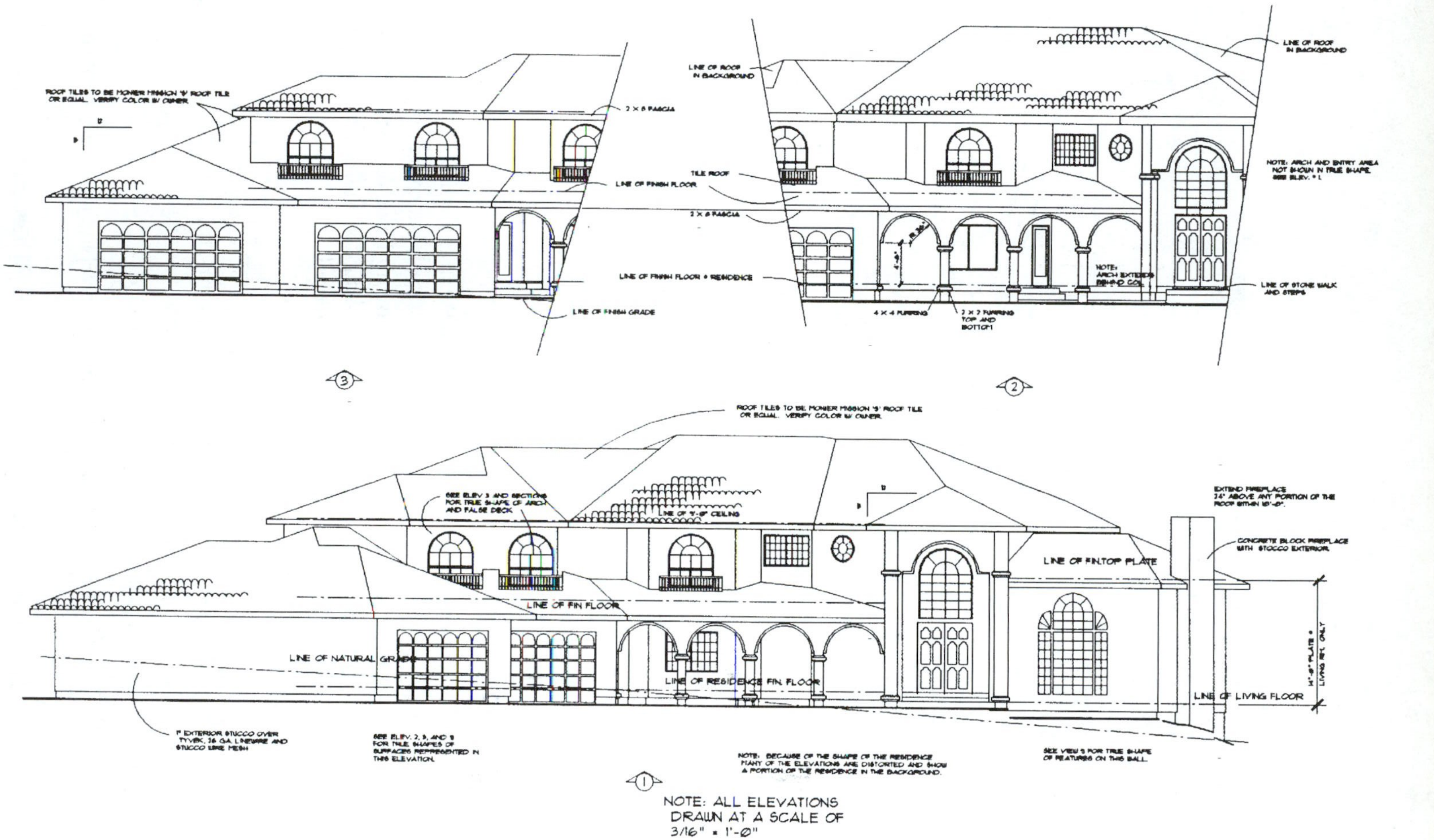

Figure 7.11 *Arcs are found throughout many structures. (Courtesy Residential Designs.)*

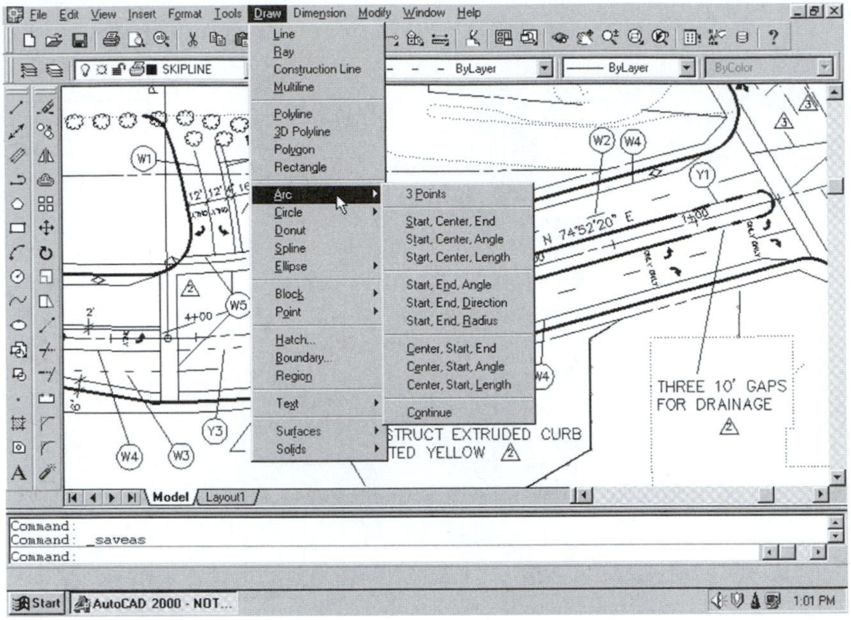

Figure 7.12 *The Arc pull-down menu. (Courtesy Lee Engineering, Inc., Oregon City, OR.)*

D: starting Direction

E: Endpoint

L: Length of chord

R: Radius

Each of the terms used to describe an arc can be seen in Figure 7.13 Notice that center describes the center of the circle that the arc is part of, not the center of the arc. When the ARC command is started from the pull-down menu, you will not have to enter any of the endpoints, center points, angles, or lengths. Pull-down arc segments are automatically provided to match the option selected.

THREE-POINT ARCS

Drawing a three-point arc is similar to drawing a 3P circle and can be seen in Figure 7.14. The first and third points are the arc endpoints; the second point is any point on the arc. The command sequence for a three-point arc is:

Command: **A** (EnterKey)

ARC Specify start point of arc or [CEnter/ENd]: *(Select any point or enter coordinate.)*

Specify second point of arc or [CEnter/ENd]: *(Select center point.)*

Specify end point of arc: *(Select the arc's endpoint or enter coordinates.)*

Command:

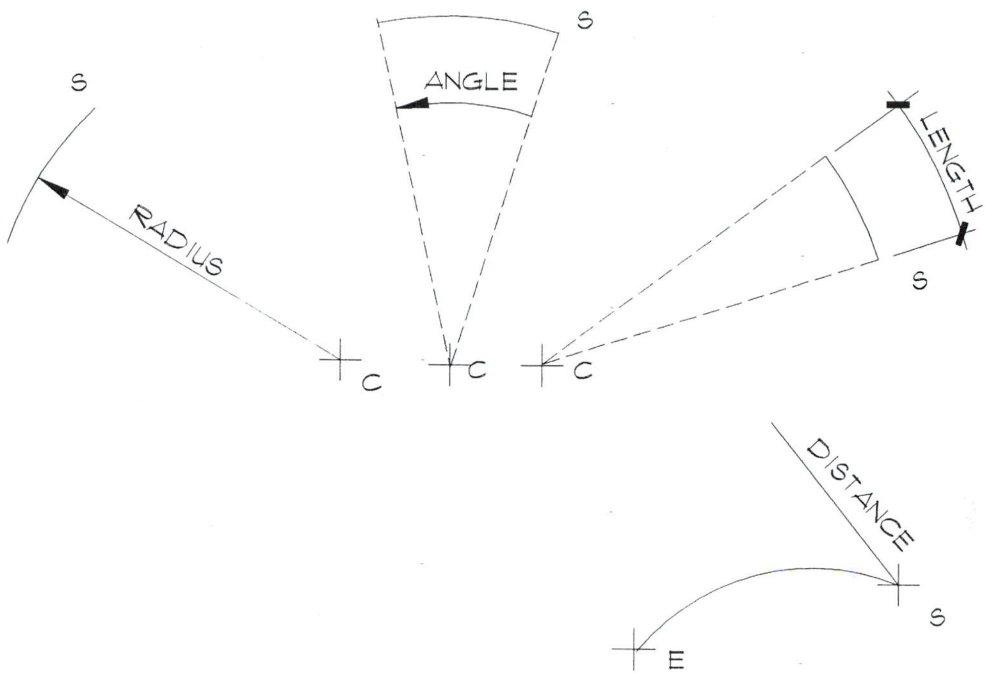

Figure 7.13 *Terms used to describe arcs are Start, Center, End, Angle, Length, Radius, and Distance.*

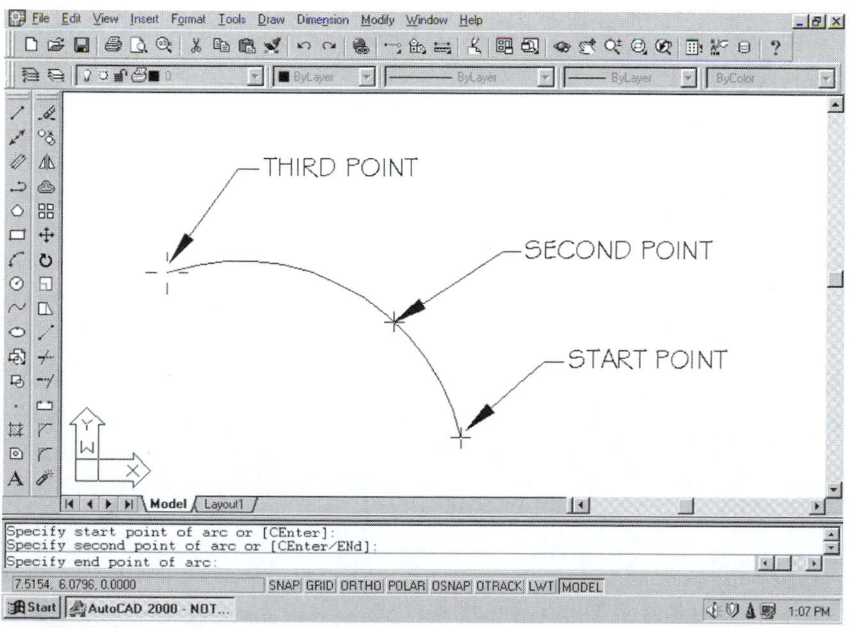

Figure 7.14 *Drawing a three-point arc is similar to drawing a three-point circle.*

Notice that once the second point is selected, the arc will drag across the screen with the crosshairs. The arc will continue to change size until the endpoint is selected.

START, CENTER, END ARCS

When the ends and center of an arch are known, the Start, Center, End arc option can be used. The stair railing in Figure 7.15 provides an example of where this arc option could be used. The command sequence is:

> Command: **A** (EnterKey)
>
> ARC Specify start point of arc or [CEnter]: *(Select any point or enter coordinate.)*
>
> Specify second point of arc or [CEnter/ENd]: *(Select second point.)*
>
> Specify end point of arc [Angle/chord Length]: *(Select the arc's endpoint or enter coordinates.)*
>
> Command:

As the last point is entered, the arc will be drawn and the command prompt will be returned. The process can be seen in Figure 7.16. Notice that as the last point is entered, the arc may not end at the point entered. The endpoint that was specified is

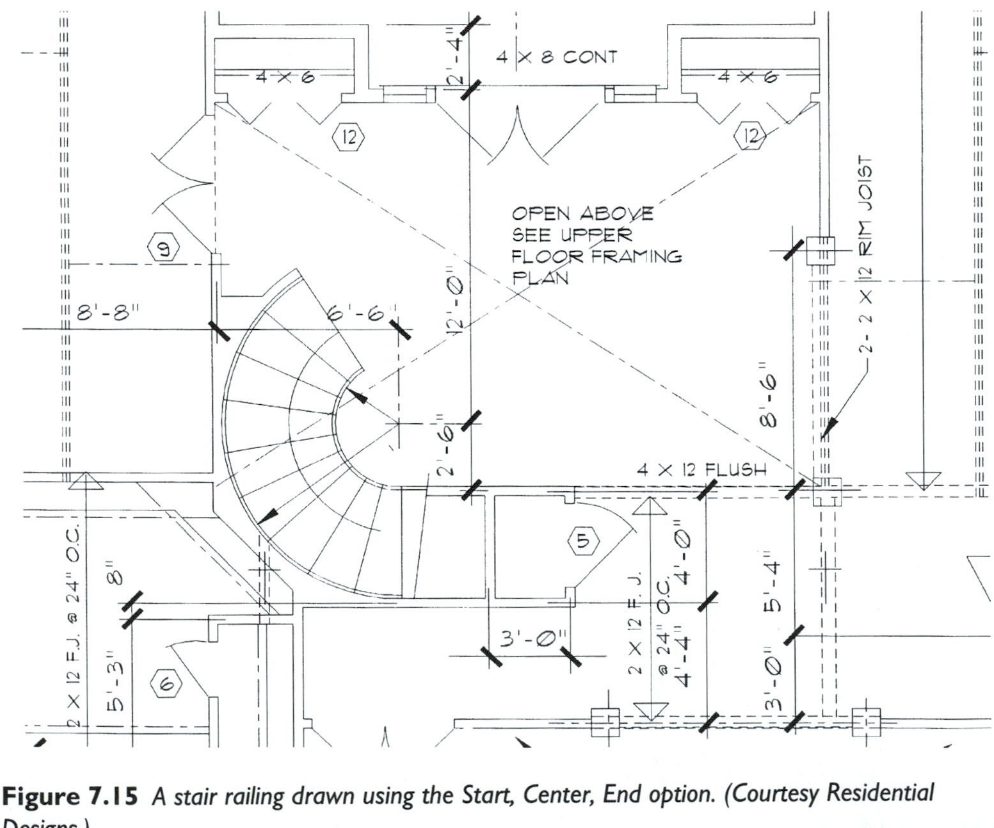

Figure 7.15 *A stair railing drawn using the Start, Center, End option. (Courtesy Residential Designs.)*

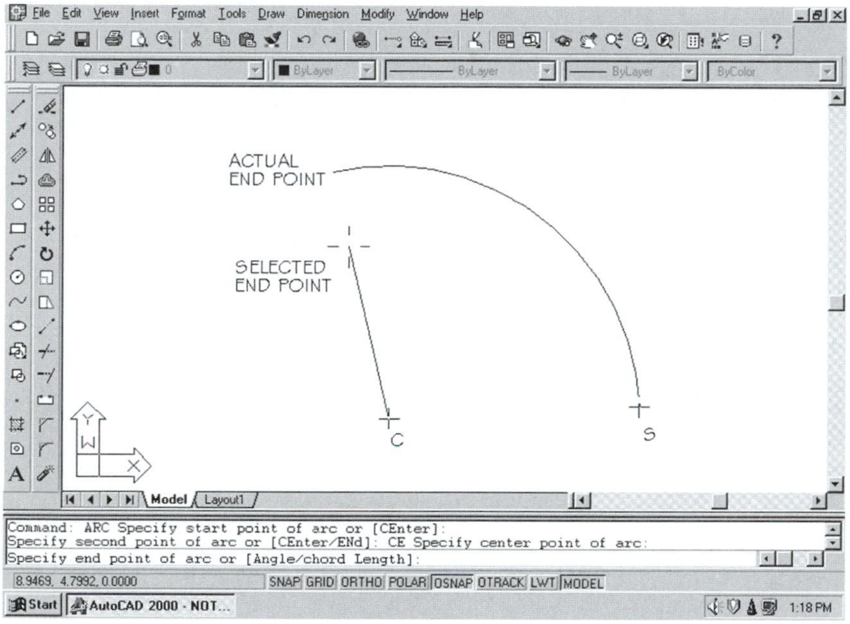

Figure 7.16 *The selected endpoint is used to determine the angle of the specified circle. The actual size of the circle is determined by the distance from the Start to the center point.*

used only to determine the angle at which the arc will end. The actual radius was determined by the Start and Center points.

START, CENTER, ANGLE ARCS

Often when doing road and utility layouts for a site plan it is necessary to draw an arc using lengths or angles. Drawing an arc using the Start, Center point, and the included Angle is ideal for this type of work. When entering the angle, if a positive value is used, the angle will be drawn in a counterclockwise direction. If a negative value is used, the angle will be made in a clockwise direction. The effects of positive and negative values can be seen in Figure 7.17. The command sequence for a Start, Center, Angle arc is:

Command: **A** (EnterKey)

ARC Specify start point of arc or [CEnter]: *(Select any point or enter coordinate.)*

Specify second point of arc or [CEnter/ENd]: **C** (EnterKey)

Specify center point of arc: *(Select center point of arc.)*

Specify end point of arc [Angle/chord Length]: **A** (EnterKey)

Specify included angle: **30** (EnterKey)

Command:

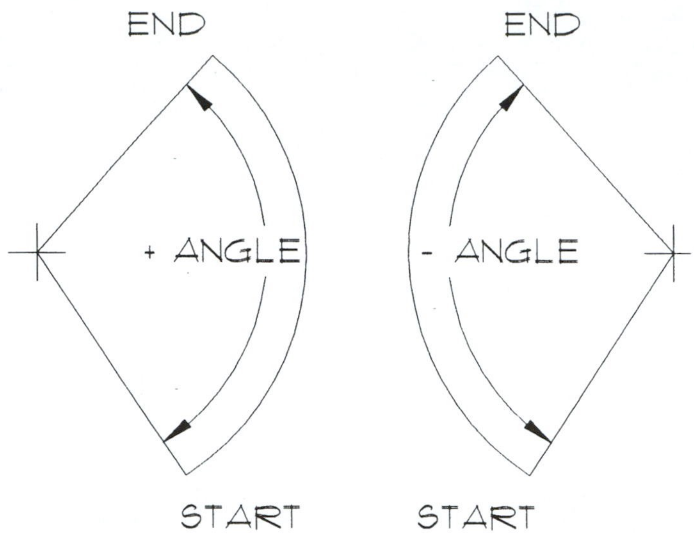

Figure 7.17 *The direction of layout is controlled by entering a positive (counterclockwise) or negative (clockwise) angle when drawing arcs with the Start, Center, Angle arc option.*

When the desired angle is entered, the arc will be drawn and the command line will return.

START, CENTER, LENGTH ARCS

A chord is the straight line connecting the endpoints of an arc. Subdivision maps and drawings relating to land often include arcs that require use of the chord length. Figure 7.18 shows an example of arcs that could be drawn using Start, Center, and Length of chord. The arc was drawn by entering the start and center points, and then supplying the chord length. A positive or negative value must be entered with the length. Figure 7.19 shows the difference between a chord of +3" or –3". The command sequence for a Start, Center, Length arc is:

Command: **A** (EnterKey)

ARC Specify start point of arc or [CEnter]: *(Select start point or enter coordinates.)*

Specify second point of arc or [CEnter/ENd]: **C** (EnterKey)

Specify center point of arc: *(Select any point or enter coordinates.)*

Specify end point of arc [Angle/chord Length]: **L** (EnterKey)

Specify length of chord: *(Select desired length.)* **3** (EnterKey)

Command:

When a length is specified or a point is selected, the arc will be drawn and the command line will return.

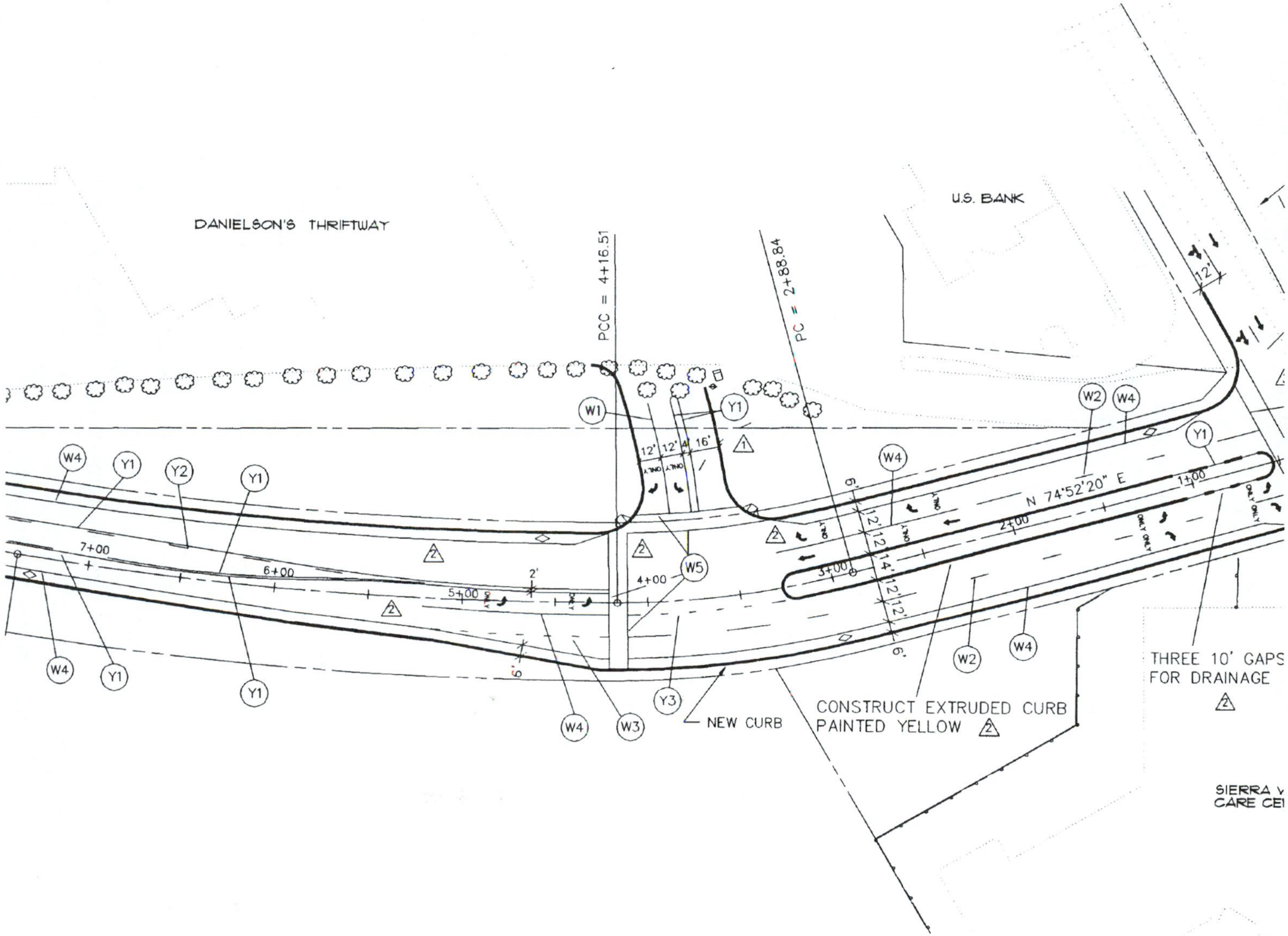

Figure 7.18 *The Start, Center, Arc option is useful when drawing parcels of land. (Courtesy Lee Engineering, Inc., Oregon City, OR.)*

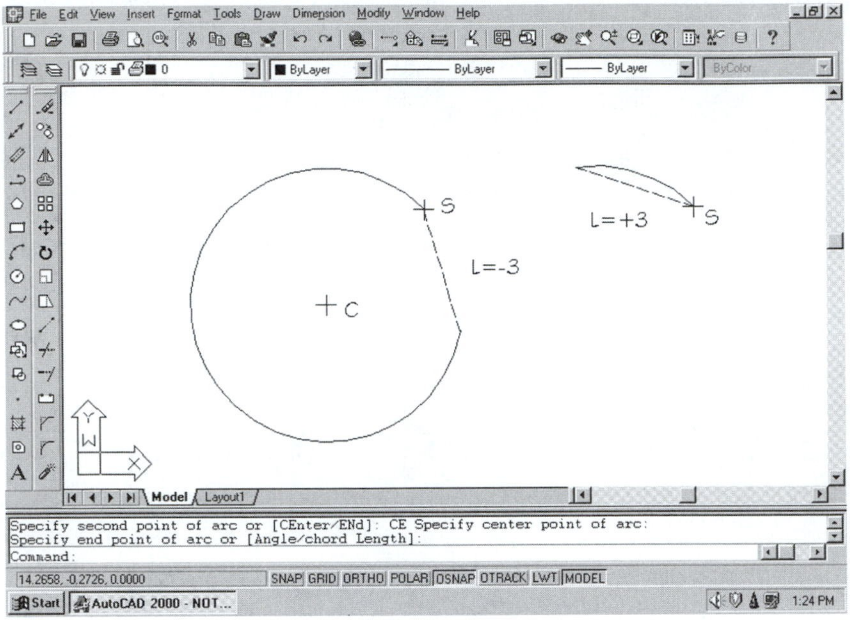

Figure 7.19 *Entering a positive or negative chord length determines which portion of the arc will be drawn using the Start, Center, Length ARC option.*

START, END, ANGLE ARCS

An arc drawn using Start, End, and Included Angle is drawn counterclockwise from the start point to the endpoint, if a positive angle value is used. If a negative angle is entered, the angle will be drawn clockwise. The results of coordinate entry can be seen in Figure 7.20. The command sequence is:

Command: **A** (EnterKey)

ARC Specify start point of arc or [CEnter]: *(Select start point or enter coordinates.)*

Specify second point of arc or [CEnter/ENd]: **E** (EnterKey)

Specify end point of arc: *(Select the desired endpoint or enter coordinates.)*

Specify center point of arc [Angle/chord Length]: **A** (EnterKey)

Specify Included angle: *(Select desired angle.)*

Command:

When a length is specified or a point is selected, the arc will be drawn and the command line will return.

START, END, DIRECTION ARCS

The Start, End, Direction option allows the direction of the arc to be selected. Major or minor arcs may be drawn in any orientation to the start point. Effects can be seen in Figure 7.21. The command sequence is:

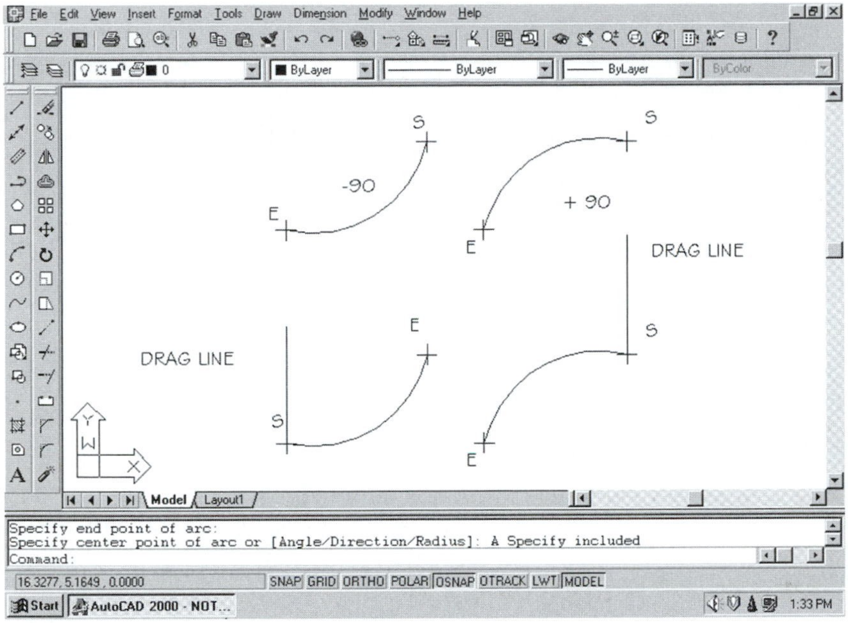

Figure 7.20 *Entering a positive or negative angle determines which direction an arc will be drawn using the Start, End, Angle ARC option.*

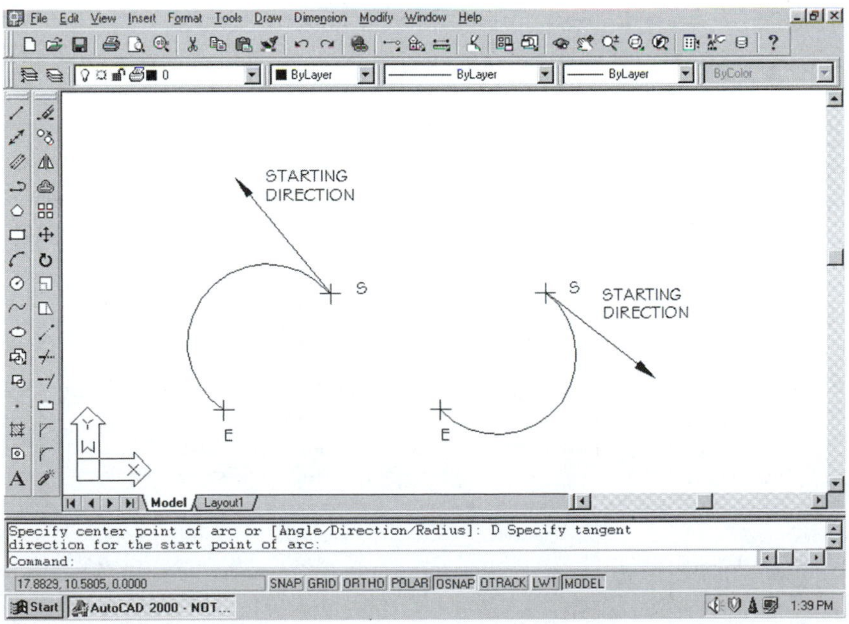

Figure 7.21 *The Start, End, Direction Arc option allows the direction of the arc to be selected.*

Command: **A** (EnterKey)

ARC Specify start point of arc or [CEnter]: *(Select start point or enter coordinates.)*

Specify second point of arc or [CEnter/ENd]: **E** (EnterKey)

Specify end point of arc: *(Select the desired endpoint or enter coordinates.)*

Specify center point of arc or [Angle/direction/Radius]: **D** (EnterKey)

Specify tangent direction for start of arc: *(Select desired angle.)*

Command:

When a length is specified or a point is selected, the arc will be drawn and the command line will return.

START, END, RADIUS ARCS

Start, End, Radius arcs can only be drawn counterclockwise. Because the variables can produce different options, the arc is drawn counterclockwise from the Start point. Entering a positive or negative value for the radius will determine if the major or minor arc will be drawn. The effects of the value can be seen in Figure 7.22. The command sequence is:

Command: **A** (EnterKey)

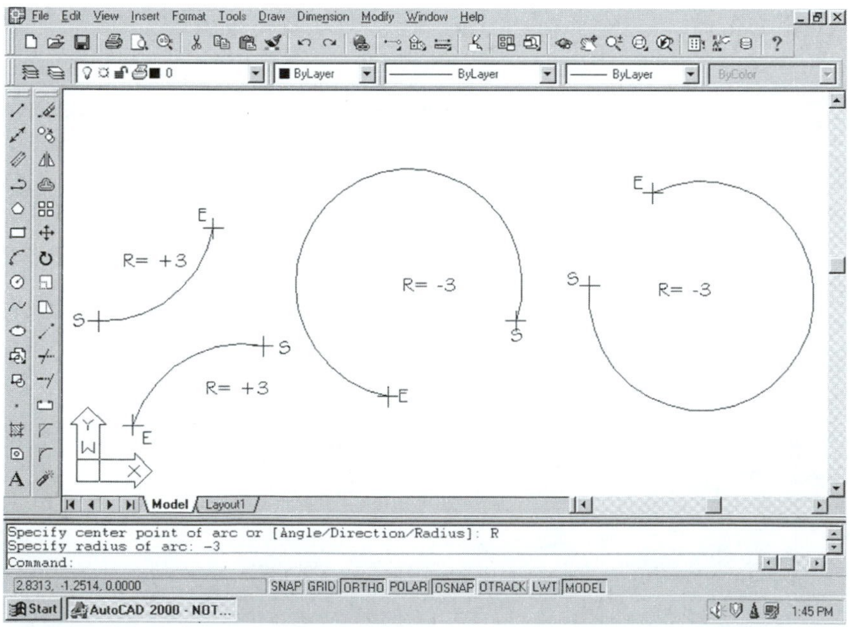

Figure 7.22 *Entering a positive radius determines which portion of the arc will be drawn using the Start, End, Radius ARC option. The direction will always be counterclockwise.*

ARC Specify start point of arc or [CEnter]: *(Select start point or enter coordinates.)*

Specify second point of arc or [CEnter/ENd]: **E** (EnterKey)

Specify end point of arc: *(Select the desired endpoint or enter coordinates.)*

Specify center point of arc or [Angle/direction/Radius]: **R** (EnterKey)

Specify radius of arc: *(Select desired radius or enter a point.)*

Command:

When a length is specified or a point is selected, the arc will be drawn and the command line will return (see Figure 7.23).

CENTER, START, END ARCS

The Center, Start, End option is similar to the Start, Center, End option. The effects of each command can be seen in Figure 7.24. The command sequence is:

Command: **A** (EnterKey)

ARC Specify start point of arc or [CEnter]: **C** (EnterKey)

Specify center point of arc: *(Select center point or enter coordinates.)*

Specify start point of arc: *(Select the start point or enter coordinates.)*

Specify end point of arc or [Angle/chord Length]: *(Select the desired endpoint or enter coordinates.)*

Command:

When a length is specified or a point is selected, the arc will be drawn and the command line will return.

CENTER, START, ANGLE ARCS

The Center, Start, Angle option is similar to the Start, Center, Angle option. The effects of each command can be seen in Figure 7.25. The command sequence for this arc is:

Command: **A** (EnterKey)

ARC Specify start point of arc or [CEnter]: **C** (EnterKey)

Specify center point of arc: *(Select center point or enter coordinates.)*

Specify start point of arc: *(Select the start point or enter coordinates.)*

Specify end point of arc or [Angle/chord Length]: **A** (EnterKey)

Specify included angle: *(Select the desired angle.)*

Command:

When a length is specified or a point is selected, the arc will be drawn and the command line will return.

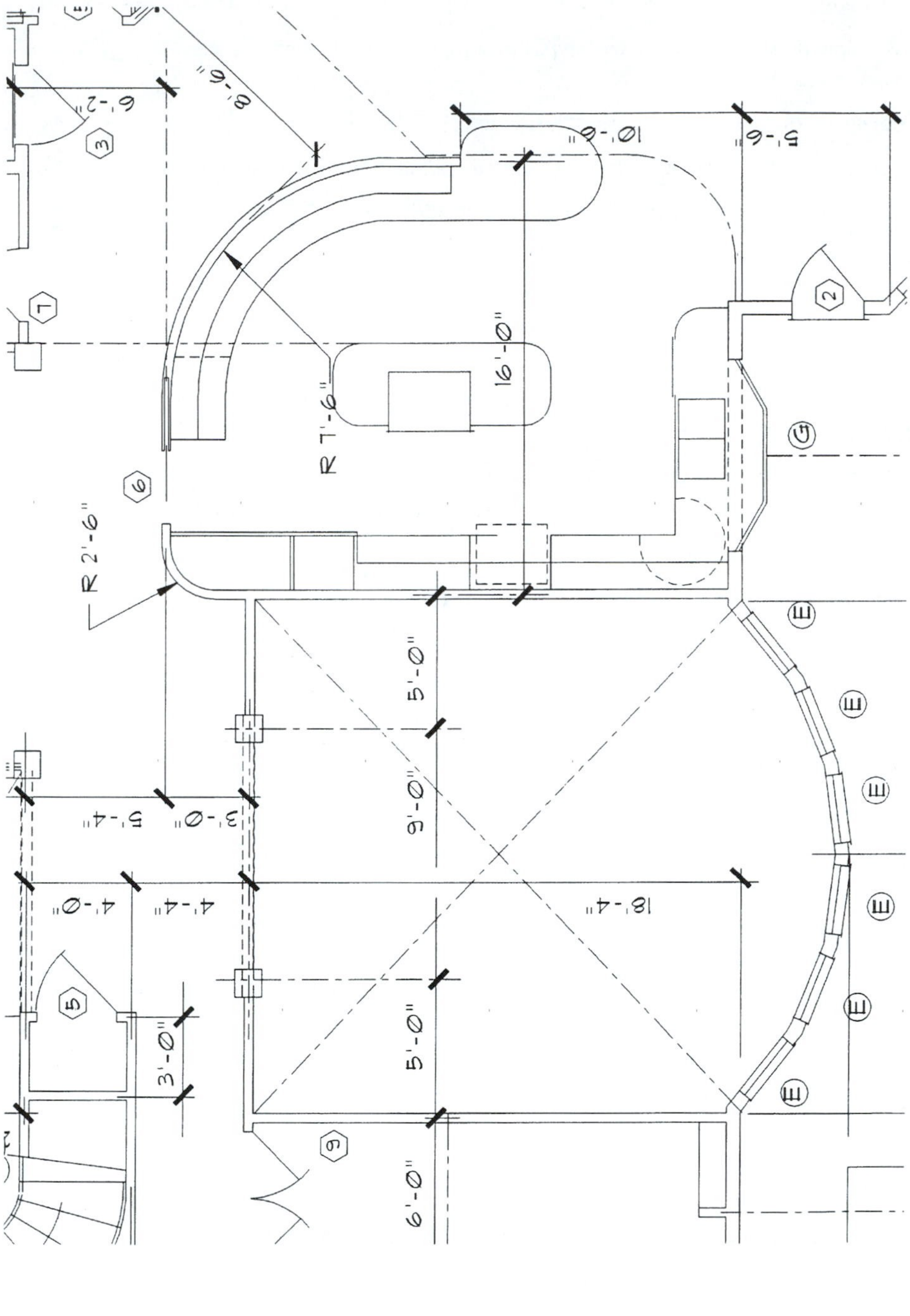

Figure 7.23 *Walls drawn using the Start, End, Radius arc option. (Courtesy Residential Designs.)*

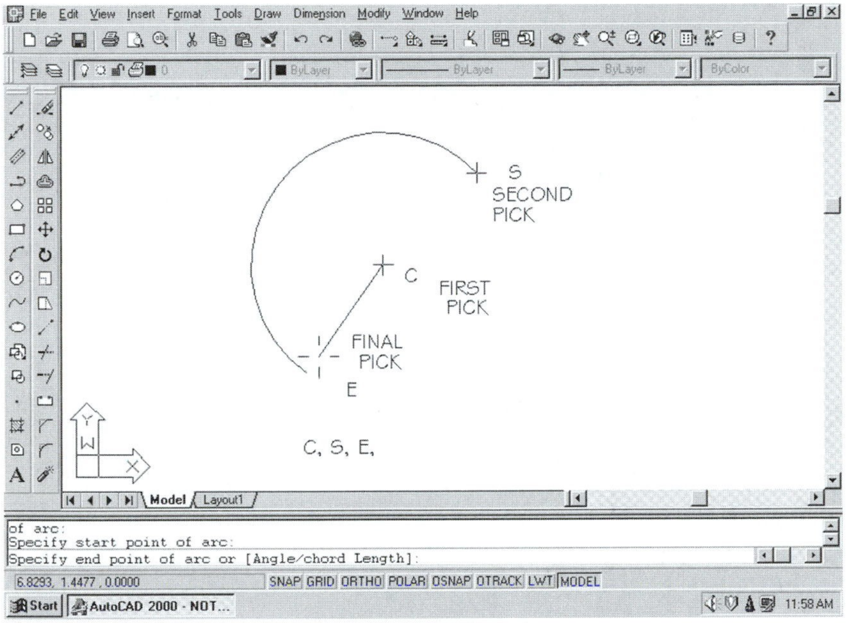

Figure 7.24 *The Center, Start, End arc option is similar to the Start, Center, End option.*

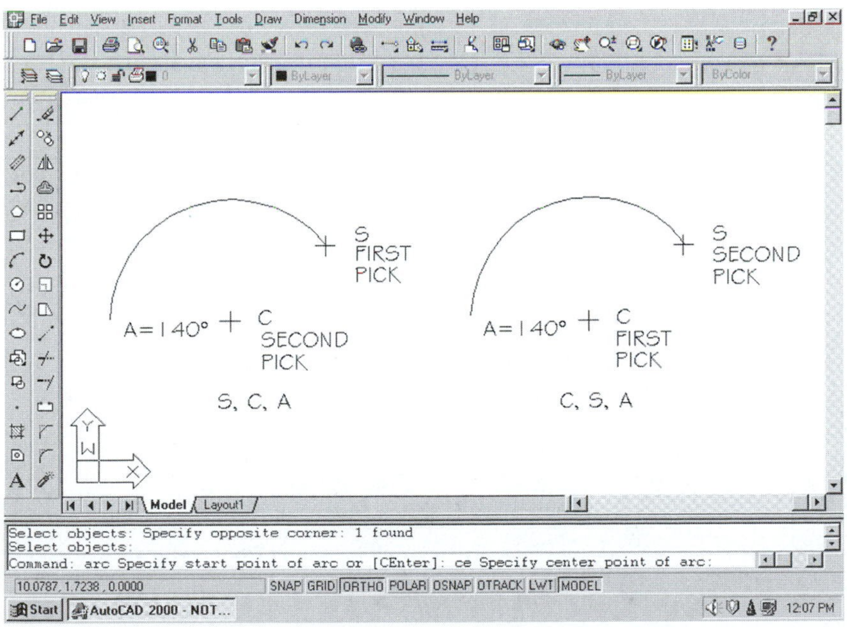

Figure 7.25 *The Center, Start, Angle ARC option produces similar results to the Start, Center, Angle option.*

CENTER, START, LENGTH ARCS

The Center, Start, Length of chord option is similar to the Start, Center, Length option. The effects of each command can be seen in Figure 7.26. The command sequence is:

Command: **A** (EnterKey)

ARC Specify start point of arc or [CEnter]: **C** (EnterKey)

Specify center point of arc: *(Select center point or enter coordinates.)*

Specify start point of arc: *(Select the start point or enter coordinates.)*

Specify end point of arc or [Angle/chord Length]: **L** (EnterKey)

Specify length of chord: *(Select desired length.)*

Command:

When a length is specified or a point is selected, the arc will be drawn and the command line will return.

CONTINUE

An arc may be continued from the previous arc. The ARC command may be continued by pressing the (EnterKey) or the space bar, or by selecting Continue from a menu in lieu of selecting the first arc entry point. By continuing from a previous arc, the current arc's start point and direction are taken from the endpoint and ending

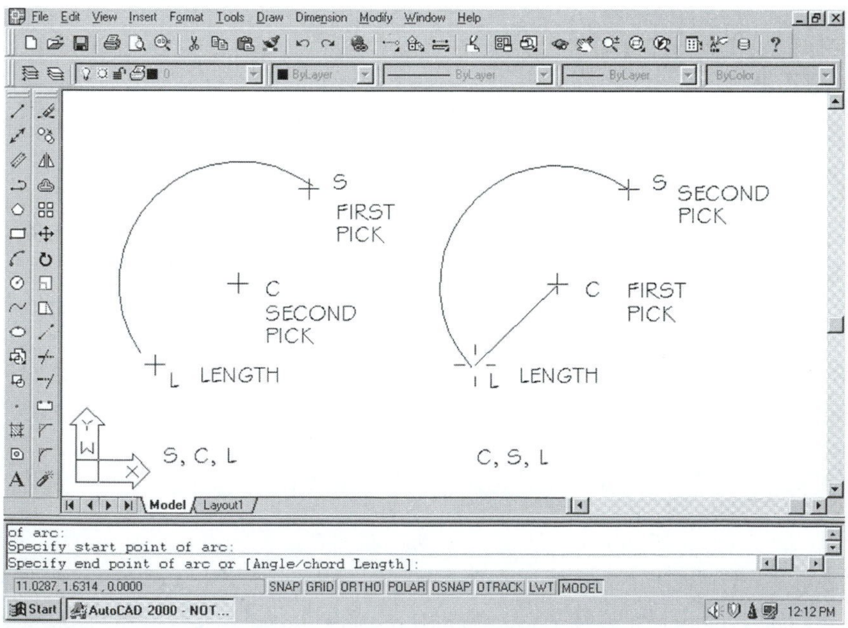

Figure 7.26 *The Center, Start, Length ARC option produces similar results to the Start, Center, Length option.*

direction of the last arc. Figure 7.27 shows an example of four continuous arcs. The command sequence is:

ARC 1

Command: **A** (EnterKey)

ARC Specify start point of arc or [CEnter/ENd]: *(Select any point or enter coordinate.)*

Specify second point of arc or [CEnter/ENd]: *(Select second point.)*

Specify end point of arc: *(Select the arc's endpoint or enter coordinates.)*

Command:

ARC 2

Command: (EnterKey) *(To resume ARC command.)*

ARC Specify start point of arc or [CEnter/End]: (EnterKey) *(To resume arc from last endpoint.)*

Specify end point of arc: *(Select the arc's endpoint or enter coordinates.)*

Command:

ARC 3

Command: (EnterKey) *(To resume ARC command.)*

ARC Specify start point of arc or [CEnter/End]: (EnterKey) *(To resume arc from last endpoint.)*

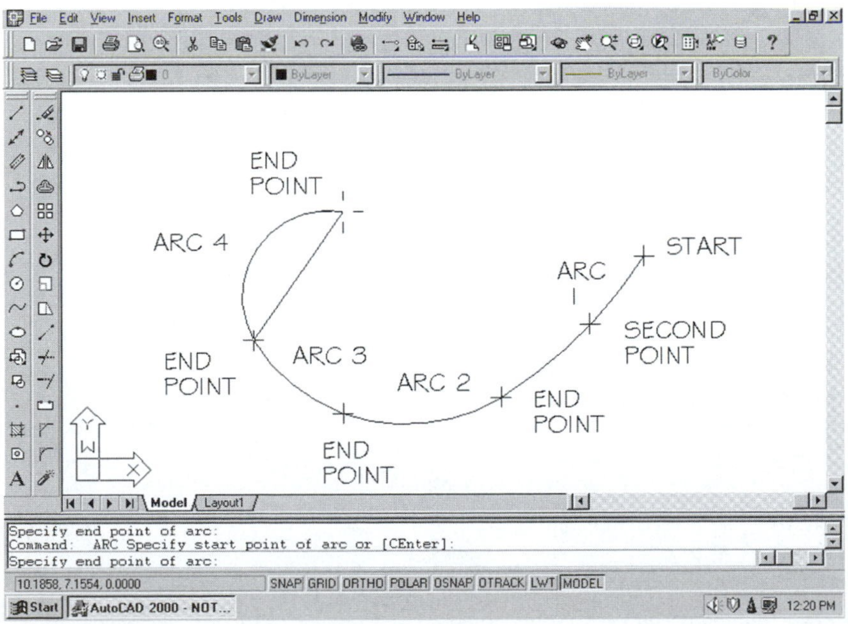

Figure 7.27 *The Continue option of ARC allows for multiple arcs to be drawn.*

Specify end point of arc: *(Select the arc's endpoint or enter coordinates.)*

Command:

ARC 4

Command: (EnterKey) (*To resume ARC command.*)

ARC Specify start point of arc or [CEnter/End]: (EnterKey) *(To resume arc from last endpoint.)*

Specify end point of arc: *(Select the arc's endpoint or enter coordinates.)*

Command:

The Continue option also can be used to extend an arc tangent to the endpoint of a line. Figure 7.28 shows an example of continuous arcs drawn from a straight line. The command sequence is:

LINE 1

Command: **L** (EnterKey)

From point: *(Select starting point or enter coordinates.)*

Specify next point or [Undo]: *(Select point or coordinates.)*

Specify next point or [Undo]: (EnterKey)

Command:

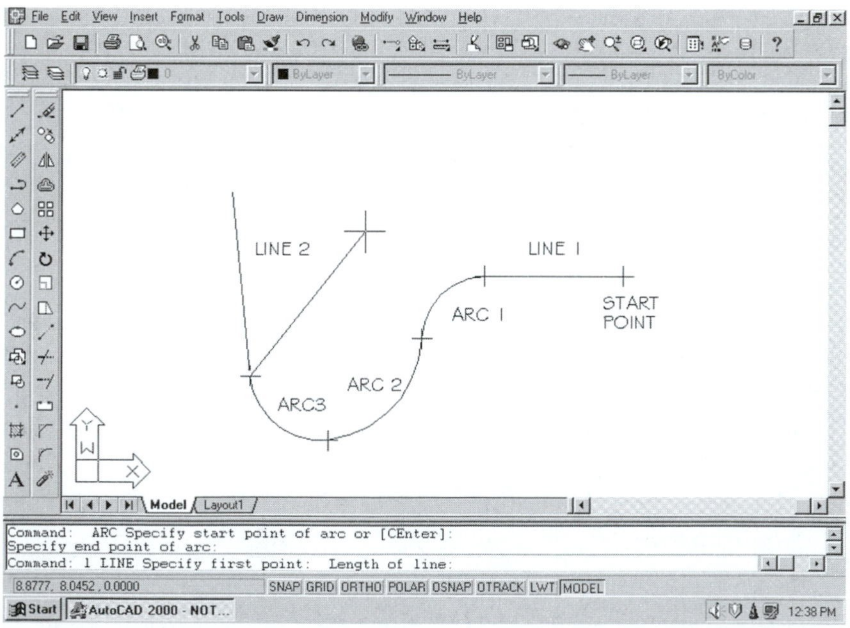

Figure 7.28 *Continue: can be used to extend an arc tangent to the last line segment.*

ARC 1

Command: **A** (EnterKey)

ARC Specify start point of arc or [CEnter]: (EnterKey) *(This will start the arc command at the line endpoint.)*

Specify endpoint of arc: *(Select the arc's endpoint or enter coordinates.)*

Command:

ARC 2

Command: (EnterKey) *(This will resume the ARC command)*

ARC Specify start point of arc or [CEnter]: (EnterKey) *(This will start the arc at the arc endpoint.)*

(EnterKey) *(This will start at the end of the last line drawn.)*

Specify endpoint of arc: *(Select the arc's endpoint or enter coordinates.)*

Command:

ARC 3

Command: (EnterKey) *(This will resume the ARC command)*

ARC Specify start point of arc or [CEnter]: (EnterKey) *(This will continue the arc from the endpoint of the last arc.)*

Specify endpoint of arc: *(Select the arc's endpoint or enter coordinates.)*

Command:

LINE 2

Command: **L** (EnterKey)

LINE Specify first point: (EnterKey)

Specify next point or [Undo]: *(Select point or coordinates.)*

Specify next point or [Undo]: (EnterKey) *(To end drawing sequence.)*

Command:

Notice in Figure 7.28 that line 2 was drawn tangent to arc 3 because Continue: was used. To place a line that continues from an arc but is not tangent to the arc will require the arc endpoint to be selected as the line "next point", using the methods described in Chapter 8.

DRAWING DONUTS

Sorry if you've started thinking about something covered with chocolate or coconut. A donut in AutoCAD is a filled circle or ring. The command can be accessed by typing **DO** (EnterKey) at the command prompt, or by selecting Donut from the Draw pull-down menu. Each can be seen in Figure 7.29. Donuts are great when drawing the end view of reinforcing steel in detail. The command sequence to create a donut is:

> Command: **DO** (EnterKey)
>
> DONUT
>
> Specify inside diameter of donut <0'-0½">**0** (EnterKey)
>
> Specify outside diameter of donut <0'-1">**.75** (EnterKey)
>
> Specify center of donut or <exit>: *(Select the desired location for the donut.)*
>
> Specify center of donut or <exit>: *(The command will continue indefinitely.)* (EnterKey)
>
> Command:

Three different options are available for drawing donuts: a ring, a filled circle that is solid black, or a circle filled with lines. Each can be seen in Figure 7.30. The fill material is determined by the status of the FILL command. Typing **FILL** (EnterKey) at the command prompt will allow for a choice to be made for DONUT and several other options. By entering OFF, filled objects would resemble the bottom circle in Figure 7.30. With FILL ON, filled objects will resemble the middle circle in Figure 7.30.

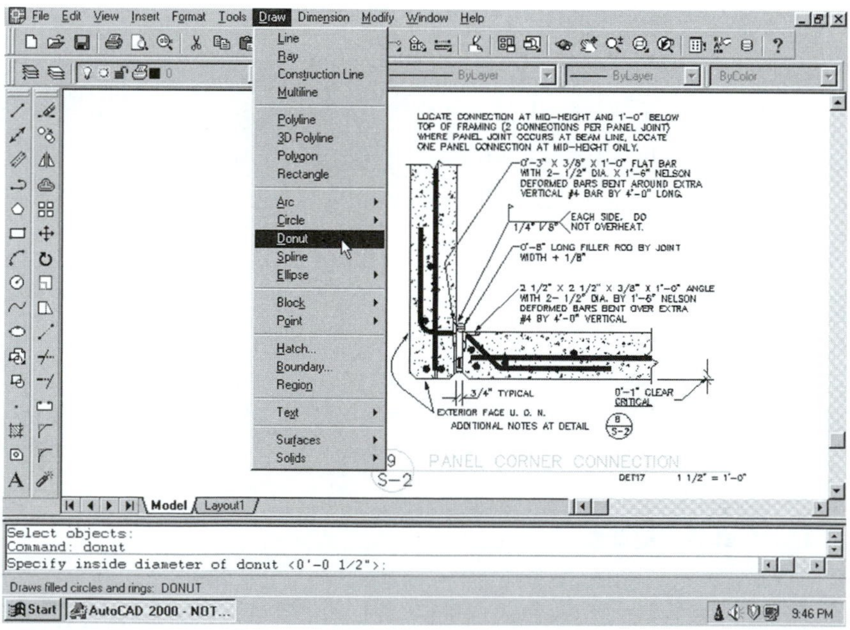

Figure 7.29 *Donuts may be drawn using the pull-down menu or from the command prompt.*

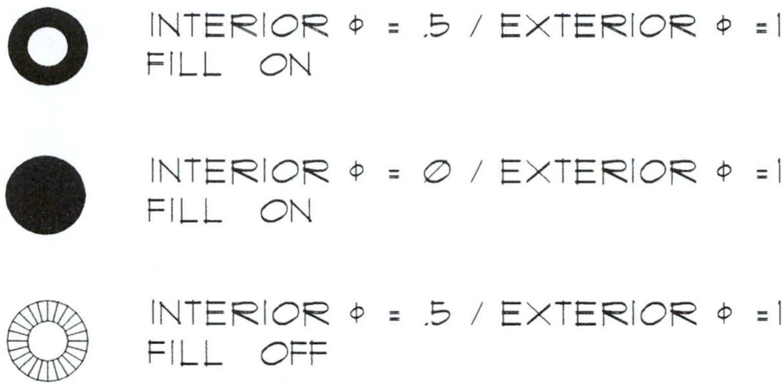

Figure 7.30 *Three options are available for the display of a donut.*

Note: FILL will affect several of the commands you will soon be using. In addition to DONUT, FILL will affect the appearance of POLYLINES, arrows created with LEADER, and the display of HATCH patterns. In older slower versions of AutoCAD, many users set FILL to OFF to a gain speed when working with large drawings. Unless you're trying to reproduce hollow objects, leave FILL in the ON setting.

Another thing to remember about FILL, is that the effect of changing the setting is usually not be seen immediately. To see the affect of FILL, enter the command, alter the setting, and press the (EnterKey). No change has occurred. Type **REGEN** (EnterKey) at the command prompt. Now the effect of FILL can be seen.

USING THE ELLIPSE COMMAND

Although not a major part of engineering drawing, ellipses sometimes are used for construction symbols. Ellipses are also used to show round objects that are not being seen in true orthographic projection, such as a round pipe that is not perpendicular to the viewing plane of the drawing. To enter the ELLIPSE command, select the Ellipse icon from the Draw toolbar, select Ellipse from the Draw pull-down menu, or type EL (EnterKey) at the command prompt. Menu options for ELLIPSE can be seen in Figure 7.31. Figure 7.32 shows the major and minor axis that will be referred to in laying out an ellipse. To draw an ellipse at the command prompt, type:

Command: **EL** (EnterKey)

ELLIPSE

Specify axis endpoint of ellipse or [Arc/Center]: *(Select a point or enter coordinates for the major axis endpoint.)*

Specify other endpoint of axis: *(Select a point or enter coordinates for the opposite major axis endpoint.)*

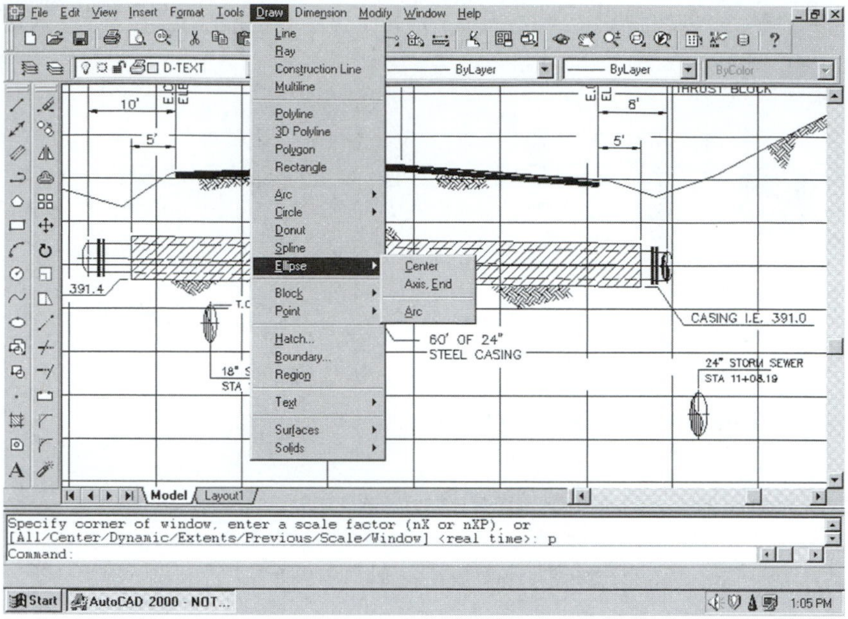

Figure 7.31 *The pull-down menu options for ELLIPSE. (Courtesy Lee Engineering, Inc., Oregon City, OR.)*

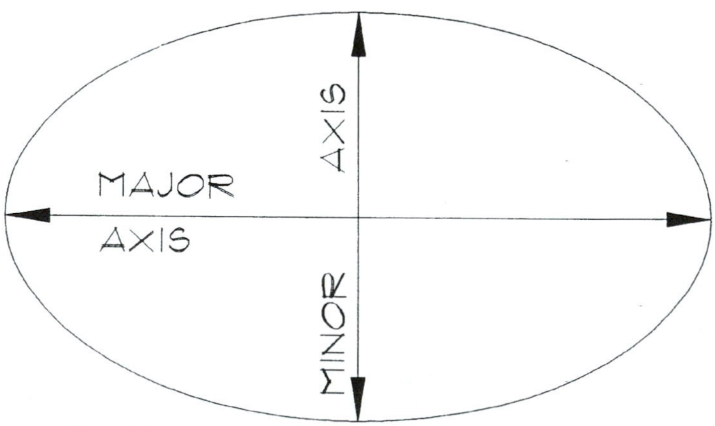

Figure 7.32 *An ellipse is defined by the major and minor axis.*

Specify distance to other axis or [Rotation]: *(Select a point or enter coordinates for the minor axis endpoint.)*

Command:

An ellipse is now shown on the screen between the two selected points. The process can be seen in Figure 7.33. AutoCAD is waiting for you to enter the axis or rotation. If the angle is not known, the crosshairs can be moved until the ellipse represents the desired shape.

If you know the ellipse represents a pipe that is at a 60° angle to the viewing plane, enter R (EnterKey). This will produce the prompt:

Command: **EL** (EnterKey)

ELLIPSE

Specify axis endpoint of ellipse or [Arc/Center]: *(Select a point or enter coordinates for the major axis endpoint.)*

Specify other endpoint of axis: *(Select a point or enter coordinates for the opposite major axis endpoint.)*

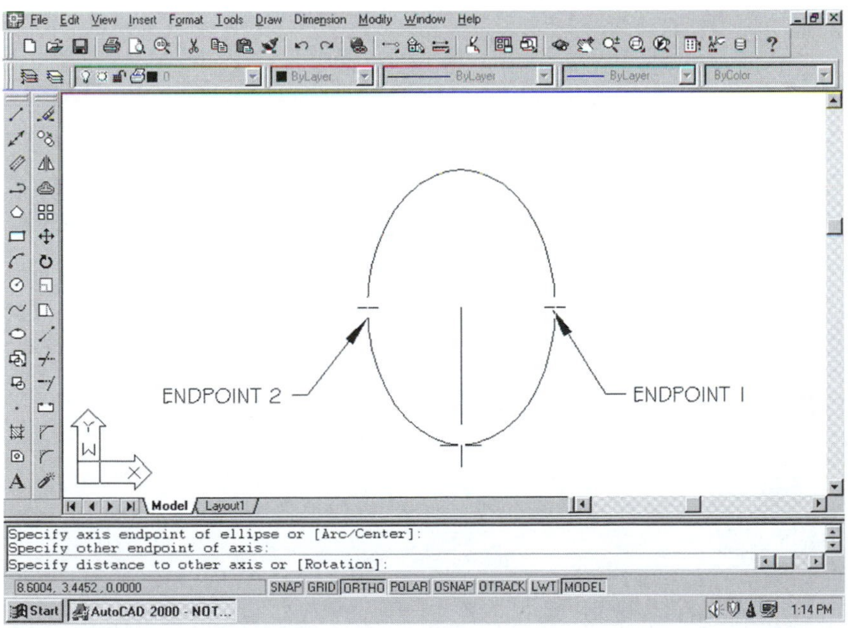

Figure 7.33 *Once the two endpoints are specified, the ellipse will rubberband as the cursor is moved.*

Specify distance to other axis or [Rotation]: **R** (EnterKey)

Specify rotation around major axis: *(Enter desired angle of rotation.)* **60** (Enter-Key)

Command:

The command sequence will produce the ellipse shown in Figure 7.34. Figure 7.35 shows the effect of rotating a circular pipe at 10° increments. An ellipse with a 90° rotation cannot be drawn using the ELLIPSE command because it would produce a single straight line.

An ellipse also can be drawn by locating the center point and then specifying an end-point for each axis. This process can be seen in Figure 7.36. The command process is:

Command: **EL** (EnterKey)

ELLIPSE

Specify axis endpoint of ellipse or [Arc/Center]: **C** (EnterKey)

Specify center of ellipse: *(Select a point or enter coordinates for the major axis center point.)*

Specify endpoint of axis: *(Select a point or enter coordinates for the endpoint of major axis.)*

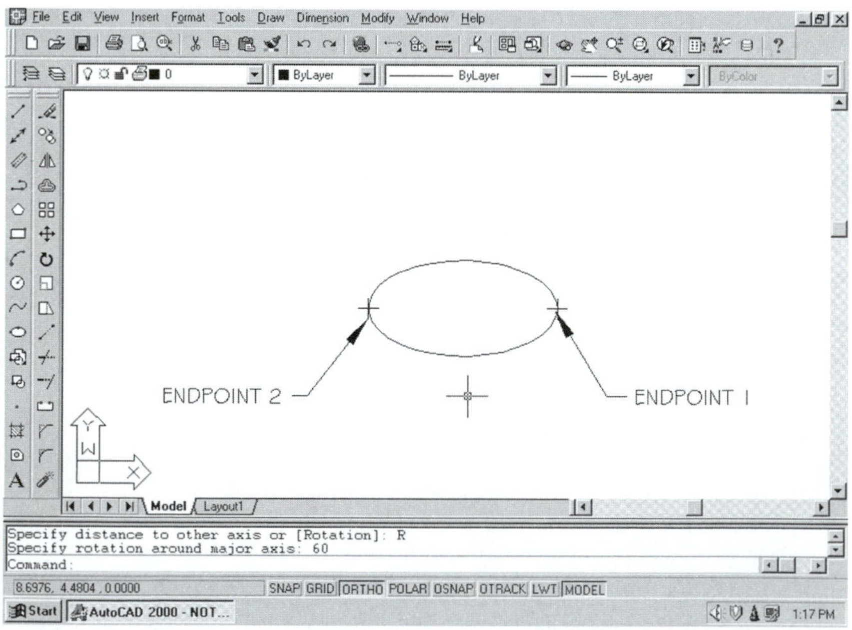

Figure 7.34 *Entering the angle of rotation of the viewing plan will produce an ellipse. Using a 60° rotation produces the ellipse shown.*

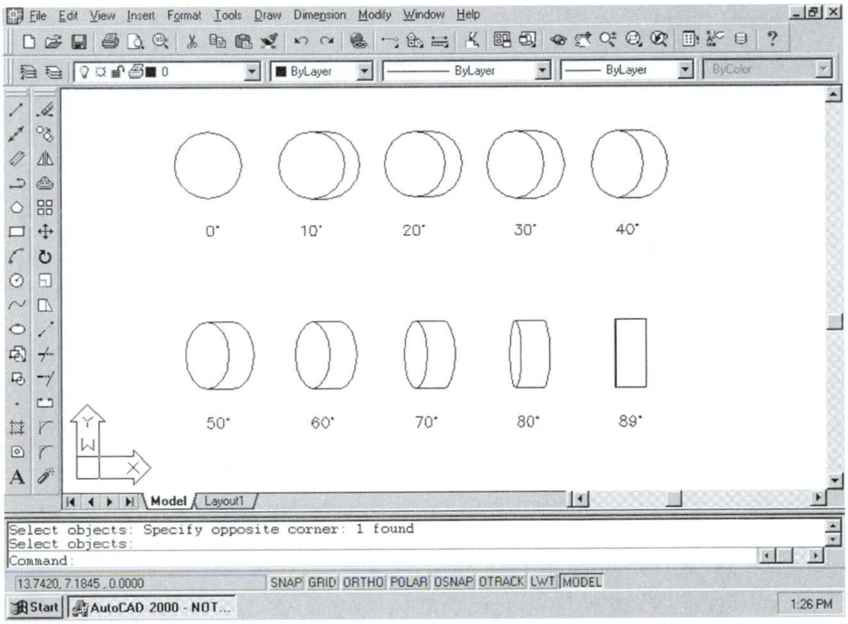

Figure 7.35 *The effects of rotating a circular object in 10° increments.*

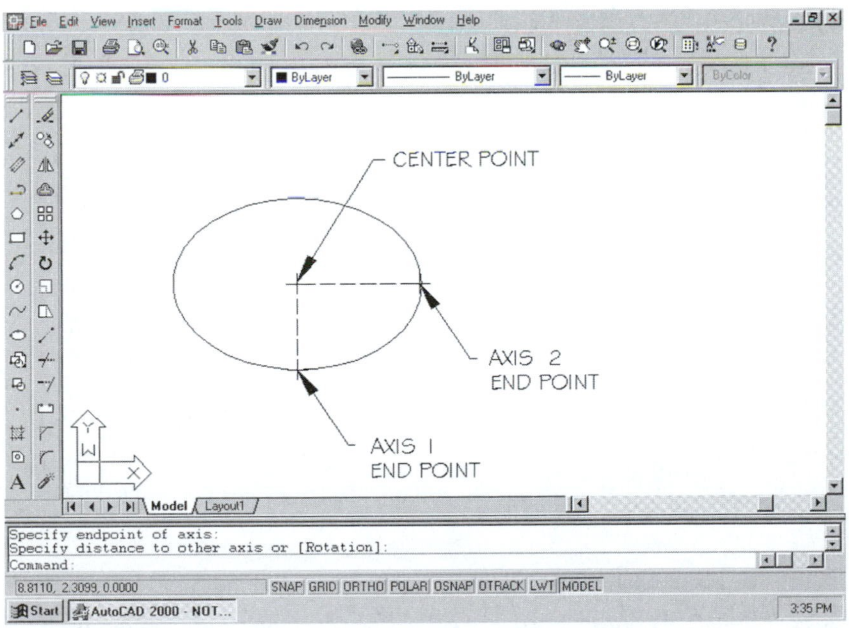

Figure 7.36 *An ellipse can be drawn by locating the center point followed by an endpoint for the major and minor axis.*

Specify distance to other axis or [Rotation]: *(Select a point or enter coordinates for the minor axis endpoint.)*

Command:

When the second axis endpoint is entered the command prompt is returned and the ellipse is drawn on the screen.

DRAWING POLYGONS

A polygon is any geometric shape lying on one plane, bound by three or more straight lines. Triangles, squares, pentagons, hexagons, and octagons are used as symbols and material shapes in construction drawings. AutoCAD draws polygons with equal length lines and angles having between three and 1,024 sides. To draw a polygon, you will be asked to choose the number of sides and whether the object is to be inscribed in or circumscribed around a circle. When the polygon is inscribed in a circle, the circle diameter provides the distance across the polygon points. When the polygon is circumscribed around a circle, the circle diameter provides the distance across the flat surfaces of the polygon. The effects of a triangle inscribed in and circumscribed around a 2.5"-diameter circle can be seen in Figure 7.37.

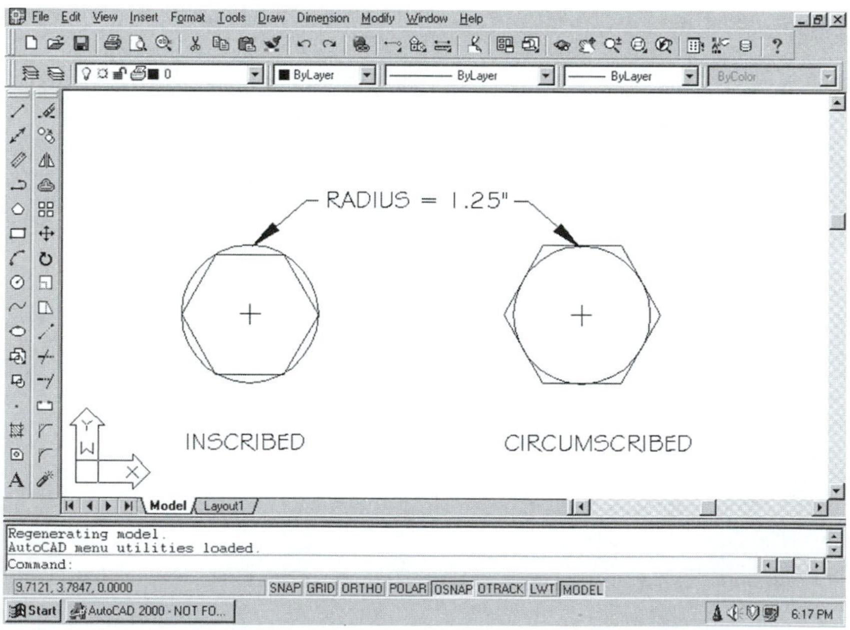

Figure 7.37 *Polygons are drawn based on a circle. Once the number of sides has been specified, you will be asked to specify whether the polygon will be placed inside or outside the circle.*

The polygon command can be accessed by selecting the Polygon icon from the Draw toolbar, by selecting Polygon from the Draw pull-down menu, or by typing **POL** (EnterKey) at the command prompt. Figure 7.38 shows the pull-down menu. Typing **POL** (EnterKey) at the command prompt will produce the following prompts:

Command: **POL** (EnterKey)

POLYGON

Enter number of sides <4>: (*Select the desired number of sides.*) **8** (EnterKey)

Specify center of polygon or [Edge]: (*Select a point or enter coordinates for the center point.*)

Enter an option [Inscribed in circle/Circumscribed about circle] <I>: *(Accept I or select C depending on your need.)* (EnterKey)

Specify radius of circle: (*Enter desired radius.*) **2.75** (EnterKey)

Command:

The results of the sequence can be seen in Figure 7.39. The setting of Ortho will affect how the polygon will be placed. When Ortho is ON, the edges of the polygon will be parallel or perpendicular to the drawing screen. When Ortho is OFF, the

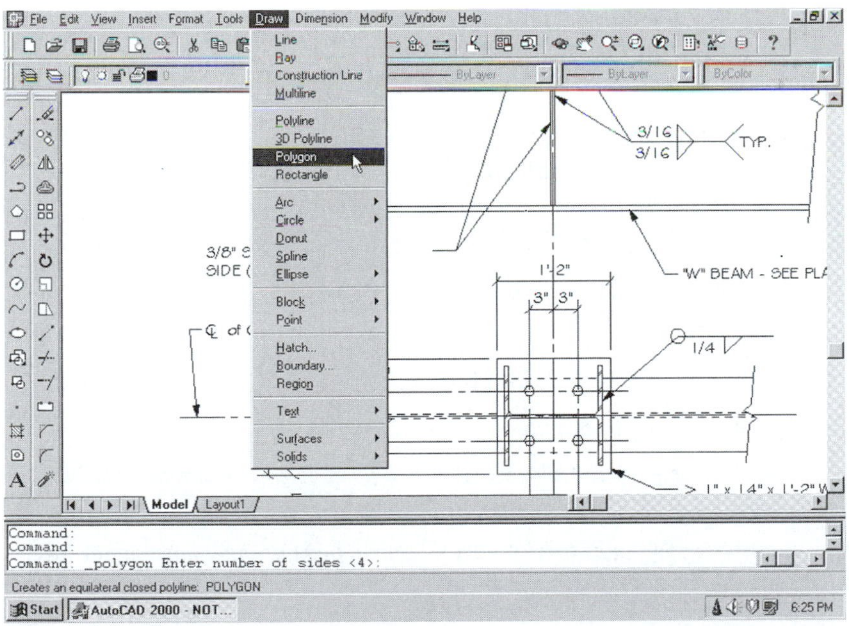

Figure 7.38 *The Pull-down menu for POLYGON. The command is useful for drawing bolt heads. (Courtesy Lee Engineering, Inc., Oregon City, OR.)*

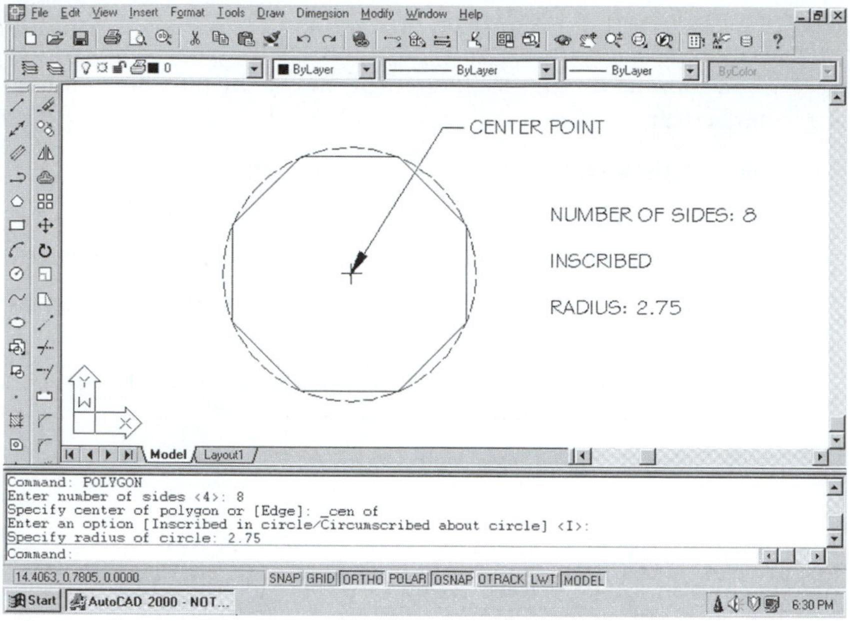

Figure 7.39 *The results of drawing an inscribed octagon with a 2.75" radius.*

edges of the polygon will rotate based on the movement of the mouse. To activate Ortho, single-click the Ortho button in the Task bar or press (F8Key).

A polygon also can be drawn based on the location of one edge, rather than the location of the center point. Once the number of sides has been provided, you will be asked for an edge or center point. By entering **E** (EnterKey) at the prompt, a polygon similar to the one in Figure 7.40 can be drawn. The command sequence is:

> Command: **POL** (EnterKey)
>
> POLYGON
>
> Enter number of sides <8>: *(Select the desired number of sides.)* **6** (EnterKey)
>
> Specify center of polygon or [Edge]: **E** (EnterKey)
>
> Specify first endpoint of edge: *(Select a point or enter coordinates for the first endpoint.)*
>
> Specify second endpoint of edge: *(Select a point or enter coordinates for the second endpoint.)*
>
> Command:

As the first point is entered, a polygon will be displayed that will vary in size as the mouse is moved to the second point. As the second point is entered, the command

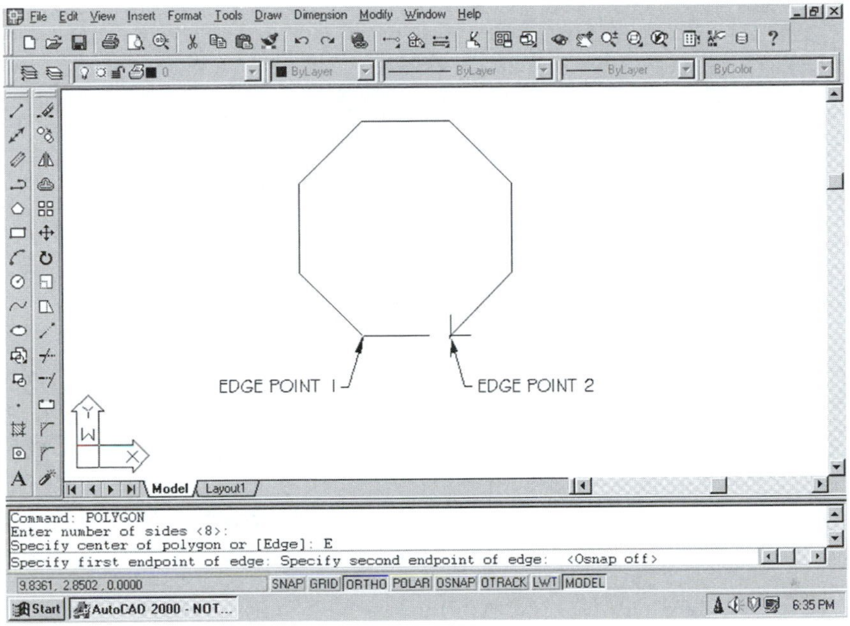

Figure 7.40 *A polygon also can be drawn by specifying the points that define one edge.*

prompt is returned and the polygon is drawn. The setting of ORTHO will affect the placement of a polygon using the Edge option.

DRAWING RECTANGLES

A square can be drawn using the POLYGON command. A rectangle is drawn by selecting the Rectangle icon from the Draw toolbar, by selecting Rectangle from the Draw pull-down menu, or by entering **REC** (EnterKey) at the command prompt. The command sequence is:

Command: **REC** (EnterKey)

RECTANGLE

Specify first corner point [Chamfer/Elevation/Fillet/Thickness/Width]: *(Select a corner of the rectangle with the mouse.)*

Specify other corner point: *(Select the opposite corner of the rectangle.)*

Command:

As the cursor is moved between the first and second corner, a rectangle will be displayed on the screen allowing the exact size to be determined. The sequence can be seen in Figure 7.41. Selecting either the Chamfer or Fillet option allows the corners of the rectangle to be altered. Selecting the Width option allows the width of the line

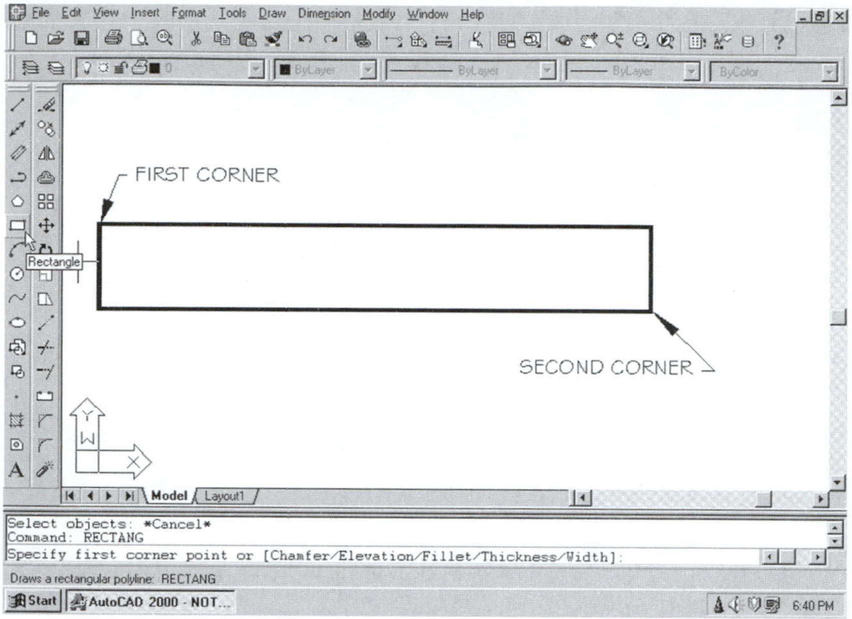

Figure 7.41 *Rectangular shapes can be drawn as one entity using the Rectangle option of the Draw toolbar, the Draw pull-down menu, or by typing* **REC** *(EnterKey) at the command line.*

to be altered. The other options only affect a rectangle drawn in 3D and will not be discussed in this chapter.

CHAMFER

A chamfer is an angled or mitered corner. Drawing a rectangle with the Chamfer option active will place a chamfer on each corner of the rectangle. Once the Chamfer option is selected, prompts will be given for the first and second chamfer distances. If the distances are equal, the result will be as shown in Figure 7.42. Providing different values will produce a rectangle similar to the example on the right in Figure 7.42. The command sequence to draw a rectangle with a chamfer is:

Command: **REC** (EnterKey)

RECTANGLE

Specify first corner point or [Chamfer/Elevation/Fillet/Thickness/Width]: **C** (EnterKey)

Select first chamfer distance for rectangles <0'-0">: *(Enter the desired size.)* **.5** (EnterKey)

Second chamfer distance for rectangles <0'-0 ½ ">: (EnterKey)

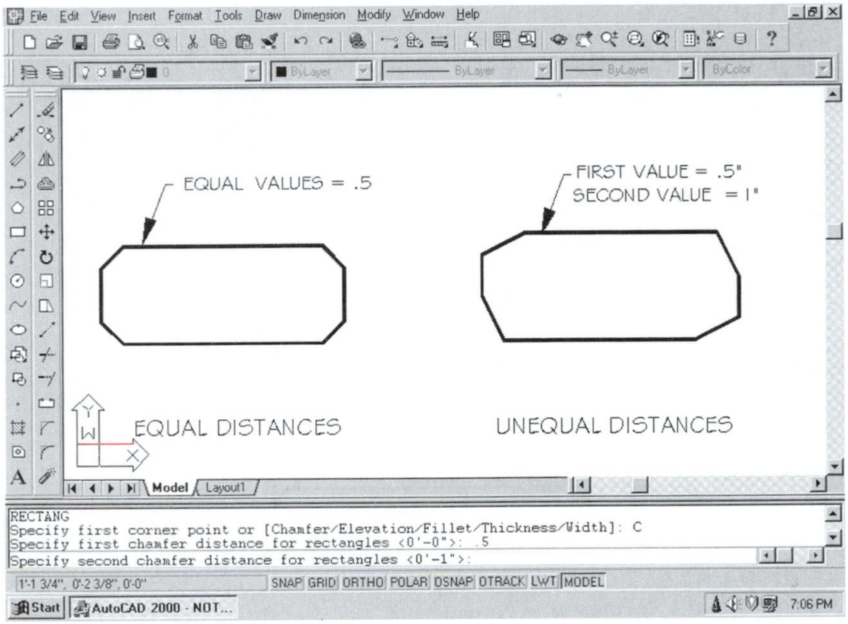

Figure 7.42 *The Chamfer option of the RECTANGLE command can be used to provide a chamfer at each corner of a rectangle. The left option shows distance 1 and 2 with a value of .5". The right option shows distance 1 with a value of .5" and distance 2 with a value of 1".*

Specify first corner point or [Chamfer/Elevation/Fillet/Thickness/Width]: *(Select a corner of the rectangle.)*

Specify other corner point: *(Select the opposite corner of the rectangle.)*

Command:

FILLET

A fillet is a rounded corner. Drawing a rectangle with the Fillet option active will place a fillet on each corner of the rectangle. Once the Fillet option is selected, a prompt will be given to provide a radius for the fillet. Figure 7.43 shows an example of a rectangle with filleted corners. The command sequence to draw a rectangle with a fillet is:

Command: **REC** (EnterKey)

RECTANGLE

Specify first corner point or [Chamfer/Elevation/Fillet/Thickness/Width]: **F** (EnterKey)

Select fillet radius for rectangles <0'-0 ½">: *(Enter the desired size.)* (EnterKey)

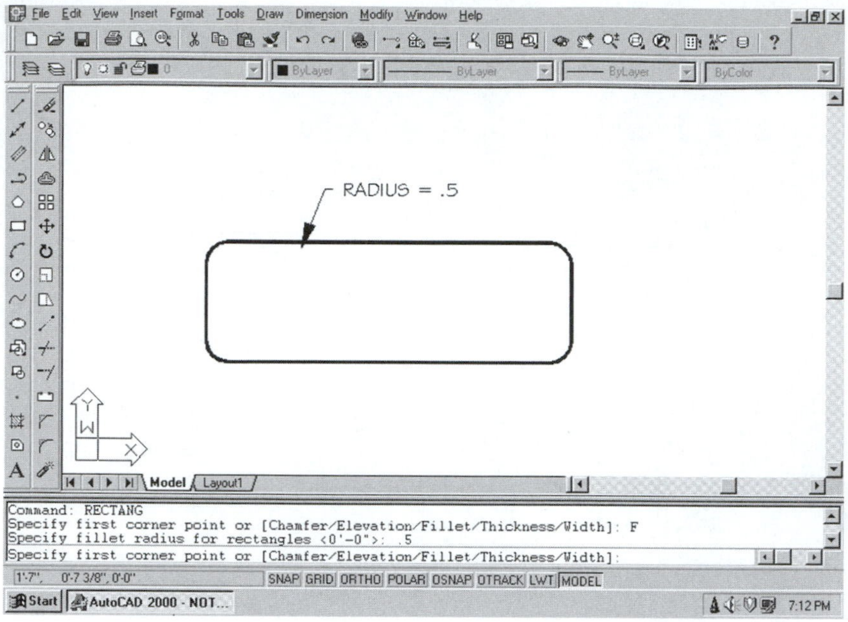

Figure 7.43 *The Fillet option of the Rectangle command can be used to provide a fillet at each corner of a rectangle. The value used for the radius will remain constant for future use until it is altered.*

Specify first corner point or [Chamfer/Elevation/Fillet/Thickness/Width]: *(Specify a corner of the rectangle.)*

Specify other corner point: *(Select the opposite corner of the rectangle.)*

Command:

Notice that the default radius value is <0'-0½">. The value that was provided for the chamfer will become the default value for all future chamfers and fillets until a new value is provided.

WIDTH

Selecting the Width option allows the line thickness to be altered. Because the fillet option was just used, setting the thickness for a rectangle will result in a rectangle with ½"-radius corners to be drawn with a thick line. To draw a rectangle with square corners, enter the RECTANGLE command and alter the value for Fillet to 0, and then set the thickness, or vise versa. The command sequence is:

Command: **REC** (EnterKey)

RECTANGLE

Specify first corner point or [Chamfer/Elevation/Fillet/Thickness/Width]: **W** (EnterKey)

Specify line width for rectangles <0'-0">: **.25** (EnterKey)

Specify first corner point or [Chamfer/Elevation/Fillet/Thickness/Width]: **F** (EnterKey)

Select fillet radius for rectangles <0'-0 ½">: **0** *(Enter the desired size)* (EnterKey)

Specify first corner point or [Chamfer/Elevation/Fillet/Thickness/Width]: *(Specify a corner of the rectangle.)*

Specify other corner point: *(Select the opposite corner of the rectangle.)*

Command:

The results of this command sequence can be seen in Figure 7.44. The width will remain the current width for all future rectangles until the value is altered. Methods of providing thickness to other entities will be discussed as Polylines are introduced.

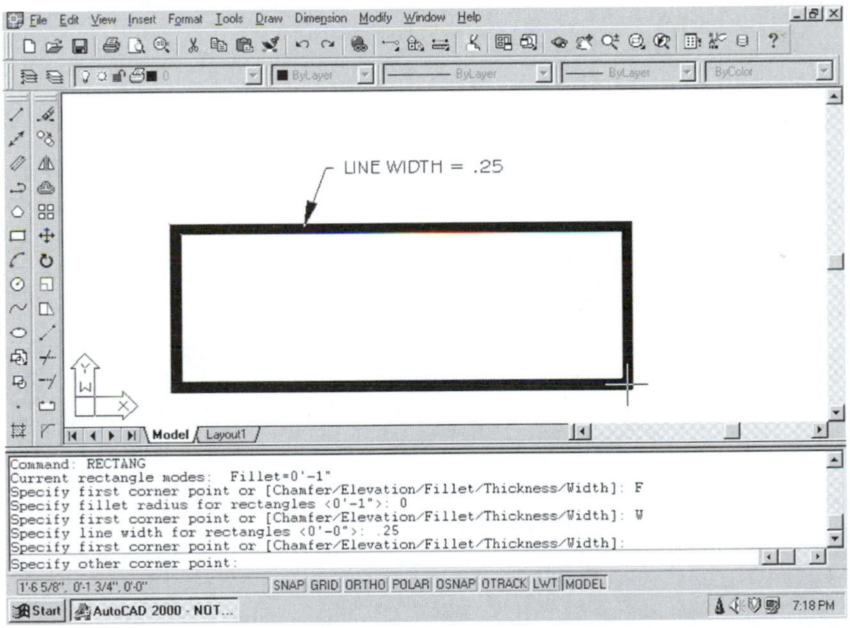

Figure 7.44 *The Width option can be used to assign a line width to the RECTANGLE command.*

CHAPTER 7 EXERCISES

Save the following exercises using the indicated filenames.

1. Draw a 3"-diameter circle, a circle with a 1" radius, and a 1"-diameter circle that is tangent to two 1"-long perpendicular lines. Make the smallest circle green with continuous lines. Make the largest circle red with hidden lines. Give one of the lines a width of .065". Save the drawing as E-7-1.
2. Draw two parallel lines that are 1" long and 1" apart, and provide an arc at each end of the lines to form a slotted hole. Draw these objects on a layer named SLOTS, using dashed blue lines, with a thin lineweight. Save the drawing as E-7-2.
3. Draw a 4"-long line. Use the continuous option to place a 1"-diameter circle at each end of the line. Save the drawing as E-7-3.
4. Draw a box formed by three 1"-long lines. Draw an arc to close the polygon. Save as drawing E-7-4.
5. Draw an arc with a start point that is 2" to the right of the endpoint with an included angle of 232°. Save as drawing E-7-5.
6. Draw an arc with an endpoint that is @2,2 from the center point with an included angle of 45°. Save as drawing E-7-6.
7. Draw an ellipse with a 3"-long axis at a 42° rotation. Save as drawing E-7-7.
8. Draw an ellipse with a center point that is 2.577 inches from the end of the major axis and 1" from the minor axis. Draw a four-sided polygon using the center of the ellipse as the center of the polygon. Make two edges of the square touch the ellipse. Save as drawing E-7-8
9. Draw the following polygons:

3 sides	inscribed	2" diameter
3 sides	circumscribed	2" diameter
4 sides	inscribed	.5" radius
4 sides	circumscribed	.5" radius
6 sides	inscribed	.75" radius
6 sides	circumscribed	.75" radius
8 sides	inscribed	1.25" radius
8 sides	circumscribed	1.25" radius

Save the drawing as E-7-9.

10. Draw a donut with an inside diameter of .25", outside diameter of .75", and FILL set at ON. Draw another donut with an inside diameter of .0" and an outside diameter of .375" with FILL set at OFF. Draw a donut with an inside diameter of 0 and an outside diameter of .5" with FILL set to ON. Save the drawing as E-7-10.

11. Draw a rectangle with filleted corners using a 2"radius. Save the drawing as E-7-11.
12. Draw a rectangle with chamfered corners using a 1" distance for both settings. Save the drawing as E-7-12.
13. Draw a rectangle with chamfered corners using a 1.5" distance for the first distance and a 2.5 value for the second setting. Assign a line width of .065 and save the drawing as E-7-13.
14. Open a new drawing and draw at least three connected combinations of lines and arcs. Save the drawing as E-7-14.
15. Draw two perpendicular lines that are two inches long and form a corner. Draw a 3"-diameter circle that is tangent to each of the lines. Draw three lines that are not connected and are neither parallel nor perpendicular to each other. Draw a circle that is tangent to all three line segments. Save this drawing as E-7-15.

CHAPTER 7 QUIZ

1. What is the maximum number of sides that can be drawn for a polygon?
2. When using the ARC command, what do the following letters represent?

    St Ce D

    Len E R

    Ang
3. What is the default method for drawing arcs from the toolbar?
4. What is the command sequence to draw three continuous arcs?
5. What command determines whether the interior of a donut will be black or hatched with lines?
6. What is the difference between inscribed and circumscribed, and what are the effects of each on a polygon?
7. What four components can be used to construct an ellipse?
8. When would TTR be used for drawing a circle?
9. What are the options available for drawing circles?
10. What command controls the rubberband effect of geometric shapes?
11. What options must be entered before a donut can be drawn?
12. What is the largest angle rotation that can be used with an ellipse?
13. What shape would a polygon with 500 sides appear as?
14. How does ORTHO affect the drawing of a polygon?
15. Use the help menu and determine the use of Elevation as it relates to a rectangle.

ENHANCING DRAWINGS

CONTROLLING DRAWING ACCURACY

DRAWING DISPLAY OPTIONS

SELECTING AND MODIFYING DRAWING OBJECTS

MODIFYING AND SELECTING DRAWING OBJECTS

POLYLINES

SUPPLEMENTAL DRAWING COMMANDS

CONTROLLING LINES AND LAYERS

PLACING PATTERNS IN DRAWING OBJECTS

INQUIRY COMMANDS

CHAPTER 8

Controlling Drawing Accuracy

Up to this point, you've been creating drawing objects with little regard to drawing accuracy. This chapter will introduce three methods of increasing the accuracy of object placement. In previous chapters, you were introduced to using the tools on the Status bar to make the crosshairs move at specific intervals. As you begin to combine geometric shapes into drawings it is often helpful to join the shapes at exact locations. Joining objects by eye is difficult and should rarely be done. Objects that appear to be tangent may overlap when the ZOOM command is used. (See Figure 8.1.) Perfect intersections can be achieved by using the object snap options. This chapter will introduce methods for controlling the aperture box, using the Drafting Settings dialog box to control object placement, and the use of object snaps for object placement.

CONTROLLING THE APERTURE BOX

The normal cursor for the drawing area is a crosshairs surrounded by a box. If you enter a drawing command such as LINE, the pick box is removed from the crosshairs as you select the First and Next points. As OSNAP is activated, an aperture box is added to the crosshairs, as seen in Figure 8.2. The marker is displayed as the cursor is moved near one of the active object snap points such as an endpoint or midpoint. The shape of the marker will vary depending on the mode being used. Various object snap modes will be introduced later in this chapter. Any object that lies within the target or aperture box is subject to the draw or edit command being used. Although objects may be added or deleted from the aperture box, this is time consuming. To obtain higher accuracy in object selection, the size of the aperture box can be adjusted.

Note: The aperture box will not be displayed unless the Display AutoSnap aperture box mode is activated. Even if it is not displayed, an invisible aperture box will still function as you draw. The aperture box is displayed on the Drafting tab of the Options dialog box. To display the dialog box, select Options... from the Tools pull-down menu.

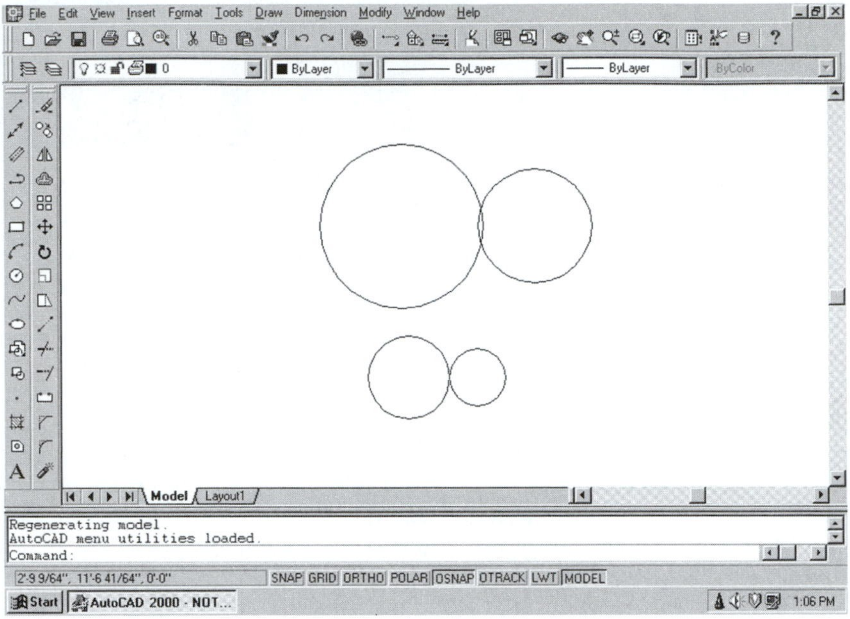

Figure 8.1 *Objects such as the lower circles that are placed by eye often appear to be perfectly placed. As the scale is changed, or the drawing is zoomed, circles that appeared tangent are crossing.*

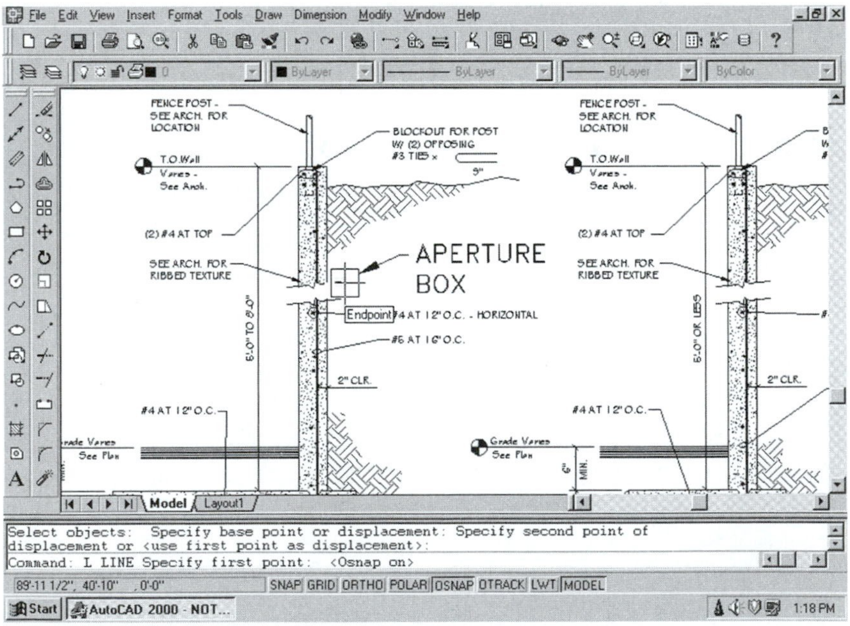

Figure 8.2 *As OSNAP is activated, a MARKER box is added to the crosshairs. The marker box is used for selecting existing drawing objects to be used as a selection point for a new line, circle, arc, or ellipse. (Courtesy Van Domelen/Looijenga/McGarrigle/Knauf Consulting Engineers.)*

SIZE

The size of the aperture box can be adjusted using the Drafting tab of the Options dialog box or by keyboard. The Options dialog box is accessed by selecting Options... of the Tools pull-down menu. This will produce the menu shown in Figure 8.3. From the drafting tab, select the slide bar in the Aperture Size edit box and hold the mouse pick button down. Dragging the bar to the left will decrease the size. Moving the bar to the right will increase the size. The size will be indicated in the Aperture Size window display.

To adjust the size of the aperture box at the command prompt type:

Command: **APERTURE** (EnterKey)

Object snap target height (1-50 pixels) <10>:

A numeric value can now be entered. The larger the number selected, the larger the box will be. Figure 8.4 shows the effect of changing the value. The value entered will now be the default for all future OSNAP commands until a new value is entered. Changing the default for the OSNAP aperture box will not affect the size of the select box used for ERASE.

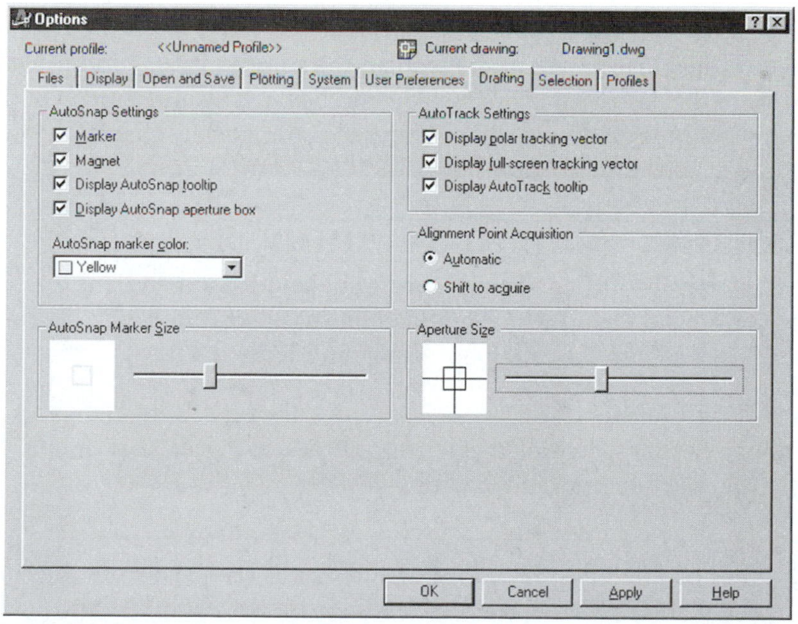

Figure 8.3 *The Options dialog box is accessed by selecting Options... of the Tools pull-down menu. The dialog box can be used to adjust the aperture box, the marker size and color, AutoSnap settings, and AutoTrack settings.*

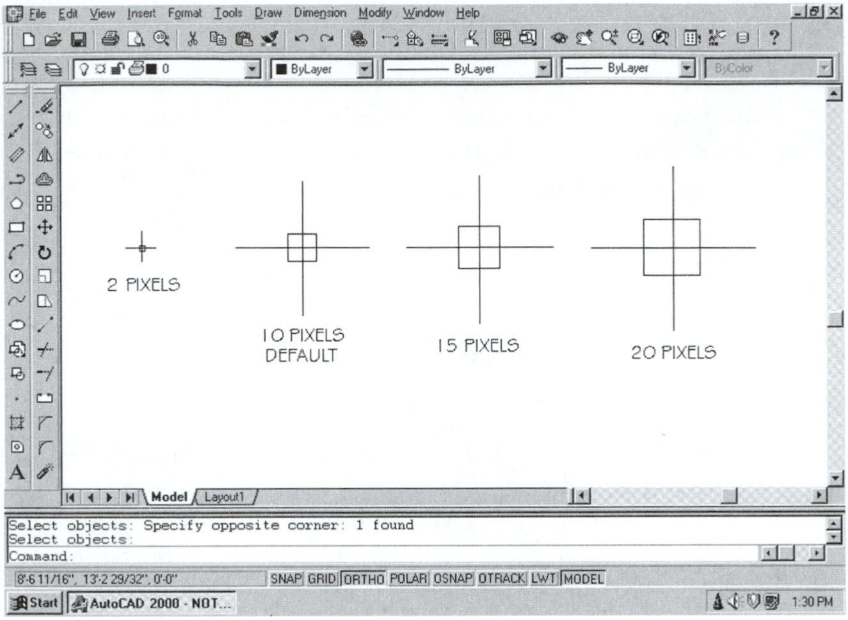

Figure 8.4 *As the value for the height of the aperture box is altered, the height is changed.*

ALTERING THE MARKER COLOR

In addition to adjusting the size of the marker, the color can also be adjusted. Selecting the arrow in the AutoSnap marker color: edit box will display a menu of the seven basic drawing colors. Selecting a color name will automatically change the marker to that color and close the color menu.

CONTROLLING DRAWING SETTINGS

To be used effectively, Osnap settings need to be understood and put to use. Major areas to be considered include the AutoSnap settings and AutoTrack Settings. Each can be set using the Options dialog box shown in Figure 8.3.

AUTOSNAP SETTINGS

The AutoSnap settings control the settings for each object snap mode. Settings include the Marker, Magnet, and Tooltip.

Marker

As the OSNAP command is used, the aperture box is removed from the cursor and replaced by the marker for the current object snap mode. AutoCAD uses a different geometric shape to represent each object snap mode. Specific markers will be introduced later in this chapter. With the Marker box activated, each time the cursor is moved to a point controlled by object snap, the appropriate object snap marker will be displayed. Similar to when you used the LINE command, each time the cursor is moved toward the end of a line, the endpoint marker is added to the cursor to indicate that the endpoint will be selected. If you deactivate this setting, the box will not be displayed, but the current object snaps will be unaffected. AutoCAD has it ON by default for a reason. It is usually very helpful for new users to be reminded of the current settings.

Magnet

The Magnet box is a toggle for the ON/OFF setting. By default, AutoCAD magnet is ON. As you've used a drawing command, the cursor often moved toward the end of a line. That is the effect of having the magnet ON. Magnet automatically moves the cursor to lock it onto the nearest object snap point as the cursor is moved. Keeping the magnet on will increase your drawing speed.

Tooltip

The Tooltip box is a toggle for the ON/OFF setting. Object snap tooltip is ON By default. Each time the cursor is moved near an active object snap mode, the object snap mode to be used will be displayed below the cursor. Figure 8.5 shows the tooltip for the endpoint object snap mode. On simple drawings, the tooltip may not seem important. On complex drawings with several object modes active, the tooltip can be helpful in determining what object snap mode will be used. Once you recognize each

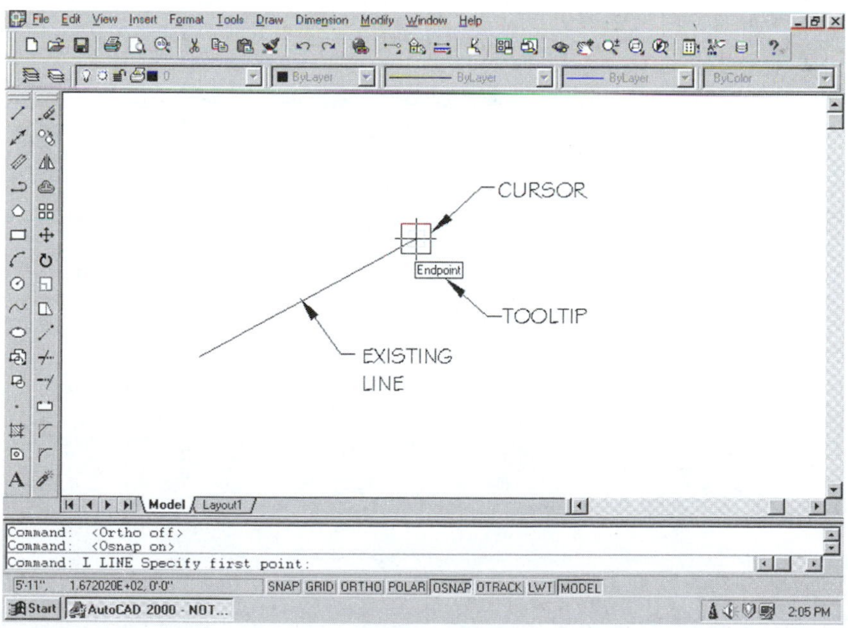

Figure 8.5 *Each time the cursor is moved near an active object snap mode, the object snap mode to be used (the tooltip) will be displayed below the cursor.*

of the object snap markers, you may find the tooltips unnecessary. As you gain confidence and drawing speed, tooltips will not even be displayed when you move from point to point.

AUTOTRACK SETTINGS

The AutoTrack settings control the settings that can enhance each object snap mode. Settings include the polar tracking vector, full-screen tracking vector, and AutoTrack tooltip displays. Each option is an ON/OFF toggle switch. By default, each is set to the ON mode.

Polar Tracking Vector

This menu option sets Polar Tracking so that lines can be drawn along angles relative to the drawing "From" or "Next" points. Angle settings must be 90° divisors, such as 15°, 30°, and 45°. Tracking increments can be set using the Polar Tracking tab of the Drafting Settings dialog box. (See Figure 8.6.) The box is accessed by selecting Drafting Settings... from the Tools pull-down menu or by typing **DSETTINGS** (EnterKey) at the command prompt. With this option ON, tracking angles will displayed similar to Figure 8.7. The display can be toggled ON/OFF using the POLAR button of the Status bar.

Full-screen Tracking Vector

The Display Full-screen Tracking vector controls the display of tracking vectors. Tracking vectors are construction lines that can be used as a base, enabling objects to be drawn at a specific angle. (See Figure 8.7.) They can also be used to draw objects

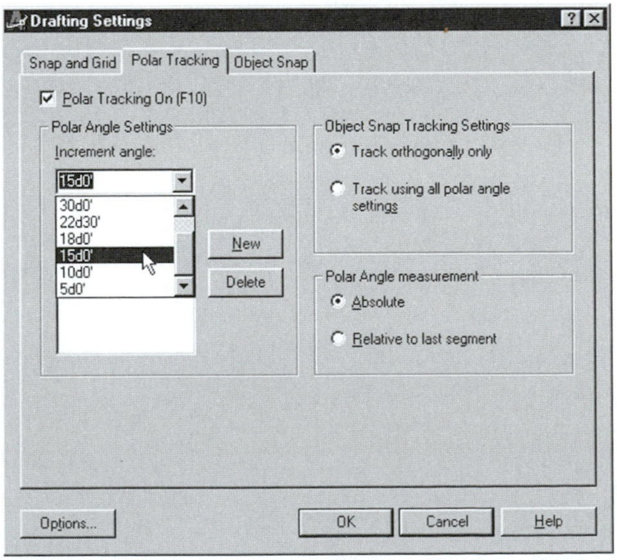

Figure 8.6 *Tracking increments can be set using the Increment angle edit box of the Polar Tracking tab of the Drafting Settings dialog box.*

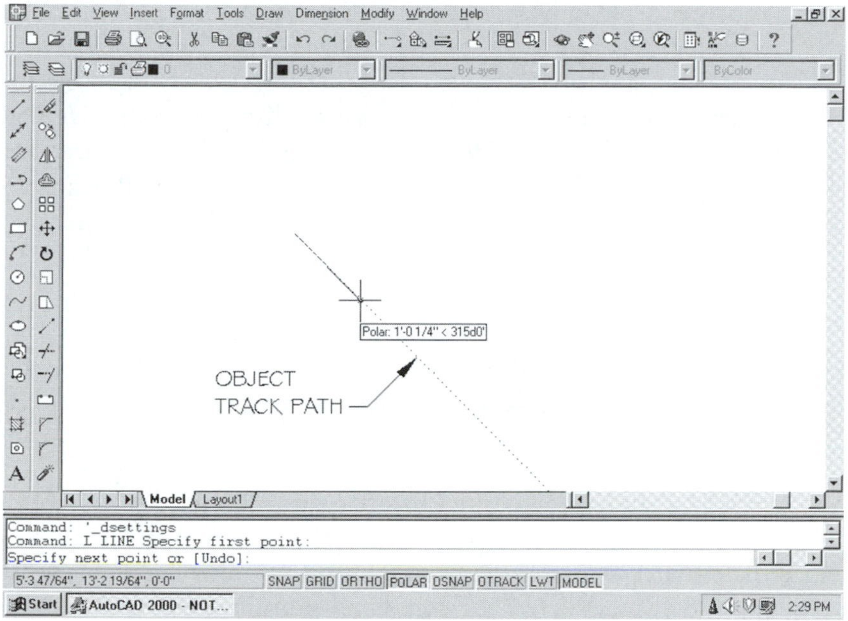

Figure 8.7 *With the Option Tracking option ON, tracking angles will displayed below the cursor.*

at a specific relationship to other objects. In their default setting, tracking vectors extend to the edge of the drawing display area. Alignment paths can be shortened or eliminated. When the setting is OFF, the path is only displayed from the object snap point to the cursor. Controls for setting Object Tracking are found in the Polar Tracking tab of the Drafting Settings dialog box.

AutoTrack Tooltip

The AutoTrack tooltip is similar to the AutoSnap tooltip. This box controls the display of the tracking coordinates.

OSNAP

Object snaps are accessed by the OSNAP command. They can be used to select an exact start or endpoint for connecting lines, rather than hoping you're close. Object snaps allow a line or other object to be snapped to a specific point. For instance, OSNAP will allow a line to be started exactly at the midpoint of an existing line, with absolutely no guessing. OSNAP has two modes of operation that include single-point override and running. The running mode allows a mode to be set and used throughout the drawing base. The override mode allows you to select an object snap mode to be used once in the middle of a command.

The desired override OSNAP can be activated by selecting its icon from the Object Snap toolbar, by selecting its icon from the Temporary Tracking flyout-out menu of the Object Properties toolbar, by pressing the (ShiftKey) and right-clicking, or by keyboard. The toolbar is activated by selecting Object Snap from Toolbars… of the View pull-down menu. Figure 8.8 shows a listing of each Object Snap mode found in the Object Snap toolbar and the marker that will be displayed with each mode. An OSNAP mode can be used in the middle of a command by entering three letters to represent the mode for the "Next" point of a command sequence. For instance, to draw a line from the endpoint of a line would require the following sequence:

Command: **L** (EnterKey)

LINE

Specify first point: **END** (EnterKey)

of: *(Select an endpoint for the starting point of the new line).*

Specify next point or [undo]:

Running Object Snap allows the desired object snap to be selected for continuous use. Running OSNAP options are set by selecting the Object Snap tab of the Drafting Settings dialog box. Choosing Drafting Settings… from the Tools pull-down menu or typing **OS** (EnterKey) at the command prompt accesses the dialog box. Running OSNAP will be discussed later in this chapter. No matter how they are accessed, using the OSNAP options will save time and greatly improve your accuracy. Once OSNAP is activated, the marker box is used to select objects to snap to. Move the crosshairs so that the desired object lies within the marker box, and then press the

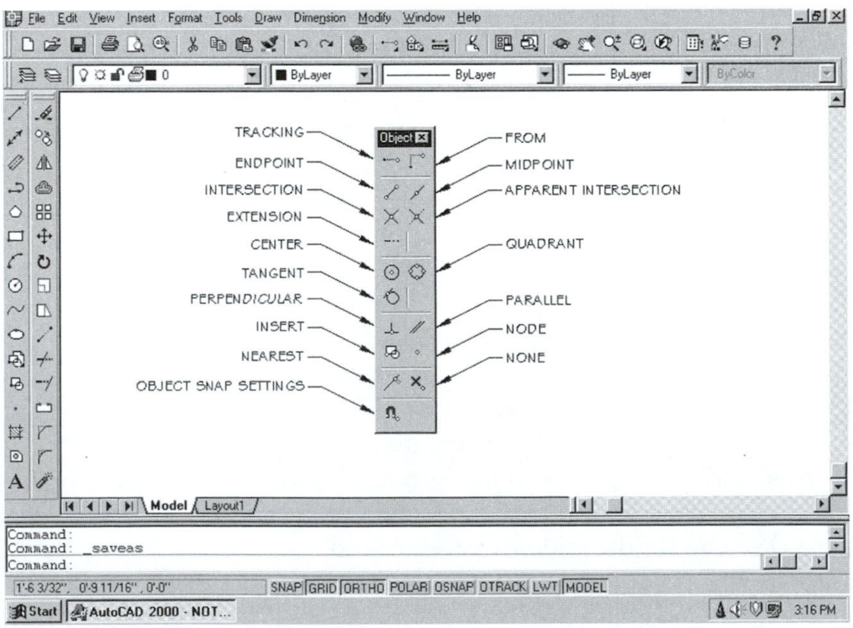

Figure 8.8 *The Object Snap toolbar and the marker box for each mode.*

pick button. If more than one object lies within the marker box, the closest object will be selected. Two options are available if no objects are found. These options will be discussed later in the chapter.

OSNAP MODES

OSNAP tools include ENDpoint, MIDpoint, INTersection, EXTension, APParent Intersection, CENter, NODe, QUAdrant, INSertion, PERpendicular, TANgent, NEArest, PARallel, QUIck, and NONe. These options can be used for drawing lines or arcs that connect at certain points to existing lines, circles, arcs, or ellipses. OSNAP snap options also can be used anytime a point needs to be specified, and for COPY, MOVE, and INSERT commands, which will be discussed in later chapters. Notice that as the options were listed, the first three letters of some of the options were written in capital letters. Typing these letters at the prompt for start point or endpoint will activate the desired object snap option.

ENDPOINT

This option allows an arc, line, or circle to be drawn to or from the endpoint of a previous line or arc by snapping to the desired endpoint. The process is started by selecting the Endpoint icon from the Object Snap toolbar when you are prompted for a "From" point. To add a line that extends between the endpoints of two existing lines, the command sequence would be:

Command: **L** (EnterKey)

From point: *(Select the Endpoint icon.)*

Now move the cursor near the end of the desired line that you would like the new line to connect to. The cursor does not have to touch the end of the line, but must be between the midpoint and endpoint so the correct end of the line is selected. As the cursor touches a line, the square marker box is displayed around the nearest endpoint. (See Figure 8.9.) Press the mouse pick button, and the new line segment will be connected to the selected endpoint. Select the Endpoint icon again at the prompt for the "Next" point, to connect the new line to the end of the second line. The new line will automatically end at the selected end of the existing line segment.

The endpoint can also be selected at the keyboard, by entering END (EnterKey) ,as the "From" point of the desired command. Move the marker box and select the desired line. To draw a line from the endpoint of an existing line to the endpoint of a second line, use the following command sequence:

Command: **L** (EnterKey)

LINE Specify first point: **END** (EnterKey)

_end of *(Select desired line.)*

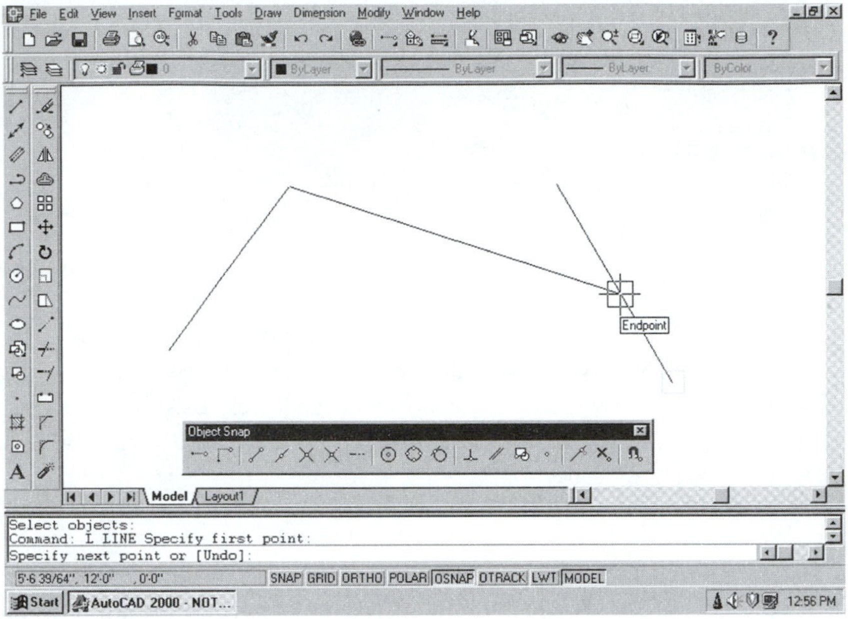

Figure 8.9 *When the End mode of OSNAP is selected, a marker is placed on the endpoint of the selected line to indicate the selected "First" or "Next" point.*

Specify next point or [undo]: **END** (EnterKey)

of *(Select desired line.)*

Specify next point or [undo]: (EnterKey)

Command:

The command sequence can be seen in Figure 8.10. If you're thinking, "This is extra work", you're right. By default, AutoCAD automatically draws lines from endpoint to endpoint. The process will work wonders as different object snap modes are entered.

MIDPOINT

Midpoint allows the selection of a line or arc at its midpoint by selecting the Midpoint icon in response to the "First" point prompt. To add a line that extends between the middle of two existing lines, the command sequence would be:

Command: **L** (EnterKey)

LINE

Specify first point: *(Select the Midpoint icon.)*

Now move the cursor near the desired line that you would like the new line to connect to. As the cursor touches a line, the marker box is displayed around the nearest mid-

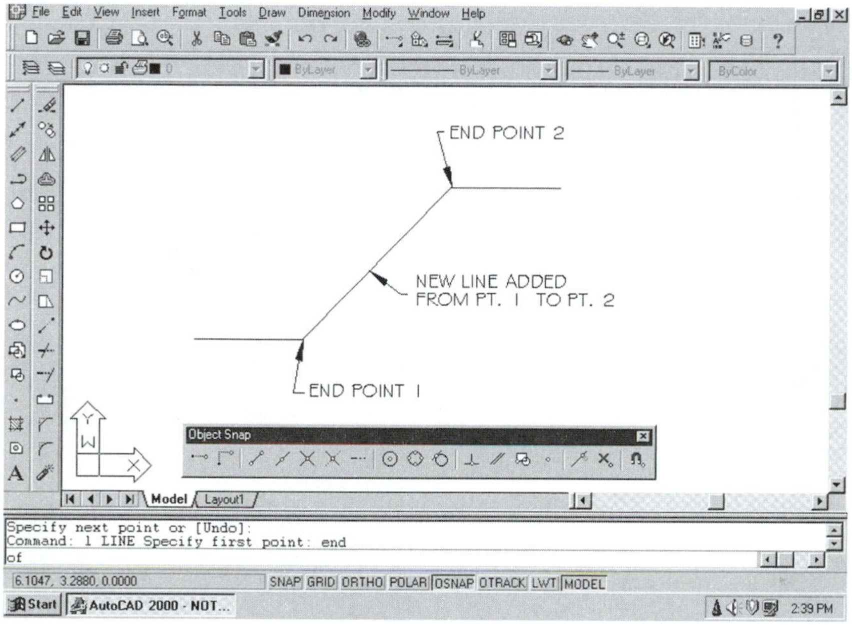

Figure 8.10 *To draw a line from the endpoint of one line to the endpoint of another, pickup the Endpoint icon or type* **END** *(EnterKey) when the "First" point and the "Next" point prompts of the LINE command are displayed.*

point. Press the pick button and the new line segment will be connected to the selected midpoint of the existing line. Select the Midpoint icon as the "Next" point is requested at the prompt and the new line will automatically end at the selected midpoint of the existing line segment.

The midpoint for an arc or line can also be selected by entering **MID** (EnterKey) as the "First" or "Next" prompt is displayed. To draw a line from the midpoint of a line to the midpoint of a second line use the following command sequence:

Command: **L** (EnterKey)

LINE

Specify first point: **MID** (EnterKey)

_mid of *(Select desired line.)*

Specify next point or [undo]: **MID** (EnterKey)

of *(Select desired line.)*

Specify next point or [undo]: (EnterKey)

Command:

The command sequence can be seen in Figure 8.11.

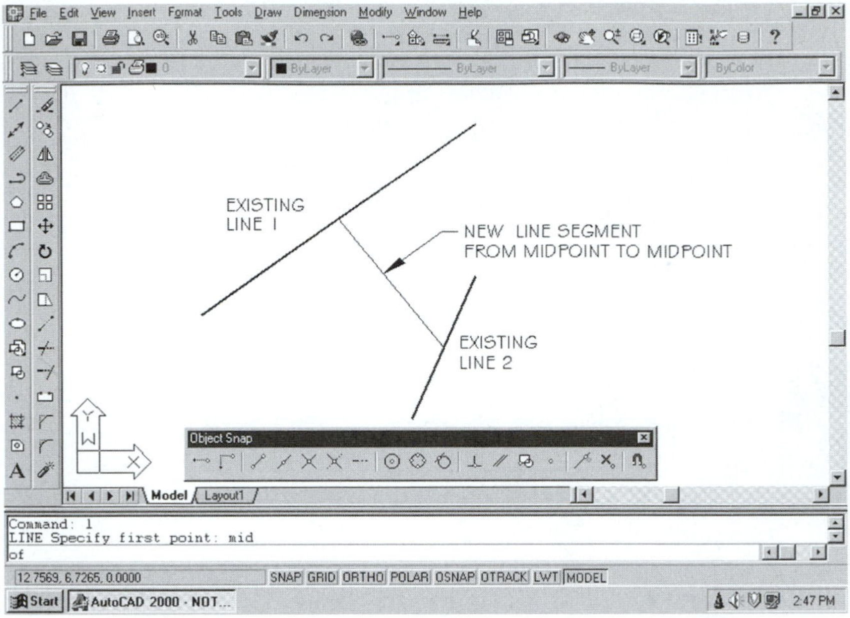

Figure 8.11 *The Midpoint option can be used to draw a line from the midpoints of one line to another by selecting the Midpoint icon or by typing* ***MID*** *(EnterKey) for the "First" and "Next" point prompts of the LINE command.*

INTERSECTION

As its name implies, Intersection allows a line or arc to be extended to or from the intersection of any lines, arcs, or circles. An example of the use of the intersection option can be seen in Figure 8.12. The command sequence is similar to other OSNAP command entries and can be seen in Figure 8.13. By selecting the intersection icon from the Object Snap toolbar or by entering INT (EnterKey) as a "First" or "Next" point the desired intersection can be snapped to. The command sequence is:

Command: **L** (EnterKey)

LINE

Specify first point: **INT** (EnterKey)

_int of *(Select desired to point.)*

Specify next point or [undo]: *(Select desired to point.)*

Specify next point or [undo]: (EnterKey)

Command:

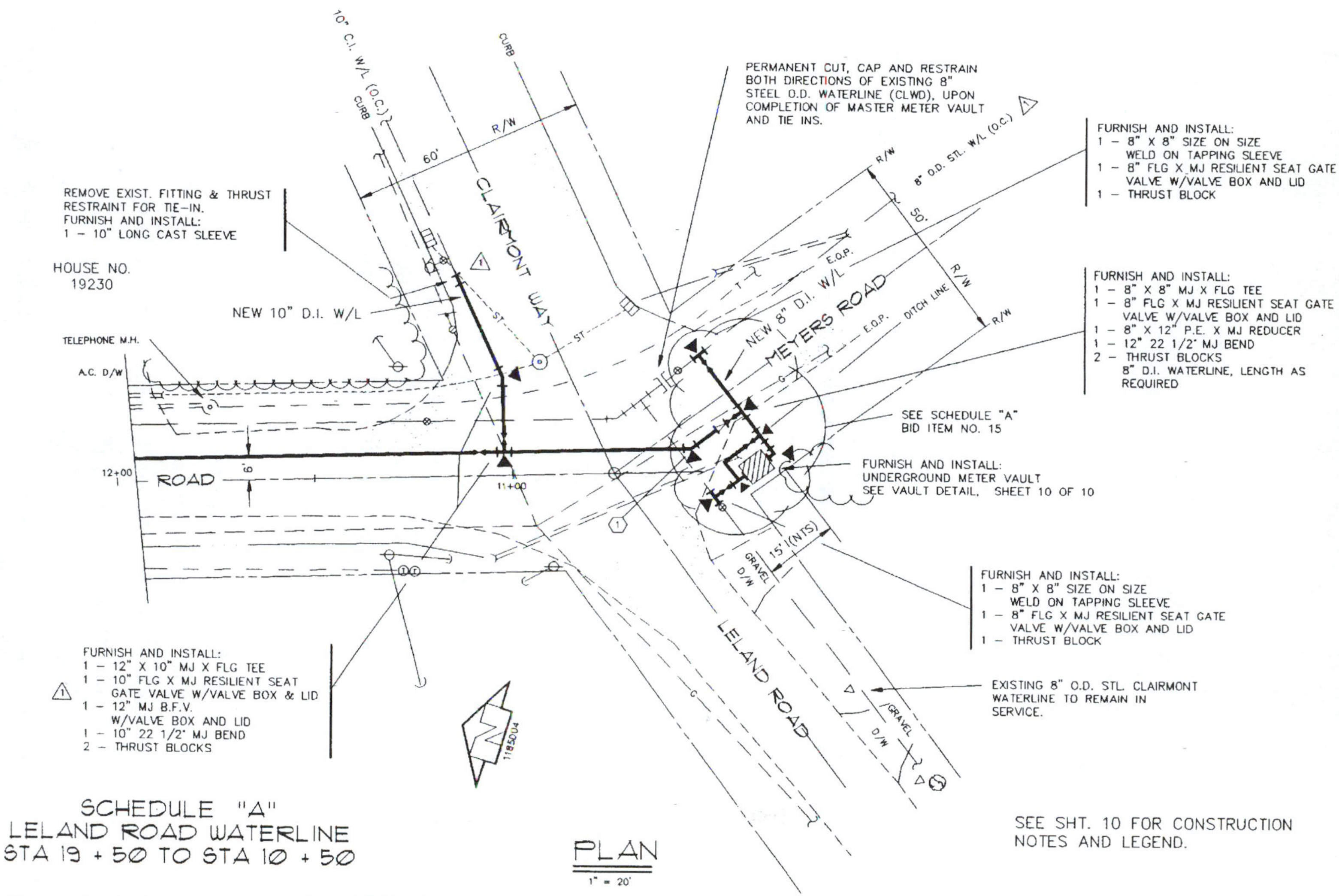

Figure 8.12 *The Intersection option of OSNAP allows lines and arcs to be connected to the intersection of other geometric shapes. (Courtesy Lee Engineering, Inc., Oregon City OR.)*

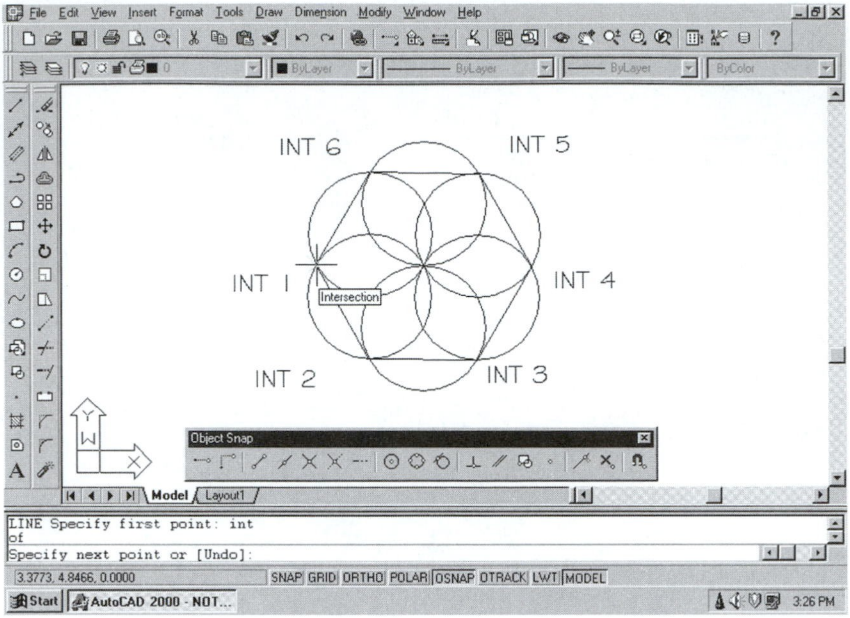

Figure 8.13 *The Intersection option of OSNAP is used to draw a line from the intersection of each circle.*

EXTENSION

The Extension mode of object snap will use a point that lies along an extension path of a line as a First, Next, or Center point for an object. This mode has two methods of operation. One method will extend a line from an existing line. (See Figure 8.14.) This method can be used to lengthen a line. To extend a line from the end of an existing line use the following procedure:

> Command: **L** (EnterKey)
>
> LINE
>
> Specify first point: **EXT** (EnterKey)
>
> _ext of *(Select desired end point.)*
>
> Specify next point or [undo]: *(Select desired to point. Be sure OTRACK is ON.)*
>
> Specify next point or [undo]: (EnterKey)
>
> Command:

The second method will use a point that lies on the path as the line is extended from the existing line. This method will work well to place a point lying on a projected

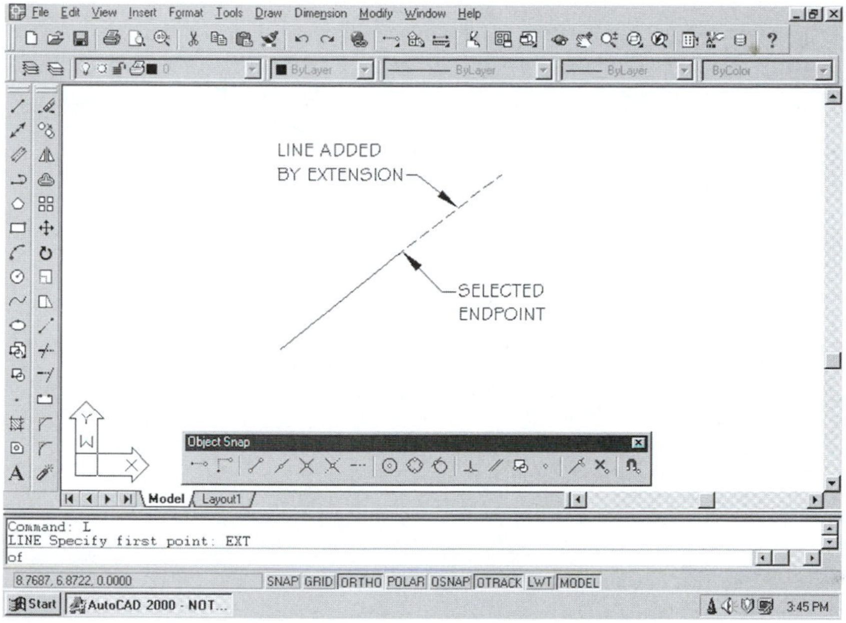

Figure 8.14 *The Extension mode of object snap will use a point that lies along an extension path of a line as a "Next" point. When the endpoint of an existing line is selected for the the extension point, the existing line will lengthened along the projected path to the "Next" point.*

path. (See Figure 8.15.) Use the following procedure to select a point that lies on an extension of an existing line:

Command: **L** (EnterKey)

LINE

Specify first point: **EXT** (EnterKey)

_ext of *(Move the cursor over the endpoint of the existing but do not pick the endpoint. Extension will automatically select the indicated endpoint and project a path. Select a point that lies on* the projected path.)

Specify next point or [undo]: *(Select desired to point.)*

Specify next point or [undo]: (EnterKey)

Command:

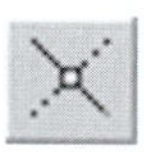

APPARENT INTERSECTION

This option finds the Apparent Intersection, or where two lines would intersect if one or both were extended. To find the apparent intersection, select the Apparent Intersection icon when prompted for the "From" point. Using the keyboard, enter

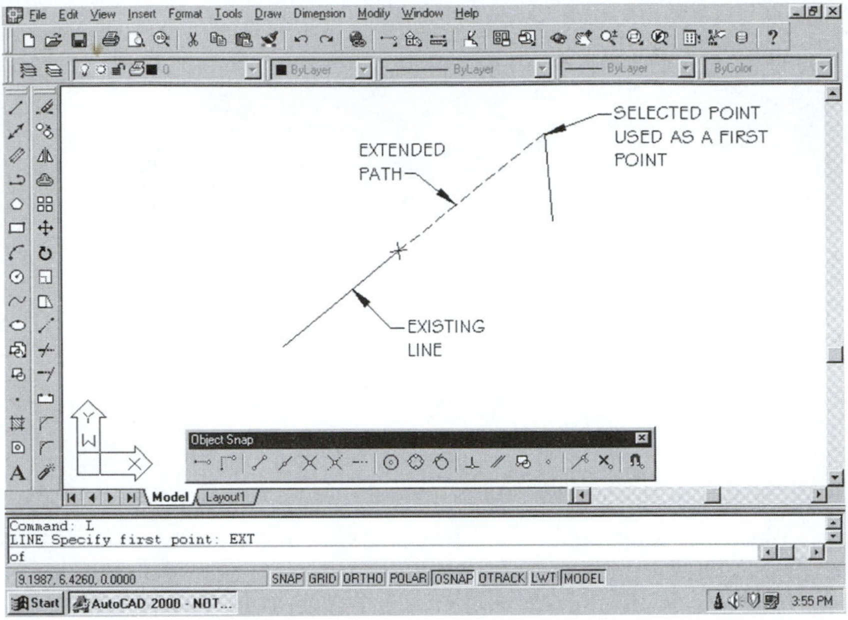

Figure 8.15 *Passing the EXT marker over the endpoint, but not selecting it as an endpoint, will use that point as a projection point for the extension path. Extension will use a point that lies on the path as the line is extended from the existing line.*

APP (EnterKey) at the "First" or "Next" prompt. Because the line will be generated from where two lines will intersect, two "First" points will need to be selected. The prompt will read " -appint of " to remind you to select two lines. The new line will be placed where the selected lines will merge. From the command prompt, use the following command sequence:

Command: **L** (EnterKey)

LINE

Specify first point : **APP** (EnterKey)

_appint of *(Select first line to be projected.)*

and *(Select second line to be projected.)*

Specify next point or [undo]: *(Select desired to point.)*

Specify next point or [undo]: (EnterKey)

Command:

Selecting (EnterKey) will extend the line. The command sequence can be seen in Figure 8.16. Chapter 10 will present two alternatives to this option.

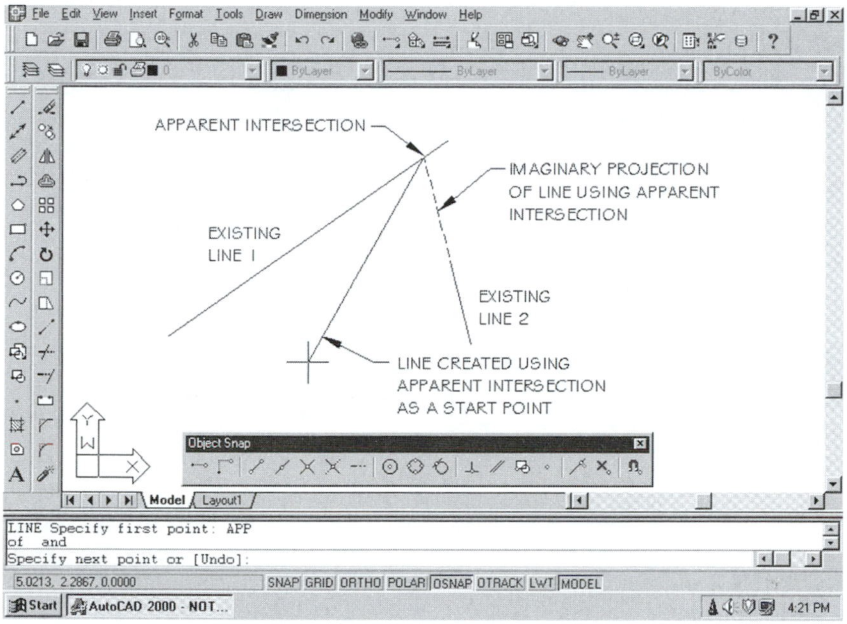

Figure 8.16 *The Apparent Intersection option of OSNAP can be used to extend a line between two lines that would intersect. The line was started by using the END option to select the end of line 2, selecting the icon from the toolbar, or by typing* **APP** *(EnterKey) for the "To" point, and then using the marker box to select line 1 as the line to intersect. The endpoint of the original line must be selected to display the new line segment.*

CENTER

The Center option can be used any time an object needs to be drawn based on the center of a circle. A line or arc often needs to be drawn from the center of an existing circle; or several circles may need to be drawn around one center point, as seen in Figure 8.17. This drawing could be easily drawn using the Center option of OSNAP.

A line or arc may be extended from the center point of a circle by selecting the Center icon from the Object Snap toolbar. The Center point option can also be selected by keyboard using the following command sequence:

Command: **L** (EnterKey)

LINE

Specify first point: **CEN** (EnterKey)

_cen of: *(Select a circle.)*

Specify next point or [undo]: *(Select the desired endpoint.)*

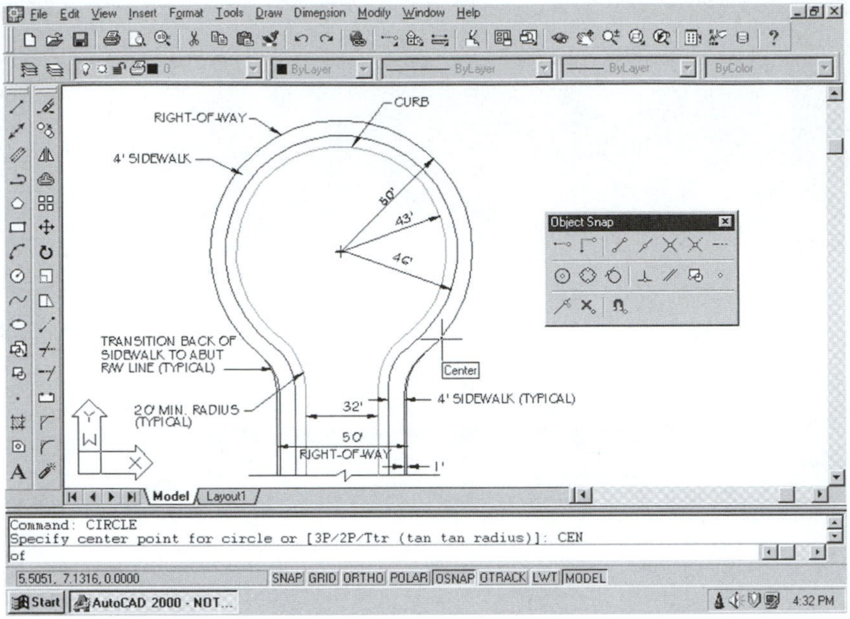

Figure 8.17 *The Center option of OSNAP can be used whenever a drawing object needs to be placed, based on the center of an existing circle. (Courtesy Lee Engineering.)*

Specify next point or [undo]: (EnterKey)

Command:

Touch any part of the circle and a rubber-band line now extends from the exact center of the circle until a "Next" point is selected. This process can be seen in Figure 8.18.

NODE

This option can be used to snap to a point. By entering Node, lines or arcs can be extended from or to predetermined points. Methods of drawing a point will be introduced in Chapter 13. This could be used, for instance, to layout points on a route survey. To activate this OSNAP option, select the Node icon or type **NOD** (EnterKey) at either the "First " or "Next" point.

QUADRANT

This OSNAP option will allow a line or arc to be snapped to the 0°, 90°, 180°, and 270° positions of a circle or arc. The command is activated by selecting the Quadrant icon from the toolbar, or by entering **QUA** (EnterKey) as the "First" or "Next" point. Place the aperture box on the circle or arc near the quadrant and the new object will

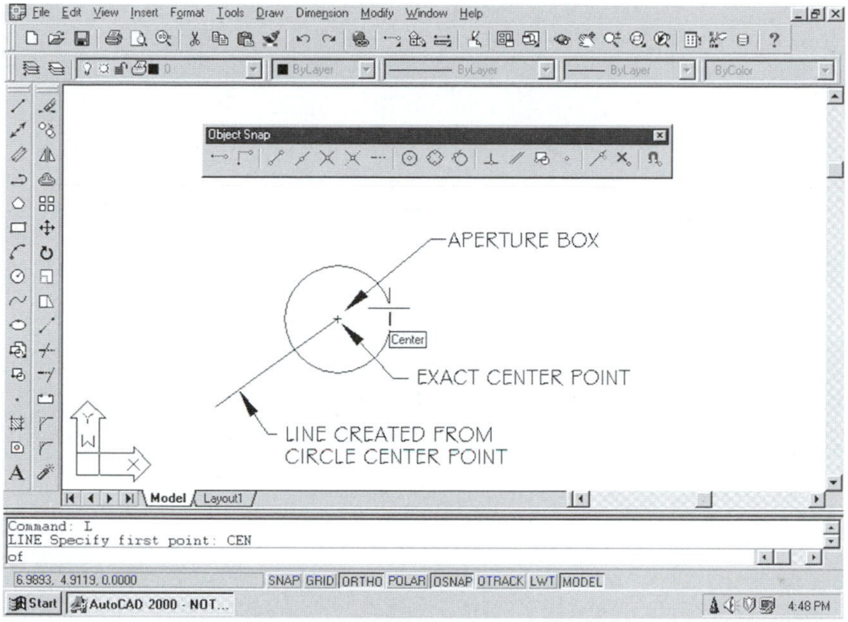

Figure 8.18 *To extend a line from the center of a circle, enter the LINE command. When prompted for a LINE "First" point, select the Center icon or type* **CEN** *(EnterKey) at the command prompt. This will produce the center marker that can be used to select the desired circle. When you place the marker anywhere on the circle and select the circle, AutoCAD will automatically use the center point of that circle as the "First" point for a line.*

automatically be joined at the quadrant, as seen in Figure 8.19. To draw a line from the quadrant point of an existing circle, use the following command sequence:

Command: **L** (EnterKey)

LINE

Specify first point: **QUA** (EnterKey)

_qua of *(Select desired circle.)*

Specify next point or [undo]: *(Select desired to point.)*

Specify next point or [undo]: (EnterKey)

Command:

INSERTION

This command will snap to the insertion point of a Shape, Text, Attribute, or Block. Meaningless stuff now, but combined with the information in future chapters it will allow information to be inserted to an exact point.

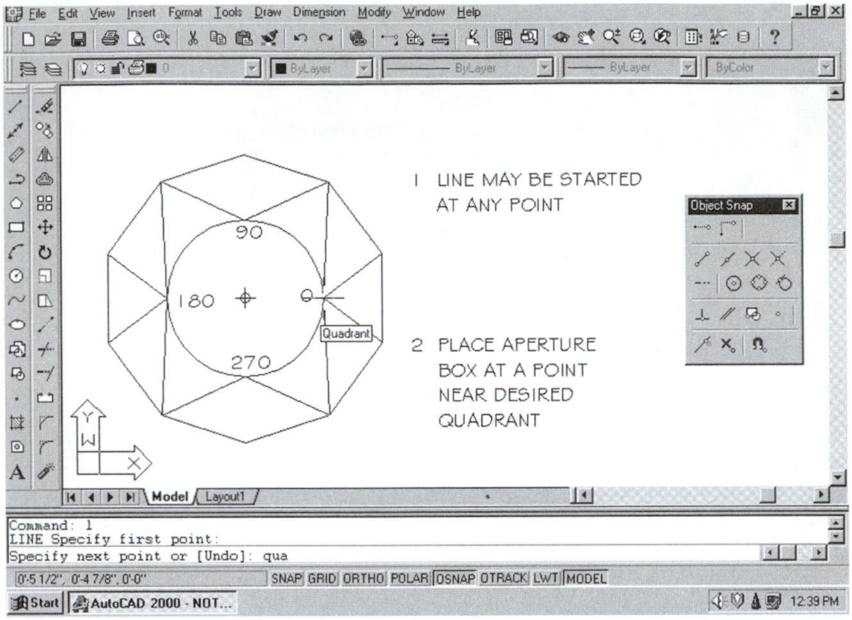

Figure 8.19 *The Quadrant option of OSNAP can be used to attach a line or arc to the quadrants of an existing circle.*

PERPENDICULAR

As seen in Figure 8.20, perpendicular lines are a basis of most construction drawings. Previous OSNAP options could be used to bring a line or an arc toward or away from an object. The Perpendicular mode is used to draw a line that is perpendicular to an existing line. The process can be seen in Figure 8.21. To draw a line from any point perpendicular to an existing line, select the Perpendicular icon when prompted for a "First" or "Next" point. Type **PER** (EnterKey) at the command prompt and use the following command sequence:

Command: **L** (EnterKey)

LINE

Specify first point: *(Select desired point.)*

Specify next point or [undo]: **PER** (EnterKey)

_per *(Select desired line.)*

Specify next point or [undo]: (EnterKey)

Command:

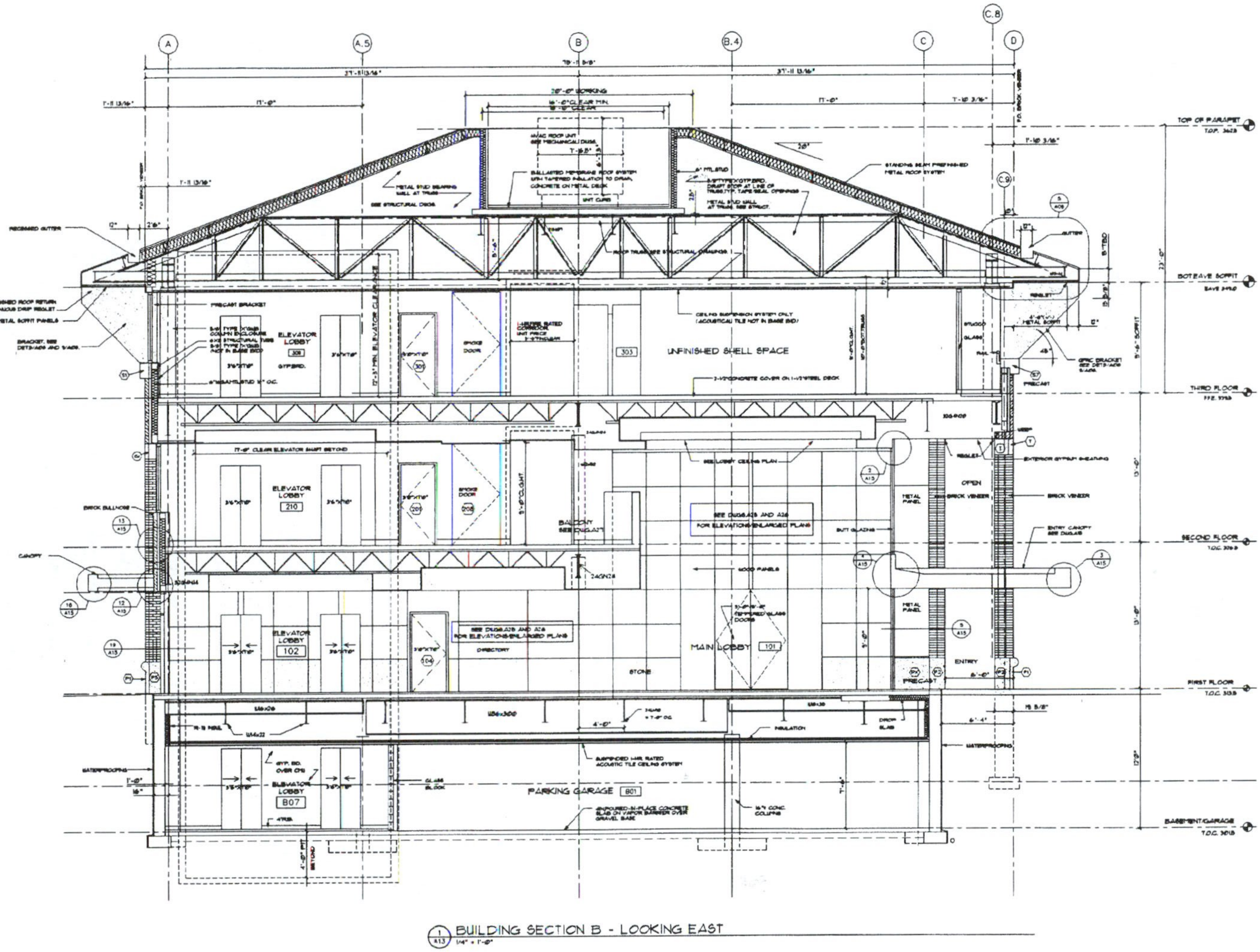

Figure 8.20 *The Perpendicular option of OSNAP can be used to draw a line perpendicular to another surface. This can be useful if the surface is not in a true vertical or horizontal allignment. (Courtesy Peck, Smiley, Ettlin Architects.)*

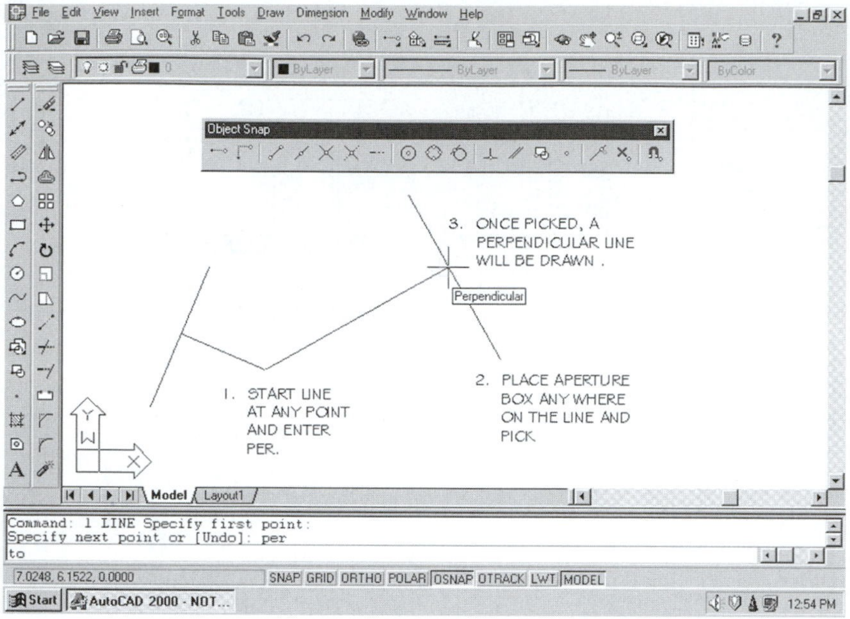

Figure 8.21 *The Perpendicular option can be used to project a new line so that it will be at a 90° angle to an existing line.*

TANGENT

This OSNAP mode will allow a line to be snapped tangent to a circle or an arc. This very common combination of geometric shapes can be seen in Figure 8.22. This mode is started by selecting the Tangent icon from the Object Snap toolbar or by typing TAN (EnterKey) at the "From" or "To" point prompts. Figure 8.23 shows an example of how Tangent can be used. To draw a line tangent to an existing circle, use the following command sequence:

Command: **L** (EnterKey)

LINE

Specify first point: *(Select desired point.)*

Specify next point or [undo]: **TAN** (EnterKey)

_tan to *(Select desired arc or circle.)*

Specify next point or [Close/undo]: (EnterKey)

Command:

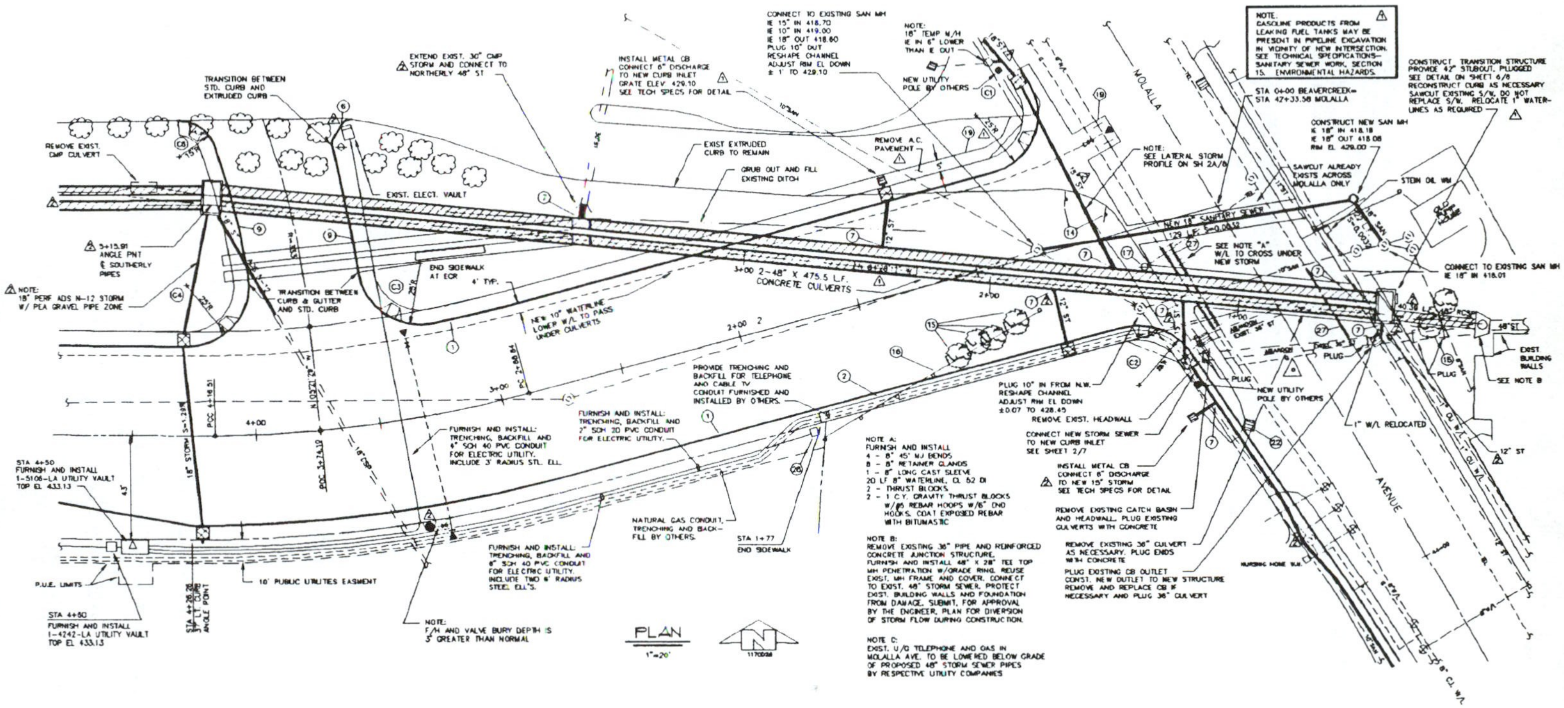

Figure 8.22 *A common drawing requirement is to draw tangent surfaces. Tangent OSNAP will allow a line or arc to be drawn tangent to an existing arc or circle. (Courtesy Lee Engineering, Inc., Oregon City, OR.)*

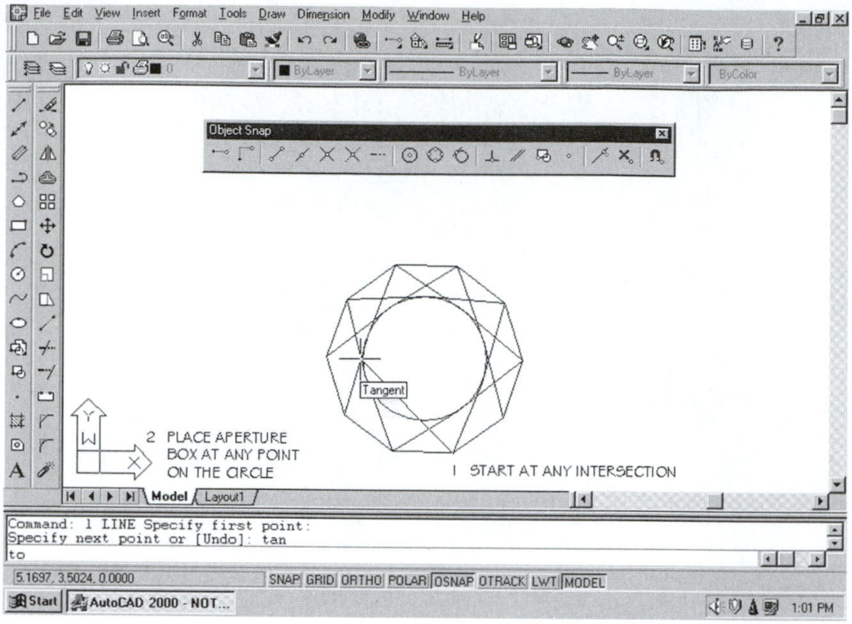

Figure 8.23 *Tangent is started by selecting the Tangent icon or by entering* **TAN** *(EnterKey) for the "First" or "Next" point of the new line or arc.*

NEAREST

This option can be used to snap to the circle, line, or arc that is nearest to the crosshairs. To activate this OSNAP option, select the Nearest icon or type **NEA** (EnterKey) at either the "First" or "Next" point. To draw a line from any point of an existing line, use the following command sequence:

Command: **L** (EnterKey)

LINE

Specify first point: **NEA** (EnterKey)

_nea of *(Select desired line.)*

Specify next point or [undo]: *(Select desired to point.)*

Specify next point or [undo]: (EnterKey)

Command:

The command sequence can be seen in Figure 8.24.

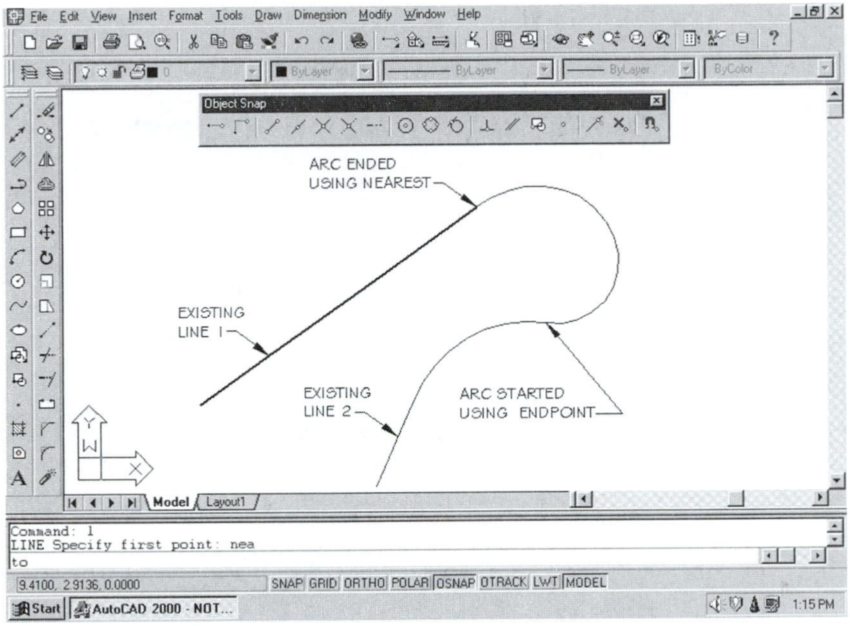

Figure 8.24 *The Nearest option can be used to select the nearest drawing object for a "First" or "Next" point when placing lines, arcs, circles, and other geometric shapes.*

PARALLEL

This mode will draw a line parallel to an existing line. Unlike other modes, parallel is specified after the "First" point is selected. Chapter 10 will explore an alternative method of placing parallel lines using the OFFSET command.

Command: **L** (EnterKey)

LINE

Specify first point: *(Select desired to point.)*

Specify next point or [undo]: **PAR** (EnterKey)

_par to *(Move the cursor over, but do not select the endpoint of the desired line. As the aperture box moves over the endpoint, the point is marked and cursor snaps to a parallel path.)*

Specify next point or [Close/undo]: (EnterKey)

Command:

The command sequence can be seen in Figure 8.25.

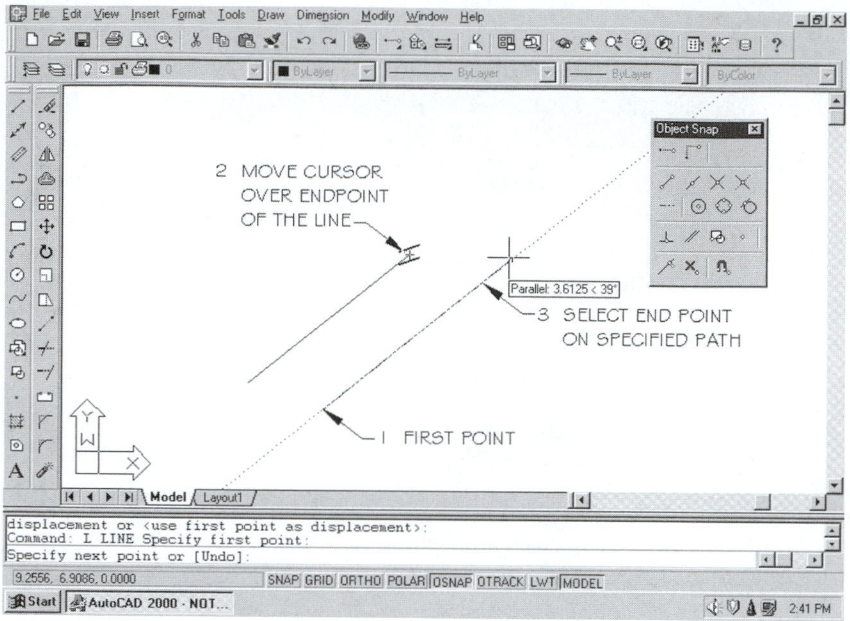

Figure 8.25 *The Parallel mode will draw a line parallel to an existing line. Unlike other modes, Parallel is specified after the "First" point is selected. Move the cursor over, but do not select the end point of the line that is to be the basis for the second line. Once the first line has been identified, specify the "Next" point for the new parallel line.*

QUICK

Although it may seem that objects are selected very quickly, AutoCAD must search each object of your drawing to find the designated objects of the aperture box. This process can be simplified by using QUICK. Quick is used in conjunction with another OSNAP option and allows for AutoCAD to stop searching as soon as it finds one object with at least one point of the specified type. OSNAP searches for all objects crossing the target and selects the closest potential snap points. If two or more objects that match the OSNAP request are in the aperture box, QUICK mode will choose the first one it sees. On a complex drawing, this may not always give the desired results. The command sequence to use QUICK and draw a line from one line endpoint to the midpoint of another line is:

Command: **LINE** (EnterKey)

LINE

Specify first point: **QUI,END** (EnterKey)

_qui,end *of (Select desired line.)*

Specify next point or [undo]: (EnterKey)

Command:

FROM

The From selection of OSNAP can be used to establish a point for a line or other drawing entity based on a known base point. The base point can be located by selecting a point with the cursor, or by providing polar or relative coordinates for the point. The From mode could be useful on site related drawings when objects are to be located based on a specific point.

Figure 8.26 shows the use of From for locating a circle from the endpoint of an existing line. The command can be entered by selecting the From icon from the Object Snap toolbar, or by entering **FRO** (EnterKey) at the command prompt. The command sequence is:

Command: **C** (EnterKey)

CIRCLE Specify center point or [3P/2P/TTR (tan tan radius)]: **FRO** (EnterKey)

Base point <offset>: **END** (EnterKey)

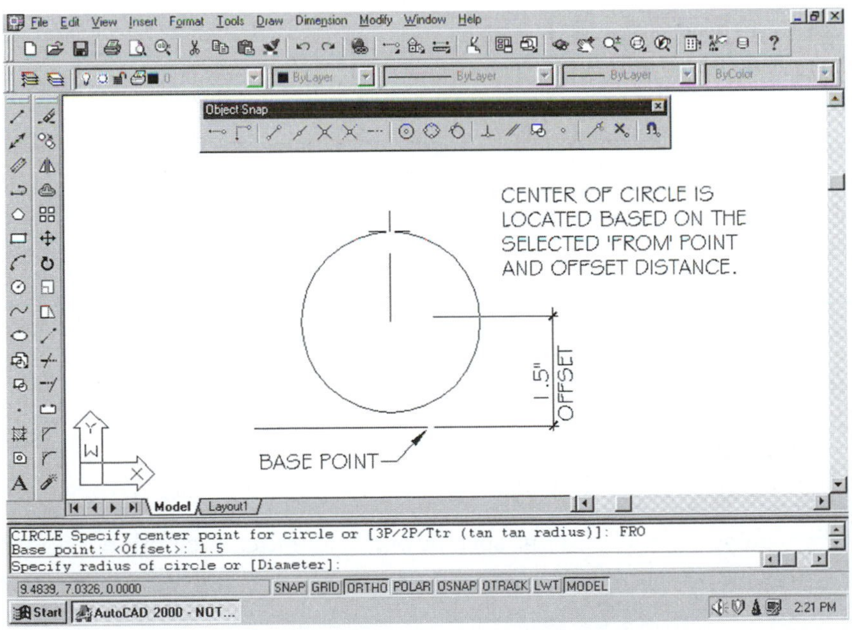

Figure 8.26 *The From setting of OSNAP allows a line, circle, or arc to be drawn from a base point by using the From icon, or by entering* **FRO** *(EnterKey) for one of the selection points for the drawing object.*

of: *(Select desired line endpoint.)*

of <offset>: *(Select location or provide coordinates from base point.)* **1.5** (Enter-Key)

Specify radius of circle or [Diameter]: *(Pick radius point or provide coordinates.)*

Command:

TRACKING

The Tracking mode can be used to find the center of a rectangle. The option functions similar to the Apparent Intersection mode. By selecting the midpoint on the bottom side of the rectangle, and then selecting the midpoint of one of the sides, a line can be drawn from the imaginary intersection of these two lines. Tracking can be started by selecting the Tracking icon from the Object Snap toolbar; or by entering **TK** (EnterKey) at the command prompt. The command sequence to draw a line from the center of a rectangle is:

Command: **L** (EnterKey)

LINE

Specify first point: **TK** (EnterKey)

First tracking point: **MID** (EnterKey)

of *(Select the bottom edge of the rectangle.)*

Next point *(Press ENTER to end tracking)*: **MID** (EnterKey)

of *(Press ENTER to end tracking.)* *(Select the edge of the rectangle.)*

Next point *(Press ENTER to end tracking)*: (EnterKey)

Specify next point or [undo]: *(Select the desired endpoint of the line.)*

Command:

As the tracking sequence is ended, the rubber-band line will now be centered in the rectangle and the prompt for a "To" point is given. The command sequence can be seen in Figure 8.27.

RUNNING OSNAP

Up to this point, OSNAP has been used in its override mode. The Osnap mode was selected by icon or keyboard for one specific point, and then the option was ended. It may be necessary to draw several lines to various endpoints of existing lines. Although you could continue to select END each time you draw a line, there is a faster way. OSNAP selection can be set to be in effect until the command is no longer needed.

Running OSNAP can be set by choosing the Object Snap Settings icon from the Object Snap toolbar, or by using the Object Snap tab of the Drafting Settings dialog

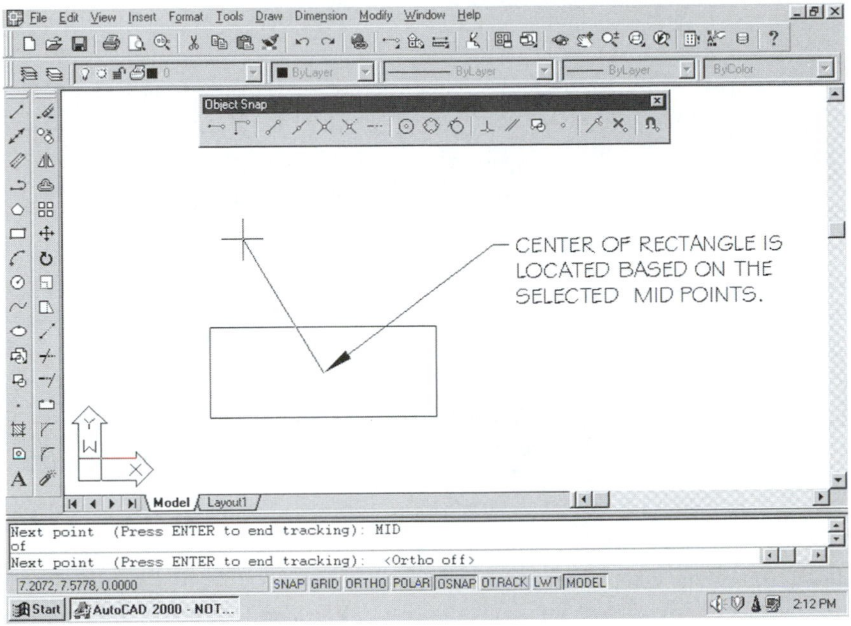

Figure 8.27 *Tracking can be used to determine the center point of a rectangle. After selecting the Tracking option, select the midpoint of the bottom and then select the midpoint of one side of the rectangle. A line or arc can now be projected from the center of the rectangle.*

box or by typing **DSETTINGS** (EnterKey) at the command prompt. The dialog box can also be displayed by selecting Drafting Settings… from the Tools pull-down menu by entering **OS** (EnterKey) at the command prompt. Each method will produce a dialog box similar to Figure 8.28.

Once the Object Snap tab is displayed, place a check in the mode to be activated, and then press the OK button. If Center is selected, a center point will automatically be selected if a circle is selected for a "First" or "Next" point each time a line or arc is drawn. Figure 8.29 shows an example where the Center option is used to place columns in the center of each pier, and lines to represent beams that start at the center of each column. The selection can be ended or altered by returning to the dialog box and altering the active modes.

OSNAP can also be set to combine multiple options. This is useful, for instance, if several lines need to be constructed from the intersection of one object to the Center of other objects, as seen in Figure 8.30. Multiple options are activated by selecting the desired options from the Object Settings dialog box. There is no limit to the number of boxes that can be selected. A temporary override of running OSNAP can be achieved by selecting the Snap to None icon from the Object Snap toolbar, or by entering **NON** (EnterKey) at the OSNAP prompt. Running OSNAP can be deactivated by picking the OSNAP button on the Status bar, or by using the F3 Key.

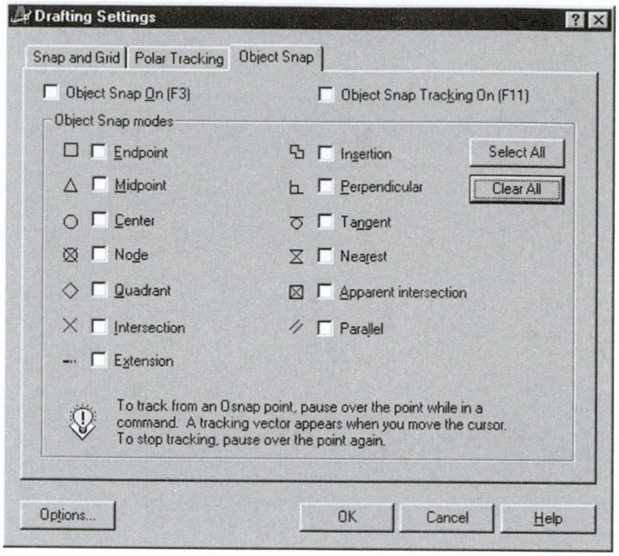

Figure 8.28 *Options for OSNAP can be adjusted by using the Object Snap dialog box. The box can be accessed by selecting the Object Snap tab after selecting Drafting Settings... from the Tools pull-down menu, or by typing* **OS** *(EnterKey) at the command prompt.*

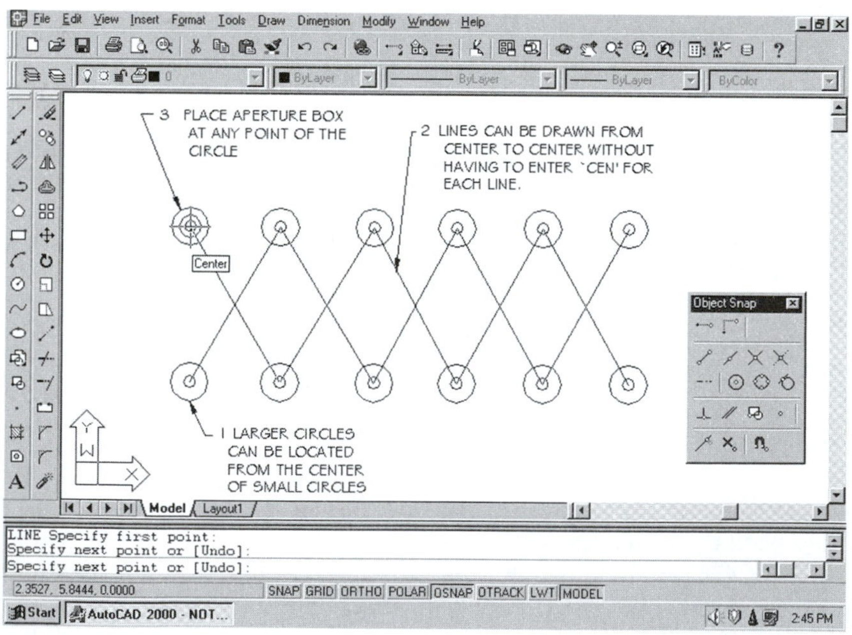

Figure 8.29 *An OSNAP mode can be used repeatedly by selecting multiple boxes from the Object Snap dialog box. Active modes can be deactivated by single-clicking the ORTHO mode of the Status bar or by using (F3Key).*

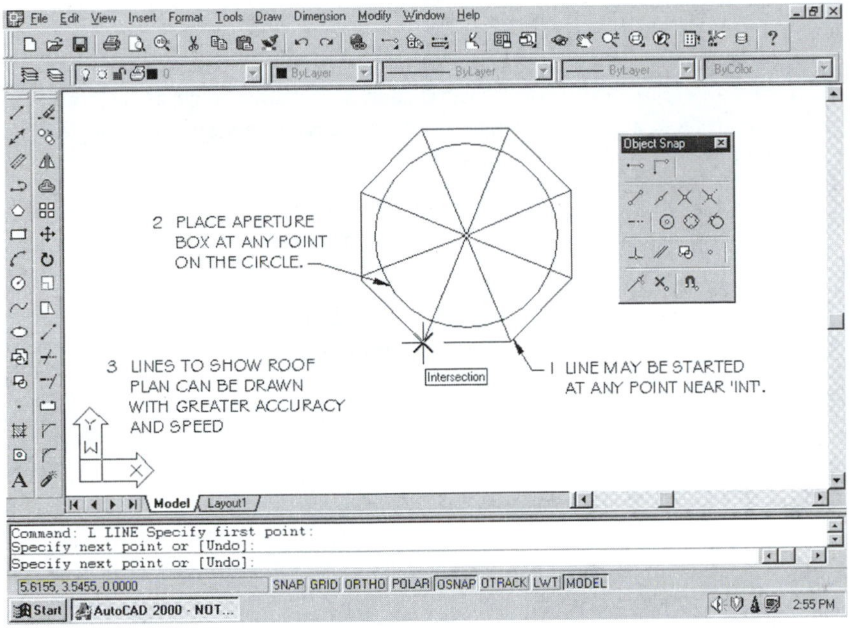

Figure 8.30 *Running OSNAP can be set to combine multiple options for "From" and "To" points.*

CHAPTER 8 EXERCISES

1. Draw an equilateral triangle inscribed in a 4"-diameter circle. Use OSNAP midpoint to form a second triangle inside the first. Draw a third triangle inside the second. Save the drawing as E-8-1.
2. Draw the base to a chimney in plan view that is 60" × 32" with 8"-wide walls. Draw an 8"-wide support wall in the center of the chimney. Save the drawing as E-8-2.
3. Use the drawing below to draw a 1"-diameter circle, a 2.25"-diameter circle, a four-sided polygon, and a triangle. Complete the drawing using the CEN, MID, INT, and TAN.

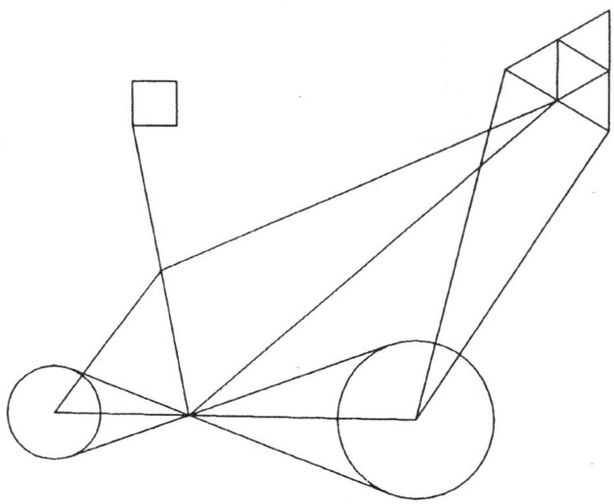

4. Set the limits and units to draw a 100' × 75' lot. Assume north to be at the top of the page. The north and south property lines will be 100' long. Five trees are located on the property. Each is located from the southwest corner with coordinates and diameters listed as follows:

	Coordinates	Trunk diameter	Branch diameter
a.	15',12'-9"	9" diameter	6' diameter
b.	30',17'	13" diameter	7.5' diameter
c.	42'-6", 19'-3"	19" diameter	9.5' diameter
d.	50', 28'-2"	15" diameter	8.5' diameter
e.	62'-0", 5'-3"	26" diameter	14' diameter

Locate each tree with a point. Use an "X" to locate each tree. Use the appropriate OSNAP options to draw the diameter of the trunk and the branch structure around each tree center. Draw a line from the southwest property corner through the center of the trees to the southeast corner. Draw another line from the southeast property corner to the northern limits and tangent to the branch diameter of each tree and ending at the southwest corner. Save the drawing as E-8-4.

5. Use the drawing below as a guide to draw a 4' × 4' window with a half-round window above. Draw all window dividers as ½"-wide. Save the drawing as E-8-5.

6. Using the drawing below as a guide, open the ARCHBASE template and draw a 32" × 21" double sink. Show the outline of the sink with rounded corners using a 2" radius. Divide the sink into two equal portions. Draw a 2"-diameter circle in the center of the right portion of the sink. Save the drawing as E-8-6SNK.

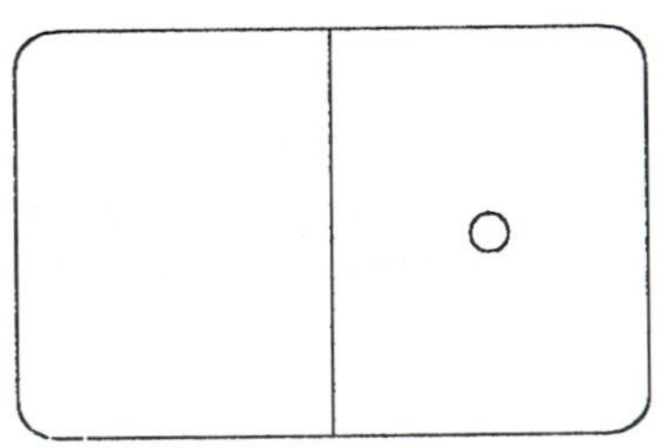

7. Using the drawing below as a guide, open the ARCHBASE template and draw a rectangle to represent a 60" × 32" tub/shower. Draw a rectangle with 3" radius rounded corners at one end and a full radius at the other end. Draw a triangle at the squared end to represent a shower head. Save the drawing as E-8-7TUB.

8. Using the drawing below as a guide, open the ARCHBASE template and draw a 48"-wide × 24"-deep, 0'-clearance fireplace. Draw the face of the fireplace as 8" wide on each side, with the interior edges 20" long, set at a 15° angle. Save the drawing as E-8-8FIRE.

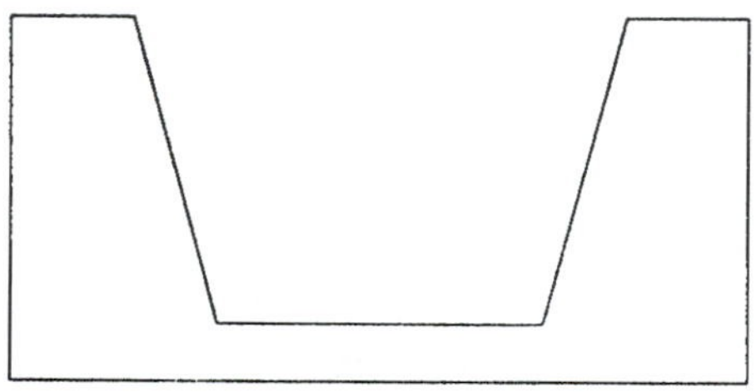

9. Open the ARCHBASE template and draw a 42" × 36" rectangle using continuous lines. Draw diagonal lines from opposite corners using hidden lines and show a 3"-diameter circle at the intersection of the interior crossing lines. Save the drawing as E-8-9SHOW.

10. Draw a plan view of a 10'-long × 24"-wide bathroom vanity. The vanity is to have a 30"-wide knee space in the center that should be represented by hidden lines. Place a 18" × 16" oval sink centered on each side of the knee space. Save the drawing as E-8-10VAN.

CHAPTER 8 QUIZ

1. What command is used to change the size of the aperture box?
2. What is the size range for the aperture box?
3. Type the command sequence to provide running snap to an endpoint.
4. List three OSNAP options for connecting a line to a circle.
5. What can be done to save time when using OSNAP on a huge drawing?
6. List three methods to stop running the OSNAP mode.
7. What steps would be needed to draw a perpendicular line to the midpoint of an existing line?
8. List the letters that are used to activate the OSNAP mode used to connect geometric shapes.
9. How can you get the dialog box for OSNAP to be displayed?
10. When using the quadrant option of OSNAP, at what degree will lines be attached to a circle?

CHAPTER 9

Drawing Display Options

In the last chapter you were introduced to drawing accurately by snapping to specific points of an object. Your drawing accuracy can also be improved by changing the size of the view of the drawing and changing the drawing resolution. The drawing view and the view size can be adjusted by the ZOOM, VIEW, PAN, and AERIAL VIEW commands. Each of these commands can be used while another command is in progress to aid viewing. Changing the view resolution can adjust the quality of the view and the speed at which the views are changed. Cleaning the screen can be accomplished by controlling the BLIPMODE, or using the REDRAW or REGEN commands.

ZOOM

Similar to the zoom lens of a camera, AutoCAD's ZOOM command has a potential zoom ratio of 10 trillion to one. "Zooming in" on a drawing can magnify a portion of a drawing to allow for better visual quality. Zoom can be used in the middle of another command sequence, which can greatly improve accuracy on detailed drawings. As the image of the object is increased, the area of the drawing that can be seen is reduced. This can be seen in Figures 9.1 and 9.2. The opposite holds true when you reduce the apparent size of an object, more of the drawing can be seen. It is important to keep in mind that the actual size of the object is not changing, only the magnification.

Zoom options can be accessed from the Zoom toolbar, by selecting one of three different icons in the Standard toolbar, from the View pull-down menu, by the right-click shortcut menu, or by keyboard. These icons include the Zoom Realtime, Zoom Flyout, and Zoom Previous and will be discussed throughout the chapter. Figure 9.3a shows the Zoom toolbar and the Zoom flyout menu found in the Standard toolbar. ZOOM can be activated by selecting Zoom from the View pull-down menu. This will produce the selections seen in Figure 9.3b. Right-clicking will produce the real-

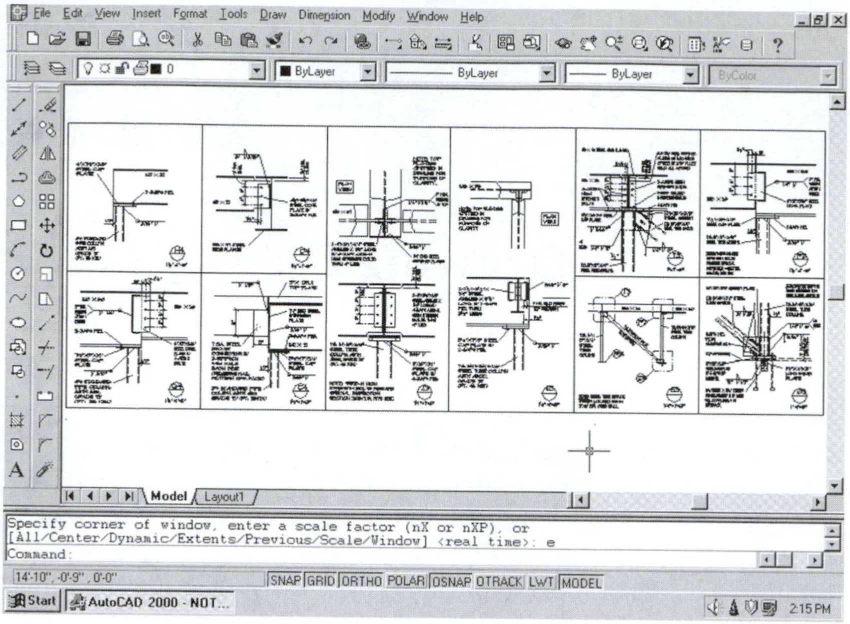

Figure 9.1 *As the drawing limits are enlarged, many features of the drawing may not be legible. (Courtesy StructureForm Masters, Inc.)*

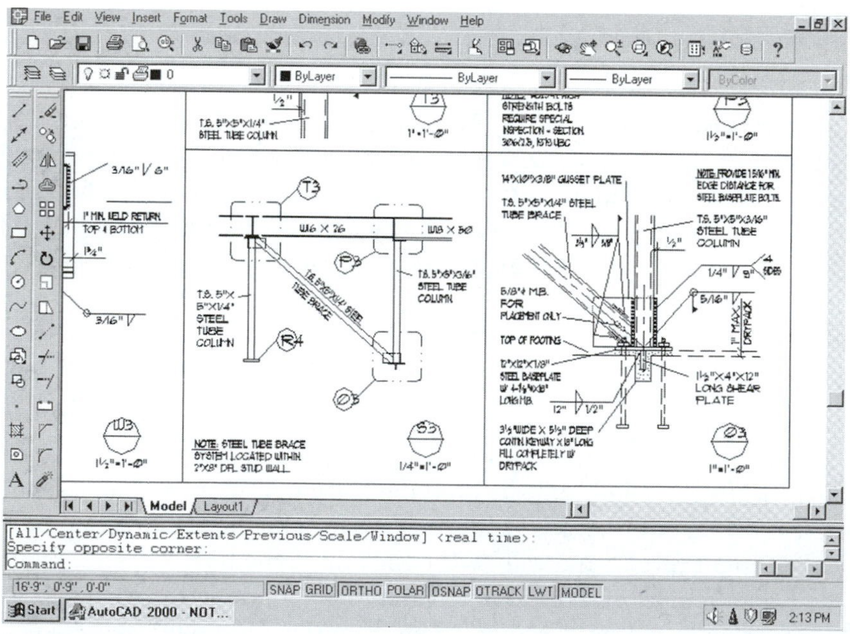

Figure 9.2 *The ZOOM command alters the position of a drawing to be viewed, giving the sense that the portion of the drawing being viewed has been enlarged. (Courtesy SctructureForm Masters, Inc.)*

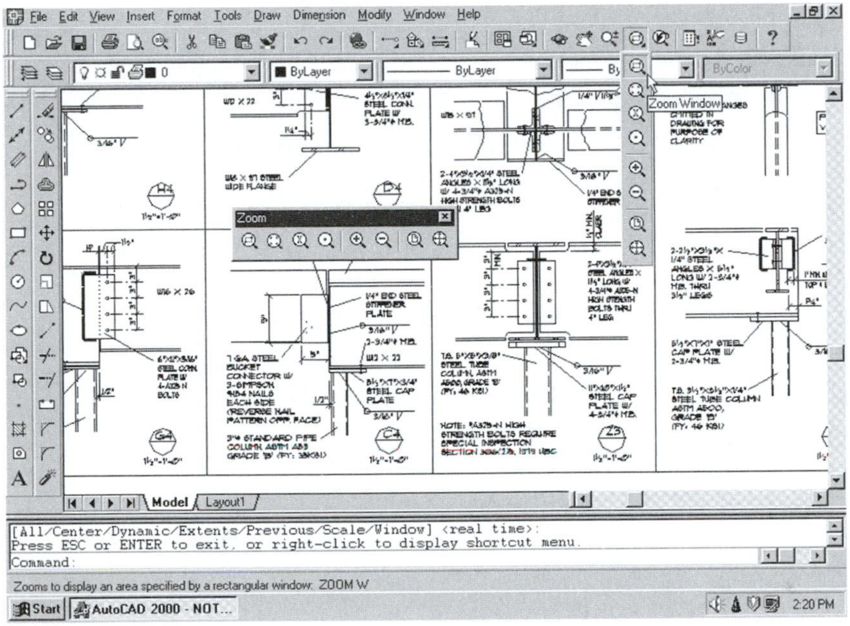

Figure 9.3a *The Zoom toolbar and the Standard toolbar flyout menu can be used to select ZOOM options.*

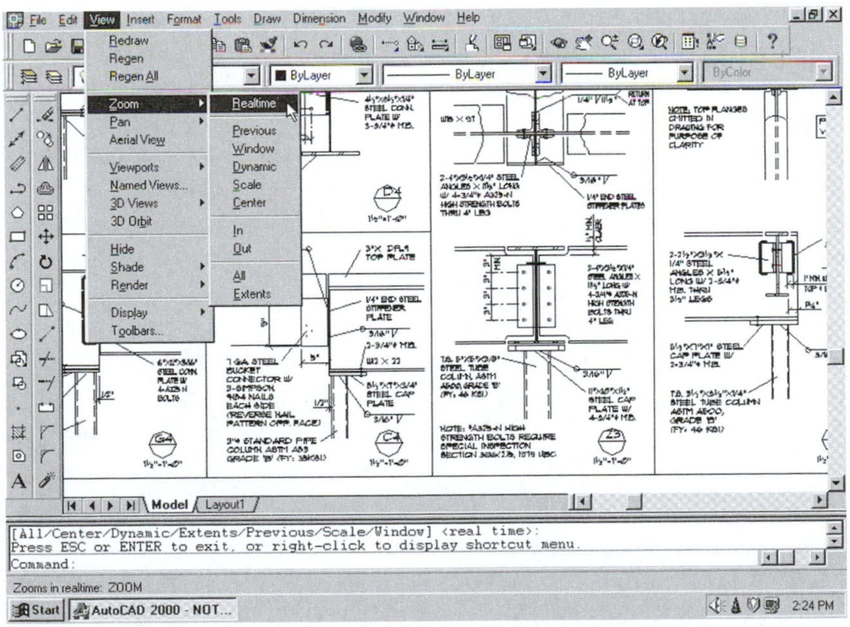

Figure 09.3b *The ZOOM command can be selected by typing* **Z** *(EnterKey) at the command prompt or by selecting the command from the View pull-down menu.*

time option of ZOOM. The command sequence can also be started by typing Z (EnterKey) at the command prompt. This will produce the prompt:

Command: **Z** (EnterKey)

[All/Center/Dynamic/Extents/Previous/Scale/Window]<Real time>:

Each of the options listed at the command prompt and in the pull-down menu can be selected from the Zoom flyout icon in the Standard toolbar.

REALTIME COMMAND

Real time zooming means that as you move the cursor, you see the results. The REALTIME command is entered by the right-click shortcut menu, by selecting the Realtime icon from the Standard toolbar, from the Zoom option of the View pull-down menu, or by entering **RTZOOM** (EnterKey) at the command prompt. The command can also be accessed by entering Z (EnterKey) at the command prompt. As the command is activated, the cursor will change to a magnifying glass with a + and – symbol.

Move the mouse around the screen and nothing happens. As you press and hold the pick button, and move the mouse toward the top of the screen, the magnification of the drawings is enlarged. Placing the magnifying glass in the middle of the screen and moving to the top will enlarge the display 100%. Pressing and holding the mouse select button and moving from the middle to the bottom of the screen will decrease the magnification 100%. Placing the magnifying glass at the bottom of the screen and moving to the top will enlarge the display 200%. Moving from the top to the bottom of the screen will decrease the magnification 200%. Each command can be repeated indefinitely, until the desired magnification is achieved. If you have trouble making Realtime Zoom alter the drawing size, remember that you must hold down the left mouse button as the mouse is moved. The command is ended by pressing (EscKey), the space bar, (EnterKey), or by pressing the right mouse button and selecting Exit. Pressing the right mouse button will produce the shortcut menu shown in Figure 9.3c. Selecting Exit will end ZOOM and return the command prompt. The menu also allows other viewing options to be selected. By alternating between the Pan and the Zoom option, a drawing can be enlarged, keeping the desired portion on the screen. Each of these options will be discussed throughout the balance of this chapter.

ZOOM WINDOW

The Window option of ZOOM is a common option for zooming into a drawing. This option lets you select the area you wish to view by providing two opposite corner locations that will form the viewing window. The size of the view box can be

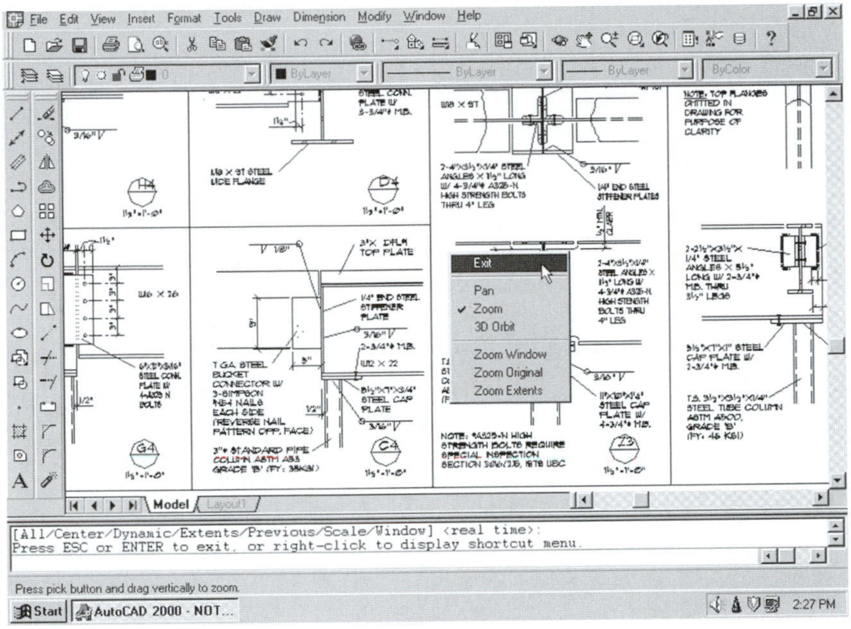

Figure 9.3c *Pressing the right-clicking while in the ZOOM command will produce the ZOOM shortcut menu.*

selected by entering coordinates or by using the mouse to select locations on the screen. The command can be accessed by toolbar, pull-down menu, from the Standard toolbar, or by entering Z (EnterKey) at the keyboard. For point selection, the command process is:

> Command: **Z** (EnterKey)
>
> [All/Center/Dynamic/Extents/Previous/Scale/Window]<Real time>: **W** (EnterKey)
>
> Specify first corner: (*Pick a corner.*)
>
> Specify other corner: (*Pick a corner.*)

This process can be seen in Figure 9.4a. Objects within the view box will be redisplayed on the screen as the new display. The process is similar when entering coordinates with a mouse. Instead of entering coordinates, you select the first corner of the zoom display by moving the cursor to the desired location. Once a corner is selected, the crosshairs will switch to a box that is enlarged or reduced as the mouse is moved across the drawing area. Figure 9.4b shows the results of the selection made in Figure 9.4a. Entering Z (EnterKey), W (EnterKey), or selecting the Zoom Window icon are the most efficient methods of starting the command.

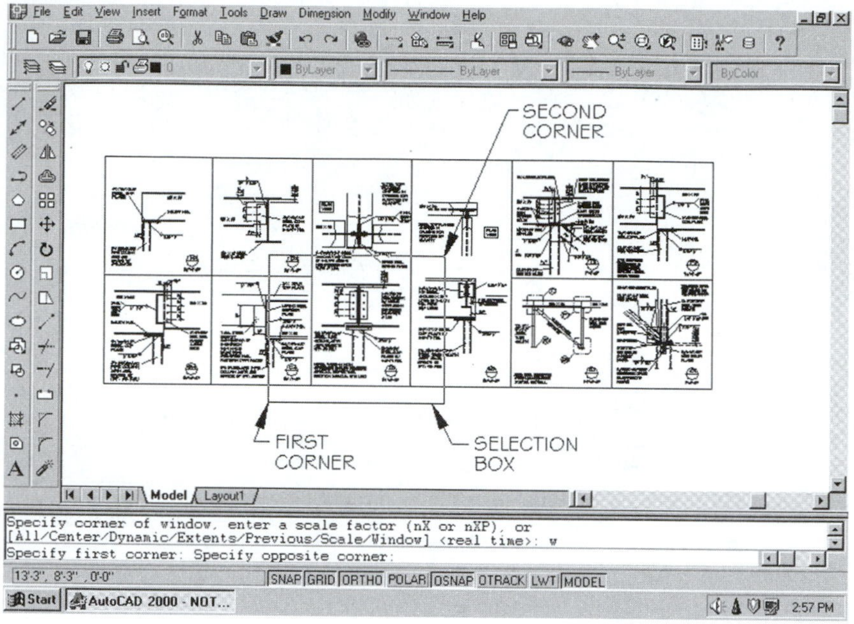

Figure 9.4a *The ZOOM window option allows the display area to be selected by specifying opposite corners of the "window." (Courtesy StructureForm Masters, Inc.)*

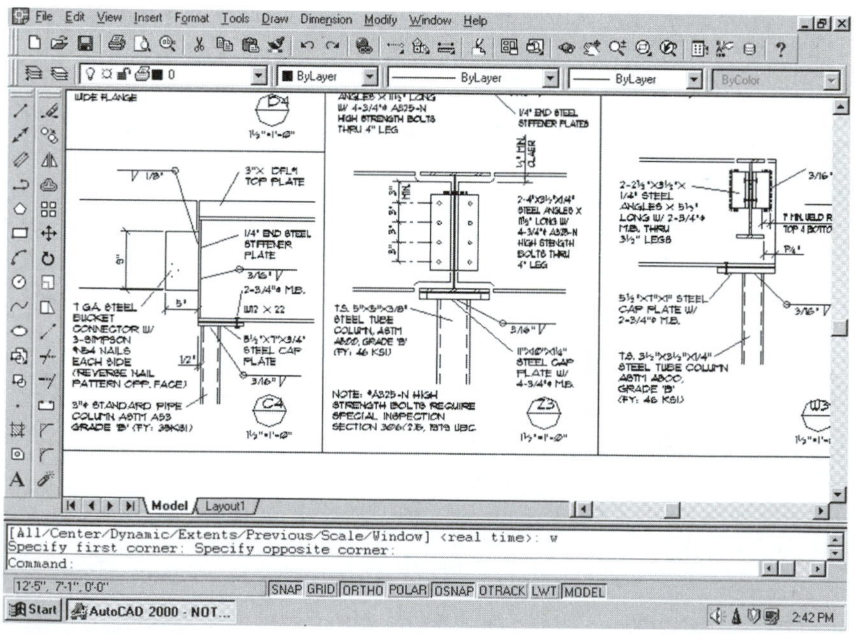

Figure 9.4b *The ZOOM window created in Figure 9.4a will now display the selected area. (Courtesy StructureForm Masters, Inc.)*

TRANSPARENT ZOOM

One of the really great features of ZOOM is that it is transparent, meaning it can be used in the middle of another command. If you start to draw a line and realize that you're straining to see the line, the screen image can be magnified using ZOOM in the middle of a LINE command sequence. Simply select the icon of the desired Zoom option or enter the command by keyboard. To use a transparent command from the keyboard, type an ' (apostrophe) prior to the command. The command sequence for using the window option of ZOOM in the middle of the LINE command would be:

Command: **L** (EnterKey)

LINE

Specify first point: *(Select first line endpoint.)*

Specify next point or [undo]: **'Z** (EnterKey)

_zoom

>>Specify corner of window, enter a scale factor (nX or nXP), or

[All/Center/Dynamic/Extents/Previous/Scale/Window]<Real time>: **W** (Enter-Key)

>>Specify first corner: *(Pick desired start point.)*

>>Specify opposite corner: *(Select opposite corner to be enlarged.)*

Resuming LINE command.

Specify next point or [undo]: *(Select desired start point.)*

Specify next point or [undo]: (EnterKey)

Command:

ZOOM could be used at any point during the LINE or any other command when a prompt is given. In the LINE command, the 'ZOOM command could have been started at the prompt for "First point" or at the "Next point" prompt. Enter the LINE, CIRCLE, or ARC commands and then use 'ZOOM to become familiar with this option of zoom. Your eyes and back will love you as you zoom more and squint less.

ZOOM ALL

This option will allow for an entire drawing to be viewed, based on the drawing limits. If the drawing extends beyond the drawing limits, the screen display will show the entire drawing. Zoom All should be used if you are zoomed into a small area of a drawing, such as Figure 9.2. Using the All option returns all of the drawing to the screen so that the display resembles Figure 9.1. Zoom All can be selected from the Zoom flyout menu in the Standard toolbar, from the right-click menu, from the pull-down menu, or by entering Z (EnterKey), A (EnterKey) at the command prompt. The command sequence is:

Command: **Z** (EnterKey)

ZOOM

Specify corner of window, enter a scale factor (nX or nXP), or

[All/Center/Dynamic/Extents/Previous/Scale/Window] <Realtime>: **A** (EnterKey)

Command:

ZOOM CENTER

This option allows a zoom to be specified by its desired center point. Once a center point is selected, a prompt will be given for the height of the display. If the current default is maintained, the drawing is redisplayed based on the new center point, but the magnification is not changed. If a smaller value for the height is selected, the display magnification will be increased. If a larger value for the height is selected, the display magnification will be reduced. The command sequence is:

Command: **Z** (EnterKey)

ZOOM

Specify corner of window, enter a scale factor (nX or nXP), or

[All/Center/Dynamic/Extents/Previous/Scale/Window]<Realtime>: **C** (EnterKey)

Specify center point: *(Enter coordinates or select a point.)*

Enter magnification or Height < 17'-4">: **10'** (EnterKey)

Command:

ZOOM DYNAMIC

This option will allow specific areas of the drawing to be selected for ZOOM viewing. The command will seem complicated at first. Once you master it, you'll be able to select any portion of the drawing for the next view without requiring a REGEN. The importance of this may seem lost on you as you compare a dynamic to a window zoom while working on the exercises. As you start to work on larger projects using hundreds of thousands of bytes of information, a Dynamic ZOOM will be well worth the effort because of the savings in time from not having to regenerate a drawing. The original view box can be enlarged or reduced and moved about the drawing screen to select the exact portion of the drawing to be zoomed. The command sequence is:

Command: **Z** (EnterKey)

ZOOM

Specify corner of window, enter a scale factor (nX or nXP), or

[All/Center/Dynamic/Extents/Previous/Scale/Window]<Realtime>: **D** (EnterKey)

The screen will be cleared, then the original view will be returned, surrounded by three color-coded boxes. Each box can be seen in Figure 9.5 and represents a different viewing area.

The first box that will be displayed is the Pan box. The Pan box allows you to drag the drawing across the screen to determine which portion of the drawing you would like to view. (See Figure 9.5.) Once the desired portion of the drawing is shown on the screen display, press the mouse pick button and the Zooming window is displayed. The view box represents the current viewport. Viewports will be discussed later. For now, you only have one viewport. The view box allows the size of the current viewport to be adjusted, either by shrinking or enlarging, or by moving it at its current size around the screen. Adjust the size of the window and press (EnterKey). Press the

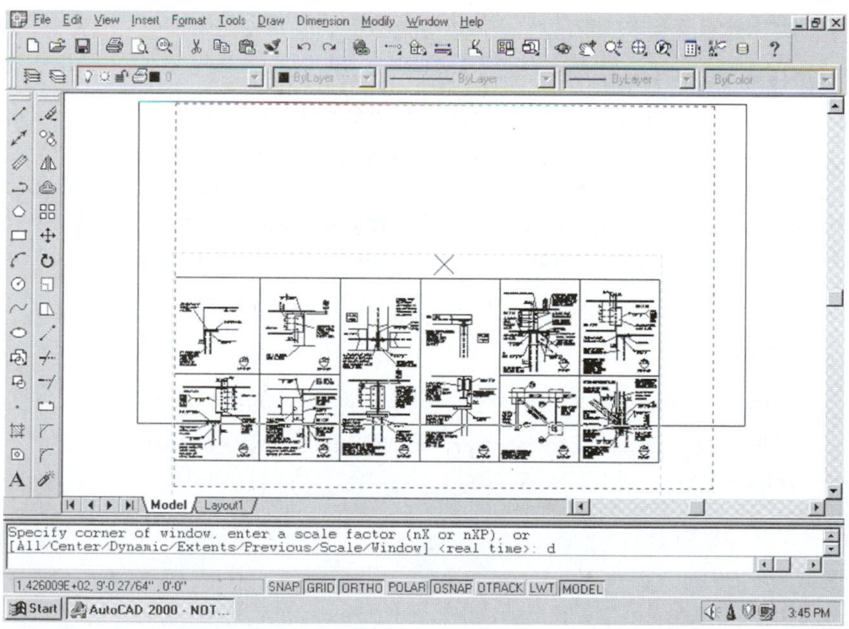

Figure 9.5 *Selecting the ZOOM Dynamic option will display the original drawing area, as well as three view boxes.*

pick button if you need to readjust the portion of the drawing to be displayed. It sounds complicated, but it's not. It may not be worth the effort either. The Dynamic Zoom was a great method of moving through a drawing, until Realtime Zoom and Pan were added to AutoCAD. With these two options added to the Standard toolbar, Zoom Dynamic has almost been rendered obsolete.

ZOOM EXTENTS

This option will display the entire drawing based on the size of the drawing—not the size of the limits that were used for Zoom All. If a small object is being examined with large drawing limits, Extents will enlarge the object to fill the drawing screen. This can be quite helpful during the initial stages of a drawing. Zoom Extents is also helpful if the extent of the drawing exceeds the drawing limits, because the option allows the entire drawing to be examined. The command sequence is:

Command: **Z** (EnterKey)

ZOOM

Specify corner of window, enter a scale factor (nX or nXP), or

[All/Center/Dynamic/Extents/Previous/Scale/Window]<Realtime>: **E** (EnterKey)

Command:

ZOOM PREVIOUS

Each of the other ZOOM options are used to enlarge an area of a drawing for easier viewing. The Previous option will allow you to go backward, so that the previous drawing area can be seen. When you use this option, the drawing displayed prior to the ZOOM will be returned to the screen. A maximum of 10 previous views can be restored using the Previous option. The command sequence for Zoom Previous is:

Command: **Z** (EnterKey)

ZOOM

Specify corner of window, enter a scale factor (nX or nXP), or

[All/Center/Dynamic/Extents/Previous/Scale/Window]<Realtime>: **P** (EnterKey)

Command:

ZOOM SCALE

The default selection for the ZOOM command is Scale or magnification factor (X). The scale may be set to be relative to the full view or to a current view.

Scale Relative to Full View

When you enter a number for the scale factor, the entire drawing will be affected. A scale factor of 1 will display the current size. A scale factor of 3 will make objects appear three times as large. By entering a number smaller than 1, you will decrease the size of the object. If you enter .5, the object will appear half as big as the full display. The command sequence is:

> Command: **Z** (EnterKey)
>
> ZOOM
>
> Specify corner of window, enter a scale factor (nX or nXP), or
>
> [All/Center/Dynamic/Extents/Previous/Scale/Window]<Realtime>: **S** (EnterKey)
>
> Enter scale factor (nX or nXP): **.5** (EnterKey)
>
> Command:

Scale Relative to Current View

If you enter a numeric value followed by an "X", the scale of the zoom will be relative to the current view, rather than the entire drawing. The command sequence is:

> Command: **Z** (EnterKey)
>
> ZOOM
>
> Specify corner of window, enter a scale factor (nX or nXP), or
>
> [All/Center/Dynamic/Extents/Previous/Scale/Window]<Realtime>: **S** (EnterKey)
>
> Enter scale factor (nX or nXP): **1.5X** (EnterKey)
>
> Command:

Scale Relative to Paper Space Units

AutoCAD thinks of space as model space and paper space. 2D and 3D drawings are typically drawn using model space. Paper space is used to plot views of the model that have been created. Zoom XP can be used to scale each space relative to paper space units. This will be helpful, for instance, in printing drawings on the same sheet that are drawn at different scales. This topic will be covered in more detail in future chapters.

ZOOM IN

Zoom In automatically scales the current drawing screen display by a scale factor of 2. Selecting this Zoom option produces the same result as selecting a scale factor of **2X** (EnterKey) from the command line. The Zoom In option is available in the

Zoom toolbar, the Zoom Flyout menu in the Standard toolbar, and from the Zoom listing in the View pull-down menu.

ZOOM OUT

Zoom Out automatically enlarges the drawing area that is displayed on the screen by a factor of 2. The drawing is, in effect, reduced by a scale factor of .5X. The Zoom Out option is available in the Zoom toolbar, the Zoom Flyout menu in the Standard toolbar, and from the Zoom listing in the View pull-down menu.

PAN

The PAN command can best be visualized by thinking of your drawing as a large sheet of paper that is rolled up. As you pan across your drawing, it's just as if you were unrolling the drawing to view a different portion. The drawing magnification is not changed, just the portion of the drawing that is displayed. This can be helpful when you've zoomed in on a drawing, similar to the one in Figure 9.2, and are looking at a specific detail. Rather than having to ZOOM All, and then ZOOM Window to a new detail, PAN can be used to bring in the desired detail. The PAN command is accessed by selecting the Pan Realtime icon from the Standard toolbar, by selecting Pan from the View pull-down menu, or by selecting Pan from the right-click menu. The command can also be accessed by entering **P** (EnterKey) by keyboard. The sub-menu for PAN can be seen in Figure 9.6.

PAN REALTIME

The Pan Realtime command is similar to Zoom Realtime. The easiest method of entering the command is to select the Pan Realtime icon in the Standard toolbar. No matter the access method, as the command is activated, the cursor will change to a hand cursor. The command prompt displays the prompt:

Press Esc or ENTER to exit, or right-click to display shortcut menu.

Move the cursor (the hand) to the desired pick point and then press and hold the pick button. This point now becomes the displacement point of the Pan Command. As the cursor is moved while the mouse pick button is being pressed, the screen display is altered. Panning can be repeated indefinitely, until the desired display is achieved. The command is ended by pressing the (EscKey) or (EnterKey) key. Figures 9.7a and 9.7b compare the effects of using the PAN Command.

Combining Realtime Zoom and Pan.

By entering either Pan Realtime or Zoom Realtime using the appropriate icon in the Standard toolbar or the right-click menu, you can easily toggle between the two options and have a convenient method of moving between different portions of a large drawing. To experiment, open a drawing and draw several lines and circles.

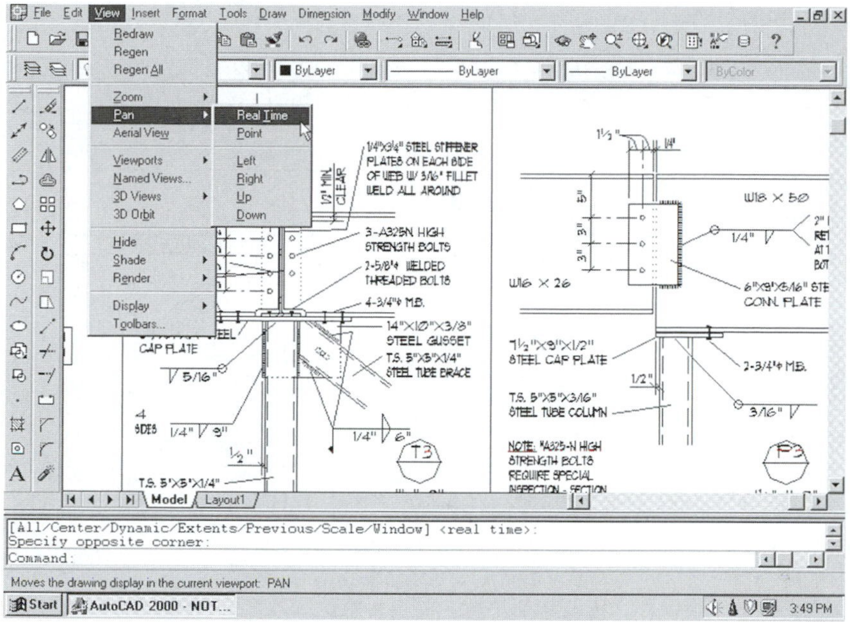

Figure 9.6 *The PAN command is accessed by selecting the Pan icon from the Standard toolbar, by selecting PAN from the View pull-down menu, or by typing* **P** *(EnterKey) at the command prompt.*

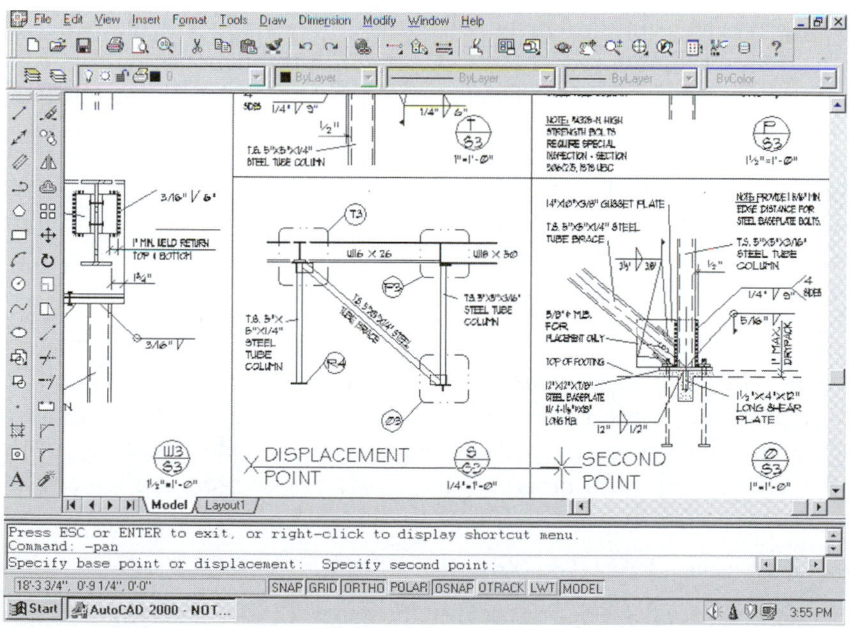

Figure 9.7a PAN allows a drawing to be "scrolled" across the screen without changing the drawing *magnification. (Courtesy StructureForm Masters, Inc.)*

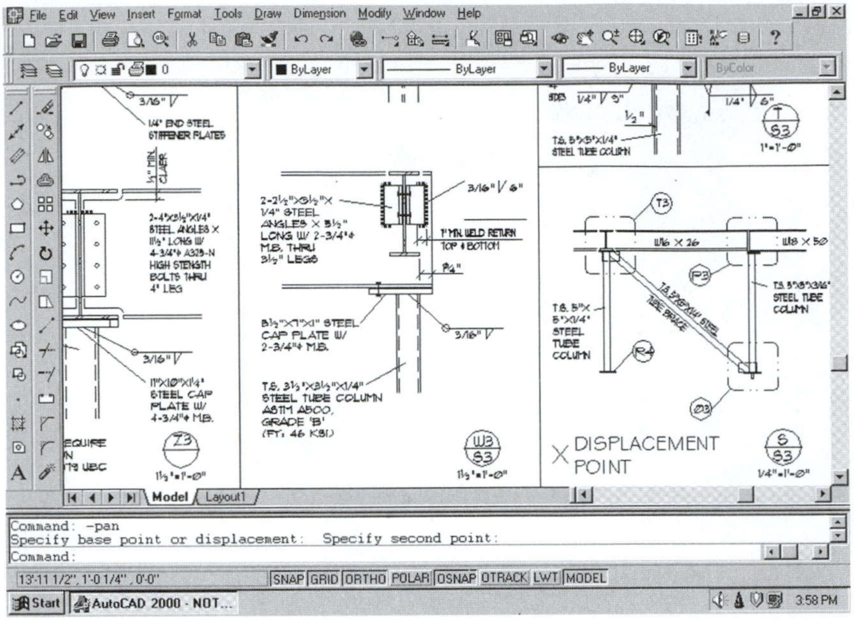

Figure 9.7b *The effects of the PAN that was started in Figure 9.7a. (Courtesy StructureForm Masters, Inc.)*

Press the Zoom Realtime icon and use the mouse to enlarge the drawing. Now press the Pan Realtime icon and the cursor will change to the hand and allow the enlarged view to be panned. Remember that by pressing the right mouse button, a pop-up menu will be displayed that allows you to toggle between the Zoom and Pan commands.

Entering Pan from the Keyboard

The PAN command can also be entered by typing **–P** (EnterKey) at the command prompt. Be sure to enter the minus sign or the command will function just as it did when the icon is selected

Once entered, the command will prompt you to select a point on the drawing screen and then ask, in effect, where would you like to move that point. Points may be entered by providing coordinates from the keyboard or by selecting a point on the screen. The drawing can be panned in greater increments by entering coordinates. Figure 9.7a shows the PAN process. The results can be seen in Figure 9.7b. The command sequence is:

Command: **–PAN** (EnterKey)

Specify base point or displacement: *(Enter point or coordinates of point to be dragged.)*

Specify second point: *(Enter point or coordinates to bring first point to.)*

For most cases, the Pan Realtime option is much quicker to enter and easier to use. The –PAN option is there to make users of older AutoCAD versions feel at home.

VIEW

An alternative to ZOOM and PAN is the use of prenamed drawing views. The VIEW command allows for views or viewports to be named and saved so that they can be restored later. Viewports can be created in model and paper space. This chapter will only address model space views. Future chapters will introduce paper space viewports. By default, AutoCAD begins each drawing as one viewport. As the limits are set, you are deciding how big the viewport is. The screen can be divided into smaller viewports for easy movement between drawings. On a drawing similar to Figure 9.1, each detail box could be named as a view so that each could be revised easily. By naming each detail as a view, you have in effect predefined the areas to be viewed. Instead of using ZOOM or PAN to select a view, you could select the desired predefined VIEW. The VIEW command may seem like a waste of time on small exercises. However, the VIEW command offers great flexibility for quickly viewing areas of a large, detailed drawing if you need to constantly move from one place to another. VIEW can be accessed by selecting the Named Views icon in the Standard toolbar, or by selecting the Named View icon from the View toolbox. Selecting Named Views... from the View pull-down menu or typing V (EnterKey) at the command prompt can also be used to work with named views. Each method will produce the View dialog box shown in Figure 9.8. Figure 9.9 shows the sheet of details that have

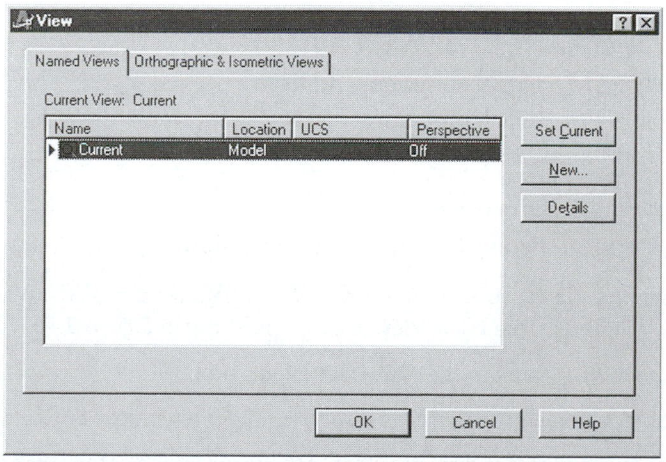

Figure 9.8 *The VIEW command offers great flexibility for quickly viewing areas of a large, detailed drawing. VIEW can be accessed by selecting the Named Views icon in the Standard toolbar, by selecting the Named View icon from the View toolbox, by selecting Named Views... from the View pull-down menu or typing **V** (EnterKey) at the command prompt.*

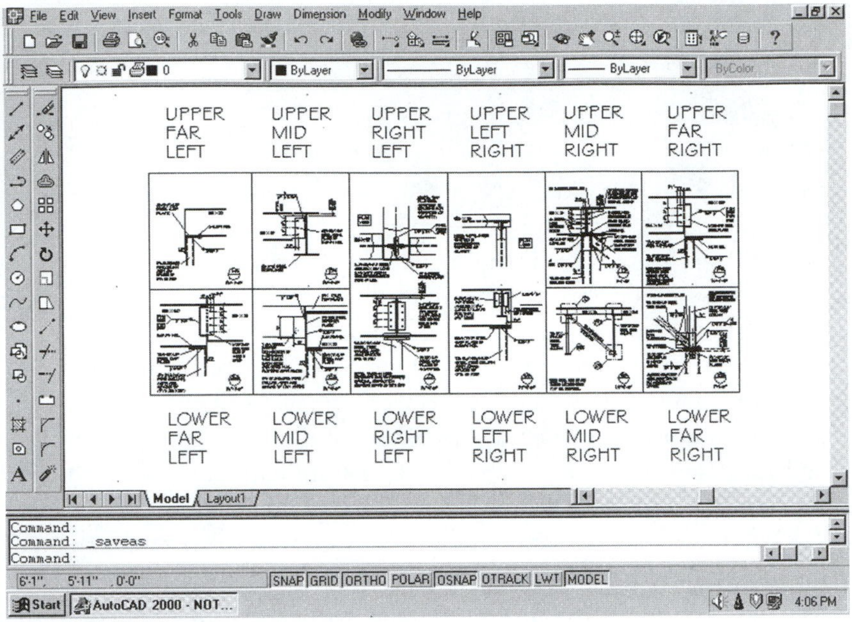

Figure 9.9 *The sheet of details that have been used throughout this chapter and the view names that will be used.*

been used throughout this chapter and the view names that will be used. The View dialog box contains the Named View and the Orthographic & Isometric Views tabs. This chapter will only explore the Named View dialog tab. Future chapters will deal with isometric views tab.

EXPLORING THE NAMED VIEWS TAB

This tab can be used to create, set, name, and delete named views. The Named Views tab displays a listing of the current drawing views. In a new drawing, the only view listed is the Current view.

Creating and Saving a Named View

The following steps can be used to create a named view.

- Select the New... button from the Named Views tab of the View dialog box. This will display the New View dialog box show in Figure 9.10.
- Assign a view name in the View name: edit box.
- Choose between the Current display or Define window buttons.
- Choose the Current display button if the view on the screen is the view that you would like to have saved.
- Choose the Define window button if you need to select the objects to be included in the view.

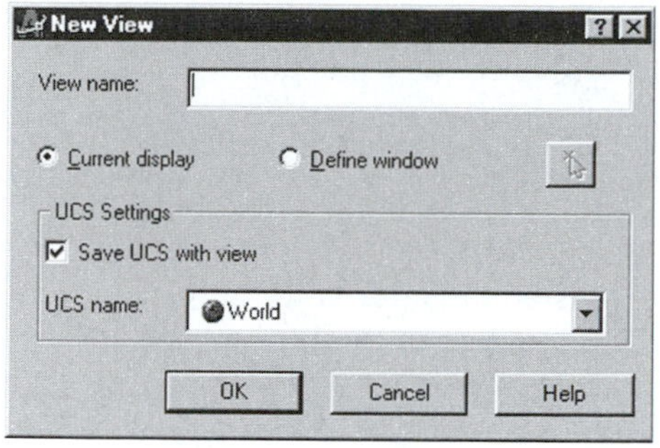

Figure 9.10 *The New View dialog box is used to assign drawing view names. The box is accessed by selecting the New... button from theView dialog box.*

- Use the view select button to define the view. Selecting this button will remove the dialog box, return the drawing area, and use the window to select the objects that will make up the view. Once the objects for the view are selected, the NewView dialog box will be restored.
- Select OK. The View dialog box will be restored, and the name of the selected view will be displayed.

Figure 9.11 shows the View dialog box with a listing of the twelve views described in Figure 9.9. The Save option saves the current screen display using a name that you supply. View titles may be up to 255 characters long, utilizing the same guidelines used to save drawing files. Views are often named by their location in the job or by

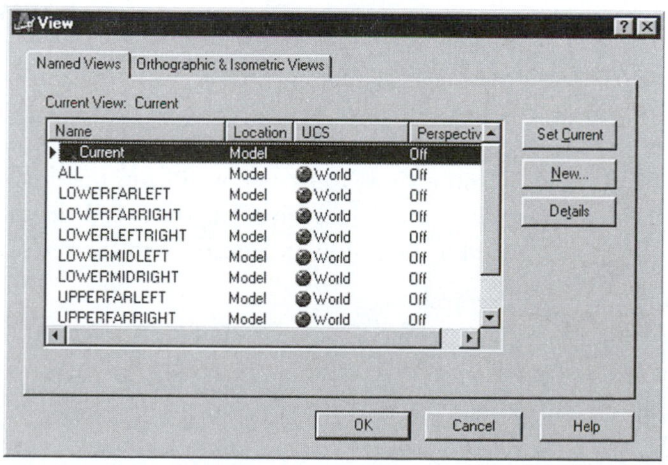

Figure 9.11 *The named views reflected in Figure 9.9 can be seen in the View dialog box. (Courtesy StructureForm Masters, Inc.)*

their specific function. If four rows of details will be drawn, each row could be given a letter and each column of details could be given a number to produce an easy grid to work with. Each room of a floor plan can also be turned into a view for easy movement through a project.

Altering the Current View

The screen display can easily be altered between named views. To make the LOWER FAR LEFT view of Figure 9.9 the current view, use the following steps:

- Enter the View dialog box. Entering **V** (EnterKey) at the command prompt is the fastest method.
- Highlight the name of the view that is to become the current view.
- Select the Set Current button.
- Pick the OK button. The dialog box will be closed and the LOWER FAR LEFT view will now occupy the screen display area.

Renaming a Named View

The name of a view can be altered in the View dialog box. Use the following steps to change the existing name of a view:

- Enter the View dialog box.
- Highlight the name of the view to be renamed.
- Right-click to produce the shortcut menu and choose Rename.
- Enter the desired name of the View.
- Select the OK button to accept the new name, close the dialog box, and return to the drawing area.

An alternative to using the shortcut menu to rename a view is to use (F2Key) once the name is highlighted. This will place the selected view name in an edit box and allow the new name to be entered.

Delete

This option will allow one or more views to be removed from the list of saved views. You're not removing the drawing, you're just removing the saved view from the list of named views. Views are deleted using a similar procedure to the rename option.

- Enter the View dialog box.
- Highlight the name of the view to be deleted.
- Right-click to produce the shortcut menu and choose Delete.
- Enter the desired name of the View to be removed from the list.
- Select the OK button to delete the view, close the dialog box and return to the drawing area.

View Details

Selecting the Details button of the View dialog box will display View Details dialog box similar to the display in Figure 9.12. This box provides information on the width and height of the selected view. Information provided includes the view name, the area of the view relative to the overall viewport, the coordinates of the view target location, and other information that is beyond the scope of entry level discussion.

AERIAL VIEW

Aerial View is a useful method of viewing different portions of a large drawing file. An overall view of the entire drawing is displayed in a small window that is overlaid over the drawing screen. This is like the picture-in-picture feature of some TVs. Portions of the drawing can be selected for viewing from this small window and redisplayed while still being able to see the entire drawing in the view window. This is especially helpful when working on a very large file. Many AutoCAD users often lose perspective of where they are in the drawing file when they have used the ZOOM command several times in the drawing. The use of Aerial View provides that perspective by displaying the overall drawing in the window. It also allows quick movement from one portion of the drawing to another, using Realtime Zoom and Pan options.

ACCESSING THE AERIAL VIEW WINDOW

The Aerial View window is displayed by selecting Aerial View from the View pull-down menu, or by typing AV (EnterKey) at the command prompt. Each method will display the window seen in Figure 9.13. As the aerial view is displayed, it will consist of the current view surrounded by a thick black box. The size of the view can be altered using either ZOOM or PAN.

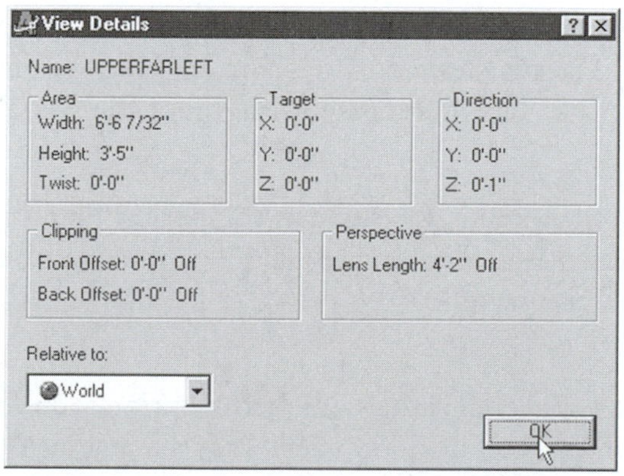

Figure 9.12 *Selecting the Details button of the View dialog box displays information including the view name; the area of the view relative to the overall viewport; and the coordinates of the view target location.*

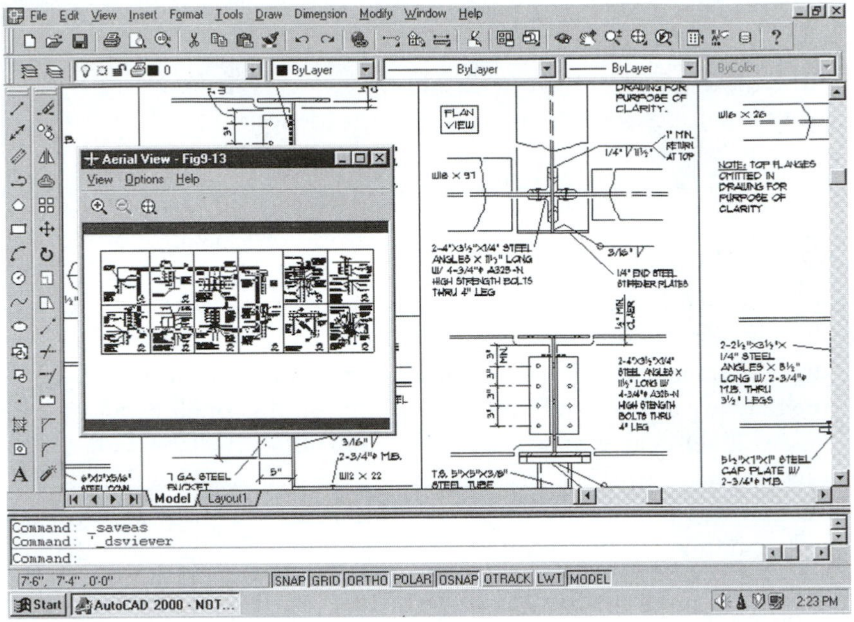

Figure 9.13 *Aerial View provides an option for moving between views of a drawing. The drawing extents are shown in the Aerial View Window box.*

AERIAL VIEW WINDOW MENUS

The two pull-down menus of Aerial View include View and Options.

View Options

The options for view include Zoom In, Zoom Out, and Global. If the entire drawing is displayed as Aerial View is entered, the Zoom In/Out options will not be active. These options are used to change the magnification of the Aerial View.

Zoom In

The Zoom In option of Aerial is the same as the option found in the Zoom option of the View pull-down menu. The option increases the magnification of the image displayed in the Aerial view window.

Zoom Out

The Zoom Out option of Aerial is the same as the option found in the Zoom option of the View pull-down menu. The option decreases the magnification of the image in the Aerial view window.

Global

Selecting this option will display the entire view in the Aerial view window.

Options Menu

The Options menu includes the options of Auto Viewport, Dynamic Update, Locator Magnification..., and Display Statistics....

Auto Viewport

This option is a toggle switch between ON/OFF. When a check is displayed, the setting is ON. In the ON mode, switching to a different tiled viewport will automatically cause the new viewport to be displayed in the aerial view window. In the OFF mode, a new viewport will not be displayed. Tiled viewports will be discussed in future chapters.

Dynamic Update

This option is a toggle switch between ON/OFF. When a check is displayed, the setting is ON. In the ON mode, the window will update the display after each change in the drawing.

Real Time Zoom

This option controls how the window will be updated. In the default setting, the window will be zoomed in realtime. By clicking this option, the view will be enlarged in Zoom rather than in Real Time Zoom.

USING THE AERIAL VIEW WINDOW

The AERIAL VIEW command allows the use of Realtime Zoom and Pan to move between areas of the drawings.

Zoom

This option changes the viewport center and view size. The Zoom mode changes the current view by allowing points to be selected that will become the new corners of the current view. Use the following steps to alter the view with the Zoom option of Aerial window:

- Display the Aerial View box.
- Right-click and select Realtime Zoom. Realtime Zoom is active when the listing is preceded by a check. Realtime Zoom can also be activated from the Options menu of the Aerial View dialog box.
- In the Aerial View window, left-click to display the pan and zoom box. The window will toggle between pan/zoom by clicking the pick button. The Pan window will have an X in the middle of the box. The Zoom window will have an arrow on the right edge of the window.
- With the Zoom option active, move the mouse to control the size of the window.
- Press escape to end Realtime Zoom.
- Left-click to toggle to Realtime Pan.

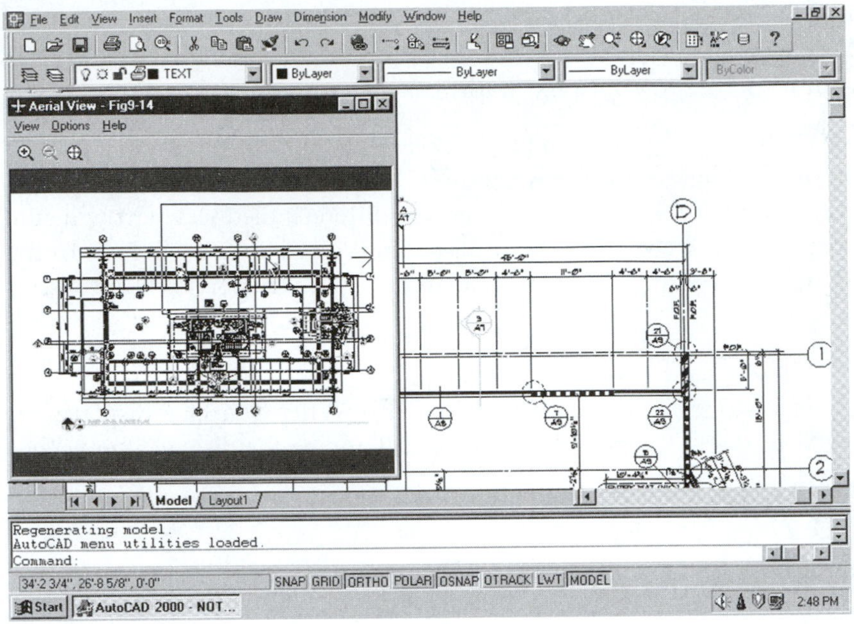

Figure 9.14 *With the Zoom option active, an aerial view window can be created. As the mouse is moved, the window size is altered and then the area and size of the drawing are altered.*

Pan

This option allows the display to be altered from side to side, or top to bottom, without changing the drawing magnification. The Aerial View provides a point of reference by showing the entire structure. By selecting the Pan option, a reference box will be displayed in the Aerial View to pan the drawing. As the reference box is moved in the aerial view, the portion of the drawing displayed in the drawing area of the screen is altered. Move the reference box, and the drawing will snap to show the area contained in the reference box. Without pressing or holding any mouse buttons, the drawing will drag in realtime to reflect the movement of the reference box. Figure 9.15 shows an example of the effect of the Pan option.

REFRESHING THE SCREEN

This chapter has introduced methods of controlling the drawing display using the Zoom, Pan, View and Aerial View commands. Each command alters the view that is placed in the drawing area. The final aspect of controlling the drawing view is by adjusting the quality of the view that is displayed. The view quality can be controlled using Blipmode, Redraw, Regeneration, and View Resolution settings.

BLIPMODE

Blips serve as temporary markers to assist in selecting objects or locations. The BLIPMODE command controls the generation of blips. In the default mode, no blips or

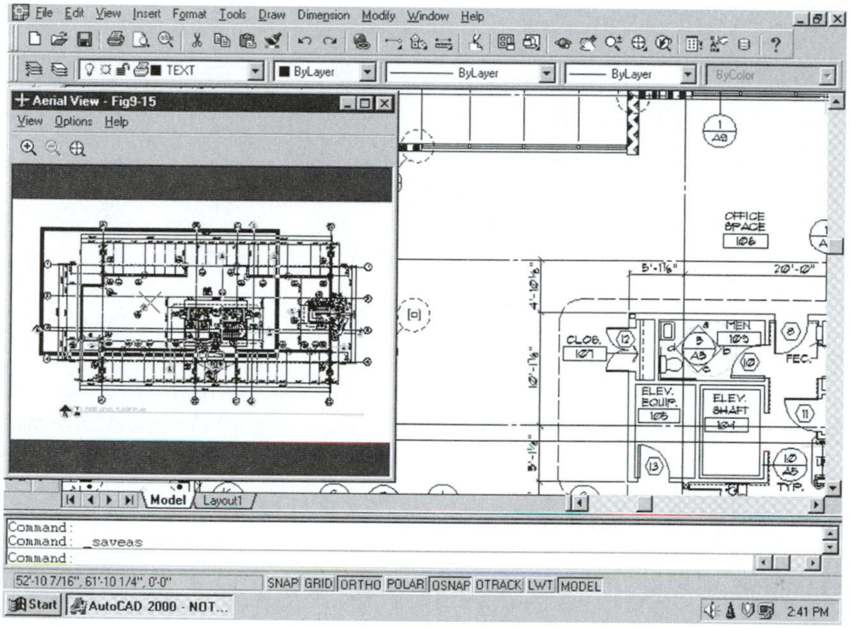

Figure 9.15 *Using the Pan option in the Aerial View to pan the drawing. As the reference box is moved in the aerial view, the portion of the drawing displayed in the drawing area of the screen is altered.*

markers will be made as lines, arcs, circles, or other drawing entities are created. These markers may be helpful for some commands, such as CIRCLE or ARC, to help you visualize the starting points. The setting may be changed throughout the life of the drawing. To change the display, type **BLIPMODE** (EnterKey) at the command prompt.

Command: **BLIPMODE** (EnterKey)

Enter mode [ON/OFF] <ON>:

With the BLIPMODE setting of ON as default, markers are displayed on the screen each time a point is designated. No markers are displayed when the OFF option is selected. For most uses, BLIPMODE can be left in the default setting of OFF. When ON, blips can be removed as desired. When ON is selected, each time an entity is drawn, a blip is created on the screen to mark the First, Next, or Center point. Blips can be removed by toggling GRID between ON/OFF with the F7 key. Blips are automatically removed from the drawing screen when the ZOOM, PAN, REDRAW, and REGEN commands are used.

REDRAW

The REDRAW command removes the current screen display momentarily, and then replaces the drawing. As the drawing is replaced, blips are removed from the drawing editor. Lines that may have been partially erased while editing other entities also will be restored. The command is accessed by selecting the Redraw All icon fromt the Standard toolbar or by typing **R** (EnterKey) at the command prompt. The same results can be achieved by selecting Redraw from the View pull-down menu. REDRAW may be used while another command is in progress if it is entered at any nontext prompt. On drawings similar in size to the exercises that you've been doing, REDRAW will seem almost instantaneous. On larger drawings, such as a subdivision map or floor plan, REDRAW can take slightly longer. Figure 9.16a and 9.16b show the effect using REDRAW.

REGEN

Similar to REDRAW is the slightly slower REGEN command. AutoCAD clears the drawing screen and then regenerates the entire drawing. AutoCAD recalculates the current screen display with 14-place accuracy, keeping track of each pixel in the X and Y direction. If you have zoomed in on a specific part of a drawing, REGEN is regenerating even the part of the drawing that cannot be seen. A screen regeneration can be started by entering **REGEN** (EnterKey) at the command prompt or by selecting REGEN from the View pull-down menu. To terminate a REGEN, press the ESC key. REGEN can also be used to improve the resolution of circular objects and polygons Figures 9.17a and 9.17b compare the effects of REDRAW and REGEN

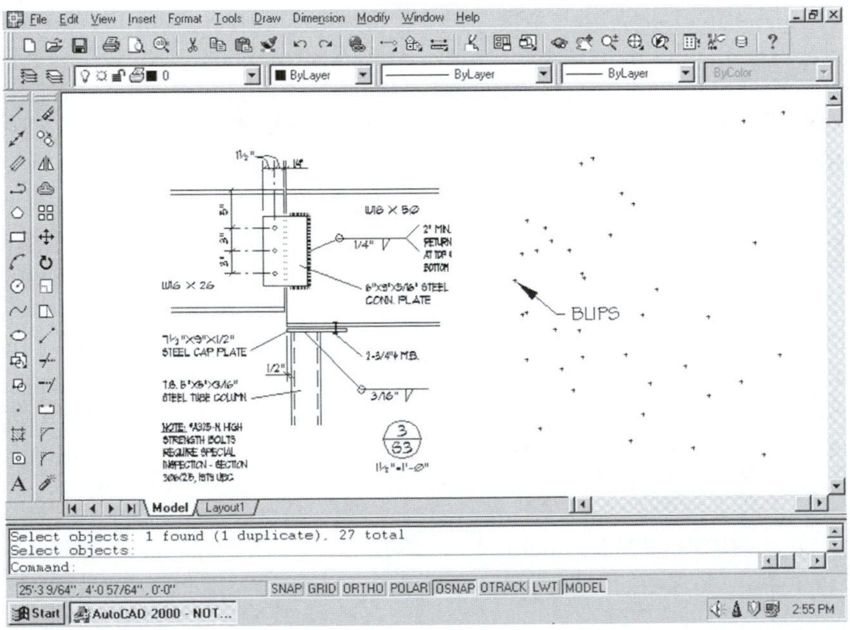

Figure 9.16a *Blips can serve as temporary markers to assist in selecting objects or locations. Blips will be added if BLIPMODE is set to ON.*

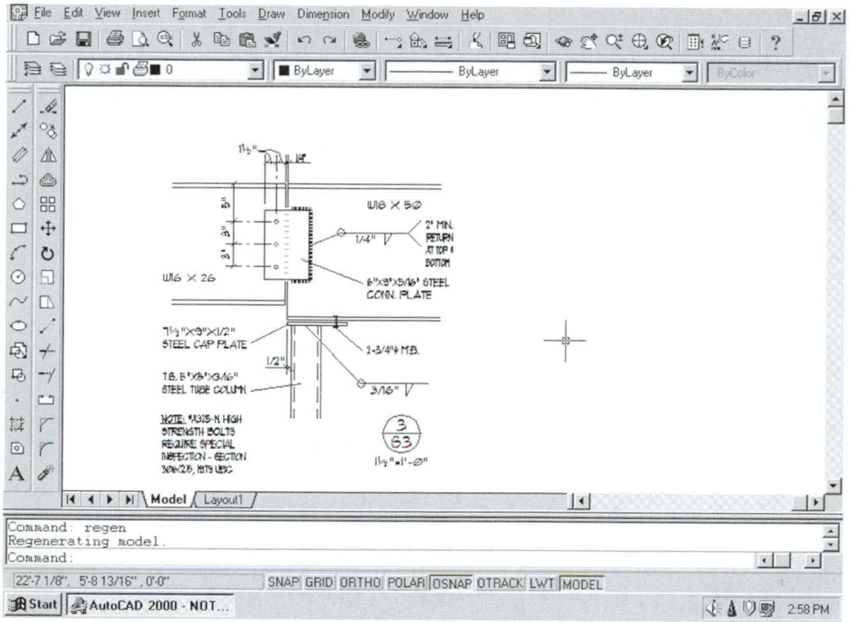

Figure 9.16b *Blips can be removed from a drawing using the REDRAW or REGEN commands.*

Figure 9.17a *Circular objects often appear as straight line segments when magnified or when a low view resolution is used. Redraw will not effect the smoothness of such features.*

Figure 9.17b *The smoothness of circular objects will be improved using the REGEN command.*

on drawing objects. You'll learn how to create viewports in future chapters. Using viewports will speed up each REGEN because only the current viewport is regenerated.

REGENALL

In future chapters you'll learn how to divide the drawing screen into different views so that specific areas of the drawing can be viewed quickly. When multiple views are created, the REGEN command will only affect the current drawing view. All other viewports will remain unaffected. The REGENALL command will regenerate all of the views, so that blips in each view will be removed and reindex the drawing database to increase objection selection. The command can be accessed by selecting Regen All from the View pull-down menu or by entering **REA** (EnterKey) at the command prompt. The significance of the command will be discussed in future chapters.

REGENAUTO

AutoCAD allows the regeneration mode to be set. On a small drawing, an automatic regeneration may not even be noticed. As the drawing size increases, using REGEN may prove to be a frustration. AutoCAD allows a prompt to be given to remind you that the drawing is about to regenerate. This aspect of REGEN can only be entered by keyboard. The command sequence is:

Command: **REGENAUTO** (EnterKey)

Enter mode [ON/OFF] <ON>: *(Enter On or Off, or press (EnterKey))*

If the ON option is retained, AutoCAD will perform an automatic regeneration when needed while altering the screen display in commands such as ZOOM. When OFF is the current mode and a regeneration is needed as the screen is altered, AutoCAD will display the prompt:

About to Regen--proceed <Y> Enter Y or N, or press ENTER

If you accept the default, the drawing regeneration will proceed allowing the ZOOM to take place. If the NO option is selected, the ZOOM will not be completed. Because of AutoCAD's increased ability to handle huge amounts of information quickly and easily, you really can leave this option in the "That's nice" file of your mind and explore other viewing options

VIEW RESOLUTION

The speed of each ZOOM or PAN can be enhanced if a REDRAW is used, rather than a REGEN. The speed of all view changes also can be influenced by the view resolution. Resolution refers to the amount of detail that is represented when arcs and circles are drawn. The higher the resolution, the more lines that are used to represent an object and the smoother the arc or circle will appear. Higher resolution also means that more time is needed for zooms and pans.

The resolution of arcs and circles can be controlled from the Options dialog box or by keyboard. To control the resolution from the dialog box, use select the Display tab of the Options dialog box. The box is displayed by selecting Options... from the Tools pull-down menu. The Arc and Smoothness option in the Display Resolution box should be considered as your drawings increase in size. The Segments in a Polyline Curve will be considered in future chapters. The other two options of this box affect 3D drawings and will not be considered in this text.

Arc and Circle Smoothness

This option controls the smoothness of circles. A value from 1 through 20,000 can be provided in the edit box, in place of the default value of 100. With a low number such as 25 or 50, circles will appear as a series of flat lines. As the value increases, the smoothness of the circle or arc is increased. Regeneration time on large drawings will slow slightly as the value is increased.

The resolution of a circular feature can also be adjusted using the VIEWRES command. The command is completed from the keyboard using the following sequence:

Command: **VIEWRES** (EnterKey)

Do you want fast zooms? [yes/no] <Y> *(a Y or N is required.)* (EnterKey)

Enter circle zoom percent (1–20000) <100>: **150** (EnterKey)

Command:

In setting VIEWRES, the first decision to be made is for fast zooms. Responding **N** (EnterKey) will cause all ZOOM, PAN, and VIEW restorations to be done using regeneration. Responding **Y** (EnterKey) will keep a large virtual screen and execute most ZOOM, PAN, and VIEW restorations at REDRAW speed. Zooming to the limits of a drawing may require regeneration, even though you have selected fast zooms. When a small circle is drawn with a low value, it will appear as a polygon. Using the Regen command will restore its circular shape.

CHAPTER 9 EXERCISES

1. Open drawing E-7-9. Use the zoom command to make the view containing the triangles the current view. Save this view as TRIANGLES. Zoom into each of the remaining shapes, and make each shape a separate saved view. Save the entire drawing as E-9-1. Write a list of each view name in the space provided.
2. Using drawing E-9-1 as a base, make the view containing the squares the current view. Draw a square in the center of one of the existing squares using midpoint to form a diagonal square. Repeat this process four additional times, using zoom as required. Inside the final square, draw a circle that is tangent to the smallest square. Save this view with the name TINY and save the entire exercise as E-9-2.
3. Load E-9-1 and enlarge the drawing to twice its current size. Save this view as VIEW2X. Zoom out so that the edges of the largest triangle touch the edge of the screen. Name this view with the name of the command required to perform the zoom. Set the drawings so that screen markers will be produced and save the drawing as E-9-3. Write the name of the views created and the commands used in the space below.

VIEWS	COMMANDS
____________	______________________________
____________	______________________________
____________	______________________________
____________	______________________________

4. Load drawing E-9-3 and use it as a base for this exercise. Change the limits of this drawing to 36,18 and set the screen display so that the total content of the limits will be displayed. The original view is now displayed:

   Use ZOOM dynamic to make the object's original view fill the screen. Save the drawing as E-9-4. Write the steps that were required to adjust the views.
5. Start a new drawing and set the limits to 96' × 144', with a grid of 24" and a snap of 6". Set units to measure in architectural, with fractions set at 16. Set angles to be measured in degrees/minutes/seconds with two-place accuracy. Set the view resolution to 150. Divide the drawing into the following seven views: All, Upper Left, Upper Cen, Upper Right, Low Right, Low Cen, and Low Left. Set All as the current view. Save the drawing as a drawing template. This drawing can be used as a future base for drawings done at ¼" = 1'-0" scale.
6. Open a file of your choice and establish a minimum of three named Views. Provide names that quickly describe the area captured in the viewport. Save the file as E-9-6.
7. Open a file of your choice and establish an aerial view window. Use the window to zoom a portion of the drawing and then save the file as E-9-7.

CHAPTER 9 QUIZ

1. What command controls the use of screen markers and how should it be set to provide markers?
2. Compare and explain the difference in VIEWRES settings between 50 and 500 and the default value.
3. Explain the difference between REGEN and REDRAW.
4. Describe the difference between ZOOM ALL and EXTENTS.
5. What commands will be needed to make a drawing exactly twice as large as the current screen display?
6. What option is used to scale space relative to paper space units?
7. Describe the meaning of "displacement point" and "second point."
8. List the steps to enlarge a portion of a drawing using the zoom command.
9. List six options for accessing the ZOOM command.
10. Describe the process that allows the ZOOM command to be used in the middle of another command?
11. What term is used to describe a command that can be used while another command is being executed?
12. What command would you use if you wanted to create a picture-in-picture of an AutoCAD drawing?
13. What must be typed at the command prompt to start Aerial View?
14. Describe how to reduce the size of the Aerial View window.
15. Other than putting you to sleep, what's the point of all of the options presented in this chapter?

CHAPTER 10

Selecting and Modifying Drawing Objects

By now you've tackled the basic components of a drawing. In this chapter you'll learn how to modify your drawing. This will include expanding your knowledge of the ERASE command and exploring four commands that can be used to alter drawings. These commands include COPY, MIRROR, OFFSET, and ARRAY. Each will reproduce all or parts of a drawing. A second group of editing commands for refining drawings and two new methods of sorting objects will be explored in the next chapter.

SELECTING OBJECTS FOR EDITING

AutoCAD provides many options for selecting objects. Individual objects can be selected using a pickbox. Groups of objects can be selected for editing using one of the following options: Window, Box, Crossing, WPolygon, CPolygon, Fence, All, Last, Add, Remove, Previous, or Multiple. As different methods of selecting objects for editing are discussed, each method will be introduced using the ERASE command. Keep in mind, these options are in no way limited to one command. These options can be used anytime a selection needs to be made for any command using any of the selection methods for the desired editing command. Selecting an icon from a toolbar or entering the command at the keyboard are generally the fastest methods. No matter which method is used to access the command, a prompt will be displayed asking you to:

Select objects:

Each of the following selection methods can be entered at the Select objects prompt for any of the editing commands.

SELECTING OBJECTS USING A PICKBOX

As you've edited your drawings, only one object at a time has been selected to be edited (ERASE). One or more objects have been selected using the pick box that is controlled by the mouse. As the prompt for the edit command is entered, the crosshairs will change to an object selection target or pickbox and the object to be edited can be selected. Any object that is included in the pickbox will be affected by the command. If the size of the pickbox is altered, it will effect the accuracy of your selections. As the size of the box is enlarged, information that is not desired in the selection set may be added. With the box too small, you may have trouble selecting information.

The size of the box can be altered to enlarge or reduce the size of the selection area. The size of the pickbox can be adjusted using the Selection tab of the Options dialog box. The dialog box is accessed by selecting Options... from the Tools pull-down menu. As the slide bar is moved to the right, the size of the pickbox is enlarged. Move the slide bar to the left to decrease the size of the pickbox. The size can also be adjusted from the keyboard. Typing **PICKBOX** (EnterKey) at the command prompt will produce a prompt of:

Command: **PICKBOX** (EnterKey)

Enter new value for PICKBOX <3>: *(Enter desired value.)* (EnterKey)

AutoCAD is now waiting for you to enter a numeric value from 0 to 50, to represent the size of the pickbox.

Selecting the Object

Objects to be edited can be selected by choosing any point of the object. Objects are selected by placing the pickbox across the desired object and pressing the mouse pick button. Objects also may be selected by entering coordinates for the location of the target. When selecting objects with a thickness, it is important to place the pickbox so that it touches an edge and not the center of the object. You also need to be careful not to select objects at their intersections.

SELECTING WITH A WINDOW

Rather than selecting objects to edit one at a time, a group of objects may be selected by surrounding them completely in a window by responding **W** (EnterKey) at the Select objects prompt. This procedure was used in Chapter 9 with the ZOOM command, and is shown in Figure 10.1a. Any object that lies entirely inside the selection window will be edited. Any object that is partially inside the window will not be edited. If the selection set in Figure 10.1a is accepted, the text that is within the selection box will be removed, and the leader lines that pass through the right side of the box will be not be affected by the edit. The result of the selection set can be seen in Figure 10.1b. The command sequence to ERASE with a window is:

Command: **E** (EnterKey)

ERASE

Select objects: **W** (EnterKey)

Specify first corner: *(Select point for first corner.)*

Specify opposite corner: *(As the first point is selected, a window will drag across the screen until a second point is selected.)*

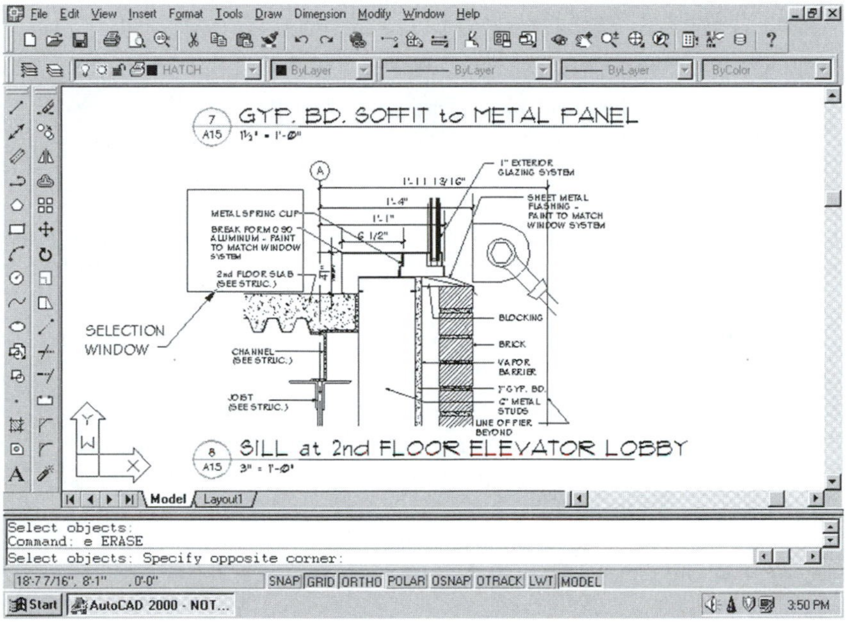

Figure 10.1a *If the Window option is used for selecting objects to be edited, all objects that lie entirely in the window will be edited. In this selection set, the text in the window will be removed, but the leader lines that cross the right side of the selection window will not be included in the selection set. (Courtesy Peck, Smiley, Ettlin Architects.)*

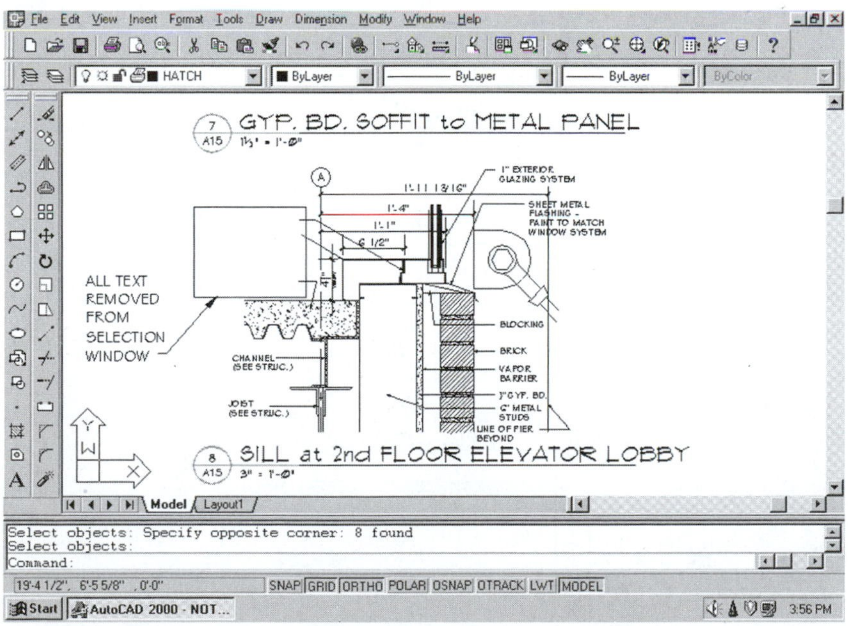

Figure 10.1b *The results of the ERASE Window option started in Figure 10.1a.*

Select objects: *(Selected objects will be highlighted. You now have an opportunity to alter the selection set. The selection set can be altered by adding additional objects using another window, or by using the add or remove selection methods. If no changes need to be made, press (EnterKey). Making changes to the selection set will be introduced later in this chapter.)*

CROSSING

Choosing an object to edit with the Crossing option is similar to the Window option. Groups of objects may be selected by surrounding them partially in a window by typing **C** (EnterKey) at the Select objects prompt. The command process will be the same as the Window option. The crossing window will appear different than the window option. The crossing window is constructed using a dashed line and a continuous line represents the window option. Any objects that lie entirely in the window will be edited. Any objects that are partially inside the window also will be edited. If the selection set in Figure 10.1a was selected using the C (Crossing) option and accepted for editing, both the text and the leader lines would be changed by the specified editing command (See Figure 10.2).

AUTO

The Auto selection mode is similar to editing using a window. The default mode for selecting objects is with the pickbox. If the pick button is pressed after the editing command is selected, the window method of selecting objects will be used. When

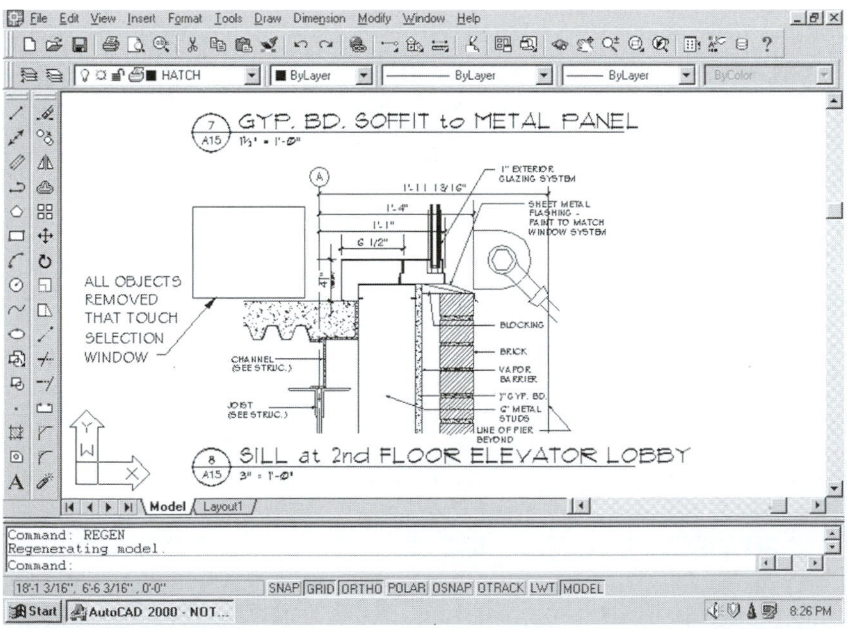

Figure 10.2 *The results of the ERASE Crossing option started in Figure 10.1a.*

prompted for the object to be erased, press the pick button. This will turn the pick box into the corner of the selection box. As the mouse is moved, the window will be formed. Any items that the window encloses will be selected. When you're satisfied with the size of the selection box, press the select button to accept the objects. To remove the selected objects from the drawing, press (EnterKey). The command process is:

Command: **E** (EnterKey)

ERASE

Select objects: *(Press the pick button to select the first corner of the selection window.)*

Specify opposite corner: *(The window will drag across the screen until a second point is selected.)*

Select objects: (EnterKey)

Command:

Implied Window Selection

The direction that is used to move the cursor as the selection window is defined will affect the selection process. If the cursor is moved from left to right when picking the selection set, only objects completely enclosed in the window will be erased. When the selection window is created by moving from right to left, any object that the window crosses will be affected by the edit command. The results of implied window selection can be seen in Figure 10.3.

WPOLYGON

The WP option is especially useful for selecting an irregular-shaped group of objects to edit. The option is similar to the window option, but you are allowed to select points to form a polygon around objects to be edited. The polygon may be any shape

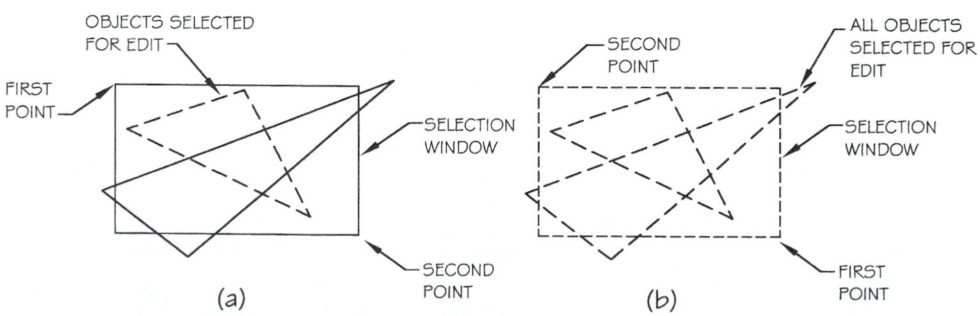

Figure 10.3 *If the cursor is moved from left to right, only objects completely enclosed will be in the selection set. When the selection window is created by moving from right to left, any object that the window crosses will be affected by the edit command.*

except a shape that crosses itself, such as a figure eight. The command sequence is started once the editing option has been entered and you are prompted to select points to form the polygon. Points to define the polygon may be entered by using a pointing device or entered as coordinates, as seen in Figure 10.4. The polygon is closed by entering (EnterKey). All objects lying totally inside the polygon will be edited. The process is completed using the following steps.

Command: **E** (EnterKey) *(Any editing command may be entered.)*

ERASE

Select objects: **WP** (EnterKey)

First polygon point: *(Enter a point.)*

Specify endpoint of line or [Undo]: *(Enter a point.)*

Specify endpoint of line or [Undo]: *(Enter a point.)*

Specify endpoint of line or [Undo]: *(Enter a point.)*

Specify endpoint of line or [Undo]: *(Continue until desired objects are selected.)* (EnterKey)

100 found *(The number will vary as AutoCAD reports the number of objects in the selection set.)*

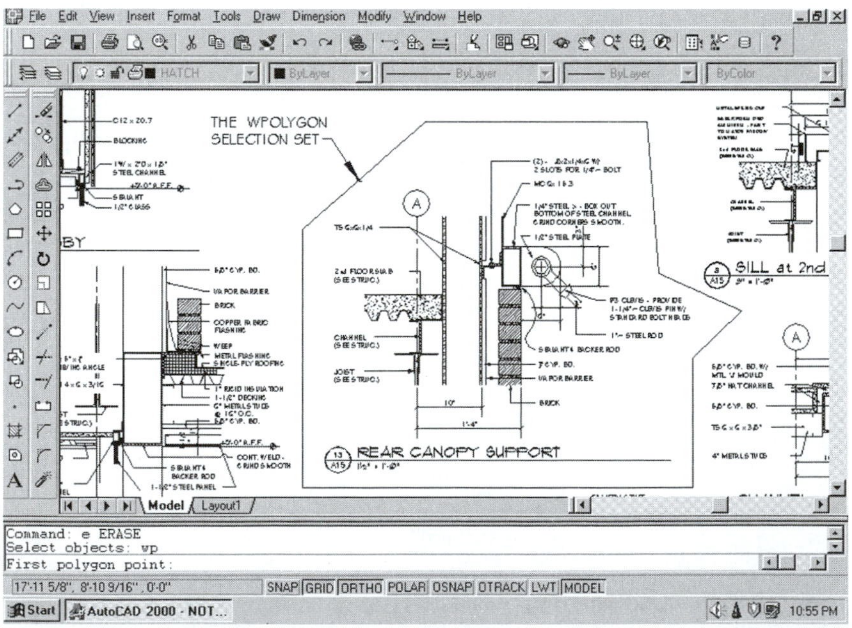

Figure 10.4 *The WP (WPolygon) option can be used as an alternative to "C" to select objects in an irregular-shaped group. Points for the polygon may be entered by selecting points or entering coordinates. (Courtesy Peck, Smiley, Ettlin Architects.)*

Select objects: (EnterKey)

Command:

CPOLYGON

Objects can also be edited by entering **CP** (EnterKey) at the Select objects: prompt. This option will be similar to the effects of combining Crossing and WPolygon, in that it will edit all objects that are in or cross the polygon. The command sequence is:

Command: **E** (EnterKey) *(Any editing command may be entered.)*

ERASE

Select objects: **CP** (EnterKey)

First polygon point: *(Enter a point.)*

Specify endpoint of line or [Undo]: *(Enter a point.)*

Specify endpoint of line or [Undo]: *(Enter a point.)*

Specify endpoint of line or [Undo]: *(Enter a point.)*

Specify endpoint of line or [Undo]: *(Continue until object set is complete.)* (EnterKey)

100 found *(The number of objects in the selection set will vary.)*

Select objects: (EnterKey)

Command:

SELECTING OBJECTS USING A FENCE

Objects can be selected for editing by Fence. This option is similar to the WPolygon and CPolygon, except that it does not close the polygon. The Fence may be a series of line segments or may form a polygon that can cross itself. This allows long rows or columns to be edited in one command sequence. As seen in Figure 10.5, the selection Fence only edits items that intersect or cross the fence. Fence points may be entered with a pointing device or by entering coordinates. The command sequence is:

Command: **E** (EnterKey) *(Any editing command may be entered.)*

ERASE

Select options: **F** (EnterKey)

First fence point: *(Enter a point.)*

Specify endpoint of line or [Undo]: *(Enter a point.)*

Specify endpoint of line or [Undo]: *(Enter a point.)*

Specify endpoint of line or [Undo]: *(Enter a point.)*

Specify endpoint of line or [Undo]: *(Continue until object set is complete.)* (EnterKey)

5 found

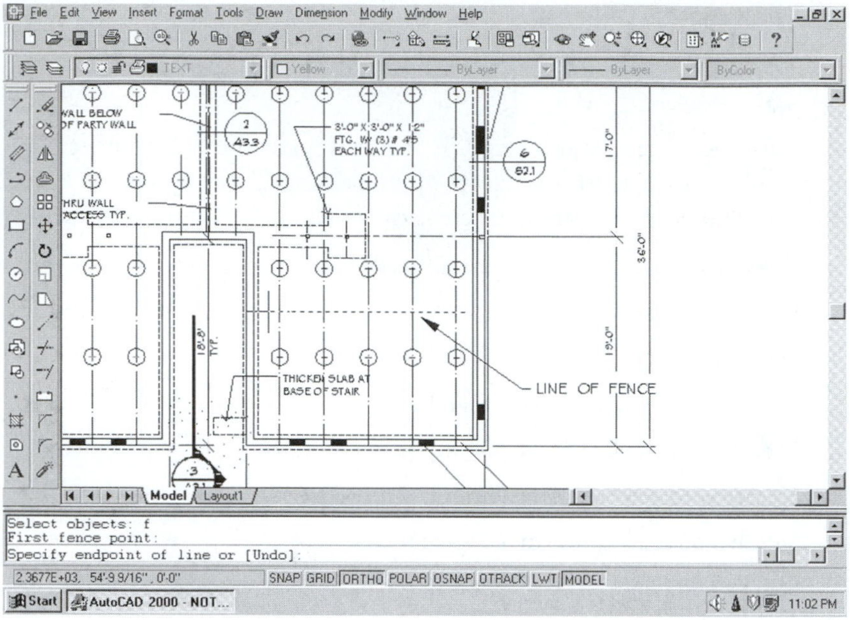

Figure 10.5 *The Fence option allows the use of straight lines to include objects in the selection set. Any objects that are touched by the Fence will be included in the selection set. (Courtesy of Scott R. Beck, Architect.)*

Select objects: (EnterKey) *(Selecting (EnterKey) will execute the editing command.)*

Command:

ALL

When this option is used, all objects that are current will be affected. You've learned how to freeze certain layers so that they cannot be seen. All objects except those on frozen or locked layers will be edited.

Command: **E** (EnterKey) *(Any editing command may be entered.)*

ERASE

Select objects: **ALL** (EnterKey) *(All items will be highlighted.)*

137 found

Select objects: (EnterKey) *(All items will be removed.)*

Command:

LAST

The LAST (L) option allows the most recently created object to be selected for editing.

Command: **E** (EnterKey) *(Any editing command may be entered.)*

ERASE

Select options: **L** (EnterKey)

1 found

Select options: (EnterKey)

Command:

ADD

When selection sets are defined, not all intended objects for editing may be selected. Entering A (EnterKey), for ADD, at the Select objects prompt will allow for additional objects to be added to the selection set. Once entered, objects may be added to the selection set individually or with any other selection method. When you are satisfied with the selection set, press (EnterKey) to edit the drawing. Objects are added to a selection set using the following steps.

Command: **E** (EnterKey) *(Any editing command may be entered.)*

ERASE

Select objects: **W** (EnterKey) *(Remember any selection method can be used.)*

Specify first corner: *(Select first corner.)*

Specify opposite corner: *(Specify second point.)*

5 found

Select objects: **A** (EnterKey)

Select objects: **C** (EnterKey) *(Remember any selection method can be used.)*

Specify first corner: *(Select first corner.)*

Specify opposite corner: *(Specify second point.)*

20 found, 25 total

Select objects:

Command:

REMOVE

This option will allow for objects to be removed from the selection set. Once objects are selected for editing, entering R (EnterKey) will allow objects to be taken out of the selected objects to edit. Objects may now be removed from the selection set individually or with any other selection method. When you are satisfied with the selection set, press the pick button to edit the drawing. The command sequence is:

Command: **E** (EnterKey) *(Any editing command may be entered.)*

ERASE

Select objects: **C** (EnterKey) *(Remember any selection method can be used.)*

Specify first corner: *(Select first corner.)*

Specify opposite corner: *(Specify second point.)*

5 found

Select objects: **R** (EnterKey)

Select objects: *(Select objects to be removed using any selection method.)*

1 found, 1 removed 4 total

Remove objects: (EnterKey)

Command:

MULTIPLE

The Multiple selection option works well when drawing objects are stacked over each other. To this point, all of your drawings have been created on one layer. In Chapter 6 you learned to place parts of your drawing on different layers. The Multiple option will allow you to edit material on different levels at one time.

DEFINING THE SELECTION MODE

The Selection tab of the Options dialog box offers six modes that will affect the formation of the selection set. Choosing Options... from the Tools pull-down menu opens the dialog box. An examination of this tab will reveal that three of the settings are active. Any combination of selection methods may be used.

NOUN/VERB SELECTION

Up to this point, as you have edited an object, you've used a process AutoCAD refers to as Verb/Noun. You've selected the verb (ERASE) and then selected the object to be edited (the NOUN). With this box active, AutoCAD allows you to reverse the process. The object can be selected first, and then the editing command.

With Noun/Verb active, any object can be selected merely by choosing it with the pickbox. Enter a drawing and select a line. Notice the line is highlighted and a box appears in the middle and at each end of the line, but the command prompt has not changed. Entering E (EnterKey) will erase the selected line. The command prompt is:

Command: *(Select object to edit using any method.)*

Command: **E** *(Enter any of the listed edit commands.)* (EnterKey)

The Noun/Verb selection can be used with the following commands to edit objects:

ARRAY	DDCHROP	MIRRORWBLOCK
BLOCK	DVIEW	MOVEEXPLODE
CHANGE	ERASE	ROTATE
CHPROP	HATCH	SCALE
COPY	LIST	STRETCH

Some commands require that the edit command be executed before the objects can be selected. These commands include:

BREAK	DIVIDE	FILLET	OFFSET
CHAMFER	EXTEND	MEASURE	TRIM

If you try to edit these commands, they will ignore any selection set created prior to starting the command.

USE SHIFT TO ADD SELECTION

This selection method controls how objects are added to an existing selection set. Open a new drawing and draw a few lines. Toggle Noun/Verb to OFF (no X) and Use Shift to Add to ON (X). Select OK and return to the drawing. Select an object to erase and it will be highlighted. Select a second object and the first object no longer will be highlighted. To retain the first object, hold the shift key down while the second object is selected. Both objects will now be edited. If Use Shift to Add is not enabled, objects are added to the selection by selecting them individually or using a selection window. To remove objects from the selection set, press (ShiftKey) while selecting the object.

PRESS AND DRAG SELECTION

This mode determines the method used for drawing a selection window. With the option active, the selection window can be made with one button selection rather than two. The selection window is created by pressing and holding the pick button and moving the cursor diagonally. The window is completed by releasing the pick button when the window reaches its desired size. When this option is not selected, objects may be selected for editing using the normal window option.

IMPLIED WINDOWING

With this option in the default mode, a selection window is automatically created when the Select object prompt is displayed. Pressing the pick button will automatically select the first corner point. A prompt will then be given requesting the other corner. If you draw the window from left to right, the window selects objects that lie entirely within the window. If the window is drawn from right to left, all objects in and touching the window will be selected. With the option off, the window must be set for each edit. In the OFF position, pressing the pick button does nothing. A mode such as Window or Crossings needs to be entered to start the creation of a window.

OBJECT GROUPING

This option determines if grouped objects will be recognized as individual objects or as a group. In the default setting of ON, grouped objects function as a group. When OFF, grouped objects can be edited individually. Grouping of objects will be discussed in future chapters.

ASSOCIATIVE HATCH

This option affects how the HATCH patterns are edited and will be discussed in Chapter 15.

MODIFYING A DRAWING USING ERASE

The ERASE command allows you to remove unwanted objects (mistakes) from a drawing. Because you're familiar with ERASE by now, the command will serve as a means of using some of the selection methods just covered. The selection prompts Point, Multiple, Window, Crossing, WPolygon, CPolygon, Fence, All, Last, Add, Remove, and Previous should start to seem more meaningful. Figure 10.6 shows a detail marker that has been selected for editing by entering a "Point". The selected object is highlighted, while AutoCAD waits for you to select other objects. When the selection process is complete, press (EnterKey) to remove the object.

The Window option can be seen in Figure 10.7. Notice that at the Select object prompt, W was entered rather than WINDOW. Remember an object must be entirely inside the window to be edited. Figure 10.8 shows the objects to be erased as highlighted. If you would like to ADD or REMOVE objects from the selection window, enter the corresponding letter at the Select objects prompt. Figure 10.9 shows the effects of the Window ERASE with no objects added or removed.

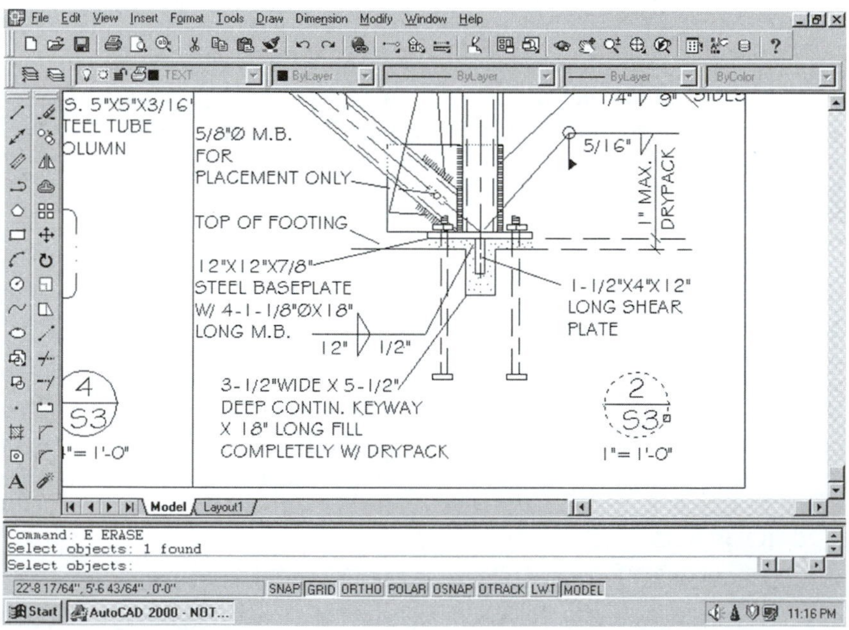

Figure 10.6 *The detail marker at the bottom of the drawing has been selected for ERASE. (Courtesy StructureForm Masters, Inc.)*

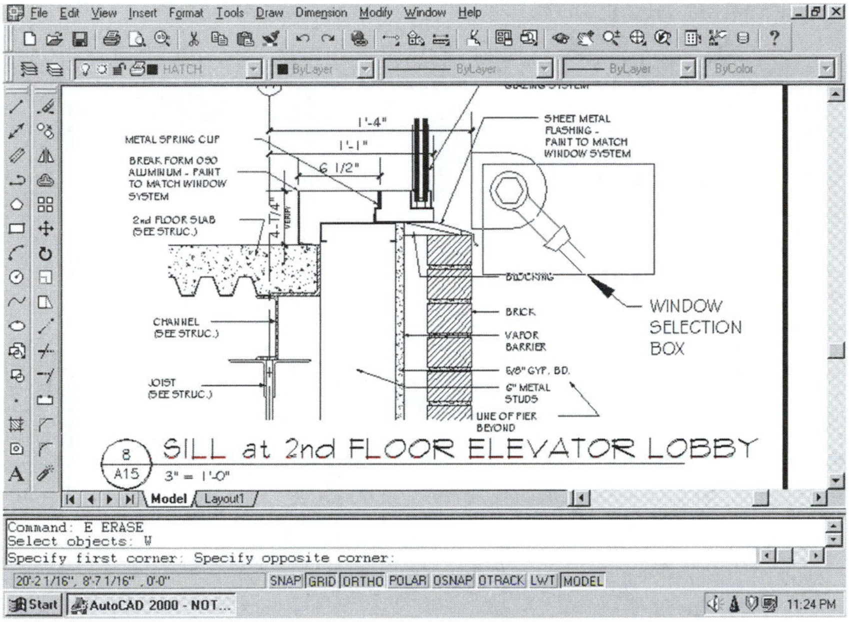

Figure 10.7 *Using the Window option to form a selection set for the ERASE command. (Courtesy Peck, Smiley, Ettlin Architects.)*

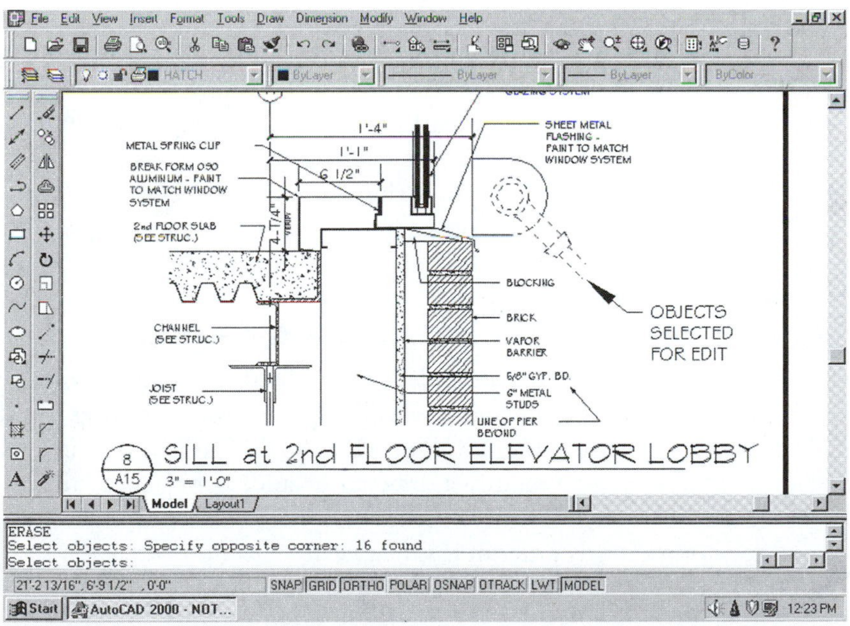

Figure 10.8 *Results of the W (Window) selection set are now displayed to confirm the choice of objects to be included in the selection set. Notice that the number of objects included in the set is displayed above the command prompt. (Courtesy Peck, Smiley, Ettlin Architects.)*

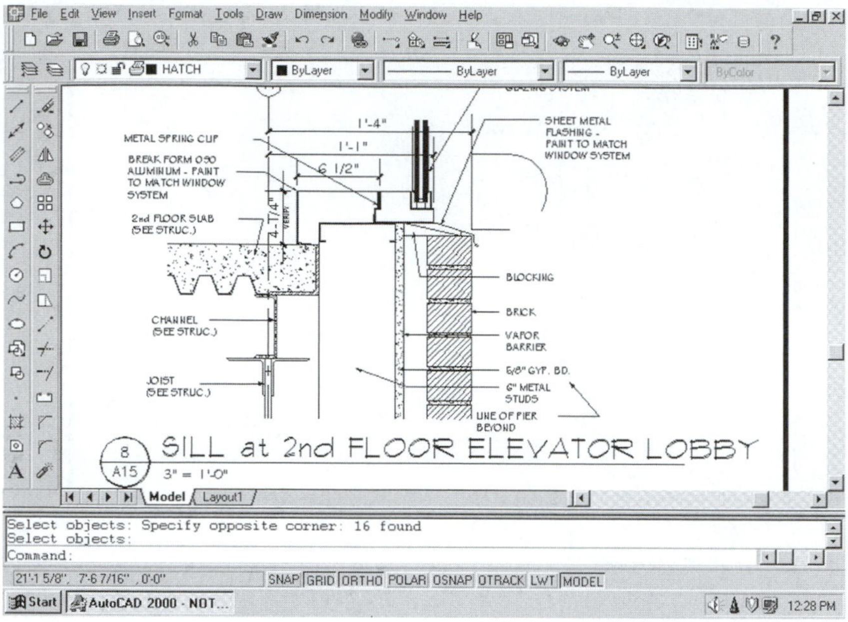

Figure 10.9 *The effect of the Window option used with the ERASE command. (Courtesy Peck, Smiley, Ettlin Architects.)*

If you type C (EnterKey) at the Select object prompt, objects that are in or cross the window are selected to be erased, as seen in Figure 10.10. The results of a Crossing window selection can be seen in Figure 10.11. Because the engineer would like to keep the beams and remove the column, the REMOVE option was used to restore the selected lines of the beams. Figure 10.12 shows the final selection set. This drawing could now be used as the basis to form a new detail of two intersecting beams of different sizes with no support column.

MODIFYING A DRAWING BY ADDING OBJECTS

The balance of this chapter will introduce the skills needed to alter a drawing by adding objects to the drawing. Future chapters will help you edit a drawing by altering existing drawing components. The COPY, MIRROR, OFFSET, and ARRAY commands can each be used to modify a drawing by adding objects to the drawing base. Objects added using one of these commands can be selected with any of the selection methods introduced earlier in this chapter.

COPY

Whenever multiple objects need to be drawn, the COPY command is a convenient method of reproducing objects. ARRAY, BLOCK, and WBLOCK are also efficient

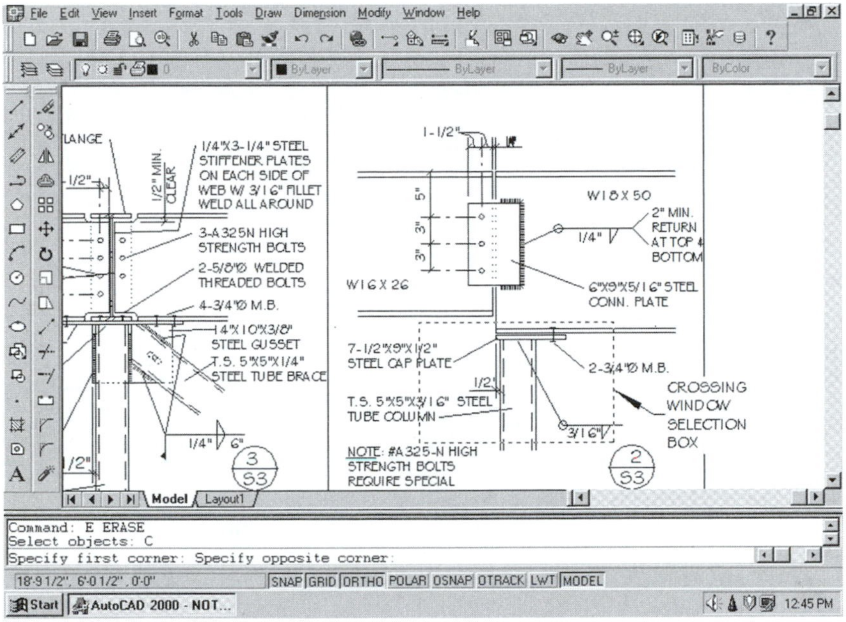

Figure 10.10 *Using the C (Crossing) option to remove the column, support plate, and related information. Any object touched by the window will be edited. (Courtesy StructureForm Masters, Inc.)*

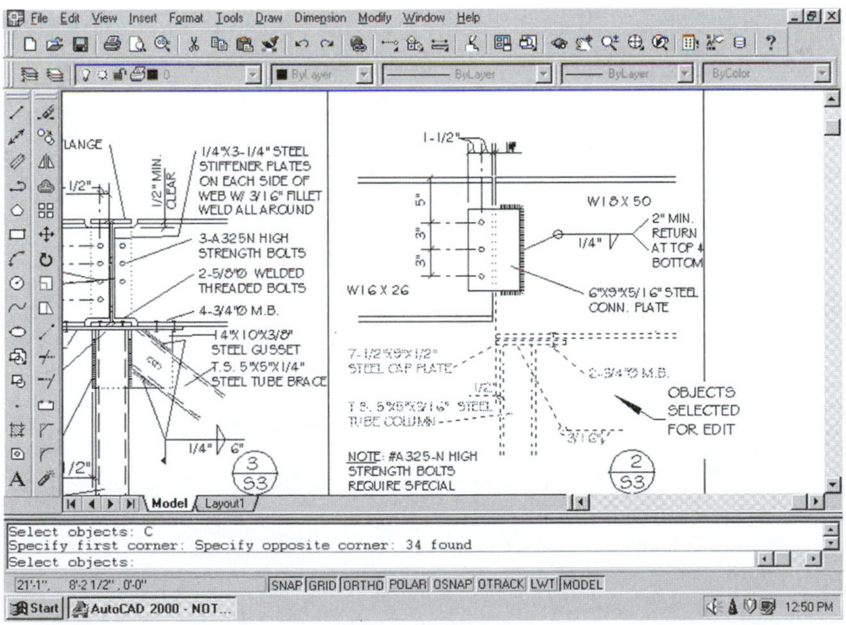

Figure 10.11 *Results of the C (Crossing Window) selection set are now displayed to confirm the choice of objects to be included in the selection set. (Courtesy StructureForm Masters, Inc.)*

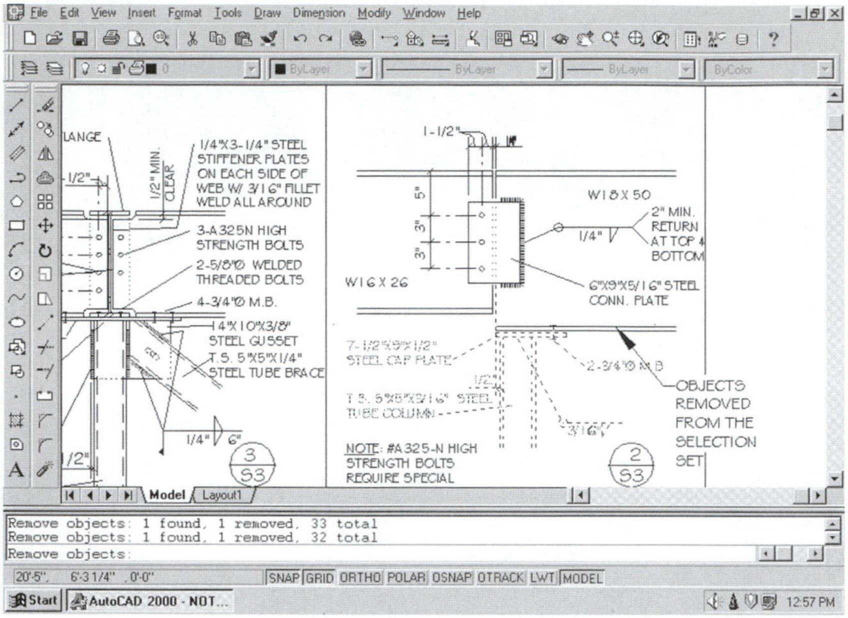

Figure 10.12 *If you would like to include more objects in the set, enter* ***A*** *(EnterKey) at the Select objects: prompt. Entering* ***R*** *(EnterKey) at the prompt will remove objects from the set. To accept the set, press (EnterKey). (Courtesy StructureForm Masters, Inc.)*

means of reproducing objects and will be covered in the coming chapters. The COPY command can be started by selecting the Copy icon from the Modify toolbar, by typing **CP** (EnterKey) at the command prompt, or by selecting Copy from the Modify pull-down menu. Figure 10.13 shows an example of a drawing created using the copy command. The drawing on the left was drawn first and copied to produce the drawing on the right. Once copied, the text was edited slightly to produce the detail. The command sequence is:

Command: **CP**

Select objects: **W** (EnterKey) *(Select objects by desired selection method.)*

Specify first corner: *(Select window corner.)*

Specify opposite corner: *(Select opposite window corner.)*

54 found

Specify base point or displacement or [Multiple]: *(Select base point or enter coordinates of object to be moved.)*

Specify second point of displacement or < use first point as displacement>: *(Select a point to relocate the base point to.)*

Command:

The copy process can be seen in Figures 10.14a through 10.14d.

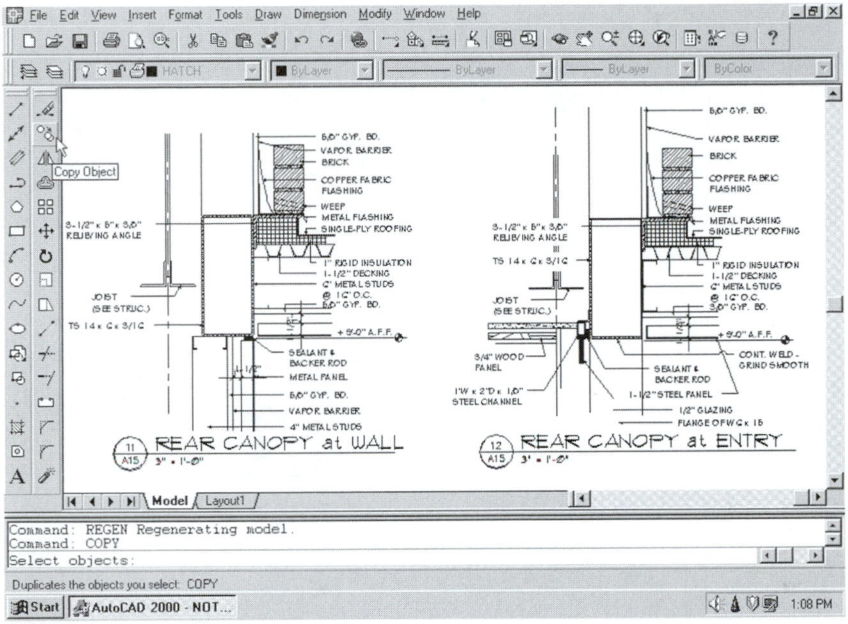

Figure 10.13 *The COPY command can be used to duplicate common features. The command can be typed, selected from the Modify toolbar, or selected from the Modify pull-down menu. (Courtesy Peck, Smiley, Ettlin Architects.)*

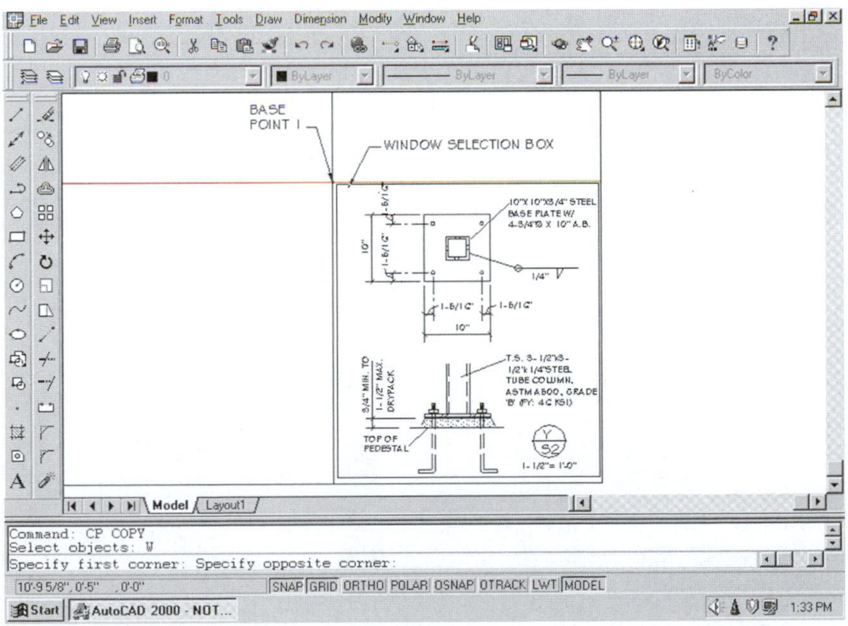

Figure 10.14a *The COPY command is started by defining the selection set. (Courtesy StructureForm Masters, Inc.)*

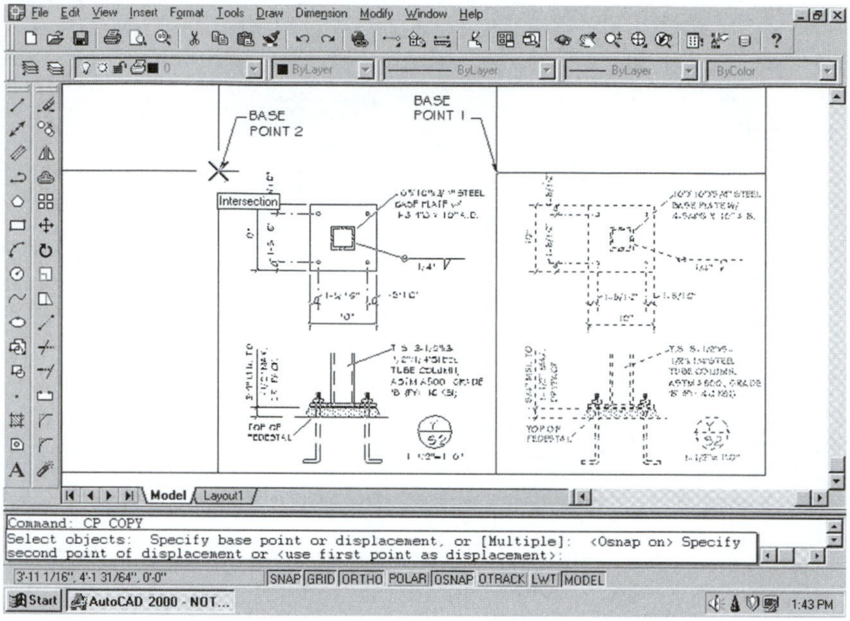

Figure 10.14b *The second step of the COPY command is to specify a base point and then specify a new location for the base point. (Courtesy StructureForm Masters, Inc.)*

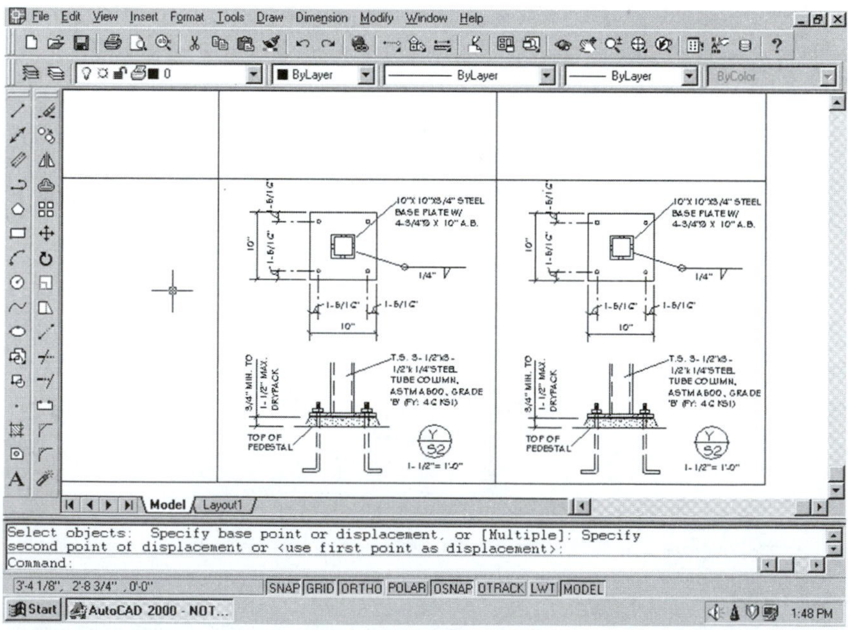

Figure 10.14c *Once (EnterKey) is pressed for the new location, the objects in the selection set will be duplicated. (Courtesy StructureForm Masters, Inc.)*

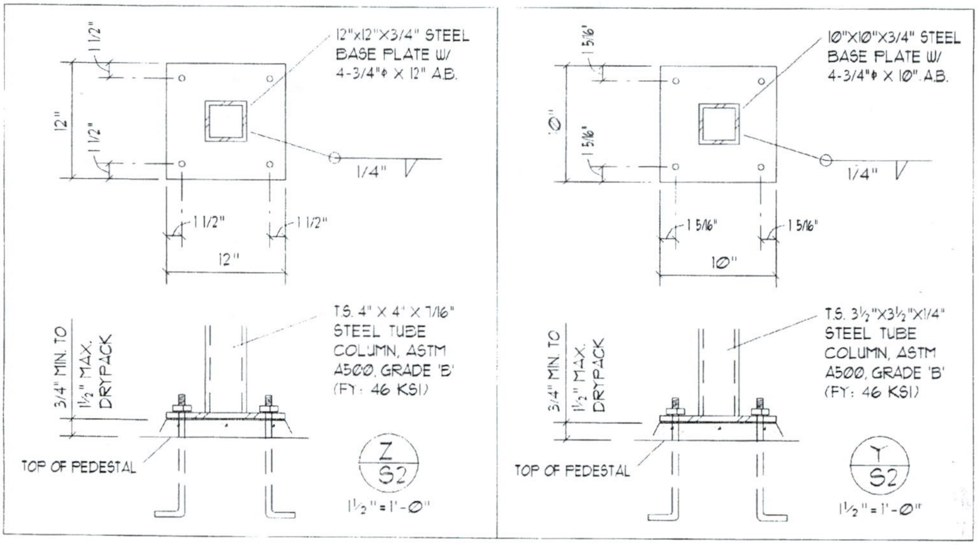

Figure 10.14d *Once copied, the detail can be edited to provide information about a similar column-plate intersection. Notice that the size of the plate and column, and the bolt locations have been edited. (Courtesy StructureForm Masters, Inc.)*

Multiple Copies

It's not uncommon to need to reproduce multiple copies of an object throughout a drawing. Multiple copies of an object or group of objects may be made once the selection set is selected. Notice in the copy command sequence that the alternative to the first point of displacement is multiple. When you enter **M** (EnterKey) at the prompt, the selection group will be placed at an unlimited amount of locations. The command sequence to make four copies of a group of objects is:

Command: **CP** (EnterKey)

COPY

Select objects: **C** (EnterKey) *(Select objects by desired selection method.)*

Specify first corner: *(Select window corner.)*

Specify opposite corner: *(Select opposite window corner.)*

25 found

Select objects: (EnterKey)

Specify base point or displacement or [Multiple]: **M** (EnterKey)

Specify base point: *(Select base point or enter coordinates of object to be moved.)*

Specify second point of displacement or < use first point as displacement>: *(Select a point to relocate the base point to for the first copy.)*

Specify second point of displacement or < use first point as displacement>: *(Select a point to relocate the base point to for the second copy.)*

Specify second point of displacement or < use first point as displacement>: *(Select a point to relocate the base point to for the third copy.)*

Specify second point of displacement or < use first point as displacement>: *(Select a point to relocate the base point to for the final copy.)*

Specify second point of displacement or < use first point as displacement>: *(EnterKey)*

Command:

Practical Use: The process can be continued until all of the desired copies are drawn. To end the sequence, choose (EnterKey) at the displacement prompt. Figure 10.15 shows the effect of the Multiple COPY command. One girder and one pier were drawn for the foundation plan. Multiple COPY was used to copy the three additional piers along the girder. Tracking mode was used for exact placement. Once the first girder was completed, Multiple COPY was used again to place the girders in the foundation plan.

(a)
ORIGINAL DRAWING

(b)
GIRDER COMPLETED WITH MULTIPLE COPY. EXACT PLACEMENT WAS OBTAINED USING TRACKING.

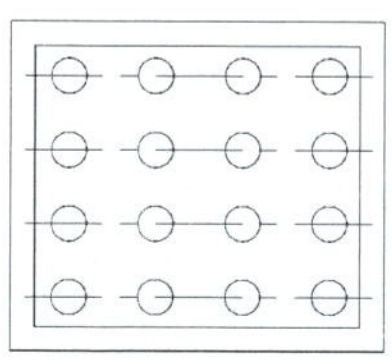

(c)
FOUNDATION PLAN COMPLETED USING MULTIPLE COPY.

Figure 10.15 *The Multiple mode of COPY allows objects to be copied several times without having to repeat the command. The drawing was started by drawing the girder and one pier. The required number of piers were copied, and finally the girder with its piers were copied to finish the drawing.*

MIRROR

Objects in a drawing often need to be reversed. Something as simple as reversing a door swing, or as complex as flipping a floor plan for an apartment unit floor plan, (Figure 10.16) can be done using the MIRROR command. The command also can be used any time objects are symmetrical, such as a fireplace or double doors. When you draw half the object, you can create the other half by using MIRROR. The MIRROR command can be started by selecting the Mirror icon from the Modify toolbar, by typing **MI** (EnterKey) at the command prompt, or by selecting Mirror from the Modify pull-down menu.

Once objects are selected to be mirrored (flipped), you will be asked to describe a "mirror line." This is done by selecting each end point of the mirror line. The process can be seen in Figure 10.17. Think of the mirror line as a fold line between the old and the new objects to be mirrored. As you pick the first point of the mirror, drag the selection set into position and alter as the mouse is moved. The use of ORTHO and

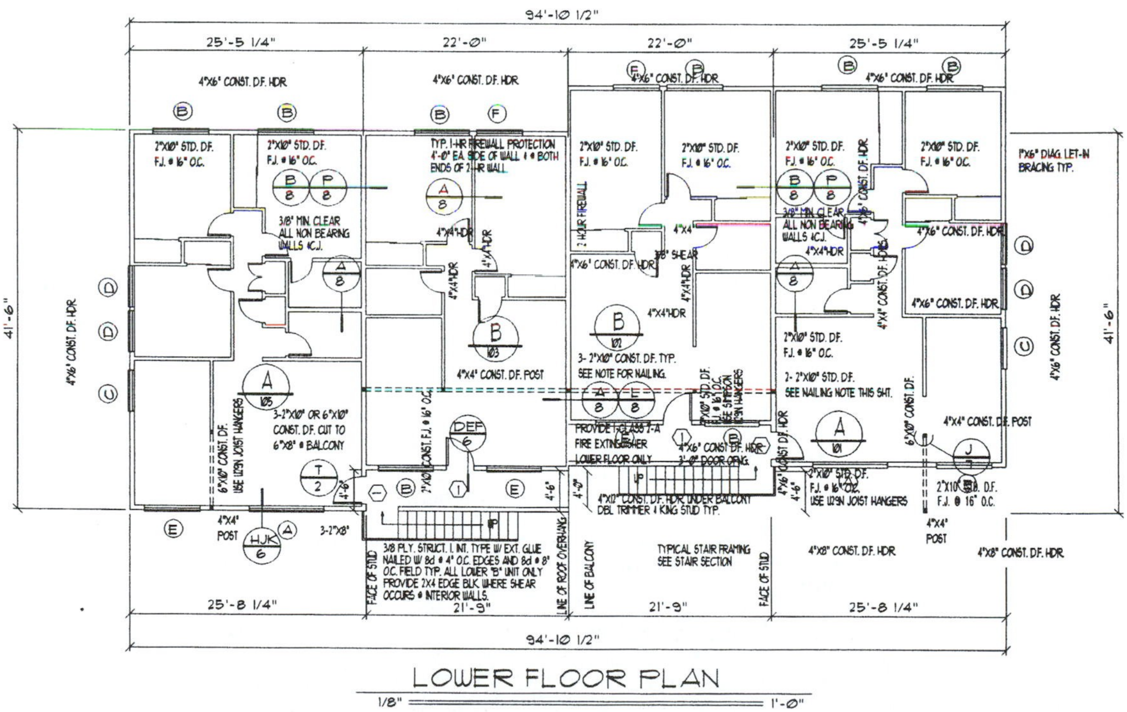

Figure 10.16 *The MIRROR command can be used to flip existing objects. The command can be selected from the Mirror icon, from the Modify toolbar, from the Modify pull-down menu, or entered by keyboard. (Courtesy StructureForm Masters, Inc.)*

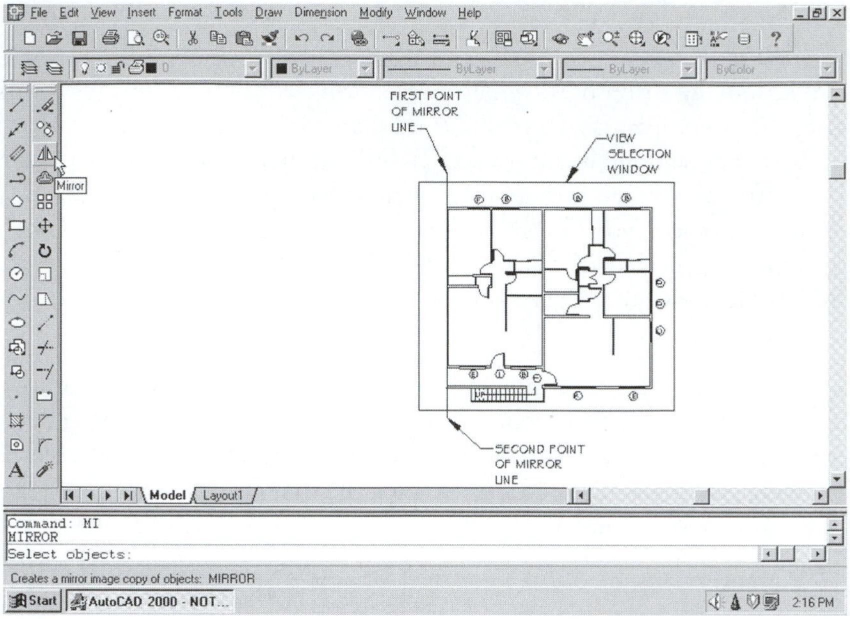

Figure 10.17 *Once the selection set has been defined, two points will be required to define the mirror line. (Courtesy StructureForm Masters, Inc.)*

Object Snaps can greatly aid in the placement of the mirrored image. The new placement can be seen in Figure 10.18.

You'll be given the option of keeping or discarding the original object prior to making the MIRROR permanent. Your choice will depend on the use of the drawing. If the object is being completed by MIRROR, don't delete the old object. The final result will look like Figure 10.19. The command sequence is:

Command: **MI** (EnterKey)

Select objects: **C** (EnterKey) *(Choose objects using the desired selection method or combination of methods.)*

Specify first corner: *(Select window corner.)*

Specify opposite corner: *(Select opposite window corner.)*

255 found

Select objects: (EnterKey)

Specify first point of mirror line: *(Select a point or enter coordinates.)*

Specify second point of mirror line: *(Select a point or enter coordinates.)*

Delete source objects? [Yes/NO] <N>: (EnterKey)

Command:

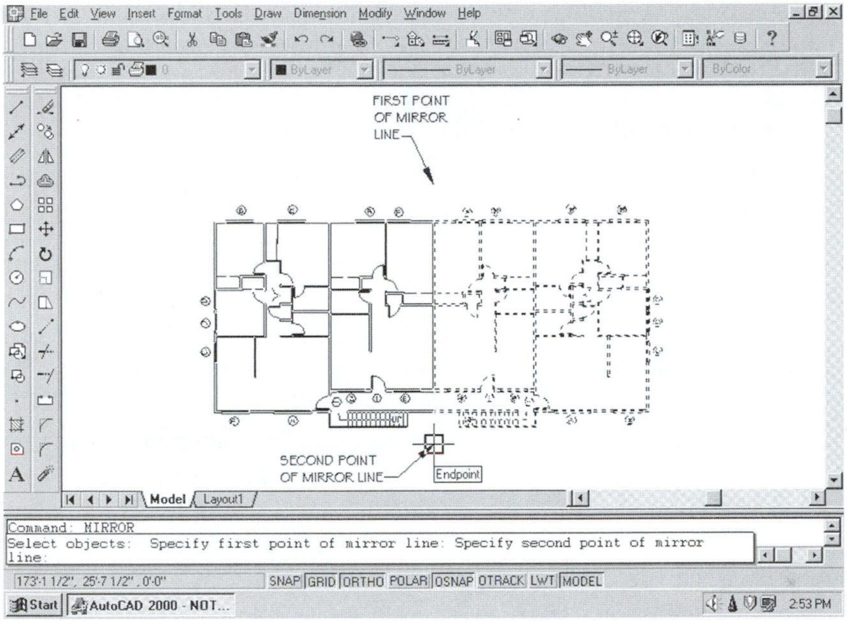

Figure 10.18 *As the second point is entered, the selection set will be mirrored into position. (Courtesy StructureForm Masters, Inc.)*

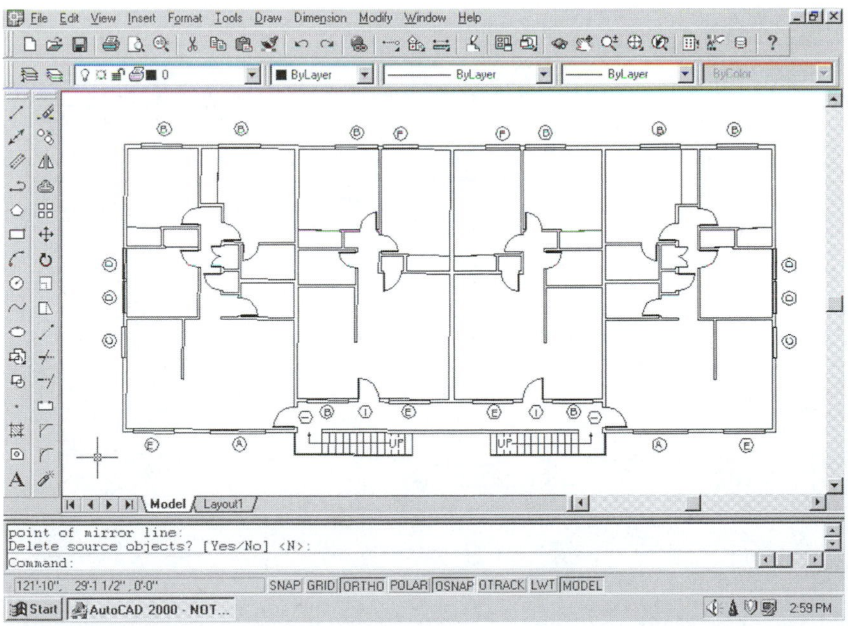

Figure 10.19 *Before the MIRROR command is complete, an option will be given to "Delete source objects?" By responding* **N** *(EnterKey), the selection set will be mirrored and the original objects will be retained. (Courtesy StructureForm Masters, Inc.)*

If the object is being flipped, you will want to remove the old object, as in Figure 10.20. The command prompts would be similar to those just noted, until the last line is reached.

Delete source objects? [Yes/NO] <N>: **Y** (EnterKey)

Typing Y (for Yes) (EnterKey) will now delete the original object and display the mirrored copy of the original object.

One of the drawbacks of the MIRROR command is how it affects text. Although you will not add text to a drawing for a few more chapters, an additional feature of MIRROR will be helpful soon. As AutoCAD is currently set, if you were to mirror a drawing with text, the text would be reversed. (See Figure 10.21). To mirror the drawing without flipping the text, adjust the MIRRTEXT value using SETVAR to zero. With MIRRTEXT set to zero, the drawing will be flopped, but the text will remain readable, as seen in Figure 10.22. This will be covered again as text is explored, but you might want to set the value in your drawing template now so that when you do add text, you won't need the UNDO or OOPS commands. The command sequence is:

Command: **MIRRTEXT** (EnterKey)

New value for MIRRTEXT <1>: **0** (EnterKey)

Command:

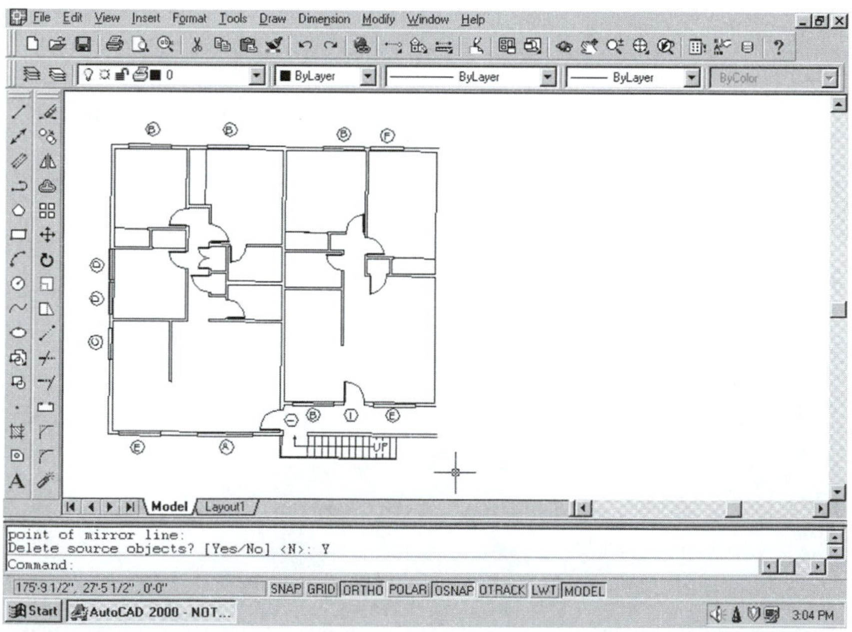

Figure 10.20 *By responding **Y** (EnterKey) to the "Delete source objects?", the selection set will be mirrored and the original objects will be removed. (Courtesy StructureForm Masters, Inc.)*

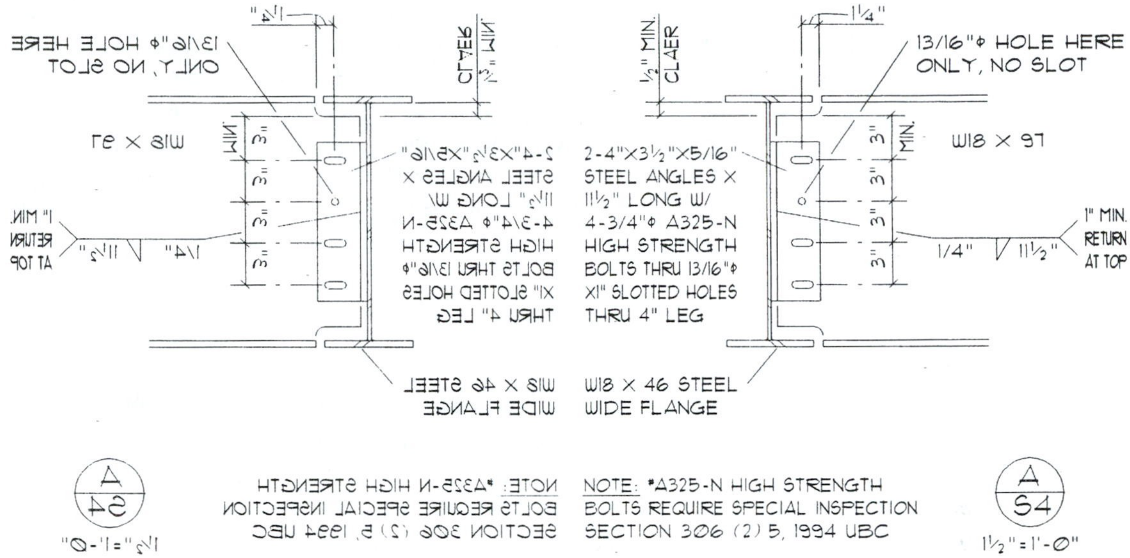

Figure 10.21 *The default setting of the MIRROR command also will reverse text. (Courtesy StructureForm Masters, Inc.)*

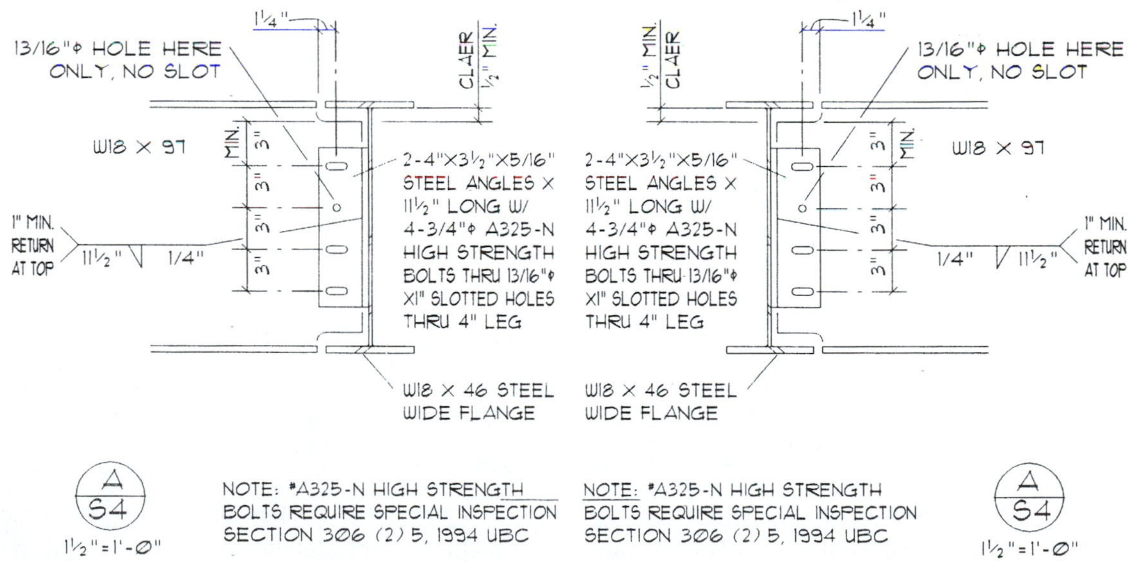

Figure 10.22 *By setting the MIRRTEXT variable set at 0, text will not be mirrored. (Courtesy StructureForm Masters, Inc.)*

Practical Uses for Mirror: Mirror is useful anytime objects need to be grouped. The girders of Figure 10.15 could have been placed using mirror. Any drawing that contains objects that lie along a centerline can be created using the MIRROR command. Common symbols that could be created with mirror for a floor plan include fireplaces, stairs, pairs of swinging doors, bay windows, sinks, convenience outlets, toilets, and tubs. Drawing half of the symbol and then using MIRROR to complete the drawing will create each of these symbols. Future chapters will explain how to make and save each symbol so that it only needs to be created once. If you're really anxious, explore BLOCKS in the Help menu.

OFFSET

The OFFSET command allows a line, circle, arc, or polygon to be copied and relocated parallel to, and at a specific distance from, the original object. The uses of this command are almost endless. A circle can be drawn and offset to represent a conduit or sewer pipe. Given a horizontal and vertical line, an entire rectangular structure can be designed by offsetting lines, and then cleaned up using other editing commands presented in future chapters.

OFFSET can be used in a manner similar to blueline layout of manual drafting. Lines can be drawn in the exact location without giving thought to their exact length. Once all of the boxes are lightly drawn to outline the desired rooms, the desired lines are darkened. Use OFFSET with the same mentality. Get the location right without worrying that the line may be too long, and then edit the length with FILLET, STRETCH, or TRIM. Each command will be introduced in Chapter 11. Figures 10.23a and 10.23b show the use of OFFSET to lay out a simple floor plan.

The command process begins by selecting the Offset icon from the Modify toolbar, by typing O (EnterKey) at the command prompt, or by selecting Offset from the Modify pull-down menu. The command sequence to offset a line 18" from the original line is:

Command: **O** (EnterKey)

OFFSET

Specify offset distance or [Through] <Through>: **1'6** (EnterKey)

A distance may be entered by providing a numeric value or by entering T (EnterKey). When the distance is entered, a new prompt will be displayed that prompts:

Select the object to offset or <exit>:

Lines, arcs, circles, ellipses, xlines, rays, and polylines can be offset. The last three commands will be introduced in future chapters. Once selected, the object will be highlighted and the prompt will read:

Specify point on side to offset:

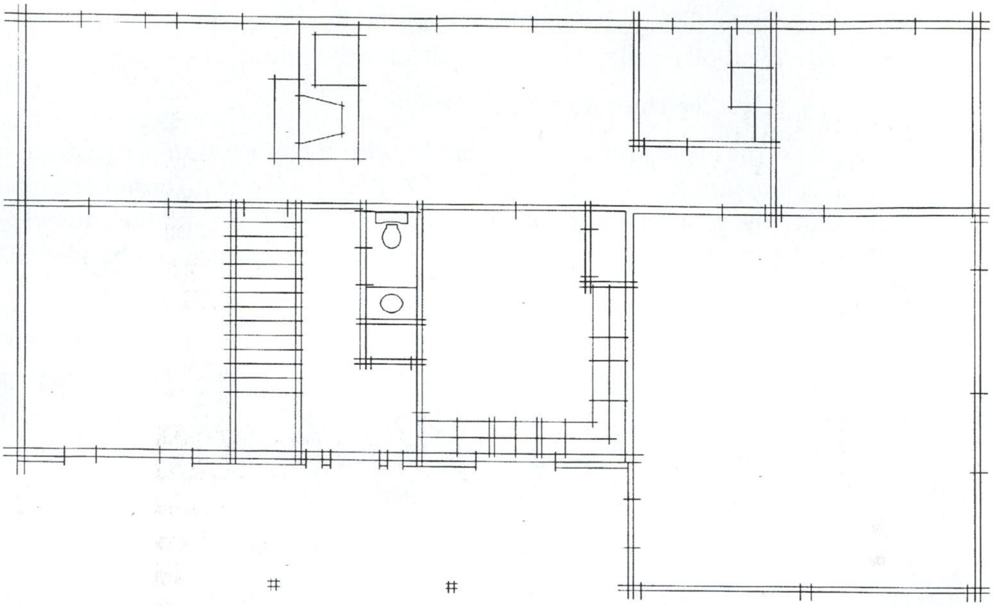

Figure 10.23a *Using one horizontal and one vertical line, the OFFSET command can be used for drawing parallel lines at a specified distance apart to create a floor plan.*

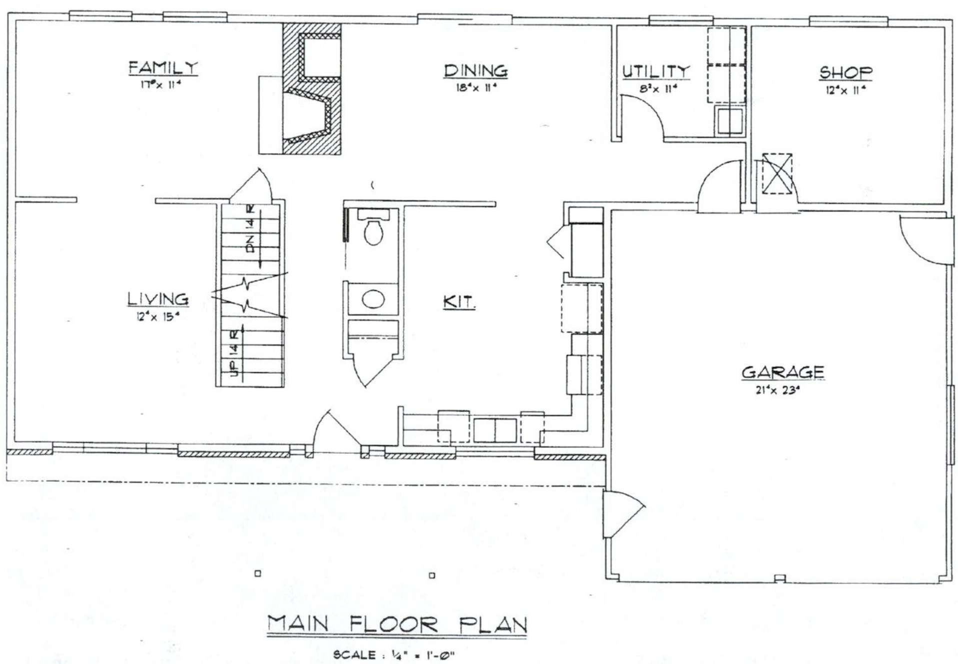

Figure 10.23b *Once the basic lines have been put in the correct position using OFFSET, other editing commands can be used to finish the drawing.*

Move the cursor slightly to the desired side and press the pick button. Pressing the pick button will offset the object and redisplay the prompt:

Select the object to offset or <exit>:

The process may be continued indefinitely, with the offset distance that was entered originally remaining as the default. The OFFSET process is terminated by (EnterKey). Once the process is terminated, a new default distance may be entered or a new command sequence may be started. The command sequence can be seen in Figure 10.24.

Entering **T** (EnterKey) at the original prompt allows a point to be entered with a pointing device, and the object to be offset will pass through the indicated point. The command sequence is:

Command: **O** (EnterKey)

OFFSET

Specify offset distance or [Through] <1'-6">: **T** (EnterKey)

Select the object to offset or <exit>: *(Select an object.)*

Specify through point: *(Move cursor to the desired point and select.)*

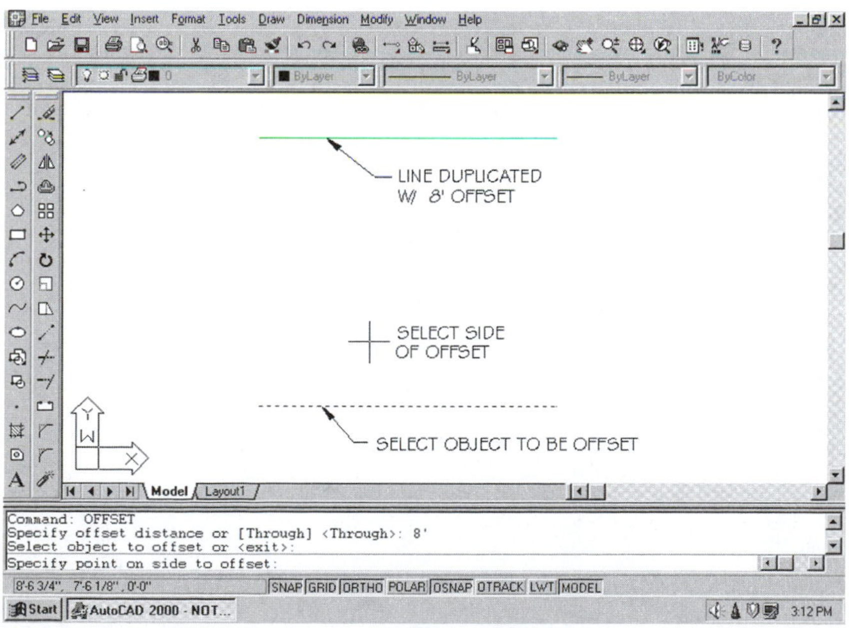

Figure 10.24 *The OFFSET command is completed by selecting the distance to be offset, the object to be offset, and the side of the original that the offset will occur on.*

Select the object to offset or <exit>:(EnterKey)

Command:

The OFFSET process can be seen in Figure 10.25.

Practical Uses for OFFSET: Open the ARCHBASE template. Now draw intersecting horizontal and vertical lines. Refer to the floor plan in Figure 10.23b. Many clients come to a designer with a floor plan that they like, along with a list of a thousand changes that they would like to make. The OFFSET command can be useful in the layout of a floor plan. Assume the two lines that have been drawn represent the outer edges of the walls in the upper right-hand corner of the plan shown in Figure 10.23b. Offset each line 6" in toward the inside of the house. Since the room in Figure 10.23b is 12'-4" wide, offset the vertical line you just created a distance of 12'-4" to the left. Offset this same line 4" to represent the wall thickness. Offset this line 8'-2", and then again 4". This will provide the wall on the left side of the Utility room.

Now go back to the inner horizontal line that you drew. Offset this line 11'-4" toward the bottom of the drawing. Offset this line 6" toward the bottom of the drawing. These two lines will represent the wall between the shop and the garage. To place the wall between the utility room and the hall, offset the inner garage wall 42". Offset this new line 4". The entire process is shown in Figure 10.26. Save this crude drawing as FLOOR10.

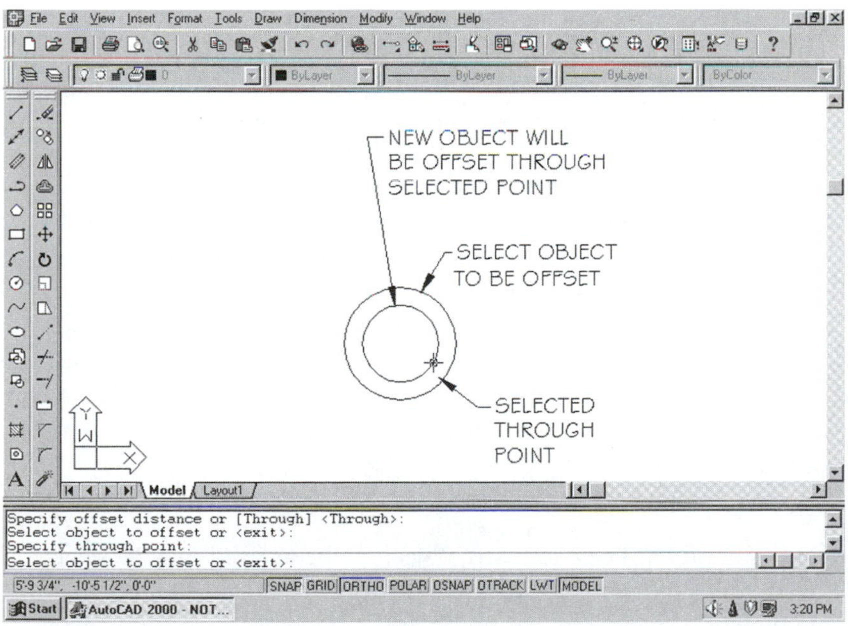

Figure 10.25 *The Through option of OFFSET allows the offset distance to be specified using the pointing device.*

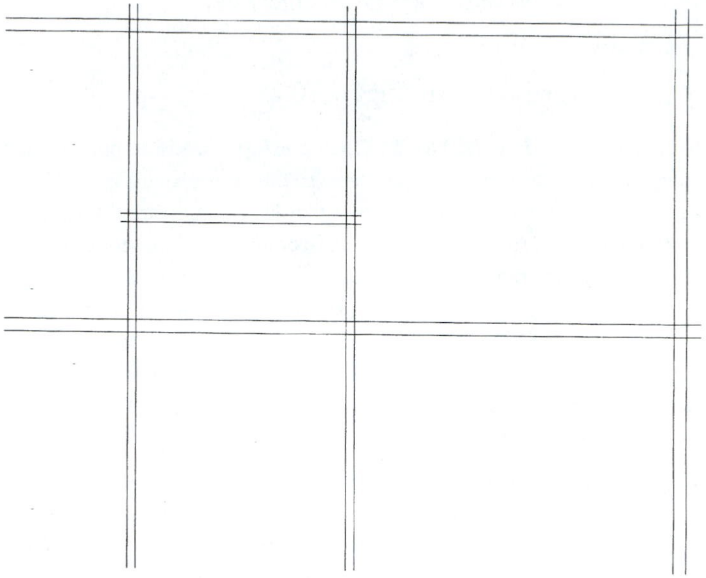

Figure 10.26 *By drawing two perpendicular lines and using the OFFSET command, a floor plan can easily be drawn.*

ARRAY

Until now, if you have wanted to reproduce a drawing element, you used the COPY command. The ARRAY command will also allow for multiple copies of an object or group of objects to be reproduced—such as the beam, post, and columns of a post-and-beam foundation. The command also provides several added features not available with COPY. ARRAY reproduces objects in rectangular or circular (polar) patterns, allows the object to be rotated as it is reproduced, and allows for easy control of the spacing of the object during the ARRAY. Figures 10.27a and 10.27b show examples of a rectangular and a polar array. The command can be accessed by selecting the Array icon from the Modify toolbar, by typing **AR** (EnterKey) at the command prompt, or by selecting Array from the Modify pull-down menu This command should not be confused with the DISARRAY option, which results from poor planning or procrastination.

Rectangular ARRAY

Once the command has been accessed, objects to be arrayed may be selected by any of AutoCAD's select features. Once selected for ARRAY, the command prompt will ask for the type of display. The prompt is:

Command: **AR** (EnterKey)

ARRAY

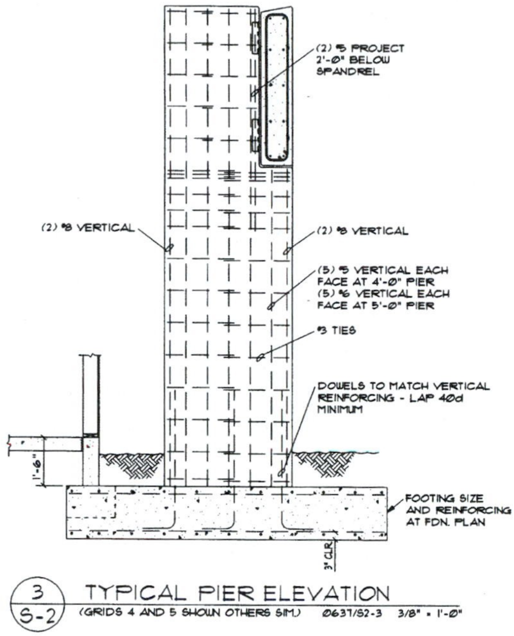

Figure 10.27a *The steel reinforcement was placed using the ARRAY command. (Courtesy Van Domelen/Looijenga/McGarrigle/Knauf Consulting Engineers.)*

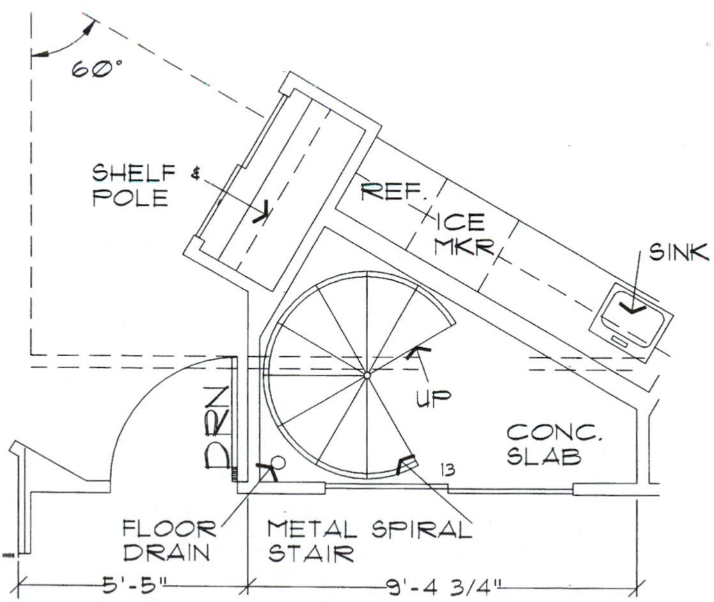

Figure 10.27b *The treads of this stair were placed using the polar option of the ARRAY command. (Courtesy Piercy & Barclay Designers, Inc.)*

Select objects: **W** (EnterKey) *(Select desired objects using any selection method.)*

Specify first corner: *(Select corner.)*

Specify opposite corner: *(Select opposite corner.)*

1 found

Select objects: (EnterKey)

Enter type of array [Rectangular/Polar] array <R>: (EnterKey)

A rectangular array will prove suitable for construction projects that are arranged in rows or columns, such as placement of posts or columns for a floor or foundation plan, or for drawing sheets of plywood for a roof-framing plan. Four specifications must be provided to perform a rectangular array. These include the number of rows, number of columns, distance between rows, and distance between columns. The effects of each can be seen in Figure 10.28.

Rows

The first prompt to respond to for a rectangular array is:

Enter the number of rows (---) <1>:

Any whole number may be entered for the value. In defining the quantity of rows, you are defining the quantity of horizontal rows to be created. Figure 10.29 shows an example of an array with two, three, and four rows.

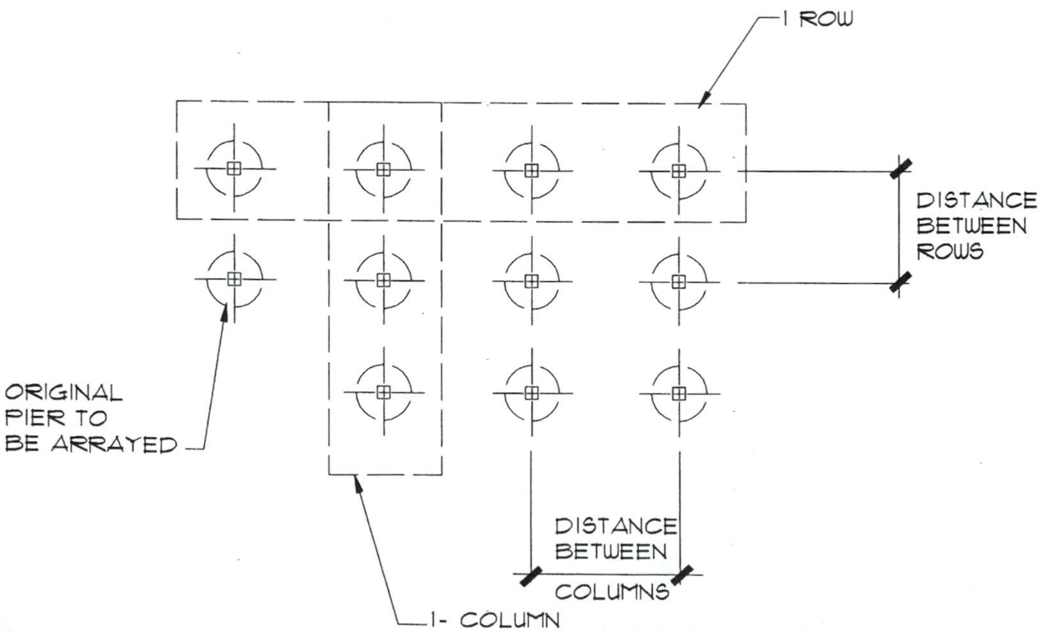

Figure 10.28 *The number of rows and columns and the spacing of each must be specified to use a rectangular ARRAY.*

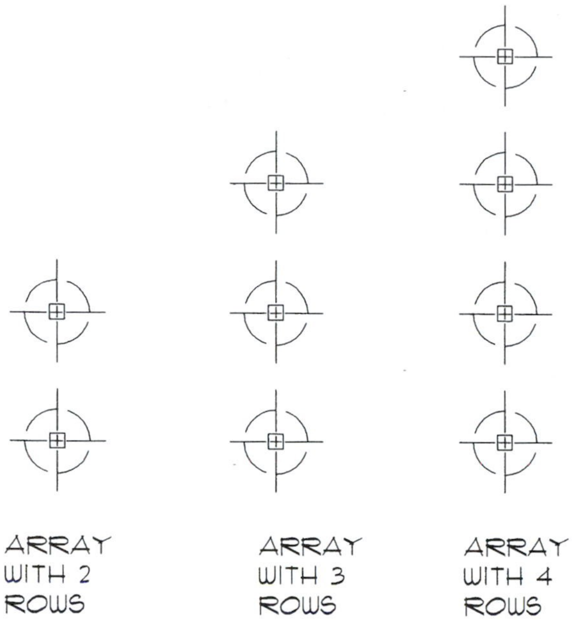

Figure 10.29 *Altering the number of rows of an ARRAY.*

Columns

Specifying the number of columns will dictate the quantity of vertical columns to be used. Any whole number may be entered for the value. If one was used for the row value, one may not be used for the column value. If it is, AutoCAD will respond with the prompt:

> One-element array, nothing to do.

and return the command prompt. Figure 10.30 shows the effect of an array using two, three, and four columns.

Distance Between Rows

The third prompt for a rectangular array is:

> Enter the distance between rows or specify unit cell (---): *(Enter desired distance between rows.)*

This prompt allows the distance to be set for row spacing, also called the unit cell. The unit cell distance includes the size of the object to be arrayed. If you are arranging 16"-wide chairs in rows, and would like 16" between rows, the spacing would require a 32" unit cell. This is shown in Figure 10.31. If you are arranging sheets of plywood, which typically are laid edge to edge, the unit spacing would be the spacing of the unit to be arrayed, in this case 48" for a sheet of plywood. This layout can be seen in Figure 10.32.

Figure 10.30 *Altering the number of columns in an ARRAY.*

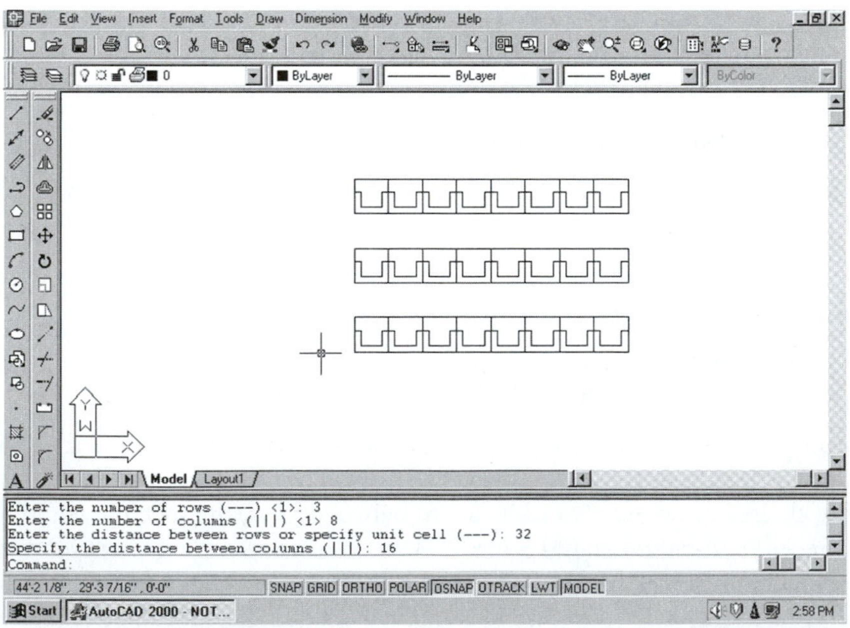

Figure 10.31 *The effect of row and column spacing in an array of three rows and eight columns.*

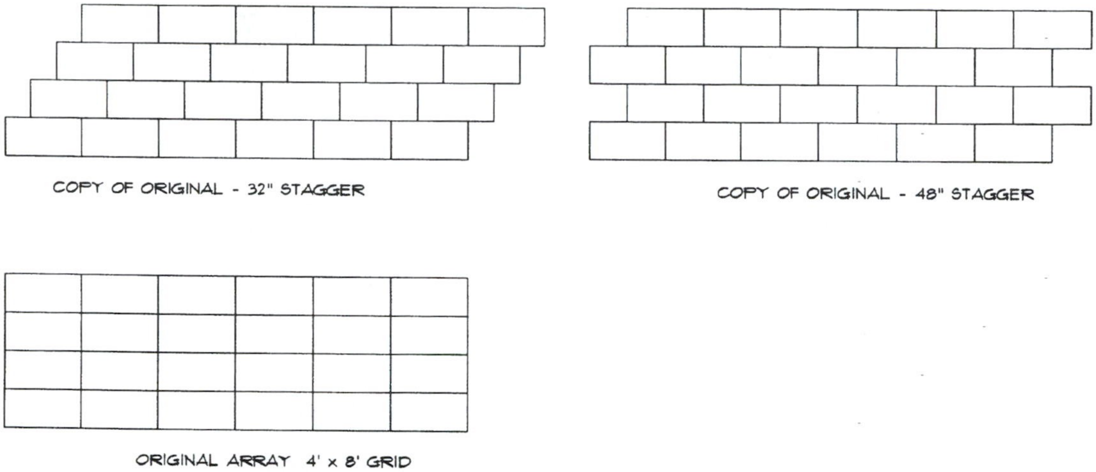

Figure 10.32 *The original array pattern at the bottom was created using four rows and six columns. The patterns at the top were altered by moving two rows.*

You also can set the distance between rows by selecting the unit cell as two opposite points of a rectangle. The rectangle will provide information for the row spacing. If more than one column is to be arrayed, the rectangle will provide information for both the row and the column distances. The unit cell spacing can be seen in Figure 10.33.

Distance Between Columns

The final prompt for a rectangular array is to provide the distance between columns. The same guidelines for spacing rows apply to spacing columns. The prompt is:

> Specify the distance between columns (| | |): *(Enter desired distance between columns.)*

Positive and Negative Values

As values are provided for rows and columns, they may be positive, negative, or a combination of both. Using positive values for both will array the objects up and to the right of the original object, as seen in Figure 10.34. If a negative value is used for the row distance, new objects will be added below the original. If negative values are used for the column distance, new objects will be added to the left of the original object, as seen in Figure 10.35. Figures 10.36 and 10.37 show the effects of combining values.

Rotated Rectangular Arrays

The array pattern is normally based on a horizontal or vertical baseline. The baseline is based on the current SNAP rotation value. By using the SNAP Rotate command, you can create a rotated array. By rotating the horizontal and vertical baselines,

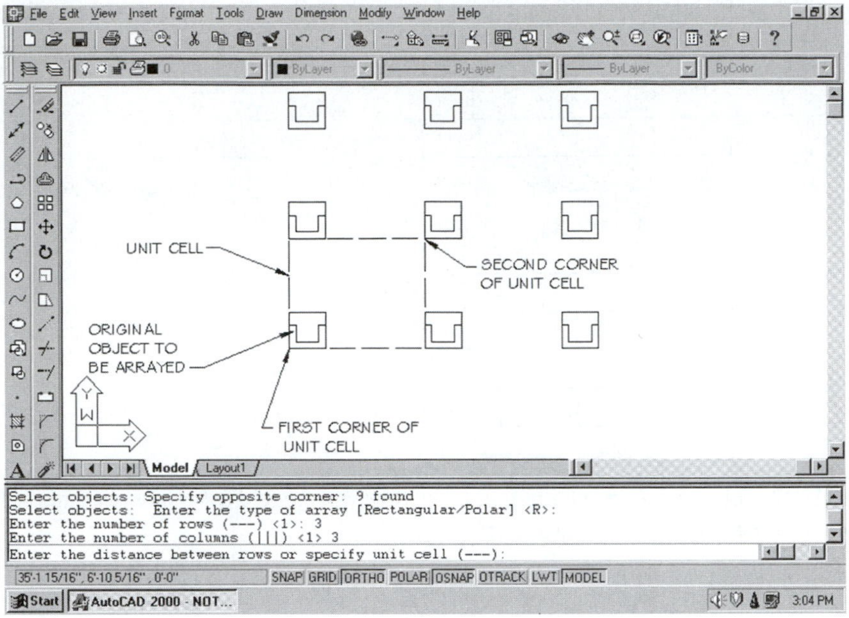

Figure 10.33 *Unit cell distance can be specified by selecting the two opposite corners of a window.*

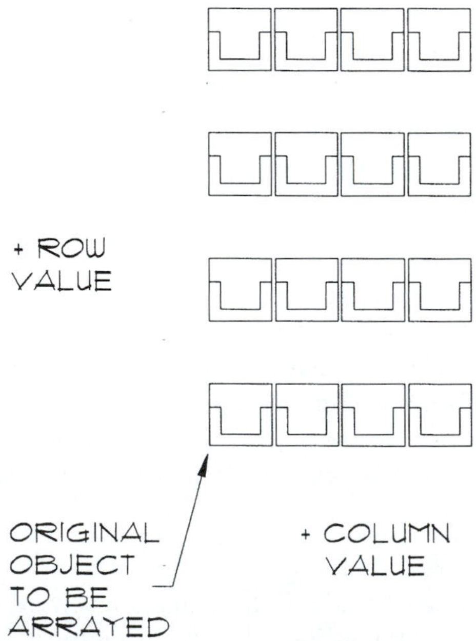

Figure 10.34 *An ARRAY pattern based on entering two positive distances.*

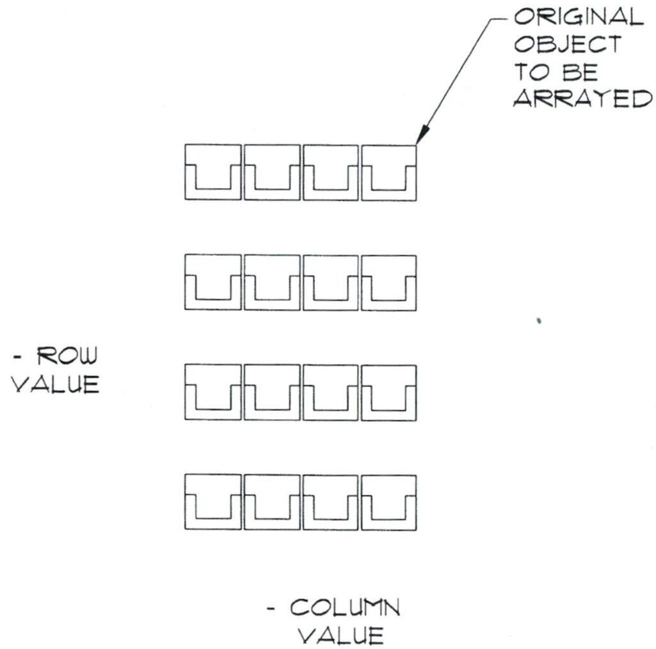

Figure 10.35 *An ARRAY pattern based on entering two negative values.*

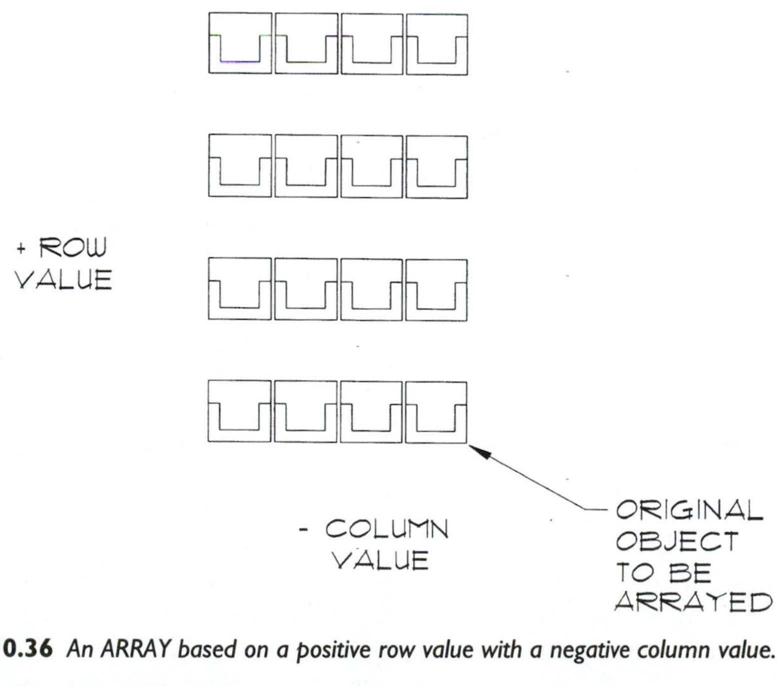

Figure 10.36 *An ARRAY based on a positive row value with a negative column value.*

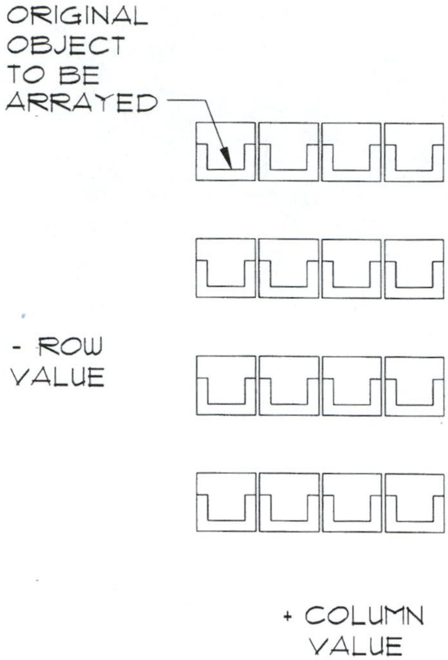

Figure 10.37 *An ARRAY based on a negative row value with a positive column value.*

objects may be arrayed along the new baseline. The SNAP angle can be adjusted using the Drafting Settings dialog box. The box is accessed using the following steps:

- Select Drafting Settings... from the Tools pull-down menu.
- Select the Snap and Grid tab.
- Enter the desired angle in the Angle: edit box.
- Choose the OK button to close the box and return to the drawing area.
- From the Command prompt the command sequence is:

  Command: **SN** (EnterKey)

  Specify snap spacing or [ON/OFF/Aspect/Rotate/Style/Type] <0'-1">: **R** (EnterKey)

  Specify base point <0'-0",0'-0">: *(Select desired base point.)*

  Specify rotation angle <0>: **45** (EnterKey)

  Command:

With the baseline rotated to 45°, the crosshairs also will be rotated. Any objects that are arrayed will be located along the selected SNAP axis. (See Figure 10.38.)

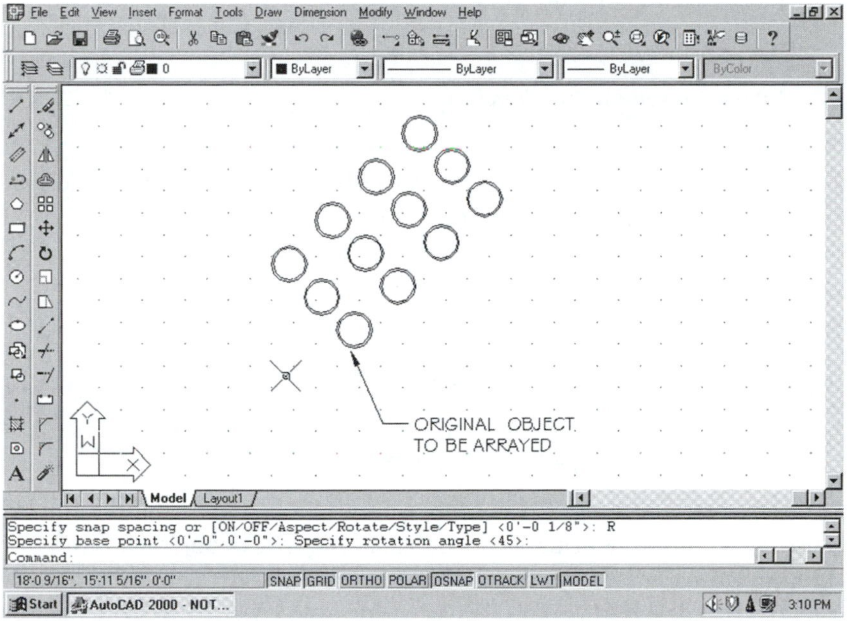

Figure 10.38 *A rotated ARRAY can be drawn by setting the SNAP spacing to ROTATE and providing the desired angle.*

Polar Array

A polar array is a circular layout of objects based on a center point. The process is started once the objects to be arrayed have been selected. The command sequence is:

Command: **AR** (EnterKey)

ARRAY

Select objects: **W** (EnterKey) *(Select desired objects.)*

Specify first corner:

Specify opposite corner:

1 found

Select objects: (EnterKey)

Enter type of array [Rectangular or Polar] <R>: **P** (EnterKey)

Specify center point of array: *(Select desired center point.)*

The center point is the point that the selected objects will be arrayed around. Four options are now presented to set parameters for the polar array. Any two of the three may be used. The parameters include:

Number of items in the array

Angle to fill (+ = CCW, – = CW) <360>

The angle between objects in the array

Rotate arrayed objects [Yes/No]

Number of Items

The number of items prompt allows you to select how many items, including the original, will be displayed around the center point. If a 0 is entered, the other two prompts will be given. If a positive value is entered, one of the two remaining prompts will be given.

Angle to Fill

This prompt controls the limits of the display around the center point. The prompt is:

Angle to fill (+ = CCW, – = CW)<360>:

The default of 360° will locate the specified number of items around a full circle. A number smaller than 360° will array the indicated number of objects around a selected center point in a specified degree range. The pattern will start with the original object and spread the balance of the pattern from its location. A positive value will produce a counterclockwise rotation. A negative value will cause a clockwise rotation. Examples of each pattern can be seen in Figure 10.39. Entering a 0 will be accepted if the

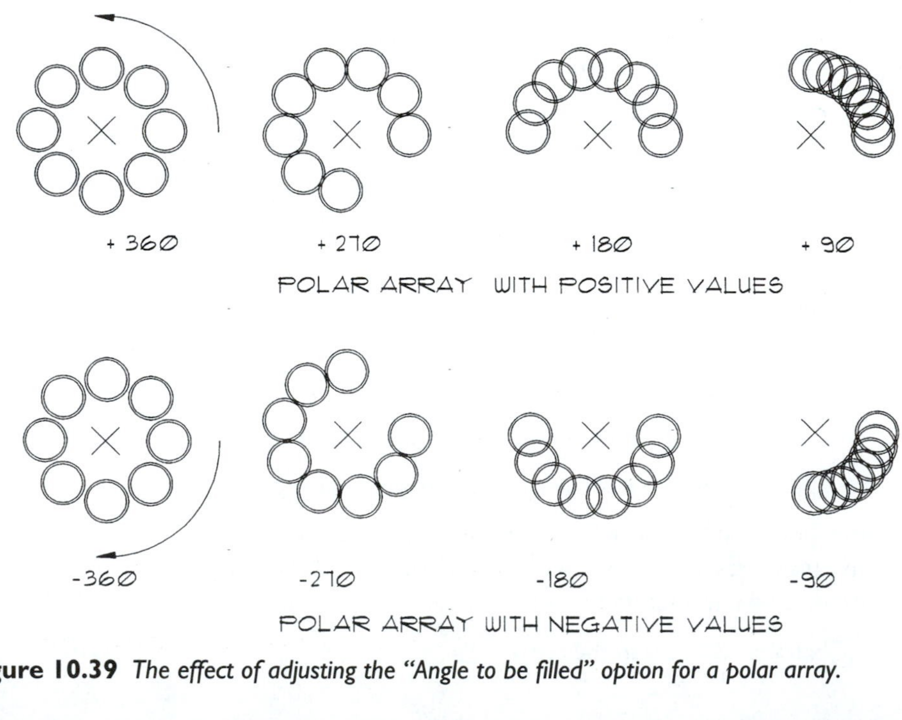

Figure 10.39 *The effect of adjusting the "Angle to be filled" option for a polar array.*

number of items has been provided. The location will now be arrayed by the final prompt, requesting the angle between items.

Angle Between Items

The third prompt that may be used for a polar array is to specify the angle between items. If the number of items and the angle to fill prompt have been specified, this prompt will not be given. The number of items to be rotated, combined with the angle distance between each item, often is used to locate bolts or other similar features for drawing connectors. (See Figure 10.40). Materials also can be arrayed over a specified degree range and intervals, such as showing pipes that intersect a cooling unit over a 75° area at each 15°, as seen in Figure 10.41.

Rotating Arrayed Objects

The final prompt for a polar array asks if the selected item to be arrayed should be rotated as it is arrayed. The default is yes, which will rotate objects as they are arrayed around the center point. If an N is entered at the prompt, the objects will not be rotated and will appear similar to the right side of Figure 10.42. The entire command sequence for a polar array of 50 segments around a 250° arc is:

Command: **AR** (EnterKey)

Select objects: *(Select desired objects.)*

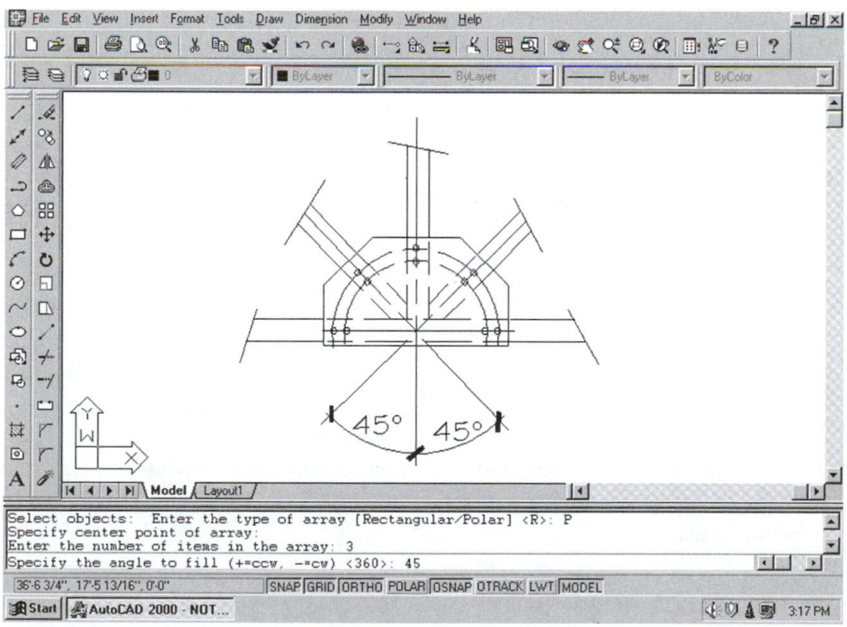

Figure 10.40 *Specifying the number of items to be rotated and the "Angle between items" is a common method for defining an array pattern.*

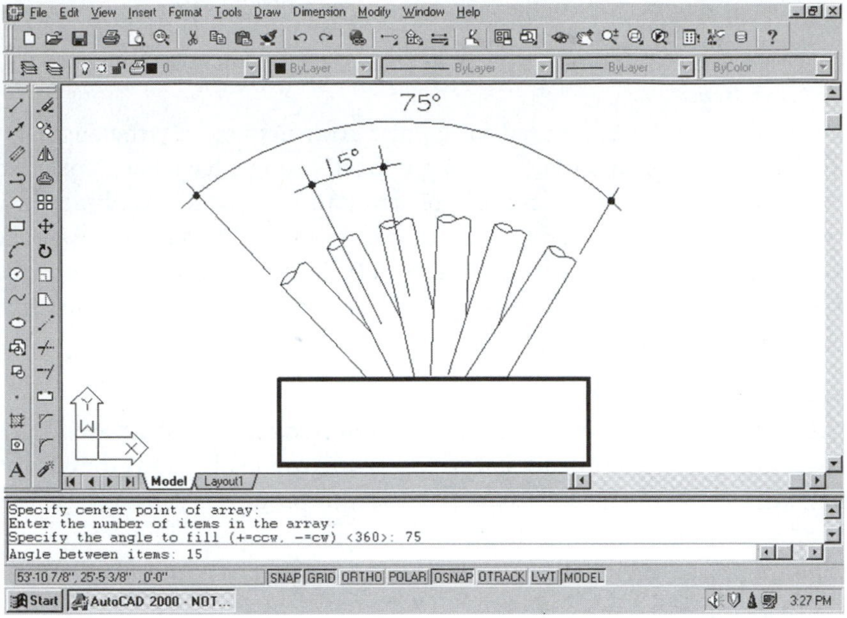

Figure 10.41 *Specifying the "Angle to fill" and the "Angle between items" is also used to define the array pattern.*

Select objects: (EnterKey)

Enter type of array [Rectangular or Polar] <R>: **P** (EnterKey)

Specify center point of array: *(Select desired center point.)*

Enter number of items in the array: **5** (EnterKey)

Angle to fill (+ = CCW, – = CW) <360>: **250** (EnterKey)

Rotate arrayed objects? [Yes/No] <Y>: (EnterKey)

Command:

Practical Uses for ARRAY: Array should be used whenever an object needs to be added to a drawing in a repetitive pattern. The girder and piers that were drawn in Figure 10.10 could have easily been placed using the ARRAY pattern. Lines to represent parking spaces on a site plan, lines to represent siding on an elevation, rectangles to represent joist or rafters in longitudinal sections, and circles to represent piers on a foundation are just a few examples of how the ARRAY command can be used to increase drawing efficiency.

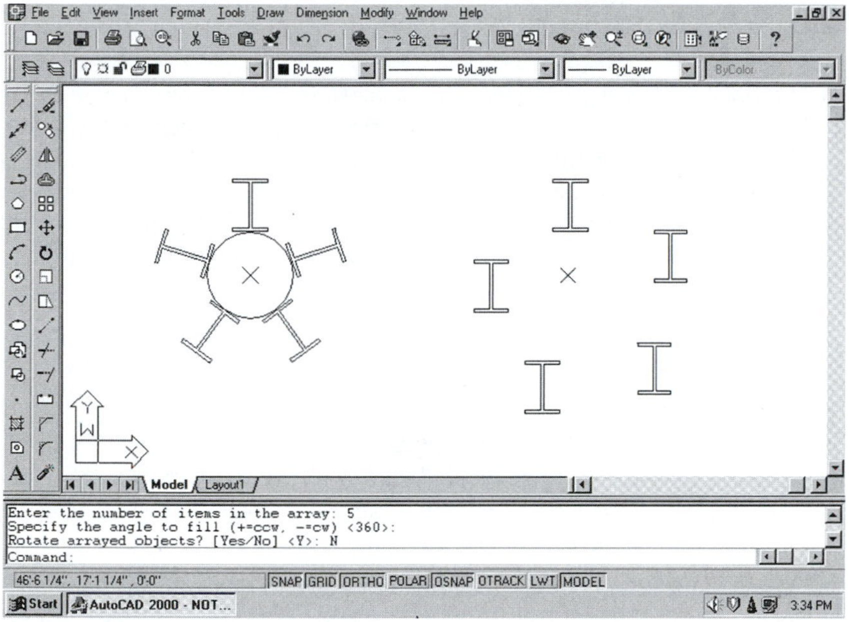

Figure 10.42 *Objects can be rotated as they are arrayed by adjusting the "Rotate objects as they are copied?" option.*

CHAPTER 10 EXERCISES

1. Start a new drawing and draw a horizontal line. Draw polygons with three, four, five, six, and eight sides with the centerline of each polygon on the horizontal line. Draw all polygons inscribed in a circle with a .5" radius. Make copies of the polygons with three, five, and eight sides and place them directly below their counterparts. Make copies of the four- and six-sided polygons and place them in a third row directly below their counterparts. Mirror the entire drawing with a horizontal mirror line, and delete the old objects. Save the drawing as E-10-1.
2. Open your existing file E-9-5. Copy the window so that a total of three pairs of windows are drawn exactly 4" apart. Align the bottoms of the windows. Save the drawing as E-10-2.
3. Draw half of a W 8"×28" (I-shaped) beam. The beam has a total height of 8.28". The top and bottom flange are .25" thick, with a total width of 5.25". The web is a total of .28" thick. Use the mirror command to complete the drawing. Save the drawing as E-10-3BM.
4. Using the attached sketch as a guide, draw an exterior swinging door that is 36" wide with a 12"-wide sidelight. Use the mirror command to draw a pair of doors with double sidelights. Save the drawing as E-10-4-DOOR.

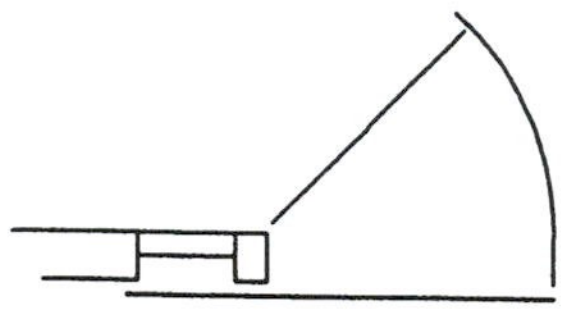

5. Using the following figure as a guide, draw a steel connector plate. The plate is 15 ½" × 10"h. Draw the bolts as ¾"-diameter hexagons, 1 ½" down and 1 ½" in from the top and ends. Bolts are 3" O.C. Each end of the plate is symmetrical. Draw a 2 ½" × 8" strap, centered in the bottom of the plate. Provide two bolts with similar spacing as the plate. Save the drawing as E-10-5.

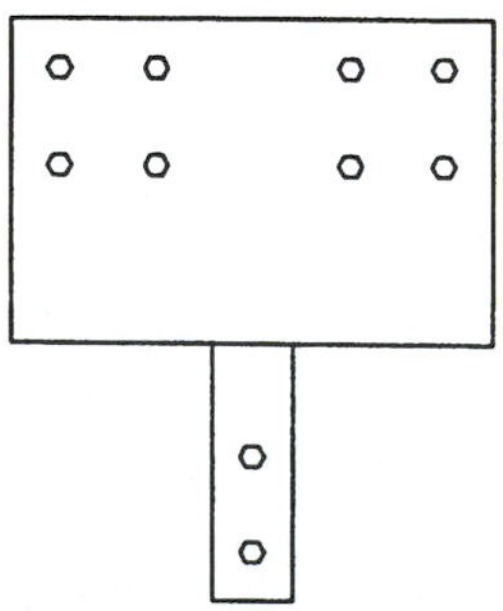

6. Use the sketch below as a guide and lay out the outline of the apartment floor plan. Draw the exterior walls as 6" wide. Use this unit plan and lay out four different building layouts with a minimum of four units per building. Save the drawing as APARTFLR.

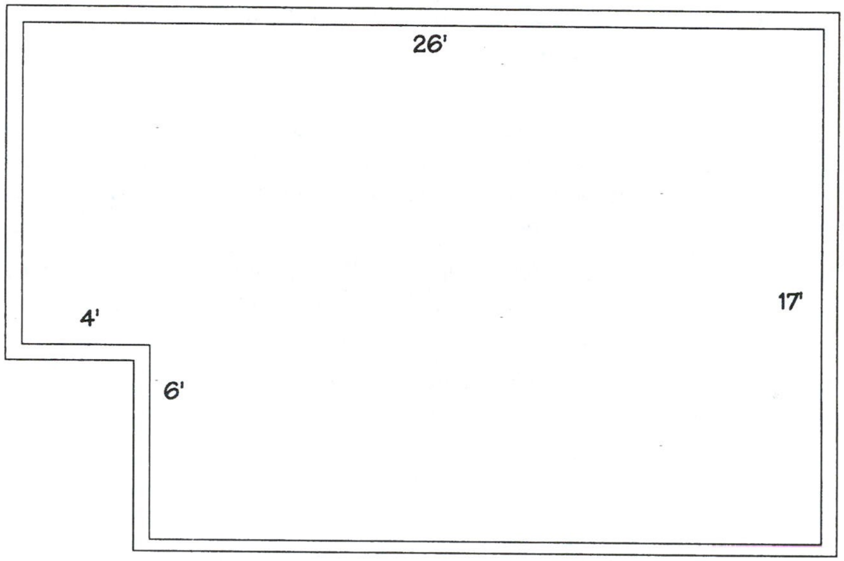

7. Draw a post-and-beam foundation plan for a 24'×16' structure. Exterior walls will be 8" thick. Piers are 18" diameter at 8' O.C. along a 4"-wide beam. Repeat the beam and piers at 4' O.C. Save this exercise as PB-FND.
8. Start a new drawing that can be used as a base for shingle siding in elevation. Draw several lines approximately 36" long, at 10" O.C. Draw shingles similar to the sketch with widths that vary from 4" through 12". Copy this row of shingles to form a minimum of five rows of shingles. Stagger each row with the rows above and below it, so that the shingle seams do not line up. Save the drawings as SHINGLE.

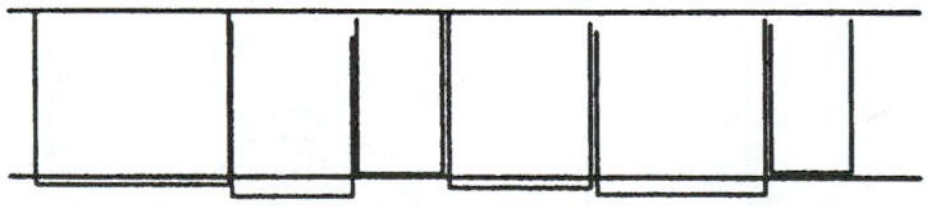

9. Extend the limits as needed to draw a 4" square centered in a 15"-diameter circle. Draw an 8" and a 12"-diameter circle with the same center point as the 15" circle. Make three additional copies of this object and move each so that the four 15" circles are tangent. Save the drawing as E-10-9.

10. Open drawing E-9-5. Your client thinks he would like a grid in each half, but isn't really sure. Copy the drawing and draw a grid in one of the windows. Provide one vertical and three horizontal dividers. Mirror the grid into the other. Save the drawing as E-10-10.
11. Draw a 60'-wide × 44'-deep structure. Use 8"-wide walls. No doors or windows need to be shown. Create a layer to represent the suspended ceiling and draw the plan for a suspended ceiling using 24" × 48" panels. Make a second copy of the plan and show the panels laid out in the opposite direction. Save the drawing as E-10-11.
12. Use the base drawing started in Figure 10.11 and draw a roof framing plan. Freeze the suspended ceiling information and create a new layer to represent roof trusses at 32" O.C. Use the attached drawing as a guide to show a portion of the roof with 4' × 8' plywood over the trusses. Save the drawing as E-10-12.
13. Draw a column that is 10'-0" tall and 12" wide. Show a 20" × 4" and a 16" × 4" corbeling 4" down from the top and 4" up from the bottom of the column. Save the drawing as E-10-13.
14. A circular structure is to be built on a hillside. Use the drawing below as a guide and draw the pier plan. Save the drawing as E-10-14.

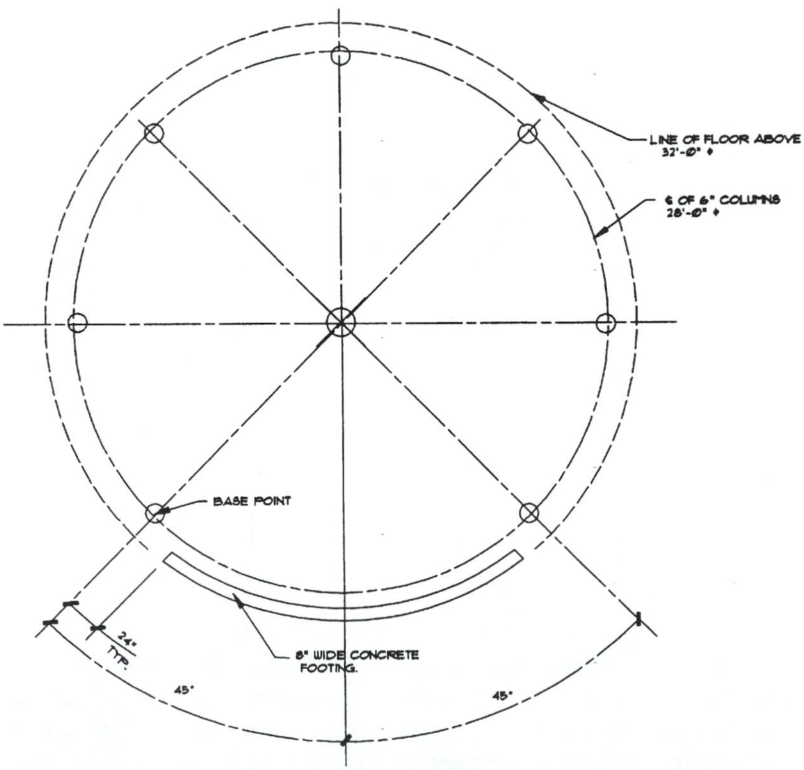

15. Draw a plan view of a 40'-diameter water reservoir. The reservoir will be supported on (12) 6"-diameter × $\frac{3}{8}$" steel columns. The columns will be equally spaced, on a 36'-diameter circle centered on an 18"-diameter × $\frac{3}{4}$"-thick vertical inlet pipe in the center of the reservoir. Assume north to be at the top of the drawing, and locate a 12"-diameter × $\frac{3}{8}$"-thick vertical outlet pipe at the outer edge of the tank, at 37° west of true north.

CHAPTER 10 QUIZ

1. Describe the process used to draw a wall that is 6" wide.
2. Several copies of an object need to be made. What option is available so that the COPY command does not have to be repeated?
3. You have selected several objects for editing when you realize you have accidentally selected an object that you do not want changed. What can you do so that you do not have to start the edit sequence over?
4. How will UNDO affect objects that have just been moved?
5. List three methods suitable for selecting irregular-shaped objects for editing.
6. When you use the Fence option, what objects will it select for editing?
7. What editing mode will allow you to select objects and then decide how to edit them?
8. What process will display the dialog box that controls selection methods?
9. List the commands in sequence required to complete Exercise 9.
10. Explain when the Through option of OFFSET could be used.
11. Open the ERASE command using each method and describe the advantages or disadvantages.
12. Experiment with the MIRROR command. Describe the effects of ORTHO on the mirror line.
13. MIRROR is being used to complete a symmetrical object. What options are available and how do they affect the mirror sequence?
14. Where is ERASE listed in the pull-down menu?
15. Describe how to change the size of the pickbox.
16. What is the affect of Base point on the COPY command, and how should it be selected?
17. List ways that the COPY command can be used to lay out a floor plan.
18. What is the benefit of having so many different methods of selecting objects for editing?
19. What are the drawbacks to enlarging the pickbox to twice its normal size, or to half its normal size?
20. Use the help menu to explain the difference between the normal setting for an editing command and the SINGLE option.
21. Explain the difference between Noun/Verb and Verb/Noun selection.
22. Explain the meaning and the effect of the following letters for selection groups: W, C, F, WP, CP, R, A, and P.
23. What effect does Press and drag have on selection?
24. Use the help menu to list the effects of the PICKADD, PICKAUTO, and PICKDRAG variables.

25. How can OBJECT SNAP be used with editing commands?
26. Explain the difference between the two types of ARRAY patterns.
27. Twelve steel columns need to be drawn 22' on center in one horizontal line. What will the array sequence be?
28. Three horizontal lines of 10 steel columns need to be drawn 18 feet on center. The horizontal lines of columns will be 32' apart. What will the array sequence be?
29. Use the drawing for Exercise 10.14 and give the command sequence to array the supports.
30. Using an illustration as an example, define a unit cell.
31. How could an array pattern for objects aligned at 45° from horizontal be drawn?
32. Ten rows and three columns are to be arrayed below and to the left of the original object at 60" O.C. How can this be done?
33. Describe how to enter information to produce a clockwise rotation.
34. Give the combination of alternatives for describing a polar array.
35. If a positive column distance and a negative row distance are entered, where will the objects be arrayed in relation to the original object?

CHAPTER 11

Modifying and Selecting Drawing Objects

In Chapter 10 you were introduced to editing commands that can affect an entire drawing. The commands discussed in this chapter can be used to modify part of an object by altering the position or size of the object. These commands include MOVE, ROTATE, ALIGN, SCALE, STRETCH, LENGTHEN, TRIM, and EXTEND. A second group of editing commands will also be introduced to edit objects by altering their length. These commands include TRIM, EXTEND, BREAK, CHAMFER, and FILLET. Mastering these commands will greatly increase your drawing ability. All of the commands can be selected from the Modify toolbar, entered by keyboard, or selected from the Modify pull-down menu. Each command can also be entered from the shortcut menu by selecting an object to edit, and then right-clicking to select the desired editing command. The final commands to be introduced will be GRIPS, SELECT, and GROUPS. These commands will expand your ability to select objects for editing.

MODIFYING DRAWINGS BY ALTERING THE POSITION AND SIZE

The Modify menus offer five options for modifying a drawing by changing the position of an object or by changing the size of the object. These commands include MOVE, ALIGN, ROTATE, SCALE, STRETCH, and LENGTHEN.

MOVE

Have you ever started a manual drawing project, only to find out that after all of your careful planning, the project no longer fits on the paper because of client changes? This common problem led to such solutions as splicing drawings into new sheets or using sepias and Mylar copies. AutoCAD has a better solution—move the object. The MOVE command allows one or more objects to be selected and moved from their current location to a new one. The command can be accessed by selecting the Move icon from the Modify toolbar, by entering **M** (EnterKey) at the command prompt, or by selecting Move from the Modify pull-down menu. Figure 11.1a and 11.1b compare the effects of the MOVE command.

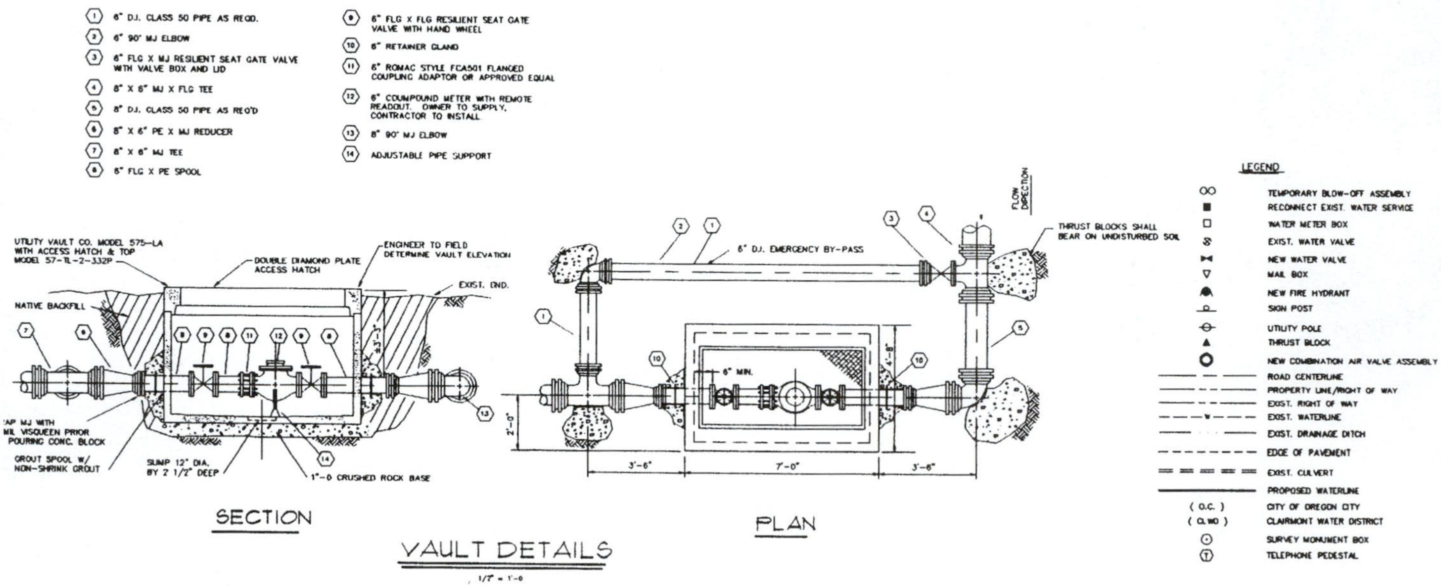

Figure 11.1a *The MOVE command can be used to alter the layout of a drawing. (Courtesy Lee Engineering, Inc., Oregon City, OR.)*

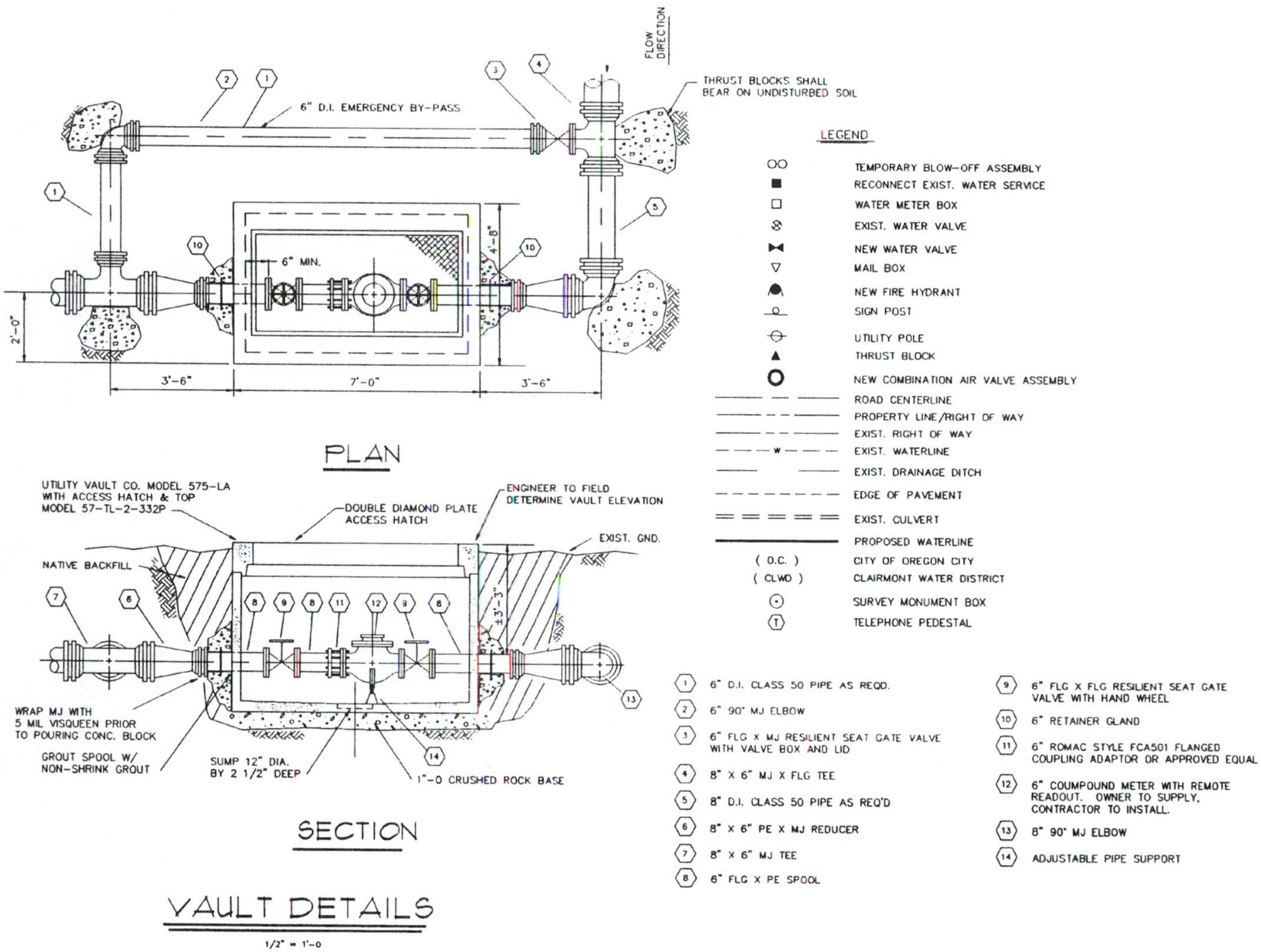

Figure 11.1b *The MOVE command was used to place the plan view of Figure 11.1a above the section drawing. (Courtesy Lee Engineering, Inc., Oregon City, OR.)*

The command sequence for MOVE is:

Command: **M** (EnterKey)

MOVE

Select objects: *(Use the desired selection method.)*

Select objects: (EnterKey)

Specify base point or displacement: *(Enter point or coordinates.)*

Specify second point of displacement or <use first point as displacement>: *(Enter point or coordinates.)*

Command:

The process can be seen in Figures 11.2a through 11.2c. Use any selection means that best suits your needs to select the objects to be moved. The Base point of displacement prompt is asking for a reference point for moving the selection set. Any point may be selected, but it may help you to visualize the results of moving by selecting a point in the drawing, such as an intersection or endpoint. This is an excellent time to make use of OSNAP.

The Second point of displacement prompt is asking where you would like to move the first point. Each point may be entered by selecting a point with a pointing device

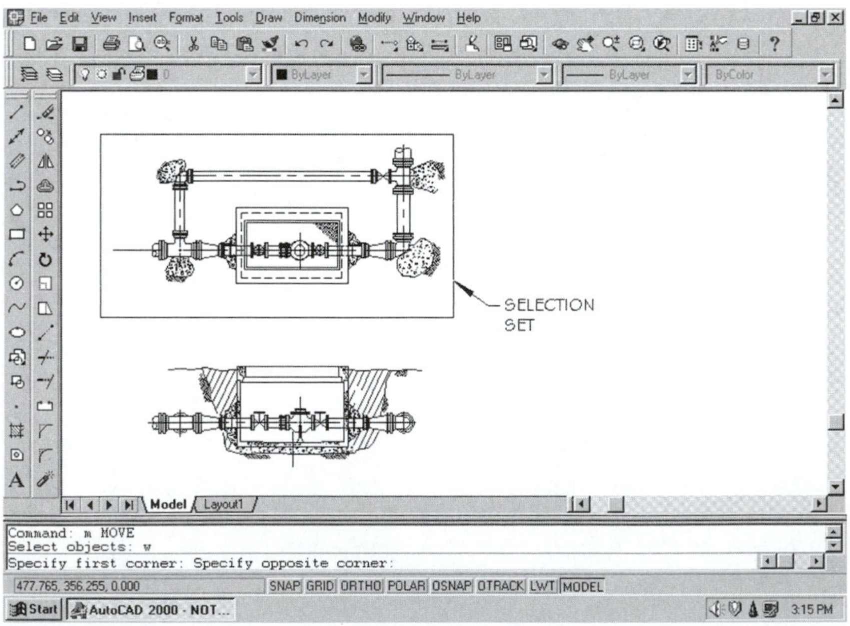

Figure 11.2a *The first step in the process for moving a drawing is to select the objects to be moved. (Courtesy Lee Engineering, Inc., Oregon City, OR.)*

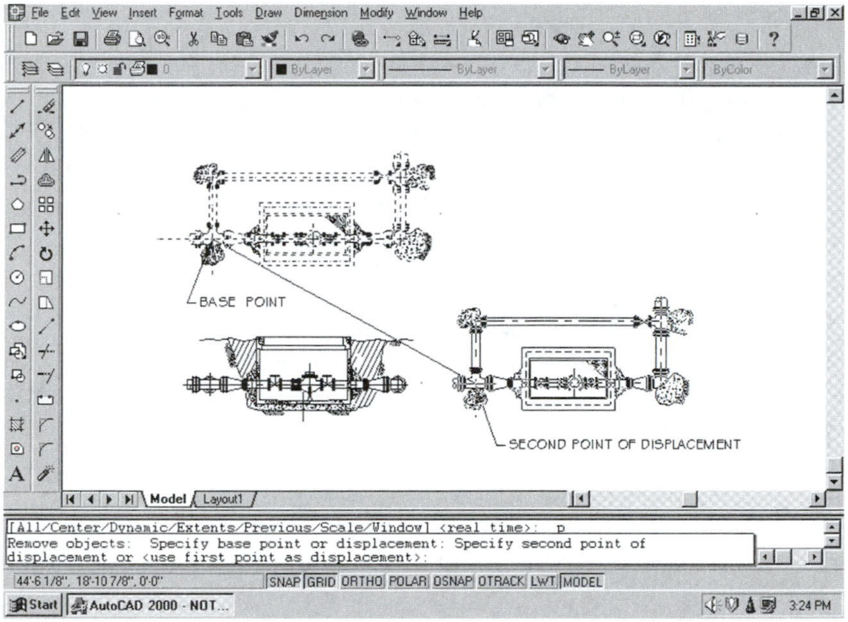

Figure 11.2b *The second step is to select the base point of displacement and the new location for this point. (Courtesy Lee Engineering, Inc., Oregon City, OR.)*

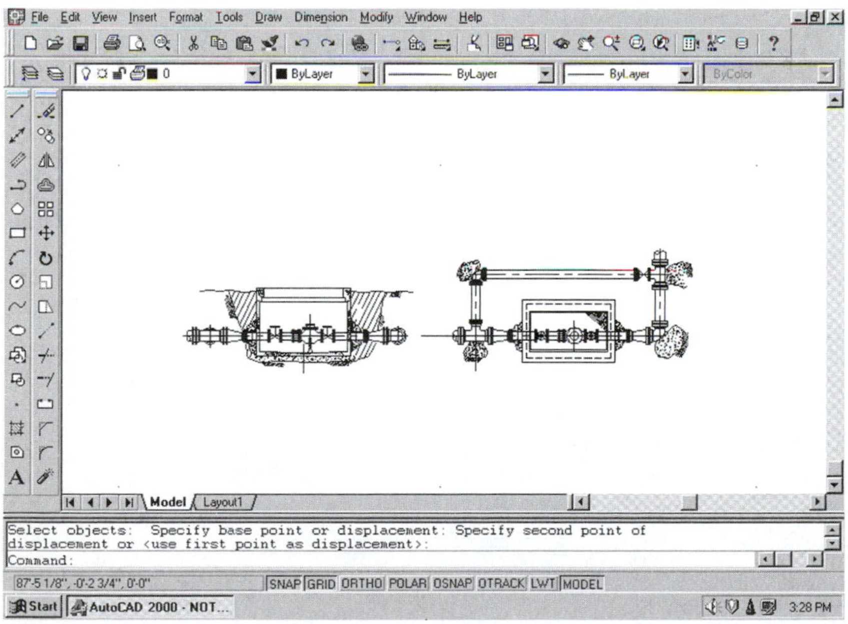

Figure 11.2c *Finally, the second point of displacement is selected and the MOVE process is complete. (Courtesy Lee Engineering, Inc., Oregon City, OR.)*

or by entering coordinates. SNAP, ORTHO, OSNAP, and ZOOM can be very helpful for aligning related objects as they are moved.

Practical Use: MOVE should give you an immense sense of freedom, but not a sense of reckless abandonment. You still need to plan your drawings, but the MOVE command offers you great flexibility in drawing arrangement. Space between views can be altered as the needs of the drawing change. On the other hand, you have to keep in mind that each of your drawings will need to fit onto a sheet of paper of a set size. For most architectural projects, this means your drawing will need to fit into a sheet of paper that is 36" × 24". If you're going to try and place ten details on this sheet, there is only so much space where the details can be moved. AutoCAD still has not incorporated the CRAM command. Planning and wise use of the editing commands are the only hope for a good solution.

ROTATE

In Chapter 10 you learned how to rotate the grid to lay out a portion of a drawing that may not be oriented to horizontal or vertical. In a similar fashion, you can draw an object or group of objects and then rotate them around a base point to change their orientation. The command can be started by selecting the Rotate icon from the Modify toolbar, by entering **RO** (EnterKey) at the command prompt, or by selecting Rotate from the Modify pull-down menu. Rotate can also be started from the shortcut menu by selecting the object, right-clicking, and then picking Rotate from the shortcut menu. Figure 11.3 shows an example of a window that was drawn for a

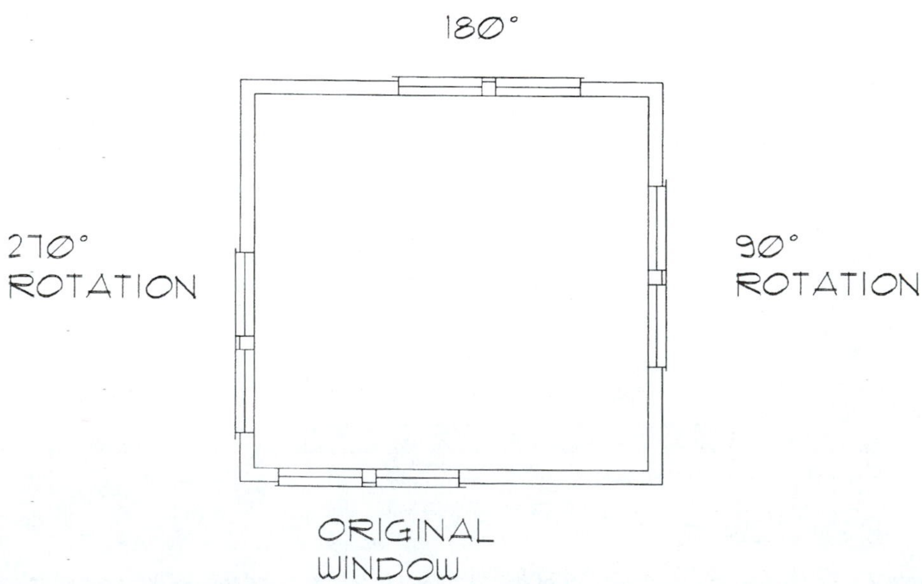

Figure 11.3 *The ROTATE command allows objects to be drawn and rotated into a new position.*

horizontal wall on a floor plan. The window was then copied and rotated 90° to be used in a vertical wall.

To complete the command, you will need to select objects to be rotated, a base point to rotate the object around, and the desired degree of rotation. Entering a positive rotation angle rotates the selected object in a counterclockwise direction. Entering a negative rotation angle produces a clockwise rotation. The base point may be anywhere on the drawing screen, but selecting one corner of the object as the base point may ease visualization of the rotation. The command sequence is:

Command: **RO** (EnterKey)

ROTATE

Current positive angle in UCS: ANGDIR = counterclockwise ANGBASE = 0

Select objects: *(Use any selection method.)*

8 found

Select objects: (EnterKey)

Specify base point: *(Select any point such as an endpoint or midpoint.)*

Specify rotation angle or [Reference]: *(Enter angle or drag into position.)*

Command:

Notice that instead of entering a precise angle, the selected objects may be dragged into the desired position as the mouse is moved. If ORTHO is toggled ON, the objects will be snapped at 90° intervals. With ORTHO off, the objects can be dragged into any desired angle. When you enter an angle, you can rotate objects to exact locations based on the current location. The process can be seen in Figure 11.4.

Objects also can be rotated based on absolute rotation angles. Suppose you've drawn the top chord of a truss at 22½° for a 5/12 pitch. The owners change their mind and desire a 6/12 pitch (27½°). You could rotate the existing top chord 5° and hope you get it close enough, or use the Reference option and place the roof exactly where it should be. The later option uses the following command sequence:

Command: **RO** (EnterKey)

ROTATE

Current positive angle in UCS: ANGDIR = counterclockwise ANGBASE = 0

Select objects: *(Use any selection method.)*

5 found

Select objects: (EnterKey)

Specify base point: *(Select any point.)*

Specify rotation angle or [Reference]: **R** (EnterKey)

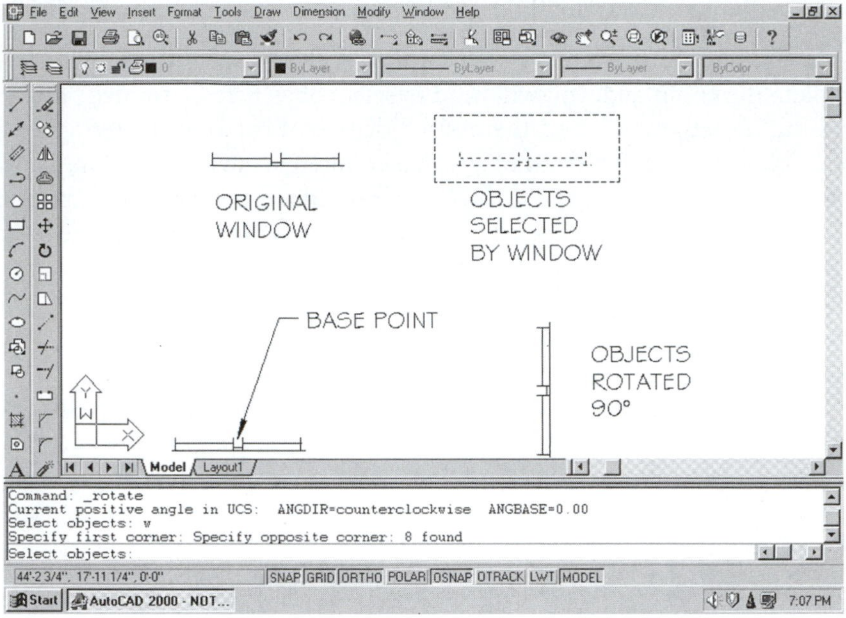

Figure 11.4 *Objects are rotated by defining the selection set, specifying a base point for the rotation, and providing the rotation angle.*

Specify the reference angle: **22.5** (EnterKey)

Specify the new angle: **27.5** (EnterKey)

Command:

The process can be seen in Figures 11.5a, 11.5b, and 11.5c.

ALIGN

Think of the ALIGN command as a combination of the MOVE and ROTATE commands. Although primarily a command for 3D drawing, it works well when an object or group of objects need to be moved and rotated. The command can be started by typing AL (EnterKey) at the command prompt, or selected from the 3D Operation flyout menu of the Modify pull-down menu. The command will prompt you for two source points and two destination points. The source points will be used to define points on the original object in its original position. The destination points represent the locations where the object will be placed. The command can be used in two ways. If one pair of source and destination points is provided, the selected object will be moved in similar fashion when the MOVE command is used. The benefit of the ALIGN command is in providing two pairs of points. If two pairs of source and destination points are provided, the selected object can be moved, rotated, and scaled

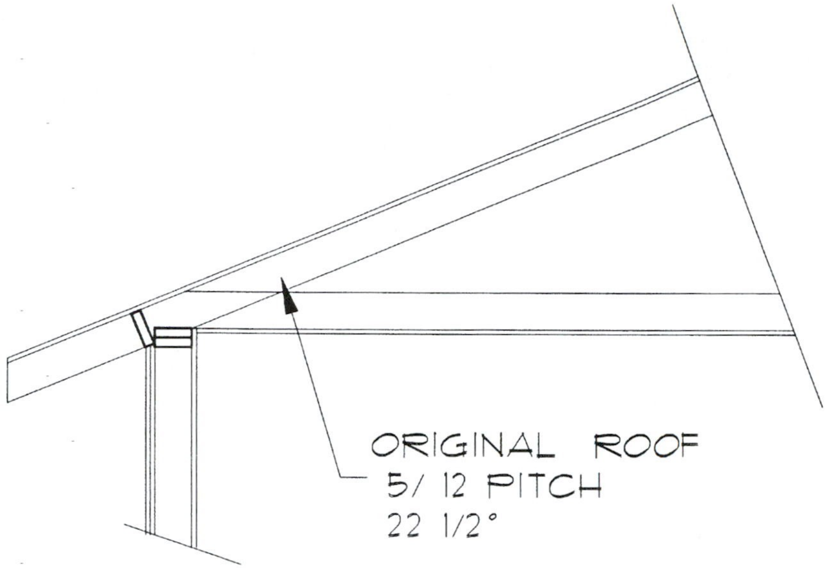

Figure 11.5a *The pitch of a roof can be altered by use of the ROTATE command.*

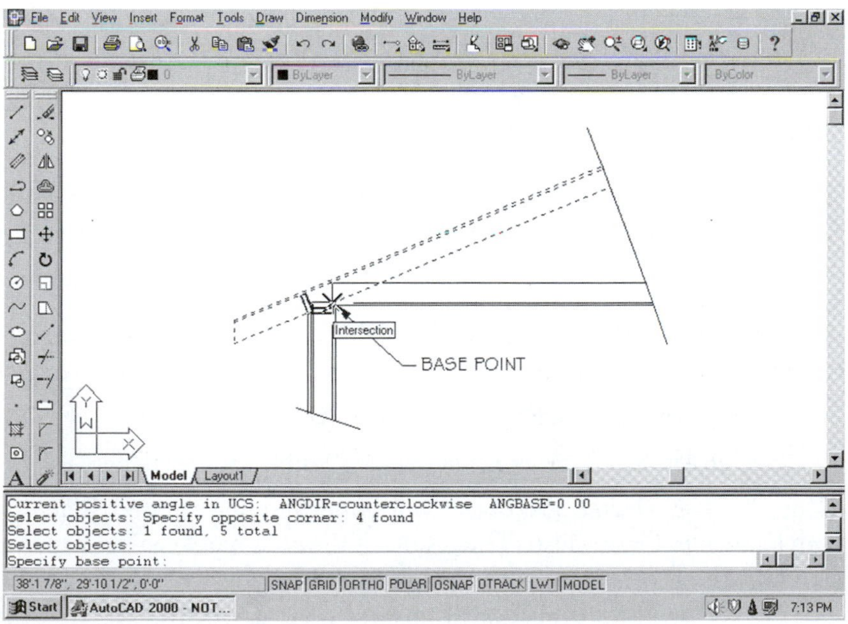

Figure 11.5b *The selection set is defined and the base point is specified.*

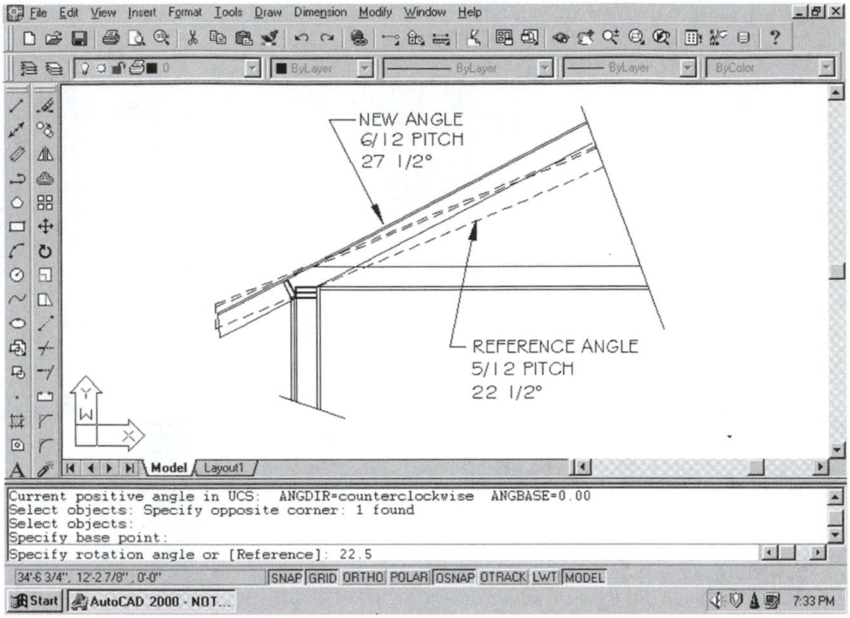

Figure 11.5c *With the reference angle specified (existing angle of 22.5°), the new angle of 27.5° is specified and the roof is rotated.*

to match other objects. This could be useful to move, rotate, and align a room with another when redesigning a floor plan. Use the following command sequence to move and rotate a kitchen from one area of a floor plan to another:

Command: **AL** (EnterKey)

ALIGN

Select object: *(Select object using any selection method.)*

Specify first source point: *(Select point with mouse or enter coordinates.)*

Specify first destination point: *(Select point with mouse or enter coordinates.)*

Specify second source point: *(Select point with mouse or enter coordinates.)*

Specify second destination point: *(Select point with mouse or enter coordinates.)*

Specify third source point or <continue>: (EnterKey)

Scale objects based on alignment points? [Yes/No] <No>: (EnterKey)

When (EnterKey) is selected the object will be moved and rotated. The selection process can be seen in Figure 11.6. The results of the command can be seen in Figure 11.7. If Y (EnterKey) is entered at the last prompt, the objects being rotated will be reduced to match the distance indicated by the first and second destination points. Figure 11.8 shows the results of scaling the objects to meet the alignment points.

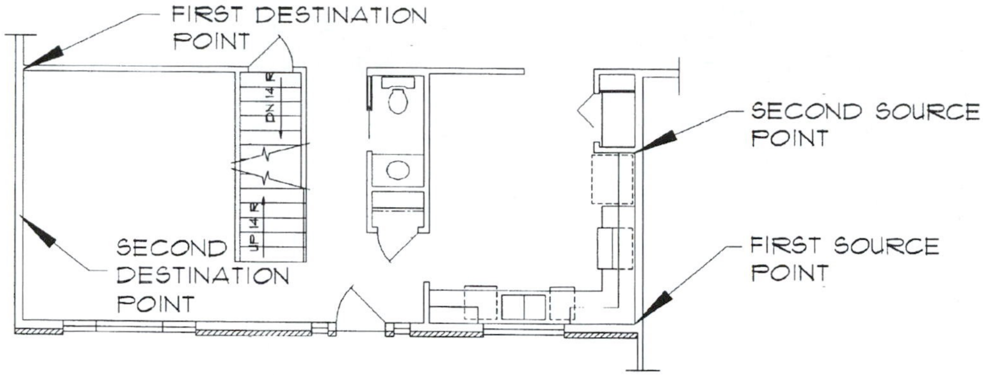

Figure 11.6 *The ALIGN command allows portions of a drawing to be moved and rotated. The command will prompt you to provide source points, and then ask where you would like that point moved to.*

SCALE

Up to this point, your drawings have been drawn at full scale. If you needed to draw a plot of land that was 100' square, you set the limits and units accordingly, zoomed out to the limits, and drew the plot of land. The whole process was done at real scale with the size of the zoom used to control the viewing screen. The SCALE command will allow you to change the size of an existing object or an entire drawing, as seen in

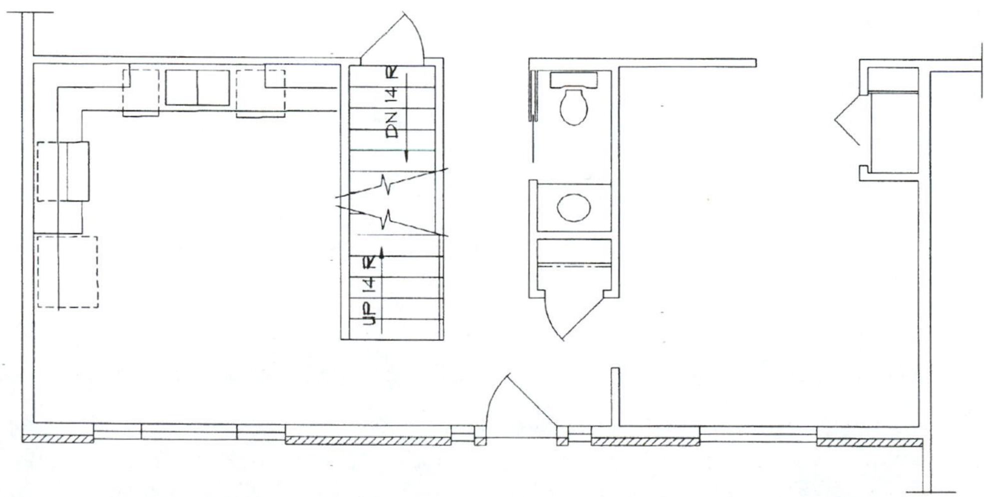

Figure 11.7 *By using the destination points shown in Figure 11.6, the kitchen was moved and rotated to a new position.*

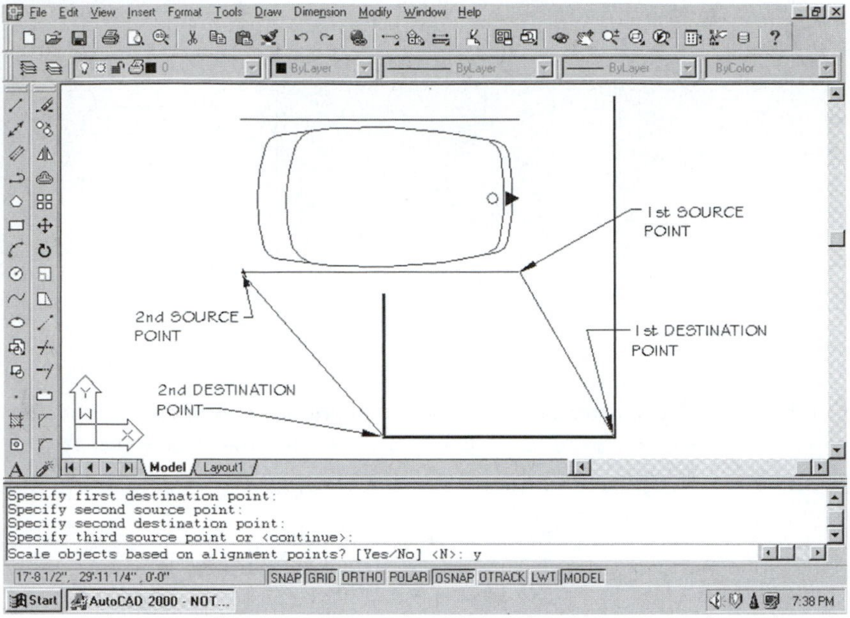

Figure 11.8a *By selecting the Scale objects to alignment points option, an object or group of objects can be moved and scaled to match the selected destination points.*

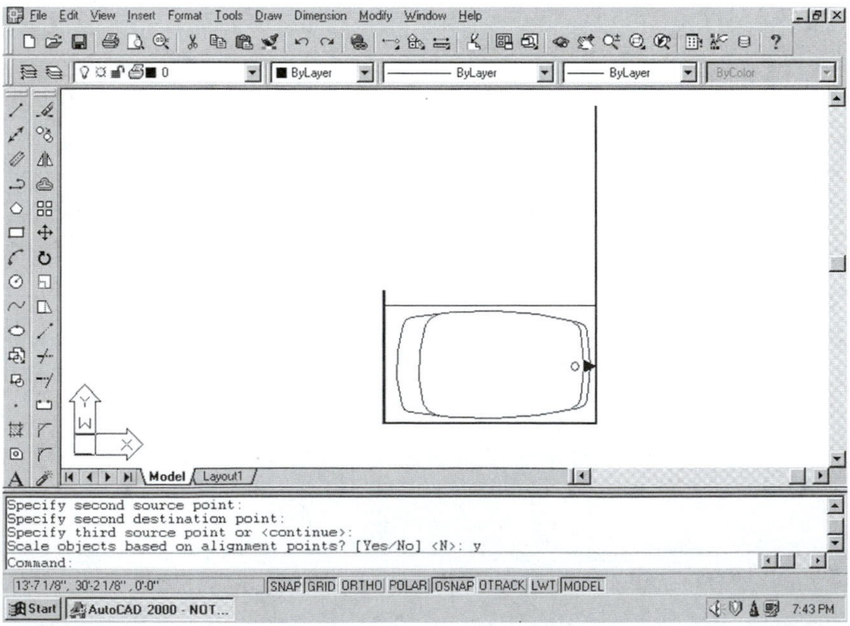

Figure 11.8b *By selecting the Scale objects to alignment points option, the 6'-long tub is scaled to fit in a 5' space.*

Figure 11.9. The command can be started by selecting the Scale icon from the Modify toolbar, by entering **SC** (EnterKey) at the command prompt, or by selecting Scale from the Modify pull-down menu. The command can be completed using the following process:

Command: **SC** (EnterKey)

Select objects: *(Select objects using any selection method.)*

1 found

Select object: (EnterKey)

Specify base point: *(Select base point.)*

Specify scale factor or [Reference]:

Command:

Scale Factor

The default for entering the effect of scaling is to enter a numeric value. To enlarge an object to twice its existing size, enter **2** (EnterKey) at the prompt. Using whole numbers will enlarge the selected objects by that factor. Enter **3** (EnterKey) at the prompt to enlarge the objects to three times their existing size. Entering a fraction, such as **.5** (EnterKey), will reduce the selected object to half of its original size.

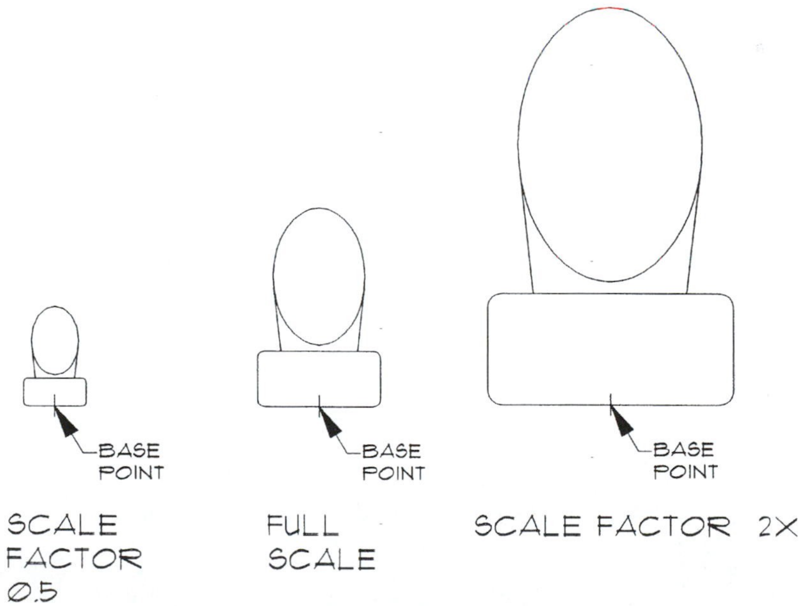

Figure 11.9 *The SCALE command allows the size of objects to be altered. In this case, the toilet on the left is half the size of the one in the middle.*

REFERENCE FACTOR

This option will allow an object to be enlarged to an absolute length rather than a relative scale. For instance, a tub that is 5' long can be scaled to 6' long. Rather than figuring the proportion of enlargement, enter **R** (EnterKey) at the command prompt. The process can be seen in Figure 11.10. The command can be completed using the following process:

> Command: **SC** (EnterKey)
>
> Select objects: *(Select objects using any selection method.)*
>
> 1 found
>
> Select object: (EnterKey)
>
> Specify base point: *(Select base point.)*
>
> Specify scale factor or [Reference]: **R** (EnterKey)
>
> Specify reference length <1>: **5'** (EnterKey)
>
> Specify new length: **6'** (EnterKey)
>
> Command:

STRETCH

The STRETCH command allows you to elongate or shrink drawing entities. This can be especially helpful when editing the shape and size of a floor plan to meet the

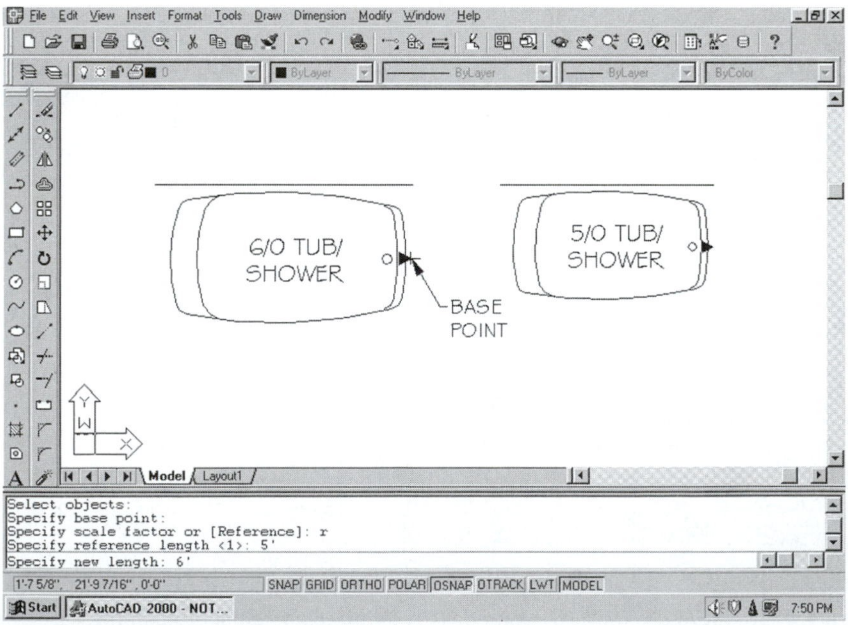

Figure 11.10 *Entering a reference factor allows an object to be scaled by an absolute length.*

changing needs of a client, as seen in Figure 11.11. Drawings made with Lines, Arcs, Traces, Solids (Chapter 15), and Polylines (Chapter 12) can all be stretched. The command can be started by selecting the Stretch icon from the Modify toolbar, by typing S (EnterKey) at the command prompt, or by selecting Stretch from the Modify pull-down menu. The command sequence is:

Command: **S** (EnterKey)

STRETCH

Select objects to stretch by crossing-window or crossing-polygon...

Select objects: *(Select object to be stretched using a window, crossing window, or a polygon options.)* **C** (EnterKey)

Specify first corner: *(Select a point for the corner of the window.)*

Specify opposite corner: *(Select the second point for the corner so that the object to be stretched is included in the window.)*

23 found

Select objects: (EnterKey)

The selection window only needs to cross a portion of the desired object because the default selection is Crossing. Objects can be added to the selection set by any of the methods described in Chapter 10. When the Window option is used to select the

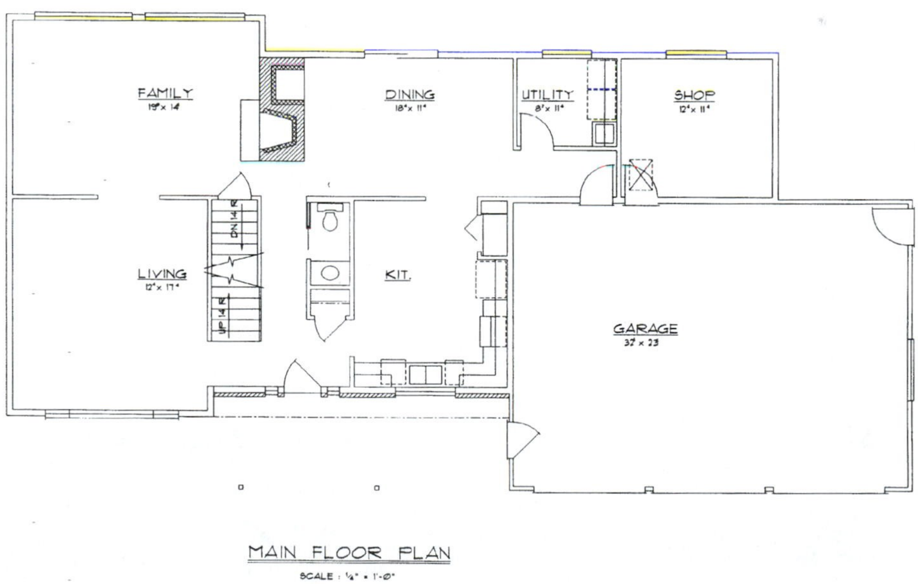

Figure 11.11 *The STRETCH command can be used to elongate an object or group of entities. Compare the garage size with the garage in Figure 10.23b.*

selection set for objects to be stretched, the desired effects are not always achieved. If too little of the group is selected, nothing will happen. If the entire object is selected, STRETCH will function like the MOVE command and move the entire object. The Crossing option can be used to consistently achieve the desired results. Once the selection set is complete, the prompt will display:

> Specify base point or displacement: *(Select a point to serve as a reference. Be sure Object Snap modes are toggled ON.)*
>
> Specify second point of displacement: *(Select the desired location that you would like to stretch the base point to.)*
>
> Command:

As the cursor is moved, the effect is displayed as the object is moved between the base and new points. The process can be seen in Figures 11.12a and 11.12b.

Not only can Stretch be used to lengthen objects, it also can be used to stretch some objects and move others, as seen in Figures 11.13a, 11.13b, and 11.13c.

LENGTHEN

The LENGTHEN command is similar to the STRETCH command. STRETCH is used to edit the length of groups of objects and closed polygons. However,

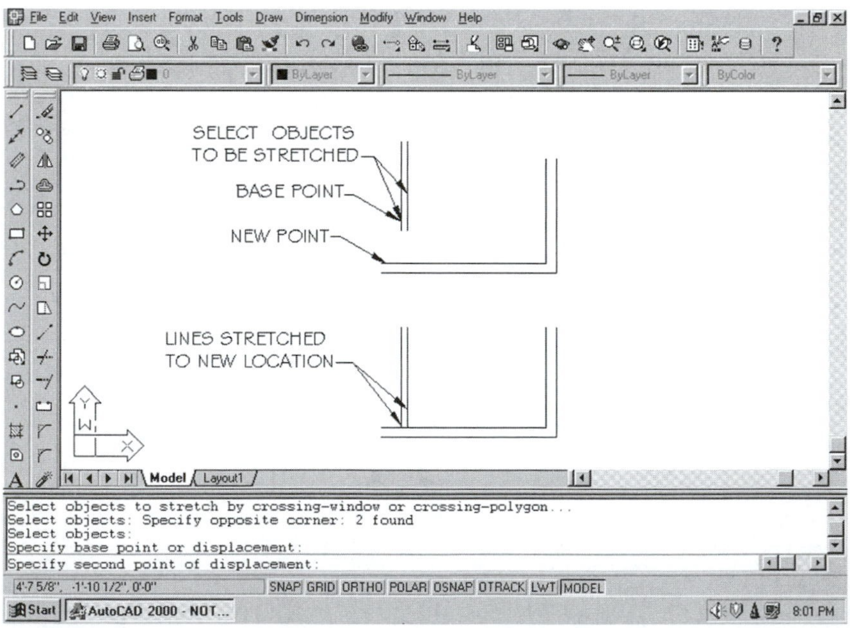

Figure 11.12a *The Window option is used to form the selection set of objects that are to be stretched. Once objects are selected, prompts will be displayed for the basepoint of the original objects and the new position they are to be stretched to.*

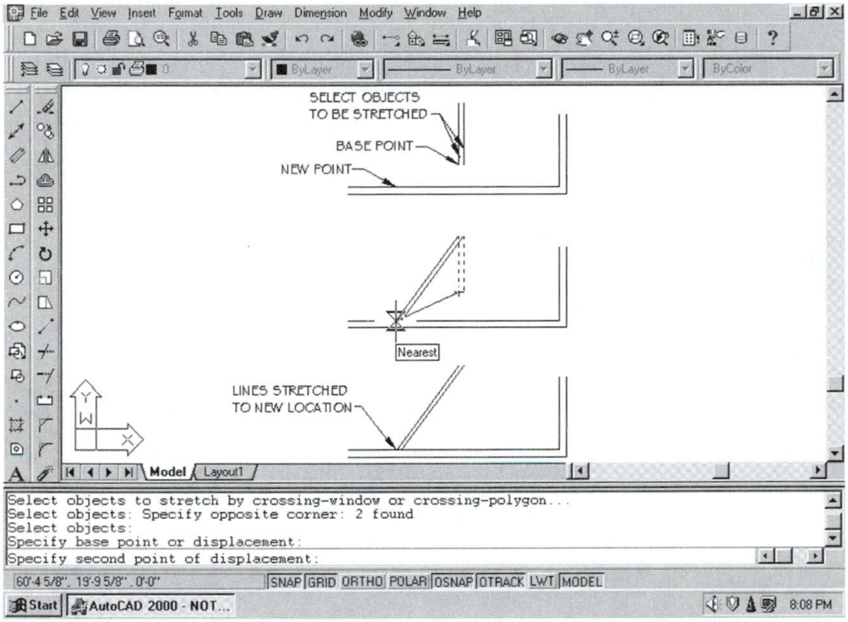

Figure 11.12b *In addition to adding length to line segments, the STRETCH command can be used to alter the angle of the selected lines.*

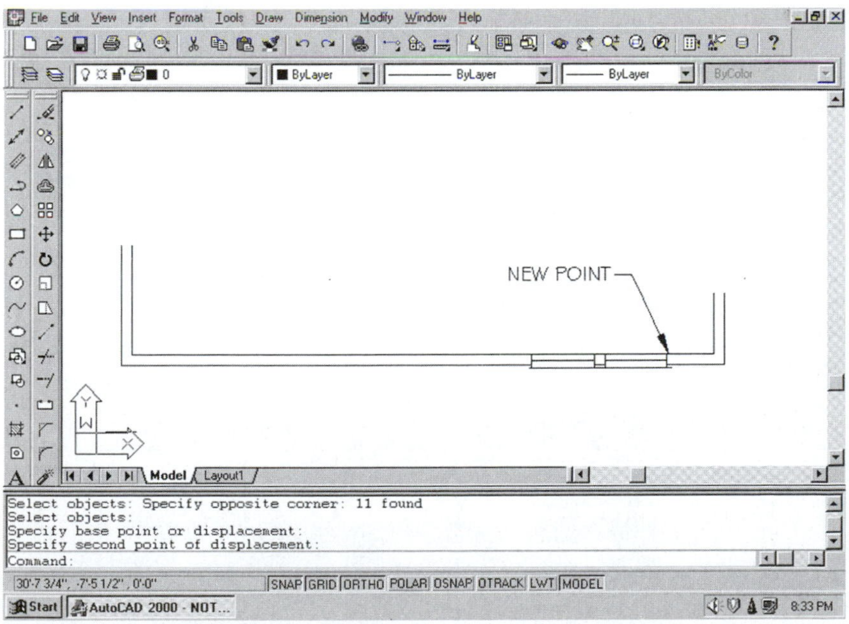

Figure 11.13a *The STRETCH command can also be used to alter the location of objects.*

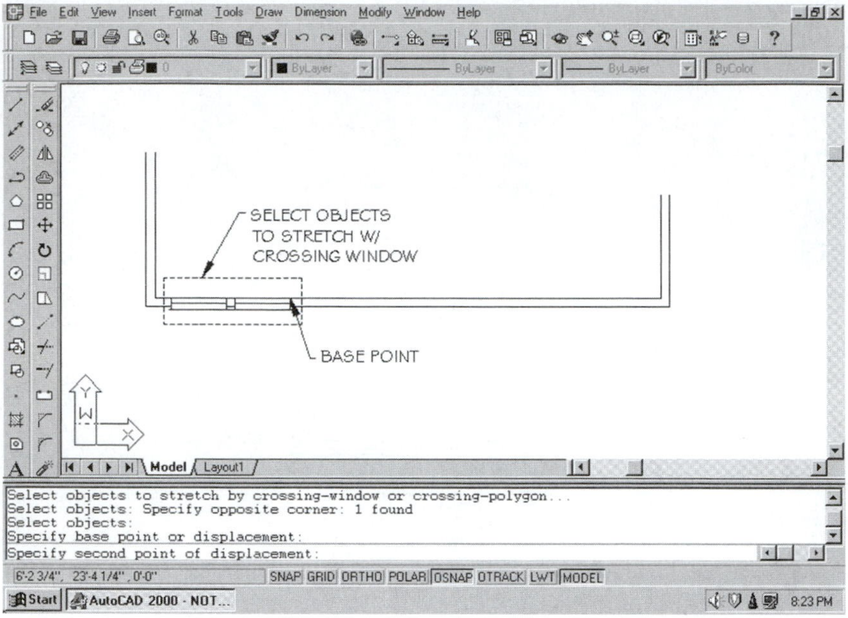

Figure 11.13b *The window and walls are stretched to alter the location of the window. Lines on the left side of the window are stretched, while the lines on the right side are shortened.*

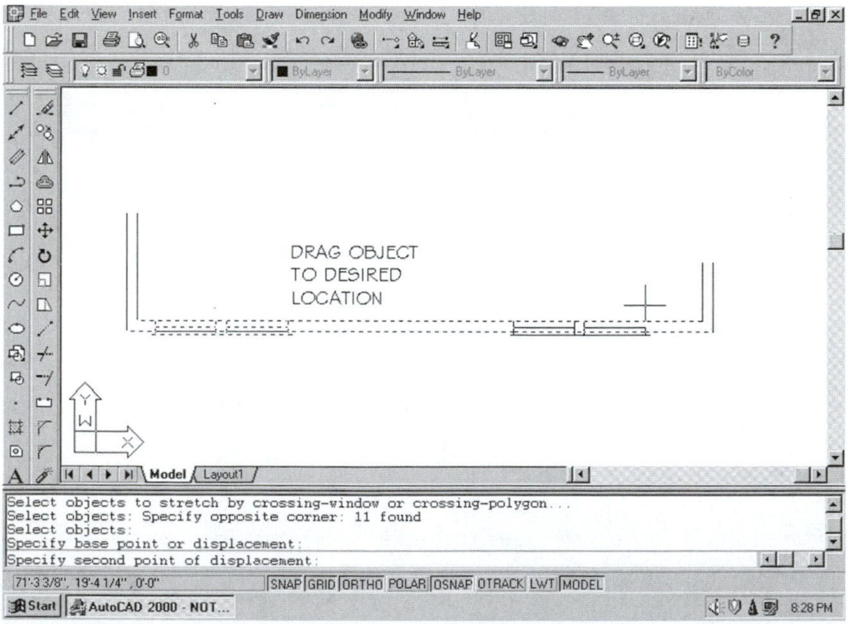

Figure 11.13c *The completed STRETCH command. This is much more efficient than moving the window, stretching the left lines, and trimming the right lines.*

LENGTHEN is used to edit the length of objects but not closed polygons. Like other edit commands, the LENGTHEN command can be started by selecting the Lengthen icon from the Modify toolbar, by typing **LEN** (EnterKey) at the command prompt, or by selecting Lengthen from the Modify pull-down menu. The command sequence is:

Command: **LEN** (EnterKey)

LENGTHEN

Select an object or [DElta/Percent/Total/DYnamic]: *(Select an object to edit.)*

Current Length: 10'-0"(Length will vary.)

Select an object or DElta/Percent/Total/DYnamic/<Select object>: **DE** (EnterKey)

Using the Delta option of Lengthen

Entering **DE** (EnterKey) at the Select object: prompt will produce the prompt:

Enter delta length or [Angle] <0'-0")>: *(Enter the desired length.)* **5'** (EnterKey)

Select an object to change [Undo]: *(Select the object again.)*

Select an object to change [Undo]: (EnterKey)

Command:

The command sequence can be seen in Figure 11.14. Notice that when the DElta was entered at the Select prompt: a choice could be made between length and angle. The default is to provide a length. Entering a positive number lengthens the line. Providing a negative response shortens the line. With each option of LENGTHEN, the end of the object that is selected will be the end of the object that is edited.

Note: The desired length entered is the length that will be added to the selected line. The value will not become the length for the total line length.

The Angle option can be used to change the included angle of an arc. The effects of this command sequence can be seen in Figure 11.15. The command sequence is:

Command: **LEN** (EnterKey)

LENGTHEN

Select an object or [DElta/Percent/Total/DYnamic]: *(Select an arc to edit.)*

Current Length: 10'-0", included angle: 90 *(Length and angle will vary.)*

DElta/Percent/Total/DYnamic/<Select object>: **DE** (EnterKey)

Enter delta length or [Angle] <0'-0">: **A** (EnterKey)

Enter delta angle<0>: **60** (EnterKey)

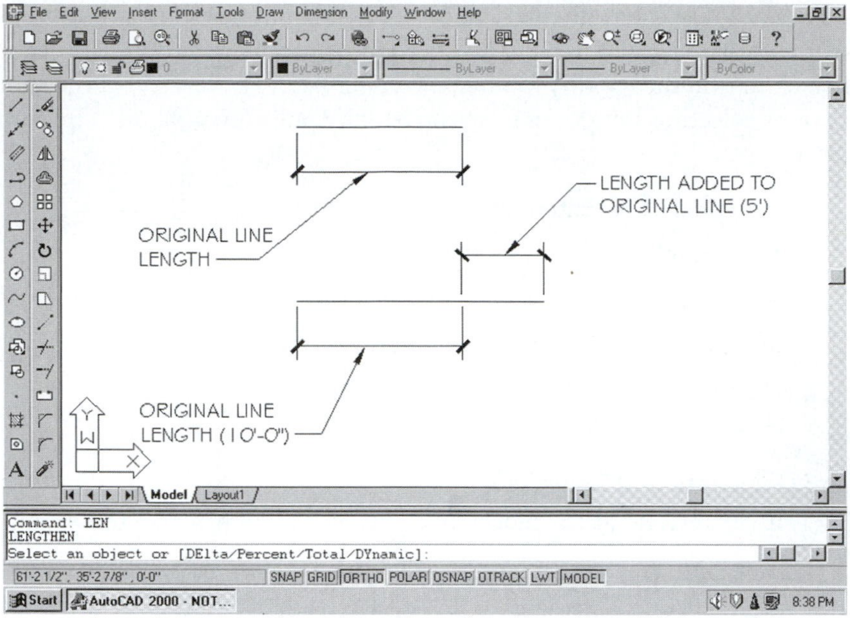

Figure 11.14 *The DElta option of the LENGTHEN command can be used to lengthen or shorten a line.*

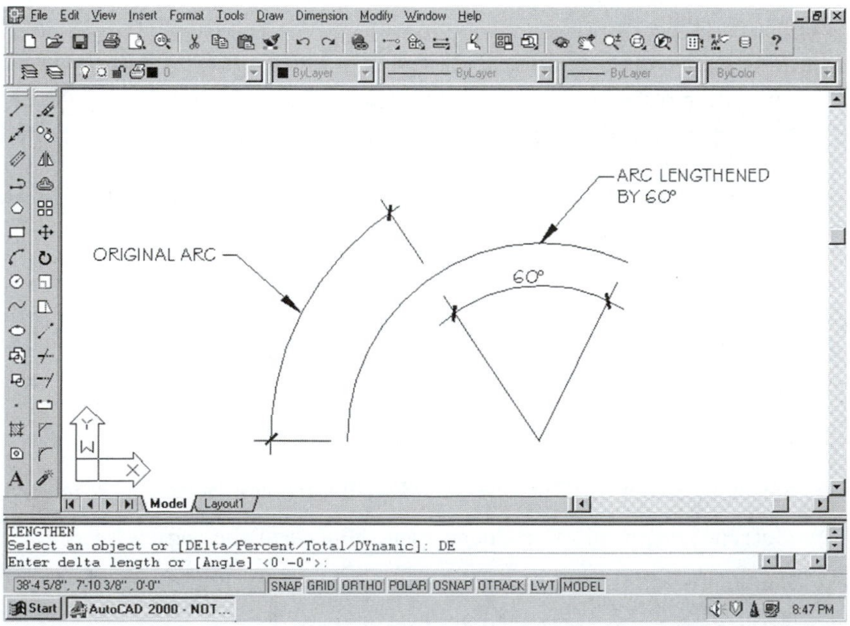

Figure 11.15 *The DElta and Angle options of the LENGTHEN command can be used to lengthen or shorten an angle.*

Select an object to change or [Undo]: *(Select desired arc to be edited.)*

Select an object to change or [Undo]: (EnterKey)

Command:

Using the Percent Option of LENGTHEN

The Percent option allows the length of a line or arc to be altered by a percentage of the existing length. The existing line length represents 100%. To lengthen the line, provide a value greater than 100%. Providing a percentage of less than 100% will shorten the line. The command sequence to lengthen a 2" line to a 3" line using the Percent option is:

Command: **LEN** (EnterKey)

LENGTHEN

Select an object or [DElta/Percent/Total/DYnamic]: *(Select an object to edit.)*

Current Length: 10'-0" *(Length will vary.)*

Select an object or [DElta/Percent/Total/DYnamic]: **P** (EnterKey)

Enter percentage length <100.0000>: **150** (EnterKey)

Select an object to change or [Undo]: *(Select desired line to be edited.)*

Select an object to change or [Undo]: (EnterKey)

Command:

The effects of the Percent option can be seen in Figure 11.16. When used with an arc, the angle of the arc will be edited based on a percentage of the total angle of the specified arc.

Using the Total Option of LENGTHEN

The Total option allows the length of a line or angle of an arc to be edited by providing the desired ending result. The command sequence to alter an existing line to a 35' line would be:

Command: **LEN** (EnterKey)

LENGTHEN

Select an object or [DElta/Percent/Total/DYnamic]: *(Select an object to edit.)*

Current Length: 10'-0" *(Length will vary.)*

Select an object or [DElta/Percent/Total/DYnamic]: **T** (EnterKey)

Specify total length or [Angle]: **35'** (EnterKey)

Select an object to change or [Undo]: *(Select desired line to be edited.)*

Select an object to change or [Undo]: (EnterKey)

Command:

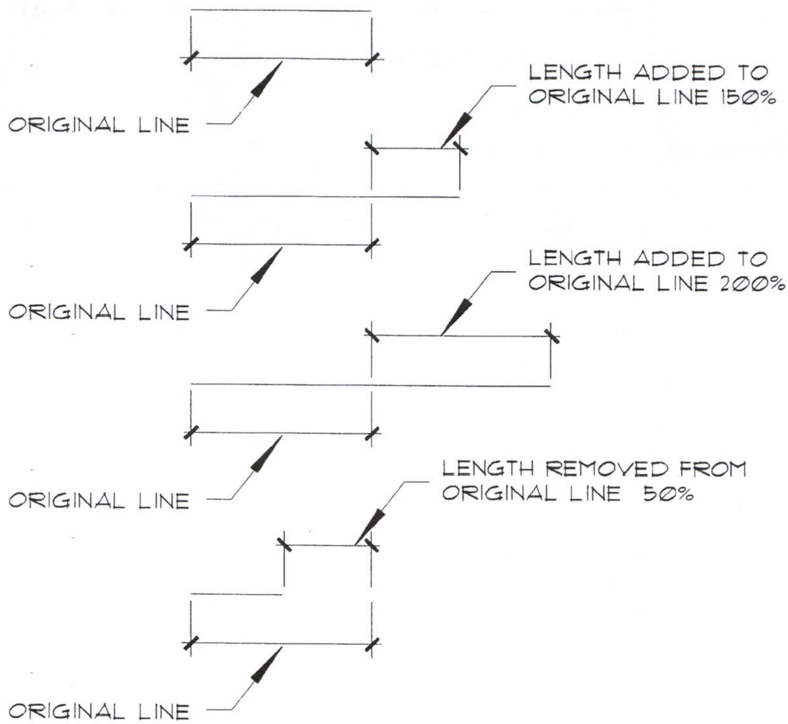

Figure 11.16 *The lengths of lines and arcs can be altered by entering a percentage of the original object. Percentages less than 100% will shorten the line or arc, and those greater than 100% will lengthen the line or arc.*

The results of this sequence can be seen in Figure 11.17. When first prompted for Select an object:, you may prefer to select an object before choosing the Total option. When a line is selected first, the length of the existing line will be displayed.

Using the DYnamic Option of LENGTHEN

The DYnamic option allows the selected line or angle to be dragged to the required length. The results of the Dynamic option can be seen in Figure 11.18.

Command: **LEN** (EnterKey)

LENGTHEN

Select an object or [DElta/Percent/Total/DYnamic]: *(Select an object to edit.)*

Current Length: 10'-0"(Length will vary.)

Select an object or [DElta/Percent/Total/DYnamic]: **DY** (EnterKey)

Select an object to change or [Undo]: *(Select desired line to be edited.)*

Specify new endpoint: *(Select new endpoint for line to be edited.)*

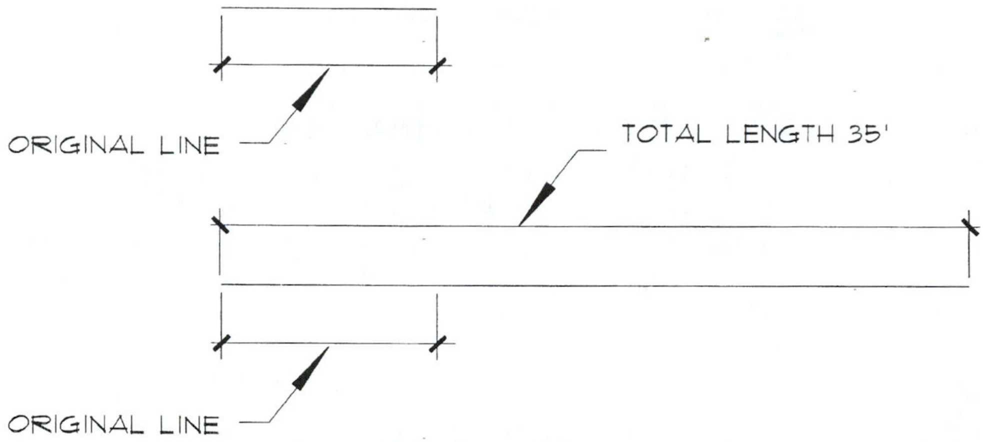

Figure 11.17 *The Total option of LENGTHEN allows a line or angle of an arc to be lengthened based on the desired total length or angle.*

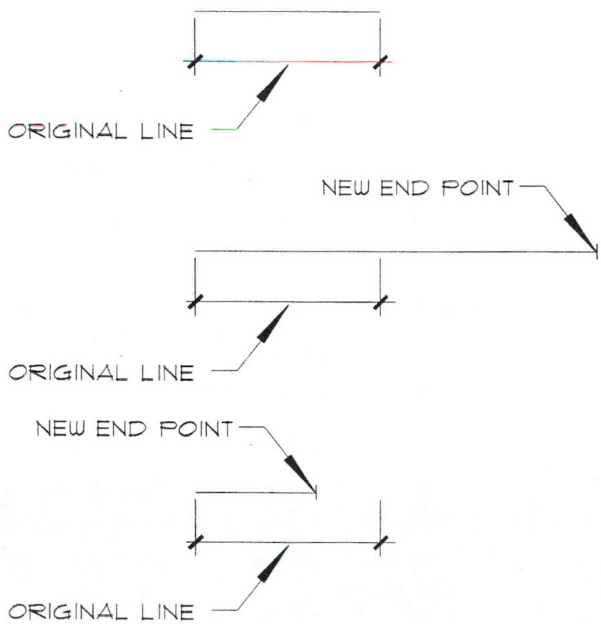

Figure 11.18 *The DYnamic option of LENGTHEN allows a line or angle of an arc to be dragged to the desired length by selecting a new ending point.*

Select an object to change or [Undo]: (EnterKey)

Command:

MODIFYING OBJECTS BY ALTERING LINES

To this point, you've learned to modify objects by making additional copies and by altering the position or size of an object. The commands contained in this portion of the Modify menu will allow you to modify a drawing by altering intersections of lines, circles or arcs, or by removing a portion of an object. The commands include TRIM, EXTEND, BREAK, CHAMFER, and FILLET.

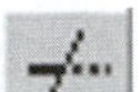

TRIM

The TRIM command can be started by selecting Trim from the Modify toolbar, by typing **TR** (EnterKey), or by selecting Trim from the Modify pull-down menu. The command allows you to edit Lines, Arcs, Circles, or Polylines using a cutting edge. Polylines will be introduced in future chapters. Any portion of the selected object extending past the designated cutting edge will be removed. Lines, Arcs, Circles, and Polylines may be used as cutting edges. TRIM can be used to clean the corners of intersecting walls in plan view, as seen in Figure 11.19, or to modify the size of objects. The command sequence is:

Command: **TR** (EnterKey)

TRIM

Current settings: Projection = UCS Edge = None

Select cutting edges... *(Select the cutting edge using any object selection method.)*

I found

Select objects: (EnterKey)

Select object to trim or [Project/Edge/Undo]: *(Select object to be trimmed.)*

Select object to trim or [Project/Edge/Undo]: *(Select another object to be trimmed or press enter when finished.)* (EnterKey)

Command:

Once the cutting edges are defined, the command will continue until (EnterKey) is selected. Any object that extends past the cutting edge can be trimmed. The process can be seen in Figure 11.20. Notice that an object can be selected to be a cutting edge and it can also be an object to be trimmed.

Multiple Trim Selections

To select more than one cutting edge, hold (ShiftKey) while selecting the cutting planes. The Window or Crossing Window options work well for selecting multiple

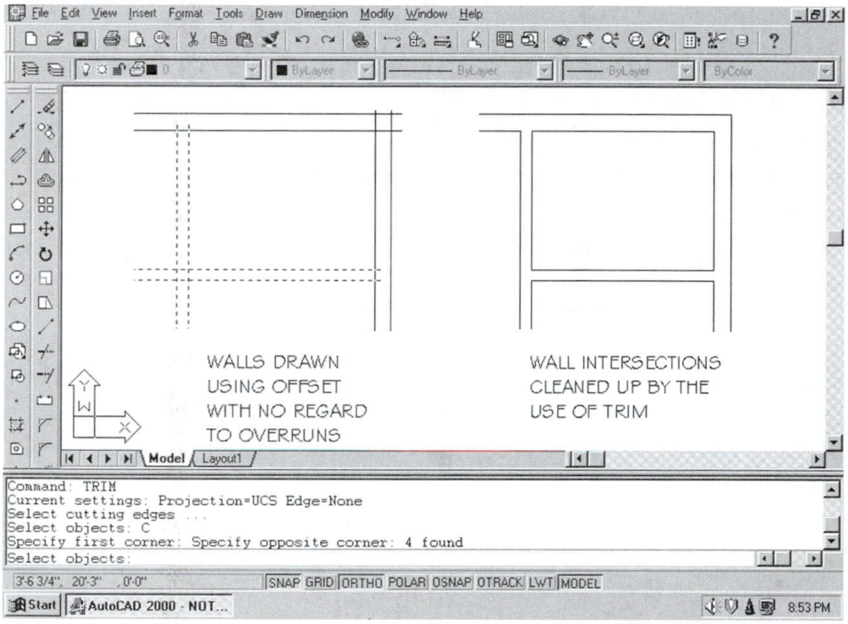

Figure 11.19 *The TRIM command can be used to eliminate line overruns that were created using OFFSET.*

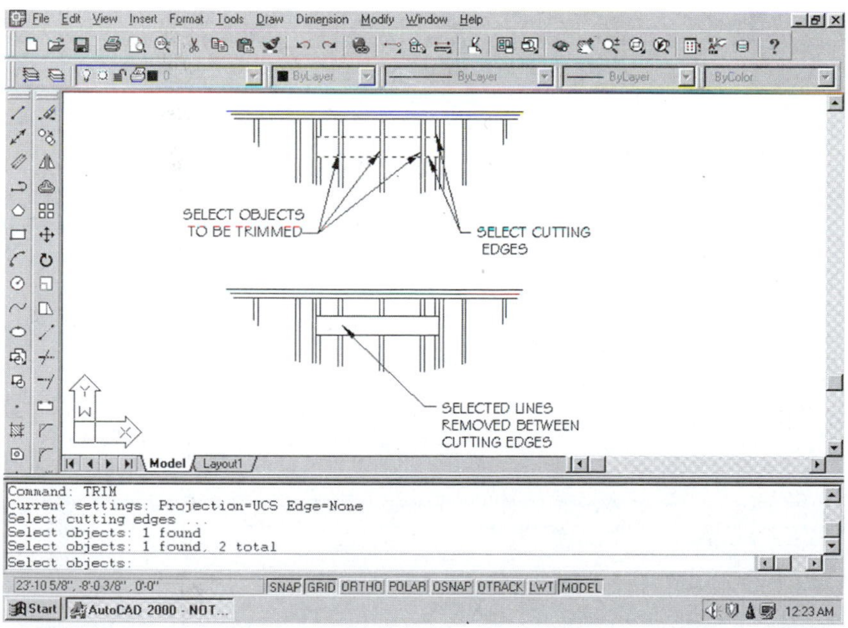

Figure 11.20 *The TRIM command is started by selecting one or several "Cutting edges". Use a Window or Crossing Window for fast selection of multiple cutting edges. Any segment extending past the selected cutting edge will be trimmed.*

objects for cutting planes. Selecting the All option will use all drawing objects as trim edges. To use the All option, you must enter **ALL** (EnterKey) (not A).

Using the TRIM Edge Option

The Edge option of TRIM can be used to trim two objects that do not actually intersect, but would intersect if one of the objects were extended. AutoCAD refers to this as an implied intersection. To use the Edge option, two objects must be selected. Once these edges have been selected, enter **E** (EnterKey) when prompted to select another object. This will now display the option of:

Extend/No extend <Extend>: (EnterKey)

When Extend is selected, AutoCAD checks the cutting edge object to verify if it will intersect when extended. If it would touch, the implied intersection is used to trim the second object. This option is shown in Figure 11.21. The command sequence is:

Command: **TR** (EnterKey)

TRIM

Current settings: Projection = UCS Edge = None

Select cutting edges... *(Select the cutting edge using any object selection method.)*

1 found

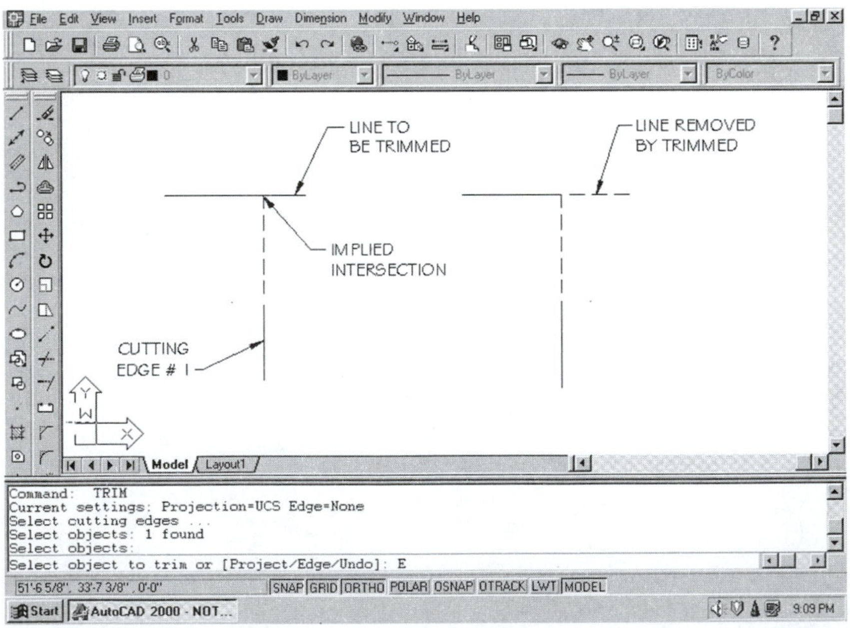

Figure 11.21 *Lines that would touch, if one were extended, can be trimmed by selecting the Edge and Extend options.*

Select objects: (EnterKey)

Select object to trim or [Project/Edge/Undo]: **E** (EnterKey)

Enter an implied edge extension mode [Extend/No Extend] <Extend: (EnterKey)

Select object to trim or [Project/Edge/Undo]: *(Select object to trim.)*

Select object to trim or [Project/Edge/Undo]: (EnterKey)

Command:

Using the TRIM Undo option

Occasionally you may trim an object and not be pleased with the results. The TRIM Undo option will allow an unsatisfactory TRIM to be undone while still in the command sequence. The command sequence is:

Command: **TR** (EnterKey)

TRIM

Current settings: Projection = UCS Edge = Extend

Select cutting edges... *(Select the cutting edge using any object selection method.)*

I found

Select objects: (EnterKey)

Select object to trim or [Project/Edge/Undo]: *(Select object to be trimmed.)*

Select object to trim or [Project/Edge/Undo]: **U** (EnterKey)

Command has been completely undone.

Select object to trim or [Project/Edge/Undo]: *(Select object to be trimmed.)*

Select object to trim or [Project/Edge/Undo]: (EnterKey)

Command:

Using the TRIM Project Option

The TRIM Project option is primarily a 3D option and that aspect of the command will not be discussed in this text. Refer to the HELP menu for further information. The option can be used just as the Edge option as a means of trimming objects that extend past a projection of the selected cutting edge. The command sequence is:

Command: **TR** (EnterKey)

TRIM

Current settings: Projection = UCS Edge = None

Select cutting edges... *(Select the cutting edge using any object selection method.)*

I found

Select objects: (EnterKey)

Select object to trim or [Project/Edge/Undo]: **P** (EnterKey)

Enter a projection option [None/Ucs/View] <Ucs> **N** (EnterKey)

Select object to trim or [Project/Edge/Undo]: *(Select object to trim)*

Select object to trim or [Project/Edge/Undo]: (EnterKey)

Command:

Practical Use of TRIM. Combining the LINE, OFFSET, and TRIM commands will allow almost any object to be drawn easily. Draw two perpendicular lines to represent the corner of an object without regard to the exact length. The exact size of the object can be determined with OFFSET or TRIM. Open the FLOOR10 file that was started earlier. With the use of TRIM this partial floor plan can be completed so that it more closely resembles the drawing in Figure 10–23b.

EXTEND

The EXTEND command can be started by selecting the Extend icon from the Modify toolbar, by typing EX (EnterKey), or by selecting Extend from the Modify pull-down menu. The command allows an object to be lengthened to an exact boundary point. The command will ask you to designate boundary edges, and then select objects to be extended to that edge. The boundary edge may be a Line, Arc, Circle, or Polyline. The process can be seen in Figure 11. 22. The command sequence is:

Command: **EX** (EnterKey)

EXTEND

Current settings: Projection = None Edge = Extend

Select boundary edges… *(Select the edge using any object selection method.)*

1 found

Select objects: (EnterKey)

Select object to extend or [Project/Edge/Undo]: *(Select object to be extended using any selection method.)*

Select object to extend or [Project/Edge/Undo]: (EnterKey)

Command:

As the object to be extended is selected, the line is extended to the other object that was selected as the boundary edge.

Using the EXTEND Edge Option

The EXTEND Edge option is similar to the Edge option presented with the TRIM command. Edge will allow lines that do not presently touch but have an implied intersection to be extended. In the default setting, the command will automatically

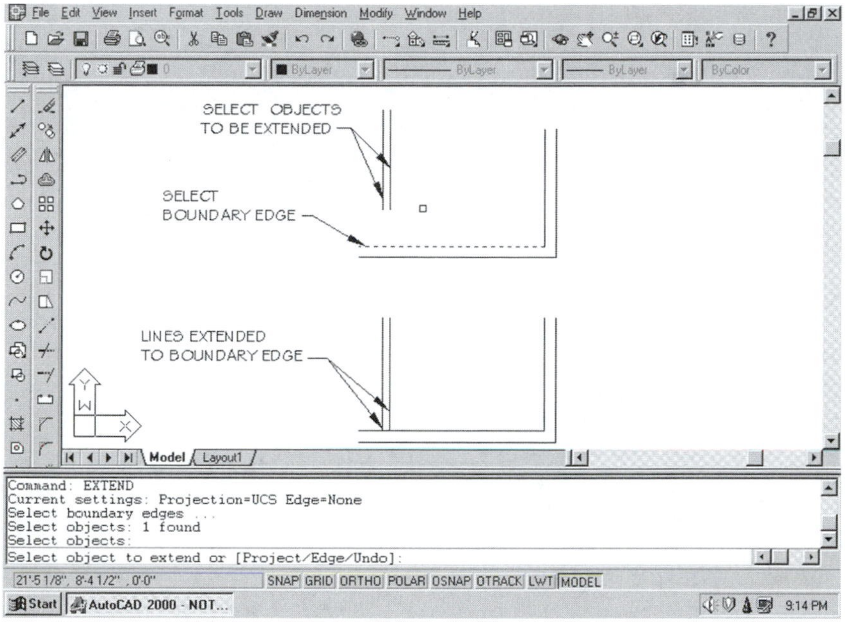

Figure 11.22 *The EXTEND command can be used to lengthen lines to a selected "Boundary edge".*

extend using an implied edge. If the Edge setting has been altered, the sequence to extend to an implied edge is:

Command: **EX** (EnterKey)

EXTEND

Current settings: Projection = None Edge = No extend

Select boundary edges… *(Select the edge using any object selection method.)*

1 found

Select objects: (EnterKey)

Select object to extend or [Project/Edge/Undo]: **E** (EnterKey)

Enter an implied edge extension mode [Extend/No Extend] <Extend>: *(Select object to be extended using any selection method.)*

Select object to extend or [Project/Edge/Undo]: (EnterKey)

Command:

The results of the command can be seen in Figure 11.23.

Using the EXTEND Undo Option

The EXTEND Undo option allows a sequence that has not produced satisfactory results to be undone while still in the command sequence.

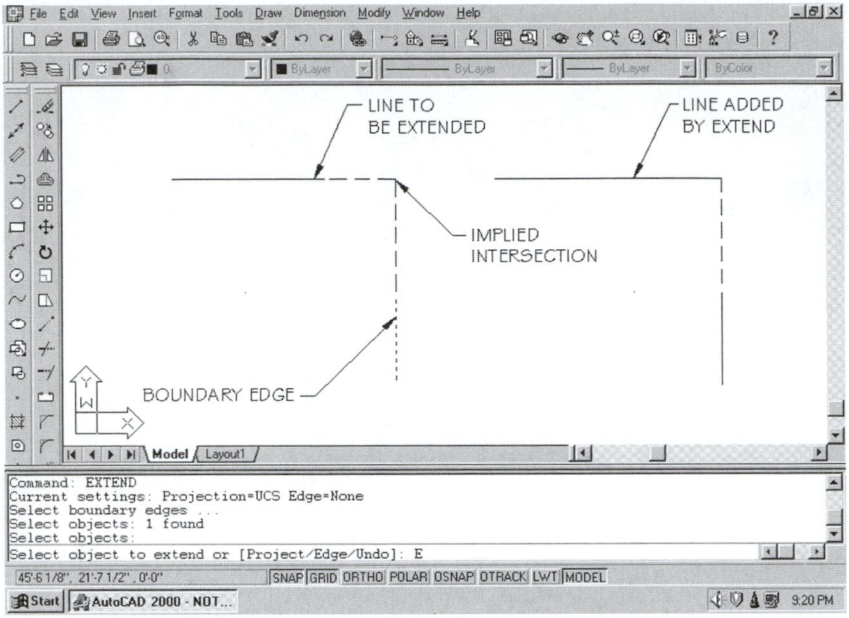

Figure 11.23 *Lines that would touch, if both were extended, can be extended by selecting the Edge option.*

Using the EXTEND Project Option.

This mode is used for 3D drawings and will not be discussed in this text. Use the HELP option for further information on this subject.

BREAK

The BREAK command can be used to remove a portion of a Line, Circle, Arc, Polyline, or Trace. Polylines and Trace lines will be introduced in future chapters. The BREAK command can be selected by selecting the Break icon from the Modify toolbar, by typing **BR** (EnterKey) at the command prompt, or by selecting Break from the Modify pull-down menu. BREAK can be completed using two different methods, but their effects are similar.

Select Object Second Point

The default option of BREAK allows you to remove a segment of an object by using two selection points. The first selection point selects the segment to be broken and specifies where the break will begin. The second point specifies where the BREAK will end. The command is:

Command: **BR** (EnterKey)

BREAK Select object: *(Select an object to break.)*

Specify second break point or [First point]: *(Select a second point.)*

Command.

Once the second point is selected, the portion of the object from the first to the second point will be removed and the command prompt will be returned. The sequence can be seen in Figure 11.24. If a circle is being edited, the break occurs in a counterclockwise direction from the first point to the second point, as seen in Figure 11.25. Enter **F** (EnterKey) when the second point is requested to allow the procedure to follow the Select Object Two Points method.

Select Object Two Points

This option allows you to remove a segment of an object by using three selection points. The first request selects the object to be broken. Once the segment has been defined for editing, the first selection point specifies where the break will begin and the second selection point specifies where the BREAK will end. The command sequence is:

Command **BR** (EnterKey)

BREAK Select object: *(Select an object to break.)*

Specify second break point or [First point]: **F** (EnterKey)

Specify first break point: *(Select a point to start the break.)*

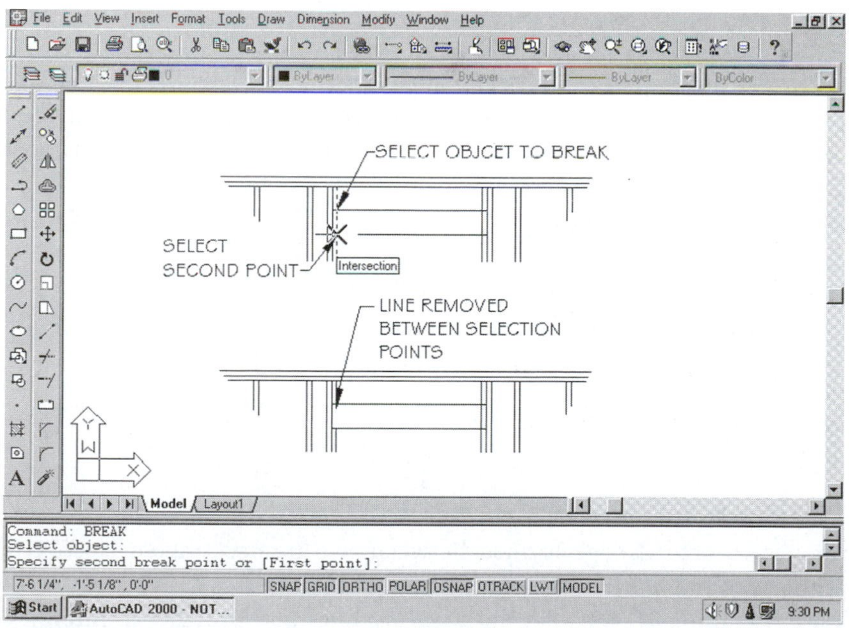

Figure 11.24 *The BREAK command's First option can be used to remove a segment from a specified line. Specified points are used to define which part of the line will be removed.*

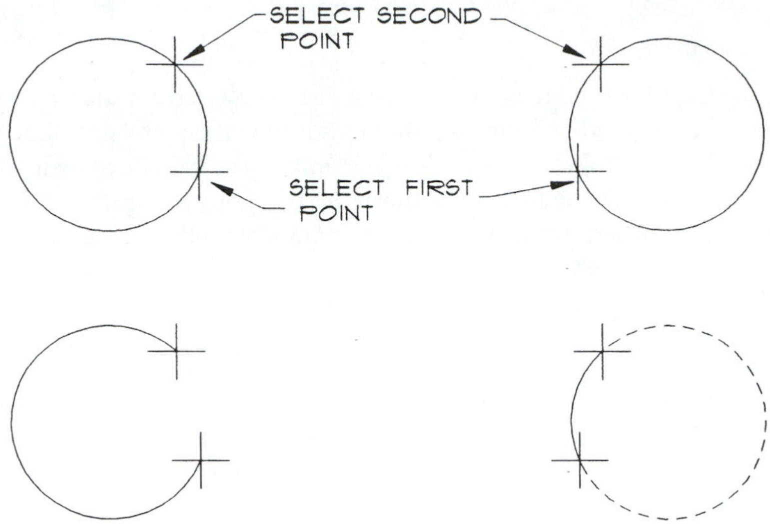

Figure 11.25 *When a break is specified for a circle, it will occur in a counterclockwise direction.*

Specify second break point: *(Select a point to end the break.)*

Command:

The command sequence can be seen in Figure 11.26.

Break @

This option will break a segment, but no apparent change will take place. BREAK @ will divide an object into two portions, but no space will be left between the portions. The @ option is useful to erase only a portion of an object by using BREAK @ and then using the ERASE command. The command sequence is:

Command: **BR** (EnterKey)

BREAK Select object: *(Select an object to break.)*

Specify second break point or [First point]: **@** (EnterKey)

Command.

The sequence is shown in Figure 11.27.

EDITING WITH THE CHAMFER COMMAND

A chamfer is an angled edge formed between two intersecting surfaces, typical of what might be found on the edge of the cabinet counter shown in Figure 11.28. The CHAMFER command trims two intersecting lines and provides a mitered corner. You'll be asked to select two lines and two distances. This will form a new line segment between the two existing line segments. The CHAMFER command can be

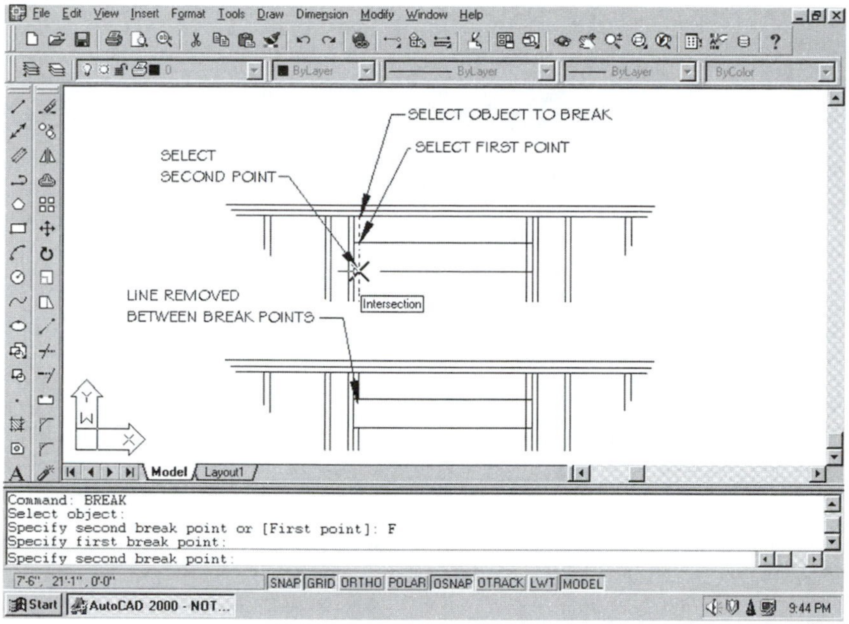

Figure 11.26 *The default option of BREAK allows for selecting the line to be broken, then specifying the segment to be removed.*

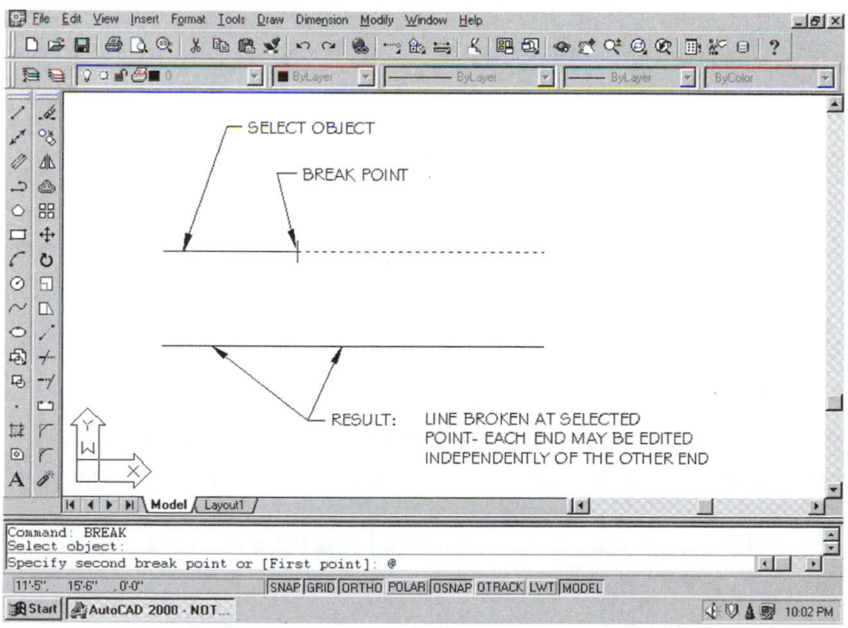

Figure 11.27 *The BREAK @ command will break a line into two segments, but no visible change will occur.*

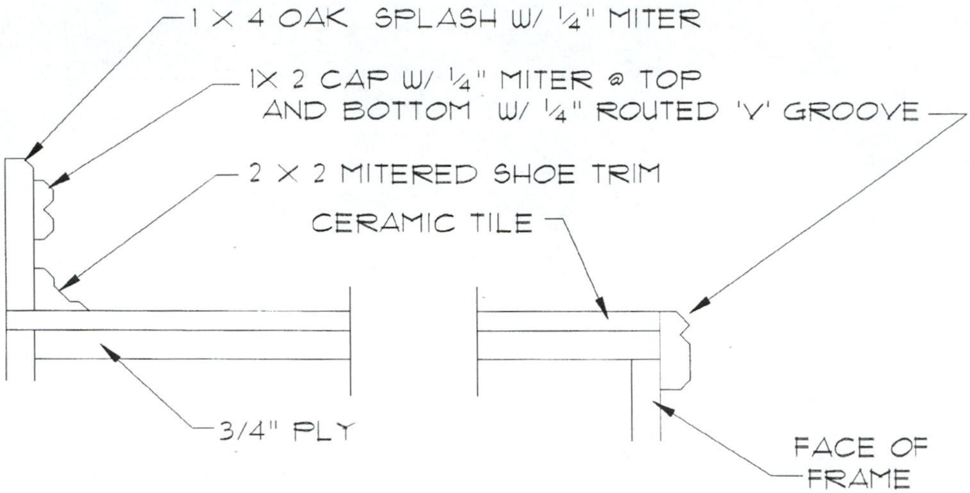

Figure 11.28 *The CHAMFER command can be used to miter a corner.*

started by selecting the Chamfer icon from the Modify toolbar, by typing **CHA** (EnterKey) at the command prompt, or by selecting Chamfer from the Modify pull-down menu. The command sequence to set the default values to 0" is:

> Command: **CHA** (EnterKey)
>
> CHAMFER
>
> (TRIM mode) Current chamfer Dist1 = 0'-0 ½", Dist2 = 0'-0 ½"
>
> Select first line or [Polyline/Distance/Angle/Trim/Method]: **D** (EnterKey)
>
> Specify first chamfer distance <0'-0 ½">: **0** (EnterKey)
>
> Specify second chamfer distance <0'-0">: (EnterKey)
>
> Command:

If the command is continued, objects can be chamfered to form square corners. Because no chamfer distance has been entered, the process and the results would resemble Figure 11.29.

To provide a mitered corner, the sequence would be similar. For the following example, a distance of 36" will be used—although any numeric value can be used, as long as the value is no longer than that of the line being chamfered. If the value exceeds the length, the following message will be displayed and the command terminated:

> Distance is too large
>
> *Invalid*

The command sequence is:

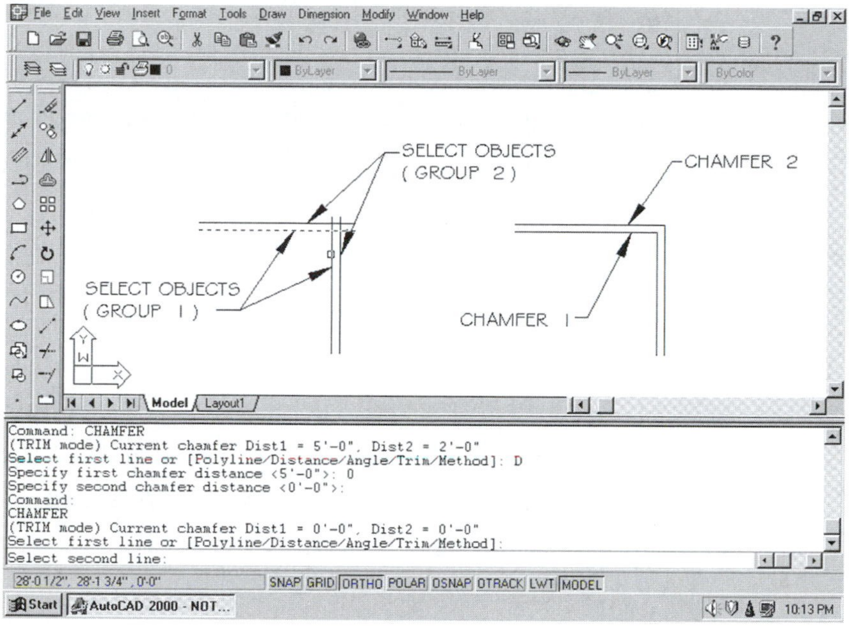

Figure 11.29 *With the value set at 0, CHAMFER can be used to trim corners.*

Command: **CHA** (EnterKey)

CHAMFER

(TRIM mode) Current chamfer Dist1 = 0'-0", Dist2 = 0'-0"

Select first line or [Polyline/Distance/Angle/Trim/Method]: **D** (EnterKey)

Specify first chamfer distance <0'-0 ½">: **36** (EnterKey)

Specify second chamfer distance <3'-0">: (EnterKey)

Command:

The first value will become the default for the second value. When the distances are equal, the results will produce a 45° miter. When the second value or the (EnterKey) is entered, the command prompt will be returned. Pressing (EnterKey) again will allow the chamfer to be drawn.

Command: (EnterKey)

CHAMFER

(TRIM mode) Current chamfer Dist1 = 3'-0", Dist2 = 3'-0"

Select first line or [Polyline/Distance/Angle/Trim/Method]: *(Select first line.)*

Specify second line <0'-0">: *(Select second line.)*

Command:

As the second line is selected, the chamfer will be performed and the command prompt returned. The effects of the command sequence can be seen in Figure 11.30. Figure 11.31 shows an example of varied distance values. CHAMFER can be used to square or chamfer two intersecting lines, as well as to square or chamfer two lines that do not touch.

Selecting the CHAMFER Angle

CHAMFER also allows the chamfer to be created using one distance and an angle. This option is started by entering A (EnterKey) when prompted for the first line. The command sequence to place a 30°-angle 4" from a corner would be:

Command: **CHA** (EnterKey)

CHAMFER

(TRIM mode) Current chamfer Dist1 = 3'-0", Dist2 = 3'-0"

Select first line or [Polyline/Distance/Angle/Trim/Method]: **A** (EnterKey)

Specify chamfer length on the first line <0'-0">: **4** (EnterKey)

Enter chamfer angle from the first line <0'-0">: **30** (EnterKey)

Command:

Command: (EnterKey)

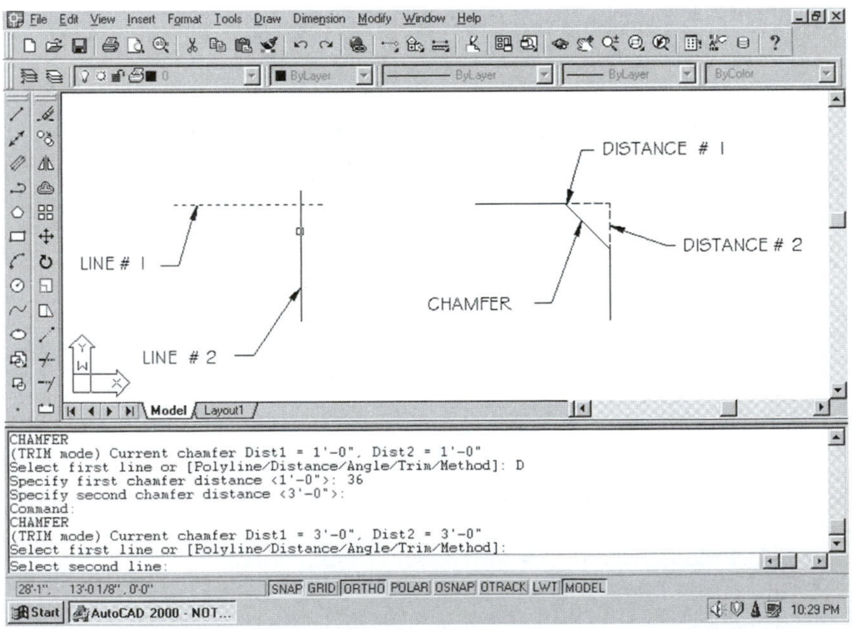

Figure 11.30 *To miter a corner, prompts will be displayed for the "first distance," and the "second distance." Once provided, a new segment will be provided between the two points.*

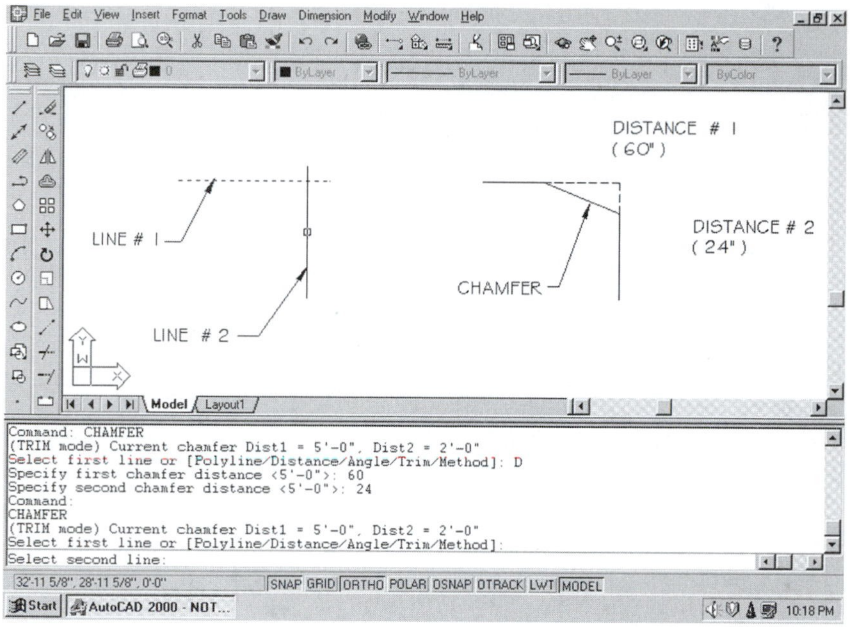

Figure 11.31 *Providing different lengths for the chamfer distances will affect the angle of the chamfer.*

CHAMFER

(TRIM mode) Current chamfer Length = 0'-4", Angle = 30

Select first line or [Polyline/Distance/Angle/Trim/Method]: *(Select first line.)*

Specify second line: *(Select second line.)*

Command:

The process can be seen in Figure 11.32.

Selecting the CHAMFER Trim Mode

The common method of using the CHAMFER command is to create the chamfer between two lines and remove a portion of the lines where the chamfer is created. Using the Method option allows for a choice of removing or not removing the lines by the chamfer. The command sequence is:

Command: **CHA** (EnterKey)

CHAMFER

(TRIM mode) Current chamfer Dist1 = 3'-0", Dist2 = 3'-0"

Select first line or [Polyline/Distance/Angle/Trim/Method]: **T** (EnterKey)

Enter Trim mode option [Trim/No trim] <Trim>: **N** (EnterKey)

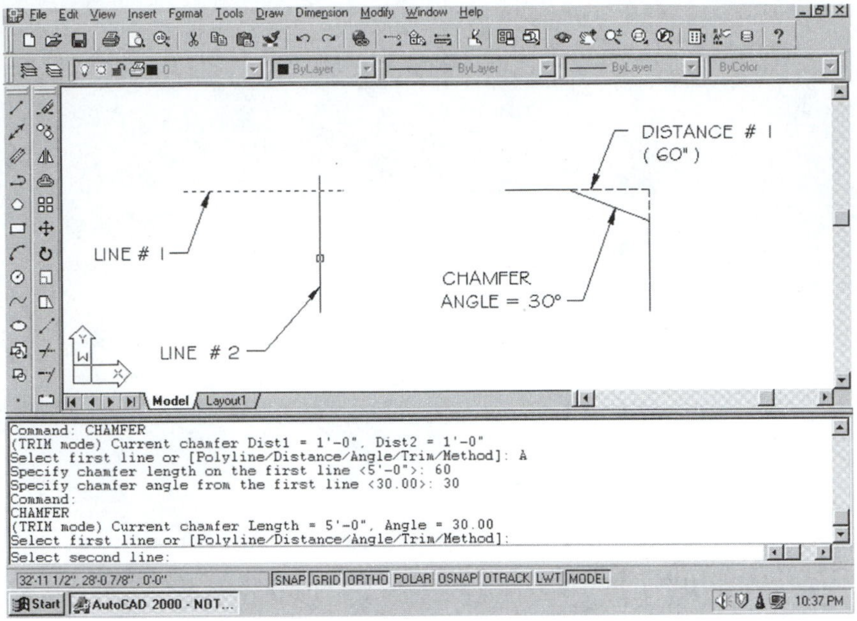

Figure 11.32 *Selecting the Angle option provides a prompt for one length and an angle. AutoCAD converts the angle to a length.*

Select first line or [Polyline/Distance/Angle/Trim/Method]: *(Select first line.)*

Specify second line: *(Select second line.)*

Command

Reminder: If you alter the trim mode of CHAMFER, the chamfer will be preformed, but no lines will be removed. The Chamfer trim mode will also affect the FILLET command. FILLET will be discussed in the next section of this chapter.

Selecting the CHAMFER Method

When a chamfer is placed using the Distance or Angle options, the values are kept as the defaults until you change them. Using the Method option, the values will not affect each other and you can switch between using either the Distance or Angle. The command sequence is:

Command: **CHA** (EnterKey)

CHAMFER

(TRIM mode) Current chamfer Dist1 = 3'-0", Dist2 = 3'-0"

Select first line or [Polyline/Distance/Angle/Trim/Method]: **M** (EnterKey)

Enter Trim mode option [Trim/No trim] <Trim>: **D** (EnterKey)

Select first line or [Polyline/Distance/Angle/Trim/Method]: *(Select first line.)*

Specify second line: *(Select second line.)*

Command: (EnterKey)

EDITING WITH THE FILLET COMMAND

You may be most familiar with a fillet as it relates to welding. It earns its name because the weld fills in a corner. The FILLET command of AutoCAD will connect two lines to form a corner, or allow two lines, arcs, or circles to be fitted together smoothly in rounded corners. The command can be started by selecting the Fillet icon from the Modify toolbar, by entering **F** (EnterKey) at the command prompt, or by selecting Fillet from the Modify pull-down menu. Once in the command, set the desired radius using a similar process to setting the Chamfer distance. The command sequence to fillet two lines using the default radius is:

Command: **F** (EnterKey)

FILLET

Current settings: Mode = TRIM Radius = 0'-0 ½"

Select first object or [Polyline/Radius/Trim/Method]: *(Select first line.)*

Select second object: *(Select second line.)*

Command:

Once the two objects are selected, the radius will be applied and the command prompt will be returned. The fillet process can be seen in Figure 11.33. If the Trim mode is set to No Trim, the results of the FILLET will not be evident. Be sure the Method Mode is set to Trim. The command can be used to form a radius between intersecting lines or two lines that do not touch.

For the command to function with crossing lines as you intend, you need to be careful which portion of the lines you select. Draw two more intersecting lines similar to those in Figure 11.33. Enter the fillet command and select the short ends of the lines. Once the FILLET command is completed, only the small portion of the lines will remain, as shown in Figure 11.34.

Altering the Fillet Radius

The Fillet radius can be altered using the Radius mode. Once the first object is selected, use the radius option to alter the radius of the FILLET. By entering **R** (EnterKey) at the first prompt, a radius may be entered for rounded corners. To transform two existing perpendicular lines into a 60"-radius corner, the command sequence is:

Command: **F** (EnterKey)

FILLET

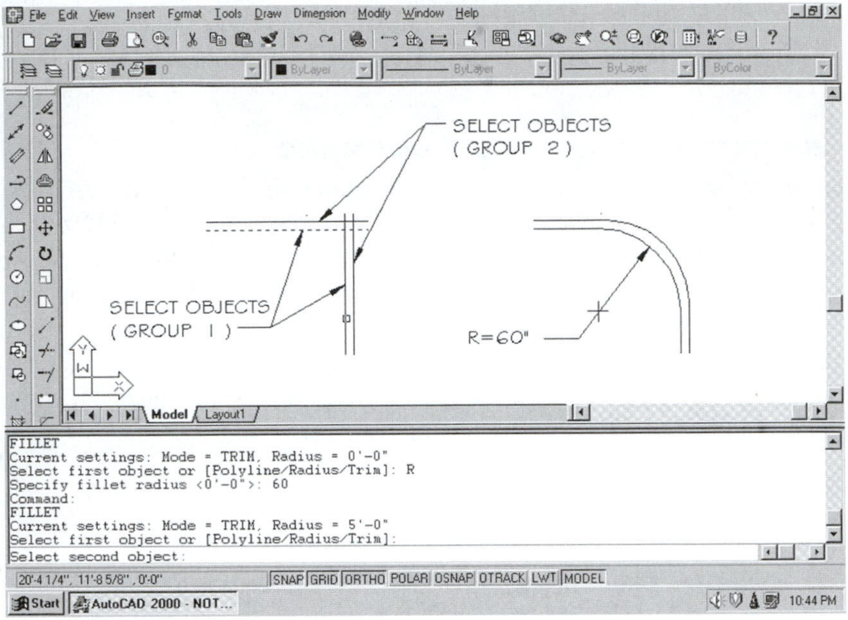

Figure 11.33 *By accepting the default or assigning a new value for the Radius, the FILLET command can be used to round square corners.*

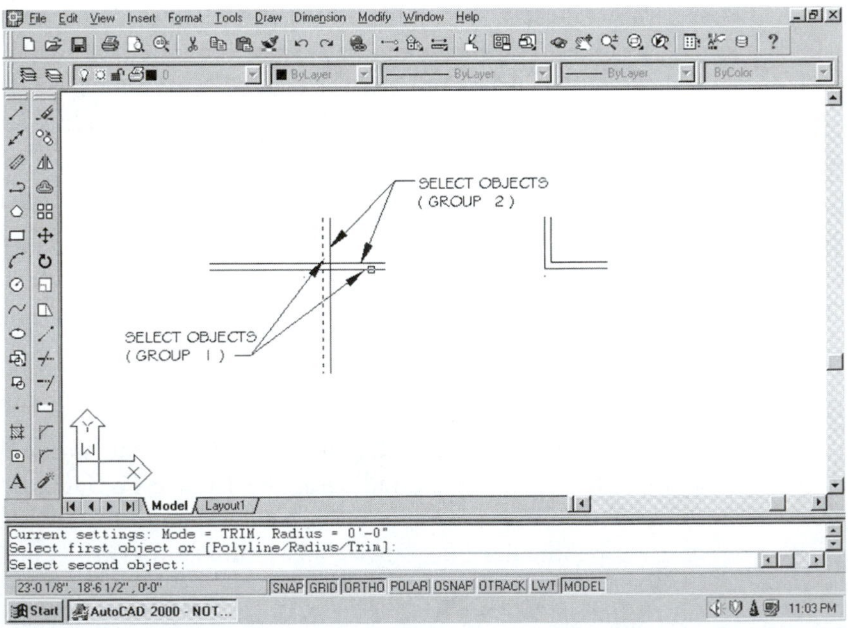

Figure 11.34 *Care must be taken when selecting objects to be edited by the FILLET command. Selecting the short ends of the lines removes that portion of the line.*

Current settings: Mode = TRIM Radius = 0'-0 ½"

Select first object or [Polyline/Radius/Trim/Method]: **R** (EnterKey)

Specify fillet radius <0'-0 ½">: **60** (or 5') (EnterKey)

Command:

Once the radius is entered, the command prompt is returned. Press (EnterKey) to reenter the command. If you continue through the command sequence, you will draw a 60" radius. The balance of the command sequence is:

Command: (EnterKey)

FILLET

Current settings: Mode = TRIM Radius = 5'-0"

Select first object or [Polyline/Radius/Trim/Method]: *(Select first line.)*

Select second object: *(Select second line.)*

Command:

As the second object is selected, the radius will be formed between the two selected lines, similar to Figure 11.33. The portion of the object that is selected will affect how the FILLET will be drawn. Examples of selections and their results can be seen in Figure 11.35. Don't speed past the command prompt. If you've entered a radius value

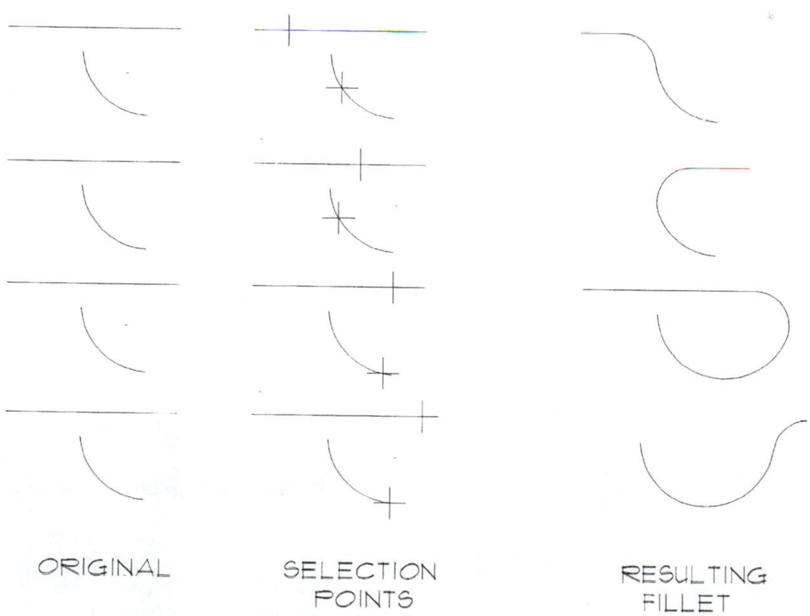

Figure 11.35 *Selecting different portions of the segment will affect the results of the FILLET.*

that exceeds the length of the lines, no change will be made to the drawing, and the following prompt will be displayed:

Radius is too large

Select second object:

Command:

Before you start thinking that you've really messed up, remember you are rarely going to damage a drawing file. Keep exploring. You just need to select a smaller radius value to make the command work.

One of the most common uses of FILLET is to form a corner with no radius. Setting the radius to 0 and then proceeding with the command will allow the command to be used to form corners. (See Figure 11.36.)

Practical Uses for Combining OFFSET, FILLET and TRIM: Open the FLOOR10 drawing that you started in Chapter 10. Since you've seen the finished floor plan in Figure 10.23b it's not too hard to visualize the lines in Figure 10.26 representing the rooms of the finished plan. Using a combination of OFFSET and FILLET will make it much easier to visualize. Figure 11.37 shows two examples of how FILLET can be used to clean up the floor plan. Notice that if FILLET is used at Fillet 3, the lower portion of the line will be removed and needs to be redrawn. This would be a good opportunity to use TRIM instead of FILLET.

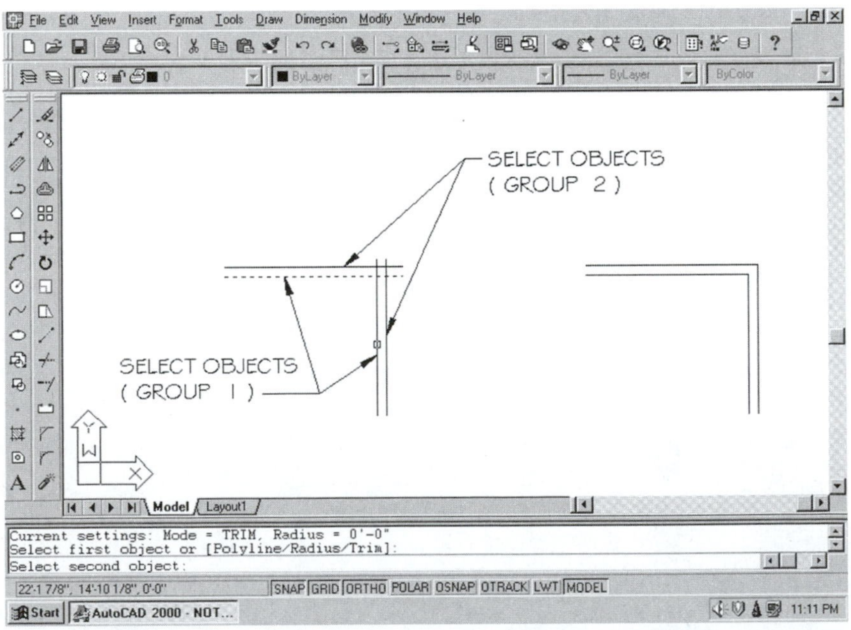

Figure 11.36 *With the value of the Radius set to 0, the FILLET command can be used to square corners of crossing lines.*

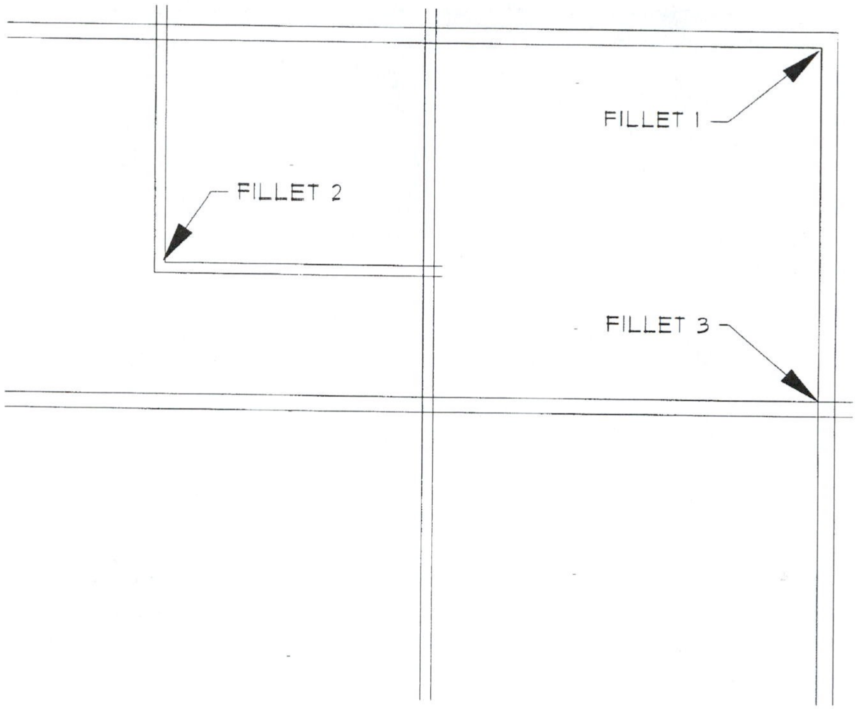

Figure 11.37 *The FILLET command was used to clean up the corners at locations 1 and 2. If FILLET had been used at location 3, the lines that will be used to form the garage would have been removed.*

SELECTING OBJECTS WITH GRIPS

In Chapter 10 you were introduced to several methods of selecting objects for editing. The GRIP command will provide an additional method of selecting objects for editing, as well as a method of combining several of the editing procedures introduced in this chapter.

Grips are similar to moving a heavy object by its handles. You can move the object without using the handles, but they make life so much easier. Grips are handles placed on objects for easy manipulation, similar to the OSNAP points, with STRETCH, MOVE, ROTATE, SCALE, and MIRROR. The default setting for Grips is ON. Grips can be toggled ON/OFF using the Selection tab of the Options dialog box. Selecting Options... from the Tools pull-down menu displays a dialog box. (See Figure 11.38.) With Enable Grips current, grips will be displayed when an edit or inquiry command is given. With the grip setting current, grip boxes will be displayed at various object-specific positions, as seen in Figure 11.39. When the cursor is moved over a grip, the cursor will snap to the grip for editing. This will save time over the use of an OSNAP location.

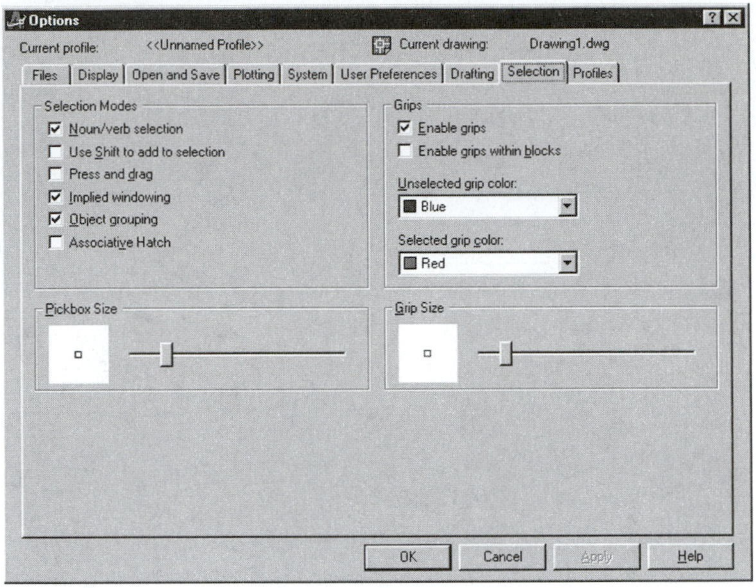

Figure 11.38 *Grips can be toggled ON/OFF using the Selection tab of the Options dialog box. The dialog box is accessed by selecting Options... from the Tools pull-down menu or by typing* ***DDGRIPS*** *(EnterKey) at the command prompt.*

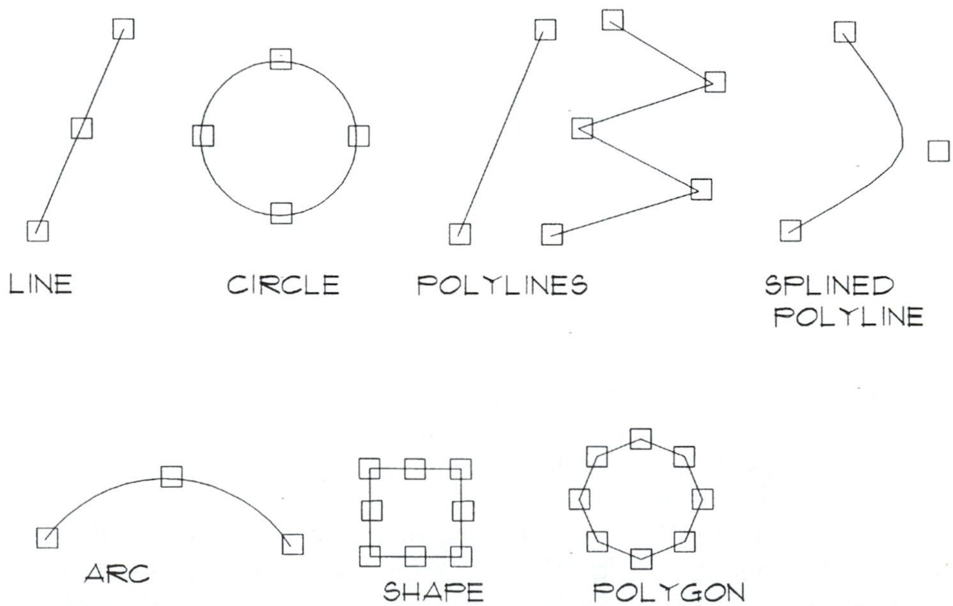

Figure 11.39 *With the default setting Enable Grips toggled to ON, Grips will be displayed for each geometric shape.*

USING GRIPS

Enter AutoCAD and draw several lines, arcs, and circles. Move the cursor so that it is over one of the lines and press the pick button. Three boxes or grips are now displayed. Move the cursor to the center grip, and press the pick button. The grip has changed from a hollow box to a solid square. This grip is now said to be *hot*. Notice that the command prompt has also changed to display the following prompt:

> ** STRETCH **
>
> Specify stretch point or [Base point/Copy/Undo/eXit]:

We'll come back to the prompt later. With the center grip hot, and the prompt reflecting STRETCH, move the cursor slightly. As the cursor moves so does the line. Pressing the pick button will move the line to this new location. The hot grip in the center of the line functions like the MOVE command. Select the same line with the cursor and make one of the end grips hot. Now the grip at the other end remains a stationary point, and as the hot grip is moved, the selected line is stretched. The effect of STRETCH on a line can be seen in Figure 11.40. GRIPS function in a similar manner with arc and circles. Figure 11.41 shows the effects of stretching each. If you decide you don't want to edit a selected line, press (EscKey) to deactivate the grips.

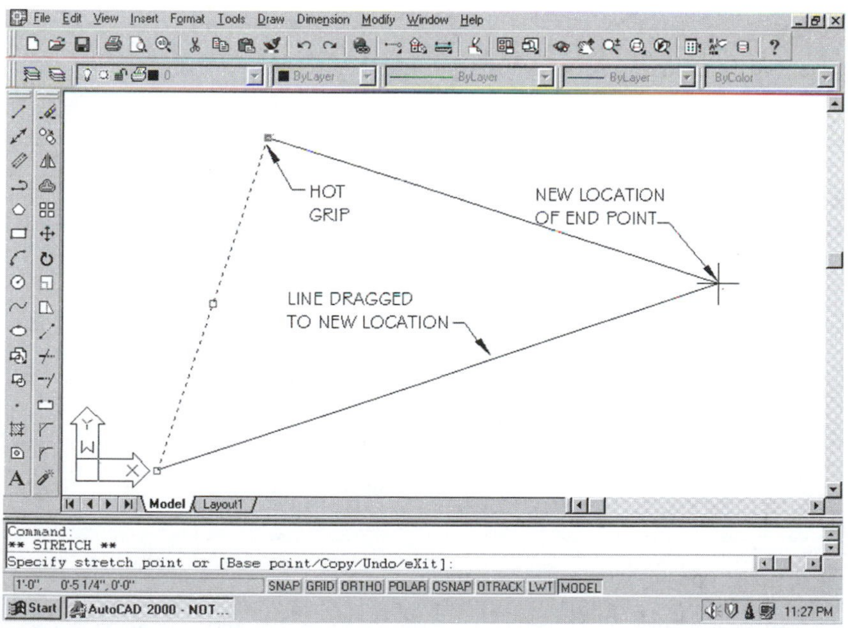

Figure 11.40 *Select a drawing object to display its grips. Moving to one of the grips and pressing the pick button will make that grip hot. The object can now be edited by STRETCH, MOVE, ROTATE, SCALE, and MIRROR.*

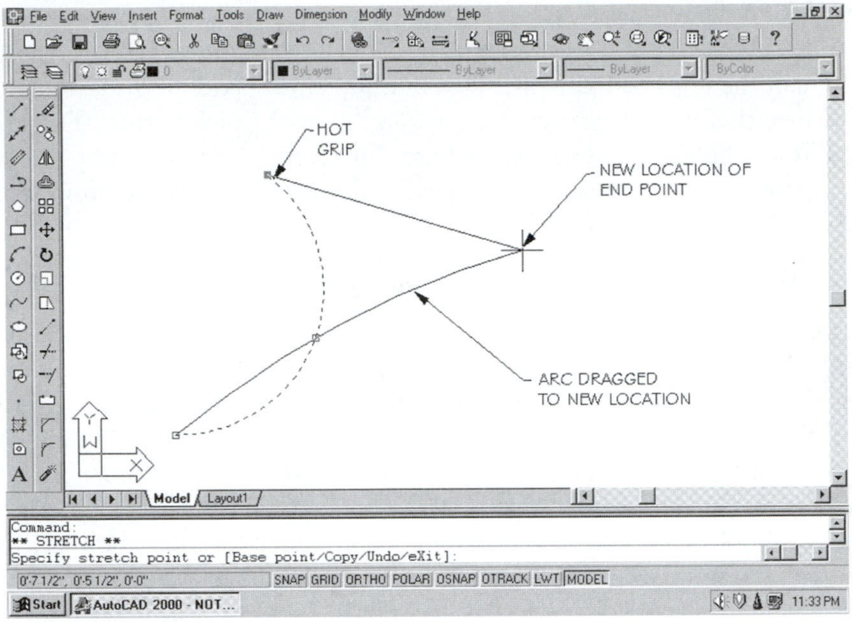

Figure 11.41a *The STRETCH command applied to an arc using the hot grip.*

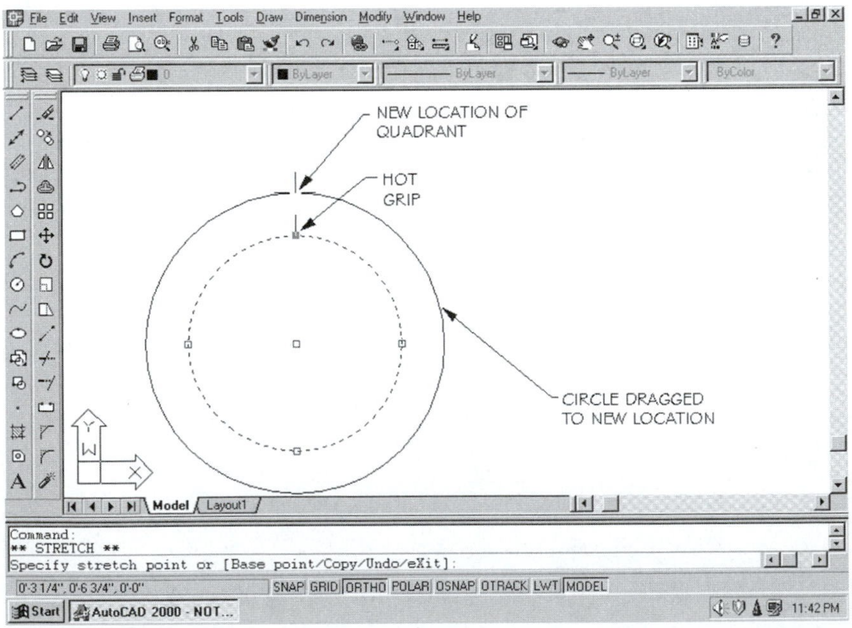

Figure 11.41b *The size of a circle can be altered using STRETCH by selecting a hot grip and a new location.*

GRIP OPTIONS

Earlier you discovered that each command displayed in grips has several options. These grip options can be used once the grips for an object have been activated. These options include:

Base Point

Typing B (EnterKey) at the prompt allows the hot box to be disregarded, and a new base point to be selected.

Copy

Typing C (EnterKey) at the prompt allows one or more copies of the selected object to be made.

eXit

Typing X (EnterKey) at the prompt terminates the GRIP edit and returns you to the command prompt.

Undo

Typing U (EnterKey) at the prompt will undo the previous selection.

CHANGING GRIP OPTIONS

Once a grip box as been selected and made hot, pressing the right mouse button will produce the menu shown in Figure 11.42. Any one of the options can be selected to alter the function to be preformed by the GRIP.

SELECTING OBJECTS FOR MULTIPLE EDITS

Often an object or group of objects may need to be edited with several different editing commands. If you were arranging units in a multi-building apartment complex, a unit may need to be mirrored, moved, and rotated to form a new building. Up to this point you would have to select the objects to be mirrored, use the MIRROR command, select the objects to be moved, use the MOVE command, select the objects to be rotated, and use the ROTATE command. It works, but you'll never impress the boss as worthy of the big bucks. The SELECT command will allow an object or group of objects to be pre-selected and used for one or more editing functions. The command sequence to mirror is:

Command: **SELECT** (EnterKey)

Select objects: *(Select objects using any of the selection methods.)* **W** (EnterKey)

First corner: Other corner: 100 found

Select objects: (EnterKey)

Command:

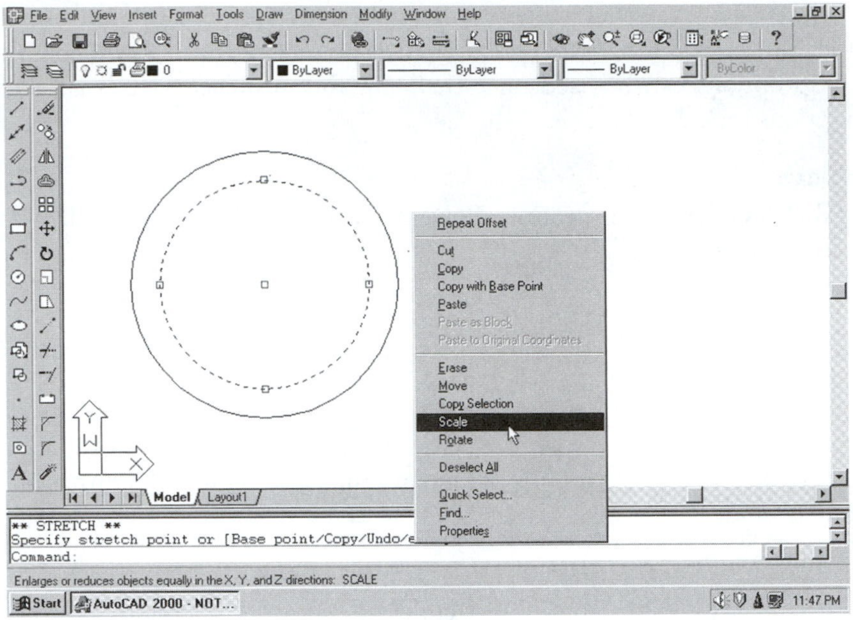

Figure 11.42 *Pressing the right mouse button will produce this menu that allows the GRIP edit option to be altered.*

The command prompt is returned but no apparent changes have occurred. The objects that were selected have been grouped as a set and will be affected by any editing command. Enter the MIRROR command and when prompted to select objects, type **P** (EnterKey) to automatically select the objects that were just placed in the set.

Type MIRROR at the command prompt. The sequence to mirror, move, and rotate the objects is:

Command: **MI** (EnterKey)

Select objects: **P** (EnterKey)

100 objects found

Select objects: (EnterKey)

Specify first point of mirror line: *(Select first point.)*

Specify second point of mirror line: *(Select second point.)*

Delete source objects? <N>: **Y** (EnterKey)

Command: **M** (EnterKey)

Select objects: **P** (EnterKey)

100 objects found

Select objects: (EnterKey)

Specify base point or displacement: *(Select base point.)*

Specify second point of displacement or <use first point as displacement>: *(Select new location for base point.)*

Command: **RO** (EnterKey)

Select objects: **P** (EnterKey)

100 objects found

Select objects: (EnterKey)

Specify base point: *(Select base point.)*

Specify rotation angle or [Reference]: **45** (EnterKey)

Command:

The GROUP command allows similar functions, but with more power.

Note: If you forget to use SELECT (EnterKey) before the editing command, don't worry. The same results can be obtained by entering the commands in the following sequence: enter **MI** (EnterKey), select your objects, select **RO** (EnterKey), and select **P** (for previous).

NAMING OBJECT GROUPS

SELECT allows a group to be created. The GROUP command allows multiple groups to be created, named, and saved. For instance, the apartment unit that was just edited could be saved as a group called Base. It could be mirrored, and both units saved as a new group called Units. A group that is part of another group is referred to as nested. The Units group could also be mirrored and saved as a third group called BLDG1. The GROUP command is started by typing **G** (EnterKey) at the command prompt. This will produce the Object Grouping dialog box shown in Figure 11.43.

CREATING AND NAMING A GROUP

In Figure 11.43 you can see that this drawing has no named groups. To create a group, enter the desired name at the keyboard. As you type, the name will be entered into the Group Name: box of the Group Identification area. Group names can be up to 31 characters long and include numbers, letters, or the character -, _, or $. Two separate words can't be used, but compound words can be joined by a dash or underline to create a name such as base_unit. As with the naming of drawing files, be sure to use names that will be easily recognized and that will clearly describe the contents of the group.

Description:

A description can be created to help you remember the contents of a specific group. To create a description, move the arrow to the Description: box and press the pick button. A description of up to 64 characters including spaces can be created.

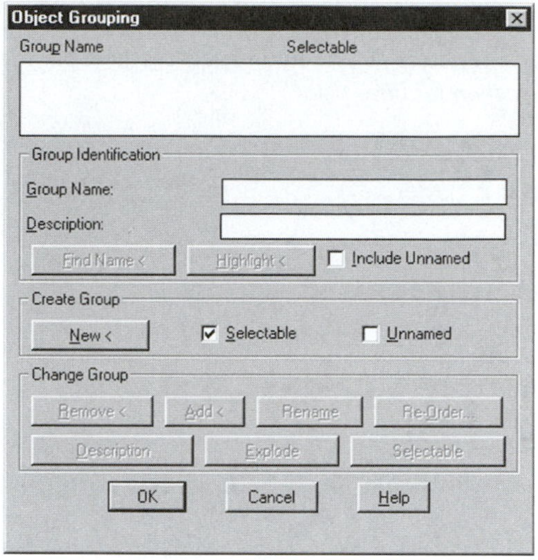

Figure 11.43 *Typing **G** (EnterKey) at the command line will display the Object Grouping dialog box. GROUP can be used to select and save drawing objects in groups for editing.*

New<

Once the desired name and description have been entered, select the New< button. This will remove the dialog box from the screen, and prompt you to Select objects. Objects to be included into the group can be selected using any of the selection methods. When the selection process is complete, the dialog box will be redisplayed and the group name will be displayed in the Group Name box. As the group is created, the word *yes* is now displayed beside the group name. If a group is selectable, selecting one object in the group selects the entire group.

Once the group is created, the other area of the dialog box can be used to alter or use the group. These options include:

GROUP IDENTIFICATION BOX

Key elements of this box can be seen in Figure 11.44. The box is used to identify named groups.

Find Name <

Selecting this button removes the dialog box from the screen and provides the prompt:

Pick a member of a group

When an object is selected, a Group Member List box similar to the display shown in Figure 11.45 is displayed. This box shows all groups where the selected object is a

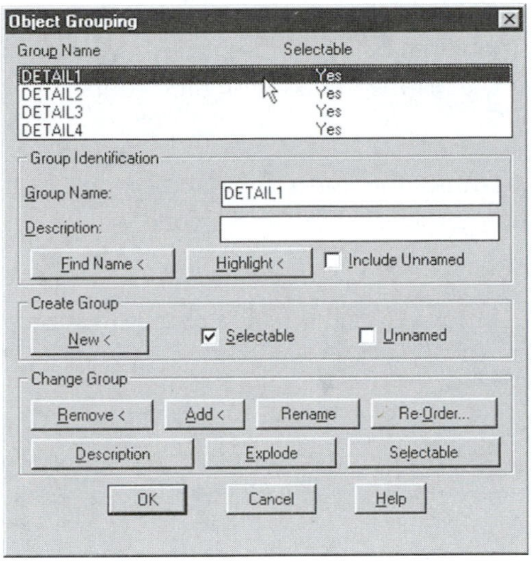

Figure 11.44 *Once groups are identified, selecting one of the groups will activate each portion of the dialog box.*

member. Selecting the OK button returns the Object Grouping dialog box. Selecting an unnamed group, and the Add< box allows a name to be provided to an unnamed group.

Highlight <

This box is not active as the dialog box is displayed. Moving the cursor to one of the named groups and pressing the pick button activates the button. Pressing the Highlight button removes the dialog box from the screen and shows the named group in dashed lines. This option is especially helpful on complicated drawings with multiple groups. Pressing the Continue button restores the Object Grouping dialog box.

Include Unnamed

Activating this box will display unnamed groups in the Group Name box. Unnamed groups can be renamed using the Rename option of Change Group.

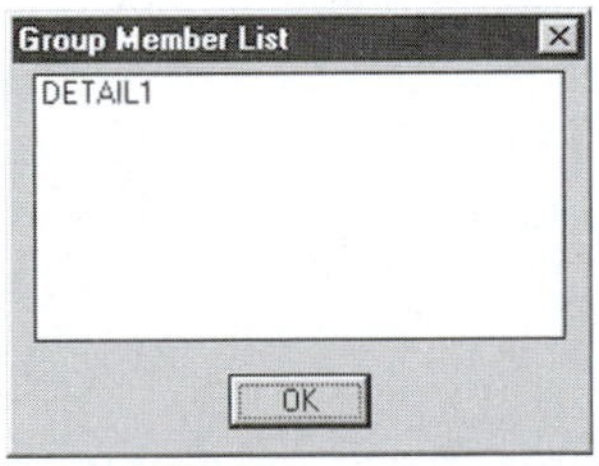

Figure 11.45 *Selecting Find Name < will display the group member List box. Select an object on the drawing and the group that it belongs to will be displayed.*

CREATE GROUP BOX

The key element of this box can be seen in Figure 11.44. The New< box was covered previously in the section, Creating and Naming a Group.

Selectable

Activating this box determines the status of named groups. Remember that selecting one object in a named group selects the entire group.

Unnamed

If this button is active, AutoCAD will provide names for unnamed groups.

CHANGE GROUP BOX

Key elements of this box can be seen in Figure 11.44. All elements except Re-Order... are inactive as the dialog box is displayed. Highlighting one of the named groups will activate all of the boxes in the Change Group area.

Add<

This option allows drawing entities to be added to an existing group. Once selected, you will be prompted to select objects. If the selected object is part of an existing group, the group name and description boxes will display the current status of the object. Choosing the OK button completes the selection process.

Description

This box allows the description to be updated and saved. To alter a description, move the arrow to the cursor box and type the desired material to be added. When complete, selecting the Description button updates the description and saves it for future reference.

Explode

The Explode box is used to remove the selected group from the effects of Group select. The objects that were in the group will not be altered, only the ability to select those objects as a group. Highlighting the group to be removed in the Group Name box, and then pressing the Explode button explodes the group.

Remove<

Selection of the Remove box allows individual objects to be removed from the grouping. Highlighting the desired group to be edited and selecting Remove will display the Remove objects: prompt. The dialog box will be removed and the objects in the group will be highlighted. Objects can be selected for removal using any of the selection methods.

Rename

This option can be used to rename an existing group. Highlight the name of an existing group so that the name and description are displayed in the Group Identification area. Edit the name as desired and select Rename to make the new name the current name.

Re-Order...

Selecting this box will produce the dialog box shown in Figure 11.46. Notice that the lower portion of the box includes three listings with numbered entries. Objects in a group are numbered in the same order that they were selected as the group was created. Re-ordering is typically used in CAD/CAM applications of manufacturing and is not often a part of the construction drawings. Options include:

Group Name

Displays the name of the group to be re-ordered.

Description

This selection displays the description of the selected group.

Remove from position (0–x):

For each listing, x will vary depending on the size of the objects in the group. This selection specifies the current numbered position of the object to be reordered.

Replace at position (0–x):

This selection specifies the new position the object will be moved to.

Number of objects (1–x):

This selection lists the range of numbers to be re-ordered.

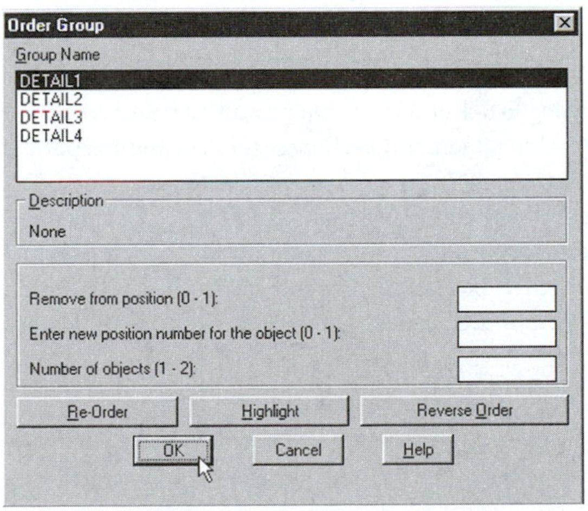

Figure 11.46 *The Order Group dialog box is displayed as the Re-Order... button is selected. Re-ordering allows the order in which objects are selected for certain functions to be altered. It is generally not used with construction drawings.*

CHAPTER 11 EXERCISES

1. Open a template and draw a 38'-long × 18 ½"-deep floor truss. Top and bottom chords will be made out of two 3 × 4s with 1"-diameter aluminum webs at approximately 45°, equally spaced throughout the truss. Save the drawing as E-11-1.
2. Draw four sets of perpendicular crossing lines. On one set of lines, use the FILLET command to square the corners. On the second set of lines, provide a 12" fillet. On the third set of lines use a 12" chamfer. On the fourth set of lines use a 12" and a 24" chamfer. Save the drawing as E-11-2.
3. Draw an 8"-wide concrete block retaining wall extending 8' above a 4"-thick concrete floor slab. Show the wall extending 8" above the finish grade. Thicken the slab to 8" thick at the edge resting on the footing. Use a 16" wide × 12" deep concrete footing, and show a 4" diameter drain on the soil side of the wall. Show a 2 × 6 top plate supporting 2 × 8 floor joists flush with the soil side of the wall. Reinforce the wall with #5 diameter @ 12" o.c. each way, 2" clear of the interior face. Use varied line weights for concrete, grade, sectioned wood, and steel. Adjust the widths to achieve good contrast. Save the drawing as E-11-3.
4. Open drawing E-10-4. Copy and rotate the pair of doors to provide doors for walls at 45°, 90°, 135°, 180°, and 270° rotations. Save the drawing as E-11-4.
5. Open drawing E-10-5 and make a copy of the column cap. Adjust the base strap so that it is 3" longer and 1" wider. Add one additional bolt. Each end of the side plate must be 2" longer (4" total) and 1 ½" higher. Save the drawing as E-11-5.
6. Open drawing E-10-7. Enlarge the foundation plan 8 feet in each direction. Add required beams and piers to meet the original criteria. Place 4' × 4' diagonal corners in the exterior walls. Save the drawing as E-11-6.
7. Open drawing E-8-1 and make three additional copies. In the original drawing, provide a 6"-long line from each intersection and perpendicular to the opposite side of the triangle. Rotate the new triangles and place a point of each new triangle tangent to the existing triangle and perpendicular to the 6" lines. Draw a 10"-diameter circle with a center point in the center of the original triangle. Trim all elements that extend beyond the circle. When complete, your drawing should resemble the attached drawing. Save the drawing as E-11-7.

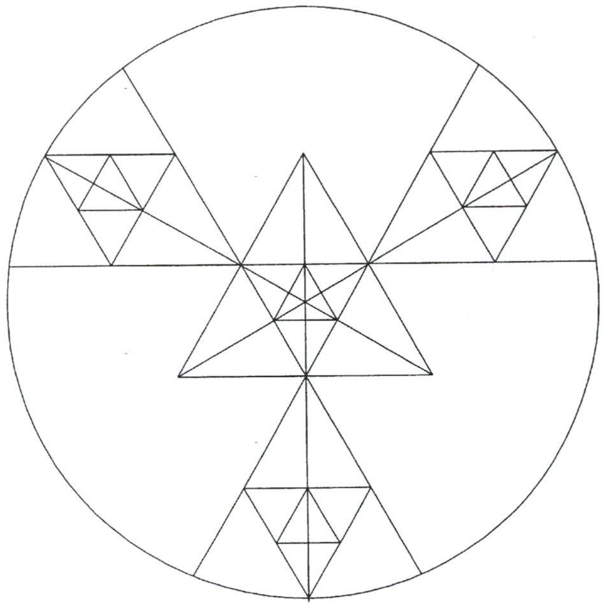

8. Start a new drawing and draw a metal chimney cap that extends a total of 12" above a 4'-wide wood chimney chase. Trim the chase with 3"-wide trim and show 6"-wide horizontal siding. The top of the chimney cap is 14" wide × 1". The main portion of the cap is 12" × 7". The chimney is 10" wide × 4" high. Use MIRROR OFFSET, TRIM, and FILLET to complete the drawing. When complete, your drawing should resemble the drawing below. Save as E-11-8.

9. Start a new drawing and draw a section of a standard one-story foundation similar to the sketch below. Copy and edit the drawing to the proper size for a two-story footing. Save the drawing as Figure E-11-9.

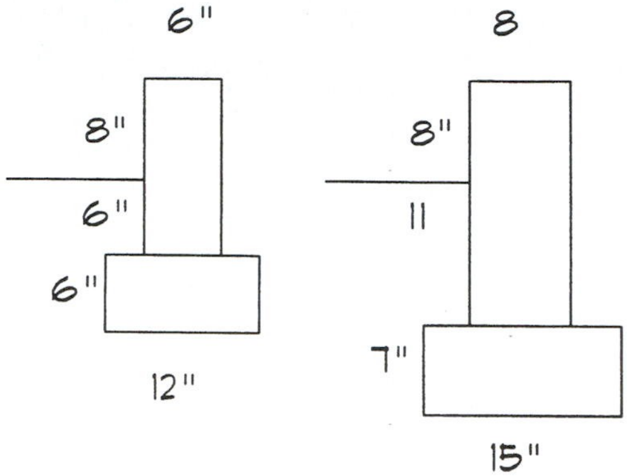

10. Open a new drawing and draw one half of a roof truss with a 5/12 pitch with a 26' span. All chords will be made of 6"-wide material; all webs will be made from 4" material. Use a 24" overhang. The webs should be spaced equally at the center of the bottom chord. As a minimum, use OFFSET, OSNAP, and TRIM. Use the MIRROR command to complete the drawing. Save the drawing as E-11-10.

11. Use the floor plan shown in Figure 10.23b and draw the walls and cabinets for this residence. No doors, windows, plumbing fixtures or appliances need to be drawn at this time. Assume:

    All exterior heated walls and toilet walls are 6" thick.

    All exterior unheated walls, interior walls are 4" thick.

    Assume the fireplace is 3'-4" wide. Determine the length by allowing for a 36"-wide hallway.

    Assume the stairway, the hall between the stair and bathroom, and the space between the stair and the front wall to be 42" wide.

    Assume the bathroom is 36" wide.

    Although door and windows symbols should not be drawn at this time, provide the opening in the wall for each door.

    Provide the outline of all cabinets. Assume 24"-wide base and 12"-wide upper cabinets. Do not draw appliances.

    Save this drawing as EFLOOR.

12. Open drawing EFLOOR. Assume North to be at the top of the drawing so that the front door is presently facing south. Make all changes so that the current wall between the family and living rooms remains straight for the entire length of the house. Alter the width of the wall as needed as it changes from interior/exterior/ toilet wall.

    Add 2'-0" to the family room in the north/south (n/s) direction.

    Add 3'-6" to the living room in the east/west (e/w) direction. Add 18" in the n/s direction.

    Enlarge the bathroom to 42" wide. (+6" e/w)

    Remove 24" e/w from the shop.

    Remove the garage doors on the south face of the garage and place them on the east wall. Remove the window from the east wall and place (2) 6'-0"-wide windows on the south wall. In addition to the 2'-8" personnel door on the east wall, enlarge the garage to accommodate (3) 8'-0"-wide doors with a 12" minimum space. Provide 24" space on each end of the garage between the door and the end of the wall. Save the drawing as E11FLOOR.

13. Draw five concentric circles with the smallest diameter of 6" and the other sizes increasing in 3" increments. Remove a 1"-wide strip going across each of the circles in both the vertical and horizontal directions. The center of the strip to be removed should be at the center of each quadrant. Rotate the remaining portions of the second and forth circles (starting at the inside) so that the remaining portions of these circles have been revolved 45° while remaining concentric. Save the drawing as E-11-13.

14. Use EFLOOR as a base. Draw the plan view of a simple tree. Draw the trees in groups of two and three with each tree in the group to have a different size. Place at least four groups of trees around the house. Design walkways, patios and a driveway.

15. Draw concentric shapes of 3, 4, 5, 6, 7, and 8 sides that are tangent to one circle. Make copies of these objects that are $\frac{1}{4}$ size, $\frac{1}{2}$ size, full size, 2 × normal size, and 90% of the normal size. Save the drawing as E-11-15.

16. Open drawing E-10-4 and make two copies of the door. Scale one door to be 2'-6" wide and the other door to be 2'-8" wide. Leave the window beside the door as is on each option. Once the sizes are altered, make copies so that double doors are available for each option. A total of 4 door arrangements should be available when complete. Save the doors as E-11-16.

17. Open Drawing E-10-10 and make two additional copies of the window. Enlarge one window to 5'-0" wide. Reduce the other window to 75% of the original size. Save the drawing as E-11-17.

18. Draw a U-shaped kitchen with a 30" × 48" island. Provide a 32" × 21" sink, a 24" × 24" dishwasher, a 34" × 24" refrigerator, and 27" × 27" oven in the kitchen, with a 30" × 22" cook top in the island. Create a second copy of the kitchen that will be flopped and set at a 45° angle to the first kitchen. Save the drawing as E-11-18.

19. Using the drawing below as a guide, lay out the outline of the office floor plan. Use 8"-wide exterior walls. Copy the plan. Save the drawing as E-11-19.

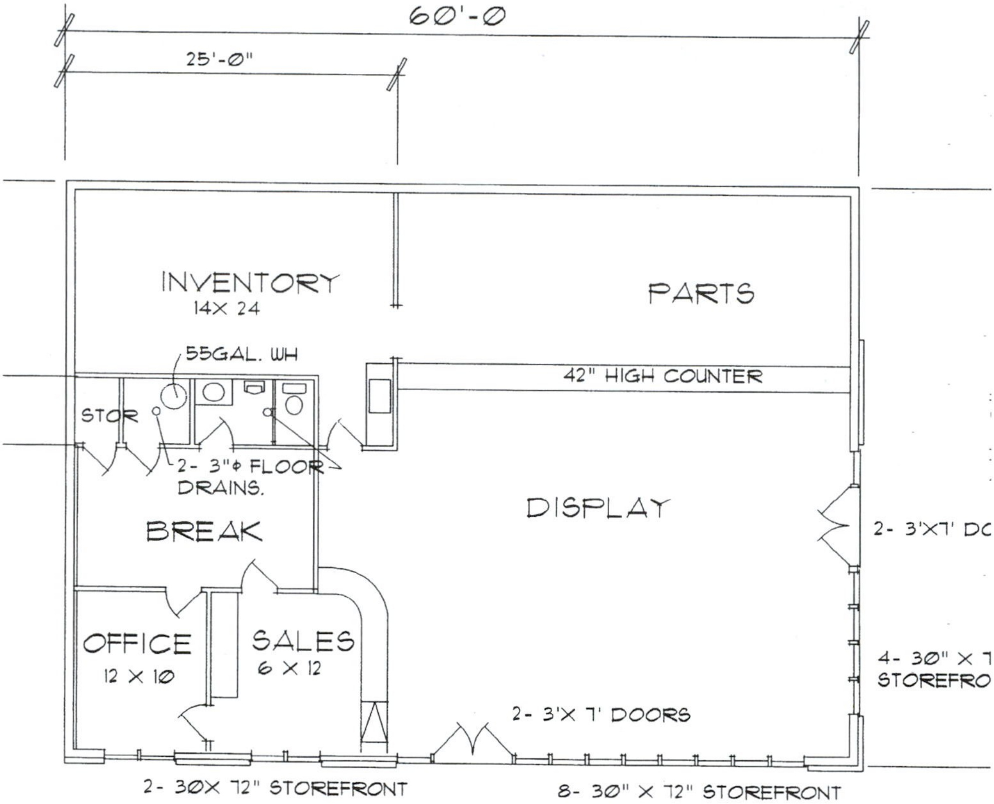

20. Draw a 3"-diameter circle, a 6"-long line, and a polygon with 8 sides which will fit in a 5 ¾"-diameter circle. Save this portion of the drawings as E-11-20a. Make enough copies of each object and the use each option of each editing function of GRIPS to edit each object. Save the drawing as E-11-20b.

21. Use E-11-20a as a template drawing to complete this assignment. Make a second copy of these objects. Use SELECT to create a group consisting of the two circles and two polygons. Use MIRROR, ROTATE, and SCALE to edit the group. Save the drawing as E-11-21.

22. Use Drawing E-11-20a as a template drawing to complete this assignment. Make a second copy of these objects and then mirror both groups so that you have a

total of four circles, lines and polygons. Use GROUP to name and save at least 6 different groups. Save the file as E-11-22.

23. Using the drawings below as a guide, redraw the plan view of the manhole cover showing 16- ½" wide support flanges equally spaced around the plate. Save the drawing as E-11-23.

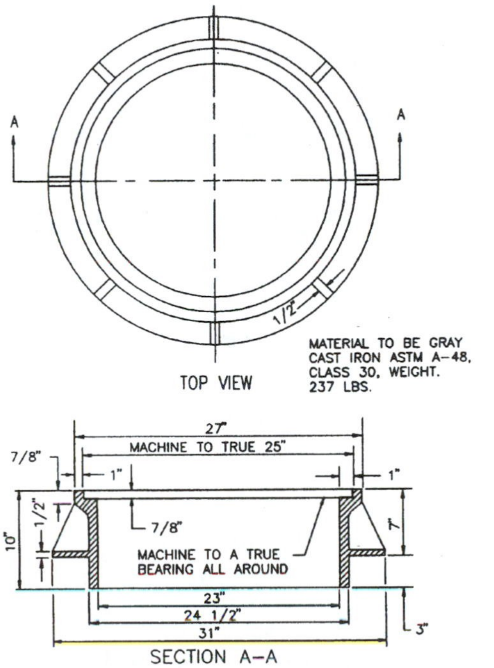

24. Using the drawing below as a guide, draw and array a pattern to represent Spanish clay tile roofing. Save the drawing as E-11-24.

25. Use a reference guide to determine the size of a W8 × 35 steel piling. Draw a plan for a 38' × 76' structure showing three rows of piling. Each of the two exterior rows will be inset 8" from the exterior face of the structure. The interior row of pilings will be 18' in from the front (the 76'-long surface) of the structure. Pilings will be placed at 19' o.c., equally spaced from each exterior wall. Save the drawing as E-11-25.

CHAPTER 11 QUIZ

1. Explain the uses of the BREAK command.
2. Describe two different uses of the FILLET command.
3. List the command sequence for ROTATE.
4. Explain the difference between MOVE and COPY.
5. List the command sequence to EXTEND a line to meet another line and sketch and label the process in the space provided.
6. Two perpendicular lines cross at a corner. List four methods to edit the lines to form a non-crossing corner.
7. A line has been drawn short of its required destination. How can the error be corrected?
8. What two options are available for offsetting an object?
9. In the CHAMFER command, what is the relationship between the distances and the selection of lines?
10. You are preparing to MOVE some objects, so you type "P" at the Select objects prompt. What will be moved?
11. Use the Help menu to determine what SETVAR controls a preset FILLET value?
12. How can an object be rotated in a clockwise direction?
13. A door is drawn at a 45° angle. It should be 49°. How can this error be corrected?
14. Six objects need to be moved. As you select the objects, an extra object is included in the selection set, and an intended object is not included. Describe the command process and prompts to move these objects.
15. List and describe the three break options.
16. Describe how each of the editing commands presented in this chapter could be used in drawing a floor plan.
17. What will happen if you try to place a 60"-radius fillet in two intersecting lines that are 48" long?
18. Explain how CHAMFER will affect two lines that do not touch if they are selected for the objects of this command.
19. What will be the effect of the Trim mode of CHAMFER on two intersecting lines?
20. What will be the effect of the TRIM edge option on two lines that do not intersect?
21. How will the Edge option of EXTEND affect two lines?
22. How will window and crossing affect objects selected for STRETCH?
23. What are the options that are given if the DElta option of LENGTHEN is selected?

24. A 6"-long line is selected to be edited using LENGTHEN. What will be the result of entering 50%? 150%? 200%?
25. How many selection points will be required when breaking an arc using BREAK@?
26. Provide an example of how the ALIGN command could save time.
27. List the commands used to create E-11-9.
28. What must be done to display GRIPS on a line?
29. What five editing functions can be executed after a grip is hot?
30. What is a hot grip and how do you create one?
31. You're exploring options and select GRIPBLOCK. What will this option do and what can you do if you don't want this option?
32. What advantage does SELECT have over GRIPS?
33. Describe the process of selecting and naming a group of objects using GROUP.
34. What is a nested group?
35. What is a selectable group?
36. You've named a group, but come to realize the name does not describe the drawing well? How do you fix the problem?
37. Save a group of objects without providing a name. How does AutoCAD keep track of this group?
38. You're exploring again and you use the Explode option of Group. What effect will this have on your computer? What effect will it have on the current drawing? What effect will it have on the selected group?
39. How do you make the options in the Change Group box of Object grouping active?
40. Use the help menu and determine what variables affect grips. What are the options and their effects?

CHAPTER 12

Polylines

One of the best ways to get objects to stand out on a complicated drawing is to vary the line thickness. In previous chapters you've learned how to add thickness to lines. Thickness can also be assigned easily with the PLINE command, which represents polylines. In plan views, walls typically are drawn thicker than all other types of lines. On site drawings, polylines are often used to represent contour elevations of specific intervals, such as every five or ten feet. Polylines are used throughout structural drawings to represent various materials. Sections and details often use several thicknesses of lines to help distinguish between concrete, steel, and wood that has been cut by the cutting plane, and materials that lie beyond the cutting plane. Figure 12.1 shows examples of polylines used to represent steel reinforcing in concrete details.

In addition to being able to control the width, Polyline has several other unique features. When the line is drawn using the LINE command, each segment is a separate object. An irregular-shaped object can be drawn with polylines, yet each individual line is part of the whole. Notice that in Figure 12.2 the horizontal line, the arc, and the vertical line forming the L-shaped rebar have been selected to be erased. Because these features were drawn using the PLINE command, they function as one object. This feature is an important editing feature when the BHATCH command is used. Features of polylines to be explored include the PLINE command, changing widths, POLYLINE arcs, undoing a PLINE, editing a PLINE, exploding a PLINE, PEDIT, and editing a vertex.

EXPLORING THE PLINE COMMAND

The PLINE command can be accessed by selecting the Pline icon from the Draw toolbar, by typing **PL** (EnterKey) at the command prompt, or by selecting Polyline from the Draw pull-down menu. The command sequence is:

Command: **PL** (EnterKey)

PLINE

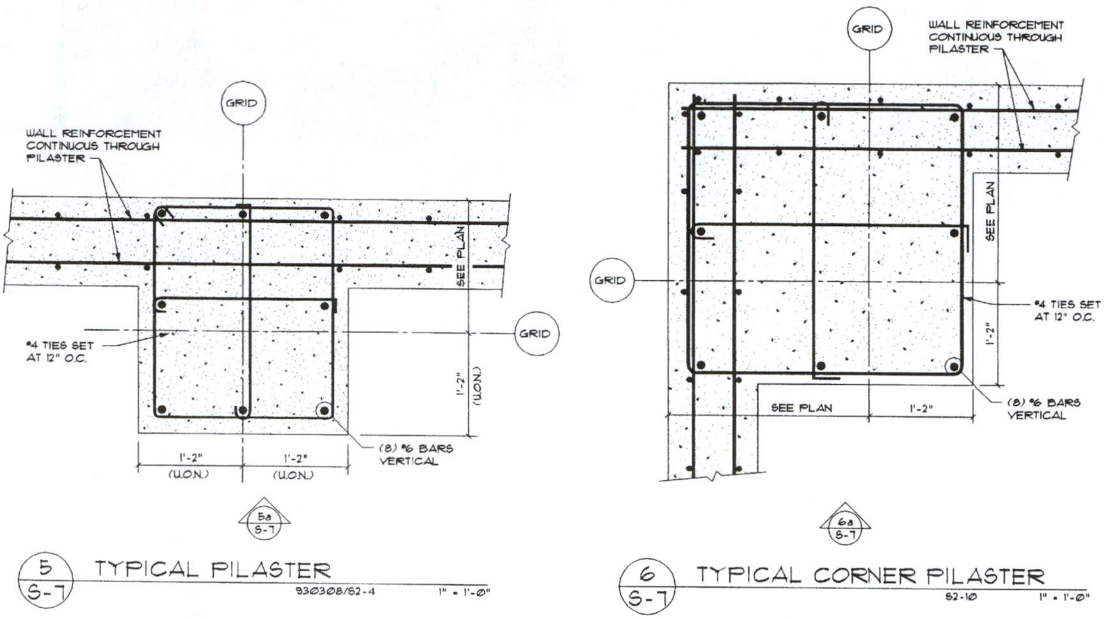

Figure 12.1 *The PLINE command can be used to distinguish between various materials. Notice that the steel is drawn with a polyline and can be easily separated from the edge of the concrete. Not only are polylines thicker, they move as one object rather than as individual lines. (Courtesy Van Domelen/Looijenga/McGarrigle/Knauf Consulting Engineers.)*

Specify start point: *(Select any point.)*

Once a beginning point has been selected, a new prompt will be displayed:

Current line–width is 0'-0"

Specify next point or [Arc/Close/Halfwidth/Length/Undo/Width]: *(Specify endpoint of line)*

This prompt will continue to be displayed each time a new endpoint is entered. Each new line segment will continue from the end of the preceding line. The command will continue until (EnterKey) is pressed.

CHANGING THE WIDTH

Enter the drawing area and draw a few polylines. Once you've drawn a polyline with several different drawing segments, select one of the segments to be erased. During editing is when you can really appreciate a polyline. Selecting one segment of a line selects the entire polyline for editing. A line can be made up of hundreds of segments, but if one is selected, they are all selected. With the line width set to "0", polylines look just like any other line. The 0 width works great if you want to outline an irregular-shaped area and find the area or the perimeter where line width is not

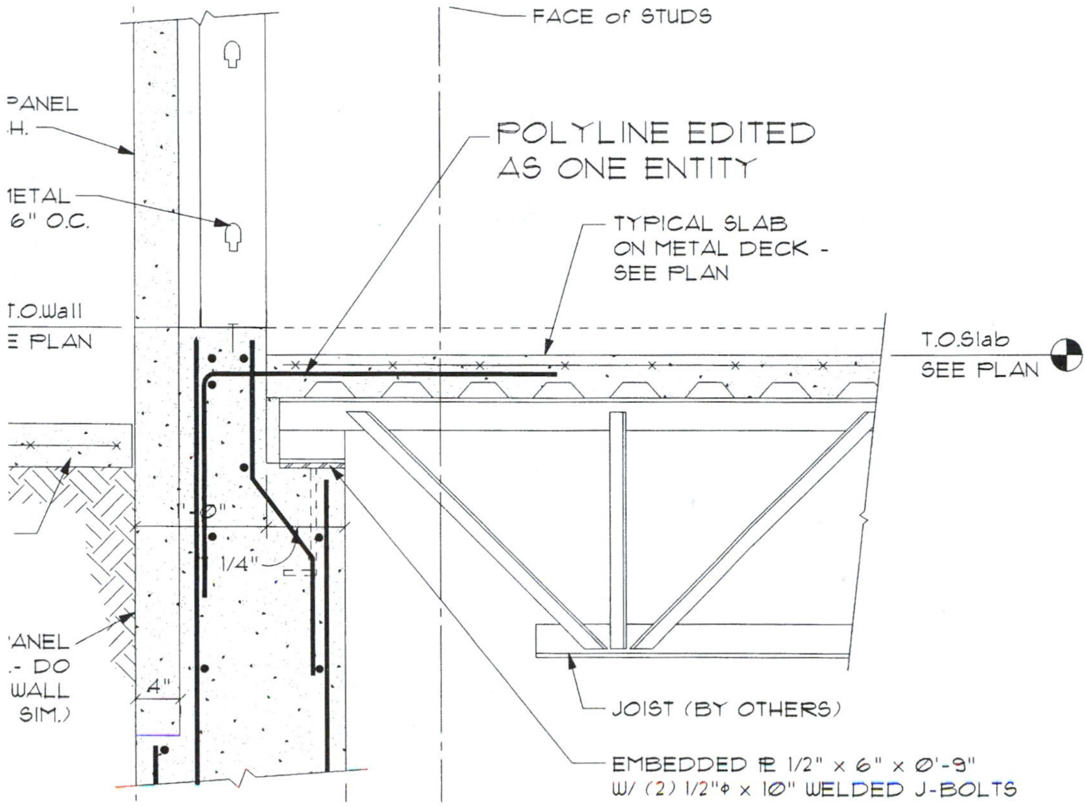

Figure 12.2 *Selecting the L-shaped rebar that was drawn using PLINE at any point will select the horizontal line, the arc, and the vertical line as one object. (Courtesy Van Domelen/Looijenga/McGarrigle/Knauf Consulting Engineers.)*

important. To change the width of a line, start the PLINE sequence and select a start point. As the option prompt is displayed, enter a **W** (EnterKey). This will allow information for a starting and ending width to be entered. The command sequence to draw a ⅛"-wide line is:

Command: **PL** (EnterKey)

PLINE

Specify start point: *(Select any point.)*

Current line–width is 0'-0"

Specify next point or [Arc/Close/Halfwidth/Length/Undo/Width]: **W** (EnterKey)

Specify starting width <0'-0">: **.125** (EnterKey)

Specify ending width <0'-0 ⅛">: (EnterKey)

AutoCAD rounds the option to the nearest fraction, depending on how the UNITS have been set. If you enter a fraction and the default returns as 0, exit PLINE and enter UNITS and set the fractional values as desired. For the examples used in this chapter, the units are set to architectural and the denominator for fractions is set to 1/8" (.125). Notice that the prompt for the Ending width shows 1/8" as a default value. By responding (EnterKey) at the prompt, the option prompt will return allowing for selection of the endpoint. With the width selected, the command sequence is:

> Specify next point [Arc/Close/Halfwidth/Length/Undo/Width]: *(Specify starting point.)*

With the starting and ending width the same, a line with uniform thickness will be drawn, as shown in Figure 12.3.

Drawing Tapered Lines

PLINE width allows the width of a Polyline to be changed for each segment or from one end of a segment to the other. A tapered polyline with a starting width of 1/8" and an ending width of 1/4" can be drawn by using the following command sequence:

> Command: **PL** (EnterKey)
>
> PLINE
>
> Specify start point: *(Select any point.)*
>
> Current line–width is 0'-0"

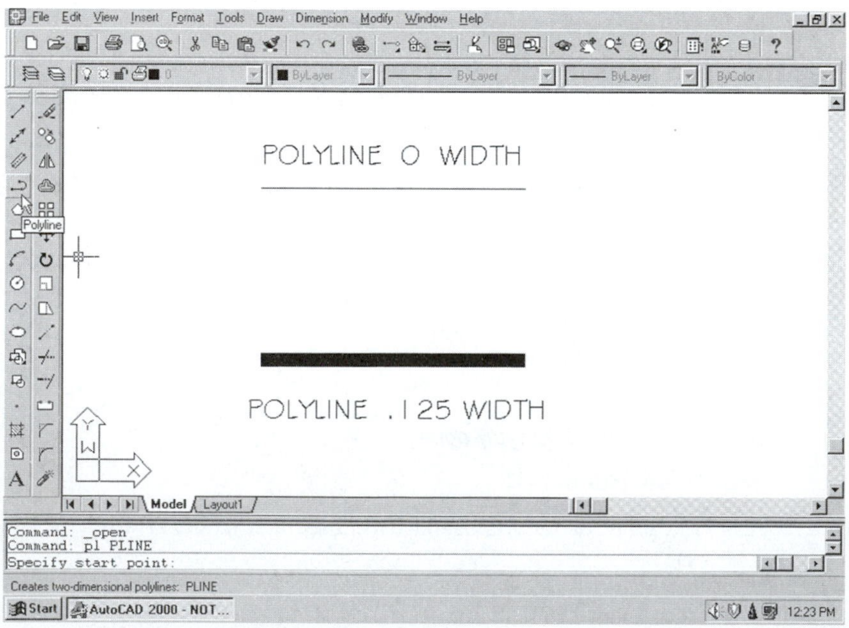

Figure 12.3 *If drawn with a width value of 0, the polyline will appear as any other line segment.*

Specify next point or [Arc/Close/Halfwidth/Length/Undo/Width]: **W** (EnterKey)

Specify starting width <0'-0">: **.125** (EnterKey)

Specify ending width <0'-0 ⅛">:**.25** (EnterKey)

Specify next point [Arc/Close/Halfwidth/Length/Undo/Width]: *(Select endpoint.)*

Specify next point [Arc/Close/Halfwidth/Length/Undo/Width]: (EnterKey)

Command:

As the endpoint is entered, the line will be drawn and the option prompt will be returned to allow continuing a polyline. Notice that if the command is continued, the default for the next line segment is the ending width. Figure 12.4 shows an example of varied line width.

Choosing a Line Width

When you experiment with PLINE, the width selection is not terribly important. As you begin to work on more involved projects that will be printed or plotted, the width selection is critical. One of the considerations in selecting a line width is the scale at which the drawing eventually will be printed. A ¼" polyline will look huge on the screen when you've zoomed in on a feature, but if printed or plotted at ⅜" = 1'-0", it will appear as a 0 width line on a drawing plotted at 1" = 30'-0".

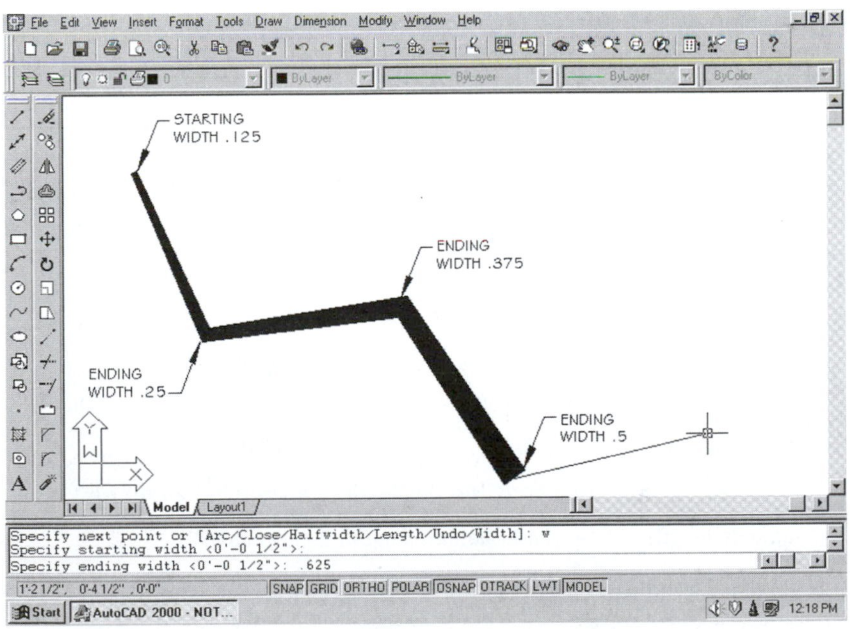

Figure 12.4 *Polylines can be used to draw a line with varied widths.*

It's also important to realize that line width can be controlled in the plotting process. This will be covered in future chapter as layers and colors are discussed, and as you explore plotting options. For now, the key is to realize that polyline width can be easily altered.

Halfwidth Option

This option allows the width of a polyline to be set from the center of the polyline to an edge and is very similar to setting the width. It's like describing the size of a circle by referring to the radius instead of the diameter. The command sequence is:

Command: **PL** (EnterKey)

PLINE

Specify start point: *(Select any point.)*

Current line–width is 0'-0 $^1/_4$"

Specify next point or [Arc/Close/Halfwidth/Length/Undo/Width]: **H** (EnterKey)

Specify starting half–width <0'-0 $^1/_8$">: (EnterKey)

Specify ending width <0'-0 $^1/_8$">: (EnterKey)

Specify next point [Arc/Close/Halfwidth/Length/Undo/Width]: *(Select endpoint.)*

Specify next point [Arc/Close/Halfwidth/Length/Undo/Width]: (EnterKey)

Command:

Length Option

This option will allow a polyline to be drawn to a specified length and parallel to the last polyline drawn. The new line will extend in the same direction as previous lines. To change the direction in which the line is drawn, enter a negative length. This process can be seen in Figure 12.5. If the last segment was an arc, a line drawn with L will be tangent to the arc.

Command: **PL** (EnterKey)

PLINE

Specify start point: *(Select any point.)*

Current line–width is 0'-0 $^1/_4$"

Specify next point or [Arc/Close/Halfwidth/Length/Undo/Width]: **L** (EnterKey)

Specify length of line: *(Enter desired length.)* **2** (EnterKey)

Specify next point or [Arc/Close/Halfwidth/Length/Undo/Width]:(EnterKey)

Command:

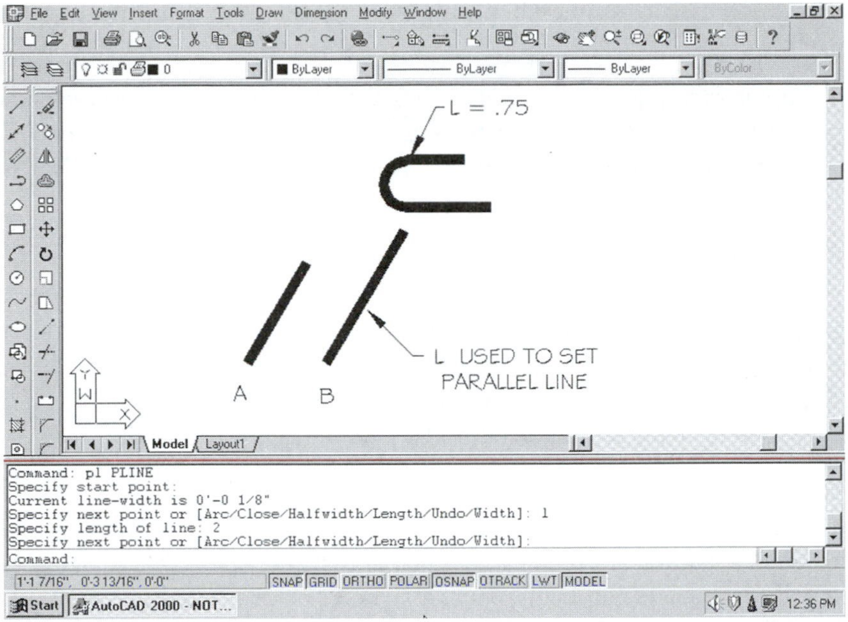

Figure 12.5 *The Length option of PLINE (L) will draw a polyline parallel to another polyline at a specified length.*

PLINE ARCS

Arc is the first option listed in the PLINE menu. When you select ARC by entering A (EnterKey) at the prompt, AutoCAD enters the ARC submenu and displays the prompt:

Command: **PL** (EnterKey)

PLINE

Specify start point: *(Select starting point.)*

Current width is 0'-0 ¼"

Specify next point or [Arc/Close/Halfwidth/Length/Undo/Width]: **A** (EnterKey)

Specify endpoint of arc or

[Angle/CEnter/CLose/Direction/Halfwidth/Line/Radius/Second pt/Undo/Width]: *(Select any point.)*

Specify endpoint of arc or

[Angle/CEnter/CLose/Direction/Halfwidth/Line/Radius/Second pt/Undo/Width]: (EnterKey)

Command:

The Arc option will continue to prompt for arc endpoints until (EnterKey) is selected.

As the submenu is displayed, an arc is started at the first point that was selected, and is dragged to the current location of the cursor. Several of the options for ARC are the same as the options for straight–line polylines. The Width option is similar, with the width for an ARC ranging from 0 up to the diameter of the arc.

Continuous Pline Arcs

If an arc is started from a previous PLINE segment, the new arc will be tangent to the existing line by default. The arc will continue in the same direction as the most recent polyline segment. An example of how the arc will be constructed can be seen in Figure 12.6. An arc can also be constructed using the Angle, CEnter, CLose, Direction, Line, Radius, and Second pt methods.

Pline Angles

This option will allow the start point, center, and angle that the arc is to span (included angle). By default, the angle will be drawn counterclockwise. By entering a negative value, the angle will be drawn clockwise. The command process to draw an arc with a 45° included angle is:

Command: **PL** (EnterKey)

PLINE

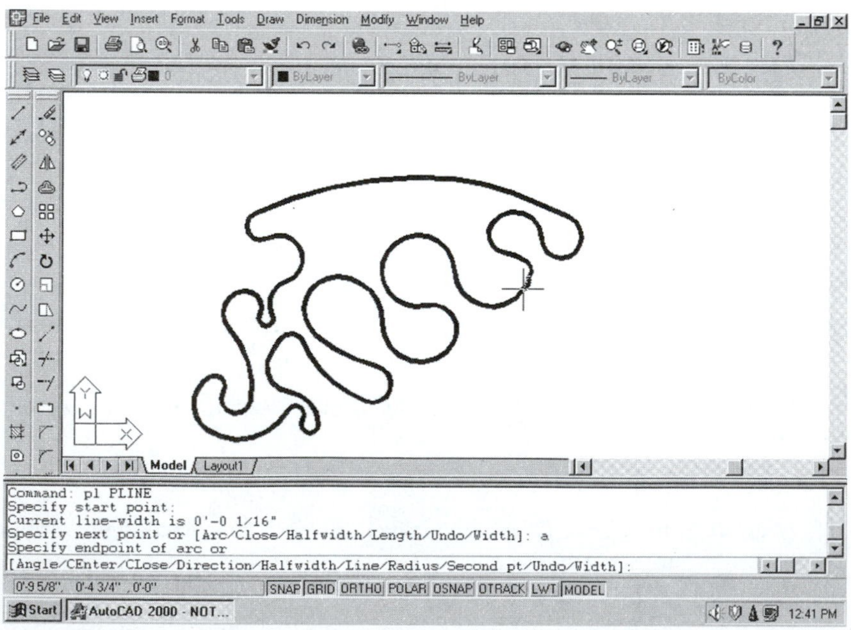

Figure 12.6 *Arcs can be constructed of polylines by using the Arc option of PLINE.*

Specify start point: *(Select starting point.)*

Current width is 0'-0 ¼"

Specify next point or [Arc/Close/Halfwidth/Length/Undo/Width]: **A** (EnterKey)

Specify endpoint of arc or

[Angle/CEnter/CLose/Direction/Halfwidth/Line/Radius/Second pt/Undo/Width]: **A** (EnterKey)

Specify included angle: **45** (EnterKey)

Specify endpoint of arc or [CEnter/Radius]: *(Select desired endpoint.)*

Specify endpoint of arc or

[Angle/CEnter/CLose/Direction/Halfwidth/Line/Radius/Second pt/Undo/Width]: (EnterKey)

Command:

As the endpoint is selected, the center is rotated as needed, based on the two endpoints. This process can be seen in Figure 12.7.

Arc Center

An arc can be drawn by specifying the endpoint and the centerpoint. When you enter **CE** (EnterKey) at the last prompt, you will be allowed to define the "first endpoint"

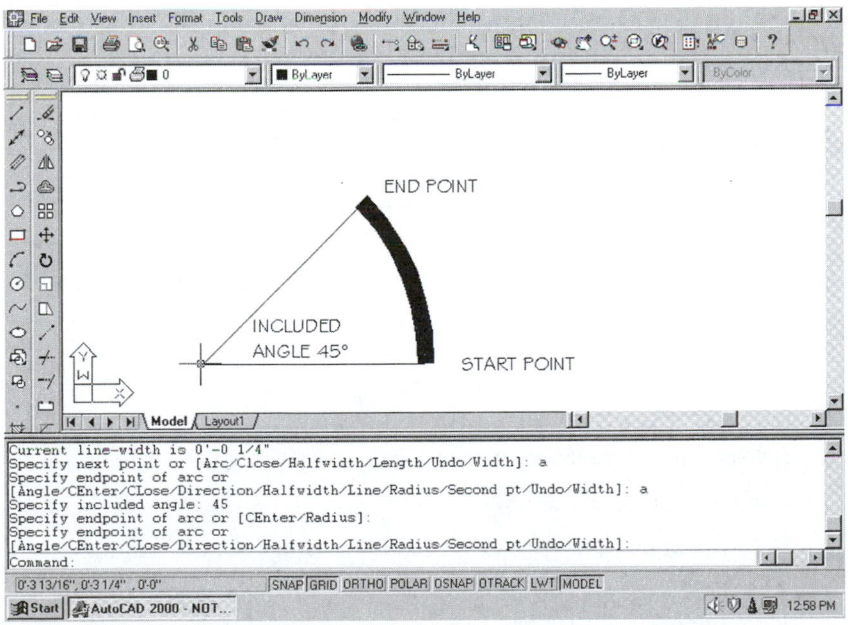

Figure 12.7 *Selecting the Angle option of Arc (A) will allow an arc to be drawn based on the provided angle.*

and the "center point" of the arc. The first steps of the command sequence are identical to other ARC commands. The command sequence for drawing an ARC by defining the center point is:

Command: **PL** (EnterKey)

PLINE

Specify start point: *(Select starting point.)*

Current width is 0'-0 ¼"

Specify next point or [Arc/Close/Halfwidth/Length/Undo/Width]: **A** (EnterKey)

Specify endpoint of arc or

[Angle/ CEnter/CLose/Direction/Halfwidth/Line/Radius/Second pt/Undo/ Width]:

CE (EnterKey)

Specify center point of arc: *(Select desired center point.)*

Specify endpoint of arc or [Angle/Length]: **A** (EnterKey)

Specify included angle: **45** (EnterKey)

Specify endpoint of arc or [CEnter/Radius]: *(Select desired endpoint.)*

Specify endpoint of arc or

[Angle/CEnter/CLose/Direction/Halfwidth/Line/Radius/Second pt/Undo/ Width]: *(Select desired endpoint)*

Specify endpoint of arc or

[Angle/ CEnter/CLose/Direction/Halfwidth/Line/Radius/Second pt/Undo/ Width]: (EnterKey)

Command:

Arc Radius

Entering R (EnterKey) will allow for specifying the first endpoint and the radius to be selected. The command sequence is:

Command: **PL** (EnterKey)

PLINE

Specify start point: *(Select starting point.)*

Current width is 0'-0 ¼"

Specify next point or [Arc/Close/Halfwidth/Length/Undo/Width]: **A** (EnterKey)

Specify endpoint of arc or

[Angle/ CEnter/CLose/Direction/Halfwidth/Line/Radius/Second pt/Undo/ Width]: **R** (EnterKey)

Specify arc radius: *(Enter desired radius.)* **3.25** (EnterKey)

Specify endpoint of arc or [Angle]: **A** (EnterKey)

Specify included angle: *(Specify desired angle)* **45** (EnterKey)

Specify direction of chord for arc <30>:

Specify endpoint of arc or

[Angle/CEnter/CLose/ Direction/Halfwidth/Line/Radius/Second pt/Undo/ Width]: (EnterKey)

Command:

By entering an angle for the chord (the line from endpoints of the arc), the arc will be drawn and the prompt will be redisplayed for continuing the arc command. Then, a new arc will be rubberbanded on the screen. Press (EnterKey) to stop the command or enter the letter representing the desired option to continue.

Arc Direction

The Direction option allows for control of the bearing of the ARC chord. The default for PLINE is to draw the arc tangent to the preceding segment. The Direction option will allow you to override this action and specify a starting direction for the arc that is not tangent to the previous line. The process can be seen in Figure 12.8. The sequence is:

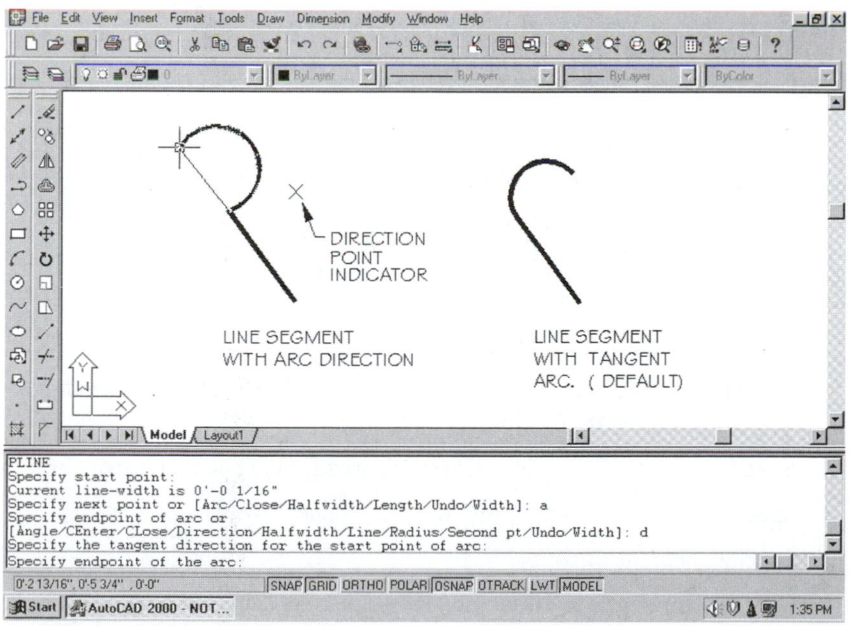

Figure 12.8 *The Direction option of Arc (D) allows an arc to be drawn that is not tangent to an existing polyline.*

Command: **PL** (EnterKey)

PLINE

Specify start point: *(Select starting point.)*

Current width is 0'-0 $^{1}/_{16}$"

Specify next point or [Arc/Close/Halfwidth/Length/Undo/Width]: **A** (EnterKey)

Specify endpoint of arc or

[Angle/CEnter/CLose/Direction/Halfwidth/Line/Radius/Second pt/Undo/Width]: **D** (EnterKey)

Specify the tangent direction for the start point of arc: *(Select a point to indicate the desired arc direction.)*

Specify endpoint of arc: *(Specify desired endpoint)*

Specify endpoint of arc or

[Angle/CEnter/CLose/Direction/Halfwidth/Line/Radius/Second pt/Undo/Width]: (EnterKey)

Command:

Three-Point Arc

An arc may be drawn by entering a second point between each endpoint. This allows for better placement of the arc. The sequence can be seen in Figure 12.9. Once in the PLINE Arc option, the command sequence is:

Command: **PL** (EnterKey)

PLINE

Specify start point: *(Select starting point.)*

Current width is 0'-0 $^{1}/_{16}$"

Specify next point or [Arc/Close/Halfwidth/Length/Undo/Width]: **A** (EnterKey)

Specify endpoint of arc or

[Angle/CEnter/CLose/Direction/Halfwidth/Line/Radius/Second pt/Undo/Width]: **S** (EnterKey)

Specify second point on arc: *(Specify desired second point for arc to pass through.)*

Specify endpoint of arc: *(Specify desired endpoint of arc.)*

Specify endpoint of arc or

[Angle/CEnter/CLose/Direction/Halfwidth/Line/Radius/Second pt/Undo/Width]: (EnterKey)

Command:

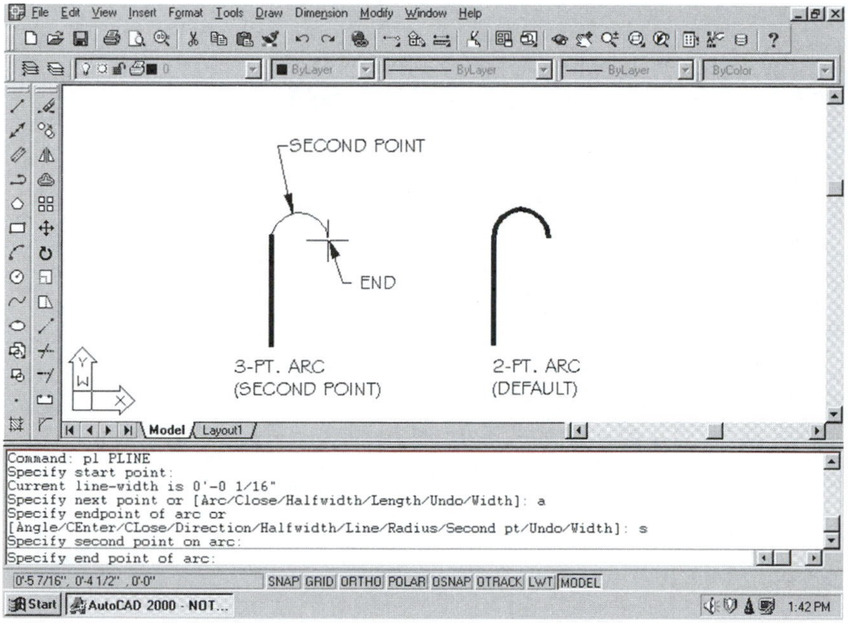

Figure 12.9 *Using the Second point option of Arc (S) draws an arc based on three points.*

CLOSING A POLYGON

As you draw with the LINE command, polygons are ended by typing **C** (EnterKey). This provides a line segment from the present location, back to the original starting point. A similar option is available with the PLINE Arc option. By entering **CL** (EnterKey) at the prompt, a series of polylines will be closed by an arc segment. Remember that CL must be entered when in the arc submenu rather than C to distinguish between CLose and CEnter. The results can be seen in Figure 12.10. The final sequence to close a polyline is:

Specify end point of arc or

[Angle/CEnter/CLose/Direction/Halfwidth/Line/Radius/Second pt/Undo/Width]: **CL** (EnterKey)

Command:

UNDOING A PLINE

As with other drawing options, PLINE allows for removing the previous segment or segments in the reverse order from the way they were drawn. This can be done with the Undo option by entering **U** (EnterKey) at the prompt.

Specify end point of arc or

[Angle/CEnter/CLose/Direction/Halfwidth/Line/Radius/Second pt/Undo/Width]: **U** (EnterKey)

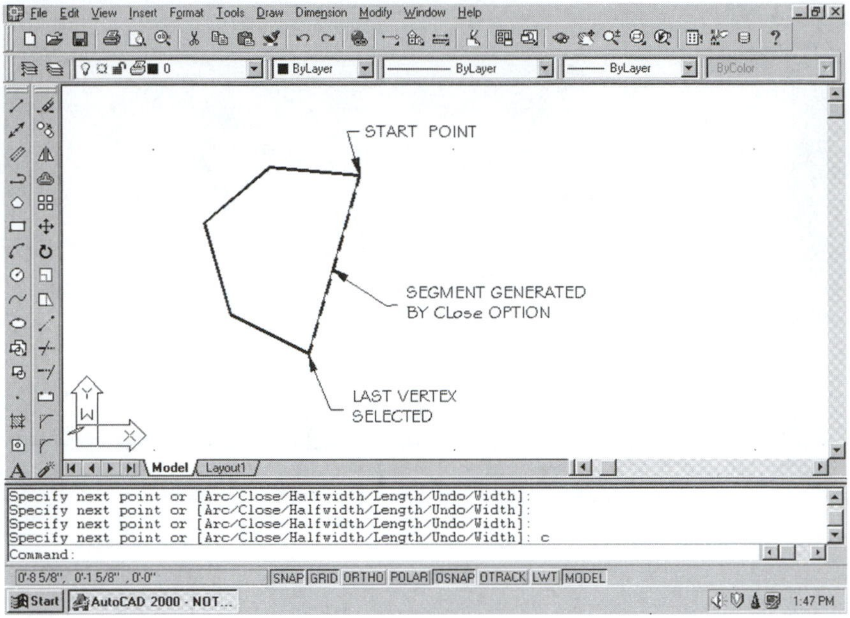

Figure 12.10 *Polygons drawn using polylines can be completed using the Close option (CL).*

This will remove the last polyline that was drawn, as seen in Figure 12.11. Undo may be used repeatedly until just one point is left in the polyline. Separate from the Undo option, the UNDO command can be used once the drawing command has been terminated. The UNDO command removes an entire command sequence and will be covered in future chapters.

EDITING A POLYLINE

Polylines may be edited with the commands introduced in previous chapters. Because polylines are segments drawn as a single object, they are edited as a single object. This can facilitate the selection process, because only one segment must be selected, rather than each of the segments that make up the entire object. This can be seen in Figure 12.12.

Most of the commands in Chapter 11 will modify a polyline in exactly the same way as another segment. FILLET and CHAMFER each have an added feature for a polyline.

ALTERING POLYLINES WITH FILLET

The FILLET command can't be used to fillet two different polylines in the same way you use it to fillet two lines. However, the command can be used on closed polygons

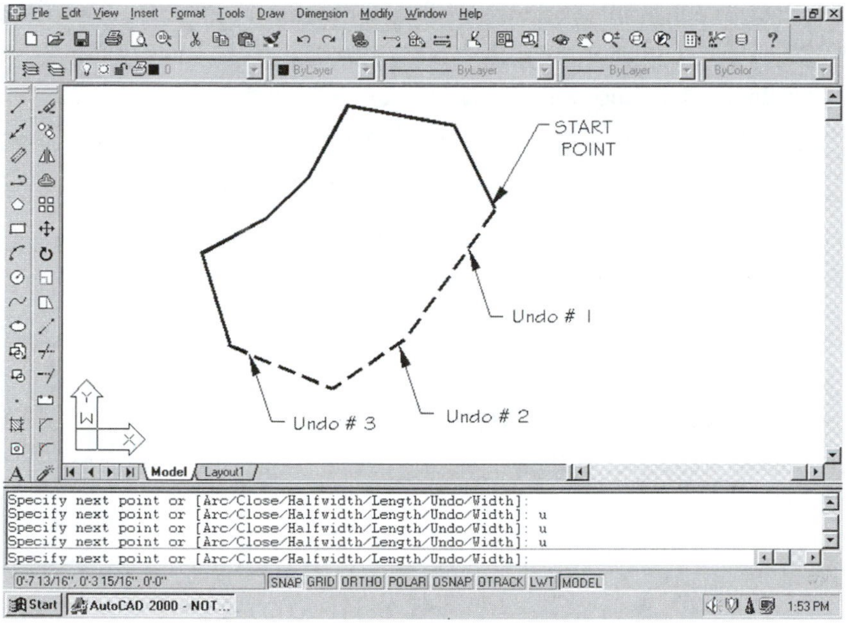

Figure 12.11 *The Undo option (U) will remove the last pline segment drawn.*

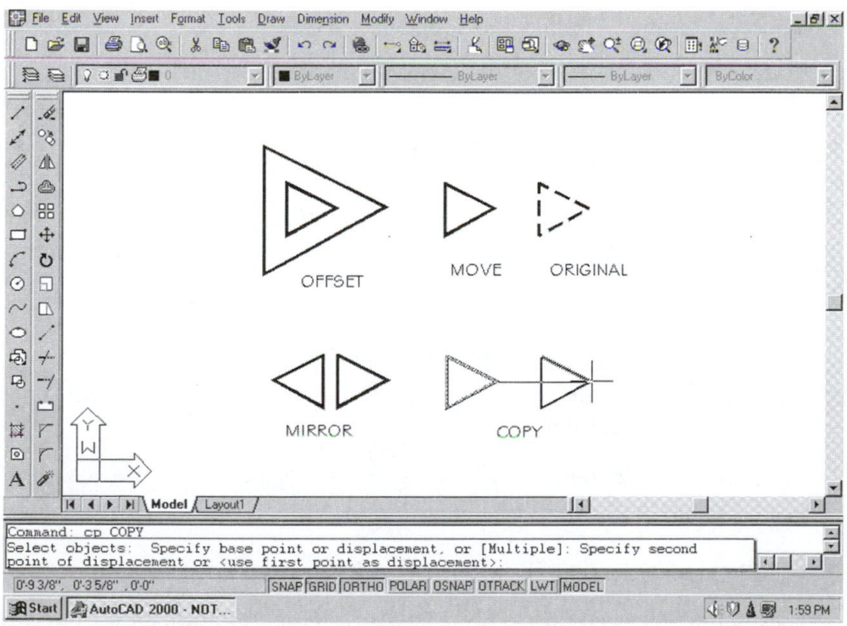

Figure 12.12 *When editing a polyline, remember that the entire object acts as one object.*

constructed of a polyline, by the use of two different methods. The command sequence to FILLET one corner of a closed polygon with a radius of 3/8" is:

> Command: **F** (EnterKey)
>
> FILLET
>
> Current settings: Mode = TRIM, Radius = 0'-0"
>
> Select first object or [Polyline/Radius/Trim]: **R** (EnterKey)
>
> Specify fillet radius <0'-0">: **.375** (EnterKey)
>
> Command: (EnterKey)
>
> FILLET
>
> Current settings: Mode = TRIM, Radius = 0'-0 3/8"
>
> Select first object or [Polyline/Radius/Trim]: *(Select edge of object to receive fillet.)*
>
> Select second object: *(Select second edge to receive fillet.)*
>
> Command:

The fillet can also be applied to all edges of the polygon in one command sequence. Once the radius is set the command sequence is:

> Command: **F** (EnterKey)
>
> FILLET
>
> Current settings: Mode = TRIM, Radius = 0'-0 3/8"
>
> Select first object or [Polyline/Radius/Trim]: **P** (EnterKey)
>
> Select 2D polyline: *(Select object to receive fillets.)*
>
> 4 lines filleted *(Quantity will vary with each object.)*
>
> Command:

The effects of the FILLET command on polylines can be seen in Figure 12.13.

ALTERING POLYLINES WITH CHAMFER

Editing polylines with CHAMFER is similar to using the FILLET command. The editing options for CHAMFER are:

> Command: **CHA** (EnterKey)
>
> CHAMFER
>
> (TRIM mode) Current chamfer Dist1 = 0'-0 3/8", Dist2 = 0'-0 3/8"
>
> Select first line or [Polyline/Distance/Angle/Trim/Method]: *(Select object to receive chamfer.)*

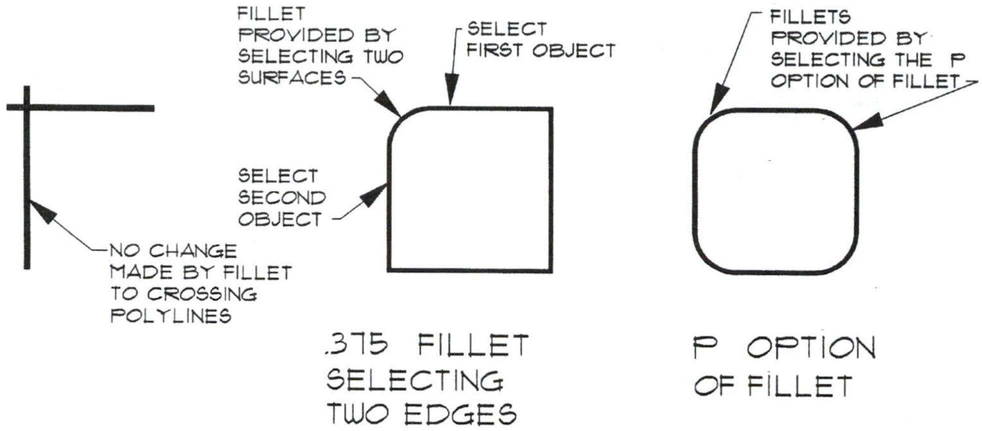

Figure 12.13 *Two separate polylines are not affected by the FILLET command. The FILLET command can be used to modify polygons constructed of polylines or the Polyline option of PLINE can be used to fillet all edges of a polygon.*

Select second line: *(Select second edge to be chamfered.)*

Command:

Once the distances are set, the command will edit a polyline as two line segments, if the default option is selected. If the Polyline option is selected, all corners of the selected polyline will receive chamfers. The command sequence is:

Command: **CHA** (EnterKey)

CHAMFER

(TRIM mode) Current chamfer Dist1 = 0'-0 $\frac{3}{8}$", Dist2 = 0'-0 $\frac{3}{8}$"

Select first line or [Polyline/Distance/Angle/Trim/Method]: **P** (EnterKey)

Select 2D polyline: *(Select object to receive chamfer.)*

4 lines were chamfered

Command:

The effect of chamfering a polygon can be seen in Figure 12.14.

EXPLODING POLYLINES

One of the benefits of a polyline is also one of its drawbacks. Whether you've drawn two or 2,000 connected polyline segments, they all function as one object. That's great if you want to move 2,000 lines at once by selecting only one option, but a hindrance if you want to change just one of the segments. To overcome this obstacle, use

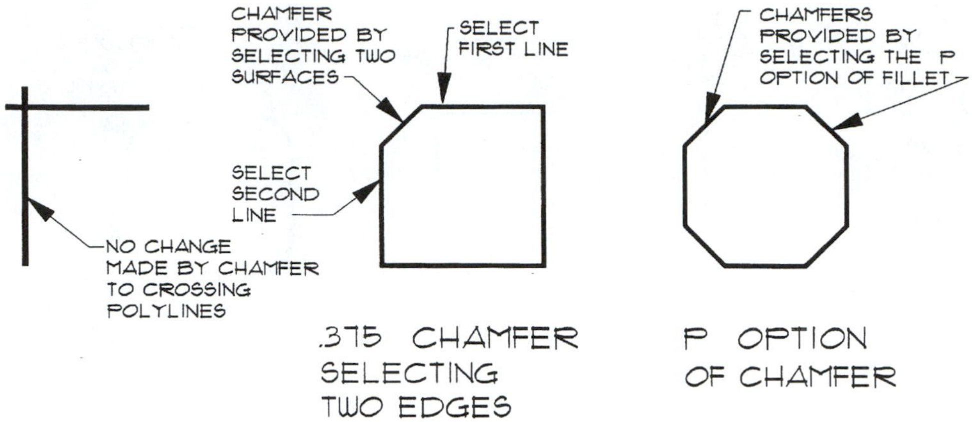

Figure 12.14 *Two separate polylines are not affected by the CHAMFER command. The CHAMFER command can be used to modify polygons constructed of polylines, or the Polyline option of PLINE can be used to chamfer all edges of a polygon.*

the EXPLODE command. This is not the long-awaited command you hope for when you're really frustrated with a computer. It only explodes groups of drawing objects, and doesn't even make a loud noise. The command can be accessed by selecting the Explode icon (the stick of dynamite) from the Modify toolbar, by entering X (EnterKey) at the command prompt, or by selecting Explode from the Modify pull-down menu. The command sequence is:

Command: **X** (EnterKey)

EXPLODE

Select objects: *(Select objects to be edited.)*

1 found

Select object: (EnterKey)

Once (EnterKey) has been pressed, the polyline will be transformed into individual line or arc segments. If the polyline had a specified width, the following message will be displayed at the command line:

Exploding this polyline has lost width information.

The UNDO command will restore it.

Command:

The effects of EXPLODE on a polyline can be seen in Figure 12.15. If you can tolerate the loss of line width, proceed. If the line width is important to the drawing, typing UNDO at the command prompt will restore the polyline to its unedited form. If you want the polyline to function as separate lines but retain their width,

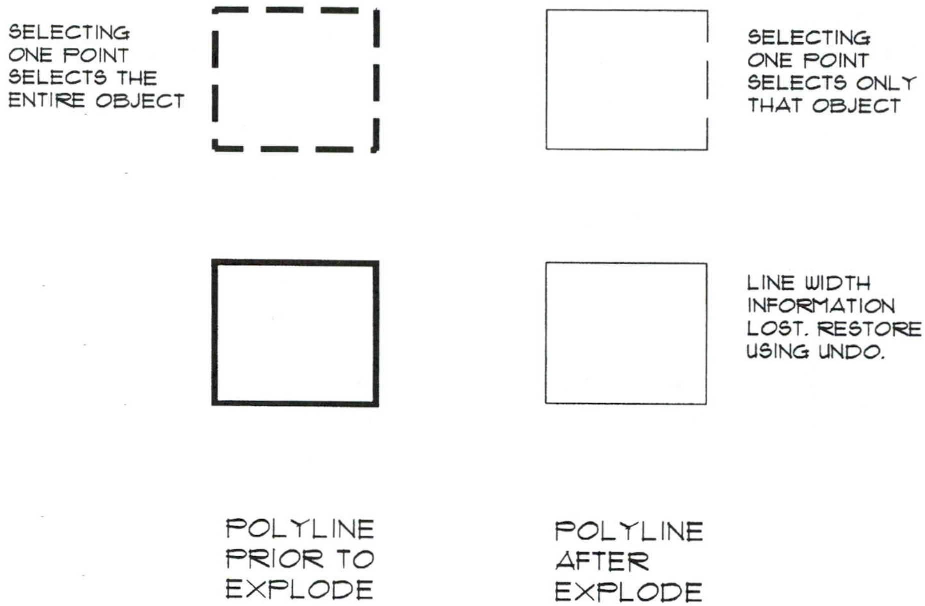

Figure 12.15 *The EXPLODE command allows segments of a polyline to be treated as individual objects.*

EXPLODE the polyline and assign width using the LINEWEIGHT and the CHANGE command. See the HELP menu or future chapters on editing.

ALTERING POLYLINES WITH PEDIT

In addition to using the standard editing options, you can edit polylines using the PEDIT command. The command can be accessed by selecting the Edit Polyline icon from the Modify II toolbar, by entering **PE** (EnterKey) at the command prompt, or by selecting Polyline from the Modify pull-down menu. Options for the command will vary slightly depending on whether the polyline forms a closed shape or a line. Options for the command, when a closed polyline is selected, are:

Command: **PE** (EnterKey)

PEDIT

Select polyline: *(Select polyline to be edited.)*

Enter an option [Open/Join/Width/Edit vertex/Fit/Spline/Decurve/Ltype gen/Undo]:

If an open polyline is selected, the Open option is replaced with the Close option. Entering (EnterKey) will exit back to the command line. Selecting objects may be done by any of the selection methods discussed in previous chapters. Notice, when you are prompted to select objects, the cursor is exchanged for a selection box to allow you to select polylines individually. The Window or Crossing methods also are excellent for selecting polylines.

Occasionally, segments might be selected for PEDIT that are not polylines. When this happens, you will be given the following prompt:

Object selected is not a polyline.

Do you want to turn it into one? <Y>

If the default is accepted, the segment will become a polyline and the PEDIT prompt will be displayed, allowing options of the new polyline to be altered. A CIRCLE can not be changed to a polyline but it can be drawn using the 360° arc of PLINE Arc option. A DONUT also could be used to draw a circle that would have similar qualities to a polyline. The editing options for a polyline include Open, Join, Width, Edit vertex, Fit curve, Spline curve, Decurve, Ltype gen, Undo, and eXit.

OPEN

The Open (O) option will remove the closing segment of a polyline. If a line is drawn back to the starting point without using the Close option, opening the polyline has no visible effect. The results can be seen in Figure 12.16. If O is entered for an open polyline, it will be closed.

CLOSE

The Close (CL) option creates a closing segment of the polyline, while in the PLINE command routine. Once you've moved on to another command sequence, the polyline must be edited with **C** (Close) option. If you select C on a closed polyline, it will be opened. Figure 12.16 shows the effects of the close command.

JOIN

This option adds Lines, Arcs, and other polylines that meet a selected polyline by typing **J** (EnterKey) at the command prompt. Its option can be useful for joining two or more individual polylines so that they function as one polyline. This option may only be used with open polylines and cannot be used to join segments that do not touch the selected polyline. Objects that cross the polyline will not be joined.

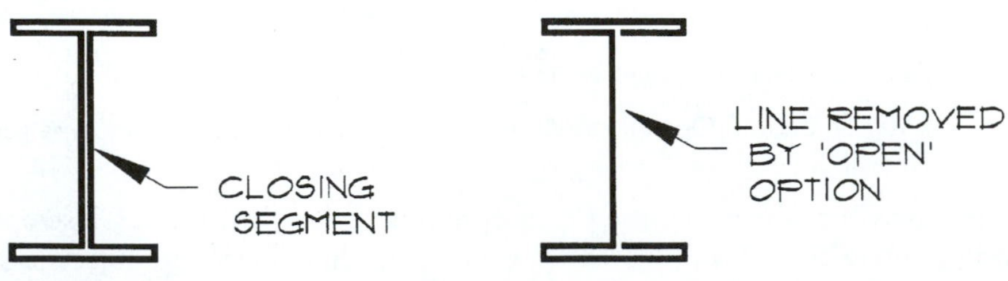

Figure 12.16 *A pline removed with the Open option.*

WIDTH

The W (Width) option will allow the current width of a polyline to be changed. Polylines that consist of segments with varying widths or tapers will be changed so that all segments are the new width. When (EnterKey) is entered at the option prompt, a new prompt will be given requesting the following:

Specify new width for all segments:

The width may be entered by keyboard or by selecting two points with the mouse. Once the width is altered, the PEDIT prompt will be displayed again, allowing for other editing to take place. Press (EnterKey) to return to the command prompt. Figure 12.17 shows the effects of the Width option.

EDIT VERTEX

Once a polyline has been formed, the shape may need to be altered. Altering the shape can be done by entering E (EnterKey) at the option prompt. An X will be placed in the first vertex and a new prompt will be displayed with the following options:

Enter an option [Open/Join/Width/Edit vertex/Fit/Spline/Decurve/Ltype gen/Undo]: **E** (EnterKey)

Enter a vertex editing option

[Next/Previous/Break/Insert/Move/Regen/Straighten/Tangent Width/eXit] <N>:

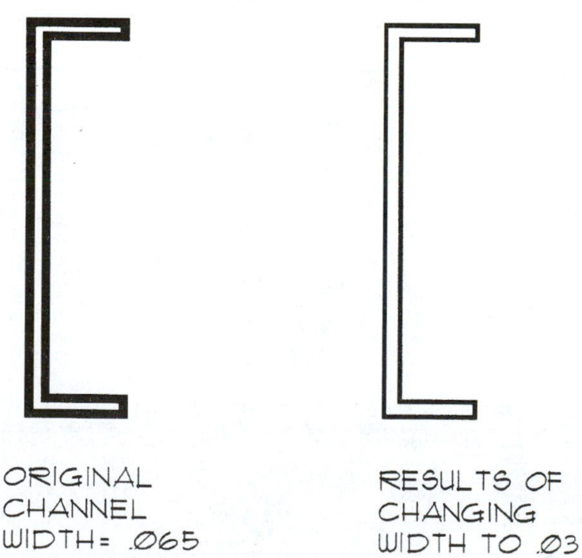

Figure 12.17 *The effects of changing the width of a polyline.*

Unless the X is placed in the location you would like to edit, use the Next or Previous options to move the X around the Polyline to select the desired editing location.

Next

The next mode (N) moves the X to the next vertex. By pressing (EnterKey), you can move the X around the object, as seen in Figure 12.18.

Previous

This mode (P) moves the X to the previous vertex. By pressing (EnterKey), you can move the X around the object in the opposite direction from the one used by Next. The option can be seen in Figure 12.18.

Break

The Break mode (B) will remove a portion of a polyline. The portion to be broken is selected by the N or P option. Once the X is in the desired position, the prompt reads:

Enter a vertex editing option [Next/Previous/Go/eXit] <N>:

As the X is now moved with N or P, any point between the original starting location and the present location will be removed. When the X is moved to the end of the segment to be used, enter **G** (EnterKey) . This will remove the desired segment. If Break is used on a closed polygon, the closing section of the polygon will be removed along

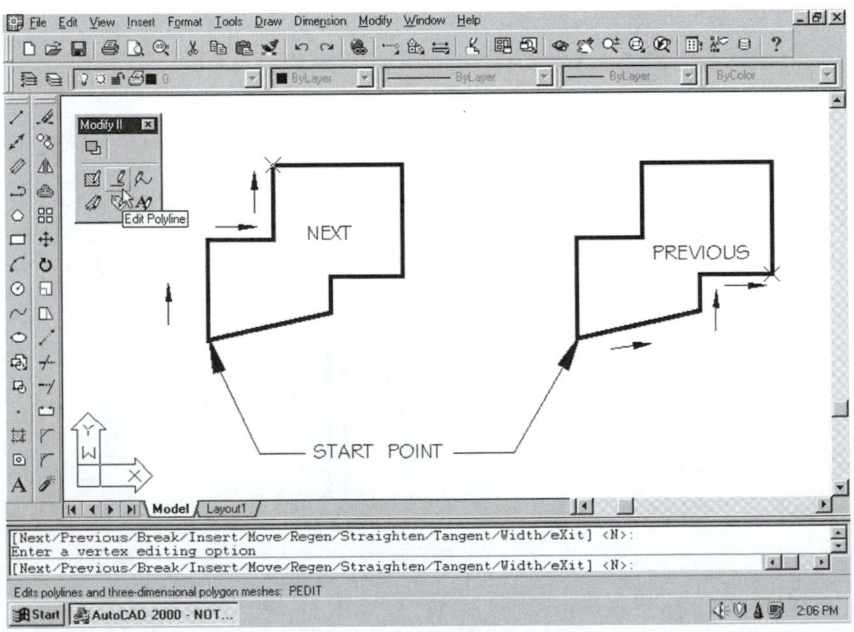

Figure 12.18 *The vertex of a polyline can be edited by moving the X using the Next or Previous option.*

with the selected portion. The process can be seen in Figure 12.19. The command sequence is:

> Enter an option [Open/Join/Width/Edit vertex/Fit/Spline/Decurve/Ltype gen/Undo]: **E** (EnterKey)
>
> Enter a vertex editing option
>
> [Next/Previous/Break/Insert/Move/Regen/Straighten/Tangent Width/eXit] <N>: **B** (EnterKey)
>
> Enter an option [Next/Previous/Go/eXit] <N>: (EnterKey)
>
> Enter an option [Next/Previous/Go/eXit] <N>: **G** (EnterKey)

Insert

Insert (I) will edit an existing polyline by adding a new vertex. The new vertex is added after the vertex marked by the X. The process can be seen in Figure 12.20. The command sequence is:

> Enter an option [Open/Join/Width/Edit vertex/Fit/Spline/Decurve/Ltype gen/Undo]: **E** (EnterKey)
>
> Enter a vertex editing option
>
> [Next/Previous/Break/Insert/Move/Regen/Straighten/Tangent/Width/eXit] <N>: (EnterKey) *(Move X to the desired location.)*

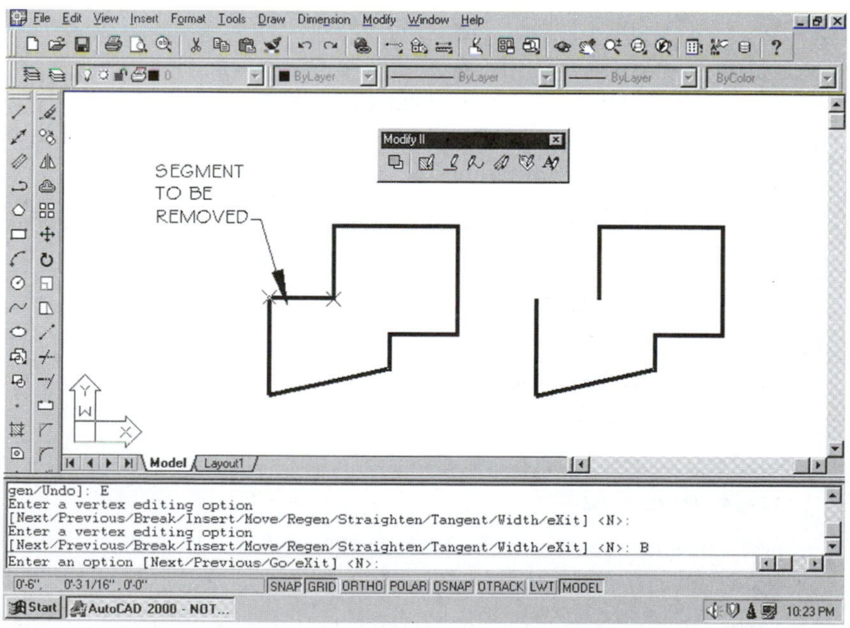

Figure 12.19 *Removing a portion of a polyline using the Break option.*

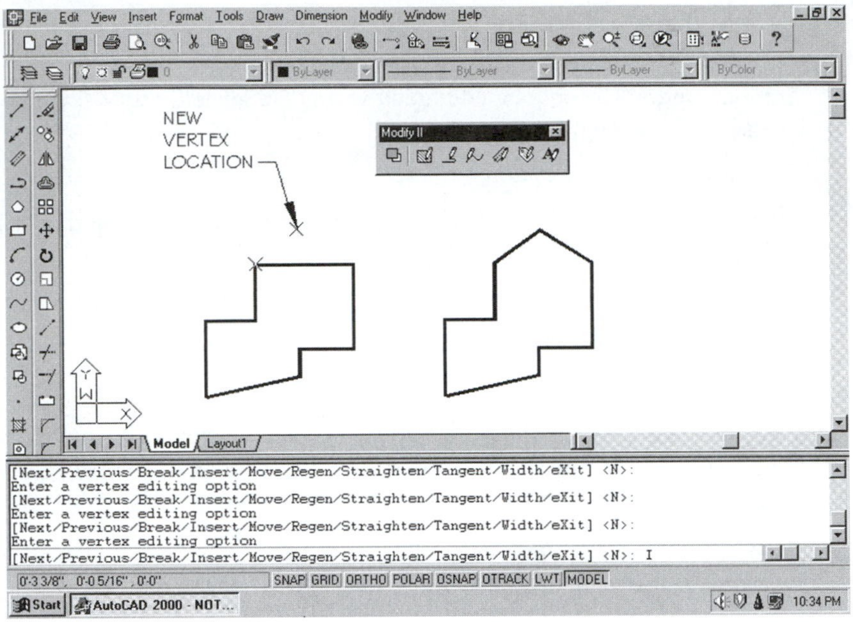

Figure 12.20 *Altering the shape of a polyline by inserting a new vertex.*

> [Next/Previous/Break/Insert/Move/Regen/Straighten/Tangent/Width/eXit] <N>: **I** (EnterKey)
>
> Specify location of new vertex: *(Select desired location.)*

Move

The Move mode (M) allows for moving an existing polyline vertex to a new location. The process can be seen in Figure 12.21. The vertex to be moved must be the one currently marked by the X. The command sequence is:

> Enter an option [Open/Join/Width/Edit vertex/Fit/Spline/Decurve/Ltype/Undo/eXit]: **E** (EnterKey)
>
> Enter a vertex editing option
>
> [Next/Previous/Break/Insert/Move/Regen/Straighten/Tangent/Width/eXit]<N>: (EnterKey) *(Move X to the desired location.)*
>
> Enter a vertex editing option
>
> [Next/Previous/Break/Insert/Move/Regen/Straighten/Tangent/Width/eXit] <N>: **M** (EnterKey)
>
> Specify new location for marked vertex: *(Select desired location.)*

Regen

Regen mode (R) will regenerate the edited polyline.

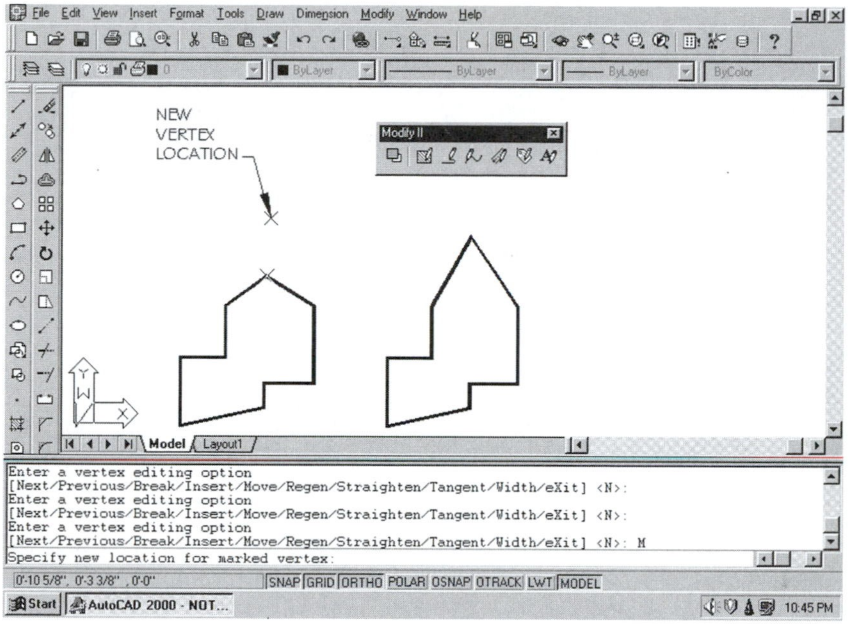

Figure 12.21 *Altering the shape of a polyline by moving an existing vertex to a new location.*

Straighten

This mode (S) will straighten existing polyline segments that lie between two selected points. Move the X to the desired location to mark the start of the edit. The X can then be moved to mark the end of the edit. Any arcs or segments between the two marks will be deleted and replaced by a straight segment. The process can be seen in Figure 12.22. The command sequence is:

Enter an option [Open/Join/Width/Edit vertex/Fit/Spline/Decurve/Ltype/Undo/eXit]: **E** (EnterKey)

Enter a vertex editing option

[Next/Previous/Break/Insert/Move/Regen/Straighten/Tangent/Width/eXit]<N>: (EnterKey) *(Move X to the desired location.)*

Enter a vertex editing option

[Next/Previous/Break/Insert/Move/Regen/Straighten/Tangent/Width/eXit] <N>: **S** (EnterKey)

Enter an option [Next/Previous/Go/eXit] <N>: (EnterKey)

Enter an option [Next/Previous/Go/eXit] <N>: (EnterKey)

Enter an option [Next/Previous/Go/eXit] <G>: (EnterKey)

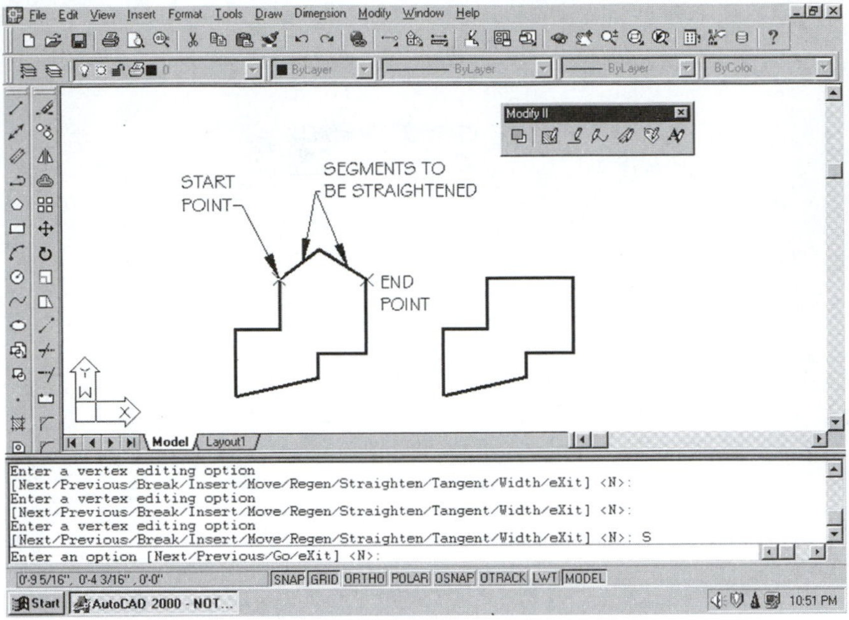

Figure 12.22 *Removing an existing polyline vertex using the Straighten option of PEDIT.*

Tangent

The tangent mode (T) will allow you to attach a tangent direction to the current vertex for later use in Curve fitting. The command sequence is:

> Enter an option [Open/Join/Width/Edit vertex/Fit/Spline/Decurve/Ltype/Undo/eXit]: **E** (EnterKey)
>
> Enter a vertex editing option
>
> [Next/Previous/Break/Insert/Move/Regen/Straighten/Tangent/Width/eXit]<N>: (EnterKey) *(Move X to the desired location.)*
>
> Enter a vertex editing option
>
> [Next/Previous/Break/Insert/Move/Regen/Straighten/Tangent/Width/eXit] <N>: **T** (EnterKey)
>
> Specify direction of vertex tangent: *(Specify a point or enter an angle.)*
>
> Enter a vertex editing option

Either a specific tangent angle may be specified from the keyboard or a point may be selected to mark the direction from the currently marked (X) vertex.

Width

The Width mode (W) changes the starting and ending width of the segment following the marked vertex. You've explored the Width option of PEDIT, which changes the entire polyline. The Width mode of the Edit Vertex options will allow only one segment to be edited. The process can be seen in Figure 12.23. The command sequence is:

> Enter an option [Open/Join/Width/Edit vertex/Fit/Spline/Decurve/Ltype/Undo/eXit]: **E** (EnterKey)
>
> Enter a vertex editing option
>
> [Next/Previous/Break/Insert/Move/Regen/Straighten/Tangent/Width/eXit]<N>: (EnterKey) *(Move X to the desired location.)*
>
> Enter a vertex editing option
>
> [Next/Previous/Break/Insert/Move/Regen/Straighten/Tangent/Width/eXit] <N>: **W** (EnterKey)
>
> Specify starting width for next segment<0'–0 ⅛"> **0** (EnterKey) *(Enter desired width.)*
>
> Specify ending for next segment width <0'-0">: **.375** (EnterKey) *(Enter desired width.)*

Exit

The Exit mode (X) exits from Vertex editing and returns you to the PEDIT prompt.

FIT CURVE

The Fit curve option (F) of PEDIT allows straight line segments to be converted into curved lines. This can be especially helpful on a topography where lines typically are drawn from elevation to elevation. Fit curve also can be used to convert zigzag lines

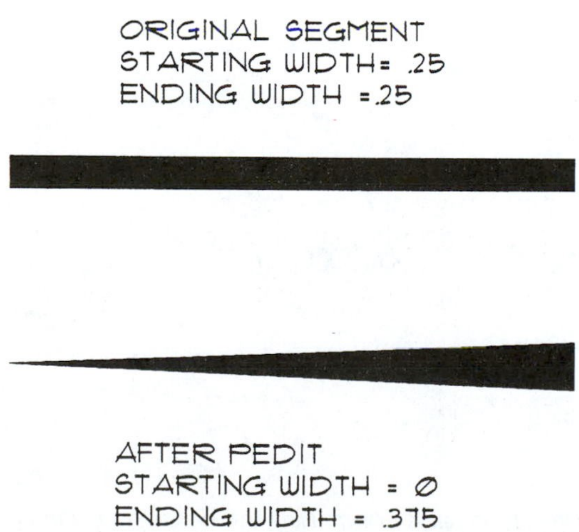

Figure 12.23 *Altering the width of an existing pline using PEDIT W mode.*

to the smooth curves often used to represent insulation in section view, as seen in Figure 12.24. The command process is:

Command: **PE** (EnterKey)

PEDIT

Select polyline:

Enter an option [Close/Join/Width/Edit vertex/Fit/Spline/Decurve/Ltype gen/Undo]: **F** (EnterKey)

As (EnterKey) is entered, Fit curve transforms the points of the original polyline into smoothed curves. The process can be seen in Figure 12.25.

SPLINE CURVE

Unlike the Fit curve that passes through all the vertices, a Spline (S) curve only passes through the first and last points of the polyline. In between, the curve will be close to each vertex, but will not pass through them. The more control points that you specify, the smoother the Spline curve will be. Figure 12.26 shows a Spline curve. The command process is:

Command: **PE** (EnterKey)

PEDIT

Select polyline:

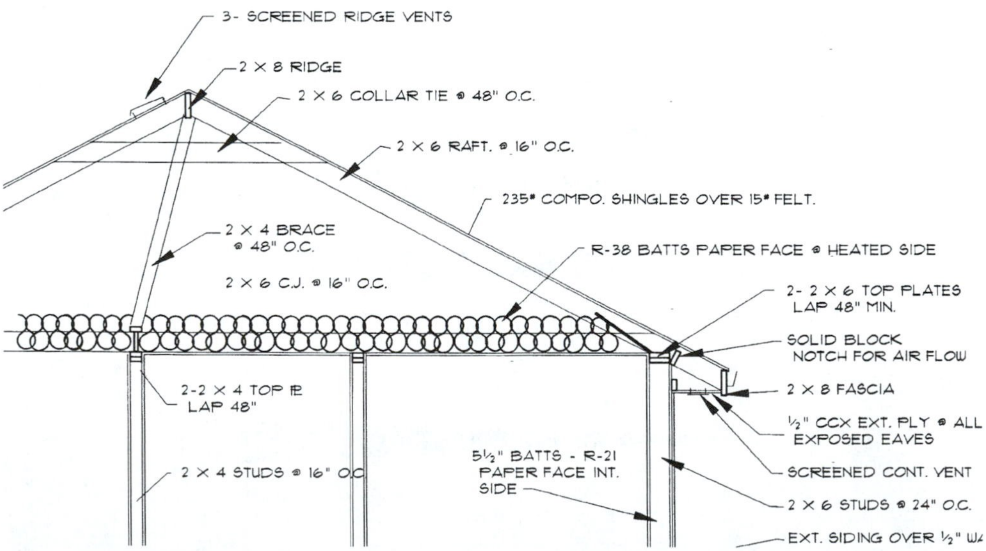

Figure 12.24 *The ceiling insulation is drawn using the Fit Curve option of PEDIT. (Courtesy Residential Designs.)*

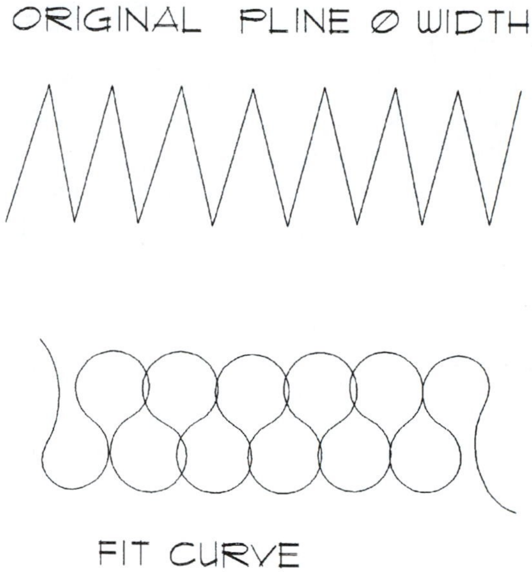

Figure 12.25 *Changing a polyline using the Fit Curve option.*

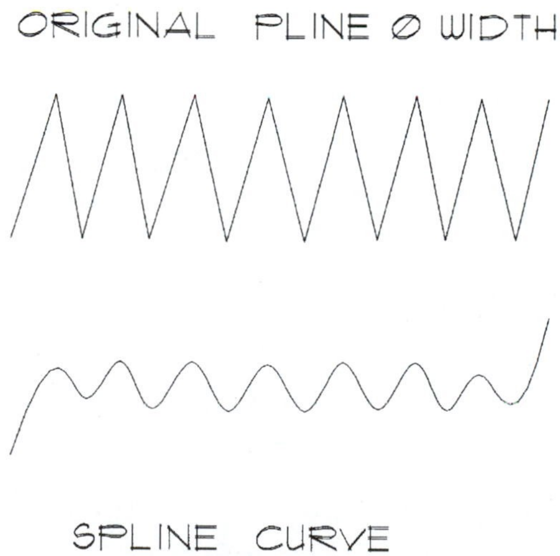

Figure 12.26 *Altering a polyline using the Spline Curve option.*

Enter an option [Close/Join/Width/Edit vertex/Fit/Spline/Decurve/Ltype gen/ Undo]: **S** (EnterKey)

As (EnterKey) is entered, the Spline curve transforms the points of polyline into smoothed curves. The Fit and Spline options are not often used on construction drawings. For further information, remember to use the HELP menu in AutoCAD. You may also want to research the commands SPLINES, SPLINETYPE, SPLINESEGS, SPLINEDIT, and NURBS for related types of drawing objects.

DECURVE

The Decurve (D) option can be used to remove any vertices that have been inserted by Fit curve or Spline curve and to straighten segments of a polyline. The effects can be seen in Figure 12.27. The command sequence is:

Command: **PE** (EnterKey)

PEDIT

Select polyline:

Enter an option [Close/Join/Width/Edit vertex/Fit/Spline/Decurve/Ltype gen/ Undo]: **D** (EnterKey)

LINETYPES

Figure 12.28 shows a polyline drawn with a centerline. Notice that the pattern extends from vertex to vertex. Each segment is, in effect, a separate centerline. The

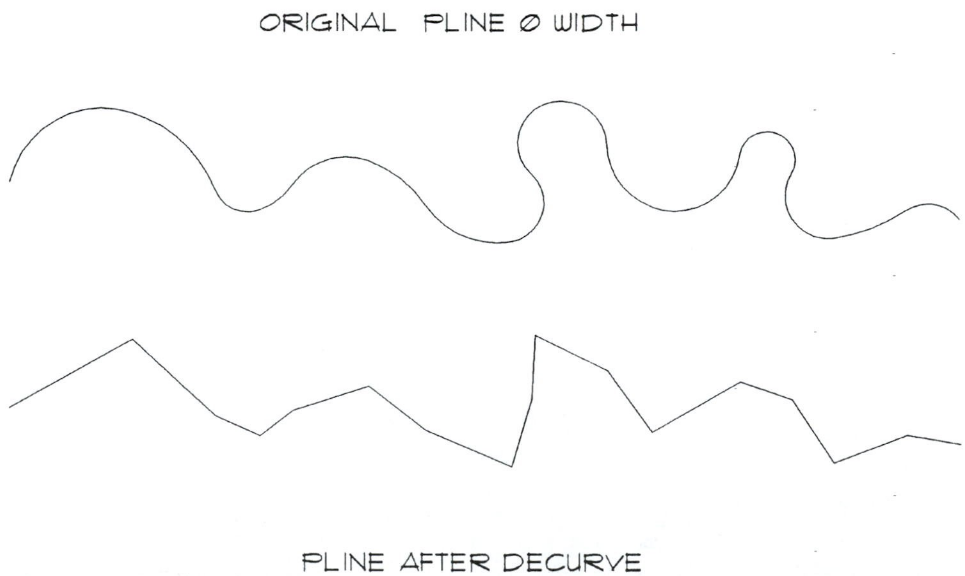

Figure 12.27 *Decurve will remove any vertices that were inserted with Fit or Spline Curve.*

polyline also can be drawn so that the pattern extends throughout the entire polyline, as seen in Figure 12.29. Ltype gen (L) mode of the PEDIT menu will adjust how the pattern is displayed. Toggled to the ON position, Ltype gen will draw line patterns in a continuous pattern from beginning to end with no consideration to vertices. In the OFF position, line patterns will be based on each vertex. Ltype gen does not affect polylines with tapered segments.

The line pattern also can be adjusted by the PLINEGEN system variable. When set to 1, PLINEGEN draws a continuous pattern. With the variable set to 0, the line pattern is based on each vertex.

UNDO

The Undo (U) option of PEDIT will undo the most recent PEDIT edit operation. Multiple entries of U will step back through the drawing, removing previous PEDIT entries.

EXIT

The eXit (X) option is the default for PEDIT. Pressing (EnterKey) will end the PEDIT command and return to the command prompt.

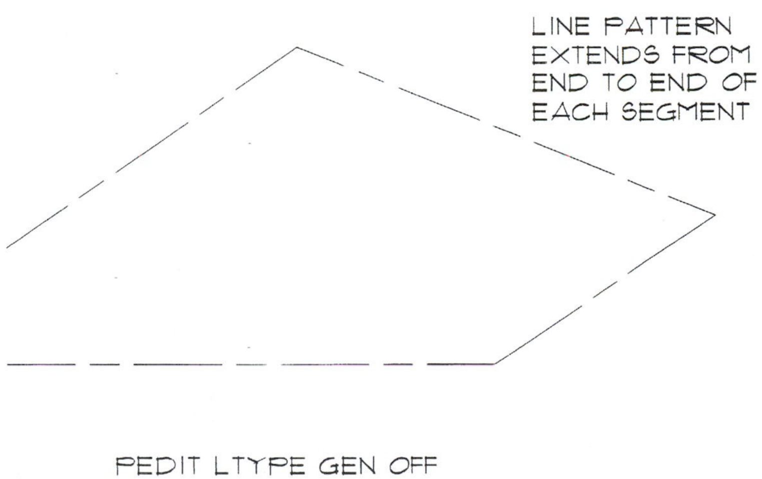

Figure 12.28 *Using the OFF setting of the Linetype option of PEDIT will draw a linetype pattern from vertex to vertex.*

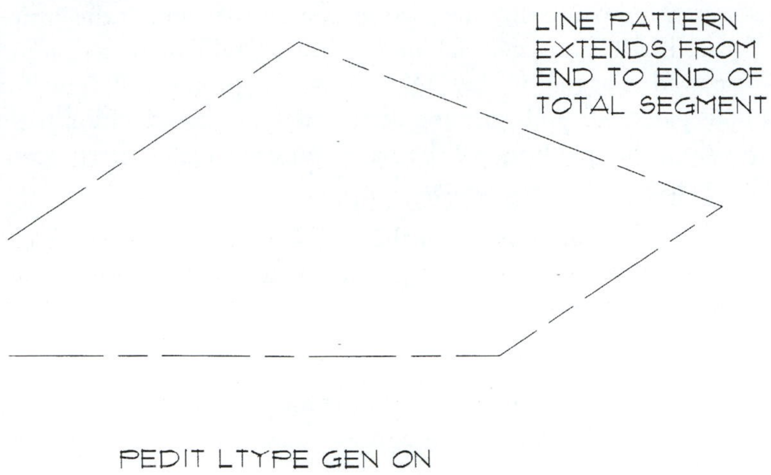

Figure 12.29 *Using the ON setting of the Linetype option of PEDIT will draw a linetype pattern from end to end of the entire line, disregarding the vertex.*

CHAPTER 12 EXERCISES

1. Open a new drawing and draw a polyline polygon consisting of five sides with a 0 width. Set the width of PLINE as 1/16" wide and draw a polygon with a minimum of four sides. Use the close option to close the polygon. Change the width to .125" and draw three open polylines. Save the drawing as E-12-1.
2. Open a new drawing and draw a polyline with a beginning width of .125", a length of 1.5", and an ending width of .25". Continue from this segment with a line that is 1.5" long and .125" wide. Draw a third segment that will close a polygon with a starting width of .25" and ending with a 0 width. Save the drawing as E-12-2.
3. Open a new drawing and draw a circle with a 3" diameter and a thickness of 1/16". Use the three-point Arc option to draw a circle with a width of 1/8" and a radius of 1.5". Save the drawing as E-12-3.
4. Open a new drawing and use a width of 0 to draw a 3" long line. Draw a .125"-wide arc with a radius of 1.75" and an angle of 60° on the right end of the original line segment. Use the right end of the straight line segment as the center of a 45° arc that ends at the left end of the arc segment. Save the drawing as E-12-4.
5. Open a new drawing and draw a four-sided polygon using polyline with a width of .125". Copy the polygon so that there are a total of five polygons. Edit one of the polygons so that the line width is .25". Edit another polygon so that the width is 0.065". Fillet two corners with a .5" radius. Fillet the remaining two corners with a .25" radius. Edit a third polygon so that the width of all lines is 0.0 wide. Provide a .4" × .25" chamfer at each corner. On the fourth polygon, add a vertex at some point so that the polygon has five segments. Save the drawing as E-12-5.
6. Open Drawing E-11-9 and convert the lines that represent the footing to polylines. Set the width as .5". Set the line that represents the finish grade as .75". Copy the one-story footing and explode the drawing. Save the drawing as E-12-6.
7. Open Drawing E-11-8. Assume the light source to be in the upper left-hand corner to shade the drawing. Use .5 width polylines to create shade on all overhanging materials. On the left side of the horizontal trim, draw a shadow that tapers from 0 up to 1" wide. Save the drawing as E-12-7.
8. Open a new drawing and draw a series of seven zigzag polyline 0.0 width lines at approximately 15° from vertical. Make the line segments 10" long. Make two additional copies. Use Spline curve to edit one set of lines and Fit curve to edit the other set. Save the drawing as E-12-8.

9. Open a drawing E-11-6 and adjust the limits as required to draw a typical wall section that is similar to the drawing below. Show:

   2 × 6 sill with a ½"×10" anchor bolt.

   2 × 8 floor joist w/ 2×8 rim joist.

   ¾" plywood sub-floor

   8'-high studs with a double top plate and a single base plate

   2 × 6 ceiling joist

   2 × 6 rafters @ a 27½° (5/12) pitch

   2 × 8 fascia and solid eave blocking

   Use .25"-wide polylines to represent any materials that would be cut by the cutting plane, such as the plate, sill, fascia, and blocking. Save the drawing as E-12-9.

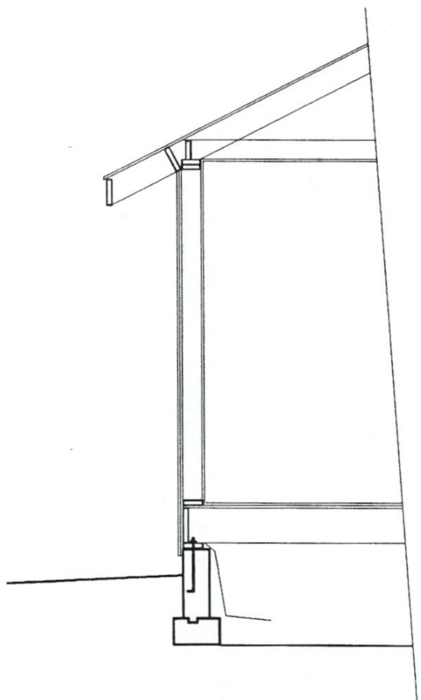

10. Open a new drawing and set the units and limits to draw the grading plan shown below. Lay out the grids and approximate the contour lines. Use a 6"-wide polyline to represent 5' intervals and a 2"-wide polyline to represent one foot intervals. Once all straight line segments have been drawn, edit the plines to provide the most accurate, smooth transitions as possible. Save the drawing as E-12-10.

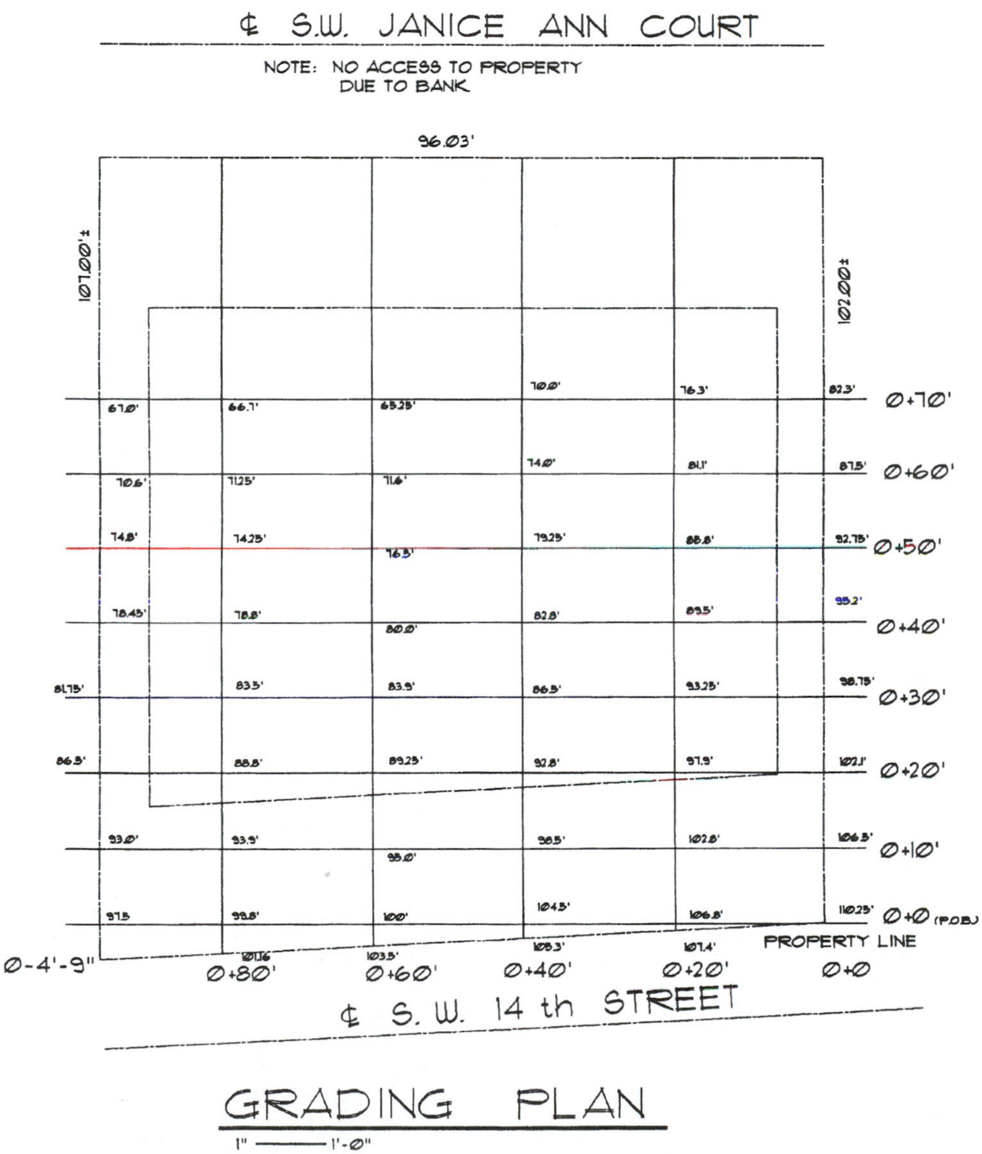

CHAPTER 12 QUIZ

1. Use the Help menu and determine what PLINEGEN controls, and what options are included.
2. Give the command sequence and show all options to draw a polyline with a width of .125".
3. List two options to control line width other than PLINE.
4. Major drawings for a residence include the site, floor, foundation plans, elevations, and sections. Find a set of professional drawings and list common ways polylines can be used to enhance these drawings.
5. What option would allow a .25"-wide line to be drawn by entering .125?
6. Explain the difference between Spline and Fit curve.
7. Explain the difference between Open and Close PEDIT options.
8. How does EXPLODE affect a polyline?
9. To edit a vertex that is formed between the third and fourth lines drawn, what two options will be needed?
10. What option will allow an additional vertex to be added to a completed polyline?
11. A polyline has been drawn, but one vertex is ½" to the left of where it belongs. What option will fix the problem?
12. What PEDIT option will remove any vertices that have been inserted by Fit or Spline curves?
13. What PEDIT option will remove and convert arcs to straight line segments?
14. What option will display an X at the start of a PLINE segment, and what three options affect the X?
15. What is the default setting for PEDIT?
16. How does the Halfwidth option differ from the Width options.
17. A friend needs to draw a 2"-long pline parallel to an existing pline. How can this be done?
18. How can a circle be drawn using the PLINE command?
19. How can a polyline arc be drawn in a clockwise direction?
20. How can the direction of an arc chord be controlled?

CHAPTER 13

Supplemental Drawing Commands

You've explored the methods for drawing and editing lines and polylines. This chapter will introduce methods of marking and dividing drawing space using the POINT, DIVIDE, and MEASURE commands. Five new methods for drawing lines will be introduced.The commands used include RAY, XLINE (Construction Line), and SKETCH. The final method of drawing lines to be presented uses the MLINES command. The command can be used to draw groups of lines of varied quantities and type. Methods for editing lines created with MLINE will also be presented. Three commands will also be presented that can be used to edit any linetype. These commands are the U, UNDO, and REDO commands.

MARKING AND DIVIDING SPACE

Three commands are available in AutoCAD for marking and dividing space. Although these commands can't be used to draw lines, each can be used to mark space in a drawing and provide a location for placing lines.

PLACING POINTS

Often, a point needs to be marked on a drawing. The loads on a column on the upper level of a structure will need to be carried down through several floors into the foundation. AutoCAD will allow you to mark these load locations on a drawing by drawing a point. To place a point in a specific location, select the Point icon from the Draw toolbar, type **PO** (EnterKey) at the command prompt, or select Point from the Draw pull-down menu. If the command is selected from the Draw pull-down menu, the command will allow for multiple points to be located. Each method will produce a prompt:

Command: **PO** (EnterKey)

POINT

Current point modes PDMODE = 0 PDSIZE = –0'–0"

Specify a point: *(Select a point location or enter coordinates.)*

The prompt is now waiting for coordinates to be typed. Points are usually selected by picking a point with the mouse.

CONTROLLING THE POINT DISPLAY

The shape that is used to identify a point can be altered using the Point Style dialog box. The box can be seen in Figure 13.1. The box is displayed by selecting Point Style… from the Format pull-down menu or by typing **DDPTYPE** (EnterKey) at the command prompt. The box allows the style and size of the point to be controlled. The active setting is shown in a black box. In this case, a point is used for the marker. The marker will be displayed on the screen but will not be plotted. Notice the second box from the left in the top row contains no marker. If this option is selected, the point will be marked with an invisible marker. Although the point can't be seen, it will be selected when the Node option of OSNAP is active.

The current point size is listed as 5% of the display screen size. Point size can be altered to meet the needs of the drawing. Just as important as the size is the display method. In the default setting, the size of the point will be displayed relative to the size of the display area. As you zoom in and out of a drawing, points will remain a set percentage of the drawing display size. If you select the display to be in absolute units, points will disappear as you zoom out, and become huge as you zoom in. Entering a

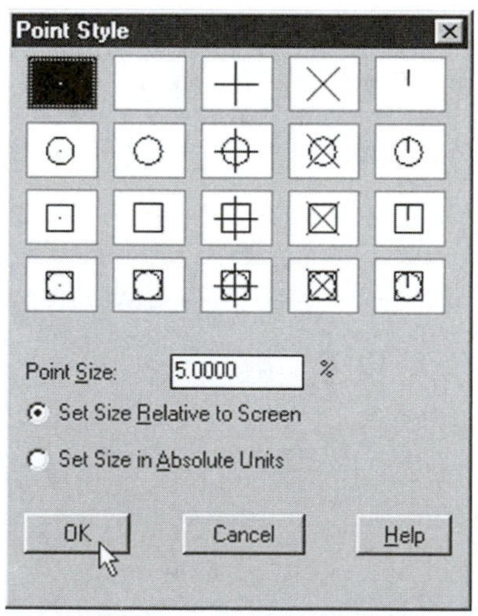

Figure 13.1 *Selecting Point Style from the Format pull-down menu or typing* ***DDPTYPE*** *(Enter-Key) at the command prompt will produce the Point Style dialog box. The box can be used to alter the size and shape of the point marker.*

positive value for the point size will display the point relative to the drawing area. Use the following procedure to display a point that is marked by crosshairs surrounded by a circle and a square, and is displayed at 10% of the screen size:

- Select Point Style... from the Format menu.
- Select the desired marker from the Point Style dialog box.
- Change the display in the Point Size edit box to read 10%.
- Select the OK button.

Note: You'll notice as you work through this chapter that several of the icons noted by commands are not found in the standard toolbars. These commands are available by selecting Customize... from the Toolbars dialog box. The dialog box can be accessed by right-clicking when the cursor is in any open toolbar, and then selecting Customize. Select the Categories edit box in the Customize Toolbars dialog box to display a listing of command categories. To display the POINT icon, select Draw from the Categories edit box. Review Chapter 2 and the Help menu to place icons in a toolbar.

EXPLORING THE DIVIDE COMMAND

Although this command does not edit an object, DIVIDE will allow an object to be divided into any number of segments of equal length for easy editing. This is done by placing markers along a line. This can be especially helpful when you are dividing space between two floors to locate where each stair tread and riser will be placed, or when locating stress points along a beam. DIVIDE will enable you to select points on a Line, Arc, Circle, or Polyline.

Before using the DIVIDE command to mark an object, you must decide how each mark will be indicated. The default marker for a point is a single point. The point style must be altered so that it can be seen on the line. Selecting the crosshairs, the X, or the crosshairs and circle, using the Point Style Dialog box, makes excellent markers (See Figure 13.1). Once the marker is set, the command sequence for DIVIDE can be started by entering **DIV** (EnterKey) at the command prompt, or by selecting Divide from the Point cascading submenu of the Draw pull-down menu. The command sequence is:

Command: **DIV**

DIVIDE

Select object to divide: *(Select the objects.)*

This is your opportunity to select a single object to divide. If you select an object that cannot be divided, an error message will be displayed and the command will be exited. Once an object is selected, another prompt will be displayed.

Enter the number of segments or [Block]: *(Enter number.)* **14** (EnterKey)

Entering a number between 2 and 32,767 and striking (EnterKey) will cause the object to be divided into the desired number of segments. The object is not physically divided into separate segments, the markers are just placed so that exact points can be selected. The process can be seen in Figure 13.2.

Another useful feature of the DIVIDE command is its ability to insert a group of objects called a *block* at a repetitive distance. Blocks will be covered in a later chapter, but the idea of inserting them is not difficult to remember. Blocks can be inserted by dividing a line into the desired number of segments and then inserting the block at each marker. This can be done if the Block option is selected. Figure 13.3 shows the effect of aligning blocks with the object. The command sequence to divide a circle and place a block named SLOT is:

Enter the number of segments or [Block]: **B** (EnterKey)

Enter name of block insert: *(Type name of desired block.)* **SLOT** (EnterKey)

Align block with object? [Yes/No]<Y>: (EnterKey)

Number of segments: *(Enter number.)* **14** (EnterKey)

Command:

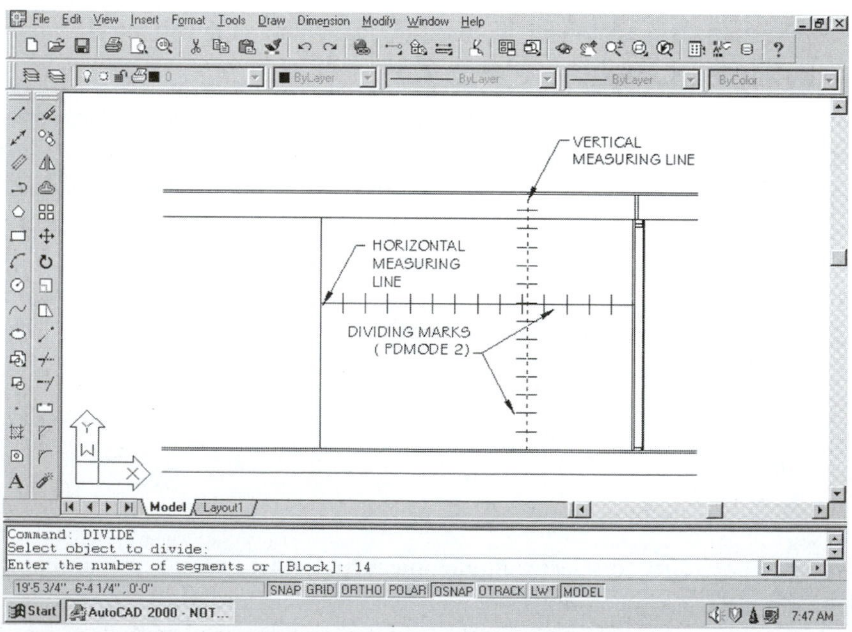

Figure 13.2 *Dividing a horizontal line into fourteen segments.*

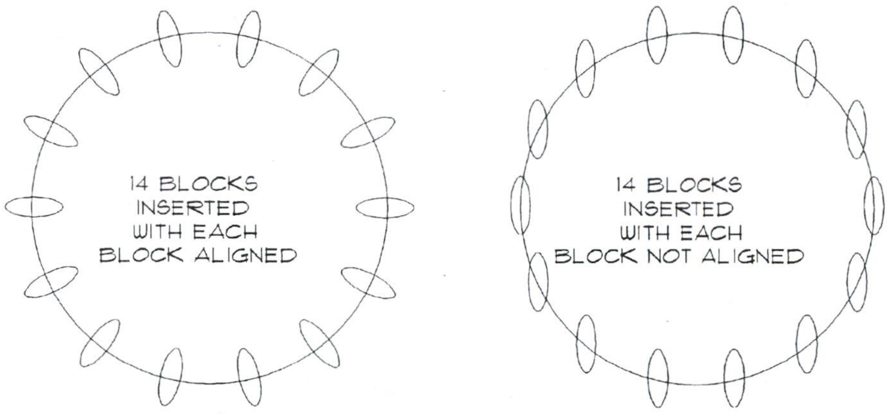

Figure 13.3 *Predetermined drawing components can be inserted into a drawing at each specified DIVIDE marker using the BLOCK option of DIVIDE.*

EXPLORING THE MEASURE COMMAND

The MEASURE command is similar to the DIVIDE command. DIVIDE places a specified number of markers at equal distances along a segment. MEASURE places markers at specific distances along a Line, Arc, Circle, or Polyline. This command can be used whenever objects occur at a repetitive spacing, such as when you are placing studs at 16"o.c. along a plate or placing a manhole at every 527.25 feet along a sewer line. Figure 13.4 shows an example of how MEASURE could be used as a drawing aid. The command can be started by entering **ME** at the command prompt, or by selecting Measure from the Point cascading menu of the Draw pull-down menu. Be sure to change the point marker to something that will be visible when placed on a line. The command sequence to measure an object is:

Command: **ME** (EnterKey)

MEASURE

Select object to measure: *(Select object.)*

This is your opportunity to select a single object to measure. If you select an object that cannot be measured, an error message will be displayed and the command will be exited. Once an object is selected, another prompt will be displayed:

Specify length of segment or [Block]: *(Enter distance.)* **16** (EnterKey)

Command:

Any specific distance may be entered with the keyboard or by specifying two points with the mouse. The markers will be placed, starting at the selected end of the line of the object to be measured. This process can be seen in Figure 13.5.

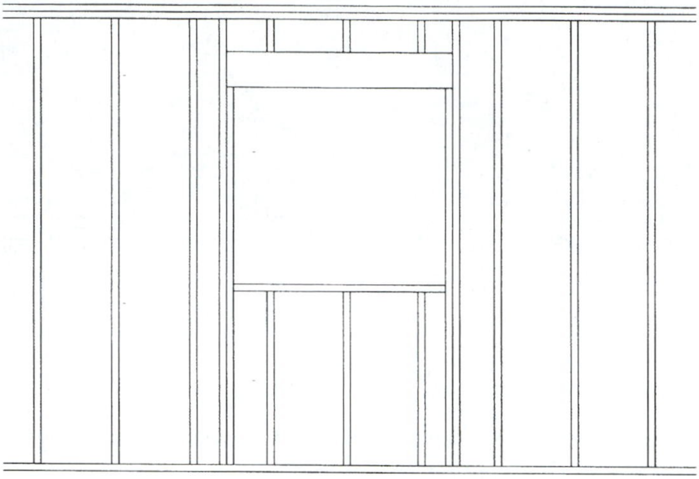

Figure 13.4 *The MEASURE command can be used to place markers at a specific distance on a line, arc, circle, or polyline, so that objects can be drawn.*

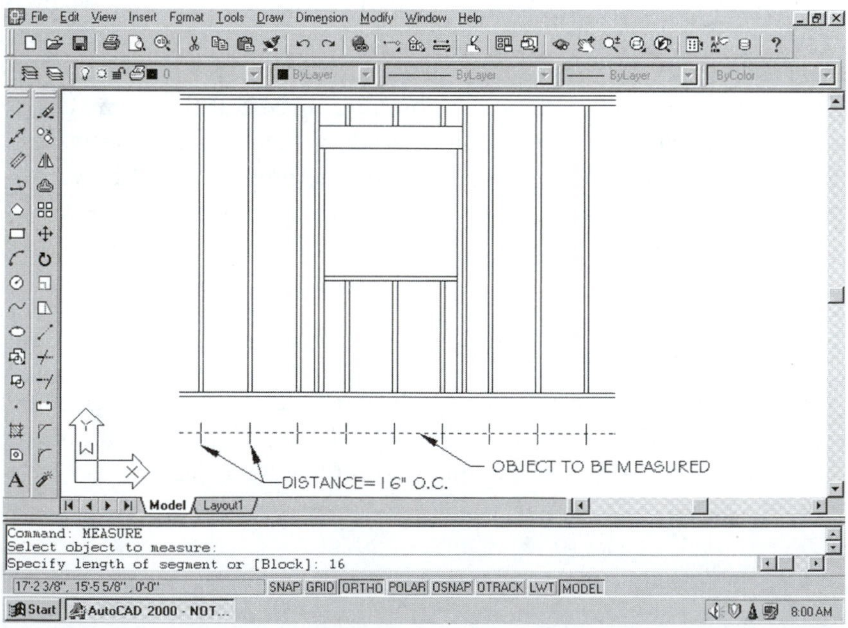

Figure 13.5 *Once a segment has been selected for measurement, the desired distance to be measured can be entered by keyboard or by selecting the distance with the pointing device.*

To use a mouse to enter the distances, move the cursor to the desired location and press the pick button. This will produce the prompt:

Command: **ME** (EnterKey)

MEASURE

Select object to measure: *(Select object.)*

Specify length of segment or [Block]: *(Select a location.)*

Second point: *(Enter a second point or coordinates.)* (EnterKey)

Command:

DRAWING LINES OF VARIED WIDTH AND LENGTHS

Each command introduced in this section can be used as an alternative to the LINE command The RAY and XLINE (Construction Line) commands allow semi-infinite and infinite lines to be added to a drawing. TRACE allows line segments to be drawn with a width similar to plines, but they function individually. The SKETCH command allows freehand sketching to be inserted into a drawing.

CREATING LINES USING THE RAY COMMAND

If you've done manual drafting, you've used light lines to plan your drawing prior to placing dark lines in graphite or ink. Those light lines are called projection lines, and are meant to be seen only by the drafter. AutoCAD allows a similar line to be drawn using the RAY command. A ray is a semi-infinite line with a defined starting point that extends into infinity without changing the drawing area. Sounds technical, but they are really very helpful.

Rays can be especially helpful in projecting features from one drawing to another. This often occurs when lines are projected from a floor, roof, or grading plan to develop exterior elevations. If you were to use a line to project from a floor plan to develop an elevation, it must have two endpoints. A zoom may be required to go from the "First" point to the "Next" point. A transparent zoom ('Z) could be used, but this is time consuming when compared to a ray. A ray will provide a guideline that will extend from the desired point indefinitely, no matter how many ZOOM Ps are required.

The RAY command can be found in the Draw pull-down menu, or executed by typing **RAY** (EnterKey) at the command prompt. The command sequence to draw a ray is:

Command: **RAY** (EnterKey)

RAY

Specify start point: *(Select desired starting point of ray.)*

Specify through point: *(Select desired direction of ray.)*

Specify through point: (EnterKey)

Command:

Figure 13.6 shows the layout of an elevation using the RAY command. A ray will show when the drawing is plotted. Rays should be kept on a separate layer from the drawing being created, so they can be easily removed from the drawing base by freezing. They can also be erased easily using the Crossing Window or Fence option of ERASE.

CREATING LINES USING THE XLINE COMMAND

A line created using the XLINE (Construction Line) command is similar to a ray, except that it is infinite in length. With RAY, you selected a starting point and the line extended from that point. With XLINE, the line extends through the point an infinite distance in two directions and offers more options than with a ray. The command can be accessed by selecting the Construction Line icon from the Draw toolbar, by typing **XL** (EnterKey) at the command prompt, or by selecting Construction

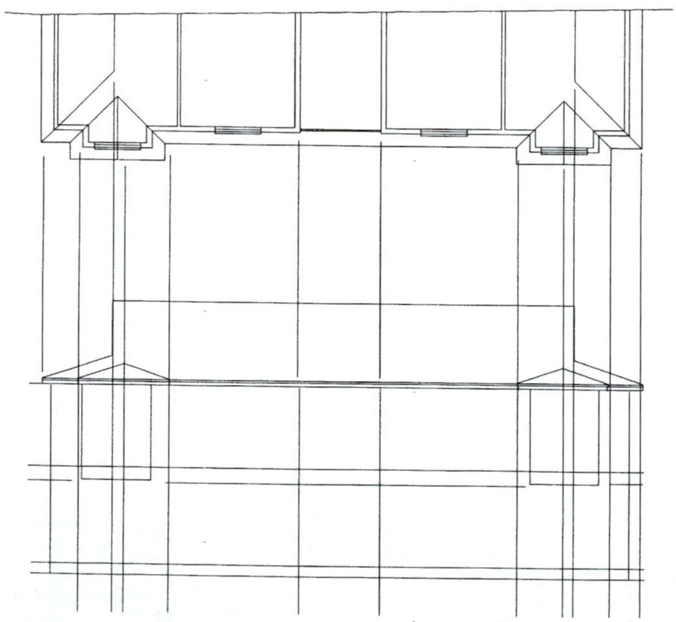

Figure 13.6 *Rays and xlines make excellent projection lines. Because of their length, ZOOMs and STRETCHs can be minimized.*

Line from the Draw pull-down menu. Unlike RAY, six options are available to place the line. The default setting is to select a point for the line to pass through.

Through Point

By selecting a through point, the line will be displayed through the point. Once the point is selected, the line will be rotated to the desired angle through the selected point. Once a line location is selected, the command sequence will allow for additional lines to be drawn through the point. The command sequence is:

Command: **XL** (EnterKey)

XLINE

Specify a point or [Hor/Ver/Ang/Bisect/Offset]: *(Select desired option.)*

Specify through point: *(Select a direction for the line to extend.)*

Specify through point: (EnterKey)

Command:

Hor Option

Choosing the Hor option will allow a horizontal line to be drawn through a specified point. The line will remain horizontal, regardless of the Ortho setting. The command sequence is:

Command: **XL** (EnterKey)

XLINE

Specify a point or [Hor/Ver/Ang/Bisect/Offset]: **H** (EnterKey)

Specify through point: *(Select the desired point for the line to pass through.)*

Specify through point: (EnterKey)

Command:

Ver Option

The Ver option will pass a vertical line through the selected point. The command functions the same as the horizontal option, once the Ver option is selected. The command sequence is:

Command: **XL** (EnterKey)

XLINE

Specify a point or [Hor/Ver/Ang/Bisect/Offset]: **V** (EnterKey)

Specify through point: *(Select the desired point for the line to pass through.)*

Specify through point: (EnterKey)

Command:

Ang Option

The Ang option allows a specific angle for the line to be specified. The selected angle is relative to horizontal. The angle option can be useful in laying out roofs or other inclined surfaces. The command sequence is:

Command: **XL** (EnterKey)

XLINE

Specify a point or [Hor/Ver/Ang/Bisect/Offset]: **A** (EnterKey)

Enter angle of xline (0) or [Reference]: **30** (EnterKey)

Specify through point: *(Select the desired point for the line to pass through.)*

Specify through point: (EnterKey)

Command:

If the Reference option of the command is selected, a line can be drawn at an angle to any line. This option could be used to draw a line 30° from a roof drawn at a 6/12 pitch—even if you can't remember what angle 6/12 represents and you don't feel like using the LIST command. Once the reference angle is selected, a location for the line to pass through needs to be indicated. The command sequence is:

Command: **XL** (EnterKey)

XLINE

Specify a point or [Hor/Ver/Ang/Bisect/Offset]: **A** (EnterKey)

Specify angle of xline (0) or [Reference]: **R** (EnterKey)

Select a line object: *(Select the line to be used as the reference.)*

Enter angle of xline <0> : **30** (EnterKey)

Specify through point: *(Select the desired point for the line to pass through.)*

Specify through point: (EnterKey)

Command:

Bisect Option

The Bisect option can be used to draw a construction line that bisects an existing angle. The line is created by selecting three points. The first point to be selected is the Angle vertex point. The second and third points are located on the lines forming the angle to be bisected. Figure 13.7 shows an xline used to bisect an angle. The command sequence to bisect an angle is:

Command: **XL** (EnterKey)

XLINE

Specify a point or [Hor/Ver/Ang/Bisect/Offset]: **B** (EnterKey)

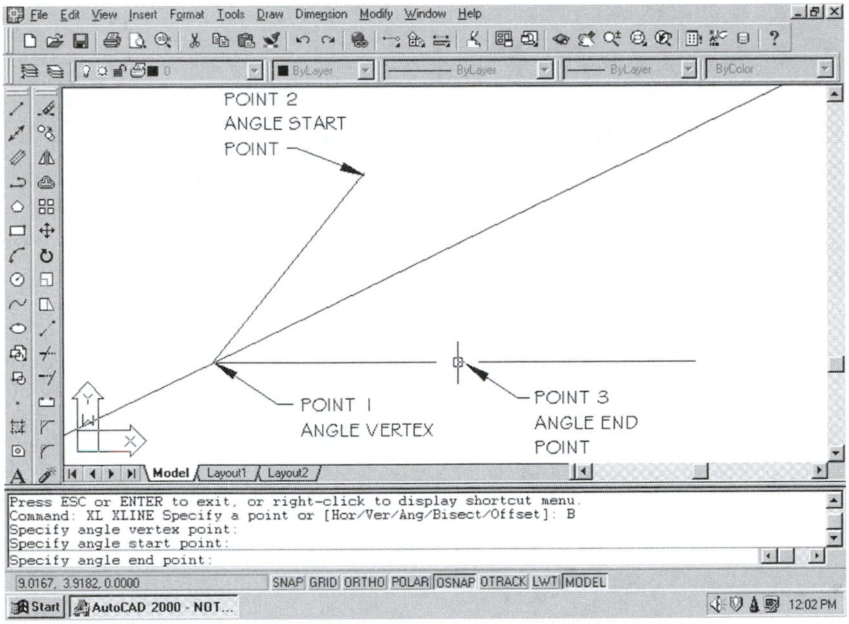

Figure 13.7 *The Bisect option of XLINE can be used to divide an angle in half, by selecting three points.*

Specify angle vertex point: *(Select the point.)* (EnterKey)

Specify angle start point: *(Select the end of one of the lines to be bisected.)*

Specify angle end point: *(Select the end of the other line to be bisected.)*

Specify angle end point: (EnterKey)

Command:

Offset Option

The Offset option draws construction lines that are parallel to the selected line, based on a specified offset distance or point. This option functions like the OFFSET command. You'll be prompted for a line, a distance, and a side to offset. Select an offset distance and then you will be prompted for a line to be offset and for a through point. The command sequence to offset a line 12" is:

Command: **XL** (EnterKey)

XLINE

Specify a point or [Hor/Ver/Ang/Bisect/Offset]: **O** (EnterKey)

Specify offset distance or [Through] <Through>: **12** (EnterKey)

Select a line object: *(Select the line to be offset.)*

Specify side to offset: *(Select the side of the offset.)*

Select a line object: (EnterKey)

Command:

CREATING LINES WITH THE SKETCH COMMAND

Using the SKETCH command allows the equivalent of freehand drawings to be generated using a computer. Line segments are entered as the mouse is moved, rather than providing a "Next" point. The command is useful for entering signatures or other irregular material on a drawing. Because each tiny segment of the sketch is recorded as a line segment, drawing size can become extremely large in a very short time. SKETCH should be kept to a minimum to avoid filling up drawing space on a disk. The size of the drawing is limited only by your storage media.

The SKETCH command is accessed by typing **SKETCH** (EnterKey) at the command prompt. Best results usually occur when ORTHO and SNAP are toggled to the OFF positions allowing greatest flexibility in line placement. With ORTHO mode ON, the SKETCH command will draw only horizontal and vertical lines. With the SNAP mode on, the record increment will equal the snap value.

Segment Length

As the command is entered, you will be prompted to provide information on the *record increment*, which is the segment length to be used while you are sketching. The increment should be set small enough to provide the needed resolution. Exceeding the resolution will only result in wasted disk space by storing lines that aren't necessary. The default line length is <0'-0⅛">. A new value can be entered by keyboard.

The increment that is chosen should be determined by the size of the reproduced drawing. If the drawing is to be reproduced at full scale on an 8½" × 11" sheet of paper, ⅛" will not produce high-quality resolution. As the mouse is moved, a line segment will be recorded for each movement greater than ⅒ of an inch. In a 10"-wide space, 100 segments could be created. By changing the length to .5", a 10"-long line could be generated with 20 segments in the same space, which would greatly decrease the resolution. However, if the drawing is going to be reproduced at a scale of ¼"=1'-0", a line segment of .5" would give extremely high resolution. A 1"-long line segment might be more appropriate. Figure 13.8 compares four record increments. The width can be entered by providing a value with the keyboard or by moving the mouse. To enter a value using the keyboard, enter the value followed by (EnterKey). The command sequence is:

Command: **SKETCH** (EnterKey)

Record increment <0'-0⅛">: **.065** (EnterKey)

Sketch. Pen eXit Quit Record Erase Connect.

AutoCAD is now ready to sketch. To set the line increment using the mouse, press the pick button when the prompt for the increment is provided. The sequence is:

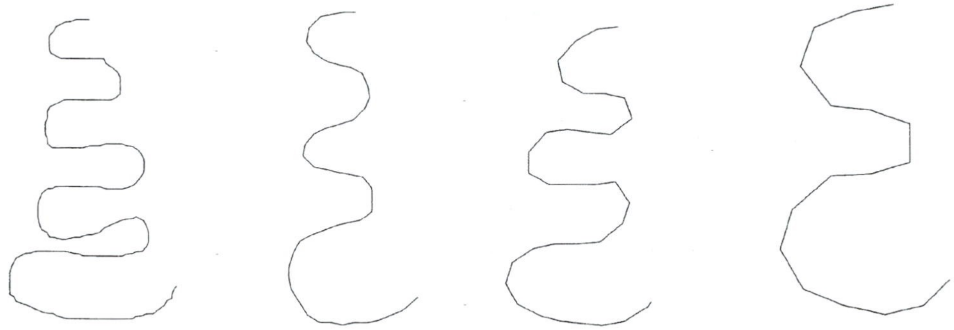

Figure 13.8 *The smaller the value for the segment length is, the smoother the resulting curves will be. A scale should be selected that provides the desired accuracy without requiring unneeded lines.*

Command: **SKETCH** (EnterKey)

Record increment <0'-0 1/8">: *(Press the pick button.)*

Second point: *(Move the mouse to the desired location and press the pick button.)*

Sketch. Pen eXit Quit Record Erase Connect.

For a drawing requiring much accuracy, enter the line increment value using the keyboard. Once the width is entered, several options will be listed in the following prompt:

Sketch. Pen eXit Quit Record Erase Connect.

Once the line length is set, move the cursor into the drawing area with the mouse. You'll notice that nothing happens as the mouse is moved throughout the drawing area.

Pen (P)

Think of the SKETCH command as sketching on paper with a pencil. The SKETCH command treats the cursor as a "pen" and allows for lifting the pen and moving it from one location to another without drawing a line. As SKETCH is started, the drawing pen is in the "up" position, allowing for selection of a starting location. Once the desired location is selected, the pen can be put down on the paper to allow for sketching. The pen is raised and lowered by pressing the mouse pick button or the P key on the keyboard. If you use the keyboard to control the pen, don't press (EnterKey) after pressing the P key. For most users, skip the keyboard once you've entered the command, and use the mouse to control the pen.

Each time the pick button is pressed, the pen is raised or lowered to allow for movement and drawing. The process can be seen in Figure 13.9.

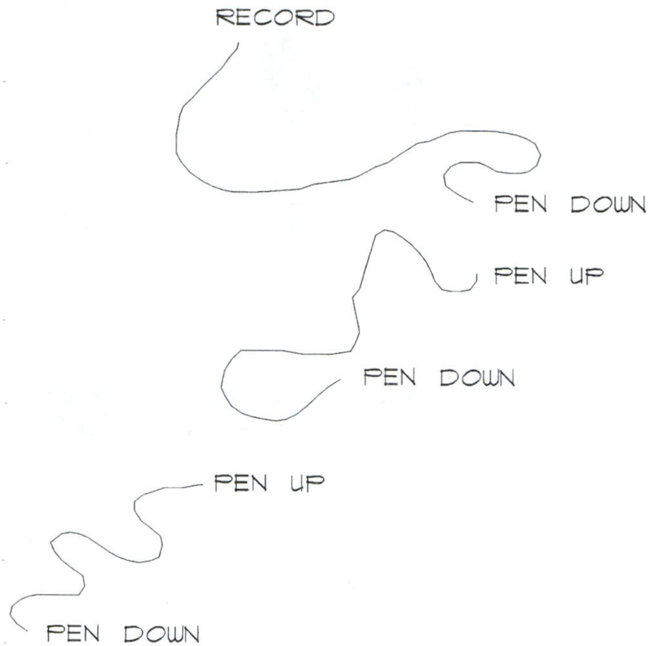

Figure 13.9 *Controlling the pen position of UP/DOWN allows breaks to be created between segments.*

. (period) Line to Point

The . (period) option can be used to draw a straight line from the endpoint of the last sketched line to the current location of the cursor. Sketch a few segments and raise the pen. Move the cursor to a new location and press the period key. This will produce a straight line from the end of the sketch segment to the new cursor location. The effect can be seen in Figure 13.10.

Record (R)

When sketches are created, they are not actually part of the drawing until they are recorded. Enter the drawing area and sketch a few lines. You'll notice that the lines are green (color will vary depending on configuration of AutoCAD), even though you've done nothing to change the drawing color. The R option records all of the temporary sketch lines as a part of the drawing base and converts them to the present drawing layer and color. Once recorded, sketch lines cannot be edited using the sketch subcommands. Sketched lines may be edited using any of the edit commands. "R" does not affect the position of the pen, and you can continue sketching after you've used it. Using (EnterKey) will record the sketch segments and exit to the command prompt.

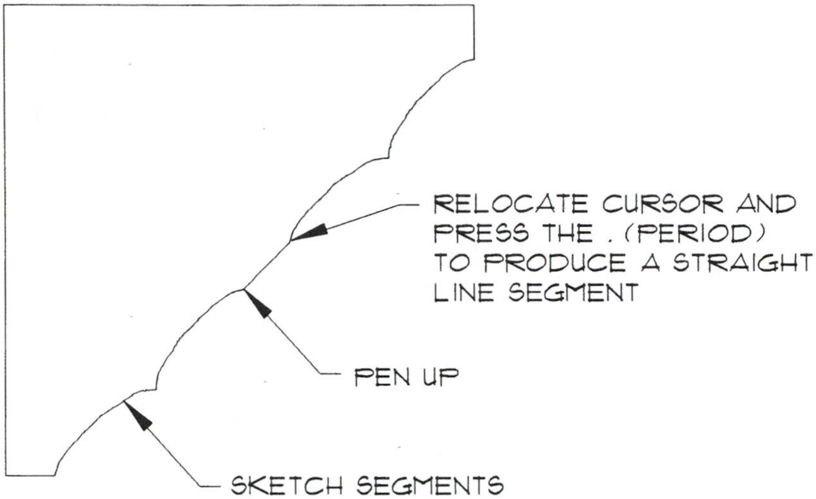

Figure 13.10 *The . (period) option of the SKETCH command allows a straight line to be drawn between the end of the last SKETCH segment and the present cursor location.*

Record and Exit (X)

Using the X key will record the sketch segments to the drawing base, return the command prompt to the screen, and re-establish the command prompt. Using this option will display the number of lines recorded just above the command line.

Quit (Q)

Pressing the Q on the keyboard will function similarly to the ERASE command. Q will remove all temporary sketch segments entered since the last Record and return you to the command prompt. This option can be used if you sketch an object and then decide that you don't need it.

Erase (E)

This option allows for part of the sketch segments to be removed prior to recording. The option may be started by the E key and will remove a portion of the sketch from the selected point to the end of the sketch segments. Once the option is started, a message is displayed:

Erase: Select end of delete. < Pen up >

The pen will be placed in the up position to allow you to select the end of a segment. Move the crosshairs to the portion of the line to be removed. The line will be displayed with the indicated portion of the line removed. If the removed section is acceptable, press the P key to erase the segment. If the selection is unacceptable, press the E key. The message "Erase aborted" will be displayed and the SKETCH prompt will be returned.

Connect (C)

Pressing the C key is similar to the END option of OSNAP when drawing lines, and will allow sketching to continue from the end of the last drawn sketch segment. When the option is selected, the following prompt will be displayed and the crosshairs will be displayed:

Connect: Move to endpoint of line

When the crosshairs are within one record increment of the end point, the pen will be placed in the down position at the end of the line. If the pen is already down when the Connect option is used, the following prompt will be displayed:

Connect Aborted

To terminate the Connect command, type **C** (EnterKey) a second time.

DRAWING WITH MULTIPLE LINES

The MLINE command allows you to draw up to 16 parallel lines at once. Each line is referred to as an element of the multiline. Drawing two to four lines will be a great aid when drawing construction-related drawings. Figure 13.11 shows an example of the use of MLINE. The command can be started by selecting the Multiline icon

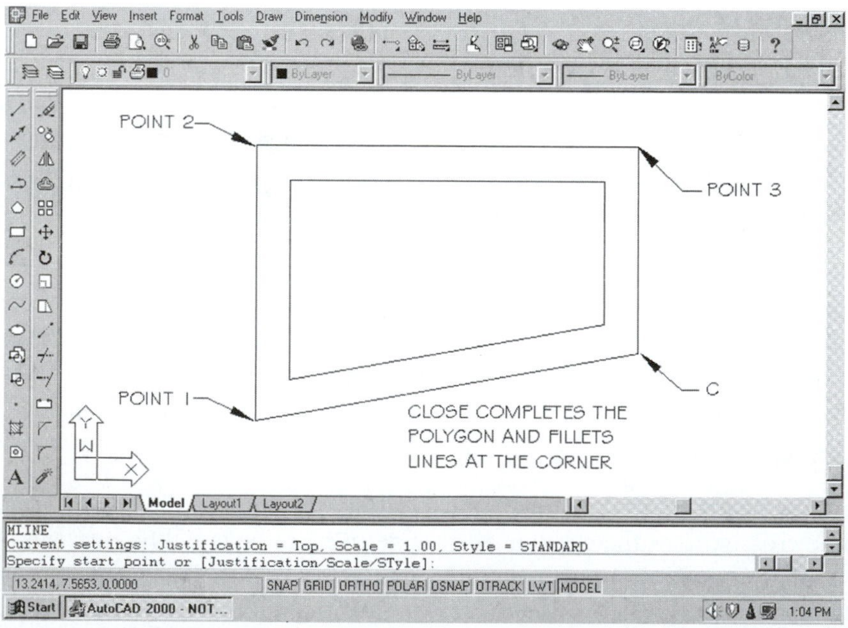

Figure 13.11 *The MLINE command allows two to sixteen lines to be drawn in a segment that functions as one line.*

from the Draw toolbar, by typing ML (EnterKey) at the command prompt, or by selecting Multiline from the Draw pull-down menu. Each of these methods will display the following prompt:

> Command: **ML** (EnterKey)
>
> MLINE
>
> Current settings: Justification = Top, Scale = 1.00, Style = STANDARD
>
> Specify start point or [Justification/Scale/STyle]: *(Select a start point or enter an option.)*

AutoCAD is waiting for you to select a point to start the command. Start the command sequence using the same procedure used with the LINE command. Enter the drawing area, enter the MLINE command, and then draw a few lines. Only two lines will be drawn as each segment is completed because the current STyle is standard (two lines). You'll notice that once a Next point is selected, the next prompt will change. The command prompts to select the four point of Figure 13.11 is:

> Specify next point: *(Select the desired point.)*
>
> Specify next point or [Undo]: *(Select another point.)*
>
> Specify next point or [Close/Undo]: *(Select another point.)*
>
> Specify next point or [Close/Undo]: *(Select another point.)* **C** (EnterKey)
>
> Command:

The Close option was used to close the object into a polygon. The Close and Undo options work with MLINE just as they do with LINE.

SETTING THE JUSTIFICATION

The term justification, as it relates to MLINE, describes where the lines will be placed relative to a base point, as the From and To points are specified. The options include Top, Zero, and Bottom, with a default setting of Top. Figure 13.12 shows the effect of each point. With Top active, the top line of the multilines will be placed

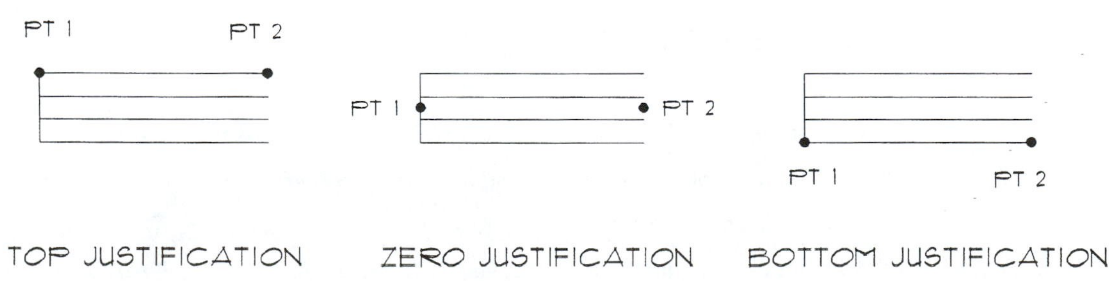

Figure 13.12 *The justification for MLINE determines where the line will be drawn relative to the selection point.*

between the selected start and next points, and other lines will be offset to go below the selected points. With Zero active, the selection points will serve as the midpoints for the multilines, with lines equally offset on each side of this theoretical center point. Setting Bottom as the active selection will form the bottom line of the multilines between the selection points, with other lines offset to the top.

The justification for multilines can only be selected once for each command sequence. The justification is altered by entering **J** (EnterKey) at the prompt for MLINE. The sequence to change to Zero is:

Command: **ML** (EnterKey)

MLINE

Current settings: Justification = Top, Scale = 1.00, Style = STANDARD

Specify start point or [Justification/Scale/Style]: **J** (EnterKey)

Enter justification type [Top/Zero/Bottom] <top>: **Z** (EnterKey)

Current settings: Justification = Zero, Scale = 1.00, Style = STANDARD

Specify start point or [Justification/Scale/STyle]: *(Select a start point.)*

Zero will now remain the default value until it is changed.

SETTING THE SCALE

Scale controls the overall width of the multilines to be drawn. The default setting is 1.00. Keep in mind that 1.00 represents 1 unit, not 1 inch or 1 foot. With Top as the current justification value, and 1 as the current scale value, a second line will be drawn 1 unit below the selection points. With the Zero option of justification current, one line will be ½ unit above and the other line will be ½ unit below for a total width of 1. Entering a value larger than 1 will enlarge the spacing between multilines. Entering a value of less than 1 will decrease the space between lines. Remember as you look at the lines in Figure 13.11, that the limits are fairly small and the setting of 1 creates what appears as wide lines. Changing the limits will allow a floor plan to be drawn, and the same lines will appear as single lines. The command sequence to adjust the scale to 6 so that 6"-wide walls can be drawn is:

Command: **ML** (EnterKey)

MLINE

Current settings: Justification = Zero, Scale = 1.00, Style = STANDARD

Specify start point or [Justification/Scale/Style]: **S** (EnterKey)

Enter mline scale <1.00>: **6** (EnterKey)

Current settings: Justification = Zero, Scale = 6.00, Style = STANDARD

Specify start point or [Justification/Scale/STyle]: *(Select a start point.)*

THE MULTILINE STYLE DIALOG BOX

A MLINE style refers to the number of lines to be drawn in the multiline pattern and how intersections and terminations will be displayed. The desired style must be created prior to using the MLINE command. The STyle option of MLINE only allows a created line style to be selected. Multiline styles can be created by selecting Multiline Style... from the Format pull-down menu, or by typing **MLSTYLE** (EnterKey) at the command prompt. Each will produce the dialog box seen in Figure 13.13. Key components and their function of the Multiline Styles dialog box include Current Style, Name, Description, and Properties.

Current

The Current edit box displays the name of the current multiline style. The current and only style at present is the default style of STANDARD. As other styles are created, the arrow can be used to scan between style names, making other styles the current selection.

Name

This box allows a name to be assigned to the style being created. The name is not specified until after the new style is created. The box can also be used to rename existing styles.

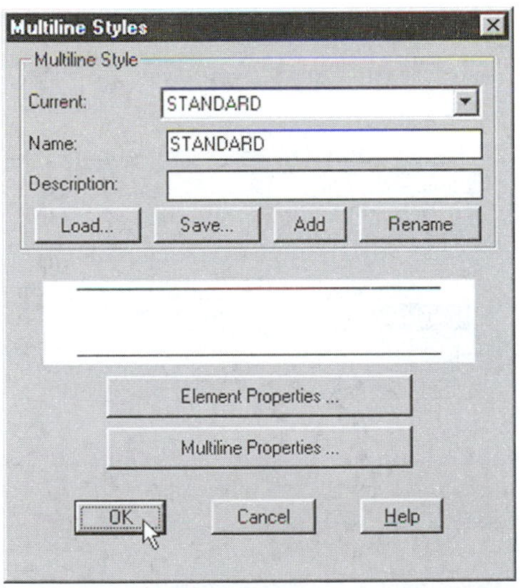

Figure 13.13 *The Multiline Styles box can be used to create, save, and load MLINE styles. Each style is comprised of the number, location, color, and linetype of each element in a style. The box is accessed by selecting Multiline Style... from the Format pull-down menu or by typing* ***MLSTYLE*** *(EnterKey) at the command prompt.*

Description

The description edit box allows a description to be entered for the style being created. Descriptions can contain up to 255 characters, including spaces between words. As with other filenames and descriptions, keep them simple.

Load....

This button will display the Load Multiline Styles dialog box seen in Figure 13.14. Multiline styles will be saved as ACAD.MLN files within the current drawing. The File... button allows you to load a multiline style that has been saved in a separate .MLN file. Selecting the File... button displays a directory of available drives for selecting stored linetype styles.

Save...

This button allows styles to be saved or copied to a new location. Typing a name in the Name: edit box and selecting this button displays a directory of available drives for saving the new file as an .MLN file.

Add

This button allows a multiline style name to be added to the current multiline file.

Rename

The Rename button allows the current style to be renamed. The Standard file can't be renamed.

Line Display Panel

This panel displays an image of the current line style.

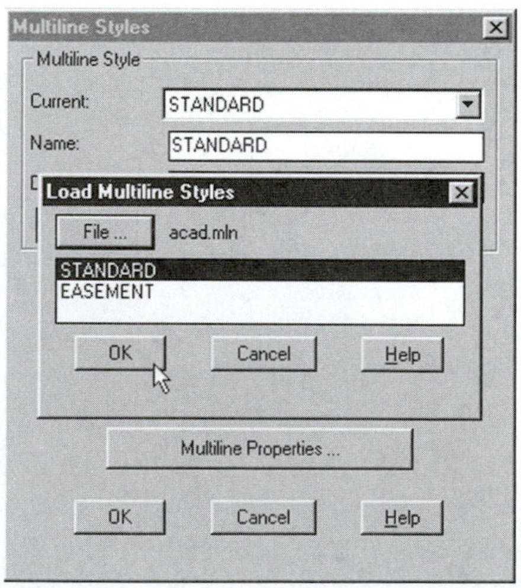

Figure 13.14 *Selecting the Load... button in the Multiline Styles edit box allows different segment styles to be loaded into a drawing file.*

Element Properties...

Selecting this option will display the Element Properties dialog box shown in Figure 13.15. This box allows lines of various linetypes and colors to be used within a style.

Multiline Properties...

Selecting this button displays the Multiline Properties dialog box shown in Figure 13.16. This box allows properties such as joints, end caps, and background fill to be set within a style. This box will be further discussed in the next section of this chapter.

ADJUSTING THE ELEMENT PROPERTIES

Pressing the Element Properties... button allows the properties of a Multiline style to be adjusted as it is being created. Elements of the box include:

Elements

This box displays the offset size, color and linetype settings of each multiline style. Lines are listed in descending order of the offset distance.

Add

This button allows new lines to be added to a multiline style.

Elements List box

Selecting Add will insert a new line with an offset distance of 0.0 to the Element Properties box. Once added, the properties of the current line can be altered.

Delete

This button allows the highlighted line style to be deleted from the Elements box.

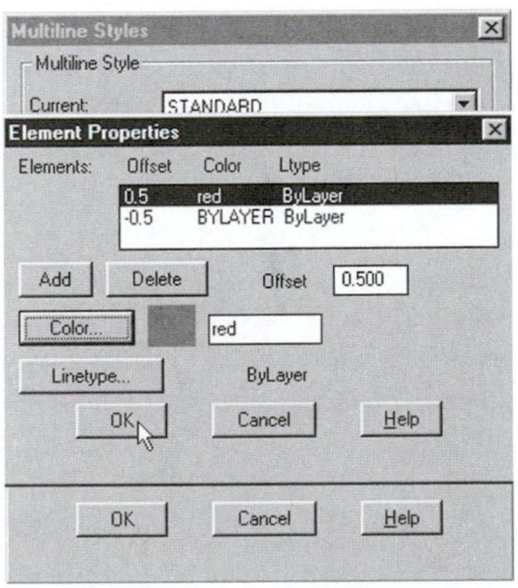

Figure 13.15 *Selecting the Element Properties... button will display the Element Properties dialog box, which allows the settings for a segment to be selected.*

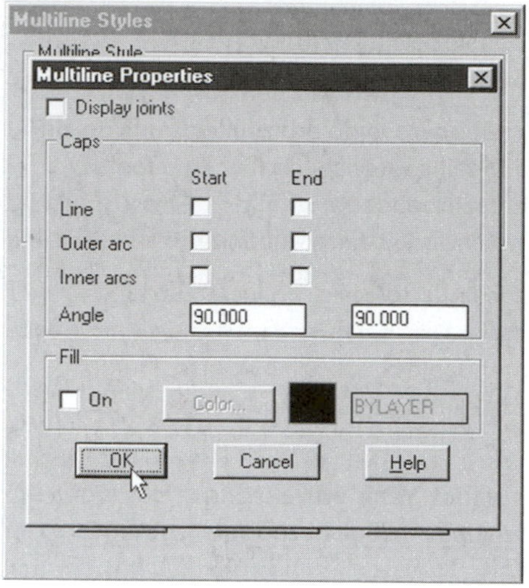

Figure 13.16 *Selecting the Multiline Properties... button in the Multiline Styles edit box allows the properties for joints, end caps, and fill to be selected for a style.*

Offset

This button allows the offset distance of the highlighted line to be altered. Offset distances are based on the location of the origin 0,0 and can be either positive or negative. To alter an offset value, select the desired line. The current offset will now be displayed in the Offset box. Use the backspace key to remove the existing value and type the desired value. The new value is added to the style by pressing the (EnterKey).

Color...

This button allows color to be assigned to a specified line. As the box is selected, a color palette will be displayed, allowing for color selection. Select the red tile in the upper left-hand corner and then select OK. The Element Properties box will be restored, and a red line will be displayed to the right of the Color... button.

Linetype...

This button allows a linetype to be assigned to the selected line of the multiline. This option allows the multiline group to have varied line patterns. The default of ByLayer, means the new line will be placed on the current layer (0) using a continuous linetype. Below the listing of the current linetype is a listing of each linetype that has been added to the drawing base (None). To draw with different linetypes, you must load the lines to be drawn. Press the Load... button and the Load or Reload Linetype dialog box will be displayed. Highlight the word CENTER by selecting it with the mouse and then scroll down. Hold the (CtrlKey) down and select HIDDEN. Select the OK box and the Linetype box will be redisplayed. The lines will be reflected in the Linetypes box.

ADJUSTING MULTILINE PROPERTIES

The Multiline Properties dialog box can be used to control how multilines will be displayed. Key components of this box include:

Display Joints

In the current setting, as multilines are drawn, the joints are off. By selecting this box, a check will be placed in the box and a line will be drawn between each line at each multiline intersection. Figure 13.17 shows the difference in line styles with joints off and on.

Caps

This option has four selections for each end of each option. The standard setting is to have no caps. In the default, all boxes are open. Highlighting a box will place a check in the selected box, indicating that the box is on. The effects of each option can be seen in Figure 13.18.

Line

With this option selected, a line will be drawn as a cap at the start or end of the multilines.

Outer Arc

This option will draw a semicircle between the endpoints of the outer lines.

Inner Arc

This option will draw a semicircle between the endpoints of the inner lines. If there is an odd number of inner lines, the centerline will not be capped.

Angle

This option allows the angle of the cap line to be adjusted. Options include angles between 10° and 170°.

Fill

This option toggles Fill ON or OFF. In the default setting OFF, multilines will be drawn as two or more parallel lines. With Fill set to ON, multilines will resemble a polyline or trace line. Once toggled ON, the color button will be activated and a color can be selected for the fill. The effect of fill on a line can be seen in Figure 13.19.

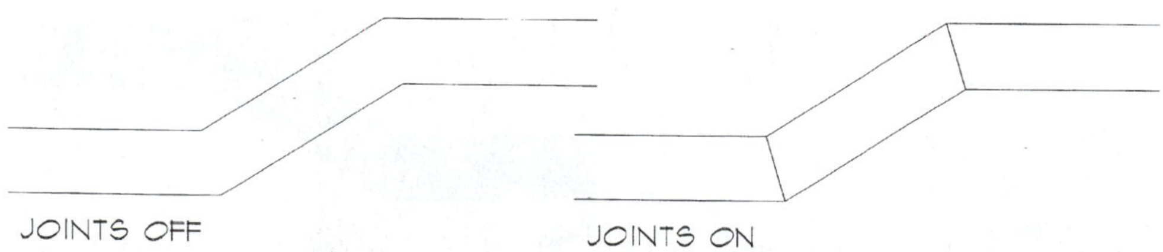

Figure 13.17 *Multiline segments can be displayed so that their intersections are OFF or ON.*

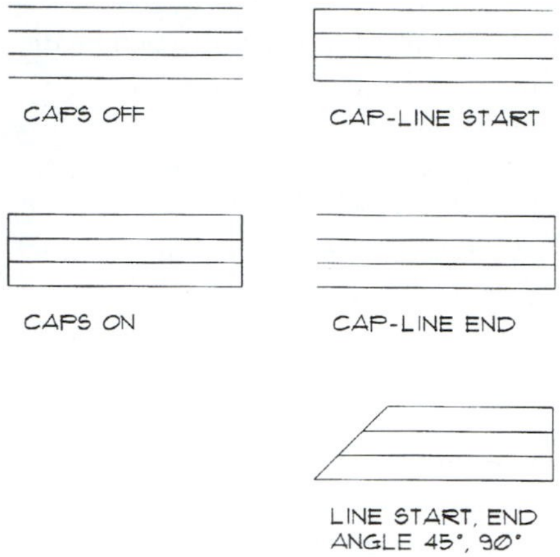

Figure 13.18 *The ends of MLINE segments can be capped or uncapped. Caps can consist of a perpendicular or inclined line, or with arcs from the outer or inner lines.*

CREATING MULTILINE STYLES

Now that the dialog box MLSTYLE has been explored, line styles that could be used on construction drawings can be created. The following section will describe how to create a multiline to represent a utility easement on a site plan. The exterior linetypes will be HIDDEN.The interior line will be a CENTERLINE. Each outer line will be red and the centerline will be black.

Open the Multiline Styles dialog box using one of the methods presented earlier in the chapter, and select the Element Properties... box. Select the desired offset values, color, and linetypes for each element of the multiline. Use offsets of .5, 0 and –.5. When the properties are acceptable, select the OK button.

Move the cursor to the Name: box and enter the desired name for the new style. For this example the name EASEMENT was used. Once typed, move the cursor to the Description: box and describe the line. A description of HIDDEN, CENTER,

Figure 13.19 *MLINE segments can be drawn as open or filled segments.*

HIDDEN was used in the example. Now move the cursor to the Save... button and save the style using the Save Multiline Style box, just as you would a drawing file. Select the Load... button to display the Load Multiline Styles dialog box and select the new style and select OK. Select OK to exit the Multiline Styles dialog box. A style named easement has now been created and loaded into the drawing file. Now the style can be used with the MLINE command.

To draw multilines using the EASEMENT style, enter the MLINE command using one of the methods described earlier. The command sequence to draw lines using WALL is:

Command: **ML** (EnterKey)

MLINE

Current settings: Justification = Zero, Scale = 1.00, Style = STANDARD

Specify start point or [Justification/Scale/Style]: **ST** (EnterKey)

Enter mline style name or [?]: **EASEMENT** (EnterKey)

Current settings: Justification = Zero, Scale = 1.00, Style = EASEMENT

Specify start point or [Justification/Scale/Style]: *(Select a starting point.)*

If you have several styles and can't remember the names of them, choosing the ? option will provide a display showing the name and description.

EDITING MULTILINES WITH GRIPS

Lines created using MLINE can't be edited using BREAK, CHAMFER, EXTEND, FILLET, or TRIM commands but are affected by the COPY, EXPLODE, MIRROR, MOVE, and STRETCH commands. Multilines can also be edited using GRIPS. Placing the cursor over a multiline object and left-clicking will produce a grip at each endpoint of each segment. The grips are located based on the justification used when the lines were drawn. Examples of grip locations can be seen in Figure 13.20.

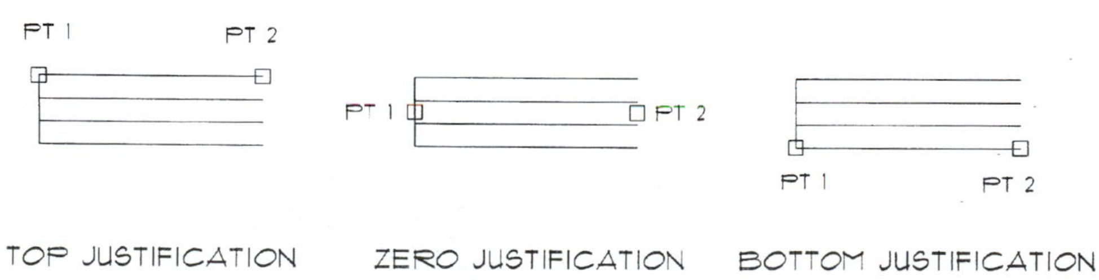

Figure 13.20 *Grips can be used to edit multiline segments. The location of the grips will depend on which line justification has been selected.*

EDITING MULTILINES USING MLEDIT

The MLEDIT command allows multilines to be edited by providing a dialog box that displays an icon of the edit to be preformed. The command can be accessed by selecting the Edit Multiline icon from the Modify II toolbar, by selecting Multiline... from the Modify pull-down menu, or by typing **MLEDIT** (EnterKey) at the command prompt. Each method will display the Multiline Edit Tools dialog box shown in Figure 13.21. This box allows editing of MLINE intersections, tees, corner joints, and lines.

EDITING MULTILINE CROSSING INTERSECTIONS

The three types of intersections using MLEDIT include closed cross, open cross, and merged cross. Each are found in the left column of the Edit Tools dialog box.

Closed Cross

This option allows one group of multilines to remain as is, while the other pair is interrupted. The order of selection determines which pair of lines will be edited. The first line selected will be trimmed, but continues to function as one set of multilines. The second pair of lines selected will remain continuous. The command sequence to provide a closed cross is:

Command: **MLEDIT** (EnterKey) *(Select the closed cross icon and OK.)*

Select first mline: *(Select the line to be trimmed.)*

Select second mline: *(Select the line to be continuous.)*

Select first mline or [Undo]: (EnterKey)

The command prompt will continue to be displayed, allowing other intersections to be edited using the closed cross option. Selecting (EnterKey) returns the command prompt. Selecting the U option allows the cross to be undone, but restores the closed crossing prompt. Selecting (EnterKey) twice restores the Multiline Edit Tools box and allows a new option to be selected. The effect of selecting the crossing opting can be seen in Figure 13.22.

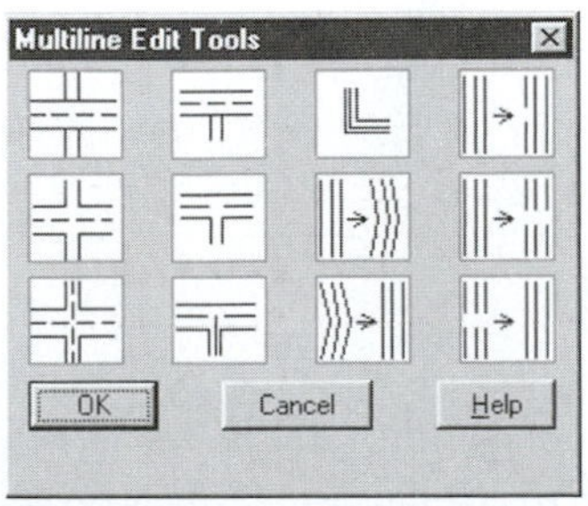

Figure 13.21 *Type **MLEDIT** (EnterKey) at the command prompt to display the Multiline Edit Tools box, which provides 12 options for editing.*

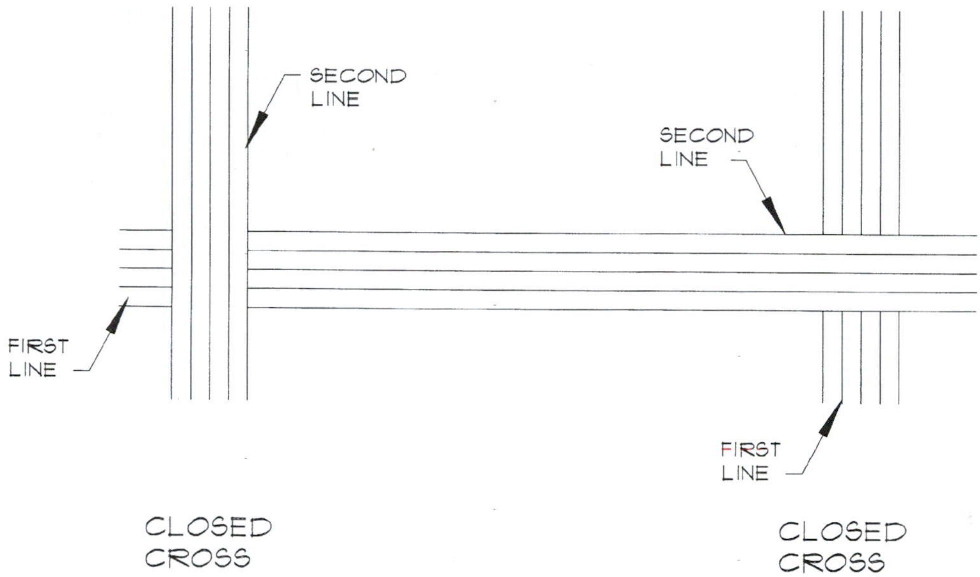

Figure 13.22 *The results of a CLOSED CROSS edit depends on which line is selected first. The segment selected second will appear unaffected.*

Open Cross

This option will trim all of the lines of the first multiline selected, and only the outer lines of the second group. The command sequence is similar to the closed cross. The results can be seen in Figure 13.23. When only two lines are used, each pair of lines will be opened.

Merged Cross

This option will trim the outer lines of both pairs of multilines and leave the inner lines to cross. The command sequence is similar to the closed cross. The results can be seen in Figure 13.24. When only two lines are used, each pair of lines will be opened.

EDITING MULTILINE TEES

Three types of tees can be completed by using the icons in the second column from the left of the Edit Tools box. These include a closed, open, and merged tee. Each command sequence is the same as its crossing counterpart. Examples of each tee can be seen in Figure 13.25.

EDITING MULTILINE CORNERS

Three types of corners can be formed using the icons in the second column from the right of the Edit Tools box. These include the top option of corner joint, the middle option of add vertex, and the delete vertex option on the bottom.

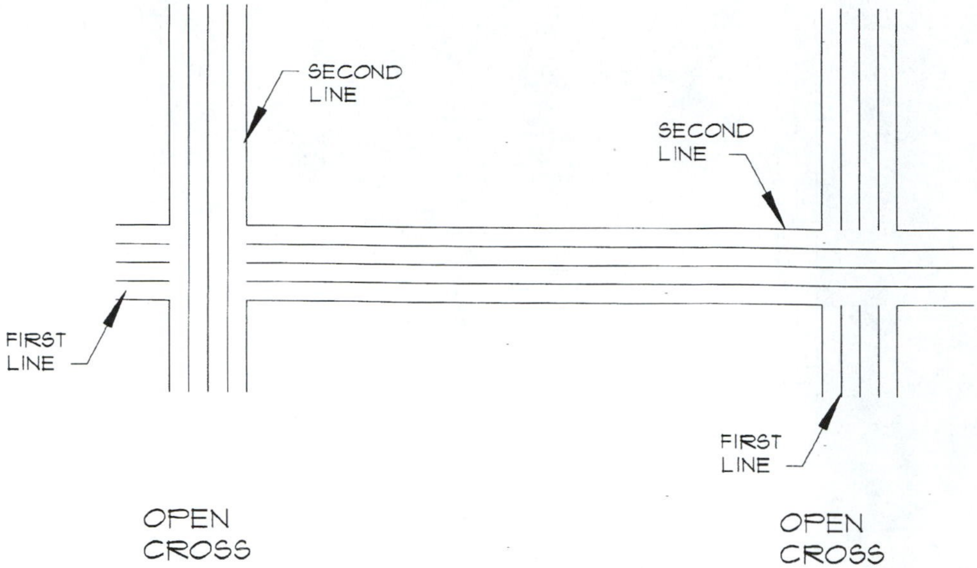

Figure 13.23 *Multilines edited using OPEN CROSS will vary, depending on the order of element selection.*

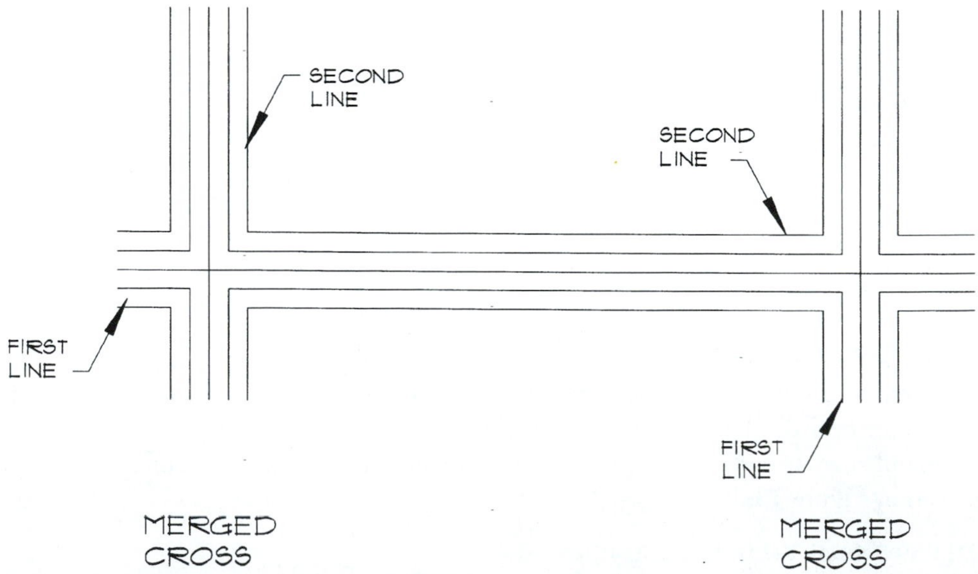

Figure 13.24 *The MERGED CROSS option allows outer and inner lines to appear as if they had been filleted.*

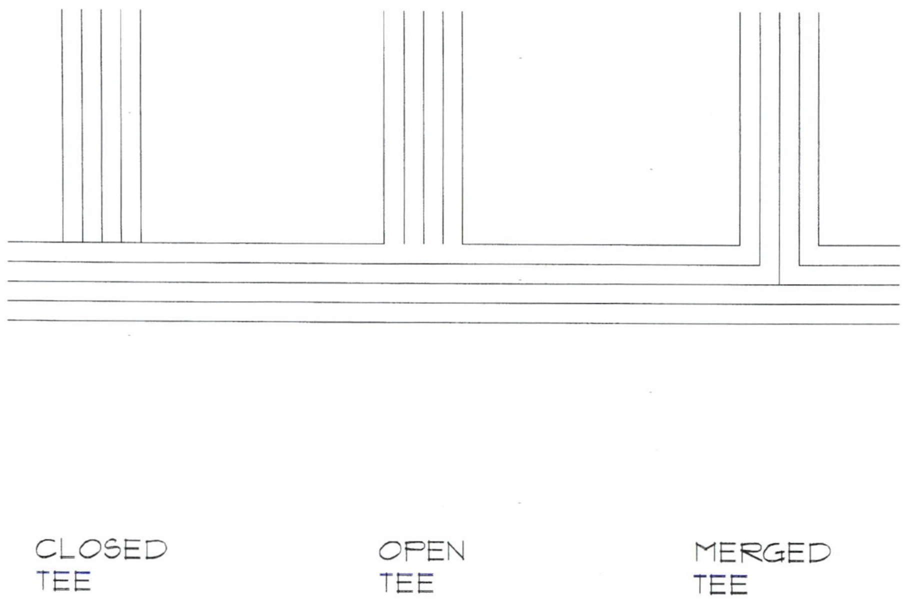

Figure 13.25 *The CLOSED, OPEN, and MERGED tee options.*

Adding a Corner

The Corner Joint forms a corner between two groups of multilines by trimming or extending each pair as required. An example can be seen in Figure 13.26.

Adding a Vertex

The middle option can be used to add a vertex to a multiline. The vertex is inserted at the position where the multiline is selected, but no apparent change is made. To see the effect of the change, move the cursor so that the grips of the desired line are displayed. Now it can be seen that a new vertex has been added along with a grip. The new grip can now be used to alter the shape of the multiline. The effect can be seen in Figure 13.27.

Delete a Vertex

A similar process can be used to remove a vertex from the selected multiline. The vertex nearest to the selection point will be removed. The effects can be seen in Figure 13.28.

EDITING THE LINES OF A MULTILINE PATTERN

Each of the three types of edit groups for cutting and welding multilines are represented in the right column of the Edit Tools box. Examples of each can be seen in Figure 13.29.

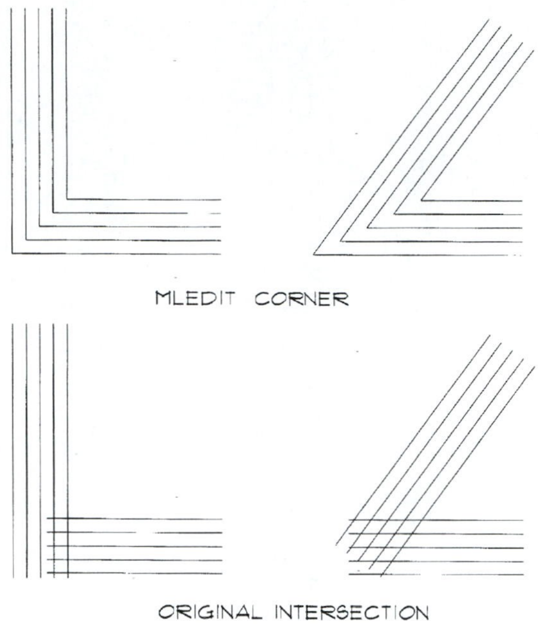

Figure 13.26 *The effects of CORNER on mlines.*

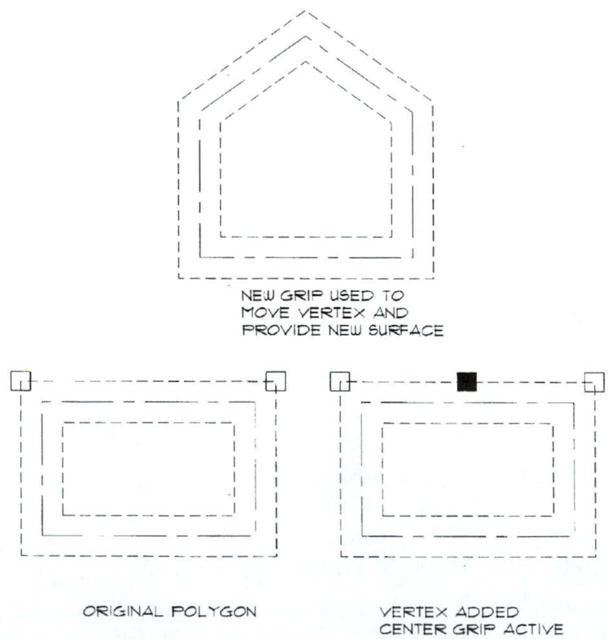

Figure 13.27 *A vertex can be added allowing the shape of mlines to be altered.*

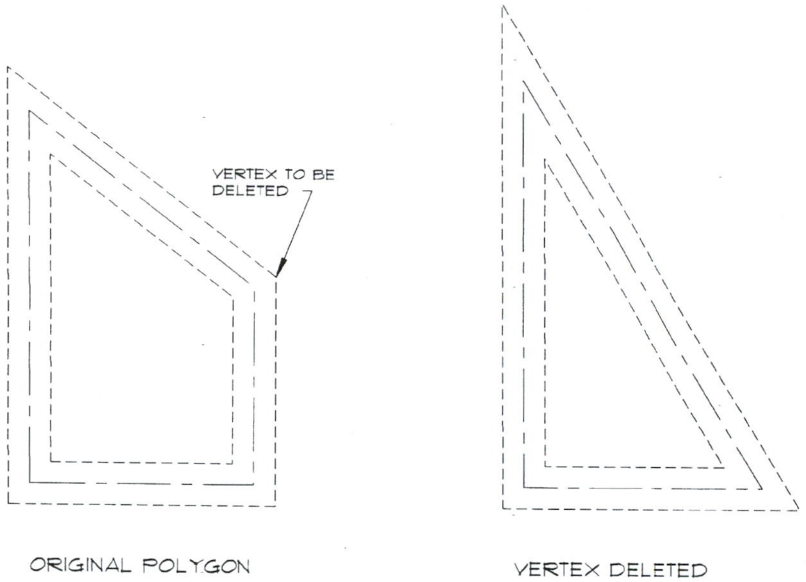

Figure 13.28 *The DELETE VERTEX can be used to remove a vertex and alter the shape of multiline segments.*

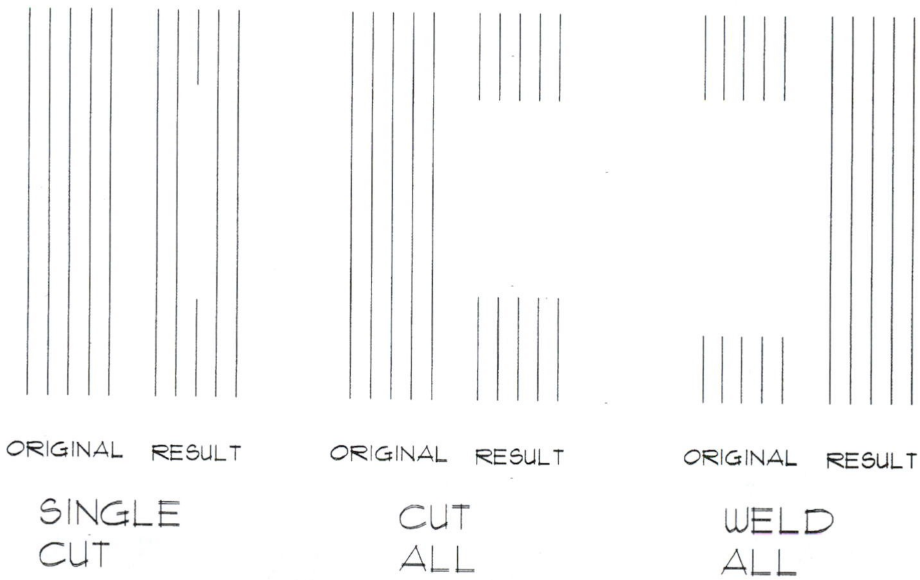

Figure 13.29 *Multiline segments can be cut or joined, using the line edit functions. The SINGLE CUT option can be used to remove a portion of one element of a segment. The option can be repeated so that more than one element of a segment can be altered. The CUT ALL option will remove a uniform length of all elements in a segment. The WELD ALL option can be used to restore lines in an MLINE segment.*

Cut Single

This option performs a function similar to BREAK. The option can be used to remove a portion from one of the lines in a multiline group. A portion of the selected line will be removed, but the balance of the elements will remain unaffected.

Cut All

This option can be used to remove a portion from all elements of a multiline group. A portion of the selected line will be removed between selected groups. Even though a portion of the line is removed, the two portions function as one multiline pattern.

Weld All

This option joins segments of a multiline that have been cut. It only restores two segments of one multiline. It can't be used to join two separate multiline groups.

EDITING MULTILINES WITH EXPLODE

Each of the twelve options in the Multiline Edit Tools dialog box allows multilines to be edited, but retains their properties. The EXPLODE command can be used to turn multiline groups into separate elements. When a multiline is exploded, all lines within the group will now function as if they had each been drawn using the line command. Color and linetype will remain unaffected, but each of the standard edit commands can be used because each element now functions as an individual object.

THE U COMMAND

You've already used the Undo option to remove a drawing object. The U option has been used in the middle of a command routine to remove an object that may not be what you intended. The U command also can be used to backtrack through a command sequence once the sequence has been ended. Use the command carefully. Once used, objects are gone and will have to be redrawn if you want them back. Enter a new drawing, draw a series of several lines, and press (EnterKey). Draw a circle, a polygon, and an ellipse in the order listed. Now if you type U (EnterKey) at the command prompt, you will see:

Command: **U** (EnterKey)

ELLIPSE GROUP

The last item drawn, the ellipse, will be removed from the screen. Notice that the most recent command (ellipse) will be listed and the command prompt will reappear. If the process is repeated, the polygon will be removed.

Command: **U** (EnterKey)

POLYGON GROUP

By continuing to enter U, the entire drawing can be removed.

Command: **U** (EnterKey)

CIRCLE

Command: **U** (EnterKey)

LINE

UNDO

The UNDO command allows several command sequences to be undone at one time and permits several operations to be carried out as the objects are undone. The UNDO command should give you the freedom to know you can try anything, and if the outcome is undesirable, enter UNDO at the command prompt to restore the drawing to its original state. The command sequence is:

Command: **UNDO** (EnterKey)

Enter the number of operations to undo or [Auto/Control/BEgin/Mark/Back]<1>:

If a number is entered, such as **1** (EnterKey), the last command will be undone. If you enter **3** (EnterKey), the last three command sequences will be removed.

AUTO

The Auto subcommand issues a prompt to toggle between ON and OFF. When Auto is ON, the default value, any group of commands that is used to insert an item, or group of items, is treated as one item and removed by U. This will be helpful later when working with blocks, wblocks, and macros. If Auto is OFF, each command in a group of commands is treated as an individual one.

CONTROL

The Control option is another method to limit the UNDO command; it also can be used to completely disable UNDO. The command sequence is:

Command: **UNDO** (EnterKey)

Enter the number of operations to undo or [Auto/Control/BEgin/Mark/Back]<1>: **C** (EnterKey)

Enter an UNDO control option [All/None/One] <All>:

If All is selected, UNDO will not be limited, and it will erase back to the last MARK. The NONE will disable U and UNDO commands so that your machine will function as if these commands have never been placed in AutoCAD. UNDO and U can be reactivated by entering UNDO and typing A (EnterKey). Selecting the One option limits U and UNDO to a single operation before returning to the command line.

BEGIN/END

The Begin option of UNDO groups a sequence of drawing operations. The group of drawing commands is defined by the use of the Begin option of UNDO. The end of the group is marked by the End option. With a specified group, AutoCAD will treat grouped commands as a single operation.

BACK

The BACK option of UNDO will remove everything in the entire drawing. Not some, not most, we're talking *everything*. Do you get the feeling you need to use this option carefully? AutoCAD will even help you ponder the use of this option by displaying the prompt:

This will undo everything. OK? <Y>

If (EnterKey) is pressed, this will accept the default yes value, and really clean up your drawing. Entering **N** (EnterKey) will ignore the Back option. If you do use the BACK option and then change your mind, use the REDO command to restore your drawing.

The BACK option also can be used with the MARK option so that the entire drawing is not erased.

MARK

This option will limit the distance that BACK will search through a drawing as it erases work. A MARK can be placed in a drawing prior to executing several commands. If the outcome of those commands is not what you expected, UNDO BACK will remove all of the commands until the MARK is reached. The command sequence is:

Command: **UNDO** (EnterKey)

Enter the number of operations to undo or [Auto/Control/BEgin/Mark/Back]<1>: **M** (EnterKey)

Command: **L** (EnterKey) *(Now draw several features in an existing drawing.)*

Command: **UNDO** (EnterKey)

Enter the number of operations to undo or [Auto/Control/BEgin/Mark/Back]<1>: **B** (EnterKey)

LINE

Mark encountered

Command:

NUMBER

The Number option of UNDO will undo the specified number of drawing operations. The effect is the same as if the U command had be used the same number of times, but only a single regeneration is required.

REDO

Occasionally you will remove an object with U or UNDO by mistake. The REDO command can be used to restore items that have been deleted. REDO will only restore objects, however, when used immediately after an UNDO or U. If U or UNDO is followed by a command such as LINE, and then REDO, REDO will have no effect on the material that was deleted prior to the UNDO. The command can be started by selecting the Redo icon from the Standard toolbar or by entering **REDO** (EnterKey) at the command prompt.

CHAPTER 13 EXERCISES

1. Start a new drawing and set the units and limits to be appropriate for this problem. Using the attached sketch as a guide, use the trace command to draw the outline of a 15' × 28' apartment unit. Make the 15' walls 6" thick, with the long walls 8" wide. Use 4"-wide lines for interior walls and lay out a 12'×15' bedroom, and a 5'-6" × 9' bathroom. No windows or doors are required at this time. Save the drawing as E-13-1.

2. Use the guidelines presented in Exercise 13-1 and draw the apartment using MLINE. Save the drawing as E-13-2.
3. Open a new drawing and create multiline segments with three, four, and five elements. Use each of the cap options to terminate each segment. Save the drawing as E-13-3.
4. Use the MLSTYLE command and create an easement similar to the style that was created in this chapter. Save the drawings as E-13-4.
5. Draw a rectangle using the MLINE command and make two additional copies. Use one copy to add a vertex and change the object into a pentagon. Use the other copy to delete a vertex and change the object into a triangle. Save the drawing as E-13-5.
6. Draw at least three different groups of multilines using three different styles of lines with at least three lines in each style. Make enough copies of these lines and use each of the intersection, tee, and corner options to edit each pair of lines. Save the drawing as E-13-6.
7. Enter a new drawing and draw a line 27'-6 ½" long. Divide the line into five equal segments. Copy the original line and place markers at 16" o.c. Save the drawing as E-13-7.
8. A stairway is to be drawn between two floors of a townhouse with a distance of 9'-1 ½" from finish floor to finish floor. Plan a set of stairs with 7 ½" maximum rise and 10 ½" minimum run per step. Determine the required number of stairs

and the total run. Draw horizontal lines to represent each floor. The upper floor level will be 13 ½" thick. Divide the space between floors into the required number of equal risers. Divide the total required run into the required individual steps. Using any editing process presented in the last two chapters, edit the grid that has been laid out to show the shape of the stairs. Draw a diagonal line that passes through the front edge of each step. Offset this line down 12" to determine the required depth of the stairs. Save the drawings as E-13-8.

Required risers. ____________ Total run. ____________________

9. Use Figure 13.4 as a guide and draw an elevation showing an 8'-high, 2 × 6 stud wall. Use (2) 2 × 6 top plates and a single base plate. Show an opening for a 6'-wide × 4'-high window. Use a 4 × 8 header over the window, and set the bottom of the header at 6'-8" above the floor level. Place the studs at 16"o.c. Save the drawing as E-13-9A. Make a copy of the elevation and show the wall at 9' high with the studs located at 24"o.c. Set the bottom of the header at 7-10" above the floor. Save the elevation as E-13-9B.
10. Use the SKETCH command and create a logo that can be used in the title block of a template drawing. Save the drawing as E-13-10.
11. Open a new drawing and draw three lines 10'-0" long. Set the point marker to show a marker with a circle with an X in the center. Measure the line into 7 equal segments. Divide the second line into 16"-long segments using a square around the crosshairs marker. Show the third line with perpendicular lines at 19.2"o.c. Save the drawing as E-13-11.
12. Use the SKETCH command and draw four curved lines using .065", .125", .25", and .5" long line segments. In at least two of the lines, include a straight line segment between the sketched segments. Save the drawing as E-13-12.
13. Open a new drawing and draw a rectangle. Place a point using the crosshairs surrounded by a circle and a square in the center of the rectangle. Draw lines of infinite length that pass through the point in a horizontal, vertical and at 30° to the horizontal and vertical planes. Trim the lines that extend beyond the rectangle. Save the drawing as E-13-13.
14. Open a new drawing and draw a circle with a semi-infinite line extending outward from each quadrant. Draw lines that extend from the center point to the edge of the circle and are placed at 15° intervals for a 360° pattern. Save the drawing as E-13-14.
15. Draw two lines that intersect. One of the lines is to be horizontal and the second line is to be 67° above the first line. Draw a line that bisects the intersection of the two lines. Save the drawing as E-13-15.

CHAPTER 13 QUIZ

Enter a new drawing and draw the following objects in the order listed. Draw 4 pairs of line segments, a three-sided polygon, a circle, a polyline with three segments with a width of .125, and an ellipse. Copy the circle, move the ellipse, and enlarge the polygon to twice its size.

1. If the U command is used after the polygon is enlarged, what will be affected?
2. If UNDO 2 is used after the polygon was drawn, what will be affected?
3. If a mark is placed after the circle is drawn, what will the effect be if the UNDO Back command is used after the circle is copied?
4. A whole drawing sequence is an experiment that you may want to delete. List the command and subcommands that would allow the entire sequence to be removed as one object. Explain when these subcommands are used.
5. If 5 is used with the UNDO Number option, what will the effect be?
6. In using the UNDO Number 5 option, one-too-many objects was removed. How can this be corrected?
7. List the keys for controlling the SKETCH subcommands.
8. Explain the difference between a TRACE and a POLYLINE line segment.
9. What is the default setting for a sketch line if the units are set to architectural units with fractions of $^1/_{16}$"?

   For engineering units with four-place decimals?
10. What is the danger of having the sketch resolution too fine?
11. If the record increment is set to 1.25", what is the effect on the sketch if the mouse is moved 1"?
12. Several lines have been drawn using SKETCH and you would like to add to the drawing in a new location. How do you move the pen to the new location without drawing?
13. You've sketched a beautiful drawing and wish to save it as part of the drawing base. How can you do this?
14. Explain how the correct option of SKETCH will affect a drawing.
15. What is a ray and how could it be used?
16. What is the difference between a ray and an xline?
17. What is the effect of HOR on a ray if Ortho is off?
18. Describe the difference between the two angle options of XLINE.
19. List and sketch an example of each of the Cap options of MLINE. Use segments with three and four elements for each option.
20. Describe the effect of each of the justifications on XLINE placement.

21. A scale of 4 has been selected to draw multiline segments. Will the segment be 4", 4', or 4 meters wide?
22. Two multiline segments need to be joined together. What command and options should be used?
23. What will the Element Properties box adjust?
24. Explain the process for selecting a color for a multiline.
25. What editing commands will affect multilines?
26. What is the difference between 5 and –5, as they relate to multilines?
27. Sketch the difference between a closed cross and an open cross in segments with three lines.
28. Sketch the difference between a merged tee and an open tee in segments with four lines each.
29. What effect will EXPLODE have on MLINE segments?
30. Explain the difference between DIVIDE and MEASURE on a line.

CHAPTER 14

Controlling Lines and Layers

This chapter will examine the various types of lines available in AutoCAD, as well as how to create your own linetypes to meet various project parameters. Once you've explored the wide variety of linetypes, this chapter will provide a method of keeping drawing information organized by the use of layers, which can be activated or made invisible. This information will be altered using the LAYER command.

CONTROLLING LINETYPES

A linetype is a specific pattern of lines of varied length and spacing presented in a uniform pattern. Varied linetypes traditionally have been used throughout the architectural and engineering communities. Chapter 6 introduced methods for loading and drawing with varied linetypes. This chapter will introduce methods for altering factors that control linetypes.

CHANGING THE LINE SEGMENT SIZE

The dashes that comprise the line pattern are currently defined as drawing units that may be unsuitable for your current drawing needs. A centerline drawn in a 12" × 9" drawing area will appear as a centerline until you zoom in on the line. If the same centerline is used in a drawing area that is 50' wide, the centerline will be displayed as a continuous line. To have the centerline appear as a centerline requires that the scale of the line be altered. The Global scale factor setting of the Linetype Manager dialog box can be used to control the length of line segments in a drawing. The dialog box is accessed by choosing Linetype... from the Format pull-down menu. Select the Show Details button to display the current line scale values.

The segment length of each pattern can also be altered by entering **LTS** (EnterKey) at the command prompt. LTS represents line type scale. No matter how the value is entered, by default AutoCAD thinks of the selected length as one unit. This unit could be in inches, feet, millimeters, kilometers, or any other unit of measurement. Altering the value will allow you to alter the selected unit length in a way similar to the way the SCALE command altered the size of objects. The distances specified in the linetype definition will be multiplied by the LTSCALE value to produce a new length for all patterns. The command sequence is:

Command: **LTS** (EnterKey)

New scale factor<1>: **2** (EnterKey)

By entering a value of two, all line segments and spaces will be doubled in size. The effect of LTSCALE on line segments can be seen in Figure 14.1. The line scale is determined by the size the drawing will be plotted at. Using an LTSCALE of 2 will make the line pattern twice as big as the line pattern created by a scale of 1. For a drawing to be plotted at ¼" = 1'-0", a dashed line that should be ⅛" when plotted, would need to be 3" long when drawn at full scale. This can be determined by multiplying .125" × 24 (the scale factor). The LTSCALE factor can be determined by using half of the scale factor. The drawing scale factor is always the reciprocal of the drawing scale. Using a drawing scale of ¼" = 1'-0" would equal .25" = 12". Dividing 12 by .25 produces a scale factor of 48 with an LTSCALE of 24. Keep in mind that when the LTSCALE is adjusted, it will affect ALL linetypes within the drawing. This is typically not a problem, unless you are trying to create one unique line. Typically, the LTSCALE factor can be preset as template drawings are set up. In most offices, template drawings are established based on the size of the paper or the scale the finished drawing will be plotted at. Common plotting scales and their respective scale factors include:

ARCHITECTURAL VALUES		**ENGINEERING VALUES**	
Drawing scale	LTSCALE	Drawing scale	LTSCALE
1" = 1'-0"	6	1" = 1'-0"	6
¾" = 1'-0"	8	1" = 10'	60
½" = 1'-0"	12	1" = 100'	600
⅜" = 1'-0"	16	1" = 20'	120
¼" = 1'-0"	24	1" = 200'	1200
3/16" = 1'-0"	32	1" = 30'	180
⅛" = 1'-0"	48	1" = 40'	240
3/32" = 1'-0"	64	1" = 50'	300
1/16" = 1'-0"	96	1" = 60'	360

CHANGING INDIVIDUAL LINE SEGMENT SIZES

Using LTSCALE changes the values of all of the lines in a drawing. Using the CELTSCALE command will allow the scale of existing lines to remain unaltered,

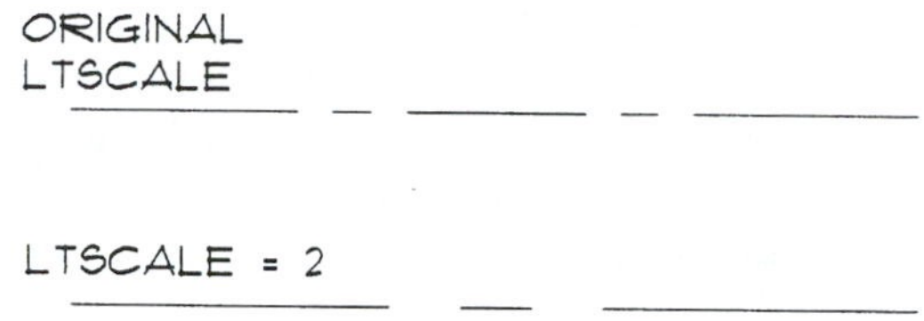

Figure 14.1 *What appears as a centerline on a 12" × 9" screen will appear as a continuous line when the LIMITS are changed to 100' × 75'. The LTSCALE command allows the length of line segments to be altered so that the pattern will still be seen.*

and change the scale of future lines. Typing **CELTSCALE** (EnterKey) at the command prompt will produce the prompt:

Enter new value for CELTSCALE <1.0000>:

The value that is assigned to CELTSCALE is used in combination with the value assigned to the LTSCALE. The CELTSCALE acts as a multiplier of the LTSCALE, producing a net scale effect on lines being drawn. With an LTSCALE of 2, and a CELTSCALE of .25, the net scale factor would be 2 × .25 = .50. Figure 14.2 shows the effects of using CELTSCALE to control the current line width.

NAMING LAYERS

Layers were introduced in Chapters 5 and 6 along with methods to create, alter, and control a layer. As your drawing ability increases, it is crucial to fully understand the importance of using layers. The use of layers will be increasingly important as the skills in the following chapters are introduced. Layers are used like separate sheets of

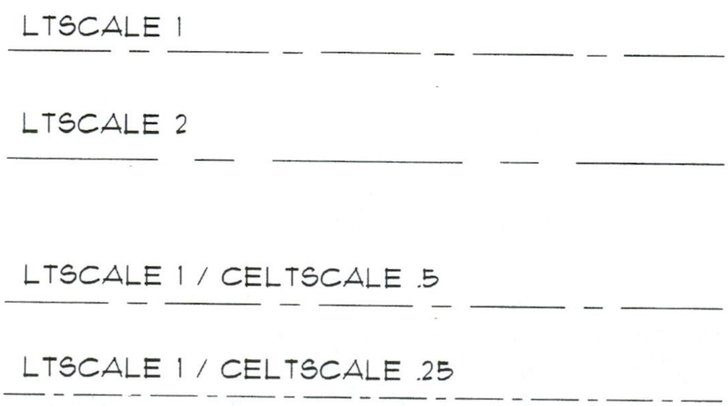

Figure 14.2 *The effect of LTSCALE and CELTSCALE on line length.*

paper to separate information within a drawing file. The base components of a floor plan can be placed on one layer; and the text, dimensions, framing information, and plumbing can be placed on separate layers to provide the information needed by various subcontractors. To help you control information, consideration must be given to layer names, controlling linetypes and colors by layer, as well as methods of displaying and hiding layer information.

LAYER NAMES

Names for layers have not become standardized throughout the architectural or engineering communities. Most offices have developed a standard method of assigning names, so that all members of the staff have easy access to information on a drawing. Even within a firm, names will vary depending on the type of project to be drawn.

As you set up names for layers, use the same guidelines for naming layers that were used to name linetypes. Use a name that will still be meaningful in six months. This might include layer names such as UWALLS, MWALLS, LWALLS, UBEAMS, MBEAMS, LBEAMS. Up to 255 characters (including spaces) may be used in naming a layer but it helps to keep layer names short and descriptive.

AIA STANDARDIZED LAYER NAMES

As with other areas of architecture and engineering, there is no one standard method of naming and using layers on construction drawings. The second edition of the AIA CAD Layer Guidelines was released in September of 1997. Copies of the standard can be obtained by calling the AIA Library at 1–800–365–2724 or from the AIA web site at www.aia.org. The guidelines are intended to coordinate with ISO (International Standards Organization) and CSI (Construction Specifications Institute) standards. The AIA has moved away from its original long and short format names, and recommends a name comprised of a discipline code and a major group name. A minor group name and a status name can also be added to the layer name, depending on the complexity of the project. The guidelines can be used to label both layers and drawing files with components of file names also used in the layer names. Because name components are common to both file and layer names, care must be taken to avoid creating project specific references within a layer name. Project designations should not be included in folder names. Use the guidelines presented in Chapter 5 for naming project folders, and the following guidelines for naming files and layers. Figure 14.3 shows examples of the new guidelines and how they can be applied to layers and files names.

Naming Model/Sheet Files

The guidelines that were introduced in Chapter 5 work well for projects that do not require input from several different subcontractors. When several firms are responsible for the project, one system that meets the needs of each discipline is required to

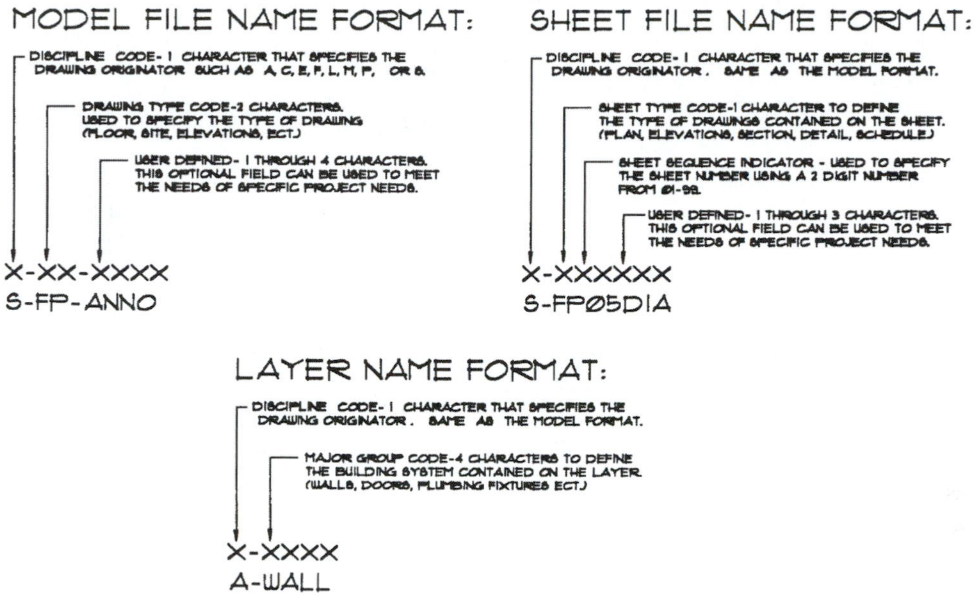

Figure 14.3 *The AIA CAD Layer Guidelines use abbreviations to name layers and drawing files. By using a uniform system to name layers and files, architects, engineers, and consultants working for several different firms can easily produce a coordinated set of drawings.*

name layers. The new AIA CAD standard can be used to name layers and drawing files. Drawing files can be classified as either a model or sheet file. Model files are the components of a set of working drawings that are drawn full scale. These are the elements—such as walls, door,s and windows—that you've been drawing in model space. Chapter 5 introduced guidelines for naming model files. A sheet file is a completed drawing, such as a floor plan, that is prepared for plotting. A sheet file may contain one or more drawings that will be plotted on one sheet of paper with a title block. Future chapters will introduce merging a model file into a template drawing using the INSERT and XREF commands. Sheet files allow for plotting in paper space at a scale of 1" = 1". Regardless of what type of file, the new CAD layer standards will provide a better means of tracking the contents of files and layers. The initial component of file and layer names is a discipline code.

Discipline Code

The discipline code is a character code that is used to identify the originator of the drawing. The discipline code will be the same for a model file, sheet file, or layer name. The code allows instant recognition of the originator of the drawing or layer. The code may be either one or two characters. If only one character is used, it is followed by a hyphen. The first character represents one of the following 16 disciplines:

Discipline Codes for the Drawing Originator

A	Architectural	M	Mechanical
C	Civil	P	Plumbing
E	Electrical	Q	Equipment
F	Fire Protection	R	Resource
G	General	S	Structural
H	Hazardous Materials	T	Telecommunications
I	Interiors	X	Other Disciplines
L	Landscape	Z	Contractor/Shop Drawings

The second character to be used should either be a hyphen or a user-defined modifier. Suggested modifiers might include G for graphics, K for kitchen, or V for audio-visual equipment. A layer or file with a discipline code of AG would represent architectural graphics, such as a rendering developed by the architectural firm.

Guidelines for Model Filenames

Figure 14.3 shows a sample of a model filename. The discipline code is followed by the drawing-type code. The AIA guideline lists eleven codes that are common to all disciplines. The codes are listed below and preceded by an asterisk that represents the space for the discipline code.

Model File Drawing Type Codes

*–FP	Floor Plan	*–SC	Section
*–SP	Site Plan	*–DT	Detail
*–DP	Demolition Plan	*–SH	Schedules
*–QP	Equipment Plan	*–3D	Isometric/3D
*–XP	Existing Plan	*–DG	Diagrams
*–EL	Elevation		

In addition to drawing-type codes that apply to all disciplines, specific codes are listed for architectural, civil, electrical, fire protection, interiors, mechanical, plumbing, structural, and telecommunications. A–CP for ceiling plans and A–EP for enlarged plans are examples of drawing types for the architectural field. S–FP for framing plans or S–NP for foundation plans are examples of structural drawing types. See the AIA CAD Layer Guidelines for a complete listing of discipline specific codes.

User-Definable Codes

The user defines the final four digits of the model filename. Guidelines for naming files and folders that were presented in Chapter 5 can be used for the final four digits. The fifth-level floor plan for a structure could be saved with a name of A–FP–2.05.

Guidelines for Naming Sheet Files

Figure 14.3 shows an example of a sheet filename. The discipline code is followed by a one-digit sheet-type code. The AIA guideline lists nine common codes that apply to all disciplines.

Sheet Type Designators

0	General (symbols, legends, notes, etc.)	5	Details
1	Plans	6	Schedules/Diagrams
2	Elevations	7	User-defined
3	Sections	8	User-defined
4	Large scale drawings that are not details	9	3D views

A two-digit sheet sequence number ranging from 01 to 99 follows the Sheet type designator. The last component of a sheet title is a three-digit or letter, user-defined code. A sheet file name of S–405CON could be used to represent a sheet of concrete details that have been assembled for plotting by the structural engineer. The sheet would be page 5 in the drawing set.

Guidelines for Naming Layer

The method for naming layers is similar to naming files. A layer name can be composed of the discipline and major group name. A minor name and status code can be added to the sequence. Figure 14.4 shows some alternatives for using components with layer names. The discipline code is the same as the one- or two-character code that is used for model and sheet file names.

Major Group Name

The major group code is a four-character code that identifies a building component specific to the defined layer. Major group layer codes are divided into the major groups of architectural, civil, electrical, fire protection, general, hazardous, interior, landscape, mechanical, plumbing, equipment, resource, structural, and telecommunication. Codes such as ANNO (annotation), EQUIP (equipment), FLOR (floor), GLAZ (glazing), and WALL (walls) are examples of major group codes that are associated with the architectural layers. A complete listing of major codes for each group can be found in the AIA Layer Guidelines.

Minor Group Name

The minor group code is an optional four-letter code that can be used to define subgroups to the major group. The code A–FLOR (architectural–floor) might include

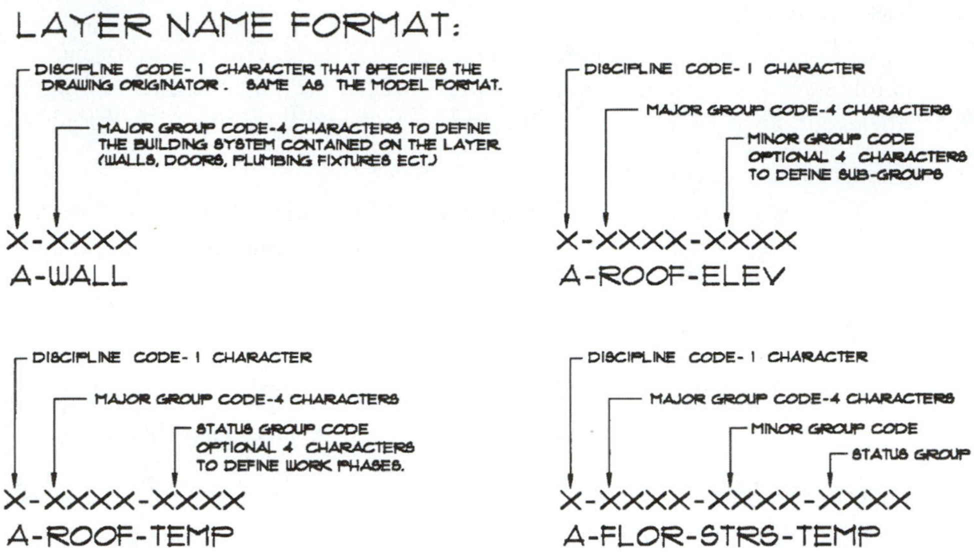

Figure 14.4 *Layer names fall into one of four categories when based on the AIA CAD Layer Guidelines.*

minor group codes for OTLN (outline), LEVL (level changes), STRS (stair treads or escalators), EVTR (elevators), or PFIX (plumbing fixtures). A layer name of A–FLOR–IDEN would contain room names, numbers, and other related titles or tags. A complete listing of minor group codes specific to each discipline is listed in the AIA guideline.

Status Code

The status code is an optional four-letter code that can be used to define a subgroup of either a major or minor group. See Figures 14.4. The code is used to specify the phase of construction. The layer names for the walls of a floor plan (A–WALL–FULL) could be further described using the status code of NEWW (new work), EXST (existing walls), or DEMO (demolition). A wall shown on the floor plan that is to be moved would be displayed on the A–WALL–FULL–MOVE layer.

CONTROLLING LAYERS

Layers can be controlled by using the Layer Properties Manager dialog box or by using the Layer Control menu display. The Layer Properties Manager dialog box shown in Figure 14.5 can be accessed by selecting the Layer icon from the Object Properties toolbar, by entering **LA** (EnterKey) at the command prompt, or by selecting Layer... from the Format pull-down menu. Previous chapters introduced each element of the Layer and Linetype Properties. This chapter will introduce methods of setting and controlling each of the layer elements.

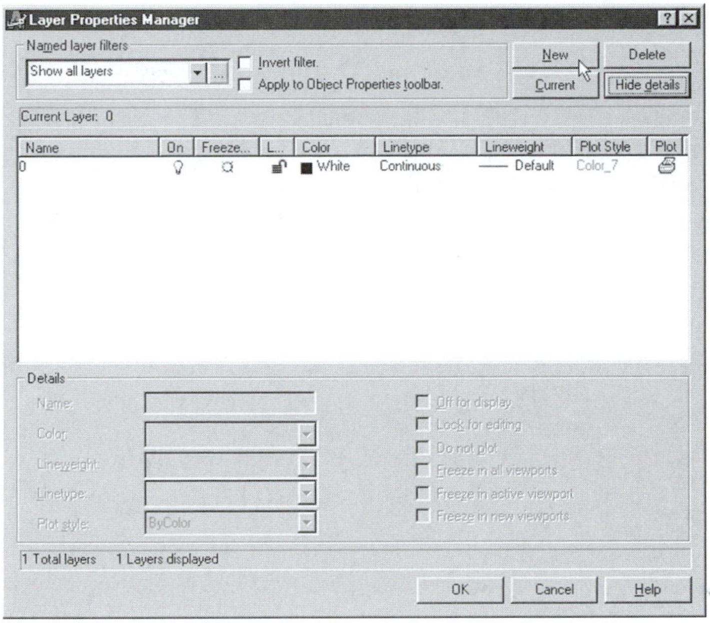

Figure 14.5 *The Layer Properties Manager dialog box can be used to control the characteristics of linetype and layers. It can be accessed by selecting the Layer icon from the Object Properties toolbar or by entering* ***LA*** *(EnterKey) at the command prompt.*

ASSIGNING COLOR BY LAYER

Colors and linetypes can be assigned by layers after a layer has been created. See Chapter 7 for a review of how to create a layer.

Figure 14.6 shows an example of the dialog box with the addition of the MWALLS, UPWALLS, and BWALLS layers. Notice that the Current Layer is still 0, but new layers have been added. As new layers are created, they are initially placed at the bottom of the layer list. As the screen is regenerated, the names will be displayed in alphabetical order. Once a layer has been created, the qualities for that layer color and linetype can be assigned.

The MWALLS layer is currently listed as black. Any lines that are created on this layer will be drawn as a black line. To alter the color of a layer, select the layer so that the layer name is highlighted. The color of each layer can be altered by selecting the color icon from the color list or by choosing Color from the Details box. The Details box is not active until a layer is highlighted and the Show Details button is selected. Selecting the color icon for the selected layer will display the color palette seen in Figure 14.7. As the desired color is selected from the palette, the name of the color will be displayed in the Details box, and the color will be displayed. The color for a layer can also be set by selecting the color icon in the Color edit box or by selecting the

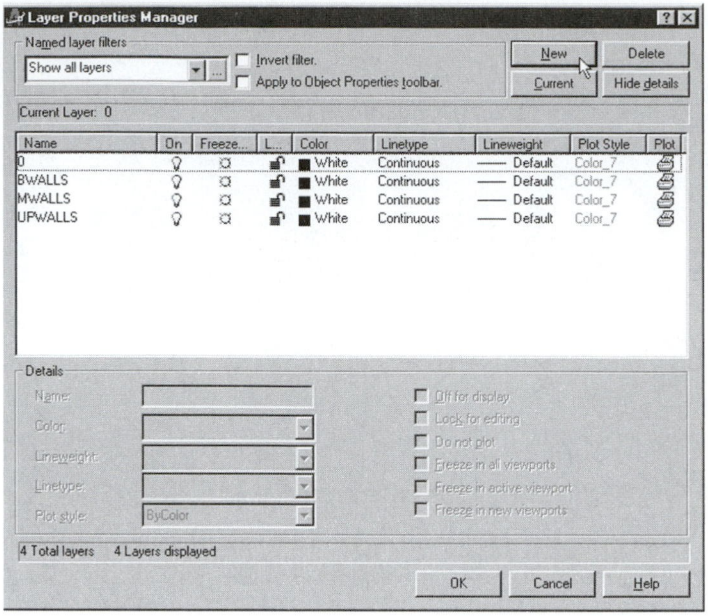

Figure 14.6 *To create new layers, press the New button. A layer with a named Layer 1 will be added. Highlight the name of Layer 1 and type in a more descriptive name.*

arrow beside the edit box in the Object Properties toolbar. Either method will produce the color menu shown in Figure 14.8. Color can be specified by selecting the specific color icon or by entering the name or number of the color. Selecting the Other option of the menu will display the color palette shown in Figure 14.7. If you want to select another color, simply select another color chip and the display will be updated. When you're satisfied with the color, select the OK button and the palette will be removed, and the Layer Control dialog box will be redisplayed. The color of individual objects on a layer can be altered by selecting the Color icon from the Object Properties toolbar. Selecting this icon will allow the color of the layer to remain unchanged, but allow all items drawn after selecting the icon to have a color different than the specified layer color.

Objects and layers may be assigned any color you desire. It may be helpful to assign like objects of different layers a different color. For instance, if all plumbing fixtures are blue, it may not be readily apparent if the UPLUMBING fixtures are ON when only the lower-layer components are desired. Assigning a different color to the LPLUMB, MPLUMB, and UPLUMB layers may help you keep track visually of a drawing as different layers are set to ON/OFF for viewing or plotting. Color, as it relates to plotting, will be introduced in future chapters.

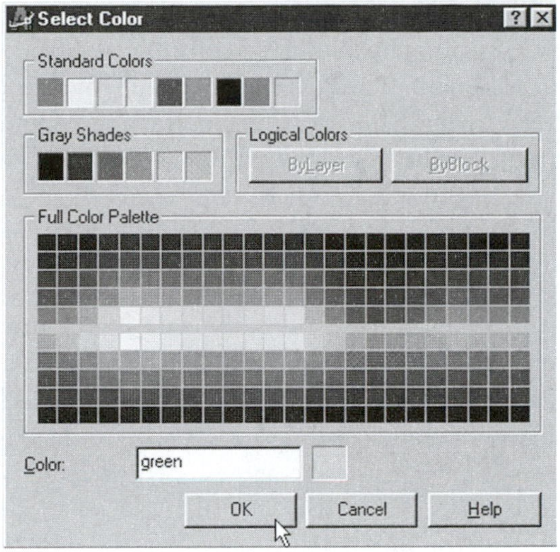

Figure 14.7 *Selecting the Color icon from the Layer Properties Manager dialog box will display the Select Color palette, allowing the current layer color to be changed.*

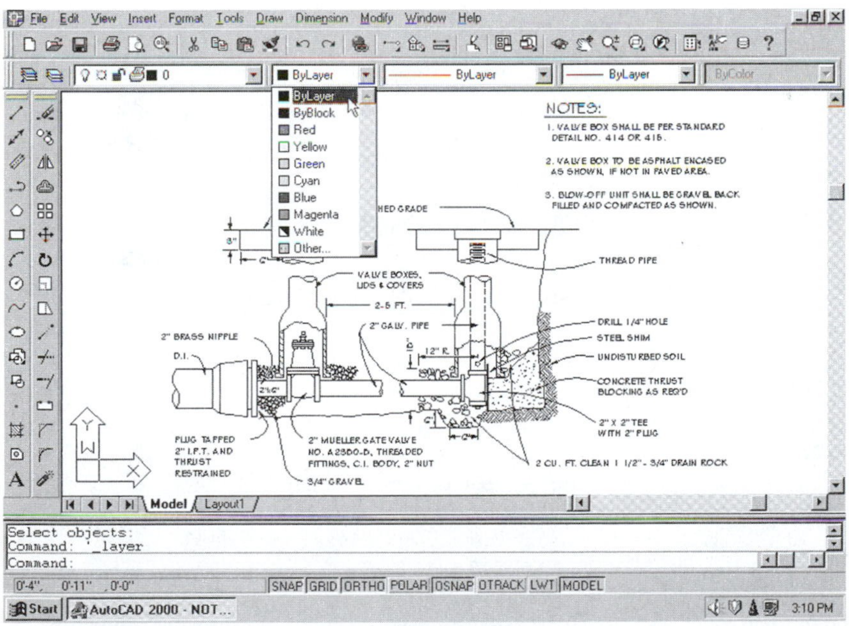

Figure 14.8 *The color for a layer can be altered by selecting the Color icon from the Object Properties toolbar or the down arrow beside the color edit box.*

ASSIGNING LINETYPES TO LAYERS

The current linetype default for each layer is a continuous line. In previous chapters, methods of drawing different linetypes were introduced. The LINETYPE command allows for setting and drawing different linetypes without regard to the layer. The Linetype option of LAYER will allow all lines on a specific layer to be drawn as one type of line. Linetypes can now be assigned to a specific layer and all lines on that layer will be drawn with the selected linetype.

The Linetype can be altered using methods similar to altering layer color. Highlighting the name of the layer to be altered, and then selecting the linetype name will display the Select Linetype dialog box shown in Figure 14.9. The dialog box can be used to load and select linetypes. Selecting the Load... button will display the Load or Reload Linetypes dialog box shown in Figure 14.10 and display a list of linetypes contained in AutoCAD. Highlighting the names of the desired linetypes to load and selecting the OK button will remove the dialog box and restore the Select Linetype dialog box. As the box is restored, it will be updated to reflect the linetypes that were just loaded. One of these linetypes can now be selected from the Select Linetype dialog box and assigned to the layer that is highlighted in the Layer control Dialog box. Linetypes are assigned by selecting the linetype and the OK button. As the button is selected, the Layer Properties Manager dialog box will be redisplayed, reflecting the selected linetype.

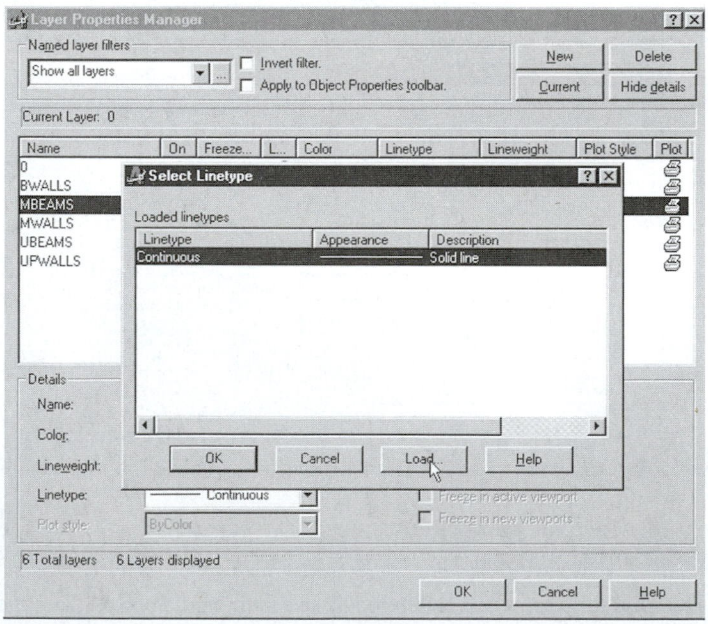

Figure 14.9 *Selecting the name of the current linetype from the Layer Properties Manager dialog box will display the Select Linetype dialog box, allowing for selection of one of the current linetypes.*

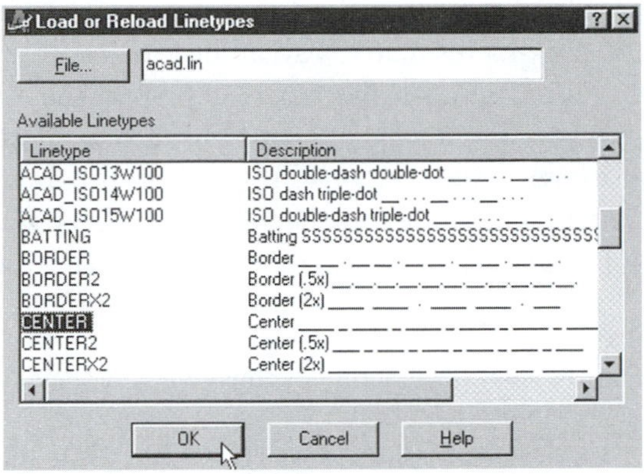

Figure 14.10 *Linetypes can be added to the drawing base by highlighting the desired layer name with the cursor and selecting the OK button.*

Note: To select multiple linetypes, hold down the (CtrlKey) while you select the linetypes. To load a sequence of linetypes, hold down the (ShiftKey) and select the first and last listing. All linetypes that are listed between the first and last listings will be selected. These two selection methods can be used anytime multiple objects are to be selected from a menu.

SETTING THE CURRENT LAYER

Up to this point, you've created a layer with the name of MFLOOR and assigned the color red and a linetype of CENTER. If you exit the dialog box and start to draw, nothing will appear to have changed. To draw red centerlines, the current layer must be set to MFLOOR. To make the MFLOOR layer current, highlight the name of the layer in the Layer Properties Manager dialog box and select the Current button. Selecting the OK button will keep this change and restore the drawing screen. As you enter the LINE command, segments will be drawn using the selected color and linetype that were just added. A new layer can also be made current by selecting the down arrow beside the Layer manager of the Object Properties toolbar and selecting the desired layer.

SETTING LAYERS ON/OFF

All of the new layers you have created up to this point are set in the ON option. If you were to switch to a layer such as UPWALLS, and then draw, you would be able to see what is drawn. Although in a multilevel structure it is sometimes helpful to view the walls of several different levels to check alignment. Typically, you will want to have only one level of a structure current. By setting a layer to OFF, objects drawn on that layer will be invisible. The information is retained, but not displayed or plotted.

The Floor Layer Table shows a listing for the layers contained in the drawing file for the plan files for a simple residence. The drawing file contains the floor, framing, and electrical plans for each level as well as the foundation, roof, and site plans. Figure 14.11 shows the floor plan of a residence with all of the nonstructural material ON, and all of the structural material OFF. Figure 14.12 shows the same floor plan with the structural material layers toggled to ON, and the architectural information set to OFF. This particular drawing file contains information for the floor plan, framing plan, electrical plan, roof plan, foundation plan, and site plan. By use of the ON/OFF option, each drawing can be displayed and edited. Each plan can be seen in Figure 14.11 through 14.16.

Drawing Layers for Electrical, Floor, Foundation, Framing and Site Plans

Base appliances	Hidden2	Frame base dimension	Continuous
Base doorwin.	Continuous	Frame beams	Center
Base plumb	Continuous	Frame exterior dimen.	Continuous
Electrical legend	Continuous	Frame joist	Continuous
Electrical symbols	Continuous	Frame interior dimen.	Continuous
Electrical text	Continuous	Frame lateral	Continuous
Electrical title block	Continuous	Frame lateral schedule	Continuous
Electrical wiring	Dashed	Frame general notes	Continuous
Floor cabinets	Continuous	Frame symbols	Continuous
Floor doorwin sch.	Continuous	Frame text	Continuous
Floor doorwin sym.	Continuous	Frame title block	Continuous
Floor general notes	Continuous	Roof beams	Center
Floor text	Continuous	Roof framing	Continuous
Floor title block	Continuous	Roof symbols	Continuous
Floor walls	Continuous	Roof text	Continuous
Foundation beams	Center	Roof title block	Continuous
Foundation ext. dimen.	Continuous	Roof walls	Dashed
Foundation footings	Hidden	Site access roads	Center
Foundation general notes	Continuous	Site dimensions	Continuous
Foundation general notes	Continuous	Site existing grades	Dashed
Foundation interior dimen.	Continuous	Site flatwork	Continuous
Foundation joist	Continuous	Site finish grades	Continuous
Foundation lateral	Continuous	Site notes	Continuous
Foundation notes	Continuous	Site property line	Phantom
Foundation piers	Hidden2	Site setbacks	Phantom2
Foundation symbols	Continuous	Site symbols	Continuous
Foundation text	Continuous	Site title block	Continuous
Foundation title block	Continuous	Site walls	Continuous
Foundation walls	Continuous		

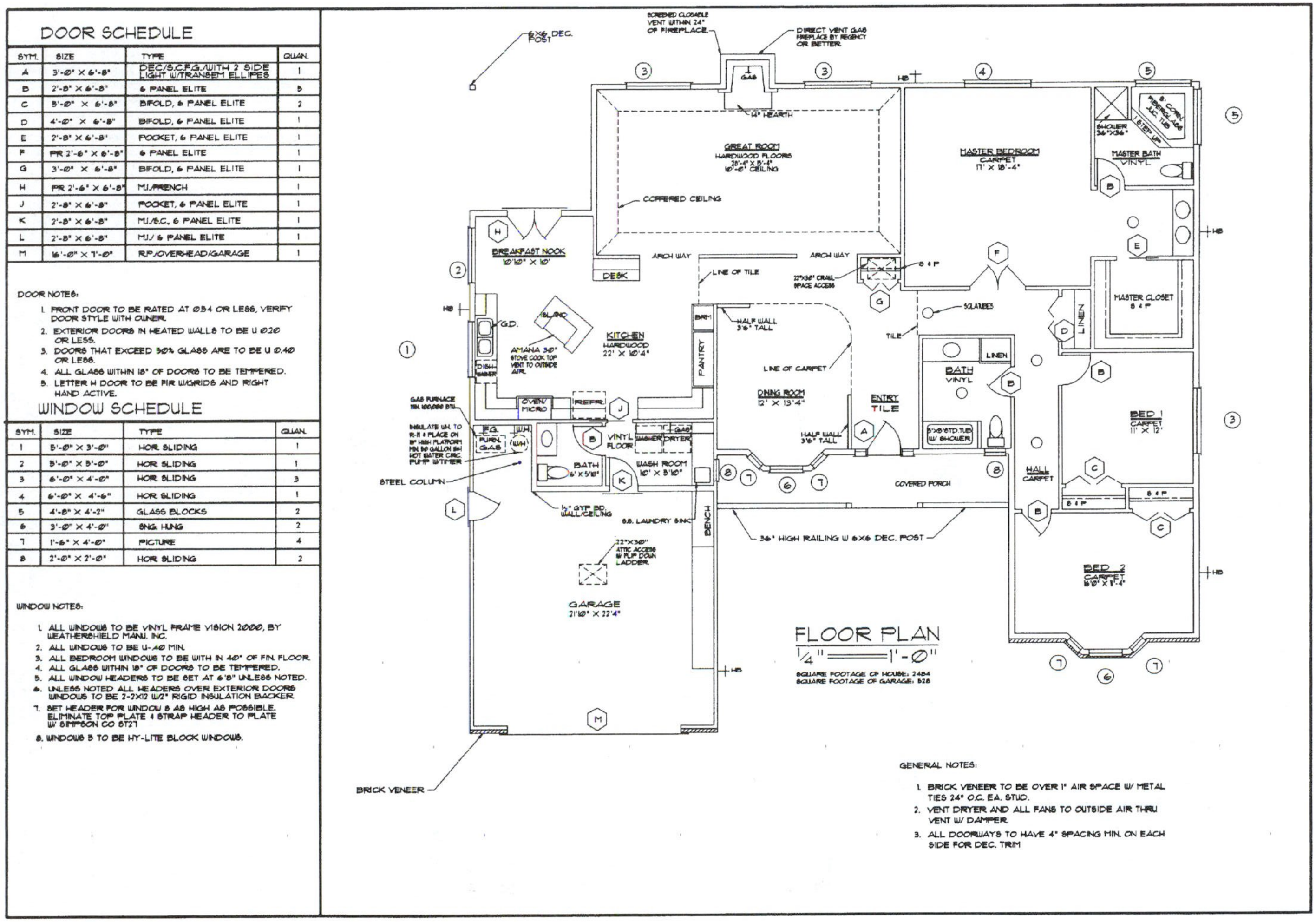

Figure 14.11 *All of the drawings seen in Figures 14.11 through 14.16 are contained in the same drawing file. Each drawing is stacked over each other using the layers. By use of a layer matrix, layers can be easily selected to display the desired drawing. (Courtesy Tereasa Jefferis.)*

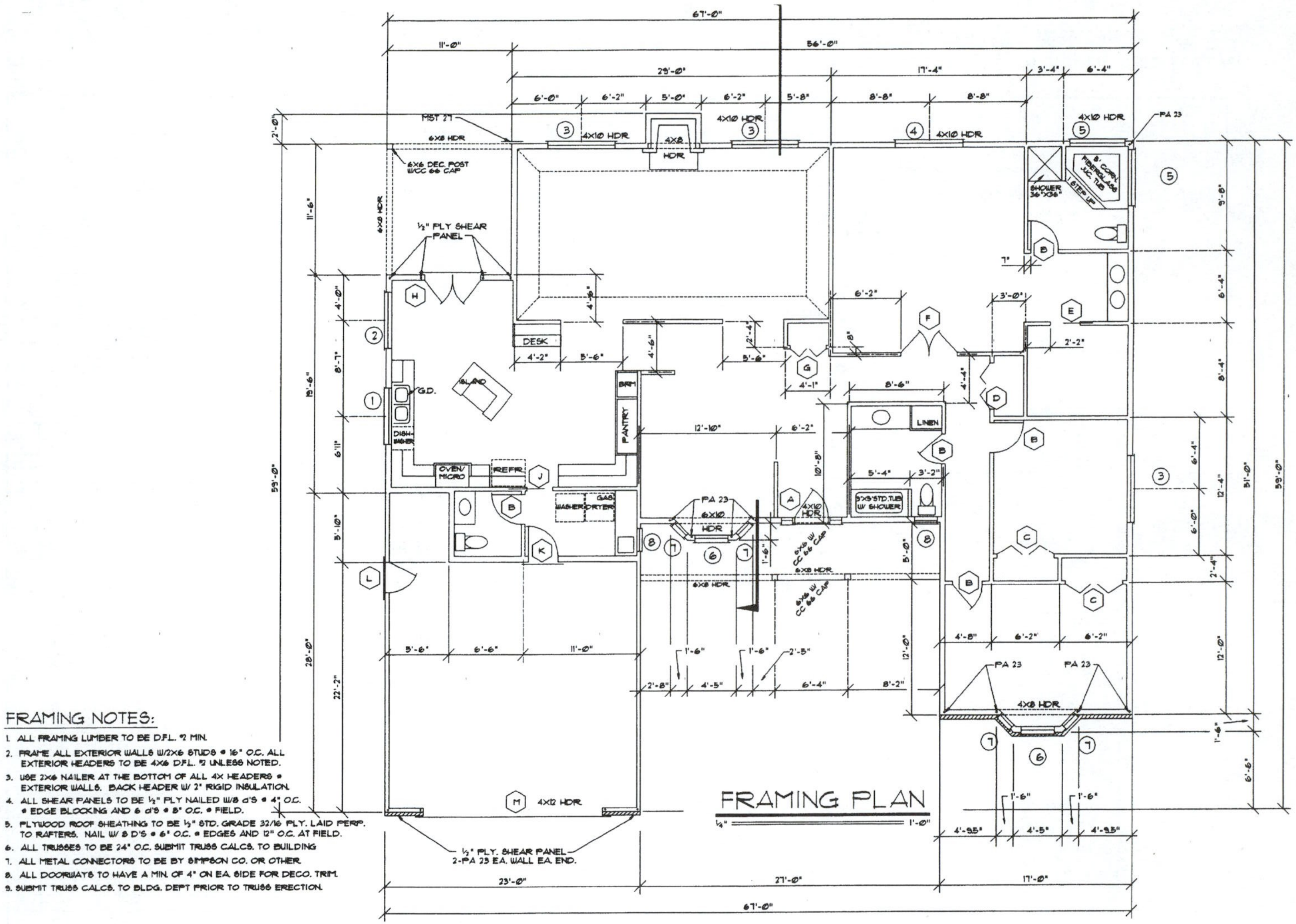

Figure 14.12 *The framing plan for the residence seen in Figure 14.11. (Courtesy Tereasa Jefferis.)*

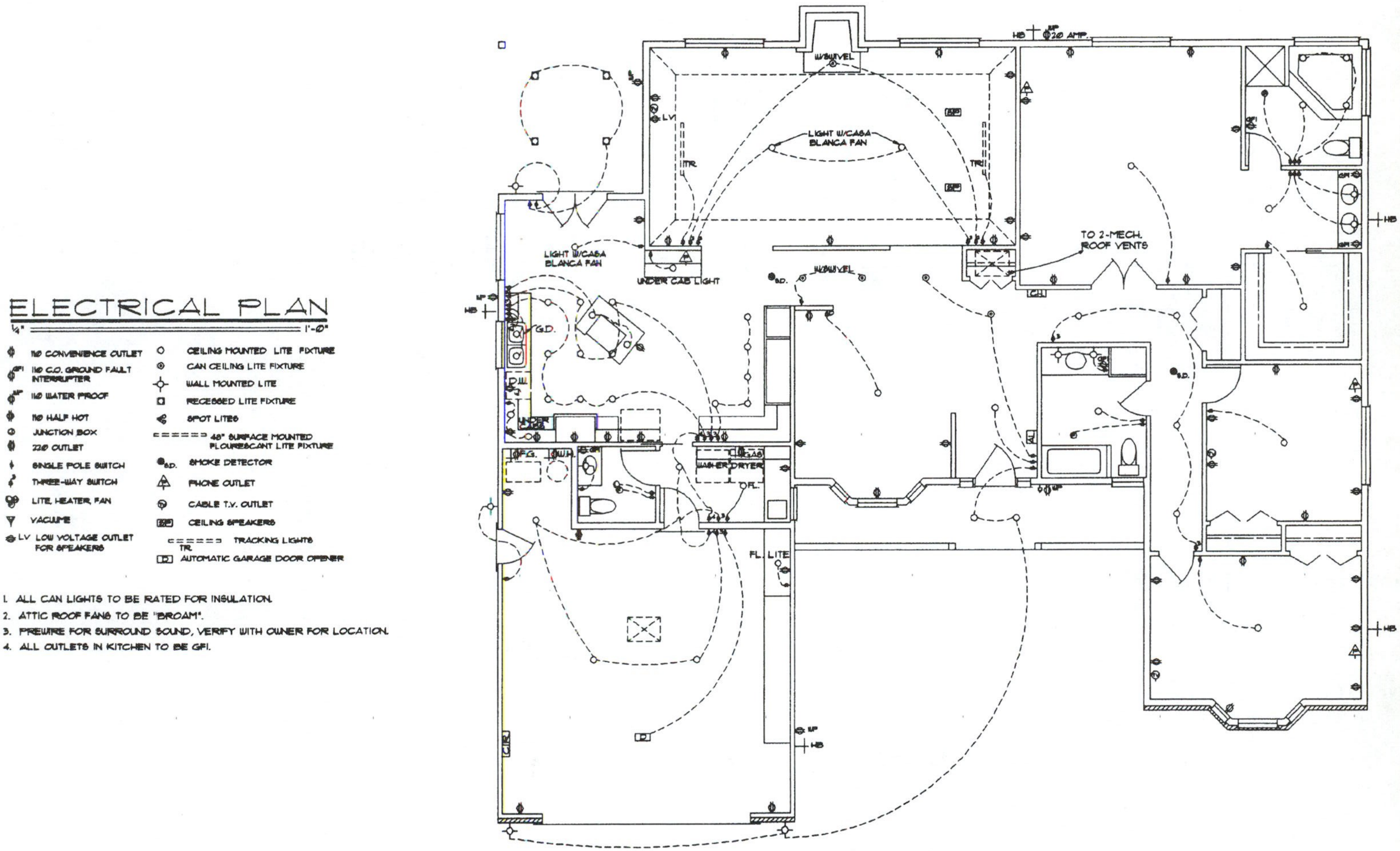

Figure 14.13 *An electrical plan uses the walls, windows, doors, and cabinets of the floor plan for a drawing base. (Courtesy Tereasa Jefferis.)*

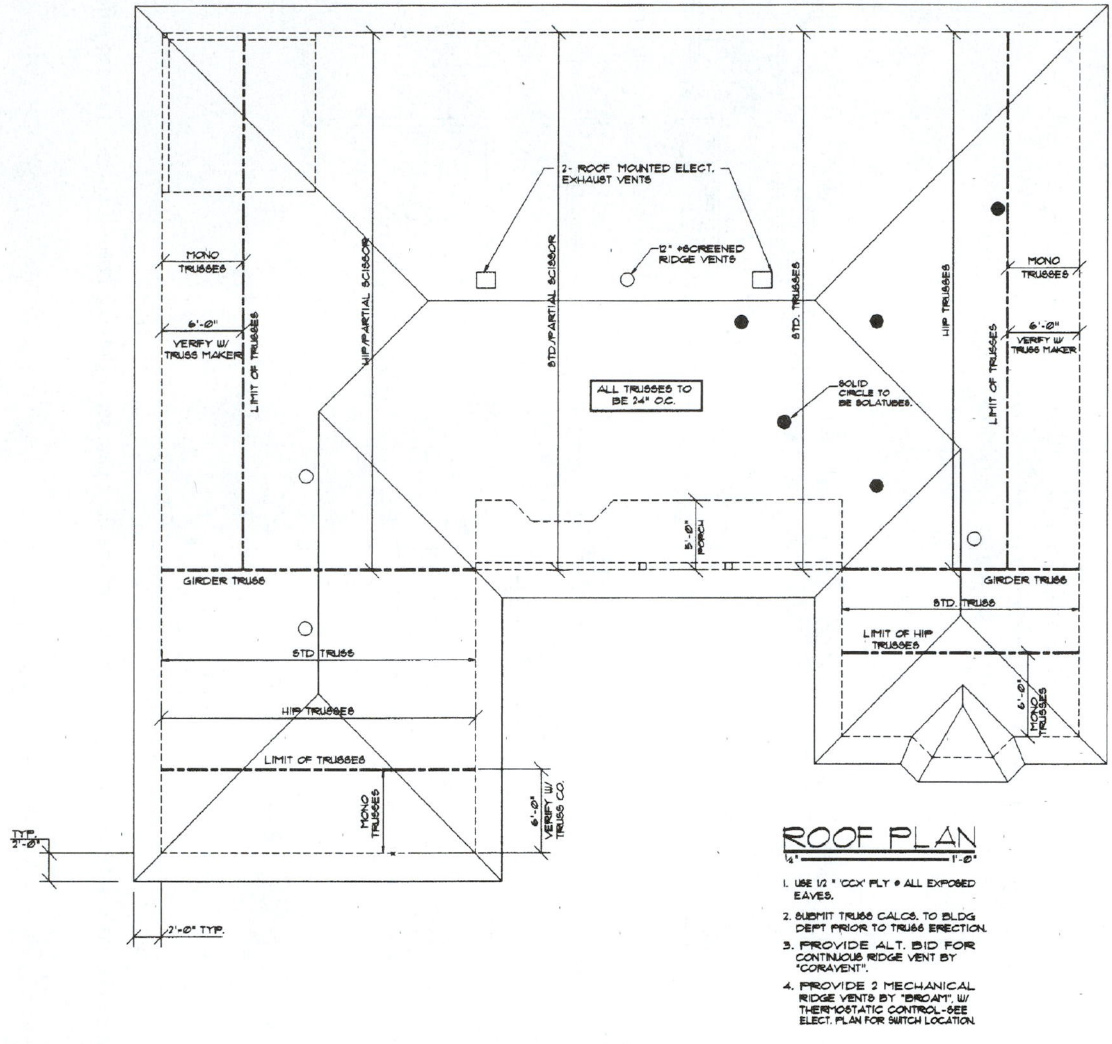

Figure 14.14 *None of the layers on the floor, framing, or electrical plans are used on the roof plan, but the walls serve as an outline for the roof. Using OSNAP and the LINE command, with the walls thawed, the outline can be quickly drawn before freezing all other layers. (Courtesy Tereasa Jefferis.)*

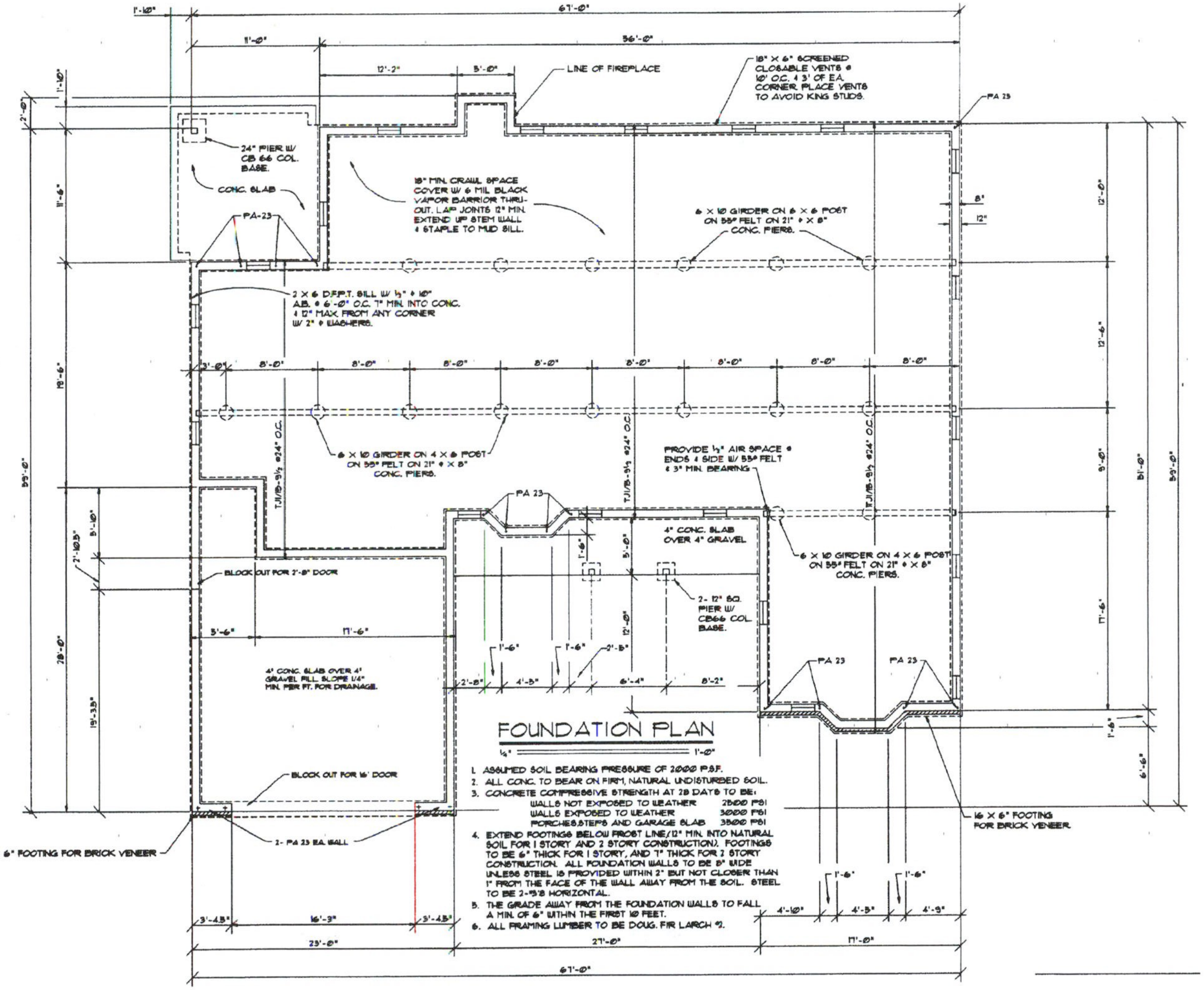

Figure 14.15 *The foundation can be drawn using the walls of the floor plan as a guide. With careful planning and good layer control, many of the dimensions on the floor plan can be used on the foundation. (Courtesy Tereasa Jefferis.)*

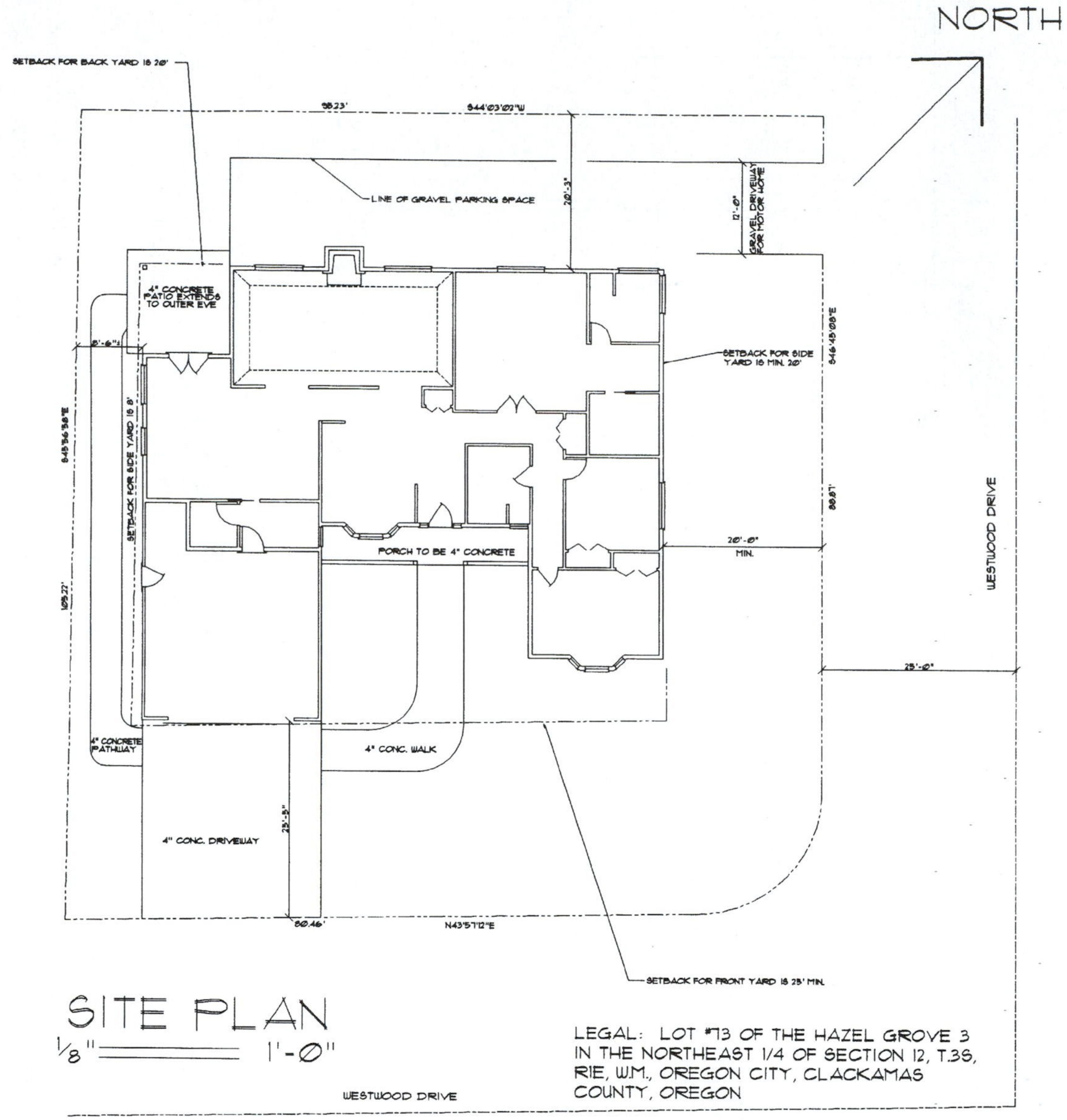

Figure 14.16 *A site plan often requires careful use of varied linetypes and CELTSCALE. (Courtesy Tereasa Jefferis.)*

Layer Off

To turn layers off, open the Layer Properties Manager dialog box. The ON/OFF icon is the light bulb—with yellow representing ON and black representing OFF. Select the ON/OFF icon for the desired layers to be turned off, followed by the OK button. The dialog box will be removed and any information that is stored on a layer that is off will be removed from the screen. If you attempt to set the current layer to be turned OFF you will be given the following prompt:

The current layer is turned off.

If you select the OK button, the dialog box is removed, and the items on the current layer are removed from the drawing screen. To restore the current layer, reenter the Layer Properties Manager dialog box and restore the current layer. If you accept the Current layer OFF, the current layer will be made invisible. Now if you draw objects, they will be drawn, but not displayed. This situation occurs more by accident than by design. If you find yourself drawing a line, but nothing is displayed, check the listing to see if the current layer has been accidently switched OFF.

Layer On

Layers that have been made invisible by the OFF option can be restored by the ON option. Open the Layer Properties Manager dialog box and highlight the desired layers to be set to ON. Selecting the ON/OFF icon followed by the OK button will restore the selected layers, remove the dialog box, and display the information on the selected layers.

THAWING AND FREEZING LAYERS

If you have set layers to OFF, you removed information from the display screen but not from the drawing base. The information merely became invisible. If you had requested a REGEN, it would have been performed with all of the information on the ON layers redisplayed. Information on layers that were set to OFF also would have to be processed which will cause a slight slowdown on large drawings. Using the Freeze/Thaw options, you will be able to control the extent of information to be scanned during the REGEN process. Layers that are in the Thaw mode are processed for a REGEN, information that is FROZEN is not processed during a REGEN, reducing the time required to REGEN. This will only be a factor on huge drawing files.

Freeze

Open the Layer Properties Manager dialog box, select the layer that you would like to Freeze, and select the Freeze button. The sun icon (Thaw) will be changed to a snowflake (Freeze). Selecting the OK button will activate the selection, restore the drawing screen, and remove the objects on the selected layer from the display. Unlike the OFF option, the current layer cannot be frozen.

Thaw

The Thaw option of the LAYER command will allow for specified frozen layers to be thawed. To thaw a layer that is frozen, open the Layer Properties Manager dialog box and select the snowflake icon of the layer to be thawed. As the snowflake is selected, it will change to the sun icon, indicating that the layer has been thawed. Selecting the OK button will activate the selection, restore the drawing screen, and display the material on the layer that had been frozen.

SAFEGUARDING LAYERS

The final pair of options of the LAYER command are Lock and Unlock. As the name implies, a lock is used to protect something. AutoCAD allows you to lock layers so that they cannot be edited accidentally. Information on a locked layer is still visible; it's just protected. To lock a layer, highlight the desired layer and select the Lock icon followed by the OK button.

To unlock a layer, highlight the desired layer and select the Unlock button, followed by the OK button. This process will reverse the consequences of the Lock. Remember that locking a layer is not a security device, but a method of protecting information on a layer from careless editing. Entities on a locked layer will not be included in a selection set of an edit command. Locking a layer will allow you to view the information on the layer, but will not affect it as you edit a drawing. You can continue to draw on a locked layer and objects can still be selected for OSNAP locations, thawed or frozen, or altered using the Linetype or Color icons. Objects on locked layers can't be edited.

FILTERS

Filters provide the ability to display certain layer names in the listing box, based on a specific attribute. Layers can be filtered based on layer name, color, lineweight, contents, locked or unlocked status, freeze or thaw status, and plot styles. There are several other filter options that will not be considered until viewports are explored in future chapters. Filters are controlled in the Layer Properties Manager Dialog box using the Named layer filters edit box arrow. Choosing the Filter edit arrow will produce the menu shown in Figure 14.17. Selecting the [...] will produce the menu shown in Figure 14.18 and allow the parameters of the filter to be set. In the example, the filter is based on the HIDDEN linetype. Selecting the Add button and then Close will shut the Named Layer Filters dialog box. Selecting the Filter Edit arrow will display the menu shown in Figure 14.17, with the HIDDEN filter shown. Selecting HIDDEN will produce a layer listing of layers created using the specified criteria. (See Figure 14.19.) All of the existing features are still displayed on the screen—they just will not be listed in the Layer display until the filter is altered. Opening the Named layer filters menu and selecting the Show All Layers option will restore the total layer listing.

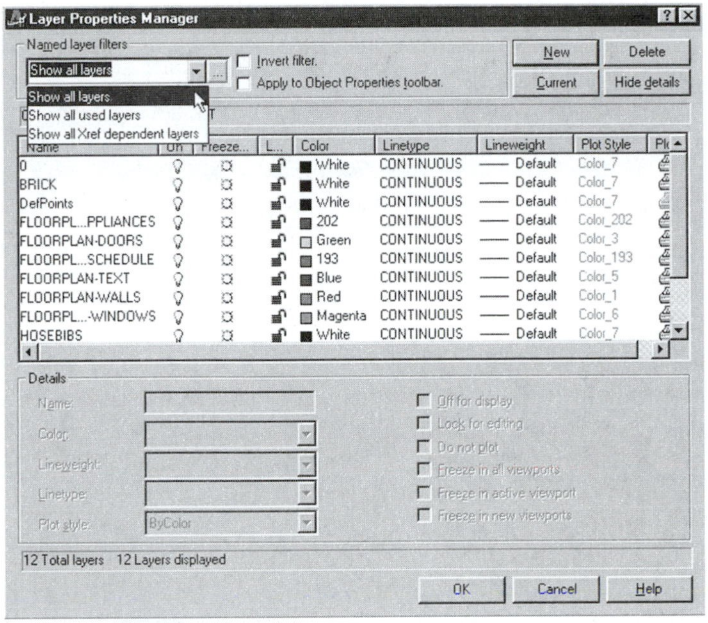

Figure 14.17 *Selecting the arrow beside the Named Layer Filters edit box will display this menu, which can be used to established filters to display.*

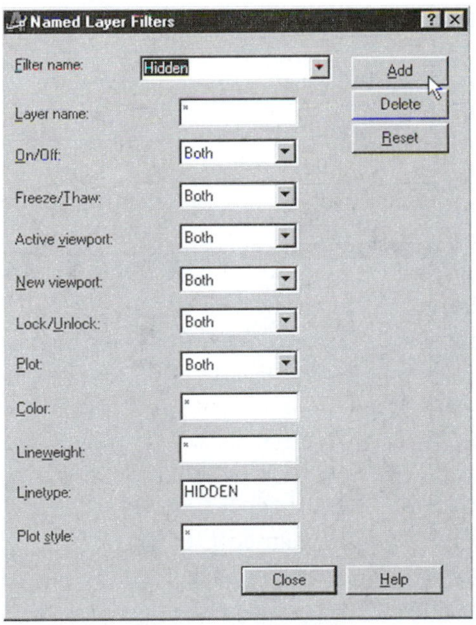

Figure 14.18 *Selecting the [...] option will produce this Named Layer Filters menu, allowing the scope of the filter to be defined.*

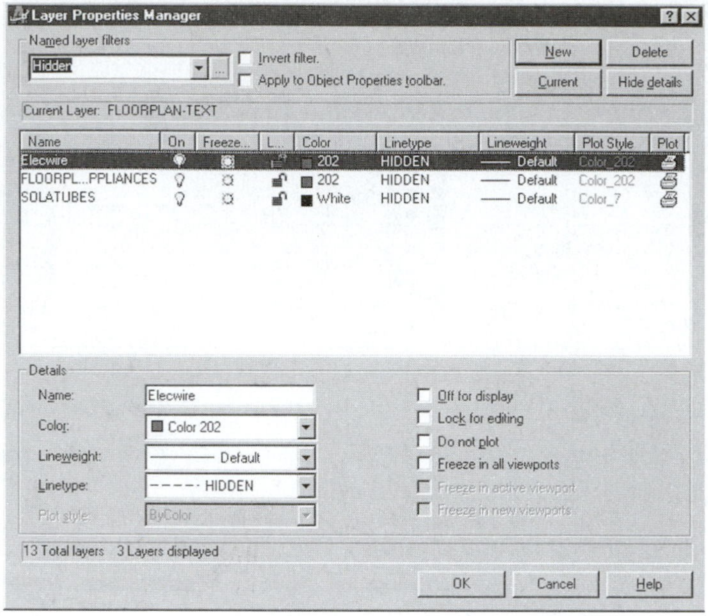

Figure 14.19 *The layers defined based on the linetype HIDDEN.*

RENAMING LAYERS

The current name of a layer can be altered to better define the use of the layer at any time throughout the life of a drawing. Only the 0 layer and XREF layers can't be renamed. XREF layers are layers that are created on a drawing that is attached to the current drawing, and will be explained in future chapters. Layers are renamed inside of the Layer Properties Manager dialog box by selecting the name of the layer to be renamed, and then selecting it again. As it is selected the second time, the name will be highlighted and a box will be placed around the original name with a blinking cursor. Pressing the backspace arrow while the name is highlighted will remove the existing name and allow a new name to be entered. Pressing (EnterKey) once will accept the name. Pressing OK removes the dialog box and restores the drawing area.

DELETING LAYERS

An unused layer can be deleted from the drawing base at any time throughout the life of the drawing—with the exception of the 0 layer or an XREF layer. Layers that are referenced to a BLOCK and a layer named DEFPOINTS can not be deleted, even if the layer is empty. XREF, DEFPOINTS, and BLOCK will be explored in future chapters. Unused layers can be deleted from inside of the Layer Properties Manager dialog box by selecting the name of the layer to be removed, and then selecting the Delete button. As the button is selected, the layer will be removed from the list.

ALTERING LINETYPE, COLOR, AND LAYERS

AutoCAD allows individual properties within a layer to be altered, as well as changing the properties of the entire layer. The Layer Control Icon from the Object Properties toolbar can be used to alter the qualities of an entire layer. Individual entities of a layer can be changed by using the Make Object's Layer Current icon, by using the MATCHPROP, Object Properties, Properties, or QSELECT commands.

LAYER CONTROL MENU

The Layer control menu shown in Figure 14.20 can be used to alter the visibility of an existing layer. The menu is displayed by selecting the Layer Control arrow on the right side of the box in the Object Properties toolbar. This menu can be used to toggle ON/OFF, Thaw/Freeze, Lock/Unlock, Plottable/Non-Plottable, and to set the current layer. The visibility of layers is adjusted by selecting the ON/OFF or Thaw/Freeze icon. As the icon for each layer is selected, the layer will be displayed or removed, depending on the setting of the icon. Layers are locked and unlocked by selecting the icon of the desired layer. The Plottable/Non–Plottable icon allows a layer to remain visible on the screen, but removed from the plotting set. In the default setting, all layers will be plotted. Selecting a layer and choosing this icon will set the

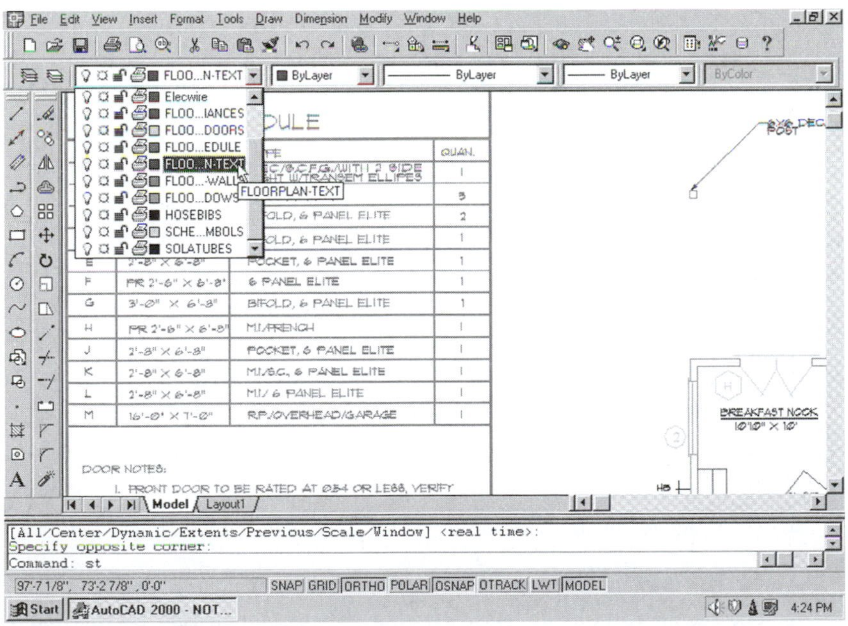

Figure 14.20 *The Layer Control menu can be accessed by selecting the Layer Control arrow in the Object Properties toolbar. The menu can be used to toggle ON/OFF, Thaw/Freeze, Lock/Unlock, and Plottable/Non-plottable.*

layer to non-plottable. The current layer is altered by selecting the name of the layer that you want to make current. As soon as the layer name is selected, the menu will be closed, the layer will be changed, and the drawing area will be restored.

Note: The Plottable/Non-plottable control can be a great asset for displaying information that you need to see, but may not be needed for others. A non-plottable layer could be created for placing questions to ask the engineer or instructor. The questions can be seen on the display, but they will not be plotted.

MAKING AN OBJECT'S LAYER CURRENT

The Make Object's Layer Current icon is located on the far-left end of the Object Properties toolbar. The icon can be used to alter the current layer. Selecting the icon displays the following prompt at the Command prompt:

Command: _ai_molc

Select object whose layer will become current:

The cursor will be changed to a pickbox as the program waits for you to select a drawing entity. Selecting a drawing object will make the layer that contains the selected object the current layer. The command prompt will then display the name of the current layer, and provide the command prompt so that drawing may continue.

ALTERING THE DISPLAY WITH MATCHPROP

The properties of one layer object can be transferred to another object by the use of the Match Properties command. The Match Properties icon is found in the Standard toolbar. Two other methods of executing the command include the Modify pull-down menu or by entering MA (EnterKey) at the command prompt. The sequence will copy the properties of the first object to the second object, including the layer, the linetype, and the color. Either method will produce the following prompt:

Command: **MA** (EnterKey)

MATCHPROP

Select Source Object: *(Select the object with the properties that you would like to copy.)*

Current active settings: Color Layer Ltype Ltscale Lineweight Thickness

PlotStyle Text Dim Hatch

Select destination object (s) or [Settings]: *(Select the object that you would like to change.)*

Select destination object (s) or [Settings]: (EnterKey)

Command:

Instead of selecting the object that you would like to change, entering **S** (EnterKey) (Settings) at the second prompt will produce the Property Settings dialog box shown in Figure 14.21. The box can be used to specify the qualities that will be copied to selected objects. One or more of the properties can be assigned including the Color, Layer, Linetype, Linetype Scale, and Thickness—as well as several other qualities that will be discussed in later chapters. Once the qualities have been selected, select the OK button and resume the command sequence by selecting the object to receive the new properties.

CHANGING OBJECT PROPERTIES

The PROPERTIES command can be used to control the properties of existing objects. The command can be started by selecting the Properties icon from the Standard toolbar, by selecting Properties... from the Modify pull-down menu, or by typing **MO** (EnterKey) or **CH** (EnterKey) at the command prompt. Each method will display the Properties dialog box similar to Figure 14.22, showing the qualities of the current layer. Object properties can be listed in alphabetical order or categorized by groups depending on which tab is selected. The dialog box allows drawing properties including the Color, Layer, Linetype, Handle, Thickness, and Ltscale to be altered. To alter a property for the selected object, choose either the name or value of the desired property. Each property will display a specific flyout menu. Figure 14.23 shows the menu for altering lineweights. Once the desired properties are altered, select the Close button to remove the dialog box and return to the drawing.

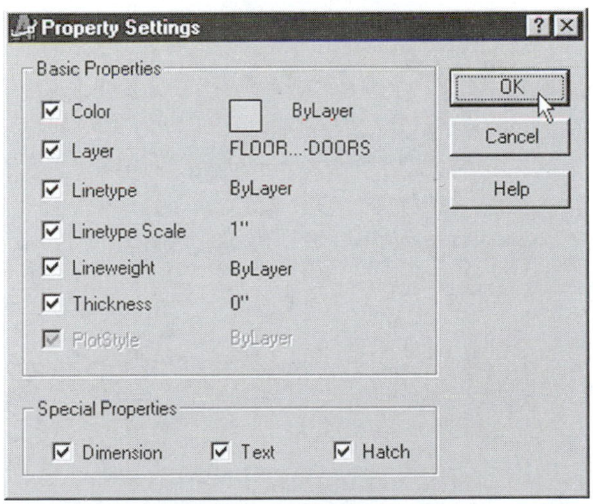

Figure 14.21 *Once a source object has been selected, entering* **S** *(EnterKey) (Settings) at the command prompt will display the Property Settings dialog box. The dialog box allows specific properties of a layer to be altered without effecting other layer qualities.*

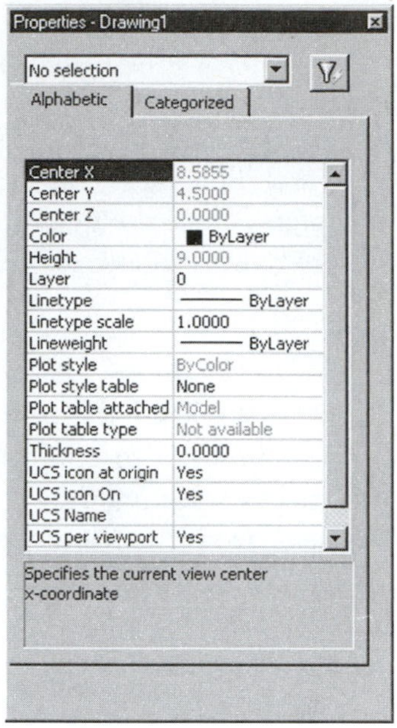

Figure 14.22 *The Properties dialog box can be used to control the properties of existing objects.*

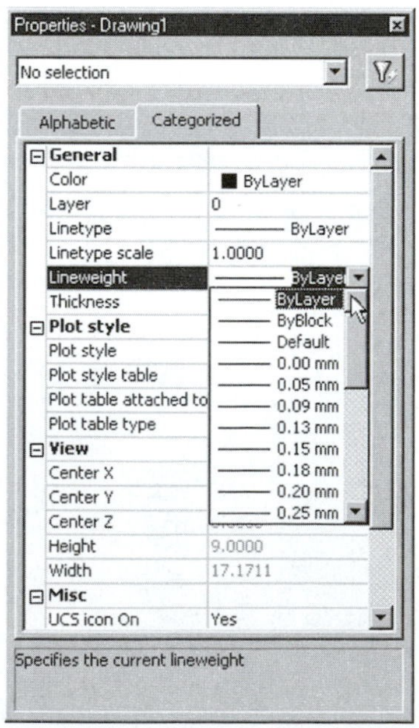

Figure 14.23 *To alter a property for the current object, choose either the name or value of the desired property. Each property will display a specific flyout menu.*

CHANGING OBJECTS WITH QSELECT

The QSELECT command can be used to create selection sets for altering the drawing objects. Selection sets can be created to either include or exclude objects matching the specified object type and criteria. For instance, QSELECT allows you to make a selection set made up of all objects that are green, or a selection set could be made to include everything except green objects. The command can be started by right-clicking and choosing Quick Select... from the shortcut menu, by selecting the Quick Select icon from the Properties dialog box, or by selecting Quick Select... from the Tools pull-down menu. The command can also be started by entering **QSELECT** (EnterKey) at the command prompt. Each method will produce the Quick Select dialog box shown in Figure 14.24. The dialog box contains seven key areas for adjusting the selection set.

- Apply to—If a selection set has been defined and is current, it will be selected automatically. If no selection set has been defined, by default the only option is to apply changes to the entire drawing. Choosing the Select Objects icon creates selection options.

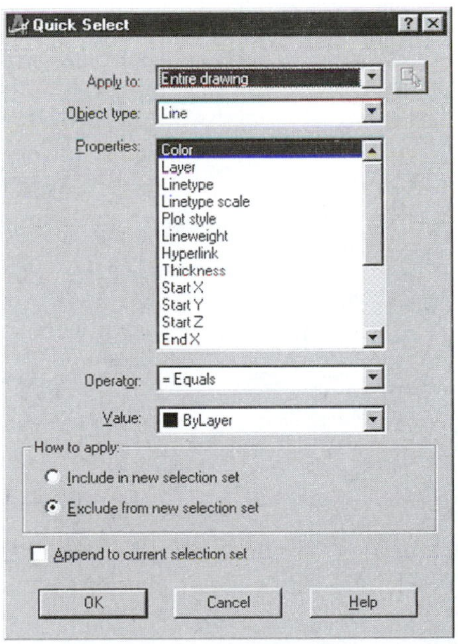

Figure 14.24 *The Quick Select dialog box is displayed by right-clicking and choosing Quick Select... from the shortcut menu, by selecting the Quick Select icon from the Properties dialog box, by selecting Quick Select... from the Tools pull-down menu, or by entering* ***QSELECT*** *(EnterKey) at the command prompt.*

- Select Objects icon—The Select Objects icon is located beside the Apply to edit box. Selecting the icon will remove the Quick Select dialog box temporarily so that items can be selected to apply the selection filters to.
- Object type—This option specifies the object type for the filter. Until a selection set is created, the Object type will only display a list of each shape included in the drawing, similar to Figure 14.25. Once a selection set is created, the list will only include the object types contained in the selection set.
- Properties—The properties displays a list of all searchable properties for the selected object to be used with the filter. The Properties display will vary depending on the object selected in the Object type edit box. If a circle is selected in the Object type edit box, a listing similar to Figure 14.26 will be displayed.
- Operator—The Operator controls the filter range and is dependent on the selected property. Operator options include: = Equals < > Not Equal > Greater than < Less than
- Value— This option allows a value for the filter to be entered. For instance, if the Property is a circle, a value of 24" could be used to include all circles equal to, not equal to, greater than, or less than 24" to define the selection set.
- How to Apply—This option contains radio buttons for including or excluding objects from the selection set. When Include in the selection set is active, a new selection set will be created composed of objects that match the filtering

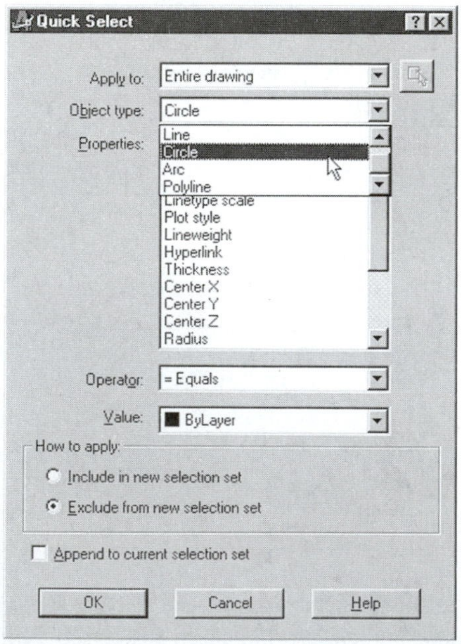

Figure 14.25 *The Object type will only display a list of each shape included in the drawing. Once a selection set is created, the list will only include the object types contained in the selection set.*

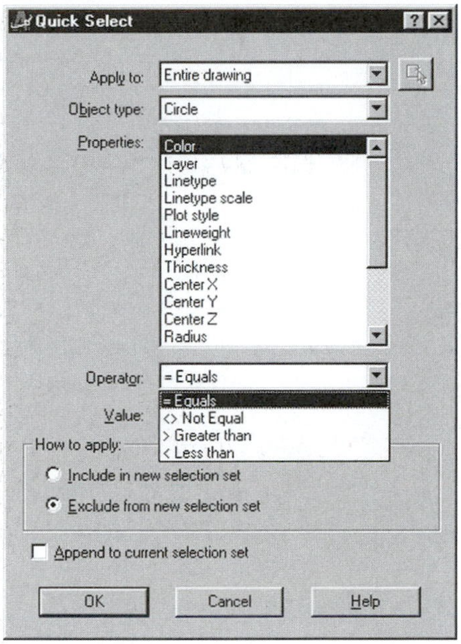

Figure 14.26 *The Properties display provides a list of all searchable properties for the selected object to be used with the filter. The display will vary depending on the object selected in the Object type edit box.*

criteria. When Exclude in the selection set is active, a selection set will be created composed of objects that do not match the filtering criteria.

- Append to the current selection set— This option determines if the selection set replaces or appends the current selection set.

Creating a Selection Set Using QSELECT

A selection set for editing all green can be created using the following steps:

- Open the Quick Select dialog box using one of the methods described earlier.
- Using the Apply to edit box, select Entire drawing.
- Select Multiple for the Object type.
- Select Color in the Properties edit box.
- Select = Equals in the Operator edit box.
- Use the down arrow beside the Value edit box, and select Green.
- Set Include in New Selection Set, in the apply box.
- Choose the OK button to close the box.
- AutoCAD will close the Quick Select edit box, and provide grips on each green object in the entire drawing. The command prompt will display a listing of the number of items in the selection set similar to Figure 14.27.

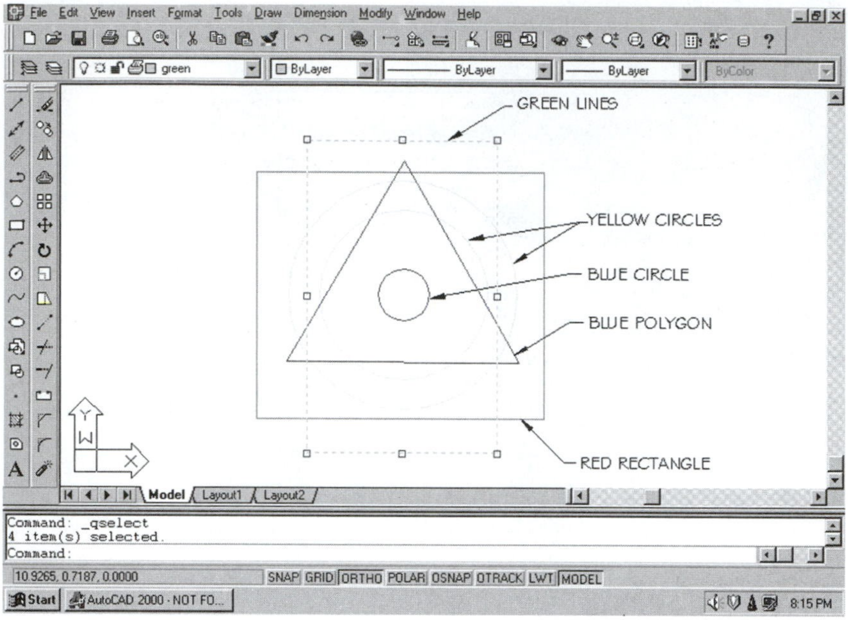

Figure 14.27 *QSELECT can be used to create a selection set made up of all objects of a specific property description. In this example, all objects that are green have be selected for editing.*

Excluding Objects from the Selection Set

A selection set excluding objects of a specific property can be created. Using the objects in Figure 14.27, a selection set excluding all circles larger than 24" could be created using the following sequence.

- Open the Quick Select dialog box using one of the methods described earlier.
- Using the Apply to edit box, choose Current Selection.
- Select Circle for the Object type.
- Select Diameter in the Properties edit box.
- Select Greater Than in the Operator edit box.
- Enter a value of 24.
- Apply the selection set to Exclude from the New Selection Set.
- Choose the OK button to close the box.
- AutoCAD will close the Quick Select edit box, and removes all circles with a diameter greater than 24" from the selection set.

CHAPTER 14 EXERCISES

1. Enter a new drawing and load CENTER, CENTER2, CENTERX2, DASHED, DASHED2, and HIDDEN. Draw a line with each linetype, long enough so that the line pattern can be seen. Save the drawing as E-14-1.
2. Open E-12-4. Create layers for WALL with red lines and DOORWIN with green lines. Move the existing entities to the appropriate layer and save the drawing as E-14-2.
3. Open drawing E-8-4. Create layers for the specified layers and options:

PROPLINE	red	Centerx2	property line
TREES	Green	Dashed	tree shape
T-LAYOUT.	7	Dot2	point/baseline

   Set the LTSCALE appropriate for future plotting at a scale of 1" = 10'-0". Save the drawing as E-14-3.
4. Open drawing E-8-1. Assign six different layers, each with a different linetype and color. Change each of the existing shapes to a different layer. Draw a different geometric shape on the remaining layers. Use PLINE and OSNAP. Save the drawing as E-14-4.
5. Open the Architectural drawing template, set the LTSCALE, and set the scale values appropriate to 1/4" = 1'-0". Set the Units to Architectural, with 1/16" fractional value. Establish layers that would be suitable for a two-level residential floor plan. Include the following: walls, dimensions, architectural notes, structural text, joist, beams, furniture, appliances, planting, plumbing, cabinets, section tags, doors and windows, decks at each level, electrical fixtures, and electrical wiring. No provisions need to be made for other drawings that typically are grouped with a floor plan. Assign colors to each layer. Assign linetypes appropriate to the layer contents. Save the drawing as QUARTER. This will become the basis for future architectural plan drawings that are to be plotted at 1/4" = 1'-0".
6. Open drawing E-11-11 and create separate layers using an AIA long format so that walls, cabinets, fireplace, doors, windows, and plumbing fixtures can be separated from each other. Assign each layer a different color. Add center and hidden lines to the drawing base. Add a dishwasher to the kitchen, using hidden lines. Save the drawing as FLRPLN.

CHAPTER 14 QUIZ

1. How many linetypes are in the ACAD.LIN library?
2. List five commands that can be used to alter the appearance of a line.
3. List four objects that are affected by a change in linetype.
4. List the command sequence to draw with a centerline and a hidden line.
5. What would be the line scale value for a line that is to be plotted at a scale of 1" = 20'?
6. When describing the STATE of a layer, what qualities are considered?
7. List the four groups of information to be used in layer names recommended by the AIA.
8. What aspects of Change Properties apply to 2D drawing?
9. Describe guidelines for naming a model file using the AIA guidelines.
10. You are working on a drawing containing several different linetypes and would like to find out their names. What command and option should be used?
11. List the preferred LTSCALE for the following scales.

Full size

3" = 1'-0"	1" = 1'-0"
3/4" = 1'-0"	3/8" = 1'-0"
1/4" = 1'-0"	1/8" = 1'-0"
1/16" = 1'-0"	1" = 10'-0"
1" = 20'-0"	1" = 60'-0"

12. What will typing DDLTYPE at the command prompt allow you to do?
13. Describe the difference between the effects of LTSCALE and CELTSCALE.
14. A layer has a title of S-ANNO-SYMB. Who created the drawing and what do you think it would contain?
15. What information is in the minor group portion of a layer name?
16. You need to erase some information that is on a layer that is currently frozen. What is your plan?
17. You're about to start the plans for a one-level residence that will require a site plan, floor plan, framing plan, electrical plan, foundation plan, exterior elevations, and sections. How many drawing files do you expect to use, and how will you create layers for these files. (It's not in this chapter, you've got to search through your cranium.)
18. What is the advantage of using QSELECT?
19. How can the Properties dialog box be accessed and what can be done with it?
20. What are the uses for the Include/Exclude buttons of quick select?

CHAPTER 15

Placing Patterns in Drawing Objects

Various materials are represented by different line or shape patterns. This is seen most often when objects are displayed in section view. Concrete material is typically filled with a pattern of dots and triangles; soil with a series of groups of perpendicular lines; and masonry with diagonal or crossing diagonal lines. These patterns are referred to as *hatch patterns* and can be created using the BHATCH command. In addition to the many patterns used in the construction industry, a solid pattern can be placed in an object using BHATCH. Using the AIA guidelines, the pattern should be placed on a layer with a minor code of PATT. Figure 15.1 shows an example of a drawing with hatch and solid patterns added to define each material. The BHATCH command adds patterns to a drawing by use of the Boundary Hatch dialog box.

HATCHING METHODS

Hatch patterns are repetitive patterns of lines, dots, and other symbols used to represent a surface or specific material. Hatch patterns are used on sections and details—although plan views and elevations also use them. Hatch patterns typically are used to distinguish between masonry and wood walls in architectural details similar to Figure 15.2. Hatch patterns can be used on elevations to represent shades and shadows, as seen in Figure 15.3. The majority of uses for hatch patterns, however, will be found in details and sections. Hatch patterns can be placed using either the BHATCH (boundary hatch) or the HATCH command. Although the patterns will look exactly alike, there is a major difference between the two commands. BHATCH creates patterns that are associative or non-associative. Associative patterns are linked to their boundaries. If the boundary is altered, the pattern is altered to match the boundary. Non-associative patterns are not affected by altering their boundaries. The other major difference between BHATCH and HATCH commands is how they are added to the drawing. BHATCH patterns are applied using the Boundary Hatch dialog box. The Boundary Hatch dialog box can be started by selecting Hatch from the Draw toolbar, by typing **BH** (EnterKey) at the command prompt, or by selecting

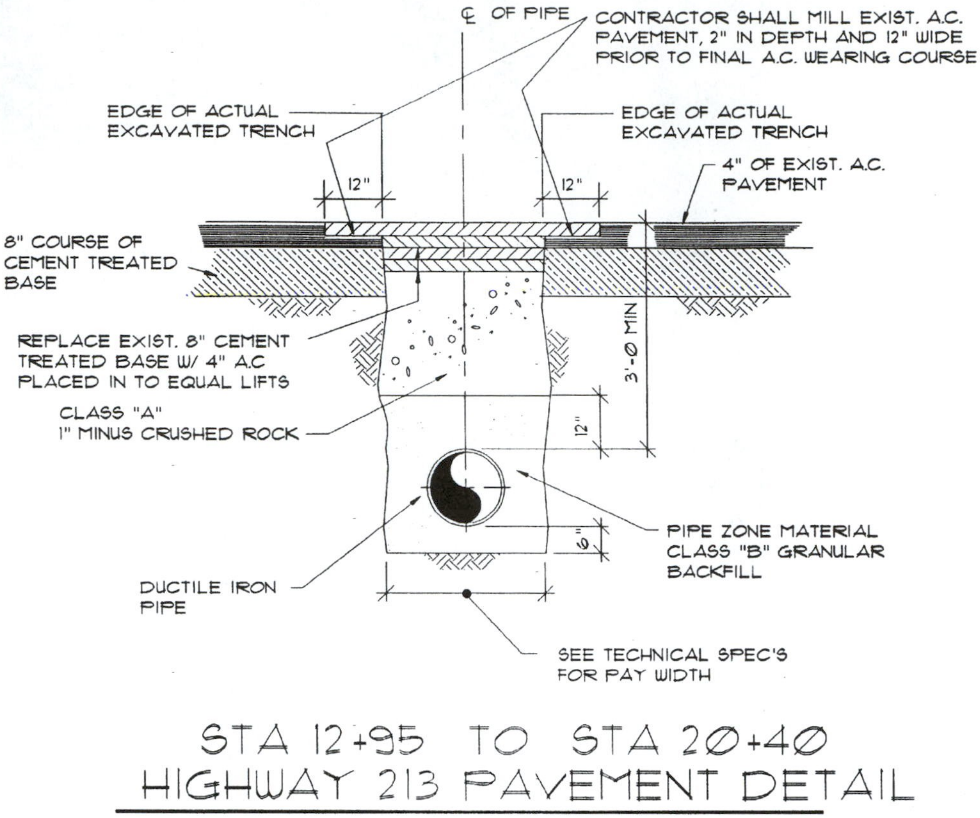

Figure 15.1 *Shading and drawing patterns to represent materials can be produced by using the BHATCH command. (Courtesy Department of Environmental Services, City of Gresham, OR.)*

Hatch... from the Draw pull-down menu. HATCH patterns are added to a drawing from the command line. HATCH will not be discussed in this text. Consult the Help menu if you're interested in using the slower, harder method.

USING THE QUICK TAB

The Boundary Hatch dialog box consists of the Quick tab and Advanced tab. The Quick tab of the Boundary Hatch dialog box is shown in Figure 15.4. The Quick tab can be used for placing patterns for many common shapes encountered on construction drawings. As you prepare to hatch an area using BHATCH, you can take several steps to speed up the process. Zooming into an area, rather than keeping the entire drawing in view, will eliminate the number of lines that must be examined to determine the boundary set that will define the hatch area. Speed also can be gained if extra layers are turned OFF or Frozen so that objects on these layers are not examined as part of the selection set.

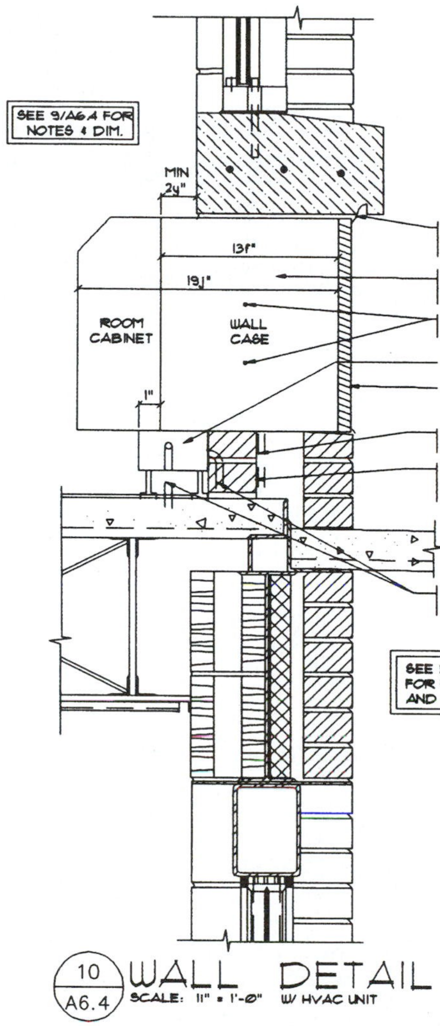

Figure 15.2 *Hatch patterns are used to distinguish various materials from each other. (Courtesy G. Williamson Archer, AIA of Archer & Archer, PA.)*

PATTERN TYPE

The Type edit box is used to set the pattern type. Options include Predefined, User defined, and Custom. The default value is a Predefined pattern. Selecting the down arrow displays the listings shown in Figure 15.5. Predefined patterns can be viewed in the Hatch Pattern Palette dialog box. The dialog box can be seen in Figure 15.6. For most uses, the predefined patterns will be all that are needed. This option allows the angle and scale of the pattern to be altered. Selecting the User Defined Patterns option creates a pattern based on the current linetype in the drawing. This option

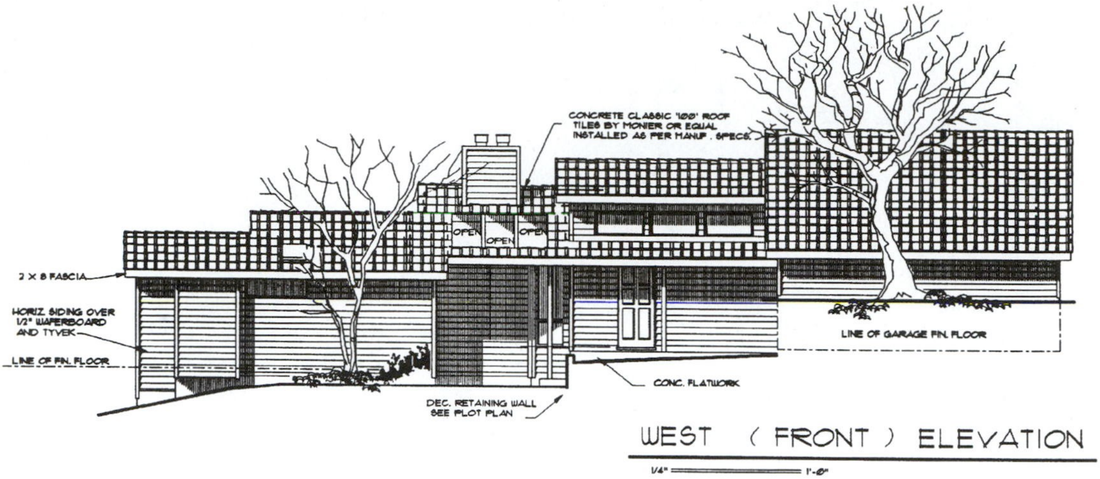

Figure 15.3 *Hatch patterns have been used to represent the tile roof, wood siding, and the shadows. (Courtesy Residential Designs.)*

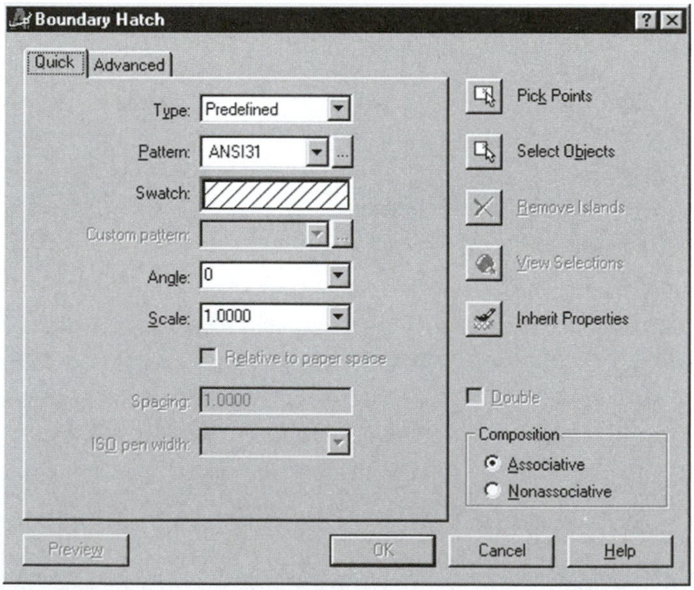

Figure 15.4 *The BHATCH command can be started by selecting Hatch from the Draw toolbar, by typing* **BH** *(EnterKey) at the command prompt, or by selecting Hatch... from the Draw pull-down menu. Each method will produce the Boundary Hatch dialog box.*

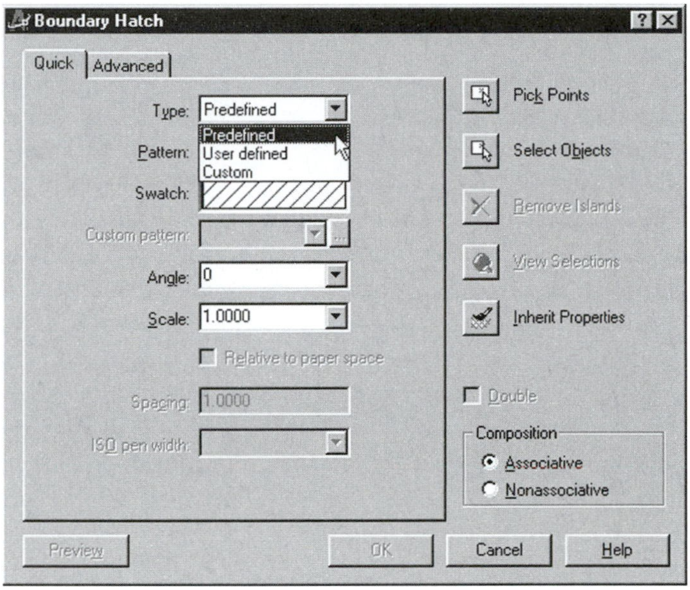

Figure 15.5 *Selecting the arrow beside the pattern edit box will display options for selecting the pattern type.*

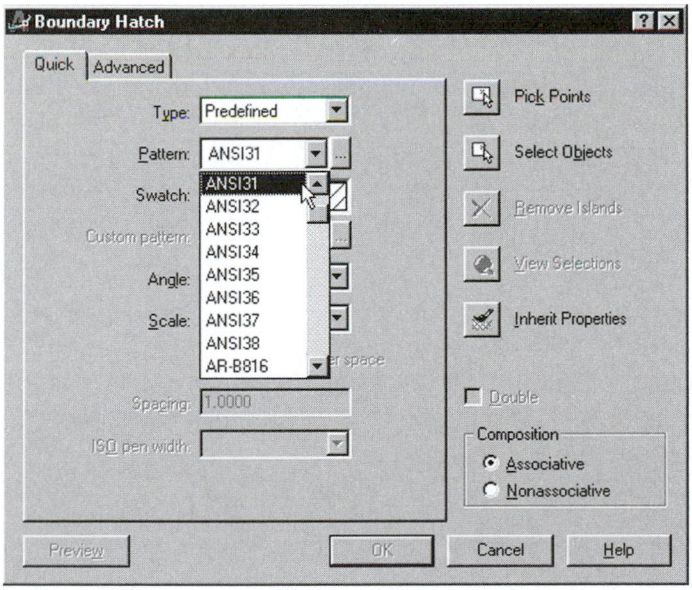

Figure 15.6 *The Hatch pattern palette can be used to select the pattern to be used for the BHATCH command.*

allows the angle and spacing of the pattern to be altered. The Custom option allows a user-created pattern to be to be added to the drawing base. Custom patterns allow drawing patterns to be saved in the acad.pat or in its own PAT file. Methods of creating patterns will be discussed later.

SELECTING A PATTERN

Notice that the ANSI31 hatch pattern is the default setting shown in the Pattern edit box. A different pattern can be selected for a BHATCH and it will remain the default until a new pattern is selected. A new pattern can be selected with the Pattern edit box using three different methods. Selecting the pattern name in the edit box or the edit box arrow will display a listing of the hatch pattern names similar to the list shown in Figure 15.6. The slide bar can be used to scroll through the list of predefined patterns. Clicking the name in the pattern box a second time will close the menu without making any changes. Selecting a name from the menu will make that pattern the current pattern and close the menu. This method of pattern selection works well if you are familiar with the names of patterns. If you're a new user, names like ANSI31 or ANSI32 will be meaningless. Select the [...] icon to see a display of each image.

Selecting the pattern icon beside the edit box [...] will display the Hatch Pattern Pallet dialog box. (See Figure 15.7a.) The ANSI and Other Predefined tabs will provide the patterns typically associated with construction drawings. (See Figure 15.7b). The slide bar can be used to move through the menu to select the desired hatch pattern.

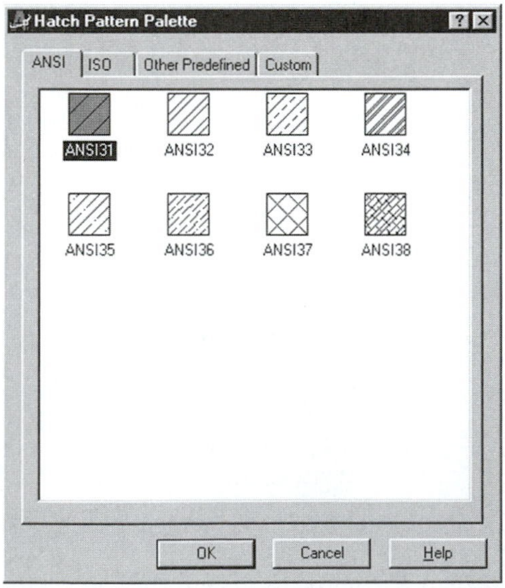

Figure 15.7a *Selecting the pattern icon beside the edit box [...] will display the Hatch Pattern Pallet dialog box.*

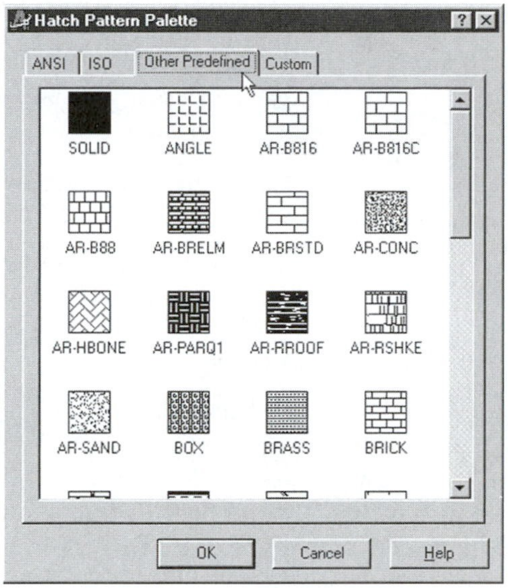

Figure 15.7b *The Other Predefined tab provides the patterns typically associated with construction drawings. The slide bar can be used to move through the menu to select the desired hatch pattern.*

Once an icon is selected for use in a drawing, press the OK button to return to the Quick tab. A third method of selecting a hatch pattern is to use the Swatch display. Clicking the Swatch pattern will display the most recently used tab of the Hatch Pattern Palette.

Note: In addition to many interesting patterns found in the Hatch Pattern Palette, several hatch patterns are specific to architectural drawings.

Pattern	Common Use
ANSI31	Masonry in plan and section view, thin veneers in section views, and shading for elevations.
ANSI32	Steel plates in section view
ANSI37	Firebrick in section view
SOLID	General usage when a surface needs to be filled solid
AR–B816 &16C	Concrete masonry units in elevation
AR–CONC	Poured concrete in section view

AR–HBONE	Flooring or brick patterns in plan and elevations
AR–PARQ1	Flooring or brick patterns in plan and elevations
AR–RROOF	General roofing in elevation (non specific to material)
AR–RSHKE	Cedar shake in elevation (typically not used for roofing)
AR–SAND	EIFS, stucco, or plaster in elevations and section
BRICK	Brick in elevation
DOTS	Shading in elevation
EARTH	Soil in section

PATTERN ANGLE

Once the pattern has been selected, the Angle edit box should be considered. By default, the pattern will be placed at an angle of 0°. By accepting the default, the pattern will be reproduced as shown in the Swatch display or the Hatch Pattern Palette. A value can be entered by keyboard or by selecting a value from the Angle menu shown in Figure 15.8. Entering a value other than zero allows the hatch pattern to be rotated. By entering an angle of 45°, pattern ANSI31 can be used to represent a shadow as seen in Figure 15.9.

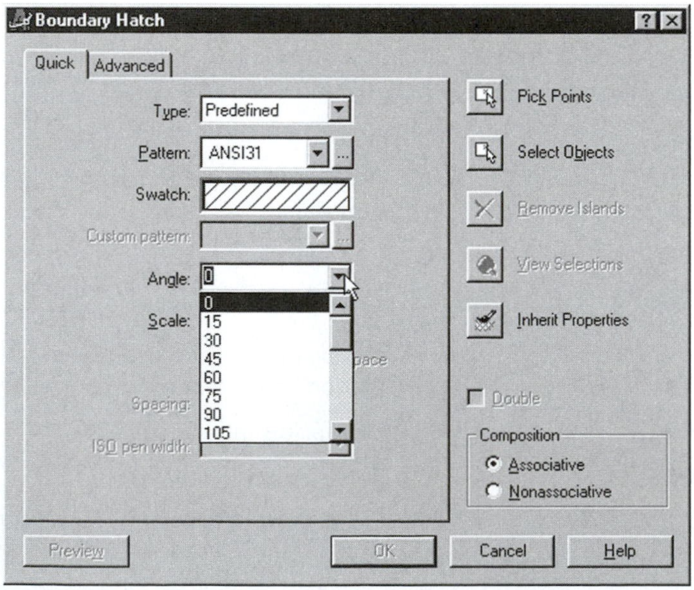

Figure 15.8 *A value for displaying the pattern can be entered by keyboard or by selecting a value from the Angle menu.*

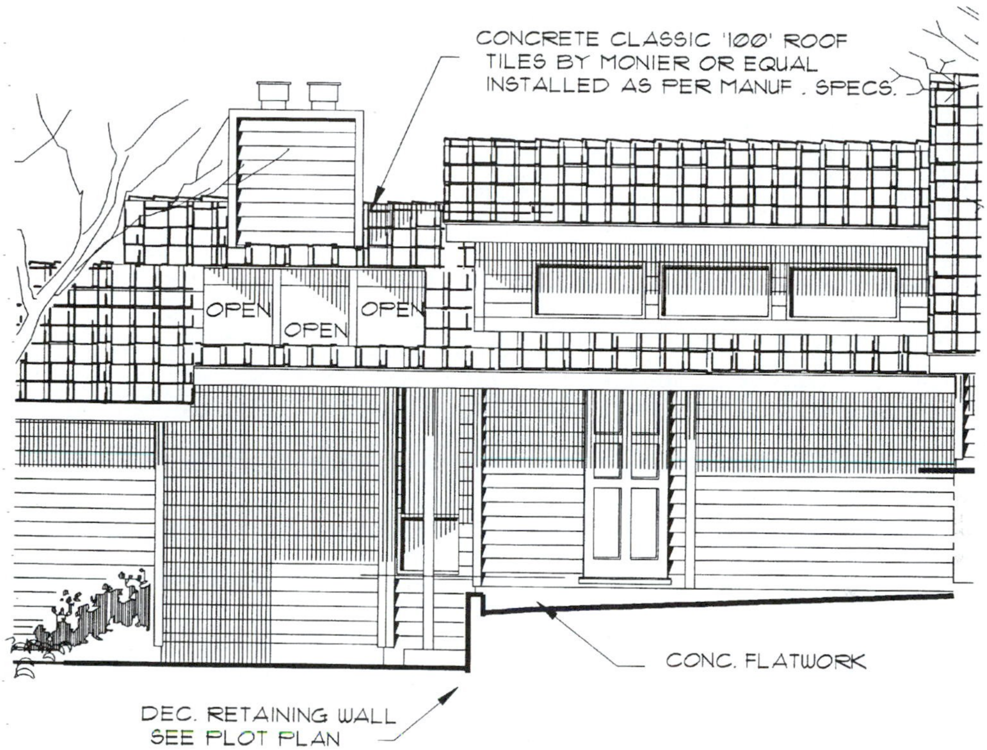

Figure 15.9 *The hatch pattern ANSI31 is drawn at a 45° angle by default. Providing a rotation angle of 45° will reproduce the pattern in a vertical position. (Courtesy Residential Designs.)*

PATTERN SCALE

This option sets the size of the pattern elements with the default of one drawing unit. The relationship of the pattern can be controlled by the drawing scale factor. The scale value is adjusted using the same method to adjust the angle value—either keyboard entry or the Scale menu. Figure 15.10 shows an example of three different scale patterns. Selecting the largest scale pattern that is still visually appealing will save REGEN time. Drawing a line the length of the desired pattern to be inserted can help to ease the scale selection. If you need a hatch pattern to represent a concrete block, draw a 16"-long line below the area to receive the hatch. Later in the command process, you'll have a chance to preview how the pattern will appear. The 16" line will serve as a reference to verify that the scale selection is accurate.

SELECTING OBJECTS TO BE HATCHED

Once the pattern, angle, and scale have been selected, use one of the methods from the right side of the dialog box to choose the area to be hatched. For now, the options are limited to Pick Points and Select Objects. If the Pick Points icon is selected, only

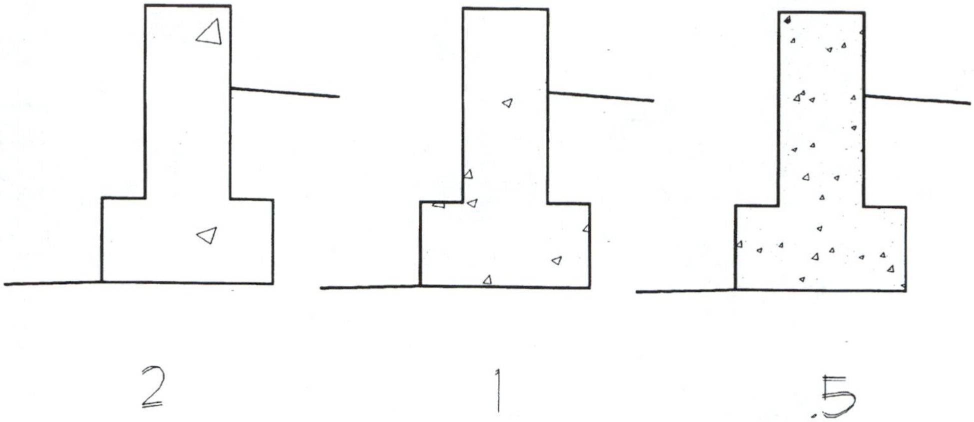

Figure 15.10 *The effect of altering the scale factor. Always use the largest size possible to save regen time.*

one point will need to be selected. If the Select Objects icon is selected, any of the selection methods introduced in previous chapters can be used to define the area. Other methods of selecting the boundary can be selected using the options from the Advanced tab, which will be discussed later in this chapter.

Using the Pick Points Option

The Pick Points option can be used to hatch drawing objects that form an enclosed boundary. Once you've selected the Pick Points icon, the dialog box will be removed and the drawing will be restored. A prompt is now displayed requesting "Select internal point." Select any point that lies within the area to be hatched. Once selected, the command line will display the prompt:

Select internal point: Selecting everything...

Selecting everything visible...

Analyzing the selected data...

Analyzing internal islands...

Select internal point: (EnterKey)

Because the Pick points icon was selected, the object to be hatched can be selected by choosing one point that lies within the object. As the point is selected, the boundary of the area to be hatched will be displayed as dashed lines. AutoCAD is waiting for you to select additional objects to be hatched. Select (EnterKey) or right-click to redisplay the Boundary Hatch dialog box. Press (EnterKey) a second time to apply the hatch pattern to the selected object. Figure 15.11 shows the resulting hatch pattern.

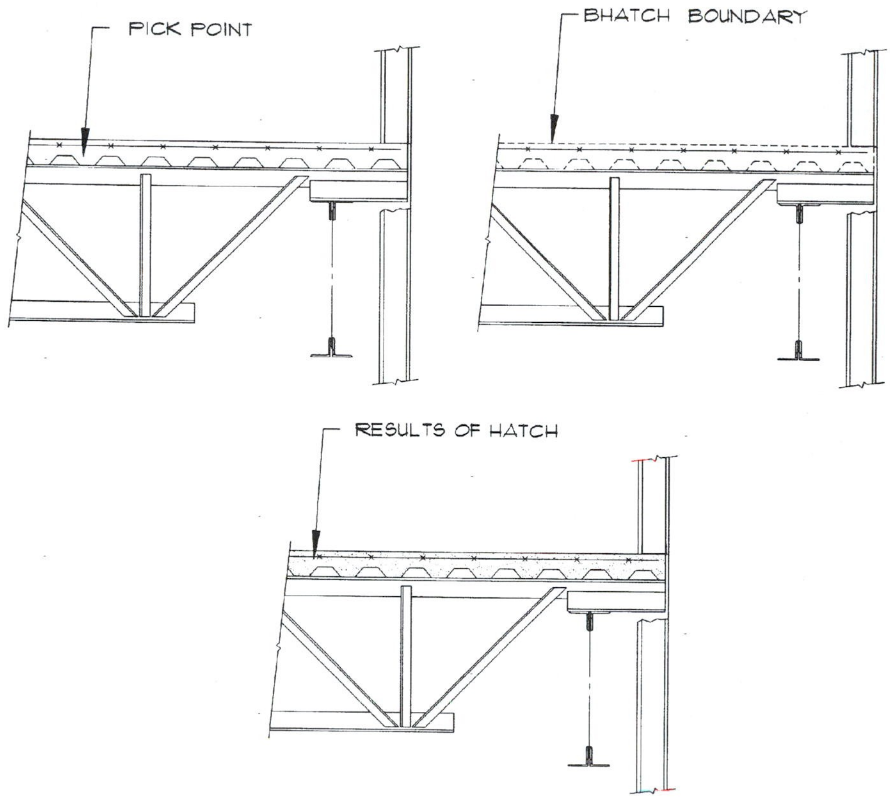

Figure 15.11 *Using BHATCH to add a pattern to a concrete floor slab. Once the area to receive the pattern is selected, the boundary will be highlighted. The AR-SAND pattern at a scale of .5 and an angle of 0 was selected. With this pattern, the angle is unimportant.*

Using the Select Objects Option.

Selecting the Select Objects icon will remove the Boundary Hatch dialog box and prompt you to select objects. Figure 15.12 shows the process for selecting an object to be hatched using the Select Objects option. In Figure 15.12a, the fireplace was selected using the Crossing Window selection option. Figure 15.12b shows the hatch pattern that would result. The object to be hatched does not need to form a closed boundary. It may contain an internal area, referred to as an *island*, that will not be hatched. Figure 15.13 shows the results of selecting an object with an island using the Window selection method.

If the object to be hatched is surrounded by a polyline, the object can be selected by picking the object at any point, as seen in Figure 15.14. The pattern will be the same as the pattern in Figure 15.12b. Once the object to be hatched is selected, press (EnterKey) to restore the Boundary Hatch dialog box. Pressing (EnterKey) a second

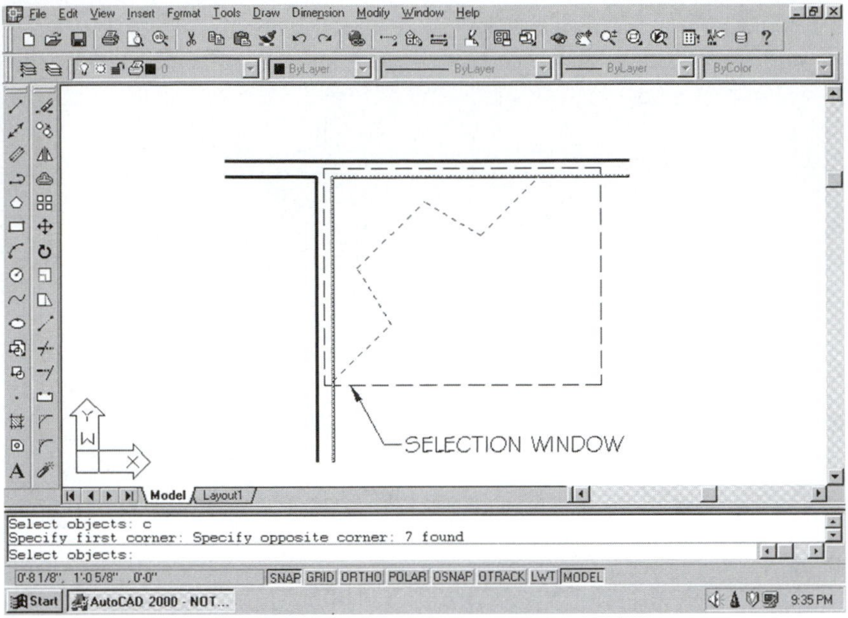

Figure 15.12a *Using the Select Object option, an object can be selected using any of the selection methods, such as Window or Crossing.*

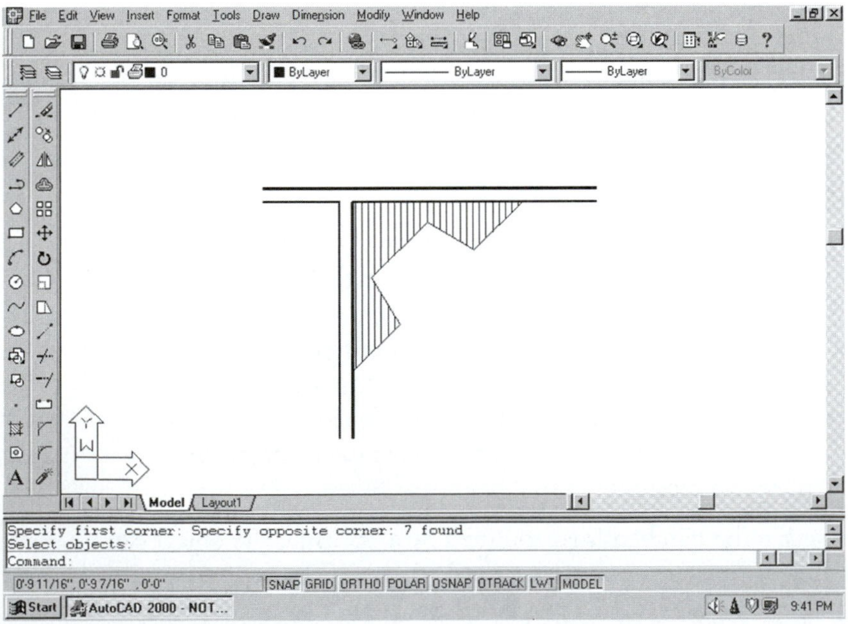

Figure 15.12b *The resulting hatch pattern for the selection set in Figure 15.12a.*

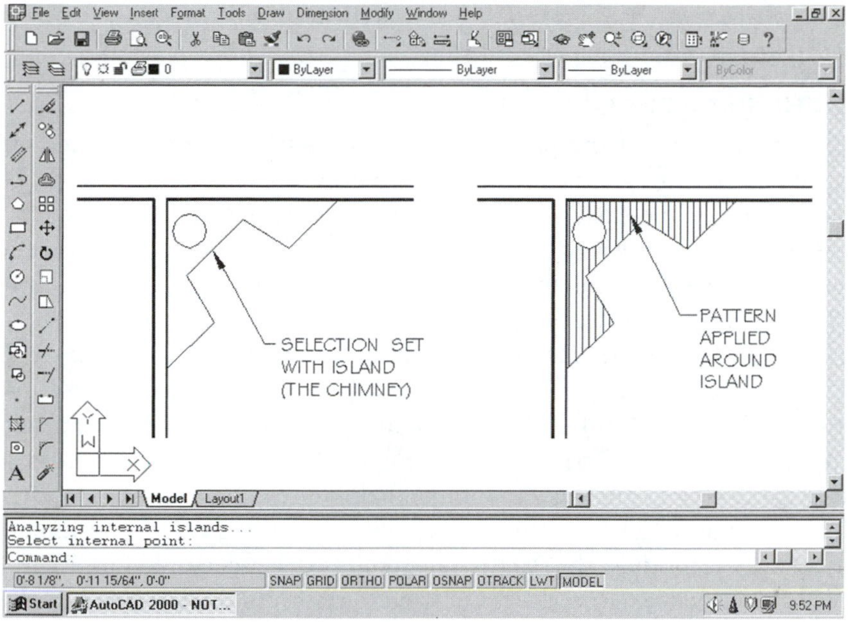

Figure 15.13 *The results of using the BHATCH Select Object option on an object with islands. The placement of the pattern can be altered by using the Style option.*

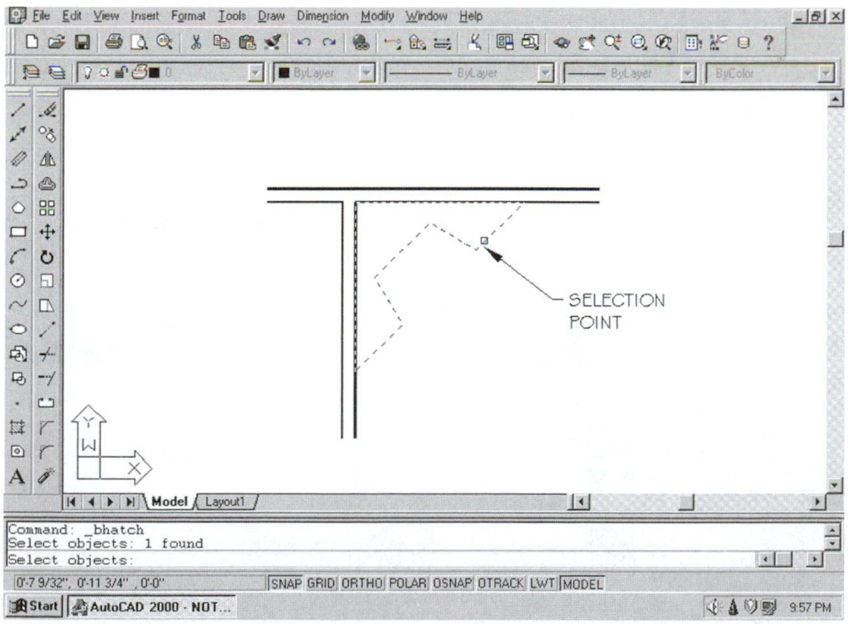

Figure 15.14 *When the boundary of an object is formed by a polyline, it can be selected by clicking on a single point. The resulting pattern will be the same as the pattern in Figure 15.12b.*

time will apply the pattern. Figure 15.15 shows the results of selecting an object with an island to be hatched by selecting an edge.

USING THE PREVIEW OPTION

If you're unsure of any of the pattern parameters, select the Preview icon prior to applying the hatch pattern. The Preview option provides a display of how the pattern will appear based on the current settings. Preview is especially helpful when a large, irregular-shaped boundary is to be hatched. When the Preview option is selected, the dialog box is removed and the object with the hatch pattern is displayed, as seen in Figure 15.16. Notice that the boundary is still dotted, which is a reminder that the hatch pattern has not yet been added to the drawing file. Select (EnterKey) or right-click to restore the Boundary Hatch dialog box. If the pattern is acceptable, pick the OK button or the (EnterKey). If the preview did not display the intended results, any of the pattern options can be altered. In Figure 15.17, a new hatch pattern was selected by returning to the Hatch options dialog box. The Preview Hatch option was selected. Using BHATCH Preview to test alternatives can go on indefinitely. The final change is to adjust the Angle. The results can be seen in Figure 15.18. Since this view was acceptable, (EnterKey) was used to return to the Boundary Hatch dialog box and the OK button was selected.

A SUMMARY OF PLACING A HATCH PATTERN

Placing the hatch pattern in Figure 15.18 requires the following steps:

- Open the Boundary Hatch dialog box by selecting Hatch from the Draw toolbar, typing **BH** (EnterKey) at the command prompt, or selecting Hatch... from the Draw pull-down menu.

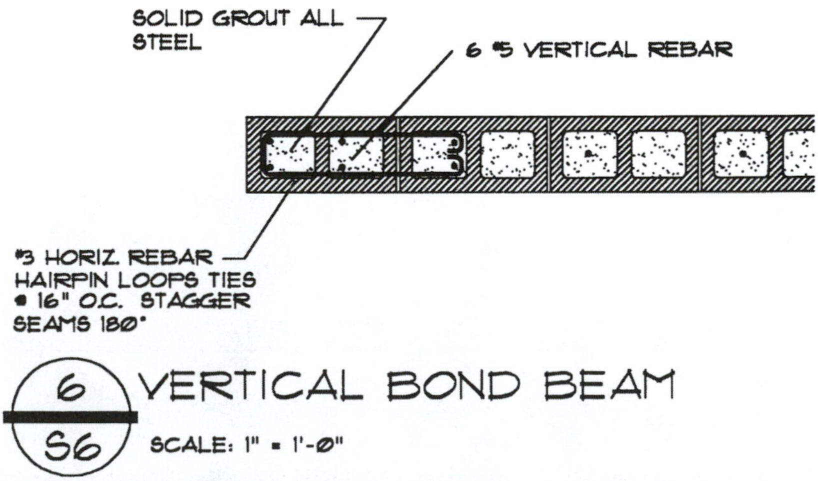

Figure 15.15 *The resulting hatch pattern for a selection set with islands.*

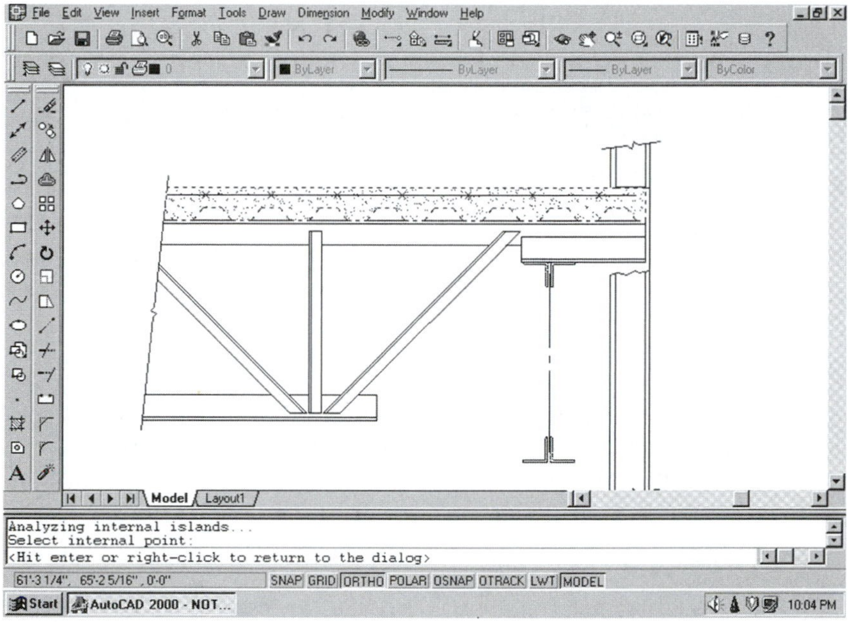

Figure 15.16 *The proposed pattern can be displayed by selecting the Preview button. If the pattern is not what you expected, the properties can be altered before applying the pattern.*

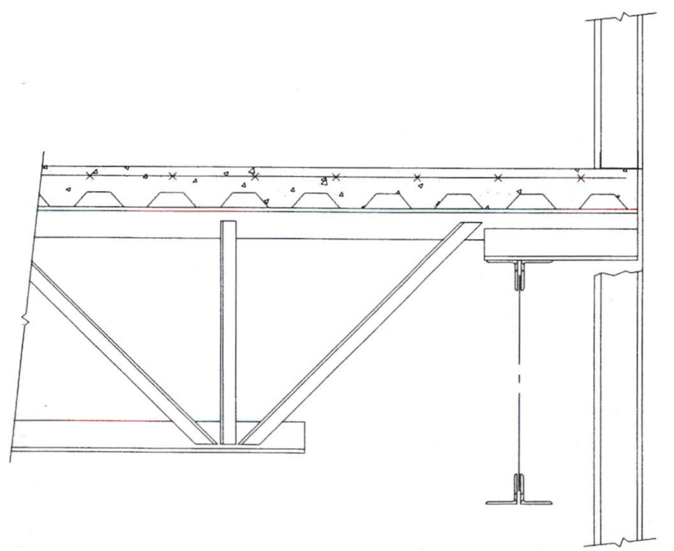

Figure 15.17 *If the Preview is acceptable, it can be added to the drawing base by selecting the OK button. A new pattern can be installed by returning to the Hatch Options dialog box.*

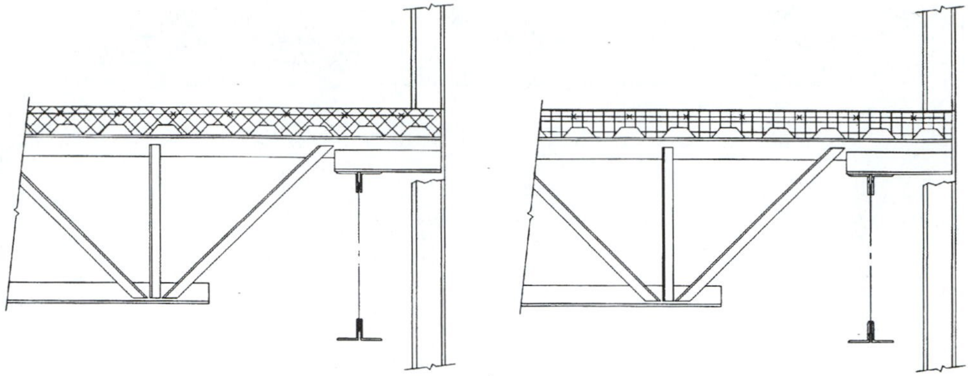

Figure 15.18 *The Preview option can be used to view the hatch pattern prior to adding it to the drawing base. If you are unhappy with the pattern, change the pattern, the scale, and the angle options, and then reuse the Preview option.*

- Choose the Quick tab.
- Select the [...] icon to display the Hatch Pattern Palette.
- Choose the ANSI tab.
- Select the ANSI37 display and select OK.
- Select Pick Points from the Boundary Hatch dialog box.
- Specify an internal point in the object to be hatched.
- Select the Preview button to display the proposed hatch pattern.

  If the pattern is not acceptable, alter the hatch properties using the Quick tab, and then use preview option again.

  If the pattern is OK, choose the OK button to apply the hatch pattern to the drawing.

REMOVING ISLANDS

Figure 15.19 shows a concrete floor slab with four islands. When the boundary is selected, each of the islands will automatically be excluded from the hatch. The hatch pattern was omitted from each of the cavities in the floor on the left side of Figure 15.19. Once the object to be hatched has been defined, selecting the Remove Islands icon allows islands to be included. An included island will receive the hatch pattern. The example on the right side of Figure 15.19 shows the results of removing two of the islands from the boundary set.

VIEW SELECTION

With all of the options available for altering the boundary or patterns used in a BHATCH, you might lose track of what options you've selected. The View Selections option will remove the dialog box and display the currently defined boundaries.

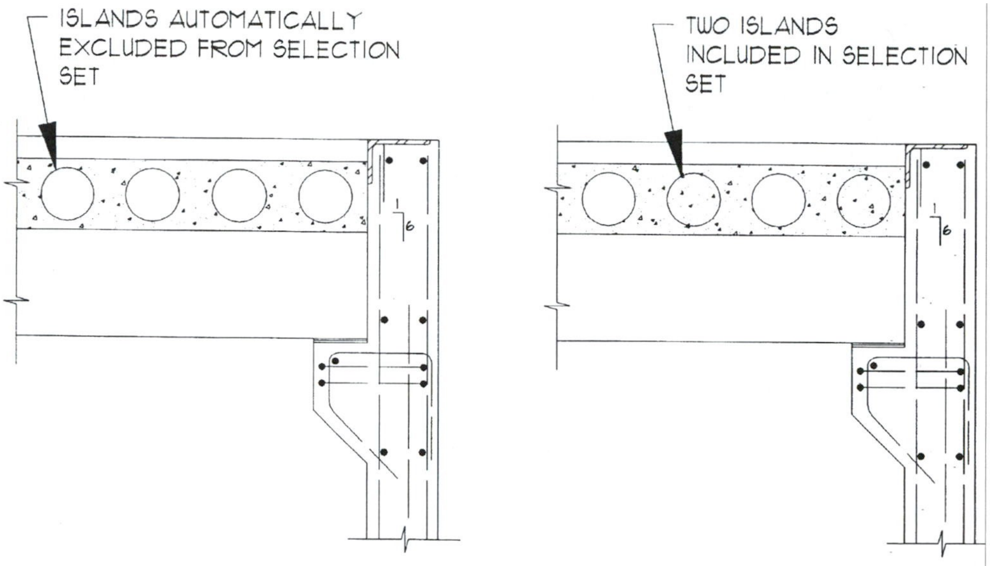

Figure 15.19 *The Remove Islands option can be used to control the placement of the pattern inside islands within the boundary.*

INHERIT PROPERTIES

This option is great to hatch an object using an existing pattern that you can't remember the qualities of. Selecting the Inherit Properties icon will close the Boundary Hatch dialog box and return the drawing area with the prompt:

> Select associative hatch object: *(Select the hatched pattern that you would like to duplicate.)*
>
> Inherited Properties: Name <ANSI31>, Scale <2.000>, Angle <0>
>
> Select internal points: (EnterKey)

Pressing (EnterKey) will restore the Boundary Hatch dialog box.

COMPOSITION

The final portion of the Boundary Hatch dialog box to be examined is the Composition box. This box contains toggles for Associative and Non-associative. With the Associative button active, as an object containing a hatch pattern is edited, so is the hatch. With Non-associative activated, if an object with a hatch pattern is altered, the object is altered, but the hatch pattern remains unaffected. Figure 15.20 shows the effect of stretching two steel beams and the affect of an associative pattern.

CONTROLLING PATTERN PLACEMENT

Although some hatch patterns such as soil, sand, or concrete can be placed without regard to the starting point of the boundary, other patterns such as brick or block should be placed relative to the hatch boundary. The SNAP command can be used to control the relationship of the hatch pattern to the boundary. Use the following command sequence to make one of the hatch boundaries the starting point for the pattern:

> Command: **SNAP** (EnterKey)
>
> Specify snap spacing or [ON/OFF/Aspect/Rotate/Style/Type] <0'-0 ½">: **R** (EnterKey)
>
> Specify base point < 0'-0",0'-0">: *(Select a corner of the pattern boundary.)*
>
> Specify rotation angle <0>: (EnterKey)
>
> Command: **BH** (EnterKey)

The pattern can now be placed using the Boundary Hatch command.

DEFINING BOUNDARY HATCH USING THE ADVANCED TAB

Although the Quick tab can be used to hatch simple objects, the Advanced tab provides options to help increase the efficiency of selecting bhatch boundaries. Unless you're working on a very large drawing, you will probably not need this option. The BHATCH command evaluates all objects displayed on the screen when you select objects to be hatched. If you are working on a very large structure, the Advanced… options can speed the boundary selection process. Key elements of the Advanced tab include controls for Island Detection Style, Object Type, Boundary Set, and Island Detection Method. The Advanced tab can be seen in Figure 15.21.

ISLAND DETECTION STYLE

The Island Detection Style option of the BHATCH command allows three styles of island detection for displaying the hatch pattern. The Normal setting is the default

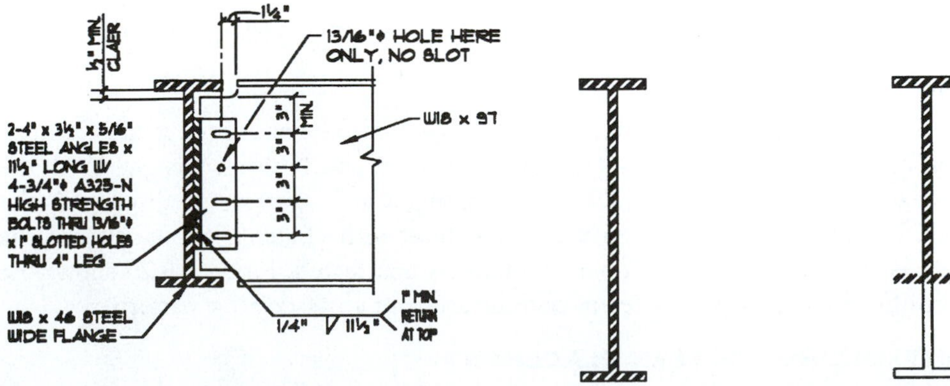

Figure 15.20 *With Associative active, stretching a steel beam will alter the hatch pattern. With the box inactive, the beam is stretched, but the hatch pattern is unaltered.*

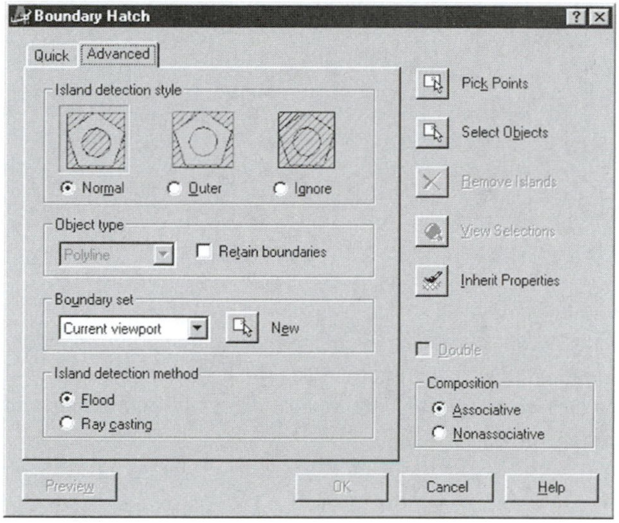

Figure 15.21 *The Advanced tab provides options to help place hatch patterns in complex shapes. Key elements of the tab include controls for Island Dection Style, Object Type, Boundary Set, and Island Detection Method.*

and the image tile reflects how the pattern will be displayed. Options may be changed by selecting one of the other radio buttons.

Normal

Figure 15.22(a) shows how hatching sets made up of multiple objects will be hatched when the default of Normal style is selected and each of the objects is selected as a boundary. Normal hatching style works inward, starting at the area boundary and proceeds until another boundary is found. The pattern is turned off until another boundary is discovered.

Outer

Figure 15.22(b) shows how hatching sets made up of multiple objects will be hatched when the Outer option is selected and each of the objects is selected as a boundary. The Outer hatching style works inward, starting at the area boundary, and proceeds until another boundary is found. The pattern is turned off at the second boundary.

Ignore

Figure 15.22(c) shows how multiple objects will be hatched when the Ignore option is selected. The Outer hatching style works inward, starting at the area boundary, and passes through all objects within that boundary.

Note: When using the Normal setting, if a hatch line encounters another shape or text, the pattern will be placed around the text. Although text has not yet been added to the drawing, it will be important to remember that text inside of a hatch boundary will still be legible. (See Figure 15.23).

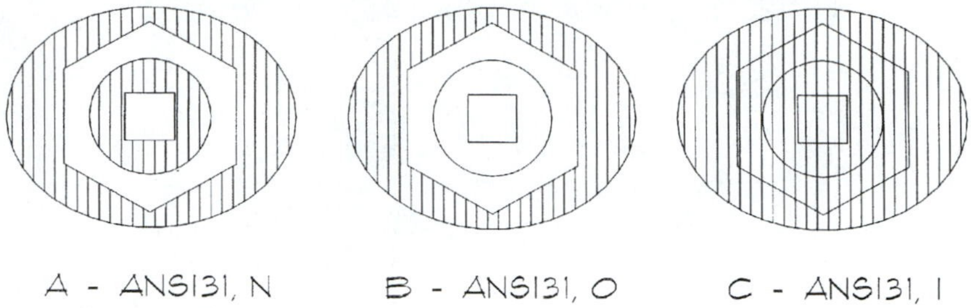

Figure 15.22 *The Island Detection Style option controls how the pattern will be displayed around multiple object sets. The N (normal) option works inward until another boundary is found. The pattern is not displayed until another boundary is found. The O (outer) option will only display the pattern in the outer boundary area. The I (ignore) option ignores all boundaries and hatches all areas within the outer boundary.*

OBJECT TYPE

The Object Type area controls boundary options for applying the hatch pattern. The options for the boundary include Retain Boundaries and Object Type. The Retain Boundaries option specifies whether or not a temporary boundary will be added to the drawing. Boundaries for hatch patterns can be specified by Object Type. These options are not active unless the Retain Boundaries option is active. Object Type controls the type of new boundary to be created. Options are Region and Polyline. You've explored polylines in a previous chapter. A closed shape comprised of a polyline can be used to form a boundary for hatch patterns.

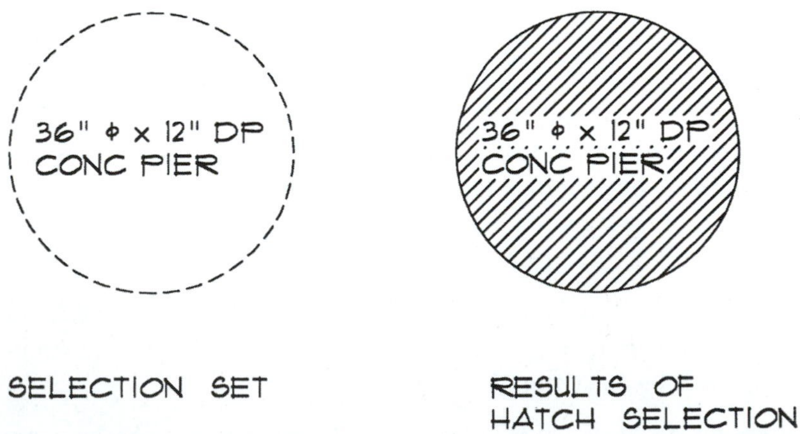

Figure 15.23 *Text will be treated like an island and be automatically be excluded from the area to be hatched.*

REGION

A region is an enclosed 2D area created from shapes called loops. Loops are a sequence of connected curves and lines that do not intersect and define an area on a plane. This sounds technical—until you realize that loops are nothing more that the basic shapes that make up a drawing. Loops can be a combination of lines, polylines, circles, arcs, elliptical arcs, splines, and solids. Loops, or regions, form the area where the hatch will be applied. Regions are created by selecting the Region icon from the Draw toolbar, by selecting Region from the Draw pull-down menu, or by typing **REG** (EnterKey) at the command prompt. Regions are created using the following command sequence.

Command: **REG** (EnterKey)

Select objects: *(Select objects to form the region using any selection method.)* **W** (EnterKey)

4 found

Select objects: (EnterKey)

1 loop extracted.

1 Region created

Command:

The selected lines will now function as a polyline, forming a boundary for future hatch patterns.

DEFINE THE BOUNDARY SET

This portion of the dialog box defines how the selection set for the hatch pattern will be selected, when the Pick Points options is selected. By fine tuning the selection set, the BHATCH command functions quicker because AutoCAD has to examine fewer objects to define the selection set. The default selection to define boundaries is Current Viewport. Selecting New will remove the dialog box from the screen and allow a new boundary to be selected. The existing boundary set will be discarded and you will be allowed to construct a new boundary set to define the area to be hatched. Once selected, the dialog box will be returned and the default setting is now the Existing Boundary Set. An unlimited number of boundary sets can be selected. The last set selected will remain the default until a new boundary set is selected. When Current Viewport is selected, the selection set will be created from everything that is visible in the current viewport. Selecting the option when there is a current boundary set will discard the current selection and place the pattern in everything in the current viewport. The normal setting is to place the pattern based on the Existing Boundary Set.

ISLAND DETECTION METHOD

The final option of the Advanced tab is the Island Detection Method box. In its current state of Flood, internal features (islands) are selected to be part of the boundary set. When the Ray Casting option is current, AutoCAD sends an invisible object called a ray to the nearest object it can find and then traces the boundary in a counterclockwise direction. You might be tempted to store this with other boring facts about AutoCAD that you wish you never knew, but it really is important when using BHATCH. If the ray hits text or an internal boundary of an object, the boundary definition error is displayed.

EDITING HATCH PATTERNS

Features of BHATCH patterns can be edited using the HATCHEDIT command. This command can be started by selecting a hatch pattern to edit and then right-clicking in the drawing area, by selecting Edit Hatch from the Modify II toolbar, by typing HE (EnterKey) at the command prompt, or by selecting Hatch... from the Modify pull-down menu. Each method will produce the prompt:

Command: **HE** (EnterKey)

Select hatch object: *(Select hatch pattern to be edited.)*

As the pattern is selected, a Hatchedit dialog box similar to the one seen in Figure 15.24 will be displayed. The exact display will vary depending on which pattern is selected. The box is very similar to the Boundary Hatch box. Each of the features of the pattern can be altered and function as they did in the Boundary Hatch box.

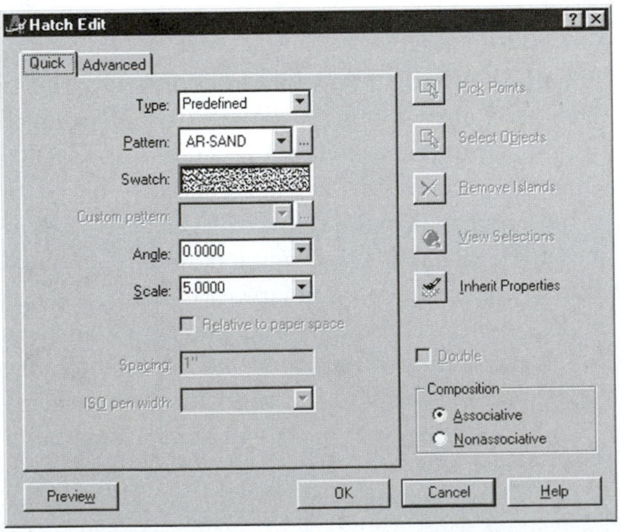

Figure 15.24 *The Hatchedit dialog box can be used to edit the pattern once it has been added to the drawing base.*

CHAPTER 15 EXERCISES

1. Enter a new drawing and use the following coordinates to draw the required object.

Start: 1,1	B. = @ 3,0	C. = @ –5,.25	D. = @ –.75,1	E. = @ –1,1
F. = @ .75,1	G. = @ .5,1	H. = @ 0,1	J. = @ .25,.25	
K. = –2.25,0	L. = @ .25,–.25	M. = @ 0,–1	N. = @ .75,–1	
P. = @ –1,–1	Q. = @ 0,–1	R. = @ 1,–.75	S. = C	

Make a copy of the completed shape. Use the SOLID pattern to fill one of the objects. Turn Fill to OFF. Use the ANSI31 hatch pattern and adjust the scale factor to a suitable scale to HATCH the object so that it appears solid. Save the drawing as E-15-1.

2. Enter a new drawing and use the drawing below as a guide to draw a 15" diameter circle. Draw a 12" diameter circle with a centerpoint 10" above the first centerpoint. Array the 12" circle so that there are a total of six circles. Copy the pattern so that you have a total of three circular patterns. Create layers to separate the object from the hatching. Use the BHATCH command to form three completely different hatching styles. Use different hatch patterns for each area to be hatched. Save the drawing as E-15-2.

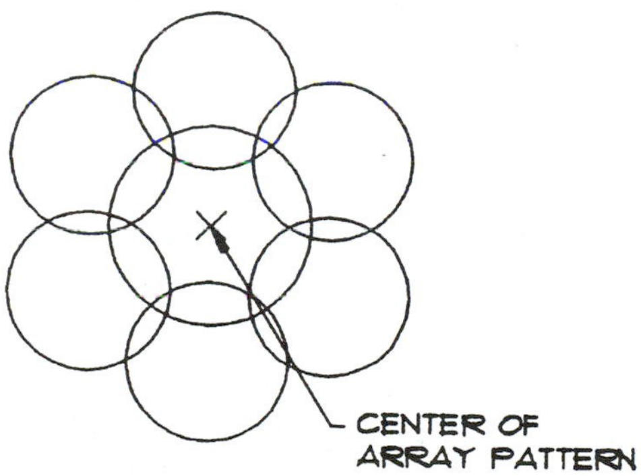

3. Enter a new drawing and draw a red square, surrounded by a blue circle, surrounded by a green octagon, surrounded by a cyan triangle. Copy this pattern so that there are a total of three copies. Create a separate layer for hatching. Use ANSI34 and create a double hatching pattern. Provide examples of Normal, Outer, and Ignore styles of hatching. Save the drawing as E-15-3.

4. Open a new drawing and draw a one-story footing. Use a 6"-wide stem wall extending from the natural grade 8". Support the wall on a 12" × 6" foundation. Show a 2 × 4 key, a 2 × 6 plate with a $\frac{1}{2}$"-diameter × 10" anchor bolt. Place wood, steel, and concrete on separate layers with different colors. Use varying widths of polylines to distinguish each material. Use EARTH to represent the soil and AR-CONC. to represent the footing and stem wall. Copy the drawing and show a 2 × 10 floor joist with $\frac{3}{4}$" plywood floor sheathing. Show the bottom of a 2 × 6 stud wall with single-wall construction. On the second copy, show a post and beam floor with 2" decking and a 6" deep beam. Save the drawing as E-15-4.

5. Draw the plan view of a 5' wide masonry fireplace. Use hatch patterns to distinguish between the firebrick of the firebox, and normal brick of the chimney. Save the drawing as E-15-5.

6. Open drawing 14-3 and create a layer for oak (green), maple (red), and Dutch elm (yellow). Change the trees from their existing layer to these new layers. Tree E is an oak, A is a maple, and the other three are Dutch elm. Hatch each type of tree with a different pattern. Save the drawing as E-15-6.

7. Open drawing E-8-5 and create a white SHADE layer. Hatch the glass area with DOTS, rotated at 45°. Adjust the scale so that the glass appears gray rather than black or white. Save the drawing as E-15-7.

8. Open drawing PB-FND (E-10-7). Create layers for FNDWALLS, FNDFOOT (dashedx2), FNDPIER (hidden), FNDCONC, FNDTXT, and FNDBEAM (center). Assign each layer a different color. Assume the wall to be made from concrete masonry units and hatch the wall with an appropriate pattern. Offset the wall lines to form a 16" wide footing centered on the wall. Trim the footings as required and change the footing lines to the appropriate layer. Save the drawing as E-15-8.

9. Open drawing E-13-9 and complete the drawing by providing the section. Use approximate sizes if dimensions are missing. Provide suitable hatch material and save as E-15-9.

10. A residence is 24' × 36' wide. The ridge is 36' long and the roof is a 6/12 pitch with 24" overhangs, with a 2 × 6 fascia. The roof has a 12" overhang at the gable end walls. The finish floor is 18" above the finish grade and the plate height for the walls is 8'-0". You are to design an elevation of the 36' front side, which must include:

    3' × 6'-8" front door (show some type of appropriate decorative pattern).

    8' × 5' sliding/picture window with grids.

    18" × 60" shutters on each side of the windows.

    12"-wide × 6"-deep stucco columns on each side of the window shutters with 6" horizontal siding above and below the window and shutters.

a 24"-wide × full-height area, of brick, on each side and above the front door (remember, bricks should be approximately 8" long).

Hatch the roof with AR-ROOF.

The area (36") between the doorway brick and the stucco columns is to be AR-RSHKE. (Rows should be approximately 10" wide.)

All other areas to have vertical siding @ 8" o.c.

Show a shadow that extends 6" below the top of the door and window, and show shade that would result from the sun being above and to the left of this gorgeous structure. (Hey! Be glad it's not in your neighborhood!).

Create a separate layer for each material and save this work of art as E-15-10.

CHAPTER 15 QUIZ

1. A circle surrounds a square, which encloses a hexagon. What style is used to hatch the circle and hexagon?
2. In the object described in question 1, what effect will Ignore have on each shape?
3. What shapes are used by the SOLID command to fill an object?
4. List the dialog boxes that are associated with BHATCH.
5. How can the Hatch Pattern Palette box be accessed?
6. How can the scale factor of the BHATCH pattern be controlled by the plotted scale factor?
7. What process is used to keep the hatch pattern from crossing text?
8. Explain the differences in using the Pick Point option and the Select Objects option.
9. What is direct hatching?
10. Describe the process to create a hatch pattern of vertical lines at 1/16" spacing.
11. What is the effect of using the ANSI32 pattern?
12. Use the Help menu to determine what effect the EXPLODE command will have on a hatch pattern?
13. What does associative have to do with BHATCH?
14. You've defined a hatch pattern, but you're not sure how it will appear. List two methods to view the pattern.
15. What is an island?

CHAPTER 16

Inquiry Commands

As the drawings you're working with get more complex, you may find it hard to keep track of some of the drawing parameters you have established. The Inquiry toolbar and the Inquiry menu in the Tools pull-down menu contain five commands used to obtain information about a drawing. These commands are DISTANCE, AREA, Mass Properties, LIST, and Locate Point (ID Point). The Inquiry menu in the Tools pull-down menu also contains the TIME and STATUS commands. In addition to these commands for determining information about the geometry of an object, the program's calculator will be examined as the CAL command is explored.

DETERMINING DISTANCE

The DIST or DISTANCE command measures the distance and angle between two selected points. This command can be especially helpful when used with SNAP or OSNAP locations to measure exact locations. DIST can be useful in determining the size of a room during the design stage, or checking an angle of an object from a baseline. The command can be accessed by selecting the Distance icon from the Inquiry toolbar, by selecting Distance from the Inquiry option of the Tools pull-down menu, or by typing **DI** (EnterKey) at the command prompt. The command sequence is:

Command: **DI** (EnterKey)

DIST: Specify first point: *(Select a point.)*

Specify second point: *(Select second point.)*

The resulting display for the points in Figure 16.1 would be:

Distance = 14'-0" Angle in XY plane = 225.0 Angle from XY plane = 0

Delta X = –9'-10 ½"   Delta Y = –9'-10 ½"   Delta Z = 0'-0"

Command:

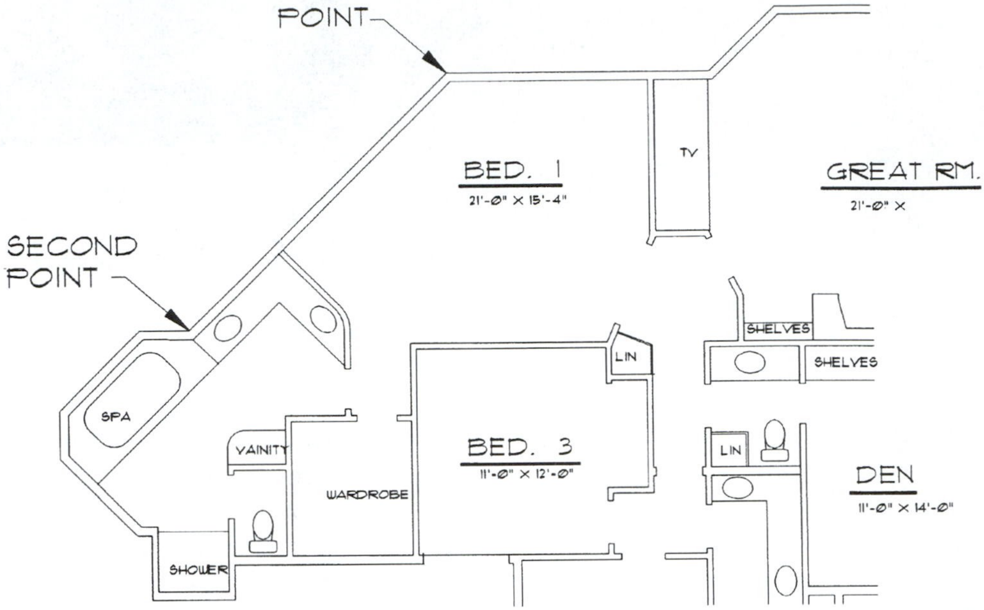

Figure 16.1 *The DISTANCE command measures the distance and angle between two selected points. (Courtesy Residential Designs)*

DETERMINING AREA

The AREA command can be used to determine the area and perimeter of an object. By using OSNAPS, the AREA command can be used to determine exact square footage sizes of structures and, at the same time, determine wall lengths. A running total can be kept and areas can be added or deleted from the total. The command is accessed by selecting the Area icon from the Inquiry toolbar, by selecting Area from the Inquiry option of the Tools pull-down menu, or by typing AA (EnterKey) at the command prompt. As the command is entered, four options are given:

> Command: **AA** (EnterKey)
>
> AREA
>
> Specify first corner point or [Object/Add/Subtract]: *(Specify a point or enter an option.)*

FIRST POINT

When you accept the AREA First point option, you will be asked to supply a series of "Next corner points" to specify the area to be defined. Points may be selected with the pointing device or selected by any of the other selection methods. Using OSNAP will greatly increase the accuracy in computing the area. Figure 16.2 shows an exam-

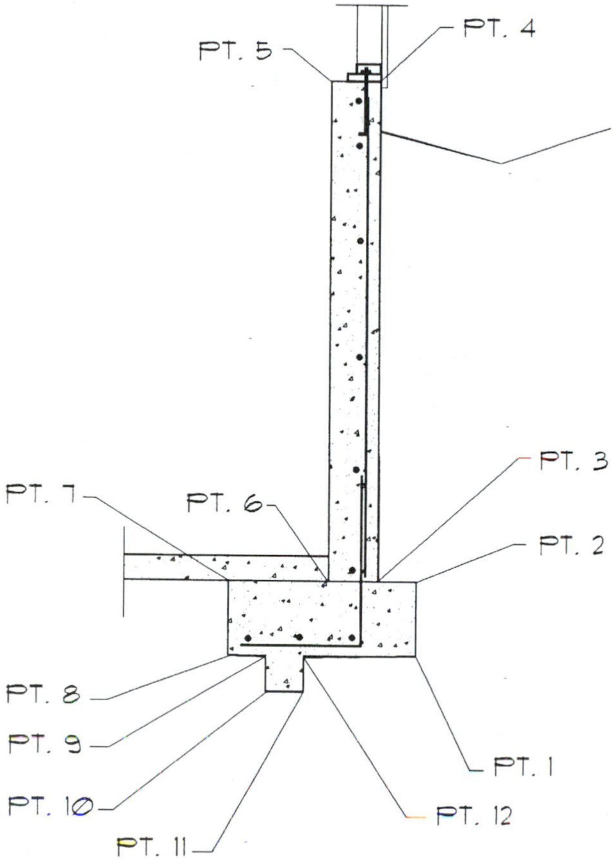

Figure 16.2 *The AREA command can be used to compute the area of a specified shape. The area and the perimeter are given at the Command Line.*

ple of computing the area of a retaining wall by using the point option. The command sequence is:

Command: **AA** (EnterKey)

AREA

Specify first corner point or [Object/Add/Subtract]: *(Specify first point.)*

Specify next corner point or press ENTER for total:

Continue to select points as required. In the example in Figure 16.2, press (EnterKey) at Point 12. This will produce a display listing the area (in square inches and square feet when units are set for architectural units) and the perimeter. The same results will occur if Point 13 (Point 1) is entered as the ending point.

OBJECT

By selecting a point, the Object option will compute the area of a circle, ellipse, spline, or an object composed of closed polylines. Selecting a circle will cause the area and the circumference to be displayed. When a closed polyline is selected, the area and perimeter will be displayed. If an open polyline is selected, the area is computed based on an imaginary line extending from each end of the polyline. The command sequence is:

Command: **AA** (EnterKey)

AREA

Specify first corner point or [Object/Add/Subtract]: **O** (EnterKey)

Select object: *(Select object.)*

Area = 27.63 square in. (0.1918 square ft.) Length 1'–1 5/8"

Command:

Once the object is selected, a display of the area and circumference will be displayed. If an open polyline is selected, AutoCAD will calculate the area as if a line were drawn between the endpoints. (See Figure 16.3.) The line that is projected by

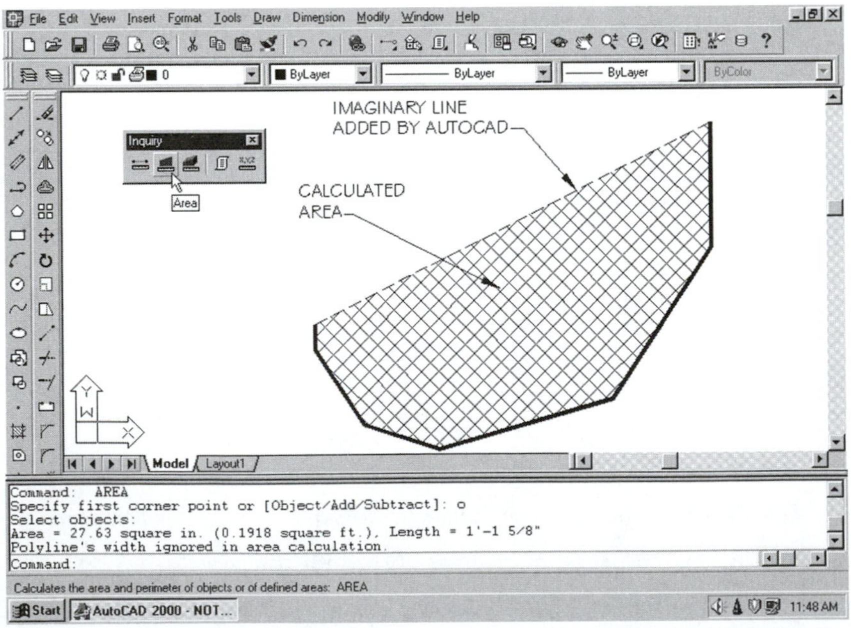

Figure 16.3 *The Object option of AREA can be used to obtain information about an open or closed polyline. AutoCAD will provide the line required to close the polygon and calculate the area without displaying the line.*

AutoCAD is not included in the calculation of the area or perimeter. The command will list the area confined by the polyline and the length of the line.

ADD/SUBTRACT

Using the Add option will allow each area measured in this mode to be displayed and added to a running total. The running total will be displayed after each addition. To begin a running total, select the Add option before selecting the objects to be added. Using the Subtract option provides the opposite effect of Add. Each area measured in the Subtract mode will be displayed and then subtracted from the running total. With both the Add and Subtract options, the prompt will be altered to reflect the active mode. The selected option will stay in effect until the command is canceled. Figure 16.4 is an example of computing the usable area of a parcel of land with the Add and Subtract options. The area of the lot was determined using running OSNAPS to select each corner of the lot, and then subtracting the area of the residence and a garage, which are both outlined by a polyline. The command sequence is:

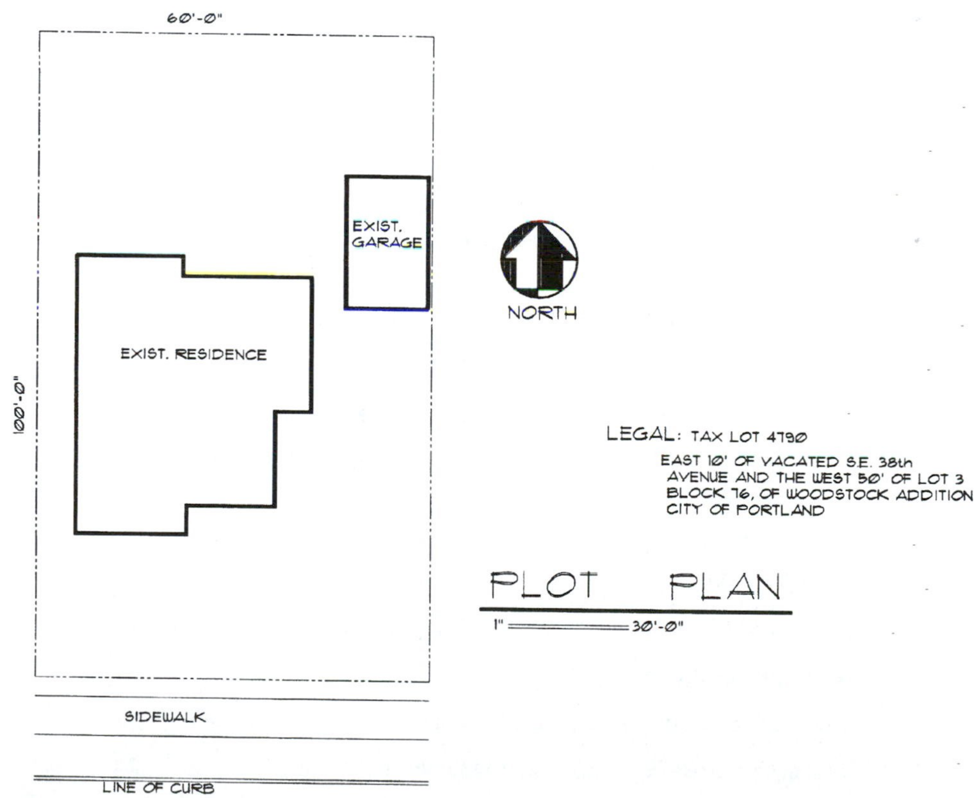

Figure 16.4 *The ADD/SUBTRACT options can be used with the AREA command to add or subtract areas from a running total. (Courtesy Residential Designs.)*

Command: **AA** (EnterKey)

AREA

Specify first corner point or [Object/Add/Subtract]: **A** (EnterKey)

Specify first corner point or [Object/Subtract]: *(Select point 1.)*

Specify first corner point or press ENTER for total (ADD mode): *(Select point 2.)*

Specify first corner point or press ENTER for total (ADD mode): *(Select point 3.)*

Specify first corner point or press ENTER for total (ADD mode): *(Select point 4.)* (EnterKey)

Area = 864000.0 square in. (6000.000 square ft.), Perimeter = 320'-0"

Total area = 864000.0 square in. (6000.000 square ft.)

Specify first corner point or [Object/Subtract]: **S** (EnterKey)

Specify first corner point or [Object/Add]: *(Select point 1.)*

Specify first corner point or press ENTER for total (SUBTRACT mode): *(Select point 2.)*

Specify first corner point or press ENTER for total (SUBTRACT mode): *(Select point 3.)*

Specify first corner point or press ENTER for total (SUBTRACT mode): (Select point 4.) (EnterKey)

Area = 189154.2 square in. (1313.571 square ft.), Perimeter = 118'-3 ½"

Total area = 674845.8 square in. (4686.429 square ft.)

Specify first corner point or [Object/add]: *(Select point 1.)*

Specify first corner point or press ENTER for total (SUBTRACT mode): *(Select point 2.)*

Specify first corner point or press ENTER for total (SUBTRACT mode): *(Select point 3.)*

Specify first corner point or press ENTER for total (SUBTRACT mode):

(Select point 4.) (EnterKey)

Area = 36365.9 square in. (252.541 square ft.), Perimeter = 65'-7"

Total area = 638479.9 square in. (4433.888 square ft.)

Specify first corner point or [Object/Add]: (EnterKey)

Command:

MASS PROPERTIES

The Mass Properties command can be used to calculate the properties of 2D and 3D objects. Although primarilary a 3D command, Mass Properties can be used to determine the area, perimeter and several other features of 2D shapes. The command can be started by selecting the Mass Properties icon from the Inquire toolbar, by selecting Mass Properties from Inquiry of the Tools pull-down menu, or by typing **MASSPROP** (EnterKey) at the command prompt. To use the Mass Properties command with a 2D object, the 2D object must first be converted into regions. If an object that is not a region is selected, the following prompt is displayed:

Command: **MASSPROP** (EnterKey)

Select Objects: *(Select desired objects using any selection method.)*

I found

No solids or regions selected.

Command

To change an object to a region, select the Region icon from the draw toolbar and select the desired objects to be converted. Mass Properties can now be used to display the properties of the selected region. The command prompt is:

Command: **MASSPROP** (EnterKey)

Select Objects: *(Select desired objects using any selection method.)*

I found

Select objects: (EnterKey)

Once objects are selected, AutoCAD will display the mass properties in the text window. The Mass Properties for the shape in Figure 16.5 are seen below.

Regions	
Area:	277.1929 sq in
Perimeter:	91.8271 in
Bounding box:	X: 23.5672 -- 46.5362 in
	Y: -3.1626 -- 18.0234 in
Centroid:	X: 39.2238 in
	Y: 9.3193 in
Moment of inertia:	X: 33838.1482 sq in sq in
	Y: 433392.8323 sq in sq in
Product of inertia:	XY: 97969.2639 sq in sq in

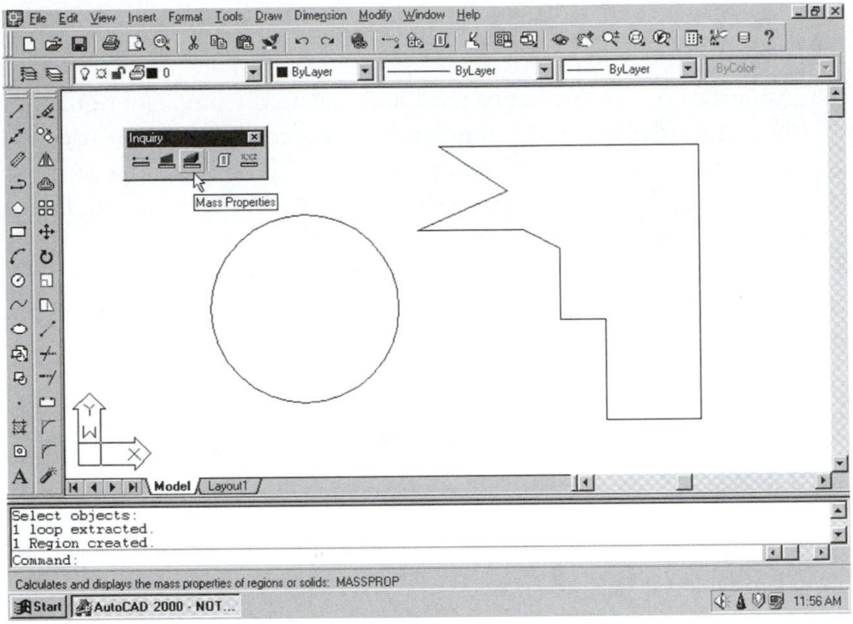

Figure 16.5 *The Mass Properties command can be used to calculate the properties of 2D objects, once they have been converted into regions. Once a region is selected, AutoCAD will display the mass properties in the text window.*

Radii of gyration: X: 11.0487 in

Y: 39.5412 in

Principal moments (sq in sq in) and X-Y directions about centroid:

I: 4704.5411 along [0.5527 -0.8334

J: 11989.3890 along [0.8334 0.5527]

Write analysis to a file? [Yes/No] <N>: *(Press ENTER to continue)* (EnterKey)

Although Mass Properties gives more information than is typically needed for most 2D situations, the command can be used to get the area and perimeter of an object. Although much of the information provided by this command is beyond the scope of this text, other information includes:

- Bounding box—the two coordinates that define the bounding box are the diagonally opposite corners that enclose the rectangle.
- Centroid—the center of the region.
- Moments of inertia—a value used in engineering formulas to compute the distributed loads.

- Product of inertia—a property used to determine forces causing motion in an object.
- Radii of gyration—a value used to determine the moments of inertia of a solid.
- Principal moments—the properties derived from the products of inertia.

LISTING DRAWING INFORMATION

The LIST command displays a listing of information about an object in the drawing. The command can be accessed by selecting the List icon from the Inquiry toolbar, by selecting List from the Inquiry option in the Tools pull-down menu, or by typing **LI** (EnterKey) at the command prompt. The command sequence is:

Command: **LI** (EnterKey)

Select objects: *(Select an object.)*

One or several entities may be selected for listing, although the information may not fit on the screen all at once. To view additional pages, press the (EnterKey) to resume the output display. Pressing (F2Key) will terminate the text box and return the command prompt. The window can also be closed by selecting the Close button in the title bar.

Figure 16.6 shows a partial floor plan with a line selected for LIST and the resulting LIST for this line. The information that is displayed depends on the type of object that is selected—but the object's type, position relative to the current UCS, the layer location, and either model or paper space are always listed. The color and linetype are listed if either is not set by layer. Four values are always given to describe a line. The X and Y values specify the horizontal and vertical distances of the "first point" and the "next point", from the 0,0 point when the limits were established. The line in Figure 16.6 is 32'-1" to the right of the X axis. Notice that distance of the from point and the to point of the Y axis is 34'-8¾" and 12'-9⅜" respectively. The difference between these two values is 21'-11⅜", which is the specified length. The angle is 270°, indicating a vertical line. Figure 16.7 shows an example of each of the values on a vertical and an inclined line.

Listing circles will provide information based on the centerpoint and radius, the area, and the circumference. Listing a circle is an easy way to determine the area of a pier for doing concrete estimates. An example of a listing for the circle in Figure 16.8 is:

CIRCLE	Layer: FNDFOOT
	Space: Model space
Handle =	2D
Center point,	X = 119'-8" Y = 103'-6" Z = 0'-0"

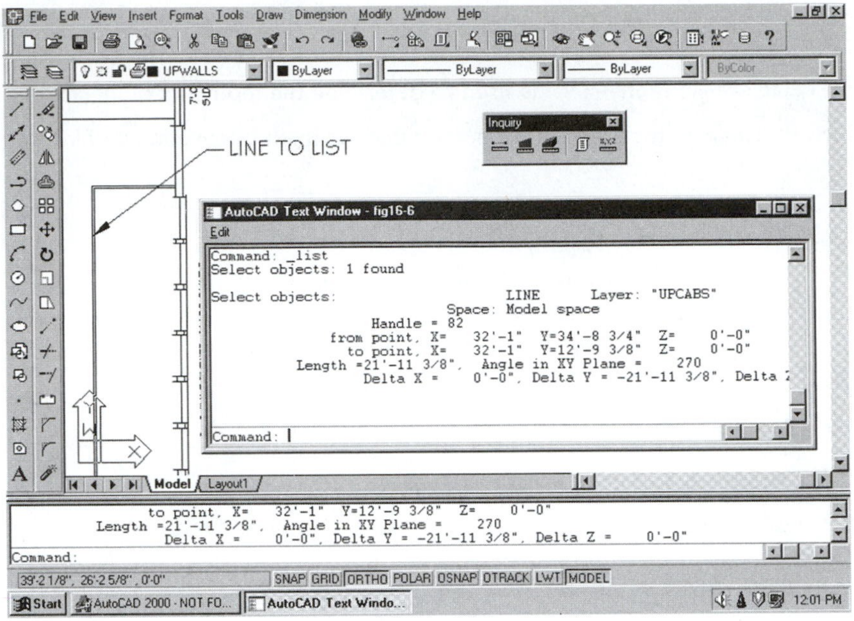

Figure 16.6 *The LIST command can be used to supply information about a drawing object.*

radius	0'-9"
circumference	4'-8 9/16"
area	254.47 sq in. (1.7671 sq ft)

The LIST command can also be used to provide information about text in the drawing base. Text will be introduced in the next chapter. The listing to describe the text for the footing in Figure 16.8 is:

TEXT	Layer: FNDTEXT
Space:	Model space
Handle =	10
Style =	ARCHSTYL Font file = archstyl
start	point, X = 27'-1 7_8" Y = 17'-3 3_4" Z = 0'-0"
height	0'-6"
text	THICKEN SLAB AT BASE POINT OF STAIR
rotation angle	0
width scale factor	1.000
obliquing angle	0
generation	normal

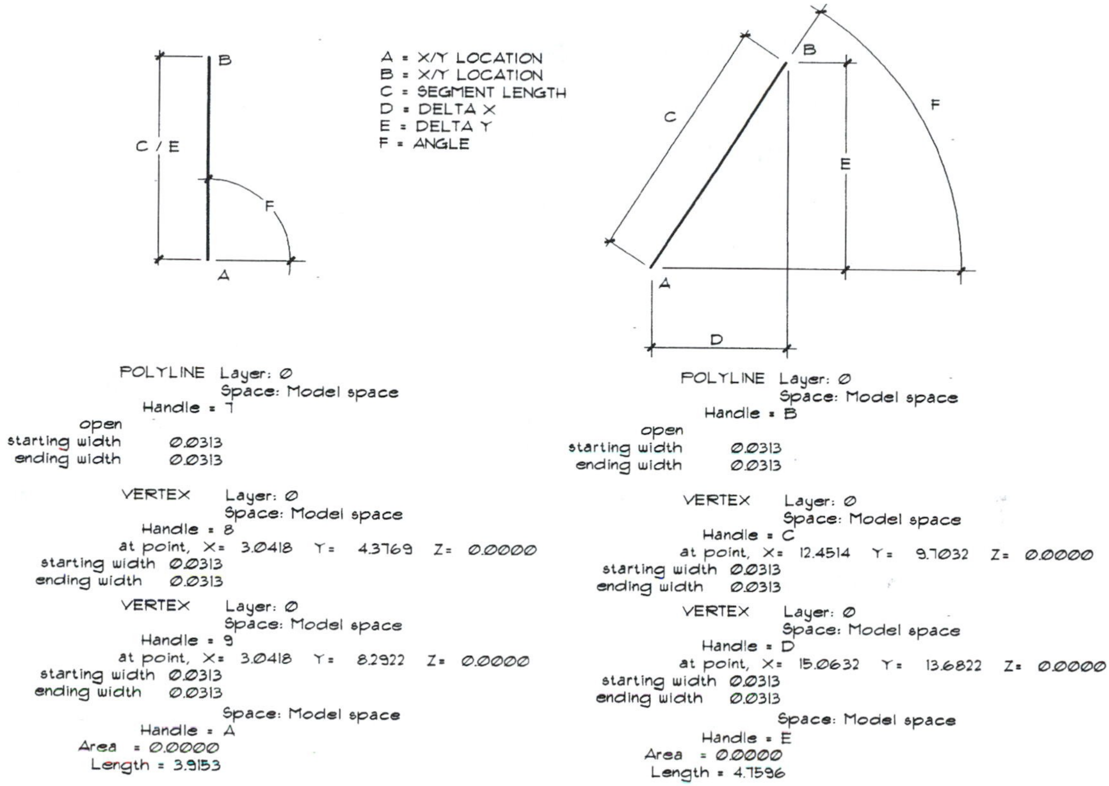

Figure 16.7 *LIST values for two lines.*

OBTAINING INFORMATION ABOUT A DRAWING

The LIST command is used to obtain information about selected objects. The DBLIST command will produce information about the entire drawing. This command can be useful when comparing drawing sequences or information contained in two similar drawings. Information is presented in the same order that is used for LIST, one page at a time. The command can only be accessed from the command line. The command sequence is:

Command: **DBLIST** (EnterKey)

Pressing (EnterKey) will advance through a listing of information for the entire drawing. Pressing (EscKey) will terminate the scrolling of the listing, and selecting (F2Key) will remove the text window and restore the drawing screen. Listings of both the DBLIST and LIST commands can be printed if your terminal is connected to a printer, by pressing the PRINT SCREEN key.

IDENTIFYNING A POINT

Information about a specific point in a drawing can be obtained by using the ID Point command. This command will allow you to determine the exact location of a specific point. The command is accessed by selecting Locate Point in the Inquiry toolbar, by typing **ID** (EnterKey) at the command prompt, or by selecting ID Point from Inquiry of the Tools pull-down menu. The sequence is:

Command: **ID** (EnterKey)

Point: *(Select a point.)*

Entering a point will display the corresponding X, Y, and Z coordinates. This can be helpful when determining the dimension of an object.

The ID command also can be used to identify a specific location in the drawing. By entering ID at the command prompt, coordinates can be supplied and that point will be identified in the drawing. This can be useful with relative coordinates entry methods to start an object at a specific location.

The ID listing for the location of intersection of the line in Figure 16.6 is:

Command: **ID** (EnterKey)

Specify point: *(Select desired point. For this example the handrail in Figure 16.6 was used.)*

X = 31'-11 ½"   Y = 34'-10 ¼"   Z = 0'-0"

TRACKING DRAWING TIME

The current time is displayed at the bottom of the drawing screen in the status bar. The TIME command uses the date and time maintained by the computer to keep track of the time related to the drawing session. The command can be accessed by selecting Time from Inquiry of the Tools pull-down menu, or by typing **TIME** (EnterKey) at the command prompt. Before the command can function properly, the date and time must be set. Using Date/Time of the Control Panel, accessed from My Computer, can set an inaccurate date or time. If you reenter a drawing, AutoCAD can now provide the current time, drawing creation time, last update time, and time elapsed while editing the drawing. Each may seem unimportant in an educational setting, but the information is highly useful in keeping track of hours spent on a project for billing purposes and minor bookkeeping functions, such as determining profit margins.

TIME DISPLAY VALUES

The time display is accessed by typing **TIME** (EnterKey) on the command line. The display will resemble:

Current time: 16 Aug, 1999 @ 11:15:25.510 AM

Times for this display:

Created:	Tuesday, Jul 20, 1999 at 3:3106.970 AM
Last updated:	Friday, Sep 10, 1999 at 2:2506.970 AM
Total editing time:	0 days 3:27:38.175
Elapsed timer (on):	0 days 1:59:51.510
Next automatic save in:	0 days 01:59.20.70

Enter option [Display/On/Off/Reset]:

Command:

Times are listed as day, month, year, hour, minute, seconds, and milliseconds.

The Current Time is the current date and time. Created time is marked when the drawing was initially started. The Last Updated time lists the last editing session when either the END or SAVE command was used with the default filename. The Total Editing time is the amount of time used editing the current drawing, since it was created until SAVE is used. If QUIT is used to end a drawing session, the time spent in that session is not added to the accumulated time. Time spent printing and plotting is excluded from the Editing time. This time is updated continuously and cannot be reset or stopped. This timer is most helpful to supervisors to track efficiency and log total hours for billing.

The Elapsed time can be reset and turned off or on. This timer is best suited for individual use each day to keep track of your time if you will be working on several different jobs. The Automatic Save time indicates when the next save is scheduled to occur (see Chapter 4).

PROMPT OPTIONS

After the time display, four options are listed on the prompt line. The Display option will redisplay the TIME display and update the current values. The ON option will activate the User Elapsed timer. The default value is set to ON when a drawing is started. The OFF option will stop the User Elapsed timer. The Reset option clears the User Elapsed timer to zero. The command line can be returned by a null response or the (F2Key).

STATUS

The STATUS command will list the current value settings of a drawing. The command can be accessed by selecting Status from Inquiry of the Tools pull-down menu or by typing **STATUS** (EnterKey) at the command prompt. The listings of the drawing shown in Figure 16.8 can be seen on page 594–595.

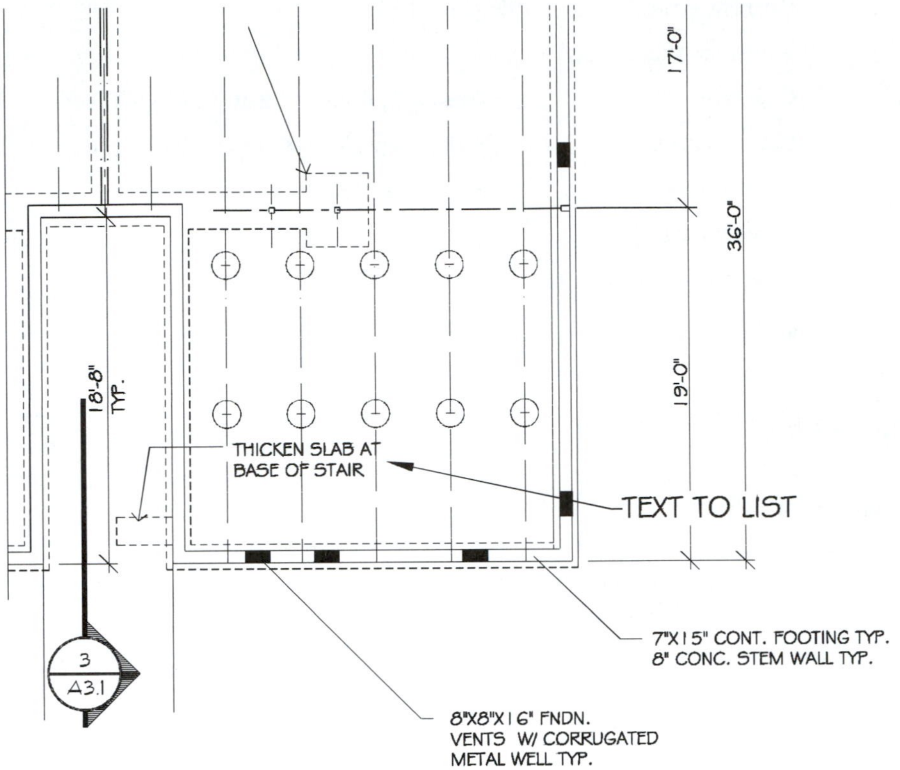

Figure 16.8 *The LIST command provides information based on the centerpoint, the radius, and the diameter. (Courtesy Scott R. Beck, Architect.)*

3931 objects in FOUND

Model Space Limits are:	X: 0'-0"	Y: 0'-0" (off)	
	X: 275'-0"	Y: 250'-0"	
Model space uses	X: 45'-3 1/8"	Y: 48'-7 1/2"	
	X:194'-11 3/8"	Y:152'-7 3/8"	
Display shows	X: 89'-5"	Y: 83'-6 3/4"	
	X: 145'-0 5/8"	Y:123'-9 3/8"	
Insertion base is	X: 0'-0"	Y: 0'-0"	Z: 0'-0"
Snap resolution is	X: 0'-1"	Y: 0'-1"	
Grid spacing is	X: 2'-0"	Y: 2'-0"	

Current space:	Model space
Current layer:	FNDFOOT
Current color:	BYLAYER—1 (red)
Current linetype:	BYLAYER—DASHED
Current lineweight:	BYLAYER—
Current plot style:	ByLayer
Current elevation:	0'-0" thickness: 0'-0"

Fill on Grid off Ortho on Qtext off Snap off Tablet off

Object snap modes: Center, Endpoint, Intersection, Node, Extension

Free dwg disk (C:) space: 2047.7 Mbytes

Free temp disk (C:) space: 2047.7 Mbytes

Free physical memory: 2.6 Mbytes (out of 31.5M)

Free swap file space 1975.7 Mbytes (out of 2016.5M)

Press ENTER to continue: (EnterKey)

Free swap file space: 1996.6 Mbytes (out of 2016.5M).

Command:

Several of the options, such as Color, Linetype, Snap, Grid, and Ortho, should be very familiar by now. Other terms may seem a little foreign. An important value to keep track of is Free Disk. The Free Disk value refers to the space that is available on the drive that contains the drawing. When this space is used up, AutoCAD will terminate the drawing session and save what has been done up to the point of termination. If you're working on the hard drive, this will rarely be a problem. Remember that much of the information provided by the STATUS command can be obtained using the Properties command.

PURGE

This command does not provide information about a drawing, but it can be a useful tool within a drawing file to eliminate unused items and minimize storage space used. Objects that have been named in a drawing session, such as Blocks, Dimstyles, Layer, Linetypes, Shapes, and Text styles, are examples of items that can be purged from a file. PURGE can be used, for instance, if a template drawing is used as a base for a new drawing. The base drawing may contain the LAYER listings for a three-story structure with prefixes such as UPPER, MAIN, and LOWER. If only a two-level structure is to be drawn, all layers with the MAIN prefix could be purged.

Items to be purged can be eliminated at any time after entering the drawing file. Purge can be started by selecting Purge from Drawing Utilities of the File pull-down menu, or by entering **PU** (EnterKey) at the command prompt. The command sequence is:

Command: **PU** (EnterKey)

Enter type of unused objects to purge

[Blocks/Dimstyles/LAyers/LTypes/PlotstylesSHapes/textSTyles/Mlinestyles/All]: *(Select one option.)* **A** (EnterKey)

Enter name(s) to purge <*>: (EnterKey)

Enter name (s) to be purged? [Yes/No]<Y>:

One object can be purged by selecting the appropriate option, or all named items can be removed by selecting the ALL option. Once an option has been entered, you can decide if you want verification of each object that is to be purged. Unused options can be removed one item at a time if you select the Y option. The command sequence is:

Command: **PU** (EnterKey)

Enter type of unused objects to purge

[Blocks/Dimstyles/LAyers/LTypes/PlotstylesSHapes/textSTyles/Mlinestyles/All]: *(Select one option.)* **A** (EnterKey)

Enter name(s) to be purged <*>: (EnterKey)

Enter name (s) to be purged? [Yes/No]<Y>: (EnterKey)

No unreferenced blocks found.

Purge Layer WOOD? <N>: (EnterKey)

Purge Layer CONC? <N>: **Y** (EnterKey)

Purge Layer STEEL? <N>: (EnterKey)

No unreferenced text styles found.

Purge shape file "ltypeshp.shx"? <N> (EnterKey)

No unreferenced dimenstyles found.

No unreferenced mlinestyles found.

No unreferenced plotstyles found.

Command:

The command will list each unused item in sequence until each unused block, dimension style, layer, linetype, shape, and plotstyle has been listed. The default will be not to purge the item. To remove the CONC layer, **Y** (EnterKey) was entered. If the **N** option is used at the "Enter name (s) to be purged? [Yes/No]<Y>:" prompt, the opportunity to verify each item to be purged will be omitted. The PURGE command will automatically remove any unused items. Press (F2Key) to restore the command line.

USING THE AUTOCAD CALCULATOR

The calculator of AutoCAD can be used just as a hand held calculator and also as a means of accurately locating points in a drawing. Typing **CAL** (EnterKey) at the command prompt starts the geometry calculator that will compute standard mathematical formulas. The prompt at the command line is:

Command: **CAL** (EnterKey)

>> Expression:

PERFORMING MATHEMATICAL CALCULATIONS

Mathematical calculations can be completed by AutoCAD but specific symbols must be included as the values are entered. Symbols recognized by AutoCAD include:

+ Add	– Subtract
* Multiply	/ Divide
^ Exponents	( ) Grouped expressions

To add 4 and 4, enter **4 + 4** (EnterKey) at the prompt. To subtract 50 from 100, enter **100 – 50** (EnterKey) at the command prompt. To multiply 50 by 50 would require **50 * 50** (EnterKey) to be entered. Dividing 100 by 50 would require an entry of **100/ 50** (EnterKey). Of course, none of these problems require a calculator, but the command is very useful for performing advanced calculations without having to clean up your work area to find a pocket calculator. By entering **CAL** (EnterKey) at the command prompt, the calculator can be used transparently. If the command line currently displays only one line of text, the answer will not be displayed.

In addition to performing simple calculations, advanced formulas can be executed using the calculator. Parentheses are used to group symbols and values of complex mathematical equations into sets. CAL evaluates expressions based on standard mathematical procedures with expressions in parentheses first, starting with the innermost set. Operations are preformed by exponents first; followed by multiplication, division, addition, and subtraction. Operations of equal precedence are done from left to right.

Entering Feet and Inches in the Calculator

Numbers are entered into the calculator just as they are at the command prompt to enter coordinates. Ten feet three inches would be entered as 10'-3" or 10'3". If you are curious about how many inches that would equal, press (EnterKey). Feet/inch values will be converted to inch values. The process is:

Command: **CAL** (EnterKey)

>> Expression: **10'3"** (EnterKey)

123.0

Command:

Entering Angles

The default units of entering angle values are decimal degrees. Angles are entered using a format of 45d30'30". The minute (') and second (") symbols can be omitted if the values are zero. If the value is less than one degree, enter the value as 0d40'10". Angles can be entered in radians if the value is followed by the letter r (5.5r), or in grads when followed by the letter g (15g). No matter how the values are entered, they will be converted to decimal degrees.

Converting Units of Measurement

As metric dimensioning becomes more important in the construction industry, the need to convert values will increase. The calculator can be used to convert units of measurement from one system to another. Enter the value in the following format: **CVUNIT** (value, from unit, to unit). The process to convert 120 inches to millimeters is:

Command: **CAL** (EnterKey)

>> Expression: **CVUNIT(120,in,mm)** (EnterKey)

3048

Command:

Additional values accepted by the calculator can be found in the CAL Help menu by researching Standard Numeric Functions.

USING THE CALCULATOR TO CREATE OR EDIT POINTS

In addition to the normal functions of a calculator, the CAL command can be used to create or edit a drawing. Listed below are common functions.

Function	Calculation to be performed
CUR	Allows a point to be selected using a cursor.
DIST	(P1,P2) Distance between point P1 and P2.
DPL	(P,P1,P2) Distance between point P and line P1-P2.
ILL	(P1,P2,P3,P4) Gives the intersection between lines P1-P2 and P3-P4.
PLD	(P1,P2,Dist) Point on a line P1-P2, that is the specified distance from P1.
PLT	(P1,P2,T) Point on line P1-P2 defined by the parameter T.
RAD	Gives the radius of a selected circle.
ROT	(P,Origin, Angle) Rotates point P through an angle about the origin.
ROT	(P, P1ax, P2ax) Rotates point P through angle using line P1-P2 as the rotation axis.

In addition to the functions listed above, some commonly used functions have shortcuts that combine some of the previous functions with the Endpoint snap mode. The shortcuts include:

Function	Shortcut for	Calculation to be performed
DEE	Dist(end,end)	Distance between two end points.
ILLE	ill(end,end,end,end)	The intersection of two lines defined by four endpoints.
MEE	(End + end)/2	Midpoint between two endpoints.
NEE	Nor(end,end)	Unit vector in the XY plane and normal to two endpoints.
VEE	Vec(end,end)	Vector from two endpoints.
VEE1	Vec1(end,end)	Unit vector from two endpoints.

To determine the length of a line, use the following sequence:

Command: **CAL** (EnterKey)

>> Expression: **DEE** (EnterKey)

>> Select one endpoint for DEE: *(Select desired endpoint.)*

>> Select another endpoint for DEE: *(Select the other endpoint.)*

10.2500 *(Length given for specified line.)*

Command:

To determine the intersection of two lines, use the following sequence:

Command: **CAL** (EnterKey)

>> Expression: **ILLE** (EnterKey)

>> Select one endpoint for ILLE: First line: *(Select first endpoint.)*

>> Select another endpoint for ILLE: First line: *(Select the other endpoint.)*

>> Select one endpoint for ILLE: Second line: *(Select first endpoint.)*

>>Select another endpoint for ILLE: Second line: *(Select the other endpoint.)*

(12.25 5.25 0.0) *(Intersection given for specified lines in XYZ coordinates.)*

Command:

To determine the radius of a circle, use the following sequence:

Command: **CAL** (EnterKey)

>> Expression: **RAD** (EnterKey)

Select circle, arc, or polyline segment for RAD function: *(Select circle.)*

1.13702 *(Radius for specified circle.)*

Command:

Using Object Snap Modes

Object snap modes can be used as part of arithmetic expressions in place of point coordinates. Prompts are automatically given for the OSNAP mode that is needed to select the specified point. Only the first three letters are required for the prompts. Prompts accepted by AutoCAD include ENDpoint, MIDpoint, INTersection, APParent intersection, EXTension, CENter, QUAdrant, TANgent, PERpendicular, PARallel, NODe, INSert, NEArest, and NONe.

Using the Calculator Effectively

So far the program calculator has been used as a tool to solve math problems. Its real value is in locating features within other features. Figure 16.9 shows an example of how the calculator can be used to place a circle at the center point of a rectangle, using the MEE function. MEE will locate the center between two endpoints. The opposite intersections will be used as endpoints. The command sequence is:

Command: **C** (EnterKey)

CIRCLE 3P/2P/TTR/<Center point>: **CAL** (EnterKey)

>> Expression: **MEE** (EnterKey)

Select one endpoint for MEE: *(Select one corner for endpoint.)*

Select another endpoint for MEE: *(Select opposite corner for endpoint.)*

Diameter/<Radius> <0'-2">: *(Select point or provide coordinates for radius.)*

Command:

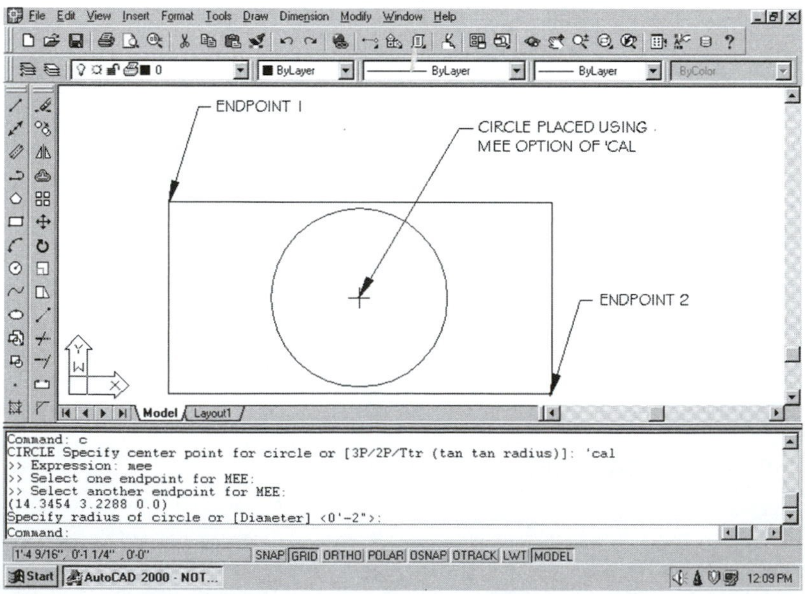

Figure 16.9 *Calculator functions can be used to locate objects in a drawing. The MEE function is used to locate the center point of a circle in the center of a rectangle.*

CHAPTER 16 EXERCISES

1. Open drawing E-15-9 and determine the area of one side of the cross-sectional view of the manhole cover. Assume all unmarked radii to be .25". Maintain a thickness of 7/8" at the cover support. Compute the following:

   Surface area of one side the cover ______________

   Sectional area ______________

   Perimeter of section ______________

   Manhole surface area ______________

   Circumference ______________

2. A steel fabricating company needs to order enough primer to cover (5,000) 21 1/2" × 8 1/2" steel plates. Each plate has (4) 3/4"-diameter holes with the center points located 2" down, 2" from plate edge and 3"o.c. The plates will be made of 1/2" steel. Draw the top, front and side views of the plate, and save the drawing as E-16-2. Use the drawing to determine the following information:

   Surface area 1 side ______________

   Surface area 2 of sides ______________

   Perimeter ______________

   Area of edges ______________

   Area of each hole ______________

   Total area of holes ______________

   Total area of 1 plate ______________

   Total area of 5,000 plates ______________

3. Enter drawing E-16-2. Copy the listed times required to complete the project.

   Created ______________

   Last updated ______________

   Total editing time ______________

   Terminate the drawing session using END or save.

4. Open E-14-3 and provide the following information.

   Drawing name ______________

   Total objects ______________

   Limits ______________ ______________

   Limits ______________ ______________

   Current layer ______________

Current linetype ______________

Free disk space ______________

Status of: Grid ______________ Snap ______________ Ortho ______________

Qtext ____________ Tablet ____________

Terminate the drawing with SAVE.

5. Open drawing E-8-5 and determine the surface area of the window.

   AREA ______________ PERIMETER ______________

   What is the distance between the end points of the inner side of the interior arch? ______________

   List the values that describe the outer arch at the top of the window.

   List the total editing time ______________

   Current drawing session time ______________

   Terminate the drawing using Save or End.

CHAPTER 16 QUIZ

1. Give four types of information that will be given if a circle is selected for a list.
2. Explain the difference between LIST and DBLIST.
3. Write the name of the command and describe two different uses for "Identifying a point."
4. What are the two methods of selecting objects to determine their area?
5. What command is best suited to determine the length of a line between two points?
6. What is the process to set the internal clock of your computer?
7. What value of the status listing keeps track of available drawing space?
8. What is the meaning of DELTA X and DELTA Y?
9. You've opened a drawing, made a few changes, and then discarded the drawing. Will the time spent in this drawing be added to the accumulated time?
10. Explain each of the categories used to measure time.
11. Describe how the AREA command can be used on an open polyline segment.
12. You need to determine the area of a rectangle. Explain the difference in selecting 3 and 4 points.
13. List the symbols used by calculator for add, subtract, multiply, divide, and to indicate a set.
14. Determine the following values for your workstation:

    Free drawing space

    Free temp disk space

    Free physical memory
15. Determine how many millimeters are in 500 inches, and how many feet are in 3000 mm. Explain the entries needed to determine the answer.
16. What is the circumference of a 4"-diameter circle to the nearest 1/6"?
17. What function of the calculator should be used to determine the radius of an existing circle?
18. Research the functions required for the calculator to determine the following functions:

    The cosine of an angle

    The square root of a positive number
19. What will the ILLE calculator function do?
20. Research and describe the calculator function to rotate a point about an axis.

SECTION 3

PLACING TEXT AND DIMENSIONS

CHAPTER 17

Placing Text on a Drawing

If you have previous drafting or design experience you will still remember the long hours of practice required to develop your lettering skills. Throw away all the templates, lettering guides, and rub-on letters of manual drafting, because AutoCAD has a better way. Figure 17.1 shows an example of stock notes drawn using AutoCAD. Instead of manual lettering, you'll be placing text using the TEXT, MTEXT, or QTEXT commands. This chapter will introduce you to the basic TEXT and MTEXT commands. Methods to change the STYLE variables, load different FONTS, change existing text styles, edit existing text, and adjust existing text for quicker regenerations will also be covered.

GENERAL TEXT CONSIDERATIONS

Before text is placed on a drawing, several important factors should be considered. Each time a drawing is started, consider who will use the drawing, the information the text is to define, and the scale factor of the text. Once these factors have been considered, the text values can often be placed in a template drawing and saved for future use. Architectural and engineering offices often use two different styles of lettering. Many engineering offices that do civil, municipal, or government projects tend to use a simple block lettering shape. Architectural and engineering offices dealing with construction projects typically use a style of lettering that features thin vertical strokes, with thicker horizontal strokes to give a more artistic flair to the drawing. Other common variations use compressed and elongated letter shapes. In addition to the basic shape of the letters, some offices use a forward or backward slant to make their office lettering more distinctive. You are unlikely to find all of these variations on one professional drawing. Figure 17.2 shows some of the variations that can be found on drawings.

TEXT SIMILARITIES

With all of the variation in styling, two major areas are typically uniform throughout architectural and engineering offices. Text height and capital letters are as close to an

WOOD:

1. MANUFACTURE OF THE STRUCTURAL GLU-LAMINATED TIMBER SHALL BE BY AN AITC APPROVED FACILITY AND SHALL BE IN CONFORMANCE WITH AMERICAN NATIONAL STANDARDS ANSI.AITC A190.1-1983. APPEARANCE GRADE SHALL BE INDUSTRIAL UNLESS NOTED OTHERWISE ON FRAMING PLANS.

2. ALL GLU-LAMINATED ADHESIVES SHALL BE FOR WET-USE CONFORMING TO ANSI. AITC A190.11983 SEC. 4.5.1.2 AND ASTM D2559-82..

3. ALL GLU-LAM BEAMS, JOIST, POST AND BLOCKING AT EDGES OF STRUCTURE SHALL BE PRESSURE TREATED W/ 0.05 PCF K-520 COPPER NAPHTHENATE IN LIGHT SOLVENT, IN ACCORDANCE WITH AWPA STANDARDS C-14, UNLESS NOTED. ALL CUTS AND BOLT HOLES SHALL BE MADE PRIOR TO TREATMENT, UNLESS NOTED. INADVERTENT FIELD DRILL HOLES AND CUTS SHALL BE REPEATEDLY FLOODED WITH 2% COPPER NAPHTHENATE, PRIOR TO FASTENER INSTALLATION IN CONFORMANCE WITH AWPA STANDARDS.

4. MINIMUM LUMBER GRADES SHALL BE AS FOLLOWS UNLESS NOTED ON PLANS:
   GLU-LAM BEAMS-SIMPLE SPANS. . . . . COASTAL D-FIR 24F-V4
   GLU-LAM CANTILEVERS . . . COASTAL D-FIR 24F-V8
   STUDS, JOIST, AND BLOCKING . . . COASTAL D-FIR # 2 AND BETTER
   4 x 6 x POST AND BEAMS . . . COASTAL D-FIR #1 AND BETTER
   DECKING . . . CEDAR #1 OR BETTER

5. PRE-MANUFACTURED FRAMING CONNECTORS SHALL BE BY SIMPSON CO. OR APPROVED EQUAL. JOIST HANGER NAILS SHALL BE COMMON. CONNECTORS SHALL BE ATTACHED TO FRAMING MEMBERS WITH THE MAXIMUM SIZE AND NUMBER OF NAILS OR BOLTS AS SPECIFIED BY SIMPSON.

Figure 17.1 *A partial listing of standard construction notes. (Courtesy Residential Designs.)*

industry standard as architects and engineers get. Text placed in the drawing area is approximately ⅛" high. Titles are usually between ¼" and 1" in height, depending on office practice. The other common feature is that text almost always comprises capital letters, although there are exceptions based on office practice. The letter "d" is always printed in lowercase when representing the pennyweight of a nail. A typical use would resemble:

USE (3) -16d @ EA. RAFT. / PL CONNECTION

ARCHITECTURAL OFFICES USE MANY STYLES OF LETTERING.
SOME TEXT IS ELONGATED.
SOME OFFICES USE A COMPRESSED STYLE.
NO MATTER THE STYLE THAT IS USED,
IT MUST BE EASY TO READ.

Figure 17.2 *Some common types of lettering found on construction drawings.*

Occasionally manufacturers use lowercase letters to represent the size of special nails or fasteners specific to their products. With the CAPS key activated, all lettering will be capitalized unless the SHIFT key is pressed. Pressing the SHIFT key will make any letters typed as lowercase letters while the key is depressed . If you have typing experience, it will take some time getting used to not having to press the SHIFT key at the start of each sentence.

Numbers are generally the same size as the text they are placed with. When a number is used to represent a quantity, it is generally separated from the balance of the note by a dash. This can be seen in the nailing specification above. Another common method of representing the quantity is to place the number in parentheses () to clarify the quantity and the size. This would resemble:

USE (3) - ¾" DIA. M.B. @ 3" O.C.

Fractions are generally placed side by side (3/4") rather than one above the other ($\frac{3}{4}$), so that larger text size can be used. When a distance such as one foot, two and one half inches is to be specified, it may be done as 1'-2½". The ' (foot) and " (inch) symbols are always used with the numbers separated with a dash. If metric sizes are to be represented on drawings, the use of a comma as a number separator is somewhat different than with English units. Metric sizes of four digits are often written with no comma. Three thousand, one hundred fifty millimeters is written as: 3150 mm. When a number five digits or larger must be specified, a space is generally placed where the comma would normally be placed. Fifty six thousand, three hundred forty five millimeters is written as: 56 345mm. Other common methods of placing numbers will be presented as dimension methods are introduced in future chapters.

TEXT PLACEMENT

Text placement refers to the location of text, relative to the drawing and within the drawing file. Care should be taken with the text orientation so that it is generally placed parallel to the bottom or parallel to the right edge of the page. When describing structural material shown in plan view, text is ideally placed parallel to the member being described. Figure 17.3 shows methods of placing text in plan views. On plan views such as Figure 17.4, text is typically placed within the drawing, but arranged so that it does not interfere with any part of the drawing. Text is generally placed within 2" of the object being described. A leader line is used to connect the text to the drawing. The leader line may be either a straight line or an arc, depending on office practice.

On details, text may be placed within the detail if large open spaces are part of the drawing. It is preferable to keep text out of the drawing. Text should be aligned to enhance clarity and may be either aligned left or right. Figure 17.5 shows an example of a detail using good text placement. The final consideration of text placement is to decide on layer placement. Text should be placed on a layer with a major group code

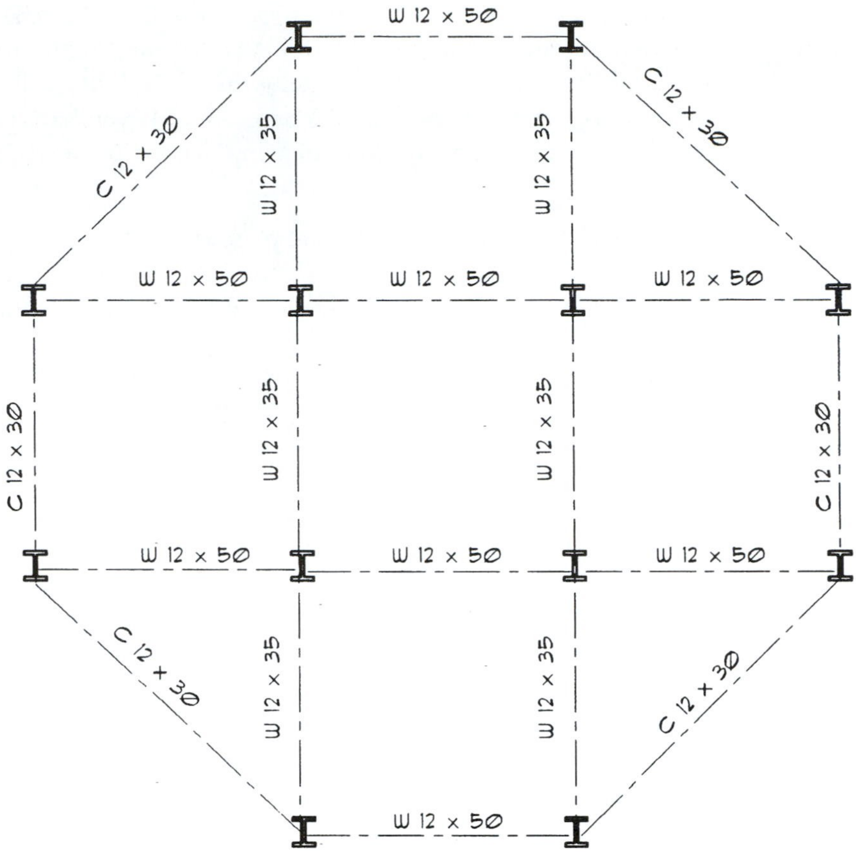

Figure 17.3 *Text should be placed so that it can be read from the bottom or right side of the drawing.*

of ANNO. Minor group codes include KEYN (key notes), LEGN (legends and schedules), NOTE, NPLT (non-plotted text), REDL (redline), REVS (revisions), SYMB (symbols) TEXT, and TTLB (border and title block).

TYPES OF TEXT

Text on a drawing is considered to be either general or local notes. General notes may refer to an entire project or to a specific drawing within a project. Local notes refer to specific areas of a project, such as the notes in Figure 17.4. Small amounts of local notes should be placed using the TEXT command. The notes shown in Figure 17.1 are general notes that specify the materials of a framing plan. The MTEXT command should be used to place large amounts of text. Practical uses for including each text type into template drawings will be discussed later in this chapter.

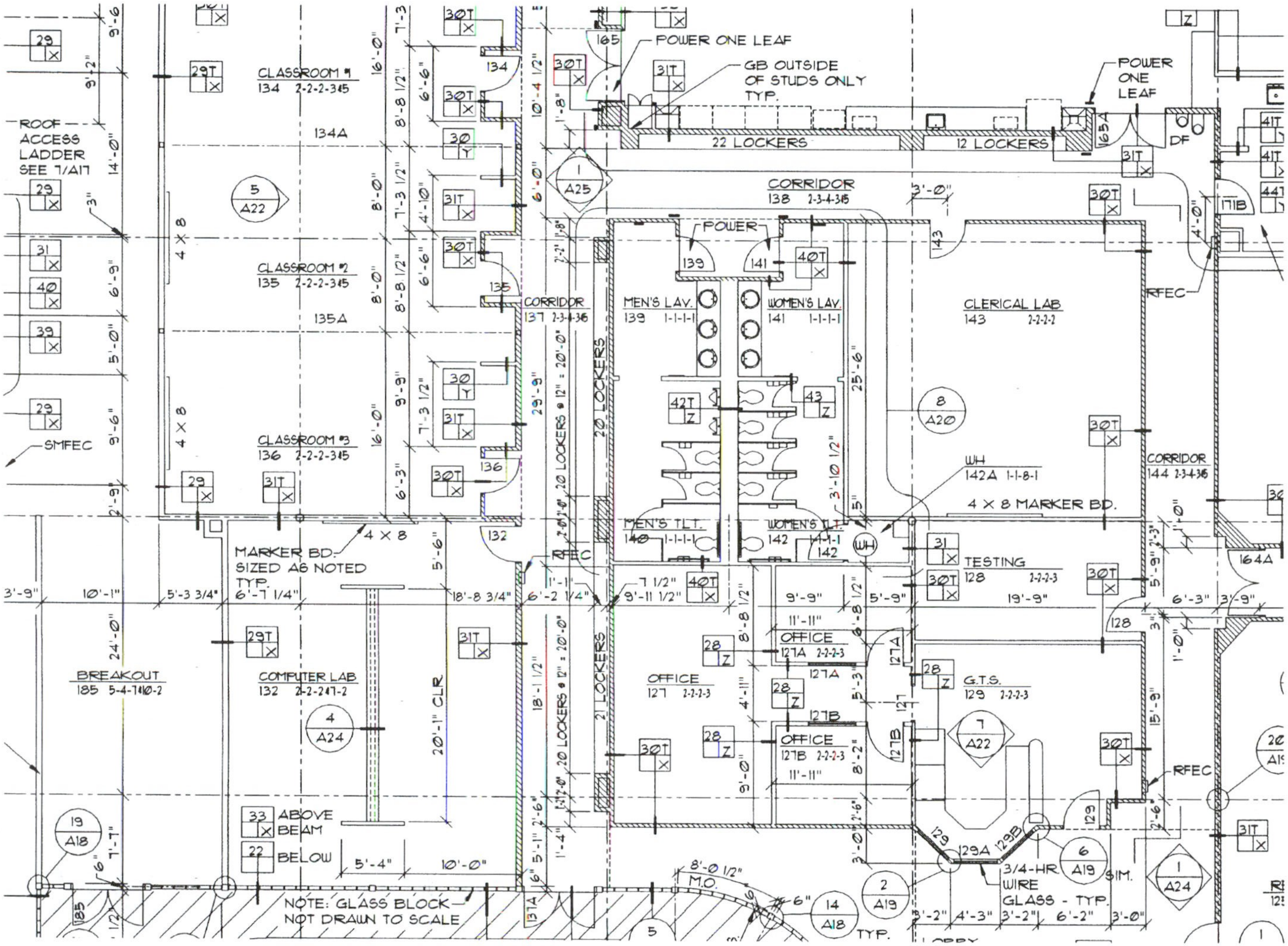

Figure 17.4 *Text can be placed within the drawing, but it should not interfere with drawing clarity. (Courtesy Tom Kuhns, Michael & Kuhns, Architects, P.C.)*

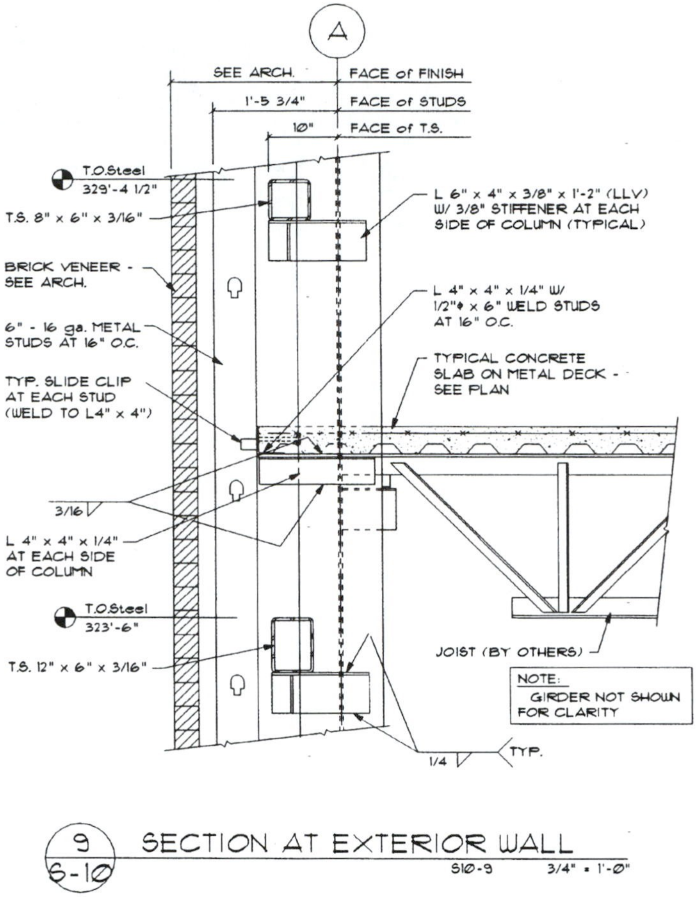

Figure 17.5 *Text should be aligned so that it can be read quickly. (Courtesy Van Domelen/Looijenga/McGarrigle/Knauf, Consulting Engineers.)*

LETTERING HEIGHT AND SCALE FACTOR

One of the first prompts you'll be given as you place text will be for the height. Manually, the height might have been ⅛"-high text, but in AutoCAD you're drawing in real (model) space. In previous chapters, the scale factor required to make line segments the desired size when plotted was considered. Similar factors must be applied to text height to produce the desired ⅛" high text when plotting is finished. Although text height can be adjusted before plotting, it is best to multiply the desired height (⅛") by the scale factor. The text scale factor is the reciprocal of the drawing scale. For drawings at ¼" = 1'-0", this would be 48. By multiplying the desired height of ⅛" (.125) × 48 (the scale factor), you see that the text should be 6" tall. Quarter-inch-high lettering should be 12" tall. You'll notice that the text scale factor is always twice the line scale factor. Other common text scale heights include:

ARCHITECTURAL VALUES		**ENGINEERING VALUES**	
Drawing scale	Text Scale	Drawing scale	Text Scale
1" = 1'-0"	12	1" = 1'-0"	12
¾" = 1'-0"	16	1" = 10'	120
½" = 1'-0"	24	1" = 100'	1200
⅜" = 1'-0"	32	1" = 20'	240
¼" = 1'-0"	48	1" = 200'	2400
1/16" = 1'-0"	64	1" = 30'	360
⅛" = 1'-0"	96	1" = 40'	480
3/32" = 1'-0"	128	1" = 50'	600
1/16" = 1'-0"	192	1" = 60'	720

Working at a scale of 1" = 20', ⅛" high lettering would be drawn as 30"-high text (.125" × 240).

USING THE TEXT COMMAND

The TEXT command creates dynamic text that can be seen on the screen as it is entered at the keyboard. Multiple lines of text can bc placed, even though the command title implies otherwise. TEXT allows some basic editing to be performed and multiple lines of text to be entered while in the same command sequence. The text command can be selected from the Draw pull-down menu by selecting the Single-Line Text option, or by typing TEXT (EnterKey) at the command prompt. The command sequence is:

Command: **TEXT** (EnterKey)

Current text style: "Standard" Text height: 0'-0 3/16"

Specify start point of text or [Justify/Style]: *(Select text starting point.)*

Specify height: <0'-0 3/16">: **6** (EnterKey)

Specify rotation angle of text <0>: (EnterKey)

Enter text: *(Enter desired text.)*

Enter text: (EnterKey)

Command:

Once you know the other options, the Text command may be started with any one of the options. Notice that the first display is the Text style. The current style is referred to as STANDARD. Methods for creating different styles will be explored after you've explored the TEXT command. Since the "Start" point is the first setting requiring input, it is a logical place to start.

START POINT

The name says it all. This is your opportunity to dictate where the start of the text will be placed. Indicating a point with the mouse will make that point the lower left corner of the text location. This type of text placement is referred to as left justified. As the prompt for Text: is displayed, the crosshairs will display the starting point at the lower left corner of the text. The relation of the text to the starting point can be seen in Figure 17.6. If you don't like the location, move the cursor to a new location and select that location with the pick button. Two other options are available instead of placing a start point. Each will be introduced after the basic sequence has been explored. Once the start point has been selected, a prompt for the height will be displayed.

HEIGHT

The second option specifies the text height, with the default of 0'-0 3/16". If you want to plot this drawing at 1/4" = 1'-0", press **6** (EnterKey) for the height. Text height can also be adjusted by dragging the pointing device until the distance between the cursor and the insertion point is the desired text height. If you have not adjusted the initial drawing limits, any text you create is going to appear to be in gigantic proportions. If the height is set to zero, the text height will be controlled by the Style height value. STYLE will be introduced later in this chapter.

ROTATION ANGLE

The third option to be adjusted prior to entering text is the rotation angle of the text. The rotation angle can be visualized by thinking of a line extending from the start point at the specified angle, with the text above this line. The rotation angle can be entered at the command prompt by entering X and Y coordinates, or by using the cursor. To use the cursor method, drag the pointing device until the angle between the cursor and the insertion point represents the desired text rotation angle. Figure

Figure 17.6 *Default placement for TEXT is to place text on the right side of the "Start" point. This is specified as* left-justified.

17.7 shows an example of text rotated at various angles. Although text typically is read looking from the bottom or right side of the drawing, this option will allow text to be aligned with specific objects of a drawing, as seen in Figure 17.8. The default of 0 rotation can be accepted by pressing (EnterKey), or a numeric value can be entered.

PLACING TEXT

The desired text can now be entered by keyboard. Be sure to have the CAP key activated. Notice that as you type, the text is displayed on the screen. Depending on the ZOOM setting, you may be unable to read it. Either cancel the command and ZOOM into the area where you will be adding the text, or read the text at the TEXT prompt line that replaces the Command line. If any special characters are used, they will show up on the screen, but not at the text prompt. When the text entry is completed, press (EnterKey) to add the text to the drawing. To add additional lines of text, press (EnterKey), and the new start position will be directly below the first line of text. The prompts for start point, height, and rotation angle will be skipped and the values for the first line of text will be reused. The spacing between lines of text will be half the height of the text. The command sequence is:

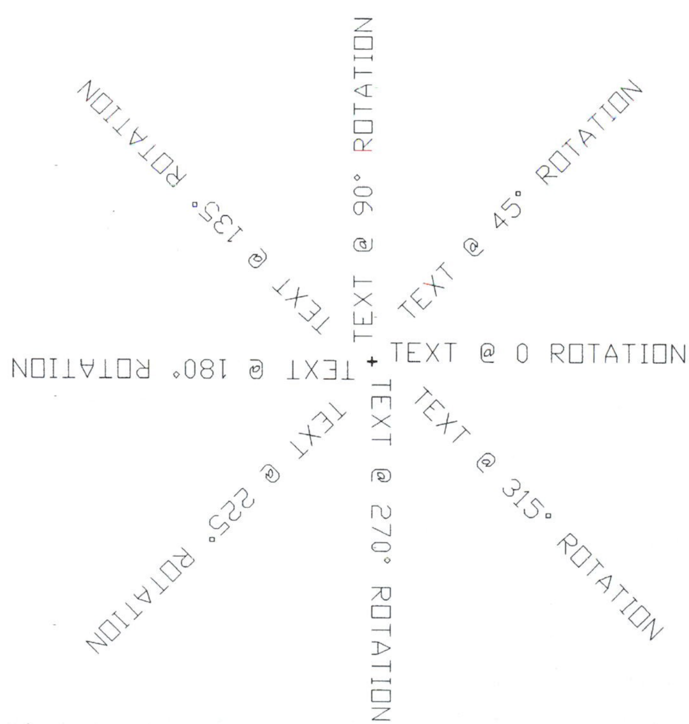

Figure 17.7 *Lines of text may be rotated so that they can be read from any angle.*

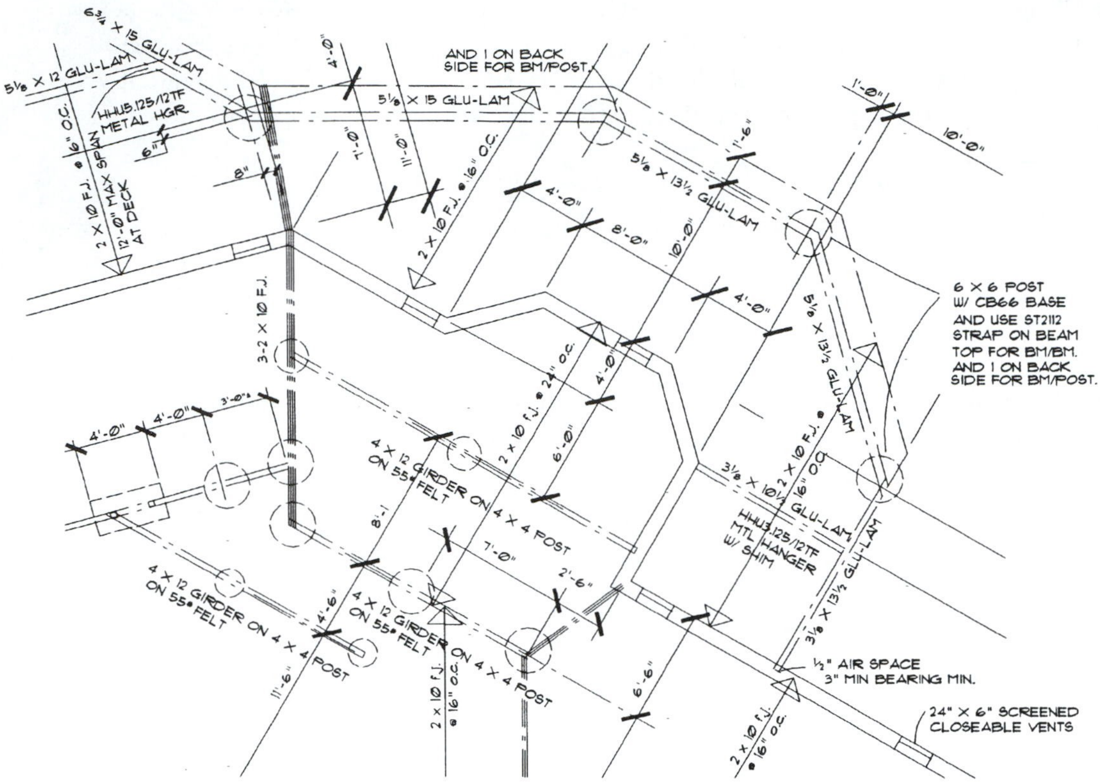

Figure 17.8 *Text should be rotated to be parallel to the structural member when a structure is an irregular shape. (Courtesy Residential Designs.)*

Command: **TEXT** (EnterKey)

Current text style: "Standard" Text height: 0'-0 3/16"

Specify start point of text or [Justify/Style]: *(Select text starting point.)*

Specify height: <0'-0 3/16">: **6** (EnterKey)

Specify rotation angle of text <0>: (EnterKey)

Enter text: **FINISH FLOOR** (EnterKey)

Enter text: ELEV. **100.00'** (EnterKey)

Enter text: (EnterKey)

Command:

Figure 17.9 shows the results of this command sequence. As you work with multiple lines of text, each succeeding line will have all of the attributes of the original text.

Figure 17.9 *The results of the TEXT command using a specified start point, a height of 6, and a rotation angle of 0°.*

Once the desired text has been entered, press (EnterKey) twice to exit TEXT. Pressing (ESCKey) will destroy any text entered during the active command sequence. The MTEXT command, which will be discussed later in this chapter, should be used to create large bodies of text.

JUSTIFYING TEXT

As the prompt for the "start" point was displayed, the options of Justify and Style were given. To justify text is to determine where and how the text will be located within a defined space. The default for placing text is for left-justified text, which places the left edge of each new line of text directly below the previous line of text. The lower left starting point is defined, and text will be placed to the right of the starting point. This method is used on construction drawings, such as sections and details, where large amounts of written information can be neatly aligned. Figure 17.10 shows an example of left-justified text.

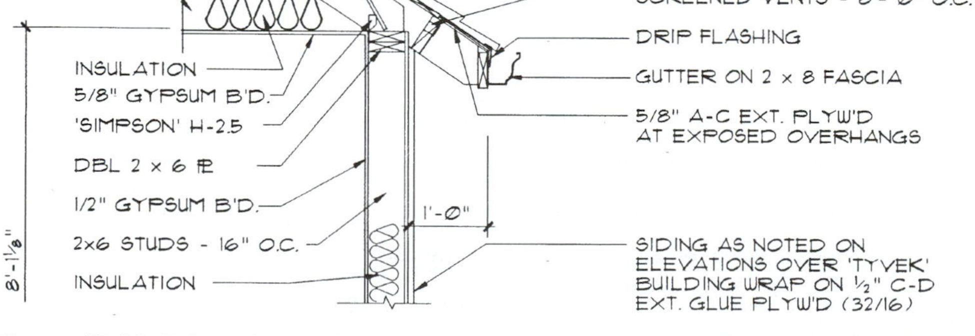

Figure 17.10 *Left-justified text helps add to the clarity of a drawing. (Courtesy Piercy & Barclay Designers, Inc., A.I.B.D.)*

Fourteen other options are available for aligning text. Each option can be accessed by the following sequence:

> Command: **TEXT** (EnterKey)
>
> Current text style: "Standard" Text height: 0'-0 3/16"
>
> Specify start point of text or [Justify/Style]: **J** (EnterKey)
>
> Align/Fit/Center/Middle/Right/TL/TC/TR/ML/MC/MR/BL/BC/BR: **R** (EnterKey)

Once you feel comfortable with each option, the option can be accessed without entering the Justify menu, by typing the letter of the option at the command prompt. The process to align text from the right can be started by the following sequence:

> Command: **TEXT** (EnterKey)
>
> Current text style: "Standard" Text height: 0'-0 3/16"
>
> Specify start point of text or [Justify/Style]: **R** (EnterKey)
>
> Specify right endpoint of text baseline:

Even though Right is not listed as one of the current options, **R** will be accepted and will change the prompt to the endpoint prompt. This prompt will then be followed by the height and rotation prompts. As text is entered using the Right option, it will be placed just as it was when left-justified was used. When (EnterKey) is selected to place the text, it will be added to the drawing as right-justified text.

Align

The aligned option will allow two useful options to be accessed. The starting and ending point can be selected so that text can be aligned with an inclined object, similar to how the Rotation option is used. This method of selection is superior to using Rotation, when the angle is not known. The second benefit of Align is the ability to select the start and endpoints of the line of text. The drawback to this option is that text height is assigned based on the amount of text to be placed between the start and endpoints. Figure 17.11 shows an example of aligned text. The command sequence is:

> Command: **TEXT** (EnterKey)
>
> Current text style: "Standard" Text height: 0'-0 3/16"
>
> Specify start point of text or [Justify/Style]: **A** (EnterKey)
>
> Specify first endpoint of text baseline: *(Select starting point.)*
>
> Specify second endpoint of text baseline: *(Select starting point.)*
>
> Specify height <0'-0 3/16"> *(Enter desired text height.)*
>
> Enter text: *(Enter desired text)*

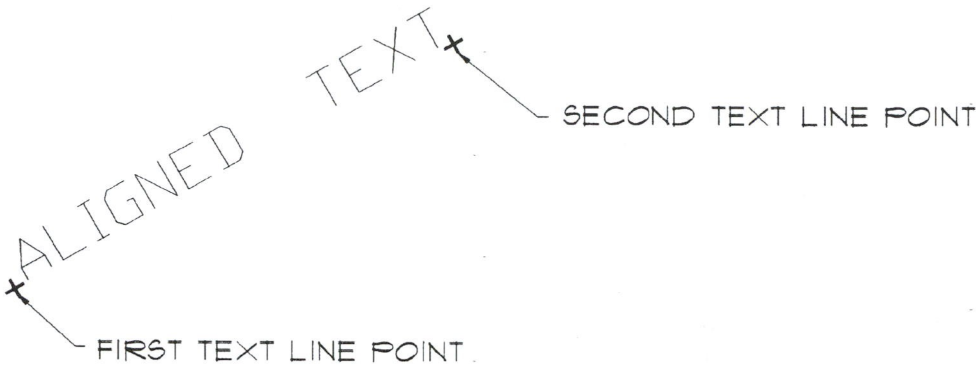

Figure 17.11 *Text can be aligned by specifying two points. The desired text will now be placed between these points. If more than one line of text is required, the height of each line may vary.*

Fit

The Fit option is very similar to the align option. Using Fit, the text height will remain constant throughout several different lines of text, but the width will be varied. This option uses a command sequence similar to Align, but allows for a height to be specified prior to placing the text. Figure 17.12 shows an example of text placed using the Fit option. The command sequence for Fit is:

Command: **TEXT** (EnterKey)

Current text style: "Standard" Text height: 0'-0 3/16"

Specify start point of text or [Justify/Style]: **F** (EnterKey)

Specify first endpoint of text baseline: *(Select starting point.)*

Specify second endpoint of text baseline: *(Select starting point.)*

Specify height <0'-0 3/16"> *(Enter desired text height.)*

Enter text: *(Enter desired text.)*

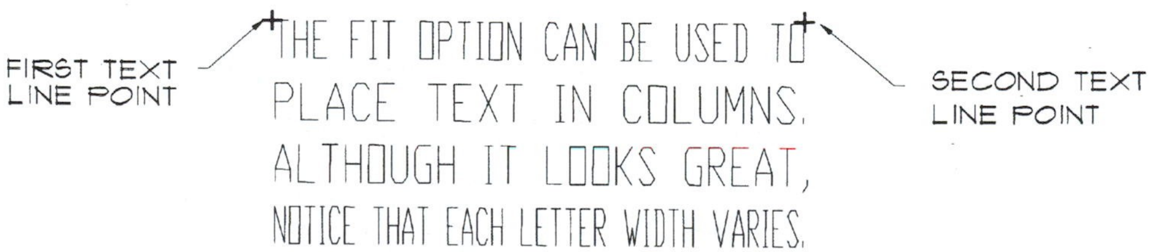

Figure 17.12 *The Fit option will place text between two specified points, but the height will remain constant if more than two lines of text are required.*

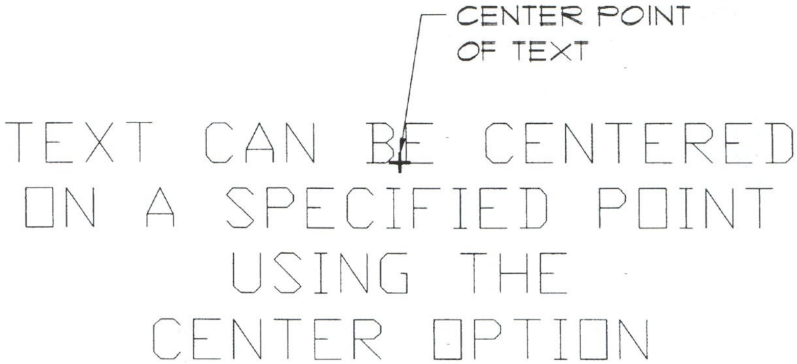

Figure 17.13 *The Center option allows text to be spaced equally around a specified center point.*

Center

The Center text option will place a line of text centered on a specified point, as seen in Figure 17.13. The command sequence is:

Command: **TEXT** (EnterKey)

Current text style: "Standard" Text height: 0'-0 3/16"

Specify start point of text or [Justify/Style]: **C** (EnterKey)

Specify center point of text: *(Select desired center point.)*

Specify height <0'-0 3/16">: *(Enter desired text height.)*

Specify rotation angle of text <0>: *(Enter desired angle.)*

Enter text: *(Enter desired text.)*

To place several lines of text using the Center option, press (EnterKey) to move the cursor to start the line below the first line. Type the second line of text and press (EnterKey) when the line is complete. This process can be continued indefinitely. Remember that MTEXT is better suited for entering large amounts of text.

Middle

This option is similar to the effects of Center, but text is centered vertically and horizontally on the start point, as seen in Figure 17.14. The command sequence is similar to the sequence used for the Center option, once the Middle point is selected.

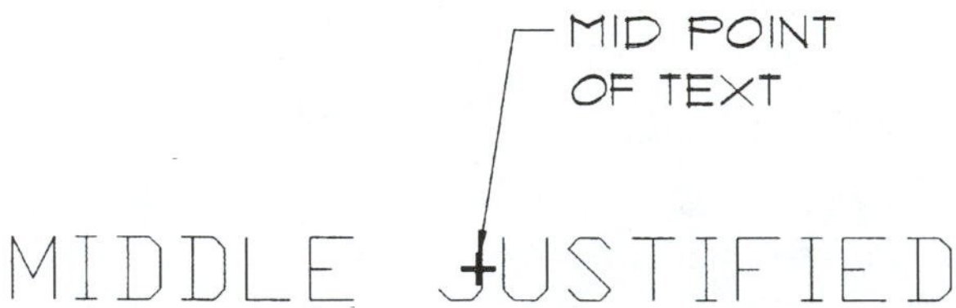

Figure 17.14 *The Middle-justified option will center text along an imaginary line.*

Figure 17.15 *The Right-justified option will align the right ends of lines of text.*

Right

This option is the reverse of the default Left-justified option. Figure 17.15 shows an example of right-justified text.

Letter Options

The remaining nine options can be entered by typing the designated letters at the Justify prompt. The letters represent:

TL-top left	TC-top center	TR-top right
ML-middle left	MC-middle center	MR-middle right
BL-bottom left	BC-bottom center	BR-bottom right

The effects of each can be seen in Figure 17.16. The command sequence is similar to the sequence used for left-justified text, although the start point is referred to by the new location. For instance, the command sequence for BR is:

Command: **TEXT** (EnterKey)

Current text style: "Standard" Text height: 0'-0$\frac{3}{16}$"

Specify start point of text or [Justify/Style]: **BR** (EnterKey)

Specify bottom right point of text: *(Select desired center point.)*

Specify height <0'-0$\frac{3}{16}$">: *(Enter desired text height.)*

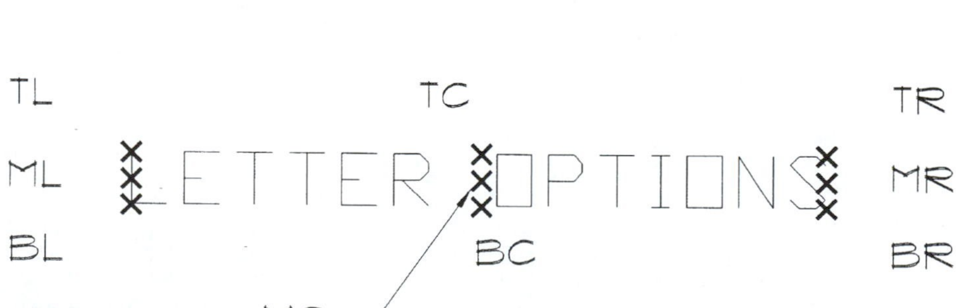

Figure 17.16 *The start points for each of the nine letter options for placing text.*

Figure 17.17 *BR (bottom right) justification is often used when notes must be placed in the left side of an object.*

Specify rotation angle of text <0>: *(Enter desired angle.)*

Enter text: *(Enter desired text.)*

The results of this sequence can be seen in Figure 17.17.

STYLE OPTION

The last option of TEXT to be considered is the Style option, which will determine the appearance of the text characters. The option is selected by entering S at the first text prompt. The sequence and options include:

Command: **TEXT** (EnterKey)

Current text style: "Standard" Text height: 0'-0 3/16"

Specify start point of text or [Justify/Style]: **S** (EnterKey)

Enter style name (or ?) <STANDARD>:

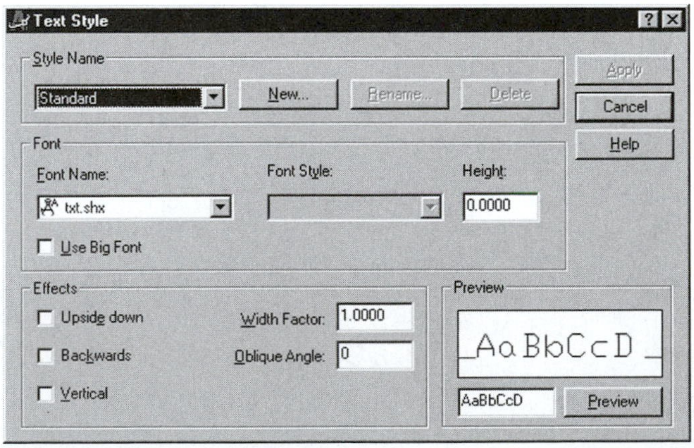

Figure 17.18 *Selecting Text Style... from the Format pull-down menu or typing* **ST***(EnterKey) at the command prompt produces the Text Style dialog box used to name and control text styles.*

Standard will remain the default style until the STYLE command is used to create other styles. The STYLE command will be explored next in this chapter. By selecting the question (?) option, the following prompt will be displayed:

> Enter text style(s) to list <*>: (EnterKey) Specify start point of text or [Justify/Style]:

WORKING WITH STYLE

The term *style* in CAD drafting describes the characteristics of a group of text—such as the font, height, width, and the angle used to display the text. One style can be used for all drawing text, a second style used for the drawing titles, and a third style used for all information in the title block. The STYLE command can be used to add different fonts to your drawing as well as change the properties of the letter shapes. As styles are created, they should be stored in drawing templates, for future use with all similar projects. Assigning a style name, selecting a font, and defining STYLE properties are the three areas to be considered when setting a new style of text.

NAMING A STYLE

The command can be accessed by typing **ST** (EnterKey) at the command prompt, or by selecting Text Style... from the Format pull-down menu. Both options will produce the dialog box shown in Figure 17.18, which can be used to name a style, assign a font, control the style options, and preview the results.

New styles can be created using the Text Style dialog box by selecting the New.... button. This will produce the New Text dialog box shown in Figure 17.19. The default style is named Style1. Accept the name by selecting the OK button. The New Text Style box will be closed and STYLE1 is now displayed as the current name in

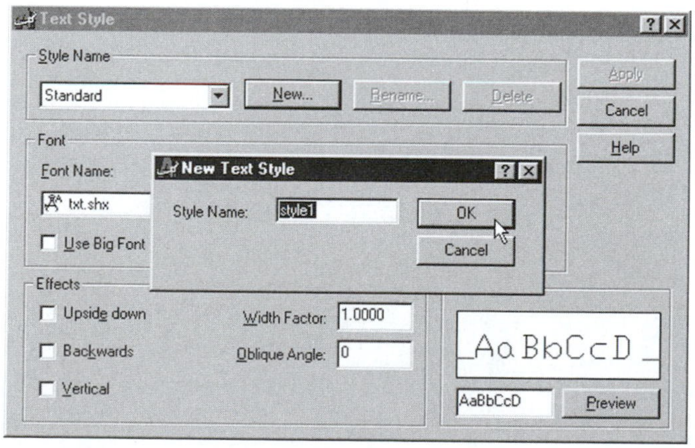

Figure 17.19 *Selecting the New... button produces the New Text dialog box, allowing a new text style to be named.*

the Style Name edit box. STYLE1 can also be altered before it is accepted as a style name by using the same procedure to rename a drawing or layer name. While the name is highlighted, type in the desired name. Enter a name of TITLES in the edit box. Selecting the OK button will remove the dialog box and change the name STYLE1 to the name provided. The name STYLE1 can also be renamed to something more meaningful by selecting the Rename... button and using the Rename Text Style dialog box seen in Figure 17.20. Selecting the Cancel button will remove the Rename Text Style box and allow a different aspect of the STYLE command to be addressed, or allow the command to be terminated. The third option in the Style Name box is the Delete… button. Selecting this button allows a style to be deleted from the current list of text styles.

Choosing the STYLE Name

The name of the text style to be created should reflect the use of the text. The name may contain up to 255 letters, numbers, spaces, and special characters. Names longer than 31 characters will spill out of the edit box. The same guidelines that were used to name linetypes and layers should be used to name text styles. Some offices use the name of the text font for the style name. Names such as TITLEBLOCK, GEN-NOTES (general notes), FNDNOTES (foundation notes), or SITENOTES are examples of names that could be used to describe different styles of text. These names could be abbreviated further as TB, GEN, FND, and ST, to speed typing text selections. Names that include the text font and usage are helpful. Much of the lettering in this text was created using a font called Archstyle that was created by a third-party vendor for use with AutoCAD. You can create styles in the template drawing, such as architl, archtxt, and italitxt. These would be used for creating titles and text using the font archstyl, and a text style using the font italic for drawing text.

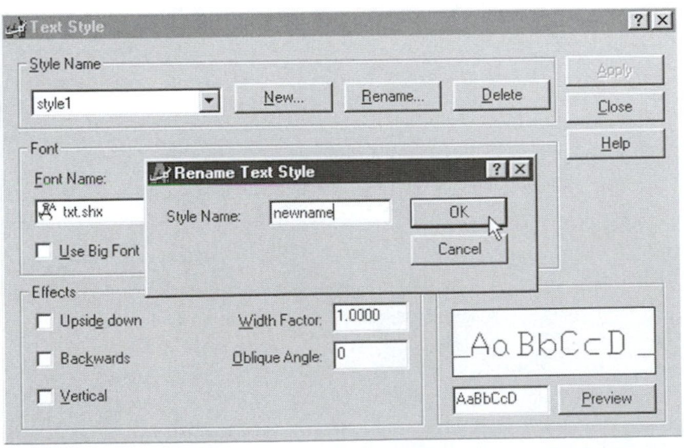

Figure 17.20 *STYLE1 can be renamed in the New text dialog box before the name is assigned, or it can be alterd by selecting the Rename… button.*

LETTERING FONTS

The word *font* is written in an italic font. The term *font* is used to describe a particular shape of text that includes all lower and upper case text and numerals. The shape of the letters is one of the defining differences between architectural and mechanical lettering. Construction drawings usually contain text fonts that are more artistic than the gothic block lettering used on mechanical drawings. Most offices use different fonts throughout a drawing to distinguish between titles and text, or to highlight specific information.

AutoCAD allows all of the fonts loaded into your computer to be accessed by the STYLE command. Many of the fonts contained in AutoCAD can be seen in Figure 17.21a and 17.21b. The illustrations in this chapter have been created using the default font of TXT. TXT is simple font design with letters consisting only of straight-line segments. Because of its simplicity it is quick to regenerate and plot. Many engineering offices use fonts such as ROMANS, ROMANC, ROMANT, or ITALICC to letter drawings. In addition to these fonts, some architectural offices use fonts that more closely resemble freehand lettering such as Stylus BT, City Blueprint, and Country Blueprint for notes on a drawing and ROM, ROMB, or SASBO for titles. Examples of these fonts can be seen in Figure 17.22. It is important to select fonts that are easy to read and quickly reproduced by a plotter. The complexity of a font also affects the amount of storage space required, the speed of regens, and the plotter speed.

Lettering Programs

One of the benefits of AutoCAD is that many third-party suppliers have developed material to supplement the standard program. Many offices add text libraries to thc basic program. You'll notice that much of the text used throughout the illustrations in this text is not found in the AutoCAD font files. This lettering is produced using material created by third-party vendors. Consult advertisements in local trade magazines for other available text libraries. The Text Style dialog box can be used to select the font used with each lettering style. The box will be discussed in detail shortly, but it's important to remember that in addition to the fonts contained in AutoCAD, TrueType fonts contained in additional programs in your computer can be used in AutoCAD drawings. Methods of accessing these files will be discussed as ways of creating text styles are introduced.

Choosing a Font

The second aspect of defining a text style is choosing a font. Once the style name has been entered, selecting the down arrow beside the Font Name: edit box will produce a menu of all of the text fonts available to you. A sample of the menu can be seen in Figure 17.23. The display shown in the menu will be altered, depending on the status of the Use Big Fonts check box. With the box inactive, all fonts in all programs on your computer will be displayed. With the box active, only the fonts contained in

Fast fonts

txt	The quick brown fox jumped over the lazy dog.	ABC123
monotxt	The quick brown fox jumped over the lazy dog.	ABC123

Simplex fonts

romans	The quick brown fox jumped over the lazy dog.	ABC123
scripts	The quick brown fox jumped over the lazy dog.	ABC123
greeks	Τηε ϑυιχκ βροων φοξ ∂υμπεδ ο∈ερ τηε λαζψ δογ.	ΑΒΧ123

Duplex font

romand	The quick brown fox jumped over the lazy dog.	ABC123

Triplex fonts

romant	**The quick brown fox jumped over the lazy dog.**	**ABC123**
italict	***The quick brown fox jumped over the lazy dog.***	***ABC123***

Complex fonts

romanc	The quick brown fox jumped over the lazy dog.	ABC123
italicc	*The quick brown fox jumped over the lazy dog.*	*ABC123*
scriptc	The quick brown fox jumped over the lazy dog.	ABC123
greekc	Τηε ϑυιχκ βροων φοξ ∂υμπεδ ο∈ερ τηε λαζψ δογ.	ΑΒΧ123
cyrillic	Узд рфивк бсоцн еоч йфмпдг охдс узд лащш гож.	АБВ123
cyriltlc	Тхе цуичк брошн фож щумпед овер тхе лазй дог.	АБЧ123

Complex fonts

romanc	The quick brown fox jumped over the lazy dog.	ABC123
italicc	*The quick brown fox jumped over the lazy dog.*	*ABC123*
scriptc	The quick brown fox jumped over the lazy dog.	ABC123
greekc	Τηε ϑυιχκ βροων φοξ ∂υμπεδ ο∈ερ τηε λαζψ δογ.	ΑΒΧ123
cyrillic	Узд рфивк бсоцн еоч йфмпдг охдс узд лащш гож.	АБВ123
cyriltlc	Тхе цуичк брошн фож щумпед овер тхе лазй дог.	АБЧ123

Gothic fonts

gothice	The quick brown fox jumped over the lazy dog.	ABC123
gothicg	The quick brown fox jumped over the lazy dog.	ABC123
gothici	The quick brown fox jumped over the lazy dog.	ABC123

Figure 17.21a *Standard AutoCAD fonts. (Courtesy Autodesk, Inc.)*

cibt	The quick brown fox jumped over the lazy dog.	ABC123
cobt	*The quick brown fox jumped over the lazy dog.*	*ABC123*
rom	**The quick brown fox jumped over the lazy dog.**	**ABC123**
romb	**The quick brown fox jumped over the lazy dog.**	**ABC123**
sas	**The quick brown fox jumped over the lazy dog.**	**ABC123**
sasb	**The quick brown fox jumped over the lazy dog.**	**ABC123**
saso	***The quick brown fox jumped over the lazy dog.***	***ABC123***
sasbo	***The quick brown fox jumped over the lazy dog.***	***ABC123***
te	**THE QUICK BROWN FOX JUMPED OVER THE LAZY DOG.**	**ABC123**
tel	THE QUICK BROWN FOX JUMPED OVER THE LAZY DOG.	ABC123
teb	**THE QUICK BROWN FOX JUMPED OVER THE LAZY DOG.**	**ABC123**

eur	The quick brown fox jumped over the lazy dog. à á â ã ä å æ ç è é ê ë ì í î ï ð ñ ò ó Û Ý ß Ø µ ¶ © ™ ® ¢ £ ¤ ¥ ¦ § ± ¼ ½ ¾ ¿ ¡	ABC123
euro	*The quick brown fox jumped over the lazy dog.* *à á â ã ä å æ ç è é ê ë ì í î ï ð ñ ò ó Ü Ý Þ ß Ø µ ¶* *© ™ ® ¢ £ ¤ ¥ ¦ § ± ¼ ½ ¾ ¿ ¡*	*ABC123*
pan	The quick brown fox jumped over the lazy dog. ə √ Æ æ Ŋ ŋ ¿ ʰ ᵊ ɗ θ ß Đ Ɖ § ¶ á ä â ǎ ā ą ã ‘a a’ Å Ǎ Ā Ä	ABC123
suf	The quick brown fox jumped over the lazy dog. £ ɗ ø ± ß « » ¶ { } ∊ Ł + ħ ʕ ¿ ª á ä â ã ‘a a’ ą Â Ã	ABC123

Figure 17.21b *Standard AutoCAD fonts. (Courtesy Autodesk, Inc.)*

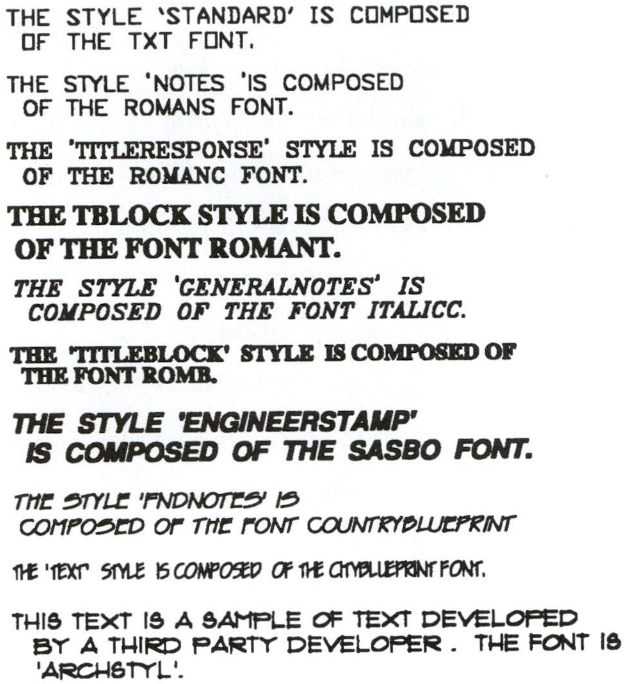

Figure 17.22 *Common fonts used in architectural and engineering offices.*

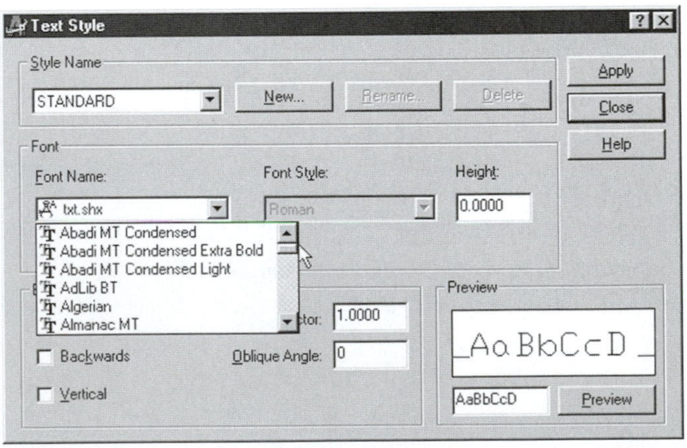

Figure 17.23 *Selecting the down arrow beside the Font Name: edit box will produce a listing of the AutoCAD and the True Type font files. Scroll through the desired list and select the name of the font to be used with the style being created.*

AutoCAD will be displayed in the menu. The AutoCAD fonts are preceded by the Divider icon and followed by .SHX. The TrueType icon precedes other fonts available to AutoCAD. Scroll through the menu and select the Stylus BT font or an architectural font of your choice. As the name of the desired font is selected from the menu, the name will be displayed in the edit box, the menu will be closed, and a sample of the font will be displayed in the Preview edit box.

DEFINING STYLE PROPERTIES

Once the Style has been named and the font has been assigned, seven options are available to further define the Style being created. These options include Font Style, Height, Upside-down, Backwards, Width Factor, Oblique Angle, and Vertical.

Font Style

So far, you've named a Style (TITLES) and assigned the STYLUS BT font to the text style. The Font Style edit box is activated once a font is named. This edit box will allow a character to be formatted with bold, italic, oblique or regular text. Options will vary, depending on the selected font.

Height

If a height is entered at this prompt, all text created using this Style of text will be the same height. If all TITLES throughout the drawing are to be the same size, the height variable of the STYLE command can be set. When height is set with the STYLE command, the variable will not have to be set each time the font is used. The height of text is best controlled with the Height option of the command used to create the actual text. Entering a height now will delete the Height prompt in the TEXT and MTEXT command sequences when this Style is used. Responding with a 0 will allow the height to be determined each time this Style is used.

Upside-Down

The default for this option will place text in the normal position. Activating this box will place text upside-down. Although this option is not typically used in construction drawings, an example of upside-down text can be seen in Figure 17.24.

Backwards

The default value for this option will produce text that is read from left to right. Activating this box will produce text that is read from right to left, or backwards. Although this option is not used often, it could be useful if a drawing is to be plotted

Figure 17.24 *Text may be printed in both normal and upside-down orientations.*

Figure 17.25 *Text can be printed in both the normal and backwards position.*

on one side of a sheet of vellum or Mylar, and text on the other side. Figure 17.25 shows an example of text printed backwards.

Width Factor

Font characters are displayed and plotted, based on a ratio of their height and width. No matter what the actual height is, the width is assigned a value of 1. The default value for the width factor is also 1, making the width equal to the height of the letter. If a value of .5 is entered, text will be half as wide as it is tall. A value of 2 will produce an elongated style of lettering. Figure 17.26 shows a comparison of various width factors.

Obliquing Angle

This option will allow text to be tilted from true vertical. The default angle of 0° will display text in the normal position, with the vertical leg of each letter in a vertical position. Entering a positive value will slant text to the right. Entering a negative number will produce a backwards slant. Figure 17.27 show examples of altering the obliquing angle.

Vertical

This option is not active if a TrueType font has been selected. The option will display .SHX fonts in the vertical position as shown in Figure 17.28.

Applying the Style

Once all of the parameters have been selected for the style, the options of Apply and Close remain to set the style. Choosing the Apply button will apply any style changes made in the dialog box to the text in that current style within this current drawing. Figure 17.29 shows an example of text created using the TXT.SHX font and the

Figure 17.26 *Adjusting the Width factor will affect the appearance of text.*

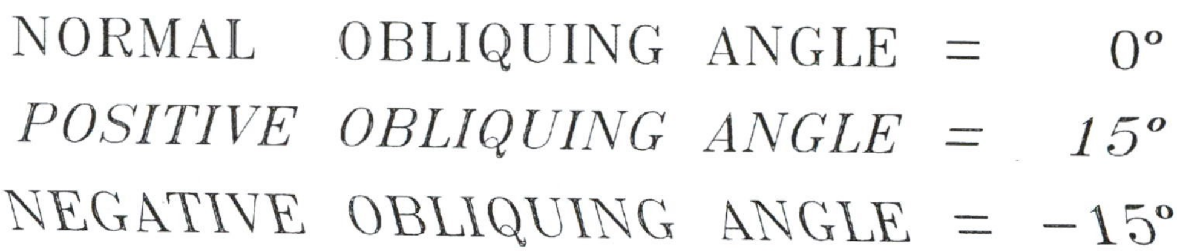

Figure 17.27 *The Obliquing angle of the STYLE command affects the angle of the vertical stroke of each letter.*

same text after the font was changed to ITALIC.SHX, with a width factor of 1.5. The text was altered by changing the appropriate values, selecting the apply button, and then selecting the Close button. If you choose the Close option without selecting Apply first, it will accept the current style values and remove the dialog box. Existing styles will remain unchanged, and new text will be added using the current settings.

SPECIAL CHARACTERS

In addition to the symbols on a keyboard, it is often necessary to print special symbols, such as a degree or diameter symbol. This can be done by typing a code letter preceded by two % (percent) symbols. Common symbols of AutoCAD include:

Figure 17.28 *Placing text in a vertical position may be required on some construction drawings.*

TEXT STYLES CAN BE CHANGED ONCE THEY HAVE
BEEN CREATED. THIS IS TXT.SHX FONT.

TEXT STYLES CAN BE CHANGED ONCE THEY
HAVE BEEN CREATED. THIS IS ITALIC WITH
A WIDTH FACTOR OF 1.5.

Figure 17.29 *Text created using the THX.SHX font was changed to ITALIC.SHX with a width factor of 1.5, by changing the appropriate values and then selecting the Apply button.*

%%C = diameter symbol

%%D = degree symbol

%%O = overscores text

%%P = plus/minus symbol

%%U = underscore text

The symbols for overscoring and underscoring text are toggles. Type **%%U** preceding the text to be underscored to activate underscoring. The symbol must also be typed at the end of the phrase to turn off underscoring. An example of the sequence would be:

%%UPROVIDE AN ALTERNATE BID FOR ALL ROOFING MATERIALS%%U

As the text line is displayed at the command line, the %%U will be displayed in addition to the desired text. When the (EnterKey) is pressed, the %%U will be removed from the screen and the text will be underscored. Although a single percent symbol can be created by using the percent key (ShiftKey) + 5, when the percent symbol must proceed another special character, %%% must be used. To write 50° +/– 5%, you must enter the following sequence:

50%%D%%P5%%%

Remember that these symbols apply to the fonts in AutoCAD and do not apply to text created by third-party vendors. It's also good to remember there is an easier method of applying special text—use the MTEXT command.

EDITING EXISTING TEXT

Text can be edited by using the backspace key and entering corrections before making the text part of the drawing base. Text can be edited once it is part of the drawing base, using the DDEDIT command. The command can be started by selecting the Edit Text icon from the Modify II toolbar, by entering **ED** (EnterKey) at the command prompt, or by selecting Text... from the Modify pull-down menu. Typing **ED**(EnterKey) will produce the following prompt:

Command: **ED** (EnterKey)

DDEDIT

Select an annotation object or [Undo]:

You may select either a line of text or an attribute definition. Editing attributes will be introduced in future chapters. For now, select the text to edit by moving the cursor to the line of text and select it with the select button. The error does not have to be selected, only the line of text containing the error. As the text is selected, a dialog box is presented containing the line of text. Figure 17.30a shows a line of text in a drawing and the line contained in the Edit Text dialog box. Because the line of text is too long to fit in the edit display, the left and right arrow keys can be used to alter the line of text that is displayed.

Once the proper portion of the text line is displayed, editing can be performed by moving the arrow to the desired location and using the pick button. In the example, the following note is used:

aLL FRAMINGG LUMBER IS TOBE D.F.L. # 2 OR BETTER UNLESSNOTED.

The "a" of "all" should be in capital letters, FRAMING is misspelled, and a space should be placed between "TO BE" and "UNLESS NOTED." To fix the "a", move

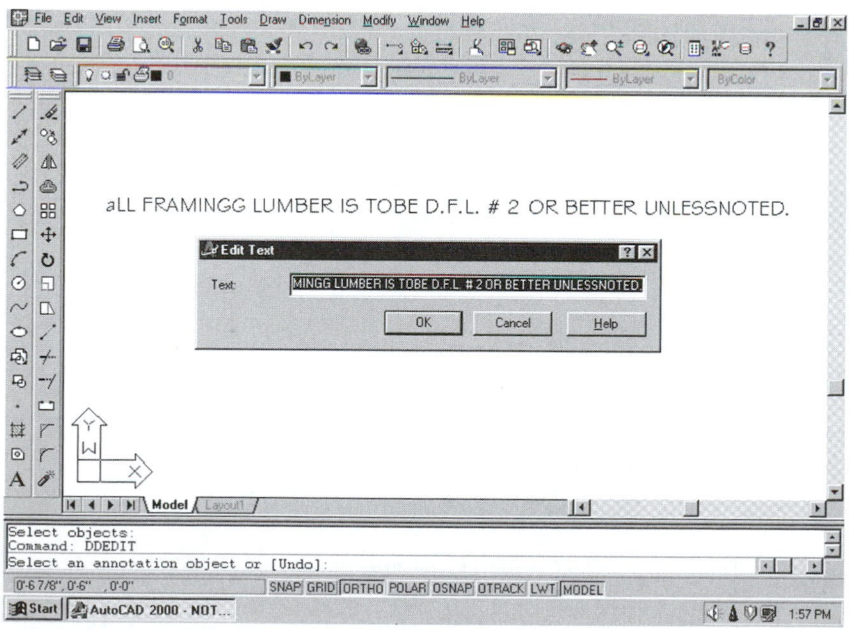

Figure 17.30a *The DDEDIT command allows a string of text to be edited. Once selected, the desired text string will be placed in the Edit Text box.*

the arrow to the right side of the "a" and press the mouse pick button. Use the back-space arrow to eliminate the "a." Without moving the cursor, the error can be fixed by typing an "A."

Use the arrow key to move the cursor to the right side of "FRAMINGG" and press the pick button. Press the backspace key, to eliminate the last "G." Now if the cursor is placed between the "O" and "B" of "TOBE," a space can be inserted by pressing the spacebar. The text should now match Figure 17.30b. "UNLESSNOTED" can be corrected in the same method by pressing the END key to scroll that area of text into the edit box. The corrected text will now read as it was intended. (See Figure 17.30c.) If the edit is not what you desired, the Undo option can be used to restore the line of text to its previous status. Once you are satisfied with the text, select the OK box to return to the drawing. The DDEDIT command is still active. If no other editing needs to be done, press the (EnterKey) to return the command line.

TEXT AND THE MIRROR COMMAND

The MIRROR command was introduced in Chapter 10. If you mirror an object that includes text, the text will also be reversed. To avoid the problems this will create use

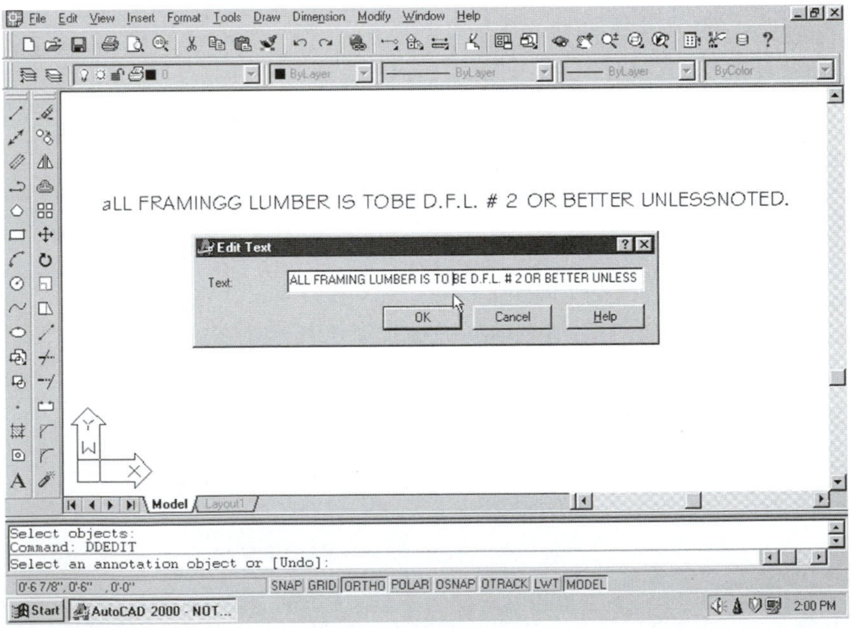

Figure 17.30b *The "a" was edited by moving the cursor to the right of the letter and pressing the pick button on the mouse. Once selected, the error can be removed with the backspace arrow, and the desired correction made at the key board. Even though the "A" and the word "FRAMING" have been corrected, these changes are not reflected in the drawing until the OK button is selected.*

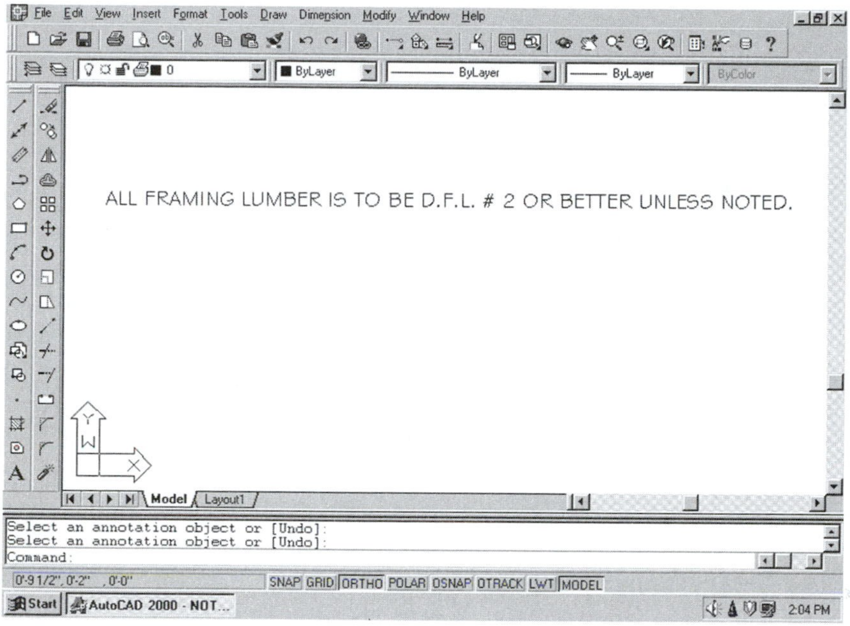

Figure 17.30c *The note shown in Figure 17.30a is now as it was intended.*

the MIRRTEXT command. The command is started by entering MIRRTEXT (EnterKey) at the command prompt. This will produce the prompt:

> Command: **MIRRTEXT**
>
> Enter new value for MIRRTEXT <1>:

Entering (EnterKey) at the prompt will accept the default value and the text will be reversed as the object is mirrored. Entering **0** (EnterKey) at the prompt will mirror the object and the text, but keep the text readable.

QUICK TEXT

Quick text is not an additional way to create text, but a very useful way to display it. As you're editing drawings containing large amounts of text, you'll find that REGEN times start to slow. The QTEXT command will display lines of text as rectangles, which are much quicker to regenerate. The height of the rectangle is the same as the text height, but the length of the rectangle is typically longer than the line of text. The command sequence is:

> Command: **QTEXT** (EnterKey)
>
> Enter mode [ON/OFF] <OFF>:

With QTEXT set to ON, text will retain its current setting until a REGEN is performed. Figure 17.31 shows an example of QTEXT in both the ON and OFF settings.

CREATING TEXT WITH MTEXT

You may be feeling overwhelmed with text by now. You've worked with TEXT, STYLE, and QTEXT and you may be wondering why you need another way to enter text into a drawing. The answer is that each method offers more options than the previous method. TEXT is ideal for placing local notes in and around the drawing. Paragraphs created in TEXT can be very frustrating to edit if they need to be moved and resized. QTEXT is a tool for displaying text and offers no benefits for creating or editing text. MTEXT or multiline text, is ideal for placing large areas of text on a drawing. MTEXT can be created in the Edit MText dialog box, which offers greater flexibility in creation and editing of the text. Using MTEXT is similar to using a word processing program within AutoCAD. MTEXT also offers advantages when editing the drawing. Lines of text placed with TEXT act as individual lines when they are moved. Text placed with MTEXT functions as one object. Multiline text also allows text to be easily underlined and varied fonts, colors, and heights to be assigned throughout the text. MTEXT is ideal for entering and editing large bodies of text, such as those seen in Figure 17.32. A paragraph can be moved by selecting one character, rather than the entire paragraph. The command can be accessed by selecting the Multiline Text icon from the Draw toolbar, by typing **T**(EnterKey) at the command prompt, or by selecting Multiline Text… from Text of the Draw pull-down menu. Each will produce the following prompt:

Command: **T** (EnterKey)

_mtext Current text style: "Stylus BT" Text height: 12"

Specify first corner

By default, AutoCAD is waiting for you to select where you would like to place the text. Once you select the first corner, you will be prompted:

Specify opposite corner or [Height/Justify/ Line spacing/Rotation/Style/Width]:

Although you can respond to the second prompt by providing a value for one or more of the options, selecting the second corner to define the text window width will produce the Multiline Text Editor box and allow each of these values, and many more, to be easily set. Notice in Figure 17.33, as the box is being dragged to indicate the second corner, an arrow is shown indicating the direction that excess text will flow out of the box. Methods of altering the direction will be introduced shortly. By selecting a second point with the cursor, AutoCAD is asking you to provide a window to place your text in. The width of the window you specify is very important. The width of the

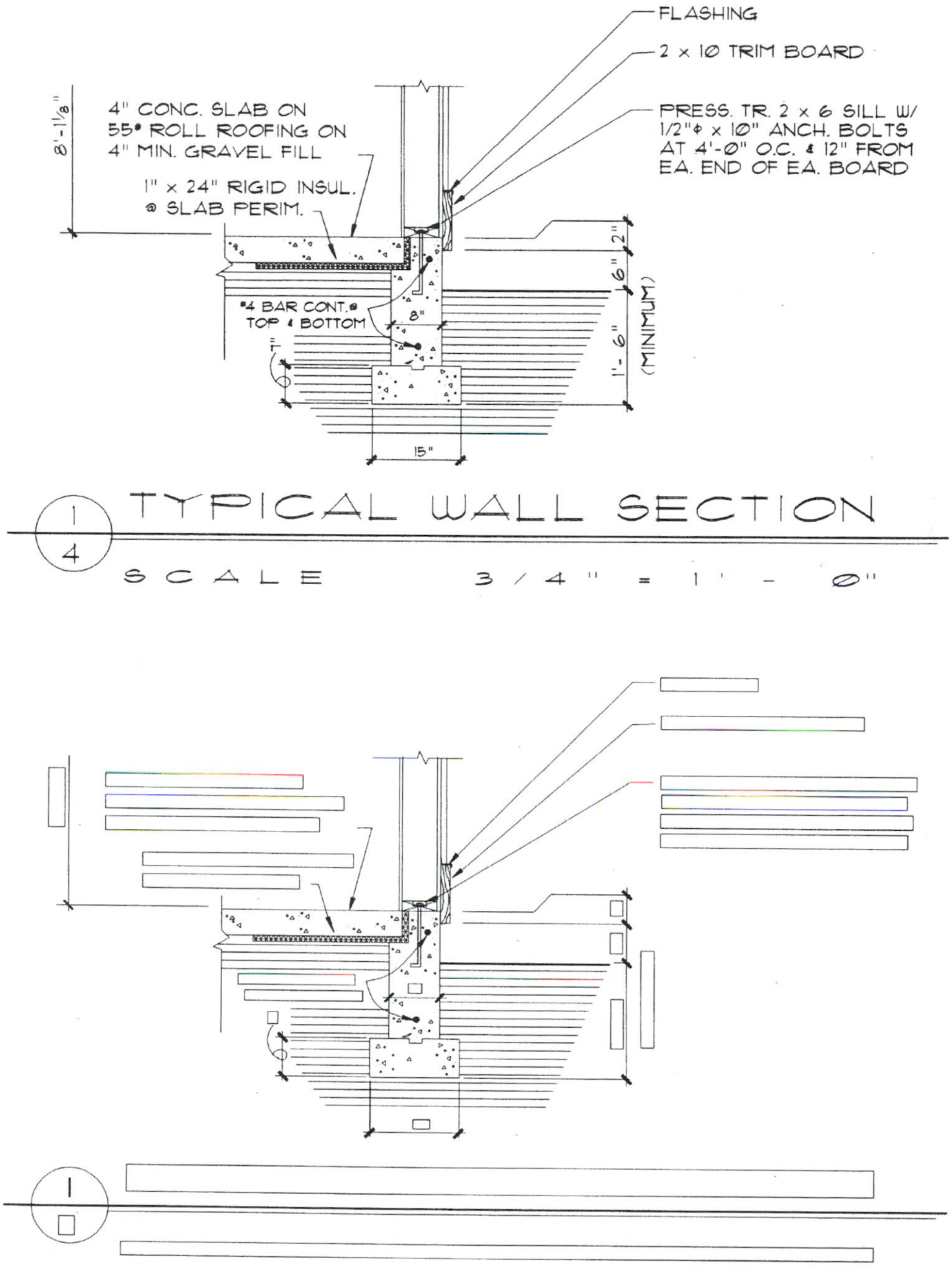

Figure 17.31 *QTEXT can be used to speed the REGEN of very large drawings. The upper drawing shows an example of normal text; the lower drawing shows the effect of QTEXT. (Courtesy Piercy & Barclay Designers, Inc., A.I.B.D.)*

STEEL JOIST
1. STEEL JOISTS SHALL BE DESIGNED, FABRICATED AND ERECTED PER SJI STANDARD SPECIFICATION.
2. CAMBER ALL JOISTS WITH STANDARD CAMBERS PER ABOVE STANDARD SPECIFICATION UNLESS NOTED ON THE DRAWINGS.
3. PROVIDE BRIDGING AS CALLED FOR BY ABOVE SPECIFICATIONS. DO NOT USE SLOTTED HOLES IN BOLTED CONNECTIONS UNLESS APPROVED BY ENGINEER.
4. PROVIDE SHOP DRAWINGS AND CALCULATIONS THAT INDICATE JOIST SIZES, LOADING, MEMBER SIZES, PANEL POINT LOCATIONS, CAMBERS, BRIDGING AND ANY OTHER INFORMATION THAT MAY BE PERTINENT TO THE JOB. SHOP DRAWINGS SHALL BE STAMPED BY AN ENGINEER (THE PRODUCT ENGINEER) REGISTERED IN THE STATE OF OREGON.
5. PROVIDE INSPECTION REPORT PREPARED BY THE PRODUCT ENGINEER CERTIFYING THAT THE JOISTS HAVE BEEN FABRICATED ACCORDING TO THE DESIGN ASSUMPTIONS AND REQUIREMENTS. REPORT IS TO ACCOMPANY THE JOISTS ON DELIVERY TO THE PROJECT.
6. JOIST ERECTOR SHALL EXERCISE EXTREME CARE DURING ERECTION OF JOISTS TO PREVENT THE JOISTS FROM BUCKLING LATERALLY. USE SPREADER BARS FOR LIFTING JOISTS AND PROVIDE LATERAL BRACING AS NECESSARY. REMOVE ANY DAMAGED JOISTS FROM THE JOB SITE. DO NOT ATTEMPT TO REINFORCE DAMAGED JOISTS.

STEEL FLOOR DECK
1. STEEL FLOOR DECK TO BE VERCO 22 GAGE 'TYPE B' FORMLOK OR EQUIVALENT, FOR UNSHORED CONSTRUCTION.
2. WELD DECK TO SUPPORTS AT ALL VALLEYS AND TO MARGINAL MEMBERS AT 24" O.C. WITH 1/2" DIA. FUSION WELDS UNLESS OTHERWISE NOTED ON DRAWINGS.
3. BUTTON PUNCH SIDE SEAMS AT 24" ON CENTER MAXIMUM.

STEEL ROOF DECK
1. STEEL ROOF DECK TO BE VERCO 20 GAGE 'TYPE HSB-36' OR EQUIVALENT.
2. WELD DECK TO SUPPORTS AT ALL VALLEYS AND TO MARGINAL MEMBERS AT 24" O.C. WITH 1/2" DIA. FUSION WELDS UNLESS OTHERWISE NOTED ON DRAWINGS.
3. BUTTON-PUNCH SIDE LAPS AT 24" ON CENTER MAXIMUM.

Figure 17.32 *MTEXT is ideal for creating large quantities of text.*

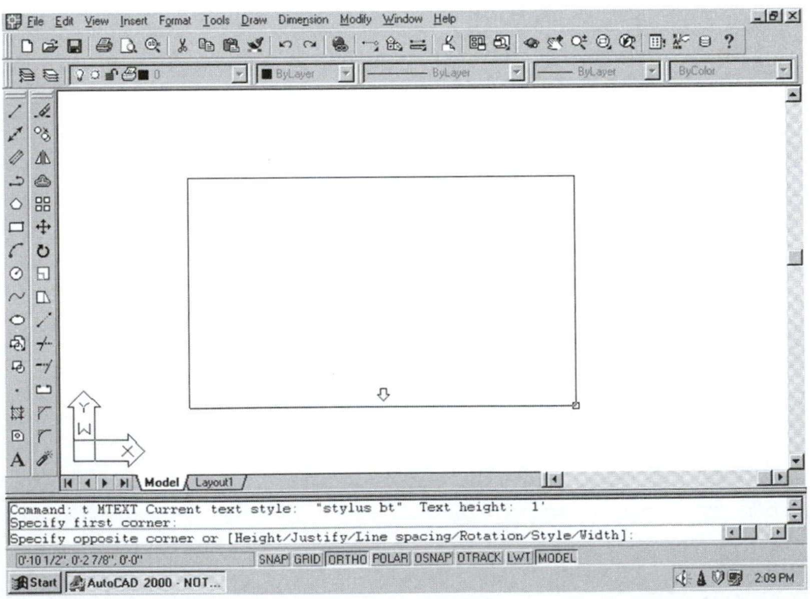

Figure 17.33 *Multiline text is started by selecting a box to define the width that the text will occupy. If the text does not fit into the box, it will overflow the bottom of the box in the current setting (the down arrow). The direction of overflow can be adjusted.*

window will control the width of your paragraph. Extra text will spill out of the top or bottom of the window, depending on the justification, so that the depth of the window is not critical. Once the other corner is selected, the Multiline Text Editor shown in Figure 17.34 will be displayed. Text can now be entered into the edit window using similar procedures of other text editors. AutoCAD allows a third-party text editor to be used by selecting the desired editor from the File tab in the Options dialog box of the Tools pull-down menu. See Chapter 9 of the User's Guide in the Help menu to select a third-party editor. The balance of this chapter will introduce the many capabilities of the AutoCAD program editor.

As text is entered at the keyboard, it will be displayed in the edit box of the text editor. Text will wrap to the next line as the desired width is reached, just as with any word processing program. Notice in Figure 17.34, that the editor contains tabs for controlling text features. The default tab controls Characters and there are also tabs for Properties, Line spacing , and Find/Replace.

CONTROLLING TEXT CHARACTERS

As with TEXT, the qualities of MTEXT can be easily altered. Qualities controlled by the Character tab include the font, height, bold, italic, underline, undo, stack,

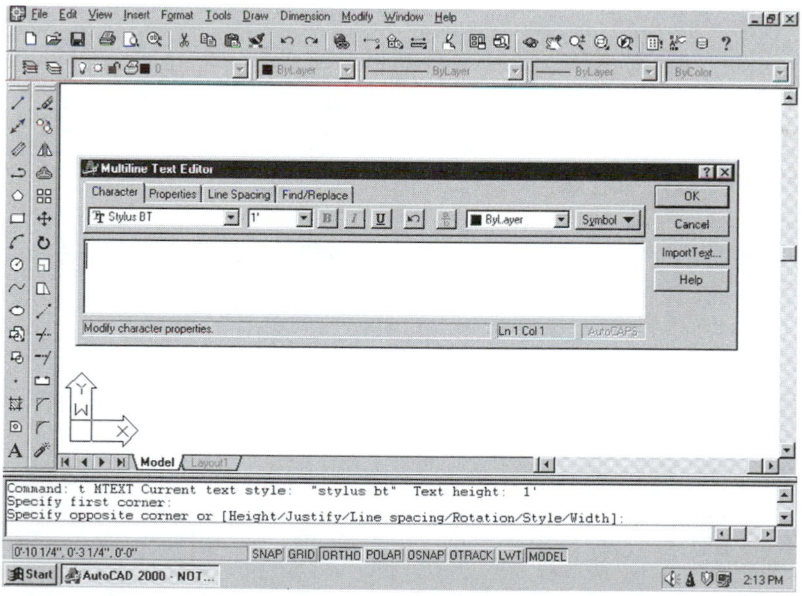

Figure 17.34 *The Multiline Text Editor is used to add large amounts of text to a drawing. By using the Character, Properties, Line spacing , and the Find/Replace tabs, text can be assigned various colors , heights, underlined, and altered. In addition to using the AutoCAD text editor, a third-part editor can be used by selecting the desired editor using the File tab of the Options dialog box. It can be selected from the Tools pull-down menu.*

color, and special symbols. This dialog box also includes the question mark button in the title bar. Pressing the ? key will produce a question mark icon beside the cursor. Now if any of the buttons are selected, an on-screen explanation will be displayed. Figure 17.35 shows an example of selecting the Stack button.

Font

Pressing the down arrow beside the Font edit box displays a menu similar to the font menu displayed with the STYLE command. The menu allows a font for new text to be selected, or changes the font of selected text. The current font for MTEXT is the same as the current font for the STYLE command. With MTEXT, the style can be altered at anytime throughout the use of the command—including mid-sentence. Figure 17.36 shows an example of text that was altered to provide different fonts within the same command sequence.

Height

Just as with the TEXT command, this option controls the height of the text entered with the MTEXT command. The default value is the same as the current style. The value for text height can be changed at any point of the command, allowing words to be printed at varied heights. Selecting the down arrow will display the heights of all current styles. To provide a new height, select the height edit box and click the cursor to highlight the current height. Remove the current height with the backspace arrow

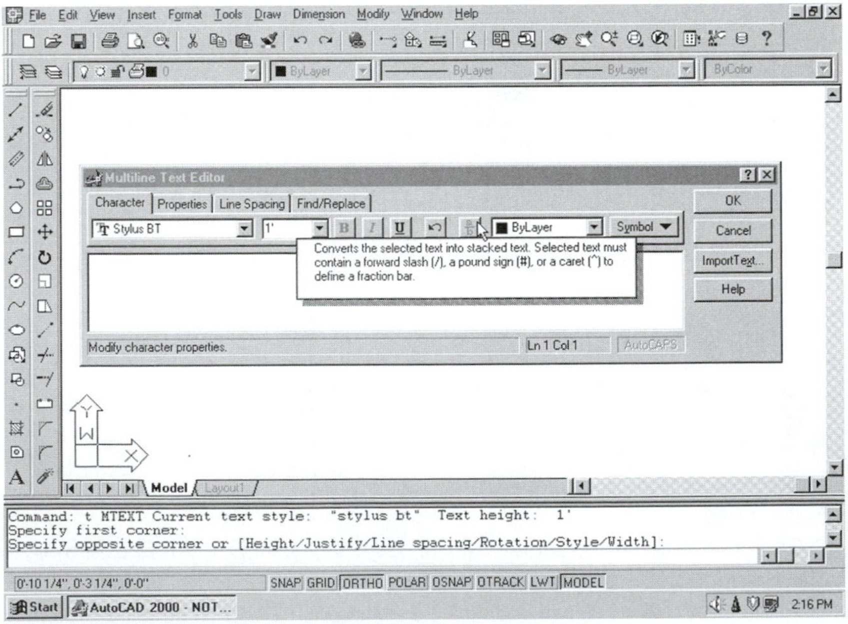

Figure 17.35 *Pressing the ? button in the Title bar, and then selecting a box or button in the Multiline Text Editor will provide a brief explanation of the function.*

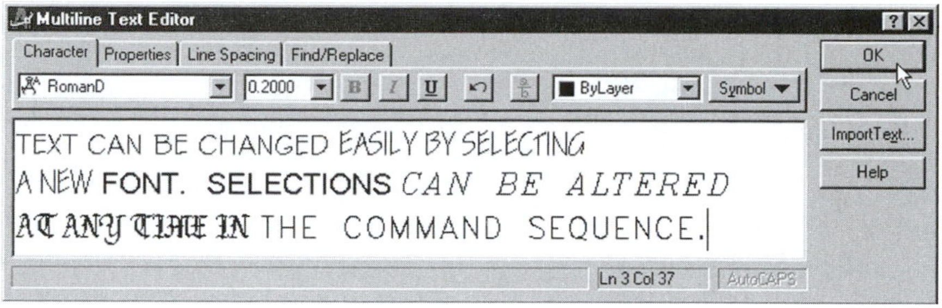

Figure 17.36 *The text font of one word or a string of words can be altered at any point of the drawing life.*

and then enter the desired height value. Text height can be altered before the text is added to the drawing by highlighting the text to be altered, and then entering the desired height in the edit box. Figure 17.37 shows an example of how text height can be altered.

Bold

Selecting the Bold button toggles between bold and normal text. The options are only active for TrueType fonts. Bold text can be created by selecting the Bold button before text is entered. Existing normal text can be changed to bold by highlighting the desired text to be changed and then pressing the Bold button. Figure 17.38 shows an example of text that was altered to bold text.

Italic

Selecting the Italic key toggles italic formatting for text. As with other options, Italic can be applied to text being created, as well as used to modify existing text. The option is only available for TrueType fonts.

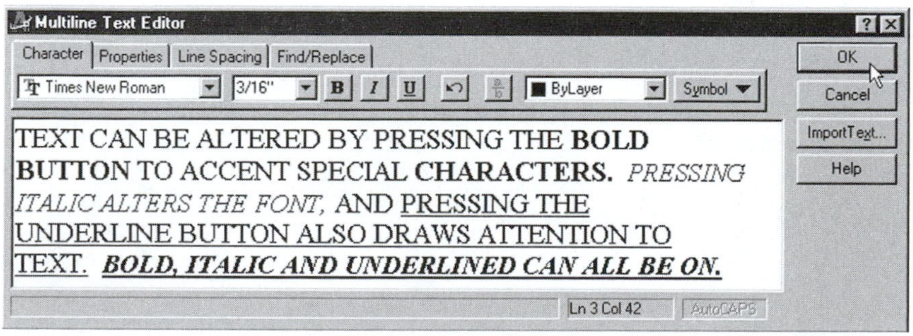

Figure 17.38 *Text can be highlighted by selecting the Bold, Italic, or Underline buttons. One or more of the features can be used in combination to affect the font.*

Underline

Selecting this button allows the underline function to be toggled ON and OFF. This option of MTEXT is so much easier than %%something that is required with the TEXT command.

Undo

This option can be used to undo the last sequence of the Text Editor. Changes can be made to the text or to one of the formatting options.

Stack

The Stack option can be used to place a portion of text over other text. This is useful to write fractions such as $\frac{1}{4}$ rather than 1/4 or to write equations. With the text written as 1/4, highlight the fraction by pressing and holding the pick button and moving the cursor over the text. Once the desired text is selected, selecting the Stack button will stack the indicated text. See Figure 17.39.

Color

Selecting the Color button allows the color of the text to be altered. By default, MTEXT will be assigned the color of the current layer. To select a new text color, select the down arrow beside the color edit box and select the desired color. Selecting the Other color option will produce the Select Color Palette and allow for a full range of colors to be selected.

Symbol

Selecting the down arrow beside the Symbol button displays the menu shown in Figure 17.40. The degree, plus/minus, and diameter are often used on construction documents. Either can be inserted into MTEXT by selecting the down arrow to display the menu, and then selecting the desired symbol. If this seems too easy, the symbols can also be inserted using the special characters that were discussed as TEXT was introduced. Selecting the Other... option will display the menu shown in Figure

Figure 17.39 *Text such as fractions can be written side by side or stacked, depending on office preference.*

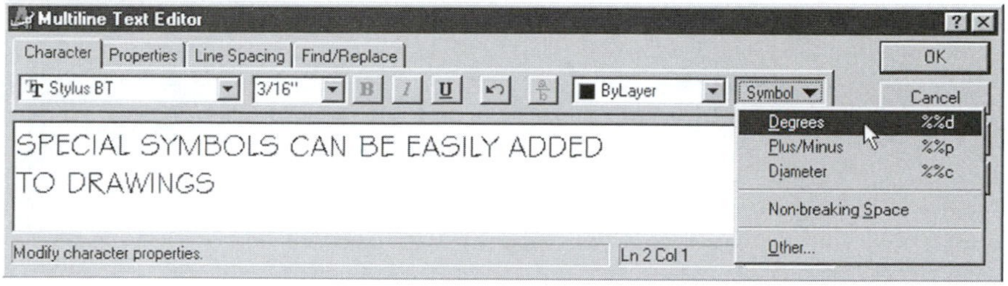

Figure 17.40 *Using MTEXT, special characters such as the degree, plus/minus, and diameter symbols can be added without using any %% special character.*

17.41, and allow one of the symbols from the Character Map to be inserted. Symbols will vary depending on the current font.

Control Buttons

As with other dialog boxes, control buttons are provided to implement the desired settings. This box features the OK, Cancel, Import Text, and Help buttons. The results of the OK, Cancel, and Help buttons should be familiar by now. The Import Text... button allows ASCII (American Standard Code for Information Interchange) and RTF (Rich Text Format) text files to be inserted into a drawing file. This may seem unimportant as you're learning AutoCAD now, but it means you can create schedules or large bodies of text up to 16K using your second favorite software, and then import the file into AutoCAD, once you feel more comfortable with MTEXT. Imported text retains it original character formatting and style properties, but it can be edited as if it had been created in MTEXT.

Note: Files larger than 16K can be inserted into AutoCAD if they are not imported directly into AutoCAD. To insert a file larger than 16K, minimize AutoCAD, and then open the file to be inserted using the word processor that was used to create the file. Now save the text file to the Clipboard. Maximize AutoCAD, and paste the file into the drawing.

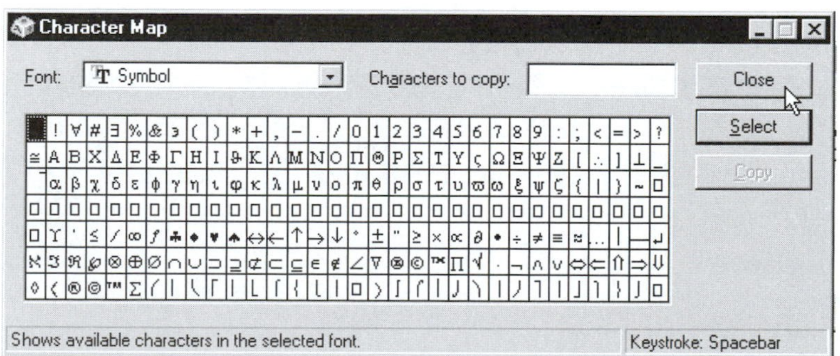

Figure 17.41 *Selecting the Other option from the Symbols menu displays the character map.*

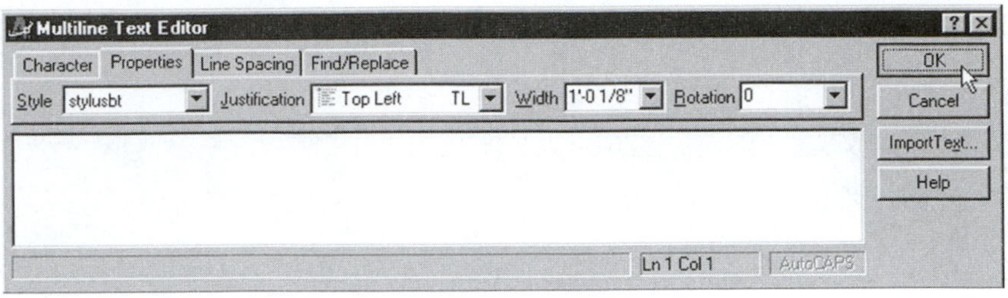

Figure 17.42 *Selecting the Properties tab displays this menu for altering the style, justification, width, and rotation of multiline text.*

ADJUSTING TEXT PROPERTIES

Selecting the Properties tab of the Multiline Text Editor will provide options for controlling text properties by changing the style, justification, width, and rotation. Figure 17.42 shows the Properties tab. The tab can be selected at any point of the MTEXT command.

Style

The Style option allows an existing text style to be applied to a new or selected text body. Character formats for font, height, bold, and italic can be overridden by applying a new style to existing multiline objects. Information can be typed in one style, and altered before the text is placed in the drawing base. To alter the current text style, select the Properties tab and then select the Style down arrow to display a listing of current text styles. Figure 17.43 shows an example of the Style menu.

Justification

Justification with MTEXT is similar to that of the TEXT command. Selecting the Justification option allows the alignment of text to be altered. The default setting places the cursor in the top left corner of the text box. Selecting the Justification down arrow will produce the menu shown in Figure 17.44. Figure 17.45 shows examples of how text will be placed based on each setting.

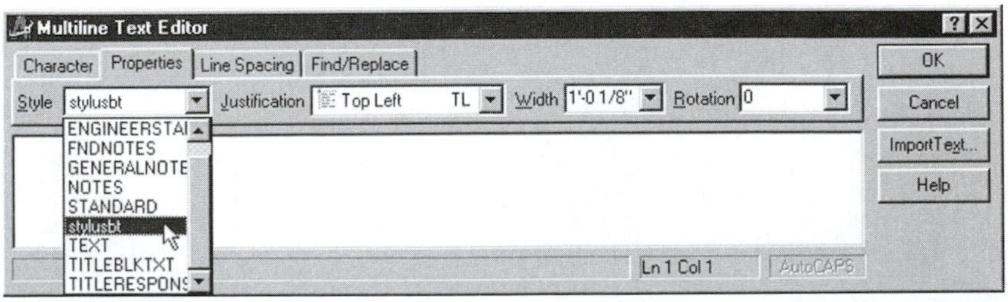

Figure 17.43 *Selecting the down arrow for Style displays a menu of current text styles. Styles for MTEXT are created using the same procedures used to create a style for TEXT.*

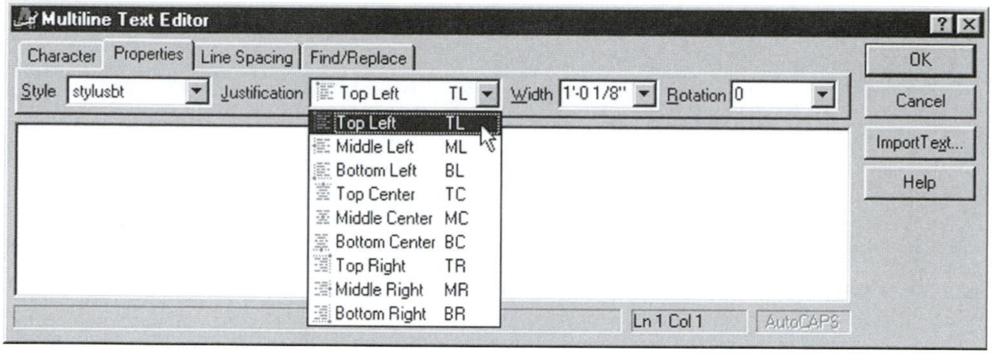

Figure 17.44 *Selecting the Justification down arrow will produce a list of options for arranging MTEXT that are similar to the TEXT justification options.*

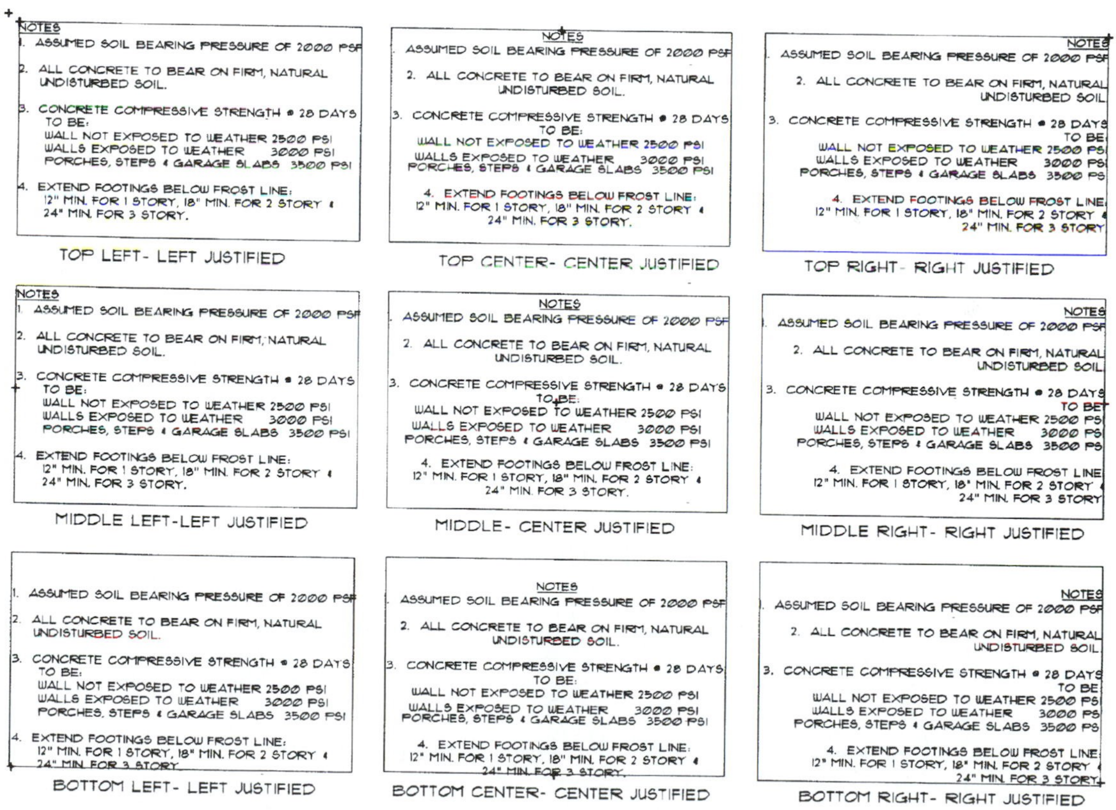

Figure 17.45 *The options for justification of MTEXT.*

Width

This option allows the width of the text to be redefined. Selecting the down arrow beside the Width edit box will display a menu of current MTEXT widths. One of the current widths can be selected, or a new value can be added to the edit box.

Rotation

This option functions just as the option did with the TEXT command—to set the text rotation angle. Selecting the Rotation option will display a listing of angles in 15° increments. One of these options can be selected from the menu, or any numerical value between 0 and 360 can be assigned. The option can be used with the current text before it is added to the drawing, or existing text can be selected and altered.

LINE SPACING

The Line Spacing tab includes two settings for controlling line spacing. These settings include the At Least/Exactly and the Single (1.0x) options.

At Least/Exactly

This setting specifies how the spacing between lines will be placed. The default setting is to allow AutoCAD to automatically add space between lines of text, based on the height of the largest character in the line of text. The Exact option sets the spacing so that it will always be the same between lines of text. The Exact option maintains the exact spacing, no matter what the variation of individual letters in each line is. The Exact option should be used when constructing tables—like a window or door schedule. Because the spacing between lines of text will always be the same, text placed with this option will have a very orderly appearance.

Single (1.0x)

This menu can be used to control the spacing increment between lines of multiline text. This setting is the equivalent to choosing between single, double, or triple spaced text with a word processor. The spacing is the vertical distance measured from bottom to bottom of lines of text. The default distance is 1.66 times the height of the text. A setting of 1 represents single spacing, and 2x represents double-spacing.

USING FIND/REPLACE

Selecting the Find/Replace tab of the Multiline Text Editor displays the options shown in Figure 17.46. The options help create search patterns for words or specified strings of text. Options include the Find box and button, the Replace box and button, Match Case, and Match Word.

Find

The Find box can be used to enter a word or string of text that you wish to find. This can be especially helpful in finding a key word in a long list of specifications. To find a word, select the Find box to place the cursor in the box, and then enter the desired word. Figure 17.47 shows an example of using Find to locate the word "JOISTS".

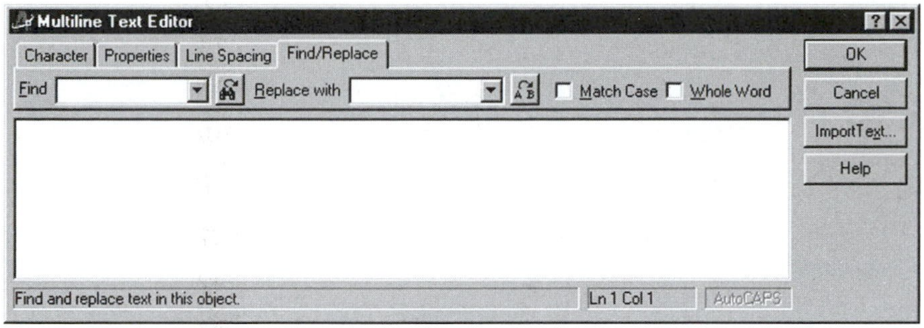

Figure 17.46 *Selecting the Find/Replace tab provide options for editing MTEXT.*

Pressing the Find button on the right end of the Find edit box starts the search for the text in the Find box. Press (EnterKey) to find the next occurrence of the selected word.

Replace

The Replace option can be used in conjunction with Find to alter a word or word string. Once a word has been found, enter a word in the Replace box and press the Replace button. Figure 17.48 shows the results of selecting "JOISTS" and replacing it with the word "trusses".

Match Case

With this box checked, only the exact word with the same case will be identified. With the box inactive, placing "TRUSS" in the find box would identify "truss" in the

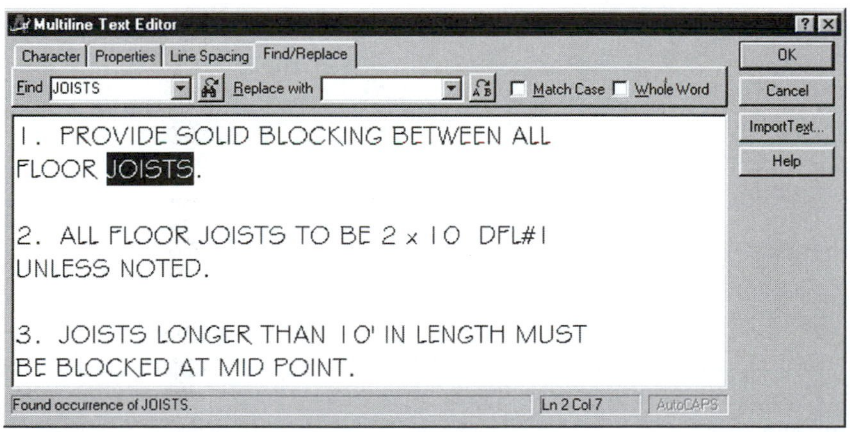

Figure 17.47 *Selecting the Find box allows a word or word string to be specified for a word search. The word "JOISTS" has been selected for the search. Pressing the Find button will highlight the first use of the specified word. Pressing (EnterKey) will highlight the next occurence of the selected word or string.*

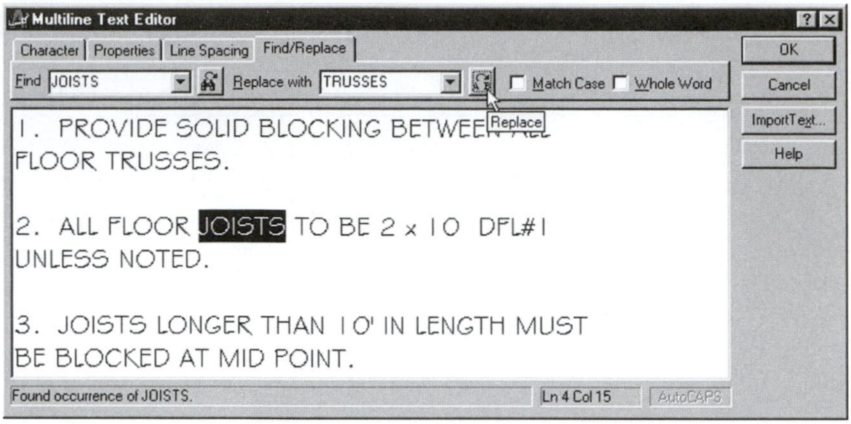

Figure 17.48 *With the word "JOISTS" as the active word for the search, the first occurence of the word was altered to read "TRUSSES" by entering "TRUSSES" in the Replace with box and pressing the Replace with button.*

body of text. With the Match Case box active, "truss" would not be identified. Only uses of the word or string that are identical will be located.

Whole Word

With this box inactive, any use of the selected word will be identified with the Find option. With "truss" indicated in the Find box, "Truss-joist" would be identified in the document. Selecting this option will identify the word only if it is a single word. When the word is part of another text string it will be ignored.

EDITING MTEXT

The contents of an MTEXT object can be edited using the COPY, ERASE, GRIPS, MIRROR, MOVE, and ROTATE commands. The text in MTEXT can be edited using either DDEDIT or DDMODIFY commands.

USING DDEDIT

To use the DDEDIT command with MTEXT, you start the same way as when editing TEXT. The quickest method of accessing the command is from the command prompt. The command can also be accessed by selecting Edit Text from the Modify II toolbar, or by selecting Text... from the Modify pull-down menu. The command sequence is:

Command: **ED** (EnterKey)

Select an annotation object or [Undo]: *(Select the desired text to be edited.)*

When the paragraph to be edited is selected, the text will be displayed in the Multiline Text Editor dialog box. Each of the four tabs that were used to create the text can be used to alter the selected text. The mouse can also be used to move the cursor to the desired location. As the cursor is moved into the edit box, it changes to a blinking cursor. To edit text, move the cursor to highlight the desired text to be changed. With the text highlighted, anyone of the features set using the text editor can be altered. When the desired changes are made, select the OK button to return to the drawing screen. As the screen is restored, the prompt for selecting additional annotation is still displayed. The program is waiting for you to select additional text to be edited. Press (EnterKey) a second time to exit DDEDIT and restore the command prompt.

In addition to using the four tabs of the Multiline Text Editor to edit text, standard Window control keys can be used to alter the text in the editor. The control keys include:

(CtrlKey) +A	Select all text in the Multiline Text Editor.
(CtrlKey) +B	Applies or removes bold format for selected text.
(CtrlKey) +C	Copies selected text to the Clipboard.
(CtrlKey) +I	Applies or removes italic format for selected text.
(CtrlKey) + (ShiftKey) +L	Converts selected text to lowercase.
(CtrlKey) +U	Applies or removes underline format for selected text.
(CtrlKey) + (ShiftKey) +U	Converts selected text to uppercase.
(CtrlKey) +V	Pastes Clipboard contents to cursor location.
(CtrlKey) +X	Cuts selected text to the Clipboard.
(CtrlKey) +SPACEBAR	Removes character formatting in selected text.
(EnterKey)	Ends the current paragraph and starts a new line.

EDITING WITH PROPERTIES

The PROPERTIES command can be used to alter the properties of multiline text. The command has been introduced earlier, but a reminder of its power to alter text should not be forgotten. The properties of multiline text can be displayed using the following steps:

- Select the body of text to be edited.
- Right-click in the drawing area to display the shortcut menu.
- Select Properties from the shortcut menu.

The command can also be started by selecting the Properties icon from the Standard toolbar, or by entering MO(EnterKey) at the command prompt. Each method will display the Properties window similar to Figure 17.49. When the properties are listed by category, seven general properties affecting the text and nine specific properties are listed for multiline text.

General Properties

Each of the seven General properties is used to display or set the specified property. For instance, if you select a block of Multiline text, and then select Color, the color property will be highlighted and the edit arrow will be displayed. Selecting the edit arrow will display the color palette, allowing the color of the text to be altered. Each of the other general properties of the text can be altered using the same procedure.

Specific Text Properties

Each Multiline text feature controlled in the Multiline Text Editor can be changed using the Properties Window. To alter the current Style, select the text to be altered. With the text highlighted, selecting Style will display a listing of the current styles. Selecting the name of the new style will automatically update the selected text to the indicated style. Each of the other properties can be altered using the same procedure. Other text specific properties include:

Figure 17.49 *The Properties window can be used to edit MTEXT. Select the Properties icon from the Standard toolbar or enter* ***MO*** *(EnterKey) at the command prompt to produce the Properties window.*

- Contents—The edit box specifies the selected text string. Only a portion of the text will be displayed for long text strings.
- Style—Specifies the Style name of the selected text body and allows a new style to be specified.
- Justify—This option specifies the attachment point of the text body, using the nine common justification options for placement.
- Direction—Allows the horizontal or vertical direction of the selected multiline text to be specified.
- Width—Allows the width of multiline text to be edited. With the text highlighted, a grip will be displayed at each corner of the text box. To alter the width of the text, select a grip to make it hot. Move the hot grip to the desired location and the text will be redisplayed to the specified width.
- Height—Specifies and alters the height of the selected text. Text can be enlarged or reduced by altering the current value.
- Rotation—This option specifies the rotation angle of the text body. Enter the desired rotation angle by keyboard and press (EnterKey). The text will be rotated around the current justification point.
- Line Space Factor—Controls the line spacing factor of the text body. Enter the desired value by keyboard and press (EnterKey). The text will be altered to display the selected spacing value.
- Line Space Style—Specifies the spacing based on text height, using the Exact and At least options.

Note: Keep in mind that the Properties box can be displayed beside the drawing area and used to alter more than just multiline text. With the box open, the properties listing will be altered as each new object is selected. The name of the listed object will be displayed in the object edit box located at the top of the dialog box.

USING THE SPELL CHECKER

In addition to using DDEDIT and PROPERTIES to edit text, the SPELL command can be used to correct spelling errors in TEXT and MTEXT. SPELL can be accessed by selecting Spelling from the Tools pull-down menu, or by entering **SP** (EnterKey) at the command prompt. Each option will prompt you to select objects to be checked for spelling. Once a word or body of text is selected, the Check Spelling dialog box similar to Figure 17.50 will be displayed. If no spelling errors are contained in the selected text, the display shown in Figure 17.51 will be displayed. Selecting ENTER will return you to the drawing. Selecting (EnterKey) (EnterKey) will remove the AutoCAD alert and reenter the SPELL command so that additional text can be checked.

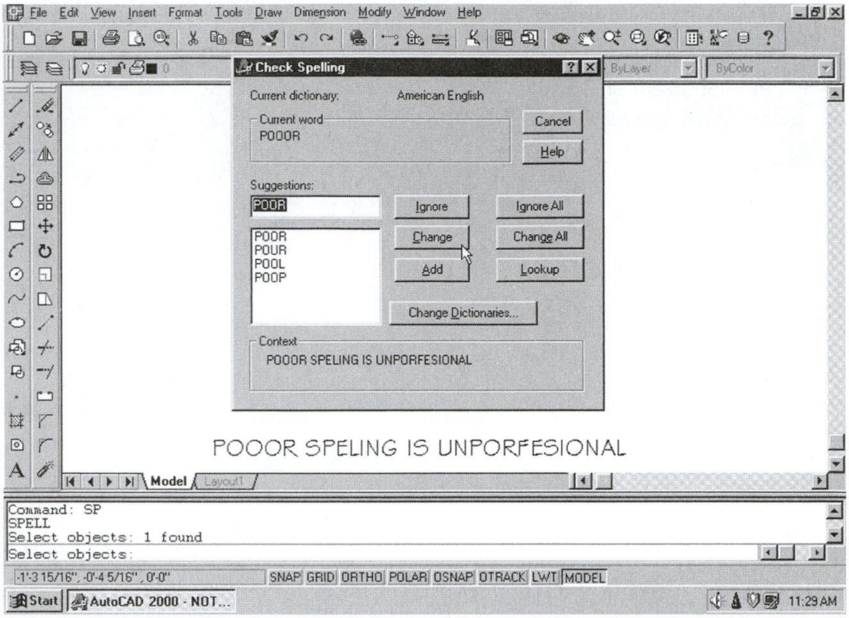

Figure 17.50 *The SPELL command can be used to check TEXT and MTEXT. The current word "pooor," four suggestions, and the current context of the word are all displayed.*

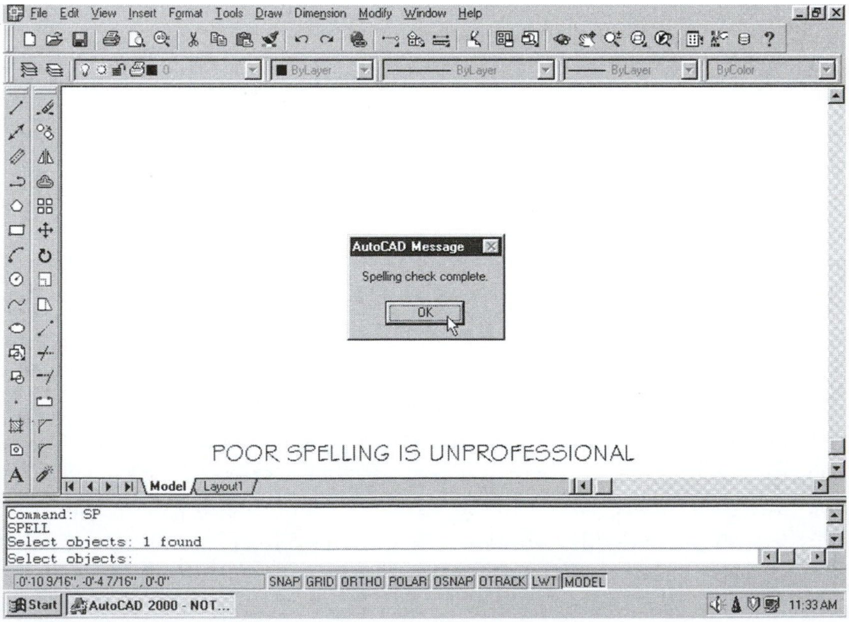

Figure 17.51 *If no errors are found, you will see this display.*

If errors are found in a selection, the line of text will be displayed in the Context box of the Check Spelling dialog box. The misspelled word will be displayed in the Current word box. A possible listing of alternatives can be found in the Suggestions: box. If the current word is as you would like it, select the Ignore option and the spell checker will move to the next word identified by the checker. If you would like to use one of the suggested words, move the cursor to highlight the word and then select Change. The editor will proceed through your text until the alert seen in Figure 17.51 is displayed.

CHANGING THE DICTIONARY

Selecting the Change Dictionaries... box will produce the dialog box shown in Figure 17.52. The Main dictionary box will allow you to change the language that is stored in the dictionary. The Custom dictionary and Custom Dictionary Words boxes can be very useful to add words, acronyms, client names, or vendor names specific to your area of expertise to the dictionary. When creating a custom dictionary, use the same guidelines that were presented in Chapter 2 for naming drawing files. An extension of .CUS must be added after the name. To enter a custom name for an architectural dictionary, type ARCHWORD.CUS in the Custom dictionary box. If several custom dictionaries have been created, the Browse... box can be used to select the desired dictionary.

To add words to a specific dictionary, select the Custom dictionary Words box. Names of up to 63 characters can be entered into the box. Multiple names can be entered at one time when they are separated by a comma. Selecting the Add box will add the names to the existing list of names contained in the dictionary. Figure 17.53

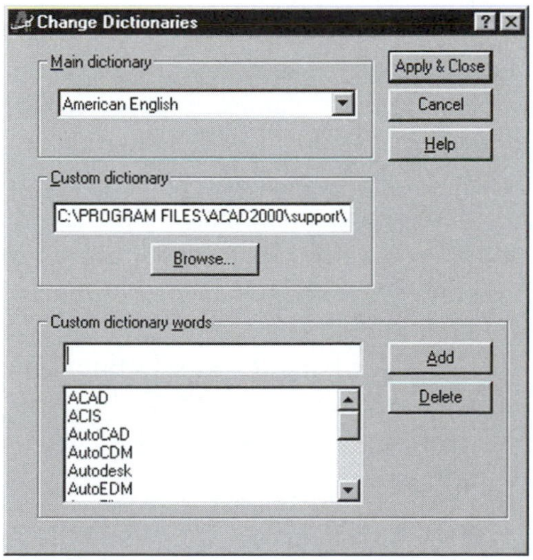

Figure 17.52 *A separate dictionary can be created to contain words specific to your field.*

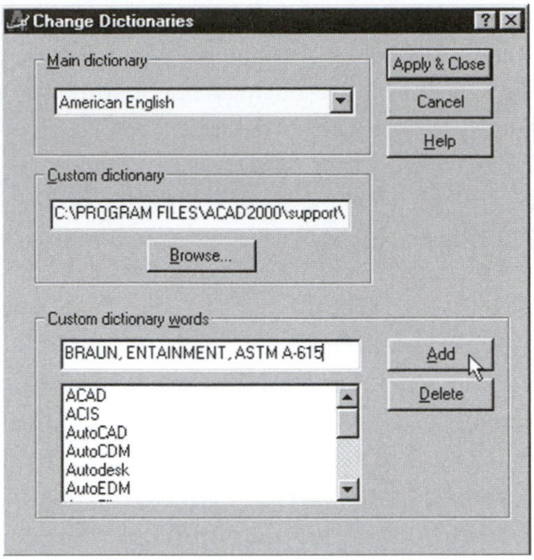

Figure 17.53 *Three words are to be added to the dictonary. A partial listing of the ARCHWORDS dictonary is shown in the lower left corner of the dialog box.*

shows a custom dialog box for the ARCHWORD.CUS dictionary with three words to be added to the existing list.

FINDING AND REPLACING TEXT

The FIND command is another tool that can be used to edit text. The command can be started by right-clicking and choosing Find... from the shortcut menu, by selecting the Find and Replace icon from the standard toolbar, or by selecting Find... from the Edit pull-down menu. Each will produce the Find and Replace dialog box similar to Figure 17.54. The Find and Replace dialog box can be used to select, find, replace, and zoom to text created by the TEXT or MTEXT commands. The dialog box can also be used to edit dimension text, block attribute text, Hyperlink descriptions, and Hyperlinks. Each of these four types of text will be introduced in the balance of the text. The command can be used in a similar method to the Find/Replace tab of the Multiline Text Editor. To find text using the dialog box, enter the text or text string that you would like to find in the Find Text String edit box. The search field can be the entire drawing or limited to the current text selection. Once the text string and the search field are defined, select the Find button. If the text is found, it will be displayed in the Search results display box. Choosing the Zoom to button will display the selected text on the screen.

In addition to locating specific text in a drawing, the edit box can be used to replace a word or phrase. After entering the text to be found, enter a phrase to be used as the

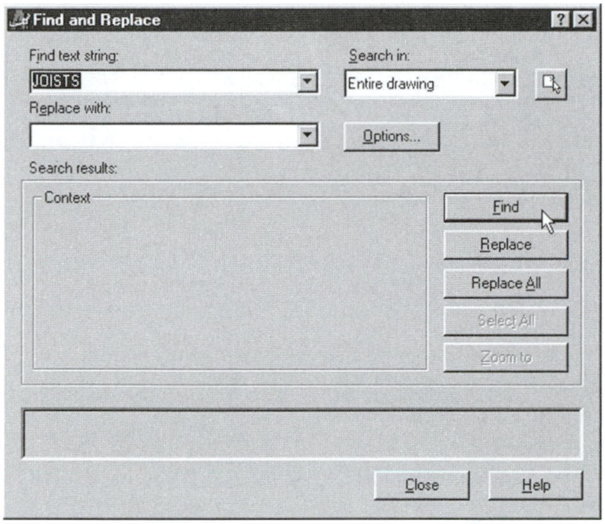

Figure 17.54 *The Find and Replace dialog box can be used to select, find, replace, and zoom to text created by the TEXT or MTEXT commands.*

replacement text in the Replace With edit box. Choosing the Replace button will replace the specified text with the replacement text. Choosing the Replace All button will replace each usage of the selected text or phrase.

INCORPORATING TEXT INTO TEMPLATE DRAWINGS

Much of the text used to describe a drawing can be standardized and placed in a template drawing. Text can also be saved as a wblock and inserted into drawings as needed. Wblocks will be discussed in future chapters. Figure 17.55 shows an example of notes that an office includes as standard notes for all site plans. These notes can be typed once and saved as a separate drawing so that they do not have to be retyped with each use. The wblock can then be either inserted into every site plan or nested in the template drawing on the SITETXT layer. The layer can be frozen, thawed when needed, and then either the DDEDIT or PROPERTIES command can be used to make minor changes based on specific requirements of the job. Notice that several notes and each of the schedules include the text "xxx". These are notes that have been set up as blocks and stored with attributes that can be altered for each usage. Attributes will be discussed later.

Many drawings, such as sections, contain basically the same notes that are displayed as local notes. Although sections for a tilt-up structure will be radically different from a section for a steel rigid frame, a template sheet can be developed for each type of construction. The actual drawing may look different, but the notes to specify stan-

GENERAL NOTES:

1. ENGINEER NOT RESPONSIBLE FOR LAND SURVEY OR TOPOGRAPHY.
2. ALL EXTERIOR SIGNS TO COMPLY W/ COUNTY SIGN ORDINANCE.
3. LOWER LEVEL FRAMED IN STEEL.
4. ALL STEEL COLUMNS TO BE 3" DIA. X 60" STEEL COLUMN FILLED W/ CONC. EMBEDDED 24" INTO GRADE.
5. SQUARE FOOTAGE OF LOT--00,000 SQ. FT.
6. SQUARE FOOTAGE OF STRUCTURE--00,000 SQ. FT. APPROXIMATELY.
7. 00.00% OF THE LOT IS COVERED.

UTILITY NOTES:

1. BALANCE OF WATER LINES TO BE SHOWN ON FLOOR PLANS.
2. CLEAN OUT @ ····

PLANTER NOTES:

1. ALL PLANTERS TO BE CONSTRUCTED W/ A SEPARATE PERMIT.
2. PLANTER ON WEST SIDE ALONG FLOYD MILLER BLVD. TO BE 44' X 7.5' X 3' CONCRETE BLOCK W/ SIMILAR PLANTER IN THE SOUTHWEST CORNER BY THE STAIRWAY.

PARKING NOTES:

1. PARKING TO BE 4" MIN. CONC. W/ 6" X 6"--4 X 4 WWM OVER 4" COMPACTED GRAVEL.
2. EVERY TWO PARKING SPACES TO HAVE A 6" X 6" WHEEL STOP W/ (2) ½" X 12" STEEL DOWELS @ EA. WHEEL STOP.
3. PARKING SPACES TO BE MARKED WITH 4" WIDE WHITE STRIPES.

PARKING SPACES		
FULL-SIZE	9' X 20'	XX
COMPACT	7'-6" X 15'	XX
HANDICAPPED	12' X 20'	XX

	OCCUPANCY OF THE STRUCTURE	MAXIMUM ALLOWABLE FLOOR AREA (SQ. FT.)	TOTAL OCCUPANTS (FOR BLDG.)
FIRST FLOOR	XX	XX,XXX	XX
SECOND FLOOR	XX	XX,XXX	XX
THIRD FLOOR	XX	XX,XXX	XX

Figure 17.55

dard materials are similar for most buildings using the same type of construction. Local notes can also be placed in the template drawing, as shown in Figure 17.56. These notes can also be stored as a wblock, moved to the desired drawing, thawed, moved into the needed position, and edited as seen in Figure 17.57. This can greatly increase drafting efficiency.

½" CD APA 32/16 PLY. ROOF SHTG. INT. GR. W/ EXT. GLUE. LAY FACE GRAIN PERP. TO JOISTS & STAGGER JOINTS. USE: 8d COM. NAILS @ 6" O.C. BOUNDARY & EDGES, 8d COM. NAILS @ 12" O.C. FIELD. SEE ROOF-FRAMING PLANS FOR SPECIAL NAILING.

'TRUS-JOIST' 16" TJI 35" @ 24" O.C.

MINERAL CAP SHEET OVER 2 LAYERS OF ASB. FELT AS PER. JOHN MANSVILLE SPEC. #406

PROVIDE ½" GYP. BD. DRAFTSTOPS FROM CEILING TO PLY ROOF SHTG. FOR EVERY 3,000 SQ. FT. MAX.

1½" 100#/FT LT. WT. CONC. OVER 15# ASPH. SATURATED FELT W/ 6" x 6" W-4" x 4" WWM IN SLAB OVER PLY. SHTG.

3/4" CD APA 42/20 T&G PLY FLR. SHTG., INT. GRD. W/ EXT. GLUE. LAY FACE GRAIN PERP. TO JOISTS & STAGGER JOINTS. USE: 10d COM. NAILS @ 6" O.C. BOUNDARY & EDGES, 10d COM. NAILS O.C. FIELD. SEE FLOOR FRAMING PLANS FOR SPECIAL NAILING.

LINE OF SUSP. CEILING (CLASS 'A')

1½" 100#/FT LT. WT. CONC.

3/4" CD APA 42/20 T&G PLY FLR. SHTG.

R-25 RATED INSULATION

PROVIDE ½" GYP. BD. DRAFT STOPS FROM CEILING TO PLY FLOOR SHTG. FOR EVERY 1,000 SQ. FT. MAX.

2-2 x 4 DFL TOP PLATES LAP 48" MIN. W/ 10-16d COM.

⅝" TYPE "X" GYP. BD.

2 x 4 STUDS @ 16" O.C. UNLESS NOTED

2-2 x 6 DFL TOP PLATES LAP 48" MIN. W/ 12-16d COM.

R-21 RATED INSULATION

2 x 6 SOLE PLATES.

R-38 RATED INSULATION

Figure 17.56

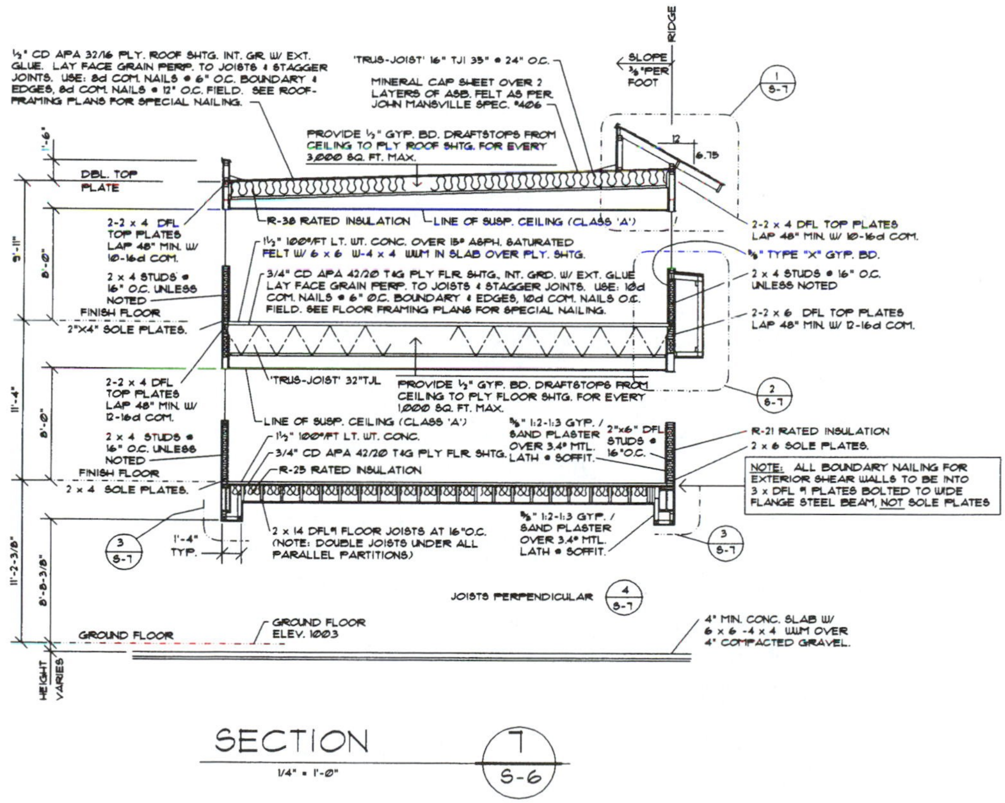

Figure 17.57

CHAPTER 17 EXERCISES

1. Open a new drawing and type the following text using left-justified text with a height of .125" and a TrueType font.

   All sheathing should be ½" standard grade 32/16-ply interior type with exterior glue. Lay perpendicular to rafters and stagger all joints. Nail with 8d's @ 6" o.c. at edge, blocking and beams, and 8d's @ 12" o.c. @ field. Use common wire nails. Save the drawing as E-17-1.

2. Open a new drawing and make ROMANS the active font, with an obliquing angle of 10° and a height of .125".

   Use 8 × 8 × 16 grade "A" concrete block units with a triple score. Use # 5 diameter rebar @ 24" o.c. each way solid-grout all steel cells. Save the drawing as E-17-2.

3. Open a new drawing and make CIBT the active font with a height of .125". Use a rotation angle of 35°. Write the following note:

   5 ⅛ × 13 ½" exposed glu-lam ridge beam f:2200

   Save the drawing as E-17-3.

4. Open a new drawing and make ROMANT the active font with a height of .125". Use a width factor of 1.25. Use the center option to type the following text:

   ATTENTION ALL CADD

   OPERATORS.

   DUE TO

   BUDGETARY PROBLEMS

   YOUR MOM DOES NOT WORK HERE.

   PLEASE CLEAN UP

   YOUR OWN MESS.

   Save the drawing as E-17-4.

5. Open a new drawing and create a style for TEXT using ROMANS as the font with a height of .125". Use a width factor of 1.00. Create a second style named TITLES using ROMANT as the font, with a height of .25 and a width factor of 1.5. Assign each style a layer named for the style and provide a separate color for each. Use MTEXT to create the following text:

   INSULATION NOTES

   Insulate all exterior walls with 5 ½" high-density fiberglass batt insulation,

   R-21 min.

   Insulate all flat ceiling joist with 12" R-38 batts (no paper facing required).

Insulate all vaulted ceiling with 10 ¼" high-density, paper-faced fiberglass R-38 min. with 2" air space above.

Insulate all wood floors with 8" fiberglass batts, R-25 with paper face. Install plumbing on heated side of insulation.

CAULKING NOTES

Caulk the following openings with expanded foam or backer rods. Elastomatic, copolymer, and siliconized acrylic latex caulks may also be used where appropriate.

Any space between window and door frames.

Between all exterior wall sole plates and plywood sheathing.

On top of rim joist prior to plywood floor application.

Wall sheathing to top plate.

Joints between wall and foundation.

Joints between wall and roof.

Joints between wall panels.

Save the drawing as E-17-5.

6. Open drawing E-17-5 and make the following changes to the insulation notes.

   Change wall insulation to reflect 6" fiberglass batt insulation, R-19 min.

   Change all flat ceiling to 12" R-38 batts with paper facing.

   Change all vaulted ceiling to 10" paper-faced fiberglass R-30 min. with 2" air space above.

   Change wood floors to 6" fiberglass batts, R-19 with paper face.

   Save the drawing as E-17-6.

7. Open drawing E-13-8. Use a text font other than TXT and label the drawing using drawing E-13-8 as a guide. Complete the drawing and add all hatch patterns. Place the required text box on a separate layer from other text. Use the LINE command to add all leader lines. Save the drawing as E-17-7.

8. Open E-15-10 and label the materials that have been drawn using a suitable font. Use QTEXT to disable the text. Create a separate layer for all text and save the drawing as E-17-8.

9. Open drawing E-15-9. Create separate layers, styles, and fonts for titles and text and then label the drawing. Use straight lines to connect text to the object being described. Save the drawing as E-17-9.

10. Open drawing FLOOR14 and design a 2"-wide title block along the right side of the page. Design space for page number, date, revision date, client information, and designer information. Since you are the designer, provide a company name,

logo, your name, address, and phone number. Create new layers for the title block lines and text that will remain constant, and a separate layer for text that will vary for each job. Use a minimum of three different text fonts. Save the drawing as FLOOR14.

11. Type the following note: PROVIDE A 5° +/–. 5° BEND AT ALL #6 DIA. STEEL REINFORCING. STEEL TO BE WITHIN 1% OF SPECIFIED LENGTH. Save it as E-17-11.

12. Create a line of text and make a copy of the text. Mirror one of the lines using a variable of 1 and one of the lines with a variable of 0. Save the drawing as E-17-12.

13. Use the MTEXT command to make a list with a minimum of three guidelines that describe the creation of MTEXT. Use a second text style and justification point and type a second list of guidelines for describing TEXT. Save the drawing as E-17-13.

14. Open drawing E-17-13 and make a copy of each set of text. Edit the new text to add additional words in the existing sentences and change the color, style, and width of text. Save the drawing as E-17-14.

CHAPTER 17 QUIZ

1. What text option will allow text to be placed between two selected points, at any angle?
2. What text option can be used to place multiple lines of text in a column?
3. If text is to be .125 high on a drawing that is to be plotted at a scale of 1" = 20'-0", what should the text height be?
4. Explain the difference between rotation angle and obliquing angle.
5. What is the difference between a style and a font?
6. What is the process to produce a degree symbol?
7. What commands are used to correct errors in text?
8. What effect will setting the style height option have on future text?
9. Explain how to correct the following problem. VERIFY EXATC hEIGHT AT JOBSITE.
10. When are lowercase letters appropriate for text on construction drawings?
11. Write the required entry to correctly specify three bolts that are three quarters of an inch in diameter at six inches on center. Include all special characters required if the style is set to be Romans.
12. A beam is located in an opening in a wall that is at 45° to horizontal. How should the text be orientated to describe the beam?
13. List and describe the two types of notes generally found on construction drawings.
14. What should be considered when choosing a font?
15. What are the effects of setting the text height in the STYLE command, as opposed to setting the height in the TEXT?
16. When would QTEXT be used?
17. What is MTEXT best suited for?
18. What should be the value for MIRRTEXT if you want to mirror the text with the object?
19. Your instructor wants you to create a body of text that starts in the upper right hand corner of a template drawing. The notes will be 4" wide and be ½" down and 4 ½" from the border. What are the commands and options, and what is the proper sequence to place these notes?
20. What is the process required to save a paragraph created using MTEXT.
21. A paragraph of MTEXT has been created using ROMANS but it should be in Italics. How can this be fixed?
22. Explain the major differences between DDEDIT and PROPERTIES on MTEXT.

23. What does the .CUS extension designate?
24. What is the advantage of placing stock notes on a template drawing?
25. A client has decided to change from floor joist to open-web trusses after all of the written specifications have been added to the drawing. How can this change be easily included?

CHAPTER 18

Introduction to Dimensioning

In addition to the visual representation and the text used to describe a feature, dimensions are needed to describe the size and location of each member of a structure. Figure 18.1 shows a floor plan for a dormitory and the dimensions used to describe the location of structural members. In this chapter you will be introduced to the basic principles of dimensioning and to the many commands that can be used to place dimensions. Several methods will be used to enter the dimension command. Because so many options are available to place dimensions, this chapter will examine only the default values for dimensioning. Chapter 19 will introduce methods of creating different styles of dimensions and describe how to adjust the variables that control the size and placement of the dimensions. Because so many options are included in the software, some options that do not apply to the construction industry will be skipped entirely. Feel free to explore these options on your own, or consult the Help menu.

BASIC PRINCIPLES OF DIMENSIONINGS

To make better use of the AutoCAD's potential, you need to understand several basic dimensioning concepts and terms before you explore the dimension commands. These principles include linetypes, placement of dimension features, and locating exterior and interior drawing features.

DIMENSIONING COMPONENTS

Dimensioning features include extension and dimension lines, text, and line terminators. Each component can be seen in Figure 18.2.

Extension Lines

Extension lines are thin lines that extend out from the object being described and set the limits for the dimensions. Extension lines usually are offset about 1/8" from the object, and extend 1/16" to 1/8" past the last dimension line. Each placement can be seen in Figure 18.3.

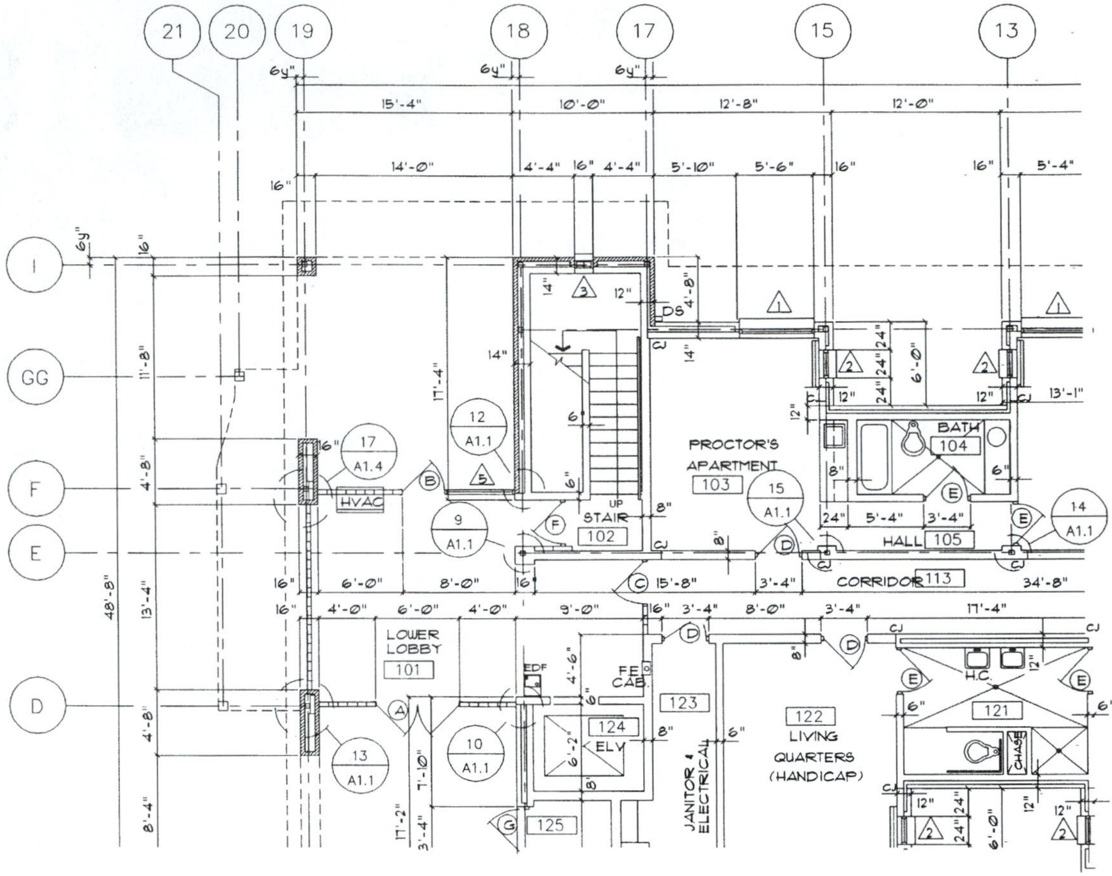

Figure 18.1 *Dimensions are used throughout construction drawings to show the location of each system to be installed. (Courtesy G. William Archer, AIA, Archer and Archer, P.A.)*

Two different types of linetypes typically are used for extension lines. Solid lines are used when dimensioning to the exterior face of an object, such as a wall or footing. A centerline is used when dimensioning to the center of a wood wall or the center of other objects. Figure 18.4 shows examples of each.

Dimension Lines

Dimension lines are thin lines used to show the extent of the object being described. Exact location will vary with each office. Dimension lines should be placed in such a way that there is room for notes, but still close enough to the features being described so that clarity will not be hindered. Guidelines for placement will be discussed later in this chapter.

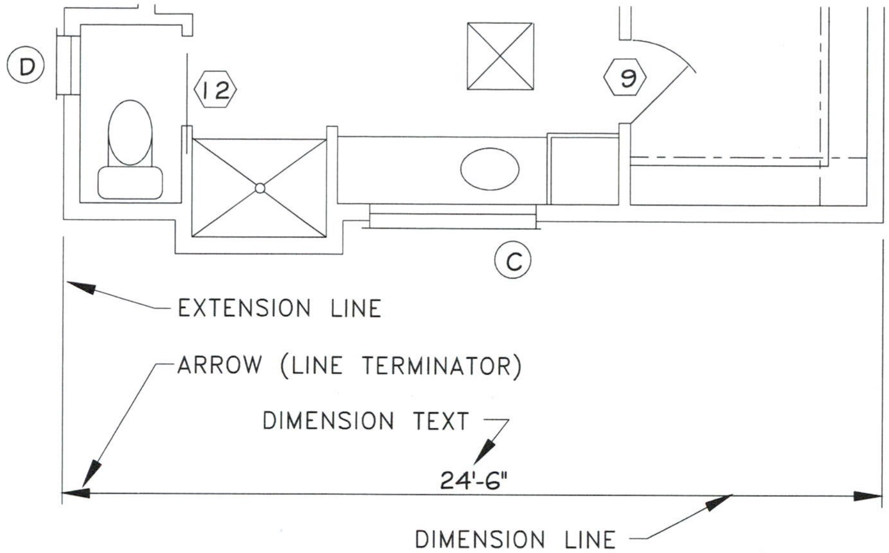

Figure 18.2 *Dimensions are composed of the dimension text, a dimension line, extension lines, and arrows (line terminators).*

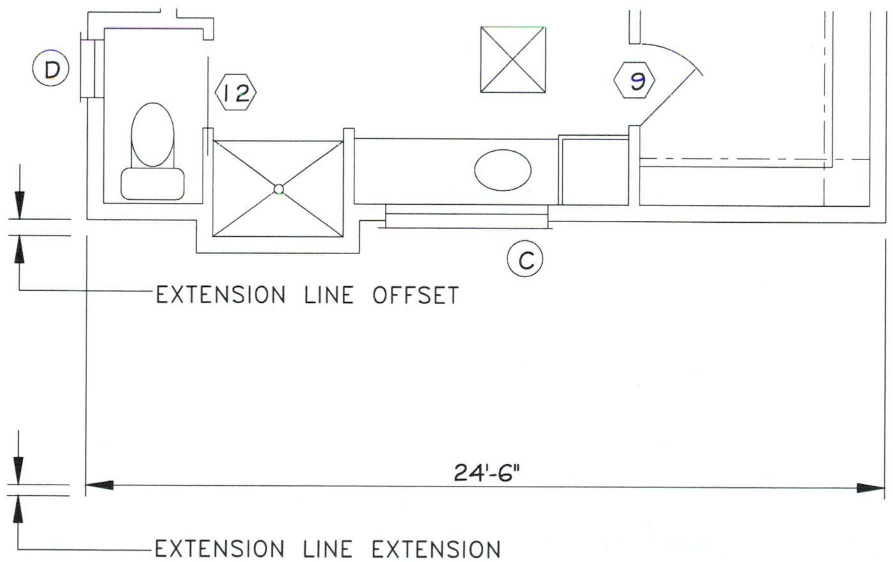

Figure 18.3 *Extension lines are typically placed about 1/8" from the object, and extend about 1/8" past the dimension line.*

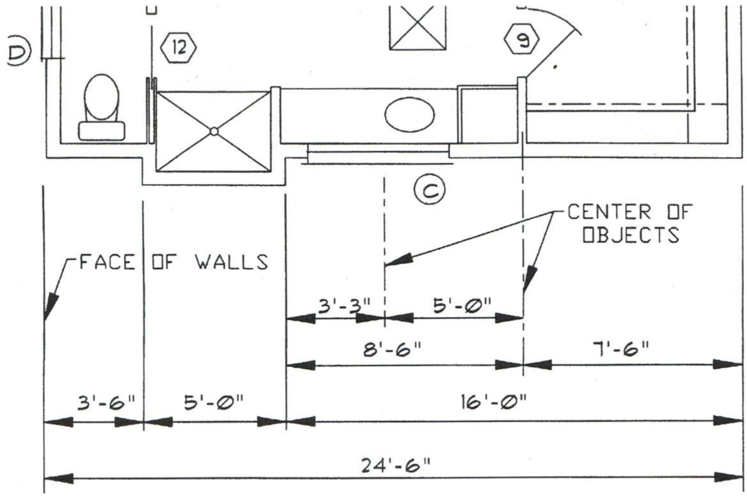

Figure 18.4 *Continuous extension lines are used to dimension to the edge of a surface; centerlines are used to dimension to the center of objects.*

Text

The text for dimensioning typically is placed above the dimension line, and centered between the two extension lines. On the left and right sides of the structure, text is placed above the dimension line and rotated so that the text can be read from the right side of the drawing page. Examples of each placement can be seen in Figure 18.5. On objects placed at an angle other than horizontal or vertical, dimension lines and text are placed parallel to the oblique object, as seen in Figure 18.6.

Often, when small areas are dimensioned, not enough space is available for the text to be placed between the extension lines. Although options vary with each office, several alternatives for placing dimensions in small spaces can be seen in Figure 18.7.

Terminators

The default method for terminating dimension lines at an extension line is with an arrowhead. Many offices use a tick mark or dot in an effort to duplicate their manual drawing styles. A third option is a thickened tick mark, which offers good contrast between lines, closely resembles its manual counterpart, and can be plotted quickly. All four options can be seen in Figure 18.8. AutoCAD also offers several other options that will be introduced in the next chapter.

DIMENSION PLACEMENT

Construction drawings requiring dimensions typically consist of plan views—such as the floor, foundation, and framing plans, and drawings showing vertical relationships—such as exterior and interior elevations, sections, details, and cabinet draw-

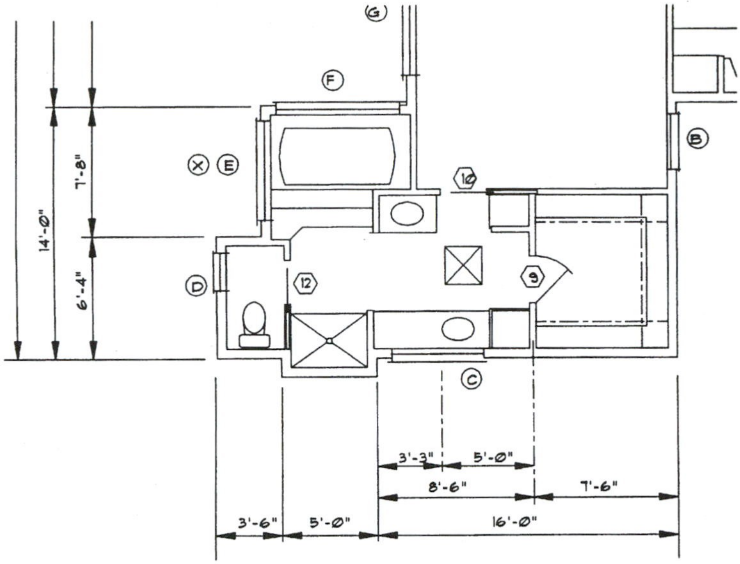

Figure 18.5 *Text should be placed above the dimension line, and read from the bottom or right side of the drawing.*

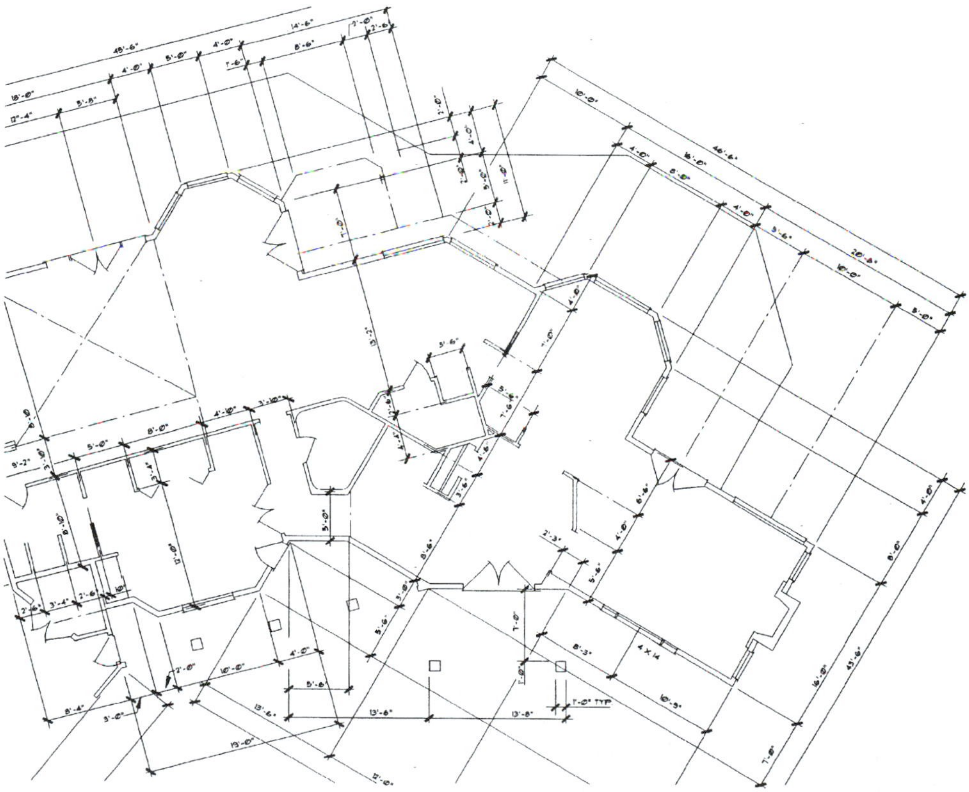

Figure 18.6 *When dimensioning objects on an angle, dimension lines and text are placed parallel to the inclined surface. (Courtesy Residential Designs.)*

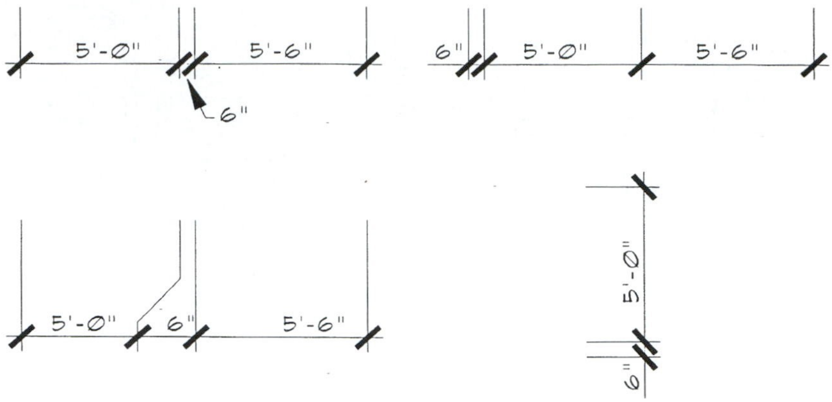

Figure 18.7 *Common professional alternatives for placing dimensions in small spaces.*

ings. Each type of drawing has its own set of challenges to be overcome in placing drawings. No matter the drawing type, dimensions should be placed on a layer titled *-ANNO-DIMS. The * represents the letter of the proper originator, ANNO represents annotation, and DIMS represents dimensions.

Note: Each of the examples that follow in this chapter has been created by professional designers and architects by altering the dimension style. This chapter introduces each of the options for placing dimension in their default settings. Chapter 19 introduces methods for altering the dimension qualities.

PLAN VIEWS

Placing dimensions in plan view can be divided into the areas of interior and exterior dimensions. Whenever possible, dimensions should be placed outside the drawing area.

Exterior Dimensions

Most offices start by placing an overall dimension, approximately 2" from the exterior wall, on each side of the structure. Dimension lines, used to describe major jogs in exterior walls, the distance from wall to wall, and the distance from wall to window or door to wall, would be placed inward at ½" intervals. Examples of placing these four dimension lines can be seen in Figure 18.9.

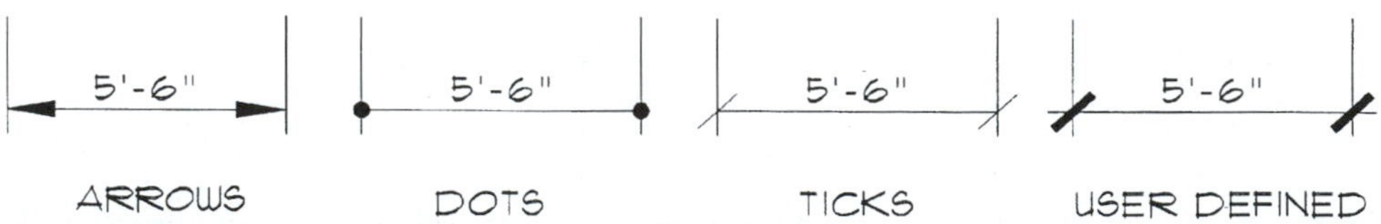

Figure 18.8 *Alternatives for terminating dimension lines.*

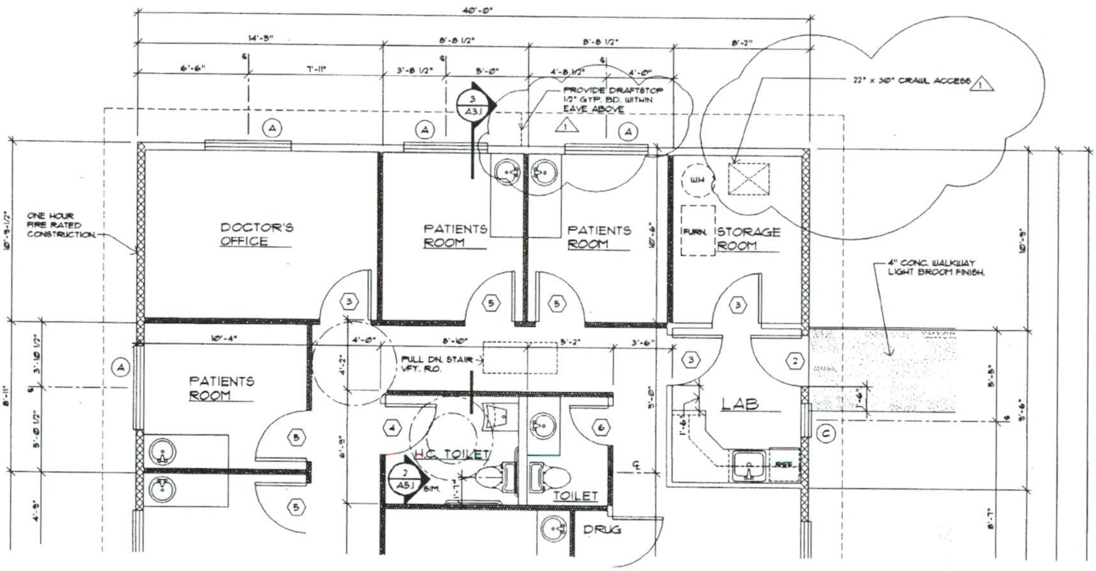

Figure 18.9 *Floor plans typically are dimensioned by placing an overall dimension, dimensions to locate major jogs, wall-to-wall dimensions, and wall-to-opening dimensions. (Courtesy Scott R. Beck, Architect.)*

Two different systems are used to represent the dimensions between exterior and interior walls. Architects tend to represent the distance of walls from edge to edge, as seen in Figure 18.10. Engineering firms tend to represent the distance from exterior edge to center of interior wood walls using methods shown in Figure 18.11. Concrete walls are dimensioned to the edge by both disciplines as seen in Figure 18.12. Concrete footings normally are dimensioned to the center. Examples of dimensioning concrete footings and walls can be seen in Figure 18.13.

Interior Dimensions

The two main considerations in placing interior dimensions are clarity and grouping. Dimension lines and text must be placed so that they can be read easily and not interfere with other information placed on a drawing. Information also should be grouped together as seen in Figure 18.14 so that construction workers can find dimensions easily.

VERTICAL DIMENSIONS

Unlike the plan views, which show horizontal relationships, the elevations, sections, and details each require dimensions that show vertical relationships as seen in Figure 18.15. Typically, these dimensions originate at a line that represents a specific point, such as the finish grade, a finish floor elevation, or a plate height.

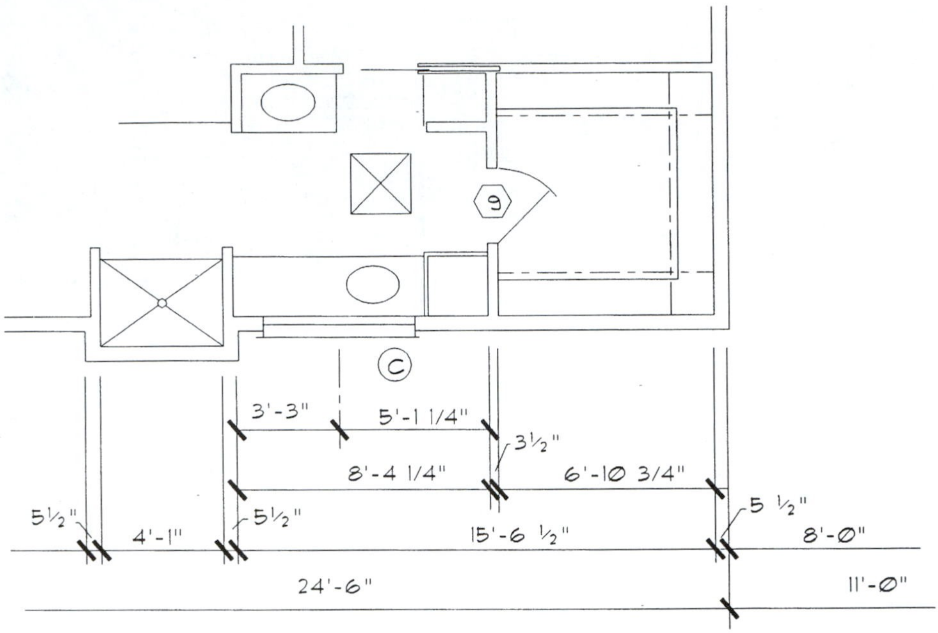

Figure 18.10 *Some architectural firms dimension interior walls from edge to edge.*

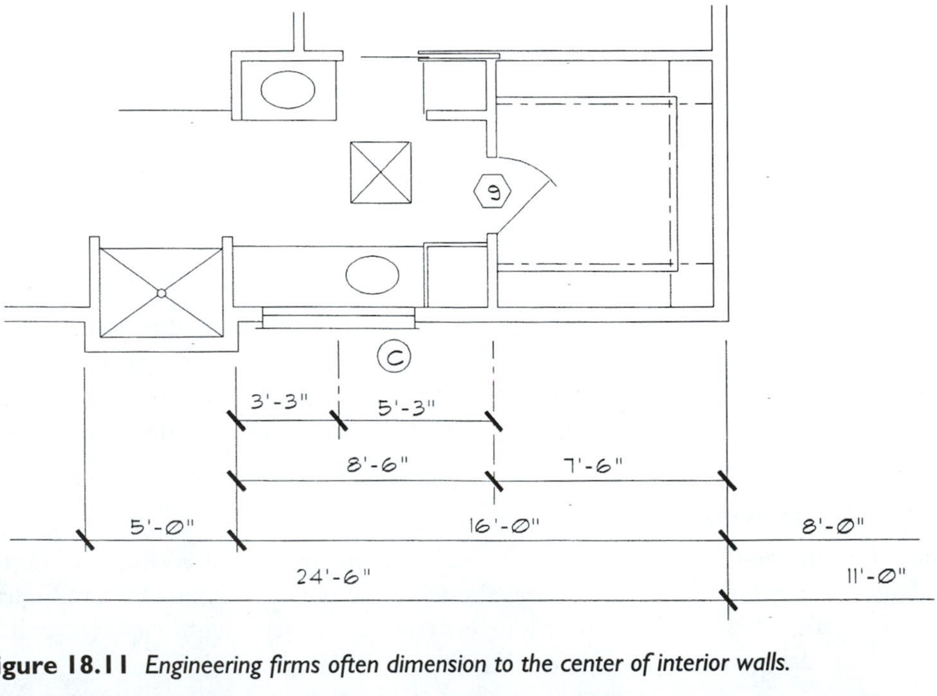

Figure 18.11 *Engineering firms often dimension to the center of interior walls.*

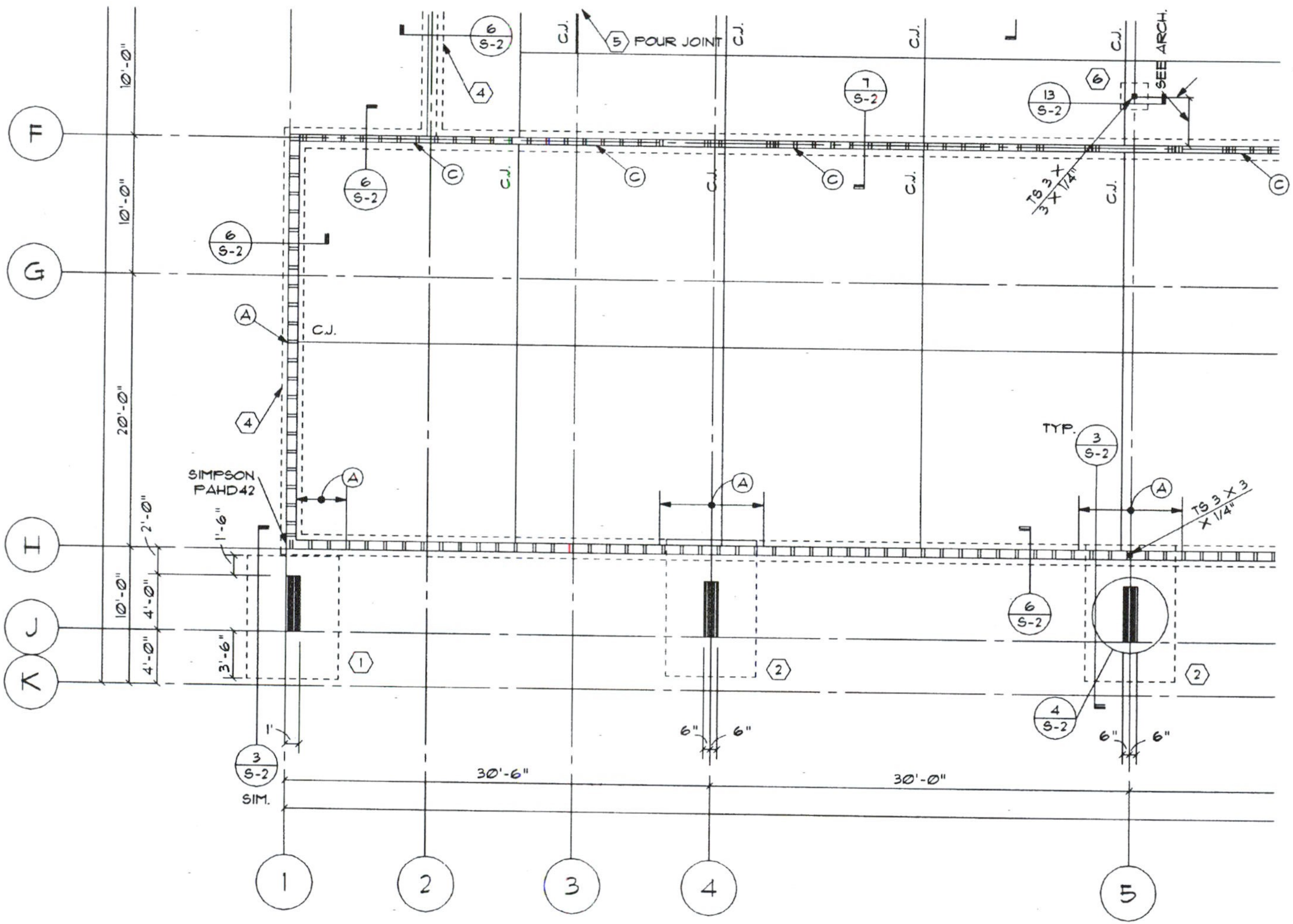

Figure 18.12 *Concrete walls are dimensioned from edge to edge. (Courtesy of Van Domelen/Looijenga/McGarrigle/Knauf Consulting Engineers.)*

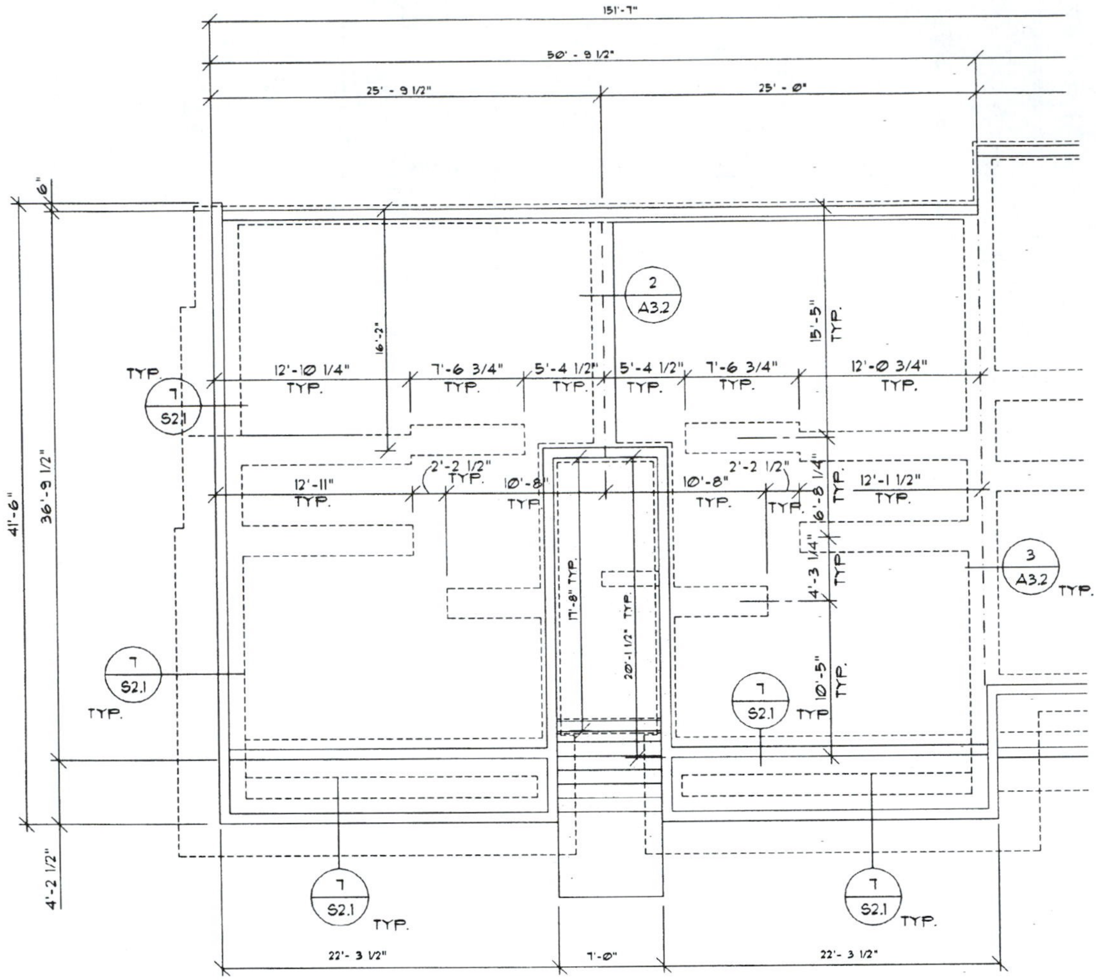

Figure 18.13 *Interior concrete piers and footings are dimensioned to their centers. (Courtesy Scott R. Beck, Architect.)*

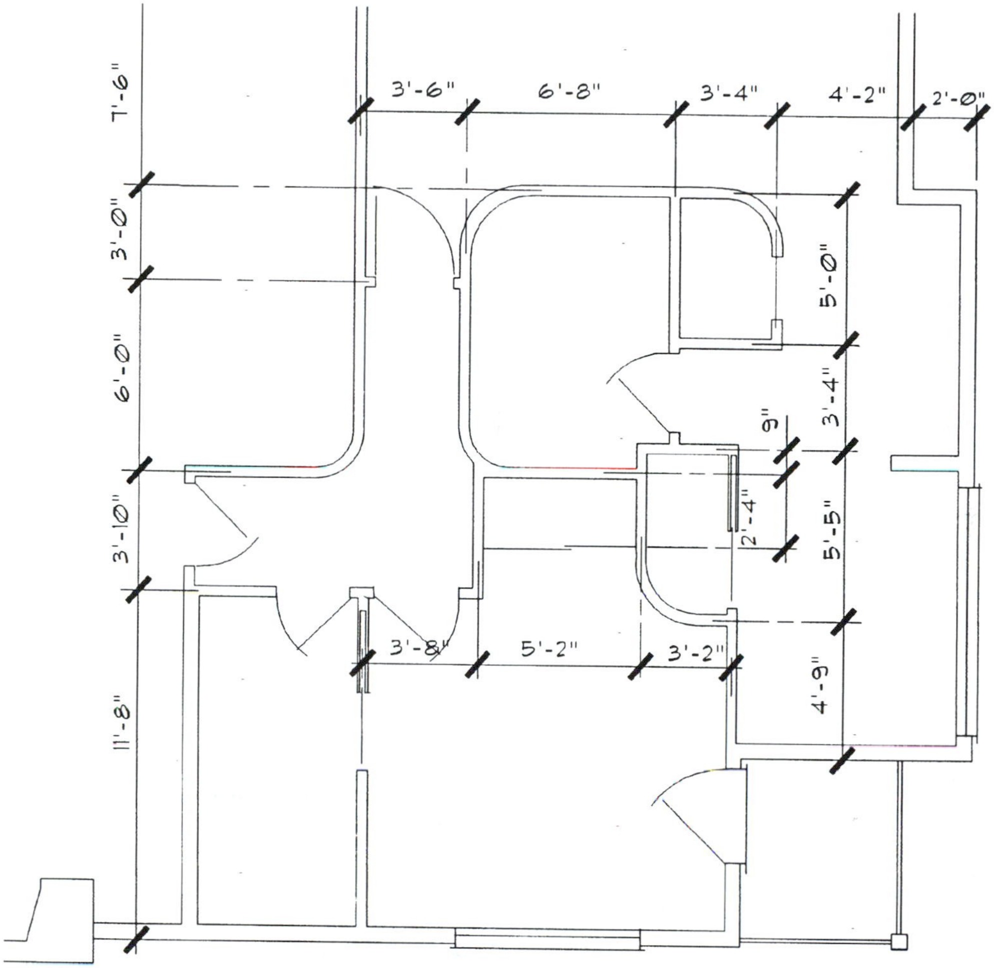

Figure 18.14 *Dimensions placed on the inside of a drawing should be grouped together and clearly visible. (Courtesy Residential Designs.)*

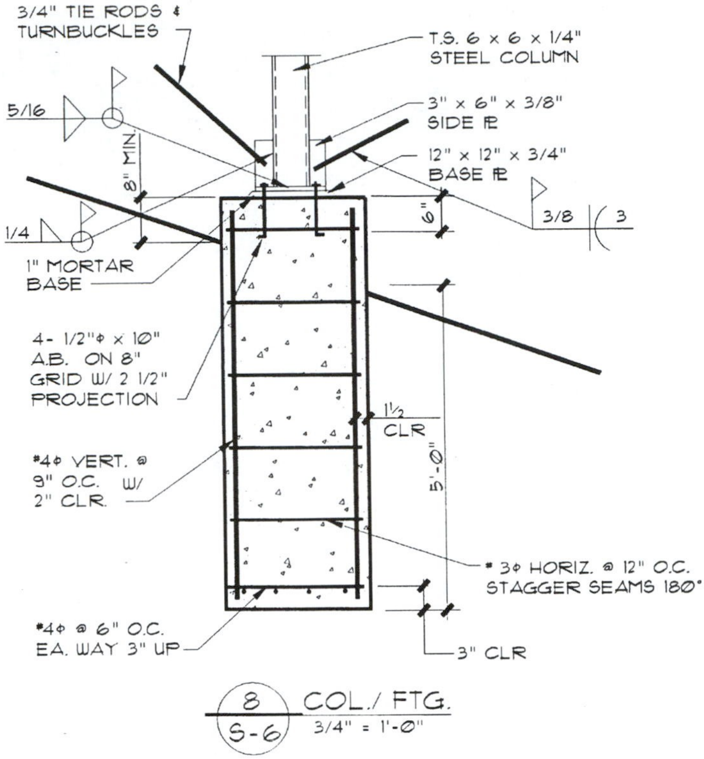

Figure 18.15 *Elevations, sections, and details each require dimensions to show vertical relationships. (Courtesy Residential Designs.)*

ACCESSING DIMENSIONING COMMANDS

Several methods are available for accessing the commands that control dimensions. Commands can be accessed by selecting icons from the Dimensioning toolbar, or by selecting an option from the Dimension pull-down menu. The pull-down menu and toolbar can be seen in Figure 18.16. Each of the options presented in these menus will be presented in this chapter. Dimensions can also be entered at the command prompt, using two different options. Menus are also available in dialog boxes, which will be presented in Chapter 19.

DIMENSION OPTIONS

Eleven methods are available for placing dimensions on a drawing. The Linear and Aligned options are used for providing linear dimensioning. Diameter, Radius, Angular, and Center Mark are used to describe circular features. Three other options,

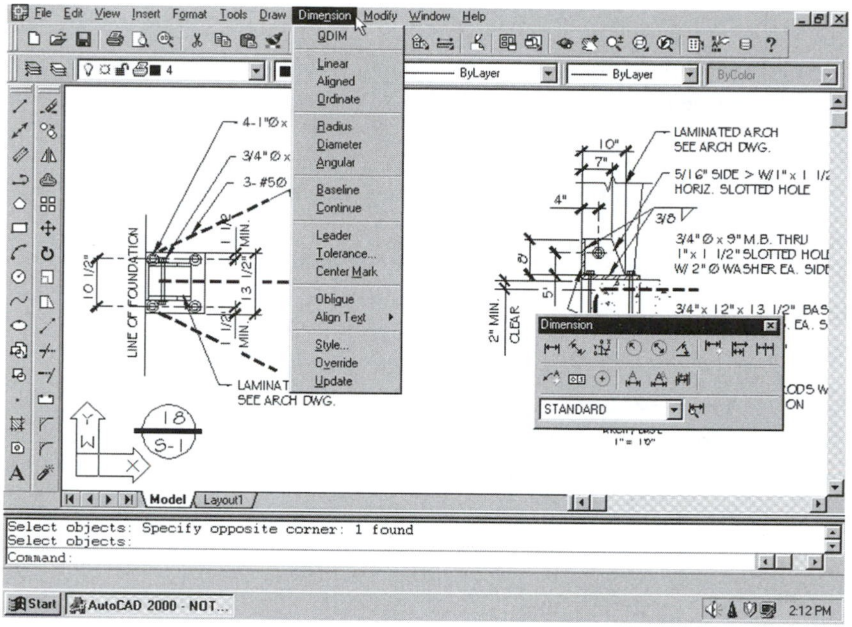

Figure 18.16 *The commands for placing dimensions can be selected from the Dimension toolbar, pull-down menus, or entered by keyboard.*

Baseline, Continue, and Ordinate, are used to place dimensions and can be combined with other commands. The LEADER command can be used for referencing text and dimensions to a drawing. The Tolerance... option can be used to access geometric dimensioning and tolerancing symbols. These symbols are rarely used and will not be covered in detail. Style can be used to access dimensioning dialog boxes that are used to save dimensioning styles. Styles will be discussed in Chapter 19.

Note: In addition to the dimensioning commands listed above, AutoCAD allows dimensions to be placed using the Quick Dimension (QDIM) command. QDIM can be used to create a series of dimensions using baseline or continued dimensions. The following portion of this chapter will introduce each dimensioning method. The final portion of this chapter will help you combine options using QDIM. Chapter 19 will introduce methods of altering the features of each dimensioning method.

LINEAR DIMENSIONS

Linear dimensions are used to dimension straight surfaces common to construction features. Selecting Linear Dimensions from the Dimension toolbar or pull-down menu will produce the command prompt, which will be discussed shortly. Linear dimensions can also be started by typing **DLI** (EnterKey) at the command prompt. This will produce the prompt:

Command: **DLI** (EnterKey)

DIMLINEAR

Specify first extension line origin or <select object>:

AutoCAD now offers two options to place a dimension. The default method is to specify two points to define the dimension. A second method allows selection of the object to be dimensioned. Once the object is selected, the end points are automatically defined.

Select Extension Line Origin

The start and end of a line or distance to be dimensioned can be specified. The command sequence for specifying each endpoint is:

Specify first extension line origin or <select object>: *(Select point.)*

Specify second extension line origin: *(Select point.)*

Specify dimension line location or

[Mtext/Text/Angle/Horizontal/Vertical/Rotated]: *(Select dimension line location.)*

The default requires you to move the cursor and select the desired location to place the dimension line. Be sure and activate OSNAP settings before selecting dimension locations. As the second extension line location is selected, the dimension and extension lines are dragged into position to allow for visual inspection.The coordinate display can be useful in spacing dimension lines. The dimension line location should be selected to allow room for all future notes. Place the dimension line in the desired location and press the pick button. The process can be seen in Figure 18.17.

Mtext

Selecting the Mtext option allows the dimension value to be replaced by a body of text in a similar manner as the Text option. Entering **M** (EnterKey) at the dimension prompt will produce a display similar to Figure 18.18 with the cursor in front of the brackets. The brackets represent the default dimension value. With the cursor preceding the brackets, text will be written followed by the dimension. An entry of F.O.S. <> would be displayed as F.O.S. 40'–0". An entry of <>F.O.S. will be displayed as 40'–0" F.O.S. Using the cursor to remove the brackets will remove the default dimension value and allow only text to be created.

Text

One of the strengths of AutoCAD is also one of its weaknesses. AutoCAD is extremely accurate. Sometimes the dimension text that will be assigned is too accurate for the drawing purpose. Chapter 19 will introduce methods of controlling dimension accurately. If you haven't been accurate as you created the drawing, the

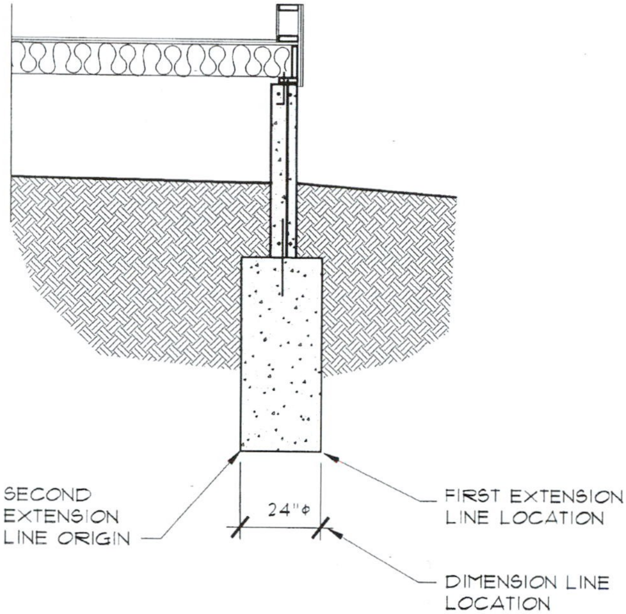

Figure 18.17 *Horizontal and vertical dimensions can be placed using the DLI (DIMLINEAR) command. The dimension is placed by selecting the starting location for each extension line and then selecting the location for the dimension line. As the location for the second extension line is selected, the dimension is displayed and can be dragged into position.*

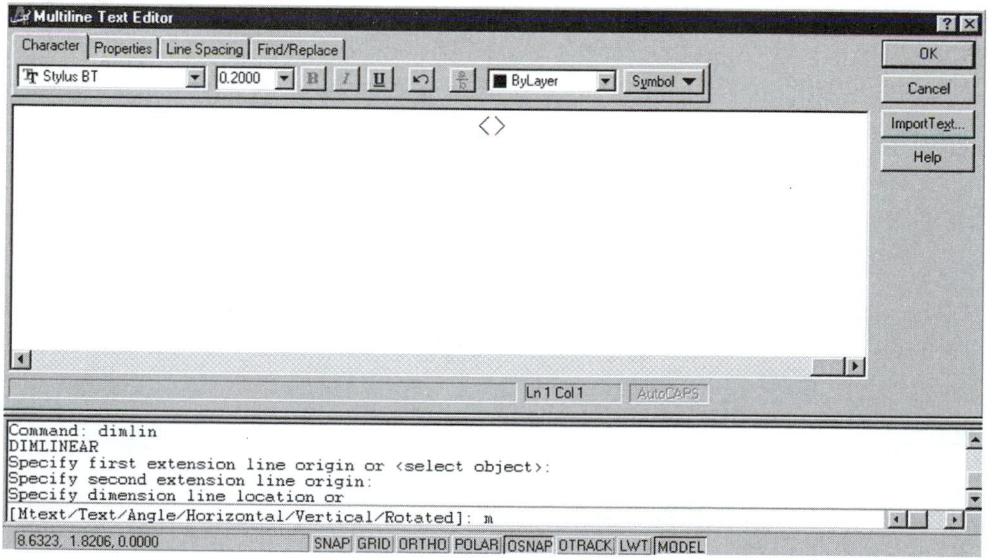

Figure 18.18 *Entering* ***M*** *(EnterKey) at the command prompt will produce the Multiline Text Editor.*

text assigned to the dimension will not produce the desired results. Choosing the Text option before the dimension line location is selected allows you to override the dimension text value by entering the desired value by keyboard. Use the following sequence to change the value from 3½" to 4":

> Specify dimension line location or
>
> [Mtext/Text/Angle/Horizontal/Vertical/Rotated]: **T** (EnterKey)
>
> Enter dimension text <3.5>: **4** (EnterKey)
>
> Specify dimension line location or
>
> [Mtext/Text/Angle/Horizontal/Vertical/Rotated]: *(Select the desired dimension location.)*
>
> Command:

Pressing (EnterKey) returns the dimension and allows it to be dragged into position. Pressing the space bar and then the (EnterKey) rather than assigning a value at the prompt for Dimension text displays the extension lines, dimension lines, and terminators—but no text is assigned. This option can be useful for specifying the limits of a specific product, as seen in Figure 18.19.

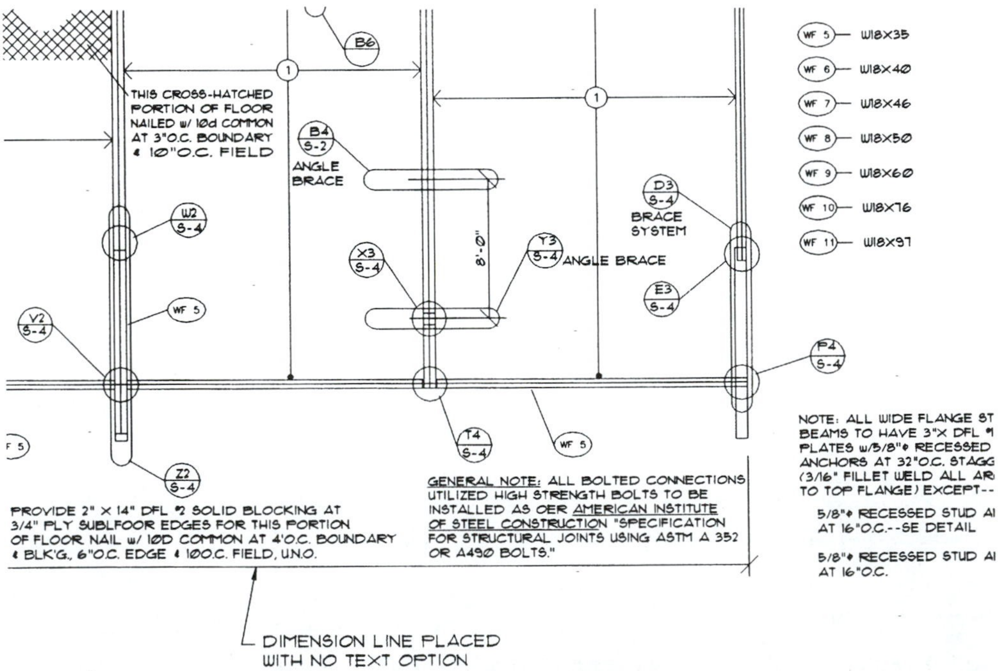

Figure 18.19 *Rather than entering a text value, pressing the Space key followed by the (EnterKey) allows the dimension line to be placed with no dimension.*

Angle

Enter **A** (EnterKey) at the line location prompt to alter the angle that the text will be displayed. Although this is not often done, it could be especially helpful to make a dimension for a small space standout. Figure 18.20 shows an example of text rotated at 45° to the dimension line. The command sequence, once the two endpoints are selected, is:

Specify dimension line location or

[Mtext/Text/Angle/Horizontal/Vertical/Rotated]: **A** (EnterKey)

Specify angle of dimension text: **45** (EnterKey)

Specify dimension line location or

[Mtext/Text/Angle/Horizontal/Vertical/Rotated]: *(Select the desired dimension location.)*

Dimension text = 20'-6"

Command:

Horizontal/Vertical

Entering **H** (EnterKey) or V (EnterKey) at the prompt allows the specified dimension direction to be entered. AutoCAD automatically places a vertical dimension if two vertical endpoints are selected, or a horizontal dimension line if two horizontal endpoints are selected.

Rotated

Entering **R** (EnterKey) at the prompt rotates the extension and dimension lines to a specified angle from the desired surface. By placing the dimension in a rotated position, this option can be used to draw attention to a dimension describing a small

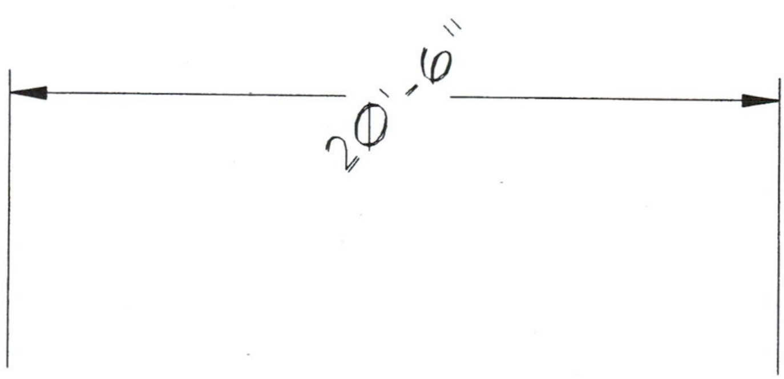

Figure 18.20 *The text angle can be rotated from the dimension line using the A option. Although not a common method of dimensioning, rotated text can be used to help draw attention to the dimension.*

space. Use this option to place the extension and dimension lines for referencing a note to a specific area. The sequence to place the lines at 20° to a vertical surface is:

Specify dimension line location or

[Mtext/Text/Angle/Horizontal/Vertical/Rotated]: **R** (EnterKey)

Specify angle of dimension line <0>: **20** (EnterKey)

Specify dimension line location or

[Mtext/Text/Angle/Horizontal/Vertical/Rotated]: *(Select the desired dimension location.)*

Dimension text = 0'-6"

Command:

Figure 18.21 shows an example of using the rotated option.

The Select Object Option

Up to this point, each of the options discussed was made available once the two origins for the extension line endpoints were selected. By choosing the Select Object option, a specific line can be selected, and each endpoint will be determined automatically. The command sequence is:

Command: **DLI** (EnterKey)

DIMLINEAR

Specify first extension line origin or <select object>: (EnterKey)

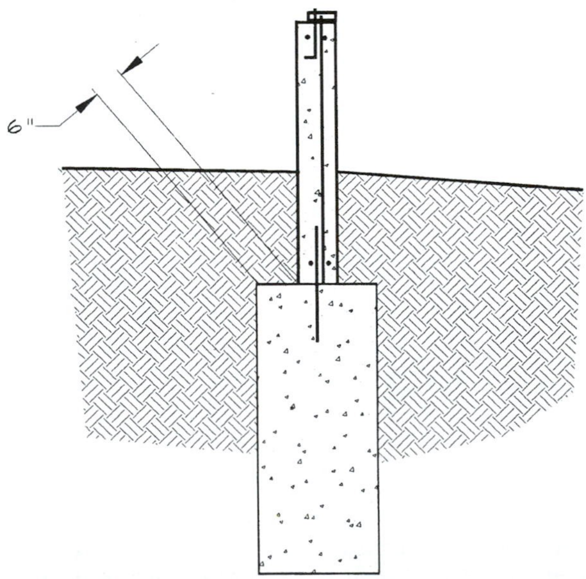

Figure 18.21 *The extension lines, dimension lines, and text can be rotated from the surface to be dimensioned using the R option.*

Once (EnterKey) is pressed, the crosshairs are changed to a pick box and a specific object may be selected:

Select object to dimension: *(Specify object to be dimensioned.)*

Specify dimension line location or

[Mtext/Text/Angle/Horizontal/Vertical/ Rotated]: *(Select the desired location.)*

Dimension text = 24

Command:

This process can be seen in Figure 18.22. The effects of the DIMLINEAR command on vertical dimensions are shown in Figure 18.23. If the area to be dimensioned is smaller than the length required to place the arrows and text, each will be placed outside the extension lines. Methods of altering placement will be discussed in Chapter 19.

ALIGNED DIMENSIONS

This method of placing dimensions is useful when the object being described is not parallel to the drawing borders. An example of aligned dimensions can be seen on the

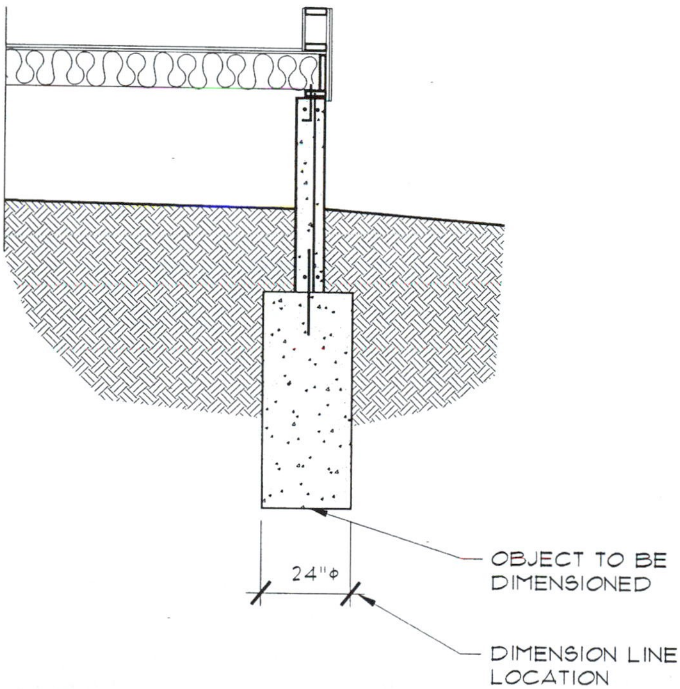

Figure 18.22 *Choosing the Select Object option rather than selecting an extension line location allows an object to be selected. Extension lines will automatically be placed at each end of the selected object.*

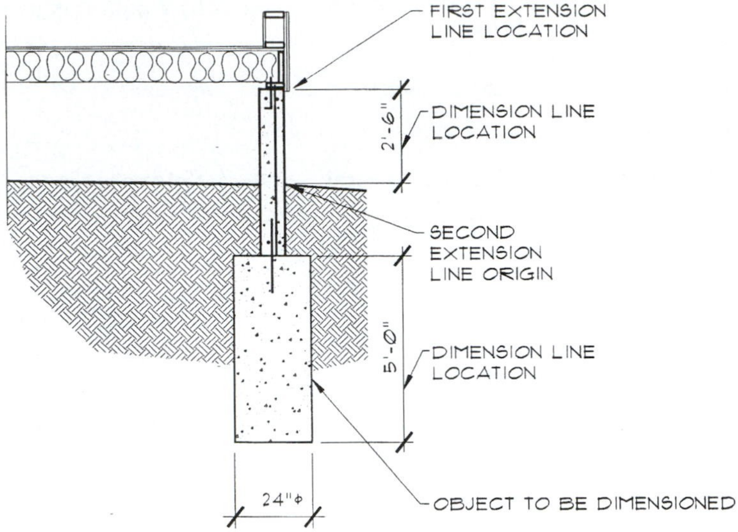

Figure 18.23 *Placement of vertical dimension lines using DIMLINEAR, with the Select Object option used to place the lower dimension and two point selection method used for the upper dimension.*

foundation plan shown in Figure 18.24. The command sequence is similar to the DIMLINEAR command. The DAL (ALIGNED) command can be selected from the Dimension toolbar, the pull-down menu, or started at the command prompt by entering **DAL** (EnterKey). The command sequence is:

Command: **DAL** (EnterKey)

DIMALIGNED

Specify first extension line origin or <Select object>: *(Select first point.)*

Specify second extension line origin: *(Select second point.)*

Specify dimension line location or

[Mtext/Text/Angle]:

Dimension text <45'-0">:

Command:

If a location is selected for the dimension line, the line will automatically be placed parallel to the surface defined by the endpoints. Selecting the Mtext or Text option allows the default text value to be altered. Selecting the Angle option allows the angle of the text to be altered. Each of the options is the same as the corresponding options for DIMLINEAR. Selecting the Select Object option automatically places the extension lines at the ends of the selected line and allows the dimension to be dragged into position.

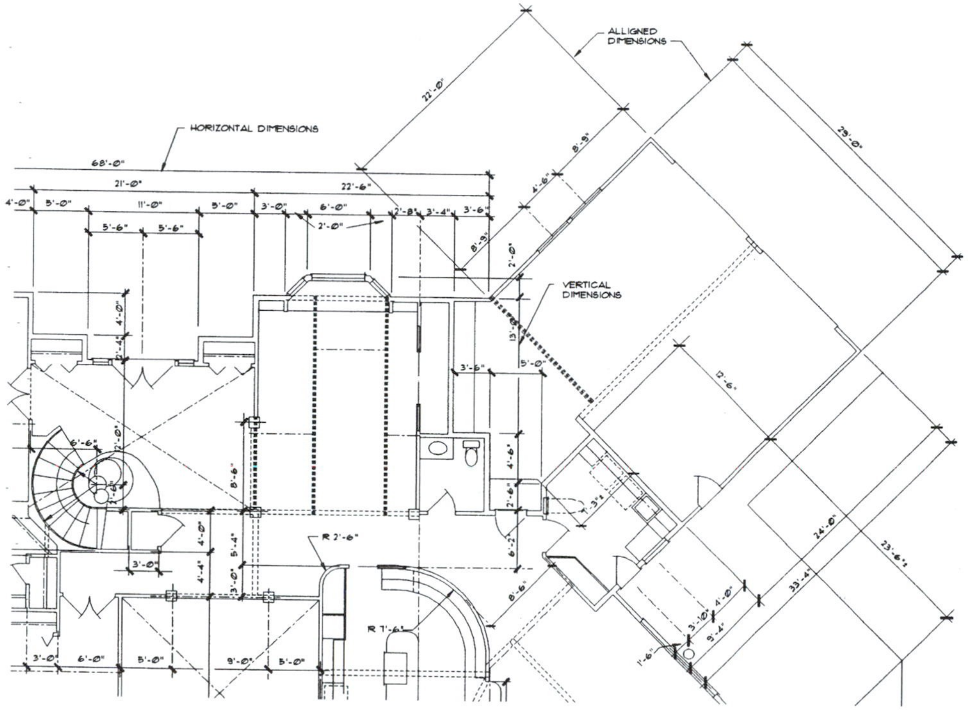

Figure 18.24 *When a surface is not horizontal or vertical, the DLI (DIMALIGNED) command can be used to place dimensions. (Courtesy Residential Designs.)*

The effects of DIMALIGNED can be seen in Figure 18.25.

ORDINATE DIMENSIONS

Ordinate dimensioning consists of placing dimensions without the use of dimension lines and arrows. An example of ordinate dimensions can be seen in Figure 18.26. They are not typically used in construction, but are occasionally used in placing vertical dimensions on details and sections. The command can be accessed from the Dimension toolbar and pull-down menu or from the command prompt by typing **DOR** (EnterKey).

Command: **DOR** (EnterKey)

DIMORDINATE

Select feature location: *(Select the surface to be dimensioned.)*

AutoCAD is prompting for the location that will serve as a base. On a section, this is often the finished floor elevation. It will be helpful to move the 0,0 location of the world coordinate system from the lower corner of the drawing to a corner of the

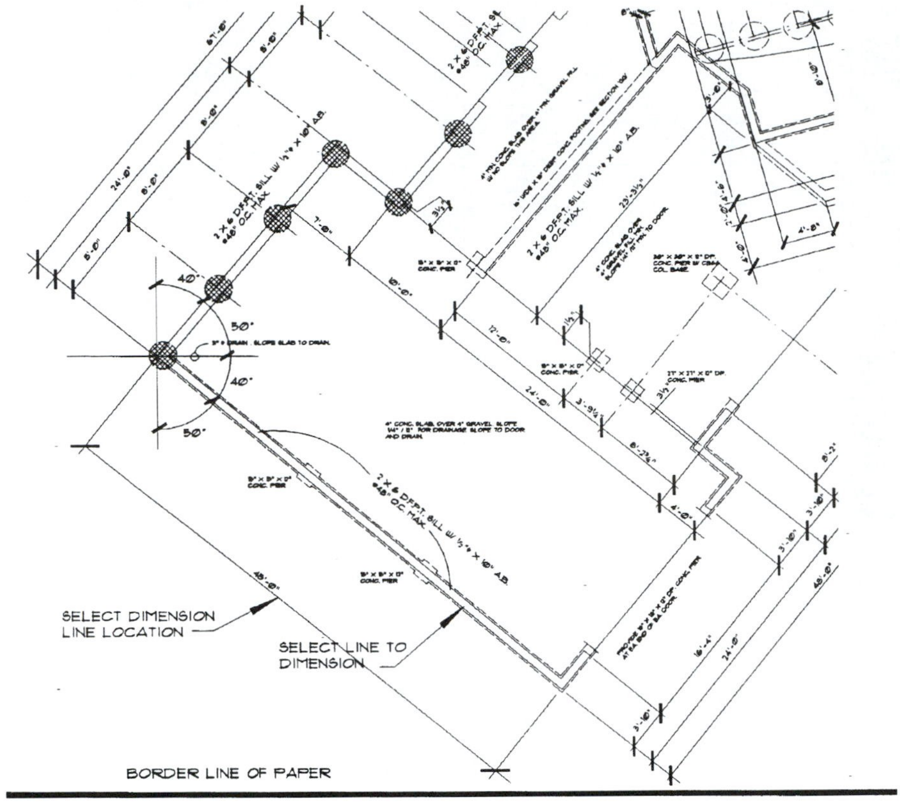

Figure 18.25 *Placing dimensions using the DAL (DIMALIGNED) command is similar to using the DIMLINEAR command.*

object being dimensioned. The Ortho mode should also be set to ON. Once the initial edge has been selected, the prompt will be altered to read:

> Specify leader endpoint or [Xdatum/Ydatum/Mtext/Text/Angle]: *(Select the endpoint of the extension line.)*
>
> Dimension text: <10'-0"> (EnterKey)

The leader is a single horizontal line when the Ydatum ordinate is used and a single vertical line for the Xdatum.

RADIUS DIMENSIONS

The DIMRAD command is useful for many engineering and architectural applications. The way the information is placed will vary depending on the size of the circle or arc to be dimensioned. Figure 18.27 shows each option. The type of dimension

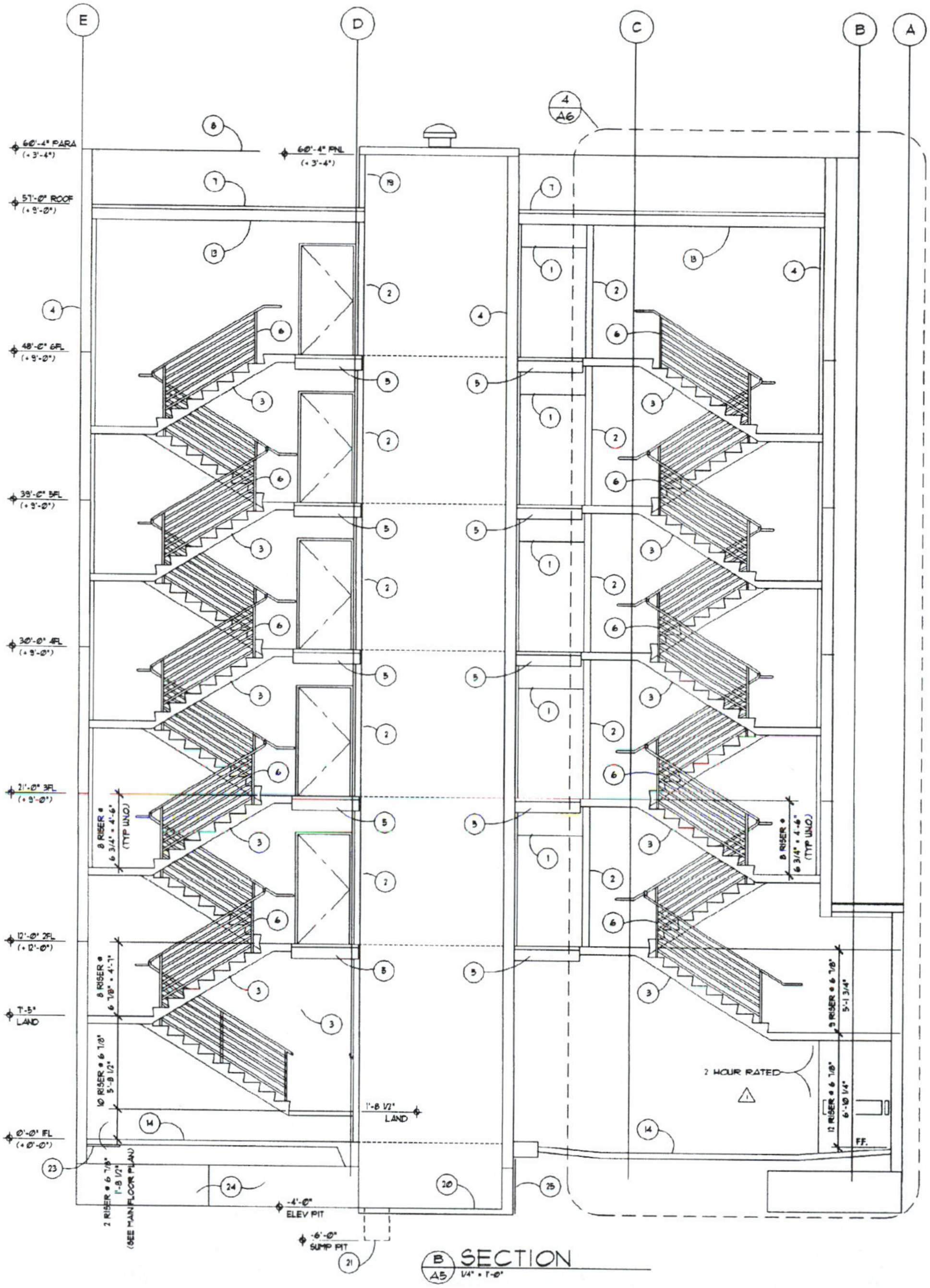

Figure 18.26 *Placement of dimensions using DOR (DIMORDINATE) command. All dimensions are placed using only a single extension line to represent each finish floor. All dimensions are referenced from the main floor level. (Courtesy Russ Hanson, HDN Architects, PC.)*

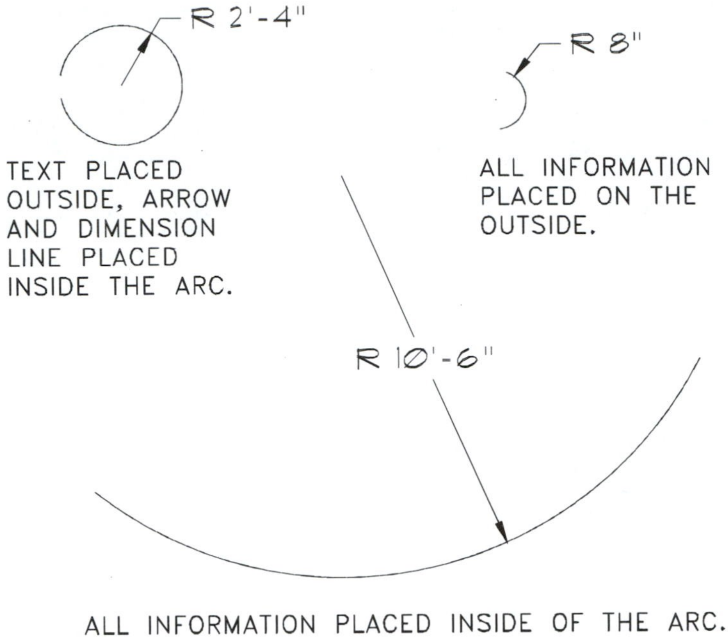

Figure 18.27 *The DRA (DIMRADIUS) command allows a radius to be placed using three different techniques.*

also depends on the DIMVAR settings. The command can be selected from the toolbar, pull-down menu, or entered by keyboard by typing **DRA** (EnterKey). The command sequence is:

Command: **DRA** (EnterKey)

DIMRADIUS

Select arc or circle: *(Select desired arc or circle.)*

Dimension text <10'-6">

Specify dimension line location or [Mtext/Text/Angle]: *(Select location for dimension.)*

Command:

When you select the circle or arc to be dimensioned, the default dimension is dragged as you move the cursor. Entering **M** (EnterKey), **T** (EnterKey), or **A** (EnterKey) allows the text or angle value to be altered. Mtext, Text, and Angle function exactly as discussed with other commands. Once the desired Text and Angle prompts have been responded to, the prompt will be returned to locate the dimension line. Figure 18.28 shows an example of the use of the DIMRADIUS command.

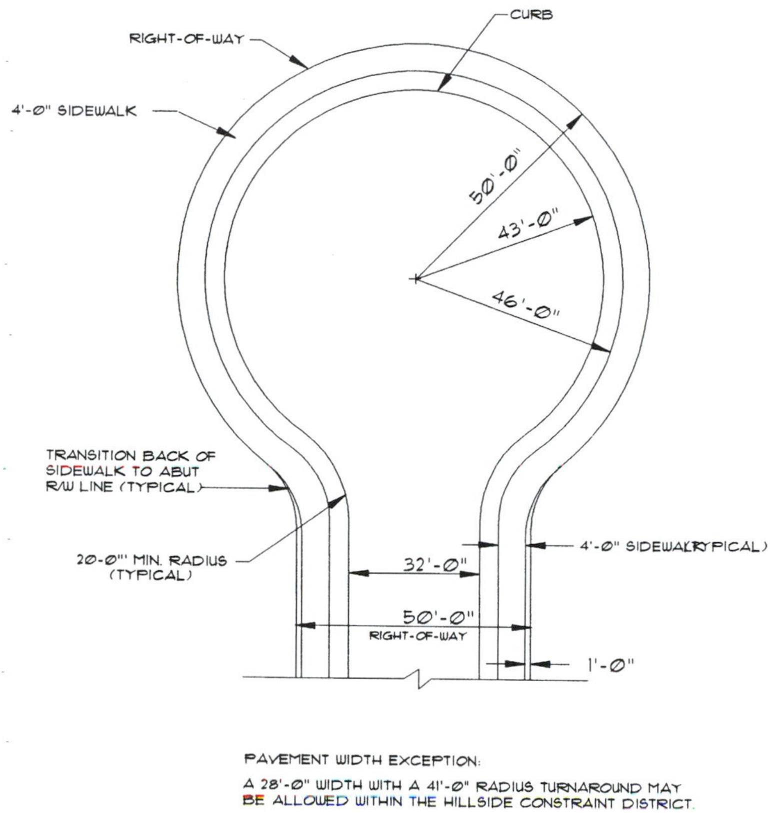

Figure 18.28 *Placing of dimensions using the DRA (DIMRADIUS) command. (Courtesy Department of Environmental Services, City of Gresham, OR.)*

DIAMETER DIMENSIONS

The DIMDIA command will allow you to dimension the diameter of circular objects. The commands can be selected from the Dimension toolbar, pull-down menu, or entered at the command prompt by typing **DDI** (EnterKey). Two methods of placing the diameter specifications on a drawing can be seen in Figure 18.29. The default method is to use a leader line to display the dimension. The command sequence is:

Command: **DDI** (EnterKey)

DIMDIAMETER

Select arc or circle:

Dimension text <24.00>

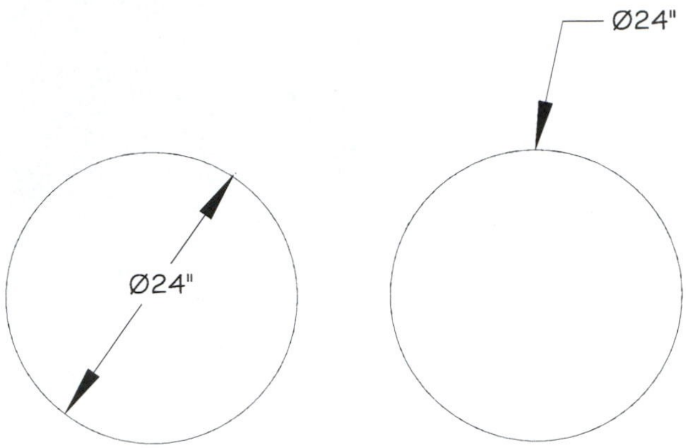

Figure 18.29 *The DDI (DIMDIAMETER) command allows the diameter to be placed using two different methods.*

Specify dimension line location or [Mtext/Text/Angle]: *(Specify a point or enter an option.)*

Command:

When you select the circle to be dimensioned, the default dimension is dragged as you move the cursor. Depending on the size of the circle, the dimension toggles between being placed inside or outside the circle. Placing the dimension inside or outside the circle depends entirely on the size of the circle and the room available. Chapter 19 will discuss methods for controlling where the text is placed. Entering **M** (EnterKey), **T** (EnterKey), or **A** (EnterKey) allows the text or angle value to be altered. Mtext, Text, and Angle function exactly as discussed earlier for the DIMLINEAR command. Once the desired prompts have been responded to, the prompt will be returned for locating the dimension line.

CREATING A CENTER MARK

The DIMCENTER (DCE) command will provide a center mark for a circle or an arc. Two commands can be used to control the center mark. The DIMCEN command allows the size of the mark to be altered. The command sequence is:

Command: **DIMCEN** (EnterKey)

Enter new value for DIMCEN <0'-0 $^1/_{16}$">: *(Provide desired value.)* **6** (EnterKey)

Command:

As the command prompt is returned, nothing is changed. Selecting Center Mark from the Dimension toolbar or pull-down menu will now allow a circle or arc to be selected. The command sequence is:

Command: **DCE** (EnterKey)

Select arc or circle: *(Select desired circle to receive center mark.)*

Command:

ANGULAR DIMENSIONS

The DAN (DIMANGLE) command can be used to dimension angles formed by selecting two straight, nonparallel lines; an arc, a circle, and another point; or three points. You can start the command by selecting the Angular Dimension icon from the Dimension toolbar, or by selecting Angular from Dimension pull-down menu. To enter the command by keyboard, type DAN (EnterKey) at the command prompt. This will produce the following prompt:

Command: **DAN** (EnterKey)

DIMANGULAR

Select arc, circle, line, or <specify vertex>:

Choosing Two Lines

One of the most common situations for dimensioning angles is describing the angle formed between two intersecting lines. Respond to the prompt by selecting a line. The command sequence continues with:

Select arc, circle, line, or <specify vertex>: *(Select first line.)*

Select second line: *(Select second point.)*

Specify dimension arc line location [Mtext/Text/Angle] *(Select location.)*

Dimension text <44.00>

Command:

As the Dimension line location (Mtext/Text/Angle): prompt is displayed, you will be able to choose the location of the dimension placement relative to the angle. Figure 18.30 shows alternatives for placing the angle dimensions.

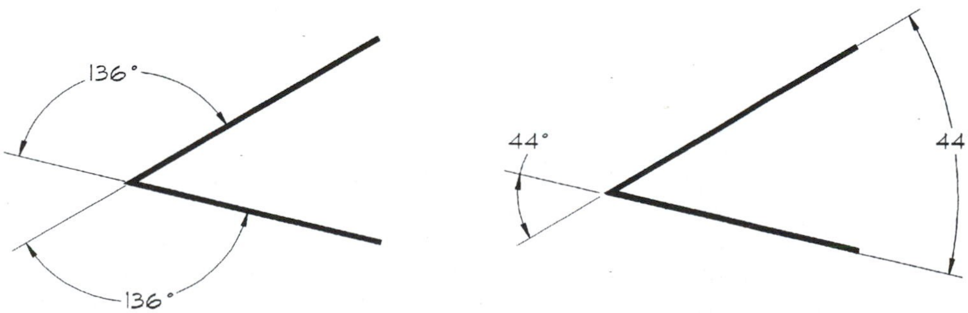

Figure 18.30 *Alternatives for placement of dimensions using the DAN (DIMANG) command.*

Angles Based on an Arc

When you respond to the initial angle prompt by choosing arc, dimensions can be placed to describe the tangent points of an arc. This could be especially useful when labeling subdivision maps or other large parcels of land. The command sequence is:

Command: **DAN**

DIMANGULAR

Select arc, circle, line, or <specify vertex>: *(Select first line.)*

Select second line: *(Select second point.)*

Specify dimension arc line location [Mtext/Text/Angle] *(Select location.)*

Dimension text <156.00>

Command:

As you select the second line, the angle is displayed. Depending on the location of the cursor, the display can be located in one of four positions. The Text and Angle options are the same as other dimensioning commands. Figure 18.31 shows an example of how text will be placed.

Angles Based on a Circle

When you respond to the initial angle prompt by choosing Circle, dimensions can be placed to describe angular patterns formed within a circle. This type of dimensioning is often used in specifying steel placement in circular columns and footings. The command sequence is:

Command: **DAN** (EnterKey)

DIMANGULAR

Select arc, circle, line, or <specify vertex>: *(Select desired arc.)*

Specify dimension arc line location [Mtext/Text/Angle] *(Select location.)*

Dimension text <67.00>

Command:

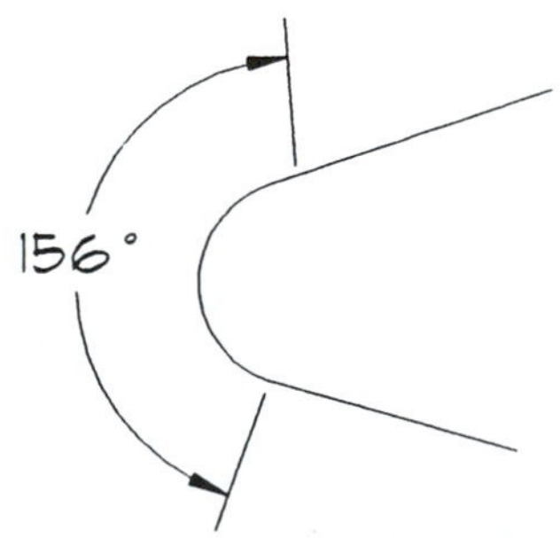

Figure 18.31 *Dimensioning of an arc using the DIMANGULAR command.*

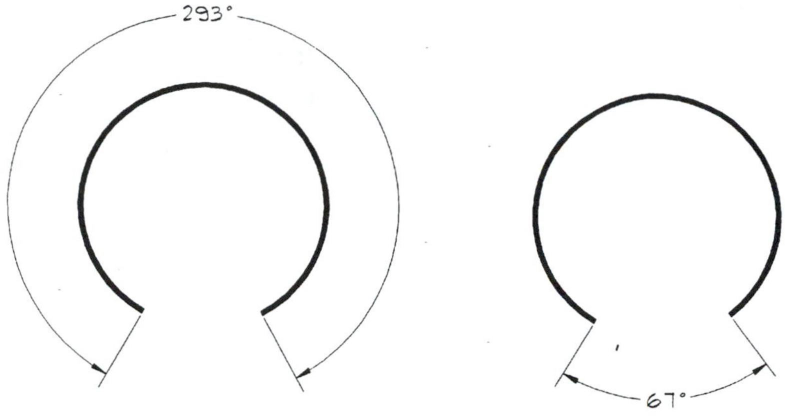

Figure 18.32 *Placement of dimensions using the circle option.*

Options for placing the dimension line can be seen in Figure 18.32.

Describing Three Points

When you respond to the original angular prompt with (EnterKey), you will be allowed to select three different points that can be used to describe an angle. This option can be used to describe the center point of bolts around a center point in a steel connector strap. The two options for placing the angle can be seen in Figure 18.33. The command sequence is:

Command: **DAN** (EnterKey)

DIMANGULAR

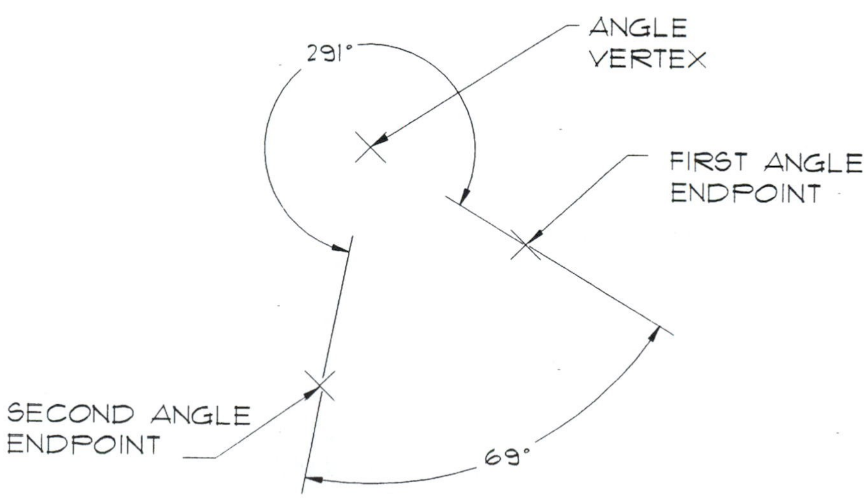

Figure 18.33 *Placement of dimensions using the three point option.*

Select arc, circle, line, or <specify vertex>: (EnterKey)

Specify angle vertex: *(Select a circle center point.)*

Specify first angle endpoint: *(Select a point.)*

Specify second angle endpoint: *(Select a point.)*

Specify dimension arc line location [Mtext/Text/Angle]: *(Select desired text location.)*

Dimension text <69>:

Command:

BASELINE DIMENSIONS

Occasionally when the spacing of objects is extremely critical, the baseline system of placing dimensions can be used. Baseline dimensions (also called datum dimensions) assume one edge to be perfect and reference all dimensions back to that surface. Figure 18.34 shows an example of baseline dimensions. Although this type of dimensioning is not used in residential design, it is common in some areas of commercial construction where a high degree of accuracy must be achieved. You start the command from the Dimension toolbar or from the Dimension pull-down menu. The command can also be started by typing **DBA** (EnterKey) at the command prompt. The DIMBASE command must be started by using a linear, angular, or ordinate

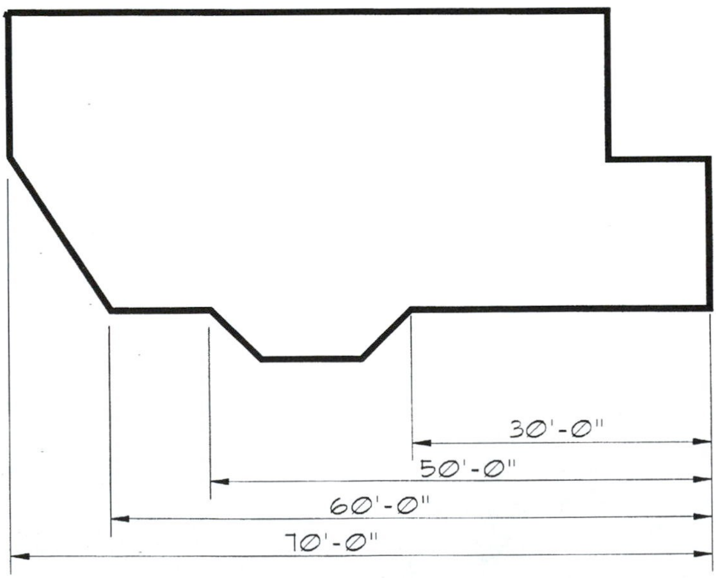

Figure 18.34 *Placement of dimensions using the DBA (DIMBASELINE) command.*

dimension. AutoCAD will use the first extension line from the existing dimension as the base point for other dimensions. The command sequence is:

Command: **DLI** (EnterKey)

DIMLINEAR

Specify first extension line origin or <select object>: (EnterKey)

Select object to dimension:

Specify dimension line location or

[Mtext/Text/Angle/Horizontal/Vertical/Rotated]: *(Select the desired dimension location.)*

Dimension text <30'-0">:

Command:

Command: **DBA** (EnterKey)

DIMBASELINE

Select base dimension: *(Select dimension to be used as the base dimension.)*

Specify a second extension line origin or [Undo/Select] <Select>: *(Select base for second point.)*

Dimension text <50'-0">

Select base dimension: *(Select dimension to be used as the base dimension.)*

Specify a second extension line origin or [Undo/Select] <Select>: *(Select base point for the next dimension.)*

Dimension text <60'-0">

Specify a second extension line origin or [Undo/Select] <Select>: *(Select base point for the next dimension.)*

Dimension text <70'-0">

Specify a second extension line origin or [Undo/Select] <Select>: (EnterKey)

Select base dimension: (EnterKey)

Command:

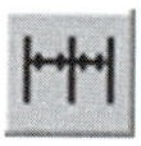

CONTINUOUS DIMENSIONS

A series of dimensions is usually needed to dimension a structure. In Figure 18.1, dimension lines extend from the exterior walls, to the interior walls, and finally to the exterior wall on the opposite side. These dimensions can be placed with the CONTINUE command. The command sequence starts by using the DIMLINEAR command to place the first dimension. To use DIMCONTINUE, it must be based on a linear, angular or ordinate dimension. You can start the CONTINUE command by selecting the Continue Dimension icon from the Dimension toolbar or by selecting Continue from the Dimension pull-down menu. You can also start the command by typing DCO(EnterKey) at the command prompt. This will produce the prompt:

Command: **DCO** (EnterKey)

DIMCONTINUE

Specify a second extension line origin or [Undo/Select] <Select>:

After you select the starting point for the second extension line, AutoCAD uses the second extension line of the DIMLINEAR command sequence as the first extension line for the DIMCONTINUE sequence. Press (EscKey) to end the DIMCONTINUE command sequence. Figure 18.35 shows an example of placing dimensions using the Continue command with the default settings. The command sequence to place the dimensions across the bottom is:

Command: **DLI** (EnterKey)

DIMLINEAR

Specify first extension line origin or <select object>: (EnterKey)

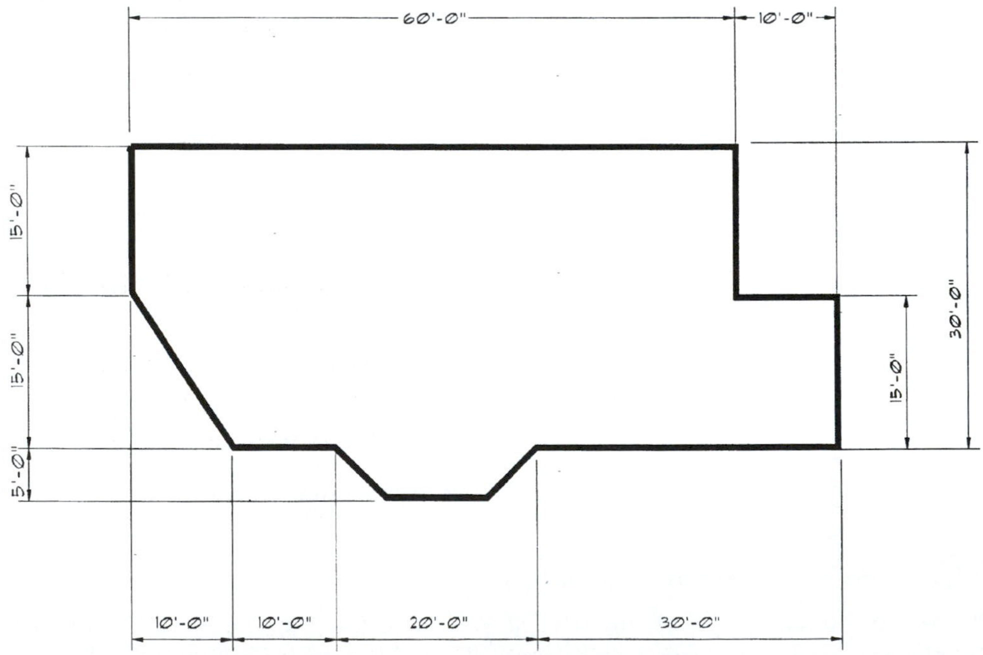

Figure 18.35 *Placing of dimensions using the DIMLINEAR and DIMCONTINUE commands. The 10'-0" dimension on the bottom left was placed using the DIMLINEAR command. The next 10'-0" dimension was placed using the DIMCONTINUE command. The DIMCONTINUE command uses the second extension line of the preceeding sequence as the first extension line of the DIMCONTINUE sequence.*

Select object to dimension:

Specify dimension line location or

[Mtext/Text/Angle/Horizontal/Vertical/Rotated]: *(Select the desired dimension location.)*

Dimension text 10'-0"

Command:

Command: **DCO** (EnterKey)

DIMCONTINUE

Select continued dimension: *(Select dimension to be used as the first dimension.)*

Specify a second extension line origin or [Undo/Select]: <Select> *(Select second point.)*

Dimension text = 10'-0"

Specify a second extension line origin or [Undo/Select]: <Select> *(Select next point.)*

Dimension text = 20'-0"

Specify a second extension line origin or [Undo/Select]: <Select> *(Select next point.)*

Dimension text = 30'-0"

Specify a second extension line origin or [Undo/Select]: <Select> (EnterKey)

Select continued dimension: (EnterKey)

Command:

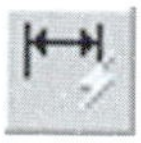

QUICK DIMENSIONS

You've been exposed to several different methods for placing dimensions throughout this chapter. Each works well for a specific need. AutoCAD 2000 provides an additional method for placing dimensions that combines features of many of the previous options. As the name implies, the QDIM command (Quick Dimension) allows a single dimension or a series of dimensions to be placed quickly. QDIM excels in placing strings of continuous or baseline dimensions with very few prompts required. The command can be accessed by selecting the QDIM icon from the dimension toolbar, by typing **QDIM** (EnterKey), or by selecting QDIM from the Dimension pull-down menu. Each will produce the following prompt at the command line:

Command: **QDIM** (EnterKey)

Select geometry to dimension: *(Select object or objects to be dimensioned.)*

AutoCAD is waiting for you to select an object to be dimensioned. Press (EnterKey) to end the selection process. The command prompt will now change to display:

Select Geometry to dimension: (EnterKey)

Specify dimension line position, or

[Continuous/Staggered/Baseline/Ordinate/Radius/Diameter/datumPoint/Edit]

<Continuous>: *(Select the desired dimension location.)*

If only one object is selected, the command will function similar to the Select Objects option of Linear Dimension. An object to be dimensioned is selected, followed by a location for the dimension to be placed. As (EnterKey) is selected, the dimension will be displayed, and can then be dragged into position as the cursor is moved. When the dimension is in the desired position, left-click to add the dimension to the drawing. The process can be seen in Figure 18.36a and 18.36b.

The best features of QDIM can be seen if multiple objects are selected to be dimensioned. In Figure 18.37a, four objects have been selected to be dimensioned. Once the selection set is completed and (EnterKey) is pressed, dimensions to describe each feature will be displayed. A dimension to describe each feature will be displayed and can be placed by identifying one location point. The process can be seen in Figure 18.37a and 18.37b. The command process to place the four dimensions is:

Command: **QDIM** (EnterKey)

Select Geometry to dimension: *(Select object or objects to be dimensioned.)*

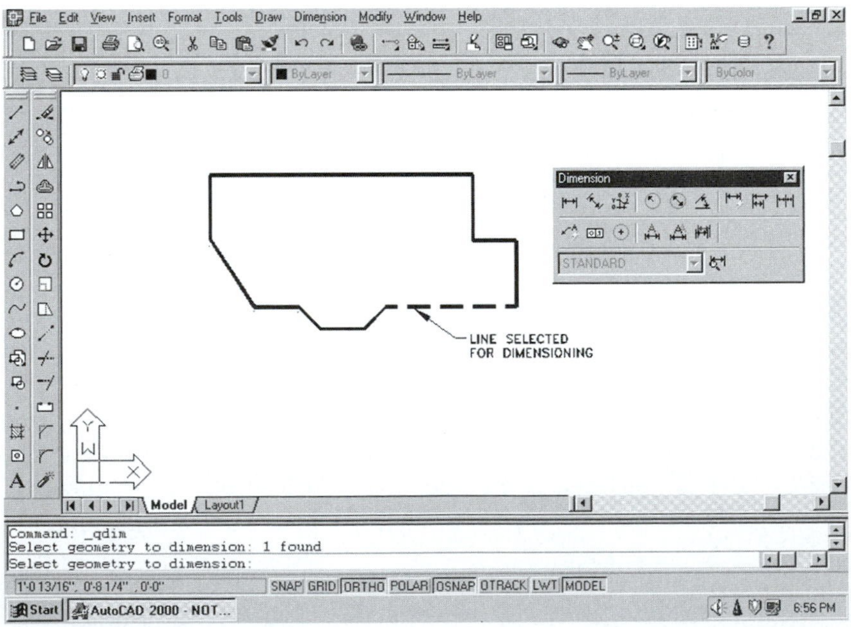

Figure 18.36a *QDIM functions similar to the Select Objects option of Linear Dimension if only one object is selected to be dimensioned.*

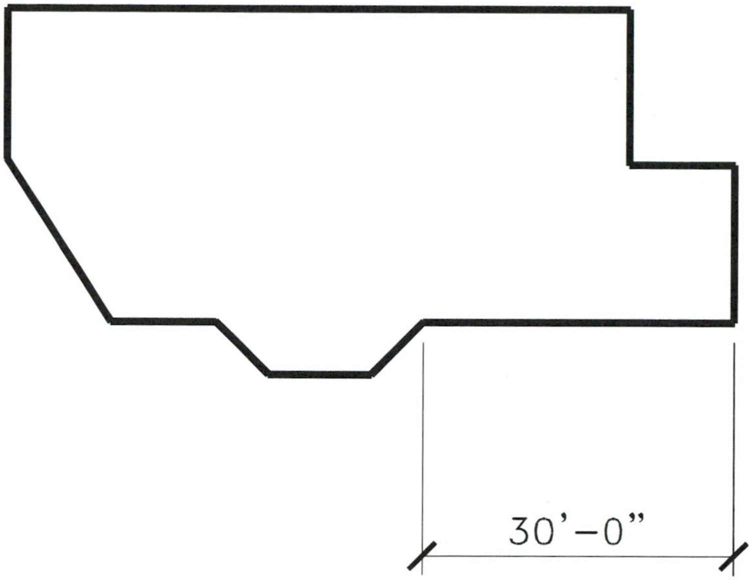

Figure 18.36b *Using the QDIM option only an object and location are required to place a dimension.*

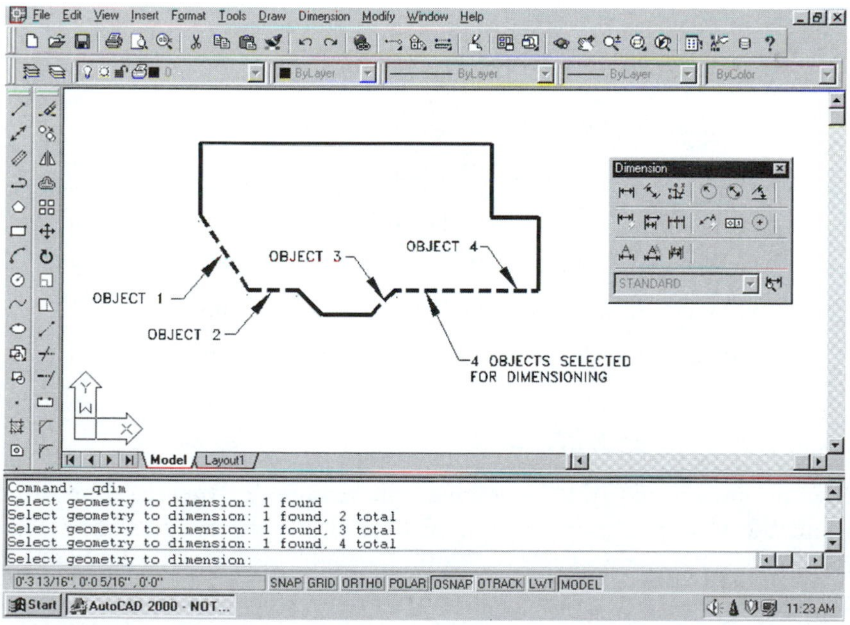

Figure 18.37a *The best features of QDIM can be seen if multiple objects are selected to be dimensioned. In Figure 18-37a, four objects have been selected to be dimensioned.*

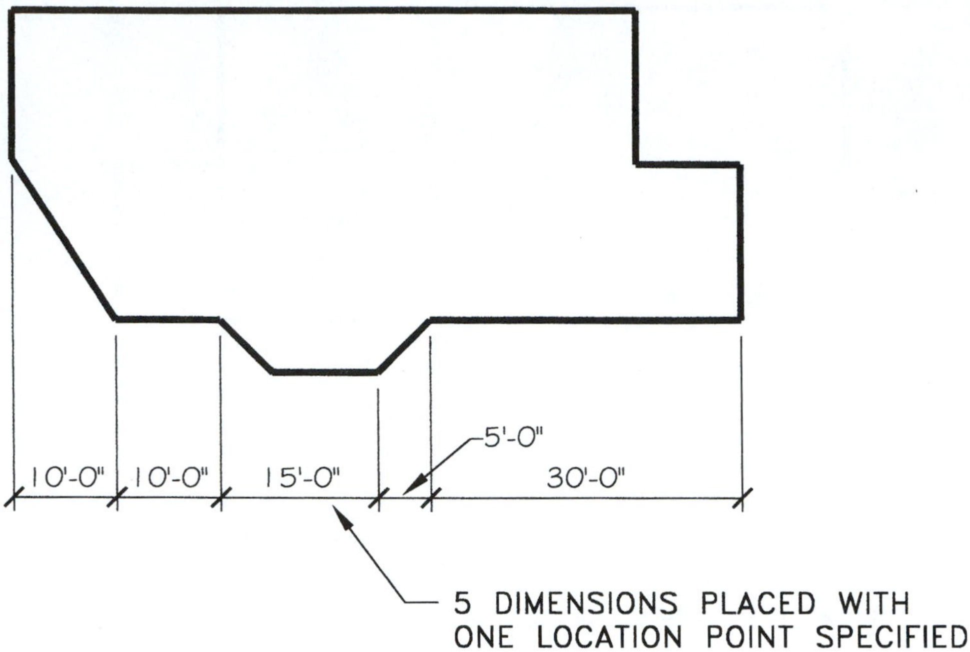

Figure 18.37b *Once the geometry to be dimensioned and the line position have been specified, the dimensions are placed.*

Select Geometry to dimension: 1 found,

Select Geometry to dimension: 1 found, 2 total

Select Geometry to dimension: 1 found, 3 total

Select Geometry to dimension: 1 found, 4 total

Select Geometry to dimension: (EnterKey)

Specify dimension line position, or

[Continuous/Staggered/Baseline/Ordinate/Radius/Diameter/datumPoint/Edit]

<Continuous>: *(Select location to place the dimension string.)*

Command:

QDIM PLACEMENT OPTIONS

The Baseline option of QDIM combines the features of the Baseline dimension option and the Continuous process to place multiple dimensions. The command functions in a similar method as the Continuous option in that each object must be selected, followed by the location. Figure 18.38 shows an example of placing dimensions using the Baseline option of QDIM. The command sequence is:

Command: **QDIM** (EnterKey)

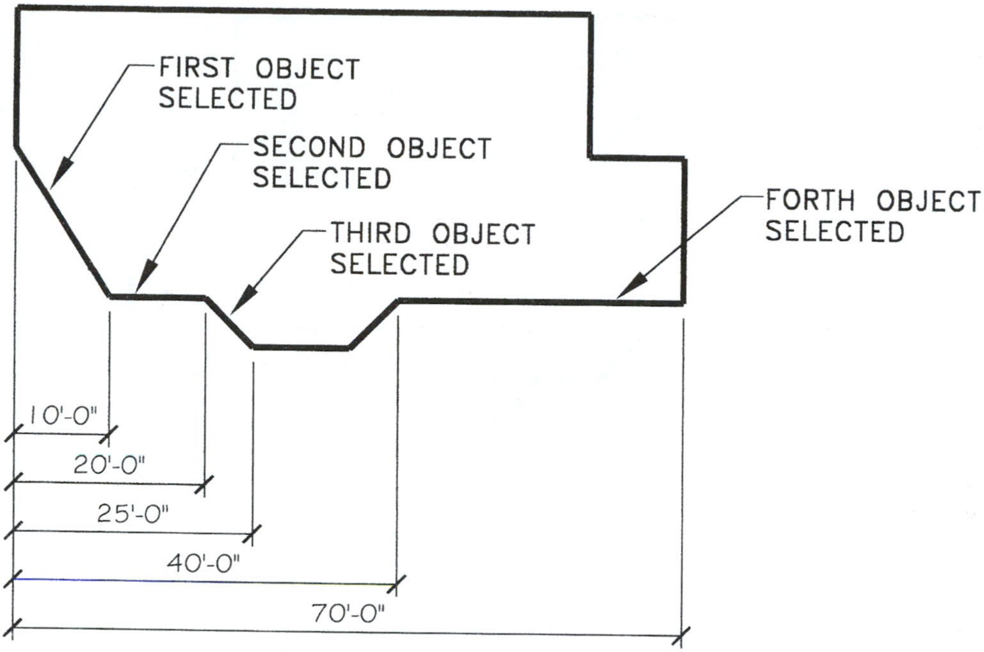

Figure 18.38 *The Baseline option of the QDIM command functions in a similar method as the Continuous option, in that each object must be selected, followed by the location.*

Select geometry to dimension: *(Select object or objects to be dimensioned.)*

Select geometry to dimension: 1 found,

Select geometry to dimension: 1 found, 2 total

Select geometry to dimension: 1 found, 3 total

Select geometry to dimension: 1 found, 4 total

Select geometry to dimension: (EnterKey)

Specify dimension line position, or

[Continuous/Staggered/Baseline/Ordinate/Radius/Diameter/datumPoint/Edit]

<Continuous>: **B** (EnterKey)

Specify dimension line position, or

[Continuous/Staggered/Baseline/Ordinate/Radius/Diameter/datumPoint/Edit]

<Baseline>: *(Select location to place the dimension string.)*

Command

The Stagger option of QDIM can be used to place staggered dimensions. Once objects to be dimensioned are selected, enter **S** (EnterKey) when prompted for an option. Figure 18.39 shows an example of staggered dimensions. Entering **O** (Enter-

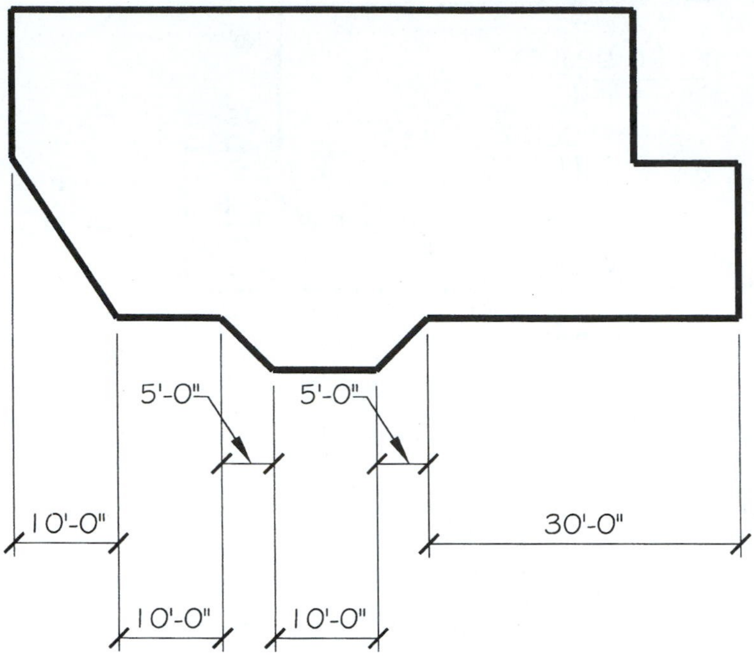

Figure 18.39 *Entering **S** (EnterKey) when prompted for an option allows staggered dimensions to be placed.*

Key) at the prompt allows ordinate dimensions to be placed using similar methods for selection as other QDIM options. The Radius and Diameter options of QDIM allow multiple circles or arcs of various sizes to be selected and dimensioned in one command sequence, rather that having to repeat a dimension sequence for each object.

The datumPoint option will place dimensions in a pattern similar to the Ordinate option. The major difference between the two options is that datumPoint allows any point to be set as the base point. This option could be used to establish a finish floor level as 00, and allow dimensions to be placed above and below the base point (the finish floor). Figure 18.40 shows an example of dimensions placed using the datumPoint option. The option is started by entering **P** (EnterKey) at the option prompt. The command sequence to place datumPoint dimensions is:

Command: **QDIM** (EnterKey)

Select Geometry to dimension: *(Select object or objects to be dimensioned.)*

Select Geometry to dimension: 1 found,

Select Geometry to dimension: 1 found, 2 total

Select Geometry to dimension: 1 found, 3 total

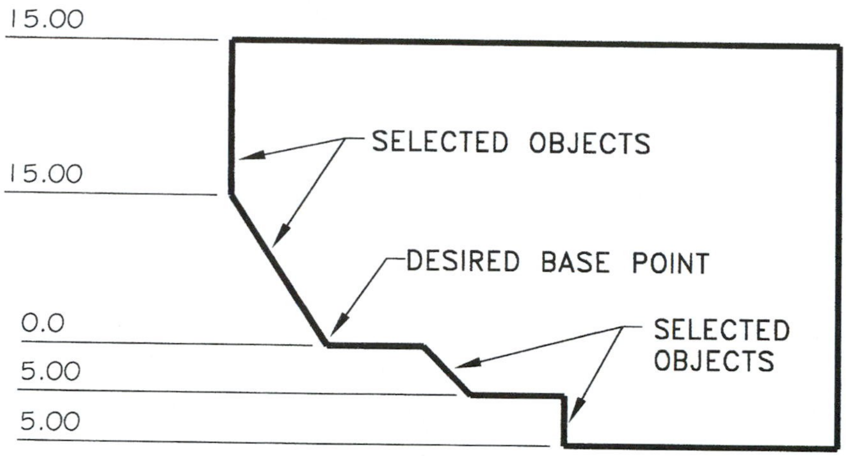

Figure 18.40 *The datumPoint option allows ordinate dimensions to be placed, using any point to be set as the base point.*

Select Geometry to dimension: 1 found, 4 total

Select Geometry to dimension: (EnterKey)

Specify dimension line position, or

[Continuous/Staggered/Baseline/Ordinate/Radius/Diameter/datumPoint/Edit]

<Continuous>: **P** (EnterKey)

Select new datum point: *(Select the point to be the base of the dimensions.)*

Specify dimension line position, or

[Continuous/Staggered/Baseline/Ordinate/Radius/Diameter/datumPoint/Edit]

<Ordinate>: *(Select location to place the dimension string.)*

Co.mmand

LEADER LINES

The QLEADER and LEADER commands can be used to place a leader line to connect notes or dimensions to a specific portion of the drawing. Figure 18.41 shows examples of the use of leader lines. Leaders are generally placed at any angle except horizontal or vertical. Some architectural offices use horizontal and vertical leaders with a custom arrow to duplicate manual practice. A horizontal leg is placed at the end of the leader line to tie the text to the leader. The horizontal leg should extend from the left side of the first line of text or from the right side of the last line of text. Figure 18.42 shows common methods of placing leader lines.

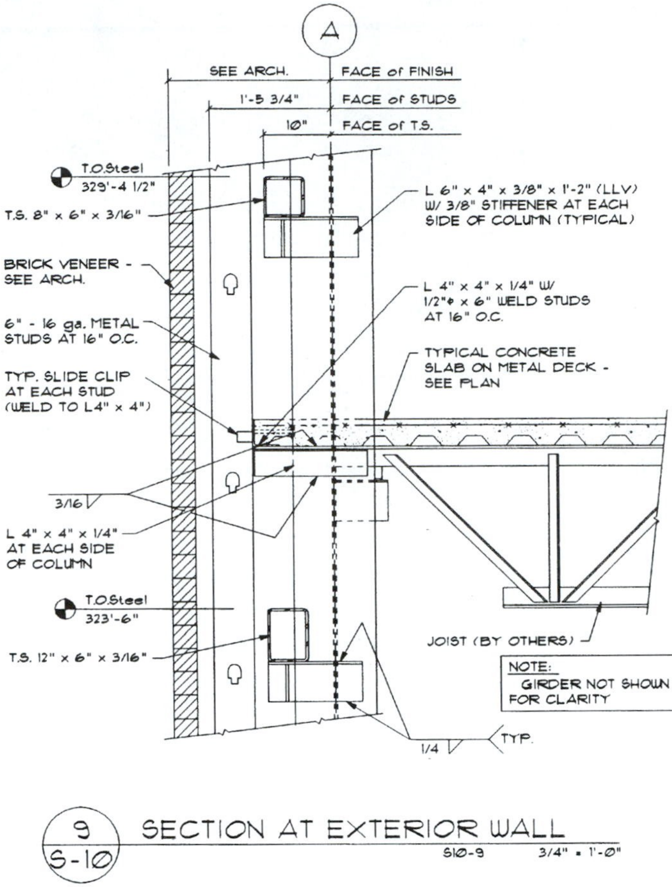

Figure 18.41 *Leader lines can be used to attach text, dimensions, and welding symbols to a specific portion of the drawing. (Courtesy of Van Domelen/Looijenga/McGarrigle/Knauf Consulting Engineers.)*

USING QUICK LEADERS

The QLEADER command should be used when several local notes need to be referenced to the drawing, or when you want to control the width of the space where the notes will be placed. The command can be accessed by selecting the Quick Leader icon from the Dimension toolbar, by typing LE (EnterKey) at the command prompt, or by selecting LEADER from the Dimension pull-down menu. Each will produce the prompt:

Command: **LE** (EnterKey)

QLEADER

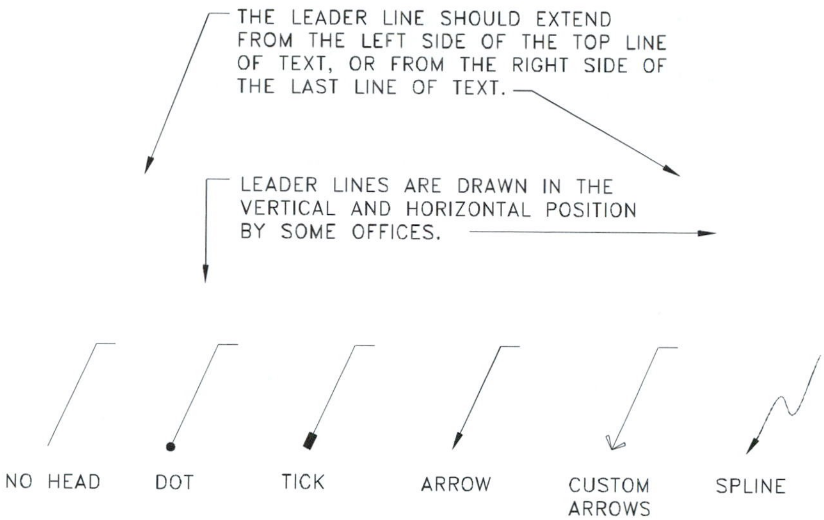

Figure 18.42 *Common methods of placing and terminating leader lines.*

Specify first leader point, or [Settings] <Settings>: *(Select desired start point for leader.)*

Specify next point: *(Select desired end point of leader.)*

Specify next point: *(Select desired end point for horizontal leader tail.)*

Specify text width <0">: (EnterKey)

Enter first line of annotation text <Mtext>:

The first option to be considered is Setting. The Setting option allows several features of the leader to be altered. Settings will be explored after the default option of QLEADER has been explored. For now, select a starting point for the leader. The "First" and "Next" point prompts are like those of the LINE command. AutoCAD will place a line with the tip of an arrow extending from the "First" point to the "Next" point. The Settings option allows alternatives to the arrow to be used. A prompt for a second "Next" point is also provided. As shown in Figure 18.43, this second "Next" point produces a horizontal line. Be sure to use ORTHO to keep the tail line horizontal.

Text Width

Once the location of the leader is specified you will be prompted to provide the Text Width. The Text Width option allows the space to be specified where the annotation will be placed. The default setting for width is 0". A line of text will be of unspecified length with this setting. Text will be placed in one line until (EnterKey) is pressed to start a new line of text. If a value is provided for the text width, that value will specify

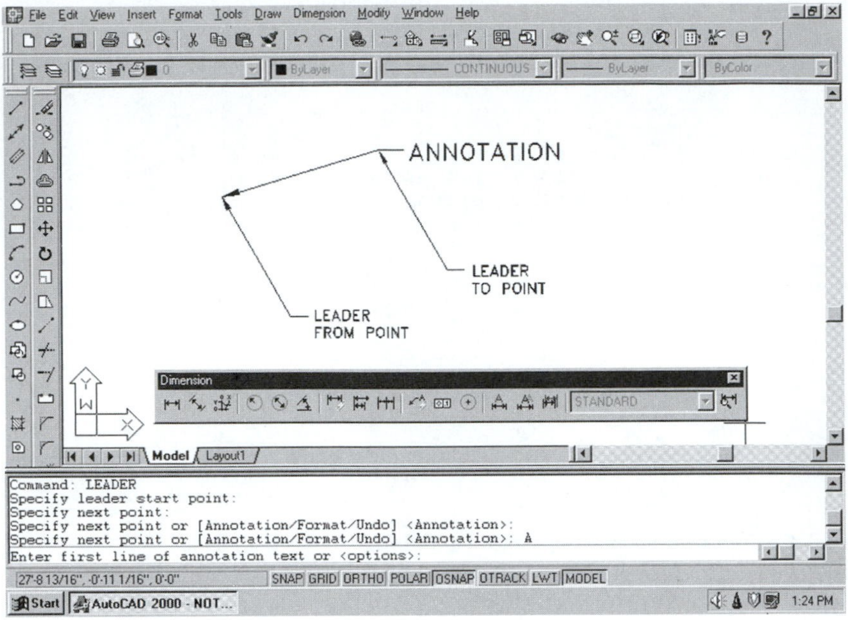

Figure 18.43 *When using QLEADER, the "First" and "Next" point prompts are like those of the LINE command. AutoCAD will place a line with the tip of an arrow extending from the "First" point to the "Next" point. The Settings option allows alternatives to the arrow to be used. A prompt for a second "Next" point is also provided. As the second "Next" point is entered, a horizontal line is produced.*

the width of each line of text. Entering **4** (EnterKey) will allow the text line to be 4" long. Once the specified width has been achieved, a new line of text will be started automatically.

Entering Annotation

The final prompt of the QLEADER command is:

> Enter first line of annotation text <Mtext>:

Three options are available for entering text beside the leader. Pressing the (EscKey) will place the leader line and close the command without placing any text beside the leader. This option works well if stock notes have been stored in a drawing template and they just need to be moved into the proper position. The notes can be placed and then leader lines can be placed to reference the note to the drawing.

A second option as the text prompt is displayed is to begin entering the desired text at the command prompt. As text is entered at the keyboard, it will not be displayed in the drawing area. Figure 18.44 shows how text will be displayed on the command prompt. Pressing (EnterKey) will display a prompt for additional lines of text. When

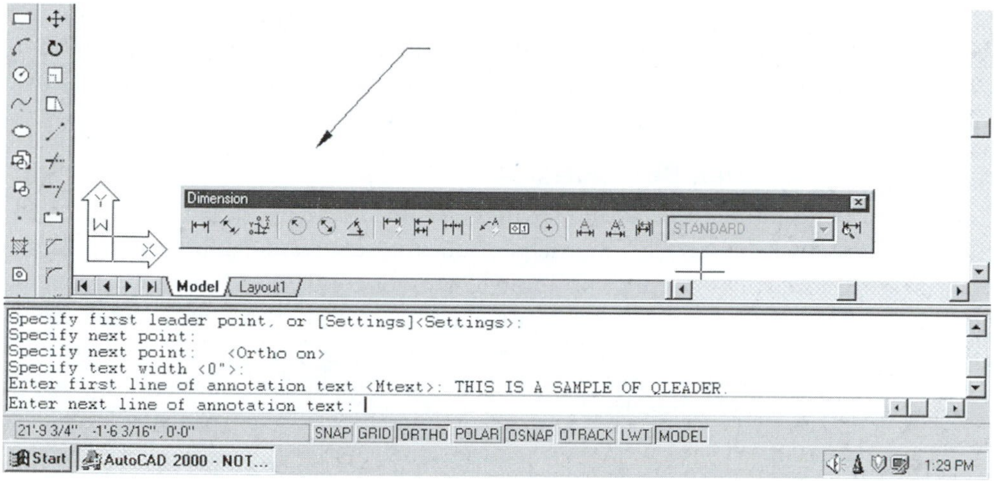

Figure 18.44 *QLEADER allows the desired text to be entered at the command prompt. As text is entered at the keyboard it will be displayed at the command prompt and will not be displayed by the leader line.*

the desired text has been entered, press (EnterKey) to end the command and display the text. Figure 18.45 shows the leader and the placement of the text.

The third option of placing text at the leader line is to press (EnterKey) at the text prompt. This option allows text, numbers, or symbols to be placed at the end of the leader line using the Multiline Text Editor edit box. This option allows multiple lines

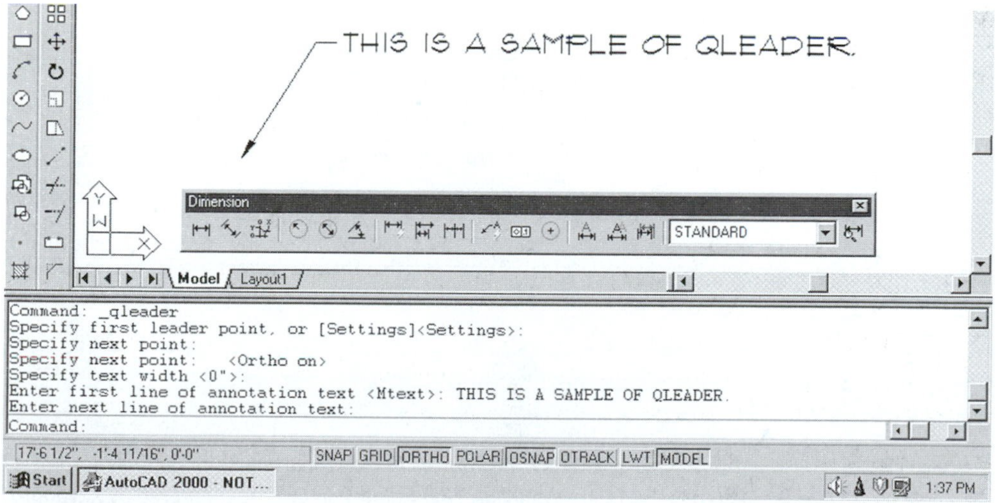

Figure 18.45 *When the desired text has been entered, pressing (EnterKey) will close the command and display the text beside the leader line.*

of text to be entered. Text entered by keyboard will be displayed in the edit box, and will not be placed by the leader line until the OK button is pressed. Figure 18.46 shows an example of the edit box. The edit box was explored as MTEXT was examined. See Chapter 17 for a review of the Multiline Text Editor.

ADJUSTING QLEADER SETTINGS

Earlier you were introduced to the Settings option of the QLEADER command. Rather than adjust the settings, methods of placing text were explored as the command was explored. The Settings option will present controls for annotation, leader line, arrow types, and methods of attaching the leader to the leader lines. QLEADER settings can be altered by pressing (EnterKey) at the first prompt.

> Command: **LE** (EnterKey)
>
> LEADER
>
> Specify first leader point, or [Settings] <Settings>: (EnterKey)

Selecting the (EnterKey) will produce the Leader Settings display shown in Figure 18.47.

Annotation Tab

The default setting of this tab will place text using MTEXT. Remember it's the default setting because it is meets a large variety of needs. Other options include:

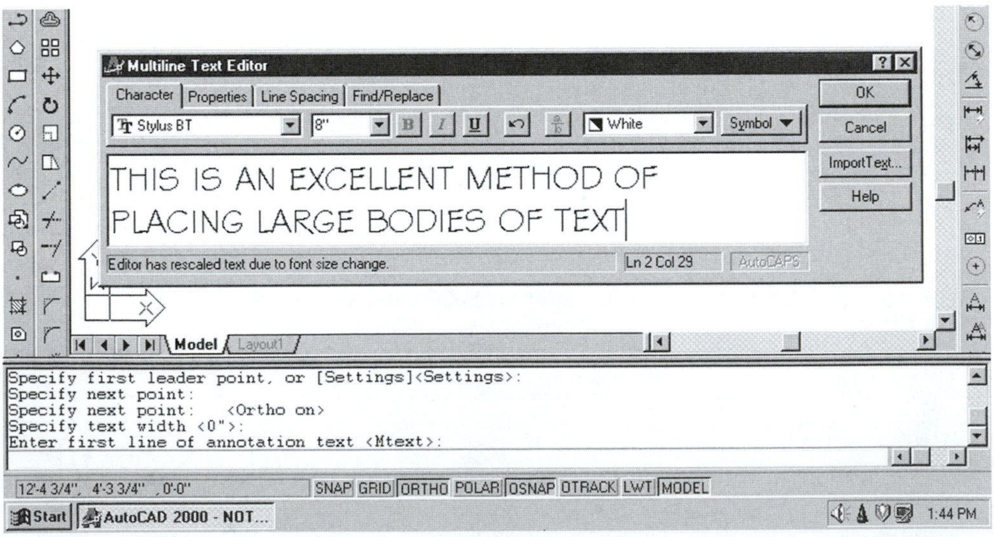

Figure 18.46 *Pressing (EnterKey) at the text prompt allows text, numbers, or symbols to be placed at the end of the leader line using the Multiline Text Editor edit box. Text entered by keyboard will be displayed in the edit box, and will not be placed by the leader line until the OK button is pressed.*

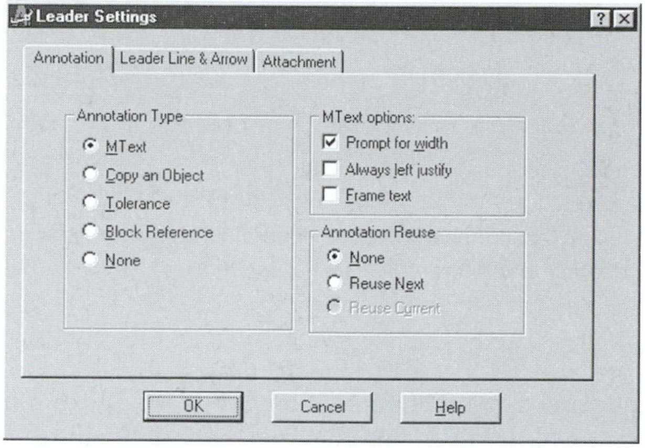

Figure 18.47 *Selecting the (EnterKey) at the first QLEADER prompt will produce the Leader Settings display.*

- The Copy option allows existing text to be copied to the new leader line. With this option active, once the leader line is placed, a prompt to select text to be copy is displayed. As the text body is selected, it will automatically be copied to the new location. This option could be used on a drawing such as the exterior elevations, where the same material is often referenced in each elevation. The COPY command can be used in the same manner.
- The Tolerance option will display the Tolerance dialog box and allow the creation of a feature control frame. These are tolerance symbols that are used to control the accuracy of a dimension in relation to a specific surface or condition. Tolerances are rarely added to architectural dimensions.
- The Block Reference option allows predefined groups of text to be inserted by the leader line. Blocks will be introduced in future chapters.
- The None option will place a leader with no text. Uses for this option have been introduced earlier in this chapter. The benefit of using this option is that it quickly produces the leader line with no prompt for text and does not require you to remember if you press the (EscKey) or (EnterKey).

MTEXT Options

The Annotation tab also allows controls for Mtext options and Annotation reuse. The Mtext options box is only activated if MTEXT is the default annotation method. The Prompt for Width option can be used to delete the width prompt. If you're placing text in a drawing and don't care how long the lines of text are, delete the prompt. Most professionals pride themselves for their neatness. For most users, aligned text will add to a professional appearance. You've explored justification as MTEXT was introduced. Future options will explore methods of altering text to leader line relationship. An option that you should explore is the Frame text option. With this option active, text will be displayed as seen in Figure 18.48.

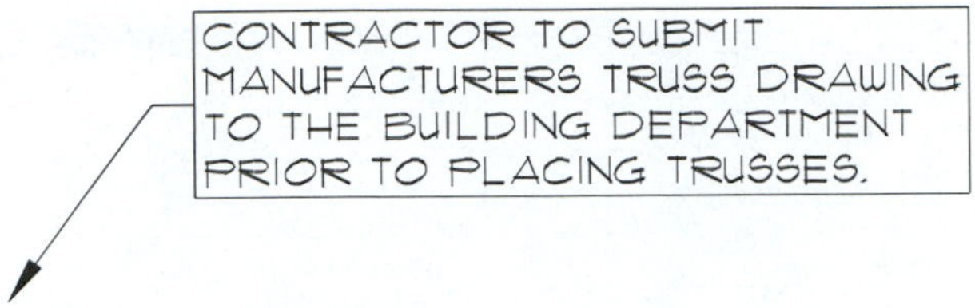

Figure 18.48 *The Frame text option will enclose leaders in a box. The option provides an excellent method of drawing attention to important notes.*

Annotation Reuse

This portion of the Annotation tab sets options for reusing text. The default option is not to reuse text at a leader line. The Reuse Next option will reuse the text that is created with the current command sequence for all future leaders. The current text will be placed by each leader until the setting is altered. The Reuse Current option will reuse current annotation. With this option active, AutoCAD will automatically select this option when you reuse text after selecting the Reuse Next option.

Leader Line and Arrow Tab

The Leader Line and Arrow tab contains options that can be used to control the appearance of the leader line and arrow. The tab can be seen in Figure 18.49.

Leader Line

Two options are available for placing the leader line. Throughout this chapter each example has been created using the Straight option. Figure 18.50 shows an example of a spline leader.

Number of Points

This option sets the number of leader points that you will be prompted to provide. With the No Limit option selected, prompts for points will be provided until (Enter-

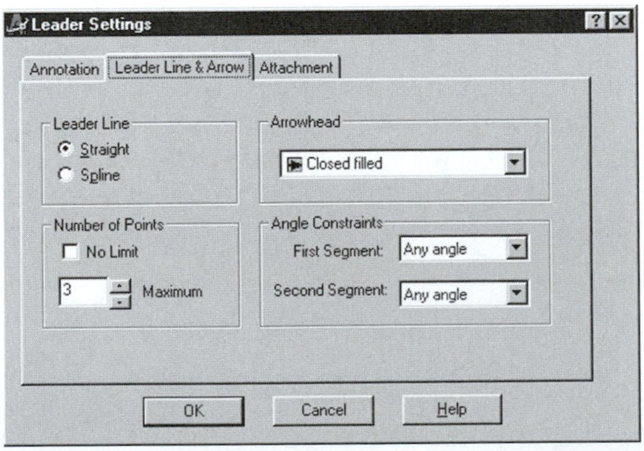

Figure 18.49 *The Leader Line and Arrow tab of the Leader Settings dialog box.*

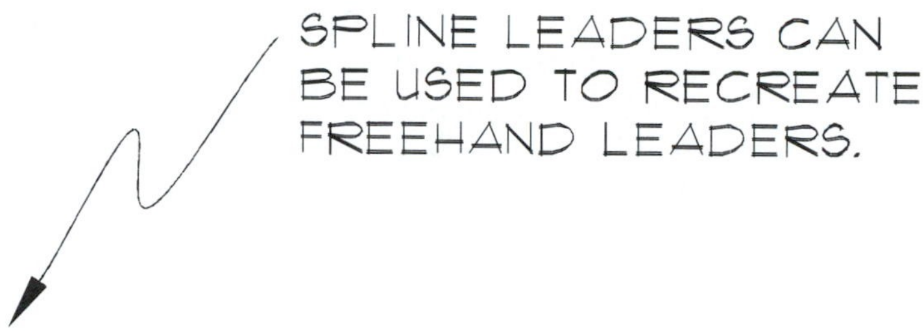

Figure 18.50 *A spline leader line.*

Key) is pressed twice. Setting a number for the maximum number of prompts will limit the number of prompts displayed as you proceed through the command. Set the value for one more than the number of leader lines you desire. In the default setting of 3, you will be prompted to place text after specifying the two lines formed by the three points.

Note: AutoCAD's option for adjusting the number of points is a powerful tool for creating spline leader lines.

Arrowhead

This portion of the tab allows the leader line terminator to be set. Figure 18.51 shows the options that are available. These options can be used to replace the filled arrow-

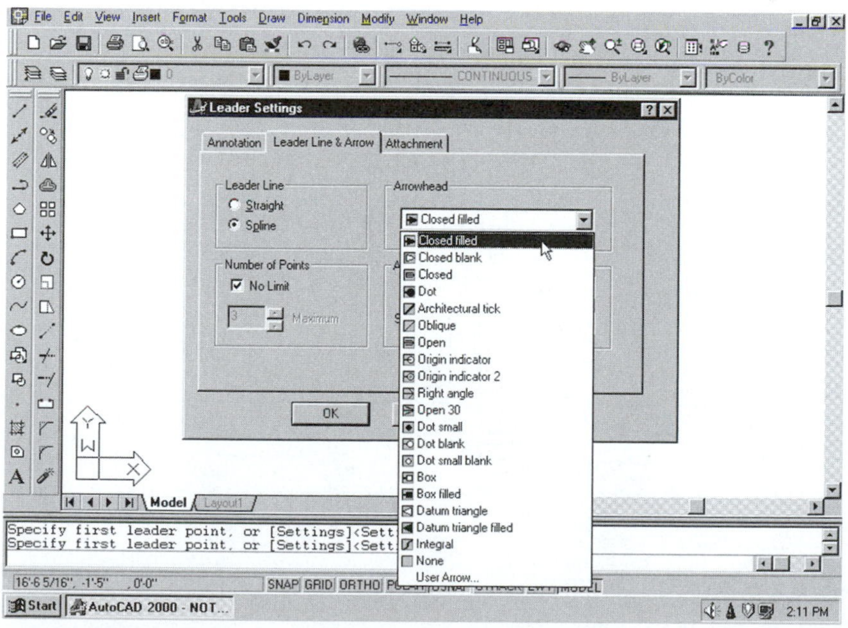

Figure 18.51 *Options that are available for leader line terminators.*

head. To replace the filled arrow, pick the desired terminator. As the selection is made, the menu will be closed and the Leader Line and Arrow tab will be displayed. Figure 18.52 shows a detail with right angle arrows. Notice that the last option is User Arrow.... This option will display the Select Custom Arrow Block prompt used to select predefined arrows that have been saved as a block. Chapter 20 will introduce the BLOCK command.

Angle Constraints

These options will allow the angle of each segment of the leader line to be controlled. Values for each option include Any Angle, Horizontal, 90°, 45°, 30° and 15°. Options

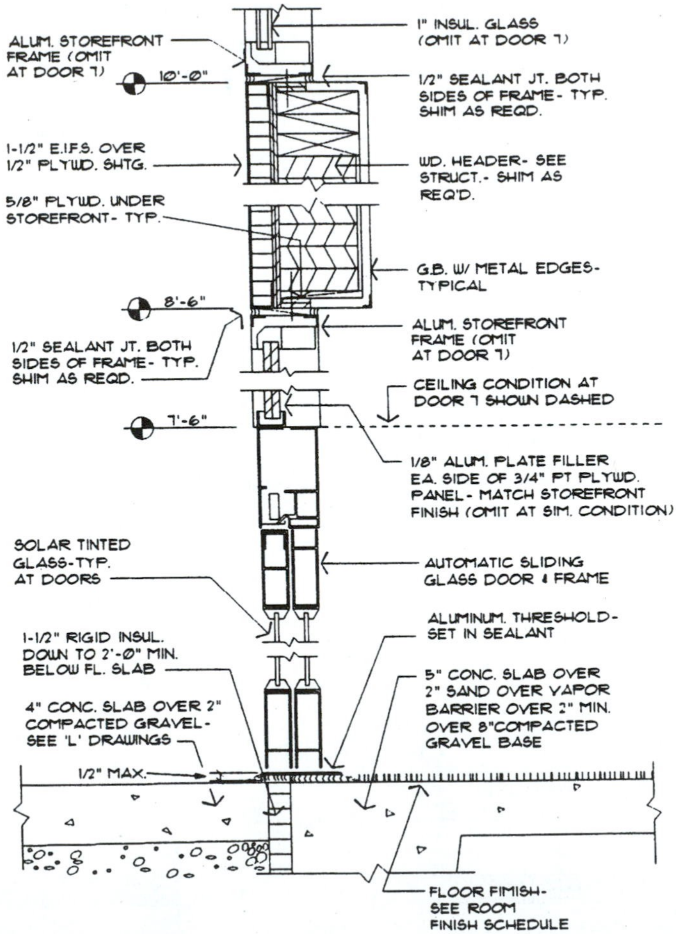

Figure 18.52 *Many architectural firms use right angle arrows. (Courtesy Barrentine, Bates and Lee, AIA.)*

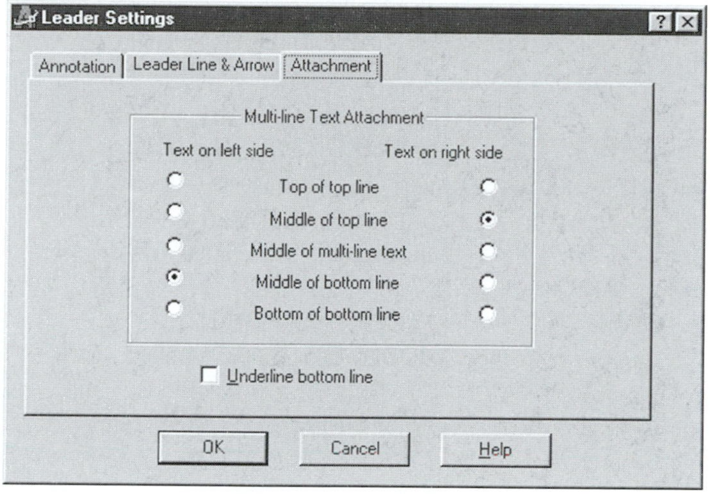

Figure 18.53 *The Attachment tab can be used to set the attachment of multiline text and the leader line.*

are provided for the first and second lines of a leader line. Using the Any Angle option for the first segment provides freedom in placing text around a drawing. Selecting an angle provides uniformity in appearance, but may make placing text more challenging. Specifying the Horizontal option for the second line will provide a horizontal tail without requiring ORTHO to be active.

Attachment Tab

The Attachment tab shown in Figure 18.53 can be used to set the attachment of multiline text and the leader line. Settings are available to control the leader to text relationship, depending on which side of the text the leader is placed. Options can be seen in Figure 18.54.

LEADER COMMAND

The LEADER command will produce a similar result as QLEADER, but it requires more effort. The command provides no dialog box as the command is executed. The command can only be accessed by keyboard. The command process is:

Command: **LEAD** (EnterKey)

LEADER

Specify leader start point: *(Select the starting point for the leader line.)*

Specify next point: *(Select the ending point for the leader line.)*

Specify next point or [Annotation/Format/Undo] <Annotation>: *(Select the ending point for the leader tail.)*

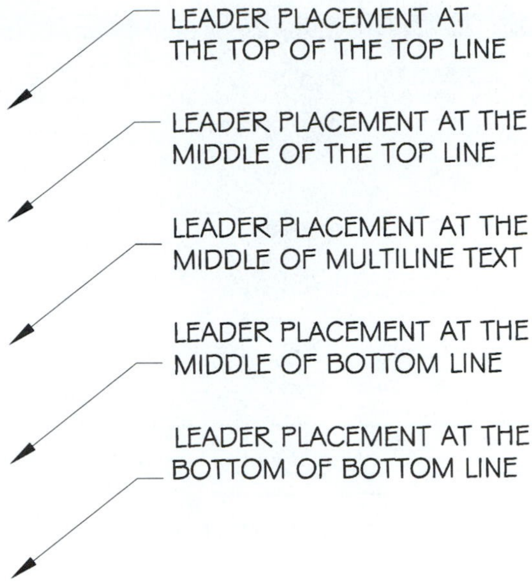

Figure 18.54 *Options for placing the leader line in relationship to the text.*

Specify next point or [Annotation/Format/Undo] <Annotation>:

The command is now waiting for you to enter the desired text or to adjust the options.

Enter first line of annotation text or < option>:

If (EnterKey) is pressed at the prompt, the prompt will be altered to display the Annotation options.

Enter an annotation option [Tolerance/Copy/Block/None/Mtext] <Mtext>:

Each option performs the same function as the QLEADER option.

Enter leader format option [Spline/STraight/Arrow/None] <Exit>

The Format option is accepted, the following prompt will be displayed. The Format option is used to determine the appearance of the leader and the arrow in a similar way that QLEADER options are controlled. Each option is selected by entering the specified letter at the prompt. Several of the options function as their counterpart in QLEADER. The None option allows a leader line to be drawn without using an arrow as a terminator. The Exit option exits the Format menu and returns to the base prompt. The Undo removes the last completed option from the command sequence. Undo functions like the Undo found in other AutoCAD commands.

CHAPTER 18 EXERCISES

1. Enter a new drawing and set the units to architectural with $\frac{1}{16}$" accuracy. Draw a hexagon with a horizontal top and base inscribed in a 4"-diameter circle. Use the DIMLIN command to provide overall dimensions. Save the drawing as E-18-1.
2. Open drawing E-18-1. Change the hexagon to a layer titled BASE. Change the dimensions to a layer titled DIMEN and turn the layer off. Create a layer titled ANGLE and make it current. Provide dimensions to indicate the angle of the sides. Save the drawing as E-18-2.
3. Enter drawing E-7-9 and dimension objects a, c, e, and g using two-place decimal dimensions. Place the remaining objects on a layer titled EXTRA, and freeze the layer. Once the dimensions are complete, freeze the existing dimensions and change the UNITS to architectural with $\frac{1}{16}$" accuracy. Dimension the objects again using alternatives to the methods already used. Save the drawing as E-18-3.
4. Open drawing E-10-3 and scale the drawing using a scale factor of .5. Create a layer for the dimensions. Completely dimension the drawing so the steel beam can be fabricated. Save the drawing as E-18-4.
5. Open drawing E-10-5 and create a layer titled DIM1. Rescale the drawing by a factor of .25 and completely dimension the drawing so that the metal hanger can be fabricated. Save the drawing as E-18-5.
6. Open drawing E-16-4 and create a layer titled DIM1. Change the lines representing the one-story footing to .5-wide polylines. Place the two-story footing on a new layer and set that layer to OFF. Rescale the drawing using a scale factor of .25. Completely dimension the one-story footing and save the drawing as E-18-6.
7. Open a new drawing and draw the following object. Starting @ 2,2:

   B. @1.5,0
   C. @ 0,1
   D. @1.25,0
   E. @ .75,.75
   F. @ 1.5,0
   G. @ 0,.75
   H. @ 1.5,0
   J. @ 0,1.75
   K. @ –2,0
   L. @ 0,–.75
   M. @ –2,0
   N. @ 0,–1.25
   P. @ –2.5,0
   Q. @ .5,–1.5
   R. AND THEN CLOSE.

   Change the line segments to polylines. Completely dimension the drawing using the Continue option where possible. Save the drawing as E-18-7.

8. Open drawing E-18-7 and create a new layer for the existing dimensions. Change the dimensions to the new layer that was just created, and set it to OFF. Create another layer for dimensions and completely dimension the object using Baseline dimensioning. Assume the lower left corner as the starting point for Horizontal and Vertical dimensions.
9. Open the architectural template, and draw an object similar to Figure 18-35. Create a layer for placing dimensions. Dimension every surface using the Continue option of QDIM. Save the drawings as E-18-9.
10. Open drawing E-18-9 and create a new layer for dimensions. Dimension each surface using the Baseline option of QDIM. Save the drawing as E-18-10.
11. Open drawing E-18-10 and create another layer for dimensions. Dimension each surface using the Staggered option of QDIM. Save the drawing as E-18-11.
12. Use Figure 15.15 as a guide and draw a detail for a concrete block wall. Use varied lineweight and colors to represent the CMUs, steel, and mortar. Create separate layers for CMU, steel, mortar, dimensions, text, and detail reference bubble. Do not place any dimension on the drawing. Place leaders so that text will be placed using an architectural font. Assume that the drawing will be plotted at a scale of ¾" = 1'-0". Save the drawing as E-18-12.
13. Use Figure 17-5 as a guide and draw a detail for a roof-to-wall connection. Use varied lineweight and colors to represent materials that have been cut by the cutting plane. Create separate layers for drawing, dimensions, and text. Do not place any dimensions on the drawing. Place leaders so that text will be placed using an architectural font in 4.5" wide columns. Use right angle arrows. Assume that the drawing will be plotted at a scale of ½" = 1'-0". Save the drawing as E-18-13.
14. Use Figure 12-1 as a guide and draw detail 5/S-7. Use varied lineweight, colors, and layers to represent the concrete, steel, dimensions, text, and detail reference bubble. Do not place any dimension on the drawing. Place leaders so that text will be placed using an architectural font. Use right angle arrows. Assume that the drawing will be plotted at a scale of 1" = 1'-0". Save the drawing as E-18-14.
15. Use Figure 12.2 as a guide and draw a detail of a truss-to-wall connection. Use varied lineweight, colors and layers to represent the concrete, steel, dimensions, text, and detail reference bubble. Do not place any dimension on the drawing. Place leaders so that text will be placed using an architectural font. Set the angle control for the first leader line segment to 90° and use a right angle arrow. Assume that the drawing will be plotted at a scale of 1" = 1'-0". Save the drawing as E-18-15.

CHAPTER 18 QUIZ

1. What are the four components of dimensioning?
2. How should text be placed on architectural drawings to ease the job of the print reader?
3. List four options typically used for the intersections of dimension and extension lines.
4. Describe how to place dimensions relative to a concrete wall.
5. List four groups of commands that relate to placing dimensions on a drawing.
6. Describe two methods of locating extension lines for linear dimensions.
7. List four methods to describe angular dimensions.
8. Enter one of the drawing exercises that was completed for this chapter, and provide the name of the layer that was added to the drawing base by AutoCAD.
9. List three methods of placing linear dimensions.
10. Explain the difference between Aligned and Rotated dimensions.
11. What is the first prompt for DIMLINEAR asking for?
12. Should the default text value always be accepted? Explain your answer.
13. What does the angle option alter in the DIMLINEAR command?
14. What is the effect of pressing the (EnterKey) at the first DIMLIN prompt rather than selecting a point?
15. Should a dimension be placed inside or outside of a circular object?
16. What is the process for placing a 6" center mark in the center of a circular steel tube?
17. List and explain the options for LEADER annotation.
18. Four lines of text need to be placed by a leader line. What option should be used?
19. Explain the options for ORDINATE dimensioning.
20. List five options of leader lines excluding the text that can be altered.

CHAPTER 19

Placing Dimensions on Drawings

In Chapter 18 you were introduced to dimension requirements of different drawings and the methods of describing linear, angular, circular, continuous, and baseline dimensions. Each system was described using the default values. If all of your drawings could fit on a computer screen, everything would be fine. To allow for drawings of various sizes, the spacing of extension and dimension lines and the size of text and arrows must be altered. In this chapter you'll learn how to alter dimension variables, establish dimension styles, and edit existing dimensions.

CONTROLLING DIMENSIONS VARIABLES WITH STYLES

As text was explained, you were introduced to the STYLE command. STYLE allows you to group variables, such as the font, text height, width, obliquing angle, and orientation, so that several different styles of text can easily be created within one drawing. AutoCAD also allows dimension styles to be created within a drawing to meet the needs of various situations. Dimensioning styles are created using the DDIM command. To save time, individual styles, as well as each dimensioning variable, should be set on template drawings on the ANNO-DIMS layer.

INTRODUCTION TO DIMENSIONING STYLES

A dimension style is a set of dimension variables that control various aspects of placing the extension lines, dimension lines, and annotation in relation to these lines and line terminators. The type of drawing you are working on will affect how the variables are to be set. Dimensions that are placed on a site plan are often written in engineering units. Dimensions that locate the property lines are placed with no extension or dimension lines. Dimensions placed to locate utilities in an easement are often dimensioned using baseline dimensions expressed in feet and inches. Dimensions written on architectural drawings are written in feet and inches, using continuous dimensions. Dimensions on elevations and sections can be placed using ordinate dimensioning. You can create styles that incorporate the requirements for each of these dimensions and save them in a template drawing. You will be introduced to creating various styles later in this chapter.

Before a style can be created, you must have a thorough understanding of the dimension variables. Dimension variables are the qualities that control how dimensions will appear. You will be introduced to dimensioning variables through the default style of STANDARD. Later, you'll be introduced to methods of creating specific styles to meet the needs of each drawing type.

CONTROLLING DIMENSION VARIABLES

Dimension variables are altered using the Dimension Style Manager dialog box. The Dimension Styles dialog box displayed in Figure 19.1 is accessed by selecting Dimension Style from the Dimension toolbar, by selecting Dimension Style... from the Format pull-down menu, by selecting Style... from the Dimension pull-down menu, or by typing **D** (EnterKey) at the command prompt.

DIMENSION STYLE DIALOG BOX

The main areas of this box include the Dimension Style box, a Preview box, and buttons for setting the Current, New, Modify, Override, and Compare.

Current Dimstyle

Notice that in Figure 19.1 the current style is named Standard. Standard is the name of the default settings for AutoCAD. As styles are created, the name of each style will

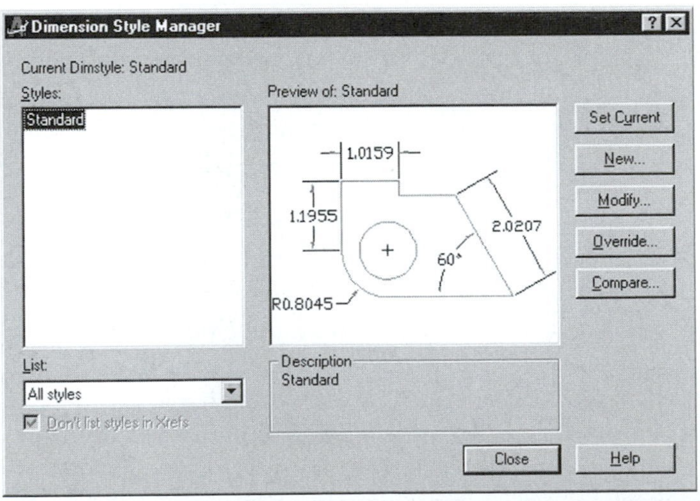

Figure 19.1 *Dimension variables are created and altered using the Dimension Style Manager dialog box. The box is accessed by selecting Dimension Style from the Dimension toolbar, by selecting Dimension Style... from the Format pull-down menu, by selecting Style... from the Dimension pull-down menu, or by typing **D** (EnterKey) at the command prompt.*

be listed in the Styles box. The current style will be highlighted. Once multiple styles have been created, a different style can be made current by highlighting the name of the desired style and then selecting the current button.

List

The list box displays the options that control the display of dimension styles. Options include All Styles and Styles in Use. With All Styles current, all of the dimension styles contained in the drawing will be displayed. With Styles in Use current, the display will only display dimension styles that are referenced by dimensions in the drawing.

Just below the List display is the option of displaying dimension styles in externally referenced drawings. XREF drawings will be discussed in future chapters.

Control Buttons

Five buttons are available for creating and altering the dimension style. Each will be introduced now and explored in detail throughout the chapter.

Set Current

This option has already been briefly introduced. Once created, a style can be set to the current style by highlighting the desired style and selecting the Set Current button. As the current style, each of the settings of the drawing style will be applied to the new dimensions.

New...

Selecting the New... button will display a Create New Dimension Style dialog box similar to Figure 19.2. The box is used to name new dimension styles, to select the style that will be used as a base to build the new style, and to indicate the dimension types that the new style will be applied to. Components include:

- New Style Name—By default, AutoCAD will assign Copy of Standard as the new style name. With the name highlighted, change the style name to provide a useful description.
- Start With—This portion of the dialog box determines what will be used as the base to form the new style. Rather than starting from scratch, AutoCAD will use the Standard style as a base and allow these options to be adjusted to meet the drawing needs. As new styles are created, each style name will be listed, allowing it to be selected to be the base for a new style.
- Use For—Selecting the Use For edit arrow will display the menu shown in Figure 19.3. The menu allows a style to be created that will only be applied to certain types of dimensions. This option could be used if you would like to use the Standard style with black text as a base, but allow all radius dimensions to be red. Selecting Standard as the Start With option and Radius Dimensions as the Use For option will begin the process of creating a sub-style of the Standard style.

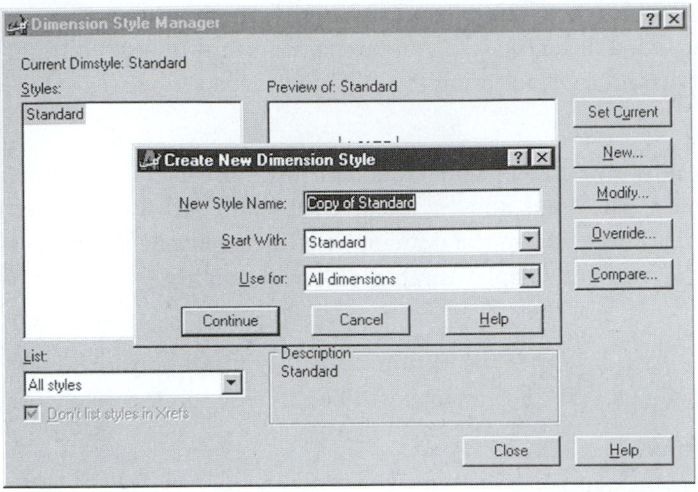

Figure 19.2 *Selecting the New... button will display a Create New Dimension Style dialog box.*

With these options active, each time the Standard style is used, radius dimensions will be red, and all other dimensions will be black. Selecting the Continue button will display the New Dimension Style dialog box. The display will be explored in the next section of this chapter.

Modify...

Selecting the Modify button from the Dimension Style Manager dialog box allows an existing style to be modified. As the Modify button is selected, the Dimension

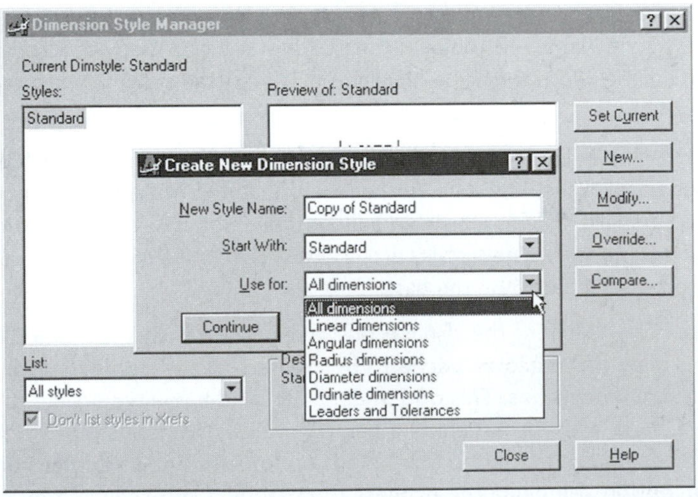

Figure 19.3 *Selecting the Use For edit arrow will display a menu of dimension types that the style will be applied to.*

Style Manager dialog box is removed and the Modify Dimension Style dialog box is displayed. The Modify dialog box is identical to the New Dimension dialog box. Methods of altering an existing style will be explored later in this chapter.

Override...

Selecting the Override button will produce the Override Dimension Style dialog box. The box is identical to the New Dimension Style dialog box. This option allows a temporary override to a dimension style. Methods of overriding an existing style will be explored later in this chapter.

Compare...

Selecting this option displays a Compare Dimension Styles dialog box similar to Figure 19.4. The contents of the display will depend on the styles being compared. This box can be very helpful in comparing the differences between two styles, or to display all of the properties of one style.

CREATING A NEW STYLE

You've just explored the Dimension Style Manager, and the Create New Dimension Style dialog boxes. Selecting the New... button of the Dimension Style Manager dialog box, and the Continue button of the Create New Dimension Style dialog box, will display a New Dimension Style dialog box similar to Figure 19.5. The exact title will vary depending on the name of the style being created. The dialog box contains six tabs for controlling the dimension style properties.

LINES AND ARROWS TAB

The Lines and Arrow tab contains five major areas for adjusting the lines and arrowheads to be used with the new style. These areas are the Display box, Dimension

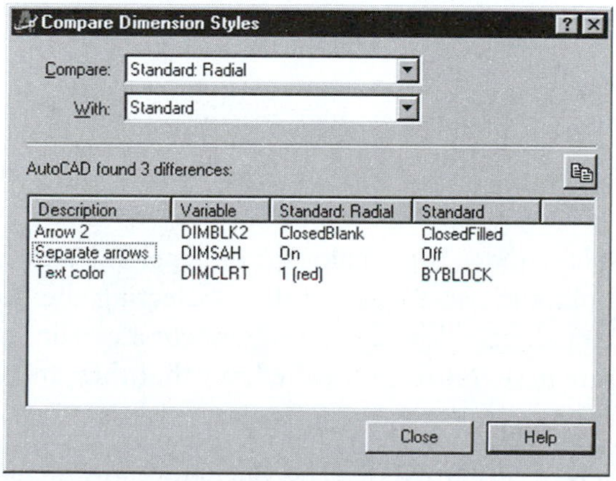

Figure 19.4 *The Compare Dimension Styles dialog box can be used to compare the differences between two styles, or to display all of the properties of one style.*

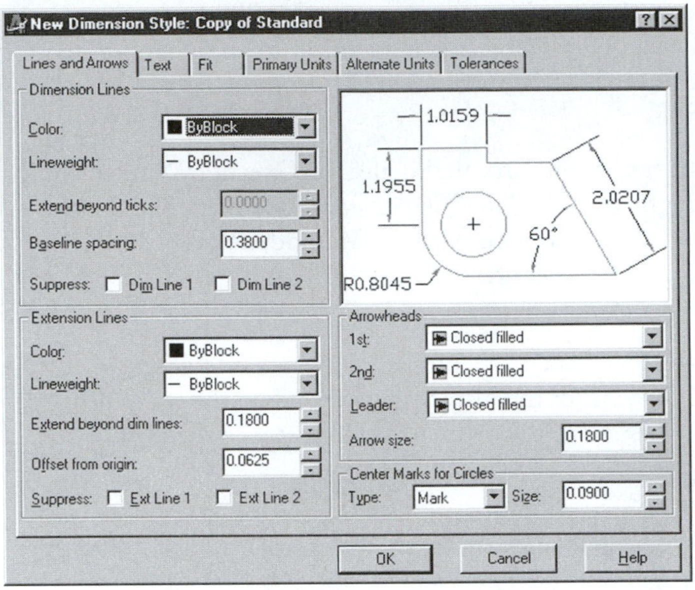

Figure 19.5 *Selecting the New... button of the Dimension Style Manager dialog box, and the Continue button of the Create New Dimension Style dialog box, will display a New Dimension Style dialog box and allow the properties of a dimension style to be set.*

Lines, Extension Lines, Arrowheads, Center Marks, and the Display tile. A display box is in each tab and will demonstrate the effect of the current tab settings.

Display Box

The Display box presents a visual representation of how the dimension properties will be displayed in the drawing area, using the current settings. Each time a property is modified in the Line and Arrows tab, the display will be updated to reflect the new property setting.

Dimension Lines

The Dimension line box contains settings to control Color, Lineweight, Extend Beyond Ticks, Baseline Spacing, and Suppress.

Color

The color menu displays and sets the color for dimension leader lines. Selecting the Color edit arrow displays the seven basic colors. Selecting the Other... option displays the Select Color palette. Although a different color can be set for each aspect of a dimension, colors are best controlled by the layer that they are to be placed on.

Lineweight

Selecting the Lineweight edit arrow displays the lineweight menu allowing the lineweight of dimension lines to be set.

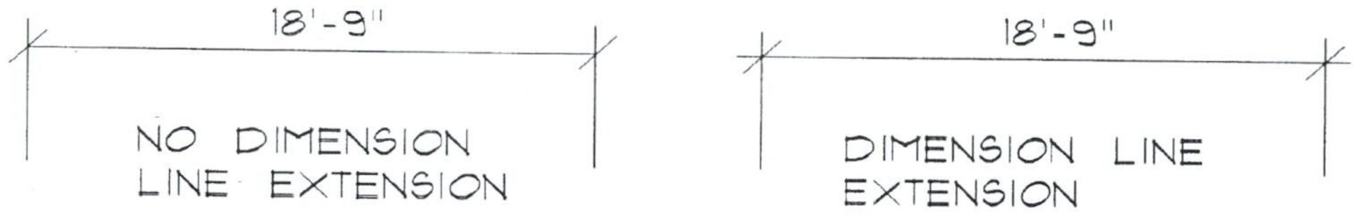

Figure 19.6 *The default placement of dimension line extension on the left is often altered so that the dimension line extends past the extension line, to simulate manual drafting methods.*

Extend Beyond Ticks

By default, this option is inactive. It will become active once the arrowhead is altered to be either an Oblique, Architectural Tick, Integral, or None. This option controls how far the dimension line will extend past the extension line. Figure 19.6 shows examples of adjusting the extension. Most architectural offices set the extension value to equal the height of the text to be used.

Baseline Spacing

Selecting the Baseline Spacing edit box allows the distance between lines of dimensions to be set. AutoCAD refers to this as the baseline setting. The default setting for architectural units is 0'-0 3/8". Most professionals use a distance of 1/2". Figure 19.7 shows an example of the baseline spacing.

Suppress

This option works as a toggle, allowing parts of the dimension line to be omitted. The Dim Line1 and Dim Line2 boxes allow the first or second portion of the dimension line to be omitted from the drawing. Figure 19.8 shows an example of the first or second dimension line suppressed. The first or second dimension line is relative to the selection of the first and second extension line. These options are generally not altered for construction drawings.

Extension Lines

This box has features similar to those in the Dimension Line box, except these features control each aspect of the extension line in relation to the object being dimensioned.

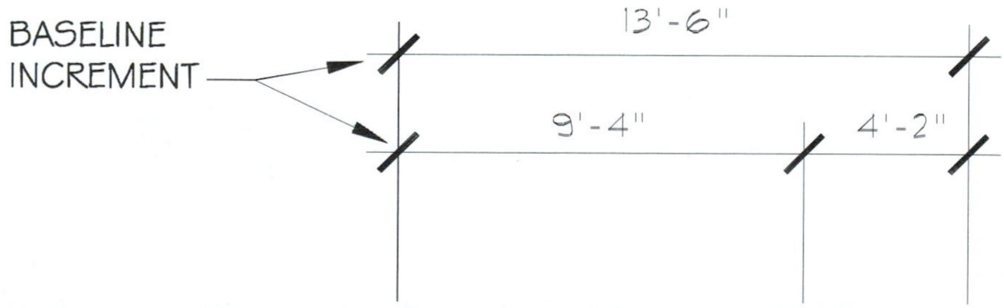

Figure 19.7 *The baseline increment controls the spacing between stacked lines of dimensions.*

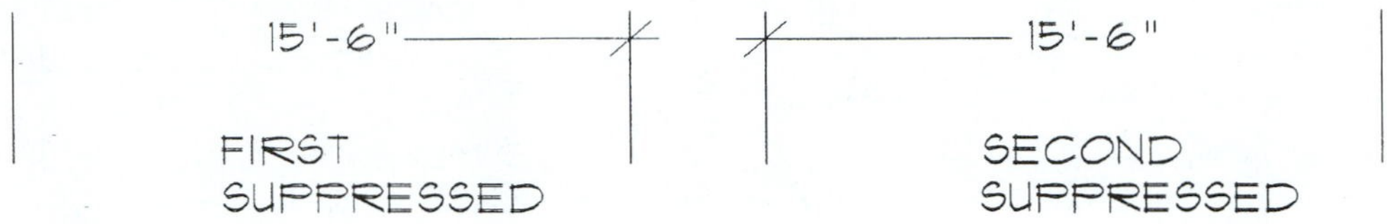

Figure 19.8 *Suppression of dimension lines is generally not required in architectural drafting.*

Setting Extension Line Color

This option can be used to set the color of the extension lines. Selecting the Color menu edit arrow displays the color palette. Although a different color can be set for extension lines, colors are best controlled by the layer that they are to be placed on.

Lineweight

Selecting the Lineweight edit arrow displays the lineweight menu, allowing the lineweight of extension lines to be set.

Extend Beyond Dim Line

The Extension: option allows the distance that the extension line extends past the dimension line to be controlled. The default value is 3/16". An extension offset of 1/8" is typical on most professional drawings. The effect of setting Extension: can be seen in Figure 19.9.

Offset From Origin

The Offset from Origin option is used to control the gap between the extension lines and the object being dimensioned. The default value is 1/16", but a space of 1/8" is common in most offices. The effect of setting the Extension line Offset is shown in Figure 19.10.

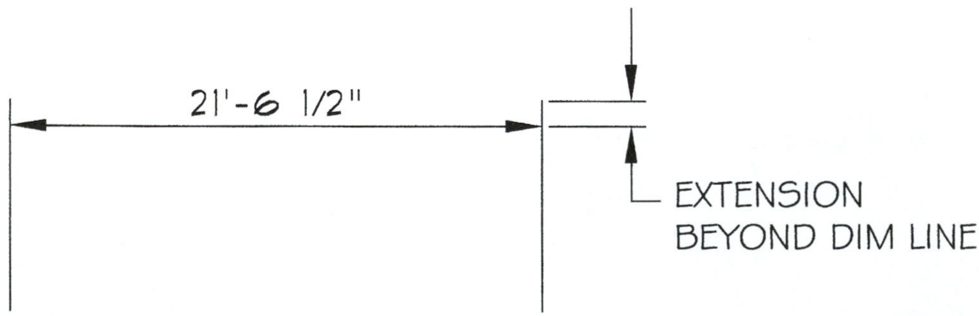

Figure 19.9 *The extension line is usually set to extend above the dimension line by a distance equal to the text height.*

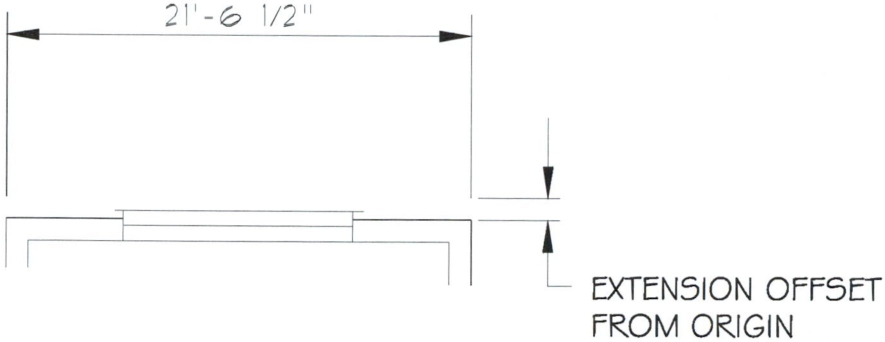

Figure 19.10 *The extension line should be offset from the object being described. A space of 1/8" is typical.*

Suppressing Extension Lines

This option allows for the suppression of the first and second extension lines, as shown in Figure 19.11. By default, extension lines will be displayed at each end of the dimension line. This option works well for establishing overall sizes. The first extension line can be omitted by selecting the Ext Line1 box. With this option active, as the location for extension lines is selected, the first location will not receive an extension line. Selecting both boxes suppresses the extension lines at each end of the dimension line. This is useful if a dimension line is to be placed between two existing extension lines. This option also works well when the object is being used as the extension lines. This is typically done when dimensioning between floor and ceiling levels on sections or elevations. Figure 19.12 shows examples of suppressing extension lines.

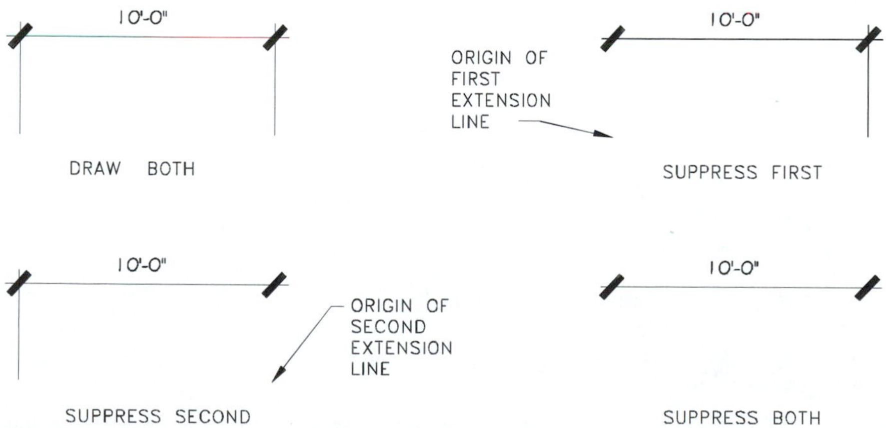

Figure 19.11 *Suppression of extension lines.*

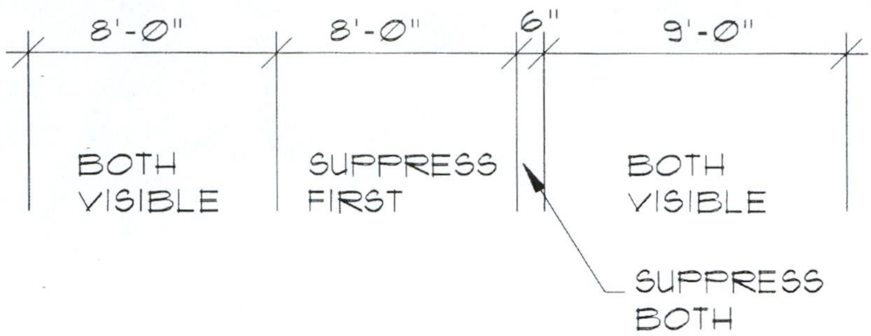

Figure 19.12 *Practical applications for supressing extension lines.*

Arrowheads

This box controls the type of dimension line terminator to be used. The current selection for the first and second terminator, the closed filled arrow, is displayed in the edit box. By default, AutoCAD uses the terminator selected for the first dimension line used for the second line. The type of terminator can be selected by selecting the icon or the arrow on the right side of the first list box. Selecting the icon or the down arrow displays the menu. Figure 19.13 shows a sample of the terminator

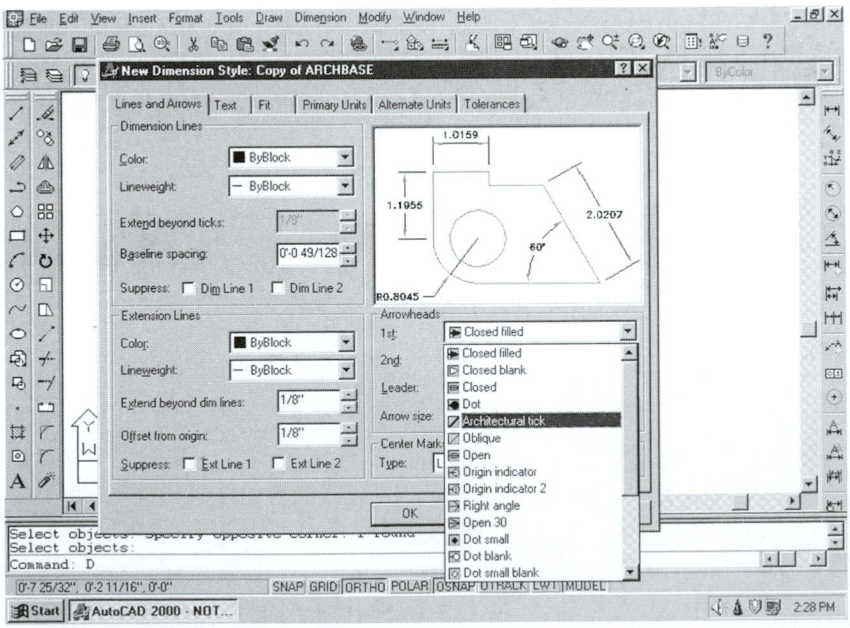

Figure 19.13 *Selecting the icon or the down arrow displays the terminator menu. The balance of the menu can be seen by using the down arrow.*

menu. Making the selection of Architectural Tick automatically changes the display of each arrow to tick marks. You can make the second terminator different from the first by using the second menu option. Rarely is this done on professional drawings, and you should not decide to start a new trend! Examples of the variations of dimension line terminators can be seen in Figure19.14.

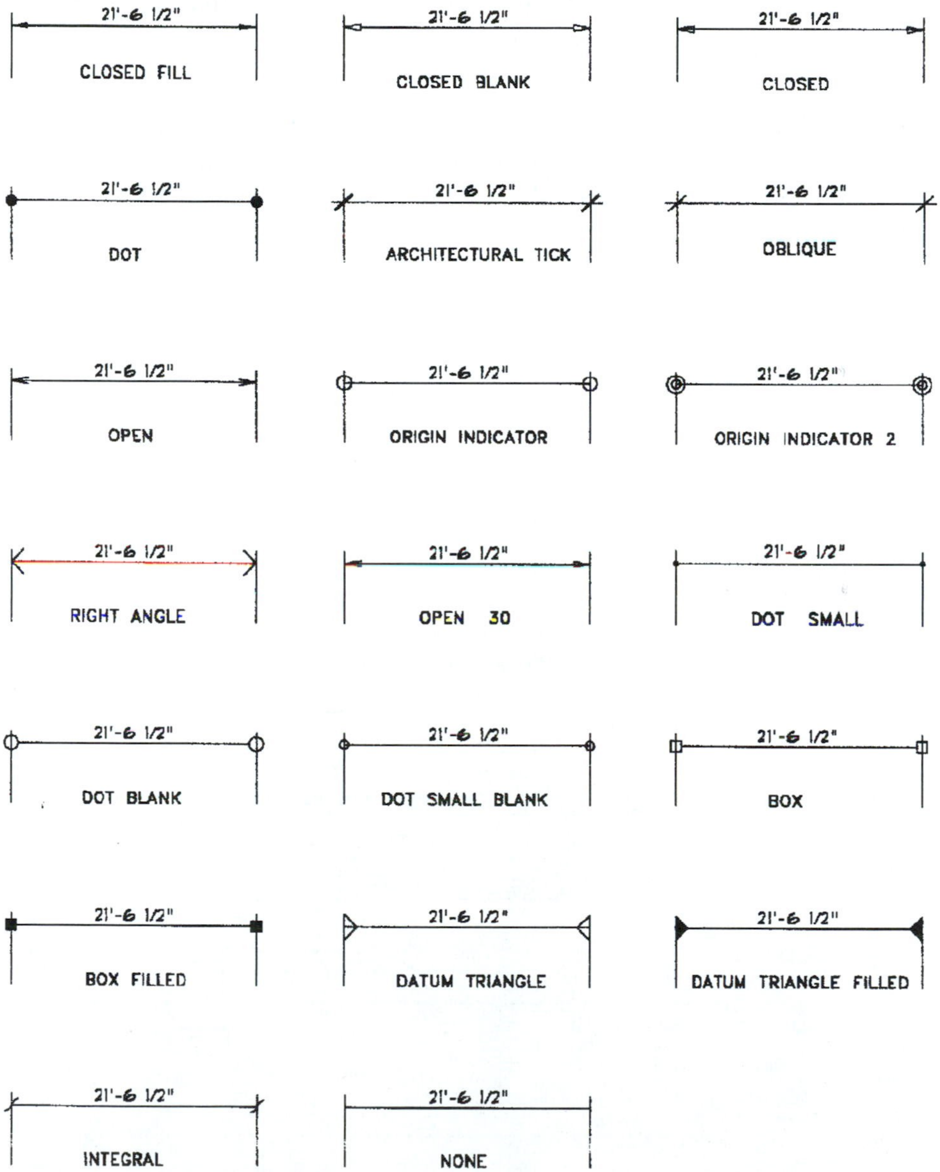

Figure 19.14 *Variations of dimension line terminators available in AutoCAD.*

User Arrow...

Most architectural and engineering offices use the Oblique, Architectural Tick, or User Arrow for a terminator. The user-defined tick can be created using the BLOCK command. BLOCK will be discussed in the Chapter 20. Once the desired block is drawn, it can be named and saved to the drawing base. Selecting User Arrow... displays the dialog box shown in Figure 19.15. The name that was assigned to the block can be assigned in the Select Custom Arrow Blocks list box. The icon tile will now display the name of the block rather than showing an icon.

Arrow Size

The Arrow Size box can be used to determine the size of the terminator, relative to other dimensioning features. Selecting a tick value that is the same height as the text is a safe guideline to follow. If 3/8"-high text is to be used, change the value to .375.

Center Marks

The Center Marks box is used to control how and when the center mark for radial dimensions will be displayed. The center mark is only drawn if the dimensions for DIMCENTER, DIMRADIUS, or DIMDIAMETER are placed outside the circle or arc being described. Figure 19.16 shows the options for placement.

Displaying the Center Mark

The default setting of Mark displays the center mark, using the center Size value. Selecting the Line option displays centerlines that project beyond the limits of the

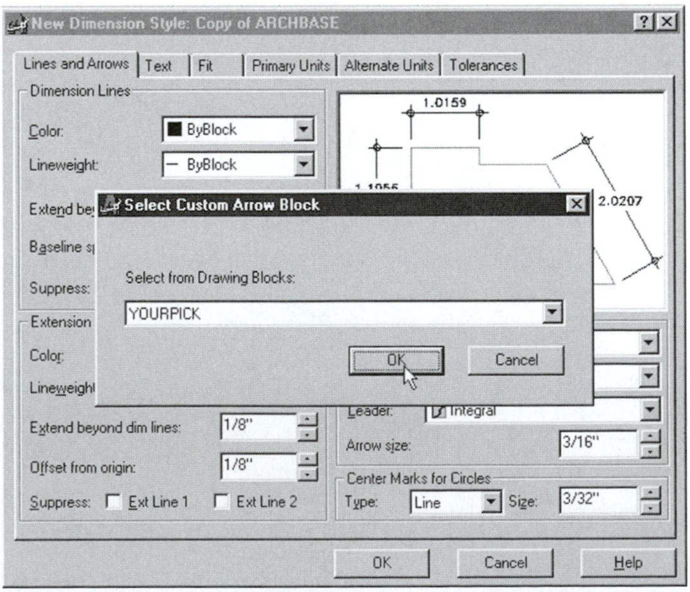

Figure 19.15 *A user-defined arrow can be created as a BLOCK and added into the drawing.*

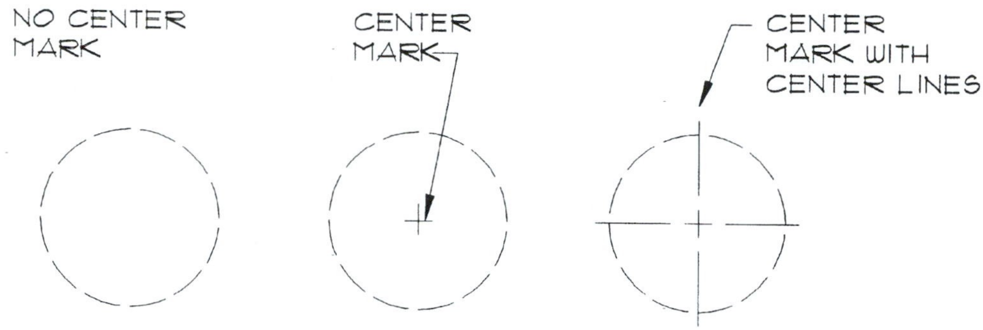

Figure 19.16 *Options for displaying center marks.*

circle. Selecting the None radio button will display no markings for the centerpoint of a circle or arc. The Size edit box can be used to control the size of the center mark. Many professionals adjust the default setting of 1⁄16" to 1⁄8".

TEXT TAB

The Text tab is used to set the appearance, placement, and alignment of the text that is used with dimensions. Each of the four major areas of the tab can be seen in Figure 19.17.

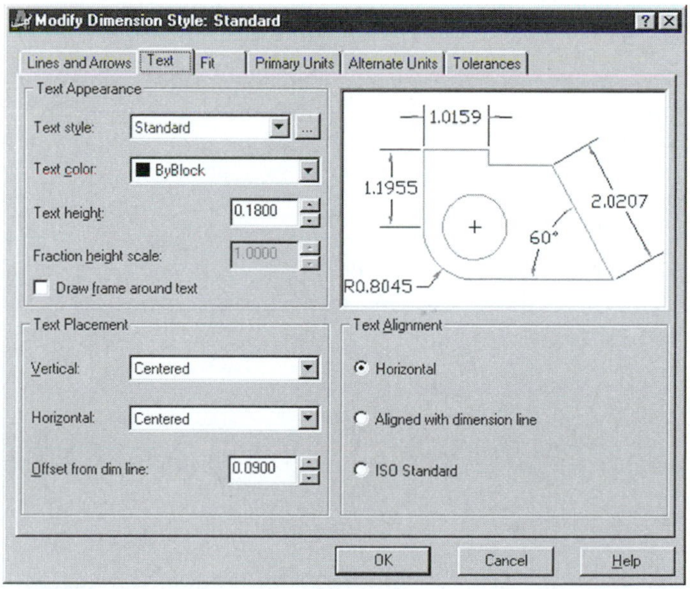

Figure 19.17 *The Text tab is used to set the appearance, placement, and alignment of the text that is used with dimensions.*

Text Appearance

The Text Appearance portion of the Text tab contains controls for the dimension text format and size. Controls include Text Style, Text Color, Text Height, Fraction Height Scale, and Frame.

Text Style

The Text Styles edit box displays the name of the current text style. It also allows one of the defined text styles to be used for dimension text. Selecting the ellipsis icon beside the edit box displays the Text Style edit box. The Text Style edit box allows a new text style to be created.

Text Color

The color menu displays and sets the color for dimension text. Selecting the Color edit arrow displays the seven basic colors. Selecting the Other… option displays the Select Color palette.

Text Height

The Text Height box displays the current text height and allows the height of the dimension text to be altered. The height entered here must be coordinated with the height that is entered in the Text Style box. If a text height is entered in the Text Styles dialog box, that height will override the text entered in the Text tab. For the value of the Text tab to be active, a text value of 0 must be entered in the Text Style dialog box.

Scale Factor for Fraction Height

This option sets the scale of fractions relative to the dimension text. In its current state, the option is inactive. For this option to be active, the Precision must be set on the Primary Units tab. Once a precision has be set, the scale factor can be altered. With the default value of 1.0000, the height of fractions will be the same as the dimension text height.

Framing Text Height

This option is currently inactive. Selecting this option will place a box around the text. With the current settings, selecting the box will do nothing. Once values are altered in the Text placement area, selecting this option box will display a box around a dimension, similar to Figure 19.18.

Text Placement

The Text Placement area of the Text tab controls the placement of the dimension text. Options include Vertical, Horizontal, and Offset.

Vertical

This option controls the vertical justification of dimension text, relative to the dimension line. Options include Centered, Above Outside, and JIS. Figure 19.19

Figure 19.18 *Once values are altered in the Text placement area, selecting the Draw Frame Around Text option box will display a box around a dimensions. This option can be used to draw attention to a dimension or to distinguish between new and existing construction.*

shows the options that relate to construction drawings. Neither the outside or the JIS (Japanese Industrial Standards) are used on construction drawings.

- Centered—This option breaks the dimension lines so that text is centered between the dimension lines.
- Above—This option is often used for architectural and engineering drawings. The option places the dimension text above the dimension line. The distance from the dimension line to the bottom of text is controlled by the Offset from Dim Line option.
- Outside—The Outside option places the dimension text on the side of the dimension farthest away from the defining point.
- JIS—This option places dimension text to conform to Japanese Industrial Standards (JIS) representation.

Horizontal

The Horizontal Justification box allows the location of text along the dimension line to be set. The box contains an icon and an edit box that can be used to control where the text is placed. Selecting the down arrow allows the options to be altered. As the

Figure 19.19 *Two common settings of the Vertical Justification menu.*

arrow is selected, the Horizontal menu is displayed. Selecting an option makes it current and removes the menu. Options include:

- Centered—The default value places text centered along the dimension line.
- At Ext Line 1—This option places the text close to the first extension line.
- At Ext Line 2—This option places the text close to the second extension line.
- Over Ext Line 1—This option places the text parallel and above the first extension line.
- Over Ext Line 2—This option places the text parallel and above the second extension line.

Examples of each option can be seen in Figure 19.20.

Offset from the Dimension Line

The Offset from Dim Line edit box displays the current gap between the text and the dimension line. The setting also allows the gap between the dimension line and the text to be altered. In construction drawings the text is placed above the dimension line. The gap value should be set to equal half of the text height value.

Text Alignment

The Text Alignment area of the Text tab can be used to control the orientation of dimension text to the dimension line. Choosing one of the inactive options alters

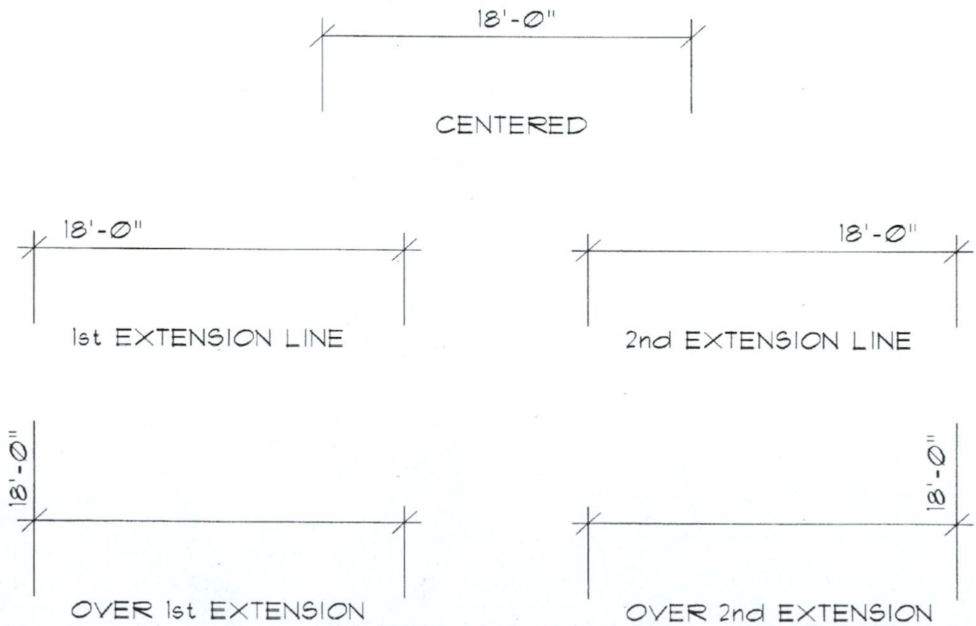

Figure 19.20 *Placement of the dimension text using the Horizontal Justification sub-menu.*

the selection. Options include the default option of Horizontal, Aligned with dimension line, and ISO Standard. Figure 19.21 shows an example of each type of text placement.

Horizontal
With this box active, dimension text is placed horizontal, regardless of the angle of the dimension line. The option should be set to OFF for most construction drawings so that the text remains parallel to the dimension line.

Aligned with Dimension Line
This option is most typically associated with construction drawings. Picking this option will align dimension text to be parallel with the dimension line.

ISO Standard
This option aligns text with the dimension line if the space between the extension lines is wide enough to display the text. Text will be placed in a horizontal position outside of the extension lines if the text will not fit between the extension lines.

FIT TAB

The Fit tab provides controls for the placement of text in relationship to the dimension line. The Fit tab can be used to set the values that control the placement of text, arrows, and leaders for dimension placement. Major areas include Fit Options, Text Placement, Scale for Dimension Features, and Fine Tuning. Figure 19.22 shows the Fit tab of the Dimension Style dialog box. Figure 19.23 shows common text placements that might be seen on a floor plan.

Fit Options

This portion of the Fit tab controls the placement of text and arrowheads if there is not enough space to place both inside of the extension lines. If you are using Architectural Ticks, you'll have more space between the extension lines than if you use arrows. Figure 19.24 shows examples of placement options.

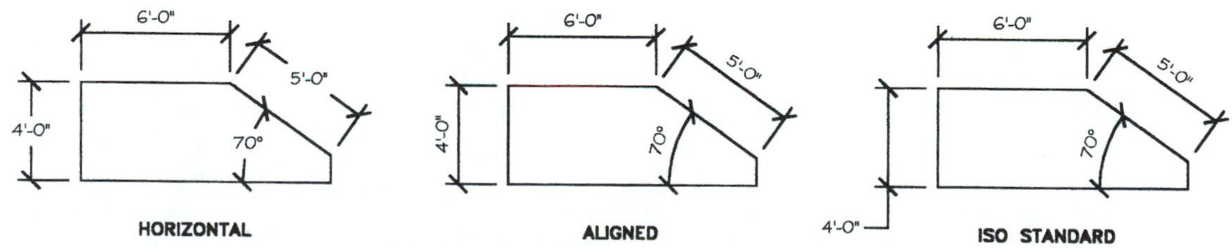

Figure 19.21 *The Text Alignment area of the Text tab can be used to control the orientation of dimension text to the dimension line. Most construction drawings use aligned dimensions.*

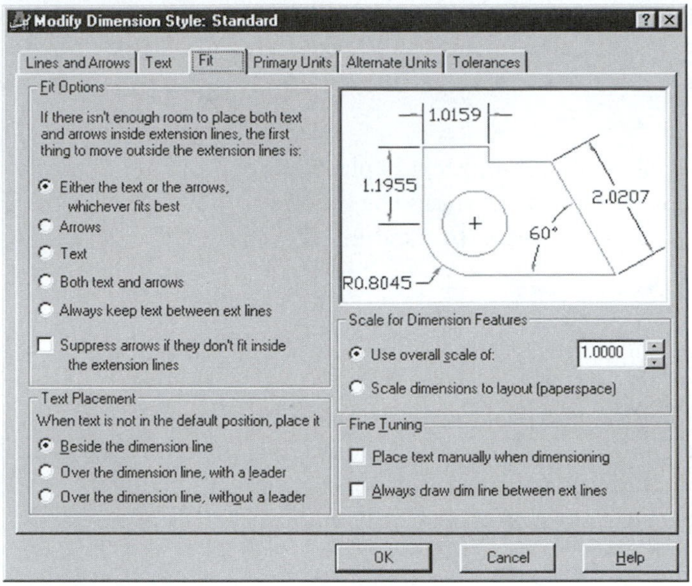

Figure 19.22 *The Fit tab of the Dimension Style dialog box.*

Either the Text or the Arrows, Whichever Fits Best

Selecting the default option will display the arrows and the text inside the extension lines, when space is available. If sufficient space is not available between the extension lines, the text is placed between the extension lines and the arrows are placed outside the extension lines. If no space is available for text, only the arrows are placed between the extension lines. If space is not available for either text or arrows, both will be placed outside the extension lines.

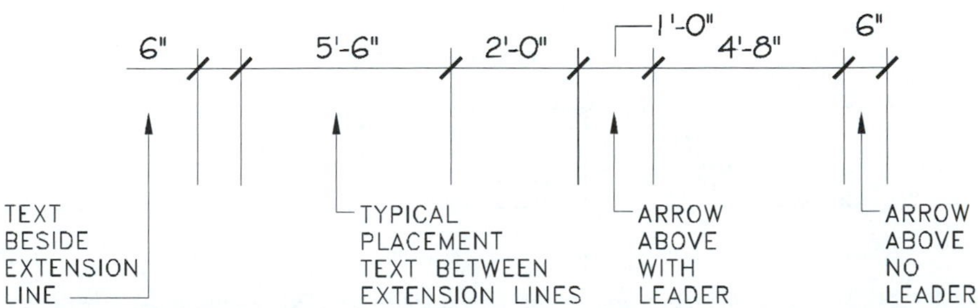

Figure 19.23 *Common dimension text placements used on floor plans.*

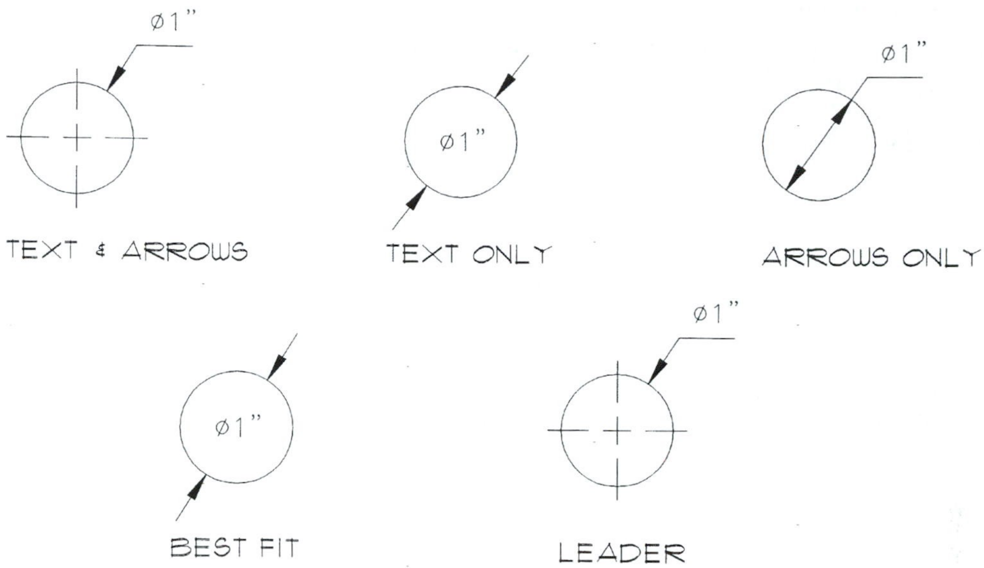

Figure 19.24 *The Fit sub-menu options. Notice that depending on the size of the space to be dimensioned, some options produce the same results.*

Arrows

This option displays the arrows and the text inside the extension lines when space is available. If both will not fit between the extension lines, the arrows will be placed inside and the text will be placed outside the extension lines. If sufficient space is not available between the extension lines, the arrows will be placed outside the extension lines.

Text

This option also displays the arrows and the text inside the extension lines when space is available. If sufficient space is not available between the extension lines, the text is placed between the extension lines and the arrows are placed outside the extension lines. If no space is available for text, both will be placed outside the extension lines.

Both Text and Arrows

This option always displays the arrows and the text outside the extension lines. This option works well with small dimensions placed at the end of a dimension string.

Always Keep Text Between Ext Lines

This option always displays the text between the extension lines. This option works well with small dimensions placed in the middle of a dimension string.

Suppressing Arrows

This option keeps the text between the extension lines and removes arrowheads when space is not available.

Text Placement

The Text Placement portion of the Fit tab allows for the control of the placement of text when it is not in the default position. Options include Beside, Over with a Leader, and Over with No Leader.

Beside the Dimension Line

This option displays the text and the arrows outside the extension lines. This option works well in combination with using the Either option of the Fit tab with small dimensions placed at the end of a dimension string.

Over with a Leader

This option displays the text above the normal position of text above and to the side of the dimension line and uses an arrow to reference the text to the dimension line. This option works well in combination with using the Either option of the Fit tab with small dimensions placed in the middle of a dimension string when several small dimensions will be place near each other.

Over without a Leader

This option displays the text above the normal position of text above the dimension line but does not use an arrow to reference the text to the dimension line. This option works well in combination with using the Either option of the Fit tab with small dimensions placed in the middle of a dimension string.

Scale for Dimension Features

Up to this point of planning dimension, each dimension will only be legible on a 12" × 9" screen. This area of the Fit tab sets the overall scale value or the paper space scaling for all dimensions. Options include Overall scale, Scale, and Layout based on Paper space.

Use Overall Scale

The Use Overall Scale Of edit box allows the scale of all dimensioning features to be increased or decreased based on the scale a drawing is to be plotted at. Typically you will be drawing in model space with actual sizes. Text that is to be ⅛" high when plotted at a scale of ¼" = 1'-0", will need to be set to 6" in height. To have legible arrows, offsets, and text you'll have to convert every variable. If you actually have a life, you may want to just set the scale factor to 48. This will control all dimension variables, and leave you a little free time. To find the desired scale factor, take the reciprocal of the drawing scale. A scale of ¼" = 1'-0" is 12/.25 = a scale factor of 48. Other common scale factors include:

ARCHITECTURAL VALUES		ENGINEERING VALUES	
Plotting Scale	Scale Factor	Plotting Scale	Scale Factor
3/4" = 1'-0"	16	1" = 1'-0"	12
1/2" = 1'-0"	24	1" = 10'-0"	120
3/8" = 1'-0"	32	1" = 100'-0"	1200
1/4" = 1'-0"	48	1" = 20'-0"	240
3/16" = 1'-0"	64	1" = 200'-0"	2400
1/8" = 1'-0"	96	1" = 30'-0"	360
3/32" = 1'-0"	128	1" = 40'-0"	480
1/16" = 1'-0"	192	1" = 50'-0"	600
		1" = 60'-0"	720

Scale Dimensions to Layout

Selecting the Scale Dimensions to Layout (Paper Space) box will apply a scaling factor to dimensions plotted in paper space. The option sets a scale factor based on the scaling between the current model space viewport and paper space.

Fine Tuning

The final portion of the Fit tab allows minor text adjustments to be made. Options include Place Text Manually, and Always Draw dimension lines.

Place Text Manually When Dimensioning

Activating this check box allows the location for dimension text to be specified as you place each dimension regardless of the justification. With this box active, AutoCAD ignores other justification settings and allows you to specify the position where the text will be placed. Adjusting the variable can be useful on radial dimensions as shown in Figure 19.25.

Always Draw Dimension Line

When a small space is dimensioned the text will be placed outside of the extension lines. Selecting the Always Draw Dim Line Between Ext Line option will place a dimension line between the extension lines even when the text and arrows are outside of the extension lines. Figure 19.26 shows the effect of using this option on a radial dimension.

PRIMARY UNITS TAB

A dimension of 14'-0" is considered a primary dimension unit. Most construction-related dimensions consist of primary units with no other information required. Occasionally a tolerance may be added to the primary units. This portion of the Dimension Style dialog box controls the display of the primary measurement units and any prefixes or suffixes for the dimension text. Selecting the Primary Units tab

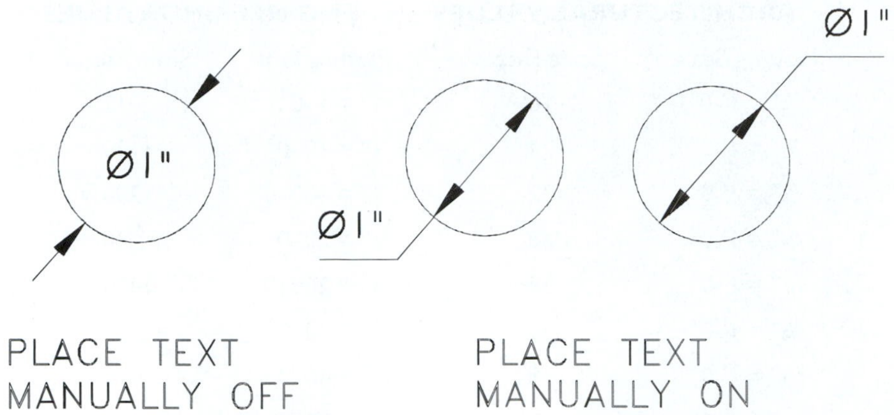

Figure 19.25 *With the Place Text Manually option active, AutoCAD ignores other justification settings and allows you to specify the position where the text will be placed.*

displays a tab similar to Figure 19.27. This tab is used to set the variables that control the appearance of text for dimension and leader line. The tab includes the areas of Linear Dimensions, Measurement Scale, and Angular Dimensions.

Linear Dimensions

This portion of the Primary Unit tab displays and sets the format and precision for linear dimensions. Key elements include the Unit Format, Precision, Fraction Format, Decimal Separator, Round off, Prefix, and Suffix.

Unit Format

Even though the Units of the drawing are set to Architectural, the default Unit Format setting is decimal. Select the down arrow to display the menu. Options include Scientific units, Decimal units, Engineering units, Architectural units, Fractional

Figure 19.26 *Selecting the Always Draw Dim Line Between Ext Line option will place a dimension line between the extension lines even when the text and arrows are outside of the extension lines.*

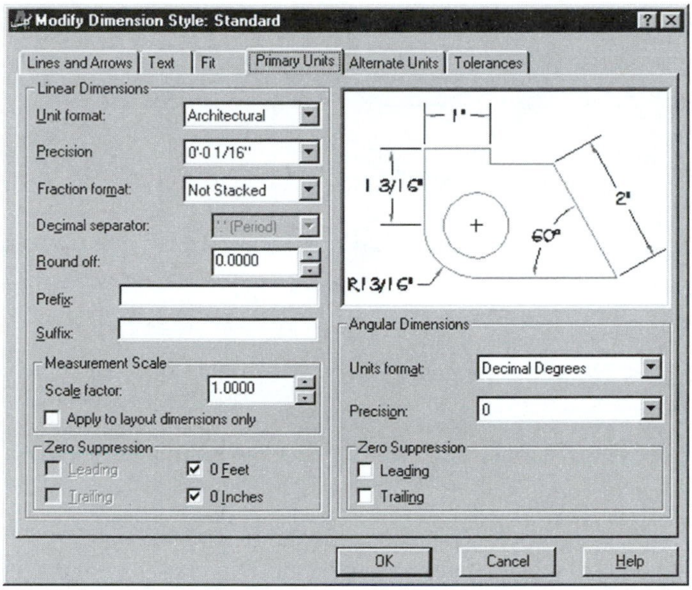

Figure 19.27 *The Primary Units tab.*

units, and Windows Desktop. The Architectural units will meet the needs for most drawings, although Engineering units are used on site-related drawings.

Precision

Selecting the down arrow of the Precision edit box displays the precision of the dimension. Options for the Architectural format are shown in Figure 19.28. For most dimensions, a setting of 0 (0'-0") will provide needed accuracy. Most engineers and architects determine the fractions based on calculations rather than relying on drawing accuracy. If a fraction needs to be added to the dimension, it can be added by editing the text once the dimension is displayed.

Fraction Format

The Fraction Format edit box sets the format for the display of fractions. Options include Diagonal, Horizontal, and Not Stacked. Each option can be seen in Figure 19.29.

Decimal Separator

This option is inactive when the Unit Format is set to Architectural units. If the Unit Format is set to Decimal, the option will be activated. Selecting the edit arrow displays the period, comma, and space options.

Round Off

This option determines the value for rounding measurements for all dimensions types except for Angular. Entering a value of 1/8 rounds all measurement values to the nearest eighth of an inch.

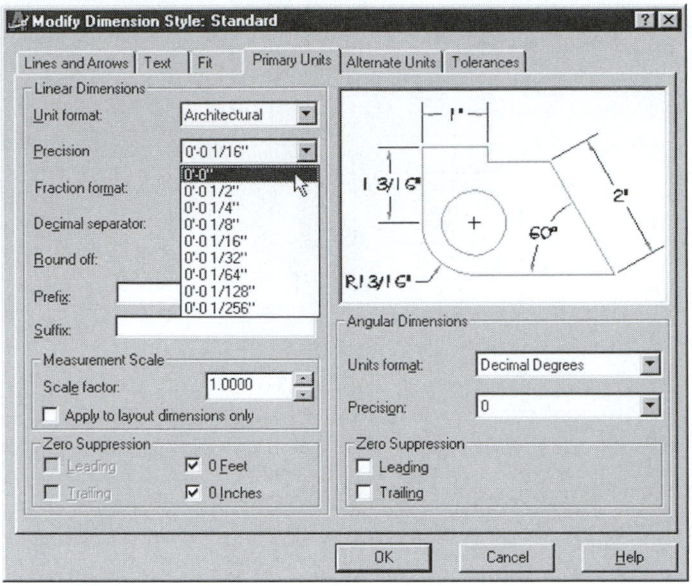

Figure 19.28 *The Precision sub-dialog box controls the accuracy assigned to a dimension.*

Prefix/Suffix

A prefix or suffix can be added by selecting the appropriate box with the cursor. Once the text is selected, it can be added to the edit box. Common suffixes might include .TYP (typical) , .MM (millimeter), or .MAX (maximum). This is one area where the control codes introduced in Chapter 17 might be useful. The %%C code can be used to enter the diameter symbol as a prefix or suffix. Other control codes include:

%%D = degree symbol %%P = plus/minus symbol

Measurement Scale

The Measurement Scale area of the Primary Units tab provides options for setting the scale factor and a method for limiting the scale values.

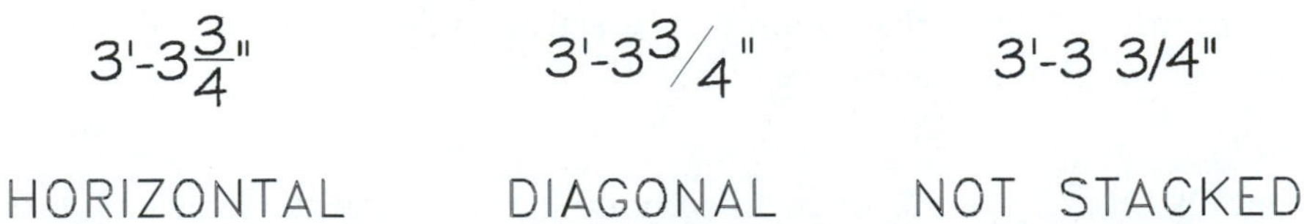

Figure 19.29 *The Fraction Format edit box sets the format for the display of fractions.*

Scale Factor

The Scale box specifies the scale factor of a dimension without affecting the component, angles, or tolerance values. The option affects all linear distances except Angular. The scale factor can be altered by using the up/down arrows, or by highlighting the existing value and then entering a new value by keyboard. A scale factor of 2 doubles the value of the dimension that is entered on the drawing. With a scale factor of 2, a dimension of 8" will be written when dimensioning a 4"-long object. For most architectural drawings the default value of 1 does not need to be altered.

Apply to Layout Dimension Only

The Apply to Layout Dimension Only check box will apply a scaling factor to dimensions plotted in paper space. The layout scale factor will be adjusted to reflect the zoom scale factor for objects created in model space viewports.

Zero Suppression

This area of the Primary Units tab controls the display of zeros for leading, trailing, feet, and inches. The Zero Suppression box controls the display of zeros. Four options are available. When an X is in the box, it is activated.

Leading

With this box active, the zero is suppressed preceding a decimal point in units less than one. Seven tenths will be written .7, rather than 0.7. This option will not be available when Architectural units are current.

Trailing

This box controls the display of zeros after a decimal point. With the box active, no zeros will be displayed after the decimal. Seven tenths will be written .7, rather than .70. This option will not be available when Architectural units are current.

0 Feet

This option will only be available when Architectural units are current. A dimension less than one foot may be displayed, as 0'-8" or 8". Professional practice is mixed and the setting for the 0 Feet box will vary from office to office. Selecting the feet box causes the zero representing feet to be suppressed when the dimension is less than twelve inches.

0 Inches

This option will only be available when Architectural units are current. Most professionals display the zero after the foot symbol. Writing 5'-0" is preferred to writing 5'. With this box active, zeros will be suppressed. This box should be inactive.

Controlling the Display of Angles

The Angular Dimensions box controls the format of angular dimensions. Options include Unit Format, Precision, and Zero Suppression. Options are similar to their angular counterpart.

Unit Format

The Unit Format for angular dimensions includes the options of Decimal Degrees, Degrees/Minutes/Seconds, Gradians, and Radians. The decimal degree option will meet the needs for most architectural drawings. The Degrees/Minutes/Seconds option will work best for most site-related drawings.

Precision

The Precision menu will vary, depending on the Unit Format setting. When units are measured in decimals, the precision will be measured in decimal places from 0 to eight places.

Zero Suppression

This option suppresses the leading or trailing zero of angular dimensions. With the leading option, 0.75 will be displayed as .75. The trailing option suppresses zeros that follow the decimal, making 7500 displayed as .75.

ALTERNATIVE UNITS TAB

This portion of the Dimension Style dialog box is not used to create a style for most construction drawings. As you enter the tab, each option is inactive. Selecting the Display Alternative Units check box activates this tab. The Alternative Units tab is shown in Figure 19.30. Major components include the Alternatives Units, Zero Suppression, and the Placement box. Features of this tab function in a similar manner to

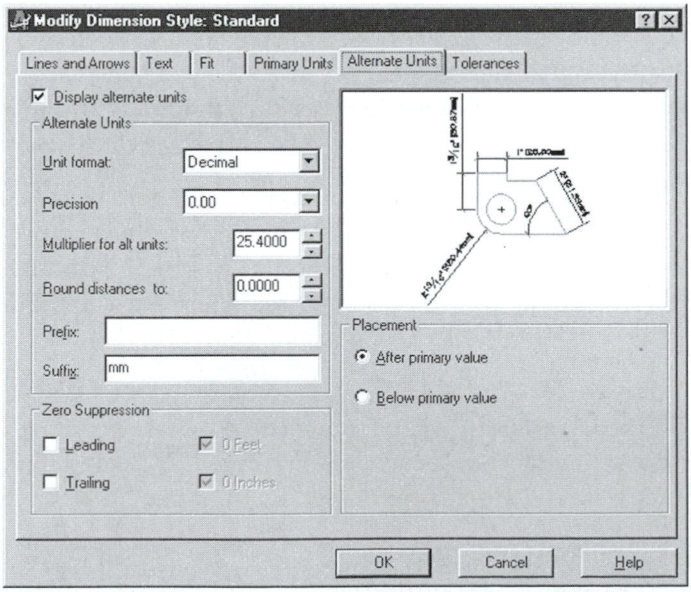

Figure 19.30 *The Alternative Units tab is inactive as it it entered. Selecting the Display Alternative Units check box activates this tab, allowing alternative units to be displayed behind the primary units.*

their counterparts on the Primary Units tab. One option that should be considered on this tab is the ability to create and display metric units behind the architectural primary units. Selecting Decimal as the Alternative unit format, a multiplier for all units of 25.4, and a suffix of mm, will display dimensions similar to Figure 19.31.

TOLERANCES TAB

A tolerance is a range that is allowed to compensate for human error. If a note specifies that bolts in a steel plate are to be 3" o.c., and the holes are placed at 3 1⁄16" o.c., should the plate be scrapped? A tolerance specification will define what is an acceptable range of variation. Most areas of construction do not use a tolerance zone. Some steel and concrete detailing may require a tolerance. The Tolerance tab of the Dimension Style dialog box controls the display and precision of the tolerance. The Tolerance tab is shown in Figure 19.32. Major components include the Tolerance Format, Zero Suppression, and the Alternate Unit Tolerance box. Features of this tab function in a similar manner to their counterparts on the Primary Units tab. With the default Method value of None, most of the box is inactive. Once a method of tolerance specification is selected, the balance of options on the tab will be activated. Figure 19.33 shows examples of the available tolerance methods.

COMBINING VARIABLES TO CREATE A STYLE

If you're feeling overwhelmed by styles, variables, and values, don't be discouraged. Remember most of these values will be set on a template drawing and left alone. Throughout this chapter, comparisons have been made between using the Dimension Style dialog box and its sub-boxes. This section will help you combine these values to create a style. Dimensioning variables can be combined to create a particular style in much the same way that options are combined with Text to create different text styles. Styles can be created, saved, modified, and quickly loaded by using the Dimension Style dialog box shown in Figure 19.1.

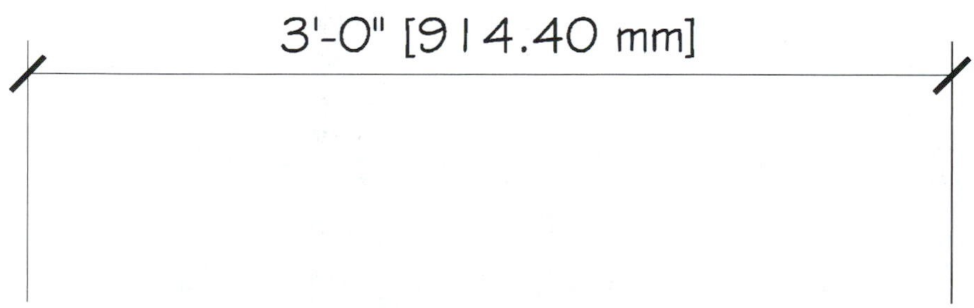

Figure 19.31 *Selecting Decimal as the Alternative unit format, a multiplier for all units of 25.4, and a suffix of mm, will display architectural units followed by metric dimensions.*

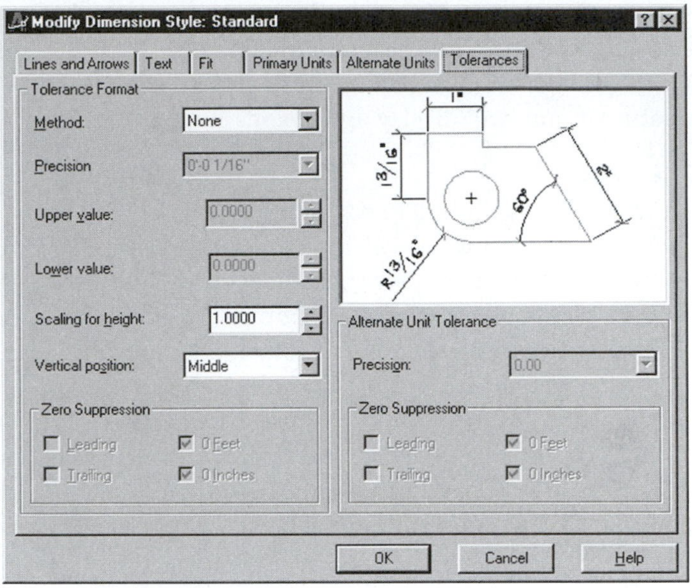

Figure 19.32 *The Tolerance tab of the Dimension Style dialog box controls the display and precision of the tolerance.*

CREATING DIMENSIONING STYLES

Dimensioning styles allow you to save groups of settings for dimension variables. If an office does both architectural and civil engineering projects, two distinct dimensioning styles could be created and saved on template drawings to speed the dimensioning process. Even in a firm that only produces architectural drawings, varied styles can be useful to group option settings. For instance, on a multilevel structure, some dimensions will be shown on each floor plan and the framing and foundation plans. These might be placed in a style titled ARCHBASE that uses a certain size, color, and style of text with two extension lines. Styles might be created for dimen-

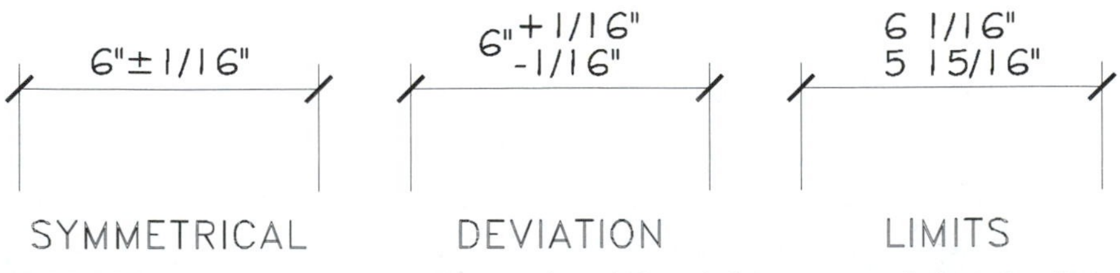

Figure 19.33 *Tolerance methods available in AutoCAD include symmetrical, deviation, limits, and basic.*

sions that will be used to dimension to the center of interior walls and have the left or right extension line suppressed, with another style created for a style that has both extension lines suppressed. Once created, these styles can be quickly loaded and interchanged. Use the following steps to create a dimension style suitable for a plan view to be plotted at a scale of ¼"=1'-0":

- Select the Dimension Style icon from the Dimension toolbar. This will display a Dimension Style Manager dialog box similar to Figure 19.1.
- Select the New button. This will display a Create New Dimension Style dialog box similar to Figure 19.2.
- Enter the desired style name. In this case PLAN VIEW BASE will be used.
- Select an existing style to use as a base. In this case the Standard style will be used.
- Select what dimensions that this style will be applied to in the Use For edit box. In this case All Dimensions will be used.
- Select the Continue button. This will display a New Dimension Style: PLAN VIEW BASE dialog box. Proceed through each tab and set the options to meet your specific situation. The next section will recommend values that you might want to consider using.
- Once each option has been selected, choose the OK button. This will return the Dimension Style Manager similar to Figure 19.34.

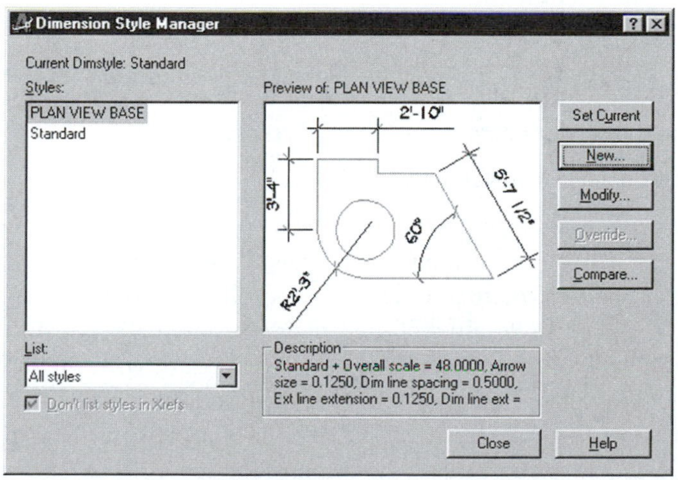

Figure 19.34 *Once each option has been selected, choosing the OK button will return the Dimension Style Manager. Select PLAN VIEW BASE from the Styles box, select the Set Current button, and then select the Close button to select, make current, and return to the drawing area.*

- Select PLAN VIEW BASE from the Styles box.
- Select the Set Current button.
- Select the Close button.
- This process will create, and set the PLAN VIEW BASE as the current dimension style.

Note: Although offices often have individual dimensioning standards, several common options can be set to meet most needs. Use the Standard style as a base to create the desired settings to update your template:

Line and Arrows Tab (Figure 19.5)

Color and Lineweight for each feature – By Block	Arrow heads – Architectural Tick
Extend beyond tick – 1/8"	Arrow size – 1/8"
Baseline Spacing – 1/2"	Center Mark – Line
Extend beyond dimen. line – 1/8"	Center Mark Size – 1/8"

Text Tab (Figure 19.17)

Style – Stylus BT	Offset from Dim Line – 1/8"
Color – ByBLock	Vertical – Above
Text Height – 1/8"	Aligned with Dimension Line – ON

Fit Tab (Figure 19.22)

Fit Option – Text	Overall Scale – 48

Always Draw dim line between ext lines – ON

Primary Units Tab (Figure 19.27)

Unit Format – Architectural	Precision – 0'-0 1/2"

Fraction Format – Not Stacked

Following the guidelines to create PLAN VIEW BASE will produce dimensions similar to Figure 19.35a. The dimension style works well for the overall dimensions, but is somewhat limited when an object has a jog in it. If the style is used to place dimension A, the desired results will be obtained, but two extension lines will be placed on the left side of space A. The style will reproduce with the desired results if you're careful to always use OSNAP when you select where the extension lines will be placed. Failure to use OSNAP may produce results similar to Figure 19.35b. Even if OSNAP or QDIM is used to place the dimensions for space B and C, the desired results will not be achieved. The extension line for B will extend to be within 1/8" of surface B, leaving no offset for surface A. The same problem will occur at B, as surface C is dimensioned.

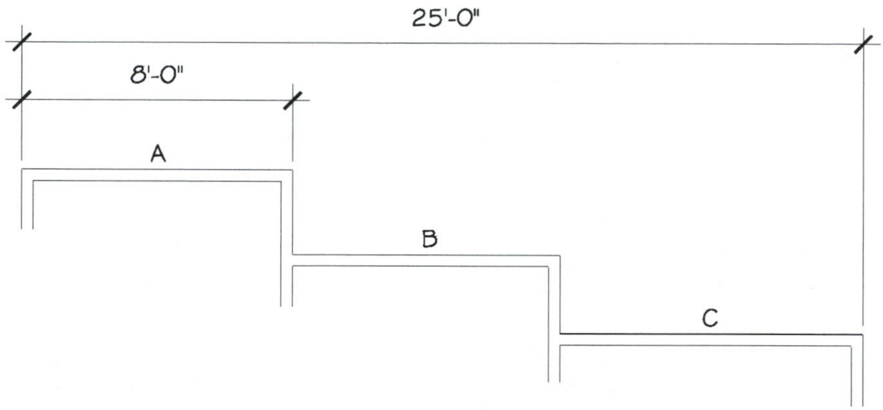

Figure 19.35a *Using the PLAN VIEW BASE style will produce dimensions suitable for many architectural applications.*

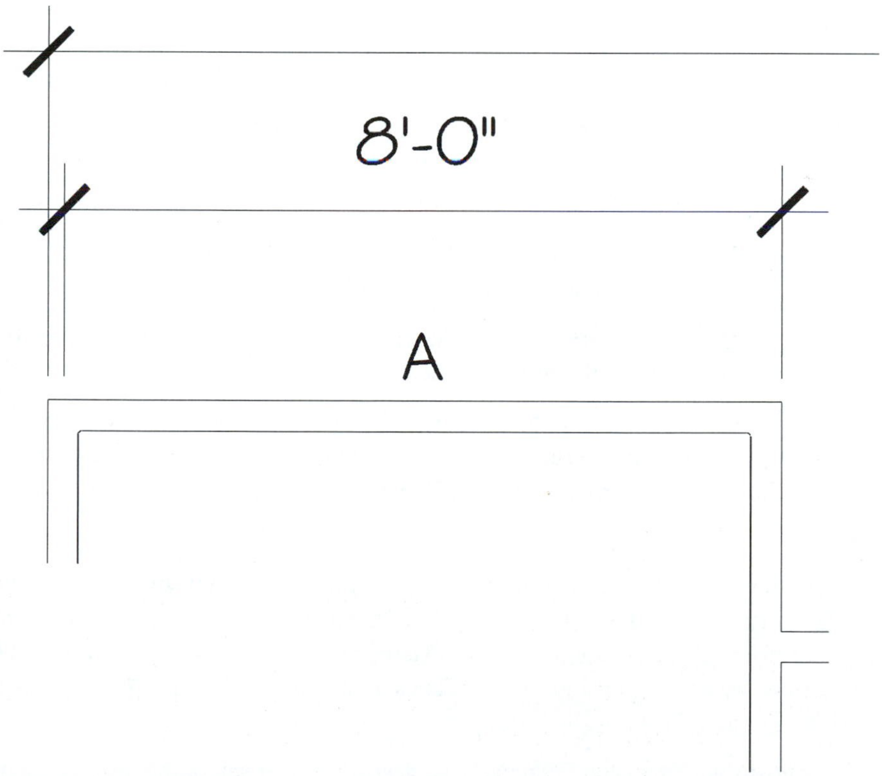

Figure 19.35b *Failure to use OSNAP may produce extension lines that do not align.*

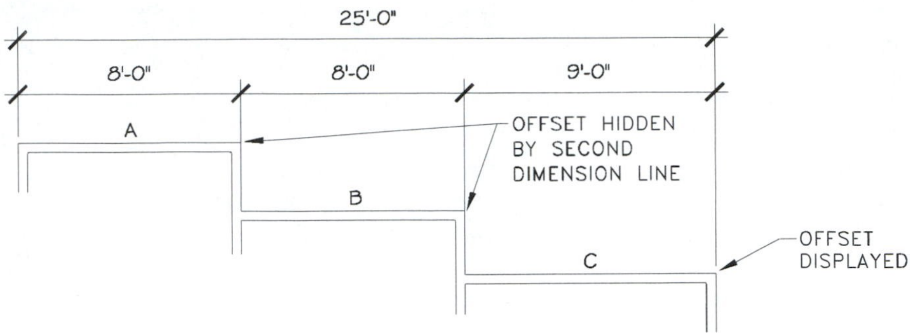

Figure 19.35c *Even if OSNAP or QDIM is used to place the dimensions for space B and C, the desired results will not be achieved. The extension line for B will extend to be within 1/8" of surface B, leaving no offset for surface A. The same problem will occur at B as surface C is dimensioned.*

CREATE A NEW STYLE BASED ON AN EXISTING STYLE

A better method of placing the interior dimensions is to create a style that will suppress one or both of the extension lines. This can easily be done using the following procedure:

- Enter the Dimension Style Manager dialog box and select the New... button.
- At the Create New Dimension Style dialog box, enter a style name of INTERIOR.
- Use PLAN VIEW BASE as the Start with style and use it with all dimensions.
- Press Continue to advance to the Dimension Style dialog Box.
- Use the Line and Arrow tab to suppress Line 1.
- Select OK, Set Current, and CLOSE buttons to set the new style as the current style and return to the drawing area.

Using the PLAN VIEW BASE and the INTERIOR styles will allow dimensions to be placed accurately for the overall length and for spaces A, B, and C. Figure 19.35d shows the result of using the PLAN VIEW BASE and the INTERIOR styles.

MODIFY A STYLE

Once a style has been created, you may find that it is not all that you hoped it would be. The Modify option of Dimension Style Manager allows an existing dimension style to be altered. The process to alter an existing style is similar to the process used to create a new style using and existing style as a base. An existing style can easily be modified using the following procedure:

- Enter the Dimension Style Manager dialog box and select the style to be modified from the Styles display box.

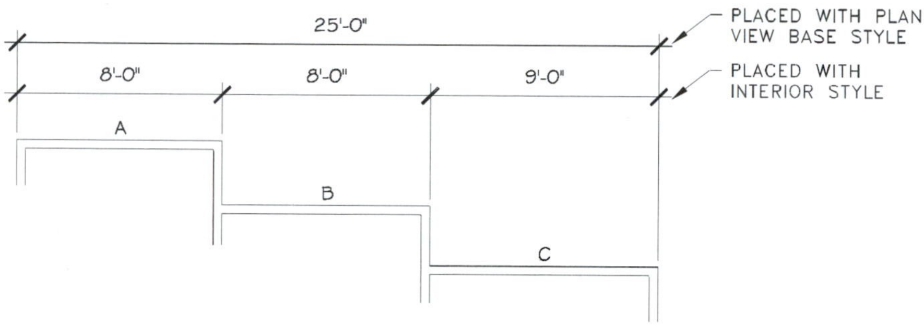

Figure 19.35d *Using the PLAN VIEW BASE and the INTERIOR styles will allow dimensions to be placed accurately for the overall length and for spaces A, B, and C.*

- With the desired style highlighted, select the Modify… button. This will display the Modify Dimension Style: dialog box. The box is exactly the same as the New Dimension Style dialog box.
- Use the necessary tabs to alter the existing values. For the example, the Line and Arrow tab was used to alter the arrowheads.

Select OK and the CLOSE buttons to modify the existing style and return to the drawing area. Using the Modify option on the INTERIOR style will alter the display to resemble Figure 19.36.

OVERRIDING EXISTING STYLES

Using the Modify option of the Dimension Style dialog box changes an entire style to the specified parameters. Using the Override option allows an existing style to stay

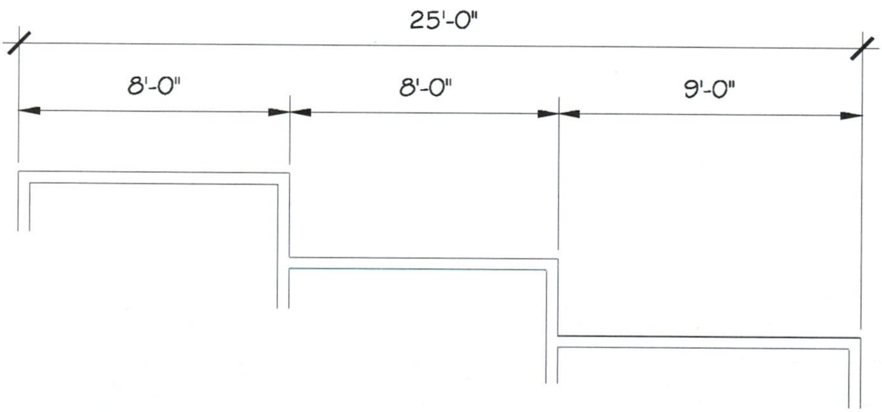

Figure 19.36 *Using the Modify option on the INTERIOR style will alter the display shown in Figure 19.35d to resemble Figure 19.36.*

intact while altering new features that are added to the style. Use the following steps to override an existing style:

- Enter the Dimension Style Manager dialog box and select the style to override from the Styles display box. In this example, the INTERIOR style will be used as the style to be altered.
- With the desired style highlighted, select the Override... button. This will display the Override Current Style: dialog box. The box is exactly the same as the New Dimension Style dialog box.
- Use the necessary tabs to alter the existing values. For the example, the Line and Arrow tab was used to alter the arrowheads and the Text tab was used to center the text between the dimension line.

Select OK and the CLOSE button to override the existing style and return to the drawing area. Using the Override option on the INTERIOR style will alter the display to resemble Figure 19.37. Styles can also be overridden at the command prompt by entering **DOV** (EnterKey). This will produce the prompt:

Enter dimension variable name to override [or Clear overrides]:

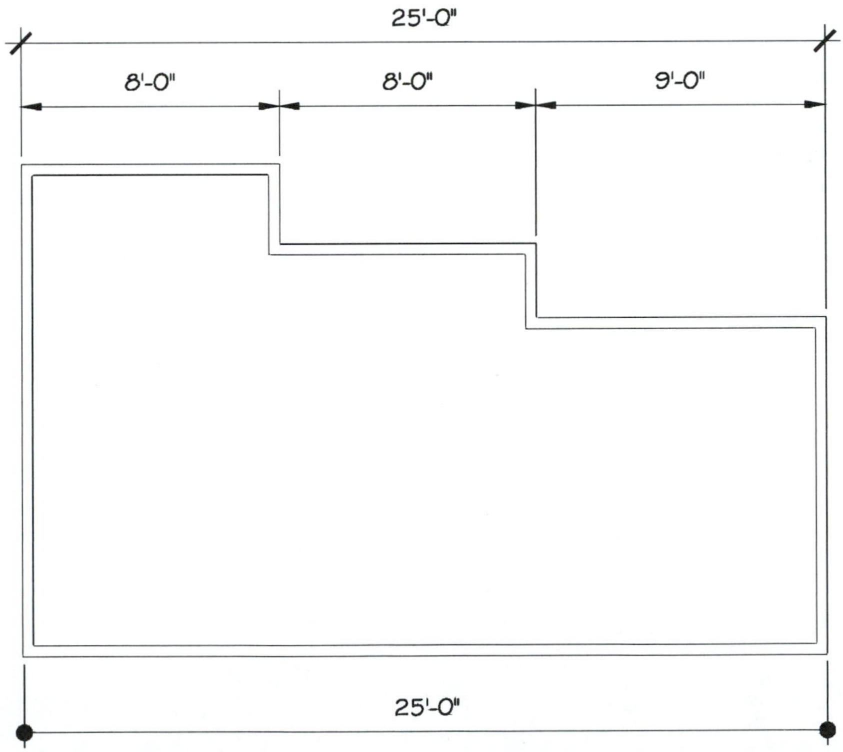

Figure 19.37 *Using the Override option on the INTERIOR style to alter the display to show dot arrowheads and centered text.*

Enter the name of the desired variable to be altered followed by the name of the new style.

COMPARING STYLES

The final option of the Dimension Style dialog box is to compare styles. This option can be used to list the properties of a style or to compare the properties of two dimension styles. The function preformed depends on the settings of the Compare and With edit boxes. To list the properties of a style, select the name of the desired style from the Compare edit box, and select <None> from the With menu. Figure 19.38 displays a partial listing of the properties for the PLAN VIEW BASE style. Entering a style name in the Compare edit box, and a different style name in the With edit box, will display a listing of the property differences similar to Figure 19.39. The list of properties and the list of differences can be copied to the Windows Clipboard. Once copied to the Clipboard, the listings can be pasted into other Windows applications.

RENAMING OR DELETING STYLES

A style can be renamed or deleted using the Style portion of the Dimension Style Manager. Use the following procedure to rename a style:

- Enter the Dimension Style Manager dialog box and move the cursor to the Styles display box.
- Highlight a style in the Style portion of the dialog box.
- With the style highlighted, right-click to produce a shortcut menu.
- Select the Rename option and enter the desired new name.
- Continue renaming other menu listings or select CLOSE to remove the dialog box and return to the drawing area.

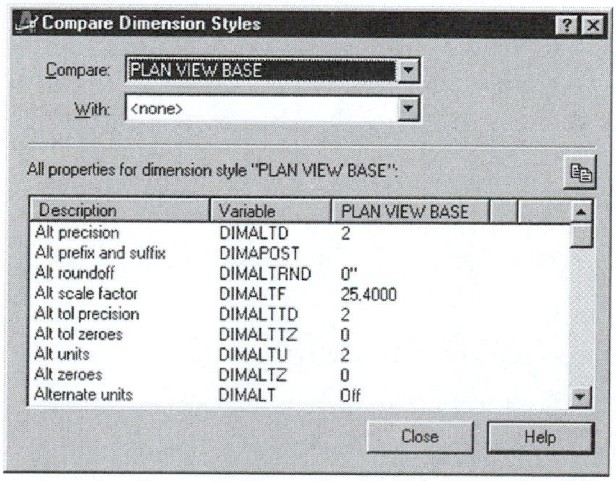

Figure 19.38 *A partial listing of the properties for the PLAN VIEW BASE style.*

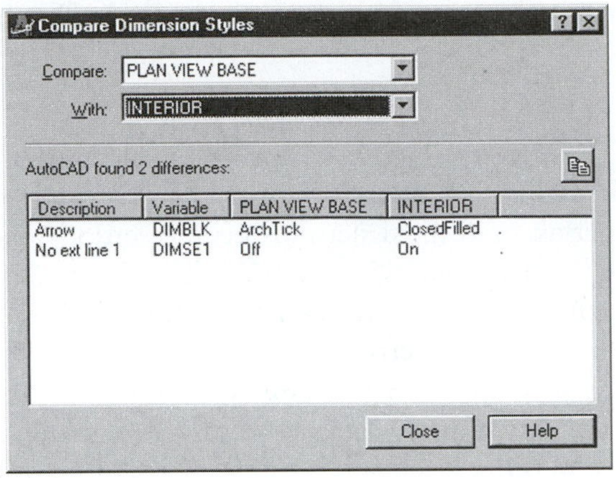

Figure 19.39 *Entering a style name in the Compare edit box, and a different style name in the With edit box, will display a listing of the property differences.*

Similarly, deleting a dimension style is done in the same dialog box, using the following procedure:

- Once the style to be deleted is highlighted, right-click and select the Delete option.
- Continue deleting other styles or select CLOSE to remove the dialog box and return to the drawing area.

APPLYING STYLES

Each dimension you create has a style. If you started to place dimensions prior to reading this chapter, the dimensions were created using the Standard style. Throughout this chapter, you've learned to create, modify, and override dimension styles. In addition to selecting the Set Current button in the Dimension Style Manager, two other methods are available to alter the dimension style. A new style can be applied by using the shortcut menu or by using the Dim Style Control edit box in the Object Properties toolbar.

Using the Shortcut Menu

Use the following steps to assign a new dimension style to an existing dimension with the shortcut menu:

- Select the desired dimension and right-click.
- Choose Dim Style from the shortcut menu.
- Choose the name of the desired style to be applied to the selected dimension and press (EnterKey).
- The process can be seen in Figure 19.40.

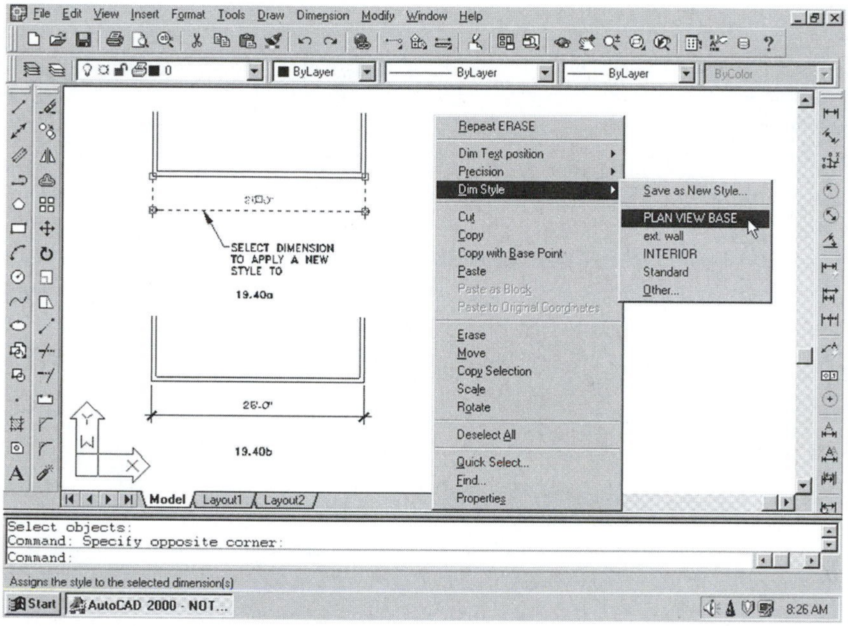

Figure 19.40 *A dimension style can be altered by **a)**. Activating the grips and right-clicking to produce the shortcut menu, and then selecting the name of the desired style. **b)**. The results of the selection.*

Using the Dim Style Control Edit Box

Use the following steps to assign a new dimension style to an existing dimension with the Dim Style edit box:

- Select the desired dimension and right-click.
- Choose the desired dimension style from the shortcut menu by right-clicking.
- Choose the name of the desired style to be applied to the selected dimension and press (EnterKey).

EDITING DIMENSION PLACEMENT

The placement of dimensions can be edited using grips or the editing commands. The editing commands can be selected from the shortcut menu or by the methods presented in earlier chapters. To use the shortcut menu, select a dimension with the cursor. Each of the grips will be activated. Right-click to obtain the editing menu. Notice in Figure 19.41 that a grip is displayed for the dimension text, at each intersection of the dimension and extension lines, and at what appears to be the end of the extension line. Closer examination will show the grips that appear to be at the end of the extension lines actually lie beyond the lines. These grips mark the definition points, or DefPoints, for the dimensions. AutoCAD automatically assigns definition points for every dimension to define the dimension location. Use the Layers icon to examine the layers for the current drawing and you'll find that the program created a

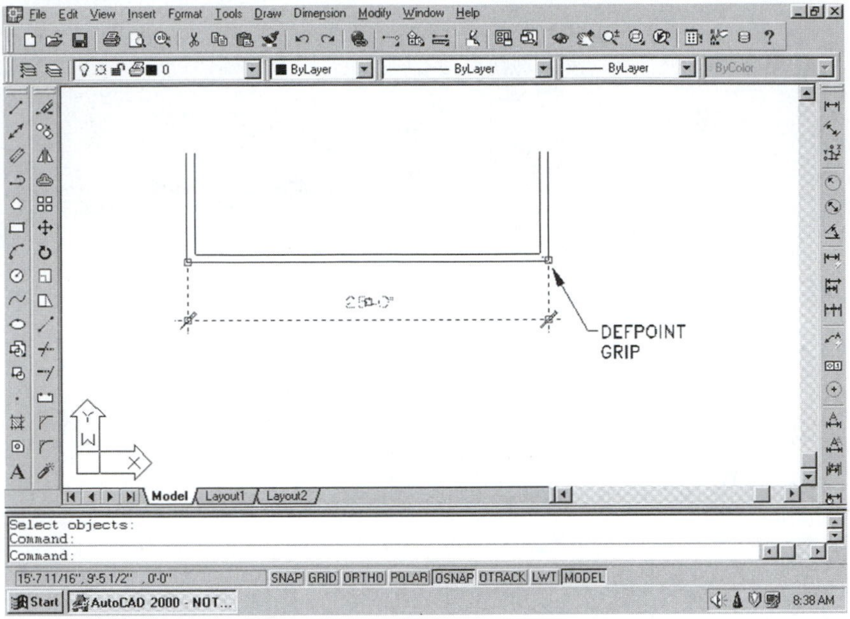

Figure 19.41 *A grip is displayed for the dimension text, at each intersection of the dimension and extension lines, and at the definition points or DefPoints for the dimensions. AutoCAD automatically assigns definition points for every dimension to define the dimension location.*

layer titled DefPoints to display the definition points. The layer will not be plotted. Dimensions can be edited using grips without knowing the location of the definition points. The DefPoints will automatically be displayed. If you alter dimensions using the edit commands, the DefPoints must be included in the selection set. Using the NODE setting of OSNAP will aid in DefPoint selection.

STRETCHING DIMENSIONS

Dimensions can be stretched to alter their position using either grips or the STRETCH command. The process of stretching a dimension can be seen in Figure 19.42. Use the following steps to alter the dimension position using grips:

- Select the dimension to be stretched.
- Make one of the grips on the dimension line the hot grip.
- Drag the grip to the desired location. OSNAP will aid in accurate placement.
- This process will stretch all features of the dimension. To move just the text, select the text grip. Make it the hot grip, and move the text to the desired location.
- The STRETCH command can be used to stretch dimensions using the same procedure that is used to move drawing objects. Use the Crossing option to select the dimension to be relocated. STRETCH can't be used to relocate the text.

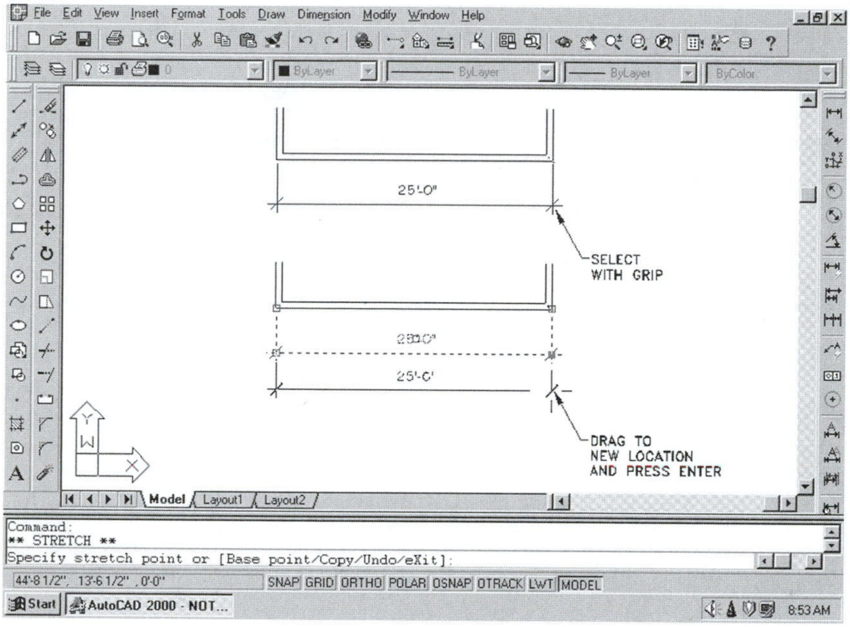

Figure 19.42 *Select the dimension to be stretched and then make one of the grips on the dimension line the hot grip. Drag the hot grip to the desired location. OSNAP will aid in accurate placement when aligning with other dimension lines.*

EDITING OBJECTS AND DIMENSIONS WITH THE MODIFY MENU

Objects and dimensions can be edited using the editing commands contained in the Modify menu, as well as the GRIP editing functions. As an object is edited, the dimensions are also edited.

STRETCH

Dimensions can be stretched in the same way that drawing objects are stretched. By selecting the dimension with the object, as seen in Figure 19.43, both will be stretched and the dimension text will be revised automatically.

EXTEND

The EXTEND command sequence also can be used to extend a dimension line and adjust the text accordingly. Figure 19.44 shows the effects of the EXTEND command on dimensions.

TRIM

The TRIM command sequence can be used to shorten a dimension line and adjust the text accordingly. Figure 19.45 shows the effects of TRIM on dimensions.

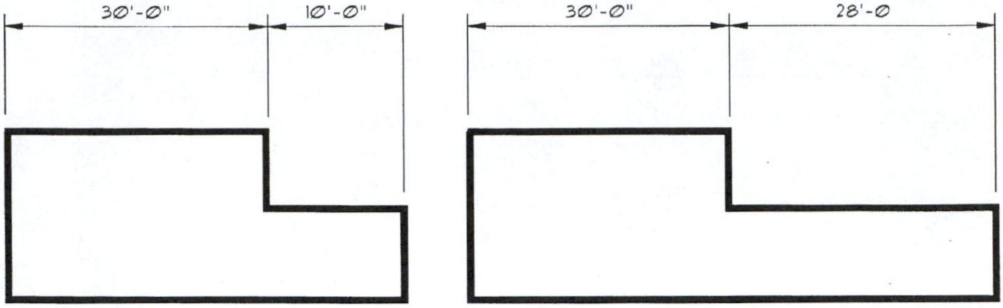

Figure 19.43 *Using the STRETCH command to alter an object with dimensions.*

EDITING DIMENSION TEXT

Dimensions placed on a drawing can be edited using the PROPERTIES, DIMEDIT, and DIMTEDIT commands.

EDITING DIMENSION TEXT USING PROPERTIES

The dimension text can easily be edited using the Properties command. Use the following steps to alter dimension text using the Properties:

- Select the dimension to be edited.
- Select the Properties icon from the Standard toolbar or by one of the other access methods.
- From the Alphabetic tab choose Text Override.
- Enter the desired dimension in the Text Override edit box.
- Continue to alter properties or close the Properties display.

Text entered in the Text Override edit box will override the measured dimension of the feature. The true measurement will continue to be displayed in the Measurement

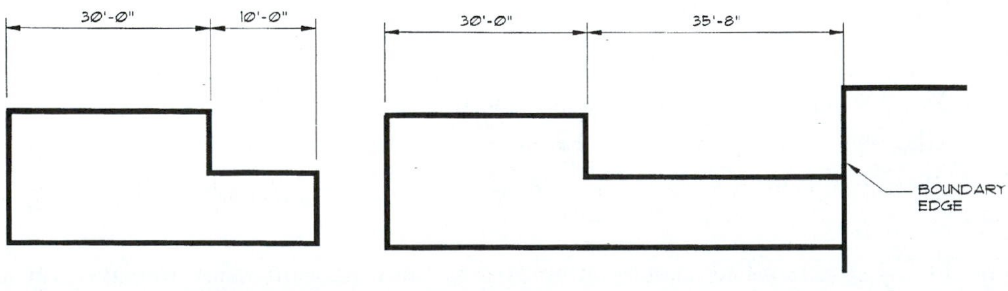

Figure 19.44 *The effect of the EXTEND command on an object with dimensions.*

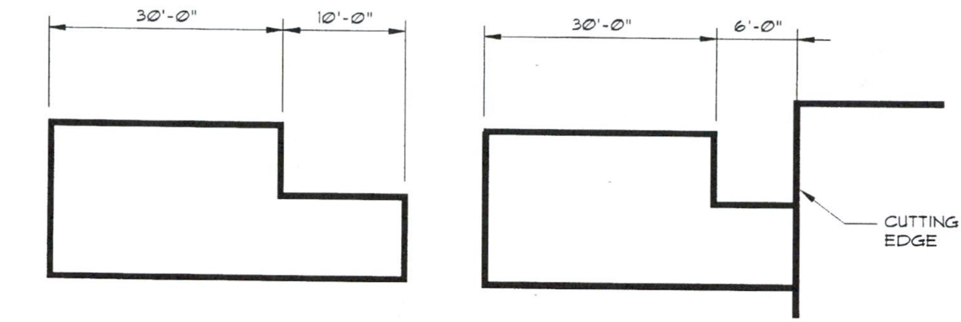

Figure 19.45 *The TRIM command also affects an object, the dimension lines, and text.*

box of Properties. Dimension Text properties can also be altered by selecting the text, right-clicking, and then picking properties from the shortcut menu. Other options for altering the location and text will be introduced later in this section.

Note: This feature can be especially helpful to dimension an object that must be a certain size, which was unknown to you when the part was drawn and dimensioned. If you know bolts are typically placed in steel plates at a 3" spacing, you can draw and dimension a detail prior to receiving the final calculation from the engineer. If the engineer determines that this application needs the bolts at 3½" spacing, the dimension can be altered without editing their location on the drawing. The bolt locations will not be drawn to scale. If the math works, so does the detail, and you've saved time.

EDITING DIMENSION TEXT USING DIMEDIT

The DIMEDIT command allows dimension text to be edited. The existing text can be edited, rotated, moved, or restored to its original position. The angle of the extension lines can also be altered from perpendicular to oblique.

The command is accessed by selecting the Dimension Edit icon from the Dimension toolbar, or by typing **DED** (EnterKey) at the command prompt using the following prompt:

Command: **DED** (EnterKey)

Enter type of dimension editing [Home/New/Rotate/Oblique] <Home>:

Changing the Dimension Text

To alter existing text select the New Option. Selecting New Option will produce the prompt:

Enter type of dimension editing [Home/New/Rotate/Oblique] <Home>: **N** (EnterKey)

This will produce the Multiline Text editor and allow the desired text to be entered. Remove the brackets to replace the existing text. If the new text is placed in front of the brackets, the new text will be placed in front of the existing dimension text. Enter the new text and select the OK button to remove the editor. The prompt will continue:

Select objects: *(Select dimension text to be altered)* I found

Select objects: (EnterKey)

Command:

The results of the command can be seen in Figure 19.46.

Rotating Existing Text

Choosing the Rotate option allows existing text to be rotated. Selecting the text in Figure 19.29 to be rotated to a 45° angle requires the following command sequence.

Command: **DED** (EnterKey)

Enter type of dimension editing [Home/New/Rotate/Oblique] <Home>: **R** (EnterKey)

Specify angle for dimension text: **45** (EnterKey)

Select object: *(Select text to be altered.)*

Select object: (EnterKey)

Command:

This produces text as shown in Figure 19.47.

Altering the Extension Line Angle

The angle of existing extension lines can be altered using the Oblique option of DIMEDIT. The command sequence to alter the angle of the extension lines in Figure 19.48 is:

Command: **DED** (EnterKey)

Enter type of dimension editing [Home/New/Rotate/Oblique] <Home>: **O** (EnterKey)

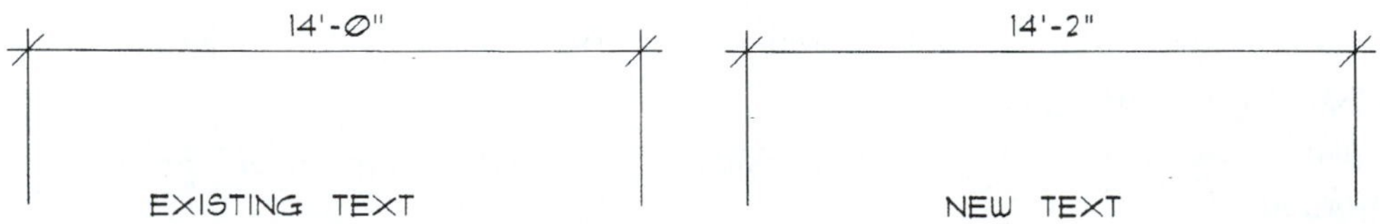

Figure 19.46 *The New option of the DED (dimension edit) command can be used to edit existing dimension text.*

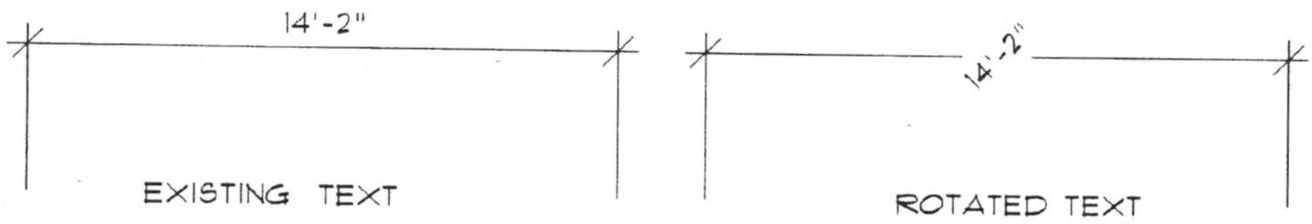

Figure 19.47 *The Rotate option of the DIMEDIT command can be used to rotate existing dimension text.*

Select object: *(Select dimension to be altered.)* 1 found

Enter obliquing angle (Press ENTER for none): **-75** (EnterKey)

Command:

The results of the sequence can be seen in Figure 19.48.

Returning Home

Using the default option of Home will return the text to its default position.

EDITING DIMENSION TEXT USING DIMTEDIT

The DIMTEDIT command is used to position the text relative to the dimension line. The command can be accessed by selecting the Dimension Text Edit icon from the Dimension toolbar, by selecting Align Text from the Dimension pull-down menu, or by typing **DIMTED** (EnterKey) at the command prompt. Either access method will prompt you to select a dimension to be edited. Once a dimension is selected, the following prompt will be displayed.

Specify new location for dimension text or [Left/Right/Center/Home/Angle]:

Choosing the Left option will move the text toward the left end of the dimension line. Choosing Right, will move the text toward the right end of the dimension line. Choosing Angle will allow the text to be rotated, choosing Center will center the text on the dimension line, and choosing Home will return the text to its default position. Each option can be seen in Figure 19.49.

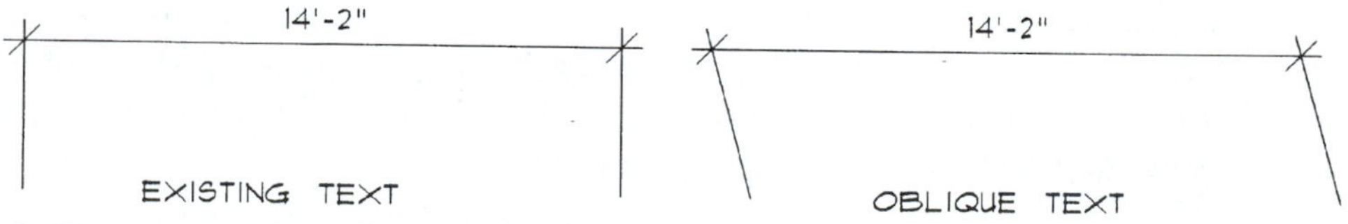

Figure 19.48 *The Oblique option of the DIMEDIT command can be used to alter the obliquing angle of existing extension lines.*

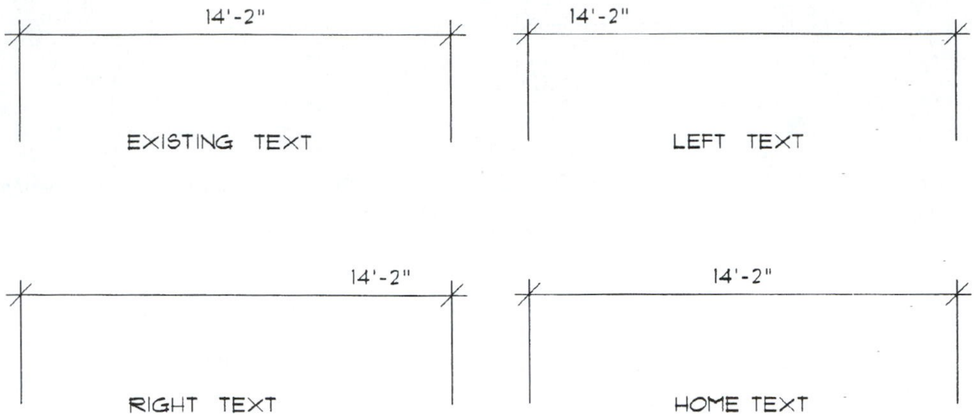

Figure 19.49 *The options of DIMTEDIT will alter the location of existing dimension text.*

WHAT IT ALL MEANS

More so than any other area of AutoCAD, dimensioning seems to cause new users the most anxiety. You've read through what must seem like an unlimited number of commands and options and may be having a hard time getting all of this information organized. The good news is that once you set the variables using the dialog boxes and store them in template drawings, most of your work is done.

ALTERNATIVE METHODS

Now that you've heard the good news, you might have guessed that there is some bad news. As good as the software is, it is really intended for mechanical drafters. They're nice people, but they do things so differently. You've probably noticed that all of the extension lines throughout the last two chapters have been created with continuous lines. This is totally unacceptable for many architectural offices. Continuous extension lines have traditionally been used to represent dimensions that extend to the face of a surface such as a stud wall. Center extension lines are traditionally used to represent the center of objects. This shortcoming can be overcome, but at a cost to other dimensioning values.

To take best advantage of the editing qualities, dimensions typically are placed as associative dimensions. Dimensions could be set as normal dimensions that would allow the extension lines to be edited using the CHANGE command. Another option is to suppress the extension lines for interior objects, and place them in the drawing using the LINE command with a centerline type.

An alternative to giving up the qualities of associative dimensions is to use the EXPLODE command. This command will be covered in detail in Chapter 20. Some layer problems will be created using EXPLODE. Basically, the command will allow you to select a group of objects that function as one object, and then allow you to separate that one group. This will allow the majority of dimensions to retain their editing qualities, while allowing the selected dimensions to be edited as individual lines, arrows, and text.

CHAPTER 19 EXERCISES

1. Open drawing Floor14. Use the appropriate dialog box and a style suitable for dimensions with both extension lines, a style with one dimension line, and a style with no extension lines. Set the text offset from the dimension line for all styles to be $\frac{1}{16}$", with a baseline increment of $\frac{1}{2}$". Assign each style a different color. Mark circles with a $\frac{1}{8}$" center mark and provide for centerlines. Use $\frac{1}{8}$" for the feature offset and extension above line settings. Use $\frac{1}{8}$"-long tick marks and a $\frac{1}{8}$"-long tick extension. Set the text height as $\frac{1}{8}$", with the text forced inside and above the extension line. Set the zero suppression controls so that a zero will be shown following feet, but not preceding inch sizes. Save these styles and then save the drawing as FLOOR14.
2. Use appropriate dimension techniques to label the floor-to-plate height of drawing E-16-8. Save the drawing as ELEV.
3. Use appropriate dimension techniques to label the base footings created for drawing E-14-4. Save the drawing as FOOTINGS.
4. Use appropriate dimension techniques to label the two-story footing created for drawing E-14-4. Show the concrete as 8" above grade, with an 8" stem wall and a 7" × 15" footing that extends 18" below the finish grade. Save the drawing as FOOTINGS.
5. Use baseline dimension methods with dot line terminators to locate each tree center. Place all dimensions separate from all other information.
6. Open drawing E-14-9 and provide the needed dimensions to describe the manhole cover created in problem E-13-9. Save the drawing as E-19-6.
7. Open drawing E-12-7 and provide the needed dimensions to describe the structure. Use the required commands to create the linetypes and text of the original problem. Hatch the 8"-wide wall with ANSI31. Save the drawing as E-19-7.
8. Open drawing E-18-7 and provide the needed dimensions and notes to describe the foundation plan created for E-9-7. Change the lines representing the beams to 3"-wide lines. Label the main structural members to represent post and beam methods of your region. Save the drawing as E-19-8.

9. Use the drawing below as a guide, draw and dimension the curb detail. Place dimensions, text, general notes, and hatching symbols on separate layers. Use different text styles for the notes and dimensions. Save the drawing as CURB.

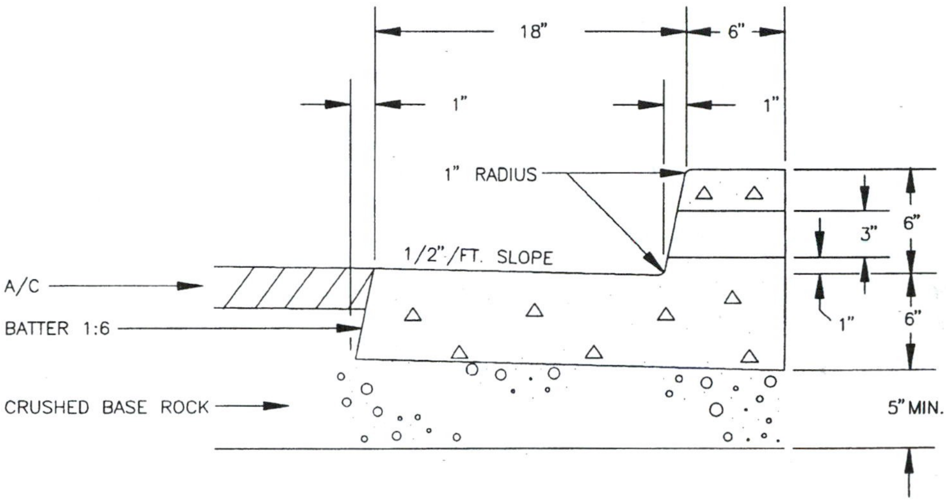

1. PCC SHALL BE 3300 PSI STRENGTH AT 28 DAYS.
2. CONTRACTION JOINTS SHALL BE PLACED AT 15' MAX. SPACING.
3. EXPANSION JOINTS SHALL BE PLACED AT 45' MAX. SPACING AND AT ALL DRIVEWAYS AND CURB RETURNS.
4. CURB EXPOSURE SHALL BE 8" AT ALL CATCH INLETS.
5. DRAIN BLOCKOUT SHALL BE SUFFICIENT FOR 3" DRAIN PIPE AND PLACED WHERE DIRECTED.
6. CRUSHED BASE ROCK SHALL BE COMPACTED TO 95% RELATIVE DENSITY PER AASHTO T-180.

10. Open a new drawing and use the drawing below as a guide to show the support for (2) 6¾" × 37½" glu-lam beams. Establish separate layers for the beam, steel, text, and dimensions. Completely label and dimension the drawing. Set the required scale factors for plotting at a scale at 1"=1'-0".

    The beams will be supported on a 5⁄16" × 6⅞" × 24" base plate with (2) 5⁄16" × 10" × 24" side plates welded to the base plate with a ¼" fillet weld. The base plate will be supported by a 6" × 6" × ¼" steel column and welded with a 5⁄16" fillet weld all around. Each beam will be bolted to the side plate with (4)- ⅝"-diameter machine bolts. Bolts will be 1½" in from the top and sides and be 3" on center.

    A second connector plate will be located with the center of the 3" × 22" × 5⁄16" steel plate 8" down from the top of the beam. Two bolts will be used to join the plate to the beam. Bolts will be 2" from the plate end, and 3" on center. Save the drawing as E-19-10

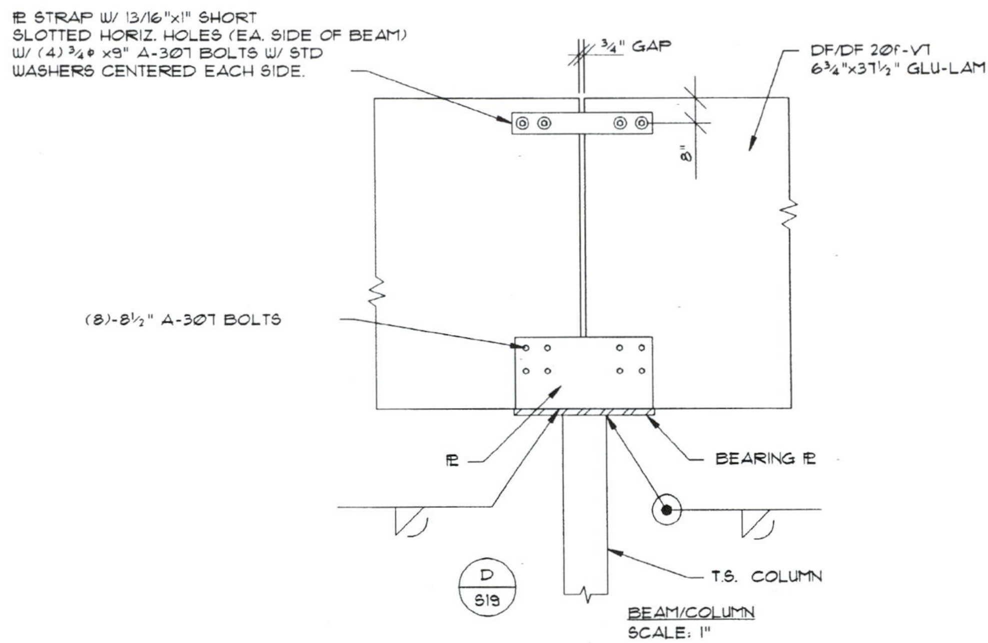

11. Open a new drawing and use the drawing below as a guide to show the column-to-baseplate connection. Establish separate layers for the steel, concrete, text, hatch, and dimensions. Completely label and dimension the drawing. Set the required scale factors for plotting at a scale of 1 ½" = 1'-0".

    The column will be a 6" × 6" × ⅜" t.s. welded to a 10" × 10" × ⅞" base plate with a $^{3}/_{16}$" fillet weld all around. Use (4) 18"-long anchor bolts set in an 8 ½" square grid around the column. Support the base plate on 1" mortar. Save the drawing as E-19-11.

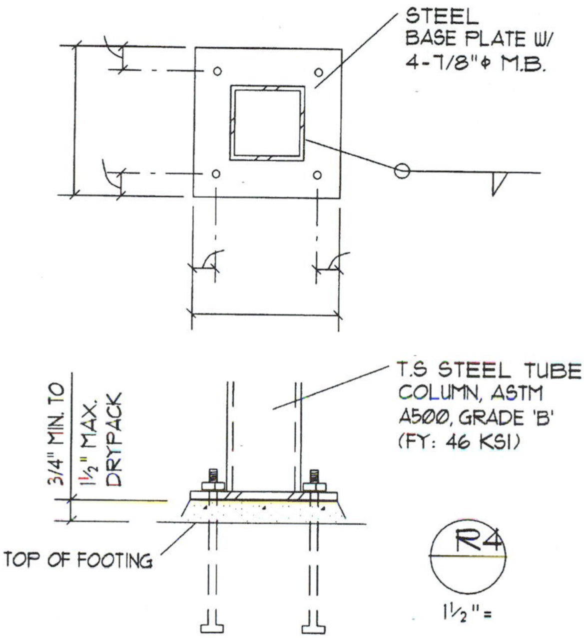

12. Open a new drawing and use the drawing on the next page as a guide to show the beam-to-beam connection. Establish separate layers for the steel, text, hatch, and dimensions. Completely label and dimension the drawing. Set the required scale factors for plotting at a scale of 1 ½" = 1'-0".

The support beam is a W18 × 76 (18 ¼" dp.) steel wide flange with a W18 × 46 (18" dp.) on the left and a W18 × 40 (17⅞" dp.) on the right. Use (2) 4" × 12" × 5/16" steel connection plates welded to the W18 × 76 with 5/16" × 12" fillet each side. Use four machine bolts through the plate and beam @ 3" o.c., 1 ½" from the edge. Save the drawing as E-19-12.

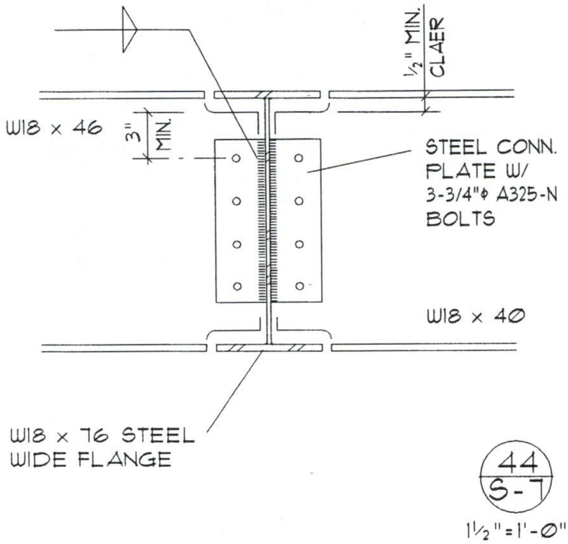

13. Open drawing FLOOR14 and use the drawing on the next page as a guide to draw the slab-on-grade plan. Establish separate layers for the concrete, text, and dimensions. Completely label and dimension the drawing. Save the drawing as E-19-13.

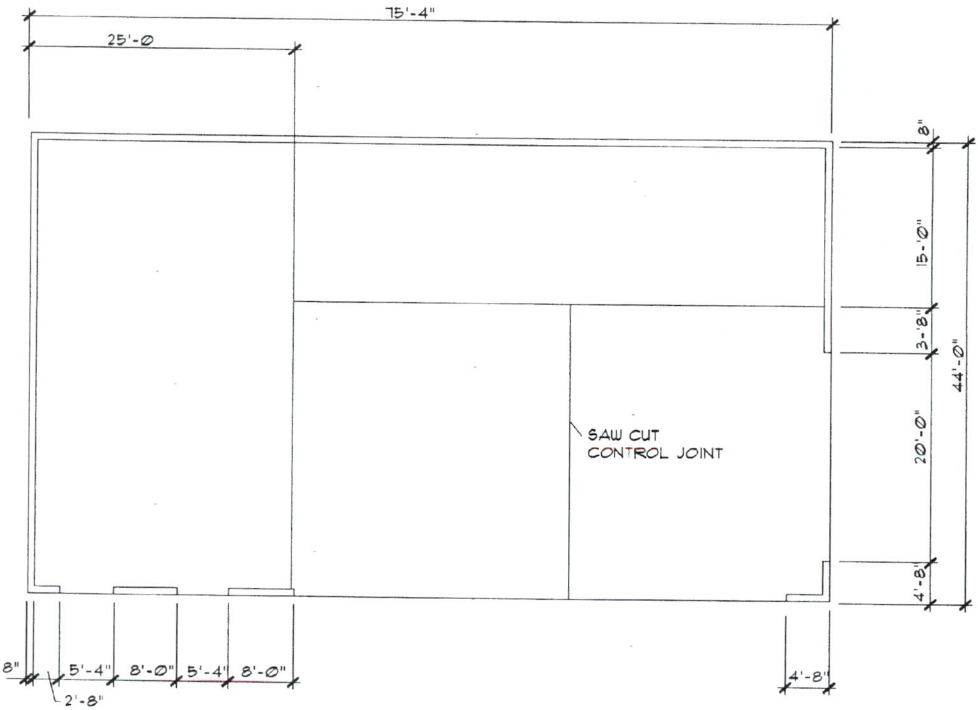

SLAB PLAN

1/8" = 1'-Ø"

14. Open a new drawing and use the drawing on the next page as a guide to steel framing elevation. Establish separate layers for the steel, text, and dimensions. Completely label and dimension the drawing. Set the required scale factors for plotting at a scale of 3/8" = 1'-0". Establish the eave height as 24' . Locate the first Z brace at 7' above the floor, with the balance at 72" o.c. Vertical supports will be set 12" in from each end of the 100'-0" structure with interior supports at 25'-0" o.c. Save the drawing as E-19-14.

15. Open a new drawing and use the drawing on the next page as a guide to draw the wall and floor detail. Establish separate layers for the wood, concrete, steel, text, and dimensions. Completely label and dimension the drawing. Set the required scale factors for plotting at a scale of 1/2" = 1'-0". Establish the upper floor 30" above the lower floor, and the bottom of the 4 × 10 that supports the lower floor is to be 18" clear to finish grade. Dimension the height of the rail as 42". Save the drawing as E-19-15.

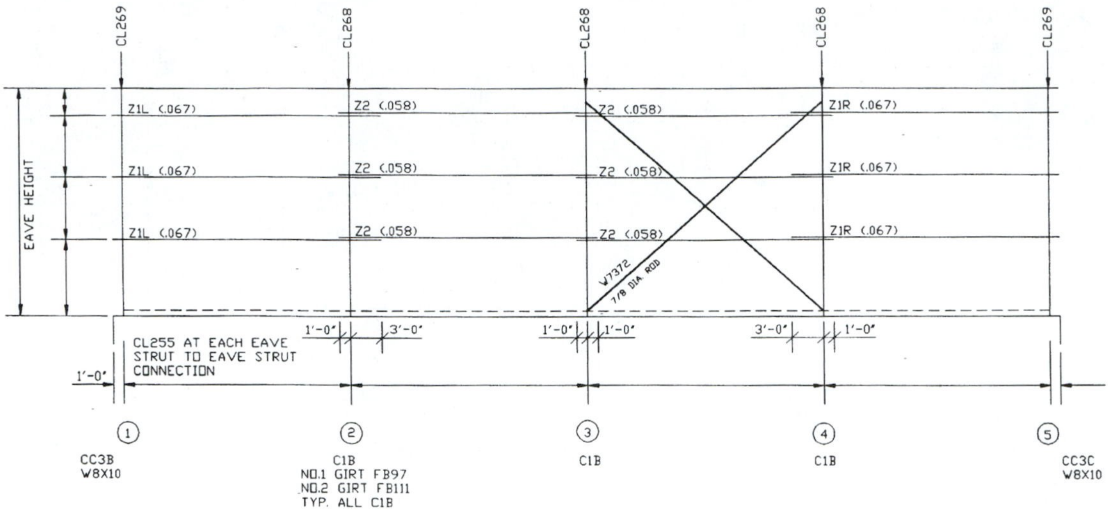
CL269
CL268
CL268
CL268
CL269
EAVE HEIGHT
Z1L (.067)
Z1L (.067)
Z1L (.067)
Z2 (.058)
Z2 (.058)
Z2 (.058)
Z2 (.058)
Z2 (.058)
Z2 (.058)
Z1R (.067)
Z1R (.067)
Z1R (.067)
W7372
7/8 DIA ROD
CL255 AT EACH EAVE STRUT TO EAVE STRUT CONNECTION
1'-0"
1'-0"
3'-0"
1'-0"
1'-0"
3'-0"
1'-0"
1
2
3
4
5
CC3B
W8X10
C1B
NO.1 GIRT FB97
NO.2 GIRT FB111
TYP. ALL C1B
C1B
C1B
CC3C
W8X10

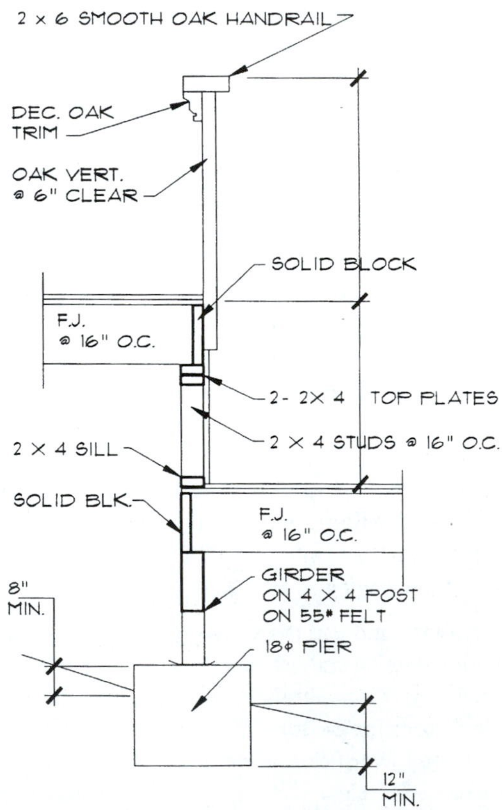
2 x 6 SMOOTH OAK HANDRAIL
DEC. OAK TRIM
OAK VERT. @ 6" CLEAR
SOLID BLOCK
F.J. @ 16" O.C.
2- 2X 4 TOP PLATES
2 X 4 STUDS @ 16" O.C.
2 X 4 SILL
SOLID BLK.
F.J. @ 16" O.C.
GIRDER ON 4 X 4 POST ON 55# FELT
18ϕ PIER
8" MIN.
12" MIN.

CHAPTER 19 QUIZ

1. What do you call a group of dimensioning controls that can be saved for future use?
2. What is used to control individual aspects of dimensioning?
3. How can a listing of dimensioning features be accessed?
4. How can styles be used in an engineering office that does both civil and structural projects?
5. What value controls the size of each dimension feature and how should this value be set for a drawing that will be reproduced at a scale of ¼" = 1'-0"?
6. What option will remove both extension lines from a dimension?
7. The spacing between two dimension lines is controlled by what option?
8. Use the Help menu and determine what the DIMGAP controls.
9. List two features that can be displayed to describe round objects.
10. List five options suitable for an architectural office for terminating a dimension line.
11. List and describe two features of the extension line as it relates to the dimension line and the object being described. Give the value that controls each.
12. What are the options for controlling the height of text?
13. Name the dialog box that can be used to make changes to dimensioning text.
14. What style will be used as a base when creating dimensioning styles?
15. What option will allow for changing an attribute without changing the entire style?
16. What does the TEDIT command affect?
17. Describe how associative dimensions will affect editing.
18. How can the LEADER command be used to provide two arrows pointing to the same text?
19. What dimensioning command is used if the text is to be placed so that it is not parallel, perpendicular, or aligned with a surface?
20. What command can be used to change existing dimensions to the current style?
21. Sketch examples of right angle and closed terminators.
22. Explain the process for creating a user-defined terminator.
23. Explain the options to display a centerline in circles to be dimensioned.
24. Give the proper scale factor for drawing a site plan to be plotted at a scale of ⅛" = 1'-0" and explain how the value is determined.
25. Explain the process to place text above the dimension line.
26. List and explain the options for controlling zeros in dimensions.

27. Explain the difference between associative and normal dimensions.
28. Explain the options of the DIMEDIT command.
29. Explain the effect of selecting an extension line to be erased.
30. Explain the effect of stretching a window that is dimensioned 3'-0" long to a length of 6'-0".

SECTION 4

GROUPING INFORMATION

CREATING BLOCKS AND WBLOCKS

ADDING ATTRIBUTES TO ENHANCE BLOCKS

COMBINING DRAWINGS USING XREF

ACCESSING DRAWING INFORMATION

PLOTTING HARDCOPIES

CHAPTER 20

Creating Blocks and Wblocks

You might have been wondering while working through the first 19 Chapters why you haven't started drawing some really neat projects. Since Chapter 4, you've had the basic drawing skills to draw most projects that you could imagine. Subsequent chapters have helped you develop skills needed to ease the drawing and design process, but still have not equipped you to easily master a complicated project . This chapter will provide the final ingredients for completing complicated projects by introducing you to the BLOCK, INSERT, MINSERT, EXPLODE, WBLOCK, and PURGE Commands.

BLOCKS

Drawings are composed of symbols. To show a door on a floor plan, you don't draw a door—instead you use a line and an arc to represent a door. A block is a group of entities that are treated as one object. The objects used to represent a door symbol can be saved as a block and be inserted throughout the drawing as one object. The BLOCK command allows symbols that are used repeatedly to be created, saved, and inserted easily in the drawing in which they were created. This provides many benefits for the user.

BENEFITS OF BLOCKS

Symbols that are used frequently can be saved as a block, which will greatly increase drawing speed, uniformity, and efficiency.

Speed

If you're drawing a typical residential floor plan, doors will need to be represented 20 or 30 times. Rather than draw each door, one door can be drawn and saved as a block. As a block, a door can be inserted quickly and the size altered or the position rotated, as seen in Figure 20.1. With a little more effort, these same qualities can also be accomplished using the COPY, ROTATE, MIRROR, ARRAY, or SCALE commands— but why use several commands when you could use one?

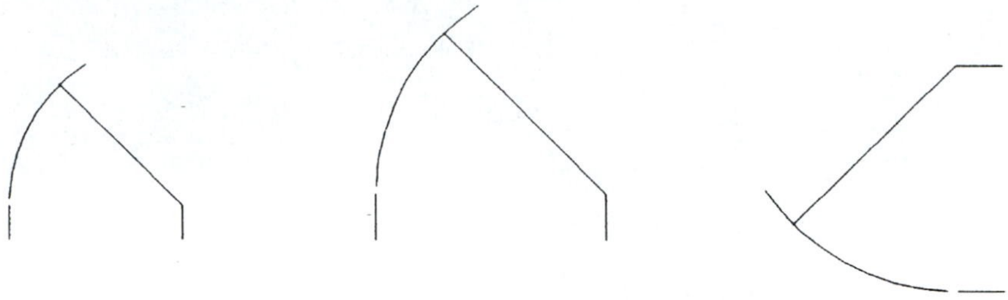

Figure 20.1 *Blocks can be inserted into a drawing, altered in size, and rotated to meet specific needs.*

Editing

Throughout the evolution of construction drawings, many changes and revisions will take place. Blocks can greatly aid in this revision process. If a client decides that all of the doors in a building are to be 6" wider, several hours might be required to change all of the doors on the various floor plans. If the doors are drawn as a block, the block can be redefined and all references to that block will be updated automatically.

Attributes

An attribute is text that is associated with a block. Many blocks, such as a door, require text to explain variations. Size, material, quantity, and supplier are examples of information that can be saved as an attribute with a door block. This text is called an attribute and can vary with each application of the block. Attributes can be displayed on the screen as text or they can be stored with the drawing but made invisible. Attributes associated with blocks can also be extracted from a drawing and transferred to a data base form, such as a door schedule.

Efficiency

In addition to saving time and providing a quick method for revisions, the BLOCK command combines the storage of groups of objects into one unit. When drawing a door, information must be stored about the start and end points of the original line and arc, the radius of the arc, the scale factor as the symbol is enlarged or reduced for various uses, and the location of each unit as it is displayed throughout the drawing. When the symbol is installed as a block, only a block reference is stored, saving valuable storage space.

USES FOR BLOCKS

Before a possible listing of blocks is considered, it should be remembered that the list is endless. Any symbol that is used repeatedly should be stored in a template drawing as a block. A sample listing for architectural blocks would include items for the site, floor and foundation plans, elevations, and sections as shown below.

Site plan: north arrow, trees, shrubs, spas, pools, fountains, curbcuts and driveways, utility symbols, and general notes. Plots for subdivisions might include structure shapes, sidewalks, and decks or patios.

Floor plans: appliances, door and window symbols, heating equipment, heating and cooling symbols, electrical symbols, and plumbing symbols.

Framing plans: joist and rafter direction indicators, section tags, and general notes.

Foundation plans: piers, post and piers, beam and column symbols, general notes, and foundation details.

Elevations: door styles, window styles, shutters, trim, eave molding, decorative posts, garage doors, gable wall vents, trees and shrubs, lighting fixtures, and people.

Sections: structural shapes such as plates, sills, double sills, beams, columns and piers, foundation components, eave components, insulation, standard dimensions, and common notes.

In addition to drawing simple symbols as a block, several individual blocks can be combined to form a larger block. Figure 20.2 shows an example of a typical truss-to-plate intersection. The block could be inserted into a section drawing, mirrored, and then the truss shape completed by the FILLET and EXTEND commands, as seen in Figure 20.3.

DRAWING OBJECTS FOR BLOCKS

Because a block is a group of objects that is saved as one object, there is no special method to drawing the block. The components that will comprise the block can be created using any of the drawing or editing commands. Some consideration should

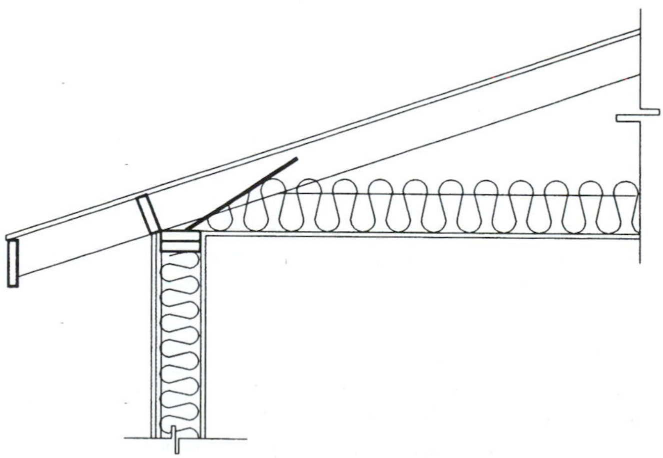

Figure 20.2 *Several objects can be combined to form a block.*

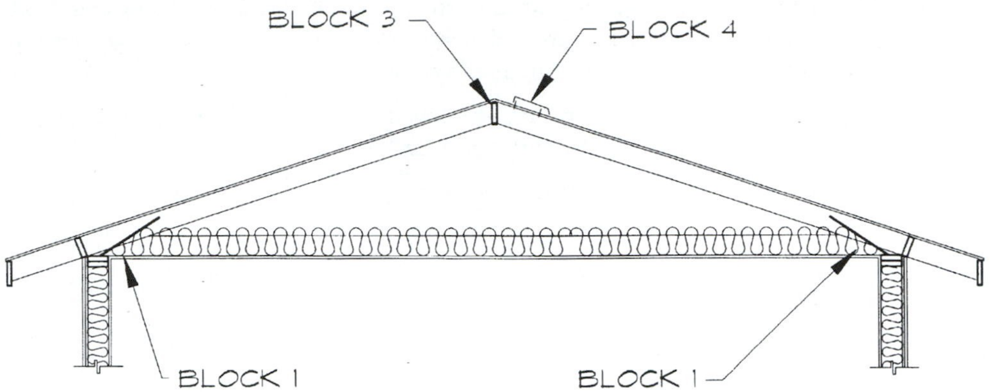

Figure 20.3 *Using the MIRROR, FILLET, and EXTEND commands, this section was created by inserting and connecting blocks.*

be given to the layer the block is created on and the size of the block while it is being constructed, to avoid some problems when the block is inserted into the drawing.

The Effects of Layer on Blocks

A block will take on the characteristics of the layer on which it is created. A block can also consist of objects that have been drawn on several layers with different linetypes and colors. When the block is inserted into a drawing, the block will retain its original characteristics. If a lavatory is created on a layer titled MPLUMB and inserted into a drawing with the UPPLUMB layer current, the lavatory will be inserted into the drawing on the MPLUMB layer. If the block was created on a layer that does not exist in the current drawing, the layer containing the block will be created. There are, however, three exceptions to this guideline.

Blocks that are created on the 0 layer will be placed on the current layer when placed into a drawing.

Blocks drawn with the color created BYLAYER are drawn with the color of the current layer, but will reflect the color of the layer into which the block is inserted.

Blocks drawn with the linetype created BYBLOCK are drawn with the linetype of the current layer, but will reflect the linetype of the layer onto which the block is inserted.

Block Size

Once the block is saved, it can be inserted throughout a drawing. When the INSERT command is introduced later in this chapter, you will have an opportunity to alter the size of the block in the X and Y direction. Altering the size of the block during the insertion process can be easier if the size of the block is considered during creation.

Some blocks, such as toilets, electrical symbols, and appliances, are a standard size. These blocks can be drawn at their actual size and inserted. Other blocks, such as doors, windows, tubs, and showers, vary in size with each application. Drawing these blocks in one-unit squares will allow the size to be adjusted each time it is inserted. Rather than having separate blocks for 4', 5', and 6' windows, one window can be created, and the X value controlling its length can be altered.

The size of the unit can vary depending on the drawing the block will be inserted in. On floor plans, you might use 1" as the unit and scale up, or use 1' as the unit and scale down. Figure 20.4 shows a window created using a 12" unit, and the effects that can be achieved with one block.

CREATING A BLOCK

Figure 20.5 shows the drawing process to create a toilet, which can then be saved as a block. Keep in mind that although the toilet has been drawn, it is not a block yet. Once the drawing is ready to be reproduced, enter the BLOCK command by selecting the Make Block icon from the Draw toolbar, by entering **BLOCK** (EnterKey) at the command prompt, or by selecting Make… from the Block flyout menu found in the Draw pull-down menu. Each method will produce the Block Definition dialog box shown in Figure 20.6.

Choosing a Name

Start the process by entering the name of the block to be created in the Name: edit box. A name of up to 255 characters can be used to describe the block. The block name can contain letters, digits, and special characters such as $, -, and _. Any character not used by Microsoft or AutoCAD, including blank spaces, can be used. Use

Figure 20.4 *BLOCKS that vary in size should be created using one unit, such as a foot, so the size of the block can be adjusted easily each time the BLOCK is inserted.*

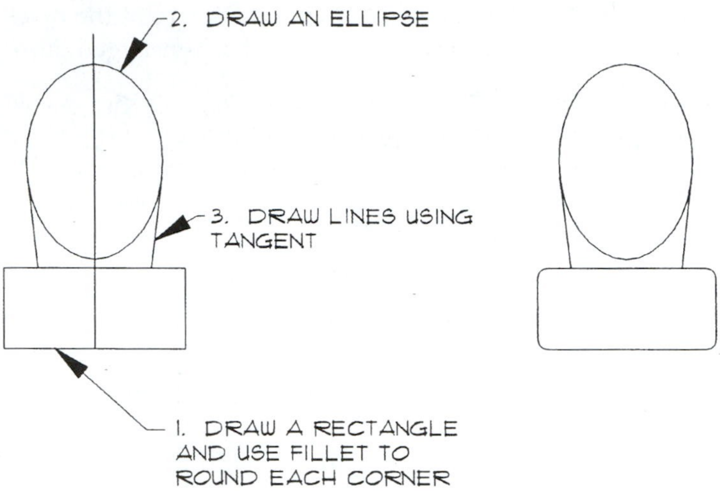

Figure 20.5 *A toilet can be created by drawing an ellipse and a rectangle and connecting the two with straight lines. The FILLET command was used to round each corner.*

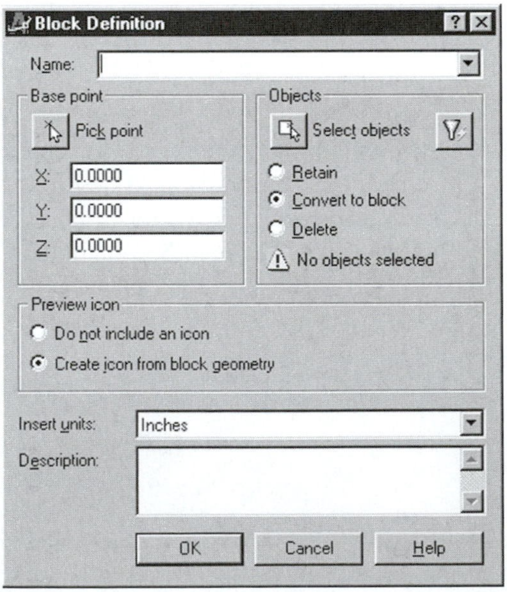

Figure 20.6 *Once the drawing is ready to be reproduced, enter the BLOCK command by selecting the Make Block icon from the Draw toolbar, by entering* **BLOCK** *(EnterKey) at the command prompt, or by selecting Make… from Block flyout menu found in the Draw pull-down menu.*

a name that will describe the contents. For instance, DOOR-SWG-32 would be an appropriate name to describe a 2'-8" swinging door. If a name for an existing block is provided for the new block, the old block will be destroyed and the new block will take its place. Before it is vaporized, you will be given an alert message by AutoCAD warning that the block is about to be redefined.

Selecting the Objects

Once the name has been provided, press either the Pick Point or Select Objects button. Each option must be selected, but the order will not effect the outcome of the command. For this discussion, the object will be selected first. Pick the Select Objects or Quick Select button to define the objects that will be in the block. The use of the Quick Select dialog box will be explained shortly. Selecting the Select Objects button temporarily removes the dialog box from the screen and allows the objects that will make up the block to be selected. As the box is removed, the command line will prompt to 'Select Objects.' Objects to comprise the block can be selected using any object selection method. Once the desired objects have been selected, pressing (EnterKey) will return the Block Definition dialog box. An image of the selected objects will be displayed in the Preview Icon display box similar to Figure 20.7. The option for no icon can be activated if you do not want to see the image you'll be using as a block.

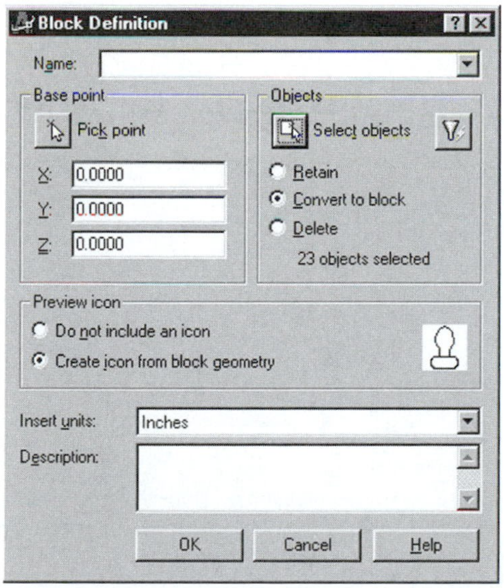

Figure 20.7 *Once the desired objects to comprise the block have been selected, pressing (EnterKey) will return the Block Definition dialog box with an image of the selected objects displayed in the Preview Icon display box.*

Note: The BLOCKICON command can be used to display images of blocks stored in drawings created using AutoCAD 2000 or previous releases. The command can be entered at the command prompt and will provide prompts for what blocks you would like to update. Pressing (EnterKey) will update all blocks.

Using Quick Select

If the Quick Select button is selected, a Quick Select dialog box similar to Figure 20.8 will be displayed. The quick select dialog box can be used to define the filtering criteria that will be used to select the objects for the block. On simple blocks such as the toilet, it can easily be selected using a window. Quick Select can be used to select complex shapes or multiple objects. Notice that the current filter is defined as the entire drawing in the Apply to edit box. If there is no current selection set, the entire drawing will be used as a block. This option will be explored later. To select object for the block, pick the Select Objects icon at the right end of the Apply to edit box. This will temporally remove the dialog box from the screen and allow the objects to be selected. The process is the same as used for the Block Definition dialog box. Once the desired objects have been selected, pressing (EnterKey) will return the Quick Select dialog box. Other key elements of Quick Select include:

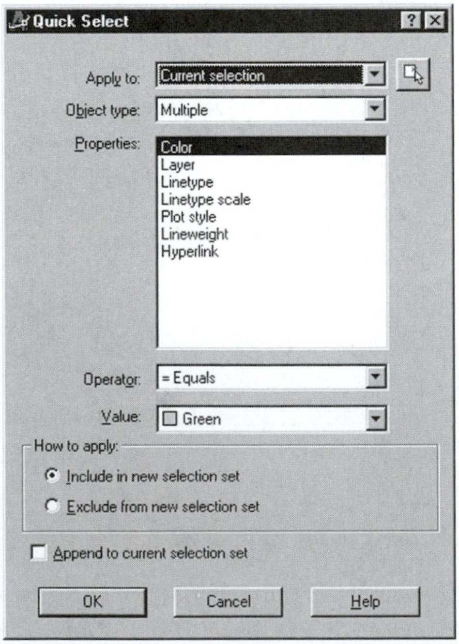

Figure 20.8 *The quick select dialog box can be used to define the filtering criteria that will be used to select the objects for the block.*

- Object Type specifies the type of object used to define the filter. The current choices are Multiple, or Block Reference. Since you're defining a block, choose Block Reference.
- The Properties box can be used to specify properties to be used as a filter. Objects to form the block can be defined by the color, lineweight, linetype, or any of the other listed properties.
- Operator controls the range of the selected property filter. Depending on the selected property, options may include Equals, Not Equal To, Greater Than, and Less Than. Selecting the edit arrow displays a menu to allow the operator value to be altered.
- Value defines the filter property value. If the selected property is a color, the value will specify a color. If the property is a layer, the 0 layer is the default value. Selecting the edit arrow displays a menu to allow the value to be altered.
- The How to Apply portion of the box specifies if the selected objects will form or be excluded from the selected set to form the block. Choosing Exclude will create a block of all objects that do not match the selection filter.

Note: To create a block such as a toilet, use Select objects. Quick select can be used to select all of the walls of a floor plan to form a block without having to select each one of them. By defining a filter set that will include all of the walls, or exclude everything except the walls, the block can be defined much faster than if each wall had to be selected.

Object Selection Options

In addition to methods for selecting objects to define the block, the Objects box contains options for Retain, Convert, and Delete the selected objects.

- Retain—Selecting Retain will keep the selected objects in their original state in the selection set. Once the object is selected to be a block, the original objects will remain in their current location.
- Convert to Block—Selecting this option will remove the selected objects from the selection set and replace them with a previously defined block.
- Delete—Selecting this option will remove the objects that define the block from the drawing. This option works well if you create the toilet (or any other object) that is to become the block in the middle of the living room. With Delete active, once the object is selected to be a block, the original object will be removed from the drawing. It will still exist as a block in the drawing base, but it can't be seen until the block is inserted.

SPECIFYING THE BASE POINT

The insertion point of the block will align with a point that will be designated as the block is inserted into other parts of a drawing. The insertion point can be selected by entering X and Y coordinates, or by selecting any point of the object. The Pick Point button temporarily removes the dialog box from the screen and allows the insertion

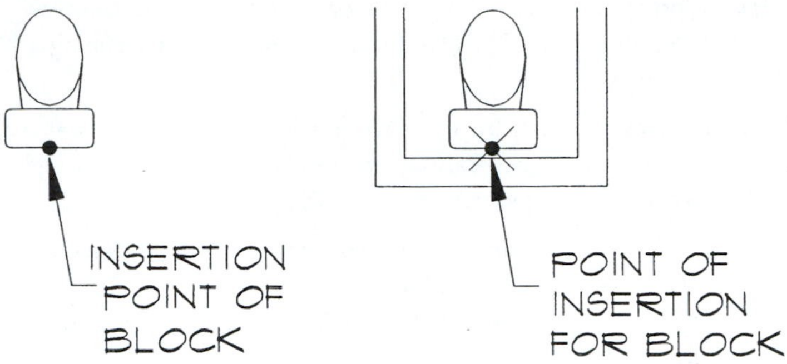

Figure 20.9 *An insertion point defines the location of the block to be used as a handle.*

point to be selected. Figure 20.9 shows the effect of the insertion point of a block with the insertion point at the destination. The insertion point should be chosen using OSNAP, based on the block's use. Blocks such as a toilet, a lavatory, or a sink will be inserted most easily by using a center point. Other objects such as a sill or beam may be best located from an edge or intersection. Figure 20.10 shows some typical insertion points for common architectural features. The insertion point also can be used as a rotation point for the block as the block is inserted into its new location. Figure 20.11 shows how a block can be rotated about its insertion point. Other options will be discussed as the INSERT command is explored. Using a center point or the lower left corner of a block will help to keep track of where to insert the block.

DESCRIPTION

The Description edit box allows a description to be assigned to the block. A block can be created without providing a description, but on a complex block, the description may help you distinguish it from a similar block.

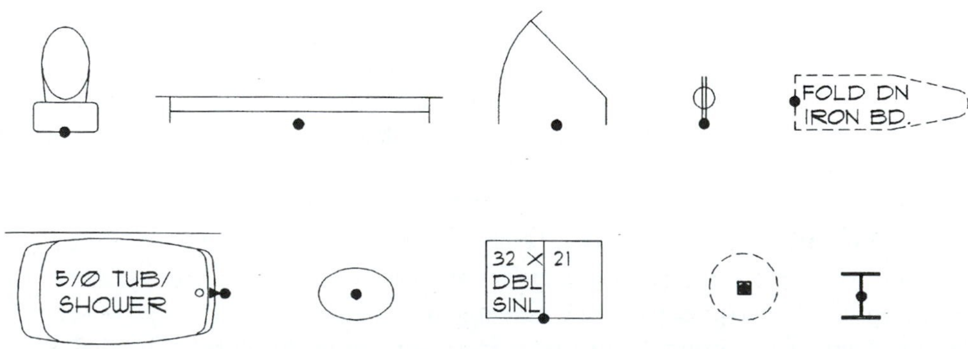

Figure 20.10 *Typical insertion points for common architectural features.*

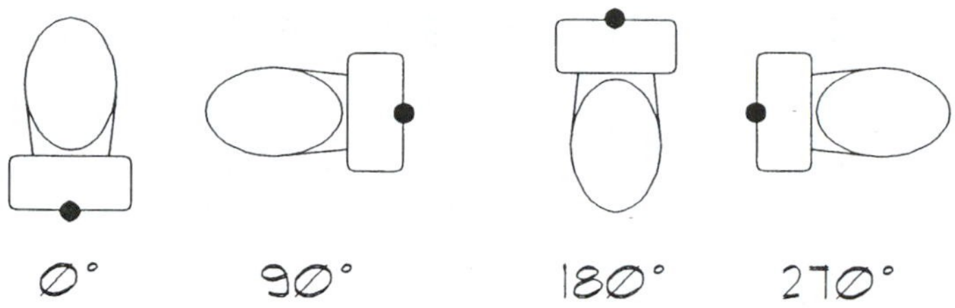

Figure 20.11 *Rotating a block around its insertion point.*

SUMMARY OF CREATING A BLOCK

It's taken several pages to explain what a block is. Remember that a block is just a collection of objects that can be used over and over. Because they are so important to your drawing future, use the summary as you create blocks. Use the following steps to create blocks:

- Enter the Block Definition dialog box by selecting the Make Block icon from the Draw toolbar, by entering **BLOCK** (EnterKey) at the command prompt, or by selecting Make… from the Block flyout menu found in the Draw pull-down menu.
- Provide a name that will define the block.
- Select the drawing objects to be reproduced as blocks.
- Select a base point to be used as the block is inserted into the drawing.
- Select modifiers in the Objects and Preview portions of the dialog box.
- Select OK to save the block to the drawing base.

The object has now been saved as part of the drawing base and is ready to be reused throughout the drawing. If another toilet, or whatever object is used as a block, needs to be added to the drawing, it can be added to the drawing by inserting the block rather than redrawing the objects.

Note: You've just read through the process to create a block. As you create a block, it may leave you waiting for something exciting to happen. The creation of a block is nothing special. It's what you can do with it that will revolutionize your drawing process.

PREPARING BLOCKS FOR INSERTION

Once a block has been created and saved, it can be placed in a drawing using the INSERT command. The command can be accessed by selecting the Insert Block icon from the Draw toolbar, by typing **I** (EnterKey) at the command prompt, or by selecting the BLOCK… option of the Insert pull-down menu. Each method will produce an Insert dialog box similar to Figure 20-12.

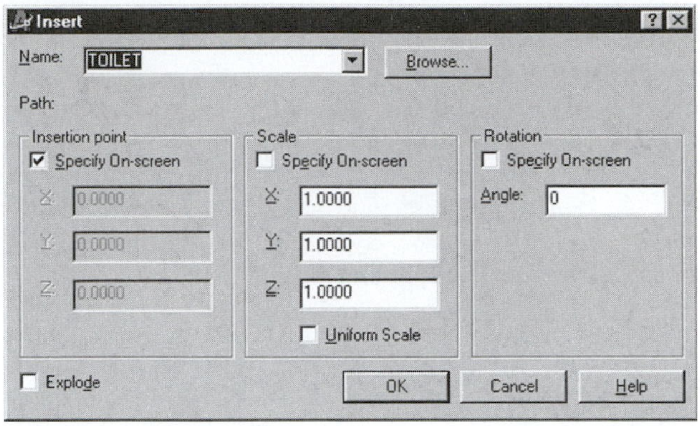

Figure 20.12 *Selecting the Insert Block icon from the Draw toolbar, typing **I** (EnterKey) at the command prompt, or selecting the BLOCK... option of the Insert pull-down menu will display the Insert dialog box.*

Choosing a Block

The Name edit box is used to select the block to be inserted. The name of the block to be inserted can be typed in the edit box, or selected from a list of blocks contained in the drawing. Selecting the edit arrow beside the box will display a listing of all blocks contained in the drawing. Selecting the Browse... button will open the Select Drawing File dialog box. The dialog box is the same as the file selection dialog boxes that you've used as drawing files are opened. The dialog allows a block or file to be selected for insertion. Inserting files will be introduced later in this chapter.

Specifying an Insertion Point

Activating the Specify Parameters On-screen box allows an insertion point to be used to specify where the block will be located. With the Specifying On-screen box inactive, the X and Y coordinate locations (Z is the vertical reference for 3D) can be used to locate the insertion point. As the block was created, an insertion point was determined. Once the desired insertion parameters are adjusted, a prompt will be displayed to provide a location of the original insertion point. Figure 20.13 shows the relationship of the block insertion point to the destination insertion point. Once the insertion point is selected, the block will be dragged into the indicated position, and the scale can be altered. The parameters to control the insertion can be adjusted before removing the dialog box to insert the block, or at the command prompt before the block is inserted.

Scale

This box allows you to define scale factors for the block to be inserted. With the Specify On-screen box active, a prompt will be displayed at the command prompt before the block is inserted. With the box inactive, the X, Y, and Z edit boxes are activated. Unless a different value is entered, the value entered for X will become the

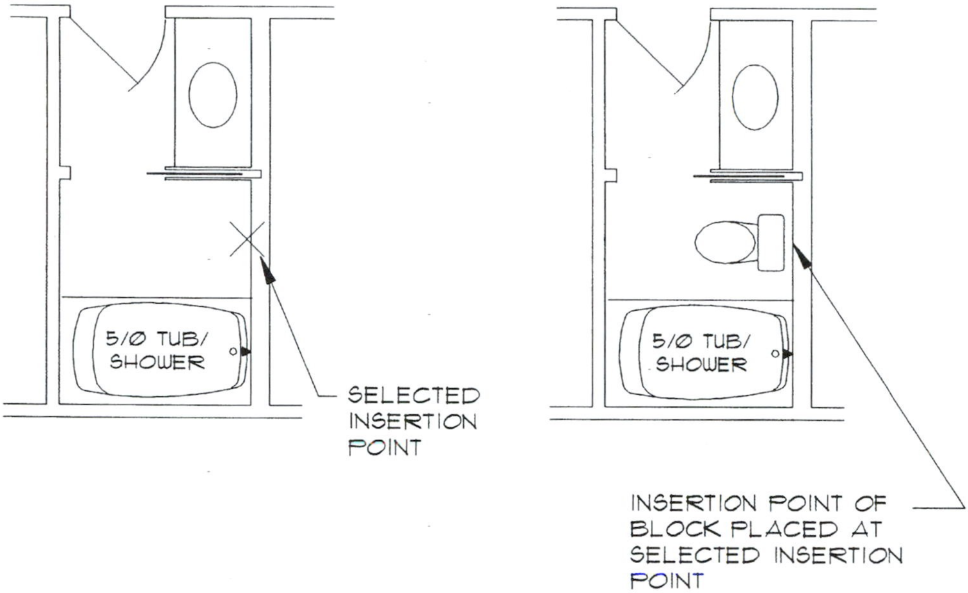

Figure 20.13 *The effect of the selected insertion point of a drawing and the insertion point of a block.*

default value for Y. Any dimensions contained in the block will be multiplied by the scale factor as the block is inserted. If the toilet in Figure 20-13 was drawn as full size, the scale for inserting the toilet would be 1. Inserting a window created as a 1' block to be a 6'-long window in a 6" exterior wall would require an X scale factor of 6, and a Y scale factor of .5. With the Uniform Scale box active, the Y value will equal the X value. The effects of the scale factor can be seen in Figure 20.14.

Negative Scale Factors

The location of a block can be altered by using negative X and Y values. Using negative values combines the INSERT and MIRROR commands. Figure 20.15 shows an example of negative value alternatives.

Rotation

The Rotation box allows you to supply a rotation angle. With the Specify On-screen box active, a prompt will be displayed at the command prompt before the block is inserted. With the box inactive, the Angle edit box is activated and an angle can be entered to control the rotation of the block at insertion. Figure 20.16 shows an object rotated around its insertion point. Any angle can be provided for the rotation angle.

Explode

Blocks function as one object when they are inserted into a drawing. If the Explode box is activated, as the block is inserted into the drawing, each object of the block will function as a separate object.

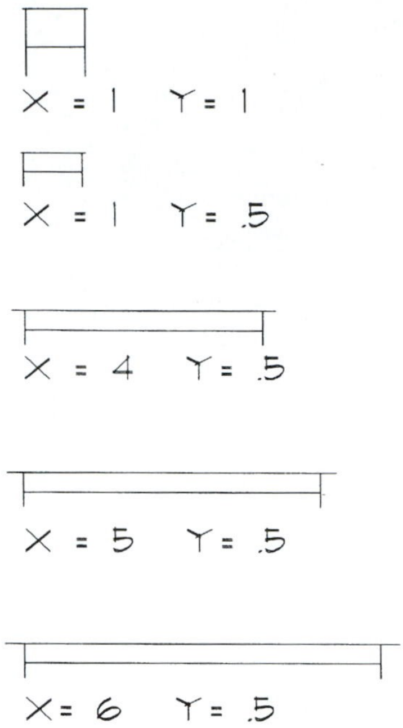

Figure 20.14 *The effect of X/Y scale factors on a block.*

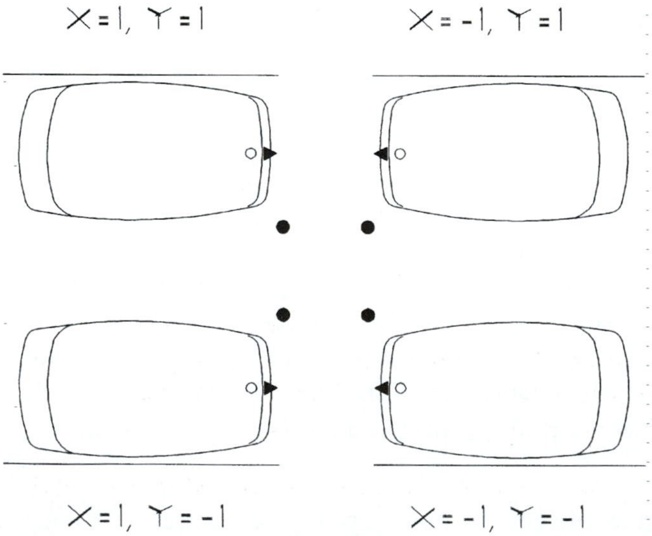

Figure 20.15 *Altering a block using negative scale factors.*

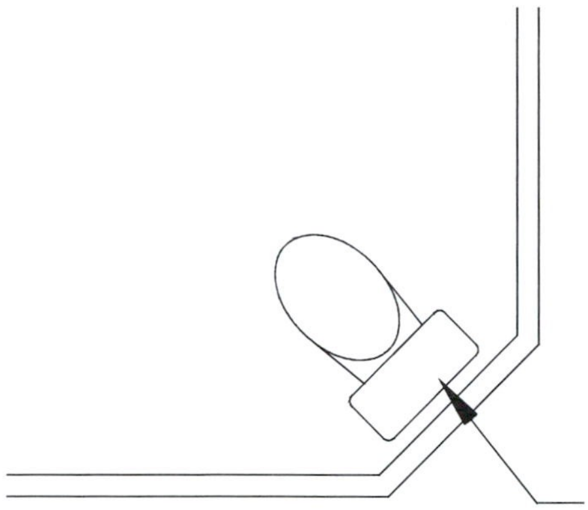

Figure 20.16 *Specifying an angle of rotation allows a block to be rotated relative to its insertion point.*

INSERTING BLOCKS

Once the parameters of the Insert dialog box have been set, selecting the OK button will close the dialog box and allow a selection point to be specified. With the Insert dialog box in the default settings, the following prompt will be displayed at the command prompt:

> Command: **I** (EnterKey) *(The Insert dialog box will be displayed. Picking the OK button will close the dialog box and continue the command prompt.)*
>
> Specify insertion point or [Scale/X/Y/Z/Rotate/PScale/PX/PY/PZ/ PRotate]:

Selecting an insertion point will insert the block to the specified parameters. The scale, insertion coordinates, and rotation values that were set in the Insert dialog box can be altered at the command prompt. Additional prompts are also shown at the command prompt.

Adjusting Insert Values from the Command Prompt

The X, Y, and Z scale factors set the scale for the block to be inserted. The PScale alters the scale of the block as it is dragged to its insertion point, but it does not control the size used to display the block, once inserted. PScale provides a preview of what could be. It's a guesstimate of how big you might want the block. Setting a PScale of 2, will double the size of the block as it is dragged to its insertion point. Once the PS option is entered at the command line, a prompt will be given to provide a scale for the display of the block. The command sequence is:

> Command: **I** (EnterKey)
>
> Specify insertion point or [Scale/X/Y/Z/Rotate/PScale/PX/PY/PZ/ PRotate]: **PS** (EnterKey)
>
> Specify preview scale factor for XYZ axes: **2** (EnterKey)

Specify insertion point: *(Select insertion point.)*

Enter X scale factor, specify opposite corner, or [corner/XYZ] <1>

This final prompt is your chance to adjust the insertion scale. Pressing (EnterKey) once allows the X value to be altered. Selecting (EnterKey) twice will insert the block based on the scale value of the Insert dialog box.

Each of the other P values function in a similar method. PX allows a preview of a variation of the X value. PY provides a similar preview of an altered Y value. PRotate allows a preview and override of the rotation angle.

Corner Specification of Scale

A final option for inserting a block is to determine the scale by picking one corner of the block and then picking the opposite corner with a window. This option works well for inserting objects, if the scale is not critical. When the prompt for X is displayed, use the cursor to indicate a point for the beginning of the box. The second location should be above and to the right of the first point to keep the block in its original orientation.

WORKING WITH BLOCKS

In addition to inserting blocks, several other options can be used to control the use of blocks. Useful tools include the MINSERT command for multiple inserts, using an entire drawing as a block, building blocks by using other blocks in a process known as nesting, exploding blocks, and controlling blocks with the Design Center

MINSERT—MAKING MULTIPLE COPIES

The MINSERT (multiple insert) command combines the features and the command sequences of INSERT and ARRAY. MINSERT will allow blocks to be arrayed in a rectangular pattern typical of what might be found in the interior column supports for multilevel structures, or the supports of a post-and-beam foundation. The MINSERT command can be started by typing **MINSERT** (EnterKey) at the command prompt. Figure 20.17 shows an example of a multiple insert. The drawing was created using the following command sequence:

Command: **MINSERT** (EnterKey)

MINSERT

Enter block name or [?]:<toilet> **PIER** (EnterKey)

Specify insertion point or [Scale/X/Y/Z/Rotate/PScale/PX/PY/PZ/ PRotate]: *(Select the insertion point.)*

Enter X scale factor, specify opposite corner, or [Corner/XYZ] <1>: (EnterKey)

Enter Y scale factor, < use X scale factor>: (EnterKey)

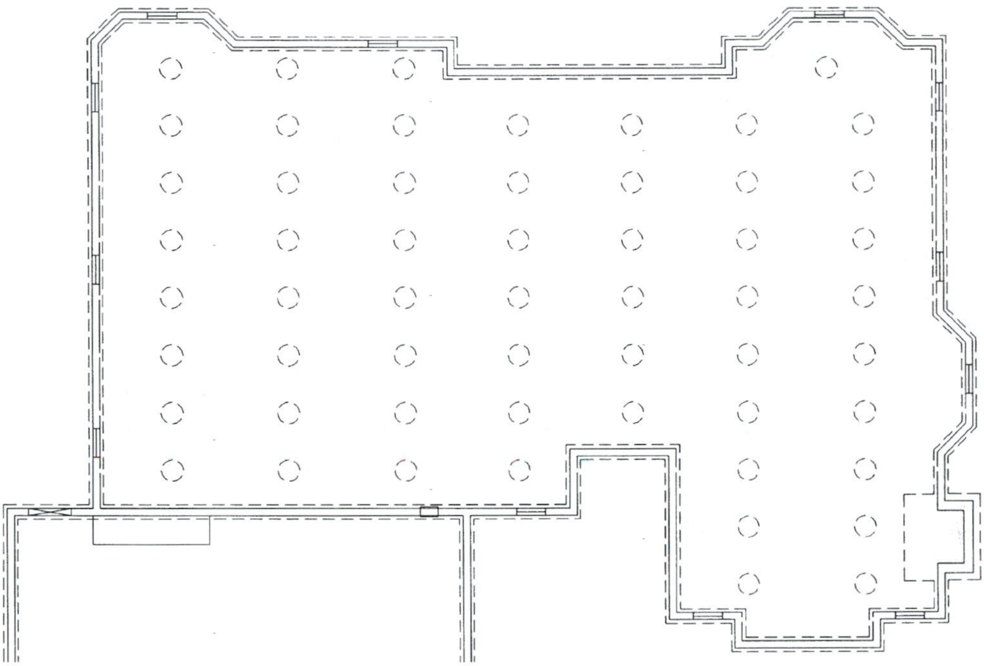

Figure 20.17 *The MINSERT command allows blocks to be arrayed in a rectangular pattern. (Courtesy Piery & Barclay Designers, Inc., A.I.B.D.)*

Specify rotation angle: <0>: (EnterKey)

Enter number of rows (---) <1>: **3** (EnterKey)

Number of columns (|||) <1>: **6** (EnterKey)

Unit of cell or distance between rows (---): **96** (EnterKey)

Distance between columns (|||): **48** (EnterKey)

Command:

The array pattern created with MINSERT has the qualities of a Block, except that it cannot be exploded.

MAKING AN ENTIRE DRAWING A BLOCK

An entire drawing can be used to form a block. When the entire drawing is to be a block, an insertion point other than 0,0 may be desired. The BASE command can be used to define a new base point. The command can be accessed by typing **BASE** (EnterKey) at the command prompt, or by selecting Base from the Block fly-out menu of the Draw pull-down menu. Each will produce the following prompt:

Command: **BASE** (EnterKey)

BASE

Enter base point: *(Select a point.)*

The base point will now become the insertion point.

A better alternative to this method of using an entire drawing as a block will be introduced later as WBLOCKS are introduced, and in Chapter 22 as externally referenced drawings (XREFS) are explored.

NESTING BLOCKS

So far you've been introduced to creating blocks to represent common symbols such as a toilet, sink, door, or any number of repetitive drawing elements. Blocks can also be combined with other objects or blocks to form a nested block. A kitchen for a large apartment complex could be created as a block and inserted in each unit. The kitchen block will contain other blocks, such as the dishwasher, sink, stove, refrigerator, and pantry. Joining blocks of individual units can create entire buildings of the complex. Uses for nested blocks are almost endless for construction drawings. Figure 20.18 shows an example of nested blocks used to create a foundation plan.

Care must be taken in the use of nested blocks to avoid confusion in naming and layering. Use the following guidelines to avoid confusion:

- Assign properties individually, rather than ByBlock, if all instances of a block require the same color, layer, linetype, and lineweight properties.
- To control the properties of each block, using the properties of the layer the block is inserted on, draw the objects that form each block, including nested blocks, on the 0 layer. Set properties such as color, layer, linetype, and lineweight properties ByLayer.
- To control the current properties of each block, using the properties of the layer the block is inserted on, draw the objects that form each block, including nested blocks, on the 0 layer. Set properties such as color, layer, linetype and lineweight properties ByBlock.

CONTROLLING BLOCKS WITH DESIGNCENTER

Earlier it was pointed out that AutoCAD 2000 is the first release of AutoCAD that allows you to work on two drawings in one drawing session of AutoCAD. One of the benefits of doing this is that you can move or copy blocks in one drawing to a different drawing. This can be done with opened or closed drawings. Use the following steps to copy blocks between two open drawings:

- Select the desired block to be copied.
- Right-click to display the shortcut menu.
- Select Copy.

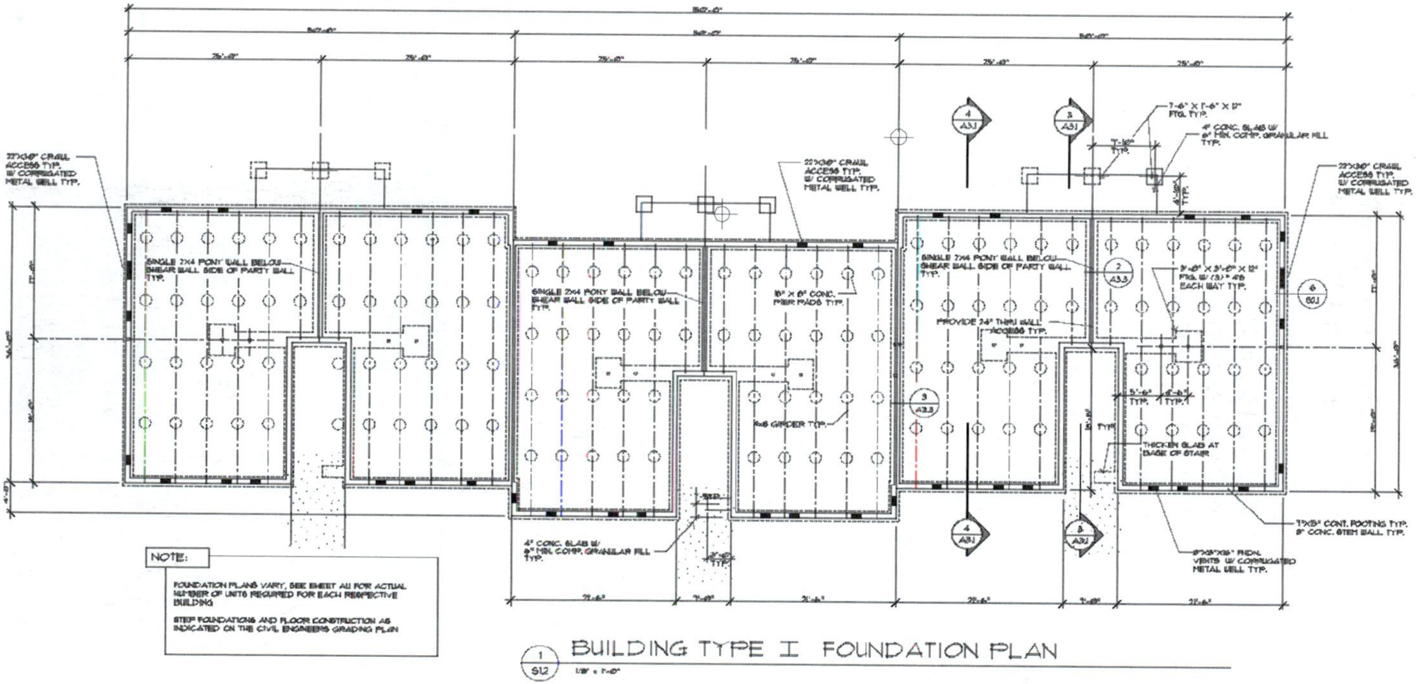

Figure 20.18 *Blocks for the vents, piers, and girders are combined to create a foundation for each unit. A block was then created for each unit and mirrored to form the adjoining unit. A block was then made of each pair of units, and then blocks of each pair were inserted to form the building foundation.*

- Move to the other drawing and make it the current drawing by picking the title bar.
- Select Paste from the Edit pull-down menu.
- Insert the block in the desired location.

Copying Blocks from Unopened Drawings

The DesignCenter allows blocks to be inserted from drawings that are not open. If you work with consulting firms, you can copy a block from a drawing in their office, if you have the proper access to their network. For now, we'll work on copying a block from one drawing to another. Future chapters will provide information on working with networked machines and consulting firms, using the Internet. To copy a block from a drawing files on your computer, open the DesignCenter by selecting the icon from the Standard toolbar. Chapter 22 will introduce additional method for opening the DesignCenter. The DesignCenter will resemble figure 20.19. The exact appearance will depend on what drawings are opened. Pick the Desktop icon in the upper left corner. This will produce a display exactly like the display produced by the Windows Explorer. Figure 20.20 shows a portion of this display. Double-clicking the listing of FLRPLAN2 will produce a display of the contents of the drawing similar to Figure 20.21. Double-clicking the Blocks listing will provide a listing of the blocks contained in the drawing, similar to Figure 20.22. Highlighting the name of the desired block displays the block and the display window, and allows that block to be dragged into the current drawing.

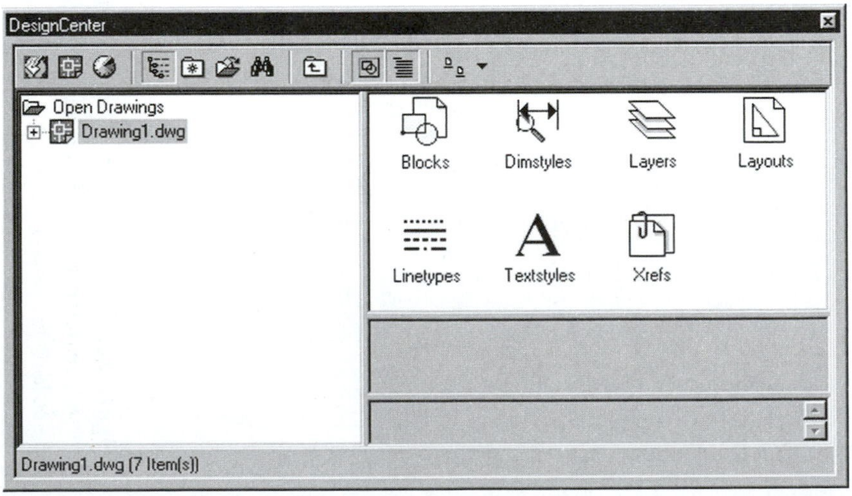

Figure 20.19 *The DesignCenter dialog box.*

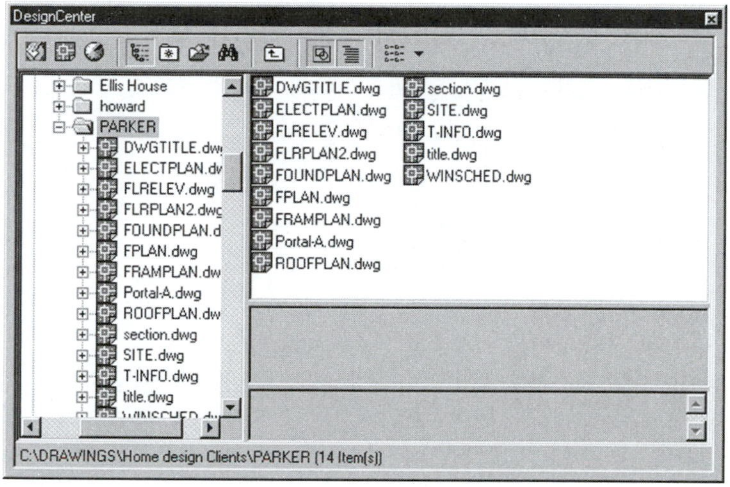

Figure 20.20 *Selecting the Desktop icon from the DesignCenter will produce a display similar to Windows Explorer.*

EDITING BLOCKS

Blocks may be edited by changing the description, by using the EXPLODE command, or by using the Redefine option of the BLOCK command.

EDITING BLOCK DESCRIPTIONS

A block description can be altered using the Block Definition dialog box. Open the dialog box by selecting the Make Block icon from the Draw toolbar or using one of

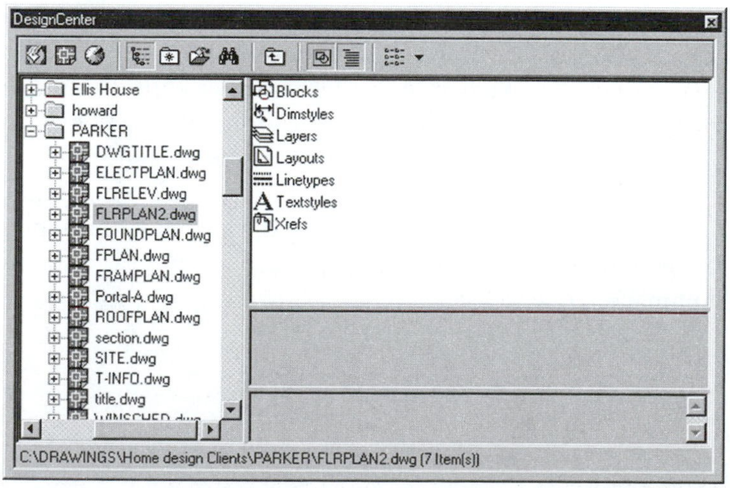

Figure 20.21 *Selecting a drawing from the menu will display a list of the contents.*

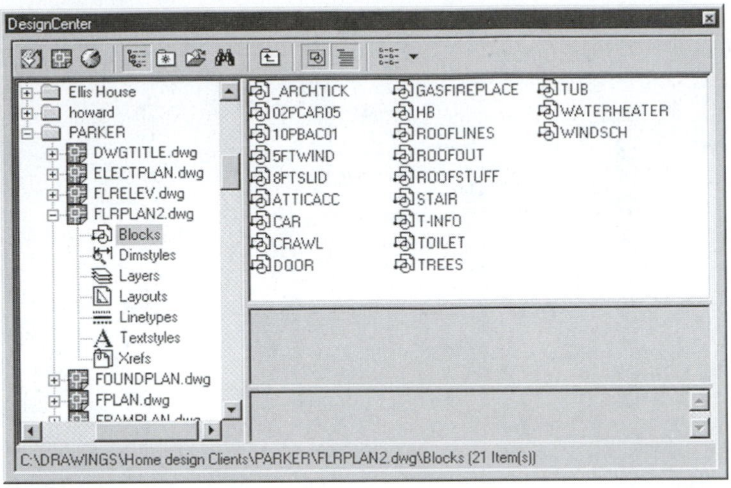

Figure 20.22 *Selecting BLOCKS from the menu in Figure 20.21 will display a list of blocks in the selected drawing. One or all of the listed blocks can be copied to a new drawing.*

the methods mentioned earlier in this chapter. Select the edit arrow beside the Name box. This will display a listing of the current blocks. Select the name of the block that will be altered. Enter the changes that will be made. Typically, this might include adding objects to the block. Choose OK when you're satisfied with the description.

EXPLODING BLOCKS

If the Explode option is activated in the Insert dialog box, each object in the block will function independently of other objects in the block. By default, as blocks are inserted into a drawing, all objects in the block will function as one object for editing purposes. Combining all of the objects of a block as one object is great if you want to MOVE, ROTATE, or COPY a block. Only one object has to be selected and the entire block will respond to the command. Using other editing commands such as ERASE to remove one object will also cause the entire block to be erased. To erase just one object in the block, you must use the EXPLODE command. The EXPLODE command was first introduced in Chapter 13. The command can be accessed by selecting the Explode icon from the Modify toolbar, by typing X (EnterKey) at the command prompt, or by selecting Explode from the Modify pull-down menu. The command sequence is:

Command: **X** (EnterKey)

Select objects: *(Select the desired blocks.)*

Select object: *(This prompt will continue until all of the desired blocks are selected.)* (EnterKey)

Command:

The block will be displayed with no apparent change. Individual components of the block can now be edited. Blocks with equal X, Y, and Z values will keep their shape when exploded. Blocks with unequal X, Y, and Z values may change shapes when exploded. Blocks that are inserted with MINSERT cannot be exploded.

REDEFINING A BLOCK

Even if a block has been inserted into a drawing in several locations, it is possible to update all of the blocks without editing them individually. Blocks are edited by redefining a copy of one of the blocks, then saving the edited copy of the block using the existing block name. It's also important to remember that this process will update *all* blocks of the specified name. There is no "I only wanted a few of them" option. Use the following sequence to redefine a block. For the following example, the TOILET block will be altered.

- Select the Edit Block or Xref icon from the Refedit toolbar, select In-place Xref and Block Edit Reference from the Modify pull-down menu, or enter **REFEDIT** (EnterKey)at the command line.
- Select the block to be edited. Selecting a block to be altered will produce a Reference Edit dialog box similar to Figure 20.23.
- Select the name of the block from the Reference Name edit box.
- Select the Enable Unique Layer and Symbol Names box.
- Select Display Attribute Definitions for Editing box.
- Choose the OK button.
- The command line will prompt to:

Select nested objects.

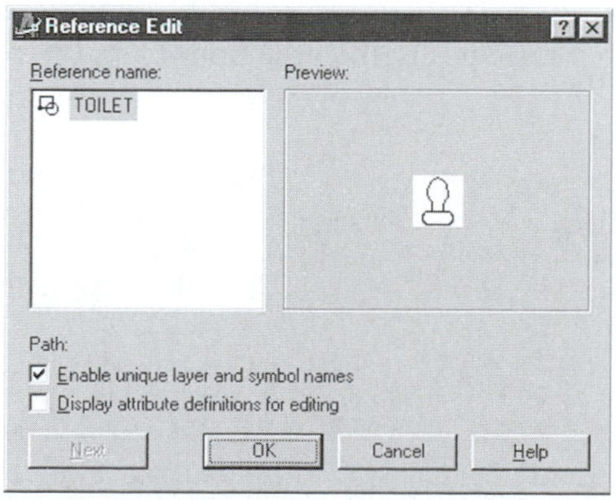

Figure 20.23 *The Reference Edit dialog box can be used to select a block to be edited.*

- Select the block to be edited. Other uses of the selected block may appear to be altered. Don't panic! The blocks will be restored shortly. As a block is selected, the Refedit tool bar will be displayed, if has not been previously opened.
- Edit the block as desired.
- Select the Save Back Changes to Reference icon from the Refedit tool bar. This will display a warning box indicating that all of the selected blocks are about to be updated.
- Select the OK button from the warning box. The box will be closed and all of the selected blocks will be redisplayed to reflect the indicated changes.

Each toilet used throughout the drawing will now be updated to reflect the current changes. Figure 20.24 shows an example of a redefined block.

WBLOCKS

Blocks are a great way to make repeated copies of a symbol throughout a drawing. A symbol that is used on a foundation for one structure can be used on the foundation of several other structures. Blocks are saved in the drawing file they are created in, but with the aid of the AutoCAD DesignCenter, a block can be used on other drawings. If a block is saved on a template drawing, the block can be used each time the template drawing is used. The WBLOCK command allows blocks to be used in multiple drawings, no matter where the drawing was created, by saving the drawing as a .DWG file. WBLOCK will be displayed as a .DWG file each time a directory of drawings is made, but will act as a block when inserted into a drawing. Figures 20.25a and 20.25b show examples of details and sections created by using blocks and combined by using the WBLOCK command.

USING WBLOCKS

Symbols and construction components that are saved as a wblock can be used to save hours of drawing time on future drawings. Using an existing drawing as a wblock can also save hours on drawings— such as elevations or sections that have features that must be located based on their position on the floor plan. By freezing all unnecessary

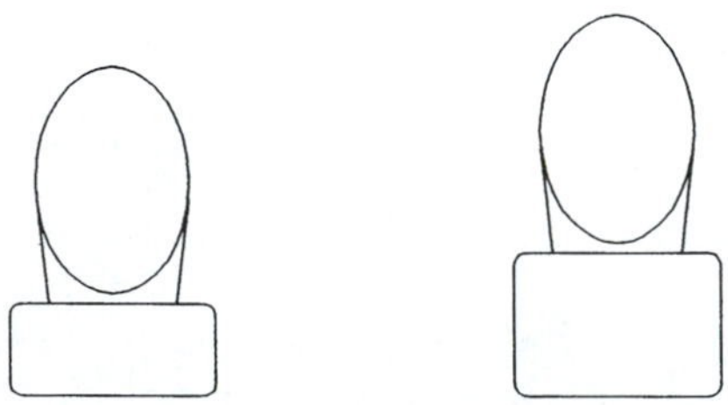

Figure 20.24 *When a block is redefined, all occurrences of the block in the drawing will be updated.*

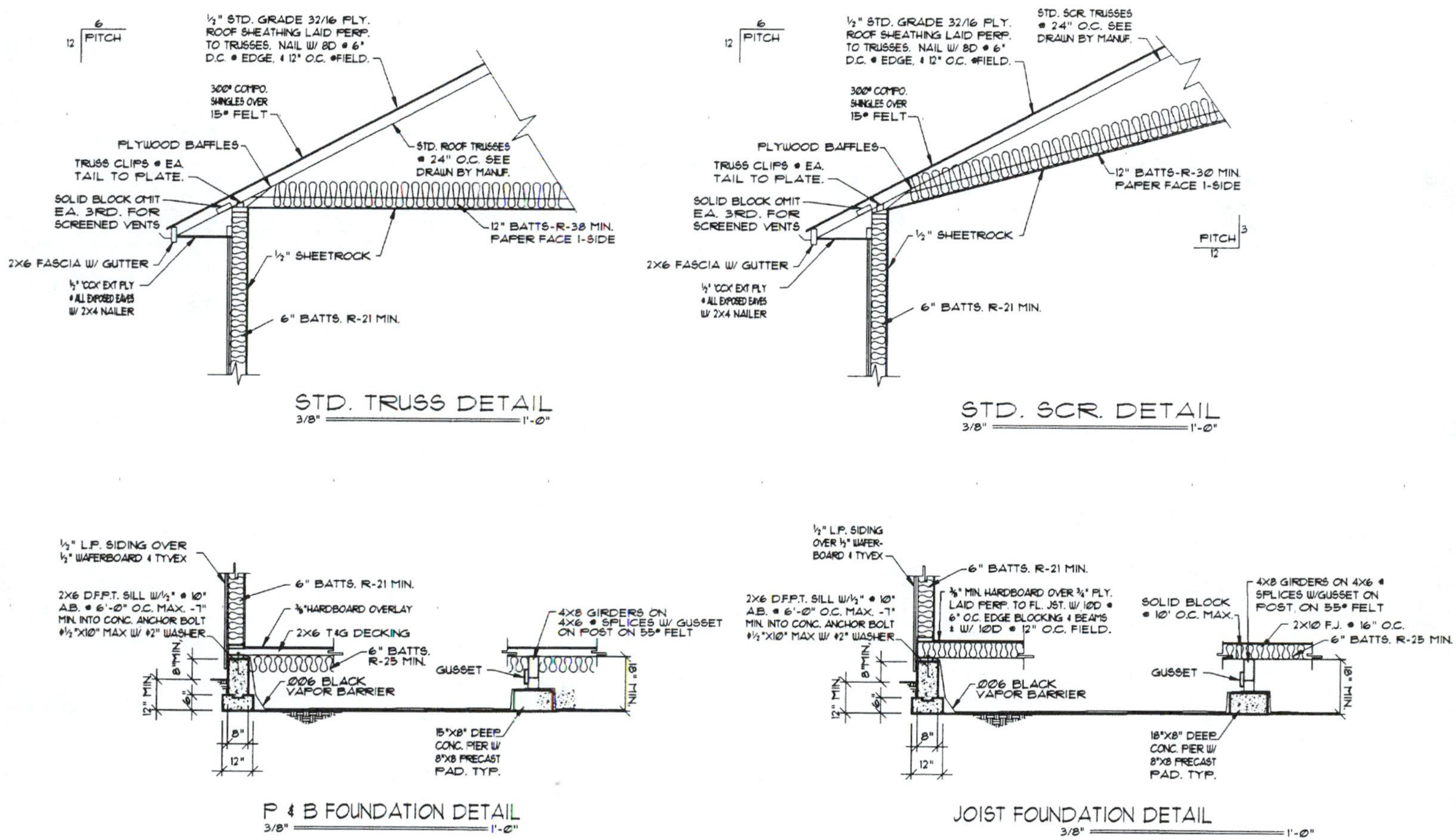

Figure 20.25a *Common details can be stored using the WBLOCK command. (Courtesy David Jefferis.)*

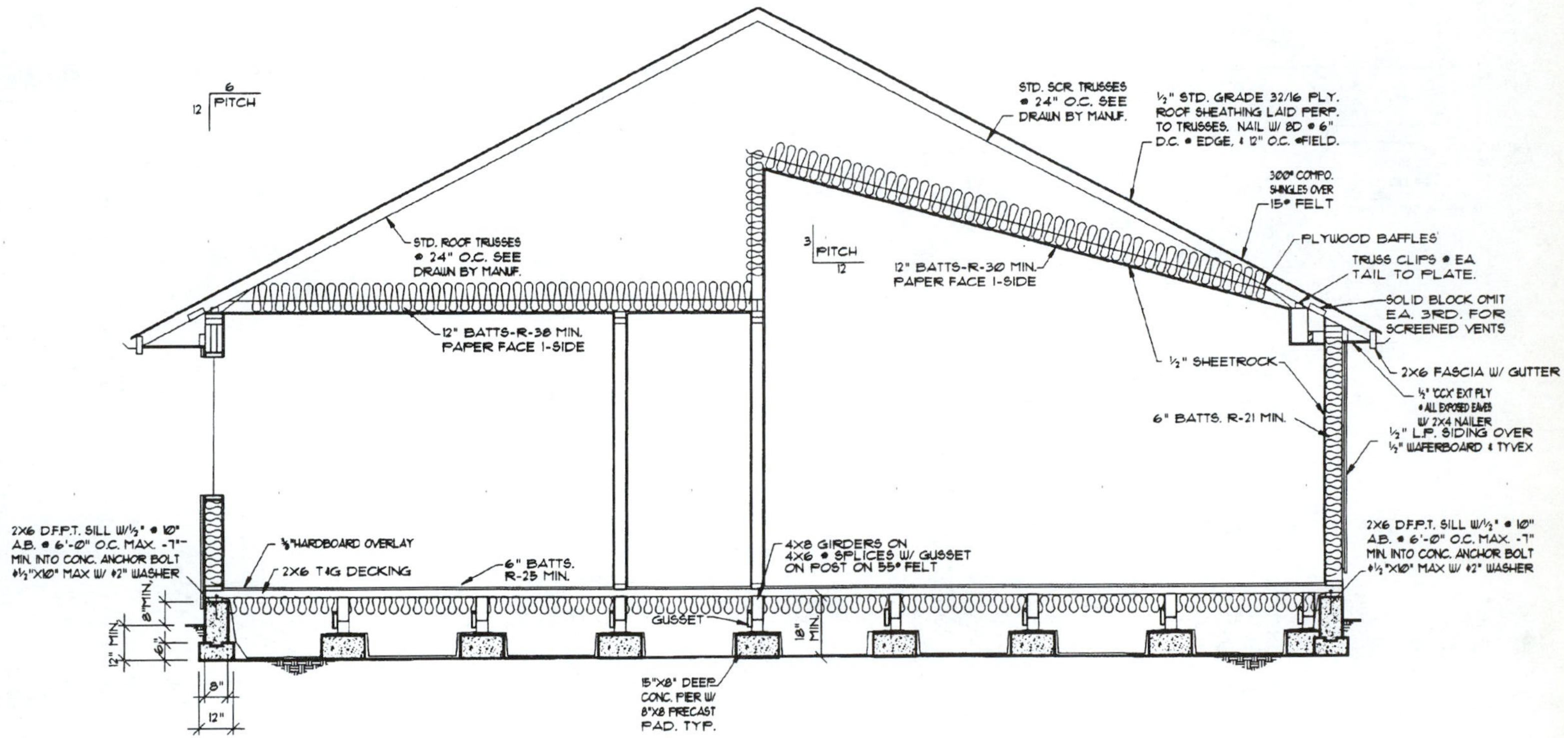

Figure 20.25b *Blocks saved using the WBLOCK command can be combined to make other drawings. (Courtesy Michael Jefferis.)*

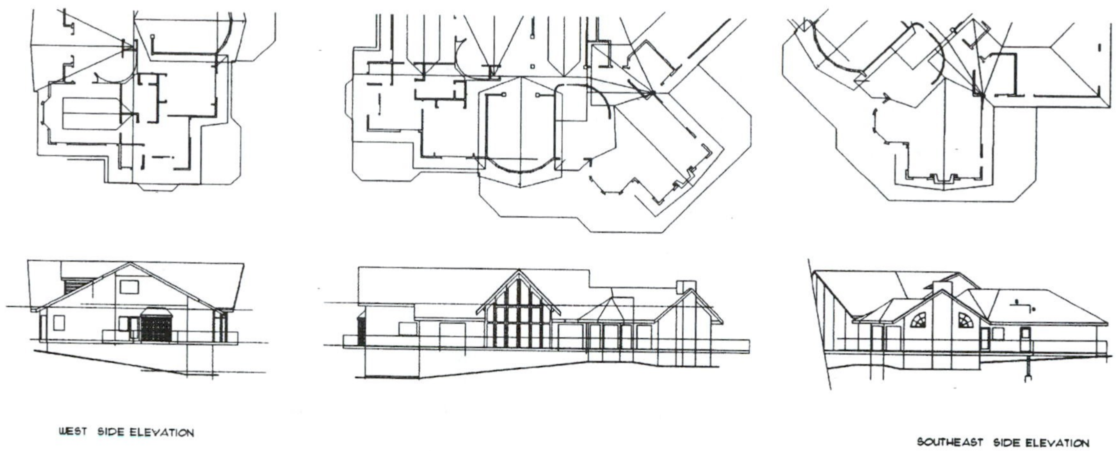

Figure 20.26 *A block was made of the floor plan, then it was reinserted and rotated to project each elevation. (Courtesy Sara Jefferis.)*

materials and only thawing materials— such as the roof outline, walls, windows and exterior doors—a wblock of the floor plan can be created and inserted into the ELEVATION.DWG file. The wblock can then be copied and rotated for use in projection of each elevation, as shown in Figure 20.20. Projecting irregular shaped structures from a wblock of the floor plan allows each surface to be accurately projected. Sizes can be projected to determine both vertical and horizontal relationships. The roof and wall locations are projected vertically from the plan to the elevation. Heights are projected from one elevation to the next.

CREATING WBLOCKS

Typing W (EnterKey) at the command prompt will start the command and display a Write Block dialog box similar to Figure 20.27. The dialog box is used to create and save blocks to a file for future use. The two major areas of the Write Block dialog box include the Source and Destination boxes. The areas that are active will depend on the selections made in the Source portion of the dialog box.

Source

The Source area of the Write dialog box selects existing blocks or objects, writes them as a file, and specifies an insertion point for when they are inserted. Options are provided for selecting an existing block, an entire drawing, or specifying objects to form the wblock.

Block

Selecting the Block option allows an existing block to be selected and stored as wblock file. With the Block setting active, the block edit box is activated and the Base Point and Objects portions of the box are inactive. Selecting the edit arrow will dis-

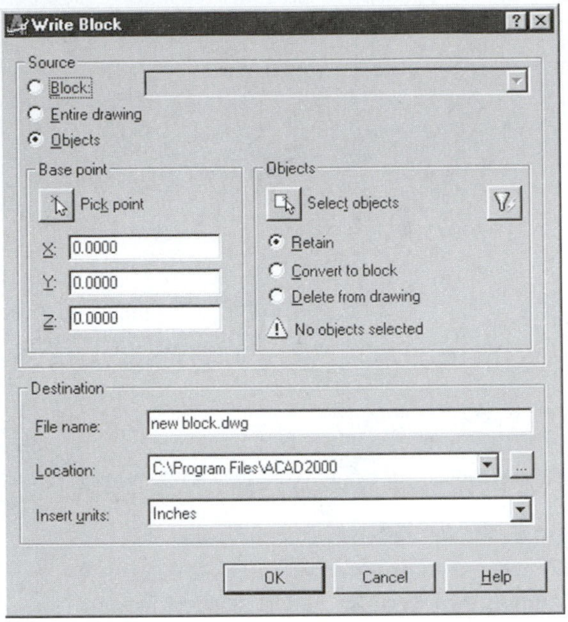

Figure 20.27 *Entering* **W** *(EnterKey) at the command prompt will display the Write Block dialog box, allowing blocks, drawings, or objects to be saved for future use.*

play a listing of blocks contained in the current drawing. Selecting a name from the menu will allow that menu to be inserted. The Destination portion of the block displays where the wblock file will be stored. If the OK button is selected, the selected block will be saved as a drawing file in the indicated destination. Methods of altering the destination will be examined in the next section of this chapter.

Entire Drawing

Selecting this portion of the drawing allows an entire drawing to be saved as a wblock. This option will have the same results as using the SAVEAS command. The benefit of using the option is that unreferenced objects, such as blocks, layers, and linetypes that were not used, will be removed from the drawing file. This is an excellent method for reducing the size of a drawing, and is an excellent method of saving a drawing at the end of a work session.

Objects

Using the default option of Objects allows drawing objects to be selected to form the wblock. The process is very similar to the procedure used to create a block with the Object button active. The Base Point and Objects portion of the dialog box are activated so that objects and a base point can be selected. The order that the Point or the Objects are selected can be altered. Each of these functions are exactly the same as their counterparts in the BLOCK command. As each icon is selected, the dialog box is removed, allowing for a selection to be made. Once the selection is made, the dialog box is restored, allowing for other options to be altered.

Destination

The box allows the name, location, and units of the drawing file to be specified.

File Name

The File Name display will vary, depending on which of the three Source options is active. If the Block option is active, the name of the block entered in the Block edit box will be displayed in the File name edit box. If the Entire Drawing box is active, the current name of the drawing will displayed in the File Name edit box. With the Objects option active, the default name of New Block.dwg will be displayed. Best results are achieved by providing a more descriptive name.

Location

Because blocks stored using the WBLOCK command are treated as a drawing file, a storage location must be given. The Location edit box displays the current drive and path that will be used to store the drawing. Selecting the Browse for Folder icon allows the drive and folder to be altered. Selecting the icon produces a display similar to the Windows Explorer. Once a folder name has been selected, the Create Drawing dialog box will allow for selection of a filename and a location for the file.

Note: Great care should be given in the naming and storing of wblocks. The name should clearly describe the contents of the drawing. Wblocks should be stored in folders and subfolders that will define the use. For instance, a name of PROTOS could be used for the folder. Subfolder names would include names such as SITE, FLOOR, ELEV, FND, or other common drawings to be used. Each subfolder would contain only wblocks for that specific type of drawing. Figure 20.28 shows a sample listing of wblocks for an architectural firm.

SUMMARY OF CREATING A WBLOCK

Remember that a wblock is just a collection of objects that can be used over and over in a multitude of drawings. Use the following steps to create wblocks:

- Enter the Write Block dialog box by entering **W** (EnterKey) at the command prompt
- Choose the source of the block to be created.
- If the wblock will be comprised of pre-existing blocks, provide a name for the block.
- Select the drawing objects to be reproduced as blocks.
- Select a base point to be used as the block is inserted into the drawing.
- Select a name to designate the file.

SITE PLAN
- NORTH ARROW
- GENERAL NOTES
- PARKING NOTES
- PAVING NOTES
- HANDIC PARKING
- PERP. SPACE
- 30° PARKING
- FIRE HYDRANT
- CURB CUT
- LIGHT STANDARD

FLOOR PLAN
- PLUMB
  - DBL. SINK
  - TRIPLE SINK
  - VEGGIE SINK
  - BATH SINK
  - LAUNDRY SINK
  - TOILET
  - BIDET
  - 60" TUB
  - 72" TUB
  - SPA
  - 36" SHOWER
  - 48" SHOWER
  - 60" SHOWER
  - WASHER
  - WATER HTR.
- ELECTRICAL
  - DISHWASHER
  - WASHER
  - WATER HTR.
  - F.A.U.
  - IRON BR.
  - VACUUM
  - RANGE
  - COOK TOP
  - DBL. OVEN
  - REFR.
  - TRASH COMP.
- DOOR
  - SINGL. SW.
  - DBL. SW.
  - SLIDING
  - FRENCH
  - BI-PASS
  - BI-FOLD
  - OVERHEAD
- WINDOW
  - STANDARD
  - BAY
  - BOW
  - GARDEN
- SKYLIGHT

FOUNDATION
- GENERAL NOTES
- VENTS
- PIER
- GIRDER
- FIREPLACE
- BAY

ELEVATIONS
- ROOF
  - VENT
  - SKYLIGHT
  - METAL CAP
  - MASON CAP
  - CUPOLA
- WALLS
  - DOOR-FR
  - DOOR-SLDG
  - DR-6 PANEL
  - DR- 10 LITE
  - DR-FRONT-1 LT
  - DR-FRT-SING LT.
  - DR-FRT-DBL. LT.
  - WIN-SLDG
  - WIN-FIXED
  - WIN-FIX/AWN
  - WIN-CASE
  - WIN-DBL. HNG
  - WALL LITE
  - SHUTTER
  - WALL VENT
  - RIDGE
  - VENT
- DECKS
  - FENCE
  - WOOD VERT
  - WOOD HORIZ
  - IRON RAIL
  - STAIR
- LANDSCAPING
  - LOW BUSH
  - BAMBOO
  - BERRY
  - VINE
  - TREE-FIR
  - TREE-OAK
  - TREE-PALM
  - TREE-ALDER
  - TREE-MAPLE

SECTIONS
- ROOF
  - TRUSS
    - STAND
    - SCISSOR
  - STICK
    - STD RAFT.
    - VAULTED
    - OPEN BM.
- UP FLOOR
  - TJI's
  - JOIST
  - TRUSS
- GND. FLOOR
  - TJI's
  - FLOOR JST.
  - POST / BM.
  - CONC. SLAB

ELECTRICAL
- PLUGS
  - 110 C.O.
  - 110 GFCI
  - 110 WP
  - 110 - 4 PLEX
  - 110- FLOOR
  - 220 C. O.
  - J-BOX
- LIGHTS
  - CEILING
  - WALL
  - CAN
  - FLOOD
  - FLUOR-SHOP
  - FLUOR-SURFACE
  - FLUOR-RECESSED
- LOW VOLT
  - YARD
  - FLOOR
  - STAIR
- SWITCHES
  - SING POLE
  - DIMMER
  - TIMER
  - THREE WAY
- SPECIALITIES
  - SMOKE DET.
  - DOOR BELL
  - CHIME
  - CIRCUIT BOX
  - CEIL. FAN
  - BATH FAN
  - PHONE
  - STEREO
  - TV
  - VACUUM
  - INTERCOM SP
- YARD

Figure 20.28 *Wblocks should be stored into folders and subfolders so that they can be easily accessed. Most offices name the folders for the drawings that will use the WBLOCK.*

- Specify the drive, folder, and sub-folder to be used as the destination.
- Select the OK button to save the block as a drawing file.

The object has now been saved as a drawing file and is ready to be reused throughout the existing drawing or in a new drawing. If the toilet from Figure 20.9 needs to be added to the existing drawing, or a totally new drawing file, it can be added to the drawing using the INSERT command.

INSERTING WBLOCKS

Once a symbol or drawing has been saved as a WBLOCK it can be used just as any other drawing. It can be retrieved using the drawing Start-up menu as a new drawing session is started, or it can be inserted into an existing drawing using the INSERT command. The INSERT command will bring an existing drawing into the current drawing from the command prompt. Each aspect of inserting a WBLOCK is the same as inserting a BLOCK.

EDITING WBLOCKS

One of the strengths of using an entire drawing as a wblock is that when it is inserted, it will move as one object. Insert a wblock of a floor/roof plan, and all of the components will move as one object. A wblock can be altered using the same commands used to edit a block.

CHAPTER 20 EXERCISES

1. Open the FOOTINGS drawing that was updated in Chapter 19. Save the drawing as a block titled FND1STORY.
2. Open drawing FLOOR14 and use the drawing to create a WBLOCK titled BASE14.
3. Open drawing E-19-8. Save the drawing as a wblock titled FNDPLN. Insert FNDPLN into drawing FND1STORY, rescale the footing, and save the entire drawing as FNDDLT.
4. Open drawing FNDDLT and use it to create a WBLOCK titled FNDPLN1. Insert FNDPLN1 into BASE14.
5. Use the attached drawings as a guide to draw the following symbols and save them as a wblock in a folder titled \PROTO\FLOOR, using the designated BLOCK name. Create all blocks on the 0 layer and use 4"-high text.

DESCRIPTION	SYMBOL	BLOCK NAME
BATH		
Draw a 60" × 32" tub		TUB60
Draw a 72" × 32" tub		TUB72
Draw a 19" × 16" lavatory		LAV
Draw a toilet		POTTY
Draw a 36" shower	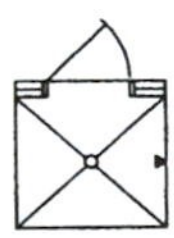	SHOWER36
Draw a 42" shower	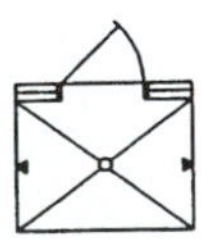	SHOWER42
Draw a 60" shower	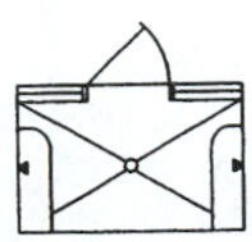	SHOWER60

DESCRIPTION	SYMBOL	BLOCK NAME
Draw a 32" × 21" double sink with a garbage disposal on either side	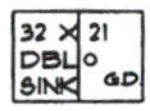	DBLSINK
Draw a 36" × 21" double sink with a garbage disposal on small side	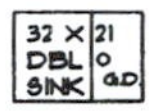	XDBLSINK
Draw a 43" × 21" triple sink with a garbage disposal in the middle		XXSINK
Draw a 16" × 16" veggie sink		VEGSINK
Draw a dishwasher		DW
Draw a trash compactor		TC
Draw a lazy susan		LS
Draw a range		RANGE
Draw a double oven		DBLOVEN
Draw an 18" pantry		PAN

DESCRIPTION	SYMBOL	BLOCK NAME
UTILITY		
Draw a washing machine		WASH
Draw a dryer		DRYER
Draw a built-in ironing board		IRON
Draw a 21"×21" laundry tray		LT
Draw a water heater		WH
Draw a hose bibb		HB

6. Use the attached drawings as a guide to draw the following symbols and save them as a wblock in a folder titled \PROTO\ELECT, using the designated BLOCK name. Use a 6"-diameter circle for all electrical symbols with no text. Use an 8"-diameter symbol with 4"-high text for symbols that require text to be placed inside the symbol. Create all blocks on the 0 layer.

DESCRIPTION	SYMBOL	BLOCK NAME
PLUGS		
Draw a 110 c.o.		110CO
Draw a multi 110 c.o.		110CO2
Draw a 220 c.o.		220CO
Draw a 110 half-hot c.o.		110HH

DESCRIPTION	SYMBOL	BLOCK NAME
Draw a 110 waterproof		110WP
Draw a 110-ground fault interrupter		110GFI
Draw a floor-mounted 110 c.o.		110F
Draw a junction box		JBOX
LIGHT FIXTURES		
Draw a ceiling-mounted light fixture		CEILLITE
Draw a wall-mounted light fixture		WALLLITE
Draw a can light fixture		CANLITE
Draw a recessed ceiling light fixture		RECESSLITE
Draw wall-mounted spotlights		WALLSPOT
Draw a 24" × 48" surface-mounted fluorescent fixture	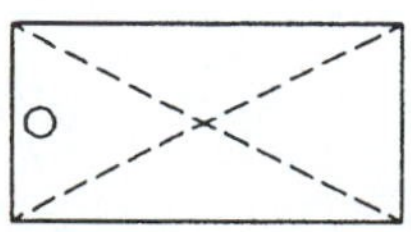	24FLUOR

DESCRIPTION	SYMBOL	BLOCK NAME
Draw a 48" × 48" surface-mounted fluorescent fixture	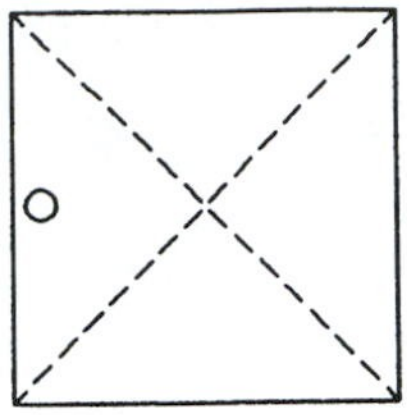	48FLUOR
Draw a 48" shop fluorescent fixture		48SHOP
Draw a 48" track light		TRACK
SWITCHES		
Draw a single-pole switch		SWITCH
Draw a three-way switch		SWITCH3
Draw a switch with dimmer		DIMSWITCH
SPECIAL SYMBOLS		
Vacuum		VACUUM
Intercom		INTERCOM
Phone	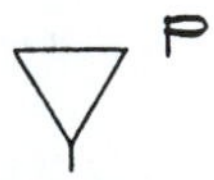	PHONE
Stereo speakers		SPEAKERS

DESCRIPTION	SYMBOL	BLOCK NAME
Smoke detector		SMOKE
Light, heat, fan		LHF

7. Use the descriptions below as a guide to draw the following symbols and save them as a wblock in a folder titled \PROTO\FLOOR\DOORS, using the designated BLOCK name. Create all blocks on the 0 layer using a 12" block unit.

DESCRIPTION	SYMBOL	BLOCK NAME
Draw an exterior door		XDOOR
Draw an exterior door with sidelight	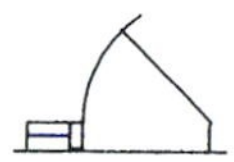	XDOOR1LITE
Draw an exterior door with two sidelights		XDOOR2LITE
Draw a pair of exterior door with 2 sidelights	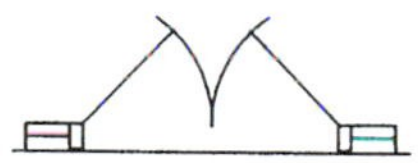	PR2DOOR2LITE
Draw an interior door		INDOOR
Draw an exterior sliding glass door		XSLDOOR
Draw an interior by-pass door		4BYPASS
Draw an interior bi-fold door		BYFOLD
Draw an interior folding door		FOLD
Draw a window		WINDOW

8. Open the appropriate template and use the drawing below to draw the bathroom. Draw the exterior windows 6' wide, except for a 2'-wide window at the toilet room, a 4-feet window over the counter, and 3'-wide windows in the bedroom. Use a pair of 2'-wide doors in the bedroom, a pair of 3'-wide doors from the bedroom to the exterior, and a 28" pocket door into the toilet room. All other doors to be 30" wide. Create the following layers: WALLS, PLUMB, APPLIANCES, DIMEN, NOTES, DOORWIN and CABS. Create and insert the required blocks to complete the drawing. Provide dimensions and label all fixtures. Save the drawing as a WBLOCK titled BATH.

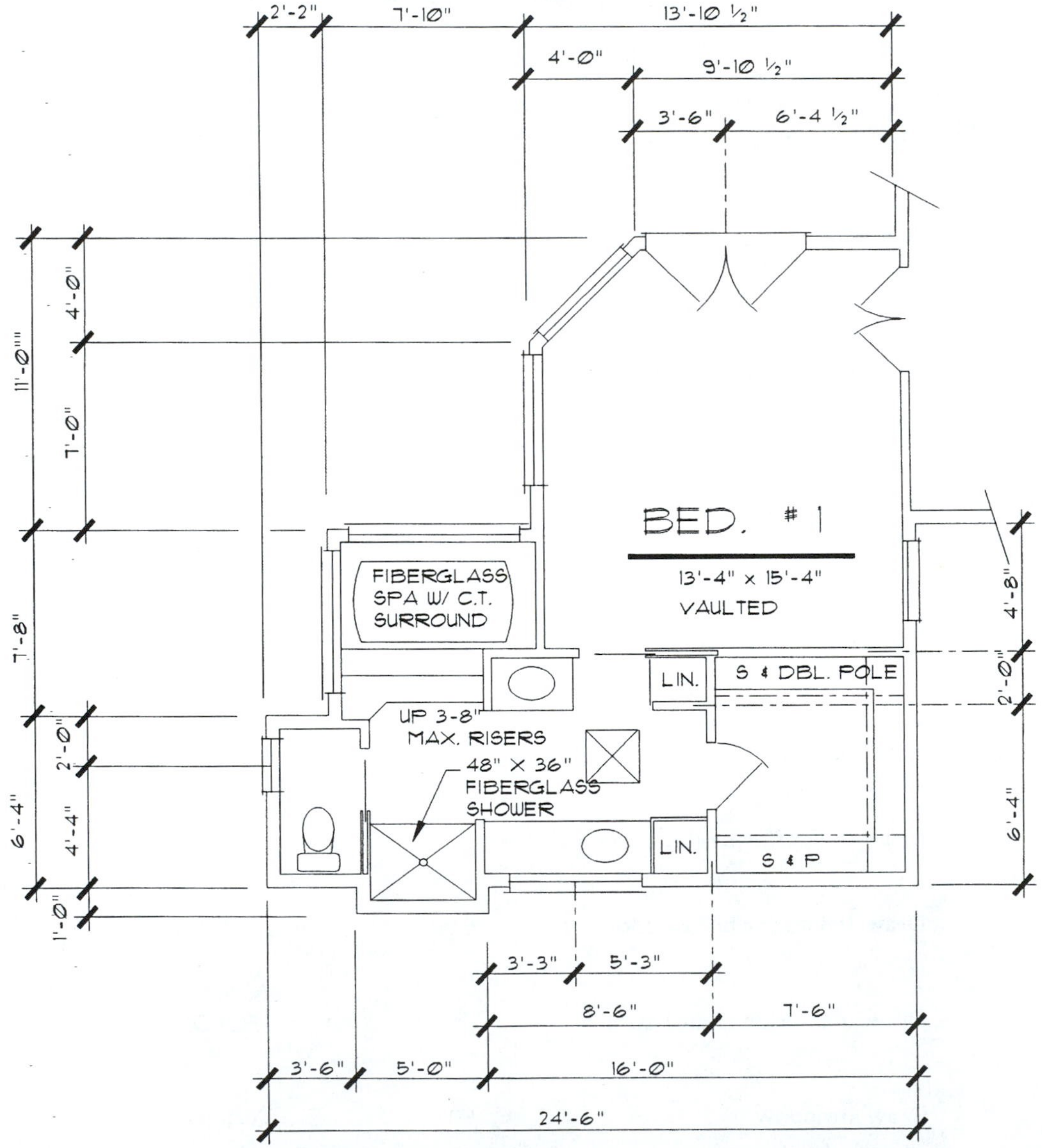

9. Open the appropriate template and use the drawing below to draw the kitchen. Design door and window sizes to fit the space provided. Create layers that will logically separate information for future printing on floor, framing, and electrical plans. Insert the required blocks to complete the drawing. Align the bathroom with the utility room and provide a 42" fiberglass shower unit. Provide dimensions, determine the square footage of each room, and label the drawing. Turn off all layers containing notes and dimensions, and design an electrical plan that

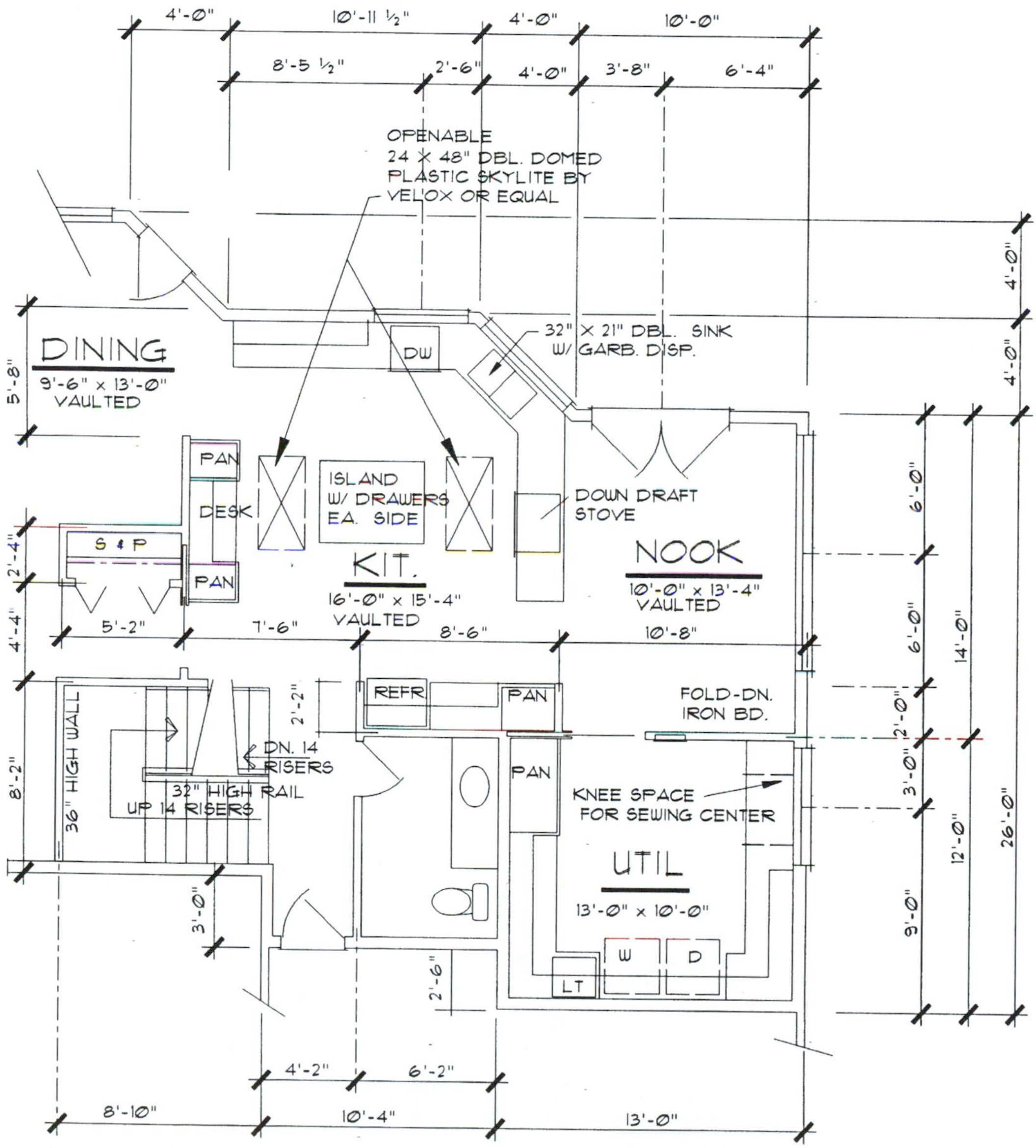

assumes all appliances are electrical. Use 2 ceiling-mounted fixtures in the nook, a surface-mounted fixture in the utility room, 8 can lights in the kitchen, 2 can lights in the hallway, a wall-mounted fixture over the bathroom sink, and a ceiling-mounted fixture in the center of the bath. Save the drawing as a WBLOCK titled KITCHEN.

10. Open the FOOTINGS drawing from Chapter 19 and use it to create a WBLOCK drawing titled BASEFTG. Save the file with an appropriate title.

CHAPTER 20 QUIZ

1. How can a wblock increase drawing speed?
2. Why is using blocks more efficient than copying an object?
3. If a block is created on the 0 layer, where will the block be located when it is inserted?
4. When the color of a block is set "BYBLOCK," how will the block be affected?
5. What is the advantage of drawing blocks such as a window in a 12" box, rather than full size?
6. List the process to create a block.
7. How does the Insertion Point specified when the block is defined affect the insertion point requested when the block is inserted?
8. What command will allow for multiple blocks to be inserted?
9. What command will allow for editing blocks?
10. What is the quickest method for editing the same mistake in several blocks?
11. Explain the difference between a block and a wblock.
12. List the process displayed for inserting a wblock into a drawing.
13. How can a block be mirrored before it is inserted?
14. What method can be used to insert a block by selecting two points?
15. What is the effect of activating the Explode box in the Insert dialog box?

CHAPTER 21

Adding Attributes to Enhance Blocks

Even with all the qualities that blocks and wblocks add to a drawing, they still can be enhanced. Many blocks need text to explain or provide a specification. Text can be added to a block using three methods. In the last chapter, text was added to the drawings using TEXT, and saved as part of the block. This works well if the text remains constant, such as the specification for a double kitchen sink. By adding the text to the block, the block must be exploded if the text is to be edited. The EXPLODE command allows the text to be edited, but characteristics of the block may be lost. Once a block has been inserted, text can be added to the drawing base to supplement the block by using the TEXT or MTEXT commands. This option works well only if a few blocks need written specifications.

This chapter will introduce you to a new, and typically the best, method of adding text to a block by using attributes. An *attribute* is a grouping of objects that contains text. An example of a block with attributes can be seen in Figure 21.1. These text groupings are similar to a block, which is then attached to a drawing block. Attributes have several features that make them superior to text created with TEXT or MTEXT and are well worth the effort to master. Once the benefits have been explored, thought must be given to defining the attribute text, controlling how the attributes will be displayed, changing the values, and extracting attributes from a drawing to compile a written listing of attributes.

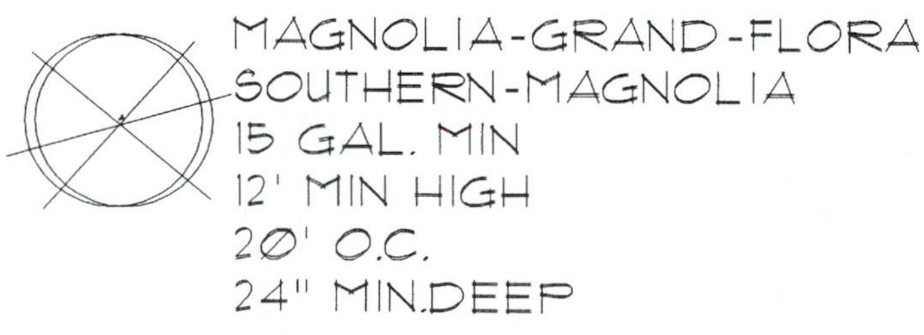

Figure 21.1 *Attributes can be added to a block to explain qualities that could vary with each insertion.*

THE BENEFITS OF ATTRIBUTES

The two major benefits of adding text through the use of attributes are control and retrieval of information.

CONTROLLING TEXT

Using attributes will allow for assigning prompts to a block to request information each time that the block is displayed. For instance, each time the door block is inserted into a drawing, the following prompts could also be provided:

> size, thickness, type, manufacturer, finish, rough opening and hardware requirements.

Default values can be provided, but altered with each insertion. Using attributes also allows for correcting the text that is displayed without altering the block characteristics. These prompts may be hidden during the display or plotting process, even though the block they are attached to is displayed.

EXTRACTING ATTRIBUTE INFORMATION

In the example above, much of the information that will comprise the attribute is displayed in a plant schedule. Attributes can be extracted from the drawing, stored in a separate file, and printed using a third party database text editor, such as the Microsoft Notepad. A list of attributes also can be compiled and printed within the drawing.

PLANNING THE DEFINITION

It is easiest to add attributes to a drawing before the drawing has been saved as a block. Once the symbols have been drawn, a decision must be made about what specifications will be associated with the block. For best efficiency this should be done in a planning session prior to entering the ATTDEF (ATTribute DEFined) command. For a block of a window, this might include the size, type, frame material, manufacturer, glazing, and rough opening. Other than the actual space within the drawing, there is no limit to the amount of information that can be associated with a block. As the attributes are being defined, an option is even available to keep the attributes invisible.

After the material to be listed with the block has been determined, decide how the prompts for each attribute will be displayed. Attribute prompts usually are in the form of a question, but statements also can be used. For instance, the prompt might read "What is the window size?" or "Provide the window size." Determine the prompts for each attribute that you will be listing.

ENTERING THE ATTRIBUTE COMMANDS

Several commands will be used as you work with attributes. Once the block has been drawn and you've decided what attributes are to be assigned, the process of attaching text to the block is started by using the ATTDEF command. Commands for controlling attributes include the ATTDEF (ATTribute DEFinition), ATTDISP (ATTribute DISPlay), ATTEDIT (ATTribute Edit), ATTREQ (ATTribute REQuest), and ATTEXT (ATTribute EXTract) commands.

CREATING ATTRIBUTES

Each of the qualities of an attribute can be controlled using the Attribute Definition dialog box or at the command line. Consult the help menu (or professional help), if you would rather use the slower, harder command prompt method. The dialog box is accessed by entering **ATTDEF** (EnterKey) at the command prompt or by selecting Define Attributes... from Block of the Draw pull-down menu. Each method will display the dialog box shown in Figure 21.2. Major elements of the dialog box include the Mode, Attribute, Insertion Point, and Text Options boxes.

ATTRIBUTE DISPLAY

The Mode box allows the display options for the attribute to be set. The initial prompt for ATTDEF provides four options to select the attributes will be displayed. These include Invisible, Constant, Verify, and Preset. Each of these options serves as a toggle switch with a default setting of Normal. If none of the options is selected the Normal setting will be kept. A Normal setting will prompt you for each attribute and the response will be displayed.

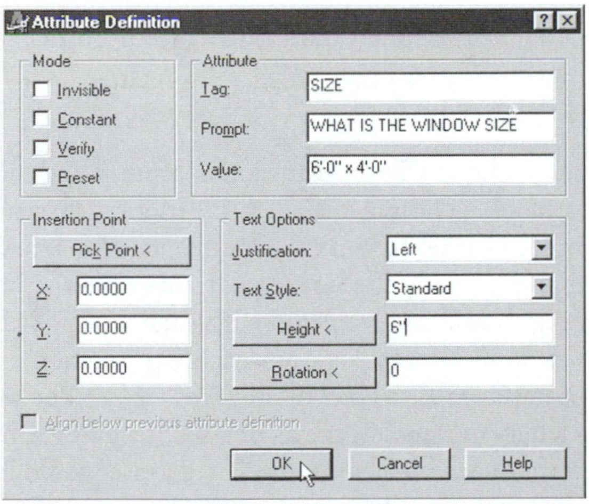

Figure 21.2 *Typing **ATTDEF** (EnterKey) at the command prompt or selecting Attributes... from Block of the Draw pull-down menu will produce the Attribute Definition dialog box. The box can be used to design the tag, prompt, value, and insertion point of each attribute as well as controlling how each will be displayed.*

Invisible

Selecting the Invisible check box will make the attributes invisible. Attributes will still be attached to the block when it is inserted, but they will not be displayed. If a separate window schedule is to be created, values can be invisible so they are not listed twice. The ATTDISP command, which will be introduced shortly, can be used to override the Invisible mode setting.

Constant

Selecting the Constant check box indicates that all uses of the block will have the same attribute value. The value is entered as the attribute is being defined and set throughout the use of the block. No prompt for new values will be given when the block is inserted with this box active. Using the default of Normal will allow variables to be changed at a later time.

Verify

By defining an attribute with this mode active, you will be given an opportunity to verify that the attribute value is correct during the insertion process.

Preset

This mode will allow you to create attributes that are variable, but are not requested each time that the block is inserted. When a block containing a preset attribute is inserted, the attribute value is not requested and is automatically set to its default value.

ASSIGNING VALUES TO ATTRIBUTES

Once all of the display options are entered, continue through the ATTDEF command prompts for defining attribute values. The three prompts are the tag, prompt, and value and they can be seen in Figure 21.3.

Attribute Tag

The tag is used to identify an attribute definition in a similar way that a specific name is used to identify a block. The tag will identify each occurrence of the attribute throughout the drawing. During the initial planning stage for the windows, you may decide that information regarding the size, type, frame material, manufacturer, glazing, and rough opening should be associated with each window block. Each of these listings can be used as an Attribute Tag, but each tag must be entered individually with the following prompts used to control only this tag. As the tag is entered, it can be composed of any letters or characters except a blank space. A hyphen can be used between words to create a space. An appropriate response would be to move the cursor to the Tag: edit box and enter a tag such as SIZE.

Attribute Prompt

The Attribute prompt is the prompt you will see when the block is inserted. As you plan the prompts, questions or statements should be used. The Attribute prompt is

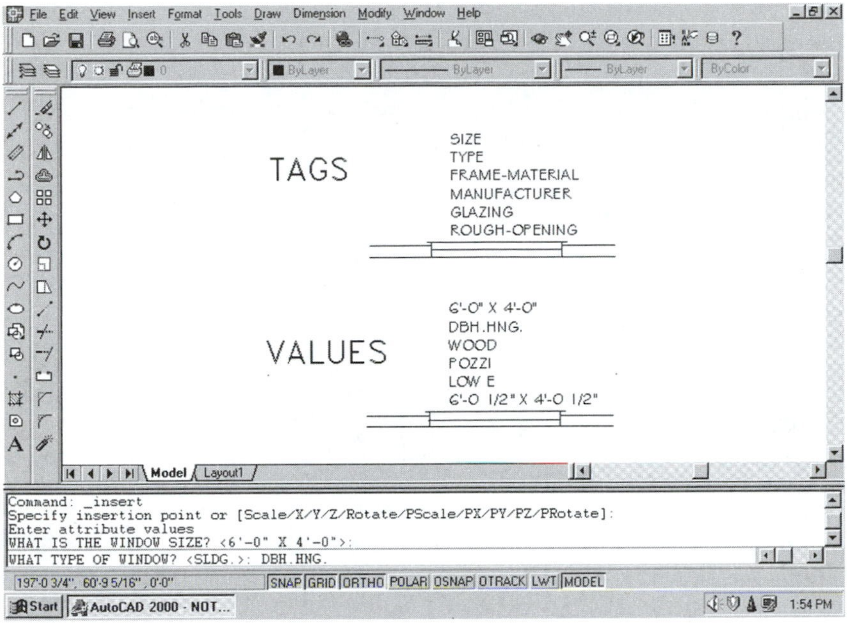

Figure 21.3 *The attribute tags are displayed as the block is being inserted. The values are displayed once the block has been inserted.*

where that prompt should be entered. An appropriate response would be to move the cursor to the Prompt: edit box and enter a prompt such as WHAT IS THE WINDOW SIZE. If the Prompt: edit box is left blank, the Tag (SIZE) will be used as the attribute prompt. If the Constant mode is active, the Prompt option will be inactive.

Attribute Value

Once the tag and the prompt have been specified, provide a value in the Value: edit box. The value that is entered at this prompt will be displayed as the default value as the block is inserted. A specific value such as 6'-0" × 4'-0" can be entered, or the null response can be used. If the Constant mode for the attribute has been selected, a request for the prompt will not be made, and the prompt "Attribute Value:" will be displayed instead. An appropriate response would be to move the cursor to the Value: edit box and enter a response such as: 6'-0" × 4'-0".

TEXT OPTIONS

Once the three Attribute prompts have been given, the prompts used to control the text can be set. Options include menus for justification and style, and edit boxes for height and rotation values.

Justification

Selecting the Justification edit arrow will display the menu shown in Figure 21.4. The options are the same as those used with the TEXT command. To replace the Left

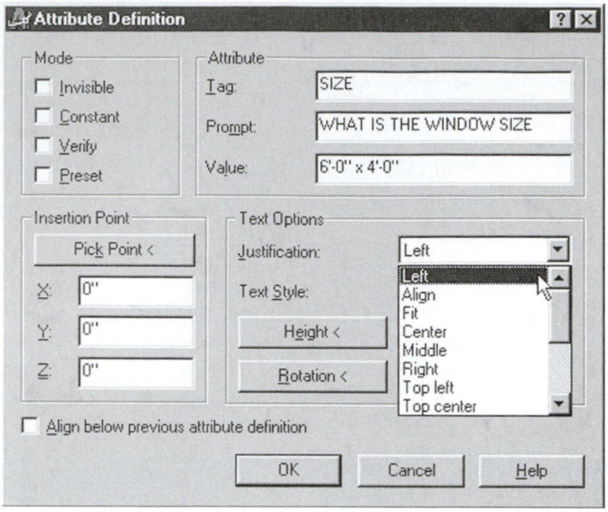

Figure 21.4 *Options for justification can be displayed by selecting the edit arrow. The options are the same as those of the TEXT command.*

option, select the desired option and the box will be closed. The name of the desired option will be placed in the text box.

Style

Selecting the style option will produce a list of current drawing styles. To alter the style, select the down arrow to display the style options and then select the name of the desired style. The selected style name will be placed in the text box, the menu will be closed, and the new style will become current.

Height

Placing the cursor in the text box and clicking with the mouse allows the attribute text height to be altered. Remove or edit the existing value and then provide the desired value by keyboard. Selecting the Height< button can also alter the height value. If the Height< button is selected, the Attribute Definition box is removed and a prompt is displayed to allow points to be selected to indicate the text height. A value can also be entered at the command line instead of selecting two points. When the points or value has been provided, the dialog box will be restored and the indicated height will be displayed in the text box.

Rotation

The rotation angle for attributes is set in the same manner as the height. Enter the desired angle in the text box or select the Rotation< button to select the angle with the mouse.

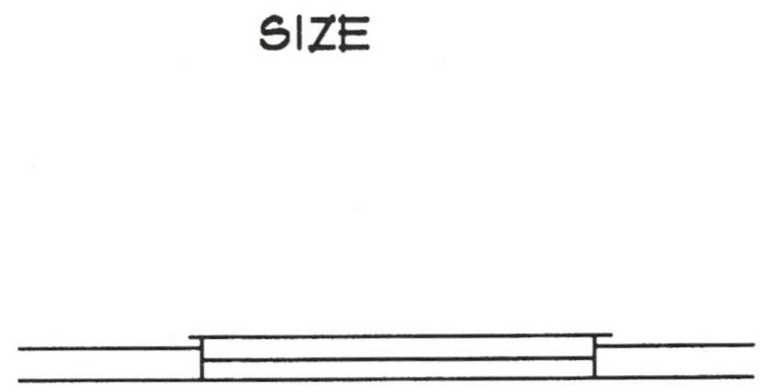

Figure 21.5 *The window block with the SIZE attribute displayed.*

SELECTING THE INSERTION POINT

The Insertion Point box allows the location of the attribute to be selected. Selecting the Pick Point< button removes the dialog box and allows the insertion point to be selected using the mouse. If the exact location is known, coordinates can be entered using the X, Y, and Z text boxes. Once all of the options have been selected, selecting the OK button will close the dialog box and display the tag at the selected insertion point, similar to the display shown in Figure 21.5. Don't be discouraged at seeing the tags displayed as the command is completed. The tags will not be displayed as the block is inserted. If you are unsatisfied with the tags, they can be altered using the PROPERTIES or DDEDIT commands.

ALIGNING ADDITIONAL ATTRIBUTES

The Align Below Previous Attribute Definition button will allow additional attributes to be placed below previous attributes, using the same justification. The option is not active when the first attribute is written. With the option active, the Text options and Insertion Point areas will be inactive.

REPEATING THE PROCESS

To add additional attributes to the block, press (EnterKey) and the entire process can be repeated. By repeating the process five more times, the tags for type, frame material, manufacturer, glazing, and rough opening can be added to the block. Activating the Align Below Previous Attribute Definition prompt will automatically place additional Tags directly below existing Tags with the proper spacing. Repeating the entire sequence will produce a display similar to the one in Figure 21.6.

ADDING ATTRIBUTES TO EXISTING BLOCKS

If a block has already been created, attributes can still be added to it. The original block name must be changed, however, to avoid an error message. If the existing POTTY block is inserted into a drawing, and attributes are added to it, it cannot be saved using the original name of POTTY. By entering a new name, the existing block

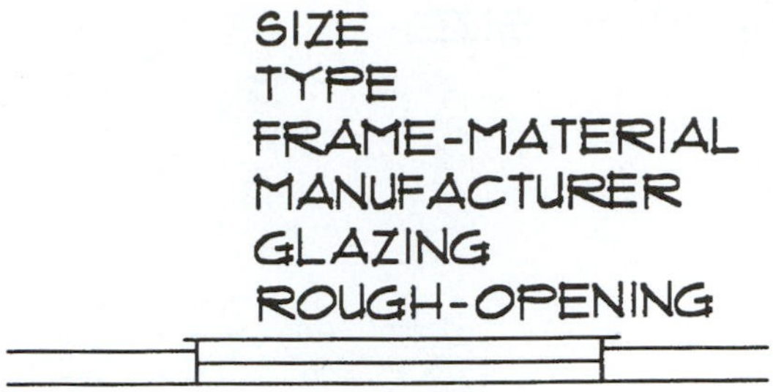

Figure 21.6 *Repeating the procedure used to create the SIZE tag allows an unlimited number of tags to be attached to a block.*

can be used with the desired attributes. Many companies will use a name such as POTTY-A, using the -A to indicate that the block contains attributes. It's much easier to add the attribute prior to saving the block.

CONTROLLING ATTRIBUTE DISPLAY

Rather than wasting disk space by storing POTTY and POTTY-A to distinguish between blocks with and without attributes, the ATTDISP (ATTribute DISPlay) command can be used to hide attributes if you decide that they should not be displayed. This command sequence is:

Command: **ATTDISP** (EnterKey)

Enter attribute visibility setting [Normal/ON/OFF] <Normal>:

Normal will display all attributes, ON will display the variables, and OFF will make all attributes invisible.

The ATTREQ (ATTribute REQuest) command will allow attributes to be displayed, but will suppress the attribute prompts. Some blocks, such as kitchen sink, typically retain their default values with very little variation. When the prompt is not required for attributes, the ATTREQ command can be toggled from 1 to 0. The 0 setting will display all attributes in their default setting.

ALTERING ATTRIBUTE VALUES PRIOR TO INSERTION

An Edit Attributes dialog box similar to Figure 21.7 can be displayed during the INSERT process by selecting the Edit Text icon of the Modify II toolbar, by typing **ED** (EnterKey) at the command prompt, or by selecting Text... from the Modify pull-down menu. This dialog box can be used to display and enter attribute values. The default values are listed to the right side of the current prompts. Displayed values can be altered using methods similar to the methods used for other dialog boxes.

The Properties box can also be used to edit attributes. The box can be displayed selecting Properties from the Object Properties toolbar, by entering **MO** at the com-

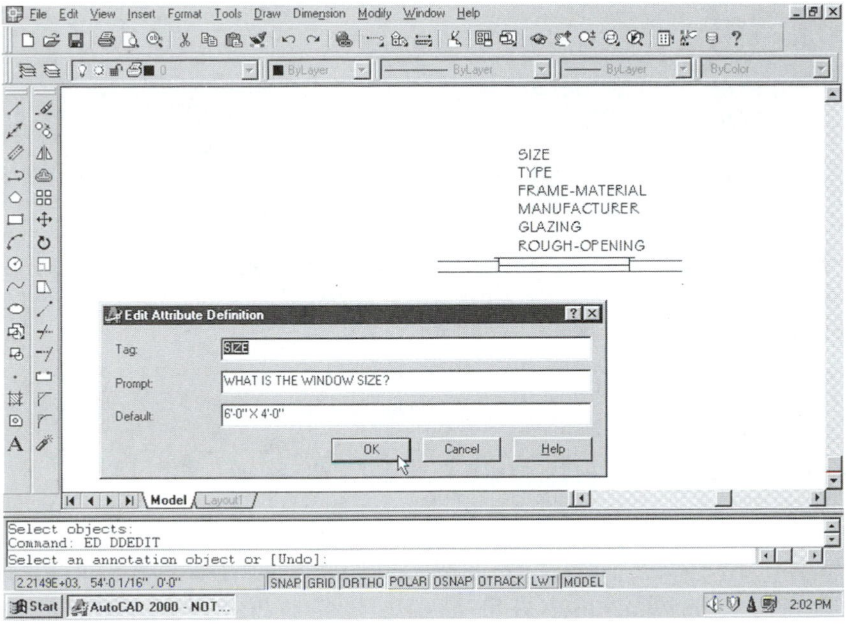

Figure 21.7 *The Edit Attribute Definition dialog box can be used to alter the tag, prompt, and default value for an attribute before it is saved with a block. The dialog box is accessed by typing* **ED** *(EnterKey) at the command prompt.*

mand prompt, or by selecting Properties… from the Modify pull-down menu. The Properties box can be used to edit the color, layer, linetype, and each of the options defined as the attribute was created.

ATTACHING ATTRIBUTES TO A BLOCK

Attributes can be attached to a block when a block is defined or redefined. Review Chapter 20. Select the attributes and the object when prompted to select the objects that will comprise the block. The order that attributes are selected when the block is created will determine the order of display when the block is inserted.

Inserting a Block with Attributes

The INSERT command introduced in Chapter 20 can be used to insert blocks with attributes into a drawing. Blocks with attributes are inserted in a similar method as blocks with no attributes. Once the block to be inserted is selected, the prompt will be given for the insertion point. Each of the options for a block can now be adjusted from the command line. The final aspect of inserting the block will be the display of the prompts defined as the attributes were assigned. Figure 21.8 shows the tags and the values for a window block.

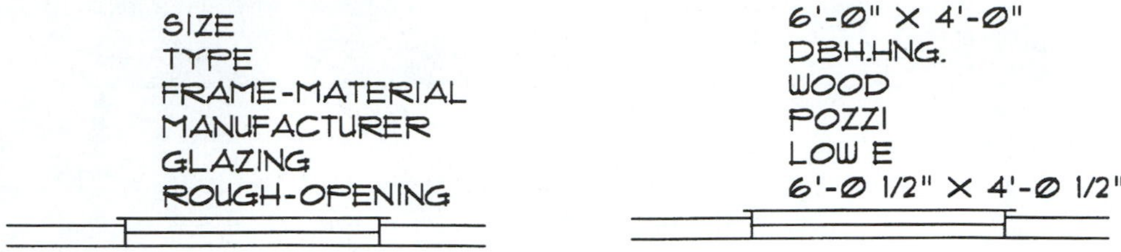

Figure 21.8 *The window on the left shows how the tags will be displayed as the block is created. The window on the right shows how the values will be displayed once the block is inserted into a drawing.*

EDITING ATTRIBUTES

Once the values for an attribute have been assigned to a block, they can be altered using the Edit Attributes dialog box. The dialog box is activated by selecting the Edit Attribute icon from the Modify II toolbar, by selecting Single… from Attribute of the Modify pull-down menu, or by typing **ATTEDIT** (EnterKey) at the command prompt. As the command is entered, you'll be prompted to select a block reference to be edited. Once the block is selected, a dialog box similar to Figure 21.9a is displayed to allow each attribute value to be altered. Text is altered by moving the cursor to the desired edit box and inserting the desired text. Selecting the OK button will alter the values displayed on the screen similar to Figure 21.9b. Selecting the Cancel button will remove the dialog box, with no changes made to the drawing base.

GLOBAL EDITING

Selecting Global... from Attribute of the Modify pull-down menu allows the same attribute to be edited in all of its applications throughout a drawing. This can be useful if you find an error in a block value that has been inserted several times. Rather than changing each attribute individually, you can change all of them using the global option. The command sequence is:

Command: *(Select Global from Modify pull-down menu.)*

Edit attributes one at a time? [Yes/No] <Y>:

Accepting the default value will allow you one last chance to verify each attribute before it is altered. Entering **N** (EnterKey) at the prompt will change all of the specified attributes, once other parameters have been set. The prompt will continue:

Preforming global editing of attribute values

Edit only attributes visible on screen? [Yes/No] <Y>:

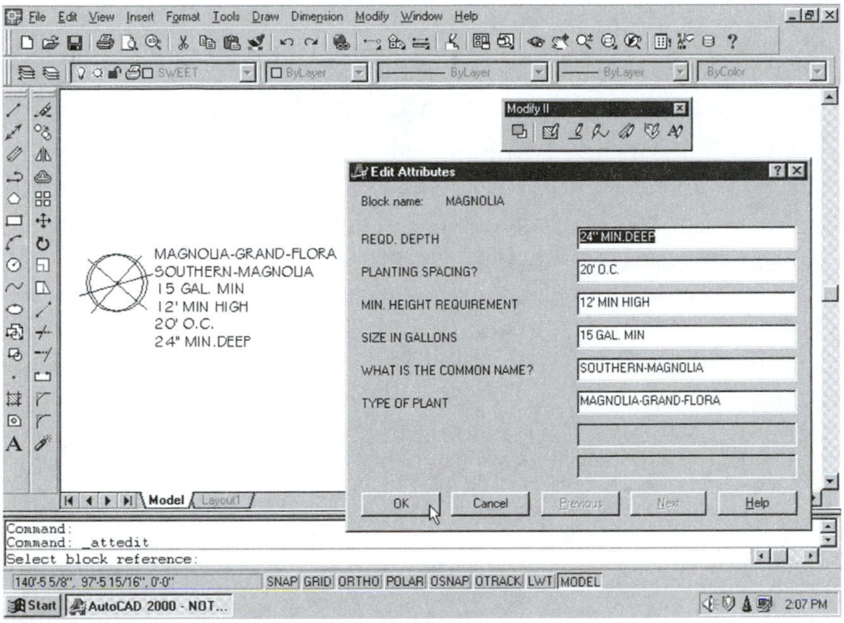

Figure 21.9a *Attributes can be edited using the Edit Attributes dialog box. The dialog box is accesed by selecting the Edit Attribute icon from the Modify II toolbar, by selecting Single... from Attribute of the Modify pull-down menu, or by typing* **ATTEDIT** *(EnterKey) at the command prompt.*

Entering an N will display the AutoCAD Text window with the following message:

Drawing must be regenerated afterward.

Any changes that are made to attributes will not be reflected until the drawing is regenerated. If AUTOREGEN is set to ON, a drawing regeneration will be performed when the editing of this command is complete.

Editing Visible Attributes

If a Y is accepted at the prompt and only the blocks shown on the screen are to be edited, four prompts will be displayed. These include:

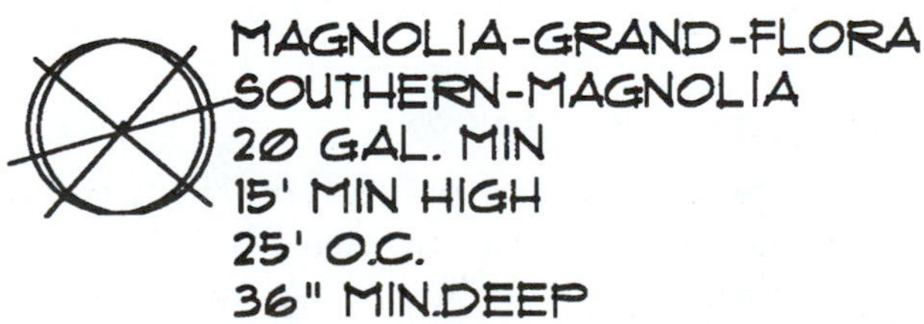

Figure 21.9b *The effects of altering the values in the Edit Attributes dialog box in Figure 21.9a.*

Enter block name specification <*>:

Enter attribute tag specification <*>:

Enter attribute value specification <*>:

Specific sets of Block names, Attribute tags, and Attribute values—including wild card characters—can be entered to define the selection set for editing. For an attribute to be selected for editing, the Block name, Attribute tag, and Attribute value must all match the name, tag, and value entered for the selection set. Be sure to match the case of type used for the selection sets to the case used when the attributes were entered originally. Matching uppercase titles with lowercase titles will not be accepted as matching sets. Attributes may be selected using Box, Window, Crossing, WPolygon, or CPolygon. Once selected, attributes will appear highlighted. When the selection process is complete, the prompt will change to:

Select attribute:

Enter string to change:

The "String to change" prompt initiates a search of all eligible attribute strings for matching strings. Each matching string will be changed to reflect the entry for the "New string" prompt. A block titled FOOTING with a tag of DIA. and a value of 17" diameter × 8" could be changed to 18" by entering:

Enter string to change: **17** (EnterKey)

Enter new string: **18** (EnterKey)

Entering * at the "String to change" prompt will cause the entry at the "New string" prompt to be placed ahead of all eligible Attribute value strings. If a block is specified as SINK with an attribute value of DBL. SINK, it can be changed to 32 × 21 DBL SINK by entering:

Enter string to change: *

Enter new string: **32 × 21** (space) (EnterKey)

Be sure to use the space bar to provide a space after the 21 or the new display will read 32 × 21DBL SINK.

Editing All Attributes

If an N is accepted at the prompt, all blocks in the drawing will be edited.

EXTRACTING ATTRIBUTES

One of the most useful benefits of using attributes is the ability to extract information stored as an attribute and assemble it in a written list. Window, door, and finish schedules typically are associated with architectural drawings. Pier, steel, and beam schedules are but a few of the types of written engineering drawings. Figure 21.10 shows portions of a floor plan and the doors that will be extracted. The doors shown

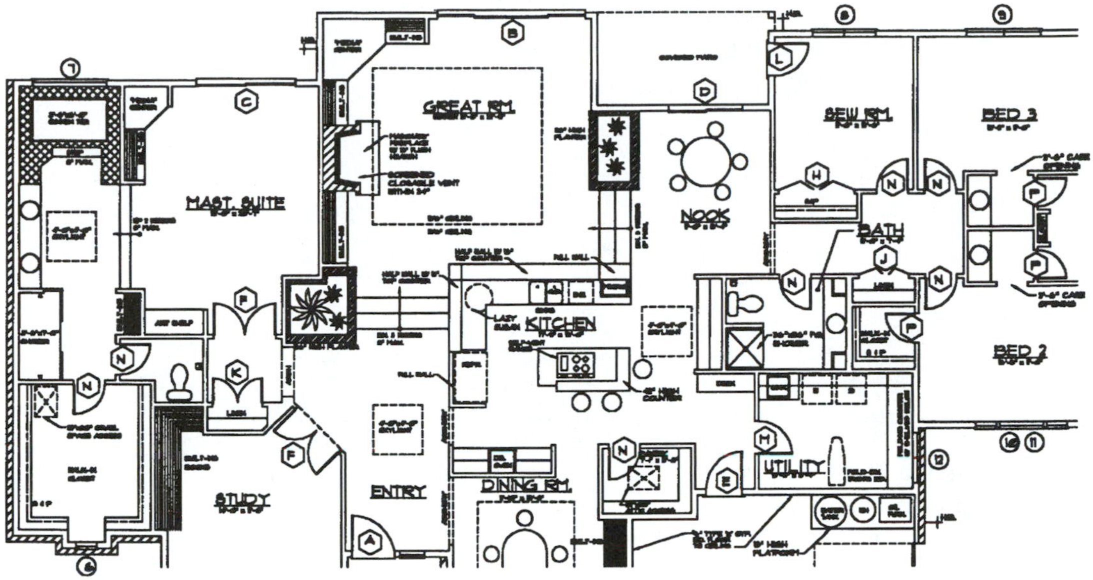

Figure 21.10 *Extracting attributes can be very useful on drawings containing multiple blocks with attributes such as this floor plan.*

are a fraction of the total number of doors that would be represented on a complete multilevel structure. Each door was inserted into the floor plan as a block with one visible and four invisible attributes. Attributes have been assigned to represent symbol, size, style, manufacturer, price, and quantity. By assigning these values as attributes, an accurate schedule can be maintained and quickly updated. By using the ATTEXT (ATTribute EXTraction) command, a complete listing of all of the doors used throughout the project can be written to file. Figure 21.11 shows an example of the Attribute Extraction dialog box. Figure 21.12 shows an example of a sample listing of door values. The heading above each column will not be listed to the extract file by the ATTEXT command. When using a program such as Notepad, the values listed under each heading will be extracted and used to sort doors by symbol, size, price, style or manufacturer.

USING A DATABASE

A database program can be used to manipulate the data stored in block attributes. A listing of doors or windows for a multilevel structure is a database that has essential data associated with it—such as size, type, rough opening, etc. Two terms generally associated with each database are record and field. A list of each of the construction components related to doors would be a record. The door, frame material, hinge type, and finishing method are each an example of a record. A door might be swinging,

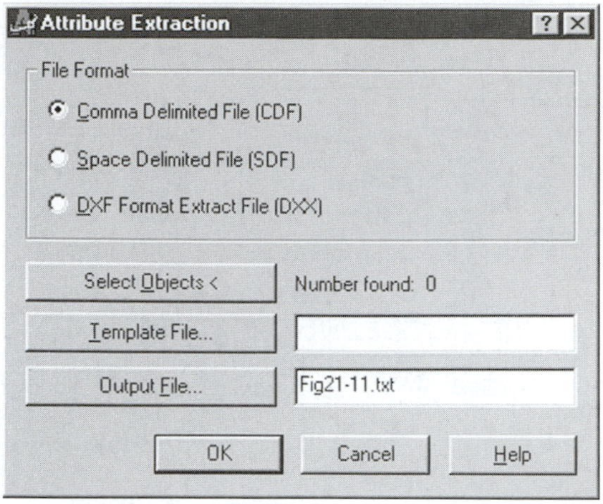

Figure 21.11 *The Attribute Extraction dialog box can be used to control the steps for extracting attributes from a drawing.*

DOOR SCHEDULE			
SYM.	SIZE	TYPE	QUAN.
A	3'-6" x 8'-Ø"	S.C.R.P., W/ 24" SIDELIGHT W/ SQ. TRANSEM ABOVE	1
B	12'-Ø" x 8'-Ø"	SL. GLASS W/ 12'-Ø" x 3'-Ø" SQ. TRANSEM ABOVE	1
C	8'-Ø" x 6'-8"	SL. FRENCH DRS. W/ 12" SQ. TRANSEM ABOVE	1
D	6'-Ø" x 6'-8"	SL. GLASS DRS. W/6'-Ø" x 3'-Ø" SQ. TRANSEM ABOVE	1
E	3'-Ø" x 6'-8"	SELF CLOSING, M.I.	1
F	PR 2'-6" X 6'-8"	HOLLOW CORE	2
G	6'-Ø" x 6'-8"	SL. MIRROR	1
H	6'-Ø" x 6'-8"	BIFOLD	1
J	4'-Ø" x 6'-8"	BIFOLD	1
K	PR 2'-Ø" x 6'-8"	HOLLOW CORE	1
L	2'-8" X 6'-8"	S.C., M.I.	1
M	2'-8" x 6'-8"	HOLLOW CORE	1
N	2'-6" x 6'-8"	HOLLOW CORE	7
P	2'-4" x 6'-8"	HOLLOW CORE	3
Q	16'-Ø" x 8'-Ø"	OVERHEAD, GARAGE	1
R	9'-Ø" x 8'-Ø"	OVERHEAD, GARAGE	1

Figure 21.12 *A list of doors values to be extracted using the ATTEXT command.*

sliding, folding, or overhead. Each listing of a door is a record. All of the doors that are swinging are a field. Swinging doors can be divided by interior, exterior, raised panel, slab, 1-lite, or 10-lite. It is possible to take a listing of doors that are listed by size, and by using a database, reorder the listing by type. The ATTEXT command of AutoCAD allows a list to be generated that contains all of the desired objects to be manipulated by the database. ATTEXT creates an extract file that allows AutoCAD to find and list attributes in the template file.

USING A TEMPLATE

A template is a file that lists the fields that specifies the tags and determines what blocks will have their attribute data extracted. The template file can be made using a database, word processor, or text-editing program. The balance of this chapter will rely on Notepad. The template must be a file on an accessible path with a .TXT extension. If a text editor or word processing program in the ASCII mode is used, you must add this extension when you name the file. A template is composed of a field name and a character-numeric value.

Field Name

The Field name must be exactly the same in every respect to the tag, in order for that tag to be extracted. If the attribute tag is SLDG, then there must be a field name in the template titled SLDG. A field name of SLDG will not have the tag of SLIDING extracted.

Character-numeric

The ATTEXT command uses the template to classify the data written to a particular field as either a numeric or as character type. Characters such as a, b, c, A, #, and " are always character type, but numbers do not have to be numeric type. Characters occupy less memory space than numeric values. Numbers that represent a width or height are better stored as characters, unless they are to be used in mathematical functions. The only significance of numbers as characters is in their order (456...) for sorting purposes by the database program. Characters have the same significance.

Template Components

The extraction template has two elements for each field. These are the field name and the character-numeric element. Numbers in strategic spaces in the character-numeric element of the template file specify the number of spaces to allow for the values to be written in the extract file. Others specify how many decimal places to carry the numeric values. The format is:

fieldname	Nwwwddd	for numeric values with decimal allowance
fieldname	Nwww000	for numeric values without decimal allowance
fieldname	Cwww000	for character values

Each line on the template is one field. The C and N's represent character or numeric information to be extracted. If a character other than a number is to be extracted, use the C. If a numeric entry, such as the price, will contain the dollar sign, use a C. The w's and d's represent the necessary digits when the template is created. The first three numbers after the letter represent the number of spaces that will be reserved for the attribute. The last three numbers specify the number of decimal spaces in the value. A listing of N010000 represents a numeric entry with 10 spaces allotted to the attribute and no space reserved for decimal points. An entry of N006002 would represent a numeric entry of six spaces with two decimal places. The order of fields listed in the template does not have to match the order that attributes tags appear in the block. Any group of fields in any order is acceptable. The only requirement for a block to be eligible for extraction by the ATTEXT command is that there is a least one tag-field match. A template file for the example in Figure 21.10 could be written as follows:

SYMBOL	C002000
TYPE	C008000
SIZE	N012000
STYLE	C008000
MANUF	C010000
PRICE	N006002
QUANTITY	C002000

From the list of doors shown in Figure 21.12, a database program can generate a sorted list with items that correspond to the selected values. To create a door schedule, listings would be sorted based on the symbol. To create a bill of materials to order doors, the listing could be sorted by the quantity or the price.

Nonattribute Fields

Additional information is automatically stored with each block as it is inserted. This can be seen by selecting a block and then entering the LIST command. Information about the block handle, insertion point for the X, Y, and Z coordinate, layer, level, name, number, orientation, scale value for X, Y, and Z, and extrusion direction in X, Y, and Zare values that can be used to manipulate attributes. In addition to listing attributes, the template file can be designed to extract block properties that include:

Handle–An identifier for a block.

Layer–The name that the layer is inserted on.

Level–Block nesting level.

Name–Block name.

Number–Counts the number of insertions.

Orient–The rotation angle of the block

X, Y, or Z. The appropriate coordinate of the insertion point.

XSCALE, YSCALE, ZSCALE–The appropriate X, Y or Z insertion scale factor.

XEXTRUDE, YEXTRUDE, ZEXTRUDE – the appropriate X, Y, or Z extrusion direction.

In creating a template, an entry might include BL:ORIENT N010000 to list all block data by orientation, allowing 10 spaces for the width of the field with no decimal places in the value. An entry of BL:PRICE C006002 will sort all blocks by price, allowing six spaces for the value to be written including two decimal places. If the value to be written exceeds the width allowed, AutoCAD truncates the written data, proceeds with the extraction, and displays the following error message:

**Field overflow in record <record number>

CREATING THE TEMPLATE

Now that the tools have been discussed, a template can be created to manipulate the attributes. Start by opening a drawing with blocks and attributes. From the Start menu, select Programs, Accessories, and then Notepad. The display will now look like Figure 21.13. Enter the required text at the flashing cursor. Use the Tab key to

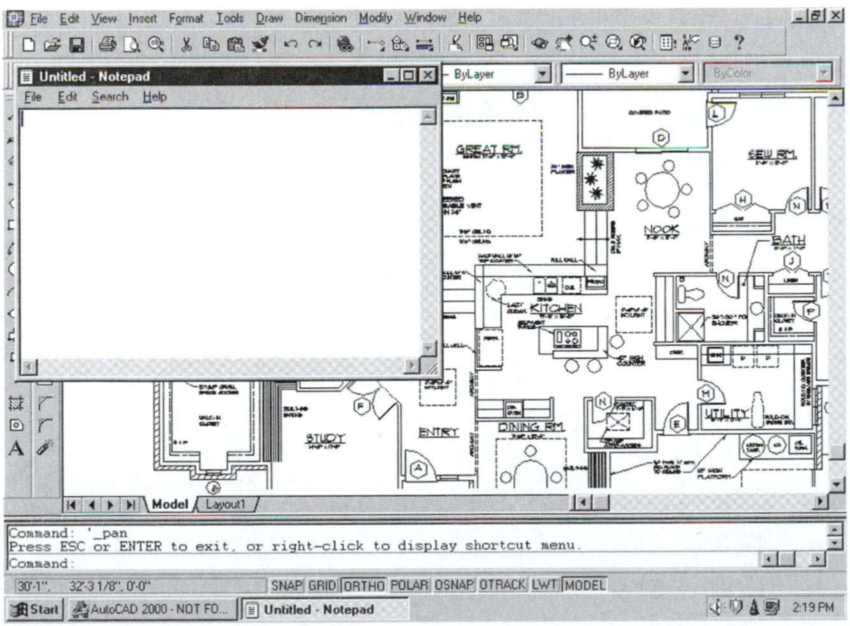

Figure 21.13 *A word processing or database program can be used to generate a template for extracting attributes. Notepad can be used by selecting the Start button, followed by selecting Programs, Accessories, and then Notepad.*

place a space between each portion of the listing. The entry will resemble Figure 21.14. When the entry is complete, use the SaveAs option from the File pull-down menu to save the file. Use the Exit option of File menu to exit the Notepad and return to AutoCAD.

EXTRACTING ATTRIBUTES

Once a template has been created, attributes can be extracted from blocks using the ATTEXT command. Entering **ATTEXT** (EnterKey) at the command prompt will produce the Attribute Extraction Dialog box that was introduced in Figure 21.11. The three major areas of the dialog box allow for selecting the file format, the template file and the Output file.

File Format

The default setting for the file format is the Comma Delimited File (CDF). Additional radio buttons allow for the selection of Space Delimited Files (SDF) and Drawing Interchange Files (DXF). CDF file format will generate a file containing one record for each block reference in the drawing that has at least one matching attribute tag in the template file. The values written under the fields in the extract files are separated by commas instead of spaces. The character fields are enclosed in single quotes. SDF format writes the values lined up in the widths allowed. It is possible for adjacent mixed fields of characters and numeric entries not to have spaces

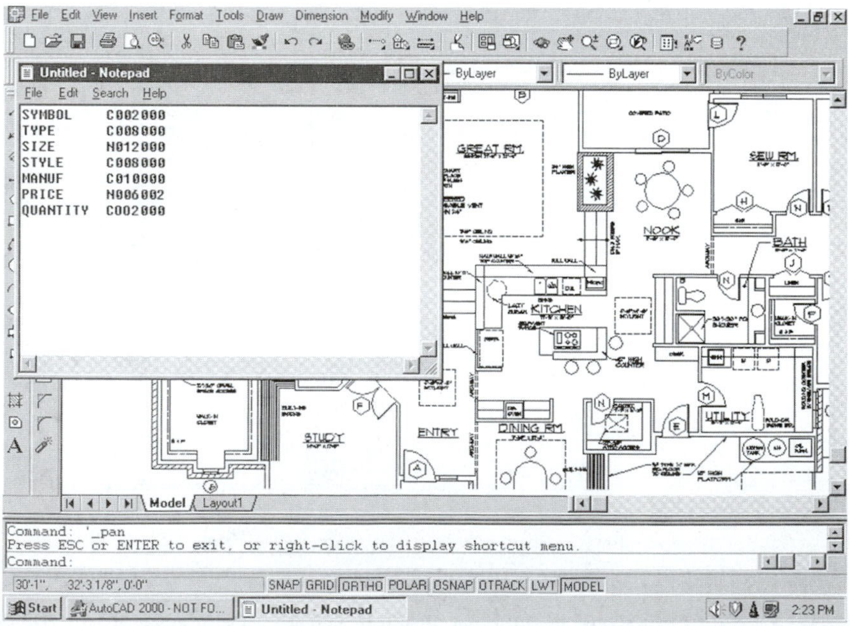

Figure 21.14 *The template for attribute extraction must include the field name and a character-numeric entry.*

between them. Dummy fields such as BLANK or SPACE can be added to provide space between fields. DXF format will generate files that contain only the block reference, the attribute, and end-of-sequence objects.

Template Files

Once the format has been selected, the template file for attribute extraction can be selected using the Selecting the Template File... button. The button will be deactivated if the DXF format has been selected. Selecting the Template File... button will display a Select Template File dialog box, allowing the DOOR.TXT template to be selected. Selecting Open will return the Attribute Extraction dialog box.

Output Files

Selecting the Output... button allows the method of output to be determined. The default file name is shown in the Output file edit box with a .TXT extension. The extension can be altered to .CON for writing to the screen, or .PRN for writing to the printer. A printer must be connected and ready for printing for the .PRN to be used.

Select Objects

With the method of output selected, selecting the Select Objects< button will remove the dialog box and return you to the drawing screen. Select the attribute objects, and the dialog box will reappear. The number of blocks selected is now displayed after the Number Found: entry. If no objects are selected, all blocks in the drawing specified by the template file will be used. With the process complete, return to Notepad to view the extract file that was created.

CHAPTER 21 EXERCISES

1. Draw an 18"-diameter footing using dashed lines and show a 4 × 4 post in the center of the pier. Use 5" high text to install the following Attributes.

TAG	PROMPT	VALUE
Dia.	What is the diameter?	18" dia.
Depth	What is the pier depth?	12" min. into grade
Strength	What is the conc. strength?	2500 p.s.i.
Rebar	Will the pier be reinforced?	10 × 10-4×4 wwm 3" up
Post	Size of post to be supported?	4 × 4 D.F.L. #1
Anchor	Will an anchor be used?	Simpson CB44 base

Save the drawing as a wblock on a floppy disk as FNDPIER.

2. Draw an elevation of a 3'-0" × 4'-6" double-hung window and use 5" high text to install the following attributes. Set the mode to invisible display.

TAG	PROMPT	VALUE
Manuf.	List the manuf.	POZZI
Frame	List the frame size	3'-0" × 4'-6"
Unit	List the unit size	UN=3'-2 5/8" × 4'-8 7/8"
R.O.	List the rough opening	R0=3'-0 1/2" × 4'-6 1/2"
Frame	Frame material	Kiln-dried fir
Exterior	Exterior finish	Prime ext. face
Sash	Sash material	1 1/16" thick sash
Glaze	Glazing material	9/16" air space
Weather	Weather stripping	foam-backed
Hardware	Hardware color	Bronze
Screen	Screen material	Alum.-bronze tint
Assembly	Where assembled	Site assembled
Shipping	Destination	Seattle, WA.

Prior to creating a wblock of this material, change the Frame material to Kiln-dried Western pine. Change the hardware to WHITE. Ship from Bend, OR. Change the Assembly point to Factory. Save the drawing as a wblock with a name of EL3-46DH.

3. Open a new drawing and draw a 24" × 24" skylight in plan view. Add the following Attribute tags, prompts, and values.

TAG	PROMPT	VALUE
Size	Unit size	24" × 24"
Manuf.	Who is the manuf?	Velux
Model	Model number	SF
Frame	Frame material	Alum.-clad frame
Finish	Finish material	Lacquered finish
Glass	Glazing	Double-tempered glazing
Flashing	Flash material	32ga. lacquered alum.

Save the drawing on a floppy disk as a wblock titled SKYLIGHT. Insert the skylight into a drawing three different times. Change the size so that the skylight will be 24 × 24, 24 × 48, and 48 × 48. Change the size value of each skylight to reflect the proper size. Use global editing to change the model value to FS, change the frame to Alum-clad Frame, and change the finish to 22ga. Save the drawing as E-21-3.

4. Open drawing E-20-5 and insert the POTTY block. EXPLODE the block and add the following Attribute tags, prompts, and values.

TAG	PROMPT	VALUE
MANUF	Who is the manuf?	Am. Std.
Cat-no	What is the Catalog number?	2006.014
Descrip.	What is the description?	Elon. Lexington
Color	What color will be used?	White
Supplier	Who will provide fixture?	General cont.
Installer	Who will install fixture?	Plumber
Hardware	Who will supply hardware?	Owner

Save the drawing as a wblock titled POTTY.

CHAPTER 21 QUIZ

1. How are attributes saved?
2. What commands can be used to add a block or wblock to a drawing?
3. What must be done to add attributes to an existing block?
4. Describe what an Attribute prompt is.
5. What is an Attribute tag?
6. List the four modes for Attribute display and give a brief explanation of each.
7. What is the initial step to assign attributes to blocks?
8. List the commands that produce dialog boxes for use with attributes.
9. List two methods to edit an attribute.
10. What ATTDIA setting is required to produce dialog box displays?
11. What two commands allow for attribute editing?
12 How can global editing of attributes be limited?
13. How do the prompts "Enter string to change" and "Enter new string" affect each other?
14. What options are provided when editing attributes individually?
15. In addition to AutoCAD, what is needed to extract attributes?

CHAPTER 22

Working with Multiple Drawings

Users of previous versions of AutoCAD have been able to work on multiple drawings. They just had to have multiple sessions of AutoCAD open and hope they had enough memory to keep their machine from crashing. AutoCAD 2000 allows multiple drawings to be opened in one drawing session, without sacrificing drawing speed. Information can be transferred using traditional Windows skills or by using the AutoCAD DesignCenter.

THE MULTIPLE DOCUMENT ENVIRONMENT

Template drawings have been used throughout this text to store frequently used styles, settings, and objects for use in similar drawings. This chapter will introduce methods for moving objects and information between two or more drawing files. Objects or drawing properties can be moved between drawings using cut/copy and paste, object drag and drop, the Property painter, and concurrent command execution.

GENERAL CONSIDERATIONS OF MULTIPLE DOCUMENT ENVIRONMENTS

AutoCAD is by default set to work with multiple documents. The toggle for switching from multiple to single document interface is the Single Drawing Compatibility Mode button in the General options area of the System tab of the Options dialog box. The dialog box is accessed from the Tools pull-down menu. With the box inactive, multiple drawings can be opened in one AutoCAD session. Don't gloss over the concept of multiple drawings and one drawing session. Just as you can have several programs open on your desktop at once, and rapidly switch from one program to another, AutoCAD now allows you to have several drawings open at the same time. Chapter 1 and 2 introduced the skills that are used to display multiple drawings. Open AutoCAD and select Open a Drawing. Select an existing drawing to be opened. While holding down (CtrlKey), select three other drawings to be opened, and then press the OPEN button. You have now used one of the most powerful tools of AutoCAD and have four drawings opened in one drawing session.

Move to Window in the Standard pull-down menu and select Cascade. Your drawing will now resemble Figure 22.1. The title bar of each drawing will be displayed. The drawing with the blue title bar on top of the stack is the active drawing. Apart from being on top of the pile, it is the only drawing that can currently be used. Return to Window in the Standard pull-down menu and select Tile Vertical. Now return to Window and select Tile Horizontal. The drawing display will now resemble Figure 22.2.With four drawings open, the horizontal and vertical displays will appear alike. With an odd number of drawings open, they will be displayed in vertical rows using the Vertical option. Important considerations to remember when having multiple drawings open include:

- In either orientation, any one of the four drawings can be made active. Notice that each small drawing area has all of the components of a normal AutoCAD drawing.
- Any one of the four drawings can be maximized or minimized to ease viewing, or all can be left open and you can switch from drawing to drawing as needed.
- You can start a command in one drawing, switch to another drawing and perform a different command, and then return to the original drawing and complete

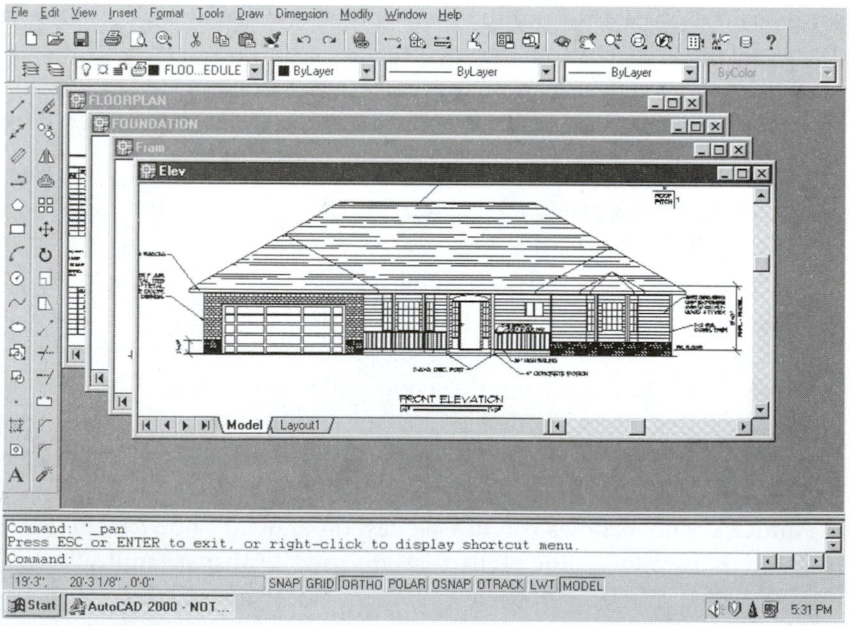

Figure 22.1 *Selecting Cascade from Window of the Standard pull-down menu will display the title bar of each drawing. The drawing with the blue title bar on top of the stack is the active drawing.*

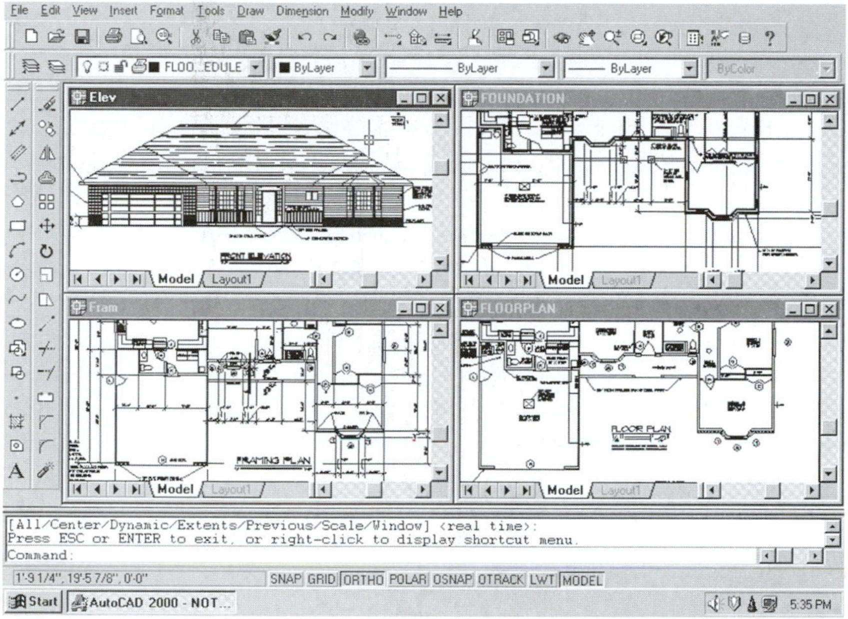

Figure 22.2 *Selecting Tile Horizontal from Window of the Standard pull-down menu will display each open drawing. In this orientation, any one of the four drawings can be made active. Notice that each small drawing area has all of the components of a normal AutoCAD drawing.*

the command in progress. As you return to the original drawing, single-click anywhere in the drawing area, and the original command will be continued.

- Each drawing is saved as a separate drawing file, independent of the other open files. You can close a file, and not affect the contents of the remaining open files.

Opening Multiple Drawings from the Windows Explorer

In addition to opening drawings from within AutoCAD, multiple drawings can also be opened from the Windows Explorer. To start the process, open the Explorer and double-click the drawing to be opened. Once the drawing is opened, reduce the size of the drawing so that the drawings in the Explorer can still be seen. With a drawing open, additional drawings can be dragged from the Explorer and dropped into the current drawing file. Files are dragged by pressing and holding (CtrlKey) and the left mouse button while the desired drawing is selected. Once the drawing icon is in the drawing area, release the (CtrlKey) and select button to open the drawing. Drawings will be placed in a cascading display similar to Figure 22.1.

Note: If a drawing is dragged from the Explorer into a current drawing, the INSERT command will be activated. Instead of opening multiple drawings, you'll be inserting a drawing into the current drawing.

USING THE CUT, COPY AND PASTE COMMANDS

In addition to standard procedures that are allowed using the Cut, Copy, and Paste commands of any Windows software, AutoCAD allows drawings objects and properties to be transferred directly from one drawing to another using these same commands. Each command can be performed using the Edit menu, by keyboard, or by shortcut menu. Cut will remove objects from a drawing and move them to the Clipboard. Once cut, the object can be placed in a new drawing using Paste. Copy allows an object to be reproduced in another location. The object to be copied is reproduced in its new location by using the Paste command. As the object is copied to the new drawing, so are any new layers, linetypes, or other properties associated the objects. Figure 22.3a and 22.3b show examples of using COPY to place the footing detail in the second drawing. The footing was copied by using the following sequence.

From the Source Drawing:

Command: *(Select Copy from the short-cut menu.)*

_copyclip

Select objects: *(Select object to be copied using any selection method.)*

Select objects: (EnterKey)

Command:

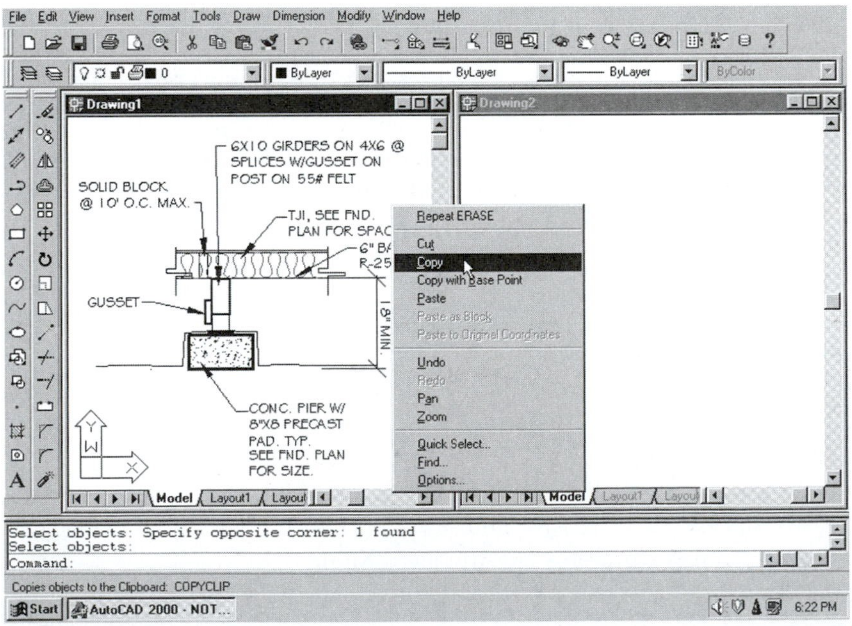

Figure 22.3a *The footing in the left drawing can be copied into the right drawing by selecting Copy from the shortcut menu.*

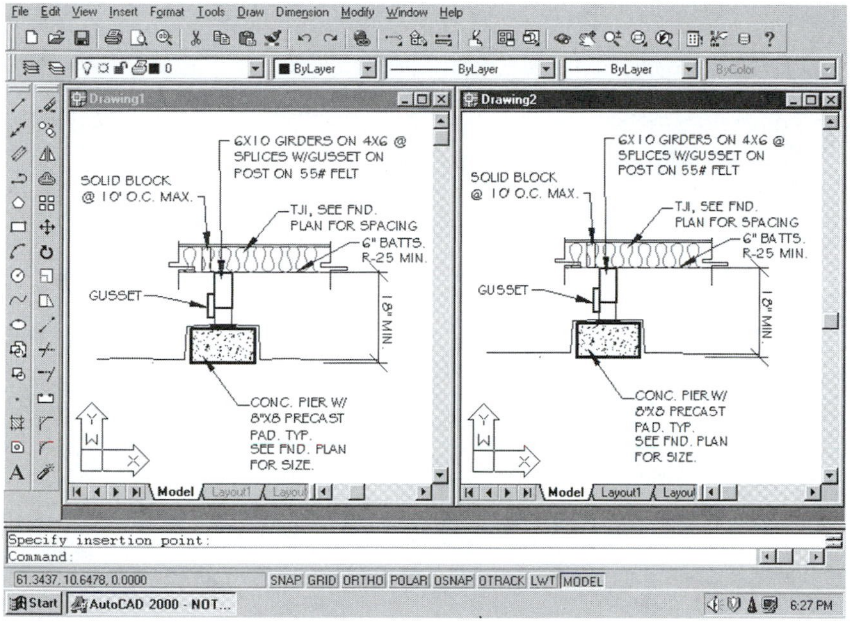

Figure 22.3b *Once selected to be copied, move the cursor to the desired drawing, and single-click to make that drawing active. Right-click and choose Paste from the shortcut menu, and then select an insertion point.*

In the Second Drawing:

Command: *(Select Paste from the short-cut menu.)*

_pasteclip

Select insertion point: *(Select the desired location for the new objects.)*

Command:

Copy With Base point

The Copy with Basepoint option of the shortcut menu is similar to the Copy option. The command can also be entered by typing COPYBASE (EnterKey) at the command prompt. With this option, before you select objects to copy, you'll be prompted to provide a base point. This option works well when objects need to be inserted accurately. The roof detail in Figure 22.4 was copied into the new drawing using the Copy to Base point option. The roof detail was copied using the following sequence:

From the Source Drawing:

Command: *(Select Copy with Base point from the short-cut menu.)*

_copybase

Specify base point: *(Select the desired base point for the objects to be copied.)*

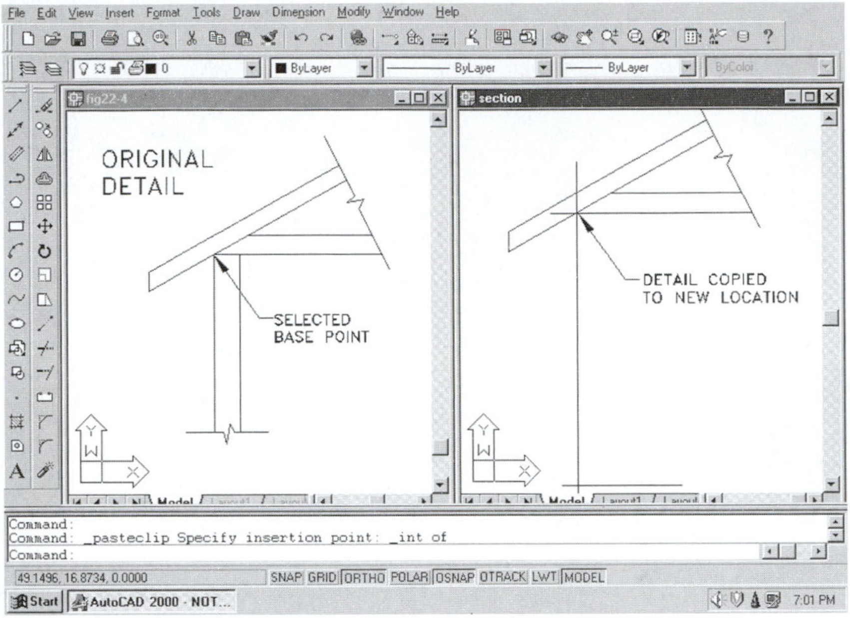

Figure 22.4 *The Copy with Base point option works well when objects need to be inserted accurately. The Copy with Base point option of the shortcut menu is similar to the Copy option. With this option, before you select objects to copy, you'll be prompted to provide a base point.*

Select objects: *(Select object to be copied using any selection method.)*

Select objects: (EnterKey)

Command:

In the Second Drawing:

Command: *(Select Paste from the short-cut menu.)*

_pasteclip

Select insertion point: *(Select the desired location for the new objects.)*

Command:

Paste as Block

The Paste as Block command can be used to paste a block into a different drawing. It functions similar to the Paste command. Selecting the objects to be copied in the original drawing and select the COPY command. Selected objects do not need to be a block, since the command will turn them into a block. Next activate the new drawing and right-click to select the Paste as Block option. The copied objects are now a block. To be edited, they must be exploded. The command can also be started by entering **PASTEBLOCK** (EnterKey) at the command prompt. The command sequence is the same as the Paste command. You will notice that blocks that are copied will be renamed. A name such as ROOFWALL will be assigned a name such as A$C2AA9F67C4 by AutoCAD. The new block can be renamed using Object Properties Manager.

Paste to Original Coordinates

This command can be used to copy an object in one drawing to the exact same location in another drawing. The command could be useful to copy an object from one apartment unit to the same location in another unit. The command can be started by using the shortcut menu or by entering **PASTEORIG** (EnterKey) with the keyboard. The command is only active when the Clipboard contains AutoCAD data from a drawing other than the current drawing. Selecting an object to be copied starts the command sequence, and is completed by making another drawing active and then selecting the Paste to Original Coordinates option.

DRAGGING AND DROPPING OBJECTS

AutoCAD allows objects to be moved between drawings using a Drag and Drop sequence. Drag and Drop allows objects:

- To be moved or copied to a new location in the same drawing.
- To be moved or copied to another drawing in the same drawing session.
- To be copied to a drawing in another session.
- To be copied to another application.
- Drag and Drop can be completed using either the left or right mouse buttons. Slightly different results and procedures will be required using each method.

Selecting Objects to Drag and Drop with Left-Click

The procedure to Drag and Drop using the left mouse button begins by selecting the objects to be moved or copied. Any object selection method can be used to select objects. Once the set is selected, move the cursor over one of the selected objects and press the left mouse button. Moving the cursor while the left button is held down will show the effect of moving the selected objects. Release the mouse button when the objects are in the desired location. If you're working with a single drawing, the selected objects will be moved. Objects can be copied in a single drawing by pressing (CtrlKey) prior to pressing the left mouse button to drag the object. Objects will always be copied when dragged into a different drawing. You are not allowed to drag objects into a drawing that has an active command. The unavailable cursor—a circle with a diagonal line through it—is displayed if you attempt to drag objects into a drawing with an active command. Figure 22.5 shows the steps to copy a detail into another drawing using the select button of the mouse.

Selecting Objects to Drag and Drop with Right-Click

The procedure to Drag and Drop using the right mouse button begins by selecting the objects to be moved or copied. Any object selection method can be used to select

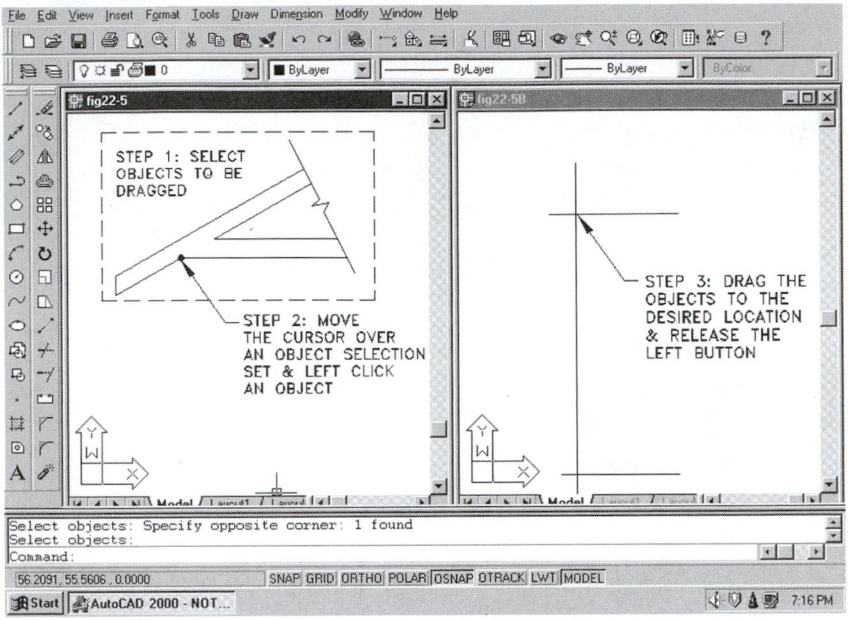

Figure 22.5 *Objects can be moved or copied easily using Drag and Drop.*

objects. Once the set is selected, move the cursor over one of the selected objects and press the right mouse button. Moving the cursor while the button is held down will show the effect of moving the selected objects. Release the mouse button when the objects are in the desired location. As the right button is released, a menu will be displayed. The menu will depend on what type of drawing you're working in. If you're working in a single drawing, the menu options will include:

Drag and Drop in the Same Drawing
- Move Here
- Copy Here
- Paste as Block
- Cancel

Drag and Drop in a Different Drawing
- Copy Here
- Paste as Block
- Paste to Orig Coords
- Cancel

Select a menu option to end the drag and drop sequence.

USING PROPERTY PAINTER

The Property Painter cannot be used to move or copy objects, but it can be used to copy properties from one object to another. The Property Painter was introduced in Chapter 14 for transferring properties within a drawing.

The command can also be used to transfer properties from an active drawing to other drawings. The process to change the layer and linetype of objects in a drawing to match another drawing can be completed using the following process:

- Select the source object using any selection method.
- Select Match Properties from the Standard toolbar.
- Move to the target drawing and make it the current drawing. The Match Properties cursor will be displayed by the selection box.
- Select the target objects. The command can be continued within the current drawing, or another drawing can be made current, and the properties of the original object can be transferred to additional objects.

EXPLORING THE AUTOCAD DESIGNCENTER

New to AutoCAD 2000 is the AutoCAD DesignCenter. DesignCenter can be used to locate, organize, and customize drawing information. The DesignCenter can be used to directly transfer data such as blocks, layers, and external references directly from one drawing to another without using the clipboard. Drawing content can be moved from other open files in AutoCAD or from files stored on your machine, on a network drive, or from an Internet site. The DesignCenter can be accessed by selecting the AutoCAD DesignCenter icon from the Standard toolbar, or by entering **ADCENTER** (EnterKey) at the command prompt. Either method will display a DesignCenter similar to Figure 22.6. The DesignCenter can also be displayed or closed using (CtrlKey) + 2.

The default position of DesignCenter is docked on the left side of the drawing screen. The DesignCenter can be resized by clicking a border, the pallet sidebar or by grabbing a corner and moving the cursor in a diagonal direction. If you are using the DesignCenter, the Properties dialog box, and a drawing, you may be tempted to shrink the size of the DesignCenter display. Resist shrinking the DesignCenter display past the point where two rows of icons can't be seen in the pallet display. The display can be reduced so no pallet is displayed, but then you've limited the usefulness of the display. Clicking and dragging the title bar will float the DesignCenter display, allowing it to be moved to any location in the drawing area. Figure 22.7 shows an example of a floating DesignCenter display. The DesignCenter display can be docked by double-clicking the title bar of the DesignCenter display or by dragging it to the left or right side of the drawing area.

FEATURES OF DESIGNCENTER

The DesignCenter provides direct access to any drawing file that you have access. Files stored on floppies, the hard drive, network or Internet can be accessed and

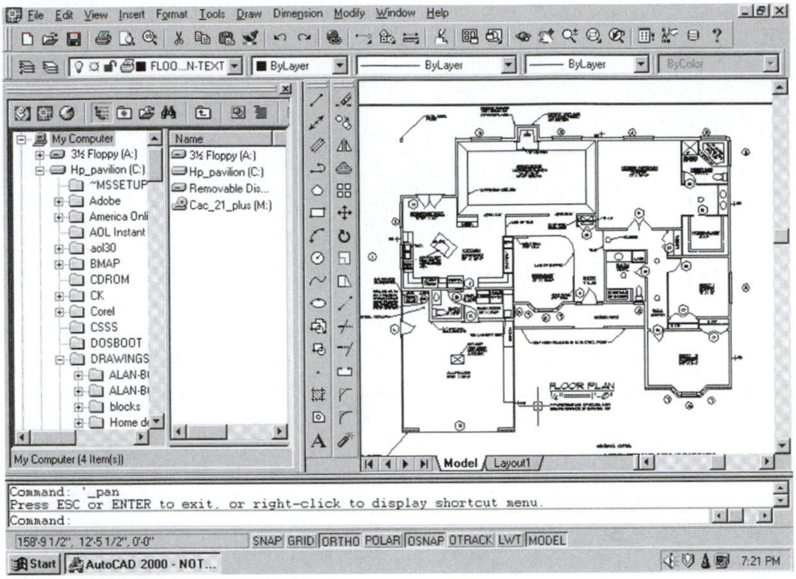

Figure 22.6 *The DesignCenter can be accessed by selecting the AutoCAD DesignCenter icon from the Standard toolbar or by entering* **ADCENTER** *(EnterKey) at the command prompt. The DesignCenter can be used to directly transfer data such as blocks, layers, and external references directly from one drawing to another.*

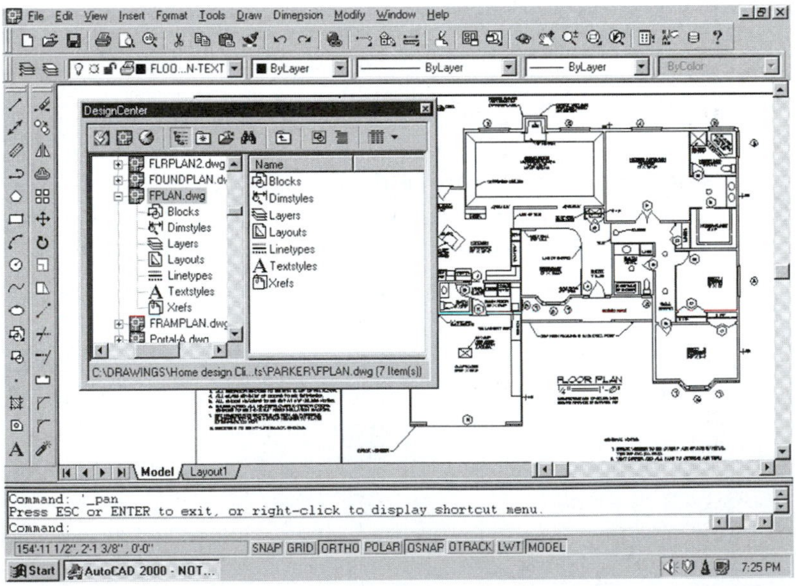

Figure 22.7 *Clicking and dragging the title bar will float the DesignCenter display, allowing it to be moved to any location in the drawing area. The DesignCenter display can be docked by double-clicking the title bar of the DesignCenter display or by dragging it to the left or right side of the drawing area.*

reused as a library for retrieving or coping blocks, referenced drawings, layouts, layers, linetypes, text styles or dimension styles. Other features of DesignCenter include:

- Create a Favorites folder to create a shortcut for frequently accessed drawings, folders, or Internet locations.
- View, attach, insert, or copy and paste layer and block definitions of any file you have access to into the current drawing. This feature could be used to update a drawing supplied by a subcontractor to match current standards.
- Open drawing files by dragging the file from the tree view into the current drawing area.
- Find drawing contents stored in any folder, drive, network, or from any Internet site that you have access to, based on the quality of the source. Search tools such as a name or date can be used to locate materials to be dragged into the current drawing file.
- Use the DesignCenter pallet to view aspects of the object, block, or aspect of the drawing content.

EXAMINING THE DESIGNCENTER TOOLBAR

The DesignCenter toolbar is located across the top of the Contents box and the Palette of the DesignCenter. Before examining the contents, you may need to adjust the size of the DesignCenter window to be sure that all of the contents are displayed. The toolbar consists of eleven icons for controlling the contents of the DesignCenter. These icons from left to right include the Desktop, Open Drawings, History, Tree View Toggle, Favorites, Load, Find, Up, Preview, Description, and Views. Each can be seen in Figure 22.8.

Desktop/Open Drawings

These icons serve as a toggle to control what is displayed in the DesignCenter. When Desktop is selected, the display will resemble the Windows Explorer. All resources available to your workstation will be displayed, including internal drives, local network, and Internet files. The exact display will depend on the contents of the selected drive and the setting for the Tree View icon. Selecting the Open Drawings icon will

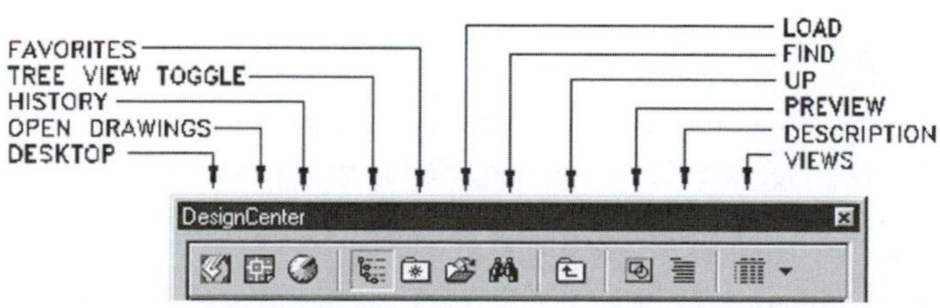

Figure 22.8 *The icons of the DesignCenter toolbar.*

display a list of drawings that are currently opened, similar to Figure 22.9. Notice that the drawings have either a + or – symbol preceding the title. The + symbol indicates that additional information can be displayed about the drawing file or folder. Double-clicking the drawing name will display the drawing contents and double-clicking a drawing with a full display (–) will close the contents of that file.

History

Selecting this icon will display a listing of the last twenty locations accessed through the DesignCenter. The history can also be displayed by right clicking in the palette and then selecting History from the shortcut menu.

Tree View Toggle

This icon displays or hides the tree view in much the same way that Details and Large Icons of the View menu of the Windows Explorer alter the display of the Explorer. Tree view displays open drawings, history, and files or folders for all of the drives available to your computer. Figure 22.9 shows a display using the tree display. Figure 22.10 shows a display with the tree view hidden. The tree view can be hidden by clicking the Tree View Toggle icon or by right-clicking in the palette and then clicking Tree from the shortcut menu. With the tree view hidden, the tree can be restored by clicking the icon or by right-clicking in the palette and then clicking Tree from the shortcut menu.

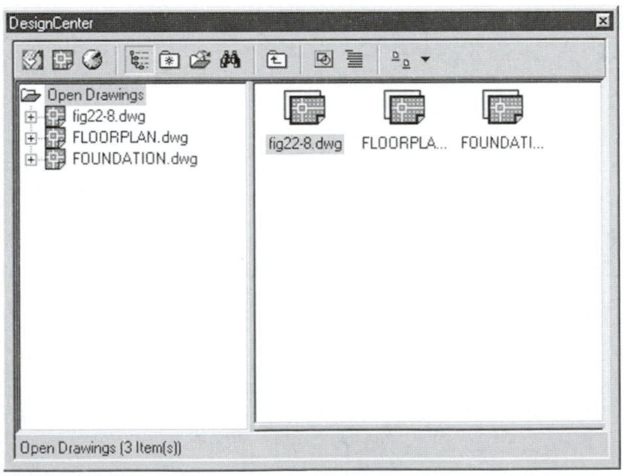

Figure 22.9 *Selecting the Open Drawings icon will display a list of drawings that are currently opened. Notice that the drawings have either a + or – symbol preceding the title. The + symbol indicates that additional information can be displayed about the drawing file or folder. Double-clicking the drawing name will display the drawing contents and double-clicking a drawing with a full display (–) will close the contents of that file.*

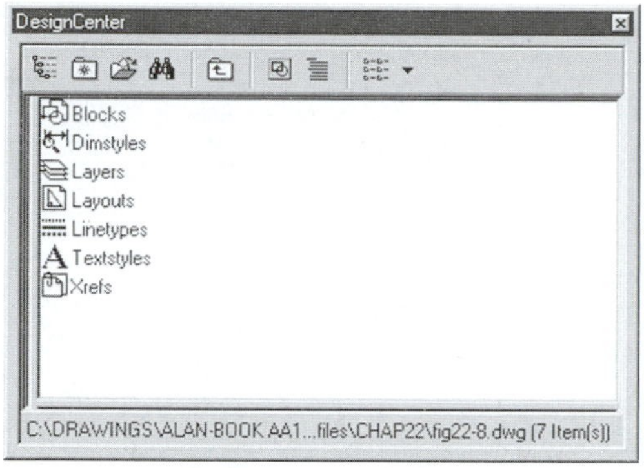

Figure 22.10 *The DesignCenter display with the tree view hidden.*

Favorites

This option of the DesignCenter displays shortcuts to files or folders that you need to access quickly on a regular basis. The tree, palette, and shortcut menus each include menus for accessing the Favorites menus.

Load

Selecting the Load icon displays the Load DesignCenter Palette dialog box. The dialog box is similar to the Select File dialog box used to open drawings. It can also be displayed by right-clicking in the pallet and then selecting Load from the shortcut menu. You can choose any available drives from the Look in edit box. The initial step in using the DesignCenter is to load the palette. Use the following steps to load the palette:

- Select the Load icon.
- In the Load DesignCenter Palette dialog box, choose one of the following options:

  Look in – this source displays the files and folders on local and network drives.

  Favorites – the Favorites icon displays the contents of the Favorites folder.

  Select the listing that is to be loaded.

Find

In addition to selecting information from the desktop, you can search for drawings or parts of drawings using Find. Selecting the Find icon displays a Find dialog box similar to Figure 22.11. The dialog box can be used to narrow the search to find drawings, blocks, layers, layouts, text styles, and referenced drawings. Currently the search will be conducted in all drawings. Selecting the Look for edit box allows the search to

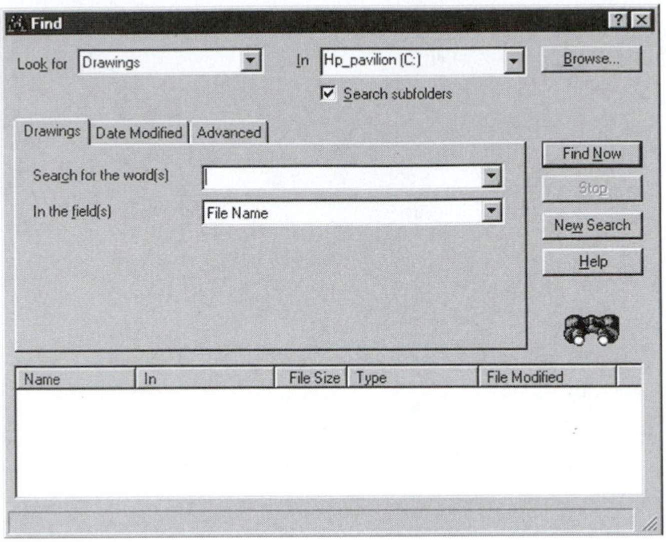

Figure 22.11 *Selecting the Find icon displays a Find dialog box. The dialog box can be used to find drawings, blocks, dimstyles, layers, layouts, linetypes, text styles, and referenced drawings.*

be limited to Blocks, Dimstyles, Drawing and Blocks, Layers, Layouts, Linetypes, Text Styles, and Xrefs. Depending on the field, the tabs in the dialog box will be altered. The tabs for finding drawings include Drawings, Date Modified, and Advanced.

Drawings

This tab can be used to specify the name or text that will be the object of the search. In Figure 22.12, the name ELECT is entered in the Search for the words edit box. The In the fields edit box can be used to limit the search criteria. In the current setting the search will be conducted in file names. Alternatives include looking in titles, subjects, author, and keywords.

Date Modified

The Date Modified tab allows a search to be conducted based on the date that the file was created or last modified. A date range specified in days or months can be used if you're not sure of the exact date the drawing was created.

Advanced

The Advanced tab allows additional search parameters to be set based on text contained in a portion of a file or by the file size. A search can be defined using the Containing edit box to search block name, block and drawing descriptions, attribute tag, or attribute value for text containing, but not limited to, a specified text. The search parameters can also be limited to looking in files of a minimum or maximum size.

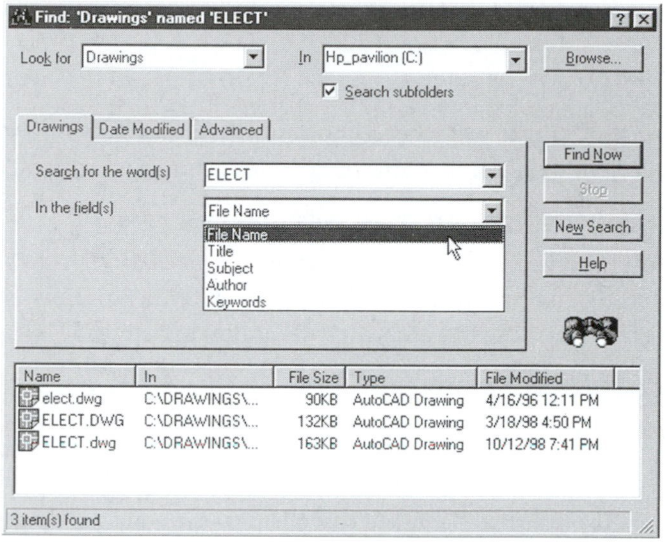

Figure 22.12 *Entering ELECT in the Search for edit box and pressing the Find Now button will conduct a search for all file names containing the word ELECT.*

Up

The Up icon is used to move through the contents of the DesignCenter. The Up icon performs the same function here as it does in the Select File dialog box. Each time the icon is selected, the display is moved up one level from the current location.

Preview

The preview window displays an image of a selected file or block.

Description

If a description was created for objects such as a block, this portion of the dialog box will display the description. The description will be given for the file or object that is displayed in the Preview window. A description can also be displayed by right-clicking in the palette, and then selecting Description from the shortcut menu.

Views

Selecting Views presents one of four options for displaying the format of the contents of the DesignCenter. Each time the icon is selected, a different method of presentation will be displayed. Selecting the edit arrow beside the icon will present a menu of the four options, so that the desired display method can be selected without having to scroll through each option. Display options include Large icons, Small icons, List, and Details.

Shortcut Menus

Right-clicking in the palette produces a shortcut menu similar to Figure 22.13. The shortcut menu contains the same menu options found in the DesignCenter toolbar.

OPENING DRAWINGS USING THE DESIGNCENTER

The DesignCenter can be used to open drawings. Open AutoCAD using any of the options presented throughout this text. The DesignCenter can now be used to drag a drawing into an empty drawing area with the following steps:

- Open the Design Center.
- Select the drawing source from the Desktop.
- Drag the icon of the drawing to be opened and drop it in the drawing area.
- Provide the insertion point and responses to the scaling prompts.
- Resize the drawing window as needed to display the drawing.
- A drawing can also be placed in another drawing by using the shortcut menu. From the DesignCenter, select the file to be opened, right-click, and select the Open to Drawing option from the shortcut menu.

ADDING CONTENT TO DRAWINGS USING THE DESIGNCENTER

One of the best uses for the DesignCenter is for adding content from the palette into a new or existing drawing. Information can be selected from the palette or from Find and dragged directly into a drawing without opening the drawing containing the

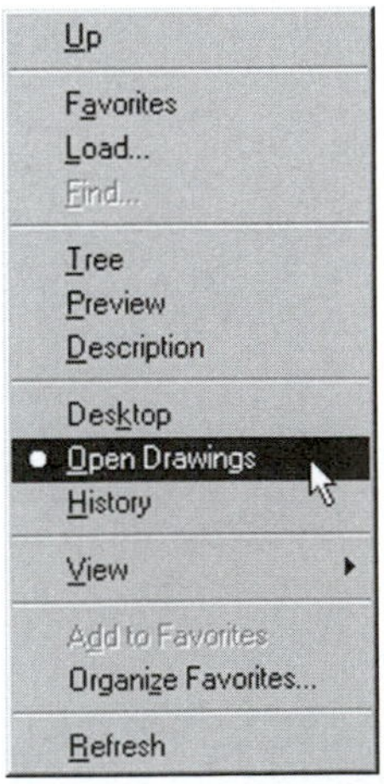

Figure 22.13 *Right-clicking in the palette produces a shortcut menu that contains the same menu options found in the DesignCenter toolbar.*

original. Chapter 20 introduced methods of moving blocks using DesignCenter. DesignCenter can also be used to attach external referenced drawings, and copy layers between drawings.

Inserting Blocks

The DesignCenter provides two methods for inserting blocks into a drawing. Blocks can be inserted using the default scale and rotation of the block, or the block parameters can be altered as the block is dragged into the new drawing.

Providing Parameters for Block Insertion

For most purposes this will be the preferred method of inserting a block using the DesignCenter. Use the following sequence to specify the insertion point, scale, and rotation angle as a block is inserted:

- Open the DesignCenter.
- Use the Palette to select the block to be copied.
- Use the right mouse button to drag the desired block to the new drawing.
- Drop the block into the drawing area.
- Release the right button. This will display a shortcut menu.
- Choose Insert Block... from the shortcut menu. This will display the Insert dialog box.
- Enter the values for each parameter or choose the Specify On-screen option for each parameter.
- Press the OK button to insert the block in the specified location.

Using the Default Scale

When blocks are inserted using the default scaling and rotation option, the block is inserted using Autoscaling. Autoscaling compares the units of the drawing with those of the block. As the block is inserted, AutoCAD scales the block as needed, based on the ratio of the block to the new drawing. Blocks inserted using this method may not retain their true size. Use the following sequence to insert a block using the default parameters:

- Open the DesignCenter.
- Use the Palette to select the block to be copied.
- Drag the desired block to the new drawing.
- Drop the block in the desired location

Attaching Referenced Drawings

The DesignCenter can be used to attach a referenced drawing using similar steps used to attach a block and provide the parameters. Use the following procedure to bring an xref into a new drawing.

- Open the DesignCenter.
- Use the Palette to select the xref to be attached.
- Use the right mouse button to drag the desired xref to the new drawing.
- Drop the xref into the drawing area.
- Release the right button. This will display a shortcut menu.
- Choose Attach Xref... from the shortcut menu. This will display the External Reference dialog box.
- Select Attach or Overlay from the Reference Type box.
- Enter the values for each parameter or choose the Specify On-screen option for each parameter.
- Select the OK button to complete the command.

Working with Layers

The DesignCenter can be used to copy layers from one drawing to another. Typically, a template drawing containing stock layers can be used when creating new drawing files. This option of the DesignCenter is useful when you're working with drawings created by a consulting firm and the drawing needs to conform to office standards. DesignCenter can be used to drag layers from a template, or any other drawing, into the new drawing and ensure drawing consistency. Layers can be dragged into a new drawing using the following command sequence:

- Open the DesignCenter.
- Use the palette or Find to select the drawing that contains the layers to be copied.
- Double-click the name of the drawing to display the options of DesignCenter.
- Double-click Layers.
- Highlight the names of the layers to be added to the new drawing.
- Press the right mouse button and drag the layer names to the new drawing.
- When the cursor is in the new drawing area, release the right button to display the shortcut menu.
- Select Add Layers(s) from the short cut menu.

ACCESSING FAVORITE CONTENTS

As with most programs, the AutoCAD DesignCenter allows frequently used sources to be stored for easy retrieval. The final section of this chapter will explore methods of adding shortcuts, displaying, and organizing your favorite drawing tools.

Adding Shortcuts to Favorites

Folders, files, drawings, or drawing content can be added to the Favorites list. Selecting a drawing or other option to be placed in the Favorites list adds a shortcut to the AutoCAD Favorites folder. The selected item is not moved to the folder, only the directions for quickly fetching the desired file. Use the following procedure for adding items to the AutoCAD favorites.

- Right-click the item to be listed as a favorite.
- Select Add to Favorites. The selected item will be listed the next time the Favorites listing is displayed.

The contents of the Favorites folder can be displayed by selecting the Favorites icon from the DesignCenter toolbar or by right clicking in the palette background and selecting Favorites. The shortcut saved in the Favorites folder can be copied, moved, or deleted, using the AutoCAD Windows Explorer. The explorer is accessed by right-clicking in the palette and choosing Organize Favorites... from the shortcut menu. Figure 22.14 shows an example of the Autodesk Explorer. Each of the four options are the same as their counterpart in Windows and AutoCAD.

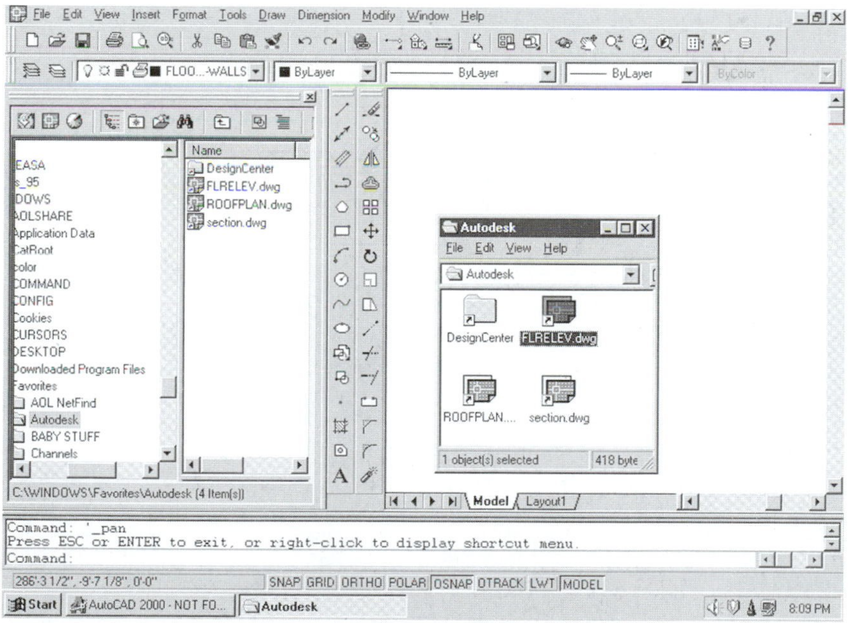

Figure 22.14 *Right-clicking in the palette and choosing Organize Favorites... from the shortcut menu will display the Autodesk Explorer. Each of the four options are the same as their counterpart in Windows and AutoCAD.*

CHAPTER 22 EXERCISES

1. Open any three drawings and copy features from one drawing into the other two. Save each drawing as E22-1a, 1b, and 1c.
2. Open the drawings created in Exercise 1. Copy one object from one of the drawings using the basepoint option so that the new objects aligns with an existing intersection. Save the drawing as E-22-2.
3. Open one of the drawings from Exercise 2. Use the Paste as Block option to place two of the blocks created in Chapter 20 into the drawing.
4. Open a template drawing and draw a 10'-long x 5'- wide bathroom. Show all walls as 4" wide. Use the DesignCenter and move blocks for a toilet, sink, and tub into the bathroom.
5. Open the floor plan that you've created and move five electrical symbols into the drawing. Save the drawing as ELECT.

CHAPTER 22 QUIZZ

1. Describe the options for inserting blocks using the DesignCenter.
2. How is AutoCAD 2000 different from previous releases, in regard to working with multiple drawings?
3. Explain the process to copy an object from one drawing to another in a multiple drawing session.
4. Explain the difference between Paste with Original Coordinates and Paste as a Block.
5. How can Property Painter be used in a multiple drawing environment?
6. List two methods for opening the DesignCenter.
7. What determines if an object will be moved or copied using drag and drop methods with the left mouse button?
8. Explain the effects of right-click on drag and drop methods.
9. How can the History icon of the DesignCenter be used?
10. Describe the four displays that can be used in the DesignCenter.
11. You need to find a site drawing created in early September. Explain the process.
12. What do the + and – symbols mean when they are displayed by a folder or file name.
13. Explain the process to place a drawing in the Favorites folder.
14. A drawing template contains five of the best layer names ever created. How can these layers be transferred into a new drawing?
15. You've come to your senses and want to remove three of the layers that were just added to the drawing. How can this be done?

CHAPTER 23

Combining Drawings Using Xref

Most of the drawings for a structure are worked on by teams of CAD drafters representing several firms. In addition to the architectural and engineering firms responsible for the design of a structure, CAD drafters often work for landscape and interior architects, as well as plumbing, mechanical, electrical, and civil firms. Each of these firms must have up-to-date drawings to ensure an error free construction process. One way to provide these drawings to subcontractors is by the use of externally referenced drawings or xrefs. The use of the XREF command is instrumental in assuring quality drawings. This chapter will introduce merging a drawing into the current drawing using the XREF, XBIND, and XCLIP commands. The XREF command can also be used to compile a drawing sheet containing details or drawings completed at varied scales. Methods for editing referenced materials will be explored using the REFEDIT, REFSET, and REFCLOSE commands.

EXTERNAL REFERENCED DRAWINGS

In Chapter 20 it was suggested that a wblock of a floor plan could be inserted into a drawing and used to form the base of that drawing. For example, the floor plan can form the base drawing for the electrical, mechanical, plumbing, and framing plans. This works well, but it does take up lots of disk space. A more efficient method than inserting a block is to use an externally referenced drawing or xref. An externally referenced drawing is similar to a wblock, in that it is displayed each time the master drawing is accessed. These drawings, however, are not stored as part of the master drawing file. Each time the base drawing is updated, all drawings that it is attached to will also be updated. Externally referenced drawings are created through the XREF command. Figure 23.1 shows a drawing that serves as the base drawing. Figures 23.2a and 23.2b show drawings that make use of the base drawings.

SAVING SPACE

An externally-referenced drawing can be brought into the current drawing for viewing, but it does not become part of the current drawing base. Depending on controls set by the originator, it may or may not be able to be altered by subcontractors work-

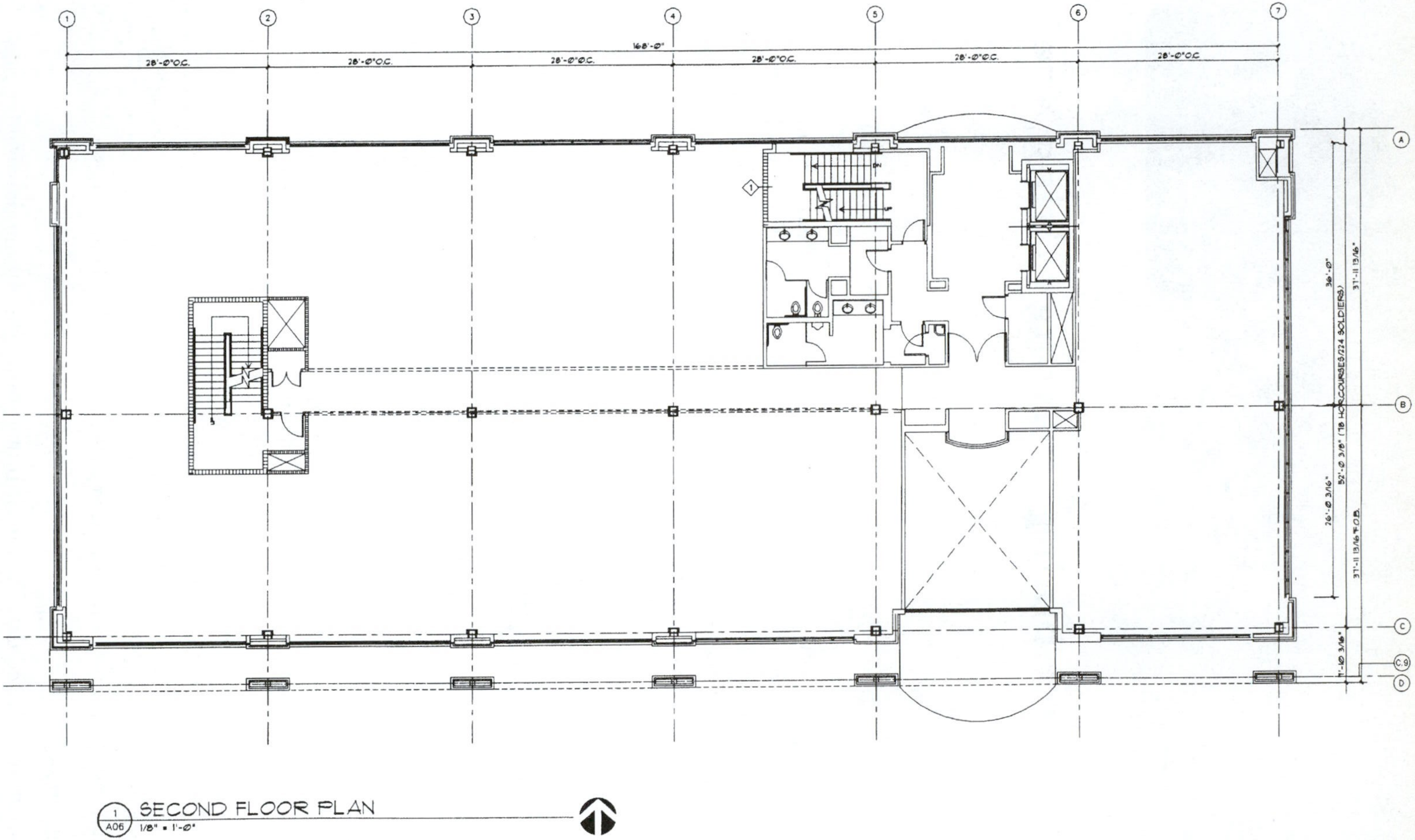

Figure 23.1 *The XREF command can be used to attach a drawing to another drawing. The information contained in the attached drawing is not stored in the current drawing. Each time the base drawing is updated, all drawings that it is attached to will also be updated. (Courtesy Peck, Smiley, Ettlin Architects.)*

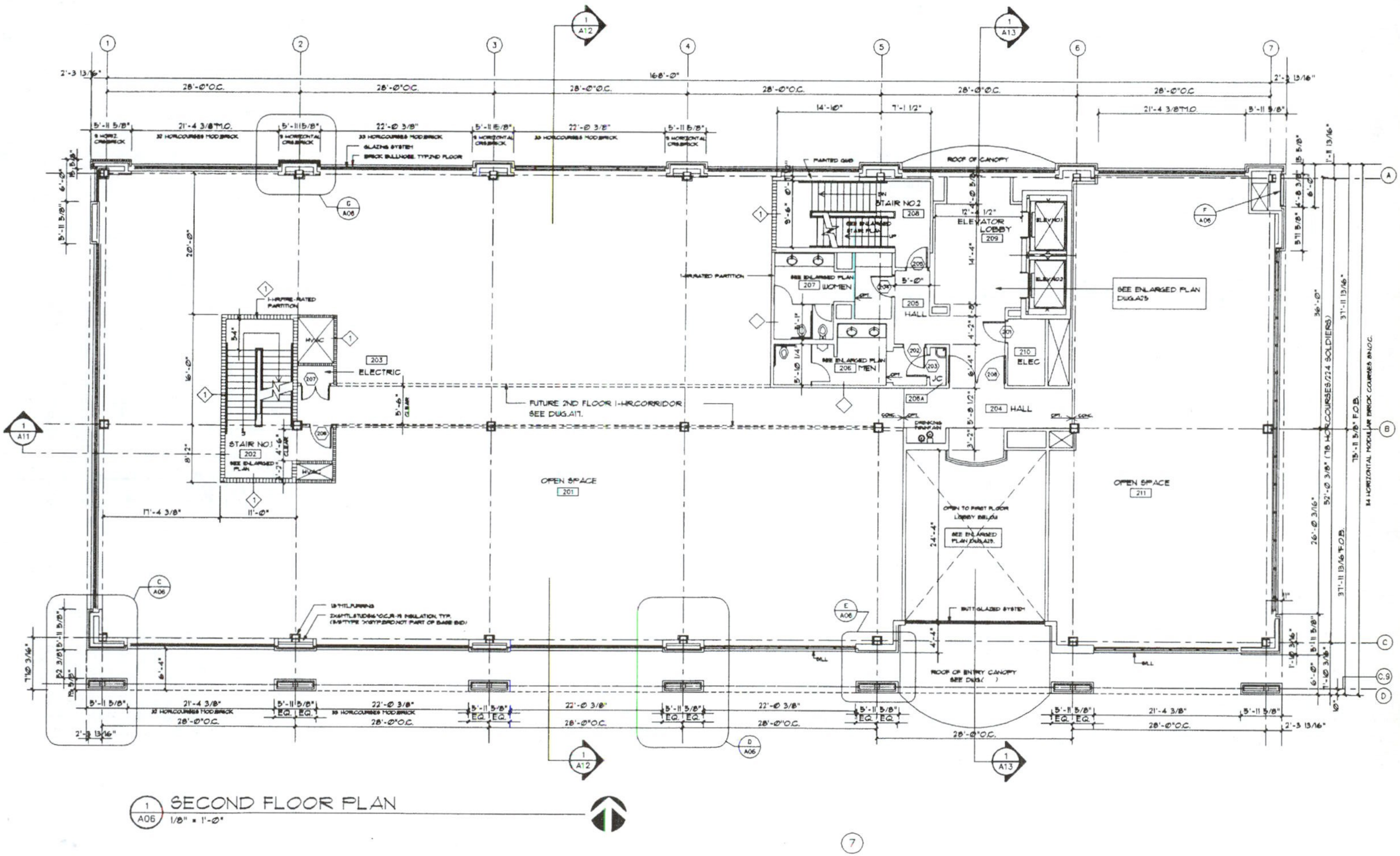

Figure 23.2a *The drawing shown in Figure 23.1 is used as the base for the floor plan shown here and the drawings prepared by the mechanical, electrical, and plumbing subcontractors. (Courtesy Peck, Smiley, Ettlin Architects.)*

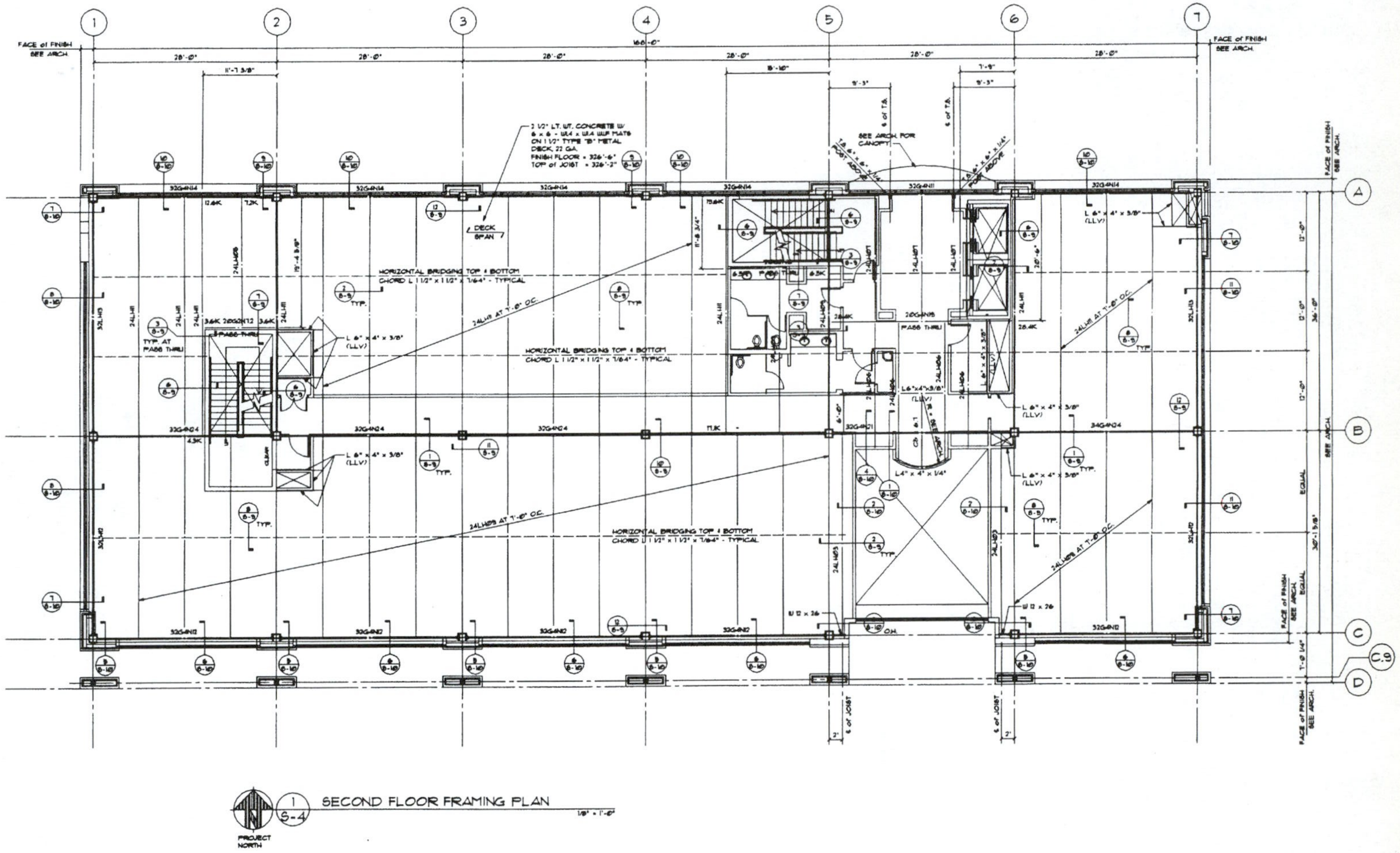

Figure 23.2b *The framing plan is completed by an engineering firm that uses the base drawing created by the architectural team. The drawing is attached using XREF, and then the needed material is added to the drawing. (Courtesy Van Domelen/Looijeng/McGarrigle/Knauf Consulting Engineers.)*

ing in the drawing. An xref is similar to a block, except that no drawing objects are recorded in the drawing base. Only the drawing name and a small amount of information needed to access the drawing are stored in the new drawing file.

AUTOMATIC UPDATING

Externally referenced drawings provide excellent benefits for drawing projects that are being developed by a team of engineers, architects, contractors, and drafters. Because of time constraints, subcontractors often need part of the drawings before they are complete. By using xref drawings, every time the drawing is accessed, the most recently saved version of the external drawing is loaded. Changes made to the xref source file will be updated automatically in every drawing where it is referenced. If one team member is working on the plan views and another is working on exterior elevations, every time the xref floor plans are loaded for reference for the elevations, they will be updated.

ASSEMBLING MULTI-SCALED DETAILS

A common use for xref drawings is to assemble details for a stock detail sheet. In most professional offices, stock libraries of details have been created for every possible construction alternative that is likely to be encountered. You can assemble a sheet of the appropriate details using externally referenced drawings with the knowledge that the most up-to-date version of the details is being provided. The next chapter will explore the process for assembling drawings to be plotted at different scales within the same drawing.

ATTACHING AN XREF TO A NEW DRAWING

Although the XREF command can be completed from the command line, using the Xref Manager dialog box will greatly aid the use of the command. The dialog box can be accessed by selecting the External Reference icon from either the Insert or Reference toolbars, or by typing **XR** (EnterKey) at the command prompt. Each method will produce the Xref Manager dialog box shown in Figure 23.3. The dialog box allows a drawing to be referenced to the current drawing. It also allows the status of each xref and the relationship of xrefs to one another to be managed. The dialog box is used to:

- Attach an xref
- Change the xref path
- Detach an existing xref
- Reload or unload an existing xref
- Change an attached xref to an overlaid reference and vice versa
- Bind an entire xref definition to the current drawing

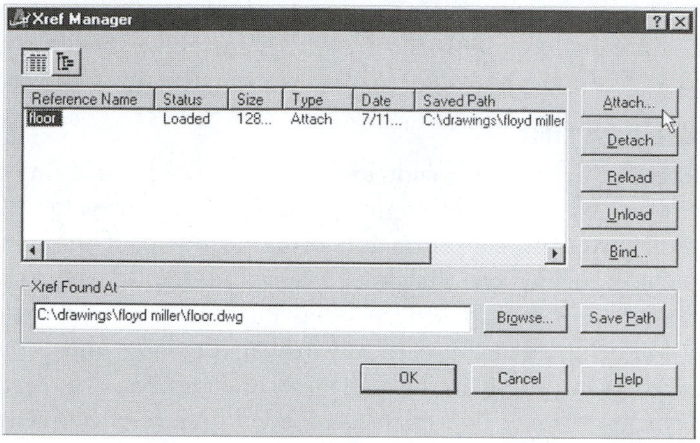

Figure 23.3 *The Xref Manager dialog box can be used to reference a drawing into a base drawing. The dialog box can be accessed by selecting the External Reference icon from either the Insert or Reference toolbar, or by typing **XR** (EnterKey) at the command prompt.*

Once a drawing has been referenced to the base drawing, the Xref Manager can be displayed using the shortcut menu. The dialog box is displayed by selecting an xref, right-clicking in the drawing area, and then selecting Xref Manager.

Selecting External Reference... from the Insert pull-down menu or selecting Attach in the Xref Manager dialog box will display the Select Reference File dialog box. Once a file to be referenced is selected, an External Reference dialog box similar to Figure 23. 4 is displayed. The dialog box allows a drawing to be added to the current file, much like the INSERT command.

Figure 23.5a shows a floor plan that will be used as the base drawing in the following example. The base drawing is titled C:\DRAWINGS\SCHOOL\ARCH \FLOOR.DWG. Figure 23.5b shows the electrical information that will be attached to the base drawing to make the electrical plan. This drawing is stored as C:\DRAWINGS\SCHOOL\ARCH\ELECT.DWG.

To attach the electrical information to the floor drawing, open the Xref Manager dialog box and select the Attach button. This will display the Select Reference Files dialog box. The box is similar to each dialog box that is used to open a new drawing. Select the desired folder and file of the drawing to be attached. With the desired file highlighted, select the Open button to display the External Reference dialog box shown in Figure 23.4. This dialog box allows you to control how and where the attached drawing will be located in the base drawing by using the Reference Type and Parameters boxes.

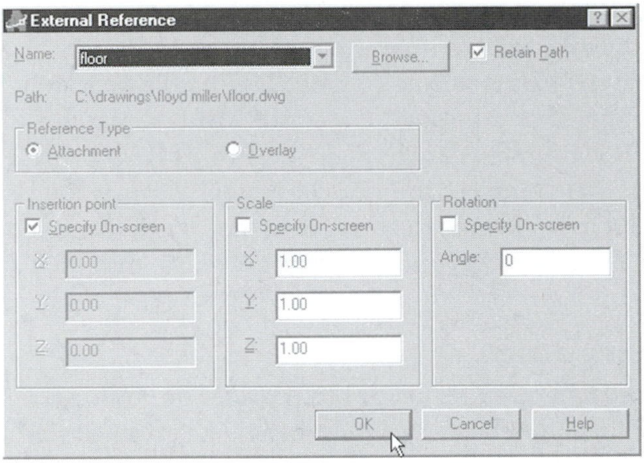

Figure 23.4 *The External Reference dialog box can be used to reference a drawing into a base drawing using a procedure similar to the INSERT command. The dialog box can be accessed by selecting External Reference... from the Insert pull-down menu.*

EXTERNAL REFERENCE DIALOG BOX OPTIONS

So far, the External Reference dialogs box shown in Figure 23.4 has only been used to attach one reference drawing to a base drawing. Several parameters can be adjusted to alter its use.

NAME AND PATH OPTIONS

The Name edit box was empty as the box was first entered. Once a drawing has been attached, the name of the drawing will be displayed. When multiple drawings are attached, selecting the edit arrow will display a list of the referenced drawings. Selecting an attached drawing from the list will display the drawing name in the Name edit box, and its location will be displayed in the Path display. Selecting Browse will display the Select a Reference File dialog box, allowing new xrefs to be selected for the current drawing.

Retain Path

The Retain Path radio button dictates if the full path of the referenced drawing will be saved. By default, this option is active so that the path to the referenced drawing will be saved in the database of the base drawing. If the option is deactivated, the referenced drawing is saved in the database without a path. AutoCAD will search paths associated with the current drawing name. The inactive option is often used when referenced drawings are shared with subcontractors who will work with the drawings. With this option inactive, as AutoCAD loads the base drawing it will only search for the referenced drawing in the path specified in the Files tab of the Options dialog box. If this box is active, it will only search through the file folders specified in the path. This will usually be the desired method. If you are working on several jobs at the same time, your drawing will not be overwritten by the wrong job. For example, if you are working on several jobs requiring an ELECT.dwg file to be attached, if you have the Retain Path box inactive, it will only search C:\PROGRAM FILES\PARKER\ELECT, rather than C:\PROGRAM FILES\ACAD2000 \ELECT.dwg. If you complete this search, AutoCAD may find several different ELECT plans, but it would grab the first file found. It is best to be as specific on the path as possible. Make sure as you create new drawings, that they are placed in the correct location.

REFERENCE TYPE BOX

This portion of the External Xref dialog box provides the options of Attachment and Overlay, for deciding how an external drawing will be attached to the base drawing. Attached drawings allow you to build a drawing using other drawings.

Attaching a Drawing

With the default setting of Attachment, selecting the OK button will remove the dialog box, return you to the drawing area, and provide a prompt for an insertion point. As the point is selected with the mouse, the drawing will be attached and the command will be closed. Figure 23.6 shows the merged drawings from Figure 23.5a and

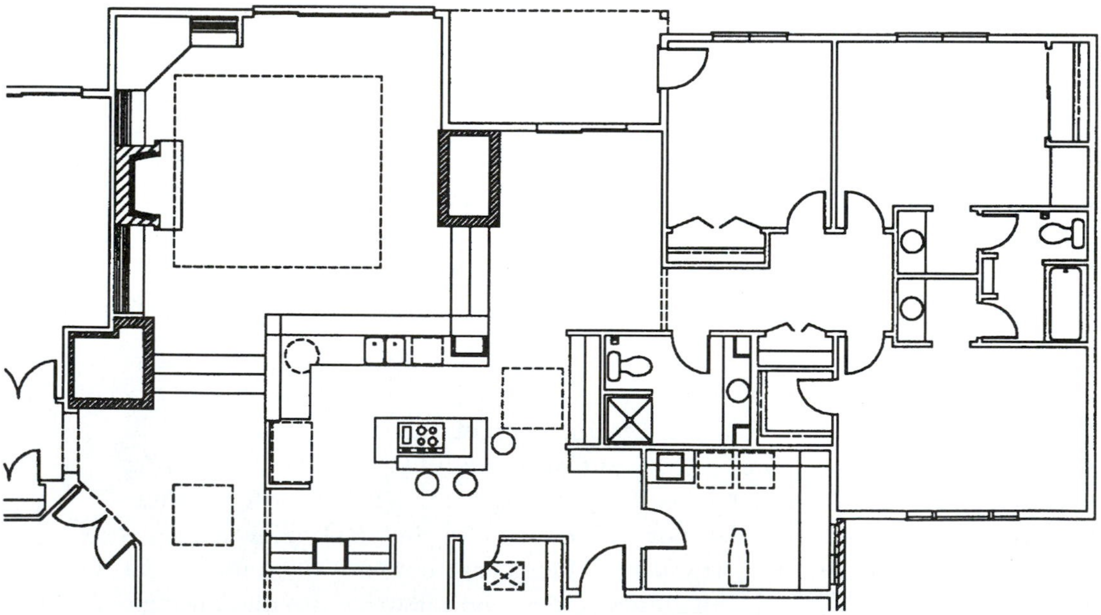

Figure 23.5a *This portion of a floor plan will serve as a base for the electrical plan shown in Figure 23.5b. The information on the referenced drawing will be displayed as the floor plan is accessed, but it will not be stored in the FLOOR.DWG drawing base. (Courtesy of Michael Jefferis.)*

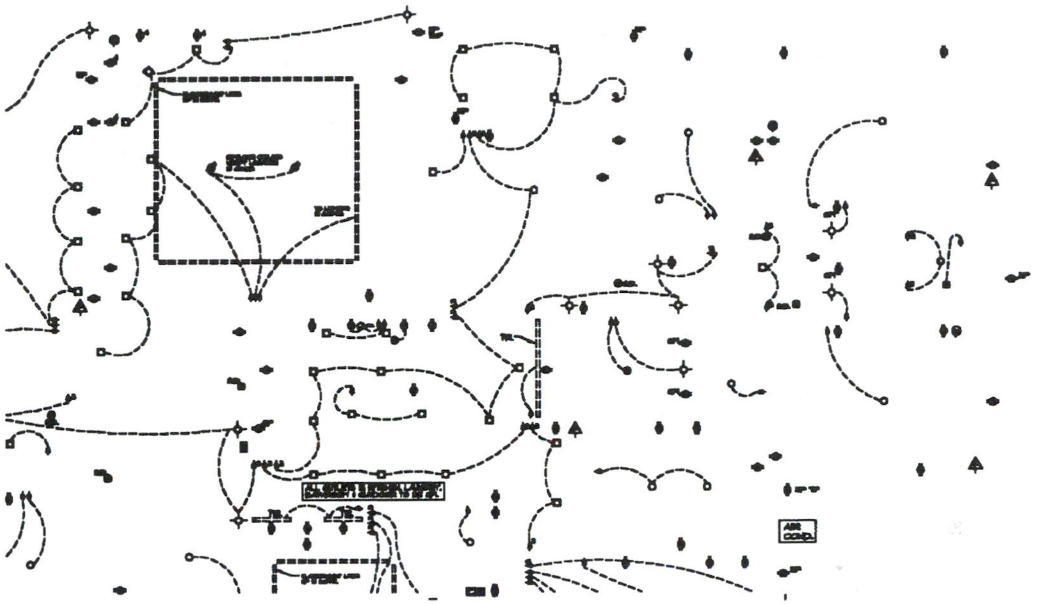

Figure 23.5b *By using referenced drawings, subcontractors from various firms can each complete their work without having to wait for others to finish. This electrical information was added to the base drawing shown in Figure 23.5a.*

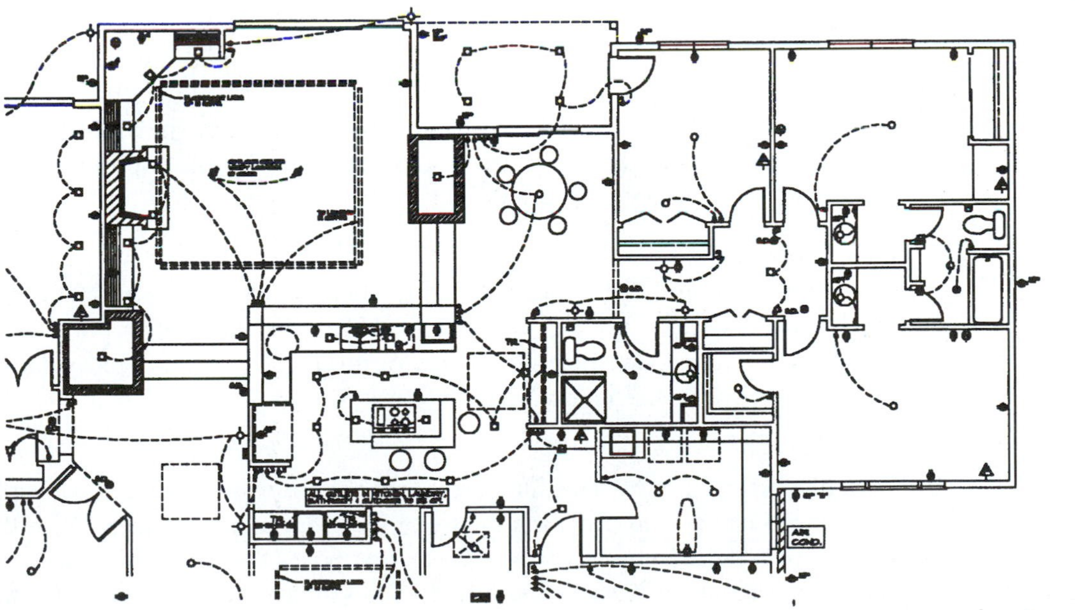

Figure 23.6 *The results of attaching the electrical drawing to the base drawing. (Courtesy of Michael Jefferis.)*

23.5b. The insertion point of the lower right corner was used with a running OSNAP of Intersection to assure proper alignment of the two drawings.

If a second drawing is to be attached to the base drawing, selecting the Attach... button will produce the Select Reference Files dialog box. Once one drawing has been selected, when the Attach... button is selected, the External Reference dialog box is redisplayed. Once a drawing is attached, AutoCAD adds dependent symbols in the xref to the base drawing. Dependent symbols are named items such as blocks, dimension styles, layers, linetypes, and text styles. Although these dependent symbols can be viewed, they can't be altered unless you preset values to allow the user to alter the drawing. This is one of the benefits of an xref—you can give a referenced drawing to a subcontractor and not have to worry about the drawing being altered. When a drawing is referenced into a base drawing, you'll notice that the layer names of the xref will be altered. Layer names for attached drawings will be displayed showing the name of the drawing, followed by a vertical bar and the layer name.

A name such as 72PLUMB|UPPLUMB would indicate a referenced drawing of 72PLUMB and a layer title of UPPLUMB.

Overlaying an External Drawing

Choosing the Overlay option will produce a drawing that looks like an attached drawing. The major difference between an attached and an overlaid drawing occurs when the referenced drawing contains nested reference drawings. A nested drawing is a drawing composed of other referenced drawings. When a nested drawing is overlaid, the nested drawings within the drawing will not be displayed. When a nested drawing is attached, all objects will be displayed. The Overlay option is typically used when you need to share information. An overlaid drawing will allow others to view how their portion of the project relates to the drawing. Rather than showing all of the drawings like the Mechanical and Electrical Xrefs, with the Overlay option you can still benefit from the automatic updated files.

SETTING PARAMETERS

As the drawing in Figure 23.5b was attached to the drawing in Figure 23.5a, the insertion point was selected prior to attachment. Using the Parameters portions of the External Reference dialog box allows the location, scale factors, and rotation angle to be controlled for a referenced drawing.

Insertion Point

In the default setting, the Specify On-screen radio button is active, allowing the insertion point to be selected using the mouse or by entering coordinates. With the Specify On-screen radio button deactivated, the text boxes for each option are activated. These coordinates allow specific X and Y coordinates to be specified for the insertion point of the referenced drawing. These options correspond to their counterparts in the Insertion Points for the INSERT command.

Note: For most firms, it is common to use an insertion point of 0,0. With an insertion point of 0,0, all drawings created by subcontractors will align perfectly. This is a much easier method than worrying about what corner of a structure may have been used as a base point.

Scale

This option specifies the scale factors for the selected referenced drawing as it is inserted into the base drawing. The default values for the X, Y, and Z scale factors are set as 1. Selecting the Specify On-screen radio button for scale factors will deactivate the scale factor values, and allow the values to be adjusted from the command prompt as the drawing is attached.

Rotation

The Rotation portion of the dialog box is similar to the Insert dialog box. Angle allows the rotation angle of the referenced drawing to be controlled as it is inserted into the base drawing. In the default setting, a rotation angle can be entered in the edit box. With the radio button active, the rotation angle can be set at the command prompt as the drawing is inserted.

XREF MANAGER DIALOG BOX OPTIONS

So far, the Xref Manager dialog box shown in Figure 23.3 has only been used to attach one reference drawing to a base drawing. This dialog box can be used to list, attach, load, and modify referenced drawings contained in the current drawing. The display provided in the box will depend on the active icon in the upper left corner of the dialog box. The icons toggle between List View and Tree View. The display can be toggled by selecting the inactive option or by pressing the F3 or F4 key.

LIST VIEW OPTIONS

In the default setting, the dialog box displays in List View, with information related to the reference name, status, size, type, date, and save path. Information can be provided in an alphabetical list of the referenced drawings contained in the current drawing. Notice in Figure 23.3, that not all of the information is visible. The path scrolls off the edge of the display field. Move the cursor to the line that divides the Reference Name area and the Status area. As the line is touched, the cursor turns to a double arrow, allowing the size of the box to be altered. This can be done to each box to allow a complete display. (See Figure 23.7a.) Click a list column heading to sort the referenced drawings by the column name. To sort referenced files by date, click the reference Date column.

Reference Name

This portion of the box lists each of the names of referenced drawings that have been stored in the block definition symbol table.

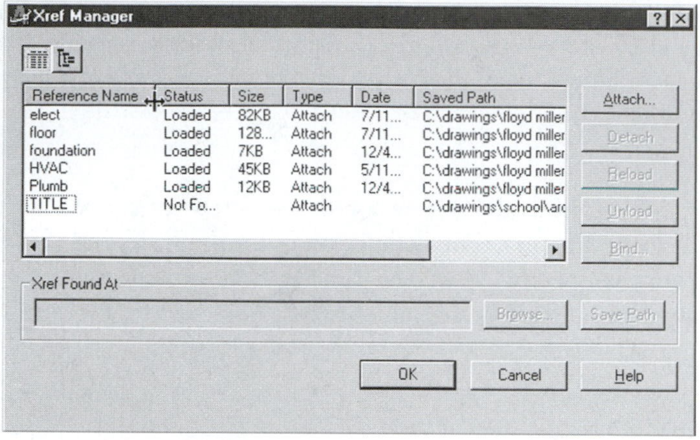

Figure 23.7a *The space between listings can be reduced to allow longer listings to be completely viewed. By placing the cursor on the lines that divide titles, the display space can be reduced or enlarged.*

Status

This box shows the current status of referenced drawings. Options include:

- Loaded—The attached drawing is displayed.
- Not Found—The referenced drawing cannot be found.
- Orphaned—The parent of the nested referenced drawing has been unloaded.
- Reload—The referenced drawing has been marked to be reloaded. The option reads and displays the most recently saved version of the drawing. A referenced drawing is loaded or unloaded after the command is closed.
- Unload—The referenced drawing has been marked to be unloaded.
- Unloaded—The referenced drawing is not displayed.
- Unrefreshed—An unrefreshed drawing is not displayed. The referenced drawing has a nested referenced drawing that is not found or it has been attached and then erased.
- Unresolved— The referenced drawing is missing from the specified path and can't be found.
- Size—The Size column displays the size of the referenced drawing.
- Type—Indicates if the referenced drawing was placed using the Attach or the Overlay option.
- Date—Lists the last date that the referenced drawing was modified.
- Saved Path—Shows the path of the referenced drawing.

TREE VIEW OPTION

With the Tree View active, a hierarchical representation of referenced drawings is shown similar to the manner that listings in the Windows Explorer are displayed. The Tree View method displays nested drawings in their relationship to the attached drawing, as well as the attachment method, and the status. Figure 23.7b shows a listing using the Tree View option.

EXTERNALLY REFERENCED DRAWING OPTIONS

Once the drawing is attached, each time the base drawing is opened, the attached drawing will be loaded and displayed as long as the path to the attached drawing is accessible. In addition to loading the referenced drawing, the other options of the External Reference dialog box will become active. Options include Attach, Detach, Reload, Unload, and Bind.

ATTACH

You've already explored the Attach option. Use the following summary of the steps to attach a drawing to another drawing:

- Choose the External Reference icon from the Insert or Reference toolbars.
- Select Attach... and choose a drawing to be inserted into the current drawing from the Select Reference File dialog box. Select the Open button to continue. Selecting Open will display the External Reference dialog box.
- Select the reference type to be used. Typically, this will be Attachment.
- Specify the insertion point, scale, and rotation angle, or select the Specify On-screen option for each parameter.
- Choose the OK button to add the referenced drawing to the current drawing.

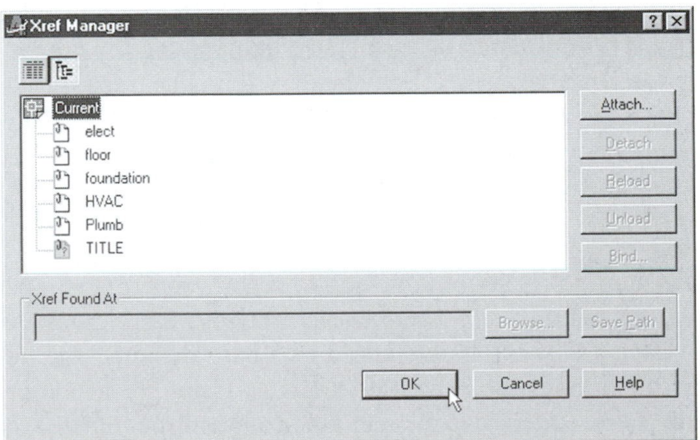

Figure 23.7b *Listings in the Xref Manager dialog box displayed using the Tree View option.*

DETACH

The Detach option will remove unneeded referenced drawings from the master drawing. Similar to how ERASE removes a block, DETACH removes copies of the referenced drawing and its definition. A drawing is detached by selecting the name of the unneeded xref. Selecting the name will highlight the listing of the drawing. Selecting the Detach button will remove the xref listing from dialog box. If the OK button is selected, the selected xref will be removed from the screen and all references to the drawing will be removed from the base drawing. Remove an attached drawing file using the following procedure:

- Choose the External Reference icon from the Insert or Reference toolbars. To use the shortcut menu, right-click in the drawing area, select the drawing to detach, and choose Xref Manager from the menu.
- Select the drawing to be removed from the base drawing.
- Select the Detach button. This will remove the drawing file from the current file.
- Select the OK button to return to the drawing area.

RELOAD

This option allows an attached referenced drawing to be reloaded and updated in the middle of a drawing session. If a coworker revises the referenced drawing while you're working on a drawing that contains the xref, your xref is out of date. The Reload option allows the most current referenced drawing to be used without having to exit the drawing session and reopening the master drawing. The process to reload a drawing is similar to the Detach option.

UNLOAD

The Unload option allows a referenced drawing to be temporarily removed from the screen display without removing the xref from the drawing base. The Unload option is similar to using the Freeze option of the LAYER command. The referenced drawing is not displayed or considered in regeneration, increasing the drawing speed. The drawing can be reactivated by using the Reload option. The process to reload a drawing is similar to the Detach option. The Unload option is very handy if you are working on a very large drawing and don't need certain information at the current time, but will need the information at a later time.

BIND

The Bind option allows a referenced drawing to become a permanent part of the base drawing. When this option is used, referenced drawings function as a wblock rather than an xref. When a drawing is referenced to a base drawing, the drawing is displayed, but you can't modify the referenced drawing. Using the Bind option allows an external drawing to be edited, but the xref will no longer be updated as the original drawing is edited. The Bind option adds dependent symbols to the drawing base so that they can be used just as any other drawing object. Layer names such as 72PLUMB|UPPLUMB in an attached drawing will be altered to read 72PLUMB0UPPLUMB. This naming system allows you to be able to quickly identify attached and bound layers. The Bind option should only be used when you know for sure that the referenced drawings has completed the review process and is no longer subject to change. Because construction drawings are subject to change, Bind may not be an acceptable option. The XBIND command will be introduced later in this chapter that will allow specific symbols of a drawing to be permanently attached.

Binding a Drawing

Selecting the Bind... button will produce the Bind Xrefs dialog box shown in Figure 23.8, providing the options of Bind and Insert.

Bind

The default option of Bind will permanently attach a drawing to the base. A drawing is bound by selecting the name of the layer to be bound so that it is highlighted. Selecting the OK button will remove the dialog box, bind the attached drawing to the base, and update the names of dependent symbols. Symbols in the bound drawing can now be edited just as any object created in the base drawing.

Insert

If the Insert option of Bind is used, an attached drawing is bound to the base drawing as if a wblock had been inserted into a drawing. Insert will bind the referenced drawing to the current drawing in a method similar to detaching and inserting the reference drawing. A drawing is inserted by selecting the name of the layer to be inserted,

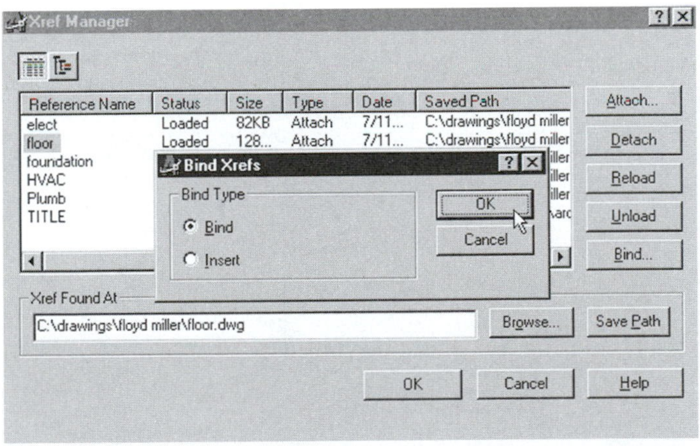

Figure 23.8 *The Bind option allows a referenced drawing to be permanently attached to a base drawing. Selecting the name of an attached drawing from the External Reference dialog box activates each of the XREF options. Selecting the Bind... option displays the Bind Xrefs dialog box.*

so that it is highlighted. Selecting the OK button will remove the dialog box, insert the attached drawing to the base, and update the names of dependent symbols. A layer of 72PLUMB|UPPLUMB would be renamed to UPPLUMB.

THE XBIND COMMAND

The bind option adds dependent symbols to the drawing base, so that they can be used. Automatic updating and other qualities of xref drawings are lost. This can be overcome using the XBIND command. XBIND allows a portion of a referenced drawing—such as a block, layer, or linetype—to be permanently added to a current drawing, with the balance of the external drawing retaining its qualities. The XBIND command can be accessed by selecting External Reference Bind from the Reference toolbar, by selecting Object, External Reference, Bind... from the Modify pull-down menu, or by typing **XBIND** (EnterKey) at the command prompt. Each option will produce the Xbind dialog box shown in Figure 23.9. The dialog box lists referenced drawings that are attached to the base drawing. In Figure 23.9, six drawings are attached to the base drawing. Each drawing is preceded by the + symbol and the AutoCAD 2000 icon. The + indicates that the listing can be expanded. Selecting one of these options and double-clicking will display the contents of the drawing. In Figure 23.10a, the FLOOR listing was expanded, revealing that the drawing contains nested Blocks, Dimstyles, Layers, and Linetypes. Notice in Figure 23.10b that the block FLOOR|10PBAC01 has been selected to be added, using the XBIND command. By selecting the desired dependent symbol to bind FLOOR|10PBAC01 and selecting the Add> button, the name of the block will be moved from the Xref listing to the Definitions to Bind display. (See Figure 23.10c.) Selecting the OK button will

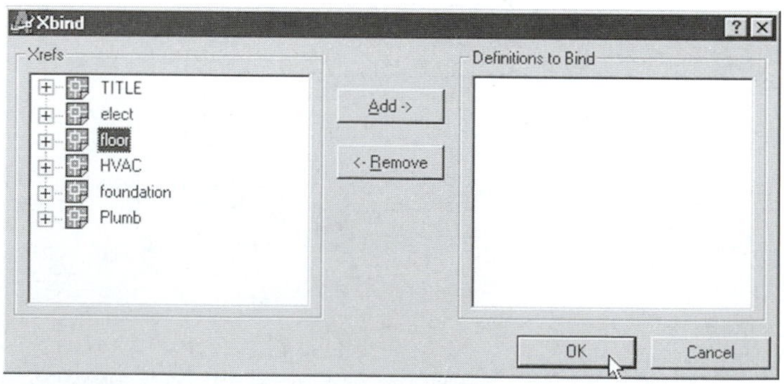

Figure 23.9 *The Xbind dialog box displays a listing of drawings that are attached to the current drawing. The + symbol preceding AutoCAD 2000 icon indicates that the listing can be expanded. Notice the Block, Dimstyle, and Textstyle of each contains additional information.*

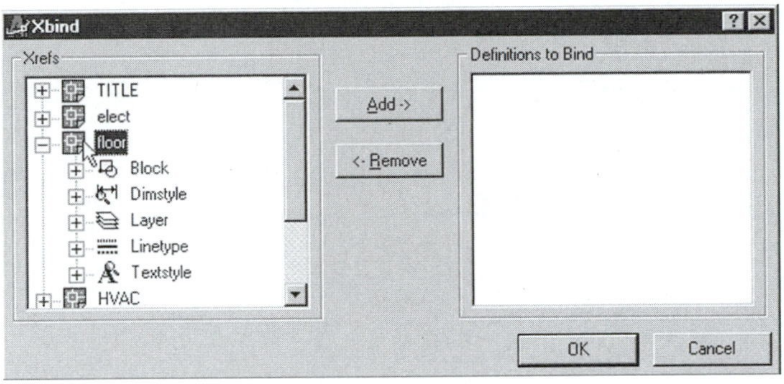

Figure 23.10a *Selecting the FLOOR listing reveals nested items in the Block, Dimstyle, and line Textstyle.*

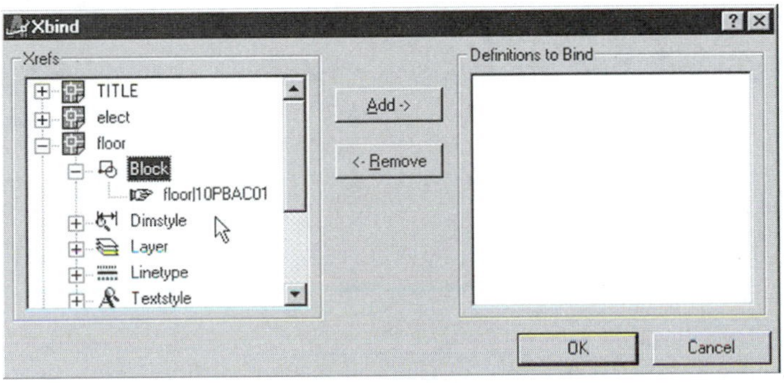

Figure 23.10b *Selecting the FLOOR|10PBAC01 block for binding to the referenced drawings.*

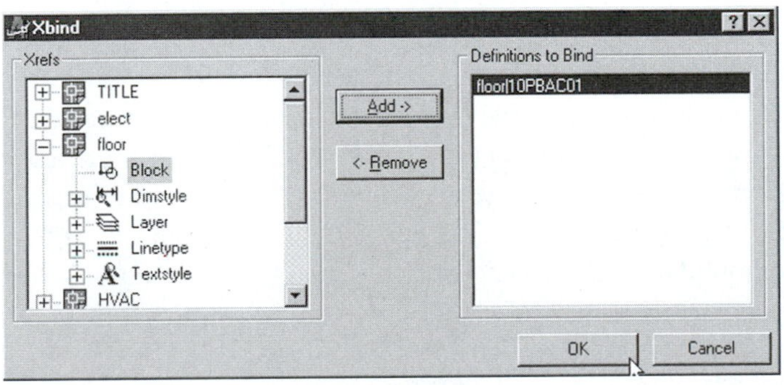

Figure 23.10c *Selecting the name and selecting the Add > button will permanently bind the selected definition to the base drawing.*

bind the block to the base drawing and close the Xbind dialog box. Selecting a listing from the Definitions to Bind display and selecting the <Remove button will unbind the definition and restore it to the Xrefs listing.

WORKING WITH XCLIP

The XCLIP command allows portions of an attached xref drawing to be hidden. The command allows a boundary to be created in a referenced drawing so that all material that is outside the boundary is invisible. The command can be accessed by selecting External Reference Clip from the Reference toolbar, or by typing **XC** (EnterKey) at the command prompt. Each access method will produce a Select object prompt. Any object selection method can be used to select xrefs to be clipped. The command sequence is:

Command: **XC** (EnterKey)

Select object: *(Select xrefs to be clipped.)*

1 found

Select objects: (EnterKey)

[ON/OFF/Clipdepth/Delete/generate Polyline/New boundary]: <New>: (EnterKey)

Accepting the default value allows the clipping boundary to be selected and produces the following prompts:

Specify Clipping boundary:

[Select polyline/Polygonal/Rectangular] <Rectangular>: (EnterKey)

Specify first corner: *(Select first window corner.)*

Specify opposite corner: *(Select opposite window corner.)*

Command:

Options for setting the boundary include the default method of forming a rectangular boundary with a selection window, selecting a polyline, or by selecting Polygonal. Selecting polygonal allows a boundary to be set by specifying points for the vertices of a polygon. Accepting the default by pressing (EnterKey) produces prompts to select the First and Other corners. As the corners are selected, objects in the xref that are outside of the boundary will become invisible. This will be helpful if there are items on certain layers that you need visible and others that need to be invisible. Rather than altering the layers, some objects can be exclude with XCLIP.

Figure 23.11a is an example of a footing detail that has been referenced into a base drawing. Figure 23.11b shows the detail on the right side of Figure 23.11a, hidden using the XCLIP command.

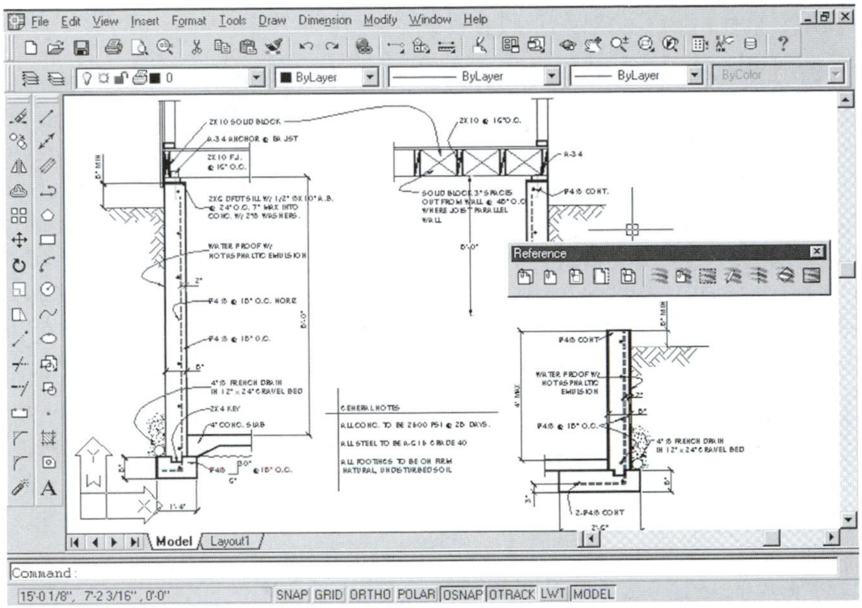

Figure 23.11a *The XCLIP command allows information that has been permanently attached to a base drawing to be hidden. The command requires that a boundary be selected. Anything outside of the boundary will be made invisible.*

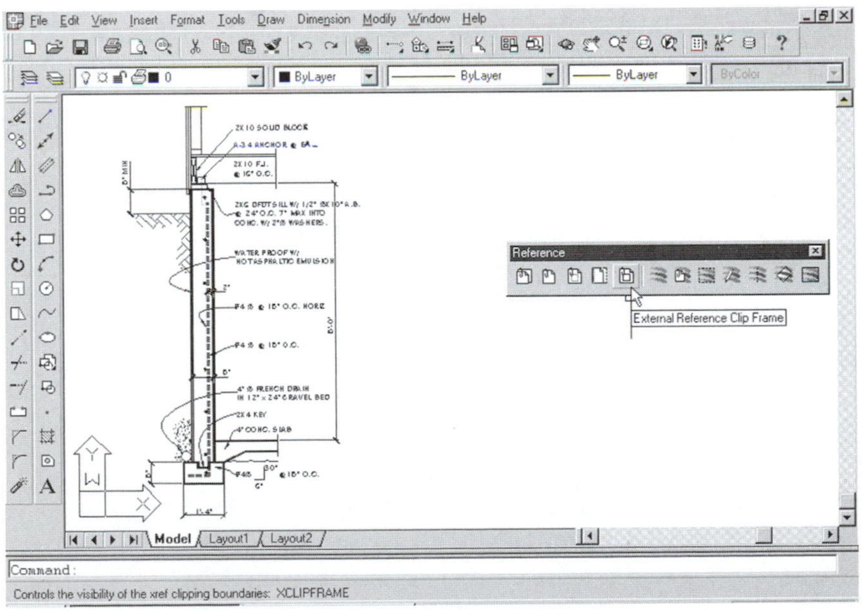

Figure 23.11b *The detail on the left side of Figure 23.11a was placed inside the boundary. Using the XCLIP command, the detail on the right side is still attached to the drawing base, but is removed from the display set.*

EDITING REFERENCED DRAWINGS

AutoCAD 2000 allows externally referenced drawings that you've created to be edited, using in-place editing. Previous releases of AutoCAD required the original drawing to be edited. In-place editing allows the referenced drawing to be revised. If the reference drawing selected for editing contains attached xrefs or block definitions, the reference and nested references are displayed in the Reference Edit dialog box. The box is displayed by selecting the Edit Block or Xref icon from the Refedit toolbar, or by selecting In-place Xref and Block Edit from the Modify pull-down menu. The toolbar and the dialog box can be seen in Figure 23.12. The toolbar presents five options for editing referenced drawings.

EDIT BLOCK OR XREF

A referenced drawing will respond like a block if you attempt to edit it. Using the REFEDIT command allows individual objects in the referenced drawing to be altered. Objects can be edited by selecting the Edit Block or Xref icon from the Refedit toolbar and then choosing the referenced drawing to be edited.

Using the Reference Edit Dialog Box

The Reference Edit dialog box allows a referenced drawing to be edited. Key elements of the dialog box include the Referenced Name box, Preview icon, Next, and editing options.

- Reference Name—This portion of the dialog box displays the reference that has been selected for in-place editing and any nested drawings or blocks within the selected drawing.
- Preview—The Preview box displays an image of the drawing selected for editing.

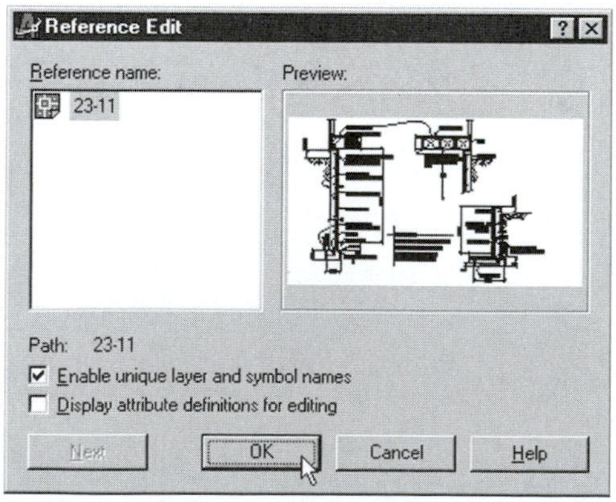

Figure 23.12 *AutoCAD 2000 allows referenced drawings to be edited using in place editing with the Reference Edit dialog box. The box is displayed by selecting the Edit Block or Xref icon from the Refedit toolbar, or by selecting In-place Xref and Block Edit from the Modify pull-down menu.*

- Next—This option can be used to cycle through selectable references.
- Path—The file location of the selected reference drawing.
- Enable Unique Layer and Symbol Names—This option controls whether layer and symbol names of objects extracted from the referenced drawing will be unique or altered. In the active mode, symbol and layer names will be displayed with a prefix of $#$. The altered prefix allows for easy identification. With the option inactive, layer and symbols will appear as they do in the original drawing.
- Display Attribute Definitions for Editing—This option controls if variable attribute definitions in a block reference will be extracted and displayed during editing. Referenced drawings and blocks without definitions are not affected by this option. When active, attributes will be invisible but the object and the attribute definitions are available for editing. If changes are saved back to the block reference, the original reference remains unchanged. The new attribute definition will affect only future insertions of the block.

Once each option is set and you've selected a reference to edit, you'll be prompted to select specific objects in the referenced drawing to edit.

Editing Referenced Drawings Using REFEDIT

The REFEDIT command is used to select and edit a referenced drawing. Use the following procedure to edit individual objects of a referenced drawing:

- Select the Edit Block or Xref icon from the Refedit toolbar, or enter **REFEDIT** (EnterKey) at the command prompt.
- Select the reference to be edited. This will highlight the referenced drawing and display the Reference Edit dialog box. If nested blocks or drawings are contained in the selected xref, all references available for selection are displayed in the dialog box.
- If multiple drawings are listed in the dialog box, highlight the drawing to be edited.
- Set the options of Enable unique layer and symbol names/Display Attribute Definitions for Editing to meet drawing needs.
- Select the OK button.
- Select objects to be edited.
- This is your opportunity to select objects in the referenced drawing that will be edited. Once selected, objects in the referenced drawing can be edited just like any other drawing.

Adding or Removing Objects from the Working Set

Objects can be added or removed from the working set as the referenced drawing is being edited using the REFSET command. The working set consists of what will be extracted, edited, and saved back to the base drawing. You'll be editing and saving changes to the original drawing—even though you're working on a referenced drawing. You can supply a referenced floor plan to the HVAC subcontractor, and they can alter the plan as needed to allow for ducts and chases to be installed. As they save changes to the plan, you can now supply an up-to-date plan to other consultants. The command can be started by selecting the appropriate icon from the Refedit toolbar, by selecting Add to Workset or Remove from Workset from In-place Xref and Block Edit from the Modify pull-down menu, or by entering **REFSET** (EnterKey) at the command prompt.

Adding Objects to the Working Set

To add objects to the working set, enter the REFSET command by selecting the Add icon from the Refedit toolbar.

This will display the prompt:

Transfer objects between the Refedit working set and host drawing...

Enter an option [Add/Remove] <Add>: _Add

Select objects: *(Select objects to be added to the working set.)*

The command will vary slightly when entered by keyboard. You'll have to select Add or Select at the prompt. When using the toolbar, you've already made the decision when the icon was selected. Once objects are selected, a prompt will be displayed giving the number of objects added to the working set. The prompt will resemble the following:

3 Added to working set.

Removing Objects from the Working Set

To remove objects to the working set, enter the REFSET command by selecting the remove icon from the Refedit toolbar.

This will display the prompt:

Transfer objects between the Refedit working set and host drawing...

Enter an option [Add/Remove] <Add>: _Rem

Select objects: *(Select objects to be removed to the working set.)*

The command will vary slightly when entered by keyboard. Once objects are selected, a prompt will be displayed giving the number of objects removed from the working set. The prompt will resemble the following:

3 Removed from working set.

Objects removed from the working set will appear in a gray tone referred to as *dithered*. The intensity of the line can be controlled using the Display tab of the Options dialog box. Options can be selected from the Tools pull-down menu. Use the Reference Edit Fading Intensity setting to control the display of removed objects. A higher value, provides greater contrast.

Saving and Discarding Changes with REFCLOSE

Up to this point, you've referenced an external drawing to a new drawing, and edited a portion of the referenced drawing. If you try to edit other portions of the referenced drawing, AutoCAD will warn you:

Command not allowed, (drawing name) already checked out for editing

This is AutoCAD's way of not allowing two subcontractors to attempt to use and update a drawing at the same time. If you save the drawing, the new drawing will still contain your revisions, but the original drawing will remain unaltered. The REFCLOSE command is used to either save back or discard changes made during the editing process of a referenced drawing. Once you've used REFCLOSE, you can edit other referenced drawings or blocks. The command can be selected by toolbar, shortcut menu, pull-down menu, or by keyboard. As the command is entered by keyboard, you'll be prompted to enter a letter for the discard or save option.

Discard

Selecting Discard will vaporize the working set so that the source drawing remains in its original state. Changes that you've made will remain in the current drawing, but the original drawing remains unchanged. You'll be given a warning, but proceeding will remove changes to the working set. Undo will restore the reference editing session if you change your mind after using Discard.

Save

Selecting the Save option will save the changes made to the working set back to the original drawing.

DENYING ACCESS TO REFERENCED DRAWINGS

Throughout the discussion of in-place editing, it has been assumed that you don't mind if someone alters your drawing. The idea of someone working for the HVAC contractor making changes to your floor plan may sound pretty good. The CAD operator for the HVAC is saving you the work of incorporating their required changes into the base drawing. If for some reason you need to provide drawings to another firm, but you want to insure accuracy, you can restrict the ability of others to edit the referenced drawing. Security can be maintained by using the Open/Save tab of the Options menu. The Options dialog box is opened by selecting Options… from the Tools pull down menu. By default, the Allow Other Users to REFEDIT Current Drawing in the External References (Xrefs) area of the dialog box is active. Making this option inactive will remove the ability to edit the source drawing.

CHAPTER 23 EXERCISES

1. Open the FOOTING drawing and use it to create an xref drawing titled BASEFTG. Open drawing FLOOR14 and attach BASEFTG to the drawing. Save the drawings as XREFTEST.
2. Open a drawing template containing the border and title block. Attach any four drawings in your library. Provide a title and the scale used by the original drawing.
3. Open any three drawings and attach them to a drawing base. Save the drawing as XREF.

CHAPTER 23 QUIZ

1. Describe how a referenced drawing can be used in an office to complete a project.
2. List and briefly define the six XREF options.
3. List the command, the options required, and the prompts to attach a drawing to a base drawing.
4. What is a dependent symbol and how does it affect a drawing?
5. Explain the effects of overlaying a drawing.
6. The bind option of XREF has been used. How will the option affect the external drawing?
7. Explain the effects of XBIND on a referenced drawing.
8. How does the XCLIP command affect an attached drawing?
9. How does the REFEDIT command affect a referenced drawing?
10. What is the difference between INSERT and XREF?

CHAPTER 24

Working with Layouts and Viewports

Chapter 3 introduced the layout tabs of the AutoCAD template drawings. A layout is a paper space drawing environment that represents how your drawing created in model space will appear on a sheet of paper when plotted. Throughout this text you've used the model space tab of the Architectural template. The template has been edited to include your favorite linetypes, lineweights, text fonts, and dimensioning styles. This chapter will introduce you to using the layout tabs to provide a predictable plotting setup. Methods of using existing layouts, and creating new layouts will be explored. By completing the steps in this chapter, you'll be able to take drawings that have been created in previous chapters, place them in a viewport within the layout, and prepare them for plotting. Methods for creating multiple viewports within a drawing will also be explored. Multiple viewports allow objects that must be displayed at different scales within the same drawing to be plotted easily.

COMPARING MODEL AND PAPER SPACE

As you begin to prepare a drawing for plotting it's important that you understand the use of the terms model space, paper space, and layout. In Chapter 3 you were introduced to drawing templates that were created in paper space. Throughout the text you've been drawing in model space. Model space is where you create the drawing, paper space is where you plot. A layout is a paper space tool that contains one or more viewports to aid in plotting. If you enter the Architectural template and select the Architectural Title Block tab, your drawing is displayed in paper space. The paper space UCS icon is displayed in the lower left corner, and the architectural title block surrounds your drawing. If you press the PAPER button below the command prompt, the UCS icon is changed to reflect model space, and is placed inside the drawing template. Your drawing is now in a viewport or window that is in model space within a paper space border. Open a template or refer to Chapter 3, and you notice that the architectural template contains a black line that overlays the border. This box is the viewport. See Figure 24.1. The drawing can be edited in the model space viewport without having to return to the model space tab. Later in this chapter

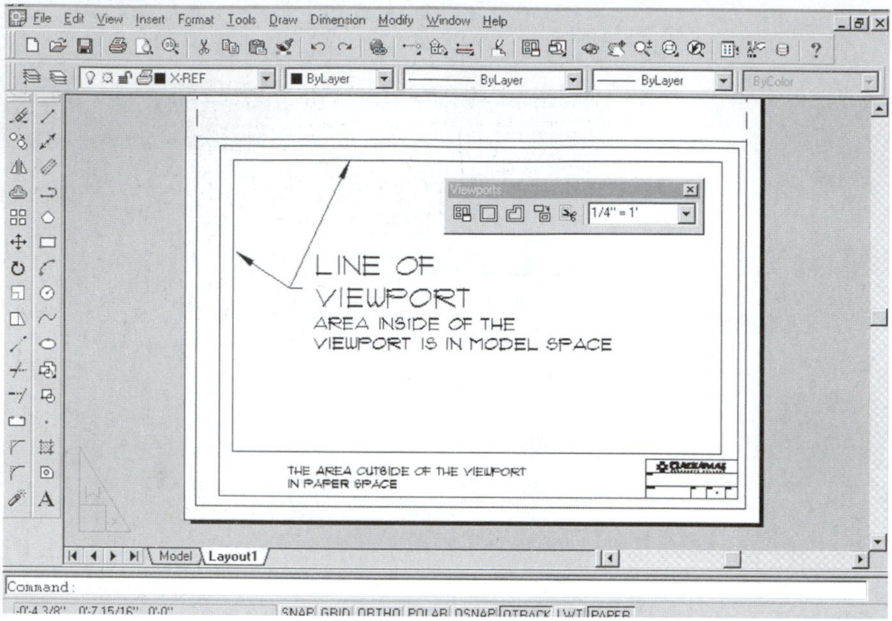

Figure 24.1 *The architectural template contains a black line that overlays the border. This box is the viewport.*

you'll be introduced to methods for dividing model space into tiled viewports to represent different views of the drawing. Switch back to paper space by pressing the MODEL button, and the drawing is displayed in the paper space layout with viewports, ready for plotting.

Note: As a general rule, NEVER alter the location of the drawing in paper space. When a drawing is altered in paper space, it may affect layouts by other subcontractors that may be working on the project. Alter the drawing in model space, and then return to paper space to display the drawing with the title block and borders.

AN OVERVIEW OF PAPER SPACE

To understand the process of viewing the floor plan drawing in Figure 24.2, visualize a sheet of 24" × 36" vellum with a title block and border printed on the sheet. Imagine a hole, the viewport, cut in the vellum that allows you to look through the paper and see the floor plan. This is a simplified version of what is required to display a drawing for plotting. Figure 24.3 shows the theory of displaying a drawing in model space to a layout in a template drawing created in paper space. The floor plan shown in Figure 24.1 is 90'-0" wide in Model space. If you were to hold a sheet of D size vellum in front of the plan, the paper would be minute. To make the floor plan fit inside the viewport on the paper, you're going to have to hold the paper a great distance away from the floor plan until the drawing is small enough to be seen through the hole.

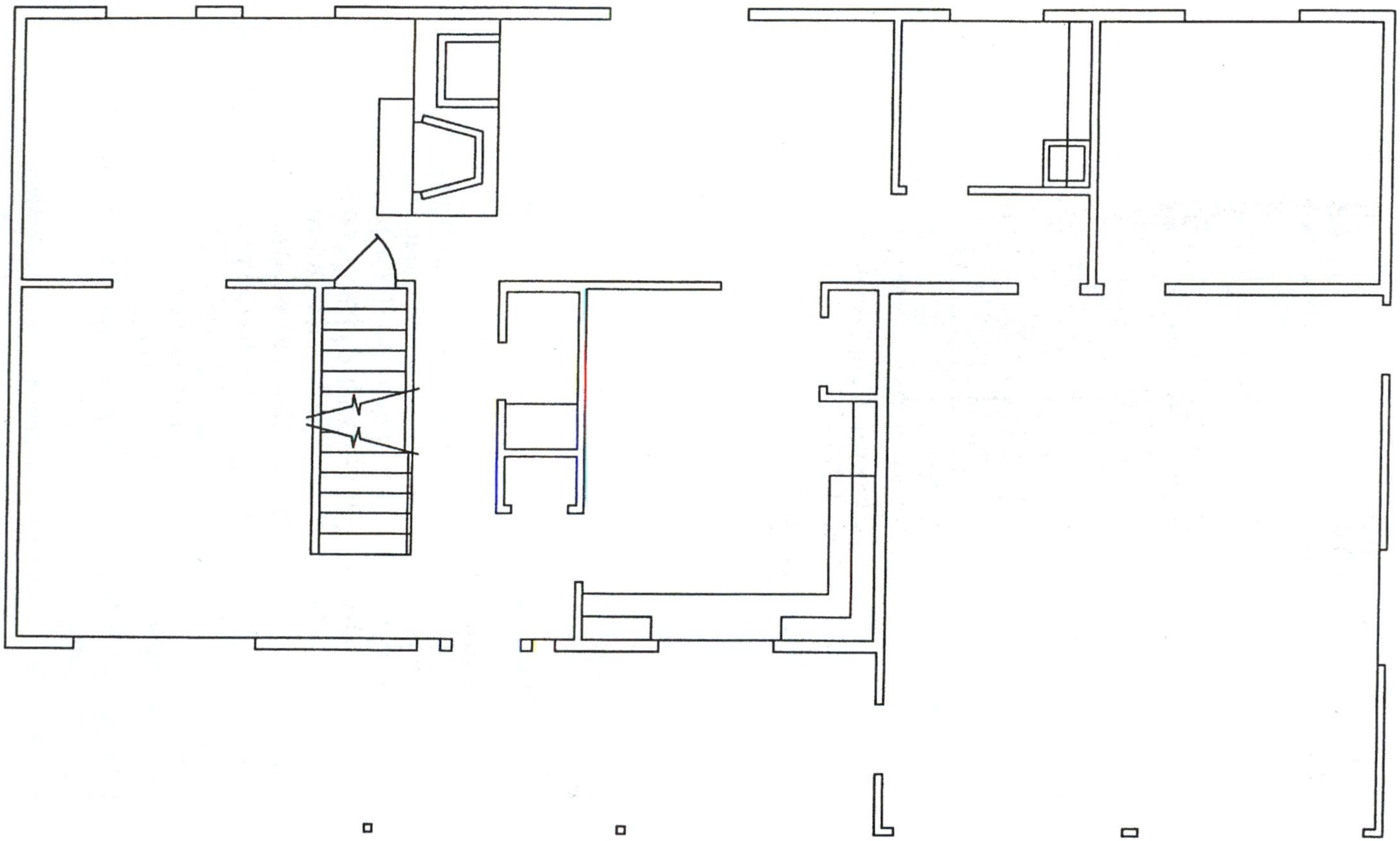

Figure 24.2 *A floor plan that is 90' long can be displayed in paper space using a paper space layout. (Courtesy Kyle Jones.)*

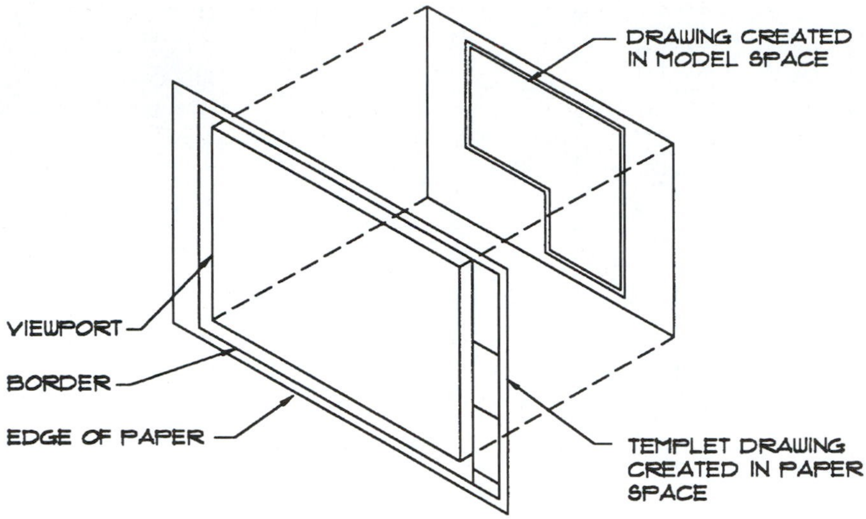

Figure 24.3 *A viewport is provided in each drawing template created in paper space. Additional layouts can be created to allow for different plotting displays.*

AutoCAD will automatically figure the distance. By entering the desired scale factor for plotting, the floor plan can be reduced to fit inside the viewport and maintain a scale typically used in the construction trade. Multiple viewports can be placed in a single layout, and multiple layouts can be created within a single drawing file

WORKING WITH LAYOUTS

Throughout this text you've been using the Architectural template or a template that you have created to meet your needs. When your drawing is complete, you can select the layout tab to begin preparing the drawing for reproduction. If you're using the Architectural template, the layout tab is titled Architectural Title Block. As the tab is entered for the first time, the Page Setup dialog box is displayed. The dialog box was introduced in earlier chapters. Chapter 25 will deal with page setup in depth. Layout settings for plotting can be set once and stored with the drawing. For now the goal is to help you create drawing displays that are ready for plotting. When your drawing is complete, use the following steps to view a drawing in the paper space:

- Select the Architectural tab.
- Select the PAPER button on the bottom of the screen to enter into model space in the viewport.
- Use the All or Extents option of Zoom to view the drawing.
- Use PAN to center the drawing in the layout.
- Select the MODEL button on the bottom of the screen to return to paper space.

- The procedure to view the drawing in paper space is simple and should be familiar by now. In addition to using layouts for plotting, several layouts can be created for the same drawing so that different parameters can be emphasized in the plotting process. Different scales and paper sizes can be assigned to different layouts, or different plotting devices can be assigned to the layouts.

CREATING NEW LAYOUTS

The existing template and layout work well for displaying a single drawing for plotting. Additional paper space layouts can be created to plot various layers, or in some other way customize a drawing for a specific plotting need. Layouts can be assigned a unique name, and multiple paper space viewports can be assigned to one or more of the layouts. Independent settings can be assigned and stored in each layout for plotter and page setup. Once created, layouts can be copied, deleted, moved, or renamed. Selecting Create Layout... from Wizards of the Tools pull-down menu allows a new layout to be created. The layout wizard contains a series of pages that walk you through the process of creating a new layout. Options are similar to the plotting options introduced in Chapter 6. Each setting will be further discussed in the next chapter. Figure 24.4 shows an example of the beginning page of the Layout Wizard. This page is used to assign a name to the layout being created. Once the desired name is entered, selecting the Next button will produce a display for choosing a printer from a list of configured printing devices. Using multiple layouts with multiple plotter configurations will allow you create check prints from a laser, or full size plots from a plotter without having to adjust plotter settings each time you make a plot.

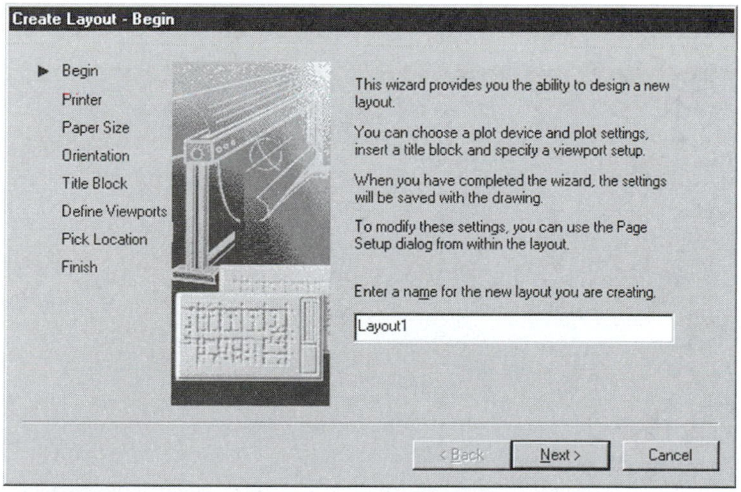

Figure 24.4 *The layout wizard contains a series of pages that walk you through the process of creating a new layout. Options are similar to the plotting options introduced in Chapter 6. The beginning page of the Layout Wizard is used to assign a name to the layout being created.*

Once the plotter to be assigned to the layout is selected, choosing Next will display a paper size display to be displayed similar to Figure 24.5. Depending on the configured plotter, this display allows the paper size and drawing units to be selected.

Selecting the Next button allows the orientation of the plot to the paper to be configured. Options include Portrait or Landscape. When complete, selecting the Next button allows a predefined template to be assigned to the layout. By default, the name of the template assigned to the previous layouts will be highlighted and displayed in the Preview area. By using multiple layouts, varied title blocks can be displayed with the drawing. This could be helpful for displaying a floor plan drawn by the architectural team, and the framing plan for the same structure drawn by the structural team.

Once a title block is selected, picking the Next button displays a Define Viewports display similar to Figure 24.6. This menu can be used to define a viewport and select the scale for presenting the information in the viewport. By default, a single viewport is displayed. Creating multiple viewports will be discussed in the next section of this chapter. The default setting for the viewport scale factor is Scaled to Fit. In this setting, a 90' structure will be crammed into the viewport. Most construction drawings need to be plotted at a measurable scale. Selecting the Next button after selecting the desired scale will display the Pick Location menu. The Pick Location menu prompts you to select the corner locations of the viewport. In the Architectural layout, the viewport edges align with the top, right, and bottom borders. This option allows a different viewport size to be selected for the new layout. The size of the viewport will depend on the size of the drawing to be displayed, the desired scale, and the number

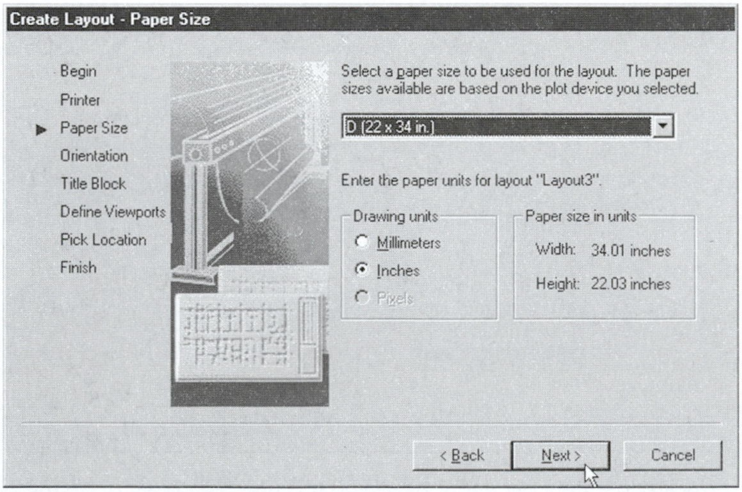

Figure 24.5 *Once the plotter to be assigned to the layout is selected, choosing Next will display a Paper Size display that allows the paper size and drawing units to be selected.*

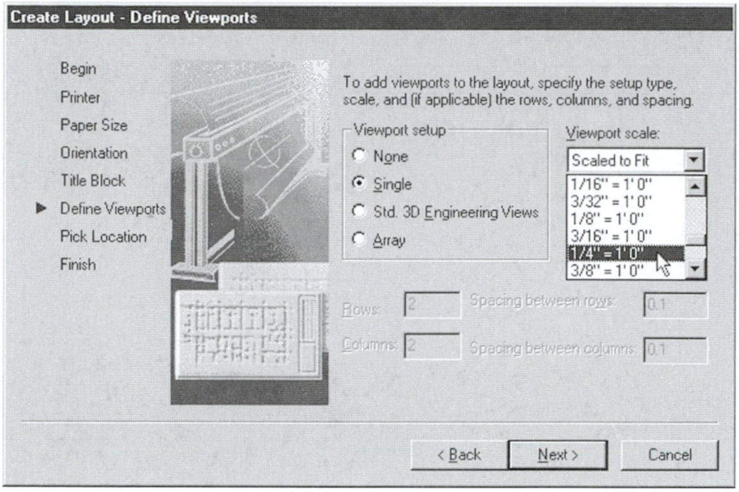

Figure 24.6 *Users of previous releases of AutoCAD will appreciate the Viewport Scale menu. The Define Viewports page allows the scale of the drawing in the viewport to be set and maintained in paper space, no matter how many times the drawing is zoomed in model space.*

of viewports to be created in the layout. As the size is selected, the final page of the wizard will be displayed. This page will display the name of the layout you've created and display methods for altering the layout. Selecting the Finish button will place the layout in the drawing base as the current layout.

In summary, the steps to creating a new layout with a wizard include:

- Select Create Layout… from Wizard of the Tools pull-down menu.
- Provide a name for the layout being created.
- Selecting the plotting device.
- Select the paper size and the drawing units.
- Select the orientation.
- Select the title block to be used.
- Select the number of viewports and the scale to be used in the viewport.
- Select the corners of the viewport.
- Press the Finish Button to return to the drawing area.

The new layout will now be displayed as the current layout similar to Figure 24.7a. Notice that the viewport that was created is not large enough to display the full floor plan. Selecting the viewport border will display the grips for the viewport. Make one of the grips hot, and then stretch the viewport to the desired size, as shown in Figure 24.7b. Pressing (EscKey) will end the STRETCH command and restore the view-

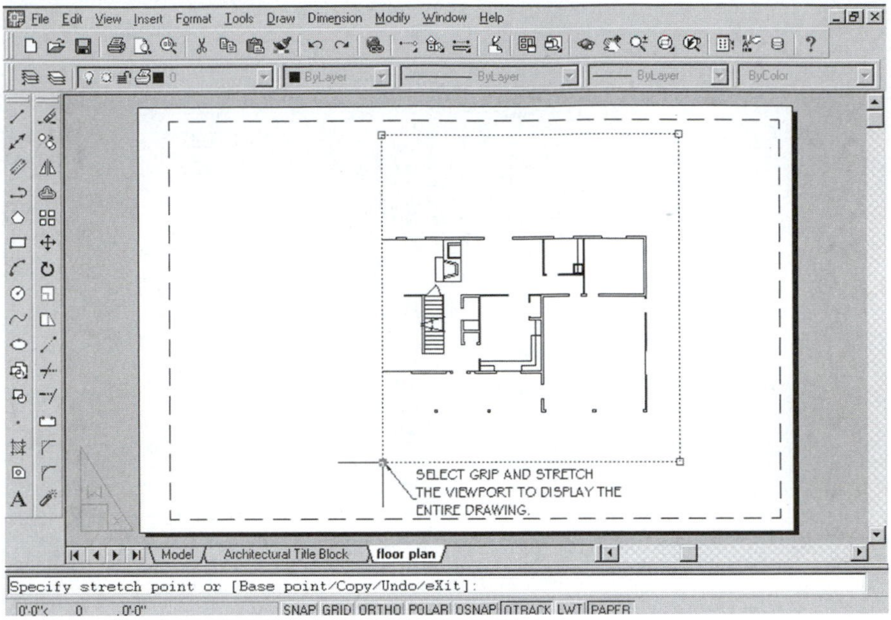

Figure 24.7a *The viewport that was created may not be large enough to display the entire drawing.*

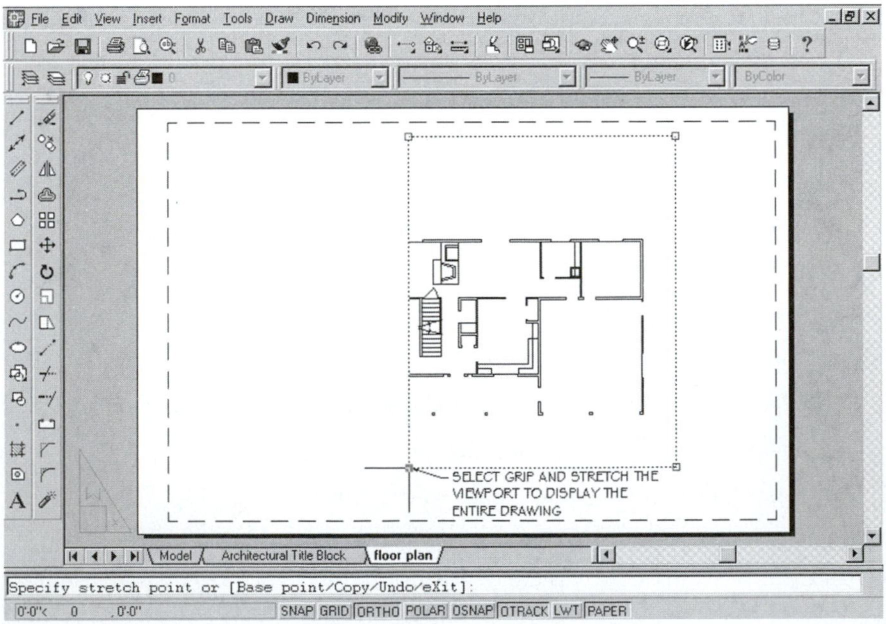

Figure 24.7b *Selecting the viewport border will display the grips for the viewport. Make one of the grips hot, and then stretch the viewport to the desired size.*

port. Figure 24.7c shows the results of stretching the viewport. Other modifications can be made to the layout by selecting the New Layout icon from the Layouts toolbar or by using the Page Setup dialog box.

Note: As several layouts are created, all of the layout names may not appear. To display more of the layout tabs, pick and hold the edge of the slide bar, and then move the cursor to the right. Figure 24.8a and 24.8b show the process.

ALTERING LAYOUTS

A new layout can be created or an existing layout can be altered, by selecting New Layouts from the Layout toolbar, or by entering **LAYOUT** (EnterKey) at the command prompt. Each method will present the following prompt at the command line:

Enter layout option [Copy/Delete/New/Template/Rename/SAve/Set/?] <Set>:

These options can be used to create, define, and save a new drawing layout.

Copy

Selecting Copy will create a new layout by making a copy of an existing layout. A prompt for a new layout name will be displayed at the command prompt. The default name will be the name of the current layout, in this case the Architectural Title Block

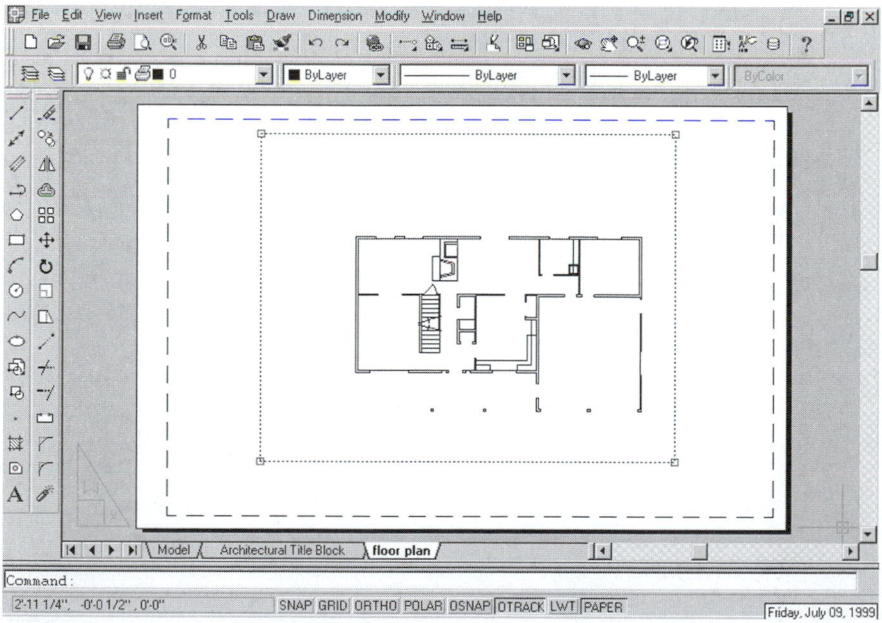

Figure 24.7c *A viewport can be stretched to display the full drawing. Pressing (EscKey) will end the STRETCH command and restore the viewport.*

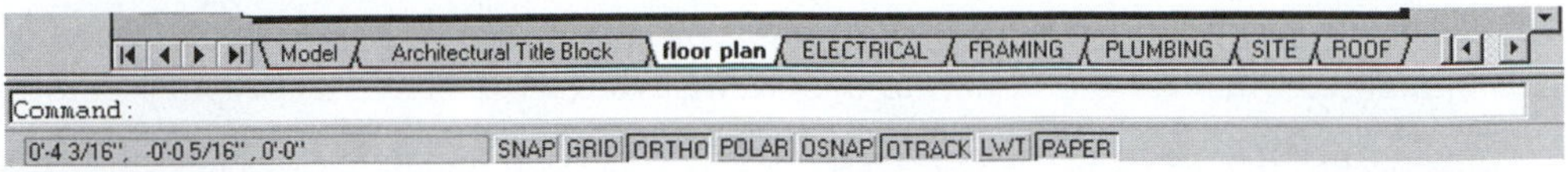

Figure 24.8a *Depending on the size of the slide bar, all of the layout tabs may not be displayed.*

is the default name. If a new layout name is not provided, the new layout assumes the name of the copied layout. An incremental number will be added to the name. The command prompt is:

> Enter layout option [Copy/Delete/New/Template/Rename/SAve/Set/?] <Set>: **C** (EnterKey)
>
> Enter layout to copy < Architectural Title Block>: (EnterKey)
>
> Enter layout name for copy < Architectural Title Block (2) >: (EnterKey)
>
> Layout "Architectural Title Block" copied to "Architectural Title Block (2)".

Delete

This option will delete a layout. By default the current layout will be deleted. The model tab cannot be deleted. One or more layouts can be deleted. If all layouts are removed, an empty layout titled Layout1 will be created and displayed.

New

This option creates a new layout tab without the use of a Layout Wizard. Selecting the New option will display the following prompts:

> Enter layout option [Copy/Delete/New/Template/Rename/SAve/Set/?] <Set>: **N** (EnterKey)
>
> Enter new Layout name <Architectural Title Block>: **COOL STUFF** (EnterKey)
>
> Command:

Template

Selecting the Template option creates a new layout based on an existing template (.DWT) or drawing file (.DWG). The layouts from an existing template or drawing file can be inserted into a new drawing. In addition to the layout, objects contained

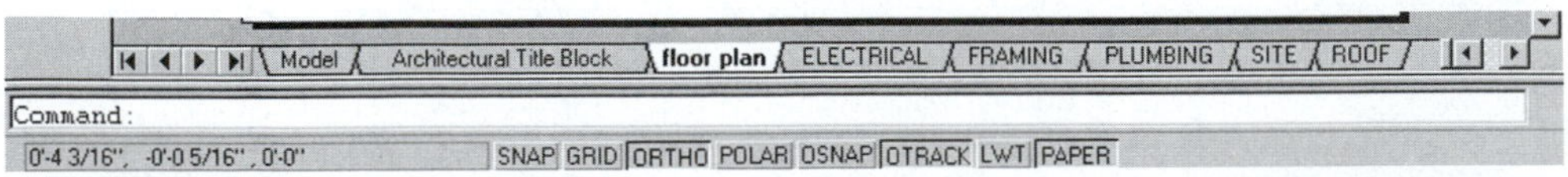

Figure 24.8b *By decreasing the size of the slide bar, more layout titles will be displayed.*

in the layout can also be made part of the layout being created. The command sequence is:

> Enter layout option [Copy/Delete/New/Template/Rename/SAve/Set/?]
> <Set>: **T** (EnterKey)

Pressing (EnterKey) will display the Select File dialog box, allowing a template to be selected. Selecting the File type edit box allows a drawing file to be selected. Once the template or file is selected, it will be displayed in the preview area. Selecting the Open button will close the Select Files dialog box and provide the Insert Layout dialog box that verifies the template to be added to the layout. Selecting OK will return the layout display with the template. A template can also be selected by choosing the Layout From Template icon from the Layout toolbar

Rename

This option can be used to rename a layout. The layout that was last used is shown as the default. Prompts will be displayed to select the layout to be renamed, and to select the new name. Prompts to rename the JUNK layout as COOL STUFF2 are:

> Enter layout option [Copy/Delete/New/Template/Rename/SAve/Set/?] <Set>: **R** (EnterKey)
>
> Enter Layout to rename <Architectural Title Block>: **JUNK** (EnterKey)
>
> Enter new layout name: **COOL STUFF2** (EnterKey)
>
> Layout "JUNK" renamed to "COOL STUFF2".
>
> Command:

Layout names are not case sensitive and can be up to 255 characters long. Only 32 characters are displayed in the layout tab.

Save

As you complete the Layout Wizard, the new layout is stored as part of the current drawing. Any existing drawing can be saved and used as a template drawing. All of the objects, properties and settings can be saved as a .DWT file (template) by selecting the SaveAs option from the LAYOUT command. New drawings can be created from drawing templates or new layouts can be created in the current drawing based on a layout in the template file.

Set

The Set option makes an existing layout the current layout. Picking the tab of a layout does the same thing.

The ? Option

Selecting this option will list all of the layouts defined in a drawing. This can be helpful when all layout names cannot be displayed below the drawing. Pressing F2 will display a list of the current layouts.

WORKING WITH VIEWPORTS

AutoCAD allows two types of viewports to be created. Floating viewports are used to create plotting layouts in paper space. Tiled viewports can be created in model space to aid in the display of the drawing. Up to four tiled viewports can be created, but only one viewport can be active at a time. Tiled viewports allow you to divide the screen into multiple areas so that you have multiple views of different parts of the same object. Figure 24.9 shows an example of a sheet of details. Creating multiple viewports would allow you to zoom in on a beam-to-column connection in one viewport, a column-to-footing detail in another viewport and a display of the a whole sheet of details in a third viewport, as shown in Figure 24.10. This multiple use of viewports allows for drawing and editing within the enlarged portions of the drawing without having to zoom in and out to move from one detail to another. It is important to remember that even though you are seeing multiple images on the screen, you're working with only database. An object that is drawn or edited in one viewport will also affect the other viewports.

DISPLAYING MULTIPLE TILED VIEWPORTS

Multiple tiled viewports can be displayed in several configurations that are controlled by the Viewports dialog box. The Viewports dialog box is displayed by selecting the Display Viewports Dialog icon from the Viewports toolbox, by selecting New View-

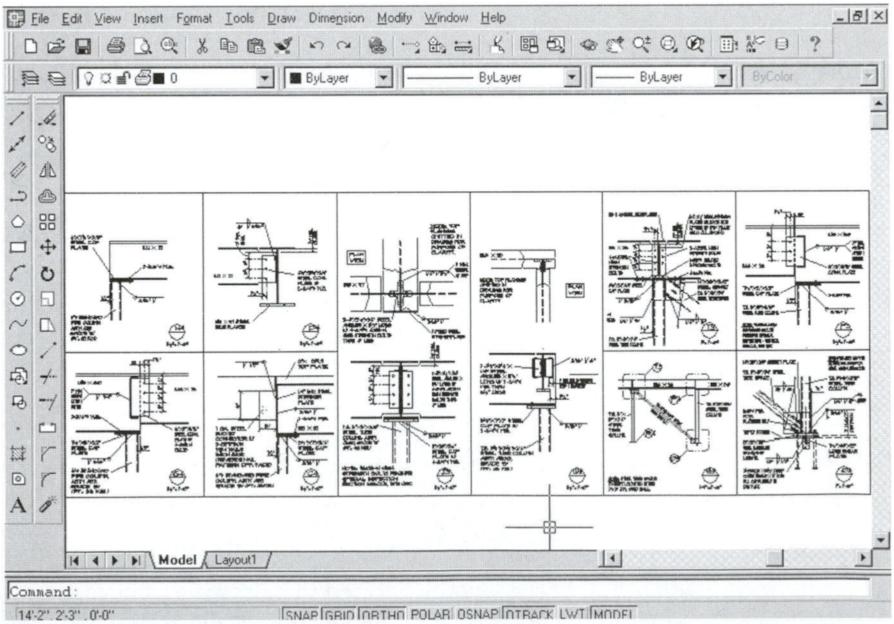

Figure 24.9 *When working on a sheet of details, it is often hard to coordinate information from one detail to another.*

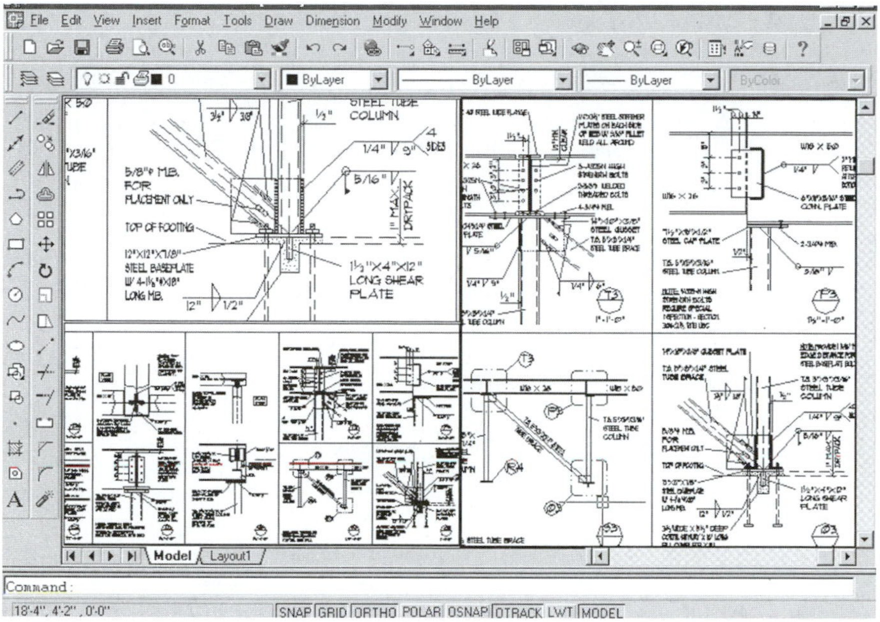

Figure 24.10 *Tiled viewports allow you to divide the screen into multiple images so that you have multiple zoom images of different parts of one drawing file. The details shown in Figure 24.9 can be easily view using multiple viewports.*

ports... from Viewports of the View pull-down menu, or by typing **VPORTS** (EnterKey) at the command prompt. Each method will produce the display shown in Figure 24.11. In the current setting, one viewport is displayed. Selecting a listing from the Viewport menu will alter the display in the preview to reflect the change. Figure 24.12 shows the display for each option. Use the following steps to display multiple tiled viewports in a new configuration:

- Open the Viewports dialog box.
- Make the New Viewports tab active.
- Select the name of the desired viewport configuration from the Standard Viewports list.
- Select Display as the current setting for the Apply to list.
- Select 2D as the current setting for the Setup list.
- Select the OK button to apply the configuration and close the dialog box.

Notice in Figure 24.10, a bold black line surrounds the viewport on the right side. This viewport also contains the crosshairs. As you move the cursor around the screen, you'll notice that the crosshairs only work in the drawing on the right side. This view is called the active viewport. You can draw or edit in this viewport just as in any other drawing. As the cursor crosses the center of the drawing into the viewport on the left,

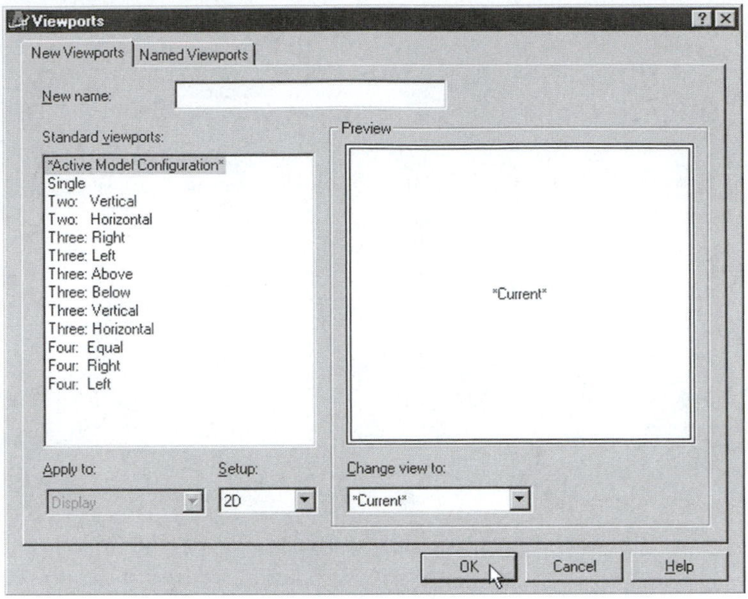

Figure 24.11 *The Viewports Layout dialog box is displayed by selecting the Display Viewports Dialog icon from the Viewports toolbox, by selecting New Viewports... from Viewports of the View pull-down menu, or by typing **VPORTS** (EnterKey) at the command prompt.*

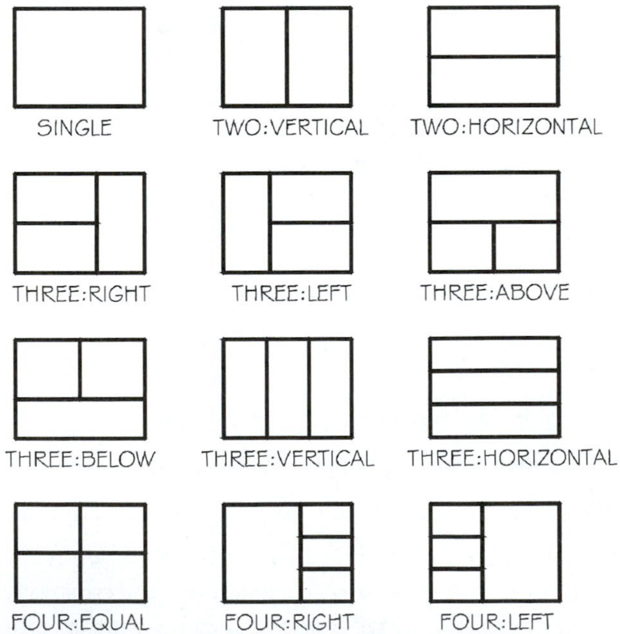

Figure 24.12 *Possible display options for tiled viewports.*

the crosshairs change to an arrow. Notice that by comparing Figure 24.13 with Figure 24.14, information related to the steel column was removed from the lower left corner of the upper left viewport, which also removed the information from the database and the total view. Although the total detail is not legible, you can see that information is missing. The fact that the information is not legible is the point. Viewports make multiple enlarged views of one drawing that are legible for easy drawing or editing, while leaving a view of the overall drawing for perspective.

To make a different viewport active, move the cursor to the desired viewport and click. To convert a drawing like Figure 24.13 to one similar to Figure 24.14, change between active viewports and use the ZOOM command. Remember, once active, all drawing and editing commands function normally. First make the upper left viewport active. Select the detail in the upper right quadrant for a ZOOM window and enlarge to fill the viewport. Finally, make the viewport on the lower left active and select the detail in the lower right quadrant for a ZOOM, using the Window option. With these details enlarged, you can make changes to each without having to zoom from one detail to another.

Working in Multiple Viewports

In addition to easing viewing, viewports can ease working. A command such as line can be started in one viewport and completed in another viewport. This could be

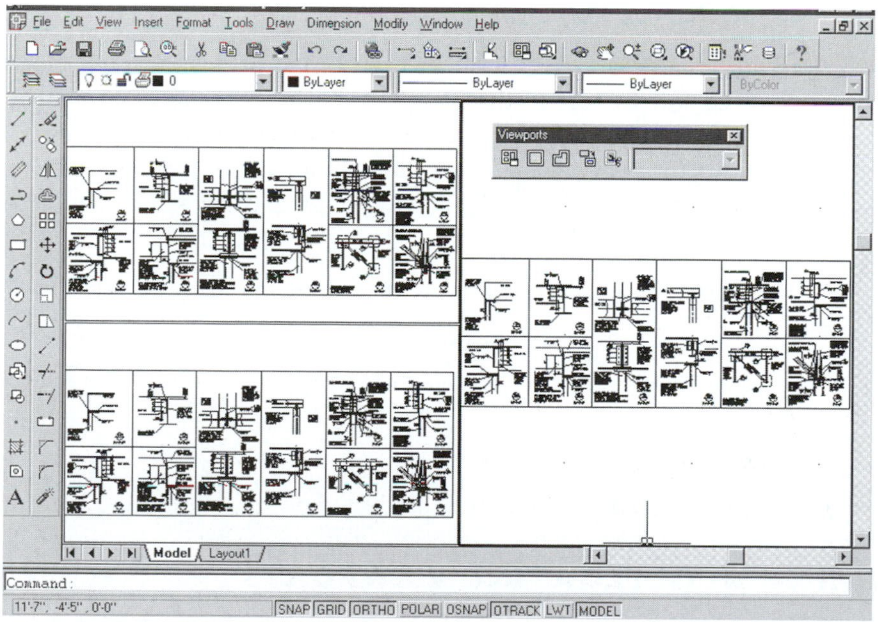

Figure 24.13 *Selecting Three:Right followed by the OK button will display the drawing shown in Figure 24.10 in three viewports. (Courtesy StructureForm Masters.)*

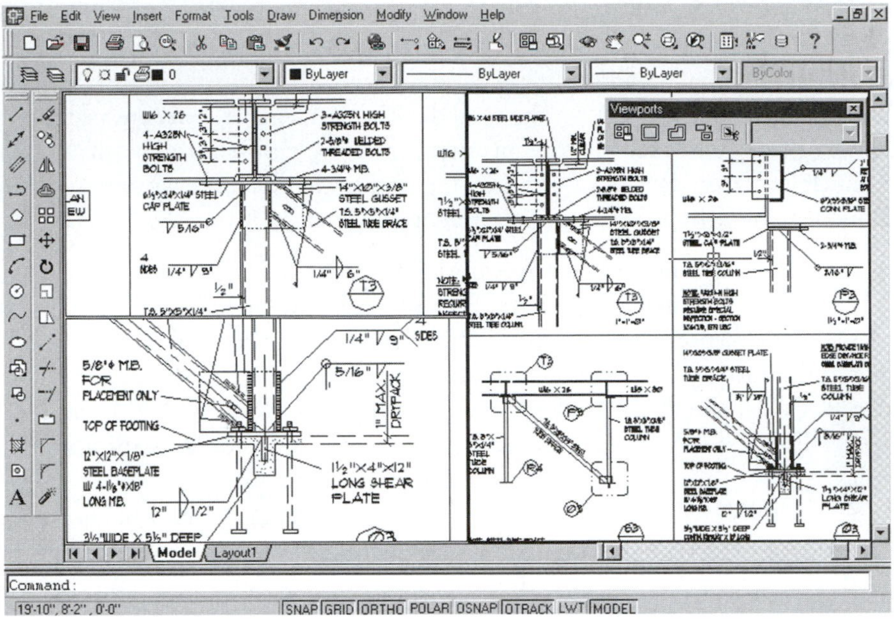

Figure 24.14 *The viewports in Figure 24.13 have each been edited to show this display. By making the upper left viewport active, the detail was enlarged using the ZOOM command to display the beam-to-beam connection. The same process was used in the lower left viewport to enlarge the column detail. (Courtesy StructureForm Masters.)*

especially helpful on large drawings, such as exterior elevations. Exterior elevations are typically created side by side so that heights can be projected from one view to another. If all views are to be displayed, each view will be tiny. Creating a viewport for each elevation allows for easy viewing. Being able to draw lines from one viewport to another will speed production. Use the following steps to draw a line between viewports:

- Select the viewport that contains the start point, making it the active viewport.
- Start the LINE command sequence.
- Select the viewport that will contain the ending point, making it the active viewport.
- Select the ending point for the line.

A line will now be displayed between the two selected points, similar to Figure 24.15.

WORKING WITH FLOATING VIEWPORTS

Each time you enter a layout, you're working with a floating viewport. The viewport is referred to as floating because it can be moved and the size can altered, just like a floating dialog box. The viewport allows you to look through the paper and see the

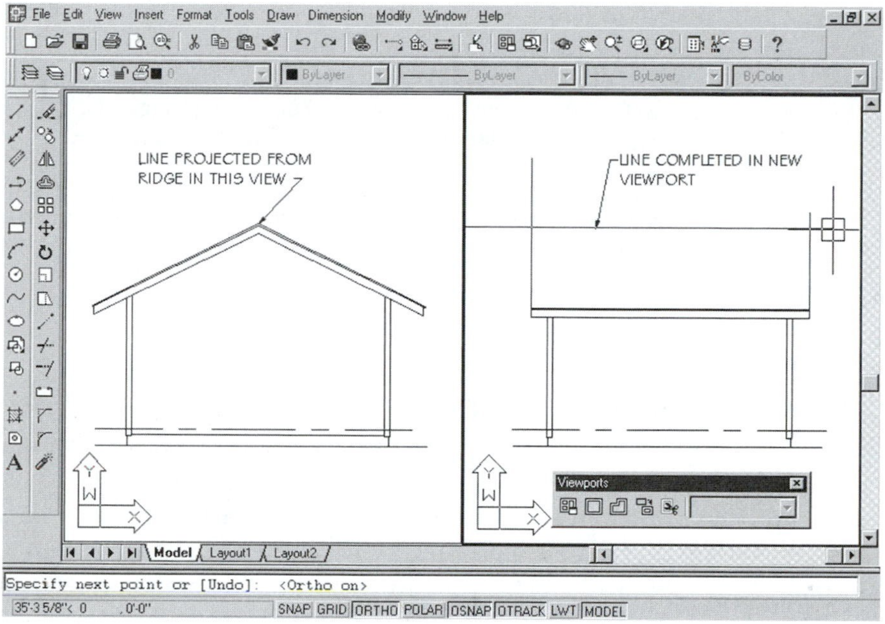

Figure 24.15 *A command can be started in one viewport and completed in another viewport. Select the viewport that contains the start point, making it the active viewport, and start the command sequence. Next select the viewport that will contain the endpoint, making it the active viewport, and select the endpoint for the line.*

drawing created in model space. The floating viewport is used to display the drawing in the layout for plotting. An unlimited amount of viewports can be created in paper space, but only 64 viewports can be visible at one time. The balance of this chapter will explore floating viewports and how they can be used to prepare a drawing layout for plotting. A floating viewport can be considered as a drawing object that provides a view into model space. Several floating viewports can be created within a layout, and the viewports can be overlapped or separated from each other. New to AutoCAD 2000, floating viewports can be an irregular shape. Since floating viewports are considered objects, drawing objects displayed in the viewport cannot be edited. To edit objects within the viewport, model space must be restored. You can toggle between model space and a floating viewport by choosing either the Model or layout tab, depending on which one is inactive. Other methods for toggling between model and paper space include:

- Double clicking in a floating viewport will switch the display to model space.

Note: Remember, while editing in model space in a layout you risk changing the drawing setup. This should only be used by experienced AutoCAD users or when in the initial setup of the view inside of the viewport.

- Select the Paper or Model button on the Status bar. Remember when you're in a layout, the page is in paper space, but the viewport can be toggled to model or paper. When Paper button is displayed, the viewport is a paper space display of the drawing in a paper layout. Selecting PAPER, will change the button to MODEL, and display the drawing in a model space viewport within a paper space layout.
- Most professionals work in model space, and arrange the final drawing for output in paper space. While working in a layout, with the viewport in paper space, any material added to the drawing will be added to the layout, but not shown in the model display. This will prove useful as the final plotting layout is constructed.

ALTERING VIEWPORT PROPERTIES

Because viewports are objects created by AutoCAD, they have properties such as color, layer, linetype, lineweight and plot style. Each of these properties can be altered using the Properties dialog box. As you create a viewport, a new layer can be created and displayed for the viewport. This layer should be frozen before the drawing is saved so that it is not plotted. Linetype and lineweight are ignored on rectangular viewports, but are displayed in nonrectangular viewports.

VIEWPORT SCALING

One of the new features of AutoCAD 2000 will greatly aid the display of drawings in a viewport. Most users of AutoCAD plot their drawings displayed in a layout at a scale of 1:1. To do this, the drawing must be inserted or attached to the viewport at a predetermined scale factor. AutoCAD has created an aid to do this. Earlier it was pointed out that the scale can be set as a drawing is placed in a viewport. The scale can be locked using the Properties dialog box. Locking the viewport scaling will prevent the scale from being altered if the drawing is zoomed once it is displayed in the viewport. The scale can be locked by:

- Enter the layout containing the viewport to be locked.
- Select the edge of the viewport so that it is highlighted.
- Open the Properties dialog box by selecting the Properties icon from the Standard toolbar.
- Select Display Locked from the Properties window.
- Select YES from the Display Locked edit menu.
- Close the Properties dialog box and return to the drawing area.
- The scale of the selected viewport is now locked. If you attempt to alter the scale while in model space, AutoCAD displays "warning: Viewport is view-locked". If you change the zoom factor in the viewport only paper space objects are affected.

Note: It's been stressed not to edit object in viewports while in paper space. Once you've placed a lock on the viewport, it is safe to edit.

CREATING MULTIPLE FLOATING VIEWPORTS

The Viewport toolbar shown in Figure 24.16 can be used to create floating viewports. Icons include Display Viewports Dialog Box, Single Viewports, Polygonal Viewports, Convert Objects to a Viewport, Clip Existing Viewport, and Viewport Scale. To this point you've used layouts with viewports that take up the entire drawing area. Floating viewports can be created in a layout that fills the entire drawing area or only take a small portion of the drawing area. Floating viewports can be created using a process that is similar to the process for creating tiled viewports. Floating viewports are created using the Viewports dialog box shown in Figure 24.17. Compare the Viewports dialog box in Figure 24.17 with the Viewports dialog box in Figure 24.11. Although each dialog box is very similar, they create different types of viewports, depending on how they are accessed. The dialog box shown in Figure 24.11 was accessed while in model space. Use the following steps to create a floating viewport.

- Enter a layout by selecting an existing layout tab or by creating a new layout.
- Open the Viewports dialog box by selecting the Display Viewports Dialog icon from the Viewports toolbox, by selecting New Viewports... from Viewports of the View pull-down menu, or by typing **VPORTS** (EnterKey) at the command prompt.
- Select the desired viewport configuration from the Standard Viewports List.
- Enter the desired spacing to be maintained between viewports in the Viewports Spacing edit box.
- Select 2D in the Setup edit box.
- Select the OK button to apply the configuration and close the dialog box.
- A process similar to placing tiled viewports in a layout can be used to place multiple floating viewports in an existing layout.

Figure 24.16 *The Viewport toolbar can be used to control floating viewports. Icons include Display Viewports Dialog Box, Single Viewports, Polygonal Viewports, Convert Object to Viewport, Clip Existing Viewport, and Viewport Scale.*

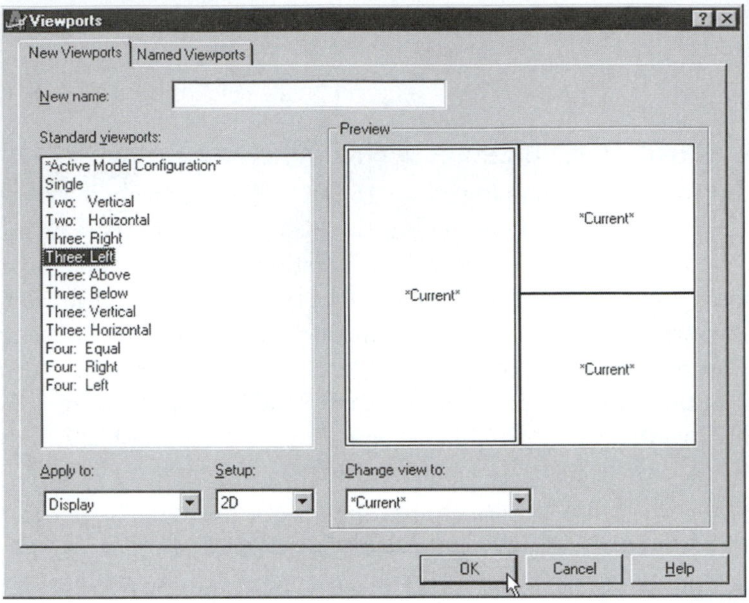

Figure 24.17 *The Viewports dialog box. Although this dialog box is very similar to the dialog box shown in Figure 24.11, they create different types of viewports, depending on how they are accessed.*

CREATING A SINGLE FLOATING VIEWPORT

Floating viewports provide an excellent means of assembling multiscaled drawings for plotting. New floating viewports can be created without using the standard viewport configurations. Figure 24.18 shows a sheet of details assembled for plotting using floating viewports. Construction projects typically comprise drawings that are drawn at a variety of scales. These details range in scale from 3/8" = 1'–0" to 1 1/2" = 1'–0". This sheet of details can not be easily drawn or plotted without the use of multiple viewports and inserting or referencing each detail to a base drawing. As with drawings where a single object is attached, the use of scale factors to control text and dimension variables will be critical to the success of the final outcome of the drawing. The process for creating multiple viewports will be similar to the creation of one viewport.

The following example will demonstrate the steps used to prepare the section and three details shown in Figure 24.19 for plotting. The section will be displayed at a scale of 1/4" = 1'-0", two of the details will be displayed at a scale of 3/4" = 1'-0", and the stair section will be displayed at a scale of 1/2" = 1'-0". For most users, drawings such as a floor plan or section will be created in a drawing template using model space. The following guidelines will walk you through the process of creating

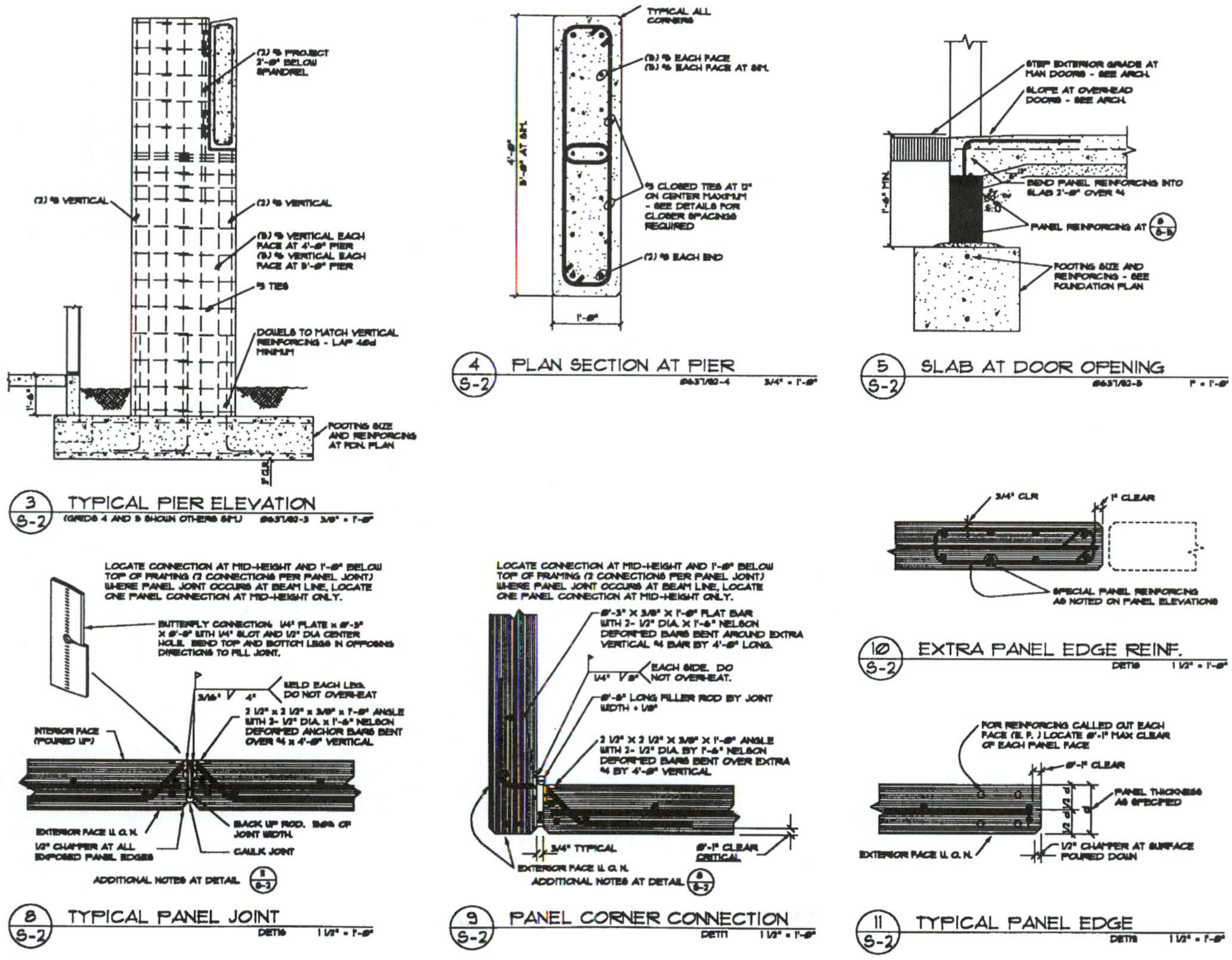

Figure 24.18 *Details drawn at varied scales can be assembled on one sheet for plotting using the XREF command. (Courtesy Van Domelen/Looijenga/McGarrigle/Knauf Consulting Engineers.)*

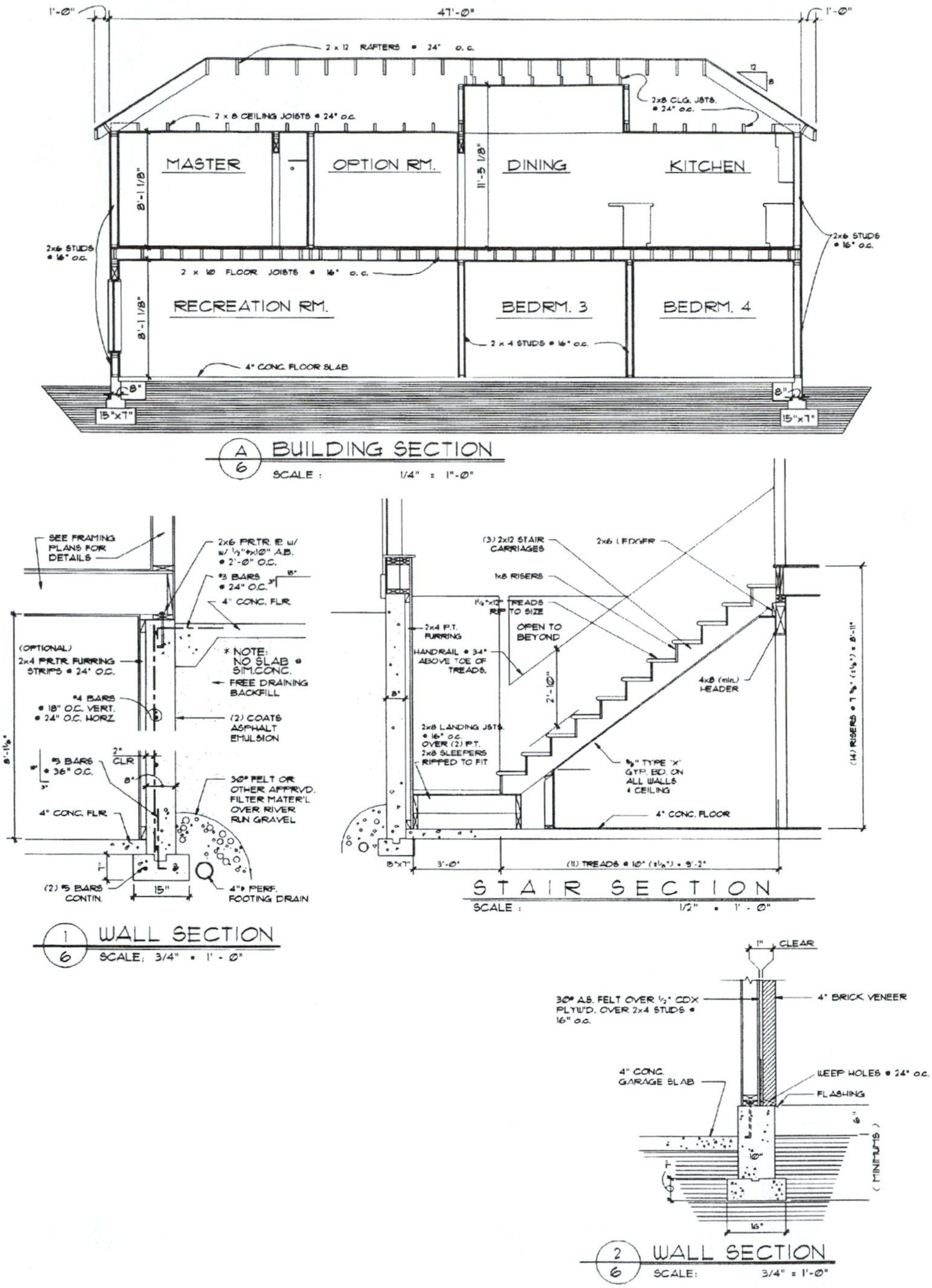

Figure 24.19 *Drawings to be attached to the basesheet. (Courtesy Piercy & Barclay Designers, Inc., A.I.B.D.)*

multiple viewports in a template and inserting or attaching multiple drawings. For this discussion, the INSERT command will be used. The drawings could also be attached using XREF command, depending on the drawings needs.

Displaying Model Space Objects in the Viewport

The easiest method to display multiple drawings at multiple scales is by inserting each of the drawings to be plotted into the template in model space. With the details arranged into one drawing similar to Figure 24.19, switch to the Architectural Title Block layout. Use the following steps to display the drawings in multiple viewports:

- In the Architectural Title Block layout, use the ZOOM All option to enlarge the area of model space to be displayed in the existing viewport. Each of the four drawings will be displayed, but each will be at an unknown scale. Your drawing will resemble Figure 24.20.
- Select what will be the largest drawing to work with first. In our example, the viewport for displaying the section will be adjusted first.
- Press the MODEL button in the status bar to toggle the drawings to paper space.
- Activate the Viewport toolbar.
- Set the scale of the viewport to the appropriate value for displaying the section using the Viewport Scale Control menu in the Viewport toolbar. For this example a scale of ¼" = 1'-0" was used.
- Once all of the drawings to be plotted have been placed in model space of the template, switch to the Architectural Title Block layout. Use the ZOOM All option to enlarge the area of model space to be displayed in the existing viewport. Each of the four drawings will be displayed, but each will be at an unknown scale.

Adjusting the Existing Viewport

Once the desired drawing is inserted or attached into the viewport, the size of the viewport can be altered if the entire drawing can't be seen. To alter the viewport size, move the cursor to touch the viewport and activate the viewport grips. Select one of the grips to make it hot and then use the hot grip to drag the window to the desired size. The Section has now been inserted into the template and is ready to be plotted at a scale of ¼" = 1'-0". The display would resemble Figure 24.21.

Note: All of the drawing commands may be used inside the viewport, but try to limit editing to changes that will only affect this plot. Changes to the drawing, which will be needed on future drawings, should be made to the original drawing. Remember, if the original is attached to this drawing by XREF, the changes made to the original drawing will be reflected in all uses of the referenced drawing. Enter Model space prior to making changes to the drawing in the current viewport. This will allow you to add detail numbers and drawing titles that change with each use of the drawing.

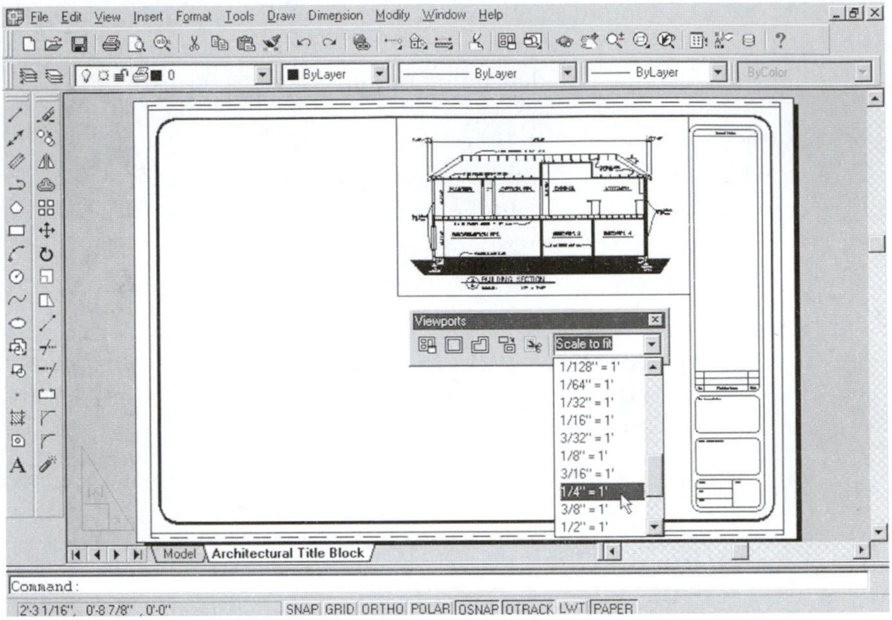

Figure 24.21 *Once the desired drawing is inserted or attached into the viewport, set the scale of the viewport to the appropriate value for displaying the drawing using the Viewport Scale Control menu in the Viewport toolbar. Next, the size of the viewport can be altered if the entire drawing can't be seen. The Section is now ready to be plotted at a scale of 1/2" = 1'-0".*

Creating Additional Viewports

Use the following steps to prepare for additional viewports:

- Set the current layer to Viewport. This layer is part of each template.
- Select Single Viewport from the Viewport toolbar.
- Select the corners for the new viewport. The coordinates for the viewport can be entered, or specified by indicating the corners of the viewport—just as if it were a window used for determining a selection set. Although you should try to size the viewport accurately, it can be stretched to enlarge or reduce its size once the scale has been set.

The results of this command can be seen in Figure 24.22.

Altering the Second Viewport

In its current state, if a new viewport is created, the display from the existing viewport will be displayed in the new viewport as well. Use the following steps to alter the display.

- Press the PAPER button in the status bar to toggle the drawings to model space.

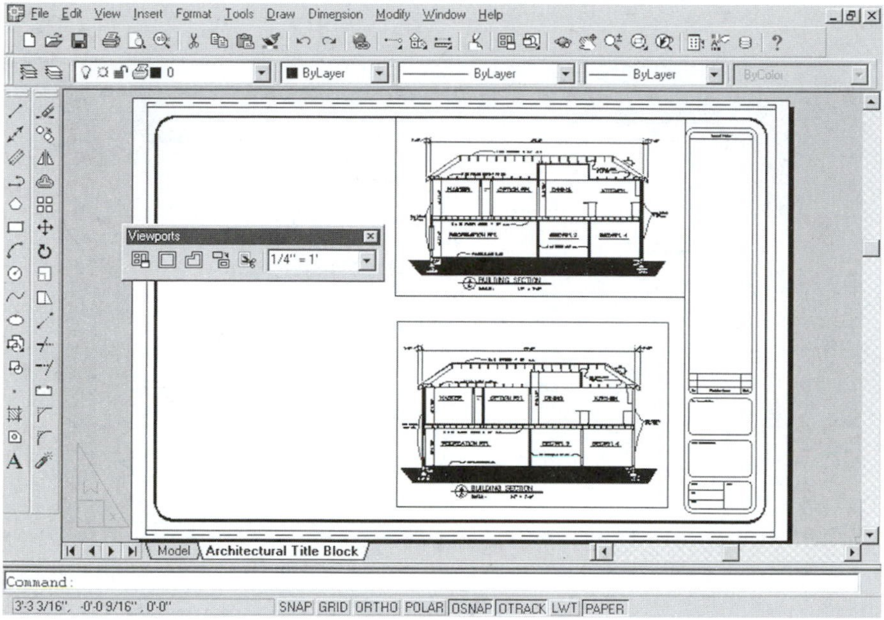

Figure 24.22 *If a new viewport is created, the display from the existing viewport will be displayed in the new viewport as well.*

- Make the new viewport the current viewport by placing the cursor in the viewport and single clicking.
- Use the ZOOM All option to display all of the model contents in the second viewport.
- Set the viewport scale to the appropriate value for displaying the section. For this example, a scale of ¾" = 1'-0" will be used to display the wall section.
- Use the PAN command to center the wall section in the viewport.
- Press the MODEL button in the status bar to toggle the drawings to paper space.
- Select the edge of the viewport to display the viewport grips.
- Make a grip hot and shrink the viewport so that only the wall section is shown.
- Use the MOVE command to move the viewport to the desired position in the template.
- Save the drawing for plotting.
- The drawing should now resemble Figure 24.23

Repeating the Process for Additional Viewports

Once the viewport for the wall section been adjusted, additional viewports can be created. The process to create the viewports to display the BRICK1ST and STAIR

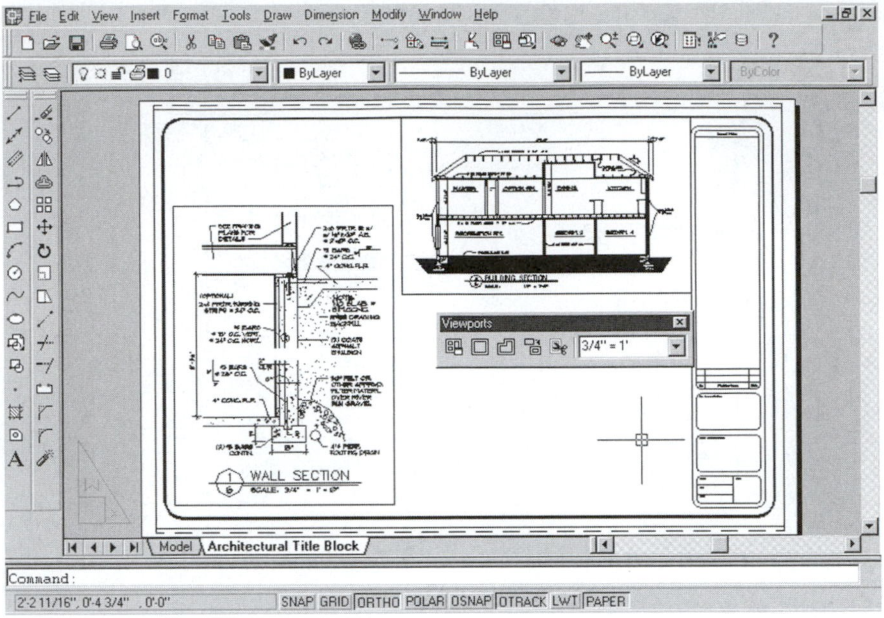

Figure 24.23 *Use the PAN command to move the drawings so that the desired information is centered in the viewport. Once centered, adjust the viewport scale to the desired value, and then adjust the size of the viewport so only the desired material is displayed.*

8FTWALL drawings is the same process used to create the viewport for the WALL SECTION drawing. If you have planned the sheet contents well, the viewports for BRICK1ST and STAIR can be created also. Placing all of the viewports at once will eliminate having to keep switching from PSPACE to MSPACE. Once all of the desired details have been added to the layout, the viewport outlines can be frozen. Enter the LAYER command and select VIEWPORT as the layer to be edited, then select the Freeze option. Freezing the viewports will eliminate the line of the viewport being produced as the drawing is reproduced. Any viewports that still exist were created on a layer other than VIEWPORT. Change any existing viewports to the correct layer. The finished drawing, with the viewports frozen will resemble Figure 24.24.

CONTROLLING THE DISPLAY IN FLOATING VIEWPORTS

As you placed drawings in viewports, the PAN and ZOOM commands were used to alter what was displayed in each viewport. AutoCAD allows displayed objects to be altered by screening, turning viewports ON or OFF, and hiding lines. Hiding lines is typically associated with 3D drawings and will not be covered in this chapter. Consult Help for additional information.

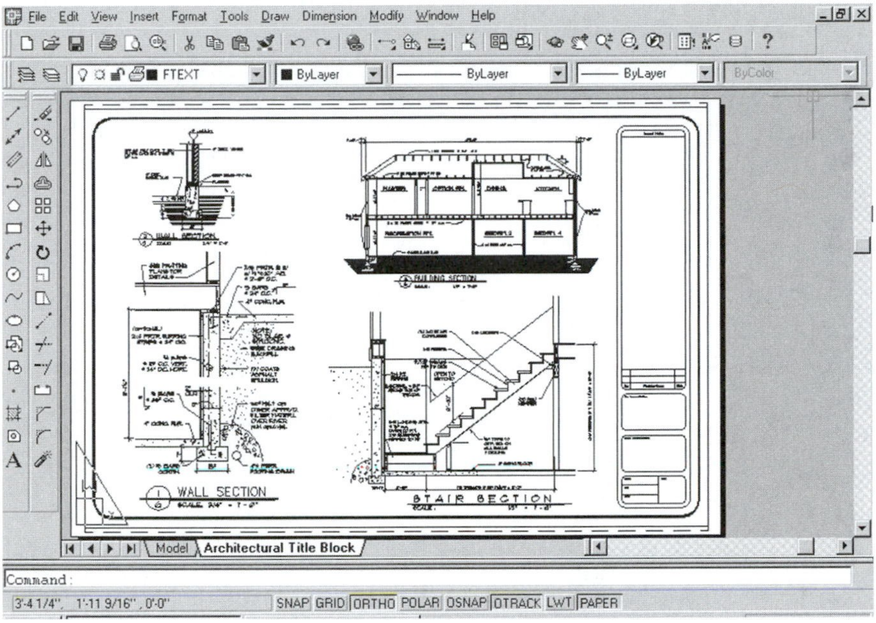

Figure 24.24 *The finished drawing, with the Viewports layer frozen, is ready for plotting.*

Screening Objects in Viewports

Screened objects use less ink when plotted, so they appear dimmer on screen and when plotted. Screening could be used to distinguish between objects, without plotting in color. On remodeling projects, existing objects can be screened to easily distinguish between new and existing work. To screen an object it must be assigned a plot style, and then the screening value is assigned to that plot style. Plot styles will be discussed in the next chapter. Values ranging from 0 to 100 can be assigned, with 100 being the default value. A value of 100 assigns no screening and will provide normal plotting values. A value of 0 would assign no ink to the object and have the same effect as freezing the viewport contents. Figure 24.25 shows an existing footing and new wood floor framing.

Displaying Floating Viewport

By default a viewport is ON as it is created. The performance of your machine can be affected if large amounts of complex viewports are created. Turning viewports OFF can save regeneration time on large drawings. Viewports can be toggled ON/OFF using the Properties dialog box by:

- Select the viewport to alter.
- Open the Properties dialog box.
- Select On from the Misc listing and then select OFF from the edit menu.

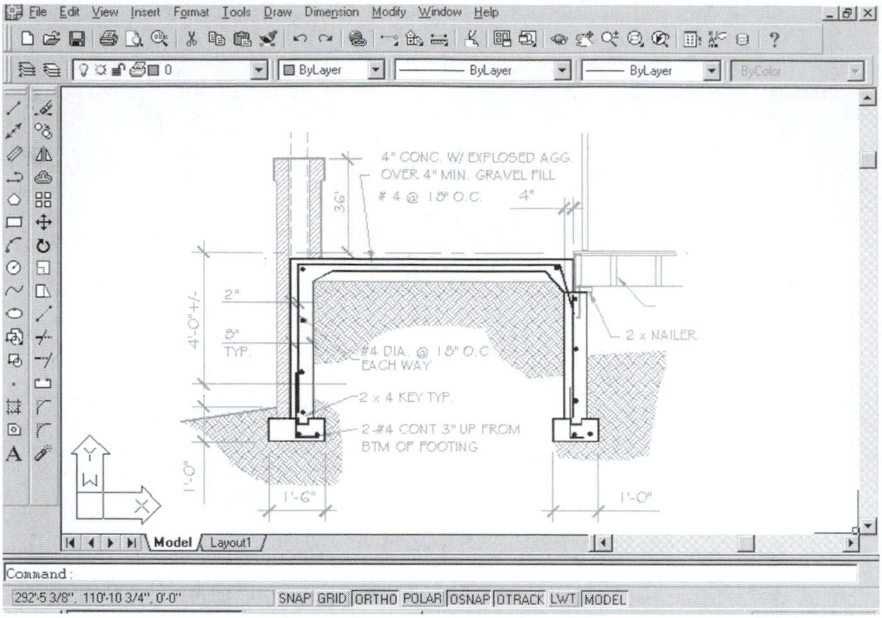

Figure 24.25 *Screened objects use less ink when plotted so they appear dimmer when plotted. To screen an object, it must be assigned a plot style, and then the screening value is assigned to that plot style. The concrete in this detail was assigned a value of 50.*

- As the viewport is toggled OFF, the contents of the viewport will be removed from the drawing display. The line representing the viewport can be removed from the display by using the Layer Properties Manager dialog box to freeze the layer. Figure 24.26 shows the drawing in Figure 24.24 with two of the viewports in the OFF setting.

Controlling Layer Display with the Layer Property Manager

The contents of specific layers in each viewport can be controlled using the Layer Properties Manager. Using the layer options works well when an object will be displayed in two or more views. Specific aspects of the drawings, such as layers containing text, can be frozen in one viewport and displayed in another. Options for freezing layers in floating viewports include Freeze in All Viewports, Freeze in Active Viewport, and Freeze in New Viewport. These options are inactive as the Layer Properties Manager is displayed, but are activated as a layer is highlighted. The Layer Properties Manager can also be used to turn plotting ON or OFF for visible layers. A layer created for interoffice notes can be displayed on the screen, but frozen for plotting. Layers are frozen for plotting by selecting the Plot icon of the desired layer.

Controlling Scaling in Viewports

Altering the size of a floating viewport does not change the scale of the drawing within the viewport. When the scale was adjusted for the viewport, you set a

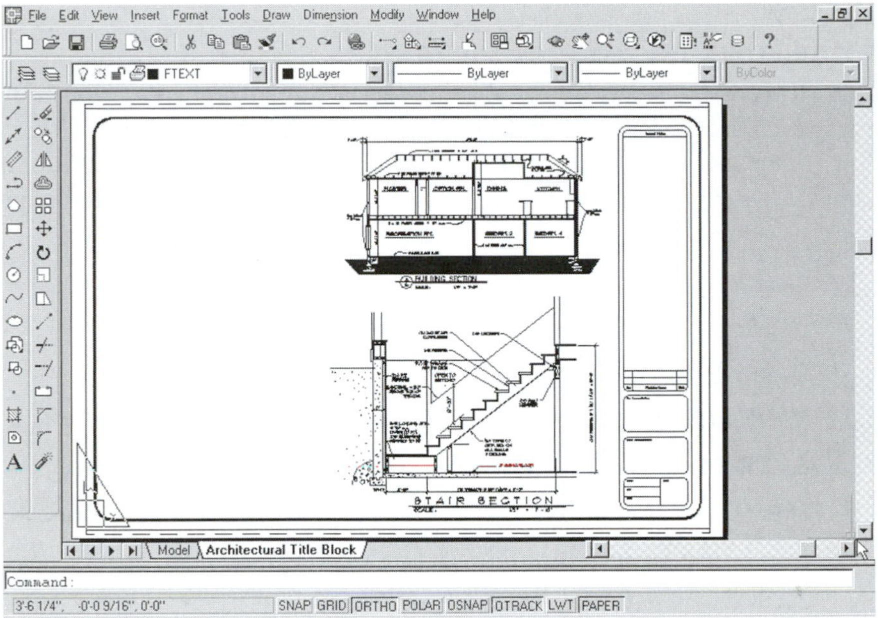

Figure 24.26 *Two of the viewports from Figure 24.24 have been toggled to the OFF setting. As the viewport is toggled OFF, the contents of the viewport will be removed from the drawing display.*

consistent scale for accurate display and plotting. Earlier in the text, selecting a scale for linetypes was discussed. If a hidden line is to be 1/8" long when plotted at 1/4" = 1'-0", a scale factor must be applied to the line to create the desired result. The scale for linetypes can be based on the drawing units of the space where the object was created, or based on paper space units. The PSLTSCALE system variable can be used to maintain a uniform linetype scaling for objects displayed at different zoom factors. The PSLTSCALE variable can be set by:

- Select Linetype from the Format pull-down menu.
- In the Linetype Manager, select Show Details.
- Enter the desired global scale factor to apply to the linetypes in the Global Scale Factor edit box.
- Select OK to return to the drawing area.

CREATING VIEWPORTS WITH IRREGULAR SHAPES

Another new feature of AutoCAD 2000 is the ability to create viewports with irregular shapes. Users of previous AutoCAD releases will appreciate that viewports are no longer limited to rectangular shapes. AutoCAD allows viewports to be created in an irregular shape that best suits the object to be displayed. Nonrectangular viewports can be created by selecting the Polygonal Viewport icon, by clipping an existing viewport, or by converting an existing object into a viewport.

CREATING POLYGONAL VIEWPORTS

A polygonal viewport is created by selecting the Polygonal Viewport icon from the Viewports toolbar, by selecting Polygonal Viewport from Viewports of the View pull-down menu, or by entering **-VPORTS** (EnterKey) at the command prompt. The command is completed by selecting a series of points using a command sequence similar to the sequence for creating a polyline. The command sequence is:

Command: *(Select Polygonal Viewport icon.)*

_-vports

Specify corner of viewport or

[ON/OFF/Fit/Hideplot/Lock/Object/Polygonal Restore/2/3/4/ <Fit>: _p

Specify start point: *(Specify viewport starting point.)*

Specify next point or [Arc/Close/Length/Undo]: *(Select points to define polygon.)*

Specify next point or [Arc/Close/Length/Undo]: *(Select points until the polygon is defined. As the polygon is closed, the command will end with the following prompt.)* Regenerating model

Command:

Figure 24.27 shows an example of an irregular shaped viewport.

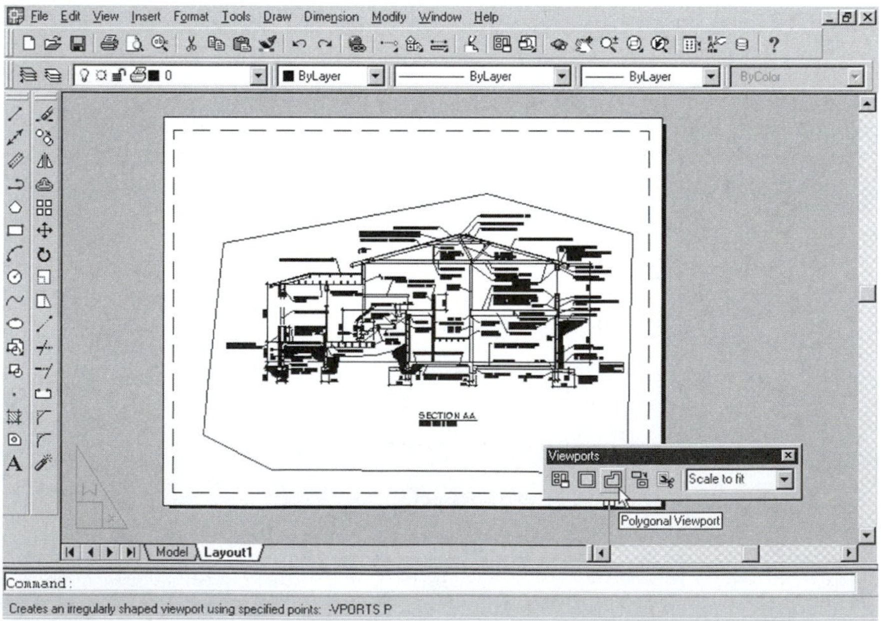

Figure 24.27 *A polygonal viewport can be created by selecting the Polygonal Viewport icon from the Viewports toolbar.*

USING VPCLIP TO ADJUST THE VIEWPORT

The VPCLIP command can be used to clip or redefine the shape of an existing viewport. The command can be accessed by selecting the Clip Existing Viewport icon from the Viewports toolbar, or by entering **VPCLIP** (EnterKey) at the command prompt. Selecting the viewport to be accessed, right-clicking to display the shortcut menu, and then selecting Viewport Clip will also access the command. The command sequence is:

> Command: *(Select Clip Existing Viewport icon.)*
>
> Select viewport to clip:
>
> Select Clipping object or [Polygonal] <Polygonal>:
>
> Specify start point: *(Specify viewport starting point)*
>
> Specify next point or [Arc/Close/Length/Undo]: *(Select a minimum of three points to define polygon.)*
>
> Specify next point or [Arc/Close/Length/Undo]: *(Select points until the polygon is defined.)*
>
> Command:

Objects that lie within the polygon will be displayed. Objects outside the polygon will be removed from the viewport and will not be displayed. The viewport in Figure 24.28 has been edited using VPCLIP. Irregular viewports can be edited by selecting the edge of the viewport. With the viewport grips displayed, the viewport can be edited in the same method any other object is altered with grips. (See Figure 24.29.)

CONVERTING EXISTING OBJECTS INTO A VIEWPORT

AutoCAD allows existing shapes to be converted into viewports. The object can be a circle, ellipse, region, spline, or closed polyline comprised of lines and arc segments. The command is accessed by selecting the Convert Object to Viewport icon from the Viewports toolbar. The command is completed using the following prompt:

> Command: (Select the Convert Object to Viewport icon from the Viewports toolbar.)
>
> Specify corner of viewport or
>
> [ON/OFF/ Fit/Hideplot/ Lock/Object/Polygonal/Restore/2/3/4] <Fit>:
>
> Select object to clip viewport:

Options include:

- ON/OFF—These serve as toggle switches to display or hide the contents of the individual viewports. Selecting either option will provide a prompt to "Select

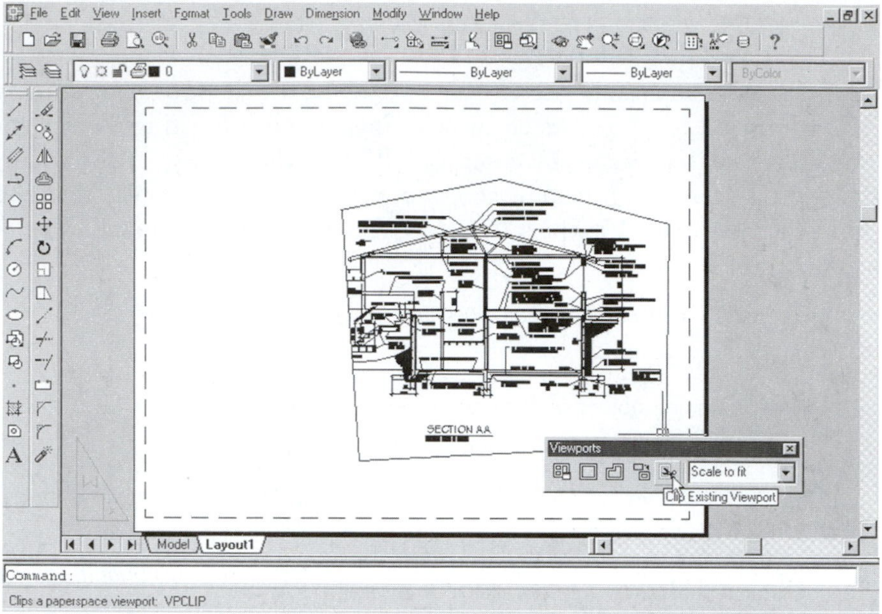

Figure 24.28 *The VPCLIP command can be used to clip or redefine the shape of an existing viewport. The command can be accessed by selecting the Clip Existing Viewport icon from the Viewports toolbar.*

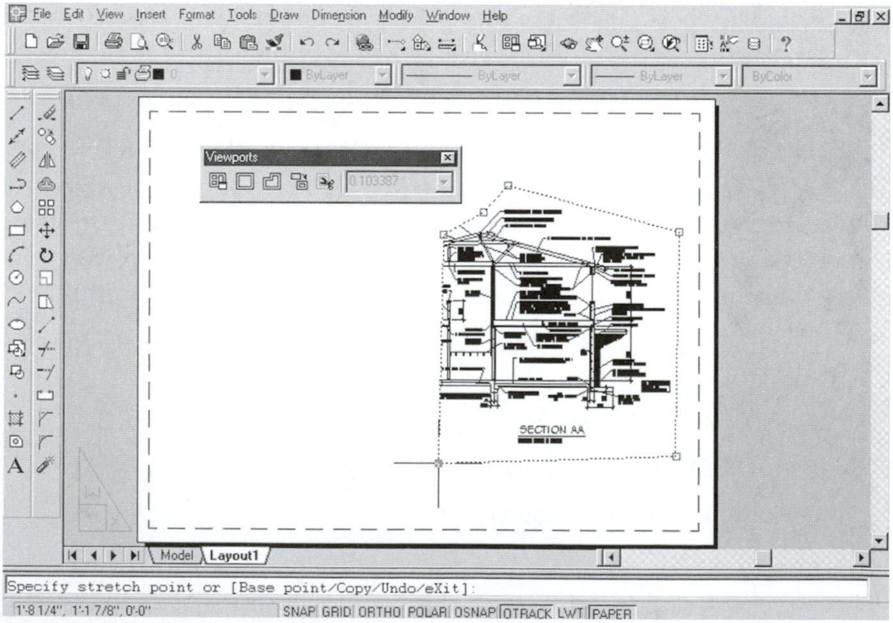

Figure 24.29 *A viewport can be edited by selecting a viewport grip.*

objects" so the desired viewport may be controlled. This option can be very useful when plotting stock sheets of details, if one or two attached details do not apply to the current use. Setting these viewports to OFF will eliminate the viewport from the plot. This option is also very useful for controlling REGEN time. Inactive viewports can be set to OFF while the current viewport is edited. If all viewports are set to OFF, you will not be allowed to work in model space of a layout until a new viewport is created.

- Fit—This option will create a viewport to fit the current screen display. The size of viewport will be determined by the dimensions of the paper space view. In the example used throughout this chapter, the objects were created first in base drawings and then drawings were inserted into a template. The FIT option will allow a viewport to be created around an existing drawing or a portion of a drawing.
- Hideplot—This option will allow hidden lines to be removed from selected viewports when plotting in paper space. This option generally is not required in architectural and engineering drawings.
- Lock—Locks the current viewport in a similar manner to locking a layer.
- Object—Selects a circle, closed polyline, ellipse, region, or spline to be used as a viewport.
- Polygonal—Creates an irregular shaped viewport using the same process that is used when the icon is selected.
- Restore—This option can be used to restore saved viewport layouts. The prompt is:

  ?/Name of window configuration to insert <default name>:

  The name of a saved view can be entered. Typing the question symbol (?) will provide a list of view names.
- 2/3/4—Selecting one of these options will create two, three, or four viewports without having to reissue the command. The required number of viewports is entered, followed by specifications for how to arrange the viewport. Each option is similar to the arrangements used for tiled viewports.

CHAPTER 24 EXERCISES

1. Open the Architectural template and create a new layout titled COOL STUFF. Use the viewports dialog box and create viewports using the "Three-above" layout. Insert a drawing into model space. Adjust each viewport so that a different aspect of the drawing is centered in each viewport. Save the drawing as E-24-1.
2. Open the Architectural template and use the layout wizard to create a new layout titled VIEWPORT TEST. Set the printers as the default system printer, and if necessary select an appropriate paper size. Set the orientation to portrait. Assign a single viewport suitable for displaying a floor plan at 1/4" = 1'-0". Select the corners of the viewport so that it will only fill half of the page. Save the drawing as E-24-2.
3. Open any three drawings and insert them to a template drawing. Create a floating viewport to display the drawings at a suitable scale. Save the drawing as E-24-3
4. Open the Architectural drawing template. Attach any four drawings from your library. Each drawing is to be shown at a different scale when plotted. Provide a title and the scale for each detail. Save the drawing as E-24-4.
5. Open drawing E-24-4 and clip two of the drawings so that they are not displayed. Save the drawing as E-24-5.

CHAPTER 24 QUIZ

1. Describe how a referenced drawing can be used in an office to complete a project.
2. Use the Help menu and determine how and what the TILEMODE setting affects.
3. What layer is added to the drawing base when working with viewports. How will it affect the display?
4. What command or process is used to create a viewport?
5. What are the benefits of using layouts?
6. How can the display in the second viewport be altered so that it does not display the same drawing as another viewport?
7. How does paper space differ from model space?
8. A drawing has been inserted into a template in model space but it can't be seen.
9. List the steps to create a layout using the Layout wizard.
10. Explain the difference between floating and tiled viewports.
11. How is a scale applied to the display in a viewport?
12. List three options for controlling the display of the viewport.
13. A drawing is displayed at the proper scale in a viewport, but the entire drawing can't be seen. Describe how the problem can be fixed.
14. Three drawings are displayed in a viewport, but you would like to hide one from a meeting with a client. Describe how to hide the drawing.
15. Describe the process to create an irregular shaped viewport.

CHAPTER 25

Controlling Output

Chapter 24 walked you through the process of taking a drawing created in model space and placing it in a layout for plotting. As computers become more common in contractor and subcontractor offices, more construction projects will be transmitted between offices electronically. AutoCAD allows an electronic plot to be created that is accessible from the World Wide Web. This allows the designers for the subcontractors to have access to areas of the project that require contributions from them. In addition to providing access to the drawings, electronic communication can save many hours by eliminating the need to produce paper copies. Unfortunately, some offices have not made the transition to CAD and still require paper copies. This chapter will introduce methods of controlling the plotting by controlling the plotting device, assigning plot styles, and configuring the drawing to be plotted. The commands to control these three areas can be accessed from the Plot dialog box. The dialog box consists of Plot Settings tabs. The Plot Device tab is used to configure plotters, and to configure plot style tables. The Plot Settings tab is used to configure plotting parameters. The Plot dialog box can be accessed by picking the Print icon from the Standard toolbar, by typing **PLOT** (EnterKey) at the command prompt, or by selecting Print... from the File pull-down menu. Each method will produce a Plot dialog box similar to Figure 25.1. The Plot dialog box can be used to control either a printer or a plotter. Printing was introduced in Chapter 6.

Each time a new layout was entered, the Page Setup dialog box was displayed. The Page Setup dialog box was first introduced when Layout tabs were first examined in Chapter 5. Although it was glossed over when the initial drawing process was started, as you gain experience it should be used to define plotting parameters. When the drawing is complete, it will be plotted based on the specified parameters during the page setup. As you gain experience, plotting parameters can be specified and saved within a layout. The Plot dialog box is similar to the Page Setup dialog box. Plotting parameters assigned in the Plot dialog box will override those set in the Page Setup dialog box. Chapter 6 introduced the skills needed to create a simple check print. This chapter will explore each of the plotting parameters so that you will be skilled at

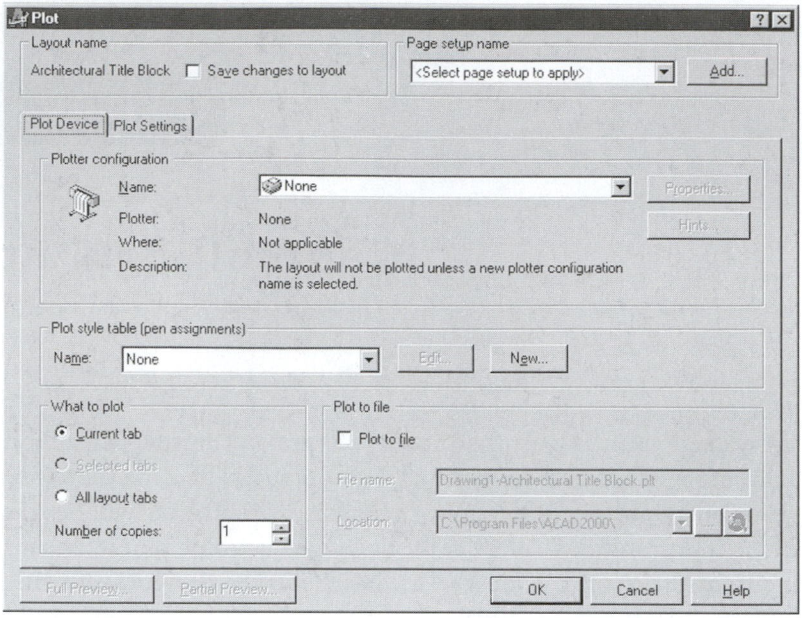

Figure 25.1 *The commands to control plotting can be accessed from the Plot dialog box. The dialog box consists of Plot Device tabs and Plot Settings tabs. The Plot Device tab is used to configure plotters, and to configure plot style tables. The Plot Settings tab is used to configure plotting parameters. The Plot dialog box can be accessed by picking the Print icon from the Standard toolbar, by typing* ***PLOT*** *(EnterKey) at the command prompt, or by selecting Print... from the File pull-down menu.*

defining plotting requirements in a page setup, and be able to plot scaled copies of the drawings contained in each layout.

Note: It's important to keep sight of the goal as you begin this chapter. Your goal is to produce a plot at a useable scale. Since Chapter 6, you've been making scaled plots using the Scaled to Fit option. This produced scaled drawings, but the scale was most likely a scale known only to AutoCAD and not the rest of the known world. In the industry most drawings are plotted at a scale such as $\frac{1}{4}$" = 1'-0". AutoCAD produces a warning as the Plot dialog box is entered that may seem a bit intimidating for new users. AutoCAD has added many aids and wizards to help make plotting easier, but at the same time has made the initial process seem more complex. Think of yourself in a new, top-of-the-line, bright red, convertible sports car. All you want to do is drive—you'll read the manual when the thrill wears off. The purpose of Chapter 6 was to get you driving. You've taken AutoCAD plotting on a test spin. This chapter will introduce the concepts needed to produce a drawing at a standard scale. Once the process to produce a plot has been explored, you'll be introduced to the many bells and whistles that have been provided by AutoCAD to enhance plotting.

AN OVER VIEW OF PLOTTING

A scaled plot can be created easily using the Plot dialog box with the same procedure that was introduced in Chapter 6. As the dialog box is entered you'll be prompted to specify the parameters that will define the plot. Once in the Plot dialog box, the plotting process will include the following steps:

- If a named page setup has been previously defined and saved, it can be selected from the Page Setup Name list.

  Using the Plot Style Tab:

  Verify the desired plotter is current.

  Verify that the correct plot style table (pen assignments) is attached to the drawing.

- Use the Plot Settings tab to define the plotting parameters. Settings include: Paper Size, Plot Area, Drawing Orientation, Plot Scale, Plot Offset, and Plot Options.
- Select a Preview Option.
- Press (EnterKey). If the preview does not provide the expected results, redefine the plotting parameters on each tab as needed and repeat the preview process.
- When the plot preview meets your expectation, select the OK button to produce the plot.

EXPLORING THE PLOT DIALOG BOX

The Plot dialog box consists of two tabs and three areas that are common to each tab. At the top of the dialog box are displays that describe the Layout and Page setup names. At the bottom of the display are buttons to control preview options and buttons to continue or cancel the plotting process. The tabs are used to control the plotting devices and the plot settings.

LAYOUT NAME DISPLAY

The Layout Name area of the Plot dialog box displays the name of the layout to be plotted. In Figure 25.1 the layout displayed is the Architectural Title Block. The selected layout names will be displayed in this area if multiple layouts are selected to be plotted. Model space will be displayed if the Model layout is the current viewport. Model space will produce a plot of the drawing, without displaying the border and title block. For best results, select a layout that will display the drawing with a title block.

Saving Plotting Specifications

AutoCAD allows the plotting parameters that are defined in the Plot dialog box to be attached to the layout. The next time you need a plot of the selected drawing, instead of setting the parameters of the two tabs, you only need to select the OK button of the Plot dialog box. With this box active, changes made to each tab will be saved in the current layout. This option is not available if multiple layouts are selected for plotting.

PAGE SETUP NAME

This portion of the Plot dialog box displays a listing of named and saved page setups. Attempting to plot a drawing displayed in the Architectural title block would include the following options: Previous Plot, Architectural Title Block, and Architectural English-no output device.

Add

Selecting the Add... button will display the User Defined Page Setups dialog box. The display is used to name the plot setup that is to be saved to the layout. In addition to naming the current setup, this option can be used to rename, delete, or import user defined setups and apply them to the current drawing.

PREVIEW OPTIONS

Once all of the plotting parameters have been set, the preview options can be used to view the outcome of the plot before the plot is produced. Options include the Full and Partial previews. Each will be explored in detail after all of the settings have been explored. A full preview will display the drawing just as it will appear when plotted. A partial preview will show a box that defines the limits of the objects to be plotted inside of a box that defines the plotting area. A display will be given in the Partial Plot Preview dialog box to define Paper Size, Printable Area, Effective Area, and Warnings. The Warnings area displays a list of warnings that will affect the effective plotting area. The other three categories can each be set using the Plot Settings tab.

EXPLORING THE PLOT DEVICE TAB

As the Plot dialog box is entered, the Plot Settings tab is displayed by default. Once a plotter is configured to your workstation, the Plot Device will not need to be accessed for each plot. The key elements of the Plot Device tab include settings for the current plotter and settings for using a plot style table. The plotting devices that are configured to your workstation or network can be used by AutoCAD to plot drawings. As you've made check prints from model space, AutoCAD has used the plotter specified in Options to determined the default plotter. As you plot using a layout, the plotting device specified on the Layout tab is used. The layout stores settings for plot devices, the plot style table, print area, rotation, paper size, and scale. These settings can be provided as the drawing is started, and stored as a named page setup. When the drawing is ready to be plotted, these values can used to control the plot production. The Plot Manager controls existing plotters, modifies existing configurations, and adds new plotter configurations. Users of previous releases of AutoCAD will appreciate the elimination of command line prompts and the addition of a Plot Configuration wizard to configure plotters.

DEFINING PLOTTING PARAMETERS

The Plot Device tab displays the settings and controls for the plotter to use, a plot style table, the layout to be plotted, and information for saving the drawing to a file. The major areas of the Plot Device tab include the Plotter Configuration, Plot Styles Table, What to Plot, Plot to File, and five buttons common to both tabs. The Plotter Configuration area of the tab displays the current plotting device and its operating parameters.

Plot Configuration

The Plotter Configuration box displays the plotting device that is currently being used and the port it is configured to. Selecting the Name edit box will display a menu similar to Figure 25.2. The list shows each available plotter that is connected to either the workstation or the network. Notice that an icon is displayed by each listing to distinguish between system devices and those stored with a PC3 file name. A system printer or plotter is a device that is configured in the Printer folder of My Computer, or added by selecting Printer from the Settings menu. Selecting the Add Printer option from the Printer Menu of Windows will produce a prompt for Add Printer

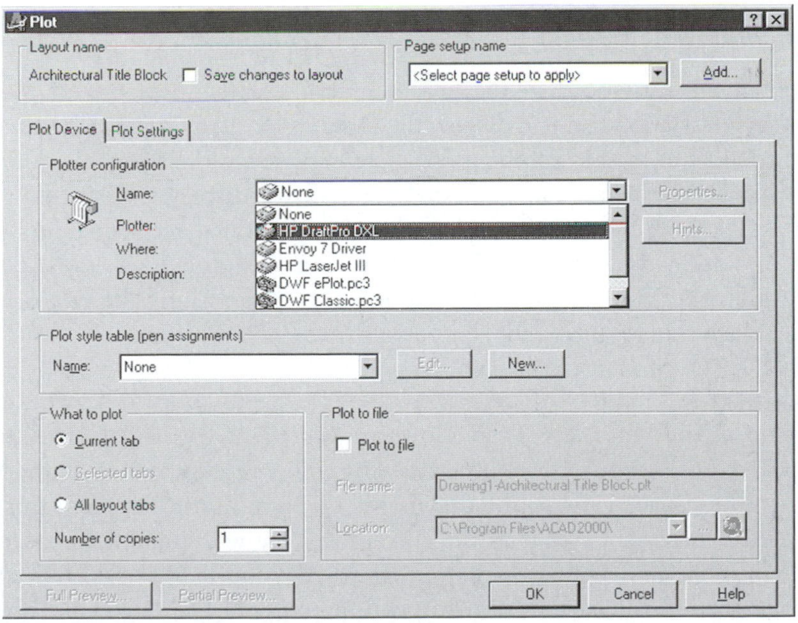

Figure 25.2 *Selecting the Name edit box will display a list of each available plotting devices that are connected to either the workstation or the network.*

Wizard and allow a printer to be added and accessed by other programs. Previously configured printers can be selected using the Device and Default Information box of the Print/Plot Configuration dialog box. Notice in Figure 25.1, that the current plotter is the DraftPro DXL. Methods of configuring a printer or plotter and adding it to this list will be introduced later in this chapter. Selecting the Properties button will display a Plotter Configuration (PC3) editor. The editor is used to modify or view the current plotter parameters. Selecting the Hints… button will display information on improving the plot quality specific to the configured plotting device.

A PC3 file is a file created by AutoCAD that is drawing independent. PC3 files are just like any other file in that they are portable, and can be shared between different workstations. A PC3 file contains all of the information on what to plot and how to plot, including the specified plotting device, paper size, and pen information. For users of previous releases of AutoCAD, PCP and PC2 files can be imported in to AutoCAD 2000. These older formats will automatically be converted into PC3 format.

Plot Style Table (Pen Assignments)

If you've used an older release of AutoCAD, part of configuring the plotter was to configure pen assignments. AutoCAD 2000 replaces pen assignments with plot styles. Just as a style can be assigned to text or dimension, AutoCAD allows a style to be assigned to each layout and model space view. This portion of the Plot Device tab is used to create, edit, or set the plot style table assigned to the current model and layout tabs. Selecting the Name edit box displays a list of available styles. Later in this chapter you'll be introduced to the procedure to create and plot with plot styles. Selecting the New… button will display the Add Color-Dependent Plot Style Table Wizard display. The wizard can be used to create a new plot style table. Once a plot style table has been created, it will be listed in the Name display. Notice that in Figure 25.3, the listing shows several plot styles, even though none have been defined. These are plot styles defined by AutoCAD. Most architectural drawings can be plotted by selecting either the Monochrome.ctb, Default R14 pen assignment.ctb, or the R14 cfg-1-Default System Printer.ctb.

What to Plot

This portion of the Plot Device tab allows you decide if you will plot using model space or one of the layout tabs of the drawing. The default setting is the Current tab. If the model space view of the drawing is displayed on the screen, this is what will be plotted. Setting a layout as the current display will make it the current tab, and select it as the drawing to be plotted. If All Layout Tabs is selected, all of the layouts contained in the current drawing will be plotted. The Selected Tabs option is currently inactive when only one tab is selected for plotting. To make this option active, hold down (CtrlKey) and select the tabs to be plotted prior to entering the

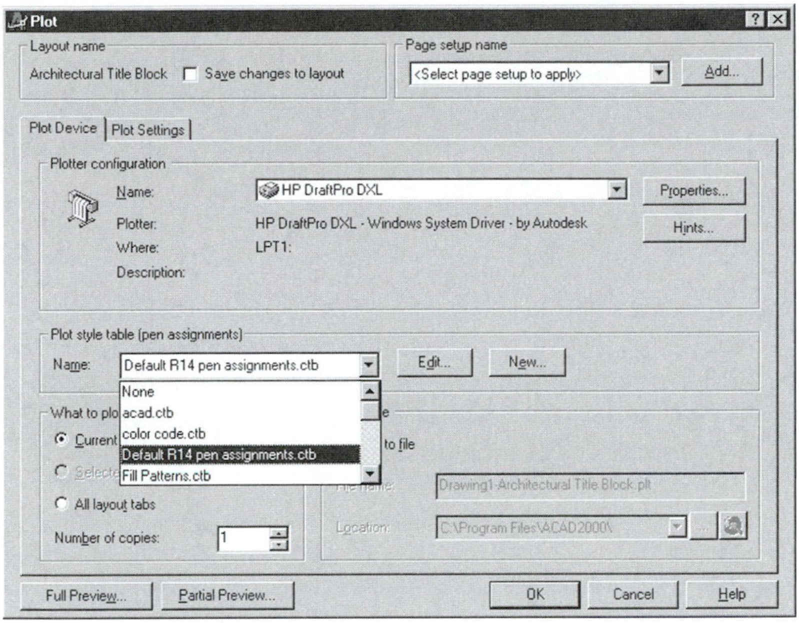

Figure 25.3 *The Plot Style Name listing shows several plot styles defined by AutoCAD. As you create new styles they will be added to the list.*

Plot dialog box. The Number of Copies setting determines the number of plots to be made of each layout.

Plot to File

When you complete all of the options on both tabs of the Plot dialog box, the information will be sent to the plotting device and the information will be processed. The time required to produce the plot will depend on the size of the drawing file and the type of plotter. If the Plot to File radio button is selected, the plot will be stored as a .PLT file instead of being sent to the plotting device. Saving a plot file allows the drawing to be plotted at a more convenient time. As the Plot to File button is selected, the other options in this box will become active. A name such as Floor-Layout1 will be displayed, indicating the drawing and layout names. The Location edit box shows the location of the current drawing. The PLT file will be stored in the same location as the drawing to be opened. Selecting the [...] icon will display the Browse for Folder dialog box with a display similar to the Explorer display. The dialog box allows the storage location of the plot file to be specified using methods similar to that used by other dialog boxes for saving files. The file will be stored to disk with the indicated name followed by the .PLT extension. Any storage device in your machine or accessible to the network can be used to store the plot file. Selecting the Browse the Web icon allows the drawing to be stored at a specified web site.

EXPLORING THE PLOT SETTING TAB

The Plot Settings tab is used to control the paper size, plot area, drawing orientation, plot scale, plot offset, and plot options. Each area can be seen in Figure 25.4.

PAPER SIZE AND PAPER UNITS

This portion of the Plot Settings tab displays the configured plotter, the current paper size, the printable area, and toggles from inch to metric. The plotter that was specified in the Plot Device tab is displayed as the Plot Device. This display indicates the standard plotting sizes that can be accommodated by the current plotting device. Selecting the edit arrow will display a menu similar to Figure 25.5. The menu will vary, based on the configured plotter. The paper size to be used for the plot can be selected from the menu. The sizes displayed are the actual paper sizes. Sizes are listed as width (X) × height (Y). Once the desired value is entered, the paper size in the Paper Area will be updated. Notice in Figure 25.4 that with 22" x 34" paper selected for plotting, an area of 31.68 x 20.62 is listed as the printable area. This is the maximum area available based on the current paper size. The final option of this area of the tab is to select the plotting units. Options include Inches and MM.

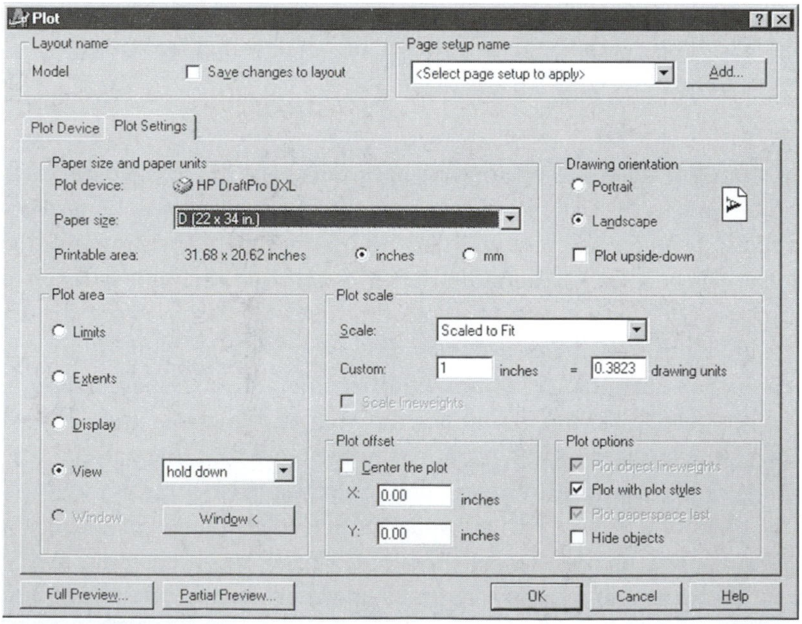

Figure 25.4 *The Plot Settings tab is used to control the paper size, plot area, drawing orientation, plot scale, plot offset, and plot options.*

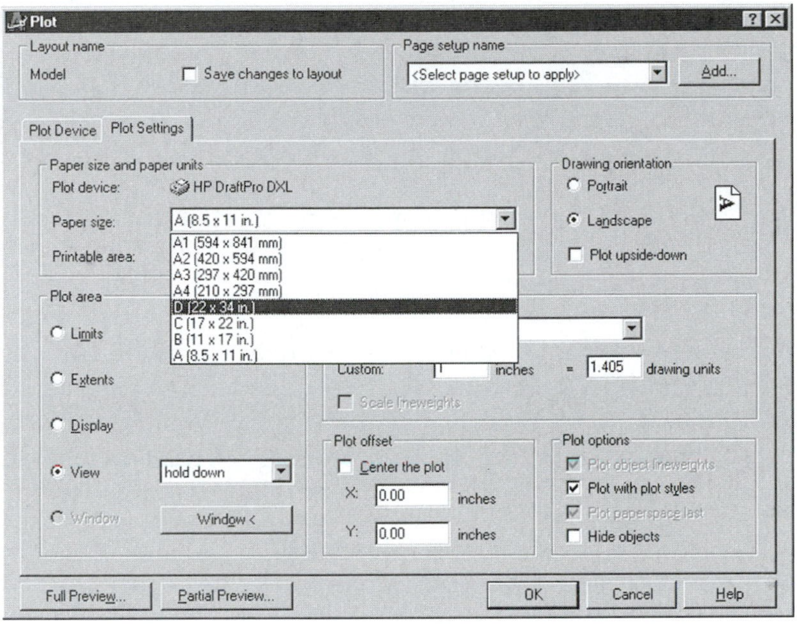

Figure 25.5 *The plotter that was specified in the Plot Device tab is displayed as the Plot Device. The Paper size menu indicates the standard plotting sizes that can be accommodated by the current plotting device. The menu will vary, based on the configured plotter.*

THE PLOT AREA

This area of the Plot Settings tab provides options for describing what portion of the current drawing will be plotted. Because the entire drawing often does not need to be plotted, five options are given to decide what portion of the drawing will be plotted.

Layout

This option will plot everything within the margins of the selected paper size. This will be the option of choice when the entire contents of the layout need to be plotted.

Extents

Choosing this option will plot the current drawing with the extent of the drawing objects as the maximum limits that will be displayed in the plot. The option is similar to the Extents option of the ZOOM command. Using this option may eliminate objects at the perimeter of the drawing because the plotter cannot reach the edge of the plotting surface.

Display

This setting plots only the material that is currently displayed on the screen. If you have zoomed into a drawing, only that portion of the drawing will be plotted.

View

If specific views have been saved using the VIEW command, the View option for plotting will be activated. This option will allow specific views to be plotted as they were saved. In the initial setting, the view radio button is inactive. When a drawing contains named views, the View button will be active and a listing showing each saved view name will be displayed. Selecting the desired view to be plotted will highlight that view, place the name in the edit box, close the menu, and allow the Plot menu to continue.

Window

This option will allow a specific area of a drawing to be defined for plotting. In the same way that a window is used to zoom or select objects for editing, a window can be specified here to define the selection plot. Initially the Window radio box is inactive. By selecting the Window< button, the Plot dialog box will be removed from the display area, and prompts for defining the window will be displayed. The prompt is:

> Specify first corner: *(Select first window corner.)*
>
> Specify other corner: *(Select opposite window corner.)*

Once the window is defined, the Plot dialog box will be redisplayed.

DRAWING ORIENTATION

This area of the Plot Settings tab allows the drawing orientation to be specified. The box on the right side of the display is the orientation icon and not a selection box. The current orientation of the box is based on the configured plotter and indicates the drawing will be plotted in the normal orientation for construction drawing. This setting is referred to as *landscape*. When landscape is used, the A will be placed on its side, indicating the print will be read from the long edge. When the plot is rotated so that it is read from the title block end, the plot is referred to as a *portrait* plot. The A will be in the vertical position indicating the print will be read looking from the title block end. A portrait plot is more likely to be used with laser printers. The orientation of the drawing to the paper can be adjusted by picking the appropriate radio button. A check box is also available for plotting the drawing upside-down. Selecting this option will rotate the portrait or landscape layout 180°.

PLOT SCALE

The Plot Scale area of the Plot Settings tab controls settings for the scale and lineweight scale. As drawings have been created throughout this text, they've been drawn at real size rather than at a reduced scale, as would be done in manual drawing. In Chapter 24 you were introduced to methods of creating layouts for plotting. If you're plotting a layout, the scale was set as the layout was created. This will allow the scale factor to be at 1:1. If you want to make a plot of a drawing created in model space, you'll need to provide a scale factor for plotting. Select the Scale edit arrow and the desired scale from the menu to provide a plotting scale. Figure 25.6 shows a portion of the scale menu that lists common scale factors that can be used to produce a plot at a given size. Dimensions and text were both multiplied by a scale factor in anticipation of the scale at which the drawing would be plotted.

An alternative to picking a precise scale is to activate the Scaled to Fit box. This option will allow the desired plot to be sized to fit within the parameters of the paper. The resulting scale will be displayed in the Custom display box. Drawings that are scaled to fit are useful as check prints to determine completeness, but are not suitable for most projects within the construction industry. The Custom box is better suited for entering a scale that is not listed in the menu. The scale is created by entering the number of inches equal to the number of drawing units.

Use the Scale Lineweights button to scale lineweights in proportion to the plotted scale. Lineweights selected during drawing construction normally specify the line width of printed objects, and are plotted with the specified width regardless of the plot scale.

PLOT OFFSET

The Plot Offset portion of the Plot Settings tab contains the Center the Plot toggle button and a display for X and Y coordinates. Activating the Center the Plot button

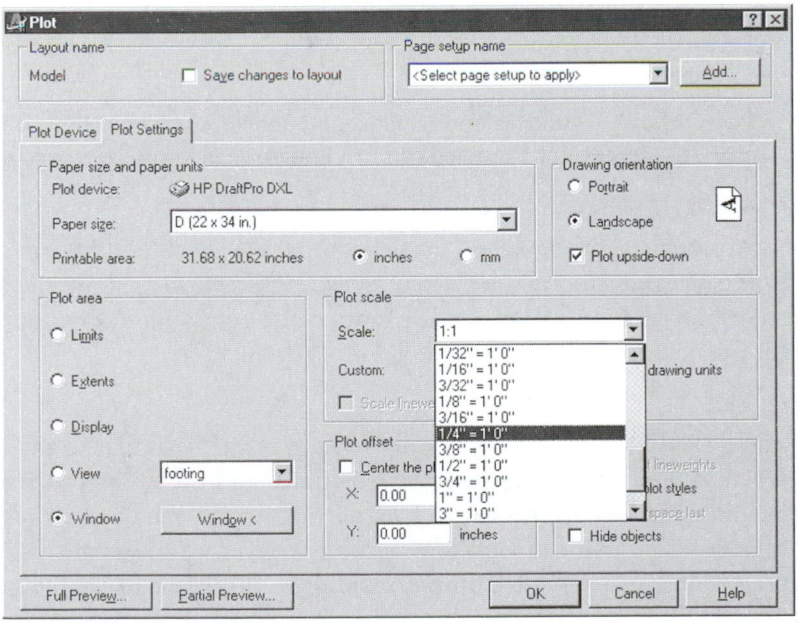

Figure 25.6 *Select the Scale edit arrow to display a list of common scale factors that can be used to produce a plot at a given size.*

will automatically center the plot in the center of the plotting paper. The Plot Offset values specify an offset distance of the drawing to be plotted from the lower left corner of the paper. If you're plotting a layout, the lower left corner of the plotting area is positioned at the lower left edge of the plotting material. The origin of the plot can be offset from the paper edge by entering a positive or negative value. The plotter unit values are in inches or millimeters on the paper. Entering a value in the X edit box specifies the plot origin in the X direction. Entering a value in the Y edit box specifies the plot origin in the Y direction. The default setting for plotting origin is X = 0 – Y = 0. For most plotters this will be the lower left corner of the paper. By selecting the X or Y origin box, new values can be entered. Entering an X value of 12 and a Y value of 10 will place the new origin 12 units to the right and 10 units above the original plotting origin.

PLOT OPTIONS

This portion of the Plot Setting tab contains four options for controlling the outcome of the plot. The options control lineweight, plot styles, and the current plot style table. Plot styles will be introduced in the next portion of this chapter.

- Plot with Lineweights—With this box active, lineweights specified in the drawing will be plotted.
- Plot with Plot Styles—With this option active, plot styles will be applied to objects that are defined in the plot style table. All style definitions with different property characteristics are stored in the plot style tables.
- Plot Paper Space Last—With this option active, model space geometry will be plotted first.
- Hide Objects—With this option active, hidden lines will be removed from layouts. Viewports can be used to remove hidden lines for model space drawings.

PREVIEWING THE PLOT

You've now explored each tab of the Plot dialog box. Once values nave been entered for each option, the drawing can be plotted. Selecting Preview... allows the outcome of the current plot settings to be viewed prior to the actual plot. Two options are available for previewing the plot: Partial or Full Preview.

Partial Preview

Selecting the Partial Preview... button will produce a display similar to the one seen in Figure 25.7. A white box represents the paper size and the limits of the effective plotting area are specified on the screen by a dashed line. A blue box represents the area that the drawing will fill. A warning also will be given of any problems that may be encountered during the plot. Common warnings displayed in the warning box include:

Effective area too small to display

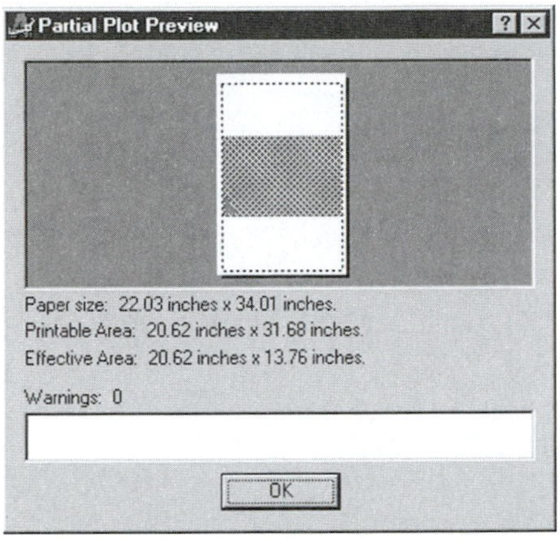

Figure 25.7 *Selecting the Partial Preview... button will produce a display of the area to be plotted. The white box represents the paper size, and the limits of the effective plotting area are specified by a dashed line. A blue box represents the area that the drawing will fill.*

Origin forced effective area off display

Plotting area exceeds plotter maximum

A small red triangular Rotation Icon is displayed in the upper left corner to represent a 0 rotation. The triangle will be moved, depending on the Landscape/Portrait/Upside-down selection. Selecting OK will restore the plot configuration dialog box.

Full Preview

Choosing this option will produce a graphic display of the drawing as it will appear when plotted. Figure 25.8 shows an example of a full preview. A few seconds will elapse between picking the Full Preview... button and the display of the proposed plot. An outline of the paper size will be drawn on the screen, which displays the portion of the drawing to be plotted. Once the drawing is displayed, the Real time Zoom icon is displayed. The Pan and Zoom option can be quite useful in viewing detailed areas of the drawing a final time prior to plotting. Once the indicated Pan or Zoom is performed, press (EnterKey) to restore the Plot Configuration dialog box.

PRODUCING THE PLOT

The two tabs of the Plot dialog box have been examined. If you change your mind on the need for a plot, select the Cancel button. The dialog box will be terminated, and the drawing will be redisplayed. If you are satisfied with the plotting parameters that have been established, select the OK box. The dialog box will be removed from the screen and the command prompt will display the effective plotting area. With older pen plotters you may be given a prompt to remind you to position paper in the plotter. Newer plotters use rolls of paper, but they also need to be reloaded occasionally.

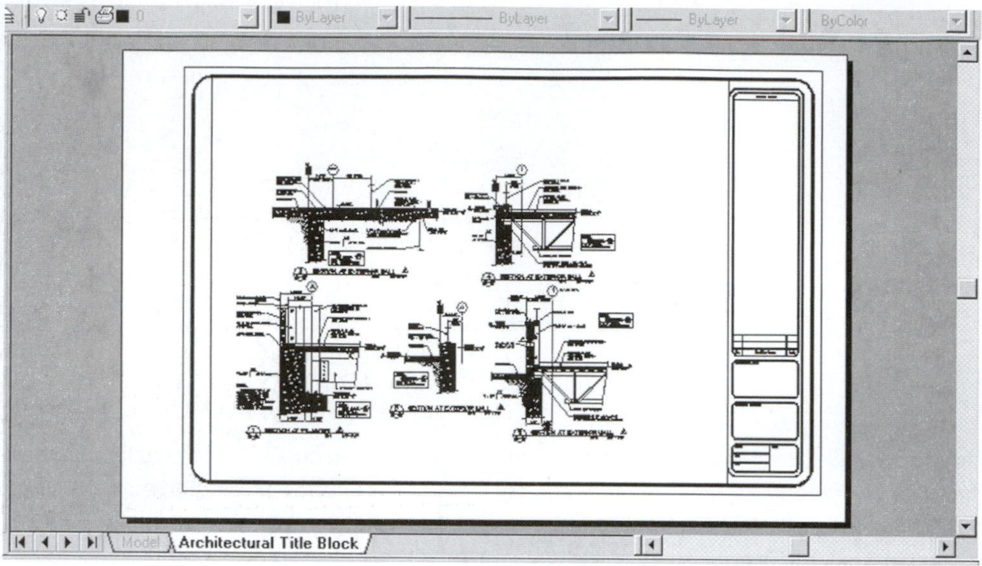

Figure 25.8 *A full preview shows each drawing object as it will appear in the finish plot.*

Don't laugh! After laboring through the Plot Configuration dialog box the first few times, simple things like turning the plotter on and installing the paper may tend to be forgotten. If you need to set up the plotter, you can do it before you press OK. Your plotter may have additional features, such as pen pressure and speed, to be adjusted. The completed plot can be seen in Figure 25.9a and 25.9b.

WORKING WITH PLOT STYLES

Rather than setting pen assignments like past releases, AutoCAD 2000 allows a plot style table to be created and saved as a file. Plot Styles are a new object property assigned to all drawing objects to control the appearance of the finished plot. Plot styles can be used as the drawing is reproduced to override the object properties assigned as the drawing was created. They can be used as overrides for color, dithering, gray scale, linetype, lineweight, pen assignments, screening, end styles, join styles, and fill styles. They can be assigned layer-by-layer or object-by-object. Each aspect of the style is defined in a plot style table that can be attached to the model or layout tab. By assigning different plot styles to the same layout, the drawing can be reproduced in different ways to meet the needs of different clients or subcontractors. The three steps required to use a plot style include:

- Define the plot style using a Plot Style Table.
- Attach the plot style table to the layout.

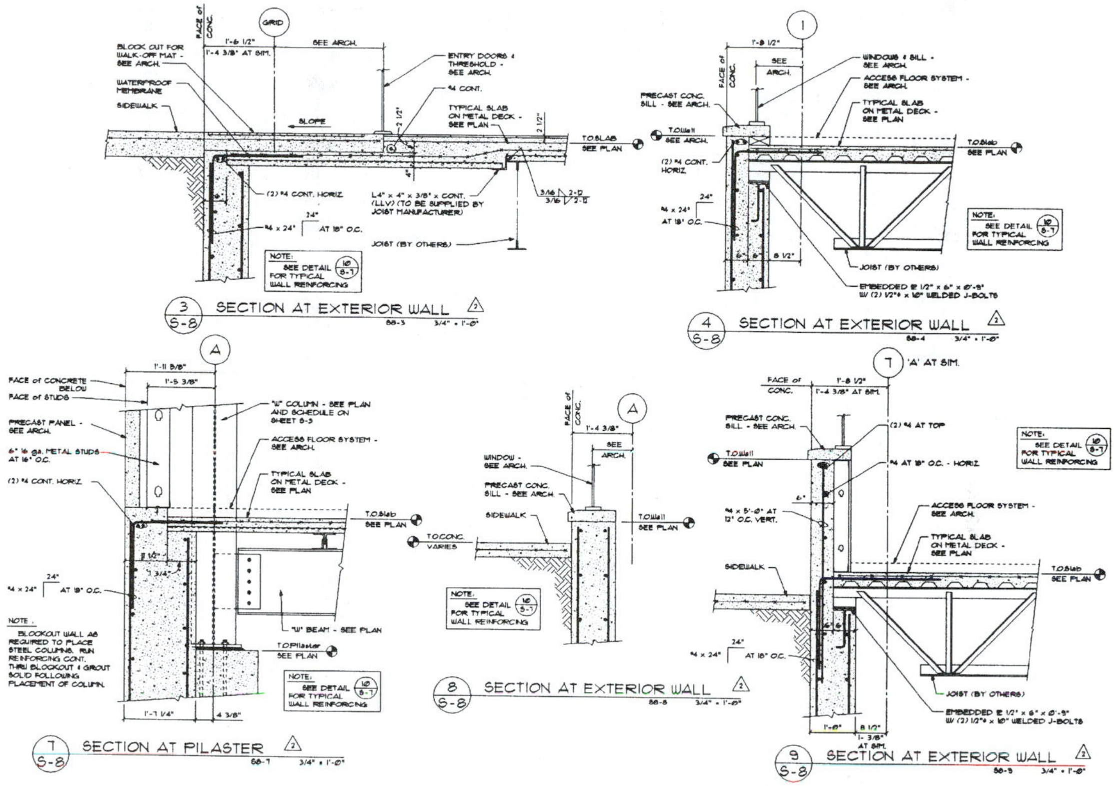

Figure 25.9a *The results of the preview shown in Figure 25.7. (Courtesy Van Domelen/Looijenga/McGarrigle/Knauf Consulting Engineers.)*

- Assign the plot style on an object or layer in the drawing.

The effects of the attached plot style can be viewed on the layout tab.

EXPLORING PLOT STYLE MODES

The two types of styles used for all drawings in AutoCAD 2000 are Color Dependent and Named plot styles. The mode is controlled using the Plotting tab of the Options dialog box. The dialog box is accessed by selecting Options... from the Tools pull-down menu. The mode is controlled using the Default Plot Style Behavior for New Drawings area of the Plotting tab. Users of previous releases of AutoCAD will be most familiar with the Color-Dependent mode.

Color-Dependent Plot Styles

A color dependent plot style controls the plotting based on the color assigned to objects in the drawing. There are 255 color-dependent plot styles based on the 255

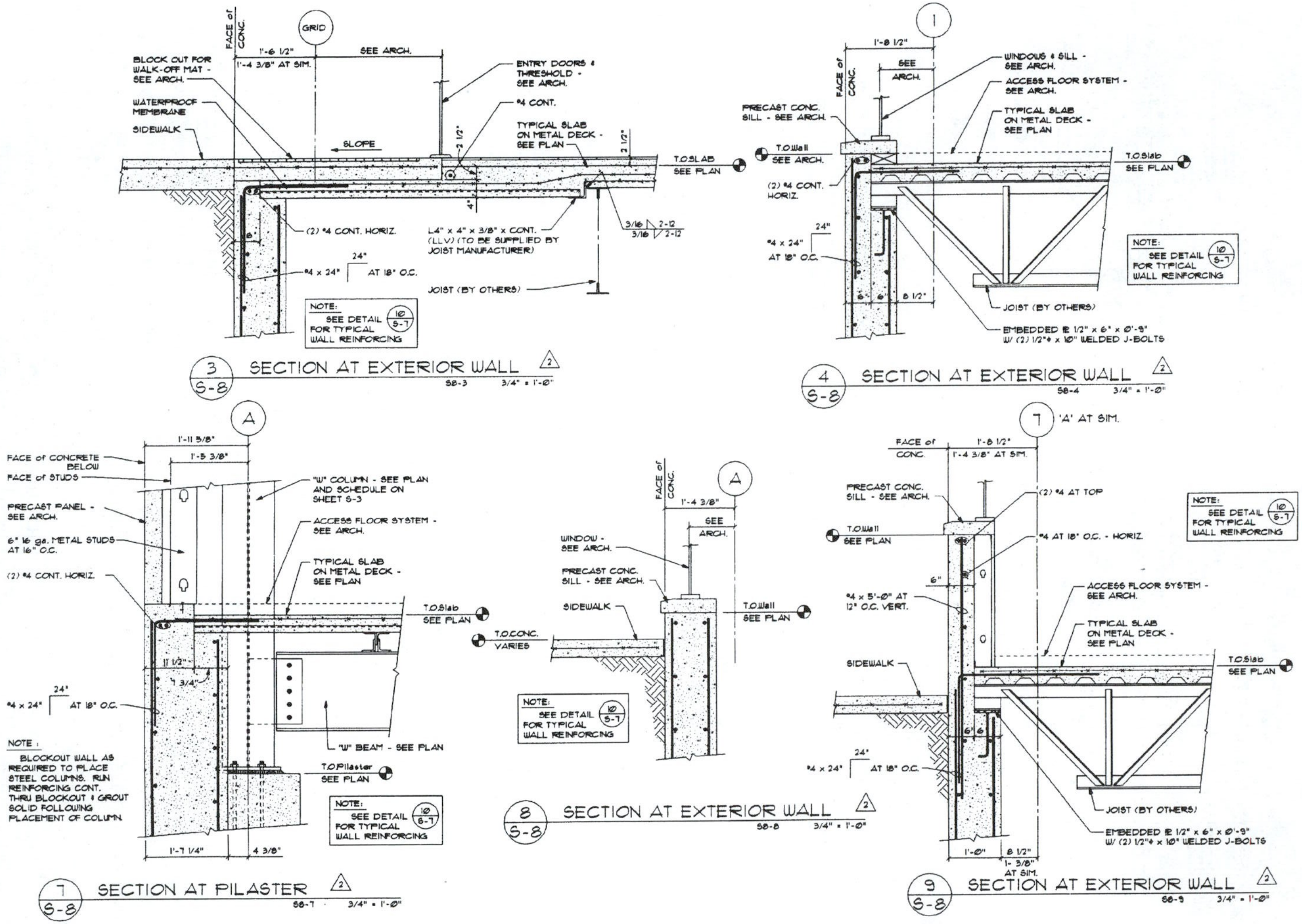

Figure 25.9b *The results of the preview shown in Figure 25.8. (Courtesy Van Domelen/ Loojenga/McGarrigle/Knauf Consulting Engineers.)*

colors in the AutoCAD color index. Color-dependent plot styles are very similar to the pen assignments introduced with AutoCAD Release 14. Key characteristics of a color dependent plot style include:

- Plot characteristics are assigned to each color to determine how the drawing will be plotted.
- Color-dependent plot styles cannot be added, deleted or renamed in a style table.

Using a color-dependent plot style, the appearance of all red drawing objects can be altered by changing the plot style that corresponds to RED. For most of the drawings that you've done in this text, a color dependent plot style will work well. All red lines could be assigned to a specific pen, all blue lines could be assigned to a different pen width. As your drawing become more complex, using the object color could limit the use of color in a drawing. By associating color with a specific pen, the ability to work with a color independent of lineweight or linetype is lost. Creating color dependent plot styles will be explored as the use of the plot wizard is explored.

Named Plot Styles

Named plot styles depends on assigned name rather than the color of an object. Named styles are independent of object colors in the style. Key characteristics of a named plot style include:

- The Normal style assigned by AutoCAD cannot be modified.
- Each defined style in the Plot Style Table has a name, and specific plotting characteristics.
- A named plot style can be assigned to layers or objects.

CREATING A PLOT STYLE TABLE

One of the most common uses of a plot style table is to plot a layout containing a wide variety of colors on a plotter using black lines. The drawing can be plotted without using a plot style table, but each line may be displayed in a gray tone rather than black. By creating a plot style table, each color can be assigned the color black for plotting, and be displayed using the same color intensity. A plot style table can be created using the STYLEMANAGER command. The command can be accessed by selecting the Plot Style Manager... from File of the Standard toolbar or by typing STYLES-MANAGER (EnterKey) at the command prompt. Each method will display the Plot Styles dialog box shown in Figure 25.10. The Plot Styles display can also be accessed by selecting the Add or Edit Plot Styles Tables... button on the Plotting tab of the Options Dialog box. Each of the available plot style tables is displayed in the Plot Style dialog box by use of a name and an icon. Selecting the Add Plot Style Table icon will produce the starting display of the wizard shown in Figure 25.11. After reading the introductory text, select the Next button.

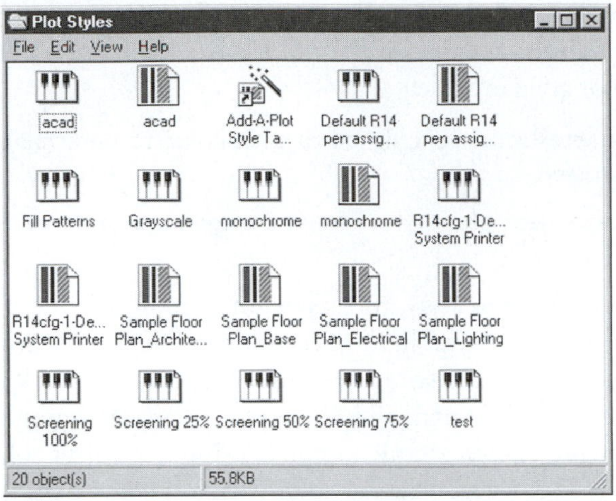

Figure 25.10 *A plot style table can be created using the STYLESMANAGER command. The command can be accessed by selecting the Plot Style Manager... from File of the Standard toolbar or by typing* ***STYLESMANAGER*** *(EnterKey) at the command prompt. Each method will display the Plot Styles dialog box.*

Figure 25.12 shows the second display prompt of the Plot Style wizard. By default, you'll be starting the plot table from scratch. The dialog box also allows options for:

- Use an Existing Plot Style Table—this option creates a new plot style using an existing style as a starting point with similar methods that are used to start a drawing, or dimension style.

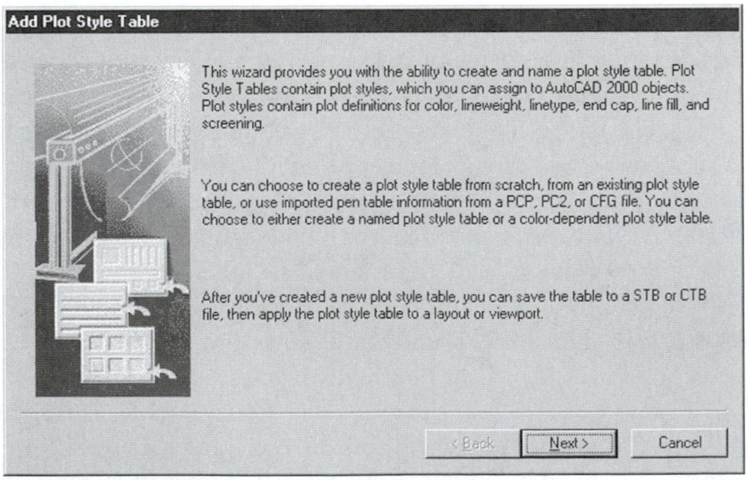

Figure 25.11 *Selecting the Add Plot Style Table icon will produce the starting display of the wizard. After reading the introductory text, select the Next button to proceed through the setup.*

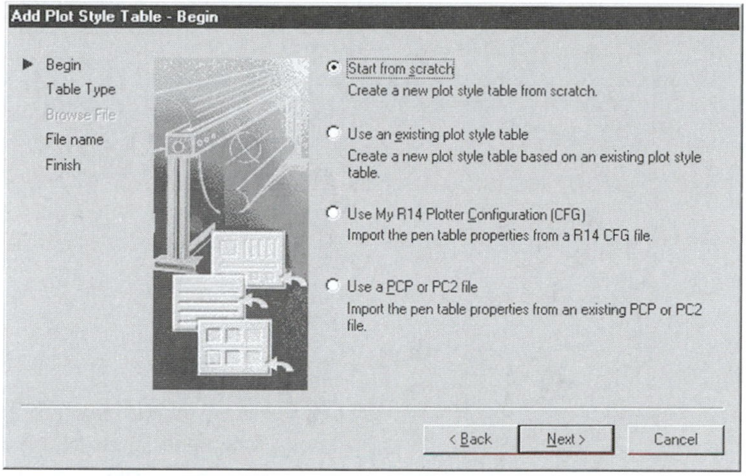

Figure 25.12 *The Begin display shows the options for selecting a starting point for creating a plot style table. By default, you'll be starting the plot table from scratch.*

- Use My R14 Plotter Configuration—this option allows a new plot style table to be created using pen assignments stored in R14 drawing, but you don't have access to the PCP or PCP2 files associated to the drawing.
- Use a PCP or PCP2 File—this option allows a new plot style table to be created using pen assignments stored in acadr14.cfg files.

As you proceed through the initial viewing of the wizard, accept the default option and select the Next button. The next display of the Plot Style wizard will display a prompt for the type of table to be used. Options include Color-dependant Plot Style Table and Named Plot Style Table. Choose the Named Plot Style Table button. Selecting the Next button will display the File Name display. Enter a name to describe the plot style in the Name edit box. For this example a name of MONO will be used. Once a name has been entered in the edit box, select the Next button to display the final display of the wizard.

To this point, you've started a plot style table named MONO. The final display of the setup wizard allows the naming process to be completed or to edit the plot style. Selecting the Plot Style Editor button will display a Plot Style Editor dialog box similar to Figure 25.13. This table can be used to alter the properties of the drawing to be plotted. Future sections of this chapter will deal with altering drawing properties. Select the Finish button to create a named plot style table and exit the wizard. Selecting Finish will create a .STB file that is listed in the Plot Style Manager. Selecting the Use This Plot Style Table for Layouts in New drawings will make the newly created plot style the default plot style for future drawings.

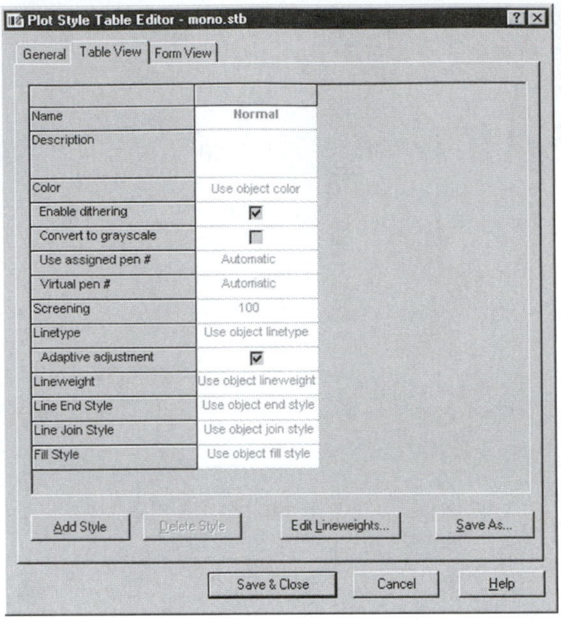

Figure 25.13 *Selecting the Plot Style Editor button will display a Plot Style Editor dialog box. This table can be used to alter the properties of the drawing to be plotted.*

Creating a Color-dependent Plot Style Table

The process to create a color-dependent plot style is similar to the process of creating a named style. The steps to create a color-dependent include:

- Select the Add Plot Style Table icon from the Plot Styles dialog box.
- Select the Next button after reading the introductory material.
- Select the option to be used to start the plotting style table and select the Next button.
- Choose the type of plot style table to be created and select the Next button.
- Enter the name of the plot style table and choose the Next button.
- Set the options on the Finish display and select the Finish button.

The color-dependent plot style will be stored as a .CTB file and listed in the plot style manager display.

ATTACHING A PLOT STYLE TABLE TO A LAYOUT

A plot style created using the plot style wizard can be attached to the model or layout tab using the Page Setup dialog box. Use the following steps to attach a page style to a layout tab.

- Select the desired layout tab to apply the plot style.
- Select Page Setup from the File pull-down menu.
- Using the Page Setup dialog box, activate the Plot Device tab.
- Select the desired plot style to be applied in the Plot Styles portion of the Plot Device tab.
- Select the OK button.
- The qualities assigned to the plot style can now be applied to the selected layout.

Viewing a Plot Style

Once a plot style has been created, it can be previewed before it is plotted. Changes made to object properties in a layout can be seen when a drawing is regenerated, or by using the Full Preview option of the Plot dialog box. Plot styles can be displayed in a drawing using the following steps:

- Select the Layout tab containing the plot style to be used.
- Select Page Setup from the File menu.
- Select Display Plot Styles from the Plot Style Table (Pen Assignments) in a Layout from the Plot Device tab.
- Select the OK button.

The properties of the plot style will now be displayed. If the changes are not displayed, enter **REGENALL** (EnterKey) at the command prompt to display plot styles in a layout.

ADDING AND DELETING NAMED PLOT STYLES

Once you've created your own plot style, it must be added to a plot style table before it can be applied to the layout to be plotted. Named plot styles are added to the plot style table using the Plot Style Table editor. Plot styles are added using the following sequence:

- Select Plot Style Manager from Files of the Standard menu.
- Use one of the following methods to display the Plot Style Table Editor:
- Double-click a .STB file.
- Right-click a .STB file and select Open from the shortcut menu.
- In the Active Plot Style Table of the Current Plot Style Table dialog box or the Select Plot Style Table, select the desired plot style and choose Editor.
- Select the Table View tab or Form View tab from the Plot Style Table Editor to view the style settings.
- Select Add Style button.

A style titled Style1 will be added to the table. While the new style is still highlighted, change the name to better describe the style contents. Adjust the options as needed, and select the Save & Close button to return to the Plot Styles dialog box. Figure 25.14 shows an example of a style that was added.

Deleting a Named Plot Style

As plotting requirements change throughout the life of a project, a plot style in a named plot style table can be deleted. The procedure to delete a style is similar to the procedure used to add the style. Use the following steps to delete a file:

- Select Plot Style Manager from File of the Standard toolbar.
- Double-click the .STB file to be edited.
- Use the Form View tab of the Plot Style Table Editor to select the plot style to be deleted.
- With the style highlighted, select the Delete Style button.
- Select the Save and Close button or the Cancel button to return to the Plot Styles display.

DEFINING A PLOT STYLE

Figure 25.14 shows an example of the MY LINES plot style. Properties of a style can be defined or edited, using similar methods used to alter a layer color in the Layer

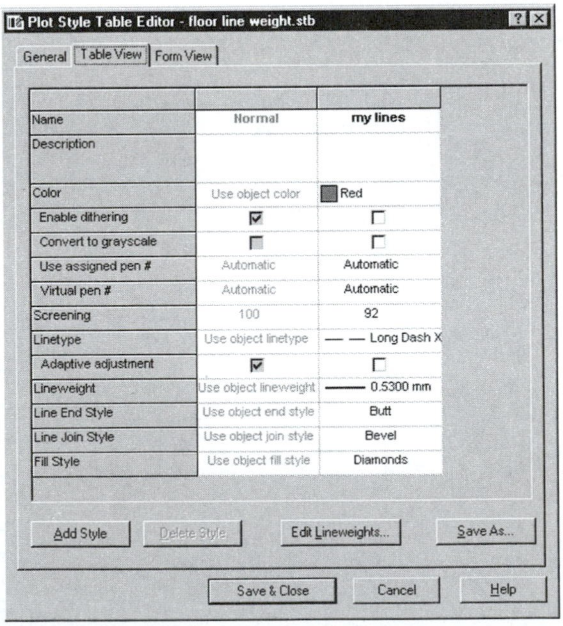

Figure 25.14 *The Table View tab of the Plot Style Table Editor can be used to add a plot style. Notice that the color, screening, linetype, lineweight, and line styles have been altered.*

Properties Manager. These settings will provide overrides to be used while plotting for values defined in the drawing. Settings can be defined using the Table or Form tabs of the Plot Style Table Editor. The Table View tab shown in Figure 25.14 works well if only a few styles are assigned to the table. Figure 25.15 shows an example of the Form View tab. The Form View tab works well if several styles have been assigned to the table. Notice that the style names are listed on the left side, and the values for the current style are listed on the right side.

To define the color for the plot style requires the following steps:

- Using the Table View tab of the Plot Style Table Editor, select the current color setting for the plot style to be altered. As Use Object Color is selected the standard color menu will be displayed.
- Select the desired color. As the color is selected, the menu will be closed allowing other options to be defined.

Other options can be defined using similar methods.

Enable Dithering

Dithering is used to approximate colors using a dot pattern. If the plotter does not support dithering, this option will be inactive. If the plotter supports dithering,

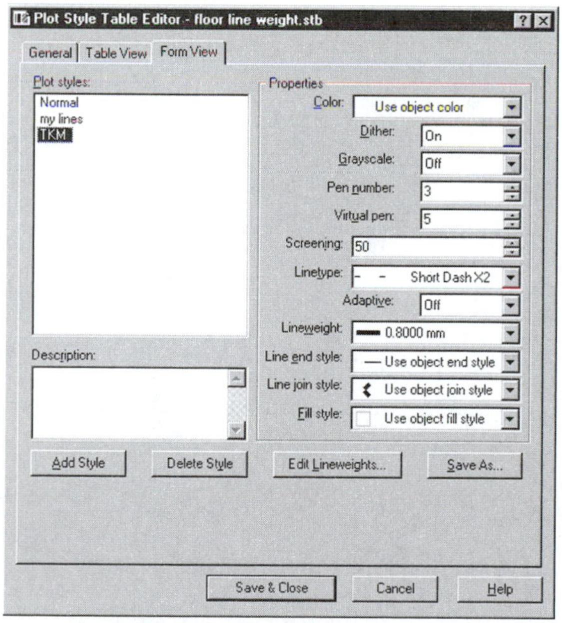

Figure 25.15 *The Form View tab works well if several styles have been assigned to the table. Notice that the style names are listed on the left side, and the values for the current style are listed on the right side.*

dithering may be turned off to provide dim colors more visibility. With dithering OFF, AutoCAD maps colors to the nearest color.

Convert to Grayscale

With this option active and a plotter that supports grayscale, AutoCAD will convert colors to grayscale.

Use Assigned Pen Numbers

This option specifies which pen will be used with each plot style. The default setting is automatic. In the Plotter Configuration Editor you can specify that a certain object will be plotted using pen #1 as black and .02" , and pen #2 as blue and .031". With this option, pen #1 can be assigned to one plot style and pen #2 can be assigned a different plot style. This can be a very useful tool if a subcontractor has used pen assignments different than the desired standard. Using the Automatic or 0 setting, AutoCAD will select the pen closest in color to the color of the object you are plotting. The value can't be altered if a color-dependent plot table is being used or if the plot style color setting of Use Object Color is active. Values can be entered by keyboard or by using the range arrow. As the box is selected, arrows are displayed to adjust the range. With a setting of 0, only the up arrow is active. A pen value of 1-32 can be specified.

Virtual Pen Number

Most new plotters do not use pens, but are capable of duplicating pen plotter output by using virtual pens. In its current setting, AutoCAD will specify a color from the AutoCAD Color Index closest to the color of the object to be plotted. A virtual pen number between 1-255 can be provided and used by plotters without pens. Using virtual pens, all other style setting are ignored and only the virtual pen settings are used. Values can be entered by keyboard or by using the range arrow. As the box is selected, arrows are displayed to adjust the range.

Screening

This setting allows a color intensity to be specified by controlling the amount of ink to be placed during the plot. The default setting of 100 plots at full intensity. A value between 0-100 can be entered. Entering 0 will have the same effect as freezing the assigned objects. Values can be entered by keyboard or by using the range arrow. As the box is selected, arrows are displayed to adjust the range. With a setting of 100, only the down arrow is active.

Linetype

Selecting the Linetype option will display a linetype menu similar to Figure 25.16. The default setting for plot style linetype is Use Object Linetype. Selecting a linetype from the menu will override the linetypes specified in the drawing.

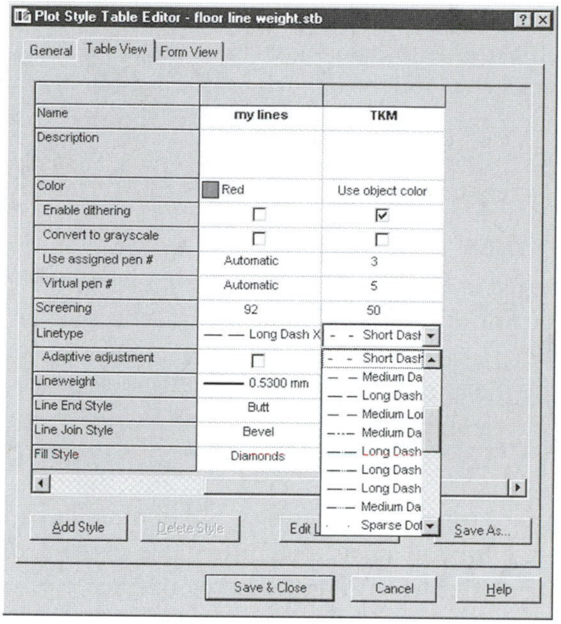

Figure 25.16 *Selecting the Linetype option will display a linetype menu. The default setting for plot style linetype is Use Object Linetype.*

Adaptive Adjustment

This setting adjust the linetype scale to be used to display the linetype pattern. This setting should be OFF when the display of the linetype is important.

Lineweight

The Lineweight field displays a sample of the selected lineweight and the numeric value of the lines to be used with the style. An override value can be selected from the drop down menu. The menu is displayed as the field is selected.

Line End Style

In addition to using the line end style specified for drawing objects, AutoCAD allows four options to be assigned to override the line end style of drawing objects. Selecting this field will produce a menu with the four options. Options include Use Object End Style, Butt, Diamond, Round, and Square.

Line Join Style

AutoCAD allows four options to be assigned to override the line join style specified for drawing objects. Selecting this field will produce a menu with the four options. Options include Use Object Join Style, Bevel, Diamond, Miter, and Round.

Assigning Fill Style

AutoCAD allows fill patterns to be assigned to override the fill style specified for drawing objects. The patterns will be applied to donuts, plines, and solids. Selecting this field will produce a menu with the options. In addition to the default of Use Object Fill Style, options include Solid, Checkerboard, Crosshatch, Diamond, Horizontal Bars, Slant Left, Slant Right, Square Dots, and Vertical Bars.

EDITING PLOT STYLES

A plot style can be edited using the same methods that were used to define it. Either the Table View or the Form View tabs can be used to edit the plot style. Qualities in a plot style can be edited using the same methods that were used to define the style. In addition to altering the qualities, the Plot Style Table Editor can be used to add, delete, copy, paste, and modify plot styles.

Copy a Plot Style

A plot style can be copied using the Table View tab of the Plot Style Table Editor. A plot style can be moved from one table to another table by using drag and drop. One or more settings from a plot style can be copied to another style using the following steps:

- Open the Plot Style Manager.
- Double-click the desired .CTB or .STB file to be opened.
- Make the Table View tab current.
- Select the plot style to be copied by selecting the gray area above the plot style name. Individual settings can be copied by selecting individual fields.
- Right-click and choose Copy from the shortcut menu.
- Switch to the plot style table to be altered.
- Select the plot style to be replaced by selecting the gray area above the plot style name.
- Right-click and choose Paste from the shortcut menu.

Renaming Plot Styles

A named plot style can be renamed using the RENAME command. Before a style can be renamed, it must exist in the plot style table associated with the layout. The new plot style name must be a name that is not currently used by any object or layer. Use the following steps to rename a named plot style:

- Select Rename... from the Format pull-down menu.
- Select Plot Style from the Named Objects list.
- Select the name of the plot style to be renamed from the Items list. The name will now be displayed in the Old Name box.

- Provide a new name in the Rename To box.
- Select the Rename To button. The new name will now be displayed in the list box.
- Select the OK button to return to the drawing area.

CHAPTER 25 EXERCISES

1. Use a laser or plotter and set the required parameters to plot E-19-8 at a scale of 1/4" = 1'-0".
2. Open drawing E-14-4. Plot the drawing at a scale of 3/8" = 1'-0".
3. Open drawing E-15-10. Use the copy command and create a second copy of the elevation. Design a second elevation using a different architectural style and materials. Save the drawing file as E-25-3. Plot the drawing containing your border and title block to a C size sheet of vellum.
4. Plot your drawing template containing the border and title block on a D size sheet of paper. Create a new base drawing that will fit within the limits of the borders. Attach any four drawings in your library. Each drawing is to be shown at a different scale when plotted. Provide a title and the scale used by the original drawing prior to plotting.

CHAPTER 25 QUIZ

1. How many colors can be configured for plotting?
2. Is it possible to have more than one plotting device connected to a single computer? Explain your answer.
3. What options are available to preview the information to be plotted prior to plotting, and how are they accessed?
4. Describe the meaning of the display used in a partial plot display.
5. What tab contains a display of the current plotting scale?
6. What can be used to control how much of the current drawing will be displayed?
7. List five options for controlling what portion of the drawing will be plotted.
8. Describe the typical origin location for a plotter and a printer.
9. What is the advantage of saving the plot information with "Plot to File"?
10. Which takes precedence, the Page setup or the Plot Style?
11. How does AutoCAD deal with pen assignments?
12. Explain the difference between portrait and landscape.
13. What plot style should be used to plot a drawing with 46 different colors using only black?
14. What command is used to create a plot style?
15. List two types of plot style tables and explain which will most typically meet your needs.
16. Explain the steps to attach a plot style to a layout.
17. What is the purpose of a plot style?
18. You've inserted five details into a drawing template. Each of the details was created for plotting at a different scale. Each was inserted into the template using the proper methods. What scale should be used to plot this drawing?
19. How are plotters configured using AutoCAD 2000?
20. What does the What to Plot area control?

CHAPTER 26

Using the Internet with AutoCAD

The Internet is an important way to exchange files and information. AutoCAD allows you to access and store drawings and related files and to interact with the Internet. The AutoCAD Internet features allow drawings to be opened and saved to an Internet location. New features in AutoCAD allow hyperlinks to be added to drawings, so that a variety of related documents can be easily accessed. AutoCAD allows drawings to be displayed in Web format (DWF) files that can be viewed with an Internet browser. DWF files can be viewed by anyone with an Internet browser using a free WHIP plug-in provided by AutoCAD. This chapter will introduce you to methods for launching a Web browser, help you use hyperlinks or Uniform Resource Locator (URL), and introduce the Drawing Web Format (DWF).

INTERNET ACCESS

You are probably familiar with the best-known uses for the Internet: email (electronic mail) and the Web (short for "World Wide Web"). Email lets users exchange messages and data at very low cost. The Web brings together text, graphics, audio, and movies in an easy-to-use format. To use the features of AutoCAD requires Internet access and an Internet browser to be installed on your workstation. To view and plot AutoCAD DWF files from an Internet browser requires the WHIP! 4.0 plug-in from http://www.autodesk.com/whip to be downloaded. Microsoft Internet Explorer must also be installed on your computer. If a full installation of AutoCAD was preformed, the required files were automatically installed. With the required software loaded, AutoCAD can open, insert, and save drawings to and from the Internet. To save files to an Internet location requires access rights to the directory where the files are stored.

AutoCAD is able to launch a Web browser from within AutoCAD using the BROWSER command. The browser is used for viewing drawings in 2D format on Web pages. The HYPERLINK command can be used to link a drawing with other

* This Chapter based on Using AutoCAD 2000, by Ralph Grabowski, ©1999, Autodesk Press.

documents on your computer or the Internet. A window or door schedule created on a spreadsheet can be linked to a floor plan created in AutoCAD. The electronic plot option of the PLOT command can create DWF files for viewing 2D drawing on Web pages. Using the OPEN, INSERT and SAVEAS commands, AutoCAD can import or export drawings to and from the Internet.

WORKING WITH UNIFORM RESOURCE LOCATORS

A URL or Uniform Resource Locator is the file naming system of the Internet. This system allows files such as a text file, a Web page, a program file, an audio file, or movie clip to be located on the Internet. A typical URL would resemble http://www.autodesk.com. The prefix http:// is typically not required to be entered. Most Web browsers automatically add in the routing prefix, which saves you a few keystrokes. Common URLs you should be familiar with include:

URL	Meaning
http://www.autodesk.com	Autodesk Primary Website.
http://data.autodesk.com	Autodesk Data Publishing Website.
news://adesknews.autodesk.com	Autodesk News Server
ftp://ftp.autodesk.com	Autodesk FTP Server
http://www.autodeskpress.com	Autodesk Press Website

URLs can access several different kinds of resources such as websites, email, or news groups. No matter the resource, the URL always take on the same general format:

scheme://netloc

The scheme accesses the specific resource on the Internet such as:

file://	File is located on your computer's hard drive or local network.
ftp://	File Transfer Protocol (used for downloading files).
http://	HyperText Transfer Protocol (the basis of websites).
mailto://	Electronic mail (email).
news://	Usenet news (news groups).
telnet://	Telnet protocol.
gopher://	Gopher protocol

The characters :// indicate a network address. Autodesk recommends these formats for specifying URL-style filenames:

Web Site	http://servername/pathname/filename
FTP Site	ftp://servername/pathname/filename
Local File	file:///drive:/pathname/filename

or	file:///drive/pathname\filename
or	file://\\localPC\pathname\filename
or	file://\\localPC\pathname\filename
or	file:////localPC/pathname/filename
Network File	file://localhost/drive:/pathname/filename
or	\\localhost\drive:\pathname\filename
or	file://localhost/drive|/pathname/filename

The following definitions will help to clarify this confusing terminology.

Term	**Meaning**
Servername	A name or location of a computer on the Internet such as: www.autodesk.com
Pathname	The same as a subdirectory or folder name.
Drive	The same as a subdirectory or folder name.
local PC	A file located on your computer.
local host	The name of the network host computer.

If you are not sure of the name of the network host computer, use Windows Explorer to check the Network Neighborhood for the network names of computers.

OPENING AND SAVING INTERNET FILES

The BROWSER command lets you start a Web browser while inside AutoCAD. The two most common browsers are the Microsoft Internet Explorer and Netscape Navigator. By default, the BROWSER command uses the Web browser program that is registered in your computer's Windows operating system. AutoCAD lists the name of the browser before prompting you for the uniform resource locator (URL). The URL is the website address, such as http://www.autodesk.com. URLs will be introduced in the next section. The BROWSER command is used to launch a Web browser from within AutoCAD and requires that Netscape Communicator or Microsoft Internet Explorer be installed on your computer. The command can also be used in scripts, toolbar or menu macros, and AutoLISP routines to automatically access the Internet. The command can be accessed using the following command sequence:

Command: **BROWSER** (EnterKey)

Enter Web location (URL) < C:PROGRAM FILES \ACAD2000\ Home.htm>: *(Enter URL)*

Selecting (EnterKey) will accept the default URL. The default URL is an HTML file added to your computer during the installation of AutoCAD. The URL is

Autodesk's own website. After you type the URL and press (EnterKey) , AutoCAD launches the Web browser and contacts the website, such as the Autodesk Web site shown in Figure 26.1

CHANGING THE DEFAULT WEBSITE

Altering the INetLocation system variable setting can change the default Web page that your browser starts with from within AutoCAD. The variable stores the URL used by the BROWSER command and the Browse the Web dialog box. Make the change as follows:

> Command: **INETLOCATION** (EnterKey)
>
> Enter new value for INETLOCATION <"C:\CAD\ACAD 2000\Home.htm">: *(Type URL.)*

DRAWINGS ON THE INTERNET

When a drawing is stored on the Internet, you access it from within AutoCAD 2000 using the standard Open, Insert, and Save commands. Instead of specifying the file's location with the usual drive- subdirectory-file name format, such as c:\acad14\file-name.dwg, use the URL format. The URL is the universal file naming system used by the Internet to access any file located on any computer hooked up to the Internet.

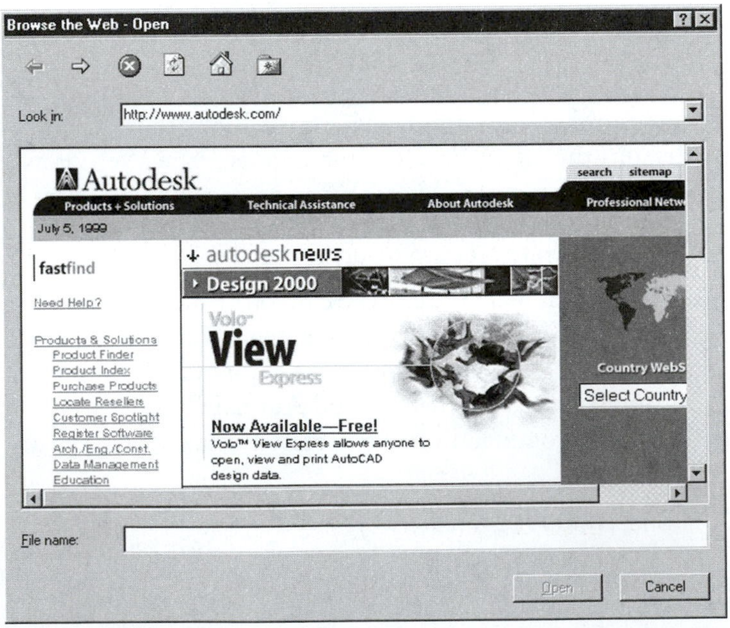

Figure 26.1 *The Autodesk Website is displayed by entering* **BROWSER** *(EnterKey) at the command prompt and then provideing the name of the URL.*

OPENING DRAWINGS FROM THE INTERNET

Use the OPEN command (File | Open) to open a drawing from the Internet. The three icons can be seen in the upper right of the Select File dialog box shown in Figure 26.2. These are specifically for use with the Internet. From left to right, they are:

Search the Web—Displays the Browse the Web dialog box, a simplified Web browser.

Look in Favorites—Opens the Favorites folder—equivalent to bookmarks that store Web addresses.

Add to Favorites—Saves the current URL (hyperlink) to the Favorites folder.

When you click the Search the Web button, AutoCAD opens the Browse the Web dialog box. This dialog box is a simplified version of the Microsoft Web browser. The purpose of this dialog box is to allow you to browse files at a website. By default, the Browse the Web dialog box displays the contents of the URL stored in the Inetlocation system variable (See Figure 26.3a). You can easily change this to another folder or website, as noted earlier. (See Figure 26.3b).

The Browse the Web dialog box has six buttons across the top to aid in working with files. These include:

Back—Go back to the previous URL.

Forward—Go forward to the next URL.

Stop—Halt displaying the Web page. This option is useful if the connection is slow or the page is very large.

Refresh—Redisplay the current Web page.

Home—Return to the location specified by the Inetlocation system variable.

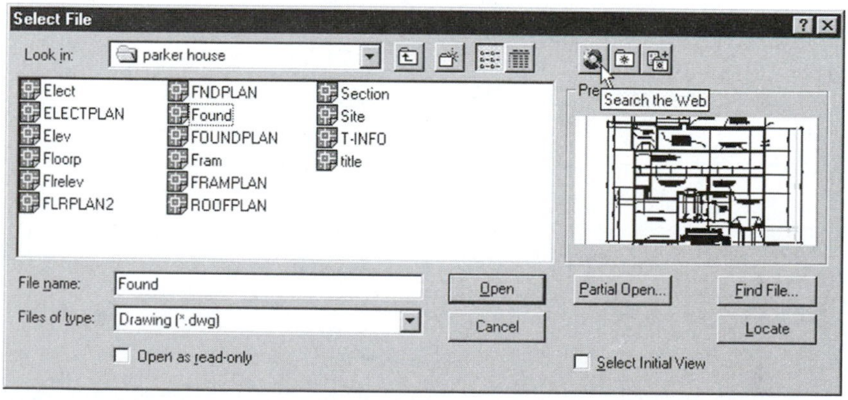

Figure 26.2 *The Select Files dialog box contains three buttons to aid in working with the Internet. These icons from left to right are the Search the Web, Look in Favorites, and Add to Favorites.*

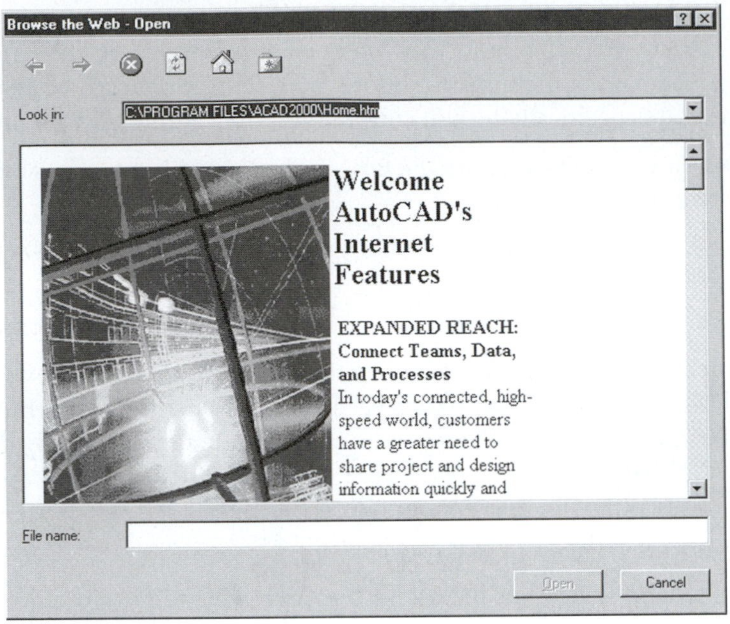

Figure 26.3a *By default, the Browse the Web dialog box displays the contents of the URL stored in the Inetlocation system variable.*

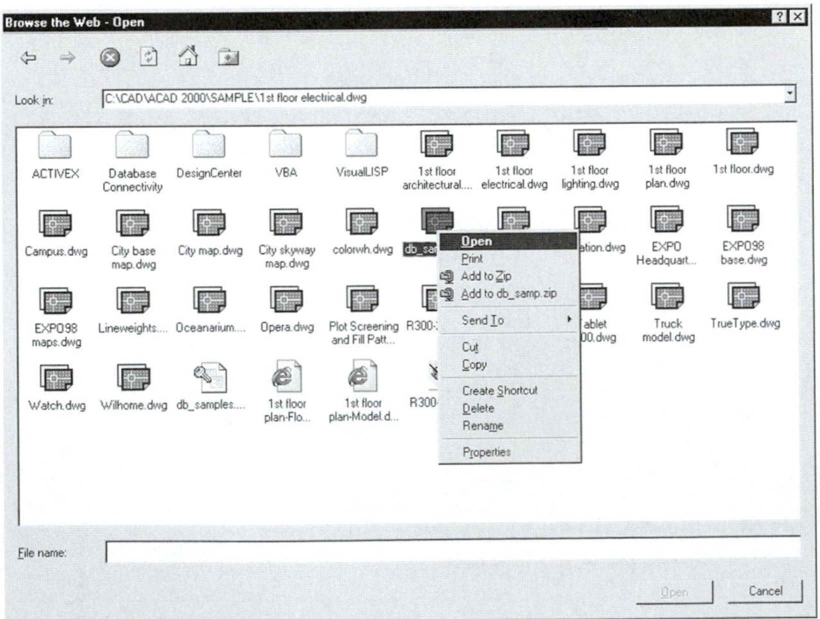

Figure 26.3b *The Browse the Web-Open dialog box.*

Figure 26.4 *The File Download display box.*

Favorites—List stored URLs (hyperlinks) or bookmarks. If you have previously used Internet Explorer, you will find all your favorites listed here. Favorites are stored in the \Windows\Favorites folder on your computer.

The Look in field allows you to type the URL. Alternatively, click the down arrow to select a previous destination. If you have stored website addresses in the Favorites folder, then select a URL from that list.

You can either double-click a filename in the window, or type a URL in the File name field. The following table gives the templates for typing the URL to open a drawing file:

Drawing Location	**Template URL**
Web or HTTP Site	http://servername/pathname/filename.dwg
	http://practicewrench.autodeskpress.com/wrench.dwg
FTP Site	ftp://servername/pathname/filename.dwg
	ftp://ftp.autodesk.com
Local File	drive:\pathname\filename.dwg
	c:\acad 2000\sample\tablet2000.dwg
Network File	\\localhost\drive:\pathname\filename.dwg
	\\upstairs\e:\install\sample.dwg

When you open a drawing over the Internet, it will probably take much longer than opening a file found on your computer. During the file transfer, AutoCAD displays a dialog box to report the progress. If your computer uses a 28.8Kbps modem, you should allow about five to ten minutes per megabyte of drawing file size. If your computer has access to a faster T1 connection to the Internet, you should expect a transfer speed of about one minute per megabyte.

The OPEN command does not copy the file from the Internet location directly into AutoCAD. Instead, it copies the file from the Internet to your computer's designated

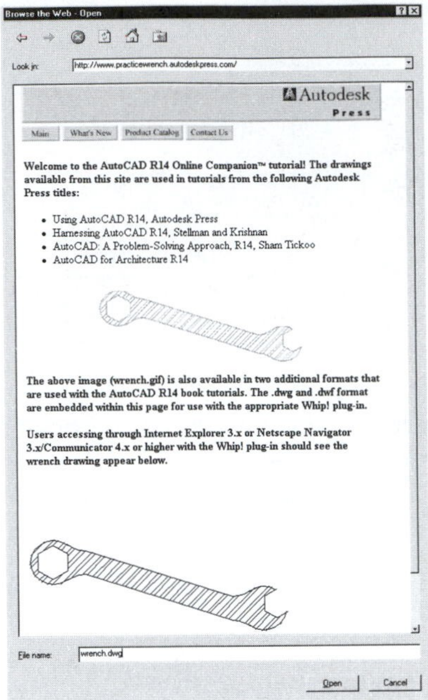

Figure 26.5 *Practice Wrench Website from autodeskpress.com.*

Temporary subdirectory, such as C:\Windows\Temp. Once in the Temporary folder, the drawing is loaded from the hard drive into AutoCAD. This is known as *caching*. It helps to speed up the processing of the drawing, since the drawing file is now located on your computer's fast hard drive instead of the relatively slow Internet.

Note: The Locate and Find File buttons in the Select File dialog box do not work for locating files on the Internet.

TUTORIAL

The Autodesk Press website has an area that allows you to practice using the Internet with AutoCAD. In this tutorial, you open a drawing file located at that website.

1. Start AutoCAD.
2. Ensure that you have a live connection to the Internet. If you normally access the Internet via a telephone (modem) connection, dial your Internet service provider now.
3. From the menu bar, select File | Open or click the Open icon on the toolbar. Notice that AutoCAD displays the Select File dialog box.
4. Click the Search the Web button. Notice that AutoCAD displays the Browse the Web dialog box.

5. In the Look in field, type:

   PRACTICEWRENCH.AUTODESKPRESS.COM (EnterKey)

   After a few seconds, the Browse the Web dialog box displays the Web site.

6. In the Filename field, type:

   WRENCH.DWG (EnterKey)

AutoCAD begins transferring the file. Depending on the speed of your Internet connection, this will take between a couple of seconds and a half-minute. Notice the drawing of the wrench in AutoCAD.

INSERTING A BLOCK FROM THE INTERNET

When a block (symbol) is stored on the Internet, you can access it from within AutoCAD using the INSERT command. When the Insert dialog box appears, click the Browse button to display the Select Drawing File dialog box. This is identical to the dialog box discussed above. After you click the select the file, AutoCAD downloads the file and continues with the Insert command's familiar prompts.

The process is identical for accessing external reference (xref) and raster image files. Other files that AutoCAD can access over the Internet include: 3D Studio, SAT (ACIS solid modeling), DXB (drawing exchange binary), WMF (Windows

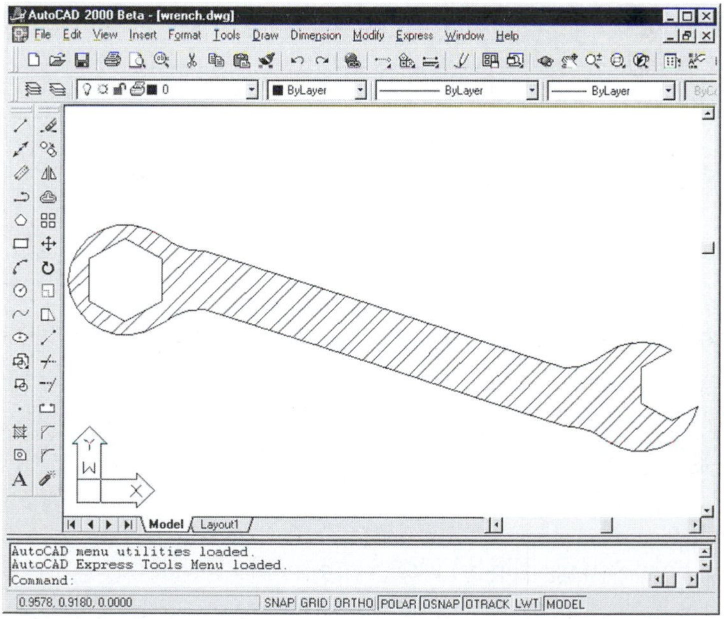

Figure 26.6 *Wrench drawing in AutoCAD.*

metafile), and EPS (encapsulated PostScript). All of these options are found in the Insert menu on the menu bar.

ACCESSING OTHER FILES ON THE INTERNET

Most other file-related dialog boxes allow you to access files from the Internet or intranet. This allows your firm or agency to have a central location that stores drawing standards. When you need to use a linetype or hatch pattern, for example, you access the LIN or PAT file over the Internet. More than likely, you would have the location of those files stored in the Favorites list.

Some examples include:

Linetypes—From the menu bar, select Linetype from the Format menu. In the Linetype Manager dialog box, click the Load, File, and Look in Favorites buttons.

Hatch Patterns—Use the Web browser to copy .PAT files from a remote location to your computer.

Multiline Styles—From the menu bar, select Multiline Style from the Format menu. In the Multiline Styles dialog box, click the Load, File, and Look in Favorites buttons.

Layer Name—From the menu bar, select Layer Properties Manager from Layers. In the Layer Manager dialog box, click the Import and Look in Favorites buttons.

LISP and ARX Applications—From the menu bar, select Load Applications from the Tools pull-down menu

Scripts—From the menu bar, select Run Scripts from the Tools pull-down menu.

Menus—From the menu bar, select Customize Menus from the Tools pull-down menu. In the Menu Customization dialog box, click the Browse and Look in Favorites buttons.

Images—From the menu bar, select View... from Image Displays of the Tools pull-down menu.

You cannot access text files, text fonts (SHX and TTF), color settings, lineweights, dimension styles, plot styles, OLE objects, or named UCSs over the Internet.

SAVING THE DRAWING TO THE INTERNET

When you are finished editing a drawing in AutoCAD, you can save it to a file server on the Internet with the SAVE command. If you inserted the drawing from the Internet using INSERT into the default Drawing.Dwg drawing, AutoCAD insists you first save the drawing to your computer's hard drive.

When a drawing of the same name already exists at that URL, AutoCAD warns you just as it does when you use the SAVEAS command. Recall from the OPEN command that AutoCAD uses your computer system's temporary subdirectory, hence the reference to it in the dialog box.

USING HYPERLINKS WITH AUTOCAD

AutoCAD 2000 allows you to employ URLs in two ways: (1) directly within an AutoCAD drawing and (2) indirectly in DWF files displayed by a Web browser. (URLs are also known as hyperlinks, the term we use from now on.) Hyperlinks are created, edited, and removed with the HYPERLINK command using a dialog box and the HYPERLINK command using prompts at the command line. The HYPERLINK command (Insert | Hyperlink) prompts you to “Select objects” and then displays the Insert Hyperlink dialog box. As a shortcut, you may press Ctrl+K or click the Insert Hyperlink button on the toolbar.

As an alternative, you may use the –HYPERLINK command. This command displays its prompts at the command line, and is useful for scripts and AutoLISP routines. The –HYPERLINK command has the following syntax:

Command: **-HYPERLINK** (EnterKey)

Enter an option [Remove/Insert] <Insert>: (EnterKey)

Enter hyperlink insert option [Area/Object] <Object>: (EnterKey)

Select objects: *(Pick an object.)*

1 found Select objects: (EnterKey)

Enter hyperlink <current drawing>: *(Enter the name of the document or website.)*

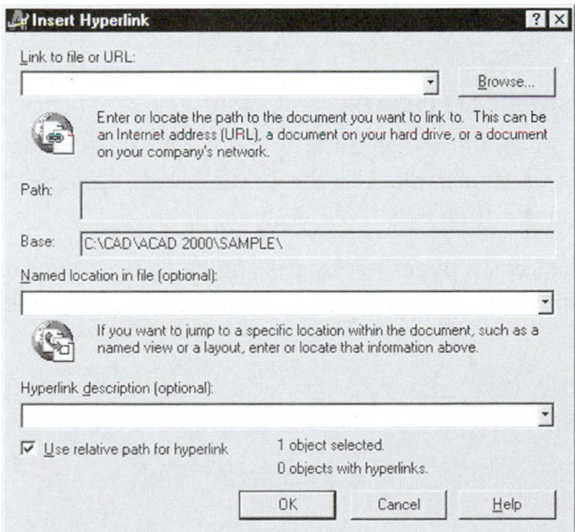

Figure 26.7 *The Insert Hyperlink Dialog Box.*

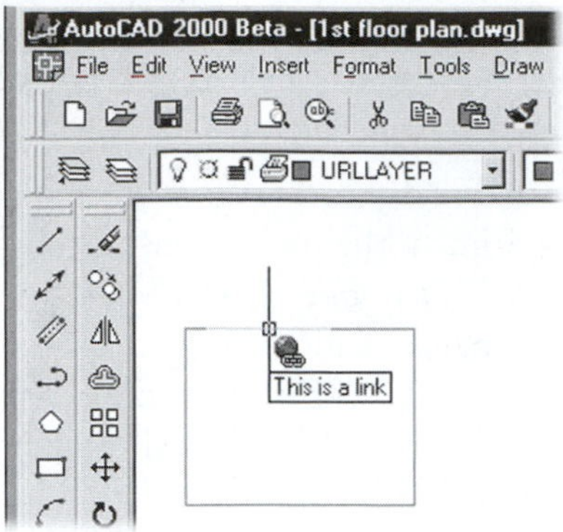

Figure 26.8 *A rectangular hyperlink area.*

Enter named location <none>: *(Enter the name of a bookmark or AutoCAD view.)*

Enter description <none>: *(Enter a description of the hyperlink.)*

Notice that the command also allows you to remove a hyperlink. It does not, however, allow you to edit a hyperlink. To do this, use the Insert option and re-specify the hyperlink data.

In addition, this command allows you to create hyperlink *areas*, which are rectangular areas that can be thought of as a 2D hyperlink. The dialog box-based Hyperlink command does not create hyperlink areas. When you select the Area option, the rectangle is placed automatically on layer URLLAYER and colored red.

In the following sections, you'll learn how to apply and use hyperlinks in an AutoCAD drawing and in a Web browser via the dialog box HYPERLINK command.

HYPERLINKS INSIDE AUTOCAD

AutoCAD allows you to add a hyperlink to any object in the drawing. An object is permitted just a single hyperlink. On the other hand, a single hyperlink may be applied to a selection set of objects.

You can tell an object has a hyperlink by passing the cursor over it. The cursor displays the "linked Earth" icon, as well as a tooltip describing the link similar to Figure 26.9.

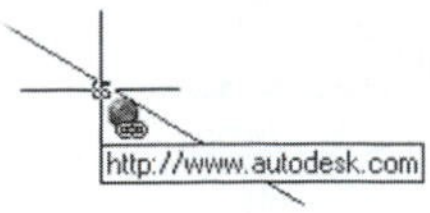

Figure 26.9 *Moving the cursor over an object reveals a hyperlink.*

If, for some reason, you do not want to see the hyperlink cursor, you can turn it off. From the menu, select Options from the Tools pull-down menu and then click the User Preferences tab. The Display hyperlink cursor and shortcut menu item toggles the display of the hyperlink cursor, as well as the Hyperlink tooltip option on the cursor menu.

TUTORIAL: ATTACHING HYPERLINKS

Let's see how that works by working through an example. In this example, we have the drawing of a floor plan. To this drawing, we will add hyperlinks to another AutoCAD drawing, a Word document, and a website. Hyperlinks must be attached to objects. For this reason, we will place some text in the drawing, then attach the hyperlinks to the text.

1. Start AutoCAD 2000.
2. Open the 1st floor plan.dwg file found in the AutoCAD 2000\Sample folder. (If necessary, click the Model tab to display the drawing in model space.)
3. Start the Text command and place the following text in the drawing:

   Command: **TEXT** (EnterKey)

   Current text style: "Standard" Text height: 0.20000

   Specify start point of text or [Justify/Style]: *(Pick a point in the drawing.)*

   Specify height <0.2000>: 2 (EnterKey)

   Specify rotation angle of text <0>: (EnterKey)

   Enter text: Site Plan (EnterKey)

   Enter text: Lighting Specs (EnterKey)

   Enter text: Electrical (EnterKey)

   Enter text: (EnterKey)

The results of this exercise can be seen in Figure 26.10

4. Click the "Site Plan" text to select it.
5. Click the Insert Hyperlink button on the toolbar. Notice the Insert Hyperlinks dialog box.
6. Next to the Link to File or URL field, click the Browse button. Notice the Browse the Web – Select Hyperlink dialog box.
7. Go to AutoCAD 2000's Sample folder and select the City base map.dwg file. Click Open. AutoCAD does not open the drawing; rather, it copies the file's name to the Insert Hyperlinks dialog box. Notice in Figure 26.11 the name of the drawing you selected in the Link to File or URL field.

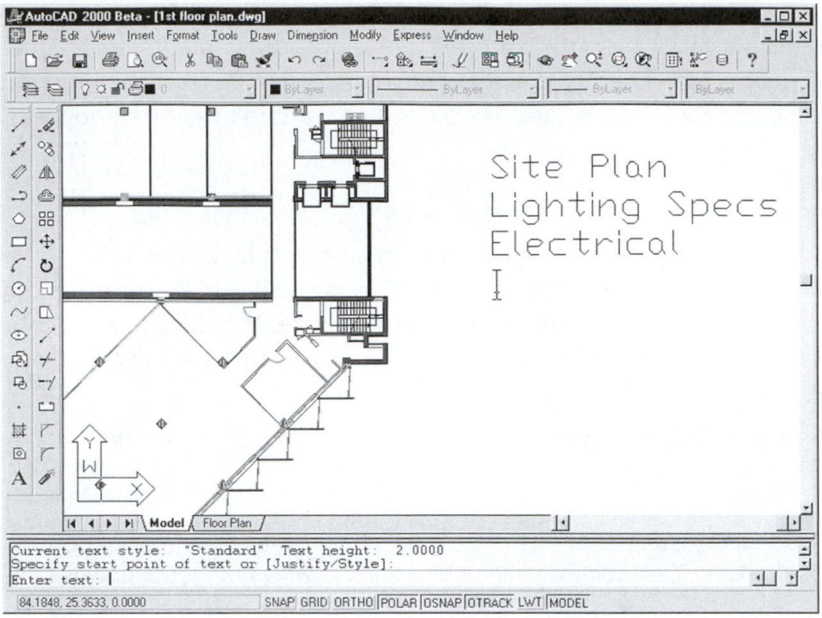

Figure 26.10 *Text placed in drawing.*

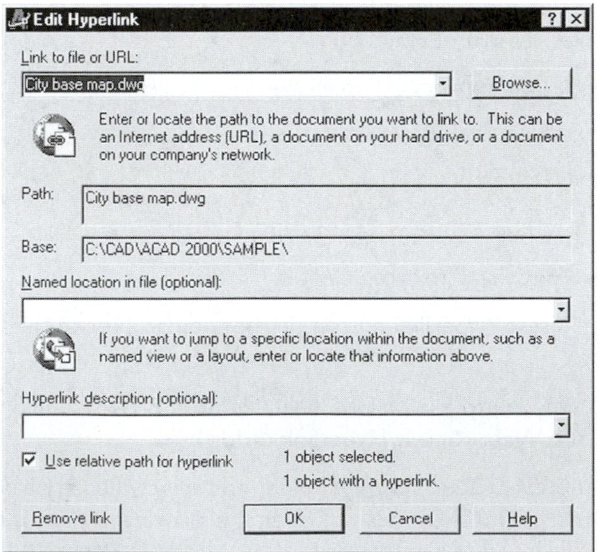

Figure 26.11 *The name selected to be opened in the Browse the Web Insert Hyperlinks dialog box.*

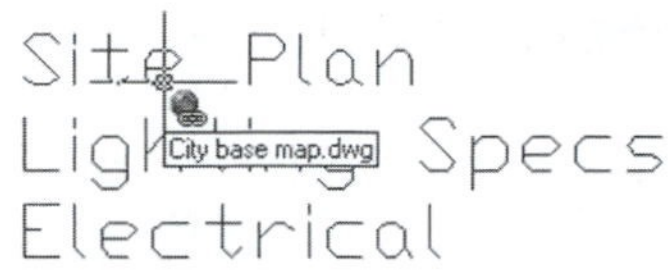

Figure 26.12 *The Hyperlink cursor and tooltip.*

You can fill in two other fields:

Named Location in File—When the hyperlinked file is opened, it goes to the named location, called a bookmark. In AutoCAD, the bookmark is a named view created with the VIEW command.

Hyperlink Description—You may type a description for the hyperlink, which is displayed by the tooltip. If you leave this blank, the URL is displayed by the tooltip.

8. Click OK to close the dialog box. Move the cursor over the Site Plan text. Notice in Figure 26.12 that the display of the "linked Earth" icon appears; a moment later, the tooltip displays "City base map.dwg".
9. Repeat the HYPERLINK command twice more, attaching these files to the drawing text:

Text	**URL**
Lighting Specs	License.rtf *(Found in the AutoCAD 2000 folder.)*
Electrical Bylaw	Home.htm *(Also found in the AutoCAD 2000 folder.)*

You have now attached a drawing, a text document, and a Web document to objects in the drawing. Let's try out the hyperlinks.

10. Click "Site Plan" to select it. Right-click and select Open "City base map.dwg" from Hyperlink of the shortcut menu. Notice in Figure 26.13 that AutoCAD opens "City base map.dwg". To see both drawings, select Tile Vertically from the Window pull-down menu.
11. Select, then right-click "Lighting Specs" hyperlink. Select Open from Hyperlink. Notice in Figure 26.14 that Windows starts a word processor and opens the License.rtf file.
12. If your word processor has a Web toolbar, open it. Click the Back button *(the back arrow)*. Notice how that action sends you back to AutoCAD.
13. Select the right-click "Electrical Bylaw" hyperlink. Select Open from Hyperlink. Notice in Figure 26.15 that Windows starts your Web browser and opens the Home.Htm file.
14. Back in AutoCAD, right-click any toolbar and select Web. Notice in Figure 26.16 that the Web toolbar has four buttons. From left to right, these are:

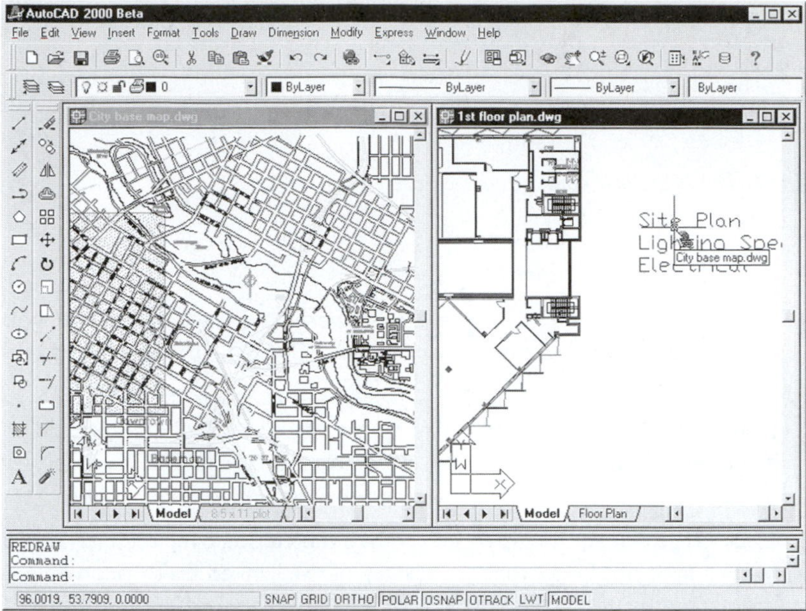

Figure 26.13 *Viewing two drawings.*

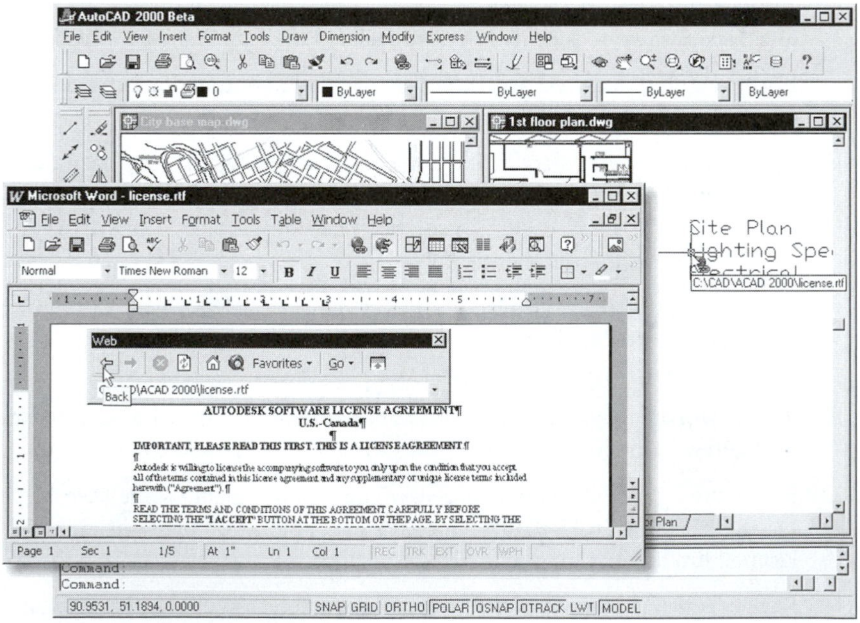

Figure 26.14 *Viewing a Word document.*

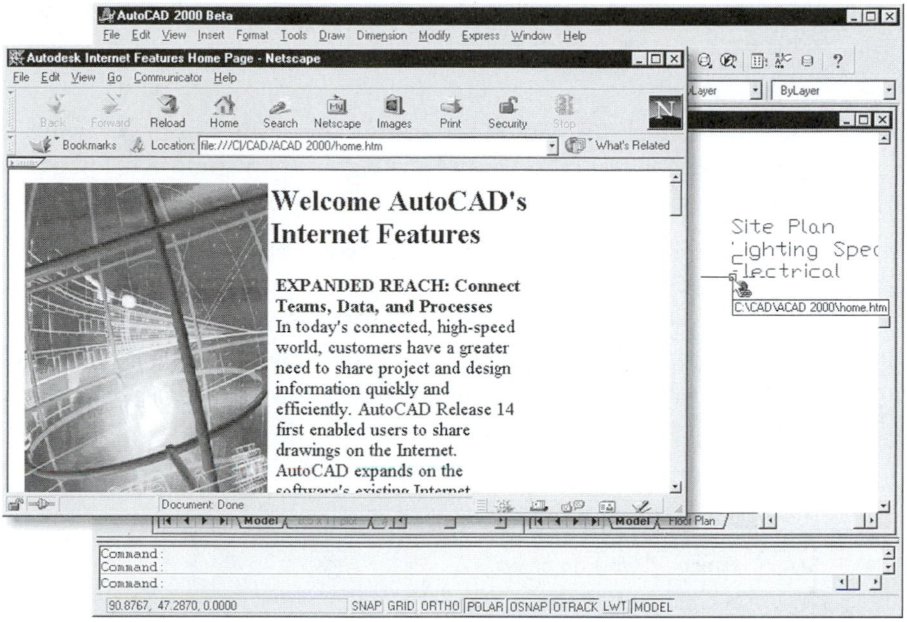

Figure 26.15 *Viewing a Website.*

Button	Command	Meaning
Go Back	HyperlinkBack	Move back to the previous hyperlink.
Go Forward	HyperlinkFwd	Move to the next hyperlink.
Stop Navigation	HyperlinkStop	Stops the hyperlink access action.
Browse the Web	Browser	Launches the Web browser.

15. Try clicking the Go Forward and Go Back buttons. Notice how these allow you to navigate between the two drawings, the RTF document, and the Web page.

When you work with hyperlinks in AutoCAD, you may come across these three limitations:

- AutoCAD does not check that the URL you type is valid.
- If you attach a hyperlink to a block, be aware that the hyperlink data is lost when you scale the block unevenly, stretch the block, or explode it.

Figure 26.16 *AutoCAD's Web Toolbar can be displayed by moving the cursor to any open toolbar, right-clicking, and selecting Web.*

- Wide polylines and rectangular hyperlink areas are only "sensitive" on their outline.

Pasting as Hyperlink

AutoCAD 2000 has a shortcut method for creating hyperlinks in the drawing. The PASTEASHYPERLINK command pastes any text in the Windows Clipboard as a hyperlink to any object in the drawing. Here is how it works:

1. In a word processor, select some text and copy it to the Clipboard using either Ctrl+C or the Edit Copy command. The text can be a URL such as http://www.autodeskpress.com or any other text.
2. Switch to AutoCAD and select Paste as Hyperlink from the Edit pull-down menu bar. Note that this command does not work (grayed out) if anything else is in the Clipboard, such as a picture.
3. Select one or more objects, as prompted:

   Command: _**PASTEASHYPERLINK** (EnterKey)

   Select objects: *(Select an object.)*

   1 found Select objects: (EnterKey)
4. Pass the cursor over the object and note the hyperlink cursor and tooltip. The tooltip displays the same text that you copied from the document.

If the text you copy to the Clipboard is very long, AutoCAD displays only portions of it in the tooltip, using ellipses (...) to shorten the text. You cannot select text in the AutoCAD drawing to paste as a hyperlink. You can, however, copy the hyperlink from one object to another. Select the object, right-click, and select Copy Hyperlink from the Hyperlink shortcut menu. The hyperlink is copied to the Clipboard. You can now paste the hyperlink into another document, or use AutoCAD's Paste as Hyperlink command to attach the hyperlink to another object in the drawing.

Note: The MATCHPROP command does not work with hyperlinks.

Highlighting Objects with URLs

Although you can see the rectangle of area URLs, the hyperlinks themselves are invisible. For this reason, AutoCAD has the QSELECT command, which highlights all objects that match specifications. From the menu bar, select Quick Select from the Tools pull-down menu. AutoCAD displays a Quick Select dialog box similar to Figure 26.17.

In the fields, enter these specifications:

Apply to: **ENTIRE DRAWING**

Object type: **MULTIPLE**

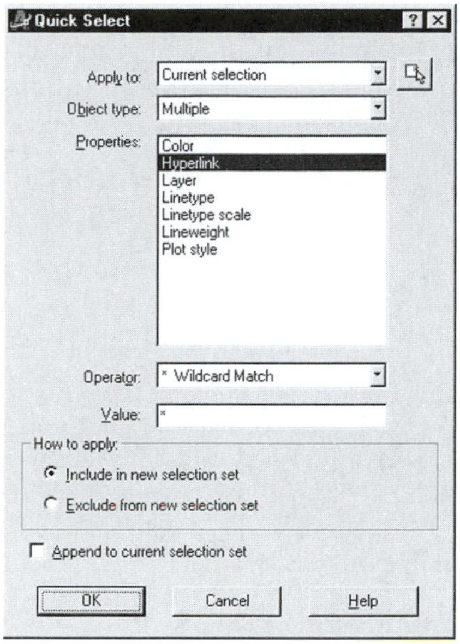

Figure 26.17 *Selecting Quick Select from the Tools pull-down menu displays the Quick Select dialog box.*

Properties: **HYPERLINK**

Operator: * **WILDCARD MATCH**

Value: *

Click OK and AutoCAD highlights all objects that have a hyperlink. Depending on your computer's display system, the highlighting shows up as dashed lines or as another color.

Editing Hyperlinks

Now that you know where the objects with hyperlinks are located, you can use the HYPERLINK command to edit the hyperlinks and related data. Select the hyperlinked object and start the HYPERLINK command. When the Edit Hyperlink dialog box appears (it looks identical to the Insert Hyperlink dialog box), make the changes and click OK.

Removing Hyperlinks from Objects

To remove a URL from an object, use the HYPERLINK command on the object. When the Edit Hyperlink dialog box appears, click the Remove Hyperlink button.

To remove a rectangular area hyperlink, you can simply use the ERASE command. Select the rectangle and AutoCAD erases the rectangle.

HYPERLINKS OUTSIDE AUTOCAD

The hyperlinks you place in the drawing are also available for use outside of AutoCAD. The Web browser makes use of the hyperlink(s) when the drawing is exported in DWF format. To help make the process clearer, here are the steps that you need:

1. Open a drawing in AutoCAD.
2. Attach hyperlinks to objects in the drawing with the HYPERLINK command. To attach hyperlinks to areas, use the –HYPERLINK command's Area option.
3. Export the drawing in DWF format using the PLOT command's "DWF ePlot PC2" plotter configuration.
4. Copy the DWF file to your website.
5. Start your Web browser with the BROWSER command.
6. View the DWF file and click on a hyperlink spot.

THE DRAWING WEB FORMAT

To display AutoCAD drawings on the Internet, Autodesk developed a new file format called drawing Web format (DWF). The DWF file has several benefits and some drawbacks over DWG files. The DWF file is compressed as much as eight times smaller than the original DWG drawing file so that it takes less time to transmit over the Internet, particularly with the relatively slow telephone modem connections. The DWF format is more secure, since the original drawing is not being displayed; another user cannot tamper with the original DWG file.

However, the DWF format has some drawbacks:

- You must go through the extra step of translating from DWG to DWF.
- DWF files cannot display rendered or shaded drawings.
- DWF is a flat 2D-file format; therefore, it does not preserve 3D data, although you can export a 3D view.
- AutoCAD itself cannot display DWF files.
- DWF files cannot be converted back to DWG format without using file translation software from a third-party vendor.
- Earlier versions of DWF did not handle paper space objects (version 2.x and earlier) or linewidths and non-rectangular viewports (version 3.x and earlier).

To view a DWF file on the Internet, your Web browser needs a *plug-in*, which is a software extension that lets a Web browser handle a variety of file formats. Autodesk makes the DWF plug-in freely available from its Web site at http://www.autodesk.com/whip. It's a good idea to regularly check for updates to the DWF plug-in, which is updated about twice a year.

OTHER DWF VIEWING OPTIONS

In addition to Whip! with a Web browser, Autodesk provides two other options for viewing DWF files.

CADViewer Light is designed to be a DWF viewer that works on all operating systems and computer hardware because it is written in Java. As along as your Windows, Mac, or Unix computer has access to Java (included with most Web browsers), you can view DWF files. Volo View Express is a stand-alone viewer that views and prints DWG, DWF, and DXF files. Both products can be downloaded free from the Autodesk website.

VIEWING DWG FILES

To view AutoCAD DWG and DXF files on the Internet, your Web browser needs a DWG-DXF plug-in. At the time of publication, these plug-ins were available from the following vendors:

Autodesk: http://www.autodesk.com/whip

SoftSource: http://www.softsource.com/

California Software Labs: http://www.cswl.com

Cimmetry systems: http://www.cimmetry.com

CREATING A DWF FILE

Use the following steps to create a DWF file from AutoCAD 2000.

1. Type the PLOT command or select Plot from the File pull-down menu. Notice that AutoCAD displays the Plot dialog box.
2. In the Name list box of the Plotter Configuration area, select "DWF ePlot pc3".
3. Click the Properties button. This will display the Plotter Configuration Editor dialog box similar to Figure 26.18.
4. Click Custom Properties in the tree view. Notice the Custom Properties button that appears in the lower half of the dialog box.
5. Click the Custom Properties button. Notice the DWF Properties dialog box similar to Figure 26.19 is displayed.

Resolution

Unlike AutoCAD DWG files that are based on real numbers, DWF files are saved using integer numbers. The Medium precision setting saves the drawing using 16-bit integers, which is adequate for all but the most complex drawings. High resolution saves the DWF file using 20-bit integers, while Extreme resolution saves using 32-bit integers. Figure 29.20 shows an extreme close-up of a drawing exported in DWF format. The portion at left was saved at medium resolution and shows some bumpiness in the arcs. The portion at right was saved at extreme resolution. More significant is

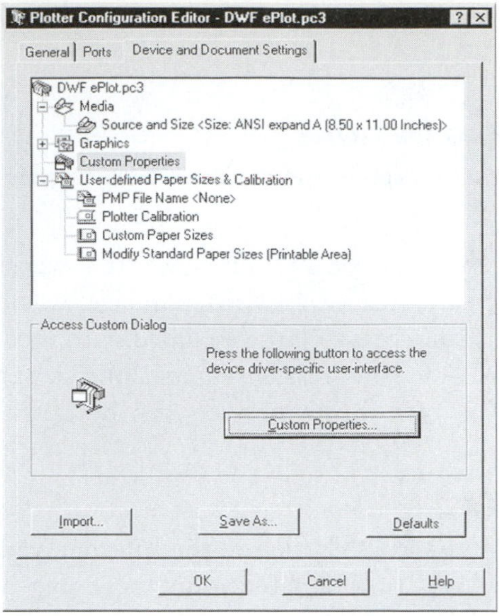

Figure 26.18 *Selecting the Properties button in the Plot dialog box will display the Plotter Configuration Editor dialog box.*

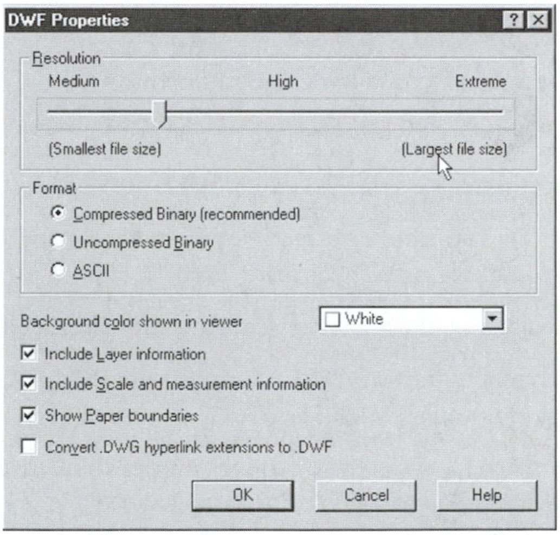

Figure 26.19 *Selecting the Custom Properties button will display the DWF Properties Dialog Box.*

Figure 26.20 *DWF Output in Medium Resolution (at left) and Extreme Resolution (at right).*

the difference in file size. When a DWF file from the "1st floor plan.dwg" file was created (342KB), the medium resolution DWF file was 27KB, while the extreme resolution file was 1,100KB—some 40 times larger. The medium resolution DWF file would transmit over the Internet 40 times faster, a significant saving in time.

Format

Compression further reduces the size of the DWF file. You should always use compression, unless you know that another application cannot decompress the DWF file. Compressed binary format is seven times smaller than ASCII format (7,700KB). Therefore, a compressed binary DWF file transmits over the Internet seven times faster than an ASCII DWF file.

Other Options

Background Color Shown in Viewer: While white is probably the best background color, you may choose any of AutoCAD's 255 colors.

- Include Layer Information—This option includes layers, which allows you to toggle layers off and on when the drawing is viewed by the Web browser.
- Include Scale and Measurement Information—This allows you to use the Location option in the Web browser's Whip plug-in to show scaled coordinate data.
- Show Paper Boundaries—Includes a rectangular boundary at the drawing's extents.
- Convert .DWG Hyperlink Extensions to .DWF—Includes hyperlinks in the DWF file.

  In most cases, you would turn on all options.

6. Once the desired options have been set, select OK to exit the dialog boxes back to the Plot dialog box.
7. Accept the DWF file name listed in the File name text box, or type a new name. If necessary, change the location that the file will be stored in. Note the two buttons: one has an ellipsis (...) that displays the Browse for Folder dialog box. The second button brings up AutoCAD's internal Web browser, a simplified version of the Microsoft browser.
8. Select OK to save the drawing in DWF format.

VIEWING DWF FILES

To view a DWF file, you need to use a Web browser with a special plug-in that allows the browser to correctly interpret the file. (Remember: you cannot view a DWF file with AutoCAD.) Autodesk named their DWF plug-in "Whip!," short for "Windows HIgh Performance."

Autodesk updates the DWF plug-in approximately twice a year. Each update includes some new features. In summary, all versions of the DWF plug-in perform the following functions:

- Views DWF files created by AutoCAD within a browser.
- Right-clicking the DWF image displays a cursor menu with commands.
- Real-time pan and zoom lets you change the view of the DWF file as quickly as a drawing file in AutoCAD.
- Embedded hyperlinks let you display other documents and files.
- File compression means that a DWF file appears in your Web browser much faster than the equivalent DWG drawing file would.
- Print the DWF file alone or along with the entire Web page.
- Works with Netscape Navigator or Microsoft Internet Explorer. A separate plug-in is required, depending upon which of the two browsers you use.
- Allows you to "drag and drop" a DWG file from a Web site into AutoCAD as a new drawing or as a block.
- Views a named view stored in the DWF file.
- Can specify a view using x, y coordinates.
- Toggles layers off and on.

If you don't know whether the DWF plug-in is installed in your Web browser, select About Plug-ins from Help in the browser's menu bar. You may need to scroll through the list of plug-ins to find a display similar to Figure 26.21. If the plug-in is not installed, or is an older version, then you need to download it from Autodesk's website at:

http://www.autodesk.com/whip

For Netscape users, the file is quite large at 3.5MB and takes about a half-hour to download using a typical 28.8Kbaud modem. The file you download from the Autodesk Web site is either a self-installing file (with a .JAR extension) or a self-extracting installation file with a name such as Whip4.Exe. After the download is complete, follow the instructions on the screen. For Microsoft Internet Explorer users, the DWF plug-in is an ActiveX control. It is automatically installed into Explorer when you set up AutoCAD 2000.

DWF PLUG-IN COMMANDS

To display the DWF plug-in commands, move the cursor over the DWF image and right-click. This displays a cursor menu with commands, such as Pan, Zoom, and Named Views similar to Figure 26.22. To select a command, move the cursor over the command name and press the left mouse button. Options include:

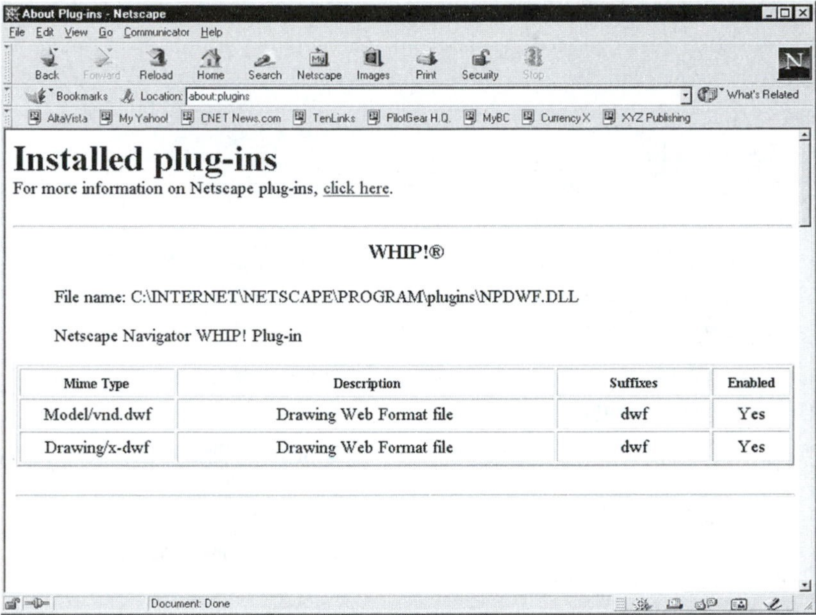

Figure 26.21 *Checking the status of browser plug-ins.*

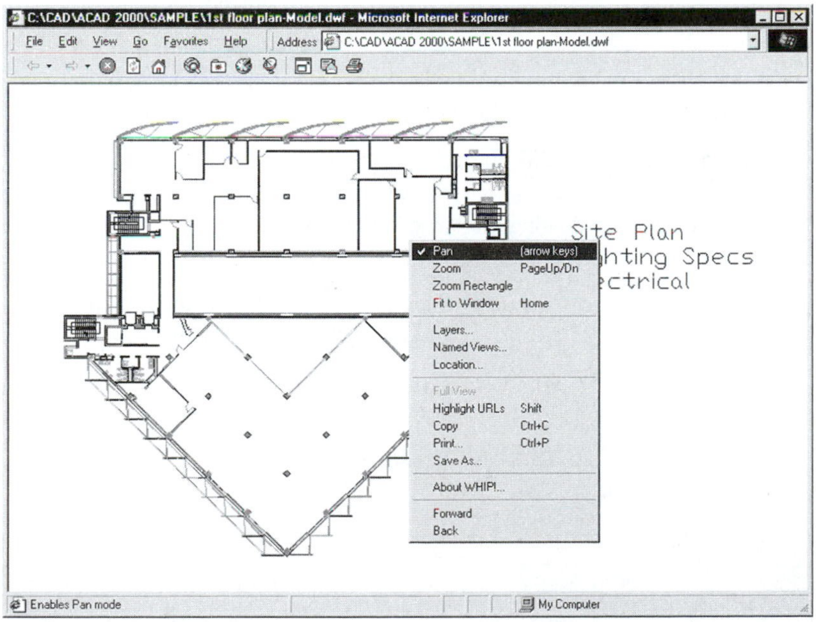

Figure 26.22 *Moving the cursor over the DWF image and right-clicking displays the Whip's shortcut menu.*

- Pan—is the default command. Press the left mouse button and move the mouse. The cursor changes to an open hand, signaling that you can pan the view around the drawing. This is exactly the same as real-time panning in AutoCAD. Panning only works when you are zoomed in; it does not work in Full View mode.
- Zoom—is like the Zoom command in AutoCAD. The cursor changes to a magnifying glass. Hold down the left mouse button and move the cursor up (to zoom in) and down (to zoom out).
- Zoom Rectangle—is the same as Zoom Window in AutoCAD. The cursor changes to a plus sign. Click the left mouse button at one corner, then drag the rectangle to specify the size of the new view.
- Fit to Window—is the same as AutoCAD's Zoom Extents command. You see the entire drawing.
- Layers—displays a non-modal dialog box similar to Figure 26.23 that lists all layers in the drawing. (A non-modal dialog box remains on the screen. Unlike AutoCAD's modal dialog boxes, you do not need to dismiss a non-modal dialog box to continue working.) Click a layer name to toggle its visibility between on (yellow lightbulb) and off (blue lightbulb).
- Named Views—works only when the original DWG drawing file contained named views created with the VIEW command. Selecting Named Views displays a non-modal dialog box that allows you to select a named view. Click a named view to see it; click the small **x** in the upper-right corner to dismiss the dialog box.
- Location—displays a non-modal dialog box that shows the x, y, z coordinates of the cursor location. Since DWF files are only 2D, the Z coordinate is always 0.
- Full View—causes the Web browser to display the DWF image as large as possible. This is useful for increasing the physical size of a small DWF image. When you have finished viewing the large image, right-click and select the BACK command or click the browser's Back button to return to the previous screen.
- Highlight URLs—displays a flashing gray effect on all objects and areas with a hyperlink. Figure 26.24 shows the text highlighted. This helps you see where the

Figure 26.23 *Selecting Layers... from the Whip shortcut menu displays Whip's Layer dialog box.*

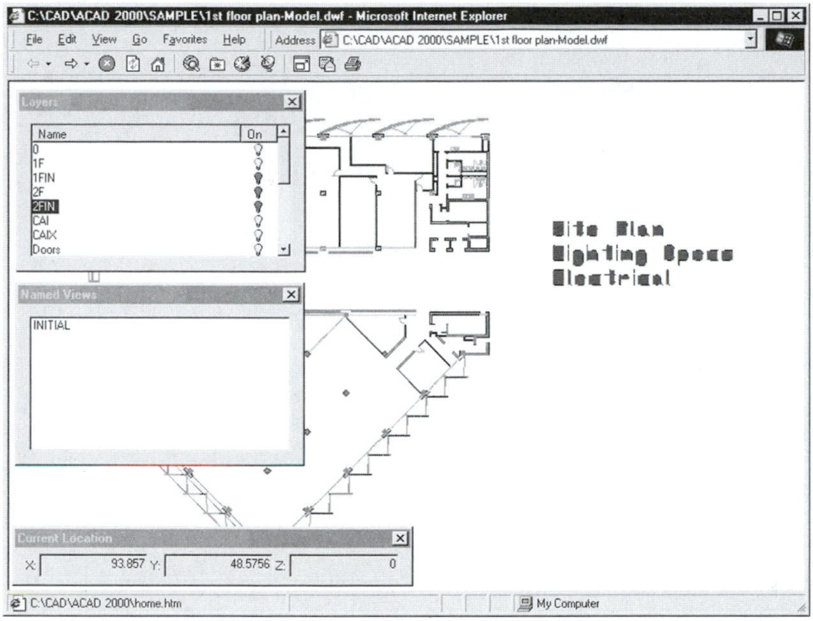

Figure 26.24 *Highlighting Hyperlinks displays a flashing gray effect on all objects and areas with a hyperlink.*

hyperlinks are. To read the associated URL, pass the cursor over a highlighted area and look at the URL on the browser's status line. A shortcut is to hold down (ShiftKey), which causes the URLs to highlight until you release the key. Clicking the URL opens the document with its associated application.

- Copy—copies the DWF image to the Windows Clipboard in EMF (Enhanced Metafile or "Picture") format.
- Print—prints only the DWF image . To print the entire Web page, including the DWF image, use the browser's Print button.
- Save As—saves the DWF file in three formats to your computer's hard drive:

  DWF

  BMP (Windows bitmap)

  DWG (AutoCAD drawing file) works only when a copy of the DWG file is available at the same subdirectory as the DWF file.

You cannot use the SAVEAS command until the entire DWF file has been transmitted to your computer.

- About WHIP...—displays information about the DWF file similar to Figure 26.25, including DWF file revision number, description, author, creator, source filename, creation time, modification time, source creation time, source

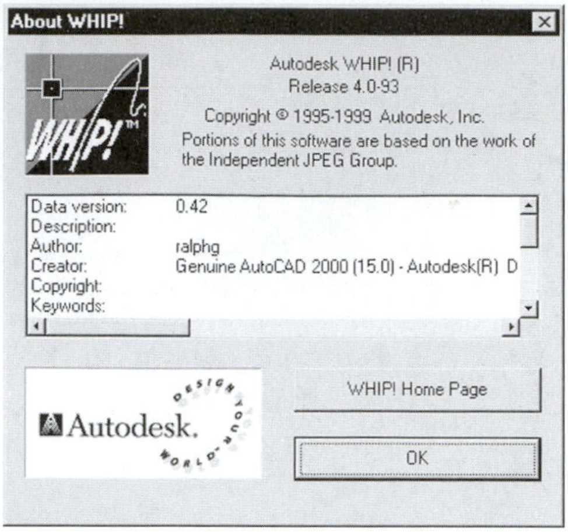

Figure 26.25 *Selecting About Whip... from the shortcut menu displays the About Whip! dialog box.*

modification time, current view left, current view right, current view bottom, and current view top.

- Forward—is almost the same as clicking the browser's Forward button. When the DWF image is in a frame, only that frame goes forward.
- Back—is almost the same as clicking the browser's Back button. It works differently when the DWF image is displayed in a frame.

Drag and Drop from Browser to Drawing

The DWF plug-in allows you to perform several "drag and drop" functions. Hold down the "(" key to drag a DWF file from the browser into AutoCAD. Remember that AutoCAD cannot translate a DWF file into DWG drawing format. For this reason, this form of drag and drop only works when the originating DWG file exists in the same subdirectory as the DWF file.

AutoCAD executes the INSERT command using the following sequence:

Command: _.-insert Enter block name or [?]: C:\CAD\ACAD 2000\SAMPLE\1st floor plan.dwg

Specify insertion point or [Scale/X/Y/Z/Rotate/PScale/PX/PY/PZ/PRotate]: **0,0** (EnterKey)

Enter X scale factor, specify opposite corner, or [Corner/XYZ] <1>: (EnterKey)

Enter Y scale factor <use X scale factor>: (EnterKey)

Specify rotation angle <0>: (EnterKey)

To see the inserted drawing, you may need to use the Extents option of the ZOOM command.

Another drag and drop function is to drag a DWF file from the Windows Explorer into the Web browser. This causes the Web browser to load the DWF plug-in, then display the DWF file. Once displayed, you can execute all of the commands listed in the previous section.

Finally, you can also drag and drop a DWF file from Windows Explorer (Windows 95/98) or File Manager (Windows NT) into AutoCAD. This causes AutoCAD to launch another program that is able to view the DWF file. This does not work if you do not have other software on your computer system capable of viewing DWF files.

EMBEDDING A DWF FILE

To let others view your DWF file over the Internet, you need to embed the DWF file in a Web page. Here are the steps for embedding a DWF file in a Web page.

1. This hypertext markup language (HTML) code is the most basic method of placing a DWF file in your Web page:

   ```
   <embed src = "filename.dwf">
   ```

   The <embed> tag embeds an object in a Web page. The src option is short for "source." Replace filename.dwf with the URL of the DWF file. Remember to keep the quotation marks in place.

2. HTML normally displays an image as large as possible. To control the size of the DWF file, add the Width and Height options:

   ```
   <embed width = 800 height = 600 src = "filename.dwf">
   ```

   The Width and Height values are measured in pixels. Replace 800 and 600 with any appropriate numbers, such as 100 and 75 for a "thumbnail" image, or 300 and 200 for a small image.

3. To speed up a Web page's display speed, some users turn off the display of images. For this reason, it is useful to include a description, which is displayed in place of the image:

   ```
   <embed width = 800 height = 600 name = description src = "filename.dwf">
   ```

   The Name option displays a textual description of the image when the browser does not load images. You might replace description with the DWF filename.

4. When the original drawing contains named views created by the VIEW command, these are transferred to the DWF file. Specify the initial view for the DWF file:

   ```
   <embed width = 800 height = 600 name = description
    namedview = "viewname" src = "filename.dwf">
   ```

The namedview option specifies the name of the view to display upon loading. Replace viewname with the name of a valid view name. When the drawing contains named views, the user can right-click on the DWF image to get a list of all named views.

As an alternative, you can specify the 2D coordinates of the initial view:

<embed width = 800 height = 600 name = description view = "0,0 9,12"
src = "filename.dwf">

The View option specifies the x, ycoordinates of the lower-left and upper-right corners of the initial view. Replace 0,0 9,12 with other coordinates. Since DWF is 2D only, you cannot specify a 3D viewpoint. You can use View or Namedview but not both.

5. Before a DWF image can be displayed, the Web browser must have the DWF plug-in called Whip!. For users of Netscape Communicator, you must include a description of where to get the Whip plug-in when the Web browser is lacking it.

   <embed pluginspage = http://www.autodesk.com/products/autocad/whip/
   whip.htm width = 800 height = 600 name = description view = "0,0 9,12"
   src = "filename.dwf">

   The plug-in space option describes the page on the Autodesk website where the Whip-DWF plug-in can be downloaded.

   The code listed above works for Netscape Navigator. To provide for users of Internet Explorer, the following HTML code must be added:

   <object classid = "clsid:B2BE75F3-9197-11CF-ABF4-08000996E931"
   codebase = "ftp://ftp.autodesk.com/pub/autocad/plugin/
   whip.cab#version = 2,0,0,0" width = 800 height = 600>

   <param name = "Filename" value = "filename.dwf">

   <param name = "View" value = "0,0 9,12">

   <param name = "Namedview" value = "viewname">

   <embed pluginspage = http://www.autodesk.com/products/autocad/whip/

   whip.htm width = 800 height = 600 name = description view = "0,0 9,12"
   src = "filename.dwf">

   </object>

   The two <object> and three <param> tags are ignored by Netscape Navigator; they are required for compatibility with Microsoft Internet Explorer. The Classid and Codebase options tell Explorer where to find the plug-in. Remember that you can use View or Namedview but not both.

6. Save the HTML file.

CHAPTER 26 QUIZ

1. Can you launch a Web browser from within AutoCAD?
2. What does DWF mean?
3. What is the purpose of DWF files?
4. What is URL short for?
5. Which of the following URLs are valid?

   www.autodesk.com

   http://www.autodesk.com

   Both of the above.

   Neither of the above.
6. What is the purpose of a URL?
7. What is FTP short for?
8. What is a "local host"?
9. Can URLs be used in an AutoCAD drawing?
10. The purpose of URLs is to let you create ______________________ between files.
11. When you attach a URL to a block, the URL data is ____________ when you scale the block unevenly, stretch the block, or explode it.
12. Can you attach a URL to rays and xlines?
13. The ATTACHURL command allows you to attach a URL to __________________ and __________________.
14. To see the location of URLs in a drawing, use the ____________________ command.
15. Rectangular URLs are stored on layer______________________________.
16. The LISTURL command tells you ______________________________________.
17. The DETACHURL command removes a __________________ from an object.
18. Compression in the DWF file causes it to take (less, more, the same) time to transmit over the Internet.
19. A DWF is created from a ________________ file using the _________________ or ________________ command.
20. What does a plug-in allow a Web browser do?
21. Can a Web browser view DWG drawing files over the Internet?
22. ________ is an HTML tag for embedding graphics in a Web page.
23 Are the Whip plug-in and the Whip display driver the same?

24. A file being transmitted over the Internet via a 28.8Kbps modem takes about ______________ minutes per megabyte.

25. To open a drawing located on the Internet, use the ____________________ command.

SECTION 5

PICTORIAL DRAWINGS

OBLIQUE AND ISOMETRIC DRAWING

CHAPTER 27

Oblique and Isometric Drawing

The drawings that have been examined throughout this text have been orthographic drawings. These are drawings that show an object by looking directly at it. Objects are assumed to be located in a glass box, with the sides of the object parallel to the sides of the glass box. Orthographic drawings comprise the majority of construction drawings. Oblique and isometric drawings are used in some disciplines of engineering and architectural drawings to show specific details. This chapter will introduce the creation of oblique and isometric drawings, the skills necessary to produce these drawings, and the specific commands that can be used to aid in drawing construction.

OBLIQUE AND ISOMETRIC DRAWINGS

Before examining what oblique and isometric drawings are, it is important to understand what they are not. These drawings appear to be three-dimensional, but are really two-dimensional drawings using lines that are created with their X and Y coordinates.

3D DRAWINGS

One of the major features of AutoCAD is the ability to create true 3D drawings. Three-dimensional drawings require creation of a 3D model. For instance, in drawing a floor plan, heights are provided in addition to the normal information that would be drawn. Once the model is created, AutoCAD can rotate the model in space, allowing the viewing point to be altered. You've most likely seen commercials on television with structures growing out of a floor plan and progressing through the entire construction process. These drawings could be created starting with 3D models. Individual points, lines, and planes comprising the drawings are stored and become part of the drawing base. Figure 27.1 shows an example of a 3D drawing created using AutoCAD and a third party software program. Although this type of drawing is both beautiful and very useful, how it is created will not be discussed in this text.

Figure 27.1 *A 3D drawing rendered on AutoCAD often resembles a photograph.*

OBLIQUE DRAWINGS

Oblique drawings show height, width, and depth of an object. Three common methods of oblique drawings include cavalier, cabinet, and general drawings. Examples of each can be seen in Figure 27.2. Each system places one or more surfaces parallel to the viewing plane, and shows the receding surface at an angle. Each drawing method may be drawn at any scale.

Cavalier Drawings

Cavalier drawings use the same scale for the front and receding surfaces. The front side is drawn parallel to the viewing plane. The receding surface is drawn at a 45° angle to show depth. See Figure 27.3.

Cabinet Drawings

Cabinet drawings are similar to cavalier drawings but show the receding surfaces using a scale that is half of the scale used on the front surfaces. The receding surfaces are also shown at a 45° angle. See Figure 27.4.

General Drawings

General drawings represent the receding surfaces using a scale that is three-quarters of the scale used to draw the front surfaces. The angle used to create the receding

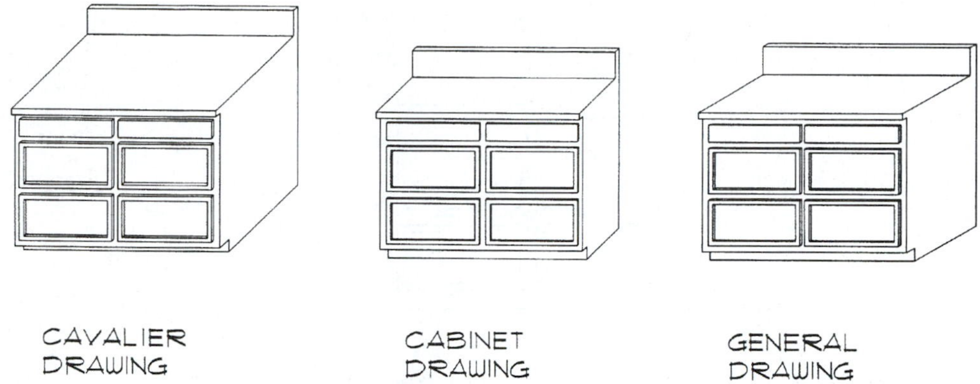

Figure 27.2 *Although lacking the details of 3D drawings, three types of oblique drawings are often used to show height, width, and depth of an object.*

edge can be any angle except 45 degrees. Figure 27.5 shows an example of a general oblique drawing.

ISOMETRIC DRAWINGS

Isometric drawings are the most common form for viewing three surfaces of an object. Rather than having one surface resting squarely on a base, the object is rotated at 45°-16' from a horizontal plane. This results in a view of the object from the front where each edge is drawn at 30° to a horizontal base. All vertical lines remain vertical,

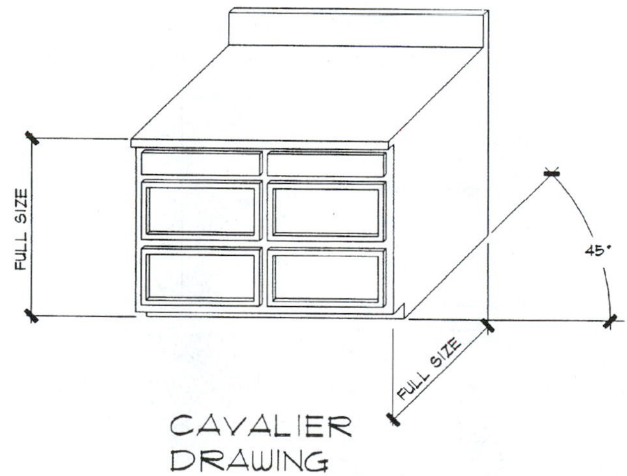

Figure 27.3 *Cavalier drawings show all surfaces in full scale, with the receding side drawn at a 45° angle.*

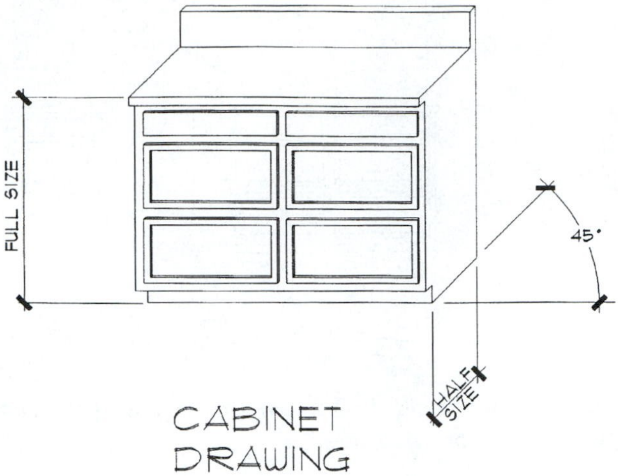

Figure 27.4 *Cabinet drawings show the front view at full scale, with the receding side shown at half scale and a 45° angle.*

and an equal scale is used to represent all surfaces. Figure 27.6 shows the basic construction methods of an isometric drawing. Lines that are parallel to one of the three axes will remain parallel. Inclined surfaces that are not parallel to an axis will need to be projected from a point that lies on a line that is parallel to an axis. Figure 27.7 shows the steps for blocking out an irregularly shaped detail.

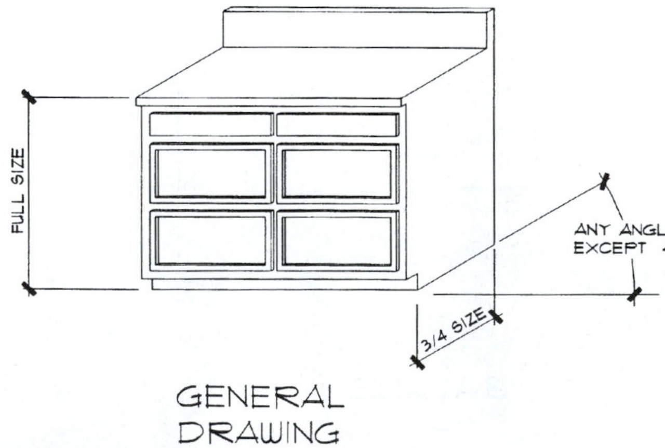

Figure 27.5 *General drawings show the front view at full scale, with the receding side shown at three-quarter scale and any angle except 45°.*

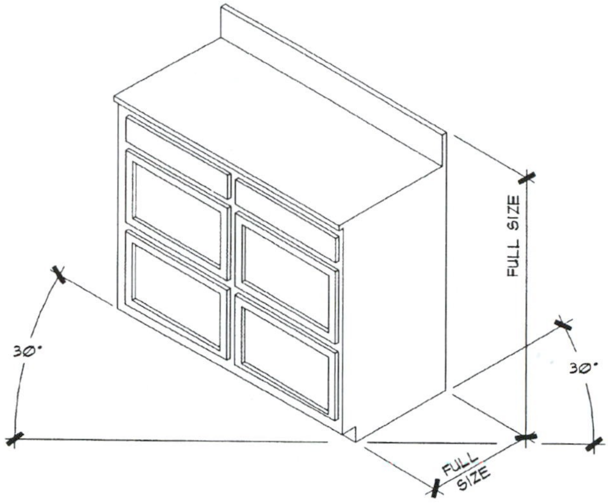

Figure 27.6 *Isometric drawings show all sides at full scale, and each side at a 30° angle.*

DRAWING TOOLS FOR PICTORIAL DRAWINGS

The drawing tools used to create two-dimensional drawings also can be used to create pictorial drawings. Commands such as SNAP, GRID, ISOPLANE, and ELLIPSE can be used to aid in drawing setup.

SNAP

Pictorial drawings can be created easily by opening the Drafting Settings dialog box from the Tools pull-down menu. Once in the menu, select Isometric Snap from the Snap Type & Style area of the Snap and Grid tab. Selecting Isometric Snap and the OK button will close the dialog box, alter the snap setting to isometric, and then restore the drawing area. Returning to the drawing display will display an isometric grid and cursor, as seen in Figure 27.8. Activating SNAP and GRID will greatly aid in creating isometric drawings.

CROSSHAIRS ORIENTATION

Three surfaces are seen in pictorial drawings and are referred to by AutoCAD as Isoplane Left, Isoplane Top, and Isoplane Right. These three orientations are also available for the crosshairs to ease drawing construction. Figure 27.9 shows the three isometric planes and the options for crosshairs positioning. The crosshairs orientation can be changed using one of several methods including a Function key, the Isoplane command or the Control keys.

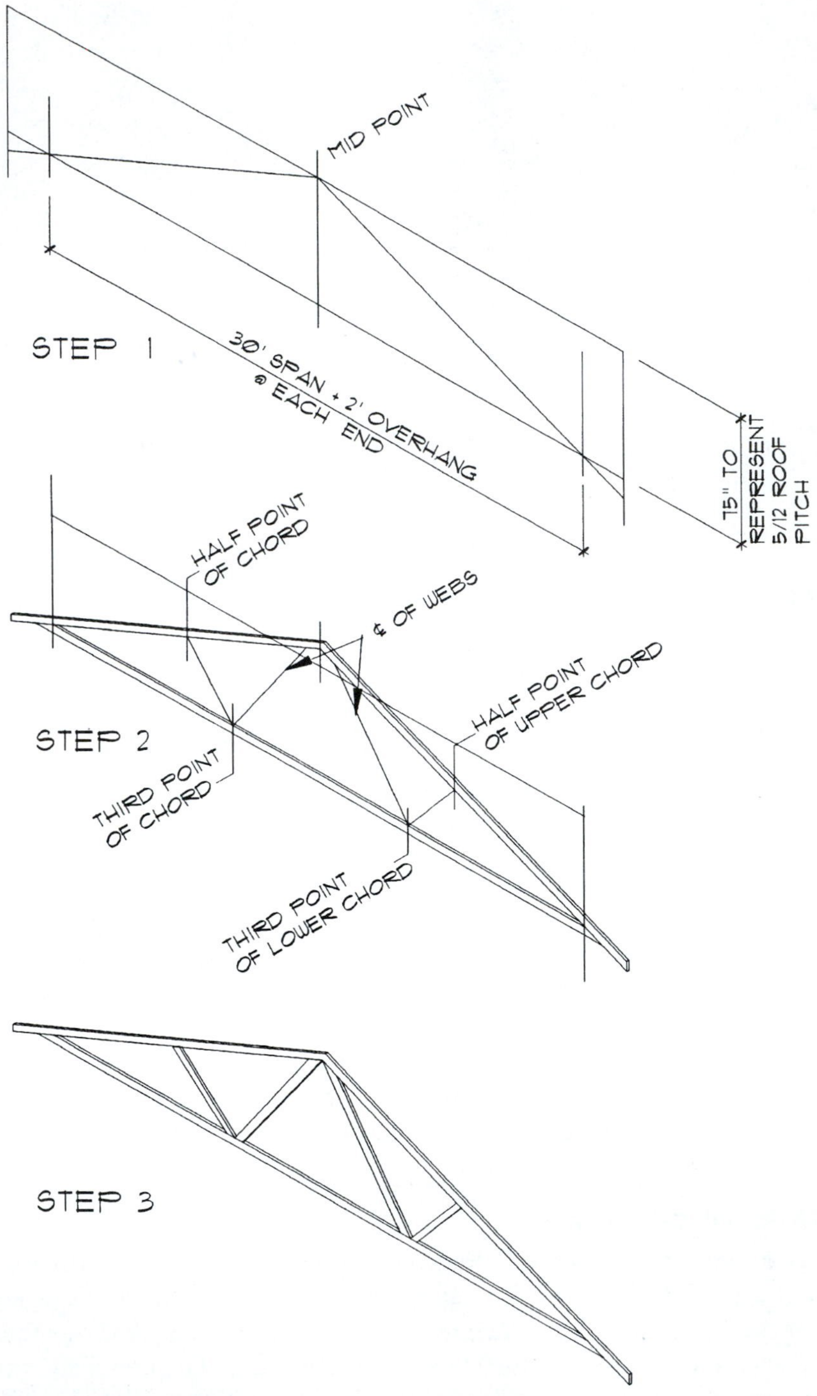

Figure 27.7 *Blocking irregular-shaped objects in rectangles will aid in drawing layout.*

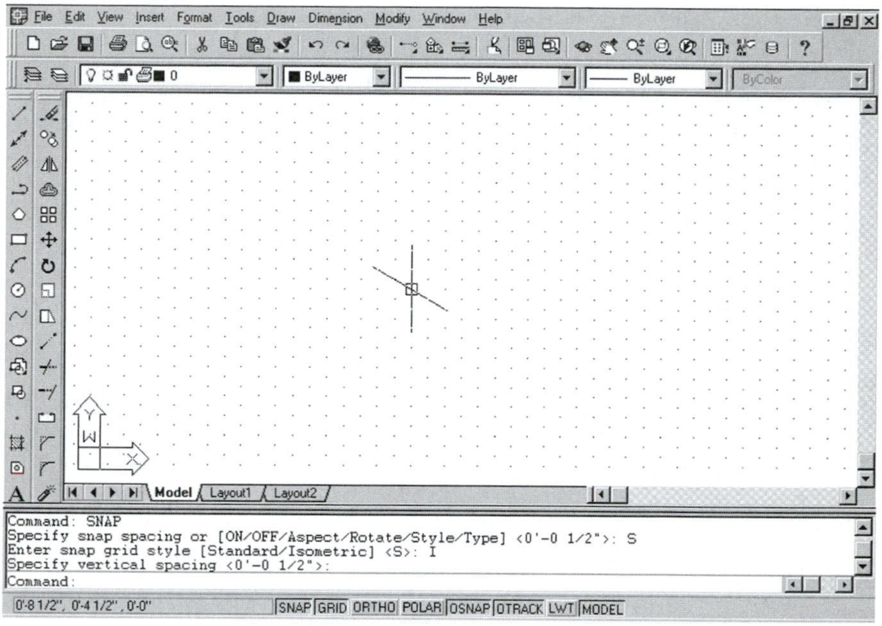

Figure 27.8 *An isometric grid can be established by using the Drafting Settings dialog box from the Tools pull-down menu. Once in the menu, select Isometric Snap from the Snap Type & Style area of the Snap and Grid tab.*

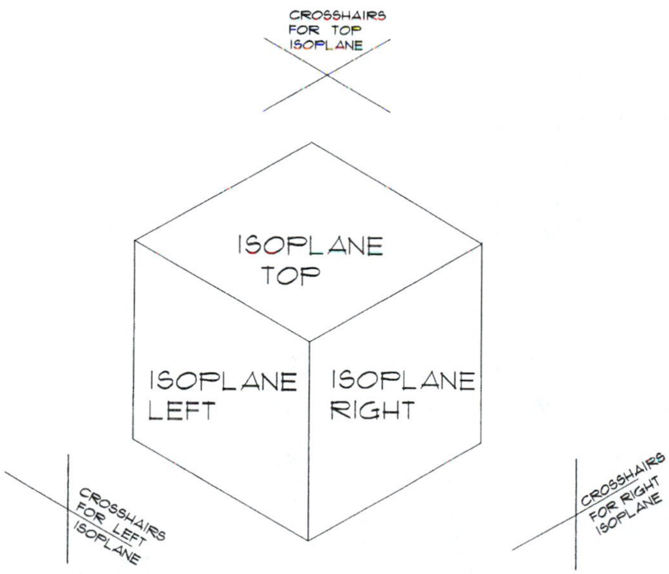

Figure 27.9 *AutoCAD refers to the three isometric surfaces as Isoplane Left, Isoplane Top, and Isoplane Right.*

Function Key

The easiest toggle between the three isoplane options is to use the F5 function key. Each time the F5 key is pressed, the crosshairs will toggle to the next isoplane and the current setting will be displayed at the command line.

The ISOPLANE Command

The crosshairs position also can be controlled using the ISOPLANE command. The command sequence is:

> Command: **ISOPLANE** (EnterKey)
>
> Enter isometric plane setting [Left/Top/Right] <top>: **L** (EnterKey) *(Enter either option.)*
>
> Current isoplane: LEFT
>
> Command:

The letter representing an option can be entered, followed by the (EnterKey). Pressing (EnterKey) at the toggle prompt will change the crosshairs automatically to the next option and restore the command prompt.

Control Key Control

The crosshairs also can be controlled by using the CTRL+E keys. Pressing CTRL+E with the crosshairs in the current setting will display the following prompt:

> Command: <Isoplane Top>:

Repeated pressing of the CTRL+E keys will toggle transparently the crosshairs between the Left, Top, and Right settings.

DRAWING ISOMETRIC CIRCLES

Circles in pictorial drawings are represented by ellipses. With the SNAP option set to ISOMETRIC, enter the ELLIPSE command and select the ISOCIRCLE option. An ellipse will be drawn automatically in the current ISOPLANE setting as seen in Figure 27.10. The command sequence is:

> Command: **EL** (EnterKey)
>
> Specify axis endpoint of ellipse or [Arc/Center/Isocircle]: **I**(EnterKey)
>
> Specify center of isocircle: *(Specify desired point.)*
>
> Specify radius of isocircle or [Diameter]: *(Specify desired diameter.)*
>
> Command:

A centerpoint may be selected or a radius or diameter may be specified. One of the drawbacks to using pictorial drawings to show construction details is that circular objects often appear distorted. Round objects will appear as an ellipse in an isometric drawing. Figure 27.11 shows the proper orientation of circles for each plane of an

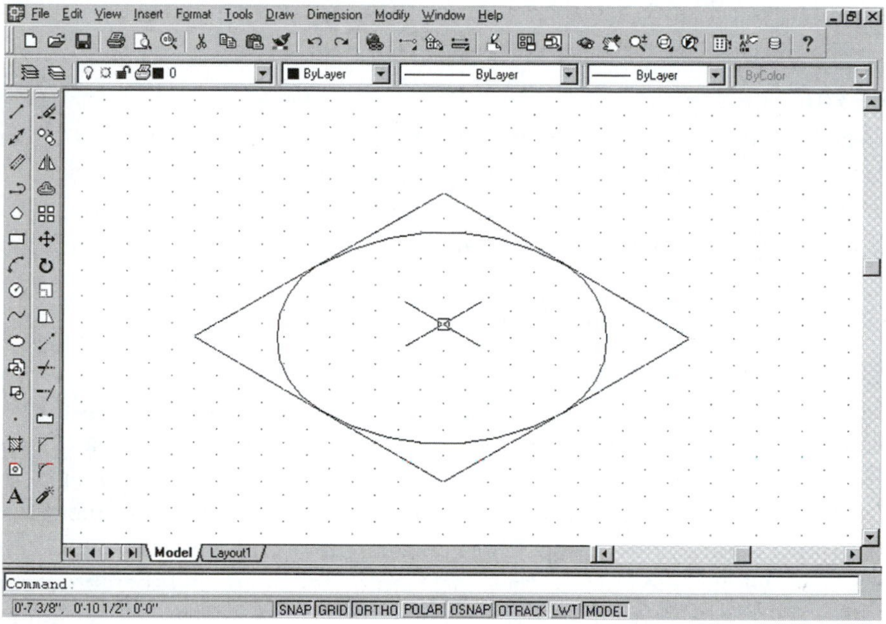

Figure 27.10 *An isometric ellipse will be drawn automatically in the current isoplane setting using the ISOCIRCLE option of the ELLIPSE command.*

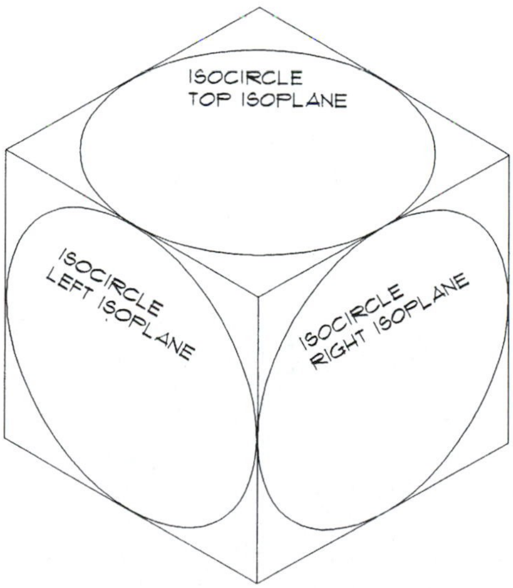

Figure 27.11 *Proper orientation of isometric circles.*

isometric drawing. Notice that the centerline of the circle is parallel to the edge of one of the three surfaces.

ISOMETRIC TEXT

No special commands are required to place text on pictorial drawings. On small details, text is best placed in its normal position and connected to the detail with leader lines. On larger drawings, text can be placed parallel to the isoplane on which it will appear by adjusting the Obliquing Angle in the TEXT command. Figure 27.12 shows an example of text placed at 30°, –30°, 90°, and –90°. This can best be done by setting up text styles for each angle.

ISOMETRIC DIMENSIONS

Although no specific command exists to create isometric dimensions, dimensions created using the Aligned option can be adjusted quickly to conform to an isometric drawing. Aligned dimensions can be placed by selecting the Aligned icon or from the keyboard. Figure 27.13 shows an example of dimensions added to an isometric drawing using the Aligned Dim option. Once created, enter Oblique from the DimEdit: prompt. You will be asked to select a dimension to be edited, and then prompted to specify an obliquing angle. Use either 30° or –30° for the angle. The command sequence is:

Command: **DIMALI** (EnterKey)

Specify first extension line origin or <select object>: *(Select first point.)*

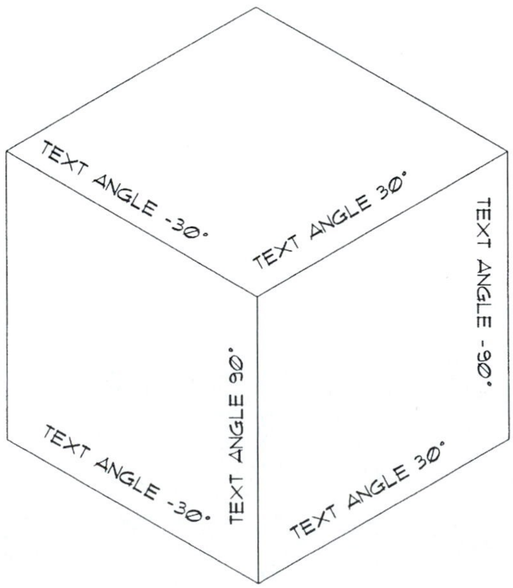

Figure 27.12 *Text can be placed in each plane of an isometric drawing using the TEXT command and rotating the angle.*

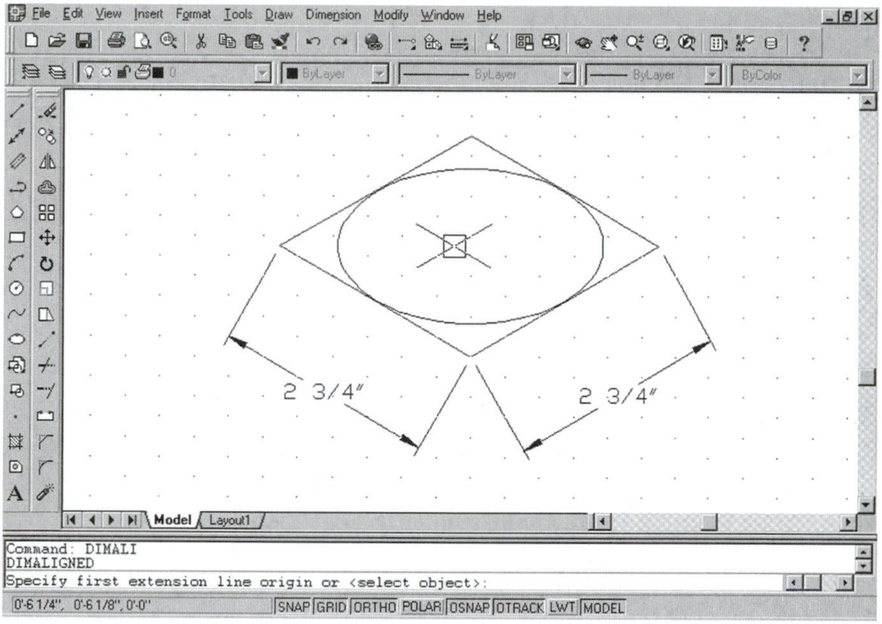

Figure 27.13 *Dimensions can be placed on an isometric drawing using the DIMALI command.*

Specify second extension line origin: *(Select second point.)*

Specify dimension line location or [Mtext/Text/Angle]: *(Select point.)*

Dimension text <2 ¾">:(EnterKey)

Command:

Command: **DIMEDIT** (EnterKey)

Enter type of dimension editing [Home/New/Rotate/Oblique] <Home>: **O** (EnterKey)

Select objects: *(Select dimension to edit.)*

1 found

Select object: (EnterKey)

Enter obliquing angle (RETURN for none): **30** (EnterKey)

Command:

The results of this sequence can be seen in Figure 27.14.

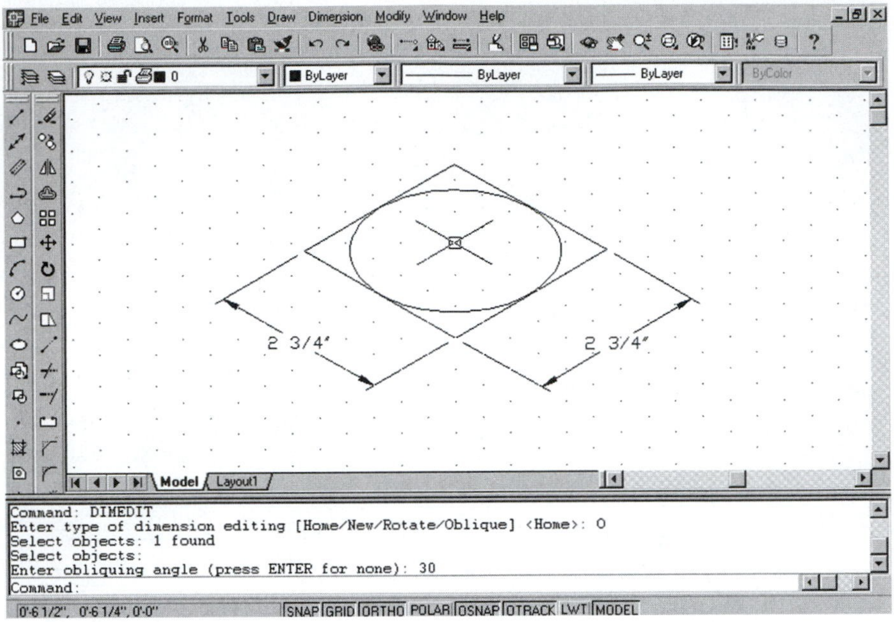

Figure 27.14 *Entering OBLIQUE at the DIMEDIT prompt will alter aligned dimensions to isometric dimensions.*

CHAPTER 27 EXERCISES

1. Open a new drawing and draw a cube that is 8' wide, 6' high, and 4'-6" deep, using each of the oblique drawing methods. Place each drawing on a separate layer. Save the drawing as E-27-1.
2. Open a new drawing and draw a cube that is 8' in each direction using isometric techniques. Place an isometric circle in each plane so that the circle is tangent to the edges of each plane. Save the drawing as E-27-2.
3. Open a new drawing and draw an isometric cube that is 6' in each direction. Show the bottom, left, and right planes. Place an isometric circle in each plane so that the circle is tangent to the edges of each plane. Save the drawing as E-27-3.
4. Use the drawing below and draw an isometric drawing showing the framing of a 5'-0" × 3'-6" window. Use 2 × 6 studs. Label the top plates, a 4 × 8 header, 2 × 6 sill, and 2 × 6 studs, king studs, trimmers, and cripples. Save the drawing as E-27-4.

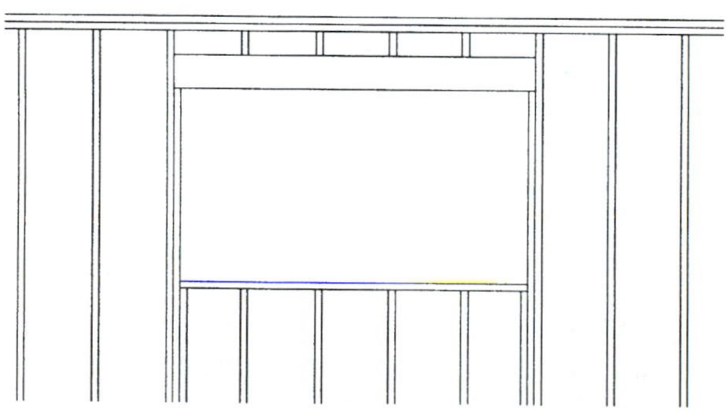

5. A window supplier would like to update its catalogue. Provide an isometric drawing of a 4'-0" wide × 3'-6"-high vinyl window with a half-round window above. Show the window with a grid in each slider, and a radial pattern in the upper window. Draw the frame as 1" aluminum, and the grid as ½" wide material. Place the grid on a separate layer from the window. Save the drawing as E-27-5.

6. Use the floor plan below and draw an isometric drawing showing the left wall with the refrigerator and double oven. Design and show drawers and cabinet drawers. Use glass in some of the doors and show the shelving. Consult vendor catalogues for specific sizes of appliances and cabinets. Omit the refrigerator. 8'-0" high ceilings. Show the drawing as E-27-6.

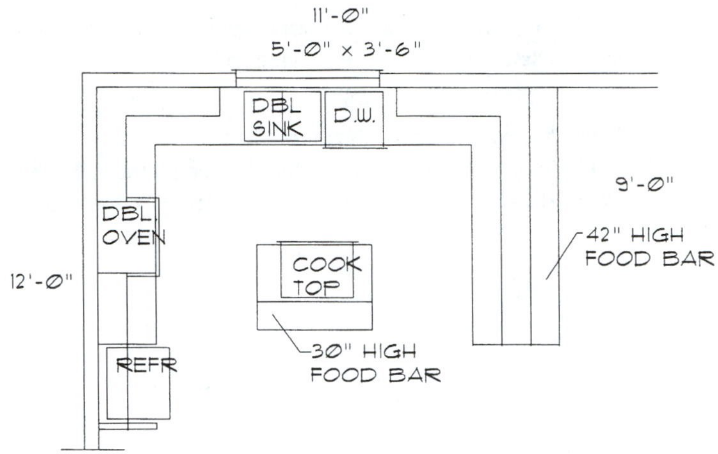

7. Use the floor plan for drawing E-27-6 and the drawing started in exercise E-24–6 to draw an isometric of the complete kitchen. Draw the cabinets over the food bar as 30" high. Save the drawing as E-27-7.

8. Use the drawing on the next page to draw an isometric drawing showing the metal beam seat. Show the bracket, but omit the beam. Make the side plates from ½" steel and the base plate from ⅞" × 7" wide steel plates. Place bolts at 2" in from each edge and at 3" o.c. Dimension the size of the steel plates. Save the drawing as E-27-8.

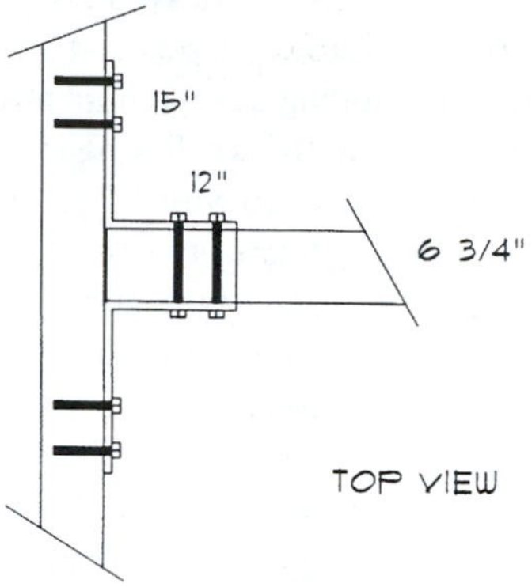

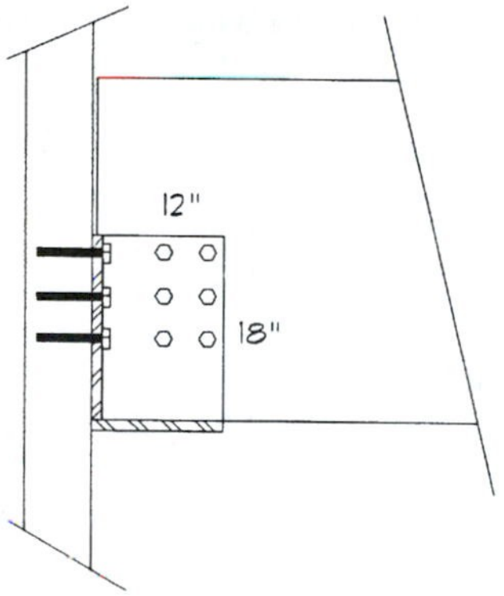

9. Use the drawing below as a guide to draw an isometric view of the truss. Assume that the top and bottom chords are made from two 2 × 3s, and the webs are 1" in diameter. Show the distance from the top of the top chord to the bottom of the metal bracket as 3 ½". Show the total depth as 36 inches, and lay out the webs at 45 degrees from the top chord. Save the drawing as E-27-9.

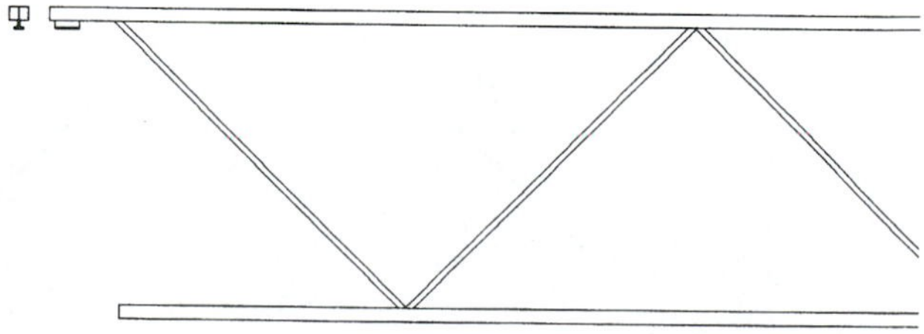

10. Use the drawing below as a guide to draw an isometric view of the 6 ¾ × 13 ½" glu-lam beam supported on two 6 × 6 posts at 10 feet o.c. on 18" diameter concrete piers. Show piers extending 8" minimum above the finish grade. Show the wall framed with 2 × 6 studs at 16" o.c. Frame the floor with 2 × 10 floor joist at 12" o.c. 12' above the grade covered with ¾" plywood floor sheathing and an 18" cantilever. Use a 13" × 8" metal hanger with 2" × 10" legs to post. Show two bolts through the hanger into each beam, and two bolts through the hanger legs and post. Place bolts 1 ½" from steel edges and 3" o.c. Use ½" steel cables with turnbuckles forming an "X" between each post. Attach the cable to each post with eye bolts 6" up and from the bottom. Place the top eyebolt 3" below the bottom hanger's leg bolt. Label major materials and dimension the cantilever and post spacing. Since your drawing will be used to assemble this project and not to manufacture each item, great accuracy is not required on items such as the turnbuckle and eyebolts. Save the drawing as E-27-10.

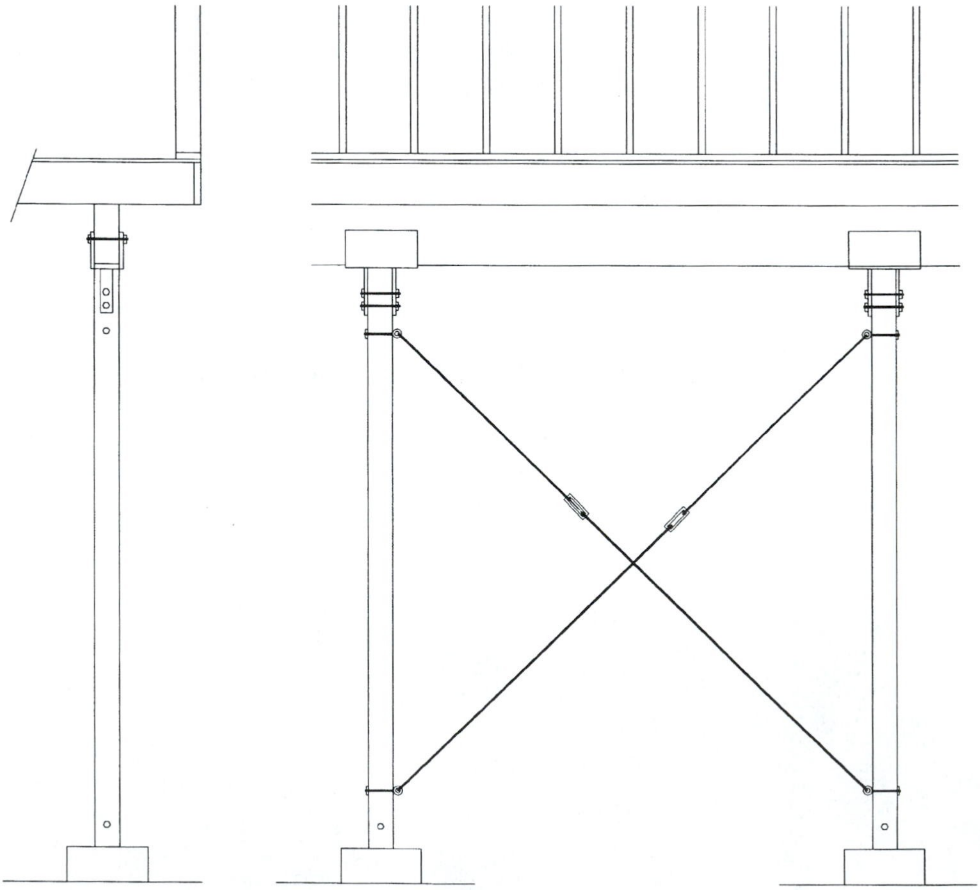

CHAPTER 27 QUIZ

1. What are the differences between cabinet and general oblique drawings?
2. Compare the differences between cavalier and cabinet drawings.
3. Describe the relationship of horizontal and vertical lines in an isometric drawing.
4. What command and option adjust the grid for isometric drawings?
5. List the three options for selecting the isometric SNAP options.
6. Explain the easiest way to construct isometric circles.
7. List common angles for placing text on isometric drawings.
8. Sketch and label the three positions of the crosshairs in isometric drawings.
9. What is the major difference between oblique and 3D drawings?
10. Explain the process for placing dimensions on an isometric drawing.

INDEX

A

B

C

D

M

N

O

P

S

T

U

V

W

X

Z